Genetics Essentials

Concepts and Connections

FOURTH EDITION

Benjamin A. Pierce
Southwestern University

w.h.freeman
Macmillan Learning
New York

Vice President, STEM: Ben Roberts
Executive Editor: Lauren Schultz
Development Editor: Maria Lokshin
Executive Marketing Manager: Will Moore
Marketing Assistant: Cate McCaffery
Director of Content: Clairissa Simmons
Content Development Manager, Biology: Amber Jonker
Lead Content Developer, Genetics: Cassandra Korsvik
Senior Media and Supplements Editor: Amy Thorne
Assistant Editor: Kevin Davidson
Director, Content Management Enhancement: Tracey Kuehn
Managing Editor: Lisa Kinne
Senior Content Project Manager: Jessica Gould
Project Management: Vanavan Jayaraman
Director of Design, Content Management: Diana Blume
Cover Design: John Callahan
Text Design: Jerry Wilke Design
Illustrations: Dragonfly Media Group
Illustration Coordinator: Janice Donnola
Photo Editor: Christine Buese
Photo Researcher: Richard Fox
Senior Production Supervisor: Paul Rohloff
Composition: Lumina Datamatics, Inc.
Printing and Binding: LSC Communications
Cover and Title Page Illustration: S. Klewitz-Seemann/AGE Fotostock

Library of Congress Control Number: 2017957157

ISBN-13: 978-1-319-10722-2
ISBN-10: 1-319-10722-2

PRINTED IN THE UNITED STATES OF AMERICA

First printing
W. H. Freeman and Company
One New York Plaza
Suite 4500
New York, NY 10004-1562
www.macmillanlearning.com

To the students who enroll in my genetics class each year and continually inspire me with their intelligence, curiosity, and enthusiasm

Brief Contents

Contents

Letter from the Author

Marlene Tyrrell

Genetics is among the most exciting and important biology courses that you will take. Almost daily, we see examples of the relevance of genetics: the discovery of genes that influence human diseases, traits, and behaviors; the use of DNA testing to trace disease transmission and to solve crimes; the use of genetic technology to develop new products. And today, genetics is particularly important to the student of biology, serving as the foundation for many biological concepts and processes. It is truly an exciting time to be learning genetics!

When writing this book, I tried to imagine that I was sitting with a small group of students, having a conversation about genetics. My goal as the author of *Genetics Essentials: Concepts and Connections* is to have that conversation with you. I want to become a trusted guide on your journey through introductory genetics. In this book, I've tried to share some of what I've learned in my 37 years of teaching genetics. I provide advice and encouragement at places where students often have difficulty, and I tell stories of the people, places, and experiments of genetics—past and present—to keep the subject relevant, interesting, and alive. My goal is to help you learn the necessary details, concepts, and problem-solving skills while encouraging you to see the elegance and beauty of the larger landscape.

Genetics Essentials: Concepts and Connections was written in response to requests from instructors and students for a more streamlined and focused genetics textbook that covers less content than a full-length genetics textbook. It has as its foundation my more comprehensive *Genetics: A Conceptual Approach*, which is now in its sixth edition.

At Southwestern University, my office door is always open, and my own students frequently drop by to share their own approaches to learning as well as their experiences, concerns, and triumphs. I would love to hear from you—by email (pierceb@southwestern.edu), by telephone (512-863-1974), or in person (Southwestern University, Georgetown, Texas).

Ben Pierce
Professor of Biology
and holder of the Lillian Nelson Pratt Chair,
Southwestern University

Preface

Welcome to the fourth edition of *Genetics Essentials: Concepts and Connections*, a brief genetics textbook designed specifically for your one-semester course. The title *Genetics Essentials: Concepts and Connections* conveys the major goals of the book: to help students uncover major concepts of genetics and make connections among those concepts so as to have a fuller understanding of genetics. The text maintains the features that made the first three editions so effective: simple and instructive illustrations, accessible writing style, a strong emphasis on problem solving, and useful pedagogical features throughout the book.

Hallmark Features

- **Key Concepts and Connections** Throughout the book, I've included features to help students focus on the major concepts of each topic.
    - ***Concepts boxes*** throughout each chapter summarize the key points of the preceding section. ***Concept Checks*** allow students to quickly assess their understanding of the material they've just read. Concept Checks are in multiple-choice or short-answer format, and their answers are given at the end of each chapter.
    - ***Connecting Concepts*** sections compare and contrast processes or integrate ideas across sections and chapters to help students see how different genetics topics relate to one another. All major concepts in each chapter are listed in the ***Concepts Summary*** at the end of the chapter.

> **CONCEPTS**
>
> The Punnett square is a shorthand method of predicting the genotypic and phenotypic ratios of progeny from a genetic cross.
>
> ✓ CONCEPT CHECK 4
>
> If the F_1 plant depicted in Figure 3.5 is backcrossed to the parent with round seeds, what proportion of the progeny will have wrinkled seeds? (Use a Punnett square.)
>
> a. $\frac{3}{4}$ b. $\frac{1}{2}$
> c. $\frac{1}{4}$ d. 0

- **Accessibility** The conversational writing style of this book has always been a favorite feature for both students and instructors. In addition to carefully walking students through each major concept of genetics, I invite them into each chapter with an **introductory story**. These stories include relevant examples of diseases or other biological phenomena, to give students a sample of what they'll be learning in a chapter. More than a third of the introductory stories in this edition are new.
- **Clear, Simple Illustration Program** The attractive and instructive figures have proved to be an effective learning tool for students throughout the past three editions and continue to be a signature feature of the new edition. Each figure has been carefully rendered to highlight main points and to step the reader through experiments and processes. Most figures include text that walks students through the graphical presentation. Illustrations of experiments reinforce the scientific method by first proposing a hypothesis, then pointing out the methods and results, and ending with a conclusion that reinforces concepts explained in the text.
- **Emphasis on Problem Solving** One of the things that I've learned in my 37 years of teaching is that students learn genetics best through problem solving. Working through an example, equation, or experiment helps students see concepts in action and reinforces the ideas explained in the text. In the book, I help students develop problem-solving skills in a number of ways. **Worked Problems** guide students through each step of a difficult concept. **Problem Links** spread throughout each chapter point to end-of-chapter problems that students can work to test their understanding of the material they have just read, all with answers in the back of the book so that students can check their

> **35.** E. W. Lindstrom crossed two corn plants with green seedlings and obtained the following progeny: 3583 green seedlings, 853 virescent-white seedlings, and 260 yellow seedlings (E. W. Lindstrom. 1921. *Genetics* 6:91–110).
>
> **a.** Give the genotypes for the green, virescent-white, and yellow progeny.
>
> **b.** Explain how color is determined in these seedlings.
>
> **c.** Does epistasis take place among the genes that determine color in the corn seedlings? If so, which gen is epistatic and which is hypostatic?

results. I provide a wide range of end-of-chapter problems, organized by chapter section and split into Comprehension Questions, Application Questions and Problems, and Challenge Questions. Some of these questions, marked by a data analysis icon, draw on examples from published, research articles.

New to the Fourth Edition

NEW **SaplingPlus for *Genetics: Essentials Concepts and Connections*** The fourth edition is now fully supported in SaplingPlus. This comprehensive and robust online teaching and learning platform incorporates online homework with the e-Book, all instructor and student resources, and powerful Gradebook functionality. Students benefit from just-in-time hints and feedback specific to common misconceptions, while instructors benefit from automatically graded homework and robust Gradebook diagnostics.

NEW **Active learning components** One of my main goals for this new edition is to provide better resources for active learning in the classroom. In this edition, I have added Think-Pair-Share questions, which require students to work, and learn, in groups. These questions not only focus on the genetics topics covered in the chapter, but also tie them to genetics in medicine, agriculture, and other aspects of human society. Found at the end of each chapter, the Think-Pair-Share questions cover both the chapter opening story and key topics throughout the chapter itself. An online instructor guide provides resources for instructors leading the in-class discussion of these questions.

THINK-PAIR-SHARE QUESTIONS

Introduction

1. Propose some possible reasons why mutations in the *RPSA* gene affect only the spleen and not other tissues where ribosomes carry out translation.
2. What are some possible reasons that researchers might be interested in identifying the gene that causes a genetic disease such as ICA? In other words, what benefits might result from this research?

- ***Think-Pair-Share Questions*** for the chapter opening story get students to discuss the chapter opening story itself and to connect it with what they know about genetics.
- ***Think-Pair-Share Questions*** on key topics provide more challenging problem solving for students to work on in groups, and encourage them to discuss the bigger-picture aspects of the material they learned in the chapter. They also allow students to connect the material they have learned to broader genetics topics.

New and Reorganized Content

The fourth edition addresses recent discoveries in genetics corresponding to our ever-changing understanding of inheritance, the molecular nature of genetic information, epigenetics, and genetic evolution. This edition also focuses on updating the new research techniques that have become available to geneticists in the past few years. For example, I have expanded coverage of CRISPR-Cas systems and reorganized the chapter on molecular genetic analysis.

New and updated content includes

- New section on bacterial defense mechanisms, including CRISPR (Chapter 7)
- Expanded discussion of replication licensing, including some of the specific molecules involved. (Chapter 9)
- New section on CRISPR RNA (Chapter 10)
- Significant reorganization to focus on methods currently in use; significant updates on new technologies; new section on CRISPR-Cas genome editing (Chapter 14)
- Updated methods in genomics (Chapter 15)

NEW Introductory Stories Each chapter begins with a brief **introductory story** that illustrates the relevance of a genetic concept that students will learn in the chapter. These stories—a favorite feature of past editions—give students a glimpse of what's going on in the field of genetics today and help to draw the reader into the chapter. Among new introductory story topics are "The Genetics of Blond Hair in the South Pacific," "The Genetics of Medieval Leprosy," "Editing the Genome with CRISPR-Cas9," "Building a Chromosome for Class," and "The Wolves of Isle Royale." End-of-chapter problems specifically address concepts discussed in many of the introductory stories, both old and new.

3 Basic Principles of Heredity

The Genetics of Blond Hair in the South Pacific

A thousand miles northeast of Australia lies an ancient chain of volcanic and coral islands known as the Solomons (Figure 3.1). The Solomon Islands were first inhabited some 30,000 years ago, when Neanderthals still roamed northern Europe. Today, the people of the Solomons are culturally diverse, but consist largely of Melanesians, a group that also inhabits other South Pacific islands. Most people from the Solomon Islands have dark skin. Remarkably, 5% to 10% also have strikingly blond hair; in fact, people of the Solomon Islands have the highest frequency of blond hair outside of Europe.

How did the Solomon Islanders get their blond hair? A number of hypotheses have been proposed over the years. Some suggested that the blond islanders had naturally dark hair that was bleached by the sun and salt water. Others proposed that the blond hair color was caused by diet. Still others suggested that it was the result of genes for blond hair left by early European explorers.

The mystery of the blond Solomon Islanders was solved in 2012 by geneticists Eimear Kenny and Sean Myles and their

Blond hair occurs in 5% to 10% of dark-skinned Solomon Islanders. Research demonstrates that blond hair in this group is a recessive trait and has a different genetic basis from blond hair in Europeans. [© Anthony Asael/Danita Delimont/

Media and Supplements

For this edition, we have thoroughly revised and refreshed the extensive set of online learning tools for *Genetics Essentials: Concepts and Connections*. All of the new media resources for this edition will be available in our new SaplingPlus system.

SaplingPlus is a comprehensive and robust online teaching and learning platform that also incorporates all instructor resources and Gradebook functionality.

Student Resources in SaplingPlus for *Genetics Essentials: Concepts and Connections*

SaplingPlus provides students with media resources designed to enhance their understanding of genetic principles and improve their problem-solving ability.

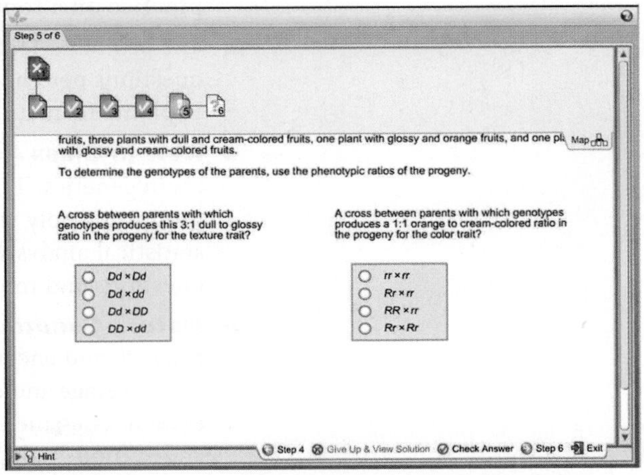

- **Detailed Feedback for Students** Homework questions include hints, wrong-answer feedback targeted to students' misconceptions, and fully worked out solutions to reinforce concepts and to build problem-solving skills.
- **The e-Book** The e-Book contains the full contents of the text as well as embedded links to important media resources (listed following).
- **Updated and New Problem-Solving Videos** offer students valuable help by reviewing basic problem-solving strategies. The problem-solving videos demonstrate an instructor working through problems that students find difficult in a step-by-step manner.

- **New Online Tutorials** identify where students have difficulty with a problem and route them through a series of steps to reach the correct answer. Hints and feedback at every step guide students along the way, as if they were working the problem with an instructor. Complete solutions are also included.

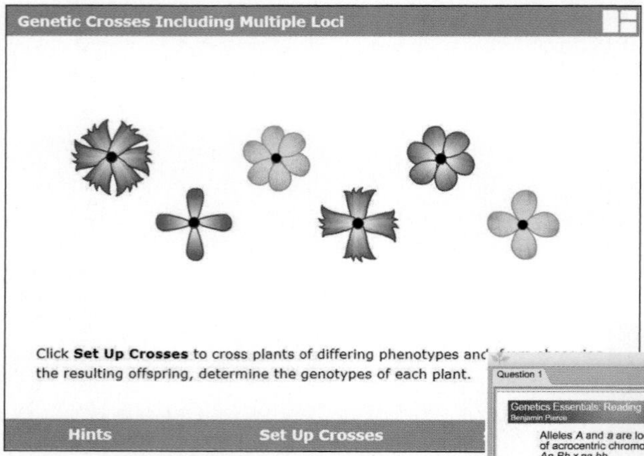

- **Updated and New Animations/Simulations** help students understand key processes in genetics by outlining them in a step-by-step manner. All of the animations and simulations include assessment questions to help students evaluate whether they understood the concept or technique they viewed.
- **Comprehensively Revised Assessment** All media resources have undergone extensive rewriting, reviewing, and accuracy checking.
- **Online Reading Quizzes,** covering the key concepts in each chapter, allow instructors to assess student preparedness before class and to identify challenging areas.

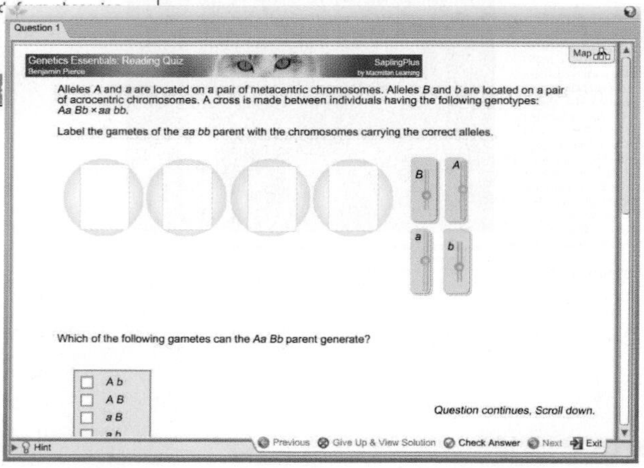

- **New Online Homework.** SaplingPlus offers robust, high-level homework questions with hints and wrong-answer feedback targeted to students' misconceptions as well as detailed worked-out solutions to reinforce concepts. Online Homework includes select end-of-chapter Application Problems from the text, converted into a variety of auto-graded formats. It also includes a variety of Sapling Genetics questions, curated for alignment with the text. These questions can also be used for quizzing or student practice. The questions are tagged by difficulty level.
- **LearningCurve** allows students to test their comprehension of the chapter concepts with questions similar to the Comprehension Questions at the end of each chapter. The system adapts to each student's individual level of preparedness by giving them questions at varying levels of difficulty depending on whether they answer a question without any help, if they need help but eventually get the question right, or if they are unable to answer the question. Links to the appropriate e-Book section, hints, and feedback help students realize where they need more practice on a topic.
- The printable **Test Bank** contains at least 50 multiple-choice and short-answer questions per chapter. The Test Bank questions are also available in a downloadable Diploma format.
- **New In-Class Activities** contribute to active learning of more challenging topics in genetics. Ten activities (15–45 minutes in length) allow students to work in groups to apply what they have learned to problems ranging from gene mapping to statistical analysis to interpreting phylogenetic trees. Each activity includes clicker questions and multiple-choice assessment questions.
- ***Nature Genetics* Articles with Assessment** engage students with primary research and encourage critical thinking. Specifically selected for both alignment with text coverage and exploration of identified difficult topics, the *Nature Genetics* articles include assessment questions that can be automatically graded. Some of the open-ended (non-multiple-choice) questions are also suitable for use in flipped classrooms and active learning discussions either in class or online.

Instructor Resources in SaplingPlus for *Genetics Essentials: Concepts and Connections*

- **Updated Clicker Questions** allow instructors to integrate active learning into the classroom and to assess students' understanding of key concepts during lectures. Available in PowerPoint format, numerous questions are based on the Concept Check questions featured in the textbook.
- **Updated Lecture PowerPoint Files** have been developed to minimize preparation time for new users of the book. These files offer suggested lectures, including key illustrations and summaries, that instructors can adapt to their teaching styles.
- **Layered PowerPoint Slides** deconstruct key concepts, sequences, and processes from the textbook illustrations, allowing instructors to present complex ideas step by step.
- **Textbook Illustrations and Tables** are offered as high-resolution JPEG files. Each image has been fully optimized to increase type sizes and adjust color saturation. These images have been tested in a large lecture hall to ensure maximum clarity and visibility. Images are presented in both labeled and unlabeled formats.
- The **Solutions and Problem-Solving Manual** (written by Jung Choi and Mark McCallum) contains complete answers and worked-out solutions to all questions and problems in the textbook. The Solutions Manual is also available in print (ISBN: 9781319200619).

Acknowledgments

Every day as I enter the classroom, I feel a rush of excitement in anticipation of exploring with my students the concepts, knowledge, and questions that constitute the field of genetics. For me, teaching genetics is never dull or routine. The field is constantly changing, with important new advances occurring weekly, and the information is interesting and relevant. But mostly, my pleasure in teaching genetics derives from the students who have filled my classes, whose energy, intelligence, curiosity, and humor have been a source of inspiration and motivation for the past 37 years. I have also learned from students worldwide who have used earlier editions of this book and kindly shared with me—through emails and phone calls—their thoughts about the book and how it could be improved. I thank my own teachers, Dr. Raymond Canham and Dr. Jeffrey Mitton, for introducing me to genetics and showing me how to teach and do scholarship.

I am indebted to Southwestern University for providing a stimulating environment in which quality teaching and research flourish. My daily life is greatly enriched by my colleagues in the Biology Department, who sustain me with their friendship, collegiality, and advice. I am grateful to Edward Burger, President of Southwestern University and Alisa Gaunder, Dean of the Faculty, for helping to create this supportive academic environment and for their friendship and encouragement.

Writing a modern science textbook requires a team effort, and I have been blessed with an outstanding team at W. H. Freeman and Macmillan Learning. Managing Director Susan Winslow has been a champion of the book for a number of years; I value her support, strategic vision, and commitment to education. Vice President for STEM Ben Roberts guided the project throughout its development. Lauren Schultz, has been a great project leader. She has been a good friend and colleague, as well as a continual source of encouragement, support, and creative ideas. I greatly enjoyed working daily with Development Editor Maria Lokshin. She has great knowledge of genetics and a passion for excellence, as well as superior organizational skills and good humor. Working with her was always fun and inspiring. I am also grateful to Lisa Samols, Director of Development, for shepherding the development of this edition and for great insight at key points.

Project manager Vanavan Jayaraman expertly managed the production of this fourth edition. I thank Dragonfly Media Group for creating and revising the book's illustrations and Janice Donnola for coordinating the illustration program. John Callahan designed the cover image. Thanks to Paul Rohloff at W. H. Freeman and Vanavan Jayaraman at Lumina Datamatics, Inc. for coordinating the composition and manufacturing phases of production. Jerry Wilke developed the book's design. I thank Christine Buese and Richard Fox for photo research. Amy Thorne, Cassandra Korsvik, Amber Jonker, Clairissa Simmons, Amanda Nietzel, and Elaine Palucki developed the excellent media and supplements that accompany the book. I am grateful to Jung Choi and Mark McCallum for writing solutions to the end-of-chapter problems. Robert Fowler, Marcie Moehnke, Ellen France, Amy McMillan, Daniel Williams, Douglas Thrower, Victor Fet, and Usha Viveganathan developed and reviewed assessment questions.

As always, I am grateful to the Macmillan Learning sales representatives, regional managers, and STEM specialists, who introduce my book to genetic instructors throughout the world. I have greatly enjoyed working with this sales staff; their expertise, hard work, and good service are responsible for the success of Macmillan books.

A number of colleagues served as reviewers of this book, kindly lending me their technical expertise and teaching experience. Their assistance is gratefully acknowledged. Any remaining errors are entirely my own.

Marlene Tyrrell—my spouse and best friend for 37 years—our children and their spouses—Sarah, Matt, Michael, and Amber—and now my F_2 progeny—Ellie, Beckett, and Caroline—provide love, support, and inspiration for everything I do.

My gratitude goes to the reviewers of *Genetics Essentials:*

Jeanne M. Andreoli
Marygrove College

Catalina Aragno
Saint Joseph's University

Nicanor Austriaco
Providence College

Lori Bergeron
New England College

Leah Brault
Becker College

Jill A. Buettner
Richland College

Alyssa Bumbaugh
Pennsylvania State University

Douglas Burks
Wilmington College

James Carpenter
Barton College

Helen Chamberlain
Ohio State University

Craig Coleman
Brigham Young University

Erin J. Cram
Northeastern University

James Daniels
Huntingdon College

Claudette Davis
George Mason University

David Donnell
The Citadel

Edward Eivers
California State University, Los Angeles

William Gilliland
DePaul University

Kurt Elliott
Northwest Vista College

Carina Enders Howell
Lock Haven University

Cheryld Emmons
Alfred University

Silviu Faitar
D'Youville College

Victor Fet
Marshall University

Christine Fleet
Emory and Henry College

Jeffrey O. French
North Greenfield University

Julie Torruellas Garcia
Nova Southwestern University

Matthew Gilg
University of North Florida

James Godde
Monmouth College

John Gray
University of Toledo

Ruth Grene
Virginia Tech

Idit Hazan
Grand View University

Xiaotang Hu
Barry University

Tina Hubler
University of North Alabama

Diana Ivankovic
Anderson University

Kristi Jones
Huntingdon College

David Kass
Eastern Michigan University

Kiran Kaur
El Centro College

Brian Kreiser
University of Southern Mississippi

Traci Lee
University of Wisconsin–Parkside

Melanie J. Lee-Brown
Guilford College

Jane Lopilato
Simmons College

Michelle Mabry
Davis and Elkins College

Clint Magill
Texas A & M University–College Station

Endre Mathe
University of Debrecen & Vasile Goldis University of Arad, Romania

Vida Mingo
Columbia College

Srinidi Mohan
University of New England

Shaheed Mukhtar
University of Alabama at Birmingham

Jessica L. Moore
West Carolina University

Richard D. Noyes
University of Central Arkansas

Gary Ogden
St. Mary's University

Henry Owen
Eastern Illinois University

Jason Rauceo
John Jay College of Criminal Justice

Jason Rawlings
Furman University

Eugenia Ribeiro-Hurley
Fordham University–Lincoln Center

Brian C. Ring
Valdosta State University

Michael Robinson
Miami University

Aaron Schirmer
Northeastern Illinois University

Brian W. Schwartz
Columbus State University

Mark Shotwell
Slippery Rock University of Pennsylvania

Wendy Shuttleworth
Lewis-Clark State College

Agnes-Southgate
College of Charleston

Walter Sotero
University of Central Florida

Ron Strohmeyer
Northwest Nazarene University

Tatiana Tatum Parker
Saint Xavier University

Fernando Tenjo
Virginia Commonwealth University

Kathleen Toedt
Housatonic Community College

Lauren Tranter
Corning Community College

Michelle Wien
Bryn Mawr College

Daniel Williams
Winston-Salem State University

Steven D. Wilt
Bellarmine University

Jacqueline Wittke-Thompson
University of St. Francis

Eric J. Yager
Albany College of Pharmacy and Health Sciences

Taek You
Campbell University

Michael Young
Concordia University

1 Introduction to Genetics

Albinism among the Hopis

A Hopi pueblo on Black Mesa. Albinism, a genetic condition, arises with high frequency among the Hopi people and occupies a special place in the Hopi culture. [Ansel Adams/National Park Archive at College Park, MD.]

Rising a thousand feet above the desert floor, Black Mesa dominates the horizon of the Enchanted Desert and provides a familiar landmark for travelers passing through northeastern Arizona. Black Mesa is not only a prominent geological feature: more significantly, it is the ancestral home of the Hopi Native Americans. Fingers of the mesa reach out into the desert, and alongside or on top of each finger is a Hopi village. Most of the villages are quite small, with only a few dozen inhabitants, but they are incredibly old. One village, Oraibi, has existed on Black Mesa since 1150 A.D. and is the oldest continually occupied settlement in North America.

In 1900, Alěs Hrdliěka, an anthropologist and physician working for the American Museum of Natural History, visited the Hopi villages of Black Mesa and reported a startling discovery. Among the Hopis were 11 white people—not Caucasians, but white Hopi Native Americans. These people had a genetic condition known as albinism (**Figure 1.1**).

Albinism is caused by a defect in one of the enzymes required to produce melanin, the pigment that darkens our skin, hair, and eyes. People with albinism either don't produce melanin or produce only small amounts of it, and consequently, have white hair, light skin, and no pigment in the irises of their eyes. Melanin normally protects the DNA of skin cells from the damaging effects of ultraviolet radiation in sunlight, and melanin's presence in the developing eye is essential for proper eyesight.

The genetic basis of albinism was first described by the English physician Archibald Garrod, who recognized in 1908 that the condition was inherited as an autosomal recessive trait, meaning that a person must receive two copies of an albino mutation—one from each parent—to have albinism. In recent years, the molecular natures of the mutations that lead to albinism have been elucidated. Albinism in humans is caused by defects in any one of several different genes that control the synthesis and storage of melanin. Many different types of mutations can occur in each gene, any one of which may lead to albinism. The form of albinism found among the Hopis is most likely oculocutaneous albinism (albinism affecting the eyes and skin) type 2, caused by a defect in the *OCA2* gene on chromosome 15.

The Hopis are not unique in having people with albinism among the members of their tribe. Albinism is found in almost all human ethnic groups and is described in ancient writings: it has probably been present since humankind's beginnings. What is unique about the Hopis is the high frequency of albinism in their population. In most

1.1 Albinism among the Hopi Native Americans. The Hopi girl in the middle of this photograph, taken about 1900, has albinism. [©The Field Museum, #CSA118. Charles Carpenter.]

human groups, albinism is rare, present in only about 1 in 20,000 people. In the villages on Black Mesa, it reaches a frequency of 1 in 200, a hundred times greater than in most other populations.

Why is albinism so frequent among the Hopis? The answer to this question is not completely known, but geneticists who have studied albinism among the Hopis speculate that the high frequency of the albino gene is at least partly related to the special place that albinism occupied in the Hopi culture. For much of their history, the Hopis considered members of their tribe with albinism to be important and special. People with albinism were considered pretty, clean, and intelligent. Having a number of people with albinism in one's village was considered a good sign, a symbol that the people of the village contained particularly pure Hopi blood. Members of the tribe with albinism performed in Hopi ceremonies and held positions of leadership, often as chiefs, healers, and religious leaders.

Hopis with albinism were also given special treatment in everyday activities. The Hopis have farmed small garden plots at the foot of Black Mesa for centuries. Every day throughout the growing season, the men of the tribe trek to the base of Black Mesa and spend much of the day in the bright southwestern sunlight tending their corn and vegetables. With little or no melanin in their skin, people with albinism are extremely susceptible to sunburn and have increased incidences of skin cancer when exposed to the sun. Furthermore, many don't see well in bright sunlight. Therefore, male Hopis with albinism were excused from farming and allowed to remain behind in the village with the women of the tribe, performing other duties.

Throughout the growing season, the men with albinism were the only male members of the tribe in the village with the women during the day, and thus they enjoyed a mating advantage, which helped to spread their albino genes. In addition, the special considerations given to Hopis with albinism allowed them to avoid the detrimental effects of albinism: increased skin cancer and poor eyesight. The small size of the Hopi tribe probably also played a role by allowing chance to increase the frequency of the albino gene. Regardless of the factors that led to the high frequency of albinism, the Hopis greatly respected and valued the members of their tribe who possessed this particular trait. Unfortunately, people with genetic conditions in many societies are often subject to discrimination and prejudice. ▶TRY PROBLEMS 1 AND 22

THINK-PAIR-SHARE Questions 1 and 2

Genetics is one of the most rapidly advancing fields of science, with important new discoveries reported every month. Look at almost any major news source and chances are that you will see articles related to genetics: the completion of another organism's genome, such as that of the monarch butterfly; the discovery of genes that affect major diseases, including multiple sclerosis, depression, and cancer; analyses of DNA from long-extinct animals such as the woolly mammoth; or the identification of genes that affect skin pigmentation, height, or learning ability in humans. Even among advertisements, you are likely to see ads for genetic testing to determine a person's ancestry, paternity, and susceptibility to diseases and disorders. These new findings and applications of genetics often have significant economic and ethical implications, making the study of genetics relevant, timely, and interesting.

This chapter introduces you to genetics and reviews some concepts that you may have encountered briefly in a preceding biology course. We begin by considering the importance of genetics to each of us, to society, and to students of biology. We then turn to the history of genetics and how the field as a whole developed. The final part of the chapter presents some fundamental terms and principles of genetics that are used throughout the book.

1.1 Genetics Is Important to Us Individually, to Society, and to the Study of Biology

Albinism among the Hopis illustrates the important role that genes play in our lives. This one genetic defect, among the 20,000 genes that humans possess, completely changes the life of a Hopi who possesses it. It alters his or her occupation, role in Hopi society, and relations with other members of the tribe. We all possess genes that influence our lives in significant ways. Genes affect our height, weight, hair color, and skin pigmentation. They influence our susceptibility to many diseases and disorders (**Figure 1.2**) and even contribute to

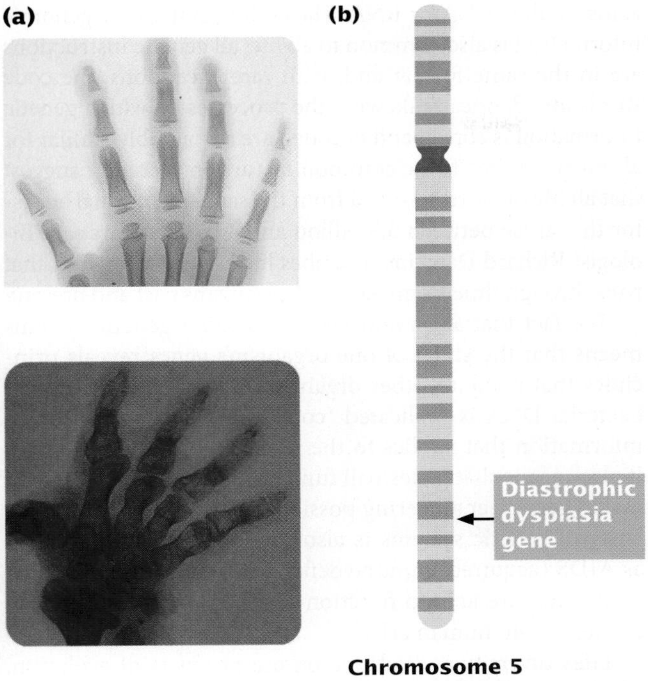

1.2 Genes influence susceptibility to many diseases and disorders. (a) An X-ray of the hand of a person suffering from diastrophic dysplasia (bottom), a hereditary growth disorder that results in curved bones, short limbs, and hand deformities, compared with an X-ray of a normal hand (top). (b) This disorder is due to a defect in the *SLC26A2* gene on chromosome 5. [Part a: (top) Biophoto Associates/Science Source; (bottom) Reprinted from Cell, 78(6) Johanna Hästbacka, et al, The diastrophic dysplasia gene encodes a novel sulfate transporter: Positional cloning by fine-structure linkage disequilibrium mapping, pp. 1073 - 1087, ©1994 with permission from Elsevier. Permission conveyed through Copyright Clearance Center, Inc. Courtesy of Prof. Eric Lander, Whitehead Institute, MIT.]

our intelligence and personality. Genes are fundamental to who and what we are.

Although the science of genetics is relatively new compared with sciences such as astronomy and chemistry, people have understood the hereditary nature of traits and have practiced genetics for thousands of years. The rise of agriculture began when people started to apply genetic principles to the domestication of plants and animals. Today, the major crops and animals used in agriculture are quite different from their wild progenitors, having undergone extensive genetic alterations that increase their yields and provide many desirable traits, such as disease and pest resistance, special nutritional qualities, and characteristics that facilitate harvest. The Green Revolution, which expanded food production throughout the world in the 1950s and 1960s, relied heavily on the application of genetics (**Figure 1.3**). Today, genetically engineered corn, soybeans, and other crops constitute a significant proportion of the food produced worldwide.

The pharmaceutical industry is another area in which genetics plays an important role. Numerous drugs and food additives are synthesized by fungi and bacteria that have been genetically manipulated to make them efficient producers of these substances. The biotechnology industry employs molecular genetic techniques to develop and mass-produce substances of commercial value. Growth hormone, insulin, clotting factor, enzymes, antibiotics, vaccines, and many drugs are now produced commercially by genetically engineered bacteria and other cells (**Figure 1.4**). Genetics has also been used to produce bacteria that remove minerals from ore, break down toxic chemicals, and inhibit damaging frost formation on crop plants.

1.3 In the Green Revolution, genetic techniques were used to develop new high-yielding strains of crops. (Left) Norman Borlaug, a leader in the development of new varieties of wheat that led to the Green Revolution. Borlaug was awarded the Nobel Peace Prize in 1970. (Right) A modern, high-yielding rice plant (left) and a traditional rice plant (right). [Left: © Bettmann/CORBIS. Right: IRRI.]

1.4 The biotechnology industry uses molecular genetic methods to produce substances of economic value. [Reuters/Jerry Lampen]

Genetics also plays a critical role in medicine. Physicians recognize that many diseases and disorders have a hereditary component, including not only rare genetic disorders such as sickle-cell anemia and Huntington disease, but also many common diseases such as asthma, diabetes, and hypertension. Advances in genetics have resulted in important insights into the nature of diseases such as cancer and in the development of diagnostic tests, including those that identify pathogens and defective genes. Gene therapy—the direct alteration of genes to treat human diseases—has now been administered to thousands of patients, although its use is still experimental and limited.

THINK-PAIR-SHARE Question 3

The Role of Genetics in Biology

Although an understanding of genetics is important to all people, it is critical to the student of biology. Genetics provides one of biology's unifying principles: all organisms use genetic systems that have a number of features in common. Genetics also undergirds the study of many other biological disciplines. Evolution, for example, is genetic change that takes place over time, so the study of evolution requires an understanding of genetics. Developmental biology relies heavily on genetics: tissues and organs develop through the regulated expression of genes (**Figure 1.5**). Even such fields as taxonomy, ecology, and animal behavior are making increasing use of genetic methods. The study of almost any field of biology or medicine is incomplete without a thorough understanding of genes and genetic methods.

Genetic Diversity and Evolution

Life on Earth exists in a tremendous array of forms and features in almost every conceivable environment. Life is also characterized by adaptation: many organisms are exquisitely suited to the environment in which they are found. The history of life is a chronicle of new forms of life emerging, old forms disappearing, and existing forms changing.

Despite their tremendous diversity, living organisms have an important feature in common: all use similar genetic systems. The complete set of genetic instructions for any organism is its **genome**. All genomes are encoded in nucleic acids—either DNA or RNA. The coding system for genomic information is also common to all life: all genetic instructions are in the same format and, with rare exceptions, the code words are identical. Likewise, the processes by which genetic information is copied and decoded are remarkably similar for all forms of life. These common features of heredity suggest that all life on Earth evolved from the same primordial ancestor that arose between 3.5 billion and 4 billion years ago. Biologist Richard Dawkins describes life as a river of DNA that runs through time, connecting all organisms past and present.

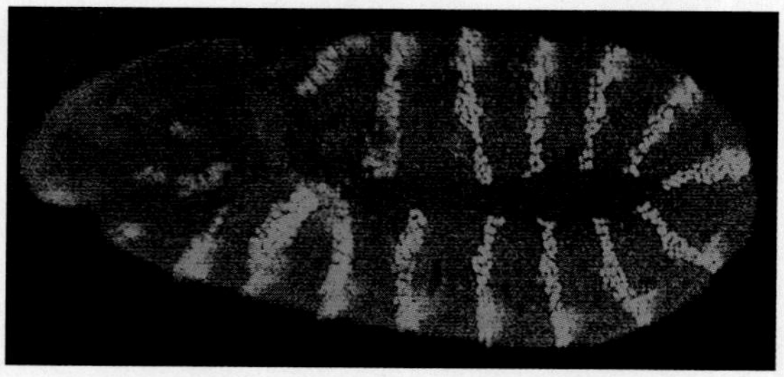

1.5 The key to development lies in the regulation of gene expression. This early fruit-fly embryo illustrates the localized expression (indicated by bright green) of the *engrailed* gene, which helps determine the development of body segments in the adult fly. [Steven Paddock.]

The fact that all organisms have similar genetic systems means that the study of one organism's genes reveals principles that apply to other organisms. Investigations of how bacterial DNA is replicated (copied), for example, provide information that applies to the replication of human DNA. It also means that genes will function in foreign cells, which makes genetic engineering possible. Unfortunately, the similarity of genetic systems is also the basis for diseases such as AIDS (acquired immune deficiency syndrome), in which viral genes are able to function—sometimes with alarming efficiency—in human cells.

Life's diversity and adaptation are products of evolution, which is simply genetic change over time. Evolution is a two-step process: first, inherited differences arise randomly, and then the proportion of individuals with particular differences increases or decreases. Genetic variation is therefore the foundation of all evolutionary change and is ultimately the basis of all life as we know it. Furthermore, techniques of molecular genetics are now routinely used to decipher evolutionary relationships among organisms; for example, recent analysis of DNA isolated from Neanderthal fossils has provided insight into the relationship between Neanderthals and modern humans, demonstrating that Neanderthals and the ancestors of modern humans probably interbred some 30,000 to 40,000 years ago. Genetics, the study of genetic variation, is critical to understanding the past, present, and future of life.

THINK-PAIR-SHARE Question 4

CONCEPTS

Heredity affects many of our physical features as well as our susceptibility to many diseases and disorders. Genetics contributes to advances in agriculture, pharmaceuticals, and medicine and is fundamental to modern biology. All organisms use similar genetic systems, and genetic variation is the foundation of the diversity of all life.

✓ CONCEPT CHECK 1

What are some of the implications of all organisms having similar genetic systems?

a. That all life forms are genetically related
b. That research findings on one organism's gene function can often be applied to other organisms
c. That genes from one organism can often exist and thrive in another organism
d. All of the above

Divisions of Genetics

The study of genetics consists of three major subdisciplines: transmission genetics, molecular genetics, and population genetics (**Figure 1.6**). **Transmission genetics** (also known as classical genetics) encompasses the basic principles of heredity and how traits are passed from one generation to the next. This subdiscipline addresses the relation between chromosomes and heredity, the arrangement of genes on chromosomes, and gene mapping. Here, the focus is on the individual organism—how an individual inherits its genetic makeup and how it passes its genes to the next generation.

Molecular genetics concerns the chemical nature of the gene itself: how genetic information is encoded, replicated, and expressed. It includes the cellular processes of replication, transcription, and translation (by which genetic information is transferred from one molecule to another) and of gene regulation (the processes that control the expression of genetic information). The focus in molecular genetics is the gene—its structure, organization, and function.

Population genetics explores the genetic composition of groups of individuals of the same species (populations) and how that composition changes over time and space. Because evolution is genetic change, population genetics is fundamentally the study of evolution. The focus of this subdiscipline is the group of genes found in a population.

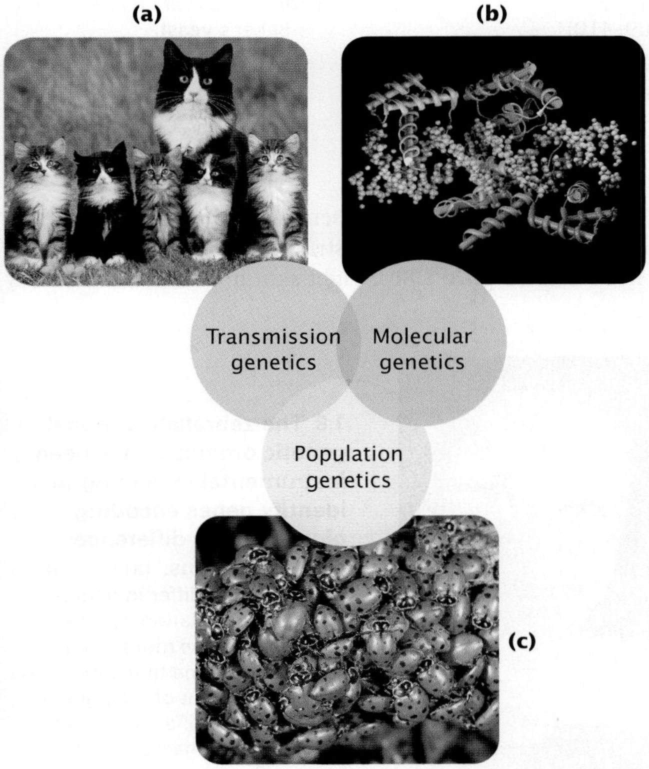

1.6 Genetics can be subdivided into three interrelated fields. [Top left: Juniors Bildarchiv/Alamy. Top right: Martin McCarthy/Getty Images. Bottom: Stuart Wilson/Science Source.]

The division of the study of genetics into these three subdisciplines is convenient and traditional, but we should recognize that these subdisciplines overlap and that each one can be further divided into a number of more specialized fields, such as chromosomal genetics, biochemical genetics, quantitative genetics, and so forth. Alternatively, the study of genetics can be subdivided by organism (fruit-fly, corn, or bacterial genetics), and each of these organisms can be studied at the level of transmission, molecular, or population genetics. Modern genetics is an extremely broad field, encompassing many interrelated subdisciplines and specializations. TRY PROBLEM 17

Model Genetic Organisms

Through the years, genetic studies have been conducted on thousands of different species, including almost all major groups of bacteria, fungi, protists, plants, and animals. Nevertheless, a few species have emerged as **model genetic organisms**: organisms with characteristics that make them particularly useful for genetic analysis and about which a tremendous amount of genetic information has accumulated. Six model organisms that have been the subject of intensive genetic study are *Drosophila melanogaster*, a fruit fly; *Escherichia coli*, a bacterium present in the gut of humans and other mammals; *Caenorhabditis elegans*, a nematode (also called a roundworm); *Arabidopsis thaliana*, the thale cress plant; *Mus musculus*, the house mouse; and *Saccharomyces cerevisiae*, baker's yeast (**Figure 1.7**). These species are the organisms of choice for many genetic researchers, and their genomes were sequenced as a part of the Human Genome Project (described in Chapter 15).

At first glance, this group of lowly and sometimes unappreciated creatures might seem to be unlikely candidates for model organisms. However, all possess traits that make them particularly suitable for genetic study, including a short generation time, large but manageable numbers of progeny, adaptability to a laboratory environment, and the ability to be housed and propagated inexpensively. The life cycles, genomic characteristics, and features that make these model organisms useful for genetic studies are included in special illustrations in later chapters for five of the six species. Other species that are frequently the subject of genetic research and considered model genetic organisms include *Neurospora crassa* (bread mold), *Zea mays* (corn), *Danio rerio* (zebrafish), and *Xenopus laevis* (clawed frog). Although not generally considered a model genetic organism, humans have also been subjected to intensive genetic scrutiny.

The value of model genetic organisms is illustrated by the use of zebrafish to identify genes that affect skin pigmentation in humans. For many years, geneticists have recognized that differences in pigmentation among human ethnic groups (**Figure 1.8a**) are genetic, but the genes causing these differences were largely unknown. The zebrafish has become an

(a) (b) (c)

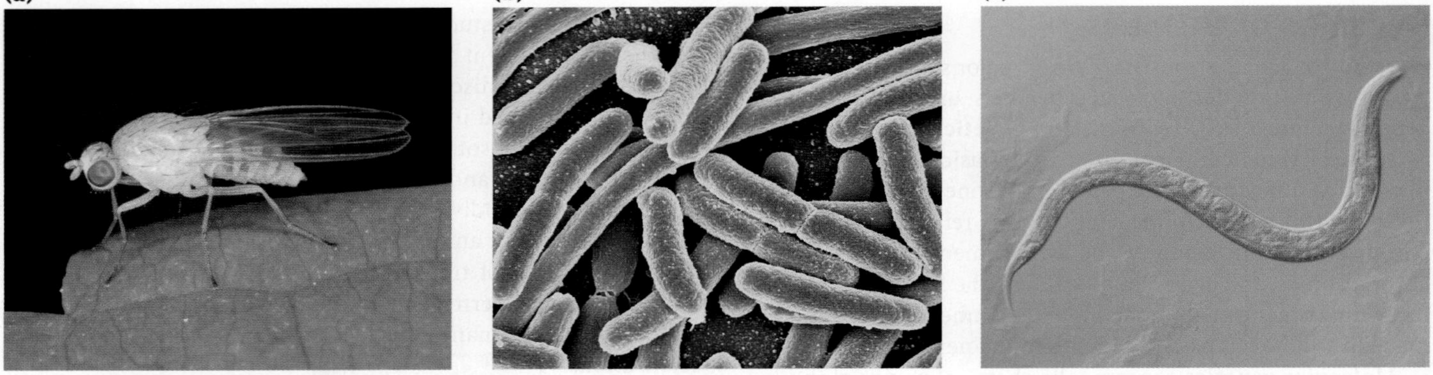

Drosophila melanogaster
Fruit fly (pp. 83–84)

Escherichia coli
Bacterium (pp. 197–198)

Caenorhabditis elegans
Nematode (pp. 291–293)

(d) (e) (f)

Arabidopsis thaliana
Thale-cress plant (pp. 343–345)

Mus musculus
House mouse (pp. 409–410)

Saccharomyces cerevisiae
Baker's yeast

1.7 Model genetic organisms are species with features that make them useful for genetic analysis. Organisms (a) through (e) are discussed in more detail on the pages referred to below each. [Part a: Alfred Schauhuber / imageBROKER / Alamy Stock Photo. Part b: Pasieka/Science Source. Part c: Sinclair Stammers/Science Source. Part d: Peggy Greb/ARS/USDA. Part e: AP Photo/Joel Page. Part f: Biophoto Associates/Science Source.]

important model in genetic studies because it is a small vertebrate that produces many offspring and is easy to rear in the laboratory. The mutant zebrafish called *golden* has light pigmentation due to the presence of fewer, smaller, and less dense pigment-containing structures called melanosomes in its cells (**Figure 1.8b**). Light skin in humans is similarly

(a)

(b)

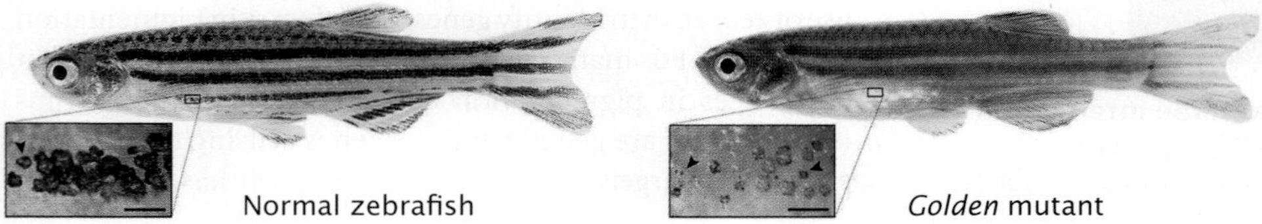

1.8 The zebrafish, a model genetic organism, has been instrumental in helping to identify genes encoding pigmentation differences among humans. (a) Human ethnic groups differ in degree of skin pigmentation. (b) The zebrafish *golden* mutation is caused by a gene that controls the amount of melanin in melanosomes. [Part a: Barbara Penoyar/Getty Images; Amos Morgan/Getty Images; Stockbyte/Getty Images. Part b: Keith Cheng/Jake Gittlen, Cancer Research Foundation Penn State College of Medicine.]

due to fewer and less dense melanosomes in pigment-containing cells.

Keith Cheng and his colleagues hypothesized that light skin in humans might result from a mutation that is similar to the *golden* mutation in zebrafish. Taking advantage of the ease with which zebrafish can be manipulated in the laboratory, they isolated and sequenced the gene responsible for the *golden* mutation and found that it encodes a protein that takes part in calcium uptake by melanosomes. They then searched a database of all known human genes and found a similar gene called *SLC24A5*, which encodes the same function in human cells. When they examined human populations, they found that light-skinned Europeans typically possess one form of this gene, whereas darker-skinned Africans, East Asians, and Native Americans usually possess a different form of the gene. Many other genes also affect pigmentation in humans, as illustrated by mutations in the *OCA2* gene that produce albinism among the Hopis (discussed in the introduction to this chapter). Nevertheless, *SLC24A5* appears to be responsible for 24% to 38% of the differences in pigmentation between Africans and Europeans. This example illustrates the power of model organisms in genetic research. However, we should not forget that all organisms possess unique characteristics and that sometimes the genetics of models do not accurately reflect the genetic systems of other organisms.

CONCEPTS

The three major divisions of genetics are transmission genetics, molecular genetics, and population genetics. Transmission genetics examines the principles of heredity; molecular genetics deals with the gene and the cellular processes by which genetic information is transferred and expressed; population genetics concerns the genetic composition of groups of organisms and how that composition changes over time and space. Model genetic organisms are species that have received special emphasis in genetic research: they have characteristics that make them useful for genetic analysis.

✓ CONCEPT CHECK 2

Would the horse make a good model genetic organism? Why or why not?

1.2 Humans Have Been Using Genetics for Thousands of Years

Although the science of genetics is young—almost entirely a product of the past 100 years or so—people have been using genetic principles for thousands of years.

The Early Use and Understanding of Heredity

The first evidence that people understood and applied the principles of heredity in earlier times is found in the domestication of plants and animals, which began between approximately 10,000 and 12,000 years ago in the Middle East. The first domesticated organisms included wheat, peas, lentils, barley, dogs, goats, and sheep (**Figure 1.9a**). By 4000 years ago, sophisticated genetic techniques were already in use in the Middle East. The Assyrians and Babylonians developed several hundred varieties of date palms that differed in fruit size, color, taste, and time of ripening (**Figure 1.9b**). Other crops and domesticated animals were developed by cultures in Asia, Africa, and the Americas in the same period.

The ancient Greeks gave careful consideration to human reproduction and heredity. Greek philosophers developed

(a) (b)

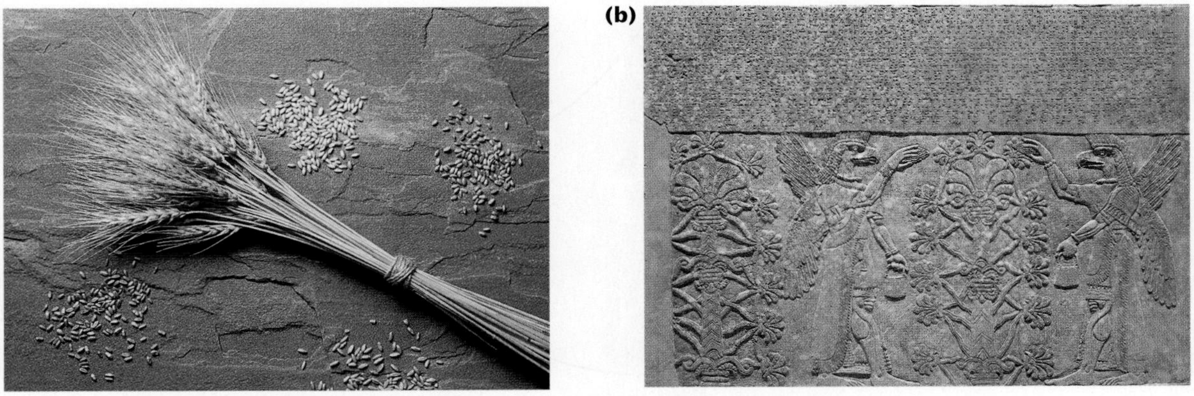

1.9 Ancient peoples practiced genetic techniques in agriculture. (a) Modern wheat, with larger and more numerous seeds that do not scatter before harvest, was produced by interbreeding at least three different wild species. (b) Assyrian bas-relief sculpture showing artificial pollination of date palms at the time of King Assurnasirpalli II, who reigned from 883 to 859 B.C. [Part a: Scott Bauer/ARS/USDA. Part b: Image copyright © The Metropolitan Museum of Art. Image source: Art Resource, NY.]

the concept of **pangenesis**. This concept suggested that specific pieces of information travel from various parts of the body to the reproductive organs, from which they are passed to the embryo (**Figure 1.10a**). Pangenesis led the ancient Greeks to propose the notion of the **inheritance of acquired characteristics**, in which traits acquired in a person's lifetime become incorporated into that person's hereditary information and are passed on to offspring; for example, people who developed musical ability through diligent study would produce children who are innately endowed with musical ability. Although incorrect, these ideas persisted through the twentieth century.

Additional developments in our understanding of heredity occurred during the seventeenth century. Dutch eyeglass makers began to put together simple microscopes in the late 1500s, enabling Robert Hooke (1635–1703) to discover cells in 1665. Microscopes provided naturalists with new and exciting vistas on life, and perhaps it was excessive enthusiasm for this new world of the very small that gave rise to the idea of **preformationism**. According to preformationism, inside the egg or sperm there exists a fully formed miniature adult, a *homunculus*, which simply enlarges during development (**Figure 1.11**). Preformationism meant that all traits were inherited from only one parent—from the father if the homunculus was in the sperm or from the mother if it was in the egg. Although many observations suggested that offspring possess a mixture of traits from both parents, preformationism remained a popular concept throughout much of the seventeenth and eighteenth centuries.

Another early notion of heredity was **blending inheritance**, which proposed that the traits of offspring are a blend, or mixture, of parental traits. This idea suggested that the genetic material itself blends, much as blue and yellow pigments blend to make green paint. It also suggested that after having been blended, genetic differences could not be separated in future generations, just as green paint cannot be separated into blue and yellow pigments. Some traits do *appear* to exhibit blending inheritance; however, thanks to Gregor Mendel's research with pea plants, we now understand that individual genes do not blend.

The Rise of the Science of Genetics

In 1676, Nehemiah Grew (1641–1712) reported that plants reproduce sexually by using pollen from the male sex cells. With this information, a number of botanists, including Gregor Mendel (1822–1884; **Figure 1.12**), began to experiment with crossing plants and creating hybrids. Mendel went on to discover the basic principles of heredity in the 1860s.

Developments in cytology (the study of cells) in the 1800s had a strong influence on genetics. Building on the work of others, Matthias Jacob Schleiden (1804–1881) and Theodor

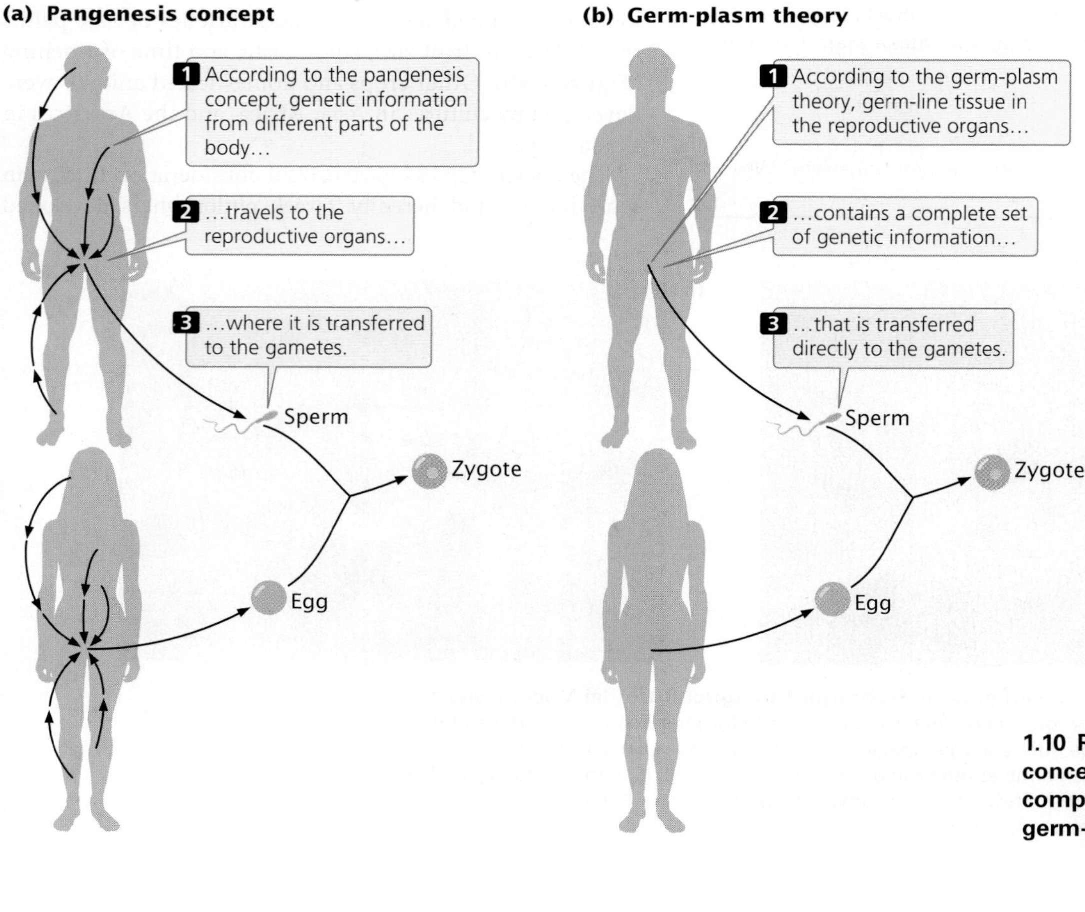

1.10 Pangenesis, an early concept of inheritance, compared with the modern germ-plasm theory.

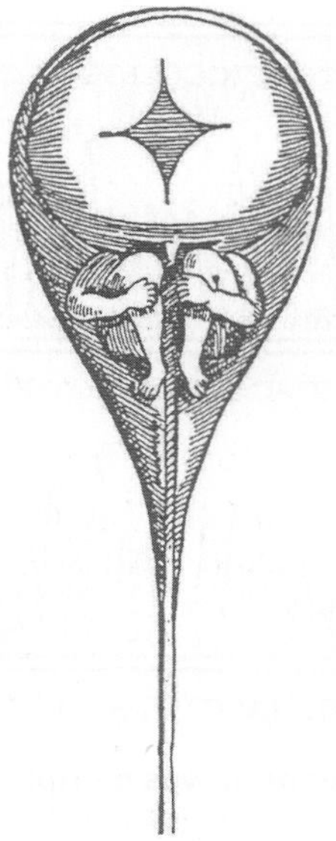

1.11 Preformationists in the seventeenth and eighteenth centuries believed that a sperm or an egg contains a fully formed human (the homunculus). Shown here is a drawing of a homunculus inside a sperm. [Science Source.]

Schwann (1810–1882) proposed the **cell theory** in 1839. According to this theory, all life is composed of cells, cells arise only from preexisting cells, and the cell is the fundamental unit of structure and function in living organisms.

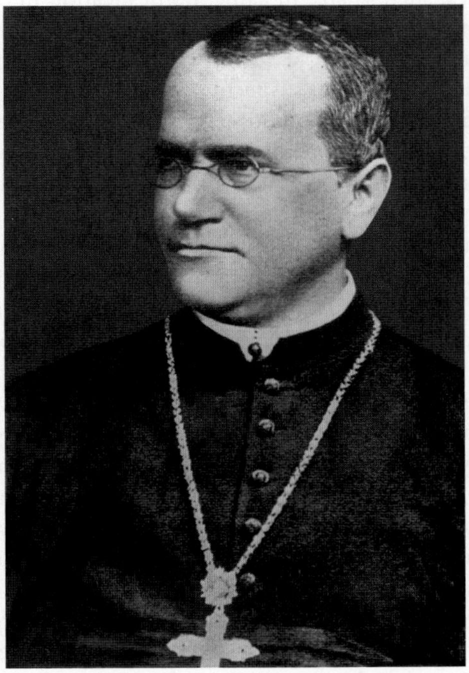

1.12 Gregor Mendel was the founder of modern genetics. Mendel first discovered the principles of heredity by crossing different varieties of pea plants and analyzing the pattern of the transmission of traits in subsequent generations. [Hulton Archive/Getty Images.]

Biologists interested in heredity began to examine cells to see what takes place in the course of cell reproduction. Walther Flemming (1843–1905) observed the division of chromosomes in 1879 and published a superb description of mitosis. By 1885, biologists generally recognized that the cell nucleus contains the hereditary information.

Charles Darwin (1809–1882), one of the most influential biologists of the nineteenth century, put forth the theory of evolution through natural selection and published his ideas in *On the Origin of Species* in 1859. Darwin recognized that heredity was fundamental to evolution, and he conducted extensive genetic crosses with pigeons and other organisms. He never understood the nature of inheritance, however, and this lack of understanding was a major omission in his theory of evolution.

In the last half of the nineteenth century, cytologists demonstrated that the nucleus had a role in fertilization. Near the close of the nineteenth century, August Weismann (1834–1914) finally laid to rest the notion of the inheritance of acquired characteristics. He cut off the tails of mice for 22 consecutive generations and showed that the tail length in descendants remained stubbornly long. Weismann proposed the **germ-plasm theory**, which holds that the cells in the reproductive organs carry a complete set of genetic information that is passed to the egg and sperm (**Figure 1.10b**). This theory, and some of the other early theories of heredity that we have discussed up to this point, are summarized in **Table 1.1.**

TABLE 1.1 Early concepts of heredity

Concept	Proposed	Correct or incorrect
Pangenesis	Genetic information travels from different parts of the body to reproductive organs.	Incorrect
Inheritance of acquired characteristics	Acquired traits become incorporated into hereditary information.	Incorrect
Preformationism	Miniature organism resides in sex cells; thus all traits are inherited from one parent.	Incorrect
Blending inheritance	Genes blend and mix.	Incorrect
Germ-plasm theory	All cells contain a complete set of genetic information.	Correct
Cell theory	All life is composed of and cells arise only from cells.	Correct
Mendelian inheritance	Traits are inherited according to specific principles proposed by Mendel.	Correct

The year 1900 was a watershed in the history of genetics. Gregor Mendel's pivotal 1866 publication on his experiments with pea plants (discussed in more detail in Chapter 3), which revealed the principles of heredity, was rediscovered. Once the significance of his conclusions was recognized, other biologists immediately began to conduct similar genetic studies on mice, chickens, and other organisms. The results of these investigations showed that many traits indeed follow Mendel's rules.

In 1902, after the acceptance of Mendel's theory of heredity, Walter Sutton (1877–1916) proposed that genes are located on chromosomes. Thomas Hunt Morgan (1866–1945) discovered the first genetic mutant of fruit flies in 1910 and used fruit flies to unravel many details of transmission genetics. The foundation for population genetics was laid in the 1930s, when geneticists begin to integrate Mendelian genetics and evolutionary theory.

Geneticists began to use bacteria and viruses in the 1940s; the rapid reproduction and simple genetic systems of these organisms allowed detailed study of the organization and structure of genes. At about the same time, evidence accumulated that DNA was the repository of genetic information. James Watson (b. 1928) and Francis Crick (1916–2004), along with Maurice Wilkins (1916–2004) and Rosalind Franklin (1920–1958), described the three-dimensional structure of DNA in 1953, ushering in the era of molecular genetics (see Chapter 8).

By 1966, the chemical structure of DNA and the system by which it determines the amino acid sequence of proteins had been worked out. Advances in molecular genetics led to the first recombinant DNA experiments in 1973, which touched off another revolution in genetic research. Methods for rapidly sequencing DNA were first developed in 1977, which later allowed whole genomes of humans and other organisms to be determined. The polymerase chain reaction, a technique for quickly amplifying tiny amounts of DNA, was developed by Kary Mullis (b. 1944) and others in 1983. This technique is now the basis of numerous types of molecular analysis. In 1990, the Human Genome Project was launched. By 1995, the first complete DNA sequence of a free-living organism—the bacterium *Haemophilus influenzae*—was determined, and the first complete sequence of a eukaryotic organism (yeast) was reported a year later. A rough draft of the human genome sequence was reported in 2000 (see Chapter 15), and the sequence was essentially completed in 2003, ushering in a new era in genetics (**Figure 1.13**). Today, the genomes of numerous organisms are being sequenced, analyzed, and compared. ▶TRY PROBLEMS 19 AND 20

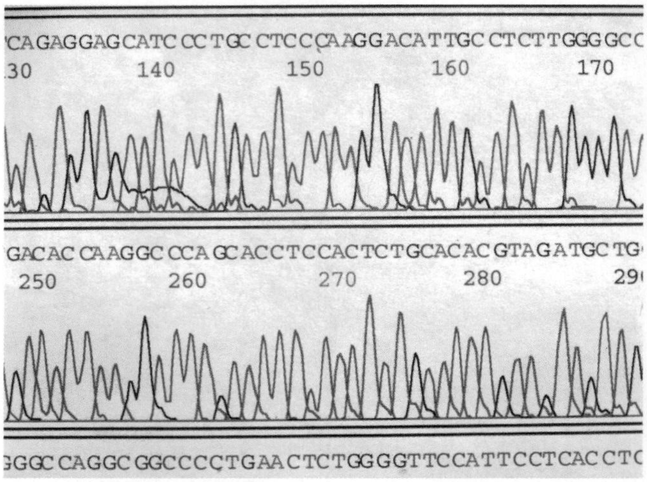

1.13 The human genome was completely sequenced in 2003. A chromatograph of a small part of the human genome. [Science Museum / Science & Society Picture Library — All rights reserved.]

The Future of Genetics

Numerous advances in genetics are being made today, and genetics remains at the forefront of biological research. New, rapid methods for sequencing DNA are being used to sequence the genomes of numerous species, from strawberries to butterflies to elephants. Recently, these methods were used to reconstruct the entire genome of an unborn fetus from fetal DNA circulating in the mother's blood, providing the potential for noninvasive prenatal genetic testing. Analysis of DNA from ancient bones demonstrates that several different species of humans roamed Earth as recently as 30,000 years ago. Powerful modern genetic techniques have been used to identify genes that influence agriculturally important characteristics, such as size in cattle, domestication in chickens, speed in racehorses, and leaf shape in corn. DNA analysis is now routinely used to identify and convict criminals or prove the innocence of suspects.

The power of new methods to identify and analyze genes is illustrated by genetic studies of myocardial infarction (heart attack) in humans. Physicians have long recognized that heart attacks run in families, but finding specific genes that contribute to an increased risk of heart attack has, until recently, been difficult. An international team of geneticists examined the DNA of 26,000 people in 10 countries for single nucleotide differences in their DNA (called single-nucleotide polymorphisms, or SNPs) that might be associated with an increased risk of heart attack. This and other similar studies identified several new genes that affect the risk of coronary artery disease and early heart attacks. These findings may make it possible to identify people who are predisposed to heart attack, allowing early intervention that might prevent the attacks. Analyses of SNPs are helping to locate genes that affect all types of traits, from eye color and height to glaucoma and cancer.

Information about sequence differences among organisms is also a source of new insights about evolution. For example, scientists analyzed DNA sequences at 26 genes to construct a comprehensive evolutionary tree of mammals. The tree revealed many interesting features of mammalian evolution; for instance, marine mammals (whales, dolphins, and porpoises) are most closely related to hippos.

In recent years, scientists have discovered that alterations to DNA and chromosome structure that do not involve the base sequence of the DNA play an important role in gene expression. These alterations, called epigenetic changes, affect our appearance, behavior, and health, and they are currently the focus of intense research. Other studies demonstrate that RNA is a key player in many aspects of gene function. The discovery in the late 1990s of tiny RNA molecules called small interfering RNAs and microRNAs led to the recognition that these molecules play central roles in gene expression and development. New genetic microchips that simultaneously analyze thousands of RNA molecules are providing information about the activities of thousands of genes in a given cell, allowing a detailed picture of how cells respond to external signals, environmental stresses, and diseases such as cancer. In the field of proteomics, powerful computer programs are being used to model the structure and function of proteins from DNA-sequence information. All this information provides us with a better understanding of numerous biological processes and evolutionary relationships. The flood of new genetic information, which requires the continuous development of sophisticated computer programs to store, retrieve, compare, and analyze genetic data, has given rise to the field of bioinformatics, a merging of molecular biology and computer science.

As the cost of sequencing decreases, the focus of DNA-sequencing efforts will shift from the genomes of different species to individual differences within species. In the not-too-distant future, each person will probably possess a copy of his or her entire genome sequence, which can be used to help assess the risk of acquiring various diseases and to tailor their treatment should they arise. The use of genetics in agriculture will continue to improve the productivity of domestic crops and animals, helping to feed the future world population. This ever-widening scope of genetics raises significant ethical, social, and economic issues.

This brief overview of the history of genetics is not intended to be comprehensive; rather, it is designed to provide a sense of the accelerating pace of advances in genetics. In the chapters to come, we will learn more about the experiments and the scientists who helped shape the discipline of genetics.

THINK-PAIR-SHARE Question 5

CONCEPTS

Humans first applied genetics to the domestication of plants and animals between 10,000 and 12,000 years ago. Developments in plant hybridization and cytology in the eighteenth and nineteenth centuries laid the foundation for the field of genetics today. After Mendel's work was rediscovered in 1900, the science of genetics developed rapidly and today is one of the most active areas of science.

✓ CONCEPT CHECK 3

How did developments in cytology in the nineteenth century contribute to our modern understanding of genetics?

1.3 A Few Fundamental Concepts Are Important for the Start of Our Journey into Genetics

Undoubtedly, you learned some genetic principles in other biology classes. Let's take a few moments to review some fundamental genetic concepts.

- **Cells are of two basic types: eukaryotic and prokaryotic.** Structurally, cells consist of two basic types, although, evolutionarily, the story is more complex (see Chapter 2). Prokaryotic cells lack a nuclear membrane and do not generally possess membrane-bounded cell organelles, whereas eukaryotic cells are more complex, possessing a nucleus and membrane-bounded organelles such as chloroplasts and mitochondria.
- **The gene is the fundamental unit of heredity.** The precise way in which a gene is defined often varies depending on the biological context. At the simplest level, we can think of a gene as a unit of information that encodes a genetic characteristic. We will expand this definition as we learn more about what genes are and how they function.
- **Genes come in multiple forms called alleles.** A gene that specifies a characteristic may exist in several forms, called alleles. For example, a gene for coat color in cats may exist as an allele that encodes black fur or as an allele that encodes orange fur.
- **Genes confer phenotypes.** One of the most important concepts in genetics is the distinction between traits and genes. Traits are not inherited directly. Rather, genes are inherited and, along with environmental factors, determine the expression of traits. The genetic information that an individual organism possesses is its genotype; the trait is its phenotype. For example, the albinism seen in some Hopis is a phenotype, and the information in *OCA2* genes that causes albinism is the genotype.
- **Genetic information is carried in DNA and RNA.** Genetic information is encoded in the molecular structure of nucleic acids, which come in two types: deoxyribonucleic acid (DNA) and ribonucleic acid (RNA). Nucleic acids are polymers consisting of repeating units called nucleotides; each nucleotide consists of a sugar, a phosphate group, and a nitrogenous base. The nitrogenous bases in DNA are of four types: adenine (A), cytosine (C), guanine (G), and thymine (T). The sequence of these bases encodes genetic information. DNA consists of two complementary nucleotide strands. Most organisms carry their genetic information in DNA, but a few viruses carry it in RNA. The four nitrogenous bases of RNA are adenine, cytosine, guanine, and uracil (U).

THINK-PAIR-SHARE Question 6

- **Genes are located on chromosomes.** The vehicles of genetic information within a cell are chromosomes (**Figure 1.14**), which consist of DNA and associated proteins. The cells of each species have a characteristic number of chromosomes; for example, bacterial cells normally possess a single chromosome; human cells possess 46; pigeon cells possess 80. Each chromosome carries a large number of genes.
- **Replicated chromosomes separate through the processes of mitosis and meiosis.** The processes of mitosis and meiosis ensure that a complete set of an organism's chromosomes exists in each cell resulting from cell division. Mitosis is the separation of replicated chromosomes in the division of somatic (nonsex) cells. Meiosis is the pairing and separation of replicated chromosomes in the division of sex cells to produce gametes (reproductive cells).
- **Genetic information is transferred from DNA to RNA to protein.** Many genes encode characteristics by specifying the structure of proteins. Genetic information is first transcribed from DNA into RNA, and then RNA is translated into the amino acid sequence of a protein.
- **Mutations are changes in genetic information that can be passed from cell to cell or from parent to offspring.** Gene mutations affect the genetic information of only a single gene; chromosome mutations alter the number or the structure of chromosomes and therefore usually affect many genes.
- **Many traits are affected by multiple factors.** Many traits are affected by multiple genes that interact in complex ways with environmental factors. Human height, for example, is affected by many genes as well as by environmental factors such as nutrition.
- **Evolution is genetic change.** Evolution can be viewed as a two-step process: first, genetic variation arises, and second, some genetic variants increase in frequency, whereas other variants decrease in frequency. ▶TRY PROBLEM 21

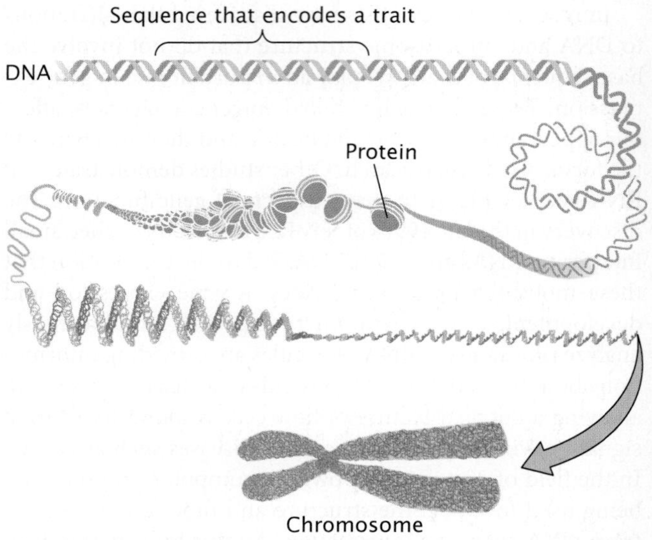

1.14 Genes are carried on chromosomes.

CONCEPTS SUMMARY

- Genetics is central to the life of every person: it influences a person's physical features, personality, intelligence, and susceptibility to numerous diseases.
- Genetics plays important roles in agriculture, the pharmaceutical industry, and medicine. It is key to the study of biology.
- All organisms use similar genetic systems. Genetic variation is the foundation of evolution and is critical to understanding all life.
- The study of genetics can be divided into transmission genetics, molecular genetics, and population genetics.
- Model genetic organisms are species about which much genetic information exists because of characteristics that make them particularly amenable to genetic analysis.
- The use of genetics by humans began with the domestication of plants and animals.
- The ancient Greeks developed the concepts of pangenesis and the inheritance of acquired characteristics, both of which were later disproved.
- Preformationism suggested that a person inherits all of his or her traits from one parent. Blending inheritance proposed that offspring possess a mixture of the parental traits. These ideas were later shown to be incorrect.
- By studying the offspring of crosses between varieties of peas, Gregor Mendel discovered the principles of heredity. Developments in cytology in the nineteenth century led to the understanding that the cell nucleus is the site of heredity.
- In 1900, Mendel's principles of heredity were rediscovered. Population genetics was established in the early 1930s, followed closely by biochemical genetics and bacterial and viral genetics. The structure of DNA

was discovered in 1953, stimulating the rise of molecular genetics.

- There are two basic types of cells: prokaryotic and eukaryotic.
- The genes that determine a trait are termed the genotype; the trait that they produce is the phenotype.
- Genes are located on chromosomes, which are made up of nucleic acids and proteins and are partitioned into daughter cells through the process of mitosis or meiosis.
- Genetic information is expressed through the transfer of information from DNA to RNA to proteins.
- Evolution requires genetic change in populations.

IMPORTANT TERMS

genome (p. 4)
transmission genetics (p. 5)
molecular genetics (p. 5)
population genetics (p. 5)
model genetic organism (p. 5)
pangenesis (p. 8)
inheritance of acquired characteristics (p. 8)
preformationism (p. 8)
blending inheritance (p. 8)
cell theory (p. 9)
germ-plasm theory (p. 9)

ANSWERS TO CONCEPT CHECKS

1. d

2. No, because horses are expensive to house, feed, and propagate, they have too few progeny, and their generation time is too long.

3. Developments in cytology in the 1800s led to the identification of parts of the cell, including the cell nucleus and chromosomes. The cell theory focused biologists' attention on the cell, eventually leading to the conclusion that the nucleus contains the hereditary information.

COMPREHENSION QUESTIONS

Answers to questions and problems preceded by an asterisk can be found at the end of the book.

Section 1.1

*1. How did Hopi culture contribute to the high incidence of albinism among members of the Hopi tribe?

2. Give at least three examples of the role of genetics in society today.

3. Briefly explain why genetics is crucial to modern biology.

4. List the three traditional subdisciplines of genetics and summarize what each covers.

5. What are some characteristics of model genetic organisms that make them useful for genetic studies?

Section 1.2

6. When and where did agriculture first arise? What role did genetics play in the development of the first domesticated plants and animals?

7. Outline the concept of pangenesis and explain how it differs from the germ-plasm theory.

8. What does the concept of the inheritance of acquired characteristics propose and how is it related to the notion of pangenesis?

9. What is preformationism? What did it have to say about how traits are inherited?

10. Define blending inheritance and contrast it with preformationism.

11. How did developments in botany in the seventeenth and eighteenth centuries contribute to the rise of modern genetics?

12. List some advances in genetics made in the twentieth century.

13. Briefly explain the contribution that each of the following people made to the study of genetics.
  a. Matthias Schleiden and Theodor Schwann
  b. August Weismann
  c. Gregor Mendel
  d. James Watson and Francis Crick
  e. Kary Mullis

Section 1.3

14. What are the two basic cell types (from a structural perspective) and how do they differ?

15. Summarize the relations between genes, DNA, and chromosomes.

For more questions that test your comprehension of the key chapter concepts, go to **LearningCurve** for this chapter.

APPLICATION QUESTIONS AND PROBLEMS

Section 1.1

*16. How are genetics and evolution related?

*17. For each of the following genetic topics, indicate whether it focuses on transmission genetics, molecular genetics, or population genetics.

a. Analysis of pedigrees to determine the probability of someone inheriting a trait.

b. Study of people on a small island to determine why a genetic form of asthma is so prevalent on the island.

c. Effect of nonrandom mating on the distribution of genotypes among a group of animals.

d. Examination of the nucleotide sequences found at the ends of chromosomes.

e. Mechanisms that ensure a high degree of accuracy during DNA replication.

f. Study of how the inheritance of traits encoded by genes on sex chromosomes (sex-linked traits) differs from the inheritance of traits encoded by genes on nonsex chromosomes (autosomal traits).

Section 1.2

*18. Genetics is said to be both a very old science and a very young science. Explain what this means.

*19. Match each description (*a* through *d*) with the correct theory or concept listed below.

a. Each reproductive cell contains a complete set of genetic information.

b. All traits are inherited from one parent.

c. Genetic information may be altered by the use of a characteristic.

d. Cells of different tissues contain different genetic information.

Preformationism

Pangenesis

Germ-plasm theory

Inheritance of acquired characteristics

*20. Compare and contrast the following ideas about inheritance.

a. Pangenesis and germ-plasm theory.

b. Preformationism and blending inheritance.

c. The inheritance of acquired characteristics and our modern theory of heredity.

Section 1.3

*21. Compare and contrast the following terms:

a. Eukaryotic and prokaryotic cells

b. Gene and allele

c. Genotype and phenotype

d. DNA and RNA

e. DNA and chromosome

CHALLENGE QUESTIONS

Introduction

*22. The type of albinism that arises with high frequency among the Hopis (discussed in the introduction to this chapter) is most likely oculocutaneous albinism type 2, which is caused by a defect in the *OCA2* gene on chromosome 15. Do some research on the Internet to determine how the phenotype of this type of albinism differs from the phenotypes of other forms of albinism in humans and the mutated genes that result in those phenotypes. Hint: Visit the website Online Mendelian Inheritance in Man and search the database for albinism.

Section 1.1

23. We now know a great deal about the genetics of humans. What are some of the reasons humans have been the focus of intensive genetic study?

24. Describe some of the ways in which your own genetic makeup affects you as a person. Be as specific as you can.

25. Describe at least one trait that appears to run in your family (appears in multiple members of the family). Do you think that this trait runs in your family because it is an inherited trait or because it is caused by environmental factors that are common to family members? How might you distinguish between these possibilities?

Section 1.3

*26. Suppose that life exists elsewhere in the universe. All life must contain some type of genetic information, but alien genomes might not consist of nucleic acids and have the same features as those found in the genomes of life on Earth. What do you think might be the common features of all genomes, no matter where they exist?

27. Choose one of the ethical or social issues in *a* through *d* below and give your opinion on the issue. For background information, you might read one of the articles on ethics listed and marked with an asterisk in the Suggested Readings for Chapter 1 in your SaplingPlus.

a. Should a person's genetic makeup be used in determining his or her eligibility for life insurance?

b. Should biotechnology companies be able to patent newly sequenced genes?

c. Should gene therapy be used on people?

d. Should genetic testing for inherited disorders for which there is no treatment or cure be made available?

***28.** A 45-year-old woman undergoes genetic testing and discovers that she is at high risk for developing colon cancer and Alzheimer disease. Because her children have 50% of her genes, they may also be at an increased risk for these diseases. Does she have a moral or legal obligation to tell her children and other close relatives about the results of her genetic testing?

***29.** Suppose that you could undergo genetic testing at age 18 for susceptibility to a genetic disease that would not appear until middle age and has no available treatment.

a. What would be some of the possible reasons for having or not having such a genetic test?

b. Would you personally want to be tested? Explain your reasoning.

THINK-PAIR-SHARE QUESTIONS

Think-Pair-Share questions are designed to be worked in collaboration with other students. First THINK about the question, then PAIR up with one or more other students, and finally SHARE your answers and work together to arrive at a solution.

Introduction

1. Albinism occupied a special place in the Hopi culture; individuals who possessed this trait were valued by members of the tribe. What are some examples of genetic traits that, in contrast, sometimes result in discrimination and prejudice?

2. Albinism in humans can be caused by mutations in any one of several different genes. This situation, in which the same phenotype may result from variation in several different genes, is referred to as genetic heterogeneity. Is genetic heterogeneity common? Are most genetic traits in humans the result of variation in a single gene, or are there many genetic traits that result from variation in several genes, as albinism does?

Section 1.1

3. Bob says that he is healthy and has no genetic diseases such as hemophilia or Down syndrome. Therefore, he says, genetics plays little role in his life. Do you think Bob is correct in his conclusion? Why or why not?

4. Are mutations good or bad? Explain your answer.

Section 1.2

5. Do you support or oppose the development of genetically engineered foods (genetically modified organisms, or GMOs)? Find someone who takes the opposite position, and discuss this question with him or her. Think about the economic and environmental benefits, health risks, ecological effects, and social impact of the use of genetically engineered foods. List some reasons for and against genetically engineering the foods we eat.

Section 1.3

6. Why do you think all organisms use nucleic acids for encoding genetic information? Why not use proteins or carbohydrates? What advantages might DNA have as the source of genetic information?

SaplingPlus Self-study tools that will help you practice what you've learned and reinforce this chapter's concepts are available online. Go to www.macmillanlearning.com/PierceGenetics6e.

2 Chromosomes and Cellular Reproduction

The Blind Men's Riddle

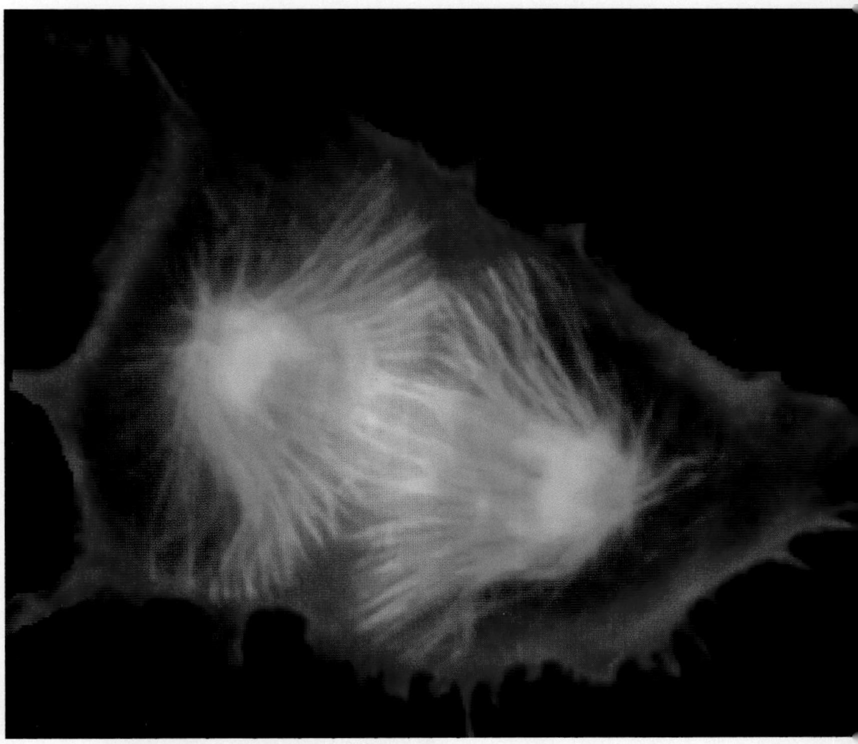

A rat kangaroo kidney cell undergoing mitosis, the process by which each new cell receives a complete copy of the genetic material. Chromosomes are shown in blue. [Courtesy of Julie Canman and Ted Salman.]

In a well-known riddle, two blind men by chance enter a department store at the same time. They both go to the same counter, and both order five pairs of socks, each pair a different color. The salesclerk is so befuddled by this strange coincidence that he places all ten pairs (two black pairs, two blue pairs, two gray pairs, two brown pairs, and two green pairs) into a single shopping bag and gives the bag of socks to one blind man and an empty bag to the other. The two blind men happen to meet on the street outside, where they discover that one of their bags contains all ten pairs of socks. How do the blind men, without seeing and without any outside help, sort out the socks so that each man goes home with exactly five different pairs of colored socks? Can you come up with a solution to the riddle?

By an interesting coincidence, cells face the same challenge that the blind men do. Most organisms possess two sets of genetic information, one set inherited from each parent. Before cell division, the DNA in each chromosome replicates; after replication, there are two copies—called sister chromatids—of each chromosome. At the end of cell division, it is critical that each of the two new cells receives a complete copy of the genetic material, just as each blind man needs to go home with a complete set of socks.

The solution to the riddle is simple. Socks are sold as pairs; the two socks of a pair are typically connected by a thread. As a pair is removed from the bag, the men each grasp a different sock of the pair and pull in opposite directions. When the socks are pulled tight, one of the men can take a pocket knife and cut the thread connecting the pair. Each man then deposits his single sock in his own bag. At the end of the process, each man's bag will contain exactly two black socks, two blue socks, two gray socks, two brown socks, and two green socks.[1]

Remarkably, cells employ a similar solution for separating their chromosomes into new daughter cells. As we will learn in this chapter, the replicated chromosomes line up at the center of a cell undergoing division and, like the socks in the riddle, the sister chromatids of each chromosome are pulled in opposite directions. Like the thread connecting two socks of a pair, a molecule called cohesin holds the sister chromatids together until it is severed by a molecular knife called separase. The two resulting chromosomes separate and the cell divides, ensuring that a complete set of chromosomes is deposited in each cell.

[1] This analogy is adapted from K. Nasmyth, Disseminating the genome: Joining, resolving, and separating sister chromatids, during mitosis and meiosis, *Annual Review of Genetics* 35:673–745, 2001.

In this analogy, the blind men and cells differ in one critical regard: if the blind men make a mistake, one man ends up with an extra sock and the other is a sock short, but no great harm results. The same cannot be said for human cells. Errors in chromosome separation, producing cells with too many or too few chromosomes, are frequently catastrophic, leading to cancer, reproductive failure, or—sometimes—a child with severe disabilities.

THINK-PAIR-SHARE Question 1

This chapter explores the process of cell reproduction and explains how a complete set of genetic information is transmitted to new cells. In prokaryotic cells, reproduction is simple because each cell possesses a single chromosome. In eukaryotic cells, multiple chromosomes must be copied and distributed to each of the new cells, making cell reproduction more complex. Cell division in eukaryotes takes place through mitosis and meiosis, processes that serve as the foundation for much of genetics.

Grasping mitosis and meiosis requires more than simply memorizing the sequences of events that take place in each stage, although these events are important. The key is to understand how genetic information is apportioned in the course of cell reproduction through the dynamic interplay of DNA synthesis, chromosome movement, and cell division. These processes bring about the transmission of genetic information and are the basis of similarities and differences between parents and progeny.

2.1 Prokaryotic and Eukaryotic Cells Differ in a Number of Genetic Characteristics

Biologists have traditionally classified all living organisms into two major groups, the prokaryotes and the eukaryotes (**Figure 2.1**). A **prokaryote** is a unicellular organism

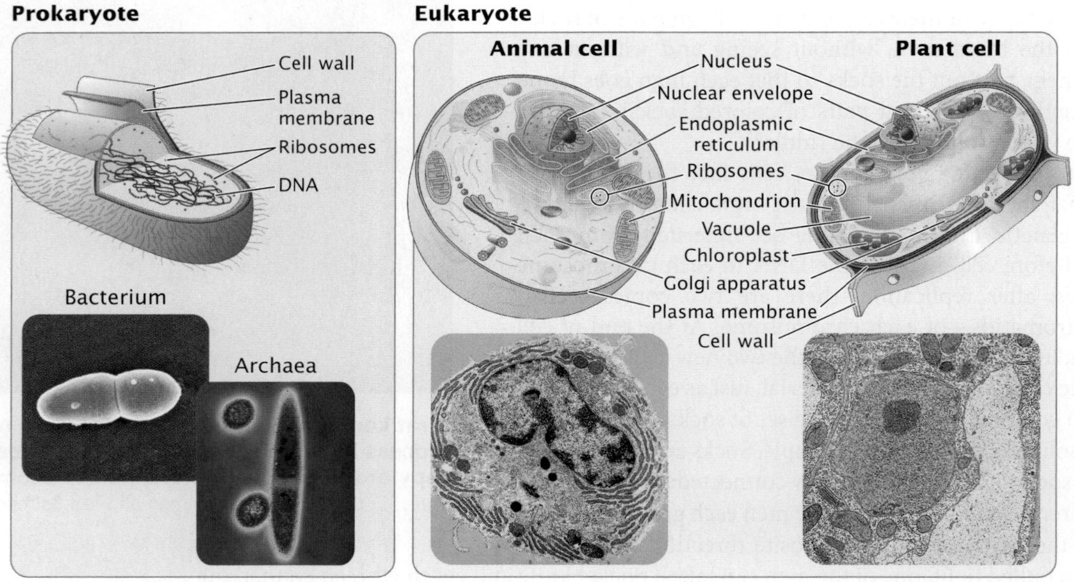

	Prokaryotic cells	Eukaryotic cells
Nucleus	Absent	Present
Cell diameter	Relatively small, from 1 to 10 μm	Relatively large, from 10 to 100 μm
Genome	Usually one circular DNA molecule	Multiple linear DNA molecules
DNA	Not complexed with histones in bacteria; some histones in archaea	Complexed with histones
Amount of DNA	Relatively small	Relatively large
Membrane-bounded organelles	Absent	Present

2.1 Prokaryotic and eukaryotic cells differ in structure. [Photographs (left to right): Dr. Gary D. Gaugler/Newscom; Dr. Kari Lounatmaa/Science Source; Dr. Gopal Murti/Phototake; Biophoto Associates/Science Source.]

with a relatively simple cell structure. A **eukaryote** has a compartmentalized cell structure with components bounded by intracellular membranes; eukaryotes may be unicellular or multicellular.

Research indicates, however, that the division of life is not so simple. Although all prokaryotes are similar in cell structure, prokaryotes include at least two fundamentally distinct types of bacteria: the **bacteria** (also called eubacteria or "true bacteria") and the **archaea** ("ancient bacteria"). An examination of equivalent DNA sequences reveals that the bacteria and the archaea are as distantly related to each other as they are to the eukaryotes. Although bacteria and archaea are similar in cell structure, some genetic processes in archaea (such as transcription) are more similar to those in eukaryotes, and the archaea are actually closer evolutionarily to eukaryotes than to bacteria. Thus, from an evolutionary perspective, there are three major groups of organisms: bacteria, archaea, and eukaryotes. In this book, the prokaryotic–eukaryotic distinction will be made frequently, but important bacterial–archaeal differences will also be noted.

From the perspective of genetics, a major difference between prokaryotic and eukaryotic cells is that a eukaryotic cell has a *nuclear envelope*, which surrounds the genetic material to form a **nucleus** and separates the DNA from the other cellular contents. In prokaryotic cells, the genetic material is in close contact with other components of the cell—a property that has important consequences for the way in which genes are controlled (**Figure 2.2**).

Another fundamental difference between prokaryotes and eukaryotes lies in the packaging of their DNA. In eukaryotes, DNA is closely associated with a special class of proteins, the **histones**, to form tightly packed chromosomes (**Figure 2.3**). This complex of DNA and histone proteins, called **chromatin**, is the stuff of eukaryotic chromosomes. The histone proteins limit the accessibility of the chromatin to enzymes and other proteins that copy and read the DNA, but they enable the DNA to fit into the nucleus. Eukaryotic DNA must separate from the histones before the genetic information in the DNA can be read. Archaea also have some histone proteins that complex with DNA, but the structure of their chromatin is different from that found in eukaryotes. Bacteria do not possess histones, so their DNA does not exist in the highly ordered, tightly packed arrangement found in eukaryotic cells. The copying and reading of DNA are therefore simpler processes in bacteria.

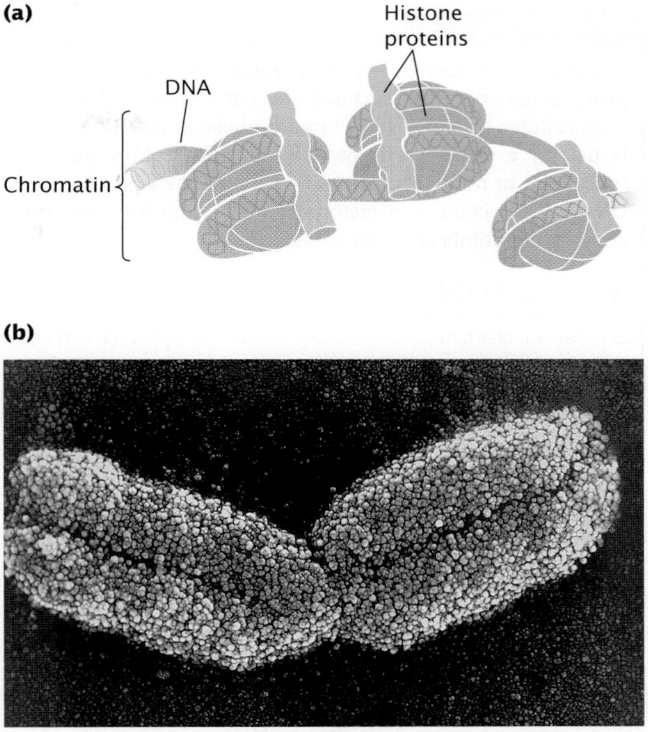

2.3 Eukaryotic DNA is closely associated with histone proteins (a) to form chromosomes—one of which is shown in (b)—that are located in the nucleus. [Part b: Biophoto Associates/Science Source.]

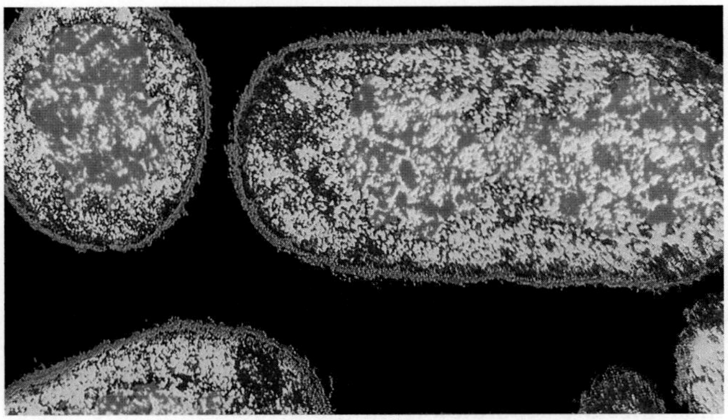

2.2 Prokaryotic DNA (red) is neither surrounded by a nuclear membrane nor associated with histone proteins. [A. Barry Dowsett/Science Source]

The genes of prokaryotic cells are generally on a single circular molecule of DNA—the chromosome of a prokaryotic cell. In eukaryotic cells, genes are located on multiple, usually linear DNA molecules (multiple chromosomes). Eukaryotic cells therefore require mechanisms that ensure that a copy of each chromosome is faithfully transmitted to each new cell. However, this generalization—a single circular chromosome in prokaryotes and multiple linear chromosomes in eukaryotes—is not always true. A few bacteria have more than one chromosome, and important bacterial genes are frequently found on other DNA molecules called *plasmids* (see Chapter 7). Furthermore, in some eukaryotes, a few genes are located on circular DNA molecules found in certain organelles (such as mitochondria and chloroplasts).

CONCEPTS

Organisms are classified as prokaryotes or eukaryotes, and the prokaryotes consist of archaea and bacteria. A prokaryote is a unicellular organism that lacks a nucleus, and its genome is usually a single chromosome. Eukaryotes may be either unicellular or multicellular, their cells possess a nucleus, their DNA is associated with histone proteins, and their genomes consist of multiple chromosomes.

✓ CONCEPT CHECK 1

List several characteristics that bacteria and archaea have in common and that distinguish them from eukaryotes.

Viruses are neither prokaryotic nor eukaryotic because they do not possess a cellular structure. Viruses are actually simple structures composed of an outer protein coat surrounding nucleic acid (either DNA or RNA; **Figure 2.4**). All known viruses can reproduce only within host cells and their evolutionary relationship to cellular organisms is uncertain. Their simplicity and small genomes make viruses useful for studying some molecular processes and for some types of genetic analyses.

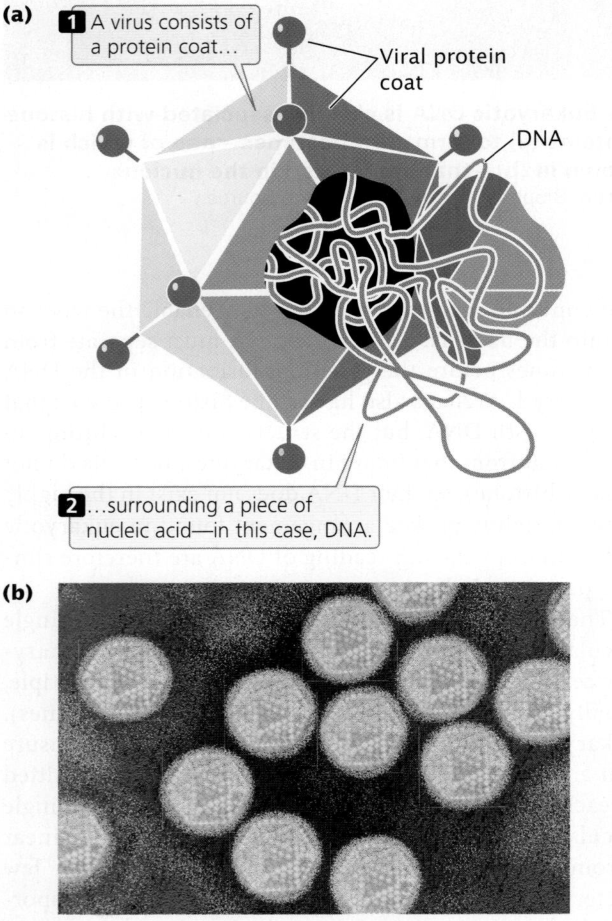

2.4 A virus is a simple replicative structure consisting of protein and nucleic acid. (a) Structure of a virus. (b) Adenoviruses are shown in the micrograph. [Part b: BSIP/Science Source.]

2.2 Cell Reproduction Requires the Copying of the Genetic Material, Separation of the Copies, and Cell Division

For any cell to reproduce successfully, three fundamental events must take place: **(1)** its genetic information must be copied, **(2)** the copies of the genetic information must be separated from each other, and **(3)** the cell must divide. All cellular reproduction includes these three events, but the processes that lead to these events differ in prokaryotic and eukaryotic cells because of their structural differences.

Prokaryotic Cell Reproduction

When a prokaryotic cell reproduces, the circular chromosome replicates and the cell divides in a process called *binary fission*. Replication usually begins at a specific place on the chromosome, called the origin of replication (or simply the *origin*). In a process that is not well understood, the origins of the two newly replicated chromosomes move away from each other and toward opposite ends of the cell. In at least some prokaryotes, proteins bind near the origins and anchor the new chromosomes to the plasma membrane at opposite ends of the cell. Finally, a new cell wall forms between the two chromosomes, producing two cells, each with an identical copy of the chromosome. Under optimal conditions, some prokaryotic cells divide every 20 minutes. At this rate, a single cell could produce a billion descendants in a mere 10 hours.

Eukaryotic Cell Reproduction

Like prokaryotic cell reproduction, eukaryotic cell reproduction requires the processes of DNA replication, copy separation, and division of the cytoplasm. However, the presence of multiple DNA molecules requires a more complex mechanism to ensure that exactly one copy of each molecule ends up in each of the new cells.

EUKARYOTIC CHROMOSOMES Eukaryotic chromosomes are separated from the cytoplasm by the nuclear envelope. The nucleus has a highly organized internal scaffolding, called the *nuclear matrix*, that consists of a

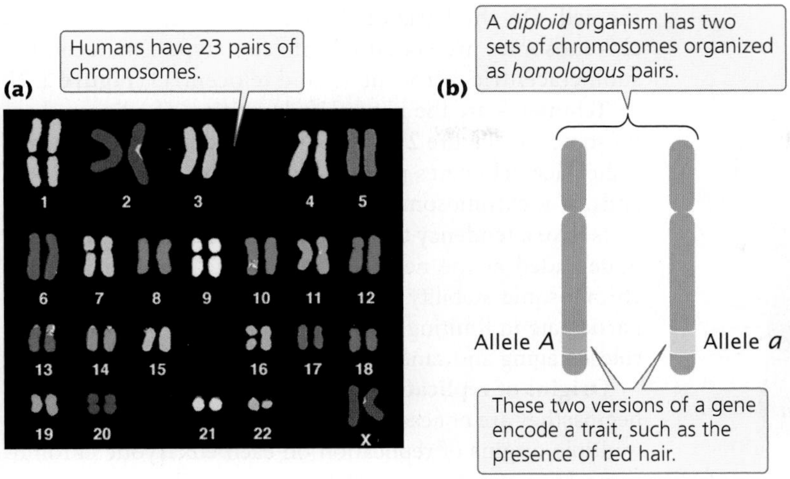

2.5 Diploid eukaryotic cells have two sets of chromosomes. (a) A set of chromosomes from a female human cell. Each pair of chromosomes is hybridized to a uniquely colored probe, giving it a distinct color. (b) The chromosomes are present in homologous pairs. Each pair consists of two chromosomes that are alike in size and structure and carry information for the same characteristics. [Part a: Courtesy of Dr. Thomas Ried and Dr. Evelin Schrock.]

network of protein fibers. The nuclear matrix maintains precise spatial relations among the components of the nucleus and takes part in DNA replication, the expression of genes, and the modification of gene products before they leave the nucleus.

Each eukaryotic species has a characteristic number of chromosomes per cell: potatoes have 48 chromosomes, fruit flies have 8, and humans have 46. There appears to be no special relation between the complexity of an organism and its number of chromosomes per cell. In most eukaryotic cells, there are two sets of chromosomes. The presence of two sets is a consequence of sexual reproduction: one set is inherited from the male parent and the other from the female parent. Each chromosome in one set has a corresponding chromosome in the other set; together, the two chromosomes constitute a homologous pair (**Figure 2.5**). Human cells, for example, have 46 chromosomes, constituting 23 homologous pairs.

The two chromosomes of a **homologous pair** are usually alike in structure and size, and each carries genetic information for the same set of hereditary characteristics (the sex chromosomes are an exception and will be discussed in Chapter 4). For example, if a gene on a particular chromosome encodes a characteristic such as hair color, another copy of the gene (each copy is called an *allele*) at the same position on that chromosome's homolog *also* encodes hair color. However, these two alleles do not need to be identical: one of them might encode brown hair and the other might encode blond hair.

Cells that carry two sets of genetic information are **diploid**. In general, the ploidy of the cell indicates how many sets of genetic information the cell possesses. Reproductive cells (such as eggs, sperm, and spores), and even nonreproductive cells in some eukaryotic organisms, contain a single set of chromosomes and are **haploid**. The cells of some organisms contain more than two sets of genetic information and are **polyploid** (see Chapter 6).

CONCEPTS

Cells reproduce by copying and separating their genetic information and then dividing. Because eukaryotic cells possess multiple chromosomes, mechanisms exist to ensure that each new cell receives one copy of each chromosome. Most eukaryotic cells are diploid, and their two chromosome sets can be arranged in homologous pairs. Haploid cells contain a single set of chromosomes.

✓ CONCEPT CHECK 2

Diploid cells have

a. two chromosomes.
b. two sets of chromosomes.
c. one set of chromosomes.
d. two pairs of homologous chromosomes.

CHROMOSOME STRUCTURE The chromosomes of eukaryotic cells are larger and more complex than those found in prokaryotes, but each unreplicated chromosome nevertheless consists of a single molecule of DNA. Although linear, the DNA molecules in eukaryotic chromosomes are highly folded and condensed; if stretched out, some human chromosomes would be several centimeters long—thousands of times as long as the span of a typical nucleus. To package such a tremendous length of DNA into the small volume of the nucleus, each DNA molecule is coiled around histone proteins and tightly packed, forming a rod-shaped chromosome. Most of the time the chromosomes are thin and difficult to observe, but before cell division they condense further into thick, readily observed structures; it is at this stage that chromosomes are usually studied.

A functional chromosome has three essential elements: a centromere, a pair of telomeres, and origins of replication. The *centromere* is the attachment point for *spindle microtubules*, which are the filaments responsible for moving chromosomes during cell division

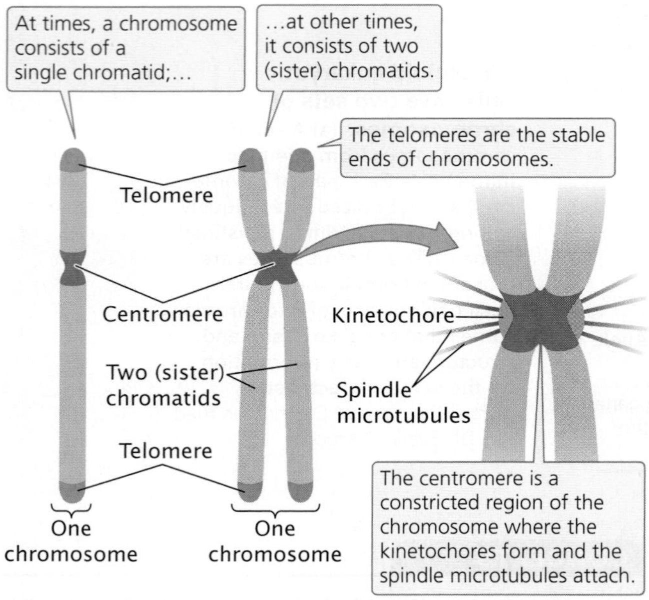

At times, a chromosome consists of a single chromatid;...
...at other times, it consists of two (sister) chromatids.
The telomeres are the stable ends of chromosomes.
Telomere
Centromere
Kinetochore
Two (sister) chromatids
Spindle microtubules
Telomere
One chromosome
One chromosome
The centromere is a constricted region of the chromosome where the kinetochores form and the spindle microtubules attach.

2.6 Each eukaryotic chromosome has a centromere and telomeres.

(**Figure 2.6**). The centromere appears as a constricted region of the chromosome. Before cell division, a multiprotein complex called the *kinetochore* assembles on the centromere; later, spindle microtubules attach to the kinetochore. Chromosomes lacking a centromere cannot be drawn into the newly formed nuclei; these chromosomes are lost, often with catastrophic consequences for the cell. On the basis of the location of the centromere, chromosomes are classified into four types: metacentric, submetacentric, acrocentric, and telocentric (**Figure 2.7**).

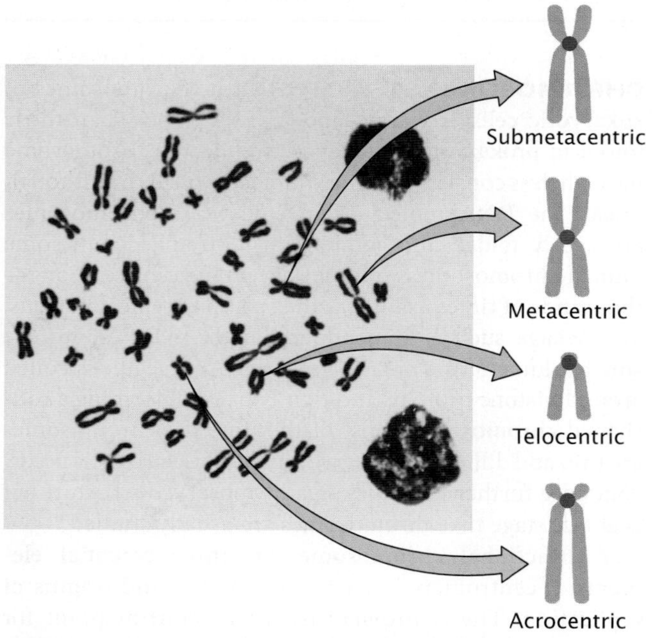

2.7 Eukaryotic chromosomes are divided into four major types based on the position of the centromere.
[Don W. Fawcett/Science Source.]

Telomeres are the natural ends, or tips, of a linear chromosome (see Figure 2.6). Like the plastic tips on the ends of a shoelace, telomeres protect and stabilize the chromosome ends. If a chromosome breaks, producing new ends, these ends have a tendency to stick together, and the chromosome is degraded at the newly broken ends. Telomeres provide chromosome stability. Research shows that telomeres also participate in limiting cell division and may play important roles in aging and cancer (as we will see in Chapter 9).

Origins of replication are the sites where DNA synthesis begins; they are not easily observed by microscopy. There are multiple origins of replication on each eukaryotic chromosome. In preparation for cell division, each chromosome replicates, making a copy of itself, as already mentioned. These two initially identical copies, called **sister chromatids**, are held together at the centromere (see Figure 2.6). Each sister chromatid consists of a single molecule of DNA.

THINK-PAIR-SHARE Question 2

CONCEPTS

Sister chromatids are copies of a chromosome held together at the centromere. Functional chromosomes contain centromeres, telomeres, and origins of replication. The kinetochore is the point of attachment for the spindle microtubules. Telomeres are the stabilizing ends of a chromosome. Origins of replication are sites where DNA synthesis begins.

✓ CONCEPT CHECK 3

What would be the result if a chromosome did not have a kinetochore?

The Cell Cycle and Mitosis

The **cell cycle** is the life story of a cell: the stages through which it passes from one division to the next (**Figure 2.8**). This process is critical to genetics because it is through the cell cycle that the genetic instructions for all characteristics are passed from parent cell to daughter cells. A new cell cycle begins after a cell has divided and produced two new cells. Each new cell metabolizes, grows, and develops. At the end of its cycle, it, too, divides to produce two cells, which can then undergo additional cell cycles. Progression through the cell cycle is regulated at key transition points called **checkpoints**, which allow or prohibit the cell's progression to the next stage. Checkpoints ensure that all cellular components are present and in good working order, and they are necessary to prevent cells with damaged or missing chromosomes from proliferating. Defects in checkpoints can lead to unregulated cell growth, as is seen in some cancers. The molecular basis of these checkpoints will be discussed in Chapter 16.

The cell cycle consists of two major phases. The first is **interphase**, the period between cell divisions, in which the

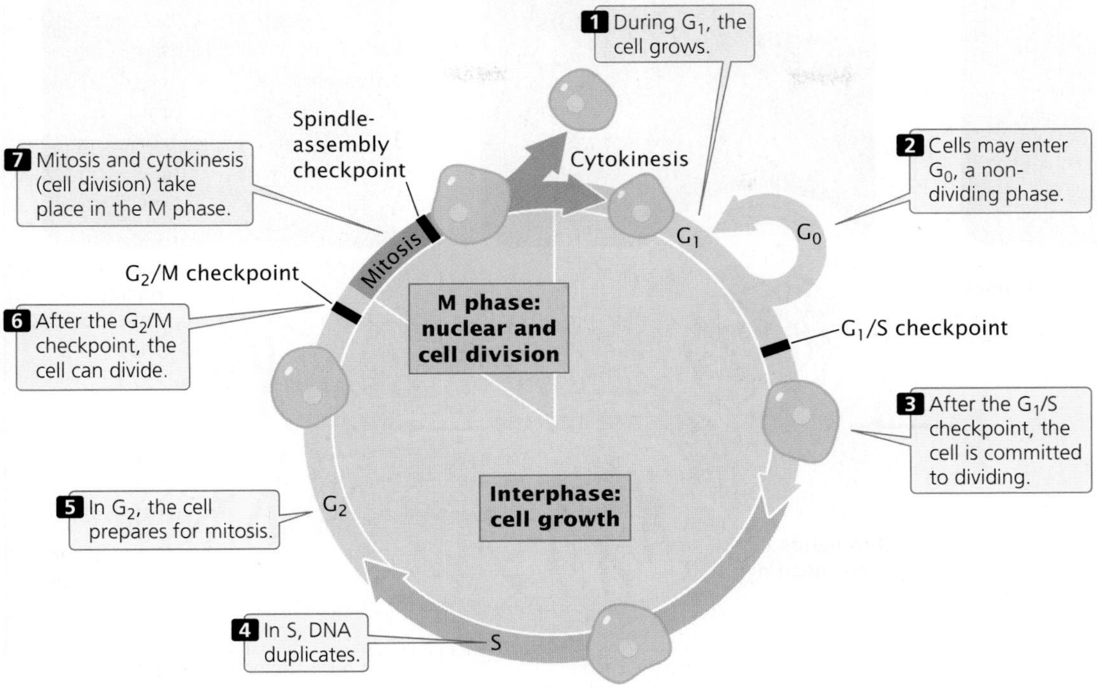

2.8 The cell cycle consists of interphase and the M phase.

cell grows, develops, and functions. During interphase, critical events necessary for cell division also take place. The second is the **M phase** (mitotic phase), the period of active cell division. The M phase includes **mitosis**, the process of nuclear division, and **cytokinesis**, or cytoplasmic division. Let's take a closer look at the details of interphase and the M phase.

INTERPHASE Interphase is the extended period of growth and development between cell divisions. Interphase includes several checkpoints.

By convention, interphase is divided into three phases: G_1, S, and G_2 (see Figure 2.8). Interphase begins with a phase called G_1 (for gap 1). In G_1, the cell grows and proteins necessary for cell division are synthesized; this phase typically lasts several hours. There is a critical point termed the *G_1/S checkpoint* near the end of G_1. The G_1/S checkpoint holds the cell in G_1 until the cell has all the enzymes necessary for the replication of DNA. After this checkpoint has been passed, the cell is committed to divide.

Before reaching the G_1/S checkpoint, cells may exit the active cell cycle in response to regulatory signals and pass into a nondividing phase called *G_0*, a stable state during which cells usually maintain a constant size. Cells can remain in G_0 for an extended period, even indefinitely, or they can reenter G_1 and the active cell cycle. Many cells never enter G_0; rather, they cycle continuously.

After G_1, the cell enters the *S* phase (for DNA synthesis), in which each chromosome is duplicated. Although the cell is committed to divide after the G_1/S checkpoint has been passed, DNA synthesis must take place before the cell can proceed to mitosis. If DNA synthesis is blocked (by drugs or by a mutation), the chromosomes will not be duplicated, and the cell will not be able to undergo mitosis. Before the S phase, each chromosome is unreplicated; after the S phase, each chromosome is composed of two chromatids (see Figure 2.6).

After the S phase, the cell enters G_2 (gap 2). In this phase, several additional biochemical events necessary for cell division take place. The *G_2/M checkpoint* is reached near the end of G_2. This checkpoint is passed only if the cell's DNA is completely replicated and undamaged. Unreplicated or damaged DNA can inhibit the activation of some proteins that are necessary for mitosis to take place. After the G_2/M checkpoint has been passed, the cell is ready to divide, and it enters the M phase. Although the length of interphase varies from cell type to cell type, a typical dividing mammalian cell spends about 10 hours in G_1, 9 hours in S, and 4 hours in G_2 (see Figure 2.8).

Throughout interphase, the chromosomes are in a relaxed, but by no means uncoiled, state, and individual chromosomes cannot be seen with a microscope. This condition changes dramatically when interphase draws to a close and the cell enters the M phase.

M PHASE The M phase is the part of the cell cycle in which the copies of the cell's chromosomes (sister chromatids) separate and the cell undergoes division. The separation of sister chromatids in the M phase is a critical process that results in a complete set of genetic information for each of the daughter cells. Biologists usually divide the M phase into six stages: the five stages of mitosis (prophase, prometaphase, metaphase, anaphase, and telophase) illustrated in **Figure 2.9**, and cytokinesis. It's important to keep

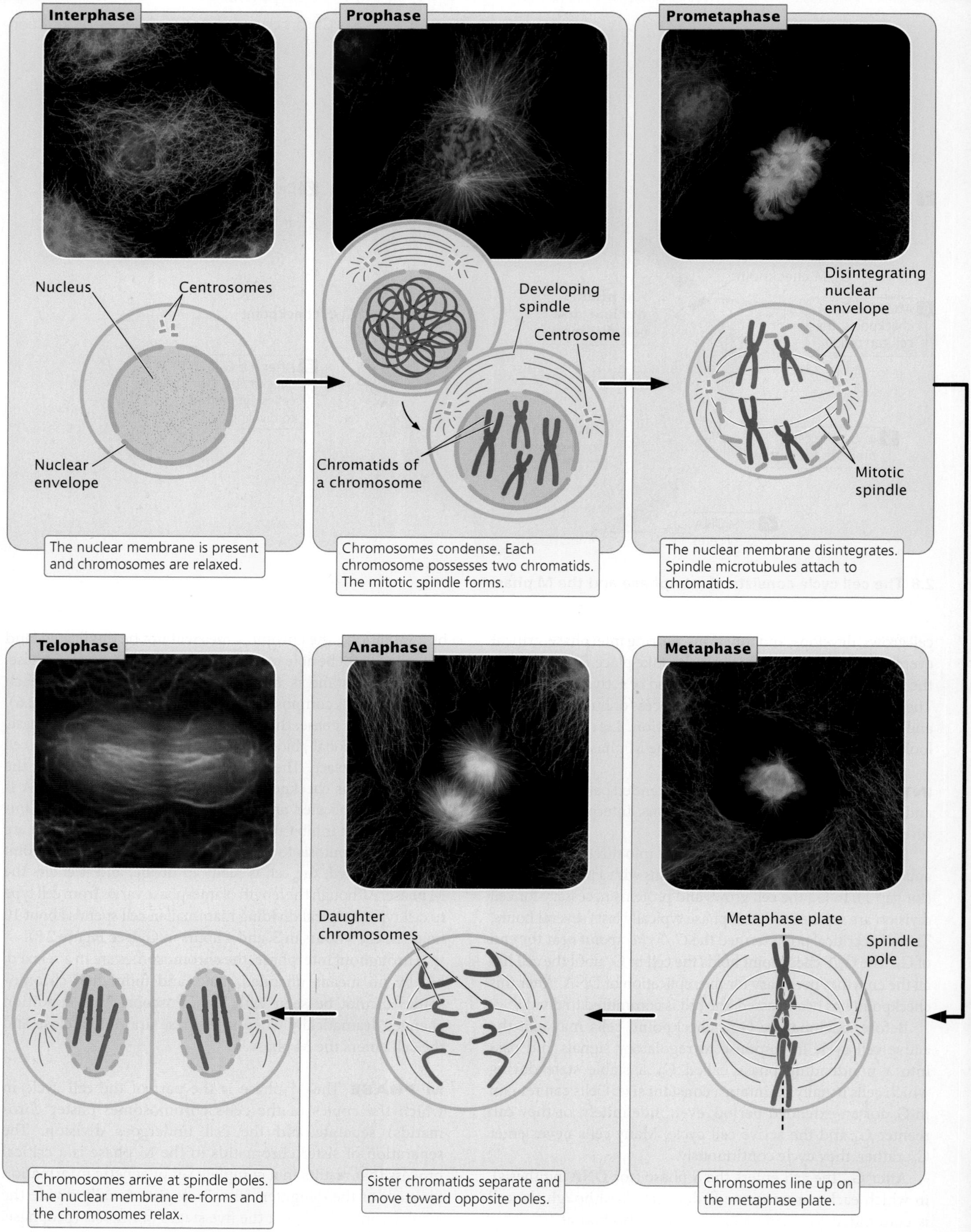

2.9 The M phase is divided into stages. [Photographs: top row and bottom row center and right - Dr. Torsten Wittmann/Science Source; bottom row left - Jennifer Waters/Science Source.]

in mind that the M phase is a continuous process and that its separation into these six stages is somewhat arbitrary.

Prophase As a cell enters prophase, the chromosomes become visible under a light microscope. Each chromosome possesses two chromatids because the chromosome was duplicated in the preceding S phase. The mitotic spindle, an organized array of microtubules that move the chromosomes in mitosis, forms. In animal cells, the spindle grows out from a pair of centrosomes that migrate to opposite sides of the cell. Within each centrosome is a special organelle, the centriole, which is also composed of microtubules. Some plant cells do not have centrosomes or centrioles, but they do have mitotic spindles.

Prometaphase Disintegration of the nuclear membrane marks the start of prometaphase. Spindle microtubules, which until now have been outside the nucleus, enter the nuclear region. When the end of a microtubule encounters a kinetochore, the microtubule becomes stabilized. Eventually, each chromosome becomes attached to microtubules from opposite poles of the spindle: for each chromosome, a microtubule from one of the centrosomes anchors to the kinetochore of *one* of the sister chromatids, and a microtubule from the opposite centrosome then attaches to the other sister chromatid, anchoring the chromosome to both of the centrosomes. This arrangement is known as chromosome bi-orientation.

Metaphase During metaphase, the chromosomes become arranged on a single plane, called the *metaphase plate*, between the two centrosomes. The centrosomes, now at opposite ends of the cell, with their microtubules radiating outward and meeting in the middle of the cell, are centered at the spindle poles. A *spindle-assembly checkpoint* ensures that each chromosome is aligned on the metaphase plate and attached to spindle microtubules from opposite poles.

Anaphase After the spindle-assembly checkpoint is passed, the connection between sister chromatids breaks down, and the sister chromatids separate. This marks the beginning of anaphase, during which the chromosomes move toward opposite spindle poles.

Telophase After the sister chromatids have separated, each is considered a separate chromosome. Telophase is marked by the arrival of the chromosomes at the spindle poles. The nuclear membrane re-forms around each set of chromosomes, producing two separate nuclei within the cell. The chromosomes relax and lengthen, once again disappearing from view. In many cells, division of the cytoplasm (cytokinesis) is simultaneous with telophase.

The major features of the cell cycle are summarized in **Table 2.1**. You can watch the cell cycle in motion by viewing

TABLE 2.1 Features of the cell cycle

Stage	Major features
G_0 phase	Stable, nondividing period of variable length.
Interphase	
G_1 phase	Growth and development of the cell; G_1/S checkpoint.
S phase	Synthesis of DNA.
G_2 phase	Preparation for division; G_2/M checkpoint.
M phase	
Prophase	Chromosomes condense and mitotic spindle forms.
Prometaphase	Nuclear envelope disintegrates, and spindle microtubules anchor to kinetochores.
Metaphase	Chromosomes align on the metaphase plate; spindle-assembly checkpoint.
Anaphase	Sister chromatids separate, becoming individual chromosomes that migrate toward spindle poles.
Telophase	Chromosomes arrive at spindle poles, the nuclear envelope re-forms, and the condensed chromosomes relax.
Cytokinesis	Cytoplasm divides; cell wall forms in plant cells.

Animation 2.1. This interactive animation allows you to determine what happens when different processes in the cycle fail. ▶TRY PROBLEM 20

Genetic Consequences of the Cell Cycle

What are the genetically important results of the cell cycle? From a single cell, the cell cycle produces two daughter cells that contain the same genetic instructions. The resulting cells are genetically identical with each other and with their parent cell because DNA synthesis in the S phase creates an exact copy of each DNA molecule, giving rise to two genetically identical sister chromatids. Mitosis then ensures that one of the two sister chromatids from each replicated chromosome passes into each new cell.

Another genetically important result of the cell cycle is that each of the cells produced contains a full complement of chromosomes: there is no net reduction or increase in chromosome number. Each cell also contains approximately half the cytoplasm and organelle content of the original parent cell, but no precise mechanism analogous to mitosis ensures that organelles are evenly divided. Consequently, not all cells resulting from the cell cycle are identical in their cytoplasmic content.

THINK-PAIR-SHARE Question 3

CONCEPTS

The active cell cycle phases are interphase and the M phase. Interphase consists of G_1, S, and G_2. In G_1, the cell grows and prepares for cell division; in the S phase, DNA synthesis takes place; in G_2, other biochemical events necessary for cell division take place. Some cells enter a quiescent phase called G_0. The M phase includes mitosis, which is divided into prophase, prometaphase, metaphase, anaphase, and telophase, and cytokinesis.

✓ CONCEPT CHECK 4

Which is the correct order of stages in the cell cycle?

a. G_1, S, prophase, metaphase, anaphase
b. S, G_1, prophase, metaphase, anaphase
c. Prophase, S, G_1, metaphase, anaphase
d. S, G_1, anaphase, prophase, metaphase

CONNECTING CONCEPTS

Counting Chromosomes and DNA Molecules

The relations among chromosomes, chromatids, and DNA molecules frequently cause confusion. At certain times, chromosomes are unreplicated; at other times, each chromosome possesses two chromatids (see Figure 2.6). Chromosomes sometimes consist of a single DNA molecule; at other times, they consist of two DNA molecules. How can we keep track of the number of these structures in the cell cycle?

There are two simple rules for counting chromosomes and DNA molecules: (1) to determine the number of chromosomes, count the number of functional centromeres; (2) to determine the number of DNA molecules, first determine if sister chromatids are present. If sister chromatids are present, then the number of DNA molecules is twice the number of chromosomes. If the chromosome is unreplicated, without sister chromatids, the number of DNA molecules is the same as the number of chromosomes.

Let's examine a hypothetical cell as it passes through the cell cycle (**Figure 2.10**). At the beginning of G_1, this diploid cell has two complete sets of chromosomes, for a total of four chromosomes. Each chromosome is unreplicated, without sister chromatids, so there are four DNA molecules in the cell during G_1. In the S phase, each DNA molecule is copied. The two resulting DNA molecules combine with histones and other proteins to form sister chromatids. Although the amount of DNA doubles in the S phase, the number of chromosomes remains the same, because the two sister chromatids are tethered together and share a single functional centromere. At the end of the S phase, this cell still contains four chromosomes, each with two chromatids; so there are $4 \times 2 = 8$ DNA molecules present.

Through prophase, prometaphase, and metaphase, the cell has four chromosomes and eight DNA molecules. At anaphase, however, the sister chromatids separate. Each now has its own functional centromere, and so each is considered a separate chromosome. Until cytokinesis, the cell contains eight unreplicated chromosomes, without sister chromatids; thus, there are still eight DNA molecules present. After cytokinesis, the eight chromosomes (and eight DNA molecules) are distributed equally between two cells; so each new cell contains four chromosomes and four DNA molecules, the number present at the beginning of the cell cycle.

In summary, the number of chromosomes increases briefly only in anaphase, when the two chromatids of a chromosome separate and become distinct chromosomes. The number of chromosomes decreases only through cytokinesis. The number of DNA molecules increases only in the S phase and decreases only through cytokinesis. ▶TRY PROBLEM 23

THINK-PAIR-SHARE Question 4

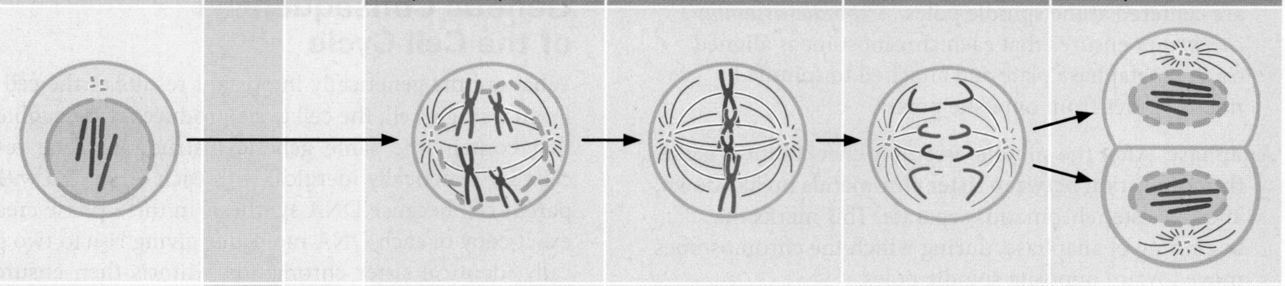

	G_1	S	G_2	Prophase and prometaphase	Metaphase	Anaphase	Telophase and cytokinesis
Number of chromosomes per cell	4	4	4	4	4	8	4
Number of DNA molecules per cell	4	4 → 8	8	8	8	8	4

2.10 The number of chromosomes and the number of DNA molecules change in the course of the cell cycle. The number of chromosomes per cell equals the number of functional centromeres. The number of DNA molecules per cell equals the number of chromosomes when the chromosomes are unreplicated (no sister chromatids are present) and twice the number of chromosomes when sister chromatids are present.

2.3 Sexual Reproduction Produces Genetic Variation Through the Process of Meiosis

If all reproduction were accomplished through mitosis, life would be quite dull because mitosis produces only genetically identical progeny. With only mitosis, you, your children, your parents, your brothers and sisters, your cousins, and many people whom you don't even know would be clones—copies of one another. Only the occasional mutation would introduce any genetic variation. All organisms reproduced in this way for the first 2 billion years of Earth's existence (and some organisms still do today). Then, about 1.5 billion to 2 billion years ago, something remarkable evolved: cells that produce genetically variable offspring through sexual reproduction.

The evolution of sexual reproduction is among the most significant events in the history of life. By shuffling the genetic information from two parents, sexual reproduction greatly increases the amount of genetic variation and allows for accelerated evolution. Most of the tremendous diversity of life on Earth is a direct result of sexual reproduction.

Sexual reproduction consists of two processes. The first is **meiosis**, which leads to *gametes* in which the number of chromosomes is reduced by half. The second process is **fertilization**, in which two haploid gametes fuse and restore the number of chromosomes to its original diploid value.

Meiosis

The words *mitosis* and *meiosis* are sometimes confused. They sound a bit alike, and both refer to chromosome division and cytokinesis. But don't be deceived. The outcomes of mitosis and meiosis are radically different, and several unique events that have important genetic consequences take place only in meiosis.

How does meiosis differ from mitosis? Mitosis consists of a single nuclear division and is usually accompanied by a single cell division. Meiosis, on the other hand, consists of two divisions. After mitosis, chromosome number in the newly formed cells is the same as that in the original cell, whereas meiosis causes chromosome number in the newly formed cells to be reduced by half. Finally, mitosis produces genetically identical cells, whereas meiosis produces genetically variable cells. Let's see how these differences arise.

Like mitosis, meiosis is preceded by interphase, which includes G_1, S, and G_2 phases. Meiosis consists of two distinct processes: *meiosis I* and *meiosis II*, each of which includes a cell division. The first division, which comes at the end of meiosis I, is termed the reduction division because the number of chromosomes per cell is reduced by half (**Figure 2.11**). The second division, which comes at the end of meiosis II, is sometimes termed the equational division. The events of meiosis II are similar to those of mitosis. However, meiosis II differs from mitosis in that chromosome number has already been halved in meiosis I, and the cell does not begin with the same number of chromosomes as it does in mitosis (see Figure 2.11).

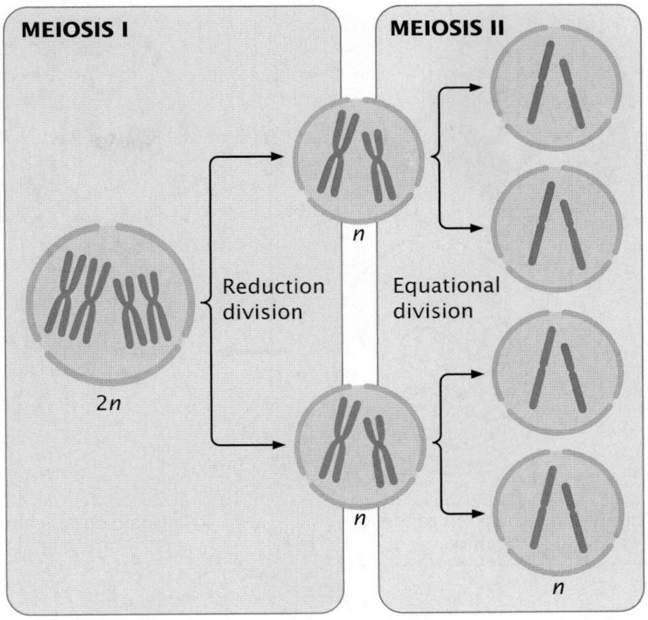

2.11 Meiosis includes two cell divisions. In this illustration, the original cell is $2n = 4$. After two meiotic divisions, each resulting cell is $1n = 2$.

MEIOSIS I The stages of meiosis are outlined in **Figure 2.12**. During interphase, the chromosomes are relaxed and visible as diffuse chromatin. **Prophase I** is a lengthy stage in which the chromosomes form homologous pairs and crossing over takes place. First, the chromosomes condense, pair up, and begin **synapsis**, a very close pairing association. Each homologous pair of synapsed chromosomes, called a **bivalent** or **tetrad**, consists of four chromatids. The chromosomes become shorter and thicker. **Crossing over**, a process in which homologous chromosomes exchange genetic information, takes place. Crossing over generates genetic variation, as we will see shortly, and is essential for the proper alignment and separation of homologous chromosomes. Each location where two chromosomes cross is called a *chiasma* (plural, *chiasmata*). The centromeres of the paired chromosomes then move apart; the two homologs remain attached at each chiasma. Near the end of prophase I, the nuclear membrane breaks down and the spindle forms.

Metaphase I is initiated when homologous pairs of chromosomes align along the metaphase plate (see Figure 2.12). A microtubule from one pole attaches to one chromosome of a homologous pair, and a microtubule from the other pole attaches to the other member of the pair. **Anaphase I** is marked by the separation of homologous chromosomes. The two chromosomes of a homologous pair are pulled toward opposite poles. Although the homologous chromosomes separate, the sister chromatids remain attached and travel together. In **telophase I**, the chromosomes arrive at the spindle poles and the cytoplasm divides.

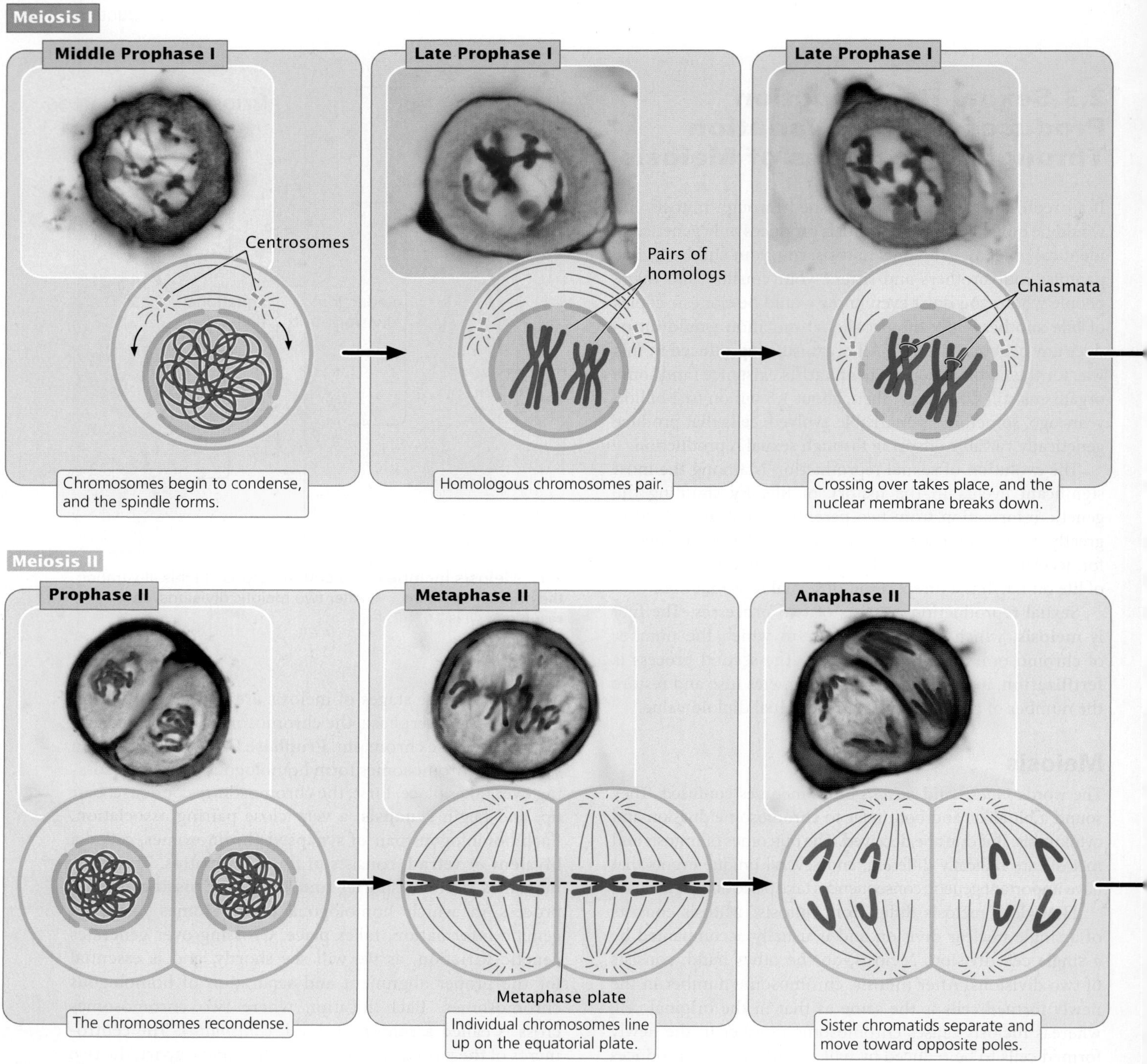

2.12 Meiosis is divided into stages. [University of Wisconsin Plant Teaching Collection, photographs by Michael Clayton.]

MEIOSIS II In the period between meiosis I and meiosis II, called **interkinesis**, the nuclear membrane re-forms around the chromosomes clustered at each pole, the spindle breaks down, and the chromosomes relax. The cells then pass through **prophase II**, in which the events of interkinesis are reversed: the chromosomes recondense, the spindle re-forms, and the nuclear envelope once again breaks down. In interkinesis in some types of cells, the chromosomes remain condensed, and the spindle does not break down. These cells move directly from cytokinesis into **metaphase II**, which is similar to metaphase of mitosis: the individual chromosomes line up on the metaphase plate, with the sister chromatids facing opposite poles.

In **anaphase II**, the kinetochores of the sister chromatids separate and the chromatids are pulled to opposite poles. Each chromatid is now a distinct chromosome. In **telophase II**, the chromosomes arrive at the spindle poles, a nuclear envelope re-forms around the chromosomes, and the cytoplasm divides. The chromosomes relax and are no longer visible. To examine the details of meiosis and the

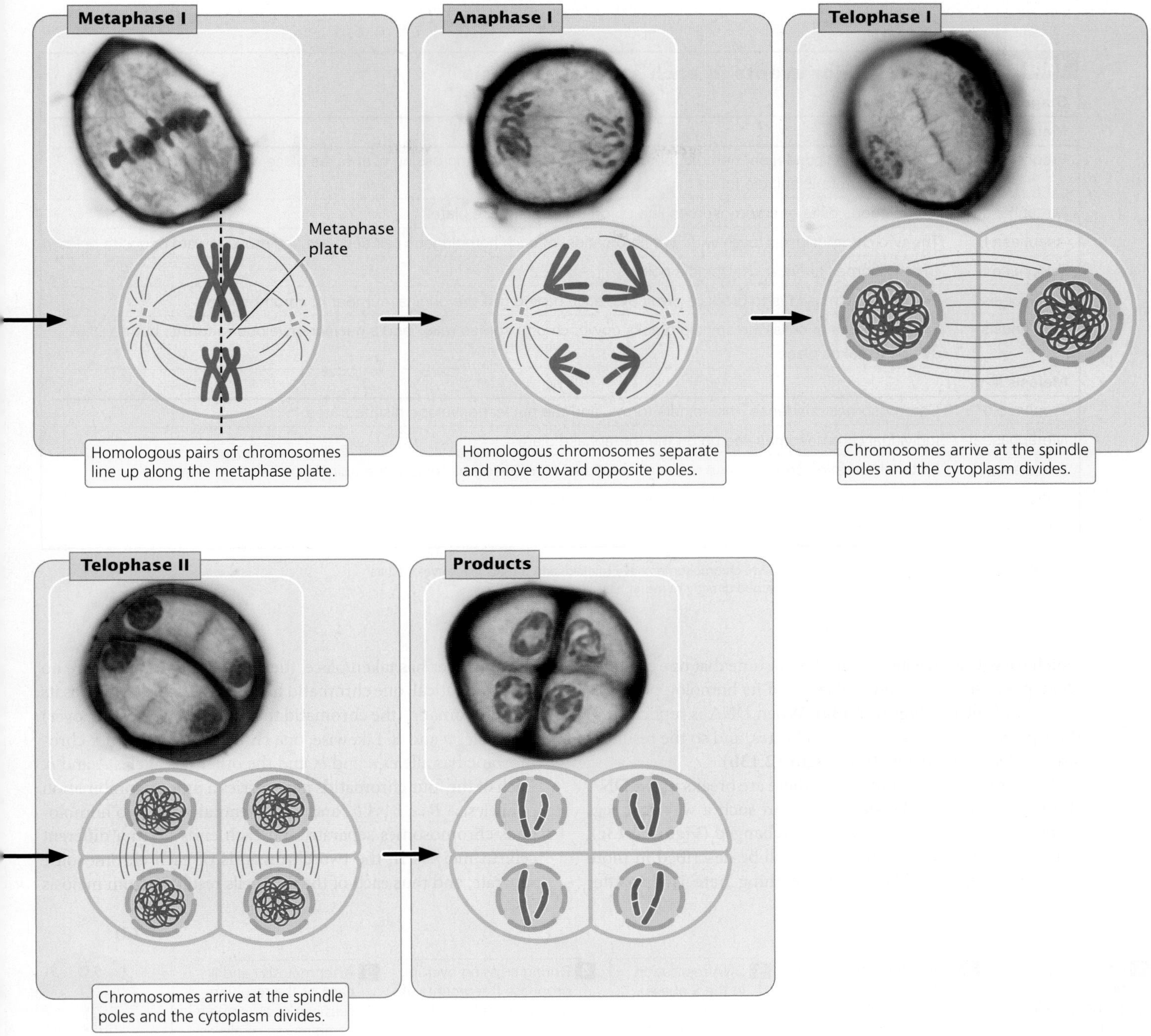

consequences of its failure, take a look at **Animation 2.2**. The major events of meiosis are summarized in **Table 2.2**.

Sources of Genetic Variation in Meiosis

What are the overall consequences of meiosis? First, meiosis comprises two divisions, so each original cell produces four cells (although there are exceptions to this generalization, as, for example, in many female animals; see Figure 2.17b). Second, chromosome number is reduced by half, so cells produced by meiosis are haploid. Third, cells produced by meiosis are genetically different from one another and from the parent cell. These genetic differences result from two processes that are unique to meiosis: crossing over and the random separation of homologous chromosomes.

CROSSING OVER Crossing over, which takes place in prophase I, refers to the exchange of genetic material between nonsister chromatids (chromatids from different homologous chromosomes). After crossing over has taken place, the sister chromatids are no longer identical. Crossing over is the basis for intrachromosomal **recombination**, creating new combinations of alleles on a chromatid. To see how crossing over produces genetic variation, consider two pairs of alleles,

TABLE 2.2 Major events in each stage of meiosis

Stage	Major events
Meiosis I	
Prophase I	Chromosomes condense, homologous chromosomes synapse, crossing over takes place, nuclear envelope breaks down, and mitotic spindle forms.
Metaphase I	Homologous pairs of chromosomes line up on the metaphase plate.
Anaphase I	The two chromosomes (each with two chromatids) of each homologous pair separate and move toward opposite poles.
Telophase I	Chromosomes arrive at the spindle poles.
Cytokinesis	The cytoplasm divides to produce two cells, each having half the original number of chromosomes.
Interkinesis	In some types of cells, the spindle breaks down, chromosomes relax, and a nuclear envelope re-forms, but no DNA synthesis takes place.
Meiosis II	
Prophase II*	Chromosomes condense, the spindle forms, and the nuclear envelope disintegrates.
Metaphase II	Individual chromosomes line up on the metaphase plate.
Anaphase II	Sister chromatids separate and move as individual chromosomes toward the spindle poles.
Telophase II	Chromosomes arrive at the spindle poles; the spindle breaks down and a nuclear envelope re-forms.
Cytokinesis	The cytoplasm divides.

*Only in cells in which the spindle has broken down, chromosomes have relaxed, and the nuclear envelope has re-formed in telophase I. Other types of cells proceed directly to metaphase II after cytokinesis.

which we will abbreviate *Aa* and *Bb*. Assume that one chromosome possesses the *A* and *B* alleles and its homolog possesses the *a* and *b* alleles (**Figure 2.13a**). When DNA is replicated in the S phase, each chromosome duplicates, and so the resulting sister chromatids are identical (**Figure 2.13b**).

In the process of crossing over, there are breaks in the DNA strands, and the breaks are repaired in such a way that segments of nonsister chromatids are exchanged (**Figure 2.13c**). The molecular basis of this process will be described in more detail in Chapter 9; the important thing here is that after crossing over has taken place, the two sister chromatids are no longer identical: one chromatid has alleles *A* and *B*, whereas its sister chromatid (the chromatid that underwent crossing over) has alleles *a* and *B*. Likewise, one chromatid of the other chromosome has alleles *a* and *b*, and the other has alleles *A* and *b*. Each of the four chromatids now carries a unique combination of alleles: $\underline{A\ B}$, $\underline{a\ B}$, $\underline{A\ b}$, and $\underline{a\ b}$. Eventually, the two homologous chromosomes separate, and each ends up in a different cell. In meiosis II, the two chromatids of each chromosome separate, and thus each of the four cells resulting from meiosis

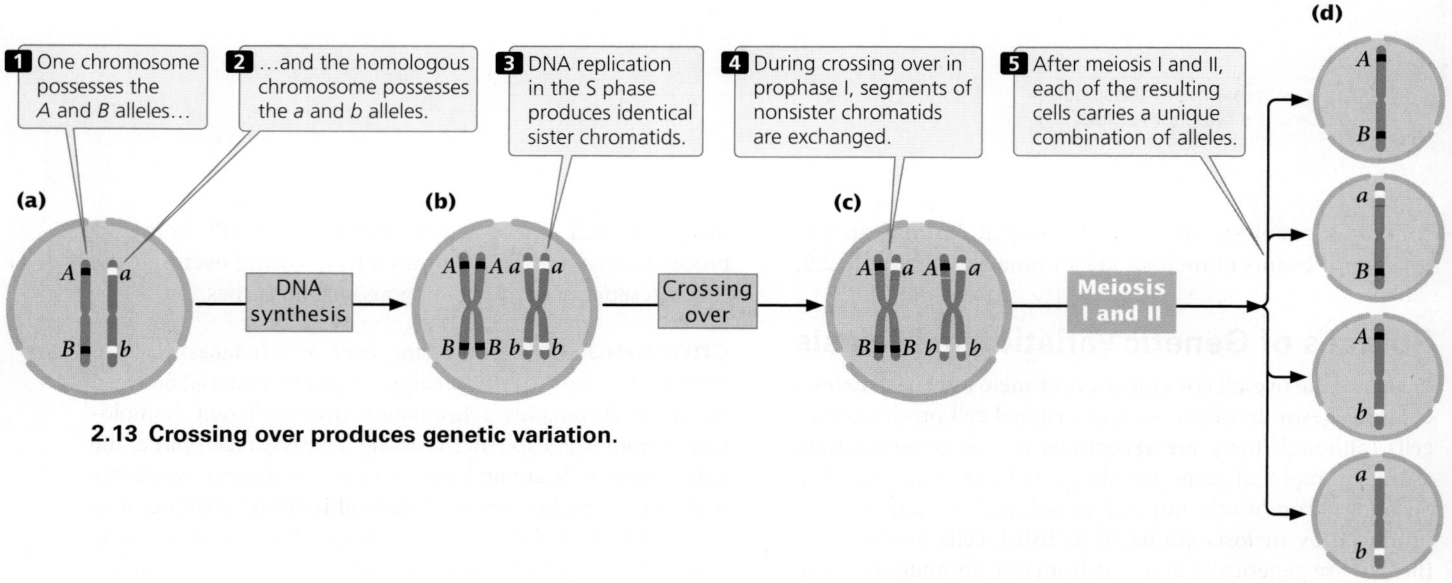

2.13 Crossing over produces genetic variation.

(a)

1 This cell has three homologous pairs of chromosomes.

2 One of each pair is maternal in origin (I_m, II_m, III_m)...

3 ...and the other is paternal (I_p, II_p, III_p).

DNA replication

4 There are four possible ways for the three pairs to align in metaphase I.

(b)

(c) Gametes

$2 \times$ $I_m\ II_m\ III_m$
$2 \times$ $I_p\ II_p\ III_p$

$2 \times$ $I_m\ II_m\ III_p$
$2 \times$ $I_p\ II_p\ III_m$

$2 \times$ $I_m\ II_p\ III_p$
$2 \times$ $I_p\ II_m\ III_m$

$2 \times$ $I_m\ II_p\ III_m$
$2 \times$ $I_p\ II_m\ III_p$

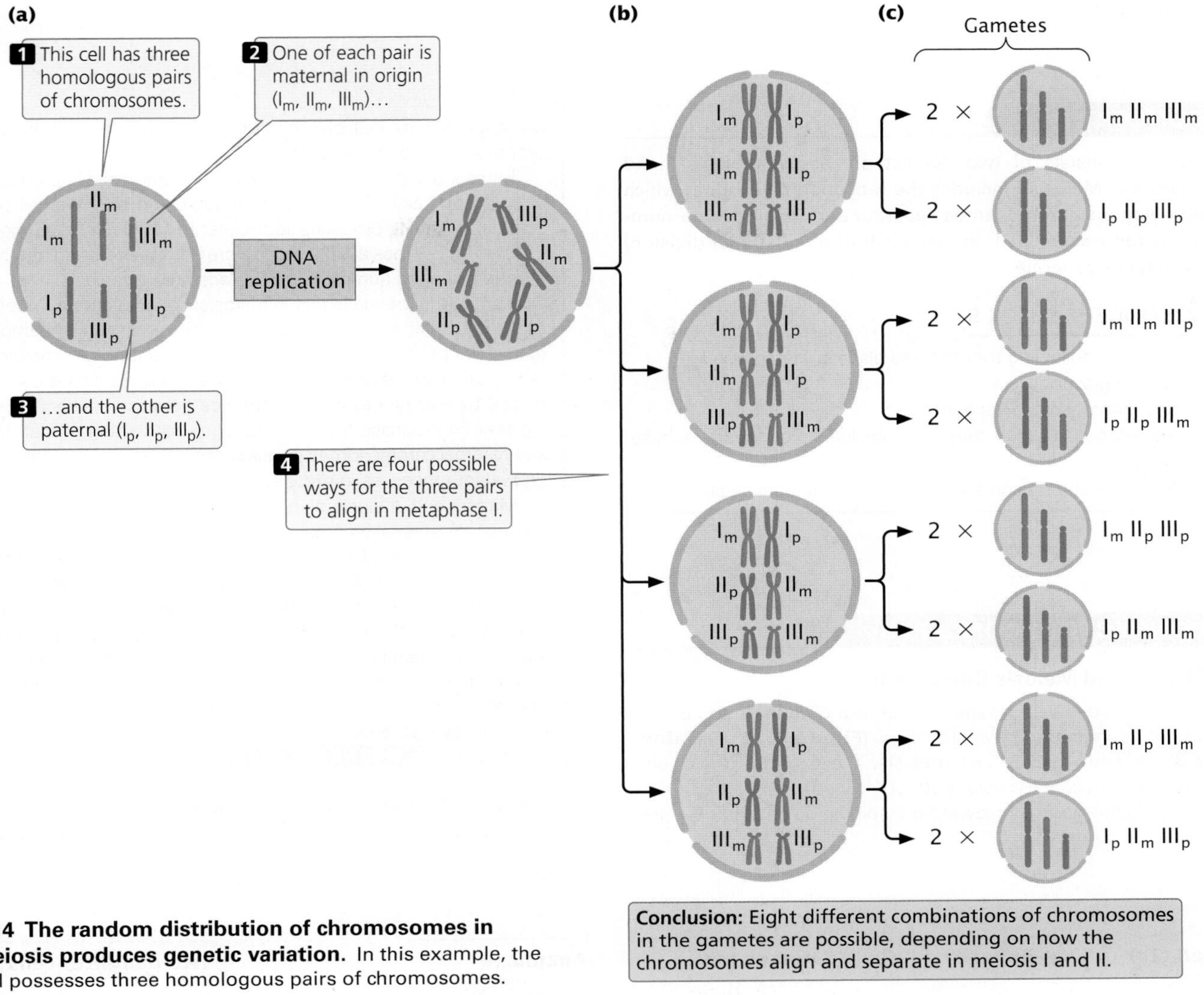

2.14 The random distribution of chromosomes in meiosis produces genetic variation. In this example, the cell possesses three homologous pairs of chromosomes.

Conclusion: Eight different combinations of chromosomes in the gametes are possible, depending on how the chromosomes align and separate in meiosis I and II.

carries a different combination of alleles (**Figure 2.13d**). You can see how crossing over affects genetic variation by viewing **Animation 2.3**.

RANDOM SEPARATION OF HOMOLOGOUS CHROMOSOMES The second process of meiosis that contributes to genetic variation is the random distribution of chromosomes in anaphase I after their random alignment in metaphase I. To illustrate this process, consider a cell with three pairs of chromosomes, I, II, and III (**Figure 2.14a**). One chromosome of each pair is maternal in origin (I_m, II_m, and III_m); the other is paternal in origin (I_p, II_p, and III_p). The chromosome pairs line up in the center of the cell in metaphase I, and in anaphase I, the chromosomes of each homologous pair separate.

How each pair of homologs aligns and separates is random and independent of how other pairs align and separate (**Figure 2.14b**). By chance, all the maternal chromosomes might migrate to one side and all the paternal chromosomes to the other. After division, one cell would contain chromosomes I_m, II_m, and III_m, and the other, I_p, II_p, and III_p. Alternatively, the I_m, II_m, and III_p chromosomes might move to one side and the I_p, II_p, and III_m chromosomes to the other. The different migrations would produce different combina-

tions of chromosomes in the resulting cells (**Figure 2.14c**). There are four ways in which the chromosomes in a diploid cell with three homologous pairs can migrate, producing a total of eight different combinations of chromosomes in the gametes. In general, the number of possible combinations is 2^n, where n equals the number of homologous pairs. As the number of chromosome pairs increases, the number of combinations quickly becomes very large. In humans, who have 23 pairs of chromosomes, 8,388,608 different combinations of chromosomes are made possible by the random separation of homologous chromosomes. You can explore the random distribution of chromosomes by viewing **Animation 2.3**. The genetic consequences of this process, termed independent assortment, will be explored in more detail in Chapter 3.

In summary, crossing over shuffles alleles on the *same* chromosome into new combinations, whereas the random distribution of maternal and paternal chromosomes shuffles alleles on *different* chromosomes into new combinations. Together, these two processes are capable of producing tremendous amounts of genetic variation among the cells resulting from meiosis. ▶TRY PROBLEMS 29 AND 30

THINK-PAIR-SHARE Question 5

CONCEPTS

Meiosis consists of two distinct processes: meiosis I and meiosis II. Meiosis I includes the reduction division, in which homologous chromosomes separate and chromosome number is reduced by half. In meiosis II (the equational division) chromatids separate.

✓ CONCEPT CHECK 5

Which of the following events takes place in metaphase I?

a. Crossing over occurs.
b. The chromosomes condense.
c. Homologous pairs of chromosomes line up on the metaphase plate.
d. Individual chromosomes line up on the metaphase plate.

CONNECTING CONCEPTS

Mitosis and Meiosis Compared

Now that we have examined the details of mitosis and meiosis, let's compare the two processes (**Figure 2.15** and **Table 2.3**). In both mitosis and meiosis, the chromosomes condense and become visible; both processes include the movement of chromosomes toward the spindle poles; and both are accompanied by cell division. Beyond these similarities, the processes are quite different.

Mitosis results in a single cell division and usually produces two daughter cells. Meiosis, in contrast, comprises two cell divisions and usually produces four cells. In diploid cells, homologous chromosomes are present before both meiosis and mitosis, but the pairing of homologs takes place only in meiosis.

Another difference is that in meiosis, chromosome number is reduced by half as a consequence of the separation of homologous pairs of chromosomes in anaphase I, but no chromosome reduction takes place in mitosis. Furthermore, meiosis is characterized by two processes that produce genetic variation: crossing over (in prophase I) and the random distribution of maternal and paternal chromosomes (in anaphase I). There are normally no equivalent processes in mitosis.

Mitosis and meiosis also differ in the behavior of chromosomes in metaphase and anaphase. In metaphase I of meiosis, *homologous pairs* of chromosomes line up on the metaphase plate, whereas *individual chromosomes* line up on the metaphase plate in metaphase of mitosis (and in metaphase II of meiosis). In anaphase I of meiosis, *paired chromosomes* (each possessing two chromatids attached at the centromere) separate and migrate toward opposite spindle poles. In contrast, in anaphase of mitosis (and in anaphase II of meiosis), *sister chromatids* separate, and each chromosome that moves toward a spindle pole is unreplicated. ▶TRY PROBLEMS 25 AND 26

THINK-PAIR-SHARE Questions 6 and 7

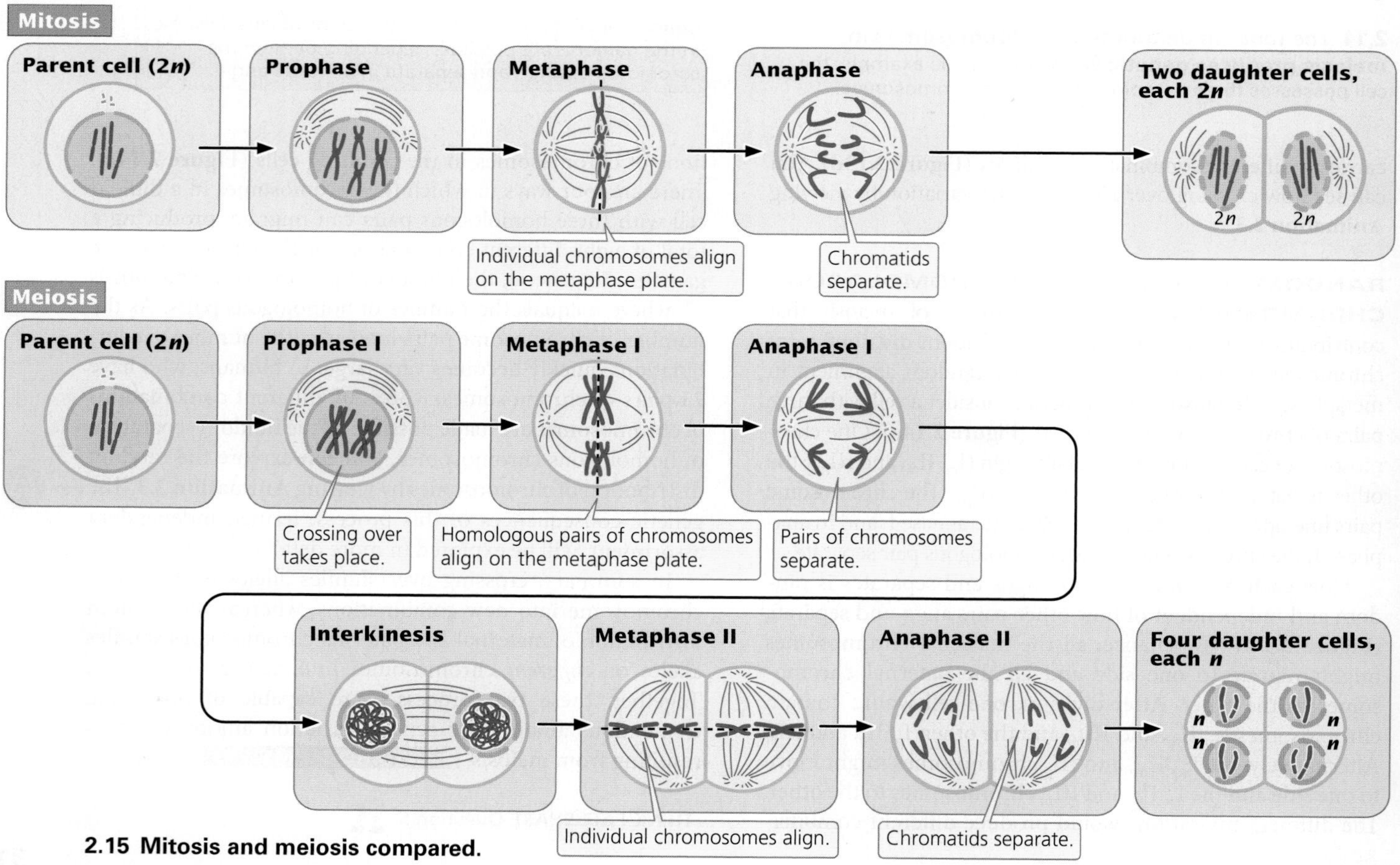

2.15 Mitosis and meiosis compared.

TABLE 2.3 Mitosis, meiosis I, and meiosis II compared			
Event	**Mitosis**	**Meiosis I**	**Meiosis II**
Cell division	Yes	Yes	Yes
Chromosome reduction	No	Yes	No
Genetic variation produced	No	Yes	No
Crossing over	No	Yes	No
Random distribution of maternal and paternal chromosomes	No	Yes	No
Metaphase	Individual chromosomes line up	Homologous pairs line up	Individual chromosomes line up
Anaphase	Chromatids separate	Homologous chromosomes separate	Chromatids separate

The Separation of Sister Chromatids and Homologous Chromosomes

In recent years, some of the molecules required for the joining and separation of chromatids and homologous chromosomes have been identified. **Cohesin**, a protein that holds chromatids together, is key to the behavior of chromosomes in mitosis and meiosis (**Figure 2.16**). The sister chromatids are held together by cohesin, which is established in the S phase and persists through G_2 and early mitosis. In anaphase of mitosis, cohesin along the entire length of the chromosome is broken down by an enzyme called separase, allowing the sister chromatids to separate.

In anaphase I of meiosis, cohesin along the chromosome arms is broken, allowing the two homologs to separate (see Figure 2.16b). However, cohesin at the centromere is protected by a protein called shugoshin, which means "guardian spirit" in Japanese. Because of this protective action by shugoshin, the

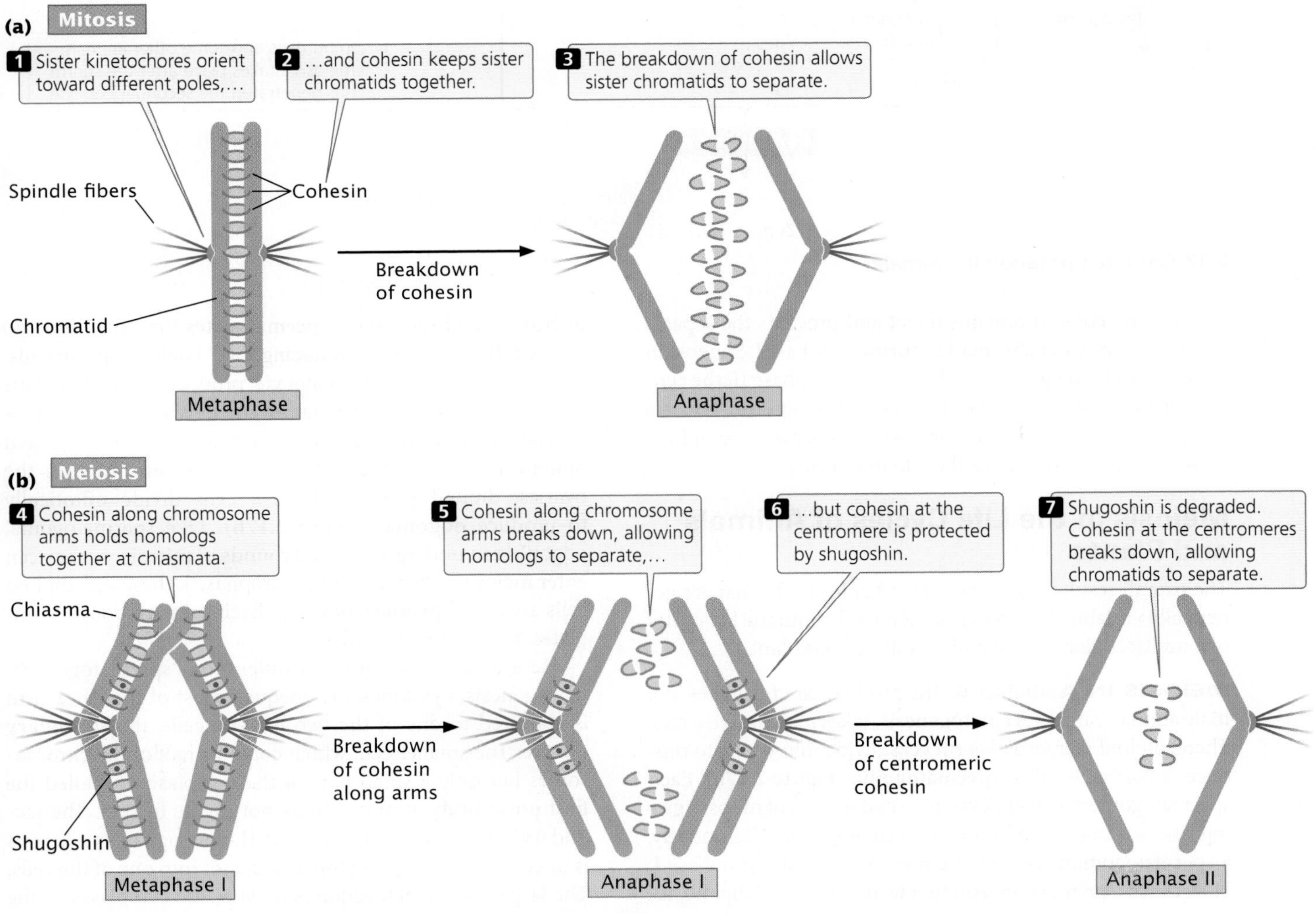

2.16 Cohesin controls the separation of chromatids and chromosomes in mitosis and meiosis.

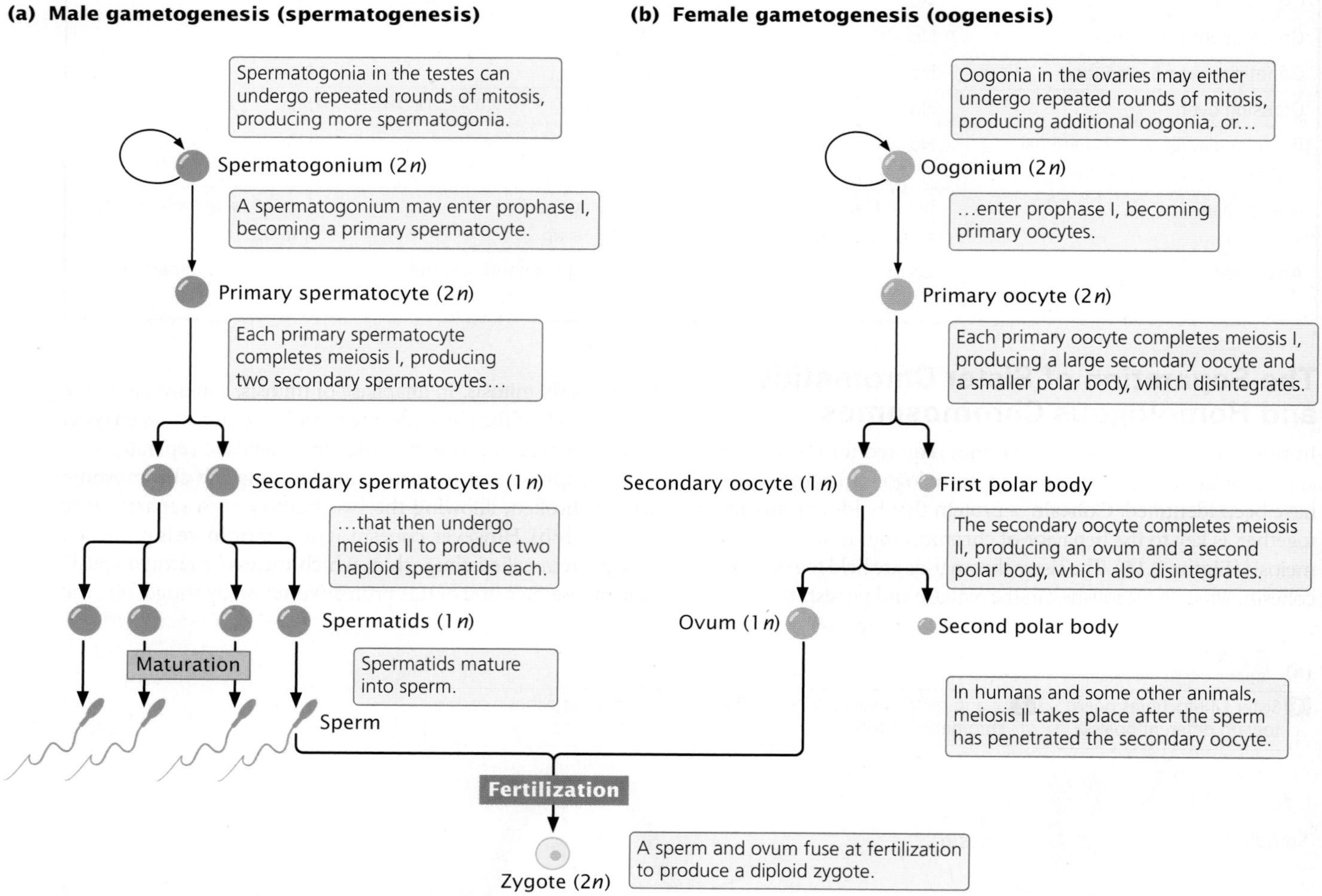

2.17 Gamete formation in animals.

centromeric cohesin remains intact and prevents the separation of the two sister chromatids during anaphase I. Shugoshin is subsequently degraded. At the end of metaphase II, the centromeric cohesin—no longer protected by shugoshin—breaks down, allowing the sister chromatids to separate in anaphase II (see Figure 2.16b), just as they do in mitosis.

Meiosis in the Life Cycles of Animals and Plants

The overall result of meiosis is four haploid cells that are genetically variable. Let's see how meiosis fits into the life cycles of a multicellular animal and a multicellular plant.

MEIOSIS IN ANIMALS The production of gametes in a male animal, called **spermatogenesis**, takes place in the testes. There, diploid primordial germ cells divide mitotically to produce diploid cells called **spermatogonia** (**Figure 2.17a**). Each spermatogonium can undergo repeated rounds of mitosis, giving rise to numerous additional spermatogonia. Alternatively, a spermatogonium can initiate meiosis and enter prophase I. Now called a **primary spermatocyte**, the cell is still diploid because the homologous chromosomes have not yet separated. Each primary spermatocyte completes meiosis I, giving rise to two haploid **secondary spermatocytes** that then undergo meiosis II, with each producing two haploid **spermatids**. Thus, each primary spermatocyte produces a total of four haploid spermatids, which mature and develop into sperm.

The production of gametes in a female animal, called **oogenesis**, begins much like spermatogenesis. Within the ovaries, diploid primordial germ cells divide mitotically to produce **oogonia** (**Figure 2.17b**). Like spermatogonia, oogonia can undergo repeated rounds of mitosis, or they can enter meiosis. When they enter prophase I, these still-diploid cells are called **primary oocytes**. Each primary oocyte completes meiosis I and divides.

Here, oogenesis begins to differ from spermatogenesis. In oogenesis, cytokinesis is unequal: most of the cytoplasm is allocated to one of the two haploid cells, the **secondary oocyte**. The smaller cell, which contains half of the chromosomes but only a small part of the cytoplasm, is called the **first polar body**; it may or may not divide further. The secondary oocyte completes meiosis II, and, again, cytokinesis is unequal: most of the cytoplasm passes into one of the cells. The larger cell, which acquires most of the cytoplasm, is the **ovum**, the mature female gamete. The smaller cell is the **second polar body**. Only the ovum is capable of being fertilized,

CONCEPTS

In the testes, a diploid spermatogonium undergoes meiosis, producing a total of four haploid sperm cells. In the ovary, a diploid oogonium undergoes meiosis to produce a single large ovum and smaller polar bodies that normally disintegrate.

✓ CONCEPT CHECK 6

A secondary spermatocyte has 12 chromosomes. How many chromosomes will be found in the primary spermatocyte that gave rise to it?

a. 6
b. 12
c. 18
d. 24

and the polar bodies usually disintegrate. Oogenesis, then, produces a single mature gamete from each primary oocyte.

MEIOSIS IN PLANTS Most multicellular plants and algae have a complex life cycle that includes two distinct structures (generations): a multicellular diploid *sporophyte* and a multicellular haploid *gametophyte*. These two generations alternate: the sporophyte produces haploid spores through meiosis, and the gametophyte produces haploid gametes through mitosis (**Figure 2.18**). This type of life cycle is sometimes called *alternation of generations*. In this cycle, the immediate products of meiosis are called *spores*, not gametes; the spores undergo one or more mitotic divisions to produce gametes. Although the terms used for this process are somewhat different from those commonly used for animals (and from some of those employed so far in this chapter), the processes in plants and animals are basically the same: in both, meiosis leads to a reduction in chromosome number, producing haploid cells.

In flowering plants, the sporophyte is the obvious, vegetative part of the plant; the gametophyte consists of only a few haploid cells within the sporophyte. The flower, which is part of the sporophyte, contains the reproductive structures. The male part of the flower, the stamen, contains diploid reproductive cells called **microsporocytes**, each of which undergoes meiosis to produce four haploid **microspores** (**Figure 2.19a**). Each microspore divides mitotically, producing an immature pollen grain consisting of two haploid nuclei. One of these nuclei, called the tube nucleus, directs the growth of a pollen tube. The other, termed the generative nucleus, divides mitotically to produce two sperm cells. The pollen grain, with its two haploid nuclei, is the male gametophyte.

The female part of the flower, the ovary, contains diploid cells called **megasporocytes**, each of which undergoes meiosis to produce four haploid **megaspores** (**Figure 2.19b**), only one of which survives. The nucleus of the surviving megaspore divides mitotically three times, producing a total of eight haploid nuclei that make up the female gametophyte, the embryo sac. Division of the cytoplasm then produces separate cells, one of which becomes the *egg*.

When the plant flowers, the stamens open and release pollen grains. Pollen lands on a flower's stigma—a sticky platform that sits on top of a long stalk called the style. At the base of the style is the ovary. If a pollen grain germinates, it grows a tube down the style into the ovary. The two sperm cells pass down this tube and enter the embryo sac (**Figure 2.19c**). One of the sperm cells fertilizes the egg cell, producing a diploid zygote, which develops into an embryo. The other sperm cell fuses with two nuclei enclosed in a single cell, giving rise to a $3n$ (triploid) endosperm, which stores food that will be used later by the embryonic plant. These two fertilization events are termed *double fertilization*.

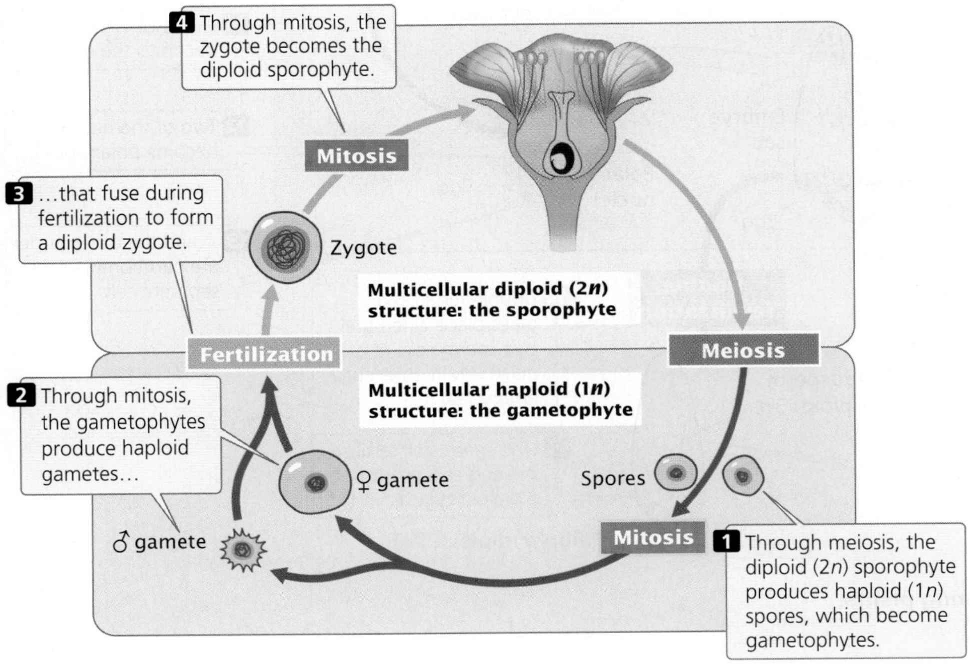

2.18 Plants alternate between diploid and haploid life stages (female, ♀; male, ♂).

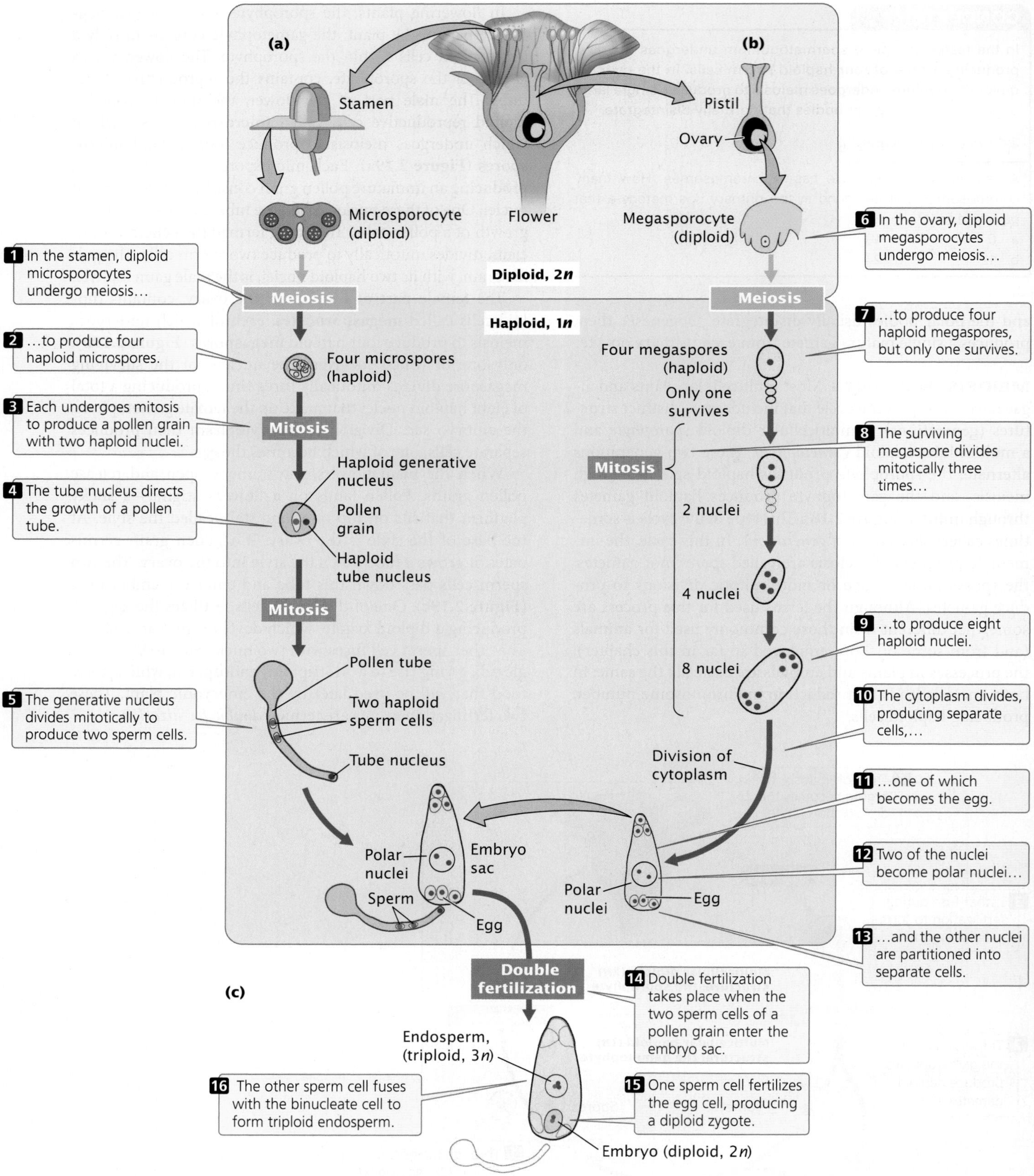
(a)
(b)
Stamen
Pistil
Ovary
Microsporocyte (diploid)
Flower
Megasporocyte (diploid)
Diploid, 2n
Haploid, 1n
1 In the stamen, diploid microsporocytes undergo meiosis…
Meiosis
Meiosis
6 In the ovary, diploid megasporocytes undergo meiosis…
2 …to produce four haploid microspores.
Four microspores (haploid)
Four megaspores (haploid)
7 …to produce four haploid megaspores, but only one survives.
Only one survives
3 Each undergoes mitosis to produce a pollen grain with two haploid nuclei.
Mitosis
Mitosis
8 The surviving megaspore divides mitotically three times…
Haploid generative nucleus
Pollen grain
Haploid tube nucleus
2 nuclei
4 The tube nucleus directs the growth of a pollen tube.
4 nuclei
Mitosis
9 …to produce eight haploid nuclei.
8 nuclei
Pollen tube
5 The generative nucleus divides mitotically to produce two sperm cells.
Two haploid sperm cells
Tube nucleus
10 The cytoplasm divides, producing separate cells,…
Division of cytoplasm
11 …one of which becomes the egg.
Polar nuclei
Embryo sac
Sperm
Egg
Polar nuclei
Egg
12 Two of the nuclei become polar nuclei…
13 …and the other nuclei are partitioned into separate cells.
(c)
Double fertilization
14 Double fertilization takes place when the two sperm cells of a pollen grain enter the embryo sac.
Endosperm, (triploid, 3n)
16 The other sperm cell fuses with the binucleate cell to form triploid endosperm.
15 One sperm cell fertilizes the egg cell, producing a diploid zygote.
Embryo (diploid, 2n)

2.19 Sexual reproduction in flowering plants.

CONCEPTS

In the stamen of a flowering plant, meiosis produces haploid microspores that divide mitotically to produce haploid sperm in a pollen grain. Within the ovary, meiosis produces four haploid megaspores, only one of which divides mitotically three times to produce eight haploid nuclei. After pollination, one sperm fertilizes the egg cell, producing a diploid zygote; the other fuses with two nuclei to form the endosperm.

✓ CONCEPT CHECK 7

Which structure is diploid?

a. Microspore c. Megaspore
b. Egg d. Microsporocyte

We have now examined the place of meiosis in the sexual cycles of two organisms, a typical multicellular animal and a flowering plant. These cycles are just two of the many variations found among eukaryotic organisms. Although the cellular processes that produce reproductive cells in plants and animals differ in the number of cell divisions, the number of haploid gametes produced, and the relative sizes of the final products, the overall result is the same: meiosis gives rise to haploid, genetically variable cells that then fuse during fertilization to produce diploid progeny. ▶ TRY PROBLEMS 32 AND 34

CONCEPTS SUMMARY

- A prokaryotic cell possesses a simple structure, with no nuclear envelope and usually a single circular chromosome. A eukaryotic cell possesses a more complex structure, with a nucleus and multiple linear chromosomes consisting of DNA associated with histone proteins.
- Cell reproduction requires the copying of genetic material, separation of the copies, and cell division.
- In a prokaryotic cell, the single chromosome replicates, the two copies move toward opposite sides of the cell, and the cell divides. In eukaryotic cells, reproduction is more complex than in prokaryotic cells, requiring mitosis and meiosis to ensure that a complete set of genetic information is transferred to each new cell.
- In eukaryotic cells, chromosomes are typically found in homologous pairs. Each functional chromosome consists of a centromere, telomeres, and multiple origins of replication. After a chromosome has been copied, the two copies remain attached at the centromere, forming sister chromatids.
- The cell cycle consists of the stages through which a eukaryotic cell passes between cell divisions. It consists of interphase, in which the cell grows and prepares for division, and the M phase, in which nuclear and cell division take place. The M phase consists of mitosis, the process of nuclear division, and cytokinesis, the division of the cytoplasm.
- Mitosis usually results in the production of two genetically identical cells.
- Sexual reproduction produces genetically variable progeny and allows for accelerated evolution. It includes meiosis, in which haploid sex cells are produced, and fertilization, the fusion of sex cells. Meiosis includes two cell divisions. In meiosis I, crossing over takes place and homologous chromosomes separate. In meiosis II, chromatids separate.
- The usual result of meiosis is the production of four haploid cells that are genetically variable. Genetic variation in meiosis is produced by crossing over and by the random distribution of maternal and paternal chromosomes.
- In animals, a diploid spermatogonium undergoes meiosis to produce four haploid sperm cells. A diploid oogonium undergoes meiosis to produce one large haploid ovum and one or more smaller polar bodies.
- In plants, a diploid microsporocyte in the stamen undergoes meiosis to produce four pollen grains, each with two haploid sperm cells. In the ovary, a diploid megasporocyte undergoes meiosis to produce eight haploid nuclei, one of which forms the egg.

IMPORTANT TERMS

prokaryote (p. 18)
eukaryote (p. 19)
bacteria (p. 19)
archaea (p. 19)
nucleus (p. 19)
histone (p. 19)
chromatin (p. 19)
homologous pair (p. 21)
diploid (p. 21)
haploid (p. 21)
polyploid (p. 21)
telomere (p. 22)
origin of replication (p. 22)
sister chromatid (p. 22)
cell cycle (p. 22)
checkpoint (p. 22)
interphase (p. 22)
M phase (p. 23)
mitosis (p. 23)
cytokinesis (p. 23)
prophase (p. 25)
prometaphase (p. 25)
metaphase (p. 25)
anaphase (p. 25)
telophase (p. 25)
meiosis (p. 27)
fertilization (p. 27)
prophase I (p. 27)
synapsis (p. 27)
bivalent (p. 27)
tetrad (p. 27)
crossing over (p. 27)
metaphase I (p. 27)
anaphase I (p. 27)
telophase I (p. 27)
interkinesis (p. 28)
prophase II (p. 28)
metaphase II (p. 28)
anaphase II (p. 28)

telophase II (p. 28)
recombination (p. 29)
cohesin (p. 33)
spermatogenesis (p. 34)
spermatogonium (p. 34)
primary spermatocyte (p. 34)
secondary spermatocyte (p. 34)
spermatid (p. 34)
oogenesis (p. 34)
oogonium (p. 34)
primary oocyte (p. 34)
secondary oocyte (p. 34)
first polar body (p. 34)
ovum (p. 34)
second polar body (p. 34)
microsporocyte (p. 35)
microspore (p. 35)
megasporocyte (p. 35)
megaspore (p. 35)

ANSWERS TO CONCEPT CHECKS

1. Bacteria and archaea are prokaryotes. They differ from eukaryotes in possessing no nucleus, a genome that usually consists of a single circular chromosome, and a small amount of DNA.

2. b

3. The kinetochore is the point at which spindle microtubules attach to the chromosome. If the kinetochore were missing, spindle microtubules would not attach to the chromosome, the chromosome would not be drawn into the newly formed nuclei, and the resulting cells would be missing a chromosome.

4. a

5. c

6. d

7. d

WORKED PROBLEM

A student examines a thin section of an onion-root tip and records the number of cells that are in each stage of the cell cycle. She observes 94 cells in interphase, 14 cells in prophase, 3 cells in prometaphase, 3 cells in metaphase, 5 cells in anaphase, and 1 cell in telophase. If the complete cell cycle in an onion-root tip requires 22 hours, what is the average duration of each stage in the cycle? Assume that all cells are in the active cell cycle (not G_0).

Solution Strategy

What information is required in your answer to the problem?

The average duration of each stage of the cell cycle.

What information is provided to solve the problem?

- The numbers of cells in different stages of the cell cycle and the time required for a complete cell cycle

For help with this problem, review:

- The Cell Cycle and Mitosis in Section 2.2.

Solution Steps

This problem is solved in two steps. First, we calculate the proportions of cells in each stage of the cell cycle, which correspond to the amount of time that an average cell spends in each stage. For example, if cells spend 90% of their time in interphase, then, at any given moment, 90% of the cells will be in interphase. The second step is to convert the proportions into lengths of time, which is done by multiplying the proportions by the total time of the cell cycle (22 hours).

STEP 1 Calculate the proportion of cells at each stage.

The proportion of cells at each stage is equal to the number of cells found in that stage divided by the total number of cells examined:

Interphase	$^{94}/_{120} = 0.783$
Prophase	$^{14}/_{120} = 0.117$
Prometaphase	$^{3}/_{120} = 0.025$
Metaphase	$^{3}/_{120} = 0.025$
Anaphase	$^{5}/_{120} = 0.042$
Telophase	$^{1}/_{120} = 0.008$

Hint: The total of all the proportions should equal 1.0.

We can check our calculations by making sure that the proportions sum to 1.0, which they do.

STEP 2 Determine the average duration of each stage.

To determine the average duration of each stage, multiply the proportion of cells in each stage by the time required for the entire cell cycle:

Hint: The total tim for all stages should equal 22 hours.

Interphase	0.783×22 hours $= 17.23$ hours
Prophase	0.117×22 hours $= 2.57$ hours
Prometaphase	0.025×22 hours $= 0.55$ hour
Metaphase	0.025×22 hours $= 0.55$ hour
Anaphase	0.042×22 hours $= 0.92$ hour
Telophase	0.008×22 hours $= 0.18$ hour

COMPREHENSION QUESTIONS

Section 2.1

1. What are some genetic differences between prokaryotic and eukaryotic cells?
2. Why are viruses often used in the study of genetics?

Section 2.2

3. List three fundamental events that must take place in cell reproduction.
4. Name three essential structural elements of a functional eukaryotic chromosome and describe their functions.
5. Sketch and identify four different types of chromosomes based on the position of the centromere.
6. List the stages of interphase and the major events that take place in each stage.
7. What are checkpoints? List some of the important checkpoints in the cell cycle.
8. List the stages of mitosis and the major events that take place in each stage.
9. What are the genetically important results of the cell cycle and mitosis?
10. Why are the two cells produced by the cell cycle genetically identical?

Section 2.3

11. What are the stages of meiosis, and what major events take place in each stage?
12. What are the major results of meiosis?
13. What two processes unique to meiosis are responsible for genetic variation? At what point in meiosis do these processes take place?
14. List some similarities and differences between mitosis and meiosis. Which differences do you think are most important and why?
15. Outline the processes of spermatogenesis and oogenesis in animals.
16. Outline the processes of male and female gamete formation in plants.

For more questions that test your comprehension of the key chapter concepts, go to **LearningCurve** for this chapter.

APPLICATION QUESTIONS AND PROBLEMS

Introduction

17. Answer the following questions about the Blind Men's Riddle, presented at the beginning of the chapter.
 a. What component of the cell cycle do the two socks of a pair represent?
 b. In the riddle, each blind man buys his own pairs of socks, but the clerk places all the pairs in one bag. Thus, there are two pairs of socks of each color in the bag (two black pairs, two blue pairs, two gray pairs, etc.). What do the two pairs (four socks in all) of each color represent?
 c. What in the riddle performs the same function as spindle fibers?

Section 2.1

18. A cell has a circular chromosome and no nuclear membrane. Its DNA is associated with some histone proteins. Does this cell belong to the bacteria, the archaea, or the eukaryotes? Explain your reasoning.

Section 2.2

19. A certain species has three pairs of chromosomes: an acrocentric pair, a metacentric pair, and a submetacentric pair. Draw a cell of this species as it would appear in metaphase of mitosis.

*20. A biologist examines a series of cells and counts 160 cells in interphase, 20 cells in prophase, 6 cells in prometaphase, 2 cells in metaphase, 7 cells in anaphase, and 5 cells in telophase. If the complete cell cycle requires 24 hours, what is the average duration of the M phase in these cells? Of metaphase?

Section 2.3

21. A certain species has three pairs of chromosomes: one acrocentric pair and two metacentric pairs. Draw a cell of this species as it would appear in the following stages of meiosis:
 a. Metaphase I
 b. Anaphase I
 c. Metaphase II
 d. Anaphase II
22. Construct a table similar to that in **Figure 2.10** for the different stages of meiosis, giving the number of chromosomes per cell and the number of DNA molecules per cell for a cell that begins with four chromosomes (two homologous pairs) in G_1. Include the following stages in your table: G_1, S, G_2, prophase I, metaphase I, anaphase I, telophase I (after cytokinesis), prophase II, metaphase II, anaphase II, and telophase II (after cytokinesis).

*23. A cell in G_1 of interphase has 12 chromosomes. How many chromosomes and DNA molecules will be found per cell when this original cell progresses to the following stages?

a. G_2 of interphase

b. Metaphase I of meiosis

c. Prophase of mitosis

d. Anaphase I of meiosis

e. Anaphase II of meiosis

f. Prophase II of meiosis

g. After cytokinesis following mitosis

h. After cytokinesis following meiosis II

24. How are the events that take place in spermatogenesis and oogenesis similar? How are they different?

*25. All of the following cells, shown in various stages of mitosis and meiosis, come from the same rare species of plant.

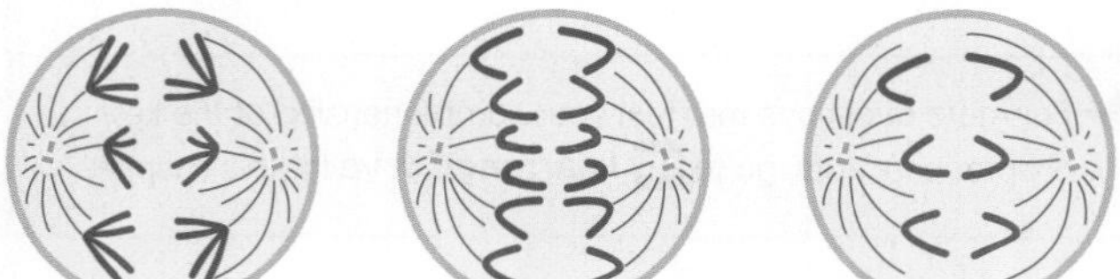

a. What is the diploid number of chromosomes in this plant?

b. Give the names of each stage of mitosis or meiosis shown.

c. Give the number of chromosomes and number of DNA molecules per cell present at each stage.

*26. The amount of DNA per cell of a particular species is measured in cells found at various stages of meiosis, and the following amounts are obtained:

Amount of DNA per cell in picograms (pg)

_____ 3.7 pg _____ 7.3 pg _____ 14.6 pg

Match the amounts of DNA above with the corresponding stages of meiosis (*a* through *f*, below). You may use more than one stage for each amount of DNA.

Stage of meiosis

a. G_1

b. Prophase I

c. G_2

d. Following telophase II and cytokinesis

e. Anaphase I

f. Metaphase II

27. A cell in prophase II of meiosis has 12 chromosomes. How many chromosomes would be present in a cell from the same organism if it were in prophase of mitosis? Prophase I of meiosis?

28. A cell has eight chromosomes in G_1 of interphase. Draw a picture of this cell with its chromosomes at the following stages. Indicate how many DNA molecules are present at each stage.

a. Metaphase of mitosis

b. Anaphase of mitosis

c. Anaphase II of meiosis

*29. The fruit fly *Drosophila melanogaster* (below left) has four pairs of chromosomes, whereas the house fly *Musca domestica* (below right) has six pairs of chromosomes. In which species would you expect to see more genetic variation among the progeny of a cross? Explain your answer.

[Left: © Graphic Science/Alamy Stock Photo. Right: © Debug/iStock Photo.]

*30. A cell has two pairs of submetacentric chromosomes, which we will call chromosomes I_a, I_b, II_a, and II_b (chromosomes I_a and I_b are homologs, and chromosomes II_a and II_b are homologs). Allele *M* is located on the long arm of chromosome I_a, and allele *m* is located at the same position on chromosome I_b. Allele *P* is located on the short arm of chromosome I_a, and allele *p* is located at the same position on chromosome I_b. Allele *R* is located on chromosome II_a and allele *r* is located at the same position on chromosome II_b.

a. Draw these chromosomes, identifying genes *M*, *m*, *P*, *p*, *R*, and *r* as they might appear in metaphase I of meiosis. Assume that there is no crossing over.

b. Taking into consideration the random separation of chromosomes in anaphase I, draw the chromosomes (with genes identified) present in all possible types of gametes that might result from this cell's undergoing meiosis. Assume that there is no crossing over.

31. A horse has 64 chromosomes and a donkey has 62 chromosomes. A cross between a female horse and a male donkey produces a mule, which is usually sterile. How many chromosomes does a mule have? Can you think of any reasons for the fact that most mules are sterile?

*32. Normal somatic cells of horses have 64 chromosomes ($2n = 64$). How many chromosomes and DNA molecules will be present in the following types of horse cells?

[© Dozornaya/iStock Photo.]

Cell type	Number of chromosomes	Number of DNA molecules
a. Spermatogonium	________	________
b. First polar body	________	________
c. Primary oocyte	________	________
d. Secondary spermatocyte	________	________

33. Indicate whether each of the following cells is haploid or diploid.

Cell type	Haploid or diploid?
Primary spermatocyte	________
Microsporocyte	________
First polar body	________
Oogonium	________
Spermatid	________
Megaspore	________
Ovum	________
Secondary oocyte	________
Spermatogonium	________

*34. A primary oocyte divides to give rise to a secondary oocyte and a first polar body. The secondary oocyte then divides to give rise to an ovum and a second polar body.

a. Is the genetic information found in the first polar body identical with that found in the secondary oocyte? Explain your answer.

b. Is the genetic information found in the second polar body identical with that in the ovum? Explain your reasoning.

CHALLENGE QUESTIONS

Section 2.3

*35. Female bees are diploid, and male bees are haploid. The haploid males produce sperm and can successfully mate with diploid females. Fertilized eggs develop into females, and unfertilized eggs develop into males. How do you think the process of sperm production in male bees differs from sperm production in other animals?

36. On average, what proportion of the genome in the following pairs of humans would be exactly the same if no crossing over occurred? (For the purposes of this question only, we will ignore the special case of the X and Y sex chromosomes and assume that all genes are located on nonsex chromosomes.)

a. Father and child

b. Mother and child

c. Two full siblings (offspring that have the same two biological parents)

d. Half siblings (offspring that have only one biological parent in common)

e. Uncle and niece

f. Grandparent and grandchild

THINK-PAIR-SHARE QUESTIONS

Introduction

1. In the blind men's riddle, two blind men must sort out ten pairs of socks so that each man gets exactly five pairs of different colored socks. In the analogy, is it important that the men are blind? In a cell, what does the blindness represent?

Section 2.2

2. A chromosome consists of two sister chromatids. Does the genetic information on the two sister chromatids come from only one parent or from both parents? Explain your reasoning.

3. Are homologous pairs of chromosomes present in mitosis? Explain your reasoning.

4. A cell has 8 chromosomes in metaphase II of meiosis. How many chromosomes and DNA molecules will be present per cell in this same organism at the following stages?
   a. Prophase of mitosis
   b. Metaphase I of meiosis
   c. Anaphase of mitosis
   d. Anaphase II of meiosis
   e. Anaphase I of meiosis
   f. After cytokinesis that follows mitosis
   g. After cytokinesis that follows meiosis II

Section 2.3

5. What is the difference between sister chromatids and homologous chromosomes?
6. List as many similarities and differences between mitosis and meiosis as you can. Which differences do you think are most important, and why?
7. Describe how and where each of the following terms applies to mitosis, meiosis, or both: (1) replication; (2) pairing; (3) separation.
8. Do you know of any genetic diseases or disorders that result from errors in mitosis or meiosis? How do errors in mitosis or meiosis bring about these diseases?

3 Basic Principles of Heredity

The Genetics of Blond Hair in the South Pacific

A thousand miles northeast of Australia lies an ancient chain of volcanic and coral islands known as the Solomons (Figure 3.1). The Solomon Islands were first inhabited some 30,000 years ago, when Neanderthals still roamed northern Europe. Today, the people of the Solomons are culturally diverse, but consist largely of Melanesians, a group that also inhabits other South Pacific islands. Most people from the Solomon Islands have dark skin. Remarkably, 5% to 10% also have strikingly blond hair; in fact, people of the Solomon Islands have the highest frequency of blond hair outside of Europe.

How did the Solomon Islanders get their blond hair? A number of hypotheses have been proposed over the years. Some suggested that the blond islanders had naturally dark hair that was bleached by the sun and salt water. Others proposed that the blond hair color was caused by diet. Still others suggested that it was the result of genes for blond hair left by early European explorers.

Blond hair occurs in 5% to 10% of dark-skinned Solomon Islanders. Research demonstrates that blond hair in this group is a recessive trait and has a different genetic basis from blond hair in Europeans. [© Anthony Asael/Danita Delimont/Alamy Stock photo.]

The mystery of the blond Solomon Islanders was solved in 2012 by geneticists Eimear Kenny and Sean Myles and their colleagues. Their research demonstrated that blond hair on the islands is, in fact, caused by a gene, but not one left by Europeans—blond hair in Solomon Islanders and in Europeans has completely separate evolutionary origins.

To search for the origin of blond hair among the people of the Solomon Islands, the geneticists collected saliva and hair samples from over 1,200 people on the islands, from which they then extracted DNA. In a type of analysis known as a genome-wide association study, they looked for statistical associations between the presence of blond hair and thousands of genetic variants scattered across the genome. Right away, they detected a strong correlation between the presence of blond hair and a particular genetic variant located on the short arm of chromosome 9. This region of chromosome 9 contains the tyrosinase-related protein 1 gene (*TYRP1*), which encodes an enzyme known to play a role in the production of melanin and to affect pigmentation in mice. The researchers found a single base difference between the DNA of islanders with blond hair and that of islanders with dark hair: the blonds had a thymine (T) base instead of cytosine (C) in their *TYRP1* gene.

Further research showed that blond hair in Solomon Islanders is a recessive trait, meaning that blonds carry two copies of the blond version of the gene (*TT*)—one inherited from each parent. Dark hair is dominant: dark-haired islanders carry either one (*CT*) or two (*CC*) copies of the dark-hair version of the gene. Thus, many dark-haired islanders are heterozygous, carrying a hidden copy of the blond gene that can be passed on to their offspring.

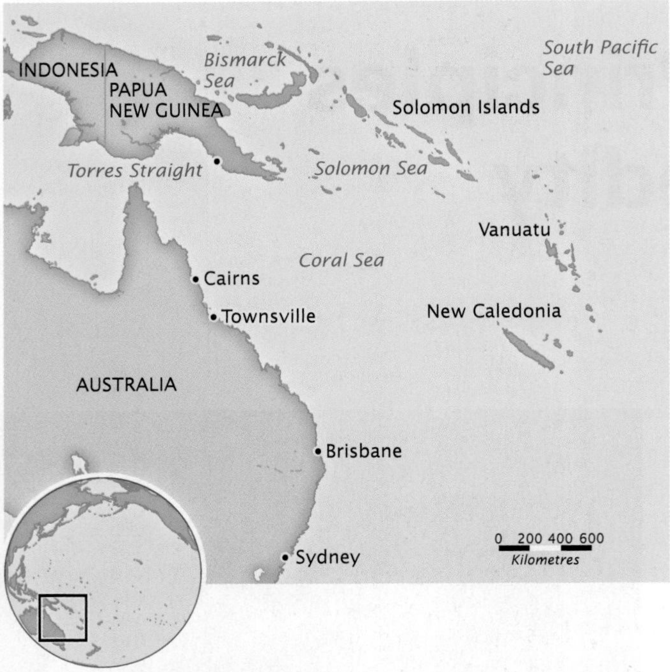

3.1 Map of the Solomon Islands.

A DNA analysis of 900 Solomon Islanders demonstrated that over 40% of dark-haired islanders carry a blond gene. Interestingly, the C to T mutation in the *TYRP1* gene that causes blond hair in Solomon Islanders is rare outside of the South Pacific, suggesting that the mutation arose independently within the Melanesian population. There is no evidence that the gene was inherited from Europeans.

THINK-PAIR-SHARE Question 1

The genetics of blond hair in Solomon Islanders differs from that in Europeans in other ways as well. In Europeans, variations in at least eight different genes have been associated with blond hair. In 2015, researchers examined one of these genes (called *KITLG*) and found that the mutation causing blond hair occurred not in the gene itself, but in a region of DNA that affects the expression of the *KITLG* gene. The *KITLG* gene produces a protein that is involved in a number of functions, including melanocyte development and melanin synthesis.

THINK-PAIR-SHARE Questions 2 and 3

This chapter is about the principles of heredity: how genes—such as the one for blond hair in Solomon Islanders—are passed from generation to generation and how factors such as dominance influence that inheritance. The principles of heredity were first put forth by Gregor Mendel, so we begin this chapter by examining Mendel's scientific achievements. We then turn to simple genetic crosses, in which a single characteristic is examined. We will consider some techniques for predicting the outcome of genetic crosses and then turn to crosses in which two or more characteristics are examined. We will see how the principles applied to simple genetic crosses and the ratios of offspring they produce can serve as the key for understanding more complicated crosses. The chapter ends with a discussion of statistical tests for analyzing crosses.

Throughout this chapter, a number of concepts are interwoven: Mendel's principles of segregation and independent assortment, probability, and the behavior of chromosomes. These concepts might at first appear to be unrelated, but they are actually different views of the same phenomenon because the genes that undergo segregation and independent assortment are located on chromosomes. This chapter aims to examine these different views and to clarify their relations.

3.1 Gregor Mendel Discovered the Basic Principles of Heredity

It was in the early 1900s that the principles of heredity first became widely known among biologists. Surprisingly, these principles had been discovered some 44 years earlier by Gregor Johann Mendel (1822–1884).

Mendel was born in what is now part of the Czech Republic. Although his parents were simple farmers with little money, he received a sound education and was admitted to the Augustinian monastery in Brno in September 1843. After graduating from seminary, Mendel became an ordained priest and was appointed to a teaching position in a local school. He excelled at teaching, and the abbot of the monastery recommended him for further study at the University of Vienna, which he attended from 1851 to 1853. There, Mendel enrolled in the newly opened Physics Institute and took courses in mathematics, chemistry, entomology, paleontology, botany, and plant physiology. It was probably there that Mendel acquired knowledge of the scientific method, which he later applied so successfully to his genetics experiments. After two years of study in Vienna, Mendel returned to Brno, where he taught school and began his experimental work with pea plants. He conducted breeding experiments from 1856 to 1863 and presented his results publicly at meetings of the Brno Natural Science Society in 1865. Mendel's paper based on these lectures was published in 1866. However, in spite of widespread interest in heredity, the effect of his research on the scientific community was minimal. At the time, no one seemed to have noticed that Mendel had discovered the basic principles of inheritance.

In 1868, Mendel was elected abbot of his monastery, and increasing administrative duties brought an end to his teaching and eventually to his genetics experiments. He died at the age of 61 on January 6, 1884, unrecognized for his contribution to genetics.

The significance of Mendel's discovery was not recognized until 1900, when three botanists—Hugo de Vries,

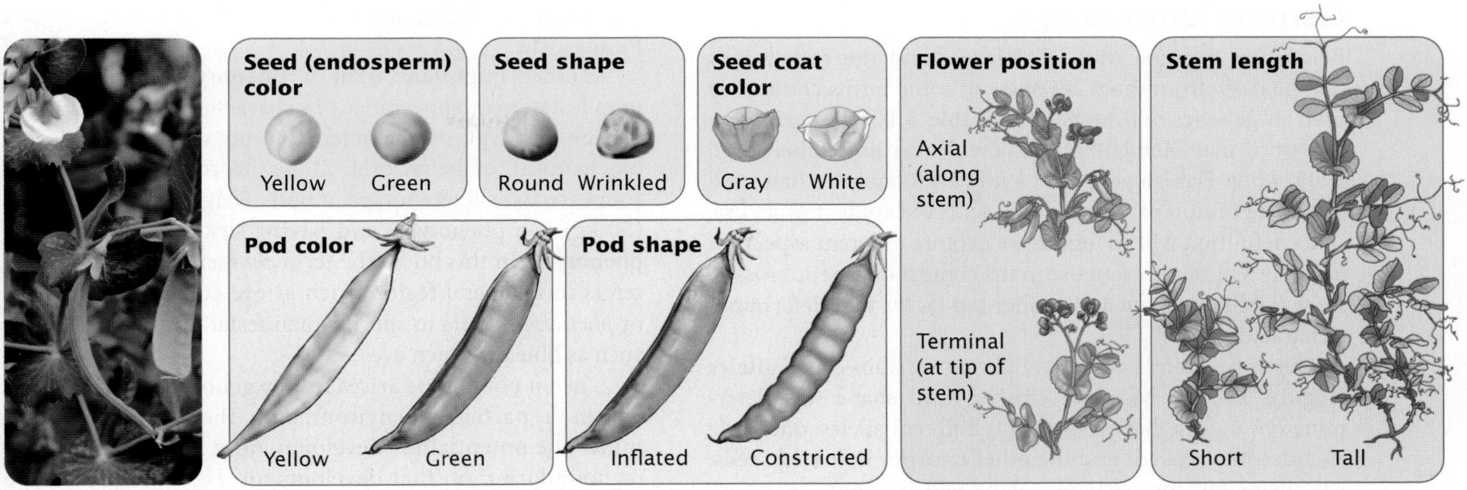

3.2 Mendel used the pea plant *Pisum sativum* in his studies of heredity. He examined seven characteristics that appeared in the seeds and in plants grown from the seeds. [© Charles Stirling/Alamy.]

Erich von Tschermak-Seysenegg, and Carl Correns—began independently conducting similar experiments with plants and arrived at conclusions similar to those of Mendel. Coming across Mendel's paper, they interpreted their results in accord with his principles and drew attention to his pioneering work.

Mendel's Success

Mendel's approach to the study of heredity was effective for several reasons. Foremost was his choice of experimental subject, the pea plant *Pisum sativum* (**Figure 3.2**), which offered clear advantages for genetic investigation. The plant is easy to cultivate, and Mendel had the monastery garden and greenhouse at his disposal. Compared with some other plants, peas grow relatively rapidly, completing an entire generation in a single growing season. By today's standards, one generation per year seems frightfully slow—fruit flies complete a generation in 2 weeks and bacteria in 20 minutes—but Mendel was under no pressure to publish quickly and was able to follow the inheritance of individual characteristics for several generations. Had he chosen to work on an organism with a longer generation time—horses, for example—he might never have discovered the basis of inheritance. Pea plants also produce many offspring—their seeds—which allowed Mendel to detect meaningful mathematical ratios in the traits he observed in the progeny. In addition, the numerous varieties of peas that were available to Mendel were crucial because these varieties differed in various traits and were genetically pure. Mendel was therefore able to begin with plants of variable, known genetic makeup.

Much of Mendel's success can be attributed to the seven characteristics of pea plants that he chose for study (see Figure 3.2). He avoided characteristics that display a range of variation; instead, he focused his attention on those that exist in two easily differentiated forms, such as white versus gray seed coats, round versus wrinkled seeds, and inflated versus constricted pods.

Finally, Mendel was successful because he adopted an experimental approach and interpreted his results by using mathematics. Unlike many earlier investigators who simply described the *results* of crosses, Mendel formulated *hypotheses* based on his initial observations and then conducted additional crosses to test his hypotheses. He kept careful records of the numbers of progeny possessing each type of trait and computed ratios of the different types. He was adept at seeing patterns in detail and was patient and thorough, conducting his experiments for 10 years before attempting to write up his results. TRY PROBLEM 14

THINK-PAIR-SHARE Question 4

CONCEPTS

Gregor Mendel put forth the basic principles of inheritance, publishing his findings in 1866. Much of Mendel's success can be attributed to the seven characteristics of pea plants that he studied.

✓ CONCEPT CHECK 1

Which of the following factors did not contribute to Mendel's success in his study of heredity?

a. His use of the pea plant
b. His study of plant chromosomes
c. His adoption of an experimental approach
d. His use of mathematics

Genetic Terminology

Before we examine Mendel's crosses and the conclusions that he drew from them, a review of some terms commonly used in genetics will be helpful (**Table 3.1**). The term *gene* is a word that Mendel never knew. It was not coined until 1909, when Danish geneticist Wilhelm Johannsen first used it. The definition of a gene varies with the context of its use, so its definition will change as we explore different aspects of heredity. For our present use in the context of genetic crosses, we will define a **gene** as an inherited factor that determines a characteristic.

Genes frequently come in different versions called **alleles** (**Figure 3.3**). In Mendel's crosses, seed shape was determined by a gene that exists as two different alleles: one allele encodes round seeds and the other encodes wrinkled seeds. All alleles for any particular gene will be found at a specific place on a chromosome called the **locus** for that gene. (The plural of locus is **loci**; it's bad form in genetics—and incorrect—to speak of "locuses.") Thus, there is a specific place—a locus—on a chromosome in pea plants where the shape of seeds is determined. This locus might be occupied by an allele for round seeds or one for wrinkled seeds. We will use the term *allele* when referring to a specific version of a gene; we will use the term *gene* to refer more generally to any allele at a locus.

The **genotype** is the set of alleles that an individual organism possesses. A diploid organism with a genotype consisting of two identical alleles is **homozygous** for that locus. One that has a genotype consisting of two different alleles is **heterozygous** for the locus.

Another important term is **phenotype**, which is the manifestation or appearance of a characteristic. A phenotype can be any type of characteristic—physical, physiological, biochemical, or behavioral. Thus, the condition of having round seeds is a phenotype, a body weight of 50 kilograms (50 kg) is a phenotype, and having sickle-cell anemia is a phenotype. In this book, the term *characteristic* or *character* refers to a general feature such as eye color; the term *trait* or *phenotype* refers to specific manifestations of that feature, such as blue or brown eyes.

A given phenotype arises from a genotype that develops within a particular environment. The genotype determines the potential for development: it sets certain limits, or boundaries, on that development. How the phenotype develops within those limits is determined by the effects of other genes and of environmental factors, and the balance between these effects varies from characteristic to characteristic. For some characteristics, the differences between phenotypes are determined largely by differences in genotype. In Mendel's peas, for example, the genotype, not the environment, largely determined the shape of the seeds. For other characteristics, environmental differences are more important. The height reached by an oak tree at maturity is a phenotype that is strongly influenced by environmental factors, such as the availability of water, sunlight, and nutrients. Nevertheless, the tree's genotype still imposes some limits on its height: an oak tree will never grow to be 300 meters (almost 1000 feet) tall, no matter how much sunlight, water, and fertilizer are provided. Thus, even the height of an oak tree is determined to some degree by genes. For many characteristics, both genes and environment are important in determining phenotypic differences.

TABLE 3.1 Summary of important genetic terms

Term	Definition
Gene	An inherited factor (region of DNA) that helps determine a characteristic
Allele	One of two or more alternative forms of a gene
Locus	Specific place on a chromosome occupied by an allele
Genotype	Set of alleles possessed by an individual organism
Heterozygote	An individual organism possessing two different alleles at a locus
Homozygote	An individual organism possessing two of the same alleles at a locus
Phenotype or **trait**	The appearance or manifestation of a characteristic
Characteristic or **character**	An attribute or feature possessed by an organism

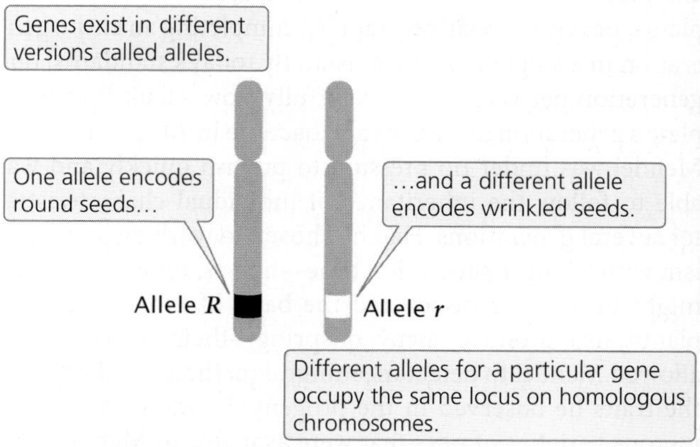

3.3 At each locus, a diploid organism possesses two alleles located on different homologous chromosomes. The alleles identified here refer to alleles in pea plants studied by Mendel.

An obvious but important point is that only the alleles of the genotype are inherited. Although the phenotype is determined, at least to some extent, by genotype, organisms do not transmit their phenotypes to the next generation. The distinction between genotype and phenotype is one of the most important principles of modern genetics. The next section describes Mendel's careful observations of phenotypes through several generations of breeding experiments. These experiments allowed him to deduce not only the genotypes of individual pea plants, but also the rules governing their inheritance.

CONCEPTS

Each phenotype results from a genotype developing within a specific environment. The alleles of the genotype, not the phenotype, are inherited.

✓ CONCEPT CHECK 2

What is the difference between a locus and an allele? What is the difference between genotype and phenotype?

3.2 Monohybrid Crosses Reveal the Principle of Segregation and the Concept of Dominance

Mendel started with 34 varieties of peas and spent 2 years selecting those varieties that he would use in his experiments. He verified that each variety was pure-breeding (homozygous for each of the traits that he chose to study) by growing the plants for two generations and confirming that all offspring were the same as their parents. He then carried out a number of crosses between the different varieties. Although peas are normally self-fertilizing (each plant mates with itself), Mendel conducted crosses between different plants by opening the buds before the anthers (male sex organs) were fully developed, removing the anthers, and then dusting the stigma (female sex organ) with pollen from a different plant's anthers (**Figure 3.4**).

Mendel began by studying **monohybrid crosses**—crosses between parents that differed in a single characteristic. In one experiment, Mendel crossed a pea plant that was pure-breeding (homozygous) for round seeds with one that was pure-breeding for wrinkled seeds (see Figure 3.4). This first generation of a cross is called the **P (parental) generation**.

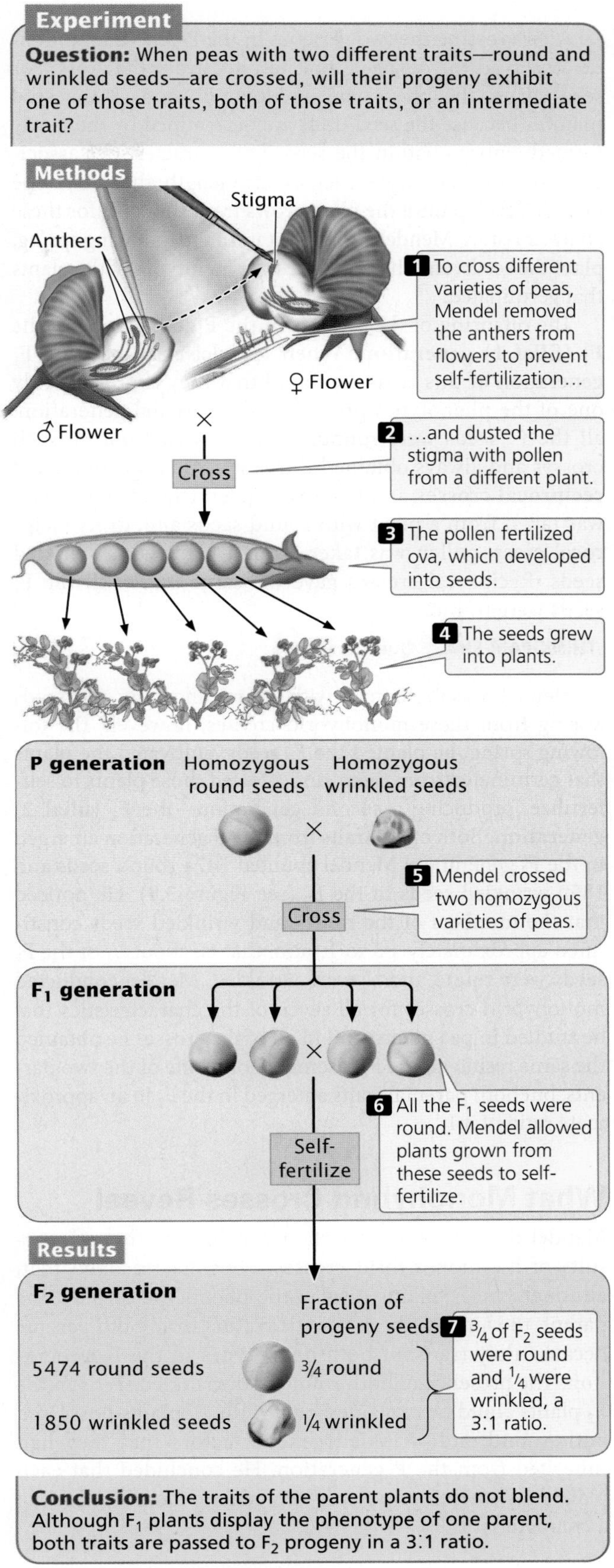

Experiment
Question: When peas with two different traits—round and wrinkled seeds—are crossed, will their progeny exhibit one of those traits, both of those traits, or an intermediate trait?
Methods
Stigma
Anthers
1 To cross different varieties of peas, Mendel removed the anthers from flowers to prevent self-fertilization…
♀ Flower
♂ Flower
×
Cross
2 …and dusted the stigma with pollen from a different plant.
3 The pollen fertilized ova, which developed into seeds.
4 The seeds grew into plants.
P generation
Homozygous round seeds
Homozygous wrinkled seeds
×
5 Mendel crossed two homozygous varieties of peas.
Cross
F1 generation
×
Self-fertilize
6 All the F1 seeds were round. Mendel allowed plants grown from these seeds to self-fertilize.
Results
F2 generation
Fraction of progeny seeds
5474 round seeds
3/4 round
1850 wrinkled seeds
1/4 wrinkled
7 3/4 of F2 seeds were round and 1/4 were wrinkled, a 3:1 ratio.
Conclusion: The traits of the parent plants do not blend. Although F1 plants display the phenotype of one parent, both traits are passed to F2 progeny in a 3:1 ratio.

3.4 Mendel conducted monohybrid crosses.

After crossing the two varieties in the P generation, Mendel observed the offspring that resulted from the cross. For seed shape, the phenotype develops as soon as the seed matures because the seed traits are determined by the newly formed embryo within the seed. For characteristics associated with the plant itself, such as stem length, the phenotype doesn't develop until the plant grows from the seed; for these characteristics, Mendel had to wait until the following spring, plant the seeds, and then observe the phenotypes of the plants that germinated.

The offspring of the parents in the P generation are the **F_1 (filial 1) generation**. When Mendel examined the F_1 generation of this cross, he found that they expressed only one of the phenotypes present in the parental generation: all the F_1 seeds were round. Mendel carried out 60 such crosses and always obtained this result. He also conducted **reciprocal crosses**: in one cross, pollen (the male gamete) was taken from a plant with round seeds and, in its reciprocal cross, pollen was taken from a plant with wrinkled seeds. Reciprocal crosses gave the same result: all the F_1 seeds were round.

THINK-PAIR-SHARE Question 5

Mendel wasn't content with examining only the seeds arising from these monohybrid crosses, however. The following spring, he planted the F_1 seeds, cultivated the plants that germinated from them, and allowed those plants to self-fertilize, producing a second generation—the **F_2 (filial 2) generation**. Both of the traits from the P generation emerged in the F_2 generation: Mendel counted 5474 round seeds and 1850 wrinkled seeds in the F_2 (see Figure 3.4). He noticed that the numbers of the round and wrinkled seeds constituted approximately a 3 to 1 ratio: that is, about $^3/_4$ of the F_2 seeds were round and $^1/_4$ were wrinkled. Mendel conducted monohybrid crosses for all seven of the characteristics that he studied in pea plants, and in all of the crosses he obtained the same result: all the F_1 resembled only one of the two parents, but both parental traits emerged in the F_2 in an approximate ratio of 3:1.

What Monohybrid Crosses Reveal

Mendel drew several important conclusions from the results of his monohybrid crosses. First, he reasoned that, although the F_1 plants display the phenotype of only one parent, they must inherit genetic factors from both parents because they transmit both phenotypes to the F_2 generation. The presence of both round and wrinkled seeds in the F_2 plants could be explained only if the F_1 plants possessed both round and wrinkled genetic factors that they had inherited from the P generation. He concluded that each plant must therefore possess two genetic factors encoding a characteristic.

The genetic factors (now called alleles) that Mendel discovered are, by convention, designated with letters: the allele for round seeds is usually represented by *R* and the allele for wrinkled seeds by *r*. The plants in the P generation of Mendel's cross possessed two identical alleles: *RR* in the round-seeded parent and *rr* in the wrinkled-seeded parent (**Figure 3.5a**).

The second conclusion that Mendel drew from his monohybrid crosses was that the two alleles in each plant separate when gametes are formed, and one allele goes into each gamete. When two gametes (one from each parent) fuse to produce a zygote, the allele from the male parent unites with the allele from the female parent to produce the genotype of the offspring. Thus, Mendel's F_1 plants inherited an *R* allele from the round-seeded plant and an *r* allele from the wrinkled-seeded plant (**Figure 3.5b**). However, only the trait encoded by the round allele (*R*) was *observed* in the F_1: all the F_1 progeny had round seeds. Those traits that appeared unchanged in the F_1 heterozygous offspring Mendel called **dominant**, and those traits that disappeared in the F_1 heterozygous offspring he called **recessive**. Alleles for dominant traits in plants are often symbolized with uppercase letters (e.g., *R*), while alleles for recessive traits are often symbolized with lowercase letters (e.g., *r*). When dominant and recessive alleles are present together, the recessive allele is masked, or suppressed. The concept of dominance was the third important conclusion that Mendel derived from his monohybrid crosses.

Mendel's fourth conclusion was that the two alleles of an individual plant separate with equal probability into the gametes. When plants of the F_1 (with genotype *Rr*) produced gametes, half of the gametes received the *R* allele for round seeds and half received the *r* allele for wrinkled seeds. The gametes then paired randomly to produce the following genotypes in equal proportions among the F_2: *RR*, *Rr*, *rR*, *rr* (**Figure 3.5c**). Because round (*R*) is dominant over wrinkled (*r*), there were three round seeds (*RR*, *Rr*, *rR*) for every one wrinkled seed (*rr*) in the F_2. This 3:1 ratio of round to wrinkled progeny that Mendel observed in the F_2 could be obtained only if the two alleles of a genotype separated into the gametes with equal probability.

The conclusions that Mendel developed about inheritance from his monohybrid crosses have been further developed and formalized into the principle of segregation and the concept of dominance. The **principle of segregation** (Mendel's first law) states that each individual diploid organism possesses two alleles for any particular characteristic, one inherited from the maternal parent and one from the paternal parent. These two alleles segregate (separate) when gametes are formed, and one allele goes into each gamete. Furthermore, the two alleles segregate into gametes in equal proportions. The **concept of dominance** states that when two different alleles are present in a genotype, only the trait encoded by one of them—the "dominant" allele—is observed in the phenotype.

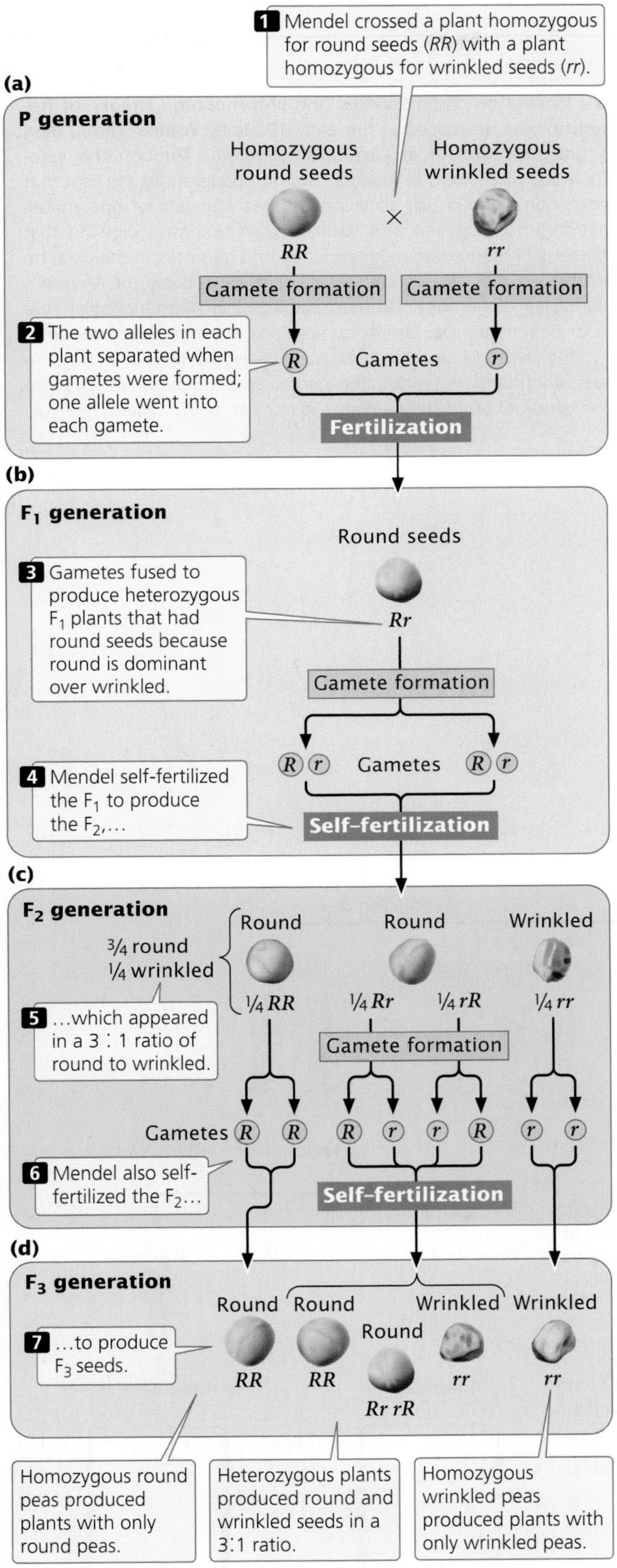

3.5 Mendel's monohybrid crosses revealed the principle of segregation and the concept of dominance.

Mendel confirmed these principles by allowing his F_2 plants to self-fertilize and produce an F_3 generation. He found that the plants grown from the wrinkled seeds—those displaying the recessive trait (*rr*)—produced an F_3 in which all plants produced wrinkled seeds. Because his wrinkled-seeded plants were homozygous for wrinkled alleles (*rr*), only wrinkled alleles could be passed on to their progeny (**Figure 3.5d**).

The plants grown from round seeds—the dominant trait—fell into two types (see Figure 3.5c). With self-fertilization, about $^2/_3$ of these plants produced both round and wrinkled seeds in the F_3 generation. These plants were heterozygous (*Rr*), so they produced $^1/_4$ *RR* (round), $^1/_2$ *Rr* (round), and $^1/_4$ *rr* (wrinkled) seeds, giving a 3:1 ratio of round to wrinkled among their progeny in the F_3. About $^1/_3$ of the F_2 plants grown from round seeds were of the second type; they produced only the round-seeded trait in the F_3. These plants were homozygous for the round allele (*RR*) and could thus produce only round-seeded offspring in the F_3 generation. Mendel planted the seeds obtained in the F_3 and carried these plants through three more rounds of self-fertilization. In each generation, $^2/_3$ of the round-seeded plants produced round and wrinkled offspring, whereas $^1/_3$ produced only round offspring. These results are entirely consistent with the principle of segregation.

CONCEPTS

The principle of segregation states that each individual organism possesses two alleles that can encode a characteristic. These alleles segregate when gametes are formed, and one allele goes into each gamete. The concept of dominance states that when the two alleles of a genotype are different, only the trait encoded by one of them—the "dominant" allele—is observed.

✓ CONCEPT CHECK 3

How did Mendel know that each of his pea plants carried two alleles encoding a characteristic?

CONNECTING CONCEPTS

Relating Genetic Crosses to Meiosis

We have now seen how the results of monohybrid crosses are explained by Mendel's principle of segregation. Many students find that they enjoy working genetic crosses but are frustrated by the abstract nature of the symbols. Perhaps you feel the same at this point. You may be asking, "What do these symbols really represent? What does the genotype *RR* mean in regard to the biology of the organism?" The answers to these questions lie in relating the abstract symbols of crosses to the structure and behavior of chromosomes, the repositories of genetic information (see Chapter 2).

In 1900, when Mendel's work was rediscovered and biologists began to apply his principles of heredity, the relation between genes and chromosomes was still unclear. The theory that genes are located on chromosomes (the **chromosome theory of heredity**) was developed in the early 1900s by Walter Sutton, then a graduate student at Columbia University. Through the careful study of meiosis in insects, Sutton documented the fact that each homologous pair of chromosomes consists of one maternal chromosome and one paternal chromosome. Showing that these pairs segregate independently into gametes in meiosis, he concluded that this process is the biological basis for Mendel's principles of heredity. German cytologist and embryologist Theodor Boveri came to similar conclusions at about the same time.

The symbols used in genetic crosses, such as *R* and *r*, are just shorthand notations for particular sequences of DNA in the chromosomes that encode particular phenotypes. The two

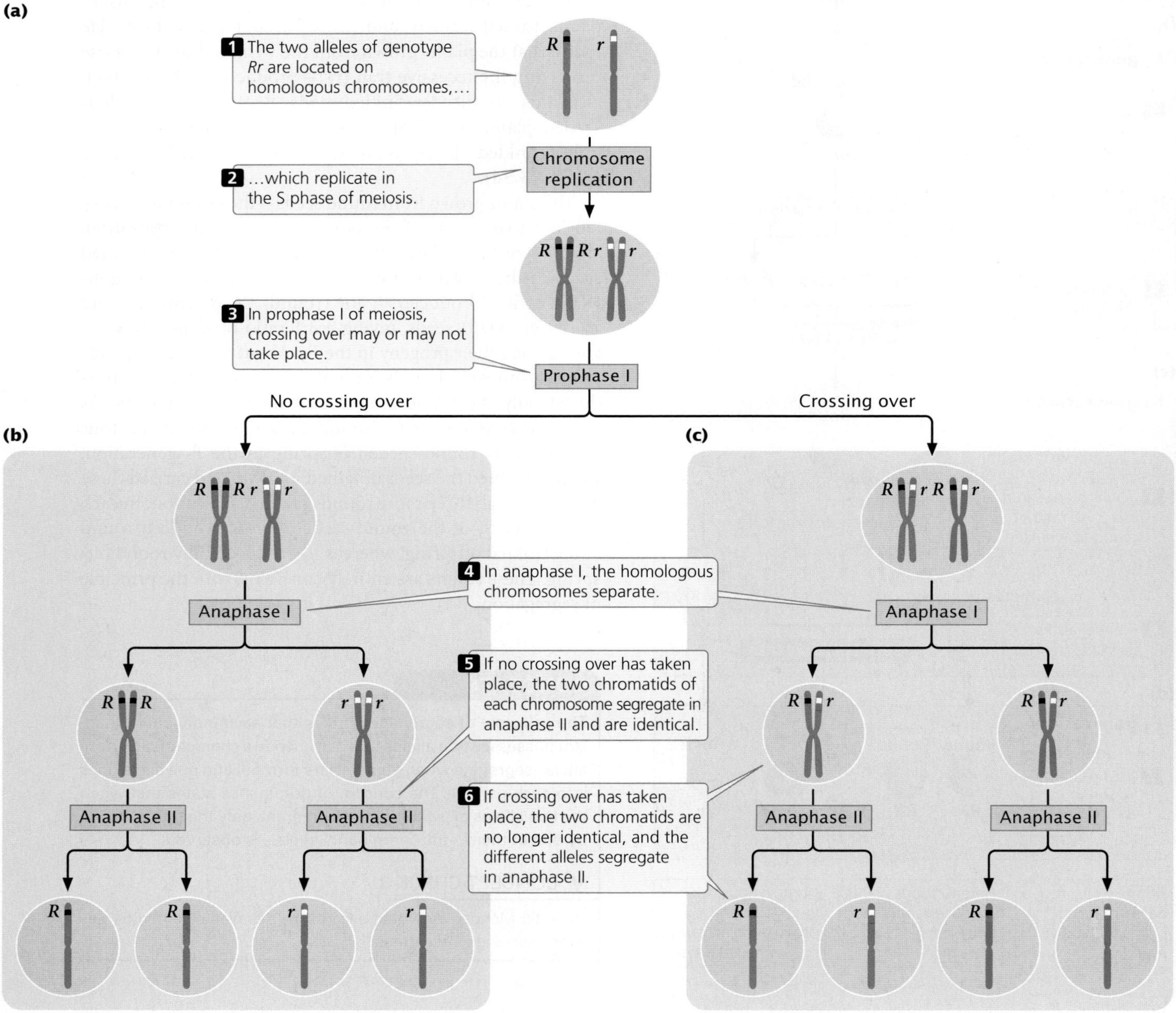

3.6 Segregation results from the separation of homologous chromosomes in meiosis.

alleles of a genotype are found on different but homologous chromosomes. One chromosome of each homologous pair is inherited from the mother and the other is inherited from the father. In the S phase of meiotic interphase, each chromosome replicates, producing two copies of each allele, one on each chromatid (**Figure 3.6a**). The homologous chromosomes segregate in anaphase I, thereby separating the two different alleles (**Figure 3.6b**). This chromosome segregation is the basis of the principle of segregation. In anaphase II of meiosis, the two chromatids of each replicated chromosome separate, so each gamete resulting from meiosis carries only a single allele at each locus, as Mendel's principle of segregation predicts.

If crossing over has taken place in prophase I of meiosis, then the two chromatids of each replicated chromosome are no longer identical, and the segregation of different alleles takes place at anaphase I and anaphase II (**Figure 3.6c**). However, Mendel didn't know anything about chromosomes; he formulated his principles of heredity entirely on the basis of the results of the crosses that he carried out. Nevertheless, we should not forget that these principles work because they are based on the behavior of actual chromosomes in meiosis. ▶TRY PROBLEM 28

The Molecular Nature of Alleles

Let's take a moment to consider in more detail exactly what an allele is and how it determines a phenotype. Although Mendel had no information about the physical nature of the genetic factors in his crosses, modern geneticists have now determined the molecular basis of these factors and how they encode a trait such as wrinkled seeds.

Alleles, such as the *R* and *r* alleles that encode round and wrinkled peas, usually represent specific DNA sequences. The locus that determines whether a pea is round or wrinkled is a sequence of DNA on pea chromosome 5 that encodes a protein called starch-branching enzyme isoform 1 (SBEI). The *R* allele, which produces round seeds in pea plants, encodes a normal, functional form of SBEI. This enzyme converts a linear form of starch into a highly branched form. The *r* allele, which encodes wrinkled seeds, is a different DNA sequence that contains a mutation or error; it encodes an inactive form of the enzyme that does not produce the branched form of starch and leads to the accumulation of sucrose within the *rr* pea. Because the *rr* pea contains a large amount of sucrose, the developing seed absorbs water and swells. Later, as the pea matures, it loses water. Because *rr* peas have absorbed more water and expanded more during development, they lose more water during maturation and afterward appear shriveled or wrinkled. The *r* allele for wrinkled seeds is recessive because the presence of a single *R* allele in the heterozygote encodes enough SBEI enzyme to produce branched starch and round seeds.

Research has revealed that the *r* allele contains an extra 800 base pairs of DNA that disrupt the normal coding sequence of the gene. The extra DNA appears to have come from a transposable element, a type of DNA sequence that has the ability to move from one location in the genome to another, which we will discuss further in Chapter 13.

Predicting the Outcomes of Genetic Crosses

One of Mendel's goals in conducting his experiments on pea plants was to develop a way to predict the outcome of crosses between plants with different phenotypes. In this section, you will first learn a simple, shorthand method for predicting the outcomes of genetic crosses (the Punnett square), and you will then learn how to use probability to predict the results of crosses.

THE PUNNETT SQUARE The Punnett square was developed by English geneticist Reginald C. Punnett in 1917. To illustrate the Punnett square, let's examine another cross carried out by Mendel. By crossing two varieties of peas that differed in height, Mendel established that tall (*T*) was dominant over short (*t*). He tested his theory concerning the inheritance of dominant traits by crossing an F_1 tall plant that was heterozygous (*Tt*) with the short homozygous parental variety (*tt*). This type of cross, between an F_1 genotype and either of the parental genotypes, is called a **backcross**.

To predict the types of offspring that will result from this backcross, we must first determine which gametes will be produced by each parent (**Figure 3.7a**). The principle of segregation tells us that the two alleles in each parent separate and that one allele passes to each gamete. All gametes from the homozygous *tt* short plant will receive a single short (*t*) allele. The tall plant in this cross is heterozygous (*Tt*), so 50% of its gametes will receive a tall allele (*T*) and the other 50% will receive a short allele (*t*).

A **Punnett square** is constructed by drawing a grid and listing the gametes produced by one parent along the upper edge and the gametes produced by the other parent down the left side (**Figure 3.7b**). Each cell (that is, each block within the Punnett square) is then filled in with an allele from each of the corresponding gametes, generating the genotype of the progeny produced by fusion of those gametes. In the upper left cell of the Punnett square in Figure 3.7b, a gamete containing *T* from the tall plant unites with a gamete containing *t* from the short plant, giving the genotype of the progeny (*Tt*). It is useful to note the phenotype expressed by each genotype; here, the progeny will be tall because the tall allele is dominant over the short allele. This process is repeated for all the cells in the Punnett square.

By simply counting, we can determine the types of progeny produced and their ratios. In Figure 3.7b, two cells contain tall (*Tt*) progeny and two cells contain short (*tt*) progeny; so the genotypic ratio expected for this cross is 2 *Tt* to 2 *tt* (a 1:1 ratio). Another way to express this result is to say that we expect $^1/_2$ of the progeny to have genotype *Tt* (and the tall phenotype) and $^1/_2$ of the progeny to have genotype *tt* (and the short phenotype). In this cross, the genotypic ratio and the phenotypic ratio are the same, but this outcome need not be the case. Try completing a Punnett square for the cross in which the F_1 round-seeded plants in Figure 3.5 undergo self-fertilization (you should obtain a phenotypic ratio of 3 round to 1 wrinkled and a genotypic ratio of 1 *RR* to 2 *Rr* to 1 *rr*).

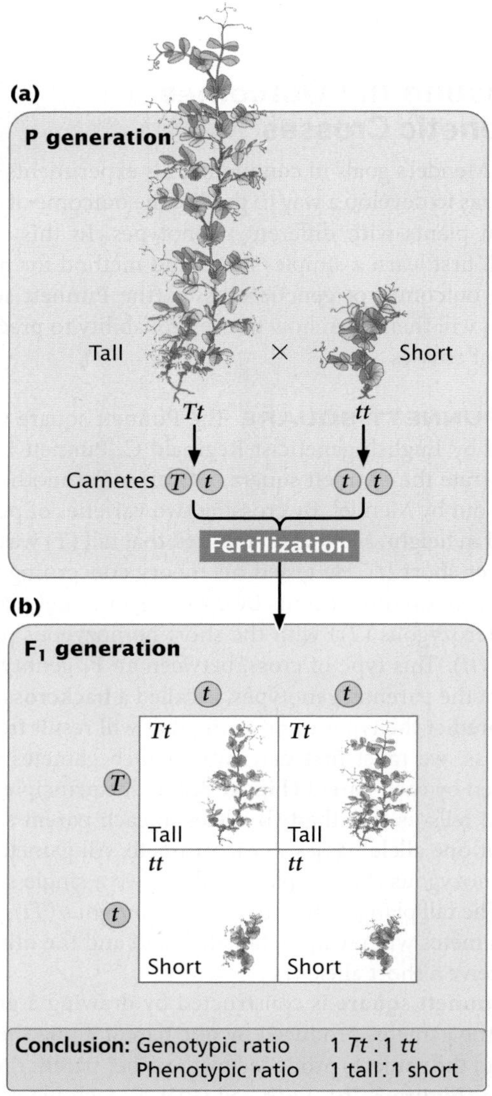

3.7 The Punnett square can be used to determine the results of a genetic cross.

CONCEPTS

The Punnett square is a shorthand method of predicting the genotypic and phenotypic ratios of progeny from a genetic cross.

✓ CONCEPT CHECK 4

If the F_1 plant depicted in Figure 3.5 is backcrossed to the parent with round seeds, what proportion of the progeny will have wrinkled seeds? (Use a Punnett square.)

a. $\frac{3}{4}$ b. $\frac{1}{2}$
c. $\frac{1}{4}$ d. 0

PROBABILITY AS A TOOL IN GENETICS Another method for determining the outcome of a genetic cross is to use the rules of probability, as Mendel did with his crosses. **Probability** expresses the likelihood of the occurrence of a particular event. It is the number of times that a particular event takes place, divided by the number of all possible outcomes. For example, a deck of 52 cards contains only one king of hearts. The probability of drawing one card from the deck at random and obtaining the king of hearts is $\frac{1}{52}$ because there is only one card that is the king of hearts (one event) and there are 52 cards that can be drawn from the deck (52 possible outcomes). The probability of drawing a card and obtaining an ace is $\frac{4}{52}$ because there are four cards that are aces (four events) and 52 cards (possible outcomes). Probability can be expressed either as a fraction ($\frac{4}{52}$ in this case) or as a decimal number (0.077 in this case).

The probability of a particular event may be determined by knowing something about *how* or *how often* the event takes place. We know, for example, that the probability of rolling a six-sided die and getting a four is $\frac{1}{6}$ because the die has six sides and any one side is equally likely to end up on top. So, in this case, understanding the nature of the event—the shape of the thrown die—allows us to determine the probability. In other cases, we determine the probability of an event by making a large number of observations. When a weather forecaster says that there is a 40% chance of rain on a particular day, this probability was obtained by observing a large number of days with similar atmospheric conditions and finding that it rains on 40% of those days. In this case, the probability has been determined empirically (by observation).

THE MULTIPLICATION RULE Two rules of probability are useful for predicting the ratios of offspring produced in genetic crosses. The first is the **multiplication rule**, which states that the probability of two or more independent events taking place together is calculated by multiplying their independent probabilities.

To illustrate the use of the multiplication rule, let's again consider the roll of a die. The probability of rolling one die and obtaining a four is $\frac{1}{6}$. To calculate the probability of rolling a die twice and obtaining two fours, we can apply the multiplication rule. The probability of obtaining a four on the first roll is $\frac{1}{6}$ and the probability of obtaining a four on the second roll is $\frac{1}{6}$; so the probability of rolling a four on both is $\frac{1}{6} \times \frac{1}{6} = \frac{1}{36}$ (**Figure 3.8a**). The key indicator for applying the multiplication rule is the word *and*; in the example just considered, we wanted to know the probability of obtaining a four on the first roll *and* a four on the second roll.

For the multiplication rule to be valid, the events whose joint probability is being calculated must be independent—the outcome of one event must not influence the outcome of the other. For example, the number that comes up on one roll of the die has no influence on the number that comes up on the next roll, so these events are independent. However, if we wanted to know the probability of being hit on the head with a hammer and going to the hospital on the same day, we could not simply apply the multiplication rule and multiply the two probabilities together, because the two events are not independent—being hit on the head with a hammer certainly influences the probability of going to the hospital.

THE ADDITION RULE The second rule of probability frequently used in genetics is the **addition rule**, which states

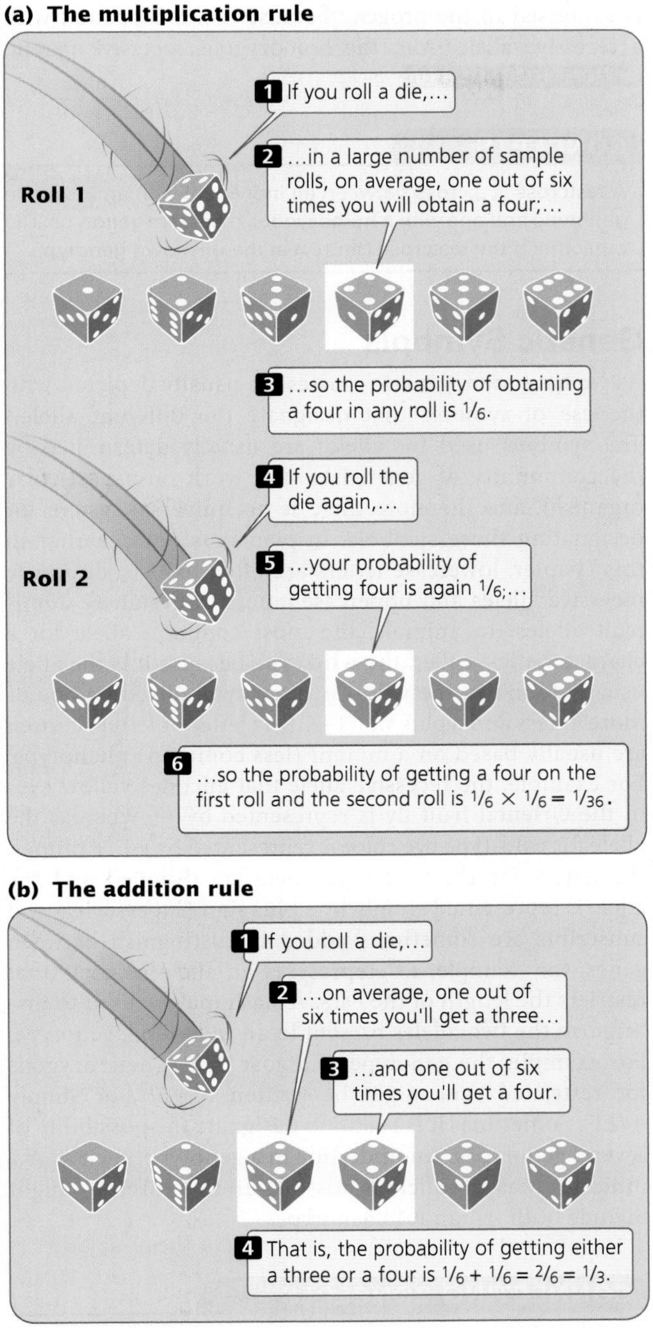

(a) The multiplication rule
1 If you roll a die,...
2 ...in a large number of sample rolls, on average, one out of six times you will obtain a four;...
Roll 1
3 ...so the probability of obtaining a four in any roll is 1/6.
4 If you roll the die again,...
5 ...your probability of getting four is again 1/6;...
Roll 2
6 ...so the probability of getting a four on the first roll and the second roll is 1/6 × 1/6 = 1/36.
(b) The addition rule
1 If you roll a die,...
2 ...on average, one out of six times you'll get a three...
3 ...and one out of six times you'll get a four.
4 That is, the probability of getting either a three or a four is 1/6 + 1/6 = 2/6 = 1/3.

3.8 The multiplication and addition rules can be used to determine the probability of combinations of events.

that the probability of any one of two or more mutually exclusive events is calculated by adding the probabilities of these events. Let's look at this rule in concrete terms. To obtain the probability of throwing a die once and rolling *either* a three *or* a four, we would use the addition rule, adding the probability of obtaining a three ($^1/_6$) to the probability of obtaining a four (again, $^1/_6$), or $^1/_6 + {}^1/_6 = {}^2/_6 = {}^1/_3$ (**Figure 3.8b**). The key indicators for applying the addition rule are the words *either* and *or*.

For the addition rule to be valid, the events whose probability is being calculated must be mutually exclusive, meaning that one event excludes the possibility of the occurrence of the other event. For example, you cannot throw a single die just once and obtain both a three and a four, because only one side of the die can be on top. These events are mutually exclusive.

CONCEPTS

The multiplication rule states that the probability of two or more independent events taking place together is calculated by multiplying their independent probabilities. The addition rule states that the probability that any one of two or more mutually exclusive events taking place is calculated by adding their probabilities.

✓ CONCEPT CHECK 5

If the probability of being blood type A is $^1/_8$ and the probability of being blood type O is $^1/_2$, what is the probability of being either blood type A or blood type O?

a. $^5/_8$ b. $^1/_2$
c. $^1/_{10}$ d. $^1/_{16}$

APPLYING PROBABILITY TO GENETIC CROSSES

The multiplication and addition rules of probability can be used in place of the Punnett square to predict the ratios of progeny expected from a genetic cross. Let's first consider a cross between two pea plants heterozygous for the locus that determines height, *Tt* × *Tt*. Half of the gametes produced by each plant have a *T* allele, and the other half have a *t* allele, so the probability for each type of gamete is $^1/_2$.

The gametes from the two parents can combine in four different ways to produce offspring. Using the multiplication rule, we can determine the probability of each possible type. To calculate the probability of obtaining *TT* progeny, for example, we multiply the probability of receiving a *T* allele from the first parent ($^1/_2$) times the probability of receiving a *T* allele from the second parent ($^1/_2$). The multiplication rule should be used here because we need the probability of receiving a *T* allele from the first parent *and* a *T* allele from the second parent—two independent events. The four types of progeny from this cross and their associated probabilities are

TT	(*T* gamete and *T* gamete)	$^1/_2 \times {}^1/_2 = {}^1/_4$	tall
Tt	(*T* gamete and *t* gamete)	$^1/_2 \times {}^1/_2 = {}^1/_4$	tall
tT	(*t* gamete and *T* gamete)	$^1/_2 \times {}^1/_2 = {}^1/_4$	tall
tt	(*t* gamete and *t* gamete)	$^1/_2 \times {}^1/_2 = {}^1/_4$	short

Notice that there are two ways for heterozygous progeny to be produced: a heterozygote can either receive a *T* allele from the first parent and a *t* allele from the second or receive a *t* allele from the first parent and a *T* allele from the second.

After determining the probabilities of obtaining each type of progeny, we can use the addition rule to determine the overall phenotypic ratios. We use the addition rule because we want to know the probability that an individual pea plant is genotype *TT*, *Tt*, or *tT*, any of which produce a tall plant. These events are mutually exclusive because an individual can have only one genotype at this locus. Using the addition rule, we find the probability of tall progeny to be $^1/_4 + ^1/_4 + ^1/_4 = ^3/_4$. Because only one genotype encodes short (*tt*), the probability of short progeny is simply $^1/_4$.

Two methods have now been introduced to predict the results of genetic crosses: the Punnett square and the probability method. At this point, you may be asking, "Why bother with probability rules and calculations? The Punnett square is easier to understand and just as quick." This is true for simple monohybrid crosses. For tackling more complex crosses involving genes at two or more loci, however, the probability method is both clearer and quicker than the Punnett square.

CONDITIONAL PROBABILITY Thus far, we have used probability to predict the chances of producing certain types of progeny given only the genotypes of the parents. Sometimes we have additional information that modifies, or "conditions," the probability, a situation termed **conditional probability**. For example, assume that we cross two heterozygous pea plants ($Tt \times Tt$) and obtain a tall offspring plant. What is the probability that this tall plant is heterozygous (*Tt*)? You might assume that the probability would be $^1/_2$, the probability of obtaining a heterozygous offspring in a cross between two heterozygotes. In this case, however, we have some additional information—the phenotype of the offspring plant—which modifies that probability. When two heterozygous individuals are crossed, we expect $^1/_4$ *TT*, $^1/_2$ *Tt*, and $^1/_4$ *tt* progeny. We know that the offspring in question is tall, so we can eliminate the possibility that it has genotype *tt*. Tall progeny must be either genotype *TT* or genotype *Tt*, and in a cross between two heterozygotes, these genotypes occur in a 1:2 ratio. Therefore, the probability that a tall offspring plant is heterozygous (*Tt*) is two out of three, or $^2/_3$.
▶TRY PROBLEMS 24 and 25

The Testcross

A useful tool for analyzing genetic crosses is the **testcross**, in which one individual of unknown genotype is crossed with another individual with a homozygous recessive genotype for the trait in question. Figure 3.7 illustrates a testcross (in this case, it is also a backcross). A testcross tests, or reveals, the genotype of the first individual.

Suppose that you were given a tall pea plant with no information about its parents. Because tallness is a dominant trait in peas, your plant could be either homozygous (*TT*) or heterozygous (*Tt*), but you would not know which. You could determine its genotype by performing a testcross. If the plant were homozygous (*TT*), a testcross would produce all tall progeny ($TT \times tt \rightarrow$ all *Tt*); if the plant were heterozygous (*Tt*), half of the progeny would be tall and half would be short ($Tt \times tt \rightarrow$ $^1/_2$ *Tt* and $^1/_2$ *tt*). When a testcross is performed, any recessive allele in the unknown genotype is expressed in the progeny because it will be paired with a recessive allele from the homozygous recessive parent.
▶TRY PROBLEM 19

CONCEPTS

A testcross is a cross between an individual with an unknown genotype and one with a homozygous recessive genotype. The outcome of the testcross can reveal the unknown genotype.

Genetic Symbols

As we have seen, genetic crosses are usually depicted with the use of symbols that designate the different alleles. The symbols used for alleles are usually determined by the community of geneticists who work on a particular organism, and therefore there is no universal system for designating those symbols. In plants, as noted earlier in this chapter, lowercase letters are often used to designate recessive alleles and uppercase letters to designate dominant alleles. In animals, the most common allele for a characteristic—called the **wild type** because it is the allele usually found in the wild—is often symbolized by one or more letters and a plus sign (+). The letter or letters chosen are usually based on a mutant (less common) phenotype. For example, the recessive allele that encodes yellow eyes in the Oriental fruit fly is represented by *ye*, whereas the allele for wild-type eye color is represented by ye^+. At times, the letters for the wild-type allele are dropped and the allele is represented simply by a plus sign. Superscripts and subscripts are sometimes added to distinguish between genes; for example, El^R represents an allele in goats that restricts the length of the ears. A slash may be used to distinguish the two alleles present in an individual genotype. For example, the genotype of a goat that is heterozygous for restricted ears might be written El^+/El^R, or simply $+/El^R$. Sometimes it is useful to designate the possibility of several genotypes. An underline in a genotype, such as *A_*, indicates that any allele is possible. In this case, *A_* might include both *AA* and *Aa* genotypes.

CONNECTING CONCEPTS

Ratios in Simple Crosses

Now that we have had some experience with genetic crosses, let's review the ratios that appear in the progeny of simple crosses in which a single locus is under consideration and one of the traits exhibits dominance. Understanding these ratios and the parental genotypes that produce them will enable you to work simple genetic crosses quickly, without resorting to the Punnett square. Later in this chapter, we will use these ratios to work more complicated crosses that include several loci.

There are only three phenotypic ratios to understand (**Table 3.2**). The 3:1 ratio arises in a simple genetic cross when both of the parents are heterozygous for a dominant trait

TABLE 3.2 Phenotypic ratios for simple genetic crosses (crosses for a single locus)

Phenotypic ratio	Genotypes of parents	Genotypes of progeny
3:1	*Aa* × *Aa*	$^3/_4$ *A*_:$^1/_4$ *aa*
1:1	*Aa* × *aa*	$^1/_2$ *Aa*:$^1/_2$ *aa*
Uniform progeny	*AA* × *AA*	All *AA*
	aa × *aa*	All *aa*
	AA × *aa*	All *Aa*
	AA × *Aa*	All *A*_

(*Aa* × *Aa*). The second phenotypic ratio is the 1:1 ratio, which results from the mating of a heterozygous parent and a homozygous parent. To obtain this 1:1 ratio, the homozygous parent in this cross (*Aa* × *aa*) must carry two recessive alleles to produce progeny of which half display the recessive trait. A cross between a homozygous dominant parent and a heterozygous parent (*AA* × *Aa*) produces progeny displaying only the dominant trait.

The third phenotypic ratio is not really a ratio: all the progeny have the same phenotype. Several combinations of parents can produce this outcome (see Table 3.2). A cross between any two homozygous parents—either between two parents of the same homozygous genotype (*AA* × *AA* or *aa* × *aa*) or between two parents with different homozygous genotypes (*AA* × *aa*)—produces progeny all having the same phenotype. Progeny of a single phenotype can also result from a cross between a homozygous dominant parent and a heterozygote (*AA* × *Aa*).

If we are interested in the ratios of genotypes instead of phenotypes, there are again only three outcomes to remember (**Table 3.3**): the 1:2:1 ratio, produced by a cross between two heterozygotes; the 1:1 ratio, produced by a cross between a heterozygote and a homozygote; and the uniform progeny produced by a cross between two homozygotes. These simple phenotypic and genotypic ratios and the parental genotypes that produce them provide the key to understanding crosses for a single locus and, as you will see in the next section, for multiple loci.

TABLE 3.3 Genotypic ratios for simple genetic crosses (crosses for a single locus)

Phenotypic ratio	Genotypes of parents	Genotypes of progeny
1:2:1	*Aa* × *Aa*	$^1/_4$ *AA*:$^1/_2$ *Aa*:$^1/_4$ *aa*
1:1	*Aa* × *aa*	$^1/_2$ *Aa*:$^1/_2$ *aa*
Uniform progeny	*AA* × *AA*	All *AA*
	aa × *aa*	All *aa*
	AA × *aa*	All *Aa*

3.3 Dihybrid Crosses Reveal the Principle of Independent Assortment

We will now extend Mendel's principle of segregation to more complex crosses that examine alleles at multiple loci. Understanding the nature of these crosses will require an additional principle: the principle of independent assortment.

Dihybrid Crosses

In addition to his work on monohybrid crosses, Mendel crossed varieties of peas that differed in *two* characteristics—that is, he performed **dihybrid crosses**. For example, he used one homozygous variety of pea with seeds that were round and yellow, and another homozygous variety with seeds that were wrinkled and green. When he crossed the two varieties, the seeds of all the F_1 progeny were round and yellow. He then allowed the F_1 to self-fertilize and obtained the following progeny in the F_2: 315 round, yellow seeds; 101 wrinkled, yellow seeds; 108 round, green seeds; and 32 wrinkled, green seeds. Mendel recognized that these traits appeared in an approximate ratio of 9:3:3:1; that is, $^9/_{16}$ of the progeny were round and yellow, $^3/_{16}$ were wrinkled and yellow, $^3/_{16}$ were round and green, and $^1/_{16}$ were wrinkled and green.

The Principle of Independent Assortment

Mendel carried out a number of dihybrid crosses for pairs of characteristics and always obtained a 9:3:3:1 ratio in the F_2. This ratio makes perfect sense in light of the principle of segregation and the concept of dominance if we add a third principle, which Mendel recognized in his dihybrid crosses: the **principle of independent assortment** (Mendel's second law). This principle states that alleles at different loci separate independently of one another.

A common mistake is to think that the principle of segregation and the principle of independent assortment refer to two different processes. The principle of independent assortment is really an extension of the principle of segregation. The principle of segregation states that the two alleles at a locus separate when gametes are formed; the principle of independent assortment states that, when these two alleles separate, their separation is independent of the separation of alleles at *other* loci.

Let's see how the principle of independent assortment explains the results that Mendel obtained in his dihybrid cross. Each plant possesses two alleles encoding each characteristic, and so the parental plants must have had genotypes *RR YY* and *rr yy* (**Figure 3.9a**). The principle of segregation tells us that the alleles for each locus separate and that one allele for each locus passes to each gamete. The gametes produced by the round, yellow parent therefore contain alleles *RY*, whereas the gametes produced by the wrinkled, green parent contain alleles *ry*. These two types of gametes unite to produce the F_1, all with genotype *Rr Yy*. Because round is dominant over wrinkled and yellow is dominant over green, the phenotype of the F_1 will be round and yellow.

When Mendel allowed the F_1 plants to self-fertilize to produce the F_2, the alleles for each locus separated, with one allele going into each gamete. This is where the principle of independent assortment becomes important. Each pair of alleles can separate in two ways: **(1)** *R* separates with *Y* and *r* separates with *y* to produce gametes *RY* and *ry* or **(2)** *R* separates with *y* and *r* separates with *Y* to produce gametes *Ry* and *rY*. The principle of independent assortment tells us that the alleles at each locus separate independently; thus, both kinds of separation take place with equal frequency, and all four types of gametes (*RY*, *ry*, *Ry*, and *rY*) are produced in equal proportions (**Figure 3.9b**). When these four types of gametes are combined to produce the F_2 generation, the progeny consist of $^9/_{16}$ round and yellow, $^3/_{16}$ wrinkled and yellow, $^3/_{16}$ round and green, and $^1/_{16}$ wrinkled and green, resulting in a 9:3:3:1 phenotypic ratio (**Figure 3.9c**).

Relating the Principle of Independent Assortment to Meiosis

An important qualification of the principle of independent assortment is that it applies to characteristics encoded by loci located on different chromosomes. Like the principle of segregation, it is based wholly on the behavior of chromosomes in meiosis. Each pair of homologous chromosomes separates independently of all other pairs in anaphase I of meiosis (**Figure 3.10**), so genes located on different pairs of homologs will assort independently. Genes that happen to be located on the same chromosome will travel together during anaphase I of meiosis and will arrive at the same destination—within the same gamete (unless crossing over takes place). Genes located on the same chromosome therefore do not assort independently (unless they are located sufficiently far apart that crossing over takes place at every meiotic division, a situation that will be discussed fully in Chapter 5).

CONCEPTS

The principle of independent assortment states that genes encoding different characteristics separate independently of one another when gametes are formed, owing to the independent separation of homologous pairs of chromosomes in meiosis. Genes located close together on the same chromosome, however, do not assort independently.

✓ CONCEPT CHECK 6

How are the principles of segregation and independent assortment related, and how are they different?

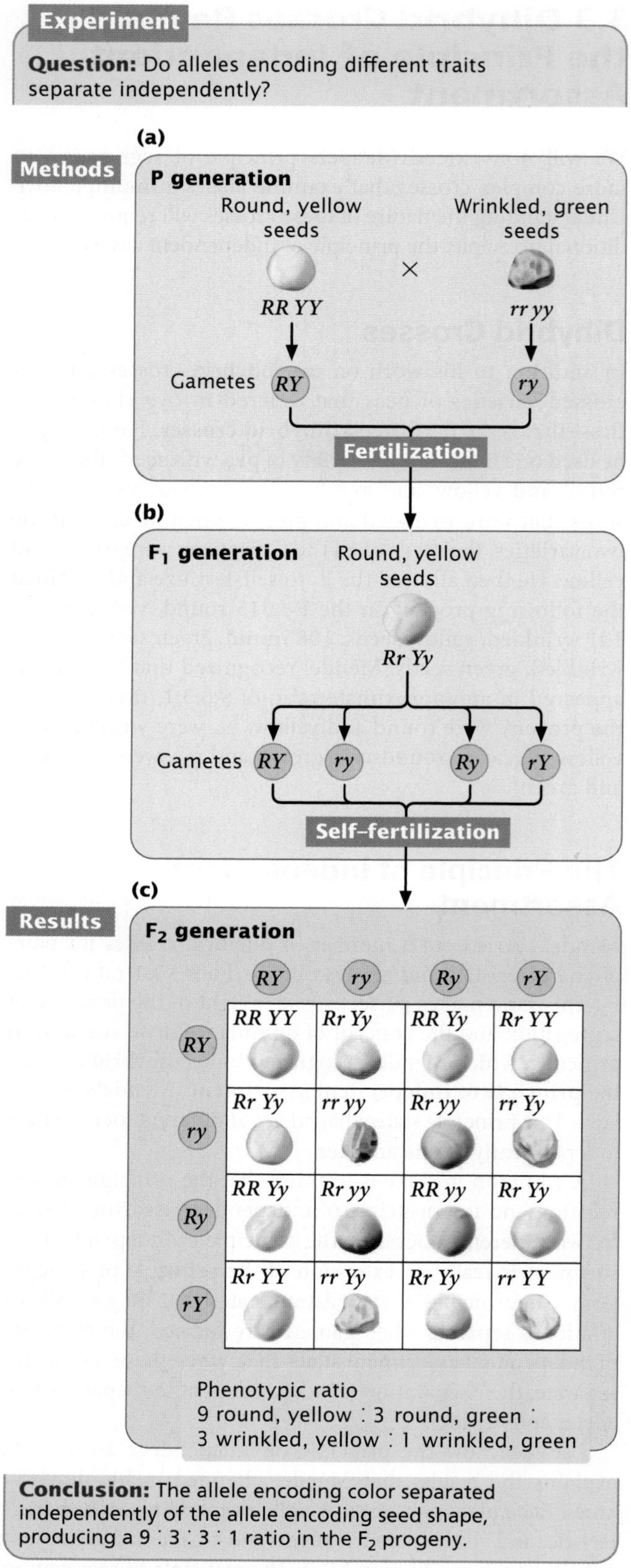

3.9 Mendel's dihybrid crosses revealed the principle of independent assortment.

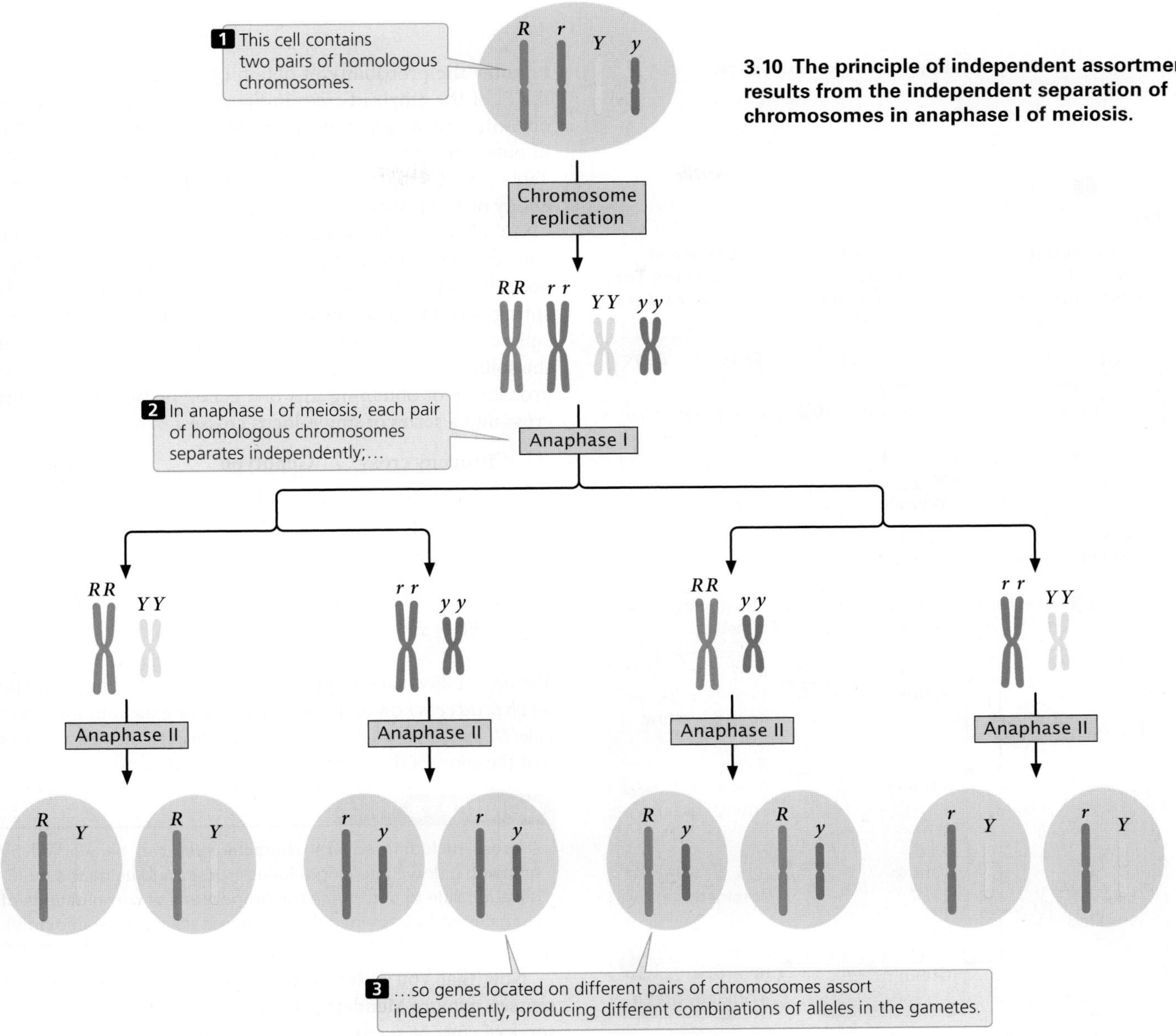

3.10 The principle of independent assortment results from the independent separation of chromosomes in anaphase I of meiosis.

Applying Probability and the Branch Diagram to Dihybrid Crosses

When the genes at two loci separate independently, a dihybrid cross can be understood as two monohybrid crosses. Let's examine Mendel's dihybrid cross (*Rr Yy* × *Rr Yy*) by considering each characteristic separately (**Figure 3.11a**). If we consider only the shape of the seeds, the cross was *Rr* × *Rr*, which yields a 3:1 phenotypic ratio ($^3/_4$ round and $^1/_4$ wrinkled progeny; see Table 3.2). Next consider the other characteristic, the color of the seed. The cross was *Yy* × *Yy*, which produces a 3:1 phenotypic ratio ($^3/_4$ yellow and $^1/_4$ green progeny).

We can now combine these monohybrid ratios by using the multiplication rule to obtain the proportion of progeny with different combinations of seed shape and color. The proportion of progeny with round and yellow seeds is $^3/_4$ (the probability of round) × $^3/_4$ (the probability of yellow) = $^9/_{16}$. The proportion of progeny with round and green seeds is $^3/_4 \times {}^1/_4 = {}^3/_{16}$; the proportion of progeny with wrinkled and yellow seeds is $^1/_4 \times {}^3/_4 = {}^3/_{16}$; and the proportion of progeny with wrinkled and green seeds is $^1/_4 \times {}^1/_4 = {}^1/_{16}$.

Branch diagrams are a convenient way of organizing all the combinations of characteristics (**Figure 3.11b**). In the first column, list the proportions of the phenotypes for one characteristic (here, $^3/_4$ round and $^1/_4$ wrinkled). In the second column, list the proportions of the phenotypes for the second characteristic ($^3/_4$ yellow and $^1/_4$ green) twice, next to each of the phenotypes in the first column: list $^3/_4$ yellow and $^1/_4$ green next to the round phenotype and again next to the wrinkled phenotype. Draw lines between the phenotypes in the first column and each of the phenotypes in the second column. Now follow each branch of the diagram, multiplying the probabilities for each trait along that branch. One branch leads from round to yellow, yielding round and yellow progeny. Another branch leads from round to green, yielding round

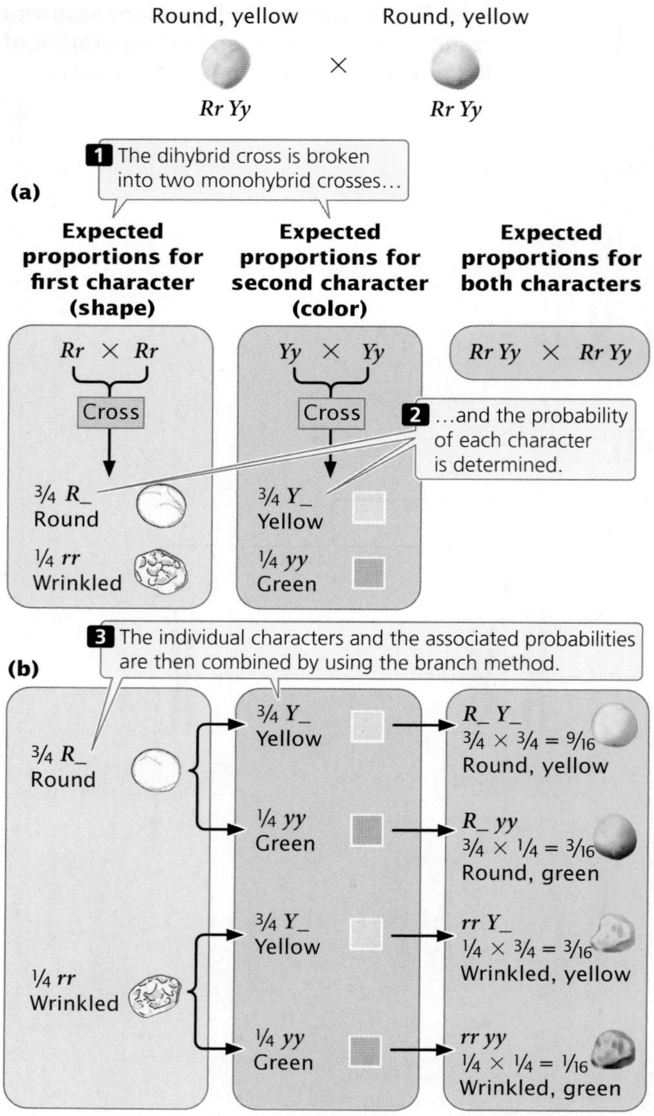

3.11 A branch diagram can be used to determine the phenotypes and expected proportions of offspring from a dihybrid cross (*Rr Yy* × *Rr Yy*).

and green progeny, and so forth. We can then calculate the probability of progeny with a particular combination of traits by using the multiplication rule, as shown above: the probability of round ($^3/_4$) and yellow ($^3/_4$) seeds, for example, is $^3/_4 \times {}^3/_4 = {}^9/_{16}$. The advantage of the branch diagram is that it helps us keep track of all the potential combinations of traits that may appear in the progeny. It can be used to determine phenotypic or genotypic ratios for any number of characteristics.

Using probability is much faster than using the Punnett square for crosses that include multiple loci. Genotypic and phenotypic ratios can be quickly worked out by combining, with the multiplication rule, the simple ratios in Tables 3.2 and 3.3. The probability method is particularly efficient if we need the probability of only a *particular* phenotype or genotype among the progeny of a cross. Suppose that we need to know the probability of obtaining the genotype *Rr yy* in the F_2 of the dihybrid cross in Figure 3.9. The probability of obtaining the *Rr* genotype in a cross of *Rr* × *Rr* is $^1/_2$, and that of obtaining the *yy* genotype in a cross of *Yy* × *Yy* is $^1/_4$ (see Table 3.3). Using the multiplication rule, we find the probability of *Rr yy* to be $^1/_2 \times {}^1/_4 = {}^1/_8$.

To illustrate the advantage of the probability method, consider the cross *Aa Bb cc Dd Ee* × *Aa Bb Cc dd Ee*. Suppose that we want to know the probability of obtaining offspring with the genotype *aa bb cc dd ee*. If we use a Punnett square to determine this probability, we might be working on the solution for months. However, we can quickly figure the probability of obtaining this one genotype by breaking this cross into a series of single-locus crosses:

Progeny cross	Genotype	Probability
Aa × *Aa*	*aa*	$^1/_4$
Bb × *Bb*	*bb*	$^1/_4$
cc × *Cc*	*cc*	$^1/_2$
Dd × *dd*	*dd*	$^1/_2$
Ee × *Ee*	*ee*	$^1/_4$

The probability of an offspring from this cross having genotype *aa bb cc dd ee* is now easily obtained by using the multiplication rule: $^1/_4 \times {}^1/_4 \times {}^1/_2 \times {}^1/_2 \times {}^1/_4 = {}^1/_{256}$. This calculation assumes that the genes at these five loci all assort independently.

CONCEPTS

A cross including several characteristics can be worked by breaking it down into single-locus crosses and using the multiplication rule to determine the proportions of combinations of characteristics (provided that the genes assort independently).

Now that you've had some experience working genetic crosses, explore Mendel's principles of heredity by setting up some of your own crosses in **Animation 3.1**.

The Dihybrid Testcross

Let's practice using the branch diagram by predicting the types and proportions of phenotypes in a dihybrid testcross between the round and yellow F_1 pea plants (*Rr Yy*) obtained by Mendel in his dihybrid cross and wrinkled and green pea plants (*rr yy*), as depicted in **Figure 3.12**. First, break the cross down into a series of single-locus crosses. The cross *Rr* × *rr* yields $^1/_2$ round (*Rr*) progeny and $^1/_2$ wrinkled (*rr*) progeny. The cross *Yy* × *yy* yields $^1/_2$ yellow (*Yy*) progeny and $^1/_2$ green (*yy*) progeny. Using the multiplication rule, we find the proportion of round and yellow progeny to be $^1/_2$ (the probability of round) × $^1/_2$ (the probability of yellow) = $^1/_4$. Four combinations of traits with the following proportions appear in the offspring: $^1/_4$ *Rr Yy*, round yellow; $^1/_4$ *Rr yy*, round green; $^1/_4$ *rr Yy*, wrinkled yellow; and $^1/_4$ *rr yy*, wrinkled green.

THINK-PAIR-SHARE Question 7

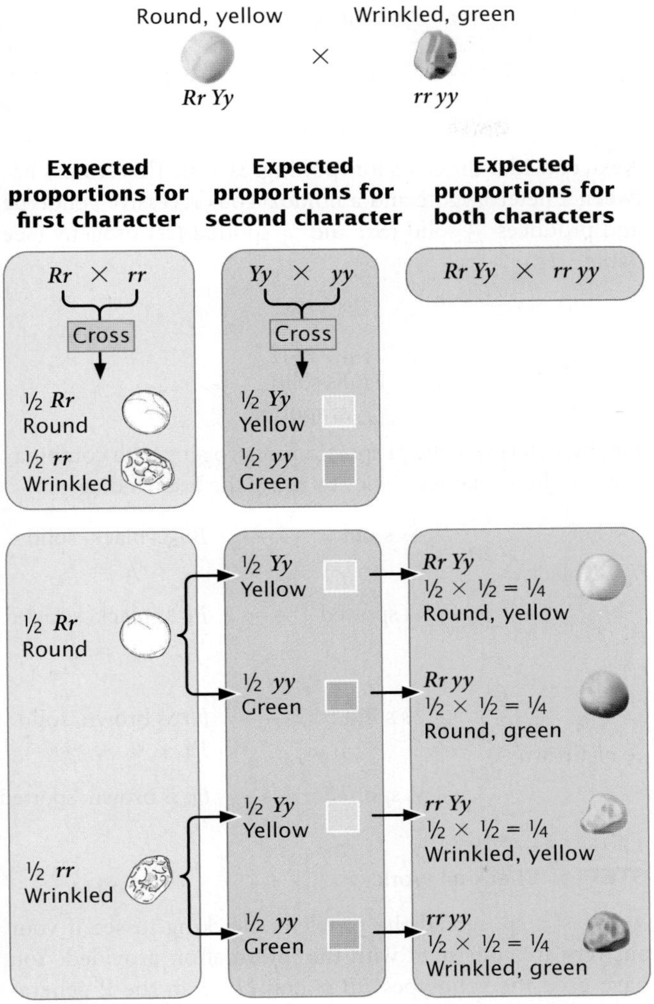

3.12 A branch diagram can be used to determine the phenotypes and expected proportions of offspring from a dihybrid testcross (*Rr Yy* × *rr yy*).

WORKED PROBLEM

The principles of segregation and independent assortment are important not only because they explain how heredity works, but also because they provide the means for predicting the outcome of genetic crosses. This predictive power has made genetics a powerful tool in agriculture and other fields, and the ability to apply the principles of heredity is an important skill for all students of genetics. Practice with genetics problems is essential for mastering the basic principles of heredity; no amount of reading and memorization can substitute for the experience gained by deriving solutions to specific problems in genetics.

You may find genetics problems difficult if you are unsure of where to begin or how to organize a solution to the problem. In genetics, every problem is different, and so no common series of steps can be applied to all genetics problems. Logic and common sense must be used to analyze a problem and arrive at a solution. Nevertheless, certain steps can facilitate the process, and solving the following problem will serve to illustrate those steps.

In mice, black coat color (*B*) is dominant over brown (*b*), and a solid pattern (*S*) is dominant over a white-spotted pattern (*s*). Color and spotting are controlled by genes that assort independently. A homozygous black, spotted mouse is crossed with a homozygous brown, solid mouse. All the F_1 mice are black and solid. A testcross is then carried out by mating the F_1 mice with brown, spotted mice.

a. Give the genotypes of the parents and the F_1 mice.

b. Give the genotypes and phenotypes, along with their expected ratios, of the progeny expected from the testcross.

Solution Strategy

What information is required in your answer to the problem?

First, determine what question or questions the problem is asking. Is it asking for genotypes, or genotypic ratios, or phenotypic ratios? This problem asks you to provide the *genotypes* of the parents and the F_1, the *expected genotypes* and *phenotypes* of the progeny of the testcross, and their *expected proportions.*

What information is provided to solve the problem?

Next, determine what information is provided that will be necessary for solving the problem. This problem gives important information about the dominance relations of the traits and the genes that encode them:

- Black is dominant over brown.
- Solid is dominant over white-spotted.
- The genes for the two characteristics assort independently.
- Symbols for the different alleles: *B* for black, *b* for brown, *S* for solid, and *s* for spotted.

It is often helpful to write down the symbols at the beginning of the solution:

B—black	*S*—solid
b—brown	*s*—white-spotted

Next, write out the crosses, as given in the problem:

P	Homozygous black, spotted	×	Homozygous brown, solid
		↓	
F_1		Black, solid	
Testcross	Black, solid	×	Brown, spotted

For help with this problem, review:

Sections 3.2 and 3.3.

Solution Steps

STEP 1: Write down any genetic information that can be determined from the phenotypes alone.

From the phenotypes and the statement that they are homozygous, you know that the P-generation mice must be *BB ss* and *bb SS*. The F_1 mice are black and solid, both dominant traits, and so the F_1 mice must possess at least one black allele (*B*) and one solid allele (*S*). At this point, you may not be certain about the other alleles; so represent the genotype of the F_1 as *B_ S_*, where _ means that any allele is possible. The brown, spotted mice used in the testcross must be *bb ss* because both brown and spotted are recessive traits that will be expressed only if two recessive alleles are present. Record these genotypes on the crosses that you wrote out in step 2:

P	Homozygous black, spotted *BB ss*	×	Homozygous brown, solid *bb SS*
		↓	
F_1		Black, solid *B_ S_*	
Testcross	Black, solid *B_ S_*	×	Brown, spotted *bb ss*

STEP 2: Break the problem down into smaller parts.

First, determine the genotype of the F_1. After this genotype has been determined, you can predict the results of the testcross and determine the genotypes and phenotypes of the progeny of the testcross. Second, because this cross includes two independently assorting loci, it can be conveniently broken down into two single-locus crosses: one for coat color and the other for spotting. Third, you can use a branch diagram to determine the proportion of progeny of the testcross with different combinations of the two traits.

STEP 3: Work the different parts of the problem.

Start by determining the genotype of the F_1 progeny. Mendel's first law indicates that the two alleles at a locus separate, one going into each gamete. Thus, the gametes produced by the black, spotted parent contain *B s* and the gametes produced by the brown, solid parent contain *b S*, which combine to produce F_1 progeny with the genotype *Bb Ss*:

P	Homozygous black, spotted *BB ss*	×	Homozygous brown, solid *bb SS*
	↓		↓
Gametes	(*Bs*)		(*bS*)
F_1		*Bb Ss*	

Use the F_1 genotype to work the testcross (*Bb Ss* × *bb ss*), breaking it into two single-locus crosses. First, consider the cross for coat color: *Bb* × *bb*. Any cross between a heterozygote and a homozygous recessive genotype produces a 1:1 phenotypic ratio of progeny (see Table 3.2):

Bb × *bb*
↓
$^1/_2$ *Bb* black
$^1/_2$ *bb* brown

Next, consider the cross for spotting: *Ss* × *ss*. This cross is between a heterozygote and a homozygous recessive genotype and produces $^1/_2$ solid (*Ss*) and $^1/_2$ spotted (*ss*) progeny (see Table 3.2):

Ss × ss
↓
$^1/_2$ *Ss* solid
$^1/_2$ *ss* spotted

Finally, determine the proportions of progeny with combinations of these characteristics by using the branch diagram.

$^1/_2$ *Bb* black	$^1/_2$ *Ss* solid	⟶	*Bb Ss* black, solid $^1/_2 \times {}^1/_2 = {}^1/_4$
	$^1/_2$ *ss* spotted	⟶	*Bb ss* black, spotted $^1/_2 \times {}^1/_2 = {}^1/_4$
$^1/_2$ *bb* brown	$^1/_2$ *Ss* solid	⟶	*bb Ss* brown, solid $^1/_2 \times {}^1/_2 = {}^1/_4$
	$^1/_2$ *ss* spotted	⟶	*bb ss* brown, spotted $^1/_2 \times {}^1/_2 = {}^1/_4$

STEP 4: Check all work.

As a last step, reread the problem, checking to see if your answers are consistent with the information provided. You have used the genotypes *BB ss* and *bb SS* in the P generation. Do these genotypes encode the phenotypes given in the problem? Are the F_1 progeny phenotypes consistent with the genotypes that you assigned? The answers are consistent with the information.

▶ Now that we have stepped through a genetics problem together, try your hand at **Problem 31** at the end of the chapter.

3.4 Observed Ratios of Progeny May Deviate from Expected Ratios by Chance

When two individual organisms of known genotype are crossed, we expect certain ratios of genotypes and phenotypes in the progeny; these expected ratios are based on the Mendelian principles of segregation, independent assortment, and dominance. The ratios of genotypes and phenotypes *actually* observed among the progeny, however, may deviate from these expectations.

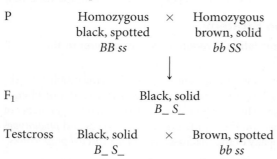

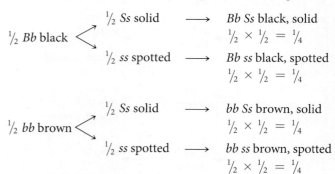

For example, in German cockroaches, brown body color (*Y*) is dominant over yellow body color (*y*). If we cross a brown, heterozygous cockroach (*Yy*) with a yellow cockroach (*yy*), we expect a 1:1 ratio of brown (*Yy*) and yellow (*yy*) progeny. Among 40 progeny, we therefore expect to see 20 brown and 20 yellow offspring. However, the observed numbers might deviate from these expected values; we might in fact see 22 brown and 18 yellow progeny.

Chance plays a critical role in genetic crosses, just as it does in flipping a coin. When you flip a coin, you expect a 1:1 ratio—$\frac{1}{2}$ heads and $\frac{1}{2}$ tails. If you flip a coin 1000 times, the proportion of heads and tails obtained will probably be very close to that expected 1:1 ratio. However, if you flip the coin 10 times, the ratio of heads to tails may be quite different from 1:1. You could easily get 6 heads and 4 tails, or 3 heads and 7 tails, just by chance. You might even get 10 heads and 0 tails. The same thing happens in genetic crosses. We may expect 20 brown and 20 yellow cockroaches, but 22 brown and 18 yellow progeny *could* arise as a result of chance.

The Chi-Square Goodness-of-Fit Test

If you expected a 1:1 ratio of brown and yellow cockroaches, but the cross produced 22 brown and 18 yellow cockroaches, you probably wouldn't be too surprised, even though it wasn't a perfect 1:1 ratio. In this case, it seems reasonable to assume that chance produced the deviation between the expected and the observed results. But if you observed 25 brown and 15 yellow cockroaches, would you still assume that this result represents a 1:1 ratio? Something other than chance might have caused this deviation. Perhaps the inheritance of this characteristic is more complicated than was assumed, or perhaps some of the yellow progeny died before they were counted. Clearly, we need some means of evaluating how likely it is that chance is responsible for the deviation between the observed and the expected numbers.

To evaluate the role of chance in producing deviations between observed and expected values, a statistical test called the **chi-square goodness-of-fit test** is used. This test provides information about how well the observed values fit the expected values. Before we learn how to use this test, however, it is important to understand what it does and does not indicate about a genetic cross. The chi-square test cannot tell us whether a genetic cross has been correctly carried out, whether our results are correct, or whether we have chosen the correct genetic explanation for those results. What it does indicate is the *probability* that the difference between the observed and the expected values is due to chance. In other words, it indicates the likelihood that chance alone could produce the deviation between the expected and the observed values.

If we expected 20 brown and 20 yellow progeny from a genetic cross, the chi-square test gives the probability that we might observe 25 brown and 15 yellow progeny simply owing to chance deviations from the expected 20:20 ratio. The hypothesis that chance alone is responsible for any deviation between observed and expected values is sometimes called the *null hypothesis*. Statistical tests such as the chi-square test cannot prove that the null hypothesis is correct, but they can help us decide if we should reject it. When the probability calculated from the chi-square test is high, we assume that chance alone produced the deviation, and we do not reject the null hypothesis. When the probability is low, we assume that some factor other than chance—some significant factor—produced the deviation. For example, the mortality rate of the yellow cockroaches might be higher than that of the brown cockroaches. When the probability that chance produced the deviation is low, we reject the null hypothesis.

To use the chi-square goodness-of-fit test, we first determine the expected results. The chi-square test must always be applied to *numbers* of progeny, not to proportions or percentages. Let's consider a locus for coat color in domestic cats, for which black color (*B*) is dominant over gray (*b*). If we crossed two heterozygous black cats ($Bb \times Bb$), we would expect a 3:1 ratio of black and gray kittens. Imagine that a series of such crosses yields a total of 50 kittens—30 black and 20 gray. These numbers are our *observed* values. We can obtain the *expected* numbers by multiplying the expected proportions by the total number of observed progeny. In this case, the expected number of black kittens is $\frac{3}{4} \times 50 = 37.5$ and the expected number of gray kittens is $\frac{1}{4} \times 50 = 12.5$.

The chi-square (χ^2) value is calculated by using the following formula:

$$\chi^2 = \Sigma\frac{(\text{observed} - \text{expected})^2}{\text{expected}}$$

where Σ means the sum. We calculate the sum of all the squared differences between observed and expected values divided by by the expected values. To calculate the chi-square value for our black and gray kittens, we would first subtract the number of *expected* black kittens from the number of *observed* black kittens ($30 - 37.5 = -7.5$) and square this value: $-7.5^2 = 56.25$. We then divide this result by the expected number of black kittens, $56.25/37.5 = 1.5$. We repeat the calculations on the number of expected gray kittens: $(20 - 12.5)^2/12.5 = 4.5$. To obtain the overall chi-square value, we add the (observed − expected)^2/expected values: $\chi^2 = 1.5 + 4.5 = 6.0$.

The next step is to determine the probability associated with this chi-square value, which is the probability that the deviation between the observed and the expected results could be due to chance. This step requires us to compare the calculated chi-square value (6.0) with theoretical values that have the same degrees of freedom in a chi-square table. The degrees of freedom represent the number of ways in which the expected classes are free to vary. For a chi-square goodness-of-fit test, the degrees of freedom are equal to $n - 1$, in which n is the number of different expected phenotypes. Here, we lose one degree of freedom because the total number of expected progeny must equal the total number of observed progeny. In our example, there are two expected phenotypes (black and gray); so $n = 2$, and the degree of freedom equals $2 - 1 = 1$.

Now that we have our calculated chi-square value and have figured out the associated degrees of freedom, we are ready to obtain the probability from a chi-square table (**Table 3.4**). The degrees of freedom are given in the left column of the table and the probabilities are given at the top;

TABLE 3.4 Critical values of the χ^2 distribution

	P								
df	**0.995**	**0.975**	**0.9**	**0.5**	**0.1**	**0.05***	**0.025**	**0.01**	**0.005**
1	0.000	0.000	0.016	0.455	2.706	3.841	5.024	6.635	7.879
2	0.010	0.051	0.211	1.386	4.605	5.991	7.378	9.210	10.597
3	0.072	0.216	0.584	2.366	6.251	7.815	9.348	11.345	12.838
4	0.207	0.484	1.064	3.357	7.779	9.488	11.143	13.277	14.860
5	0.412	0.831	1.610	4.351	9.236	11.070	12.832	15.086	16.750
6	0.676	1.237	2.204	5.348	10.645	12.592	14.449	16.812	18.548
7	0.989	1.690	2.833	6.346	12.017	14.067	16.013	18.475	20.278
8	1.344	2.180	3.490	7.344	13.362	15.507	17.535	20.090	21.955
9	1.735	2.700	4.168	8.343	14.684	16.919	19.023	21.666	23.589
10	2.156	3.247	4.865	9.342	15.987	18.307	20.483	23.209	25.188
11	2.603	3.816	5.578	10.341	17.275	19.675	21.920	24.725	26.757
12	3.074	4.404	6.304	11.340	18.549	21.026	23.337	26.217	28.300
13	3.565	5.009	7.042	12.340	19.812	22.362	24.736	27.688	29.819
14	4.075	5.629	7.790	13.339	21.064	23.685	26.119	29.141	31.319
15	4.601	6.262	8.547	14.339	22.307	24.996	27.488	30.578	32.801

P, probability; df, degrees of freedom.
*Most scientists assume that when $P < 0.05$, a significant difference exists between the observed and expected values in a chi-square test.

within the body of the table are chi-square values associated with these probabilities. First, we find the row for the appropriate degrees of freedom; for our example with 1 degree of freedom, it is the first row of the table. Then we find where our calculated chi-square value (6.0) lies among the theoretical values in this row. The theoretical chi-square values increase from left to right, and the probabilities decrease from left to right. Our chi-square value of 6.0 falls between the value of 5.024, associated with a probability of 0.025, and the value of 6.635, associated with a probability of 0.01.

Thus, the probability associated with our chi-square value is less than 0.025 and greater than 0.01. So there is less than a 2.5% probability that the deviation that we observed between the expected and the observed numbers of black and gray kittens could be due to chance.

Most scientists use the 0.05 probability level as their cutoff value: if the probability of chance being responsible for the deviation between the observed and the expected values is greater than or equal to 0.05, they assume that chance is responsible for that deviation. When the probability is less than 0.05, scientists assume that chance is not responsible and that a significant difference from the expected values exists. The expression *significant difference* means that a factor other than chance is responsible for the difference between the observed values and the expected values. In regard to the kittens, perhaps one of the genotypes had a greater mortality rate before the progeny were counted, or perhaps other genetic factors skewed the observed ratios.

In choosing 0.05 as their cutoff value, scientists have agreed to assume that chance is responsible for deviations between observed and expected values unless there is strong evidence to the contrary. Bear in mind that even if we obtain a probability of, say, 0.01, there is still a 1% probability that the deviation between the observed and the expected numbers is due to nothing more than chance. Calculation of the chi-square value is illustrated in **Figure 3.13**. See **Animation 3.2** for an example of working a chi-square goodness-of-fit test.

TRY PROBLEM 35

THINK-PAIR-SHARE Question 8

CONCEPTS

Differences between observed and expected numbers among the progeny of a cross can arise by chance alone. The chi-square goodness-of-fit test can be used to evaluate whether deviations between observed and expected values are likely to be due to chance or to some other significant factor.

✓ CONCEPT CHECK 7

A chi-square test comparing observed and expected numbers of progeny is carried out, and the probability associated with the calculated chi-square value is 0.72. What does this probability represent?

a. Probability that the correct results were obtained
b. Probability of obtaining the observed numbers
c. Probability that the difference between observed and expected numbers is significant
d. Probability that the difference between observed and expected numbers could be due to chance

Phenotype	Observed	Expected
Purple	105	$\frac{3}{4} \times 150 = 112.5$
White	45	$\frac{1}{4} \times 150 = 37.5$
Total	150	

The expected values are obtained by multiplying the expected proportion by the total,…

$$\chi^2 = \sum \frac{(O-E)^2}{E}$$

$$\chi^2 = \frac{(105-112.5)^2}{112.5} + \frac{(45-37.5)^2}{37.5}$$

$$\chi^2 = \frac{56.25}{112.5} + \frac{56.25}{37.5}$$

$$\chi^2 = 0.5 + 1.5 = 2.0$$

…and then the chi-square value is calculated.

Degrees of freedom = $n-1$
Degrees of freedom = $2-1=1$
Probability (from Table 3.5)
$0.1 < P < 0.5$

The probability associated with the calculated chi-square value is between 0.10 and 0.50, indicating a high probability that the difference between observed and expected values is due to chance.

Conclusion: No significant difference between observed and expected values.

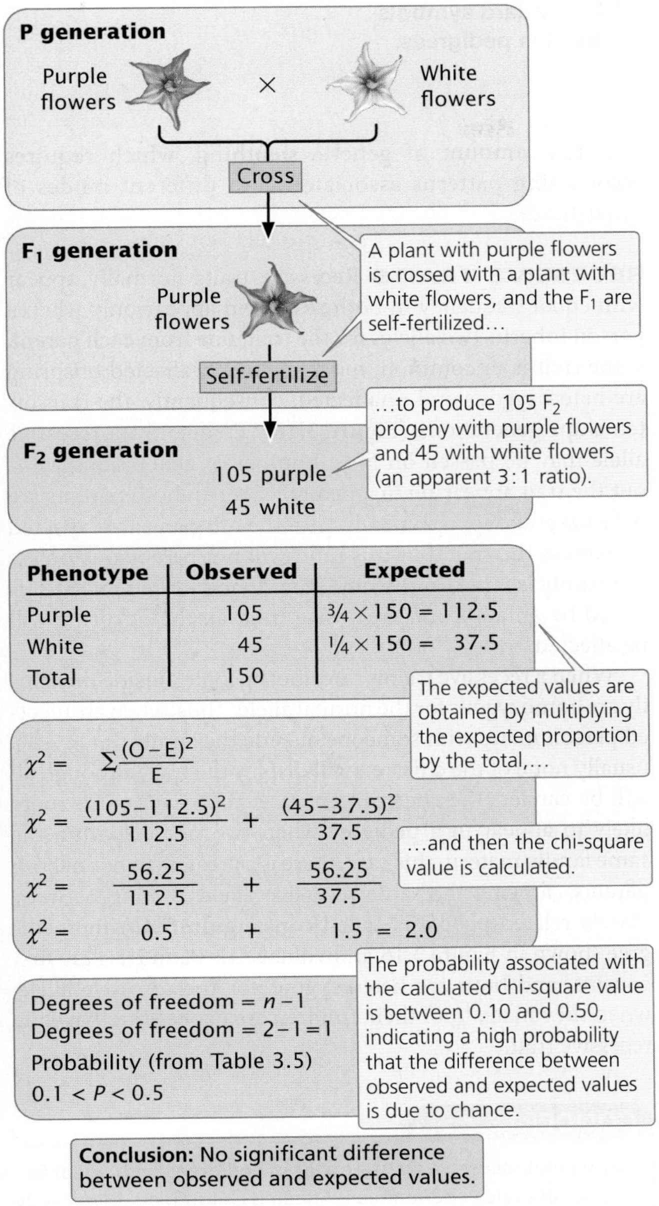

3.13 A chi-square goodness-of-fit test is used to determine the probability that the difference between observed and expected values is due to chance.

3.5 Geneticists Often Use Pedigrees to Study the Inheritance of Human Characteristics

The study of human genetic characteristics presents some major obstacles. First, controlled matings are not possible. With other organisms, geneticists can carry out specific crosses to test their hypotheses about inheritance. Unfortunately

(for geneticists at least), matings between humans are more frequently determined by romance, family expectations, and—occasionally—accident than by the requirements of geneticists. Other obstacles are the long generation time and generally small family size of our species. To overcome these obstacles, geneticists have developed special techniques for studying human inheritance that are uniquely suited to human biology and culture.

One technique used by geneticists to study human inheritance is the analysis of pedigrees. A **pedigree** is a pictorial representation of a family history, essentially a family tree that outlines the inheritance of one or more characteristics. When a particular characteristic or disease is observed in a person, a geneticist often studies the family of this affected person by drawing a pedigree.

Symbols Used in Pedigrees

The symbols commonly used in pedigrees are summarized in **Figure 3.14**. Males are represented by squares, females by circles. A horizontal line drawn between two symbols representing a man and a woman indicates a mating; children are connected to their parents by vertical lines extending below the parents. The pedigree shown in **Figure 3.15a** on p. 65 illustrates a family with Waardenburg syndrome, an autosomal dominant type of deafness that may be accompanied by fair skin, a white forelock, and visual problems (**Figure 3.15b**). An autosomal trait is one that is encoded on an autosome (nonsex chromosome). In Chapter 4, we will consider the inheritance of sex-linked traits, those encoded by genes on the sex chromosomes.

Persons in a pedigree who exhibit the trait of interest are represented by filled circles and squares; in Figure 3.15a, the filled symbols represent members of the family who have Waardenburg syndrome. Unaffected members are represented by open circles and squares. The person from whom the pedigree is initiated is called the **proband** and is usually designated by an arrow and sometimes with the letter P (IV-2 in Figure 3.15a).

Let's look closely at Figure 3.15 and consider some additional features of a pedigree. Each generation in a pedigree is identified by a roman numeral; within each generation, family members are assigned arabic numerals, and children in each family are listed in birth order from left to right. Person II-4, a man with Waardenburg syndrome, mated with II-5, an unaffected woman, and they produced five children. The oldest of their children is III-8, a male with Waardenburg syndrome, and the youngest is III-14, an unaffected female.

Analysis of Pedigrees

The limited number of offspring in most human families means that clear Mendelian ratios are usually impossible to discern in a single pedigree. Pedigree analysis requires

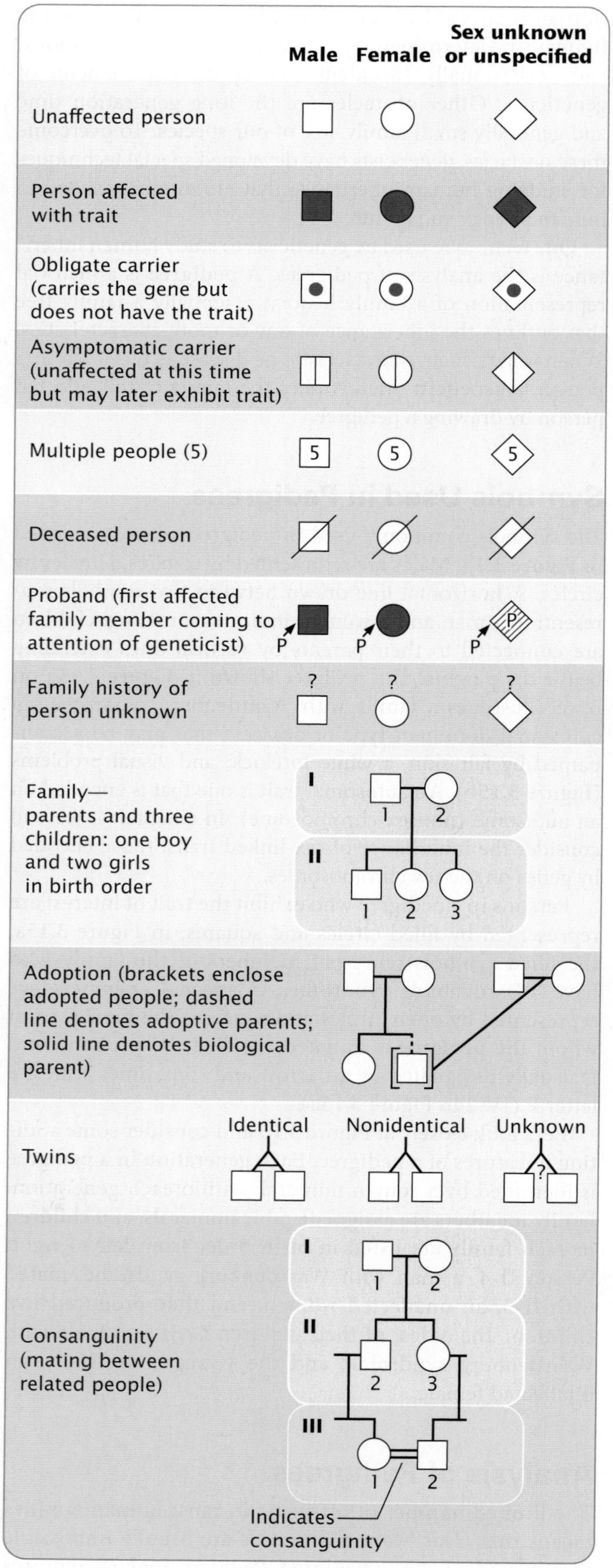

3.14 Standard symbols are used in pedigrees.

a certain amount of genetic sleuthing, which requires recognizing patterns associated with different modes of inheritance.

RECESSIVE TRAITS Recessive traits normally appear with equal frequency in both sexes and appear only when a person inherits two alleles for the trait, one from each parent. If the trait is uncommon, most parents of affected offspring are heterozygous and unaffected; consequently, the trait often skips generations (**Figure 3.16**). Frequently, a recessive allele may be passed on for a number of generations without the trait appearing in a pedigree. When both parents are heterozygous, approximately $^1/_4$ of the offspring are expected to express the trait, but this ratio will not be obvious unless the family is large. In the rare event that both parents are affected by an autosomal recessive trait, all the offspring will be affected.

When a recessive trait is rare, most people outside the family are homozygous for the normal allele. Thus, when an affected person mates with someone outside the family ($aa \times AA$), usually none of the children will display the trait, although all will be carriers (i.e., heterozygous). A recessive trait is more likely to appear in a pedigree when two people within the same family mate; in this case, there is a greater chance of both parents carrying the same recessive allele. Mating between closely related people is called **consanguinity**. In the pedigree shown in Figure 3.16, individuals III-3 and III-4 are first cousins, and both are heterozygous for the recessive allele; when they mate, $^1/_4$ of their children are expected to have the recessive trait.

CONCEPTS

Autosomal recessive traits appear in pedigrees with equal frequency in males and females. Affected children are commonly born to unaffected parents who are heterozygous carriers of the gene for the trait, and the trait often skips generations. Recessive traits appear in pedigrees more frequently among the offspring of consanguineous matings. Autosomal dominant traits also appear in both sexes with equal frequency. An affected person has an affected parent, and the trait does not skip generations. Unaffected persons do not transmit the trait.

✓ CONCEPT CHECK 8

Recessive traits often appear in pedigrees in which there have been consanguineous matings because these traits

a. tend to skip generations.
b. appear only when both parents carry a copy of the allele for the trait, which is more likely when the parents are related.
c. usually arise in children born to parents who are unaffected.
d. appear equally in males and females.

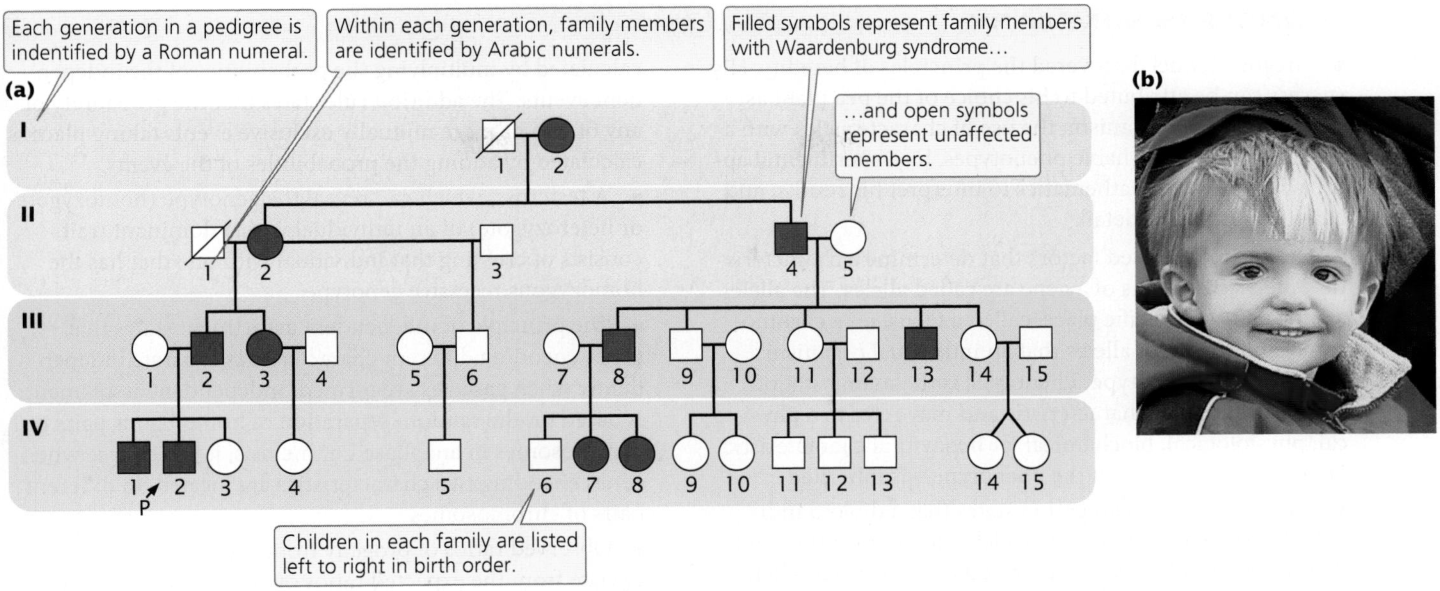

3.15 Waardenburg syndrome is (a) inherited as an autosomal dominant trait and (b) characterized by deafness, fair skin, visual problems, and a white forelock. The proband (P) is the person from whom this pedigree is initiated. [Photograph Courtesy of Guy Rowland.]

DOMINANT TRAITS Autosomal dominant traits normally appear in both sexes with equal frequency, and both sexes are capable of transmitting these traits to their offspring. Every person with a dominant trait must have inherited the allele from at least one parent (unless the person carries a new mutation), so autosomal dominant traits do not skip generations (**Figure 3.17**).

Sex-linked traits also have a distinctive pattern of inheritance, which we will consider in Chapter 4. See the analysis of a pedigree in **Animation 3.3**. A

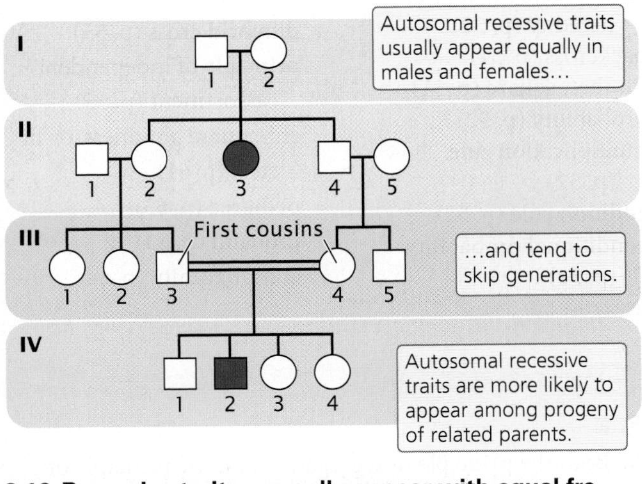

3.16 Recessive traits normally appear with equal frequency in both sexes and often skip generations. The double line between III-3 and III-4 represents consanguinity (mating between related persons).

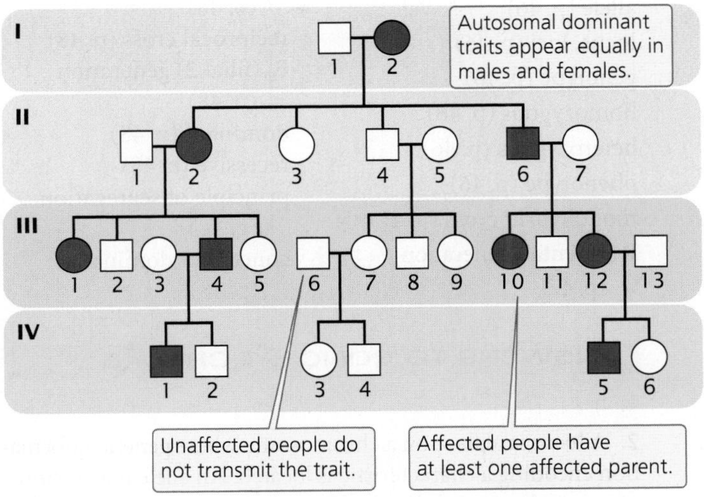

3.17 Dominant traits normally appear with equal frequency in both sexes and do not skip generations.

CONCEPTS SUMMARY

- Gregor Mendel discovered the principles of heredity. His success can be attributed to his choice of the pea plant as an experimental organism, the use of characteristics with a few, easily distinguishable phenotypes, his experimental approach, the use of mathematics to interpret his results, and careful attention to detail.
- Genes are inherited factors that determine a characteristic. Alternative forms of a gene are called alleles. The alleles are located at a specific place, called a locus, on a chromosome, and the set of alleles that an individual organism possesses is its genotype. Phenotype is the manifestation or appearance of a characteristic and may refer to a physical, physiological, biochemical, or behavioral characteristic. Only the genotype—not the phenotype—is inherited.
- The principle of segregation states that a diploid individual organism possesses two alleles encoding a trait and that these two alleles separate in equal proportions when gametes are formed.
- The concept of dominance indicates that when two different alleles are present in a heterozygote, only the trait of one of them, the dominant allele, is observed in the phenotype. The other allele is said to be recessive.
- The two alleles of a genotype are located on homologous chromosomes. The separation of homologous chromosomes in anaphase I of meiosis brings about the segregation of alleles.
- Probability is the likelihood that a particular event will take place. The multiplication rule states that the probability of two or more independent events taking place together is calculated by multiplying the probabilities of the independent events. The addition rule states that the probability of any of two or more mutually exclusive events taking place is calculated by adding the probabilities of the events.
- A testcross, which can reveal the genotype (homozygote or heterozygote) of an individual with a dominant trait, consists of crossing that individual with one that has the homozygous recessive genotype.
- The principle of independent assortment states that genes encoding different characteristics separate independently when gametes are formed. Independent assortment is based on the random separation of homologous pairs of chromosomes in anaphase I of meiosis; it takes place when genes encoding two characteristics are located on different pairs of chromosomes.
- Observed ratios of progeny from a genetic cross may deviate from the expected ratios owing to chance. The chi-square goodness-of-fit test can be used to determine the probability that a difference between observed and expected numbers is due to chance.
- Pedigrees are often used to study the inheritance of traits in humans. Autosomal recessive traits typically appear with equal frequency in both sexes and often skip generations. They are more likely to appear in families with consanguinity (mating between closely related persons). Autosomal dominant traits also appear with equal frequency in both sexes but do not skip generations. Unaffected people do not normally transmit an autosomal dominant trait to their offspring.

IMPORTANT TERMS

gene (p. 46)
allele (p. 46)
locus, loci (p. 46)
genotype (p. 46)
homozygous (p. 46)
heterozygous (p. 46)
phenotype (p. 46)
monohybrid cross (p. 47)
P (parental) generation (p. 47)
F_1 (filial 1) generation (p. 48)
reciprocal cross (p. 48)
F_2 (filial 2) generation (p. 48)
dominant (p. 48)
recessive (p. 48)
principle of segregation (p. 48)
concept of dominance (p. 48)
chromosome theory of heredity (p. 50)
backcross (p. 51)
Punnett square (p. 51)
probability (p. 52)
multiplication rule (p. 52)
addition rule (p. 52)
conditional probability (p. 54)
testcross (p. 54)
wild type (p. 54)
dihybrid cross (p. 55)
principle of independent assortment (p. 55)
chi-square goodness-of-fit test (p. 61)
pedigree (p. 63)
proband (p. 63)
consanguinity (p. 64)

ANSWERS TO CONCEPT CHECKS

1. b

2. A locus is a place on a chromosome where genetic information encoding a characteristic is located. An allele is a version of a gene that encodes a specific trait. A genotype is the set of alleles possessed by an individual organism, and a phenotype is the manifestation or appearance of a characteristic.

3. The traits encoded by both alleles appeared in the F_2 progeny.

4. d

5. a

6. Both the principle of segregation and the principle of independent assortment refer to the separation of alleles in anaphase I of meiosis. The principle of segregation says that these alleles separate, and the principle of independent assortment says that they separate independently of alleles at other loci.

7. d

8. b

WORKED PROBLEMS

Problem 1

In corn, purple kernels are dominant over yellow kernels, and full kernels are dominant over shrunken kernels. A corn plant with purple and full kernels is crossed with a plant with yellow and shrunken kernels and the following progeny are obtained:

purple, full	112
purple, shrunken	103
yellow, full	91
yellow, shrunken	94

What are the most likely genotypes of the parents and the progeny? Test your genetic hypothesis with a chi-square test.

Solution Strategy

What information is required in your answer to the problem?

The genotypes of parents and progeny. A chi-square test comparing the observed and expected results.

What information is provided to solve the problem?

- Purple kernels are dominant over yellow kernels and full kernels are dominant over shrunken kernels.
- The phenotypes of the parents.
- The phenotypes and numbers of the different types of progeny of the cross.

For help with this problem, review:

Sections 3.3 and 3.4.

Solution Steps

The best way to begin this problem is to break the cross down into simple crosses for a single characteristic (seed color or seed shape):

P	purple × yellow	full × shrunken
F_1	112 + 103 = 215 purple	112 + 91 = 203 full
	91 + 94 = 185 yellow	103 + 94 = 197 shrunken

... good ... n a cross ... multiple ... istics is to ... he results ... characteris- ... tely.

In this cross, purple × yellow produces approximately $\frac{1}{2}$ purple and $\frac{1}{2}$ yellow (a 1:1 ratio). A 1:1 ratio is usually the result of a cross between a heterozygote and a homozygote. Because purple is dominant, the purple parent must be heterozygous (*Pp*) and the yellow parent must be homozygous (*pp*). The purple progeny produced by this cross will be heterozygous (*Pp*), and the yellow progeny must be homozygous (*pp*).

Now let's examine the other character. Full × shrunken produces $\frac{1}{2}$ full and $\frac{1}{2}$ shrunken, or a 1:1 ratio, so these progeny phenotypes are also produced by a cross between a heterozygote (*Ff*) and a homozygote (*ff*); the full-kernel progeny will be heterozygous (*Ff*) and the shrunken-kernel progeny will be homozygous (*ff*).

Now combine the two crosses and use the multiplication rule to obtain the overall genotypes and the proportions of each genotype:

Recall: The multiplication rule states that the probability of two or more independent events occurring together is calculated by multiplying their independent probabilities.

P purple, full × yellow, shrunken
Pp Ff *pp ff*

F_1 *Pp Ff* = $\frac{1}{2}$ purple × $\frac{1}{2}$ full = $\frac{1}{4}$ purple, full
Pp ff = $\frac{1}{2}$ purple × $\frac{1}{2}$ shrunken = $\frac{1}{4}$ purple, shrunken
Pp Ff = $\frac{1}{2}$ yellow × $\frac{1}{2}$ full = $\frac{1}{4}$ yellow, full
Pp ff = $\frac{1}{2}$ yellow × $\frac{1}{2}$ shrunken = $\frac{1}{4}$ yellow, shrunken

Our genetic explanation predicts that from this cross, we should see $\frac{1}{4}$ purple, full-kernel progeny; $\frac{1}{4}$ purple, shrunken-kernel progeny; $\frac{1}{4}$ yellow, full-kernel progeny; and $\frac{1}{4}$ yellow, shrunken-kernel progeny. A total of 400 progeny were produced; so $\frac{1}{4} \times 400 = 100$ of each phenotype are expected. Therefore, the observed numbers do not fit the expected numbers exactly.

Could the difference between what we observe and what we expected be due to chance? If the probability is high that chance alone is responsible for the difference between observed and expected values, we will assume that the progeny have been produced in the 1:1:1:1 ratio predicted for the cross. If the probability that the difference between observed and expected values is due to chance is low, we will assume that the progeny are not really in the predicted ratio, and that some other, *significant* factor must be responsible for the deviation from our expectations.

The observed and expected numbers:

Phenotype	Observed	Expected
purple, full	112	$\frac{1}{4} \times 400 = 100$
purple, shrunken	103	$\frac{1}{4} \times 400 = 100$
yellow, full	91	$\frac{1}{4} \times 400 = 100$
yellow, shrunken	94	$\frac{1}{4} \times 400 = 100$

To determine the probability that the difference between the observed and expected numbers is due to chance, we calculate a chi-square value using the formula

Hint: See Figure 3.13 for help on how to carry out a chi-square test.

$$\chi^2 = \Sigma \frac{(\text{observed} - \text{expected})^2}{\text{expected}}$$

$$\chi^2 = \frac{(112-100)^2}{100} + \frac{(103-100)^2}{100} + \frac{(91-100)^2}{100} + \frac{(94-100)^2}{100}$$

$$= \frac{12^2}{100} + \frac{3^2}{100} + \frac{9^2}{100} + \frac{6^2}{100}$$

$$= \frac{144}{100} + \frac{9}{100} + \frac{81}{100} + \frac{36}{100}$$

$$= 1.44 + 0.09 + 0.81 + 0.36 = 2.70$$

Now that we have the chi-square value, we must determine the probability that this chi-square value is due to chance.

To obtain this probability, we first calculate the degrees of freedom, which for a chi-square goodness-of-fit test are $n - 1$, where n equals the number of expected phenotypic classes. In this case, there are four expected phenotypic classes, so the degrees of freedom equal $4 - 1 = 3$. We must now look up the chi-square value in a chi-square table (see Table 3.4). We select the row corresponding to 3 degrees of freedom and look along this row to find our calculated chi-square value. The calculated chi-square value of 2.7 lies between 2.366 (a probability of 0.5) and 6.251 (a probability of 0.1). The probability (P) associated with the calculated chi-square value is therefore $0.5 < P < 0.1$. This is the probability that the difference between what we observed and what we expected is due to chance, which in this case is relatively high, and so chance is probably responsible for the deviation. We can conclude that the progeny *do* appear in the 1:1:1:1 ratio predicted by our genetic explanation.

Problem 2

Joanna has "short fingers" (brachydactyly). She has two older brothers who are identical twins; both have short fingers. Joanna's two younger sisters have normal fingers. Joanna's mother has normal fingers, and her father has short fingers. Joanna's paternal grandmother (her father's mother) has short fingers; her paternal grandfather (her father's father), who is now deceased, had normal fingers. Both of Joanna's maternal grandparents (her mother's parents) have normal fingers. Joanna marries Tom, who has normal fingers; they adopt a son named Bill, who has normal fingers. Bill's biological parents both have normal fingers. After adopting Bill, Joanna and Tom produce two children: an older daughter with short fingers and a younger son with normal fingers.

a. Using standard symbols and labels, draw a pedigree illustrating the inheritance of short fingers in Joanna's family.

b. What is the most likely mode of inheritance for short fingers in this family?

c. If Joanna and Tom have another biological child, what is the probability (based on your answer to part *b*) that this child will have short fingers?

Solution Strategy

What information is required in your answer to the problem?

a. A pedigree to represent the family, drawn with correct symbols and labeling.

b. The most likely mode of inheritance for short fingers.

c. The probability that Joanna and Tom's next child will have short fingers.

What information is provided to solve the problem?

- The phenotypes of Joanna and Tom and their family members.

For help with this problem, review:

The information on pedigrees in Section 3.5.

Solution Steps

a. In the pedigree for the family, use filled circles (females) and filled squares (males) to represent persons with the trait of interest (short fingers). Connect Joanna's identical twin brothers to the line above by drawing diagonal lines that have a horizontal line between them. Enclose Bill, the adopted child of Joanna and Tom, in brackets; connect him to his biological parents by drawing a diagonal line and to his adopted parents by a dashed line.

Hint: See 3.14 for a r symbols use pedigree.

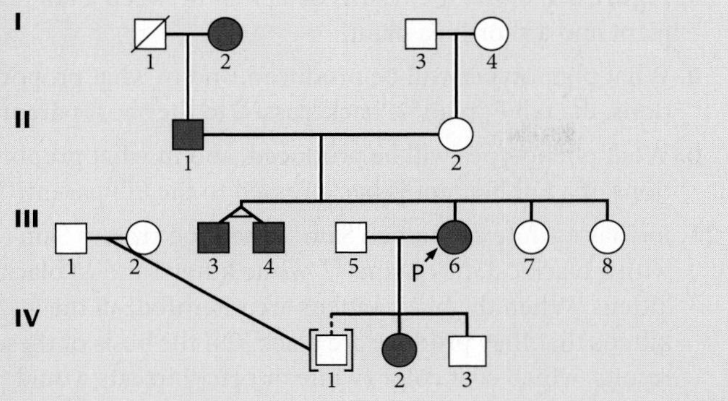

b. The most likely mode of inheritance for short fingers in this family is dominant. The trait appears equally in males and females and does not skip generations. When one parent has the trait, it appears in approximately half of that parent's sons and daughters, although the number of children in the families is small.

Hint: See Analysis of Pedigrees for a review of characteristics of dominant and recessive traits in pedigrees.

c. If having short fingers is a dominant trait, Tom must be homozygous (*bb*) because he has normal fingers. Joanna must be heterozygous (*Bb*) because she and Tom have produced both short-fingered and normal-fingered offspring. In a cross between a heterozygote and a homozygote, half the progeny are expected to be heterozygous and half homozygous ($Bb \times bb \rightarrow \frac{1}{2}\ Bb,\ \frac{1}{2}\ bb$); so the probability that Joanna's and Tom's next biological child will have short fingers is $\frac{1}{2}$.

COMPREHENSION QUESTIONS

Section 3.1

1. Why was Mendel's approach to the study of heredity so successful?
2. What is the difference between genotype and phenotype?

Section 3.2

3. What is the principle of segregation? Why is it important?
4. How are Mendel's principles different from the concept of blending inheritance discussed in Chapter 1?
5. What is the concept of dominance?
6. What are the addition and multiplication rules of probability and when should they be used?
7. Give the genotypic ratios that may appear among the progeny of simple crosses and the genotypes of the parents that may give rise to each ratio.
8. What is the chromosome theory of heredity? Why was it important?

Section 3.3

9. What is the principle of independent assortment? How is it related to the principle of segregation?
10. In which phases of mitosis and meiosis are the principles of segregation and independent assortment at work?

Section 3.4

11. How is the chi-square goodness-of-fit test used to analyze genetic crosses? What does the probability associated with a chi-square value indicate about the results of a cross?

Section 3.5

12. What features are exhibited by a pedigree of a recessive trait? What features are exhibited if the trait is dominant?

For more questions that test your comprehension of the key chapter concepts, go to LearningCurve for this chapter.

APPLICATION QUESTIONS AND PROBLEMS

Introduction

13. If blond hair in the Solomon Islanders had originated from early European explorers, what would you predict the researchers would have found when they conducted their genetic study of the islanders?

Section 3.1

*14. What characteristics of an organism would make it suitable for studies of the principles of inheritance? Can you name several organisms that have these characteristics?

Section 3.2

15. In cucumbers, orange fruit color (*R*) is dominant over cream fruit color (*r*). A cucumber plant homozygous for orange fruit is crossed with a plant homozygous for cream fruit. The F_1 are intercrossed (F_1 individuals crossed with F_1 individuals) to produce the F_2.
    - **a.** Give the genotypes and phenotypes of the parents, the F_1, and the F_2.
    - **b.** Give the genotypes and phenotypes of the offspring of a backcross between the F_1 and the orange-fruited parent.
    - **c.** Give the genotypes and phenotypes of a backcross between the F_1 and the cream-fruited parent.
16. **Figure 1.1** (p. 2) shows three girls, one of whom has albinism. Could the three girls shown in the photograph be sisters? Why or why not?

17. J. W. McKay crossed a stock melon plant that produced tan seeds with a plant that produced red seeds and obtained the following results (J. W. McKay. 1936. *Journal of Heredity* 27:110–112).

Cross	F_1	F_2
tan ♀ × red ♂	13 tan seeds	93 tan, 24 red seeds

a. Explain the inheritance of tan seeds and red seeds in this plant.

b. Assign symbols for the alleles in this cross and give genotypes for all the individual plants.

***18.** White (*w*) coat color in guinea pigs is recessive to black (*W*). In 1909, W. E. Castle and J. C. Phillips transplanted an ovary from a black guinea pig into a white female whose ovaries had been removed. They then mated this white female with a white male. All the offspring from the mating were black (W. E. Castle and J. C. Phillips. 1909. *Science* 30:312–313).

a. Explain the results of this cross.

b. Give the genotype of the offspring of this cross.

c. What, if anything, does this experiment indicate about the validity of the pangenesis and the germ-plasm theories discussed in Chapter 1?

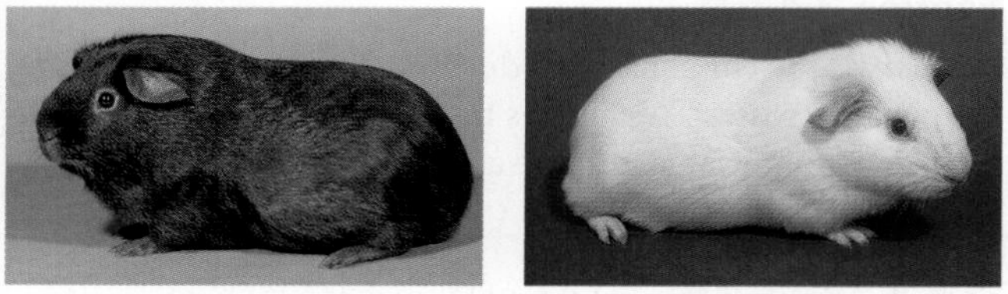

[Left: © Wegner/ARCO/Nature Picture Library. Right: © Nigel Cattlin/Alamy.]

***19.** In cats, blood type A results from an allele I^A, which is dominant over an allele i^B that produces blood type B. (There is no O blood type, as there is in humans.) The blood types of male and female cats that were mated and the blood types of their kittens are presented in the following table. Give the most likely genotypes for the parents of each litter.

	Male parent	Female parent	Kittens
a.	A	B	4 kittens with type A, 3 kittens with type B
b.	B	B	6 kittens with type B
c.	B	A	8 kittens with type A
d.	A	A	7 kittens with type A, 2 kittens with type B
e.	A	A	10 kittens with type A
f.	A	B	4 kittens with type A, 1 kitten with type B

20. **Figure 3.7** shows the results of a cross between a tall pea plant and a short pea plant.

a. What phenotypes will be produced, and in what proportions, if a tall F_1 plant is backcrossed to the short parent?

b. What phenotypes will be produced, and in what proportions, if a tall F_1 plant is backcrossed to the tall parent?

21. Joe has a white cat named Sam. When Joe crosses Sam with a black cat, he obtains $^1/_2$ white kittens and $^1/_2$ black kittens. When the black kittens are interbred, all the kittens that they produce are black. On the basis of these results, which coat color (white or black) in cats would you conclude is a recessive trait? Explain your reasoning.

***22.** Alkaptonuria is a metabolic disorder in which affected persons produce black urine. Alkaptonuria results from an allele (*a*) that is recessive to the allele for normal metabolism (*A*). Sally has normal metabolism, but her brother has alkaptonuria. Sally's father has alkaptonuria, and her mother has normal metabolism.

a. Give the genotypes of Sally, her mother, her father, and her brother.

b. If Sally's parents have another child, what is the probability that this child will have alkaptonuria?

c. If Sally marries a man with alkaptonuria, what is the probability that their first child will have alkaptonuria?

23. Hairlessness in American rat terriers is recessive to the presence of hair. Suppose that you have a rat terrier with hair. How can you determine whether this dog is homozygous or heterozygous for the hairy trait?

***24.** What is the probability of rolling one six-sided die and obtaining the following numbers?

a. 2

b. 1 or 2

c. An even number

d. Any number but a 6

***25.** What is the probability of rolling two six-sided dice and obtaining the following numbers?

a. 2 and 3

b. 6 and 6

c. At least one 6

d. Two of the same number (two 1s, or two 2s, or two 3s, etc.)

e. An even number on both dice

f. An even number on at least one die

26. Phenylketonuria (PKU) is a disease that results from a recessive gene. Two normal parents produce a child with PKU.

a. What is the probability that a sperm from the father will contain the PKU allele?

b. What is the probability that an egg from the mother will contain the PKU allele?

c. What is the probability that their next child will have PKU?

d. What is the probability that their next child will be heterozygous for the PKU gene?

***27.** In German cockroaches, a curved wing (*cv*) is recessive to a normal wing (cv^+). A homozygous cockroach that has normal wings is crossed with a homozygous cockroach that has curved wings. The F_1 are intercrossed to produce the F_2. Assume that the pair of chromosomes containing the locus for wing shape is metacentric. Draw this pair of chromosomes as it would appear in the parents, the F_1, and each class of F_2 progeny at metaphase I of meiosis. Assume that no crossing over takes place. At each stage, label a location for the alleles for wing shape (*cv* and cv^+) on the chromosomes.

***28.** In guinea pigs, the allele for black fur (*B*) is dominant over the allele for brown fur (*b*). A black guinea pig is crossed with a brown guinea pig, producing five F_1 black guinea pigs and six F_1 brown guinea pigs.

a. How many copies of the black allele (*B*) will be present in each cell of an F_1 black guinea pig at the following stages: G_1, G_2, metaphase of mitosis, metaphase I of meiosis, metaphase II of meiosis, and after the second cytokinesis following meiosis? Assume that no crossing over takes place.

b. How many copies of the brown allele (*b*) will be present in each cell of an F_1 brown guinea pig at the same stages as those listed in part *a*? Assume that no crossing over takes place.

Section 3.3

29. In watermelons, bitter fruit (*B*) is dominant over sweet fruit (*b*), and yellow spots (*S*) are dominant over no spots (*s*). The genes for these two characteristics assort independently. A homozygous plant that has bitter fruit and yellow spots is crossed with a homozygous plant that has sweet fruit and no spots. The F_1 are intercrossed to produce the F_2.

a. What are the phenotypic ratios in the F_2?

b. If an F_1 plant is backcrossed with the bitter, yellow-spotted parent, what phenotypes and proportions are expected in the offspring?

c. If an F_1 plant is backcrossed with the sweet, non-spotted parent, what phenotypes and proportions are expected in the offspring?

30. Figure 3.9 shows the results of a dihybrid cross involving seed shape and seed color.

a. What proportion of the round and yellow F_2 progeny from this cross is homozygous at both loci?

b. What proportion of the round and yellow F_2 progeny from this cross is homozygous at least at one locus?

***31.** In cats, curled ears result from an allele (*Cu*) that is dominant over an allele for normal ears (*cu*). Black color results from an independently assorting allele (*G*) that is dominant over an allele for gray (*g*). A gray cat homozygous for curled ears is mated with a homozygous black cat with normal ears. All the F_1 cats are black and have curled ears.

a. If two of the F_1 cats mate, what phenotypes and proportions are expected in the F_2?

b. An F_1 cat mates with a stray cat that is gray and possesses normal ears. What phenotypes and proportions of progeny are expected from this cross?

[Jean-Michel Labat/Science Source]

***32.** The following two genotypes are crossed: *Aa Bb Cc dd Ee* × *Aa bb Cc Dd Ee*. What will the proportion of the following genotypes be among the progeny of this cross?

a. *Aa Bb Cc Dd Ee* **b.** *Aa bb Cc dd ee*

c. *aa bb cc dd ee* **d.** *AA BB CC DD EE*

33. In cucumbers, dull fruit (*D*) is dominant over glossy fruit (*d*), orange fruit (*R*) is dominant over cream fruit (*r*), and bitter cotyledons (*B*) are dominant over non-bitter cotyledons (*b*). The three characteristics are encoded by genes located on different pairs of chromosomes. A plant homozygous for dull, orange fruit and bitter cotyledons is crossed with a plant that has glossy, cream fruit and non-bitter cotyledons. The F_1 are intercrossed to produce the F_2.

a. Give the phenotypes and their expected proportions in the F_2.

b. An F_1 plant is crossed with a plant that has glossy, cream fruit and non-bitter cotyledons. Give the phenotypes and expected proportions among the progeny of this cross.

***34.** Alleles *A* and *a* are located on a pair of metacentric chromosomes. Alleles *B* and *b* are located on a pair of acrocentric chromosomes. A cross is made between individuals having the following genotypes: *Aa Bb* × *aa bb*.

a. Draw the chromosomes as they would appear in each type of gamete produced by the individuals of this cross.

b. For each type of progeny resulting from this cross, draw the chromosomes as they would appear in a cell at G_1, G_2, and metaphase of mitosis.

Section 3.4

*35. J. A. Moore investigated the inheritance of spotting patterns in leopard frogs (J. A. Moore. 1943. *Journal of Heredity* 34:3–7). The pipiens phenotype has the normal spots that give leopard frogs their name. In contrast, the burnsi phenotype lacks spots on its back. Moore carried out the following crosses, producing the progeny indicated.

DATA ANALYSIS

Parent phenotypes	Progeny phenotypes
burnsi × burnsi	39 burnsi, 6 pipiens
burnsi × pipiens	23 burnsi, 33 pipiens
burnsi × pipiens	196 burnsi, 210 pipiens

a. On the basis of these results, what is the most likely mode of inheritance of the burnsi phenotype?

b. Give the most likely genotypes of the parent in each cross (use *B* for the burnsi allele and B^+ for the pipiens allele).

c. Use a chi-square test to evaluate the fit of the observed numbers of progeny to the number expected on the basis of your proposed genotypes.

*36. In the California poppy, an allele for yellow flowers (*C*) is dominant over an allele for white flowers (*c*). At an independently assorting locus, an allele for entire petals (*F*) is dominant over an allele for fringed petals (*f*). A plant that is homozygous for yellow and entire petals is crossed with a plant that is white and fringed. A resulting F_1 plant is then crossed with a plant that is white and fringed, and the following progeny are produced: 54 yellow and entire, 58 yellow and fringed, 53 white and entire, and 10 white and fringed.

a. Use a chi-square test to compare the observed numbers with those expected for the cross.

b. What conclusion can you make from the results of the chi-square test?

c. Suggest an explanation for the results.

Section 3.5

37. Many studies have suggested a strong genetic predisposition to migraine headaches, but the mode of inheritance is not clear. L. Russo and colleagues examined migraine headaches in several families, two of which are shown in the following pedigree (L. Russo et al. 2005. *American Journal of Human Genetics* 76:327–333). What is the most likely mode of inheritance for migraine headaches in these families? Explain your reasoning.

DATA ANALYSIS

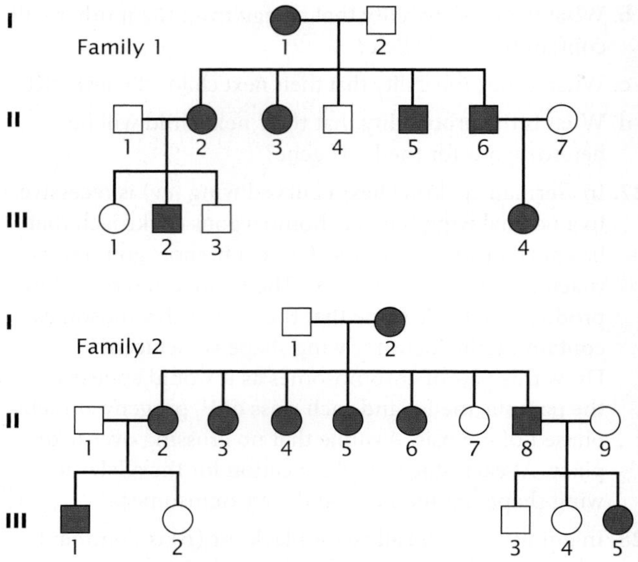

38. For each of the following pedigrees, give the most likely mode of inheritance, assuming that the trait is rare. Carefully explain your reasoning.

a.

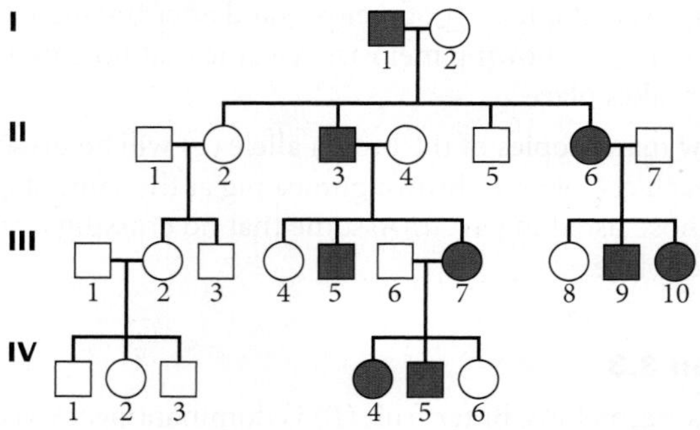

b.

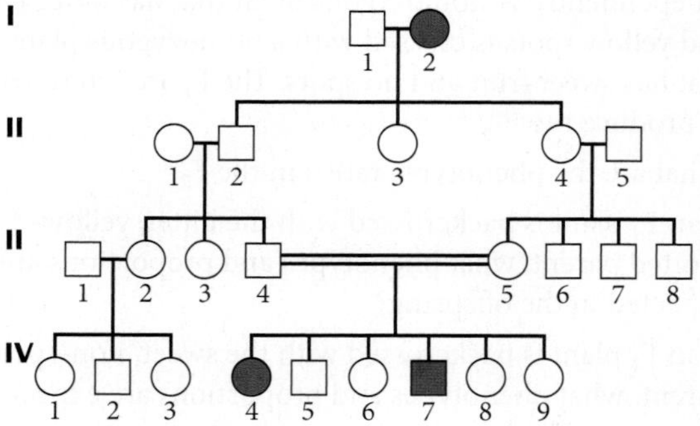

39. Ectrodactyly is a rare condition in which the fingers are absent and the hand is split. This condition is usually inherited as an autosomal dominant trait. Ademar Freire-Maia reported the appearance of ectrodactyly in a family in São Paulo, Brazil, whose pedigree is shown here. Is this pedigree consistent with autosomal dominant inheritance? If not, what mode of inheritance is most likely? Explain your reasoning.

DATA ANALYSIS

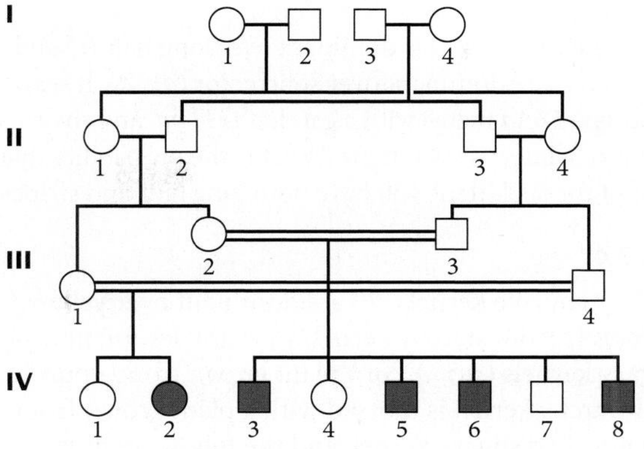

[Data from A. Freire-Maia, *Journal of Heredity* 62:53, 1971.]

CHALLENGE QUESTIONS

Section 3.2

*__40.__ A geneticist discovers an obese mouse in his laboratory colony. He crosses this mouse with a normal mouse. All the F_1 mice from this cross are normal in size. When he crosses two F_1 mice, eight of the F_2 mice are normal in size and two are obese. The geneticist then crosses two of his obese mice, and he finds that all the progeny from this cross are obese. These results lead the geneticist to conclude that obesity in mice results from a recessive allele.

A second geneticist at a different university also discovers an obese mouse in her laboratory colony. She carries out the same crosses as the first geneticist and obtains the same results. She also concludes that obesity in mice results from a recessive allele. One day the two geneticists meet at a genetics conference, learn of each other's experiments, and decide to exchange mice. They both find that, when they cross two obese mice from the different laboratories, all the offspring are normal; however, when they cross two obese mice from the same laboratory, all the offspring are obese. Explain their results.

41. Albinism in humans is a recessive trait (see the introduction to Chapter 1). A geneticist studies a series of families in which both parents are normal and at least one child has albinism. The geneticist reasons that both parents in these families must be heterozygotes and that albinism should appear in $^1/_4$ of the children of these families. To his surprise, the geneticist finds that the frequency of albinism among the children of these families is considerably greater than $^1/_4$. Can you think of an explanation for the higher-than-expected frequency of albinism among these families?

THINK-PAIR-SHARE QUESTIONS

Introduction

1. About 40% of Solomon Islanders carry a gene for blond hair, and yet only 5% to 10% of these people actually have blond hair. Why is the proportion of people with blond hair only 5% to 10% when so many people carry the genes for blond hair?
2. Why is knowing the genetic basis of a trait such as blond hair important? Why would scientists go to the trouble to investigate the genetic basis of blond hair in Solomon Islanders?
3. If a blond-haired person from northern Europe mated with a blond Solomon Islander, what proportion of their offspring would be expected to have blond hair? Explain your reasoning.

Section 3.1

4. Why was Mendel's success dependent on his studying characteristics that exhibit only two easily distinguished phenotypes, such as white versus gray seed coats and round versus wrinkled seeds? Would he have been less successful if he had instead studied traits such as seed weight or leaf length, which vary much more in their phenotypes? Explain your answer.

Section 3.2

5. Geneticists often carry out reciprocal crosses when they are studying the inheritance of traits. Why do geneticists use reciprocal crosses?

Section 3.3

6. Are Mendel's principles of segregation and independent assortment even relevant today in the age of genomics, when it is possible to sequence an organism's entire genome and determine all of its genetic information? Why is it important to study these principles, and how can they be used?

7. In cats, short hair (L) is dominant over long hair (l) and stripes (A) are dominant over solid color (a). A cat with genotype *Ll Aa* mates with a cat that is *Ll aa*, and they produce a litter of six kittens. What is the probability that four of the six kittens will have both long hair and stripes?

Section 3.4

8. In corn, purple kernels (P) are dominant over yellow kernels (p) and starchy kernels (Su) are dominant over sugary kernels (su). A corn plant grown from a purple and starchy kernel is crossed with a plant grown from a yellow and sugary kernel, and the following progeny (kernels) are produced:

Phenotype	Number
purple, starchy	150
purple, sugary	142
yellow, starchy	161
yellow, sugary	115

Formulate a hypothesis about the genotypes of the parents and offspring in this cross. Perform a chi-square goodness-of-fit test comparing the observed progeny with the numbers expected based on your genetic hypothesis. What conclusion can you draw based on the results of your chi-square test? Can you suggest an explanation for the observed results?

SaplingPlus Self-study tools that will help you practice what you've learned and reinforce this chapter's concepts are available online. Go to www.macmillanlearning.com/PierceGenetics6e.

4 Extensions and Modifications of Basic Principles

The Odd Genetics of Left-Handed Snails

The direction of shell coiling in *Lymnaea* snails is determined by a genetic maternal effect. Shown here is *Lymnaea stagnalis,* a snail with a left-handed (sinistral) shell on the left and a snail with a right-handed (dextral) shell on the right. [Courtesy of Dr. Reiko Kuroda.]

At the start of the twentieth century, Mendel's work on inheritance in pea plants became widely known (see Chapter 3), and a number of biologists set out to verify his conclusions by conducting crosses with other organisms. Biologists quickly confirmed that Mendel's principles applied not just to peas, but also to corn, beans, mice, guinea pigs, chickens, humans, and many other organisms. At the same time, biologists began to discover exceptions—traits whose inheritance was more complex than the simple dominant and recessive traits that Mendel had observed. One of these exceptions involved the spiral of a snail's shell.

The direction of coiling in snail shells is called chirality. Most snail shells spiral downward in a clockwise or right-handed direction; these are termed dextral shells. A few snails have shells that coil in the opposite direction, spiraling downward in a counterclockwise or left-handed direction; these are termed sinistral shells. The shells of most snail species are all dextral or all sinistral; only in a few rare instances do both dextral and sinistral shells coexist in the same species.

In the 1920s and 1930s, Arthur Boycott of the University of London investigated the genetics of shell coiling in *Lymnaea peregra*, a common pond snail in Britain. In this species, most snails are dextral, but a few sinistral snails occur in some populations. Boycott learned from amateur naturalists of a pond near Leeds, England, where an abnormally high number of sinistral snails could be found. He obtained four sinistral snails from this location and began to investigate the genetics of shell chirality.

Boycott's research was complicated by the fact that these snails are hermaphroditic, meaning that a snail can self-fertilize, or *self* (mate with itself). If a suitable partner is available, the snails are also capable of *outcrossing*—mating with another individual. Boycott found that if he isolated a newly hatched snail and reared it alone, it would eventually produce offspring, so he knew that it had selfed. But when he placed two snails together and one produced offspring, he had no way of knowing whether it had mated with itself or with the other snail. Boycott's research required rearing large numbers of snails in isolation and in pairs, raising their offspring, and determining the direction of shell coiling for each of the offspring. To facilitate the work, he enlisted the aid of several amateur scientists. One of his assistants was Captain C. Diver, a friend who worked as an assistant for the British Parliament. Since Parliament met for only part of the year, Diver had time on his hands and eagerly enlisted to assist with the research. Together, Boycott, Diver, and other assistants carried out numerous breeding experiments, selfing and crossing snails and raising the progeny in jam jars. They eventually raised more than 6000 broods and determined the direction of coiling in a million snails.

Initially, their results were puzzling: shell coiling did not appear to conform to Mendel's principles of heredity. They eventually realized that dextral was dominant to sinistral, but with a peculiar twist: the phenotype of a snail was determined *not* by its own genotype, but by the genotype of its mother. This phenomenon—a phenotype that is influenced by the genotype of the mother—is called a genetic maternal effect. Genetic maternal effects often arise because the maternal parent produces a substance, encoded by her own genotype, that is deposited in the cytoplasm of the egg and that influences early development of the offspring.

The substance that determines the direction of shell coiling in snails has never been isolated. However, in 2009, Reiko Kuroda and her colleagues demonstrated that the direction of coiling in *Lymnaea* snails is determined by the orientation of cells when the embryo is at an early developmental stage—specifically, the eight-cell stage. By gently pushing on the cells of eight-cell embryos, they were able to induce offspring whose mother's genotype was dextral to develop as sinistral snails; similarly, they induced the offspring of mothers whose genotype was sinistral to develop as dextral snails by pushing on the cells in the opposite direction.

THINK-PAIR-SHARE Questions 1 and 2

Boycott's research on the direction of coiling in snails demonstrated that not all characteristics are inherited as simple dominant and recessive traits like the shapes and colors of peas that Mendel described. This demonstration doesn't mean that Mendel was wrong; rather, it indicates that Mendel's principles are not, by themselves, sufficient to explain the inheritance of all genetic characteristics. Our modern understanding of genetics has been greatly enriched by the discovery of a number of modifications and extensions of Mendel's basic principles, which are the focus of this chapter.

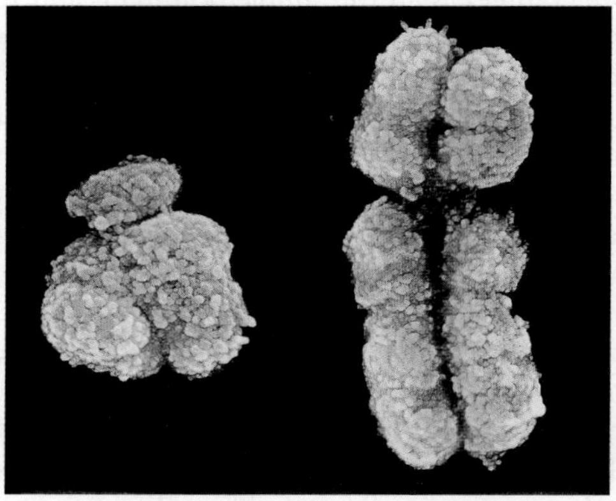

4.1 The sex chromosomes of human males (Y, at the left) and females (X, at the right) differ in size and shape. [Biophoto Associates/Science Source.]

4.1 Sex Is Determined by a Number of Different Mechanisms

One of the first extensions of Mendel's principles is the pattern of the inheritance of characteristics encoded by genes located on the sex chromosomes, which often differ in males and females (**Figure 4.1**). To understand the inheritance of sex-linked characteristics, we must first know how sex is determined—why some members of a species are male and others are female.

Sexual reproduction is the formation of offspring that are genetically distinct from their parents; most often, two parents contribute genes to their offspring, and those genes are assorted into new combinations through meiosis. Among most eukaryotes, sexual reproduction consists of two processes that lead to an alternation of haploid and diploid cells: meiosis produces haploid gametes (or spores in plants), and fertilization produces diploid zygotes (**Figure 4.2**).

The term **sex** refers to sexual phenotype. Most organisms have only two sexual phenotypes: male and female. The fundamental difference between males and females is

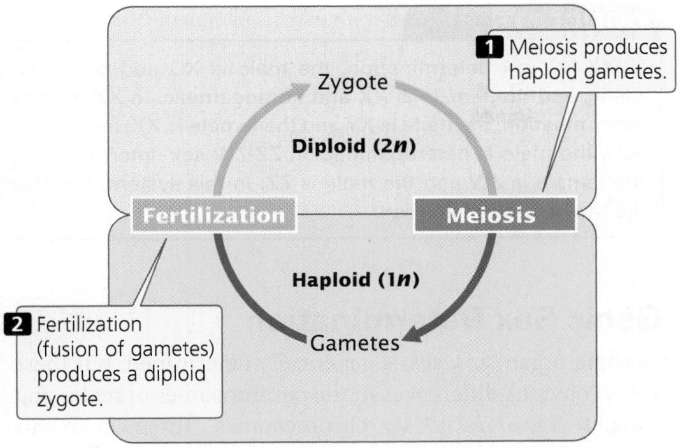

4.2 In most eukaryotic organisms, sexual reproduction consists of two processes that lead to an alternation of haploid (1*n*) and diploid (2*n*) cells.

gamete size: males produce small gametes, called sperm; females produce relatively large gametes, called eggs (**Figure 4.3**).

We define the sex of an individual organism in reference to its phenotype. Sometimes an individual organism has chromosomes or genes that are normally associated with one sex but an anatomy corresponding to the opposite sex. For instance, the cells of female humans normally have two X chromosomes, and the cells of males have one X chromosome and one Y chromosome. A few rare persons have male anatomy although their cells each contain two X chromosomes. Even though these people are genetically female, we refer to them as male because their sexual phenotype is male. (As we will see later in the chapter, these XX males usually have a small piece of the Y chromosome that is attached to another chromosome.) The mechanism by which sex is established is termed **sex determination**. Sex may be determined by differences in specific chromosomes, by genotypes, or by environmental factors.

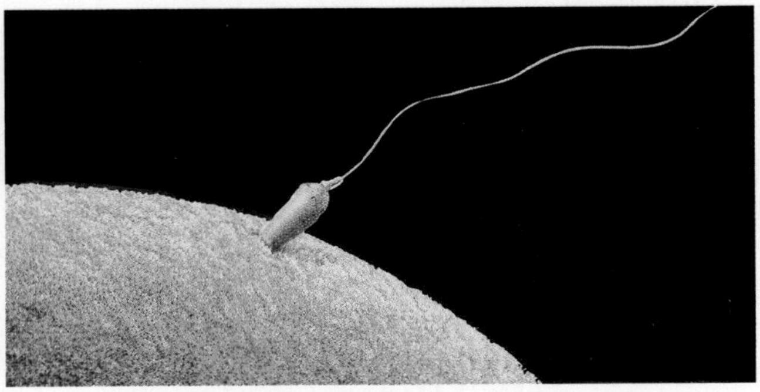

4.3 Male and female gametes differ in size. In this photograph, a human sperm (with flagellum) penetrates a human egg cell. [Francis Leroy, Biocosmos/Science Source.]

CONCEPTS

In sexual reproduction, two parents contribute genes to produce an offspring that is genetically distinct from both parents. In most eukaryotes, sexual reproduction consists of meiosis, which produces haploid gametes (or spores), and fertilization, which produces a diploid zygote.

✓ **CONCEPT CHECK 1**

What process causes the genetic variation seen in offspring produced by sexual reproduction?

Chromosomal Sex-Determining Systems

Sex in many organisms is determined by a pair of chromosomes, the **sex chromosomes**, which differ between males and females. The nonsex chromosomes, which are the same for males and females, are called **autosomes**. We think of sex in these organisms as being determined by the presence of the sex chromosomes, but in fact, the individual genes located on the sex chromosomes, in conjunction with genes on the autosomes, are usually responsible for the sexual phenotypes.

XX-XO SEX DETERMINATION In some insects, sex is determined by the XX-XO system. In this system, females have two X chromosomes (XX), and males possess a single X chromosome (XO). There is no O chromosome—the letter O signifies the absence of a sex chromosome.

In meiosis in females, the two X chromosomes pair and then separate, with one X chromosome entering each haploid egg. In males, the single X chromosome segregates in meiosis to half the sperm cells; the other half receive no sex chromosome. Because males produce two different types of gametes with respect to the sex chromosomes, they are said to be the **heterogametic sex**. Females, which produce gametes that are all the same with respect to the sex chromosomes, are the **homogametic sex**. In the XX-XO system, the sex of an individual is therefore determined by which type of male gamete fertilizes the egg. X-bearing sperm unite with X-bearing eggs to produce XX zygotes, which develop into females. Sperm lacking an X chromosome unite with X-bearing eggs to produce XO zygotes, which develop into males.

XX-XY SEX DETERMINATION In many species, the cells of males and females have the same number of chromosomes, but the cells of females have two X chromosomes (XX) and the cells of males have a single X chromosome and

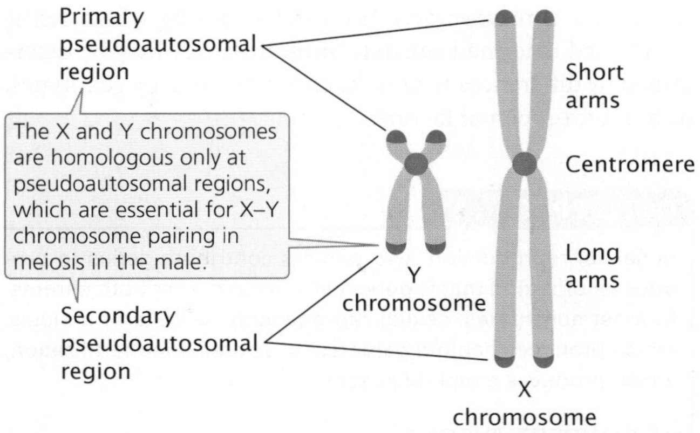

4.4 The X and Y chromosomes in humans differ in size and genetic content. They are homologous only at the pseudoautosomal regions.

a smaller sex chromosome called the Y chromosome (XY). In humans and many other organisms, the Y chromosome is acrocentric (**Figure 4.4**)—not Y-shaped, as is often assumed. In this sex-determining system, males are the heterogametic sex—half of the gametes produced by a male have an X chromosome and half have a Y chromosome. Females are the homogametic sex—all the egg cells produced by a female contain an X chromosome. A sperm containing a Y chromosome unites with an X-bearing egg to produce an XY male, whereas a sperm containing an X chromosome unites with an X-bearing egg to produce an XX female, which accounts for the 50 : 50 sex ratio observed in most organisms (**Figure 4.5**). Many organisms, including some plants, insects, reptiles, and mammals (including humans), have the XX-XY sex-determining system. Other organisms have odd variations of the XX-XY system of sex determination, including the duck-billed platypus, in which females have five pairs of X chromosomes and males have five pairs of X and Y chromosomes.

Although the X and Y chromosomes are not generally homologous, they do pair and segregate into different cells in meiosis. They can pair because these chromosomes are homologous at small regions called the **pseudoautosomal regions** (see Figure 4.4), in which they carry the same genes. In humans, there are pseudoautosomal regions at both tips of the X and Y chromosomes.

THINK-PAIR-SHARE Question 3

ZZ-ZW SEX DETERMINATION In ZZ-ZW sex determination, females are heterogametic and males are homogametic. To prevent confusion with the XX-XY system, the sex chromosomes in this system are labeled Z and W, but the chromosomes do not resemble Zs and Ws. Females in this system are ZW; after meiosis, half of the eggs have a Z chromosome and the other half have a W chromosome. Males are ZZ; all sperm contain a single Z chromosome. The ZZ-ZW system is found in birds, snakes, butterflies, some amphibians, and some fishes.

THINK-PAIR-SHARE Question 4

CONCEPTS

In XX-XO sex determination, the male is XO and heterogametic, and the female is XX and homogametic. In XX-XY sex determination, the male is XY and the female is XX; in this system, the male is heterogametic. In ZZ-ZW sex determination, the female is ZW and the male is ZZ; in this system, females are the heterogametic sex.

Genic Sex Determination

In some organisms, sex is genetically determined, but there are no obvious differences in the chromosomes of males and females: there are no sex chromosomes. These organisms have **genic sex determination**: genotypes at one or more loci determine the sex of an individual. Scientists have observed genic sex determination in some plants, fungi, protozoans, and fishes.

It is important to understand that even in chromosomal sex-determining systems, sex is actually determined by individual genes. In mammals, for example, a gene (*SRY*,

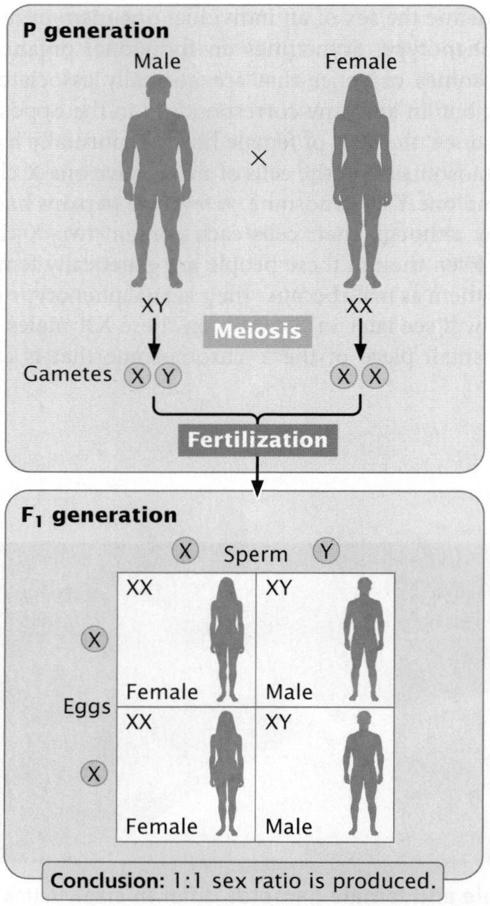

4.5 The inheritance of sex in organisms with X and Y chromosomes results in equal numbers of male and female offspring.

discussed later in this chapter) located on the Y chromosome determines the male phenotype. In both genic sex determination and chromosomal sex determination, sex is controlled by individual genes; the difference is that with chromosomal sex determination, the sex chromosomes that carry those genes look different in males and females.

Environmental Sex Determination

Genes have had a role in all the mechanisms of sex determination discussed thus far, but in a number of organisms, sex is determined fully or in part by environmental factors. Environmental factors are important in determining sex in some reptiles; the sexual phenotype of many turtles, crocodiles, alligators, and a few birds is affected by temperature during embryonic development. In turtles, for example, warm incubation temperatures produce more females during certain times of the year, whereas cool temperatures produce males. In alligators, the reverse is true. In some species, sex chromosomes usually determine whether individuals are male or female, but environmental factors can sometimes override this chromosomal sex determination. For example, bearded dragon lizards are normally ZZ when male and ZW when female, but when the eggs are incubated at high temperatures, ZZ individuals develop into phenotypic females.

Now that we have surveyed some of the different ways that sex can be determined, we will examine one mechanism (the XX-XY system) in detail. Both fruit flies and humans possess XX-XY sex determination, but as we will see, the way in which the X and Y chromosomes determine sex in these two organisms is quite different. ▶ TRY PROBLEM 1

CONCEPTS

In genic sex determination, sex is determined by genes at one or more loci, but there are no obvious differences in the chromosomes of males and females. In environmental sex determination, sex is determined fully or in part by environmental factors.

✓ CONCEPT CHECK 2

How do chromosomal, genic, and environmental sex determination systems differ?

Sex Determination in *Drosophila melanogaster*

The fruit fly *Drosophila melanogaster* has eight chromosomes: three pairs of autosomes and one pair of sex chromosomes. Usually, females have two X chromosomes and males have an X chromosome and a Y chromosome. In the 1920s, Calvin Bridges proposed that sex in *Drosophila* was determined not by the number of X and Y chromosomes, but rather by the balance of female-determining genes on the X chromosome and male-determining genes on the autosomes. He suggested that a fly's sex is determined by the so-called X:A ratio, the number of X chromosomes divided by the number of haploid sets of autosomes. Normal flies possess two haploid sets of autosomes and either two X chromosomes (females) or one X chromosome and a Y chromosome (males). Bridges proposed that an X:A ratio of 1.0 produces a female fly and an X:A ratio of 0.5 produces a male fly. He also suggested that an X:A ratio between 1.0 and 0.5 produces an intersex fly, with a mixture of male and female characteristics. An X:A ratio of less than 0.5 or greater than 1.0 produces developmentally abnormal flies called metamales and metafemales, respectively. When Bridges and others examined flies with different numbers of sex chromosomes and autosomes, the X:A ratio appeared to correctly predict the phenotypic sex of the flies (**Table 4.1**).

Although the X:A ratio correctly *predicts* the sexual phenotype, recent research suggests that the *mechanism* of sex determination is not a balance between X-linked genes and autosomal genes, as Bridges proposed. Researchers have located a number of genes on the X chromosome that affect sexual phenotype, but few autosomal sex-determining genes (required for the X:A ratio hypothesis) have been identified. New evidence suggests that genes on the X chromosome are the primary sex determinant. The influence of the number of autosomes on sex is indirect, affecting the timing of developmental events and therefore how long sex-determining genes on the X chromosome are active. For example, XX flies with three autosomal sets (XX, AAA) have an X:A ratio of 0.67 and develop an intersex phenotype. In these flies, the presence of three autosomal sets causes a critical developmental stage to shorten, not allowing female factors encoded on the X chromosomes enough time to accumulate, with the result that the flies end up with an intersex phenotype. The number

TABLE 4.1 Chromosome complements and sexual phenotypes in *Drosophila*

Sex-chromosome complement	Haploid sets of autosomes	X:A ratio	Sexual phenotype
XX	AA	1.0	Female
XY	AA	0.5	Male
XO	AA	0.5	Male
XXY	AA	1.0	Female
XXX	AA	1.5	Metafemale
XXXY	AA	1.5	Metafemale
XX	AAA	0.67	Intersex
XO	AAA	0.33	Metamale
XXXX	AAA	1.3	Metafemale

of autosomes influences sex determination in *Drosophila*, but not through the action of autosomal genes, as envisioned by Bridges.

CONCEPTS

Although the sexual phenotype of a fruit fly is predicted by the X:A ratio, sex is actually determined by genes on the X chromosome.

Sex Determination in Humans

Humans, like *Drosophila*, have XX-XY sex determination, but in humans, maleness is primarily determined by the presence of a gene (*SRY*) on the Y chromosome. The phenotypes that result from abnormal numbers of sex chromosomes, which arise when the sex chromosomes do not segregate properly in meiosis or mitosis, illustrate the importance of the Y chromosome in human sex determination.

TURNER SYNDROME People who have **Turner syndrome** are female and often have underdeveloped secondary sex characteristics. This syndrome is seen in 1 of 3000 female births. Affected women are frequently short and have a low hairline, a relatively broad chest, and folds of skin on the neck. Their intelligence is usually normal. Most women who have Turner syndrome are sterile. In 1959, Charles Ford used new techniques to study human chromosomes, and discovered that cells from a 14-year-old girl with Turner syndrome had only a single X chromosome; this chromosome complement is usually referred to as XO.

There are no known cases in which a person is missing both X chromosomes, an indication that at least one X chromosome is necessary for human development. Presumably, embryos missing both Xs are spontaneously aborted in the early stages of development.

KLINEFELTER SYNDROME People who have **Klinefelter syndrome**, which has a frequency of about 1 in 1000 male births, have cells with one or more Y chromosomes and multiple X chromosomes. The cells of most males who have this condition are XXY, but cells of a few males with Klinefelter syndrome are XXXY, XXXXY, or XXYY. Persons with this condition are male and often have small testes and reduced facial and pubic hair. They are often taller than normal and sterile; most have normal intelligence.

POLY-X FEMALES In about 1 in 1000 female births, the infant's cells possess three X chromosomes, a condition often referred to as **triple-X syndrome**. These individuals have no distinctive features other than a tendency to be tall and thin. Although a few are sterile, many menstruate regularly and are fertile. The incidence of intellectual disability among triple-X females is slightly greater than that in the general population, but most XXX females have normal intelligence. Much rarer are females whose cells contain four or five X chromosomes. These women usually have normal female anatomy but are intellectually disabled and have a number of physical problems. The severity of intellectual disability increases as the number of X chromosomes increases beyond three.

XYY MALES Males with an extra Y chromosome (XYY) occur with a frequency of about 1 in 1000 male births. These individuals have no physical differences other than a tendency to be several inches taller than the average of XY males. Their IQ is usually within the normal range, although some studies suggest that learning difficulties may be more common than in XY males. The effect of the number of sex chromosomes on human development is summarized in **Table 4.2**.

THE MALE-DETERMINING GENE IN HUMANS The phenotypes associated with sex-chromosome anomalies show that the Y chromosome in humans, and all other mammals, is of paramount importance in producing a male phenotype. However, scientists have discovered a few rare XX males whose cells apparently lack a Y chromosome. For many years, these males presented an enigma: How could a male phenotype exist without a Y chromosome? Close examination eventually revealed a small part of the Y chromosome attached to another chromosome, usually the X. This finding indicates that it is not the entire Y chromosome that determines maleness in humans; rather, it is a gene on the Y chromosome.

Early in development, all humans possess undifferentiated gonads and both male and female reproductive ducts. Then, about 6 weeks after fertilization, a gene on the Y chromosome becomes active. This gene causes the neutral gonads to develop into testes, which begin to secrete two hormones:

TABLE 4.2 Sex chromosomes and associated sexual phenotypes in humans

Sex chromosomes	Phenotype	Characteristics
XX	Female	Female traits
XY	Male	Male traits
XXY, XXYY, XXXY	Klinefelter syndrome	Male traits, tall, small testes, reduced facial and pubic hair
XO	Turner syndrome	Female traits, short, low hairline, broad chest, neck folds
XXX, XXXX, XXXXX	Poly-X females	Female traits, tall and thin
XYY	XYY males	Male traits, tall

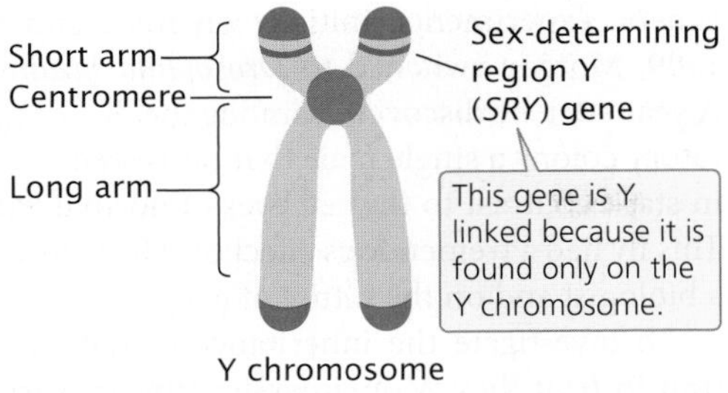

4.6 The *SRY* gene is on the Y chromosome and causes the development of male characteristics.

testosterone and Mullerian-inhibiting substance. Testosterone induces the development of male characteristics, and Mullerian-inhibiting substance causes the degeneration of the female reproductive ducts. In the absence of this male-determining gene, the neutral gonads become ovaries, and female features develop.

The male-determining gene in humans, called the **sex-determining region Y (*SRY*) gene**, was discovered in 1990 (**Figure 4.6**). This gene is found in XX males; it is also found on the Y chromosome of other mammals. Definitive proof that *SRY* is the male-determining gene came when scientists placed a copy of this gene in XX mice by means of genetic engineering. The XX mice that received this gene, although sterile, developed into anatomical males. Although *SRY* is the primary determinant of maleness in humans, other genes (some X linked, others Y linked, and still others autosomal) also play a role in fertility and the development of sexual phenotypes.

THINK-PAIR-SHARE Question 5

CONCEPTS

The presence of the *SRY* gene on the Y chromosome causes a human embryo to develop as a male. In the absence of this gene, a human embryo develops as a female.

✓ CONCEPT CHECK 3

In humans, what will be the phenotype of a person with XXXY sex chromosomes?

a. Klinefelter syndrome
b. Turner syndrome
c. Poly-X female

4.2 Sex-Linked Characteristics Are Determined by Genes on the Sex Chromosomes

In Chapter 3, we learned several basic principles of heredity that Mendel discovered from his experiments with pea plants. A major extension of these Mendelian principles is the pattern of inheritance exhibited by **sex-linked characteristics**, characteristics determined by genes located on the sex chromosomes. Genes on the X chromosome determine **X-linked characteristics**; those on the Y chromosome determine **Y-linked characteristics**. Because the Y chromosome of many organisms contains little genetic information, most sex-linked characteristics are X linked. Males and females differ in their sex chromosomes, so the pattern of inheritance for sex-linked characteristics differs from that exhibited by genes located on autosomes.

X-Linked White Eyes in *Drosophila*

The first person to explain sex-linked inheritance was American biologist Thomas Hunt Morgan (**Figure 4.7**). Morgan began his career as an embryologist, but the discovery of Mendel's principles inspired him to begin conducting

(a)

(b)

4.7 Thomas Hunt Morgan's work with *Drosophila* helped unravel many basic principles of genetics, including X-linked inheritance. (a) Morgan. (b) The Fly Room, where Morgan and his students conducted genetic research. [Part a: AP/Wide World Photos. Part b: American Philosophical Society.]

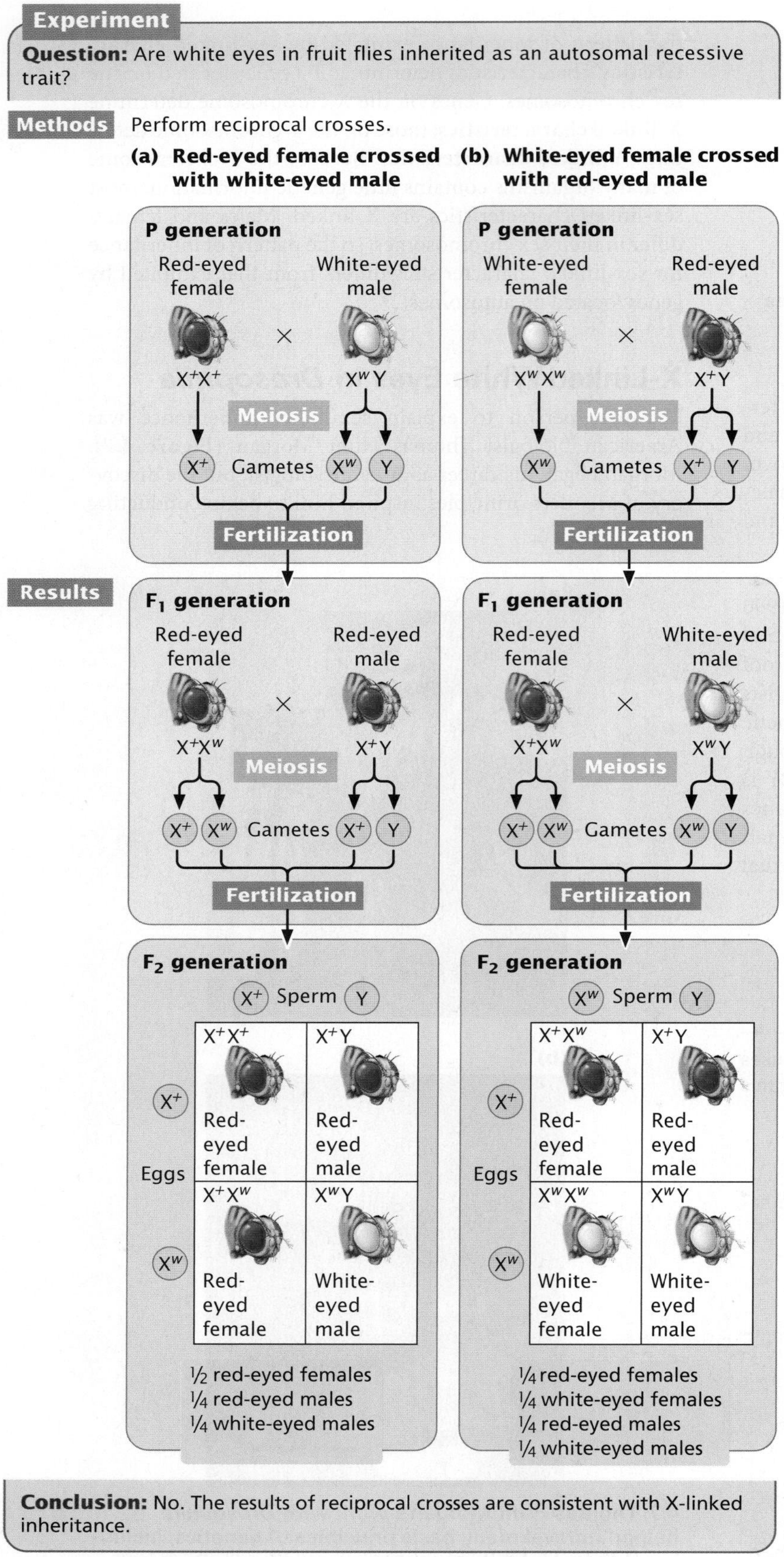

4.8 Morgan's X-linked crosses for white eyes in fruit flies. (a) Original and F_1 crosses. (b) Reciprocal crosses.

genetic experiments, initially on mice and rats. In 1909, Morgan switched to *Drosophila melanogaster*. A year later, he discovered among the flies of his laboratory colony a single male that possessed white eyes, in stark contrast to the red eyes of normal fruit flies. This fly had a tremendous effect on Morgan's career as a biologist and on the future of genetics.

To investigate the inheritance of the white-eyed trait in fruit flies, Morgan systematically carried out a series of genetic crosses. First, he crossed pure-breeding red-eyed females with his white-eyed male, producing F_1 progeny that had red eyes (**Figure 4.8a**). Morgan's results from this initial cross were consistent with Mendel's principles: a cross between a homozygous dominant individual and a homozygous recessive individual produced heterozygous offspring exhibiting the dominant trait. These results suggested that white eyes are a simple recessive trait. When Morgan crossed the F_1 flies with one another, however, he found that all the female F_2 flies possessed red eyes, but that half the male F_2 flies had red eyes and the other half had white eyes. This finding was clearly not the expected result for a simple recessive trait, which should appear in one-fourth of both male and female F_2 offspring.

To explain this unexpected result, Morgan proposed that the locus affecting eye color is on the X chromosome (i.e., eye color is X linked). He also recognized that the eye-color alleles are present on the X chromosome only; no homologous allele is present on the Y chromosome. Because the cells of females possess two X chromosomes, females can be homozygous or heterozygous for the eye-color alleles. The cells of males, on the other hand, possess only a single X chromosome and can carry only a single eye-color allele. Males, therefore, cannot be homozygous or heterozygous, but are said to be **hemizygous** for X-linked loci.

To verify his hypothesis that the white-eye trait is X linked, Morgan conducted additional crosses. He predicted that a cross between a white-eyed female and a red-eyed male would produce all red-eyed females and all white-eyed males (**Figure 4.8b**). When Morgan performed this cross, the results were exactly as predicted. Note that this cross is the reciprocal of the original cross and that the two reciprocal crosses produced different results in the F_1 and F_2 generations. Morgan also crossed the F_1 heterozygous females with their white-eyed father, the red-eyed F_2 females with white-eyed males, and white-eyed females with white-eyed males. In all these crosses, the results were consistent with Morgan's conclusion that the white-eye trait is an X-linked characteristic. You can view the results of Morgan's crosses in **Animation 4.1**.

Model Genetic Organism
The Fruit Fly *Drosophila melanogaster*

Drosophila melanogaster, a fruit fly (**Figure 4.9**), was among the first organisms used for genetic analysis, and today it is one of the most widely used and best known genetically of all eukaryotic organisms. It has played an important role in studies of linkage, epistasis, chromosomal genetics, development, behavior, and evolution. Because all organisms use a common genetic system, understanding a process such as replication or transcription in fruit flies helps us to understand these same processes in humans and other eukaryotes.

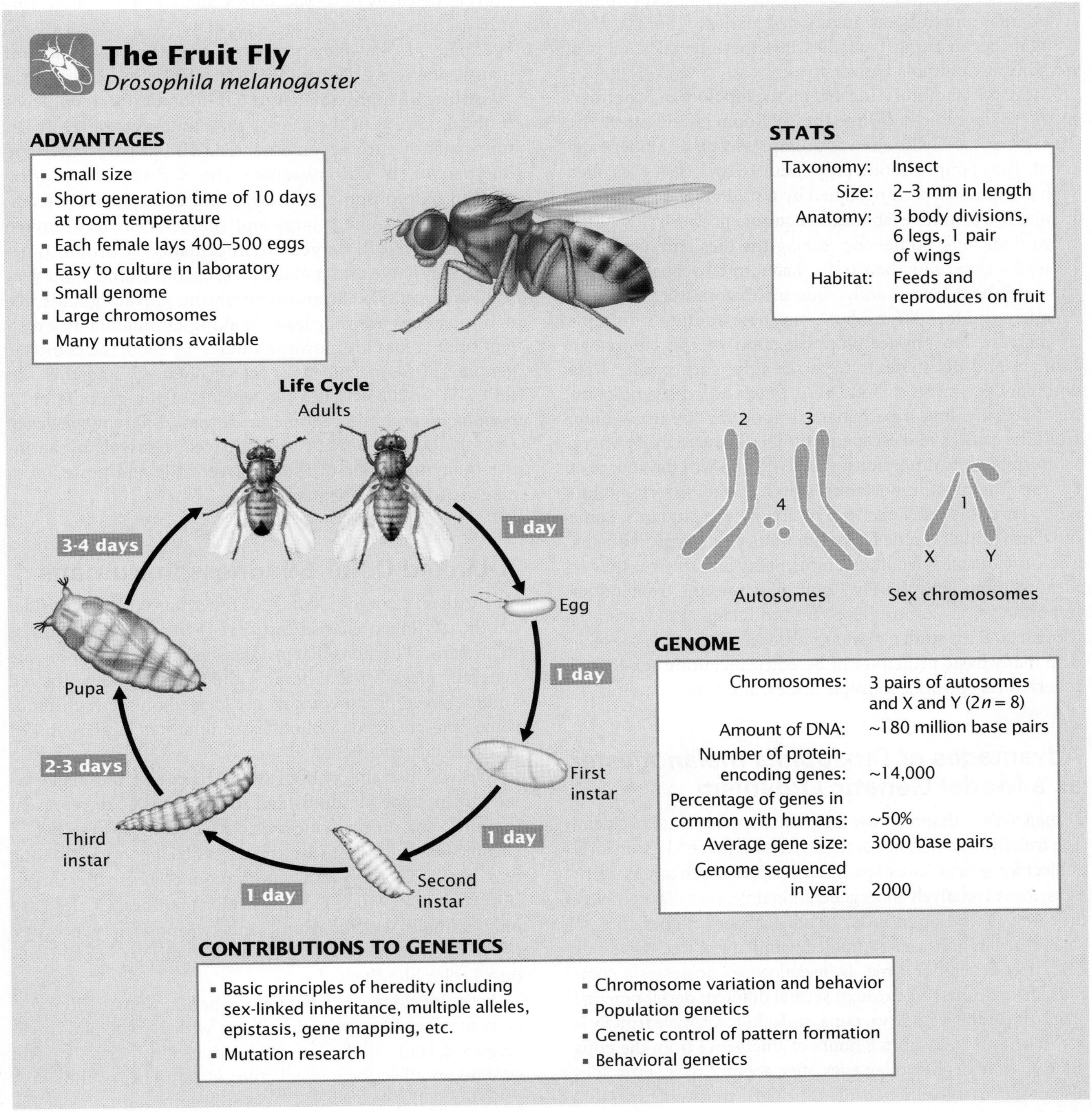
The Fruit Fly
Drosophila melanogaster
ADVANTAGES
Small size
Short generation time of 10 days at room temperature
Each female lays 400–500 eggs
Easy to culture in laboratory
Small genome
Large chromosomes
Many mutations available
STATS
Taxonomy: Insect
Size: 2–3 mm in length
Anatomy: 3 body divisions, 6 legs, 1 pair of wings
Habitat: Feeds and reproduces on fruit
Life Cycle
Adults
1 day
Egg
1 day
First instar
1 day
Second instar
1 day
Third instar
2-3 days
Pupa
3-4 days
2
3
4
1
X
Y
Autosomes
Sex chromosomes
GENOME
Chromosomes: 3 pairs of autosomes and X and Y (2n = 8)
Amount of DNA: ~180 million base pairs
Number of protein-encoding genes: ~14,000
Percentage of genes in common with humans: ~50%
Average gene size: 3000 base pairs
Genome sequenced in year: 2000
CONTRIBUTIONS TO GENETICS
Basic principles of heredity including sex-linked inheritance, multiple alleles, epistasis, gene mapping, etc.
Mutation research
Chromosome variation and behavior
Population genetics
Genetic control of pattern formation
Behavioral genetics

4.9 ***Drosophila melanogaster*** **is a model genetic organism.**

Drosophila is a genus of more than a thousand described species of small flies (about 2 to 3 mm in length) that frequently feed and reproduce on fruit, although they rarely cause damage and are not considered economic pests. The best known and most widely studied of the fruit flies is *D. melanogaster*, but genetic studies have also been extended to many other species of the genus.

D. melanogaster first began to appear in biological laboratories about 1900. After first taking up breeding experiments with mice and rats, as mentioned earlier, Thomas Hunt Morgan began using fruit flies in experimental studies of heredity at Columbia University.

Morgan's laboratory, located on the top floor of Schermerhorn Hall, became known as the Fly Room (see Figure 4.7b). To say that the Fly Room was unimpressive is an understatement. The cramped room, only about 16 by 23 feet, was filled with eight desks, each occupied by a student and his experiments. The primitive laboratory equipment consisted of little more than milk bottles for rearing the flies and hand-held lenses for observing their traits. Later, microscopes replaced the hand-held lenses, and crude incubators were added to maintain the fly cultures, but even these additions did little to increase the physical sophistication of the laboratory. Morgan and his students were not tidy: cockroaches were abundant (living off spilled *Drosophila* food), dirty milk bottles filled the sink, ripe bananas—food for the flies—hung from the ceiling, and escaped fruit flies hovered everywhere.

In spite of its limitations, the Fly Room was the source of some of the most important research in the history of biology. There was daily excitement among the students, some of whom initially came to the laboratory as undergraduates. The close quarters facilitated informality and the free flow of ideas. Morgan and the Fly Room illustrate the tremendous importance of "atmosphere" in producing good science. Morgan and his students eventually used *Drosophila* to elucidate many basic principles of heredity, including sex-linked inheritance, epistasis, multiple alleles, and gene mapping.

Advantages of *Drosophila melanogaster* as a Model Genetic Organism

Drosophila's widespread use in genetic studies is no accident. The fruit fly has a number of characteristics that make it an ideal subject for genetic investigations. Compared with other organisms, it has a relatively short generation time: fruit flies complete an entire generation in about 10 days at room temperature, so several generations can be studied within a few weeks. Despite this short generation time, *D. melanogaster* possesses a complex life cycle, passing through several different developmental stages, including egg, larva, pupa, and adult. A female fruit fly is capable of mating within 8 hours of emergence from the pupa and typically begins to lay eggs after about 2 days. Fruit flies also produce large numbers of offspring, laying as many as 400 to 500 eggs in a 10-day period. Thus, large numbers of progeny can be obtained from a single genetic cross.

Another advantage is that fruit flies are easy to culture in the laboratory. They are usually raised in small glass vials or bottles on easily prepared, pastelike food consisting of bananas or corn meal and molasses. Males and females are readily distinguished and virgin females are easily isolated, which facilitates genetic crosses. The flies are small, requiring little space—several hundred can be raised in a half-pint bottle—but they are large enough for many mutations to be easily observed with a hand lens or a dissecting microscope.

Finally, *D. melanogaster* is an organism of choice for many geneticists because it has a relatively small genome consisting of approximately 180 million base pairs of DNA (only about 5% of the size of the human genome). It has four pairs of chromosomes: three pairs of autosomes and one pair of sex chromosomes. The X chromosome (designated chromosome 1) is large and acrocentric, whereas the Y chromosome is large and submetacentric, although it contains very little genetic information. Chromosomes 2 and 3 are large and metacentric; chromosome 4 is a very small acrocentric chromosome. In the salivary glands, the chromosomes are very large, making *Drosophila* an excellent subject for chromosome studies. In 2000, the complete genome of *D. melanogaster* was sequenced, and this was followed in 2005 by the sequencing of the genome of *D. pseudoobscura*. The genomes of several different species of *Drosophila* have now been sequenced. *Drosophila* continues today to be one of the most versatile and powerful of all genetic model organisms. ■

X-Linked Color Blindness in Humans

To further examine X-linked inheritance, let's consider another X-linked characteristic: red–green color blindness in humans. Mutations that produce defective color vision are generally recessive, and because the genes encoding the red and the green photoreceptors are located on the X chromosome, red–green color blindness is inherited as an X-linked recessive characteristic.

We will use the symbol X^c to represent the allele for red–green color blindness and the symbol X^+ to represent the allele for normal color vision. Females possess two X chromosomes, so there are three possible genotypes among females: X^+X^+ and X^+X^c, which produce normal color vision, and X^cX^c, which produces color blindness. Males have only a single X chromosome and two possible genotypes: X^+Y, which produces normal vision, and X^cY, which produces color blindness.

Let's consider what happens when a woman homozygous for normal color vision mates with a color-blind man (**Figure 4.10a**). All the gametes produced by the woman will contain an allele for normal color vision. Half of the man's gametes will receive the X chromosome with the color-blindness allele, and the other half will receive the Y chromosome, which carries no alleles affecting color vision. When an

X^c-bearing sperm unites with the X^+-bearing egg, a heterozygous female with normal vision (X^+X^c) is produced. When a Y-bearing sperm unites with the X-bearing egg, a hemizygous male with normal vision (X^+Y) is produced.

In the reciprocal cross between a color-blind woman and a man with normal color vision (**Figure 4.10b**), the woman produces only X^c-bearing gametes. The man produces some gametes that contain the X chromosome and others that contain the Y chromosome. Males inherit the X chromosome from their mothers; because both of the mother's X chromosomes bear the X^c allele in this case, all the male offspring will be color blind. In contrast, females inherit an X chromosome from both parents; thus all the female offspring of this reciprocal cross will be heterozygous with normal vision. Females are color blind only when color-blindness alleles have been inherited from both parents, whereas a male only needs to inherit a color-blindness allele from his mother to be color blind; for this reason, color blindness and most other rare X-linked recessive characteristics are more common in males than in females.

In these crosses for color blindness, notice that an affected woman passes the X-linked recessive trait to her sons, but not to her daughters, whereas an affected man passes the trait to his grandsons through his daughters, but never to his sons. X-linked recessive characteristics may therefore appear to alternate between the sexes, appearing in females in one generation and in males in the next generation.

THINK-PAIR-SHARE Questions 6 and 7

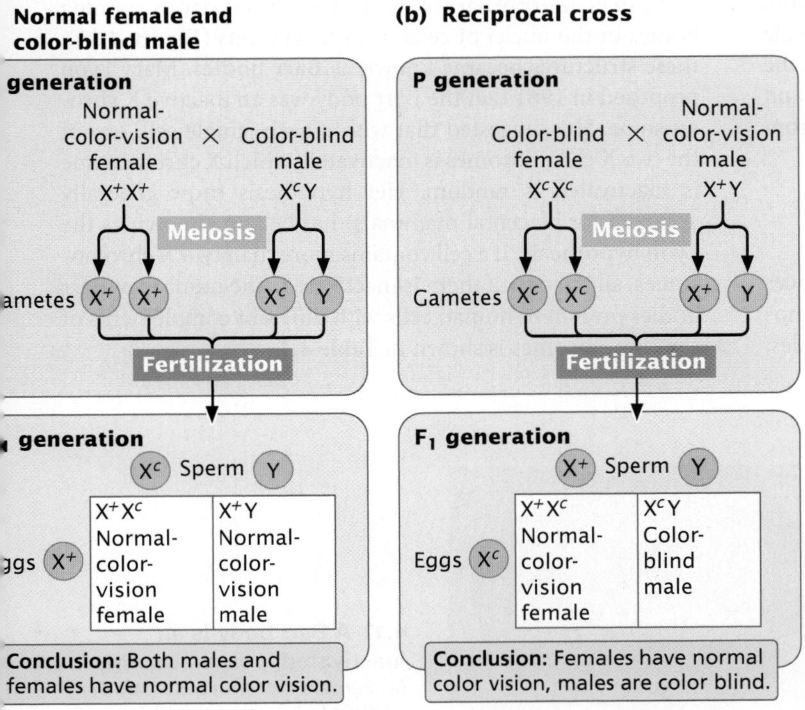

4.10 Red–green color blindness is inherited as an X-linked recessive trait in humans.

WORKED PROBLEM

Now that we understand the pattern of X-linked inheritance, let's apply our knowledge to answer a specific question.

Betty has normal vision, but her mother is color blind. Bill is color blind. If Bill and Betty marry and have a child together, what is the probability that the child will be color blind?

Solution Strategy

What information is required in your answer to the problem?

The probability that Bill and Betty's child will be color blind.

What information is provided to solve the problem?

The phenotypes of Betty, Betty's mother, and Bill.

Solution Steps

Because color blindness is an X-linked recessive trait, Betty's color-blind mother must be homozygous for the color-blindness allele (X^cX^c). Females inherit one X chromosome from each of their parents, so Betty must have inherited a color-blindness allele from her mother. Because Betty has normal color vision, she must have inherited an allele for normal vision (X^+) from her father; thus Betty is heterozygous (X^+X^c). Bill is color blind. Because males are hemizygous for X-linked alleles, he must be (X^cY). A mating between Betty and Bill is represented as

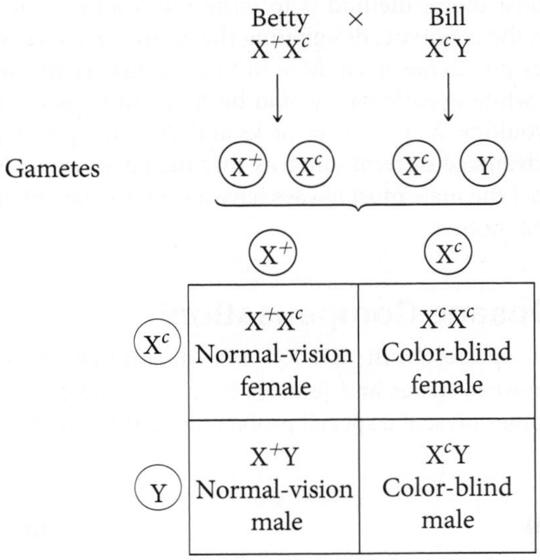

Thus, the overall probability that their child will be color blind is $^1/_2$.

THINK-PAIR-SHARE Question 8

▶ Get some additional practice with X-linked inheritance by working **Problem 15** at the end of this chapter.

CONCEPTS

Characteristics determined by genes on the sex chromosomes are called sex-linked characteristics. In organisms with XX-XY sex determination, diploid females have two alleles at each X-linked locus, whereas diploid males possess a single allele at each X-linked locus. Females inherit X-linked alleles from both parents, but males inherit a single X-linked allele from their mothers.

✓ CONCEPT CHECK 4

Hemophilia (reduced blood clotting) is an X-linked recessive disease in humans. A woman with hemophilia mates with a man who exhibits normal blood clotting. What is the probability that their child will have hemophilia?

Symbols for X-Linked Genes

There are several different ways to record genotypes for X-linked traits. Sometimes these genotypes are designated in the same way as for autosomal characteristics, but the hemizygous males are simply given a single allele: for example, the genotype of a female *Drosophila* with white eyes would be *ww*, and the genotype of a white-eyed hemizygous male would be *w*. Another method is to include the Y chromosome, designating it with a diagonal slash (/). With this method, the white-eyed female's genotype would still be *ww*, and the white-eyed male's genotype would be *w*/. Perhaps the most useful method is to write the X and Y chromosomes in the genotype, designating the X-linked alleles with superscripts, as has been done in this chapter. With this method, a white-eyed female would be X^wX^w and a white-eyed male would be X^wY. The use of Xs and Ys in the genotype has the advantage of reminding us that the genes are X linked and that the male must always have a single allele, inherited from the mother.

Dosage Compensation

In species with XX-XY sex determination, differences between males and females in their number of X chromosomes present a special problem in development. In females, there are two copies of the X chromosome and two copies of each autosome, so genes on the X chromosomes and on autosomes are "in balance." In males, however, there is only a single X chromosome, while there are two copies of every autosome. Because the amount of a protein produced is often a function of the number of gene copies encoding that protein, males are likely to produce less of a protein encoded by X-linked genes than of a protein encoded by autosomal genes. This difference can be detrimental because protein concentration often plays a critical role in development.

Some animals have overcome this problem by evolving mechanisms to equalize the amount of protein produced by the single X and two autosomes in the heterogametic sex. These mechanisms are referred to as **dosage compensation**. In fruit flies, dosage compensation is achieved by a doubling of the activity of the genes on the X chromosome of males, but not of females. In placental mammals, the expression of dosage-sensitive genes on the X chromosomes of both males and females has been increased, and one of the X chromosomes is inactivated in females, so that the expression of X-linked and autosomal genes is balanced in both males and females.

For unknown reasons, the presence of sex chromosomes does not always produce problems of gene dosage, and dosage compensation of X-linked genes is not universal. A number of animals do not exhibit obvious mechanisms of dosage compensation, including butterflies and moths, birds, some fishes, and even the duck-billed platypus. As we will see in the next section, even in placental mammals a number of genes escape dosage compensation.

In 1949, Murray Barr observed condensed, darkly staining bodies in the nuclei of cells from female cats (**Figure 4.11**); these structures became known as **Barr bodies**. Mary Lyon proposed in 1961 that the Barr body was an inactive X chromosome. She suggested that within each female cell, one of the two X chromosomes is inactivated; which X chromosome is inactivated is random. Her hypothesis (now generally accepted for placental mammals) has become known as the **Lyon hypothesis**. If a cell contains more than two X chromosomes, all but one of them is inactivated. The number of Barr bodies present in human cells with different complements of sex chromosomes is shown in **Table 4.3**.

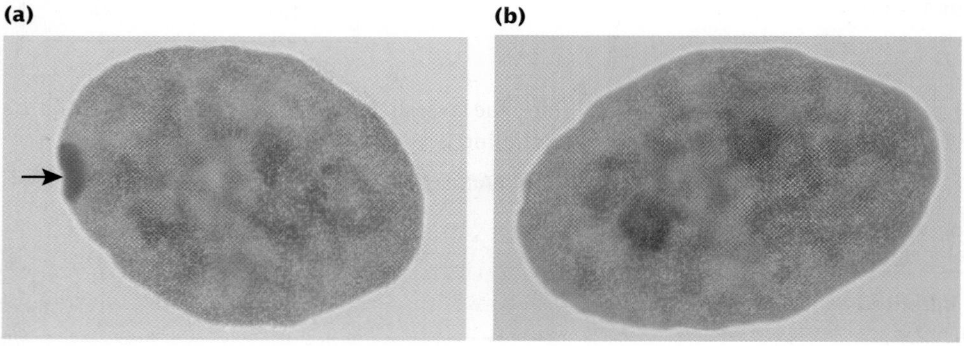

4.11 A Barr body is an inactivated X chromosome. (a) Female cell with a Barr body (indicated by arrow). (b) Male cell without a Barr body. [Chris Bjornberg/Science Source.]

TABLE 4.3 Number of Barr bodies in human cells with different complements of sex chromosomes

Sex chromosomes	Syndrome	Number of Barr bodies
XX	None	1
XY	None	0
XO	Turner	0
XXY	Klinefelter	1
XXYY	Klinefelter	1
XXXY	Klinefelter	2
XXXXY	Klinefelter	3
XXX	Triple-X	2
XXXX	Poly-X female	3
XXXXX	Poly-X female	4

As a result of X inactivation, female placental mammals are functionally hemizygous at the cellular level for X-linked genes. In females that are heterozygous at an X-linked locus, approximately 50% of the cells express one allele and 50% express the other allele; thus, proteins encoded by both alleles are produced, although not within the same cell. This functional hemizygosity means that cells in females are not identical with respect to the expression of the genes on the X chromosome; females are mosaics for the expression of X-linked genes.

Random X inactivation takes place early in development—in humans, within the first few weeks of development. After an X chromosome has become inactive in a cell, it remains inactive in that cell and in all somatic cells that descend from that cell. Thus, neighboring cells tend to have the same X chromosome inactivated, producing a patchy (mosaic) pattern of expression of X-linked characteristics in heterozygous females.

This patchy distribution of gene expression can be seen in tortoiseshell (**Figure 4.12**) and calico cats. Although many genes contribute to coat color and pattern in domestic cats, a single X-linked locus determines the presence of orange color. There are two possible alleles at this locus: X^+, which produces nonorange (usually black) fur, and X^o, which produces orange fur. Males are hemizygous and thus may be black (X^+Y) or orange (X^oY), but not black and orange. (Rare tortoiseshell males can arise from the presence of two X chromosomes, X^+X^oY.) Females may be black (X^+X^+), orange (X^oX^o), or tortoiseshell (X^+X^o), with the tortoiseshell pattern arising from a patchy mixture of black and orange fur. Each orange patch is a clone of cells derived from an original cell in which the black allele was inactivated, and each black patch is a clone of cells derived from an original cell in which the orange allele was inactivated.

4.12 The patchy distribution of color on tortoiseshell cats results from the random inactivation of one X chromosome in females. [Robert Adrian Hillman/Shutterstock.]

The Lyon hypothesis suggests that the presence of variable numbers of X chromosomes should not affect the phenotype in mammals because any X chromosomes in excess of one should be inactivated. However, persons with Turner syndrome (XO) differ from XX females, and those with Klinefelter syndrome (XXY) differ from XY males. The phenotypes associated with these disorders probably arise because some X-linked genes escape inactivation.

CONCEPTS

Dosage compensation ensures that the amount of sex-linked gene product is balanced with the amount of autosomal gene product. In placental mammals, all but one X chromosome is inactivated in each cell; which of the X chromosomes is inactivated is random and varies from cell to cell.

✓ CONCEPT CHECK 5

How many Barr bodies will a male with XXXYY chromosomes have in each of his cells? What are those Barr bodies?

Y-Linked Characteristics

Y-linked traits—also called holandric traits—exhibit a distinct pattern of inheritance. These traits are present only in males, because only males possess a Y chromosome, and are always inherited from the father. Furthermore, all male offspring of a male with a Y-linked trait will display the trait because every male inherits his Y chromosome from his father.

THE USE OF Y-LINKED GENETIC MARKERS DNA sequences in the Y chromosome undergo mutation with the passage of time and thus vary among individual males. These mutations create variations in DNA sequence—called genetic markers—that, like Y-linked traits, are passed from father to son and can be used to study male ancestry. Genetic markers can also occur on other chromosomes and are frequently used in mapping genes (see pp. 140–141). Although the markers themselves do not encode any physical traits, they can be detected with the use of molecular methods. Much of the Y chromosome is nonfunctional, and so mutations readily accumulate. Many of these mutations are unique; they arise only once and are passed down through the generations without undergoing recombination. Individual males possessing the same set of mutations are therefore assumed to be related, and the distribution of these genetic markers on Y chromosomes provides clues about the genetic relationships of present-day people. Y-chromosome sequences have also been used extensively to examine past patterns of human migration and the genetic relationships among different human populations.

Y-linked genetic markers have been used to study the offspring of Thomas Jefferson, principal author of the Declaration of Independence and third president of the United States. In 1802, a political enemy accused Jefferson of fathering a child by his slave Sally Hemings, but the evidence was circumstantial. Hemings, who worked in the Jefferson household and accompanied Jefferson on a trip to Paris, had five children. Jefferson was accused of fathering the first child, but rumors about the paternity of the other children circulated as well. Descendants of Hemings's children maintained that they were related to the Jefferson line, but some Jefferson descendants refused to recognize their claim.

To resolve this long-standing controversy, geneticists examined markers from the Y chromosomes of male-line descendants of Hemings's first son (Thomas Woodson), her last son (Eston Hemings), and a paternal uncle of Thomas Jefferson with whom Jefferson had Y chromosomes in common (descendants of Jefferson's uncle were used because Jefferson himself had no verified male descendants). Geneticists determined that Jefferson possessed a rare and distinctive set of genetic markers on his Y chromosome. The same markers were also found on the Y chromosomes of the male-line descendants of Eston Hemings. The probability of such a match arising by chance is less than 1%. The markers were not found on the Y chromosomes of the descendants of Thomas Woodson. Together with the circumstantial historical evidence, these matching markers suggest that Jefferson was the father of Eston Hemings, but not Thomas Woodson.

CONCEPTS

Y-linked characteristics exhibit a distinct pattern of inheritance: they are present only in males, and all male offspring of a male with a Y-linked trait inherit the trait.

CONNECTING CONCEPTS

Recognizing Sex-Linked Inheritance

What features should we look for to identify a trait as sex linked? A common misconception is that any genetic characteristic in which the phenotypes of males and females differ must be sex linked. In fact, the expression of many *autosomal* characteristics differs between males and females. The genes that encode these characteristics are the same in both sexes, but their expression is influenced by sex hormones. The different sex hormones of males and females cause the same genes to generate different phenotypes in males and females.

Another misconception is that any characteristic that is found more frequently in one sex than in the other is sex linked. A number of autosomal traits are expressed more commonly in one sex. These traits are said to be sex influenced. Some autosomal traits are expressed in only one sex; these traits are said to be sex limited. Both sex-influenced and sex-limited characteristics will be considered in more detail later in the chapter.

Several features of sex-linked characteristics make them easy to recognize. Y-linked traits are found only in males, but this fact does not guarantee that a trait is Y linked because some autosomal characteristics are expressed only in males. Y-linked traits are unique, however, in that all the male offspring of an affected male express the father's phenotype, and Y-linked traits can be inherited only from the father's side of the family. Thus, a Y-linked trait can be inherited only from the paternal grandfather (the father's father), never from the maternal grandfather (the mother's father).

X-linked characteristics also exhibit a distinctive pattern of inheritance. X linkage is a possible explanation when the results of reciprocal crosses differ. If a characteristic is X linked, a cross between an affected male and an unaffected female will not give the same results as a cross between an affected female and an unaffected male. For almost all autosomal characteristics, the results of reciprocal crosses are the same. We should not conclude, however, that when the reciprocal crosses give different results, the characteristic is X linked. Other sex-associated forms of inheritance, described later in the chapter, also produce different results in reciprocal crosses. The key to recognizing X-linked inheritance is to remember that a male always inherits his X chromosome from his mother, not from his father. Thus, an X-linked characteristic is not passed directly from father to son; if a male clearly inherits a trait from his father—and his mother is not heterozygous—it cannot be X linked. TRY PROBLEM 14

4.3 Additional Factors at a Single Locus Can Affect the Results of Genetic Crosses

In Chapter 3, we learned that the principle of segregation and the principle of independent assortment enable us to predict the outcomes of genetic crosses. Here, we examine several factors acting at individual loci that may alter the phenotypic ratios predicted by Mendel's principles.

Types of Dominance

One of Mendel's important contributions to the study of heredity is the concept of dominance—the idea that although an individual organism possesses two different alleles for a characteristic, the trait encoded by only one of the alleles is observed in the phenotype. With dominance, the heterozygote possesses the same phenotype as one of the homozygotes.

Mendel observed dominance in all the traits he chose to study extensively, but he was aware that not all characteristics exhibit dominance. He conducted some studies of the length of time that pea plants take to flower. When he crossed two homozygous varieties that differed in their flowering time by an average of 20 days, the length of time taken by the F_1 plants to flower was intermediate between those of the two parents. When the heterozygote has a phenotype intermediate between the phenotypes of the two homozygotes, the trait is said to display incomplete dominance.

COMPLETE AND INCOMPLETE DOMINANCE Dominance can be understood in regard to how the phenotype of the heterozygote relates to the phenotypes of the two homozygotes. In the example presented in the upper panel of **Figure 4.13**, potential flower color ranges from red to white. One homozygous genotype, A^1A^1, produces red pigment, resulting in red flowers; another, A^2A^2, produces no pigment, resulting in white flowers. Where the heterozygote falls in the range of phenotypes determines the type of dominance. If the heterozygote (A^1A^2) produces the same amount of pigment as the A^1A^1 homozygote, resulting in red flowers, then the A^1 allele displays **complete dominance** over the A^2 allele; that is, red is dominant over white. If, on the other hand, the heterozygote produces no pigment, resulting in flowers with the same color as the A^2A^2 homozygote (white), then the A^2 allele is completely dominant, and white is dominant over red.

When the phenotype of the heterozygote falls in between the phenotypes of the two homozygotes, dominance is incomplete (see the bottom panel of Figure 4.13). With **incomplete dominance**, the heterozygote need not be exactly intermediate (pink in our example) between the two homozygotes; it might be a slightly lighter shade of red or a slightly pink shade of white. As long as the heterozygote's phenotype can be differentiated and falls within the range between the two homozygotes, dominance is incomplete. The important thing to remember about dominance is that it affects the way that genes are *expressed* (the phenotype), but not the way that genes are *inherited*.

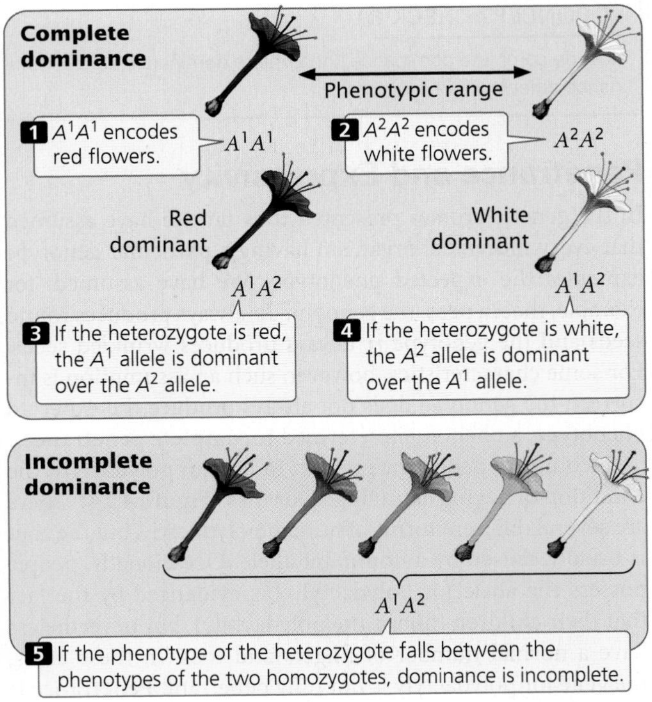

4.13 The type of dominance exhibited by a trait depends on how the phenotype of the heterozygote relates to the phenotypes of the homozygotes.

CODOMINANCE Another type of interaction between alleles is **codominance**, in which the phenotype of the heterozygote is not intermediate between the phenotypes of the homozygotes; rather, the heterozygote simultaneously expresses the phenotypes of both homozygotes. An example of codominance is seen in the MN blood types of humans.

The *MN* locus encodes one of the types of antigens on the surface of red blood cells. Unlike antigens of the ABO and Rh blood groups (which also encode red-blood-cell antigens), MN antigens do not elicit a strong immunological reaction, and therefore MN blood types are not routinely considered in blood transfusions. At the *MN* locus, there are two alleles: the L^M allele, which encodes the M antigen; and the L^N allele, which encodes the N antigen. Homozygotes with genotype L^ML^M express the M antigen on the surface of their red blood cells and have the M blood type. Homozygotes with genotype L^NL^N express the N antigen and have the N blood type. Heterozygotes with genotype L^ML^N exhibit codominance and express both the M and the N antigens; they have blood type MN.

Some students might ask why the pink flowers illustrated in Figure 4.13 exhibit incomplete dominance: Why is this trait not an example of codominance? In this situation, the heterozygote is not producing both red and white pigments, which then combine to produce a pink phenotype. The heterozygote produces only red pigment, but the amount produced by the heterozygote is less than the amount produced by the A^1A^1 homozygote. So here, the alleles clearly exhibit incomplete dominance, not codominance. The differences between

TABLE 4.4 Differences between dominance, incomplete dominance, and codominance

Type of dominance	Definition
Complete dominance	Phenotype of the heterozygote is the same as the phenotype of one of the homozygotes
Incomplete dominance	Phenotype of the heterozygote is intermediate (falls within the range) between the phenotypes of the two homozygotes
Codominance	Phenotype of the heterozygote includes the phenotypes of both homozygotes

complete dominance, incomplete dominance, and codominance are summarized in **Table 4.4**. TRY PROBLEM 26

LEVEL OF PHENOTYPE OBSERVED MAY AFFECT DOMINANCE Many phenotypes can be observed at several different levels, including the anatomical level, the physiological level, and the molecular level. The type of dominance exhibited by a characteristic depends on the level of the phenotype examined. This dependency is seen in cystic fibrosis, a common genetic disorder in Caucasians that is usually considered to be a recessive disease. People who have cystic fibrosis produce large quantities of thick, sticky mucus, which plugs up the airways of the lungs and clogs the ducts leading from the pancreas to the intestine, causing frequent respiratory infections and digestive problems. Even with medical treatment, patients with cystic fibrosis suffer chronic, life-threatening medical problems.

The gene responsible for cystic fibrosis resides on the long arm of chromosome 7. It encodes a protein termed *cystic fibrosis transmembrane conductance regulator* (CFTR), which acts as a gated channel in the cell membrane and regulates the movement of chloride ions into and out of the cell. Persons with cystic fibrosis have a mutated, dysfunctional form of CFTR that causes the channel to stay closed, so chloride ions build up in the cell. This buildup causes the formation of thick mucus and produces the symptoms of the disease.

Most people have two copies of the normal allele for CFTR and produce only functional CFTR protein. Those with cystic fibrosis possess two copies of the mutated *CFTR* allele and produce only the defective CFTR protein. Heterozygotes, who have one normal and one defective *CFTR* allele, produce both functional and defective CFTR protein. Thus, at the molecular level, the alleles for normal and defective CFTR are codominant because both alleles are expressed in the heterozygote. However, because one functional allele produces enough functional CFTR protein to allow normal chloride-ion transport, heterozygotes exhibit no adverse effects, and the mutated *CFTR* allele appears to be recessive at the physiological level. The type of dominance expressed by an allele, as illustrated in this example, is a function of the phenotypic aspect of the allele that is observed.

THINK-PAIR-SHARE Question 9

CHARACTERISTICS OF DOMINANCE Several important characteristics of dominance should be emphasized. First, dominance is a result of interactions between genes at the same locus (allelic genes); in other words, dominance is *allelic* interaction. Second, dominance does not alter the way in which the genes are inherited; it influences only the way in which they are expressed as a phenotype. The allelic interaction that characterizes dominance is therefore interaction between the *products* of the genes. Finally, the type of dominance frequently depends on the level at which the phenotype is observed. As seen for cystic fibrosis, an allele may exhibit codominance at one level and be recessive at another level.

CONCEPTS

Dominance entails interactions between genes at the same locus (allelic genes) and is an aspect of the phenotype; dominance does not affect the way in which genes are inherited. The type of dominance exhibited by a characteristic frequently depends on the level at which the phenotype is examined.

✓ **CONCEPT CHECK 6**

How do complete dominance, incomplete dominance, and codominance differ?

Penetrance and Expressivity

In the genetic crosses presented thus far, we have assumed that every individual organism having a particular genotype expresses the expected phenotype. We have assumed, for example, that in peas, the genotype *Rr* always produces round seeds and the genotype *rr* always produces wrinkled seeds. For some characteristics, however, such an assumption is incorrect: the genotype does not always produce the expected phenotype, a phenomenon termed **incomplete penetrance**.

Incomplete penetrance is seen in human polydactyly, the condition of having extra fingers or toes (**Figure 4.14**). There are several different forms of human polydactyly, but the trait is usually caused by a dominant allele. Occasionally, people possess the allele for polydactyly (as evidenced by the fact that their children inherit the polydactyly), but nevertheless have a normal number of fingers and toes. In these cases, the gene for polydactyly is not fully penetrant. **Penetrance** is defined as the percentage of individuals having a particular genotype that express the expected phenotype. For example, if we examined 42 people having an allele for polydactyly and found that only 38 of them were polydactylous, the penetrance would be $^{38}/_{42} = 0.90$ (90%).

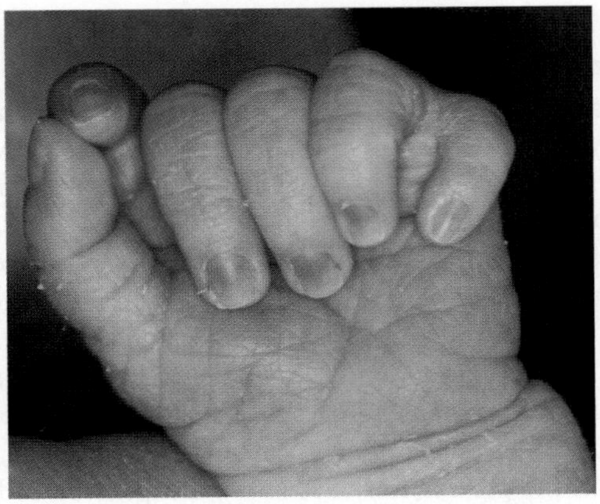

4.14 Human polydactyly (extra digits) exhibits incomplete penetrance and variable expressivity. [SPL/Science Source.]

A related concept is that of **expressivity**, the degree to which a trait is expressed. In addition to exhibiting incomplete penetrance, polydactyly exhibits variable expressivity. Some polydactylous people possess extra fingers or toes that are fully functional, whereas others possess only a small tag of extra skin.

Incomplete penetrance and variable expressivity are due to the effects of other genes and environmental factors that can alter or completely suppress the effect of a particular gene. For example, a gene may encode an enzyme that produces a particular phenotype only within a limited temperature range (see also Environmental Effects on the Phenotype, p. 103). At higher or lower temperatures, the enzyme does not function, and the phenotype is not expressed; the allele encoding such an enzyme is therefore penetrant only within a particular temperature range. Many characters exhibit incomplete penetrance and variable expressivity; thus, the mere presence of a gene does not guarantee its expression. ▶TRY PROBLEM 28

CONCEPTS

Penetrance is the percentage of individuals having a particular genotype that express the associated phenotype. Expressivity is the degree to which a trait is expressed. Incomplete penetrance and variable expressivity result from the influence of other genes and environmental factors on the phenotype.

✓ CONCEPT CHECK 7

How does incomplete dominance differ from incomplete penetrance?

a. Incomplete dominance refers to alleles at the same locus; incomplete penetrance refers to alleles at different loci.
b. Incomplete dominance ranges from 0% to 50%; incomplete penetrance ranges from 51% to 99%.
c. In incomplete dominance, the heterozygote is intermediate between the homozygotes; in incomplete penetrance, heterozygotes express phenotypes of both homozygotes.
d. In incomplete dominance, the heterozygote is intermediate between the homozygotes; in incomplete penetrance, some individuals do not express the expected phenotype.

Lethal Alleles

A **lethal allele** causes death at an early stage of development—often before birth—so that some genotypes do not appear among the progeny. An example of a lethal allele is one that determines yellow coat color in mice. A cross between two yellow heterozygous mice produces an initial genotypic ratio of $\frac{1}{4}$ *YY*, $\frac{1}{2}$ *Yy*, and $\frac{1}{4}$ *yy*, but the homozygous *YY* mice die early in development and do not appear among the progeny, resulting in a 2:1 ratio of *Yy* (yellow) to *yy* (nonyellow) in the offspring (**Figure 4.15**). A 2:1 ratio is almost always produced by a recessive lethal allele, so observing this ratio among the progeny of a cross between individuals with the same phenotype is a strong clue that one of the alleles is lethal.

Another example of a lethal allele, originally described by Erwin Baur in 1907, is found in snapdragons. The *aurea* strain in these plants has yellow leaves. When two plants with yellow leaves are crossed, $\frac{2}{3}$ of the progeny have yellow leaves and $\frac{1}{3}$ have green leaves. When green is crossed with green, all the progeny have green leaves; when yellow is crossed with green, however, $\frac{1}{2}$ of the progeny have green leaves and $\frac{1}{2}$ have yellow leaves, confirming that all yellow-leaved snapdragons are heterozygous.

In this example, like that of yellow coat color in mice, the lethal allele is recessive because it causes death only in homozygotes. Unlike its effect on *survival*, the effect of the allele on *color* is dominant; in both mice and snapdragons, a single copy of the allele in heterozygotes produces a yellow color. These examples illustrate the point made earlier that the type of dominance depends on the aspect of the phenotype examined.

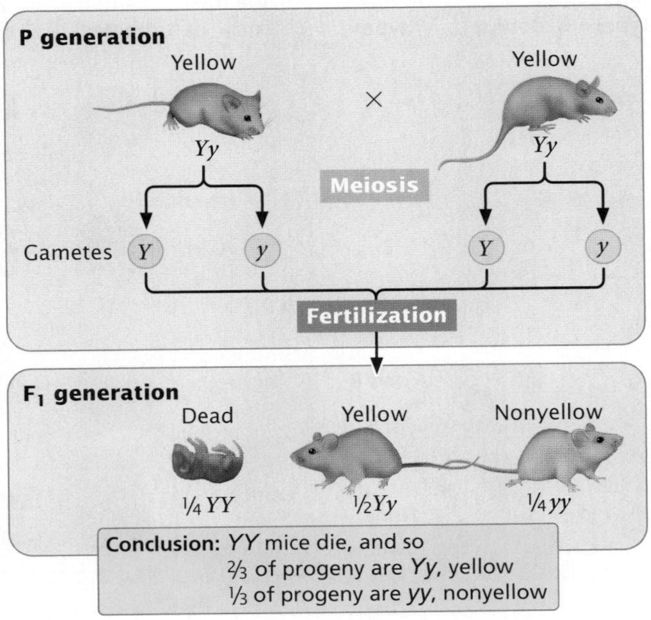

4.15 The 2:1 ratio produced by a cross between two yellow mice results from a lethal allele.

Many lethal alleles in nature are recessive, but lethal alleles also can be dominant; in this case, homozygotes and heterozygotes for the allele die. Truly dominant lethal alleles cannot be transmitted unless they are expressed after the onset of reproduction. ▶TRY PROBLEM 29

CONCEPTS

A lethal allele causes death, frequently at an early developmental stage, so that one or more genotypes are missing from the progeny of a cross. Lethal alleles therefore modify the ratios of progeny resulting from a cross.

Multiple Alleles

Most of the genetic systems that we have examined so far consist of two alleles. In Mendel's peas, for instance, one allele encoded round seeds and another encoded wrinkled seeds; in cats, one allele produced a black coat and another produced a gray coat. For some loci, more than two alleles are present within a group of individuals—the locus has **multiple alleles** (which may also be referred to as an *allelic series*). Although there may be more than two alleles present within a *group* of organisms, the genotype of each individual diploid organism still consists of only two alleles. The inheritance of characteristics encoded by multiple alleles is no different from the inheritance of characteristics encoded by two alleles, except that a greater variety of genotypes and phenotypes is possible.

THE ABO BLOOD GROUP A multiple-allele system is found at the locus for the ABO blood group, which determines your ABO blood type. This locus, like the *MN* locus, encodes antigens on the surface of red blood cells. The three common alleles for the ABO blood group locus are I^A, which encodes the A antigen; I^B, which encodes the B antigen; and i, which encodes no antigen (O). We can represent the dominance relations among the ABO alleles as follows: $I^A > i$, $I^B > i$, $I^A = I^B$. The I^A and the I^B alleles are dominant over i and are codominant with each other; the AB phenotype is due to the presence of an I^A allele and an I^B allele, which results in the production of A and B antigens on red blood cells. A person with genotype ii produces neither antigen and has blood type O. The six common genotypes at this locus and their phenotypes are shown in **Figure 4.16a**.

The body produces antibodies against any foreign antigens (see Figure 4.16a). For instance, a person with blood type A produces anti-B antibodies because the B antigen is foreign to that person. A person with blood type B produces anti-A antibodies, and a person with blood type AB produces neither anti-A nor anti-B antibodies because neither A nor B antigen is foreign to that person. A person with blood type O possesses no A or B antigens; consequently, that person produces both anti-A antibodies and anti-B antibodies. The presence of antibodies against foreign ABO antigens means that successful blood transfusions are possible only between persons with certain compatible blood types (**Figure 4.16b**).

(a)

Phenotype (blood type)	Genotype	Antigen type	Antibodies made by body
A	I^AI^A or I^Ai	A	Anti-B
B	I^BI^B or I^Bi	B	Anti-A
AB	I^AI^B	A and B	None
O	ii	None	Anti-A and Anti-B

(b)

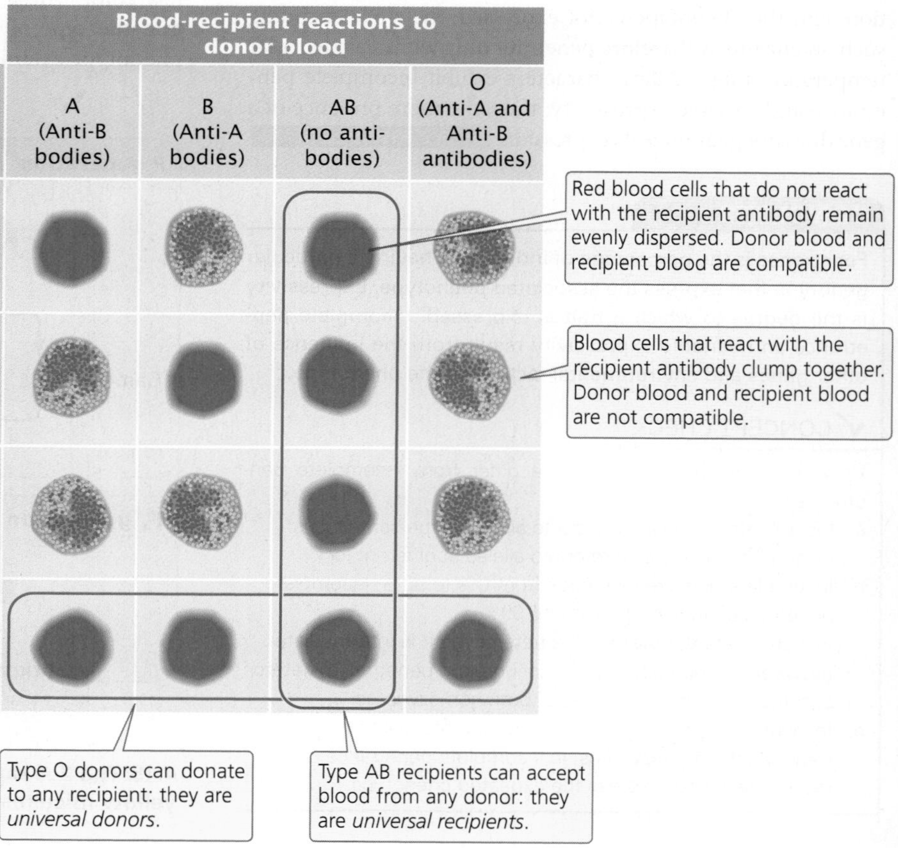

4.16 ABO blood types and possible blood transfusions.

The inheritance of alleles at the ABO locus is illustrated by a paternity suit against the movie actor Charlie Chaplin. In 1941, Chaplin met a young actress named Joan Barry, with whom he had an affair. The affair ended in February 1942, but 20 months later, Barry gave birth to a baby girl and claimed that Chaplin was the father. Barry then sued for child support. At this time, blood typing had just come into widespread use, and Chaplin's attorneys had Chaplin, Barry, and the child blood typed. Barry had blood type A, her child had blood type B, and Chaplin had blood type O. Could Chaplin have been the father of Barry's child?

Your answer should be no. Joan Barry had blood type A, which can be produced by either genotype I^AI^A or genotype I^Ai. Her baby possessed blood type B, which can be produced by either genotype I^BI^B or genotype I^Bi. The baby could not have inherited the I^B allele from Barry (Barry could not carry an I^B allele if she were blood type A); therefore, the baby must have inherited the *i* allele from her. Barry must have had genotype I^Ai, and the baby must have had genotype I^Bi. Because the baby girl inherited her *i* allele from Barry, she must have inherited the I^B allele from her father. Having blood type O, produced only by genotype *ii*, Chaplin could not have been the father of Barry's child. Although blood types can be used to exclude the possibility of paternity (as in this case), they cannot prove that a person is the parent of a child, because many different people have the same blood type.

In the course of the trial to settle the paternity suit against Chaplin, three pathologists testified that it was genetically impossible for Chaplin to have fathered the child. Nevertheless, the jury ruled that Chaplin was the father and ordered him to pay child support and Barry's legal expenses.

▶TRY PROBLEM 32

COMPOUND HETEROZYGOTES Different alleles often give rise to the same phenotype. For example, cystic fibrosis, as we saw earlier in this chapter, arises from defects in alleles at the *CFTR* locus, which encodes a protein that controls the movement of chloride ions into and out of the cell. Over a thousand different alleles at the *CFTR* locus that can cause cystic fibrosis have been discovered worldwide. Because cystic fibrosis is an autosomal recessive condition, one must normally inherit two defective *CFTR* alleles to have cystic fibrosis. In some people with cystic fibrosis, these two defective alleles are identical, meaning that the person is homozygous. Other people with cystic fibrosis are heterozygous, possessing two different defective alleles. An individual who carries two different alleles at a locus that result in a recessive phenotype is referred to as a **compound heterozygote**.

CONCEPTS

More than two alleles (multiple alleles) may be present at a locus within a group of individuals, although each individual diploid organism still has only two alleles at that locus. A compound heterozygote possesses two different alleles that result in a recessive phenotype.

4.4 Gene Interaction Takes Place When Genes at Multiple Loci Determine a Single Phenotype

In the dihybrid crosses that we examined in Chapter 3, each locus had an independent effect on the phenotype. When Mendel crossed a homozygous pea plant that produced round and yellow seeds (*RR YY*) with a homozygous plant that produced wrinkled and green seeds (*rr yy*) and then allowed the F_1 to self-fertilize, he obtained F_2 progeny in the following proportions:

$^9/_{16}$	*R_ Y_*	round, yellow
$^3/_{16}$	*R_ yy*	round, green
$^3/_{16}$	*rr Y_*	wrinkled, yellow
$^1/_{16}$	*rr yy*	wrinkled, green

In this example, the two loci showed two kinds of independence. First, the genes at each locus were independent in their *assortment* in meiosis, which produced the 9 : 3 : 3 : 1 ratio of phenotypes in the progeny, in accord with Mendel's principle of independent assortment. Second, the genes were independent in their *phenotypic expression*; the *R* and *r* alleles affected only the shape of the seed and had no influence on the color of the seed; the *Y* and *y* alleles affected only color and had no influence on the shape of the seed.

Frequently, genes exhibit independent assortment but do not act independently in their phenotypic expression; instead, the effects of genes at one locus depend on the presence of genes at other loci. This type of interaction between the effects of genes at different loci (genes that are not allelic) is termed **gene interaction**. With gene interaction, the products of genes at different loci combine to produce new phenotypes that are not predictable from the single-locus effects alone. In our consideration of gene interaction, we'll focus primarily on interactions between the effects of genes at two loci, although interactions among genes at three, four, or more loci are common.

CONCEPTS

In gene interaction, genes at different loci contribute to the determination of a single phenotypic characteristic.

Gene Interaction That Produces Novel Phenotypes

Let's begin by examining gene interaction in which genes at two loci interact to produce a single characteristic. Fruit color in the pepper *Capsicum annuum* is determined in this way. Certain types of pepper plants produce fruits in one of four colors: red, peach, orange (sometimes called yellow), and cream (or white). If a homozygous plant with red peppers is crossed with a homozygous plant with cream peppers, all the F_1 plants have red peppers (**Figure 4.17a**). When the F_1 are crossed with each other, the F_2 show a ratio of 9 red : 3 peach : 3 orange : 1 cream (**Figure 4.17b**). This dihybrid ratio (see Chapter 3) is produced by a cross between two plants that are both heterozygous for two loci ($Y^+y\ C^+c \times Y^+y\ C^+c$). In this example, the *Y* locus and the *C* locus interact to produce a single phenotype—the color of the pepper:

Genotype	Phenotype
$Y^+\_\ C^+\_$	red
$Y^+\_\ cc$	peach
$yy\ C^+\_$	orange
$yy\ cc$	cream

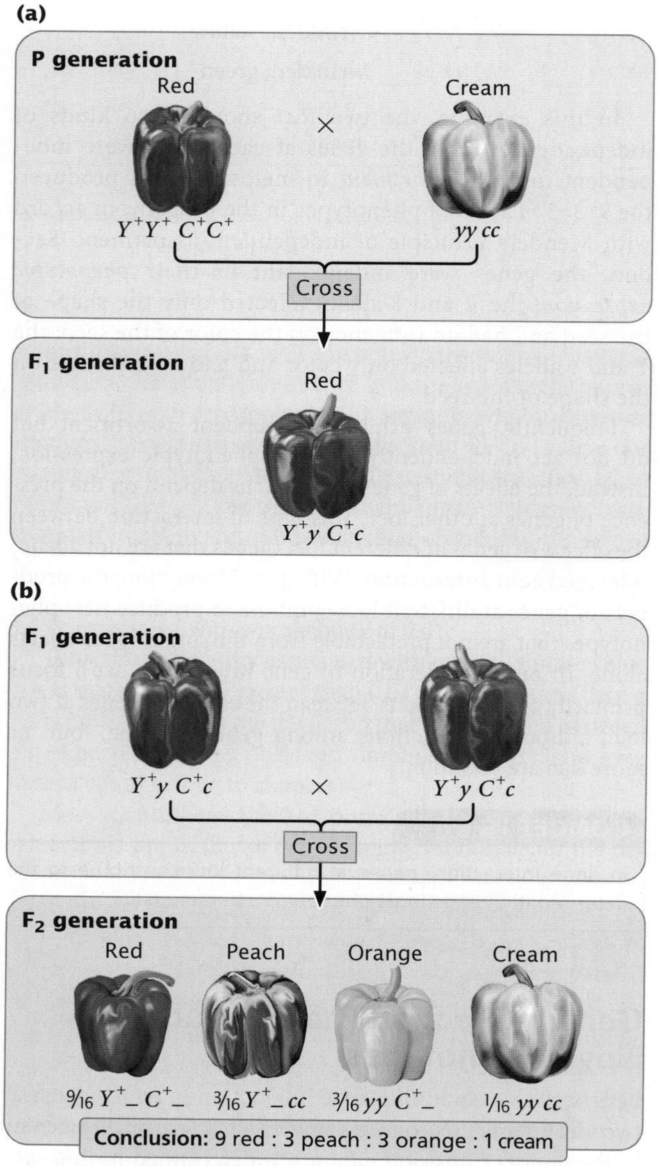

4.17 Interaction between genes at two loci determines a single characteristic, fruit color, in the pepper *Capsicum annuum*.

To illustrate how Mendel's rules of heredity can be used to understand the inheritance of characteristics determined by gene interaction, let's consider a testcross between an F_1 plant from the cross in Figure 4.17 ($Y^+y\ C^+c$) and a plant with cream peppers (*yy cc*). As outlined in Chapter 3 for independent loci, we can work this cross by breaking it down into two simple crosses. At the first locus, the heterozygote Y^+y is crossed with the homozygote *yy*; this cross produces $\frac{1}{2}\ Y^+y$ and $\frac{1}{2}\ yy$ progeny. Similarly, at the second locus, the heterozygous genotype C^+c is crossed with the homozygous genotype *cc*, producing $\frac{1}{2}\ C^+c$ and $\frac{1}{2}\ cc$ progeny. In accord with Mendel's principle of independent assortment, these single-locus ratios can be combined by using the multiplication rule: the probability of obtaining the genotype $Y^+y\ C^+c$ is the probability of Y^+y ($\frac{1}{2}$) multiplied by the probability of C^+c ($\frac{1}{2}$), or $\frac{1}{2} \times \frac{1}{2} = \frac{1}{4}$. The probabilities of each progeny genotype resulting from the testcross are

Progeny genotype	Probability at each locus		Overall probability	Phenotype
$Y^+y\ C^+c$	$\frac{1}{2} \times \frac{1}{2}$	=	$\frac{1}{4}$	red peppers
$Y^+y\ cc$	$\frac{1}{2} \times \frac{1}{2}$	=	$\frac{1}{4}$	peach peppers
$yy\ C^+c$	$\frac{1}{2} \times \frac{1}{2}$	=	$\frac{1}{4}$	orange peppers
$Yy\ cc$	$\frac{1}{2} \times \frac{1}{2}$	=	$\frac{1}{4}$	cream peppers

When you work problems with gene interaction, it is especially important to determine the probabilities of single-locus genotypes and to multiply the probabilities of *genotypes*, not phenotypes, because the phenotypes cannot be determined without considering the effects of the genotypes at all the contributing loci. ▶TRY PROBLEM 33

Gene Interaction with Epistasis

Sometimes the effect of gene interaction is that one gene masks (hides) the effect of another gene at a different locus, a phenomenon known as **epistasis**. In the examples of genic interaction that we have already examined, genes at different loci interacted to determine a single phenotype, but one gene did not mask the effect of a gene at another locus, and there was no epistasis. Epistasis is similar to dominance, except that dominance entails the masking of genes at the *same* locus (allelic genes). In epistasis, the gene that does the masking is called an **epistatic gene**; the gene whose effect is masked is a **hypostatic gene**. Epistatic genes may be recessive or dominant in their effects.

RECESSIVE EPISTASIS Recessive epistasis is seen in the genes that determine coat color in Labrador retrievers. These dogs may be black, brown (frequently called chocolate), or yellow; their different coat colors are determined by interactions between genes at two loci (although a number of other loci also help to determine coat color). One locus determines the type of pigment produced by the skin cells: a dominant allele *B* encodes black pigment, whereas a recessive allele *b* encodes brown pigment. Alleles at a second locus affect the *deposition* of the pigment in the shaft of the hair: allele *E* allows dark pigment (black or brown) to be deposited, whereas a recessive allele *e* prevents the deposition of dark pigment, causing the hair to be yellow. The presence of genotype *ee* at the second locus therefore masks the expression of the black and brown alleles at the first locus. The genotypes that determine coat color and their phenotypes are

Genotype	Phenotype
*B_ E_*	black
*bb E_*	brown
B_ ee	yellow
bb ee	yellow

If we cross a black Labrador that is homozygous for the dominant alleles (*BB EE*) with a yellow Labrador that is homozygous for the recessive alleles (*bb ee*) and then intercross the F_1, we obtain progeny in the F_2 in a 9:3:4 ratio:

P *BB EE* × *bb ee*
Black Yellow

↓

F1 *Bb Ee*
Black

↓ Intercross

F2 $^9/_{16}$ *B_ E_* black
$^3/_{16}$ *bb E_* brown
$^3/_{16}$ *B_ ee* yellow } $^4/_{16}$ yellow
$^1/_{16}$ *bb ee* yellow }

Notice that yellow dogs can carry alleles for either black or brown pigment, but these alleles are not expressed in their coat color. In this example of gene interaction, allele *e* is epistatic to *B* and *b* because *e* masks the expression of the alleles for black and brown pigments, and alleles *B* and *b* are hypostatic to *e*. In this case, *e* is a recessive epistatic allele because two copies of *e* must be present to mask the expression of the black and brown pigments.

Another example of an epistatic gene is the gene that determines the Bombay phenotype; this gene masks the expression of alleles at the ABO locus. In most people, a dominant allele (*H*) encodes an enzyme that makes H, a molecule necessary for the production of ABO antigens. People with the Bombay phenotype are homozygous for a recessive mutation (*h*) that encodes a defective enzyme. That defective enzyme is incapable of making H, and because H is not produced, no ABO antigens are synthesized. People with the genotype *hh* who would normally have A, B, or AB blood types do not produce antigens and therefore express an O phenotype. In this example, the alleles at the ABO locus are hypostatic to the recessive *h* allele.

DOMINANT EPISTASIS In recessive epistasis, the presence of two recessive alleles (the homozygous genotype) inhibits the expression of an allele at a different locus. In dominant epistasis, only a single copy of an allele is required to inhibit the expression of an allele at a different locus.

Dominant epistasis is seen in the interaction of two loci that determine fruit color in summer squash, which is commonly found in one of three colors: yellow, white, or green. When a homozygous plant that produces white squash is crossed with a homozygous plant that produces green squash and the F_1 plants are crossed with each other, the following results are obtained:

P Plants with white squash × Plants with green squash

↓

F_1 Plants with white squash

↓ Intercross

F_2 $^{12}/_{16}$ plants with white squash
$^3/_{16}$ plants with yellow squash
$^1/_{16}$ plants with green squash

How can gene interaction explain these results?

In the F_2, $^{12}/_{16}$, or $^3/_4$, of the plants produce white squash and $^3/_{16} + {}^1/_{16} = {}^4/_{16} = {}^1/_4$ of the plants produce colored squash. This outcome is the familiar 3:1 ratio produced by a cross between two heterozygotes, which suggests that a dominant allele at one locus inhibits the production of pigment, resulting in white progeny. If we use the symbol *W* to represent the dominant allele that inhibits pigment production, the genotype *W_* inhibits pigment production and produces white squash, whereas *ww* allows pigment production and results in colored squash.

Among those *ww* F_2 plants with pigmented fruit, we observe $^3/_{16}$ yellow and $^1/_{16}$ green (a 3:1 ratio). This observation suggests that a second locus determines the type of pigment produced in the squash, with yellow (*Y_*) dominant over green (*yy*). This locus is expressed only in *ww* plants, which lack the dominant inhibitory allele *W*. We can assign the genotype *ww Y_* to plants that produce yellow squash and the genotype *ww yy* to plants that produce green squash. The genotypes and their associated phenotypes are

*W_ Y_*	white squash
W_ yy	white squash
*ww Y_*	yellow squash
ww yy	green squash

Allele *W* is epistatic to *Y* and *y*: it suppresses the expression of these pigment-producing genes. Allele *W* is a dominant epistatic allele because, in contrast with *e* in Labrador-retriever coat color, a single copy of the allele is sufficient to inhibit pigment production.

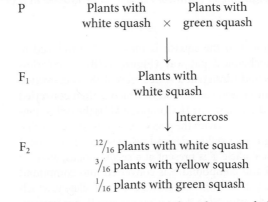

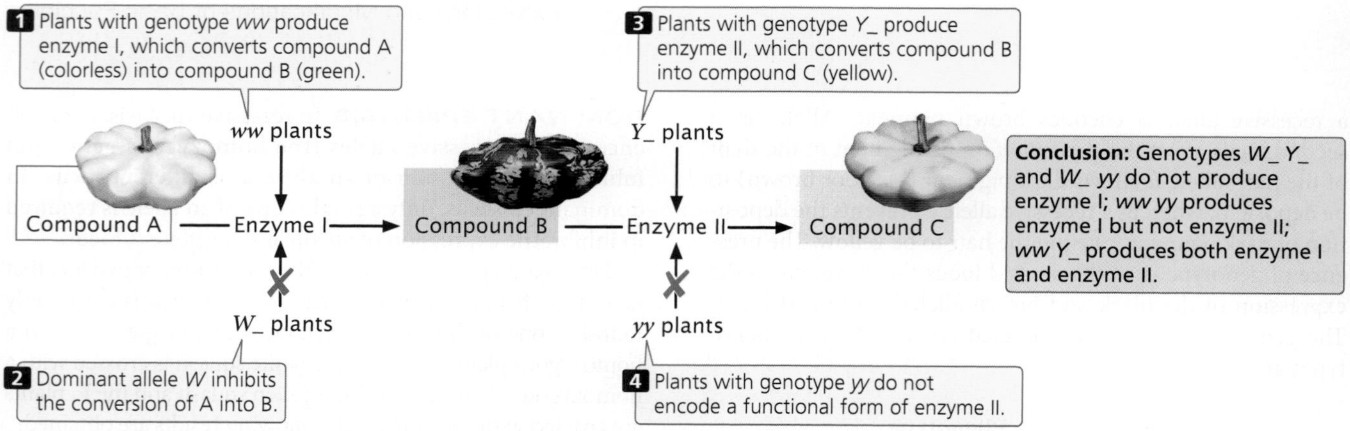

4.18 Yellow pigment in summer squash is produced in a two-step pathway.

Yellow pigment in the squash is most likely produced in a two-step biochemical pathway (**Figure 4.18**). A colorless (white) compound (designated A in Figure 4.18) is converted by enzyme I into green compound B, which is then converted into compound C by enzyme II. Compound C is the yellow pigment in the fruit. Plants with the genotype *ww* produce enzyme I and may be green or yellow, depending on whether enzyme II is present. When allele *Y* is present at a second locus, enzyme II is produced and compound B is converted into compound C, producing a yellow fruit. When two copies of allele *y*, which does not encode a functional form of enzyme II, are present, squash remain green. The presence of *W* at the first locus inhibits the conversion of compound A into compound B; plants with genotype *W_* do not make compound B, and their fruit remains white, regardless of which alleles are present at the second locus. View an explanation of epistasis in **Animation 4.2**.

A

CONCEPTS

Epistasis is the masking of the expression of one gene by another gene at a different locus. The epistatic gene does the masking; the hypostatic gene is masked. Epistatic alleles can be dominant or recessive.

✓ CONCEPT CHECK 8

A number of all-white cats are crossed, and they produce the following types of progeny: $^{12}/_{16}$ all-white, $^{3}/_{16}$ black, and $^{1}/_{16}$ gray. What is the genotype of the black progeny?

a. *Aa*
b. *Aa Bb*
c. *A_ B_*
d. *aaB_*

CONNECTING CONCEPTS

Interpreting Phenotypic Ratios Produced by Gene Interaction

A number of modified phenotypic ratios that result from gene interaction are shown in **Table 4.5**. Each of these examples represents a modification of the basic 9 : 3 : 3 : 1 dihybrid ratio. In interpreting the genetic basis of such modified ratios, we should keep several points in mind. First, the inheritance of the genes producing these characteristics is no different from the inheritance of genes encoding simple genetic characters. Mendel's principles of segregation and independent assortment still apply; each individual possesses two alleles at each locus, which separate in meiosis, and genes at the different loci assort independently. The only difference is in how the *products* of the genotypes interact to produce the phenotype. Thus, we cannot consider the expression of genes at each locus separately; instead, we must take into consideration how the genes at different loci interact.

A second point is that, in the examples that we have considered, the phenotypic proportions were always in sixteenths, because in all the crosses, pairs of alleles segregated at two independently assorting loci. The probability of inheriting one of the two alleles at a locus is $^1/_2$. Because there are two loci, each with two alleles, the probability of inheriting any particular combination of genes is $(^1/_2)^4 = {^1/_{16}}$. For a trihybrid cross, the progeny proportions should be in sixty-fourths, because $(^1/_2)^6 = {^1/_{64}}$. In general, the progeny proportions should be in fractions of $(^1/_2)^{2n}$, where n equals the number of loci with two alleles segregating in the cross.

Crosses rarely produce exactly 16 progeny; therefore, modifications of the dihybrid ratio are not always obvious. Modified dihybrid ratios are more easily seen if the number of individuals of each phenotype is expressed in sixteenths:

$$\frac{x}{16} = \frac{\text{number of progeny with a phenotype}}{\text{total number of progeny}}$$

where $x/16$ equals the proportion of progeny with a particular phenotype. If we solve for x (the proportion of the particular phenotype in sixteenths), we have

$$x = \frac{\text{number of progeny with a phenotype} \times 16}{\text{total number of progeny}}$$

For example, suppose that we cross two homozygotes, interbreed the F_1, and obtain 63 red, 21 brown, and 28 white F_2 individuals. Using the preceding formula, we find the phenotypic ratio in the F_2 to be red = (63 × 16)/112 = 9; brown = (21 × 16)/112 = 3; and white = (28 × 16)/112 = 4. The phenotypic ratio is 9 : 3 : 4.

A final point to consider is how best to assign genotypes to the phenotypes in modified ratios that result from gene interaction. Don't try to memorize the genotypes associated with all the modified ratios in Table 4.5. Instead, practice relating modified ratios to known ratios, such as the 9 : 3 : 3 : 1 dihybrid ratio. Suppose that we obtain $^{15}/_{16}$ green progeny and $^1/_{16}$ white progeny in a cross between two plants. If we compare this 15 : 1 ratio with the standard 9 : 3 : 3 : 1 dihybrid ratio, we see that $^9/_{16} + {^3/_{16}} + {^3/_{16}}$ equals $^{15}/_{16}$. All the genotypes associated with these proportions in the dihybrid cross (*A_ B_*, *A_ bb*, and *aa B_*) must give the same phenotype, the green progeny. Genotype *aa bb* makes up $^1/_{16}$ of the progeny in a dihybrid cross, the white progeny in this cross.

TABLE 4.5 Modified dihybrid phenotypic ratios due to gene interaction

Ratio*	Genotype $A\_ B\_$	$A\_ bb$	$aa B\_$	$aa bb$	Type of interaction	Example discussed in text
9:3:3:1	9	3	3	1	None	Seed shape and seed color in peas
9:3:4	9	3	4		Recessive epistasis	Coat color in Labrador retrievers
12:3:1	12		3	1	Dominant epistasis	Color in squash
9:7	9	7			Duplicate recessive epistasis	—
9:6:1	9	6		1	Duplicate interaction	—
15:1	15			1	Duplicate dominant epistasis	—
13:3	13		3		Dominant and recessive epistasis	—

*Each ratio is produced by a dihybrid cross ($Aa\ Bb \times Aa\ Bb$). Shaded bars represent combinations of genotypes that give the same phenotype.

In assigning genotypes to phenotypes in modified ratios, students sometimes become confused about which letters to assign to which phenotype. Suppose that we obtain the following phenotypic ratio: $^9/_{16}$ black : $^3/_{16}$ brown : $^4/_{16}$ white. Which genotype do we assign to the brown progeny, $A\_\ bb$ or $aa\ B\_$? Either answer is correct because the letters are just arbitrary symbols for the genetic information. The important thing to realize about this ratio is that the brown phenotype arises when two recessive alleles are present at one locus.

WORKED PROBLEM

A homozygous strain of yellow corn is crossed with a homozygous strain of purple corn. The F_1 are intercrossed, producing an ear of corn with 119 purple kernels and 89 yellow kernels (the progeny). What is the genotype of the yellow kernels?

Solution Strategy

What information is required in your answer to the problem?

The genotype of the yellow kernels.

What information is provided to solve the problem?

- A homozygous yellow corn plant is crossed with a homozygous purple corn plant.
- The numbers of purple and yellow progeny produced by the cross.

Solution Steps

We should first consider whether the cross between yellow and purple strains might be a monohybrid cross for a simple dominant trait, which would produce a 3:1 ratio in the F_2 ($Aa \times Aa \rightarrow {}^3/_4\ A\_$ and $^1/_4\ aa$). Under this hypothesis, we would expect 156 purple progeny and 52 yellow progeny:

Phenotype	Genotype	Observed number	Expected number
purple	$A\_$	119	$^3/_4 \times 208 = 156$
yellow	aa	89	$^1/_4 \times 208 = 52$
Total		208	

We see that the expected numbers do not closely fit the observed numbers. If we performed a chi-square test (see Chapter 3), we would obtain a calculated chi-square value of 35.08, which has a probability much less than 0.05, indicating that it is extremely unlikely that, when we expect a 3:1 ratio, we would obtain 119 purple progeny and 89 yellow progeny. Therefore, we can reject the hypothesis that these results were produced by a monohybrid cross.

Another possible hypothesis is that the observed F_2 progeny are in a 1:1 ratio. However, we learned in Chapter 3 that a 1:1 ratio is produced by a cross between a heterozygote and a homozygote ($Aa \times aa$), and in this cross, both original parental strains were homozygous. Furthermore, a chi-square test comparing the observed numbers with an expected 1:1 ratio yields a calculated chi-square value of 4.32, which has a probability of less than 0.05.

Next, we should look to see if the results can be explained by a dihybrid cross ($Aa\ Bb \times Aa\ Bb$). A dihybrid cross results in phenotypic proportions that are in sixteenths. We can apply the formula given earlier in the chapter to determine the number of sixteenths for each phenotype:

$$x = \frac{\text{number of progeny with a phenotype} \times 16}{\text{total number of progeny}}$$

$$x_{(\text{purple})} = \frac{119 \times 16}{208} = 9.15$$

$$x_{(\text{yellow})} = \frac{89 \times 16}{208} = 6.85$$

Thus, purple and yellow appear in an approximate ratio of 9:7. We can test this hypothesis with a chi-square test:

Phenotype	Genotype	Observed number	Expected number
purple	?	119	$^9/_{16} \times 208 = 117$
yellow	?	89	$^7/_{16} \times 208 = 91$
Total		208	

$$\chi^2 = \Sigma\frac{(\text{observed} - \text{expected})^2}{\text{expected}} = \frac{(119 - 117)^2}{117} + \frac{(89 - 91)^2}{91}$$

$$= 0.034 + 0.044 = 0.078$$

$$\text{Degree of freedom} = n - 1 = 2 - 1 = 1$$
$$P > 0.05$$

The probability associated with the chi-square value is greater than 0.05, indicating that there is a good fit between the observed results and a 9:7 ratio.

We now need to determine how a dihybrid cross can produce a 9:7 ratio and what genotypes correspond to the two phenotypes. A dihybrid cross without epistasis produces a 9:3:3:1 ratio:

$$Aa\ Bb \times Aa\ Bb$$
$$\downarrow$$
$$A\_\ B\_\ {}^{9}/_{16}$$
$$A\_\ bb\ {}^{3}/_{16}$$
$$aa\ B\_\ {}^{3}/_{16}$$
$$aa\ bb\ {}^{1}/_{16}$$

Because ${}^{9}/_{16}$ of the progeny from the corn cross are purple, purple must be produced by genotypes *A_ B_*; in other words, individual kernels that have at least one dominant allele at the first locus and at least one dominant allele at the second locus are purple. The proportions of all the other genotypes (*A_ bb*, *aa B_*, and *aa bb*) sum to ${}^{7}/_{16}$, which is the proportion of the progeny in the corn cross that are yellow, and so any individual kernel that does not have a dominant allele at both the first and the second locus is yellow.

▶ Now test your understanding of epistasis by working Problem 34 at the end of this chapter.

Complementation: Determining Whether Mutations Are at the Same Locus or at Different Loci

How do we know whether different mutations that affect a characteristic occur at the same locus (are allelic) or occur at different loci? In fruit flies, for example, *white* is an X-linked recessive mutation that produces white eyes instead of the red eyes found in wild-type flies; *apricot* is an X-linked recessive mutation that produces light orange eyes. Do the *white* and *apricot* mutations occur at the same locus or at different loci? We can use the complementation test to answer this question.

To carry out a **complementation test** on recessive mutations (*a* and *b*), parents that are homozygous for different mutations are crossed, producing offspring that are heterozygous. If the mutations are allelic (occur at the same locus), then the heterozygous offspring have only mutant alleles (*a b*) and exhibit a mutant phenotype:

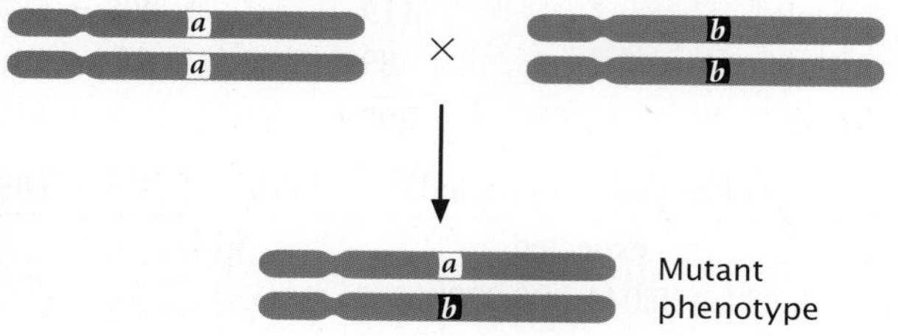

If, on the other hand, the mutations occur at different loci, each of the homozygous parents possesses wild-type genes at the other locus ($aa\ b^+b^+$ and $a^+a^+\ bb$); so the heterozygous offspring inherit a mutant allele and a wild-type allele at each locus. In this case, the presence of a wild-type allele complements the mutation at each locus, and the heterozygous offspring have the wild-type phenotype:

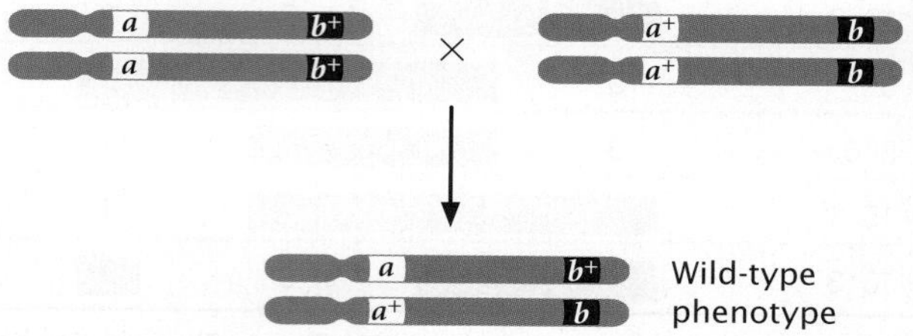

Complementation has taken place if an individual possessing two recessive mutations has a wild-type phenotype, indicating that the mutations are nonallelic genes. There is a lack of complementation when two recessive mutations occur at the same locus, producing a mutant phenotype.

When the complementation test is applied to *white* and *apricot* mutations, all the heterozygous offspring have light-colored eyes, demonstrating that white eyes and apricot eyes are produced by mutations that occur at the same locus and are allelic.

THINK-PAIR-SHARE Question 10

CONCEPTS

A complementation test is used to determine whether two mutations occur at the same locus (are allelic) or at different loci.

4.5 Sex Influences the Inheritance and Expression of Genes in a Variety of Ways

In Section 4.2, we considered characteristics encoded by genes located on the sex chromosomes (sex-linked traits) and how their inheritance differs from the inheritance of traits encoded by autosomal genes. X-linked traits, for example, are passed from father to daughter, but never from father to son, and Y-linked traits are passed from father to all sons. Here, we will examine additional influences of sex, including the effect of an individual's sex on the expression of genes on autosomes, on characteristics determined by genes located in the cytoplasm, and on characteristics for which the genotype of only the maternal parent determines the phenotype of the offspring. Finally, we'll look at situations in which the expression of genes on autosomes is affected by the sex of the parent from whom they are inherited.

Sex-Influenced and Sex-Limited Characteristics

Sex-influenced characteristics are determined by autosomal genes and are inherited according to Mendel's principles, but they are expressed differently in males and females. In this case, a particular trait is more readily expressed in one sex; in other words, the trait has higher penetrance in one of the sexes.

For example, the presence of a beard on some goats is determined by an autosomal gene (B^b) that is dominant in males and recessive in females. In males, a single bearded allele is required for the expression of this trait: both the homozygote (B^bB^b) and the heterozygote (B^bB^+) have beards, whereas the B^+B^+ male is beardless. In contrast, females require two bearded alleles in order for this trait to be expressed: the homozygote B^bB^b has a beard, whereas the heterozygote (B^bB^+) and the other homozygote (B^+B^+) are beardless. The key to understanding the expression of the bearded gene is to look at the heterozygote. In males (for which the presence of a beard is dominant), the heterozygous genotype produces a beard, but in females (for which the absence of a beard is dominant), the heterozygous genotype produces a goat without a beard.

An extreme form of sex-influenced characteristic, a **sex-limited characteristic**, is encoded by autosomal genes that are expressed in only one sex; the trait has zero penetrance in the other sex. In domestic chickens, for example, some males display a plumage pattern called cock feathering. Other males and all females display a pattern called hen feathering. Cock feathering is an autosomal recessive trait that is limited to males. Because the trait is autosomal, the genotypes of males and females are the same, but the phenotypes produced by these genotypes are different in males and females. Only homozygous recessive males have the cock-feathering phenotype.

THINK-PAIR-SHARE Questions 11, 12, 13

CONCEPTS

Sex-influenced characteristics are encoded by autosomal genes that are more readily expressed in one sex. Sex-limited characteristics are encoded by autosomal genes whose expression is limited to one sex.

✓ CONCEPT CHECK 9

How do sex-influenced and sex-limited characteristics differ from sex-linked characteristics?

Cytoplasmic Inheritance

Mendel's principles of segregation and independent assortment are based on the assumption that genes are located on chromosomes in the nucleus of the cell. For most genetic characteristics, this assumption is valid, and Mendel's principles allow us to predict the types of offspring that will be produced in a genetic cross. Not all the genetic material of a cell is found in the nucleus, however. Some characteristics are encoded by genes located in the cytoplasm, and these characteristics exhibit **cytoplasmic inheritance**.

A few organelles, notably chloroplasts and mitochondria, contain DNA. The human mitochondrial genome contains 16,569 nucleotides of DNA, encoding 37 genes. Compared with the nuclear genome, which contains some 3 billion nucleotides encoding some 20,000 genes, the size of the mitochondrial genome is very small; nevertheless, mitochondrial and chloroplast genes encode some important characteristics.

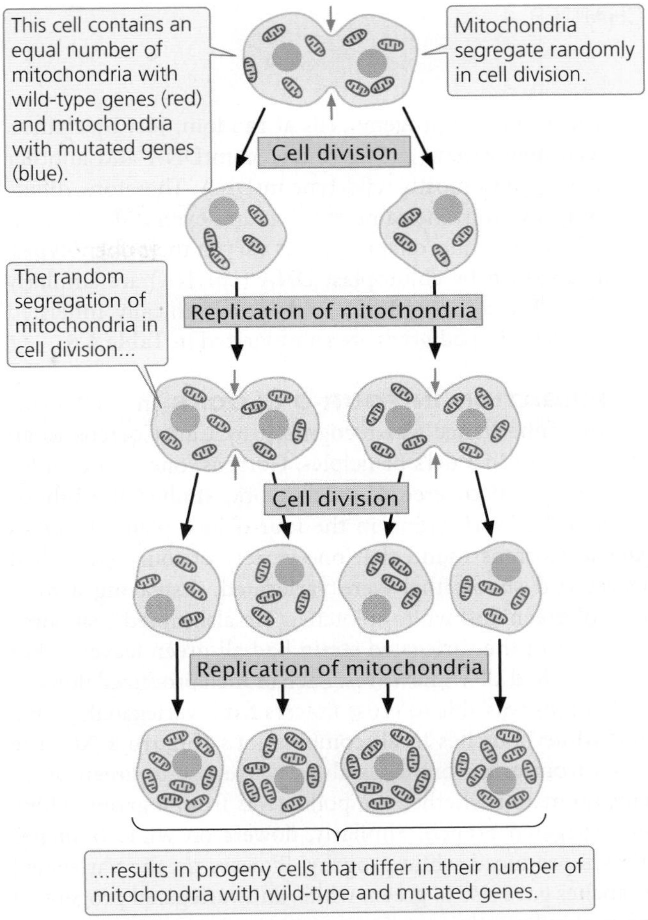

4.19 Cytoplasmically inherited characteristics frequently exhibit extensive phenotypic variation because different cells and different individual offspring may contain different proportions of cytoplasmic genes.

Cytoplasmic inheritance differs from the inheritance of characteristics encoded by nuclear genes in several important respects. A zygote inherits nuclear genes from both parents, but typically, all its cytoplasmic organelles, and thus all its cytoplasmic genes, come from only one of the gametes, usually the egg. A sperm from the male parent generally contributes only a set of nuclear genes. Thus, most cytoplasmically inherited traits are present in both males and females and are passed from mother to offspring, never from father to offspring. Reciprocal crosses, therefore, give different results when cytoplasmic genes encode a trait. In a few organisms, however, cytoplasmic genes are inherited from the male parent only or from both parents.

Cytoplasmically inherited characteristics frequently exhibit extensive phenotypic variation because no mechanism analogous to mitosis or meiosis ensures that cytoplasmic genes are evenly distributed during cell division. Thus, different cells and different individual offspring will contain various proportions of cytoplasmic genes.

Consider mitochondrial genes. Most cells contain thousands of mitochondria, and each mitochondrion contains from 2 to 10 copies of mitochondrial DNA (mtDNA). Suppose that half of the mitochondria in a cell contain a normal wild-type copy of mtDNA and the other half contain a mutated copy (**Figure 4.19**). During cell division, the mitochondria segregate into the progeny cells at random. Just by chance, one cell may receive mostly mutated mtDNA and another cell may receive mostly wild-type mtDNA. Therefore, different progeny from the same mother, and even different cells within an individual offspring, may vary in their phenotypes. Traits encoded by chloroplast DNA (cpDNA) are similarly variable. The characteristics that cytoplasmically inherited traits exhibit in pedigrees are summarized in **Table 4.6.**

VARIEGATION IN FOUR-O'CLOCKS In 1909, cytoplasmic inheritance was recognized by Carl Correns as an exception to Mendel's principles. Correns, one of the biologists who rediscovered Mendel's work, studied the inheritance of leaf variegation in the four-o'clock plant, *Mirabilis jalapa*. Correns found that one variety of four-o'clock had leaves and shoots that were variegated, displaying a mixture of green and white splotches. He also noted that some branches of the variegated strain had all-green leaves; other branches had all-white leaves. Each branch produced flowers, so Correns was able to cross flowers from variegated, green, and white branches in all combinations (**Figure 4.20**). The seeds from green branches always gave rise to green progeny, no matter whether the pollen was from a green, white, or variegated branch. Similarly, flowers on white branches always produced white progeny. Flowers on the variegated branches gave rise to green, white, and variegated progeny in no particular ratio.

Correns's crosses demonstrated the cytoplasmic inheritance of variegation in four-o'clocks. The phenotypes of the offspring were determined entirely by the maternal parent, never by the paternal parent (the source of the pollen). Furthermore, the production of all three phenotypes by flowers on variegated branches is consistent with cytoplasmic inheritance. The white color in these plants is caused by a defective gene in cpDNA, which results in a failure to produce the green pigment chlorophyll. Cells from green branches contain normal chloroplasts only, cells from white branches contain abnormal chloroplasts only, and cells from variegated branches contain a mixture of normal and abnormal chloroplasts. In the flowers from variegated branches, the random segregation of chloroplasts in the course of oogenesis produces some egg cells with normal cpDNA, which develop into green progeny; other egg cells with only abnormal cpDNA, which develop into white progeny; and, finally, still other egg cells with a mixture of normal and abnormal cpDNA, which develop into variegated progeny.

Experiment

Question: How is stem and leaf color inherited in the four-o'clock plant?

Pollen plant (♂)

Seed plant (♀)

Results

Conclusion: The phenotype of the progeny is determined by the phenotype of the branch from which the seed originated, not from the branch on which the pollen originated. Stem and leaf color exhibit cytoplasmic inheritance.

MITOCHONDRIAL DISEASES A number of human diseases (mostly rare) that exhibit cytoplasmic inheritance have been identified. These disorders arise from mutations

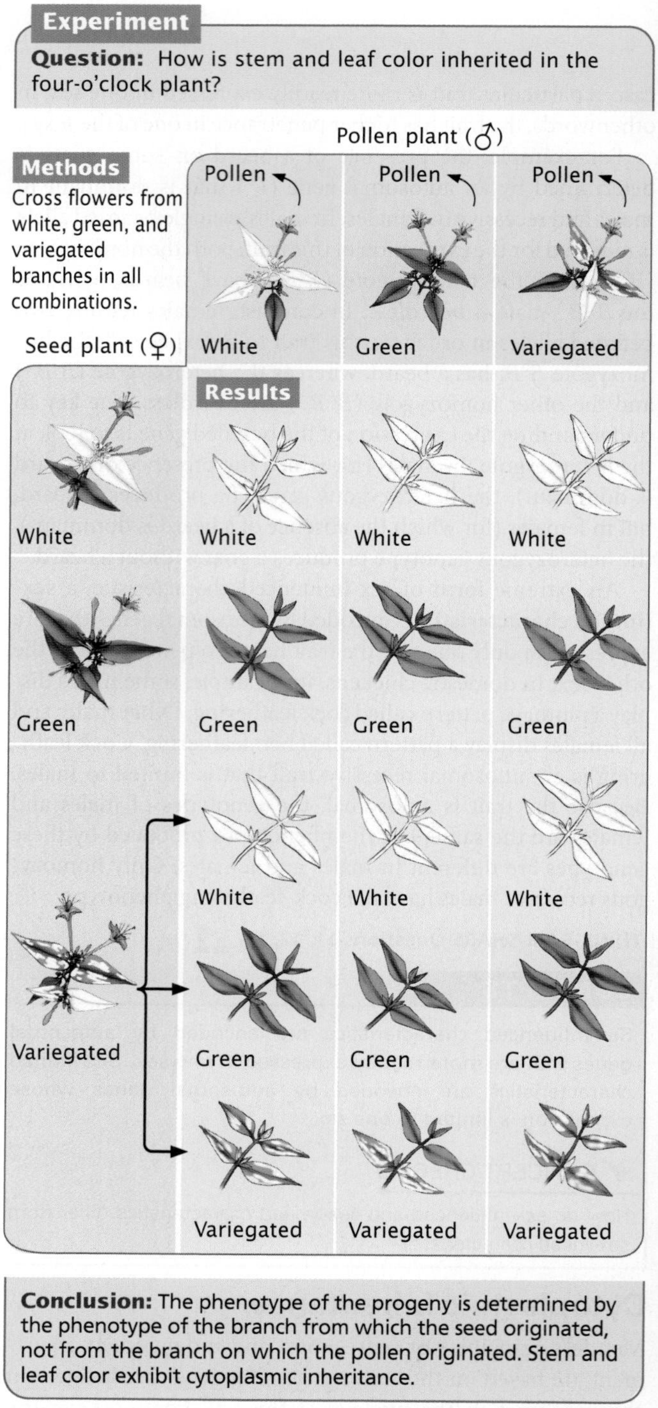

4.20 Crosses for leaf type in four-o'clocks illustrate cytoplasmic inheritance.

TABLE 4.6 Characteristics of cytoplasmically inherited traits
1. Present in males and females.
2. Usually inherited from one parent, typically the maternal parent.
3. Reciprocal crosses give different results.
4. Exhibit extensive phenotypic variation, even within a single family.

in mtDNA, most of which occur in genes encoding components of the electron-transport chain, which generates most of the ATP (adenosine triphosphate) in aerobic cellular respiration. One such disease is Leber hereditary optic neuropathy (LHON). Patients who have this disorder experience rapid loss of vision in both eyes, resulting from the death of cells in the optic nerve. This loss of vision typically occurs in early adulthood (usually between the ages of 20 and 24), but it can occur any time after adolescence. There is much clinical variation in the severity of the disease, even within the same family. Leber hereditary optic neuropathy exhibits cytoplasmic inheritance: the trait is always passed from the mother to all children, sons and daughters alike.

Genetic Maternal Effects

A genetic phenomenon that is sometimes confused with cytoplasmic inheritance is **genetic maternal effect**, in which the phenotype of the offspring is determined by the genotype of the mother. In cytoplasmic inheritance, the genes for a characteristic are inherited from only one parent, usually the mother. In genetic maternal effects, the genes are inherited from both parents, but the offspring's phenotype is determined not by its own genotype, but by the genotype of its mother.

Genetic maternal effects frequently arise when substances present in the cytoplasm of an egg (encoded by the mother's nuclear genes) are pivotal in early development. An excellent example is the shell coiling of the snail *Lymnaea peregra* (**Figure 4.21**), described in the introduction to this chapter. In *Lymnaea peregra*, the direction of coiling is determined by a pair of alleles: the allele for dextral (right-handed) coiling (s^+) is dominant over the allele for sinistral (left-handed) coiling (s). However, the direction of coiling is determined not by a snail's own genotype, but by the genotype of its *mother*. The direction of coiling is affected by the way in which the egg divides soon after fertilization, which in turn is determined by a substance produced by the mother and passed to the offspring in the cytoplasm of the egg.

If a male homozygous for dextral alleles (s^+s^+) is crossed with a female homozygous for sinistral alleles (ss), all the F_1 are heterozygous (s^+s) and have a sinistral shell because the genotype of the mother (ss) encodes sinistral coiling (see Figure 4.21). If these F_1 snails self-fertilize, the genotypic ratio of the F_2 is 1 s^+s^+ : 2 s^+s : 1 ss. Notice, however, that the phenotype of all the F_2 snails is dextral, regardless of their genotypes. The F_2 offspring are dextral because the genotype of their mother (s^+s), which encodes a right-coiling shell, determines their phenotype. With genetic maternal effects, the phenotype of the progeny is not necessarily the same as the phenotype of the mother because the progeny's phenotype is determined by the mother's *genotype*, not her phenotype. Neither the male parent's nor the offspring's own genotype has any role in the offspring's phenotype. However, a male does influence the phenotype of the F_2 generation: by contributing to the genotypes of his daughters, he affects the phenotypes of their offspring. Genes that exhibit genetic maternal effects are therefore transmitted through males to future generations. In contrast, genes that exhibit cytoplasmic inheritance are transmitted through only one of the sexes (usually the female). ▶TRY PROBLEM 38

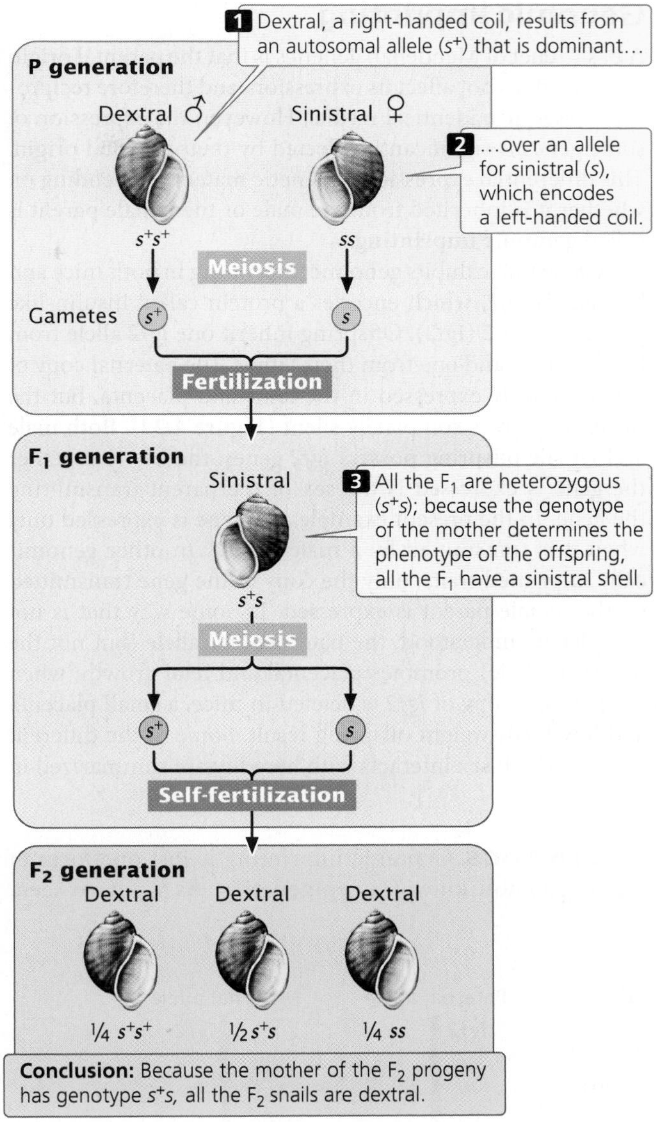

4.21 In genetic maternal effects, the genotype of the maternal parent determines the phenotype of the offspring. The shell coiling of a snail is a trait that exhibits a genetic maternal effect.

CONCEPTS

Characteristics exhibiting cytoplasmic inheritance are encoded by genes in the cytoplasm and are usually inherited from one parent, most commonly the mother. In genetic maternal effects, the genotype of the mother determines the phenotype of the offspring.

Genomic Imprinting

A basic tenet of Mendelian genetics is that the parental origin of a gene does not affect its expression, and therefore reciprocal crosses give identical results. However, the expression of some genes *is* significantly affected by their parental origin. This differential expression of genetic material depending on whether it is inherited from the male or the female parent is called **genomic imprinting**.

A gene that exhibits genomic imprinting in both mice and humans is *Igf2*, which encodes a protein called insulin-like growth factor 2 (Igf2). Offspring inherit one *Igf2* allele from their mother and one from their father. The paternal copy of *Igf2* is actively expressed in the fetus and placenta, but the maternal copy is completely silent (**Figure 4.22**). Both male and female offspring possess *Igf2* genes; the key to whether the gene is expressed is the sex of the parent transmitting the gene. In the present example, the gene is expressed only when it is transmitted by a male parent. In other genomically imprinted traits, only the copy of the gene transmitted by the female parent is expressed. In some way that is not completely understood, the paternal *Igf2* allele (but not the maternal allele) promotes placental and fetal growth; when the paternal copy of *Igf2* is deleted in mice, a small placenta and low-birth-weight offspring result. Some of the different ways in which sex interacts with heredity are summarized in **Table 4.7**.

TABLE 4.7 Influences of sex on heredity

Genetic phenomenon	Phenotype determined by
Sex-linked characteristic	Genes located on the sex chromosome
Sex-influenced characteristic	Genes on autosomes that are more readily expressed in one sex
Sex-limited characteristic	Autosomal genes whose expression is limited to one sex
Genetic maternal effects	Nuclear genotype of the maternal parent
Cytoplasmic inheritance	Cytoplasmic genes, which are usually inherited entirely from only one parent
Genomic imprinting	Genes whose expression is affected by the sex of the transmitting parent

EPIGENETICS Genomic imprinting is just one form of a phenomenon known as **epigenetics**. As we have seen, most traits are encoded by genetic information that resides in the sequence of nucleotide bases of the DNA—the so-called genetic code, which will be considered in more detail in Chapter 11. Some traits, however, may be caused

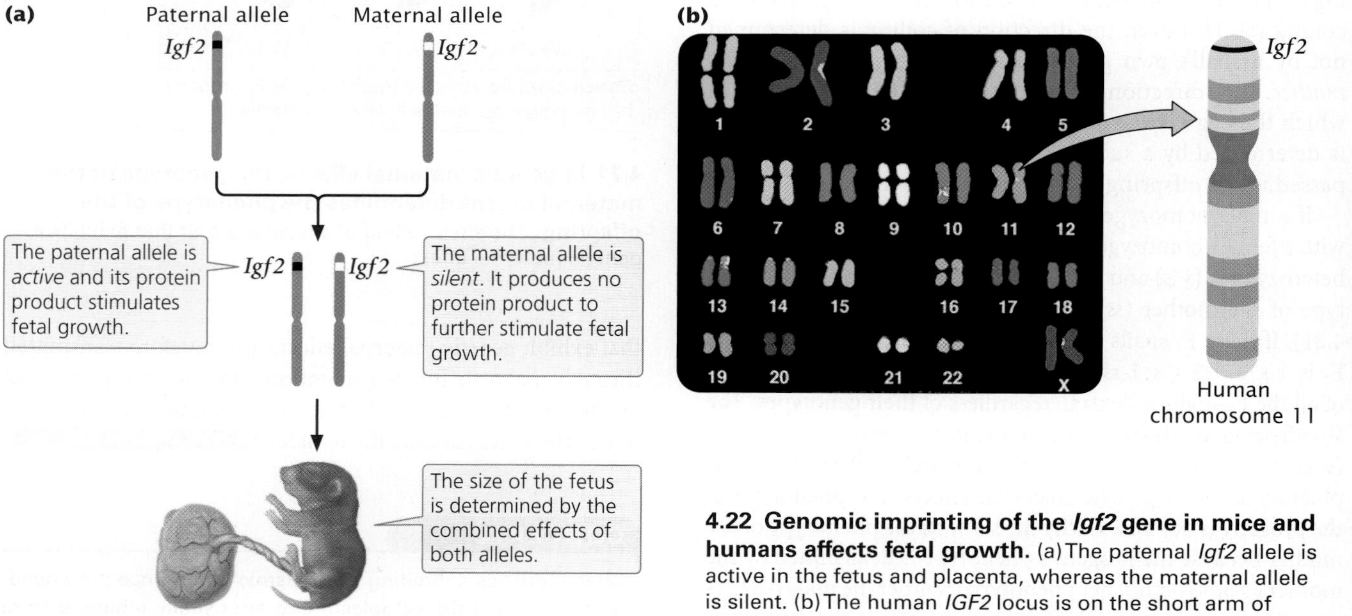

4.22 Genomic imprinting of the *Igf2* gene in mice and humans affects fetal growth. (a) The paternal *Igf2* allele is active in the fetus and placenta, whereas the maternal allele is silent. (b) The human *IGF2* locus is on the short arm of chromosome 11; the *Igf2* locus in mice is on chromosome 7. [Courtesy of Dr. Thomas Ried and Dr. Evelin Schrock.]

by alterations to the DNA that do not affect the DNA base sequence, but affect the way in which the DNA sequences are expressed. An example of this type of alteration is the addition of methyl groups to some of the DNA bases (DNA methylation). These changes are often stable and heritable in the sense that they are passed from one cell to another.

In genomic imprinting, whether the gene passes through the egg or the sperm determines how much methylation of the DNA takes place. The pattern of methylation on a gene is copied when the DNA is replicated and therefore remains on the gene as it is passed from cell to cell through mitosis. However, the methylation may be modified or removed when the DNA passes through a gamete, so a gene that is methylated in sperm may be unmethylated when it is eventually passed down to a daughter's egg. Ultimately, the amount of methylation determines whether the gene is expressed in the offspring.

These types of reversible changes to DNA that influence the expression of traits are termed epigenetic marks. The inactivation of one of the X chromosomes in female mammals (discussed earlier in this chapter) is another type of epigenetic change. We will consider epigenetic changes in more detail in Chapter 12.

CONCEPTS

In genomic imprinting, the expression of a gene is influenced by the sex of the parent that transmits the gene to the offspring. Epigenetic marks are reversible changes to DNA that do not alter the base sequence, but may affect how a gene is expressed.

4.6 The Expression of a Genotype May Be Influenced by Environmental Effects

In Chapter 3, we learned that each phenotype is the result of a genotype that develops within a particular environment; each genotype may produce several different phenotypes, depending on the environmental conditions in which development takes place. For example, a fruit fly that is homozygous for the vestigial mutation (*vg vg*) develops reduced wings when raised at a temperature below 29°C, but the same genotype develops much longer wings when raised at 31°C.

For most of the characteristics we have discussed so far, the effect of the environment on the phenotype has been slight. Mendel's peas with genotype *yy*, for example, developed green seeds regardless of the environment in which they were raised. Similarly, persons with genotype I^AI^A have the A antigen on their red blood cells regardless of their diet, socioeconomic status, or family environment. For other phenotypes, however, environmental effects play a more important role.

Environmental Effects on the Phenotype

The phenotypic expression of some genotypes critically depends on the presence of a specific environment. For example, the *himalayan* allele in rabbits produces dark fur at the extremities of the body—on the nose, ears, and feet (**Figure 4.23**). The dark pigment develops, however, only when the rabbit is reared at a temperature of 25°C or less; if a Himalayan rabbit is reared at 30°C, no dark patches develop. The expression of the *himalayan* allele is thus temperature dependent; an enzyme necessary for the production of dark pigment is inactivated at higher temperatures. The pigment is restricted to the nose, feet, and ears of a Himalayan rabbit because the animal's core body temperature is normally above 25°C and the enzyme is functional only in the cells of the relatively cool extremities. The *himalayan* allele is an example of a **temperature-sensitive allele**, an allele whose product is functional only at certain temperatures.

Environmental factors also play an important role in the expression of a number of human genetic diseases. Glucose-6-phosphate dehydrogenase is an enzyme that helps to supply energy to the cell. In humans, there are a number of genetic

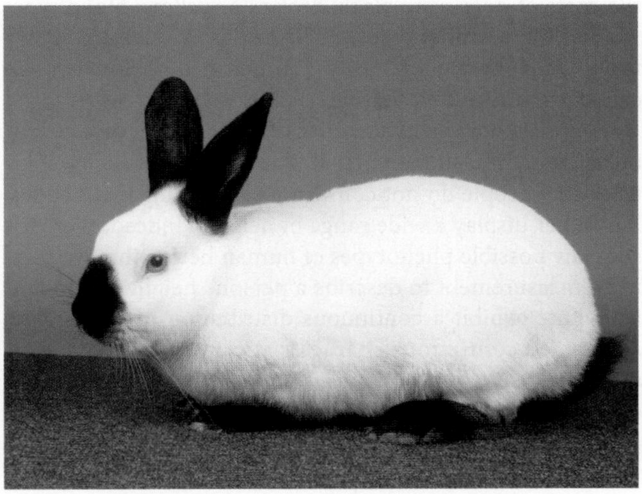

Reared at 25°C or lower

4.23 The expression of the *himalayan* allele is temperature dependent. This rabbit was reared below 25°C. Its pigment is restricted to the extremities where the body temperature falls below 25°C and the enzyme that produces pigment is functional. [Petra Wegner/Alamy.]

variants of glucose-6-phosphate dehydrogenase, some of which destroy red blood cells when the body is stressed by infection or by the ingestion of certain drugs or foods. The symptoms of the genetic disease, called glucose-6-phosphate dehydrogenase deficiency, appear only in the presence of these specific environmental factors.

These examples illustrate the point that genes and their products do not act in isolation; rather, they frequently interact with environmental factors. Occasionally, environmental factors alone can produce a phenotype that is the same as the phenotype produced by a genotype; such a phenotype is called a **phenocopy**. In fruit flies, for example, the autosomal recessive mutation *eyeless* produces greatly reduced eyes. The eyeless phenotype can also be produced by exposing the larvae of normal flies to sodium metaborate.

THINK-PAIR-SHARE Question 14

CONCEPTS

The expression of many genes is modified by the environment. A phenocopy is a trait produced by environmental effects that mimics the phenotype produced by a genotype.

The Inheritance of Continuous Characteristics

So far, we've dealt primarily with characteristics that have only a few distinct phenotypes. In Mendel's peas, for example, the seeds were either smooth or wrinkled, yellow or green; the coats of dogs were black, brown, or yellow; blood types were of four distinct types, A, B, AB, or O. Such characteristics, which have a few easily distinguished phenotypes, are called **discontinuous characteristics**.

However, many characteristics do not exhibit discontinuous phenotypes. Human height is an example of such a characteristic; people do not come in just a few distinct heights, but rather display a wide range of heights. Indeed, there are so many possible phenotypes of human height that we must use a measurement to describe a person's height. Characteristics that exhibit a continuous distribution of phenotypes are termed **continuous characteristics**. Because such characteristics have many possible phenotypes and must be described in quantitative terms, continuous characteristics are also called **quantitative characteristics**.

Continuous characteristics frequently arise because genes at many loci interact to produce the phenotypes. When a single locus with two alleles encodes a characteristic, there are three genotypes possible: *AA*, *Aa*, and *aa*. With two loci, each with two alleles, there are $3^2 = 9$ genotypes possible. The number of genotypes encoding a characteristic is $3n$, where n equals the number of loci with two alleles that influence the characteristic. For example, when a characteristic is determined by eight loci, each with two alleles, there are $3^8 = 6561$ different genotypes possible for this characteristic. If each genotype produces a different phenotype, many phenotypes will be possible. The slight differences between the phenotypes will be indistinguishable, and the characteristic will appear continuous. Characteristics encoded by genes at many loci are called **polygenic characteristics**.

The converse of polygeny is **pleiotropy**, in which one gene affects multiple characteristics. Many genes exhibit pleiotropy. Phenylketonuria (PKU) is a genetic disease that results from a recessive allele; persons homozygous for this allele, if untreated, exhibit intellectual disability, blue eyes, and light skin color. The lethal allele that causes yellow coat color in mice is also pleiotropic. In addition to its lethality and its effect on coat color, the gene causes a diabetes-like condition, obesity, and an increased propensity to develop tumors.

Frequently, the phenotypes of continuous characteristics are also influenced by environmental factors. In this situation, each genotype is capable of producing a range of phenotypes, and the particular phenotype that results depends on both the genotype and the environmental conditions in which the genotype develops. For example, there may be only three genotypes at a single locus that encode a characteristic, but because each genotype produces a range of phenotypes associated with different environments, the phenotype of the characteristic exhibits a continuous distribution. Many continuous characteristics are both polygenic and influenced by environmental factors; such characteristics are called **multifactorial characteristics** because many factors help determine the phenotype.

The inheritance of continuous characteristics may appear to be complex, but the alleles at each locus follow Mendel's principles and are inherited in the same way as alleles encoding simple, discontinuous characteristics. However, because many genes participate, because environmental factors influence the phenotype, and because the phenotypes do not sort out into a few distinct types, we cannot observe the distinct ratios that have allowed us to interpret the genetic basis of discontinuous characteristics. To analyze continuous characteristics, we must employ special statistical tools, which will be discussed in Chapter 17.

CONCEPTS

Discontinuous characteristics exhibit a few distinct phenotypes; continuous characteristics exhibit a range of phenotypes. A continuous characteristic is frequently produced when genes at many loci and environmental factors combine to determine a phenotype.

✓ CONCEPT CHECK 10

What is the difference between polygeny and pleiotropy?

CONCEPTS SUMMARY

- Sexual reproduction is the production of offspring that are genetically distinct from their parents. Most organisms have two sexual phenotypes—males and females. Males produce small gametes; females produce large gametes.
- The mechanism by which sex is specified is termed sex determination. Sex may be determined by differences in specific chromosomes, genotypes, or environment.
- Sex chromosomes differ in number and appearance between males and females. The homogametic sex produces gametes that are all identical with regard to sex chromosomes; the heterogametic sex produces gametes that differ in their sex-chromosome composition.
- In the XX-XO system of sex determination, females possess two X chromosomes and males possess a single X chromosome. In the XX-XY system, females possess two X chromosomes and males possess a single X and a single Y chromosome. In the ZZ-ZW system, males possess two Z chromosomes and females possess a Z and a W chromosome.
- In some organisms, environmental factors determine sex.
- In *Drosophila melanogaster*, sex is predicted by the X : A ratio, but is primarily determined by genes on the X chromosome.
- In humans, sex is ultimately determined by the presence or absence of the *SRY* gene located on the Y chromosome.
- Sex-linked characteristics are determined by genes on the sex chromosomes. X-linked characteristics are encoded by genes on the X chromosome, and Y-linked characteristics are encoded by genes on the Y chromosome.
- A female inherits X-linked alleles from both parents; a male inherits X-linked alleles from his female parent only.
- The fruit fly *Drosophila melanogaster* has a number of characteristics that make it an ideal model organism for genetic studies, including a short generation time, large numbers of progeny, small size, ease of rearing, and a small genome.
- In placental mammals, one of the two X chromosomes in females normally becomes inactivated. Which X chromosome is inactivated is random and varies from cell to cell.
- Y-linked characteristics are found only in males and are passed from a father to all of his sons.
- Dominance is an interaction between genes at the same locus (allelic genes). Dominance is complete when a heterozygote has the same phenotype as one homozygote. It is incomplete when the heterozygote has a phenotype intermediate between those of two parental homozygotes. In codominance, the heterozygote exhibits traits of both parental homozygotes.
- Penetrance is the percentage of individuals having a particular genotype that exhibit the expected phenotype. Expressivity is the degree to which a character is expressed.
- Lethal alleles cause the death of an individual possessing them, usually at an early stage of development, and may alter phenotypic ratios among the offspring of a cross.
- The existence of multiple alleles refers to the presence of more than two alleles at a locus within a group of individuals. The presence of multiple alleles increases the number of genotypes and phenotypes that are possible.
- Gene interaction refers to the interaction between genes at different loci to produce a single phenotype. An epistatic gene at one locus suppresses or masks the expression of hypostatic genes at other loci. Gene interaction frequently produces phenotypic ratios that are modifications of dihybrid ratios.
- Sex-influenced characteristics are encoded by autosomal genes that are expressed more readily in one sex. Sex-limited characteristics are encoded by autosomal genes that are expressed in only one sex.
- In cytoplasmic inheritance, the genes for the characteristic are found in the organelles and are usually inherited from a single (typically maternal) parent. In genetic maternal effects, an offspring inherits genes from both parents, but the nuclear genes of the mother determine the offspring's phenotype.
- Genomic imprinting refers to characteristics encoded by genes whose expression is affected by the sex of the parent transmitting the genes. Epigenetic effects such as genomic imprinting are caused by alterations to DNA—such as DNA methylation—that do not affect the DNA base sequence.
- Phenotypes are often modified by environmental effects. A phenocopy is a phenotype produced by an environmental effect that mimics a phenotype produced by a genotype.
- Continuous characteristics are those that exhibit a wide range of phenotypes; they are frequently produced by the combined effects of many genes and the environment.

IMPORTANT TERMS

sex (p. 76)
sex determination (p. 77)
sex chromosome (p. 77)
autosome (p. 77)
heterogametic sex (p. 77)
homogametic sex (p. 77)
pseudoautosomal region (p. 78)
genic sex determination (p. 78)
Turner syndrome (p. 80)
Klinefelter syndrome (p. 80)
triple-X syndrome (p. 80)

sex-determining region Y (*SRY*) gene (p. 81)
sex-linked characteristic (p. 81)
X-linked characteristic (p. 81)
Y-linked characteristic (p. 81)
hemizygosity (p. 82)
dosage compensation (p. 86)
Barr body (p. 86)
Lyon hypothesis (p. 86)
complete dominance (p. 89)
incomplete dominance (p. 89)
codominance (p. 89)
incomplete penetrance (p. 90)
penetrance (p. 90)
expressivity (p. 91)
lethal allele (p. 91)
multiple alleles (p. 92)
compound heterozygote (p. 93)
gene interaction (p. 93)
epistasis (p. 94)
epistatic gene (p. 94)
hypostatic gene (p. 94)
complementation test (p. 98)
complementation (p. 98)
sex-influenced characteristic (p. 98)
sex-limited characteristic (p. 99)
cytoplasmic inheritance (p. 99)
genetic maternal effect (p. 101)
genomic imprinting (p. 102)
epigenetics (p. 102)
temperature-sensitive allele (p. 103)
phenocopy (p. 104)
discontinuous characteristic (p. 104)
continuous characteristic (p. 104)
quantitative characteristic (p. 104)
polygenic characteristic (p. 104)
pleiotropy (p. 104)
multifactorial characteristic (p. 104)

ANSWERS TO CONCEPT CHECKS

1. Meiosis

2. In chromosomal sex determination, males and females have chromosomes that are distinguishable. In genic sex determination, sex is determined by genes, but the chromosomes of males and females are indistinguishable. In environmental sex determination, sex is determined by environmental effects.

3. a

4. All of their male offspring will have hemophilia, and none of their female offspring will have hemophilia, so the overall probability of hemophilia in their offspring is $\frac{1}{2}$.

5. Two Barr bodies. Each Barr body is an inactive X chromosome.

6. With complete dominance, the heterozygote expresses the same phenotype as that of one of the homozygotes. With incomplete dominance, the heterozygote has a phenotype that is intermediate between the two homozygotes. With codominance, the heterozygote has a phenotype that simultaneously expresses the phenotypes of both homozygotes.

7. d

8. d

9. Both sex-influenced and sex-limited characteristics are encoded by autosomal genes whose expression is affected by the sex of the individual who possesses the gene. Sex-linked characteristics are encoded by genes on the sex chromosomes.

10. Polygeny refers to the influence of multiple genes on the expression of a single characteristic. Pleiotropy refers to the effect of a single gene on the expression of multiple characteristics.

WORKED PROBLEMS

Problem 1

In *Drosophila melanogaster*, forked bristles are caused by an allele (X^f) that is X linked and recessive to an allele for normal bristles (X^+). Brown eyes are caused by an allele (b) that is autosomal and recessive to an allele for red eyes (b^+). A female fly that is homozygous for normal bristles and red eyes mates with a male fly that has forked bristles and brown eyes. The F_1 are intercrossed to produce the F_2. What will the phenotypes and proportions of the F_2 flies from this cross be?

Solution Strategy

What information is required in your answer to the problem?

Phenotypes and proportions of the F_2 flies.

What information is provided to solve the problem?

- Forked bristles are X-linked recessive.
- Brown eyes are autosomal recessive.
- Phenotypes of the parents of the cross.
- The F_1 are intercrossed to produce the F_2.

For help with this problem, review:

X-linked Color Blindness in Humans in Section 4.2.

Section 3.3 in Chapter 3.

Solution Steps

This problem is best worked by breaking the cross down into two separate crosses, one for the X-linked genes that determine the type of bristles and one for the autosomal genes that determine eye color.

For problems … multiple loci, … cross down … separate

Let's begin with the autosomal characteristics. A female fly that is homozygous for red eyes (b^+b^+) is crossed with a male with brown eyes. Because brown eyes are recessive, the male fly must be homozygous for the brown-eye allele (*bb*). All the offspring of this cross will be heterozygous (b^+b) and will have red eyes:

P	b^+b^+ Red eyes	×	bb Brown eyes
Gametes	(b^+)		(b)
F_1	b^+b Red eyes		

The F_1 are then intercrossed to produce the F_2. Whenever two individual organisms heterozygous for an autosomal recessive characteristic are crossed, $^3/_4$ of the offspring will have the dominant trait and $^1/_4$ will have the recessive trait; thus, $^3/_4$ of the F_2 flies will have red eyes and $^1/_4$ will have brown eyes:

F_1	b^+b Red eyes	×	b^+b Red eyes
Gametes	(b^+) (b)		(b^+) (b)
F_2	$^1/_4$ b^+b^+ red		
	$^1/_2$ b^+b red		
	$^1/_4$ bb brown		
	$^3/_4$ red, $^1/_4$ brown		

Next, we work out the results for the X-linked characteristic. A female that is homozygous for normal bristles (X^+X^+) is crossed with a male that has forked bristles (X^fY). The female F_1 from this cross are heterozygous (X^+X^f), receiving an X chromosome with a normal-bristle allele from their mother (X^+) and an X chromosome with a forked-bristle allele (X^f) from their father. The male F_1 are hemizygous (X^+Y), receiving an X chromosome with a normal-bristle allele from their mother (X^+) and a Y chromosome from their father:

…: Females … X-linked … ut males only … ingle X-linked

P	X^+X^+ Normal bristles	×	X^fY Forked bristles
Gametes	(X^+)		(X^f) (Y)
F_1	$^1/_2$ X^+X^f normal bristles		
	$^1/_2$ X^+Y normal bristles		

When these F_1 are intercrossed, $^1/_2$ of the F_2 will be normal-bristle females, $^1/_4$ will be normal-bristle males, and $^1/_4$ will be forked-bristle males:

F_1	X^+X^f	×	X^+Y
Gametes	(X^+) (X^f)		(X^+) (Y)

F_2

	(X^+)	(X^f)
(X^+)	X^+X^+ Normal female	X^+X^f Normal female
(Y)	X^+Y Normal male	X^fY Forked-bristle male

$^1/_2$ normal female, $^1/_4$ normal male, $^1/_4$ forked-bristle male

Recall: The multiplication rule states that the probability of two independent events occurring together is the multiplication of their independent probabilities.

To obtain the phenotypic ratio in the F_2, we now combine these two crosses by using the multiplication rule and the branch diagram:

Hint: The branch diagram is a convenient way of keeping up with all the different combinations of traits.

Eye color	Bristle and sex	F_2 phenotype	Probability
red ($^3/_4$)	normal female ($^1/_2$)	→ red normal female	$^3/_4 \times {}^1/_2 = {}^3/_8 = {}^6/_{16}$
	normal male ($^1/_4$)	→ red normal male	$^3/_4 \times {}^1/_4 = {}^3/_{16}$
	forked-bristle male ($^1/_4$)	→ red forked-bristle male	$^3/_4 \times {}^1/_4 = {}^3/_{16}$
brown ($^1/_4$)	normal female ($^1/_2$)	→ brown normal female	$^1/_4 \times {}^1/_2 = {}^1/_8 = {}^2/_{16}$
	normal male ($^1/_4$)	→ brown normal male	$^1/_4 \times {}^1/_4 = {}^1/_{16}$
	forked-bristle male ($^1/_4$)	→ brown forked-bristle male	$^1/_4 \times {}^1/_4 = {}^1/_{16}$

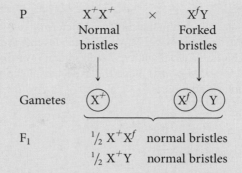

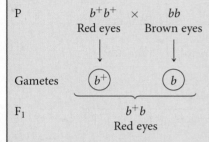

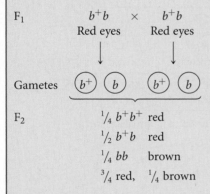

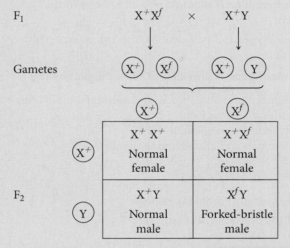

Problem 2

The type of plumage found in mallard ducks is determined by three alleles at a single locus: M^R, which encodes restricted plumage; M, which encodes mallard plumage; and m^d, which encodes dusky plumage. The restricted phenotype is dominant over mallard and dusky; mallard is dominant over dusky ($M^R > M > m^d$). Give the expected phenotypes and proportions of offspring produced by the following crosses:

a. $M^RM \times m^dm^d$

b. $M^Rm^d \times Mm^d$

c. $M^Rm^d \times M^RM$

d. $M^RM \times Mm^d$

Solution Strategy

What information is required in your answer to the problem?

The phenotypes and proportions of progeny expected for each cross.

What information is provided to solve the problem?

- Plumage in mallard ducks is determined by multiple alleles at a single locus.
- Symbols for the alleles (M^R, M, m^d).
- Dominance relations among the alleles ($M^R > M > m^d$).
- Genotypes of the parents in each cross.

For help with this problem, review:

Multiple Alleles in Section 4.3.

Solution Steps

Hint: The Punnett square provides an easy way to determine the genotypes of the progeny of each cross.

We can determine the phenotypes and proportions of offspring by **(1)** determining the types of gametes produced by each parent and **(2)** combining the gametes of the two parents with the use of a Punnett square.

a.

Parents: $M^RM \times m^dm^d$

Gametes: (M^R) (M) $\quad$ (m^d)

	M^R	M
m^d	M^Rm^d restricted	Mm^d mallard

1/2 restricted, 1/2 mallard

Hint: The phenotypes of the progeny can be determined by looking at the dominance relations of the alleles in the progeny's genotype.

b.

Parents: $M^Rm^d \times Mm^d$

Gametes: (M^R) (m^d) $\quad$ (M) (m^d)

	M	m^d
M^R	M^RM restricted	M^Rm^d restricted
m^d	Mm^d mallard	m^dm^d dusky

1/2 restricted, 1/4 mallard, 1/4 dusky

c.

Parents: $M^Rm^d \times M^RM$

Gametes: (M^R) (m^d) $\quad$ (M^R) (M)

	M^R	M
M^R	M^RM^R restricted	M^RM restricted
m^d	M^Rm^d restricted	Mm^d mallard

3/4 restricted, 1/4 mallard

d.

Parents: $M^RM \times Mm^d$

Gametes: (M^R) (M) $\quad$ (M) (m^d)

	M	m^d
M^R	M^RM restricted	M^Rm^d restricted
M	MM mallard	Mm^d mallard

1/2 restricted, 1/2 mallard

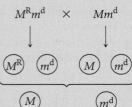

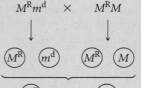

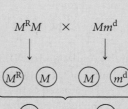

Problem 3

In some sheep, horns are produced by an autosomal allele that is dominant in males and recessive in females. A horned female is crossed with a hornless male. One of the resulting F_1 females is crossed with a hornless male. What proportion of the male and female progeny from this cross will have horns?

Solution Strategy

What information is required in your answer to the problem?

Proportions of male and female progeny that have horns.

What information is provided to solve the problem?

- The presence of horns is due to an autosomal gene that is dominant in male and recessive in females.
- A horned female is crossed with a hornless male. A resulting F_1 female is crossed with a hornless male to produce progeny.

For help with this problem, review:

Sex-Influenced and Sex-Limited Characteristics in Section 4.5.

Solution Steps

The presence of horns in these sheep is an example of a sex-influenced characteristic. Because the phenotypes associated with the genotypes differ between the two sexes, let's begin this problem by writing out the genotypes and phenotypes for each sex. We will let H represent the allele that encodes horns and H^+ represent the allele that encodes hornless. In males, the allele for horns is dominant over the allele for hornless, which means that males homozygous (HH) and heterozygous (H^+H) for this gene are horned. Only males homozygous for the recessive hornless allele (H^+H^+) are hornless. In females, the allele for horns is recessive, which means that only females homozygous for this allele (HH) are horned; females heterozygous (H^+H) and homozygous (H^+H^+) for the hornless allele are hornless. The following table summarizes the genotypes and their associated phenotypes:

...Write out the ...es and the ...ed phenotype ... sex.

...: When a ...ominant, ...e homozygote ... heterozygote ... the trait in ...enotypes.

Genotype	Male phenotype	Female phenotype
HH	horned	horned
HH^+	horned	hornless
H^+H^+	hornless	hornless

In the problem, a horned female is crossed with a hornless male. From the preceding table, we see that a horned female must be homozygous for the allele for horns (HH), and that a hornless male must be homozygous for the allele for hornless (H^+H^+); so all the F_1 will be heterozygous; the F_1 males will be horned and the F_1 females will be hornless, as shown in the following diagram:

P $\quad H^+H^+ \quad \times \quad HH$

↓

F_1 $\quad H^+H$

Horned males and hornless females

A heterozygous hornless F_1 female (H^+H) is then crossed with a hornless male (H^+H^+):

H^+H (Hornless female) $\times$ H^+H^+ (Hornless male)

↓

	Males	Females
$\frac{1}{2}\ H^+H^+$	hornless	hornless
$\frac{1}{2}\ H^+H$	horned	hornless

Therefore, $\frac{1}{2}$ of the male progeny will be horned, but none of the female progeny will be horned.

COMPREHENSION QUESTIONS

Section 4.1

*1. How does sex determination in the XX-XY system differ from sex determination in the ZZ-ZW system?

2. What is meant by genic sex determination?

3. How is sex determined in humans?

Section 4.2

4. What characteristics are exhibited by an X-linked trait?

5. Explain why tortoiseshell cats are almost always female and why they have a patchy distribution of orange and black fur.

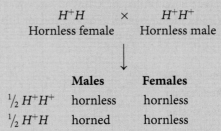

6. What is a Barr body? How is it related to the Lyon hypothesis?

Section 4.3

7. How do incomplete dominance and codominance differ?
8. What is incomplete penetrance and what causes it?

Section 4.4

9. What is gene interaction? What is the difference between an epistatic gene and a hypostatic gene?
10. What is a complementation test and what is it used for?

Section 4.5

11. What characteristics are exhibited by a cytoplasmically inherited trait?

For more questions that test your comprehension of the key chapter concepts, go to LearningCurve for this chapter.

APPLICATION QUESTIONS AND PROBLEMS

Section 4.1

12. If nondisjunction of the sex chromosomes takes place in meiosis I in the male in **Figure 4.5**, what sexual phenotypes and proportions of offspring will be produced?
13. What will be the phenotypic sex of a human with the following genes or chromosomes or both?
    a. XY with the *SRY* gene deleted
    b. XY with the *SRY* gene located on an autosome
    c. XX with a copy of the *SRY* gene on an autosome
    d. XO with a copy of the *SRY* gene on an autosome
    e. XXY with the *SRY* gene deleted
    f. XXYY with one copy of the *SRY* gene deleted

Section 4.2

*14. Joe has classic hemophilia, an X-linked recessive disease. Could Joe have inherited the gene for this disease from the following persons?

	Yes	No
a. His mother's mother	______	______
b. His mother's father	______	______
c. His father's mother	______	______
d. His father's father	______	______

*15. In *Drosophila*, yellow body color is due to an X-linked gene that is recessive to the gene for gray body color.

[Courtesy Dr. Masa-Toshi Yamamoto, Drosophila Genetic Resource Center, Kyoto Institute of Technology.]

    a. A homozygous gray female is crossed with a yellow male. The F_1 are intercrossed to produce F_2. Give the genotypes and phenotypes, along with the expected proportions, of the F_1 and F_2 progeny.
    b. A yellow female is crossed with a gray male. The F_1 are intercrossed to produce the F_2. Give the genotypes and phenotypes, along with the expected proportions, of the F_1 and F_2 progeny.

16. Coat color in cats is determined by genes at several different loci. At one locus on the X chromosome, one allele (X^+) encodes black fur; another allele (X^o) encodes orange fur. Females can be black (X^+X^+), orange (X^oX^o). or a mixture of orange and black called tortoiseshell (X^+X^o). Males are either black (X^+Y) or orange (X^oY), Bill has a female tortoiseshell cat named Patches. One night Patches escapes from Bill's house, spends the night out, and mates with a stray male. Patches later gives birth to the following kittens: one orange male, one black male, two tortoiseshell females, and one orange female. Give the genotypes of Patches, her kittens, and the stray male with which Patches mated.

17. Red–green color blindness in humans is due to an X-linked recessive gene. Both John and Cathy have normal color vision. After 10 years of marriage to John, Cathy gave birth to a color-blind daughter. John filed for divorce, claiming that he is not the father of the child. Is John justified in his claim of nonpaternity? Explain why. If Cathy had given birth to a color-blind son, would John be justified in claiming nonpaternity?

18. DATA ANALYSIS The following pedigree illustrates the inheritance of Nance–Horan syndrome, a rare genetic condition in which affected persons have cataracts and abnormally shaped teeth.

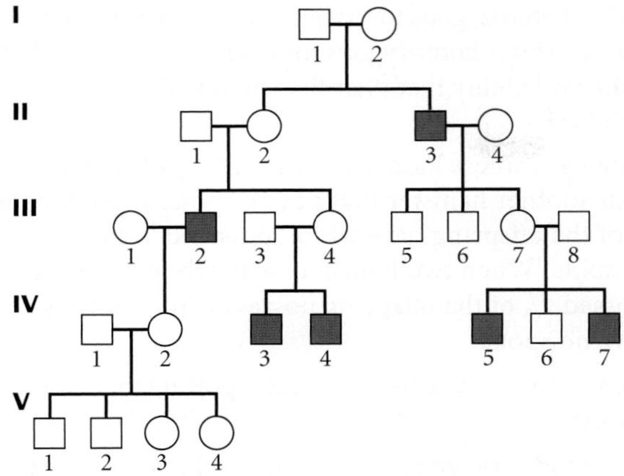

[Pedigree after D. Stambollan et al., *American Journal of Human Genetics* 47:15, 1990.]

a. On the basis of this pedigree, what do you think is the most likely mode of inheritance for Nance–Horan syndrome?

b. If couple III-7 and III-8 have another child, what is the probability that the child will have Nance–Horan syndrome?

c. If III-2 and III-7 were to mate, what is the probability that one of their children would have Nance–Horan syndrome?

*__19.__ Bob has XXY chromosomes (Klinefelter syndrome) and is color blind. His mother and father have normal color vision, but his maternal grandfather is color blind. Assume that Bob's chromosome abnormality arose from nondisjunction in meiosis. In which parent and in which meiotic division did nondisjunction take place? Assume that no crossing over has taken place. Explain your answer.

20. The Talmud, an ancient book of Jewish civil and religious laws, states that if a woman bears two sons who die of bleeding after circumcision (removal of the foreskin from the penis), any additional sons that she has should not be circumcised. (The bleeding is most likely due to the X-linked disorder hemophilia.) Furthermore, the Talmud states that the sons of her sisters must not be circumcised, whereas the sons of her brothers should be. Is this religious law consistent with sound genetic principles? Explain your answer.

21. DATA ANALYSIS Craniofrontonasal syndrome (CFNS) is a birth defect in which premature fusion of the cranial sutures leads to abnormal head shape, widely spaced eyes, nasal clefts, and various other skeletal abnormalities. George Feldman and his colleagues looked at several families in which CFNS occurred and recorded the results shown in the following table (G. J. Feldman. 1997. *Human Molecular Genetics* 6:1937–1941).

Family number	Parents		Offspring			
			Normal		CFNS	
	Father	Mother	Male	Female	Male	Female
1	normal	CFNS	1	0	2	1
5	normal	CFNS	0	2	1	2
6	normal	CFNS	0	0	1	2
8	normal	CFNS	1	1	1	0
10a	CFNS	normal	3	0	0	2
10b	normal	CFNS	1	1	2	0
12	CFNS	normal	0	0	0	1
13a	normal	CFNS	0	1	2	1
13b	CFNS	normal	0	0	0	2
7b	CFNS	normal	0	0	0	2

a. On the basis of the families given, what is the most likely mode of inheritance for CFNS?

b. Give the most likely genotypes of the parents in families 1 and 10a.

22. How many Barr bodies would you expect to see in a human cell containing the following chromosomes?

a. XX **d.** XXY **g.** XYY

b. XY **e.** XXYY **h.** XXX

c. XO **f.** XXXY **i.** XXXX

23. What is the most likely sex and genotype of the cat shown in **Figure 4.12**?

24. Red–green color blindness is an X-linked recessive trait in humans. Polydactyly (extra fingers and toes) is an autosomal dominant trait. Martha has normal fingers and toes and normal color vision. Her mother is normal in all respects, but her father is color blind and polydactylous. Bill is color blind and polydactylous. His mother has normal color vision and normal fingers and toes. If Bill and Martha marry, what types and proportions of children can they produce?

*__25.__ Miniature wings in *Drosophila* result from an X-linked gene (X^m) that is recessive to an allele for long wings (X^+). Sepia eyes are produced by an autosomal gene (s) that is recessive to an allele for red eyes (s^+).

a. A female fly that has miniature wings and sepia eyes is crossed with a male that has normal wings and is homozygous for red eyes. The F_1 are intercrossed to produce the F_2. Give the phenotypes and their proportions expected in the F_1 and F_2 flies from this cross.

b. A female fly that is homozygous for normal wings and has sepia eyes is crossed with a male that has miniature wings and is homozygous for red eyes. The F_1 are intercrossed to produce the F_2. Give the phenotypes and their proportions expected in the F_1 and F_2 flies from this cross.

Section 4.3

*26. Palomino horses have a golden yellow coat, chestnut horses have a brown coat, and cremello horses have a coat that is almost white. A series of crosses between the three different types of horses produced the following offspring:

Cross	Offspring
palomino × palomino	13 palomino, 6 chestnut, 5 cremello
chestnut × chestnut	16 chestnut
cremello × cremello	13 cremello
palomino × chestnut	8 palomino, 9 chestnut
palomino × cremello	11 palomino, 11 cremello
chestnut × cremello	23 palomino

a. Explain the inheritance of the palomino, chestnut, and cremello phenotypes in horses.

b. Assign symbols for the alleles that determine these phenotypes, and list the genotypes of all parents and offspring given in the preceding table.

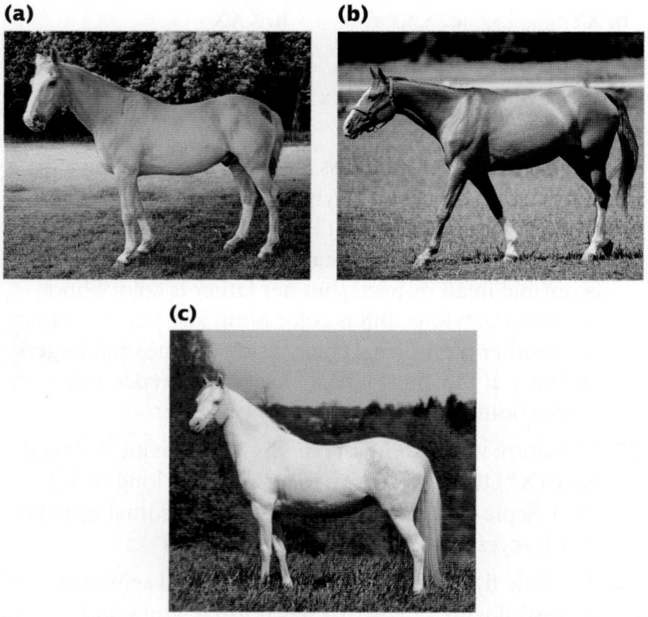

Coat color: (a) palomino, (b) chestnut, (c) cremello. [Part a: Keith J. Smith/Alamy. Part b: jirijura/iStockphoto.com. Part c: Olga_i/Shutterstock.]

27. The L^M and L^N alleles at the MN blood group locus exhibit codominance. Give the expected genotypes and phenotypes and their ratios in progeny resulting from the following crosses.

a. $L^ML^M \times L^ML^N$

b. $L^NL^N \times L^NL^N$

c. $L^ML^N \times L^ML^N$

d. $L^ML^N \times L^NL^N$

e. $L^ML^M \times L^NL^N$

*28. Assume that long earlobes in humans are an autosomal dominant trait that exhibits 30% penetrance. A person who is heterozygous for long earlobes mates with a person who is homozygous for normal earlobes. What is the probability that their first child will have long earlobes?

*29. When a Chinese hamster with white spots is crossed with another hamster that has no spots, approximately $^1/_2$ of the offspring have white spots and $^1/_2$ have no spots. When two hamsters with white spots are crossed, $^2/_3$ of the offspring possess white spots and $^1/_3$ have no spots.

a. What is the genetic basis of white spotting in Chinese hamsters?

b. How might you go about producing Chinese hamsters that breed true for white spotting?

30. In the early 1900s, Lucien Cuénot, a French scientist working at the University of Nancy, studied the genetic basis of yellow coat color in mice (discussed on p. 91). He carried out a number of crosses between two yellow mice and obtained what he thought was a 3 : 1 ratio of yellow to gray mice in the progeny. The following table gives Cuénot's actual results, along with the results of a much larger series of crosses carried out by William Castle and Clarence Little (W. E. Castle and C. C. Little. 1910. *Science* 32:868–870).

Investigators	Yellow progeny	Nonyellow progeny	Total progeny
Cuénot	263	100	363
Castle and Little	800	435	1235
Both combined	1063	535	1598

a. Using a chi-square test, determine whether Cuénot's results are significantly different from the 3 : 1 ratio that he thought he observed. Are they different from a 2 : 1 ratio?

b. Determine whether Castle and Little's results are significantly different from a 3 : 1 ratio. Are they different from a 2 : 1 ratio?

c. Combine the Castle and Little results with those of Cuénot and determine whether they are significantly different from a 3 : 1 ratio and a 2 : 1 ratio.

d. Offer an explanation for the different ratios obtained by Cuénot and by Castle and Little.

31. In rabbits, an allelic series helps to determine coat color: C (full color), c^{ch} (chinchilla, gray color), c^h (Himalayan, white with black extremities), and c (albino, all white). The C allele is dominant over all others, c^{ch} is dominant over c^h and c, c^h is dominant over c, and c is recessive to all the other alleles. This dominance hierarchy can be summarized as $C > c^{ch} > c^h > c$. The rabbits in the list on the following page are crossed and produce the progeny shown. Give the genotypes of the parents for each cross.

	Phenotypes of parents	Phenotypes of offspring
a.	full color × albino	$^1/_2$ full color, $^1/_2$ albino
b.	Himalayan × albino	$^1/_2$ Himalayan, $^1/_2$ albino
c.	full color × albino	$^1/_2$ full color, $^1/_2$ chinchilla
d.	full color × Himalayan	$^1/_2$ full color, $^1/_4$ Himalayan, $^1/_4$ albino
e.	full color × full color	$^3/_4$ full color, $^1/_4$ albino

***32.** In this chapter, we considered Joan Barry's paternity suit against Charlie Chaplin and how, on the basis of blood types, Chaplin could not have been the father of her child.

a. What blood types are possible for the father of Barry's child?

b. If Chaplin had possessed one of these blood types, would that prove that he fathered Barry's child?

Section 4.4

***33.** In chickens, comb shape is determined by alleles at two loci (*R*, *r* and *P*, *p*). A walnut comb is produced when at least one dominant allele *R* is present at one locus and at least one dominant allele *P* is present at a second locus (genotype *R_ P_*). A rose comb is produced when at least one dominant allele is present at the first locus and two recessive alleles are present at the second locus (genotype *R_ pp*). A pea comb is produced when two recessive alleles are present at the first locus and at least one dominant allele is present at the second (genotype *rr P_*). If two recessive alleles are present at the first locus and at the second locus (*rr pp*), a single comb is produced. Progeny with what types of combs and in what proportions will result from the following crosses?

Comb shape: (a) walnut, (b) rose, (c) pea, (d) single. [Parts a and d: Robert Dowling/Corbis. Part b: Robert Maier/Animals Animals. Part c: Dapne Godfrey Trust/Animals Animals.]

a. *RR PP* × *rr pp*

b. *Rr Pp* × *rr pp*

c. *Rr Pp* × *Rr Pp*

d. *Rr pp* × *Rr pp*

e. *Rr pp* × *rr Pp*

f. *Rr pp* × *rr pp*

34. DATA ANALYSIS Tatuo Aida investigated the genetic basis of color variation in the medaka (*Aplocheilus latipes*), a small fish found in Japan (T. Aida. 1921. *Genetics* 6:554–573). Aida found that genes at two loci (*B*, *b* and *R*, *r*) determine the color of the fish: fish with a dominant allele at both loci (*B_ R_*) are brown, fish with a dominant allele at the *B* locus only (*B_ rr*) are blue, fish with a dominant allele at the *R* locus only (*bb R_*) are red, and fish with recessive alleles at both loci (*bb rr*) are white. Aida crossed a homozygous brown fish with a homozygous white fish. He then backcrossed the F_1 with the homozygous white parent and obtained 228 brown fish, 230 blue fish, 237 red fish, and 222 white fish.

a. Give the genotypes of the backcross progeny.

b. Use a chi-square test to compare the observed numbers of backcross progeny with the number expected. What conclusion can you make from your chi-square results?

c. What results would you expect for a cross between a homozygous red fish and a white fish?

d. What results would you expect if you crossed a homozygous red fish with a homozygous blue fish and then backcrossed the F_1 with a homozygous red parental fish?

35. DATA ANALYSIS E. W. Lindstrom crossed two corn plants with green seedlings and obtained the following progeny: 3583 green seedlings, 853 virescent-white seedlings, and 260 yellow seedlings (E. W. Lindstrom. 1921. *Genetics* 6:91–110).

a. Give the genotypes for the green, virescent-white, and yellow progeny.

b. Explain how color is determined in these seedlings.

c. Does epistasis take place among the genes that determine color in the corn seedlings? If so, which gene is epistatic and which is hypostatic?

***36.** A summer-squash plant that produces disc-shaped fruit is crossed with a summer-squash plant that produces long fruit. All the F_1 have disc-shaped fruit. When the F_1 are intercrossed, F_2 progeny are produced in the following ratio: $^9/_{16}$ disc-shaped fruit; $^6/_{16}$ spherical fruit; $^1/_{16}$ long fruit. Give the genotypes of the F_2 progeny.

37. Some sweet-pea plants have purple flowers and others have white flowers. A homozygous variety of sweet pea that has purple flowers is crossed with a homozygous variety that has white flowers. All the F_1 have purple flowers. When these F_1 are self-fertilized, the F_2 appear in a ratio of $^9/_{16}$ purple to $^7/_{16}$ white.

a. Give genotypes for the purple and white flowers in these crosses.

b. Draw a hypothetical biochemical pathway to explain the production of purple and white flowers in sweet peas.

Section 4.5

***38.** The direction of shell coiling in the snail *Lymnaea peregra* (discussed in the introduction to this chapter) results from a genetic maternal effect. An autosomal allele for a right-handed shell (s^+), called dextral, is dominant over the allele for a left-handed shell (s), called sinistral. A pet snail called Martha is sinistral and reproduces only as a female (the snails are hermaphroditic). Indicate which of the following statements are true and which are false. Explain your reasoning in each case.

a. Martha's genotype *must* be *ss*.

b. Martha's genotype *cannot* be s^+s^+.

c. All the offspring produced by Martha *must* be sinistral.

d. At least some of the offspring produced by Martha *must* be sinistral.

e. Martha's mother *must* have been sinistral.

f. All of Martha's brothers *must* be sinistral.

39. If the F_2 dextral snails with genotype S^+s in **Figure 4.21** undergo self-fertilization, what phenotypes and proportions are expected to occur in the progeny?

Section 4.6

40. Which of the following statements is an example of a phenocopy? Explain your reasoning.

a. Phenylketonuria results from a recessive mutation that causes light skin as well as intellectual disability.

b. Human height is influenced by genes at many different loci.

c. Dwarf plants and mottled leaves in tomatoes are caused by separate genes that are linked.

d. Vestigial wings in *Drosophila* are produced by a recessive mutation. This trait is also produced by high temperature during development.

e. Intelligence in humans is influenced by both genetic and environmental factors.

41. Match each of the following terms with its correct definition (parts *a* through *h*).

phenocopy	sex-limited trait
pleiotropy	genetic maternal effect
polygenic trait	genomic imprinting
penetrance	sex-influenced trait

a. The percentage of individuals with a particular genotype that express the expected phenotype

b. A trait determined by an autosomal gene that is more easily expressed in one sex

c. A trait determined by an autosomal gene that is expressed in only one sex

d. A trait that is determined by an environmental effect and has the same phenotype as a genetically determined trait

e. A trait determined by genes at many loci

f. The expression of a trait affected by the sex of the parent that transmits the gene to the offspring

g. A gene affecting more than one phenotype

h. The influence of the genotype of the maternal parent on the phenotype of the offspring

CHALLENGE QUESTION

Section 4.2

42. A geneticist discovers a male mouse with greatly enlarged testes in his laboratory colony. He suspects that this trait results from a new mutation that is either Y-linked or autosomal dominant. How could he determine which it is?

THINK-PAIR-SHARE QUESTIONS

Introduction

1. The introduction to this chapter discusses the genetic basis of chirality in snails and the research of Arthur Boycott, whose work established the mode of inheritance for this trait. In the course of his research, Boycott enlisted the aid of several amateur scientists—men who were not trained as scientists and had other jobs. The research would have been impossible without the aid of these individuals. In the past, amateur scientists such as these often made important contributions to science, but the practice is less frequent today. Discuss some possible reasons for the decline in contributions to research by amateur scientists. Are there any areas where amateur scientists still actively contribute?

2. Genetic maternal effect is often seen in mammals. For example, research shows that the maternal genotype influences adult body size in mice. Why might these types of genetic effects be more common in mammals than in other organisms such as fishes, amphibians, or reptiles?

Section 4.1

3. The duck-billed platypus has a unique mechanism of sex determination: females have five pairs of X chromosomes ($X_1X_1X_2X_2X_3X_3X_4X_4X_5X_5$) and males have five pairs of X and Y chromosomes ($X_1Y_1X_2Y_2X_3Y_3X_4Y_4X_5Y_5$). Do you think each of the X and Y chromosome pairs in males assorts independently of other X and Y pairs during meiosis? Why or why not?

4. Most organisms with XX-XY sex determination have pseudoautosomal regions, portions of the X and Y chromosomes that are homologous. Would you predict that organisms with ZZ-ZW sex determination have pseudoautosomal regions of homology between Z and W chromosomes? Explain your answer.

5. Both men and women produce testosterone, but concentrations of testosterone in the blood are generally higher in men than in women. However, the testosterone levels of some XX females fall within the range of testosterone levels of XY men. This overlap has created controversy within women's sports. Testosterone is known to increase muscle mass and enhance some types of athletic performance, so some people have suggested that women with naturally high testosterone levels have an unfair competitive advantage. In 2011, the International Association of Athletics Federations (IAAF) adopted a policy that limits levels of testosterone in female athletes, saying that female athletes must not have a blood testosterone concentration greater than 10 nanomoles per liter (nmol/L), a level typically seen in men. Some elite female athletes have natural testosterone levels above this limit and have challenged the policy. Do you think that it is fair for XX females with naturally high testosterone (levels typically found in XY males) to compete in women's sports? Do they have an unfair advantage in competition with other women? What about male athletes with naturally high levels of testosterone? Do they have an unfair advantage over other males? In general, what role does genetics play in athletic competition—do some individuals have genes that give them an unfair advantage in competition?

Section 4.2

6. How is the inheritance of X-linked traits different from the inheritance of autosomal traits? How is the inheritance of X-linked and autosomal traits similar? List as many differences and similarities as you can.

7. On average, what proportion of X-linked genes in the first individual, in the following examples, are the same (inherited from a common ancestor) as those in the second individual?

   a. A male and his mother
   b. A female and her mother
   c. A male and his father
   d. A female and her father
   e. A male and his brother
   f. A female and her sister
   g. A male and his sister
   h. A female and her brother

8. Red–green color blindness is an X-linked recessive trait. Susan has normal color vision, but her father is color-blind. Susan marries Bob, who has normal color vision.

   a. What is the probability that Susan and Bob will have a color-blind son?

   b. Susan and Bob have a daughter named Betty, who has normal color vision. If Betty marries a man with normal color vision, and they have a son, what is the probability that the son will be color-blind?

Section 4.3

9. Over a thousand different alleles at the *CFTR* locus have been discovered that can cause cystic fibrosis. What difficulties might the presence of so many different alleles at this locus create for the diagnosis and treatment of cystic fibrosis?

Section 4.4

10. Could you carry out a complementation test on two dominant mutations? Why or why not?

Section 4.5

11. In goats, a beard is produced by an autosomal allele that is dominant in males and recessive in females. Is it possible to cross two bearded goats and obtain a beardless male offspring? Why or why not? What about a bearded female offspring?

12. In chickens, cock feathering is an autosomal recessive trait that is limited to males. Is it possible to cross a cock-feathered male with a hen-feathered female and get male progeny that are cock-feathered? Explain your reasoning.

13. Suppose you observed a new mutant phenotype, notched ears, which appears only in male mice. How might you go about determining whether notched ears is a Y-linked trait or a sex-limited trait? What crosses would you carry out to distinguish between these two modes of inheritance?

Section 4.6

14. Phenylketonuria (PKU) is an autosomal recessive disease that results from a defect in an enzyme that normally metabolizes the amino acid phenylalanine; when this enzyme is defective, high levels of phenylalanine cause brain damage. In the past, most children with PKU became intellectually disabled. Fortunately, intellectual disability can be prevented in these children by carefully controlling the amount of phenylalanine in the diet. The diet is usually applied during childhood, when brain development is taking place. As a result of this treatment, many people with PKU now reach reproductive age. Children born to women with PKU (who are no longer on a phenylalanine-restricted diet) frequently have low birth weight, developmental abnormalities, and intellectual disabilities. However, children of men with PKU do not have these problems.
    a. Provide an explanation for these observations.
    b. What type of genetic effect is this? Explain your reasoning.

5 Linkage, Recombination, and Eukaryotic Gene Mapping

Linked Genes and Bald Heads

For many, baldness is the curse of manhood. Twenty-five percent of men begin balding by age 30, and almost half are bald to some degree by age 50. In the United States, baldness affects more than 40 million men, and hundreds of millions of dollars are spent each year on hair-loss treatment. Baldness is not just a matter of vanity: it is associated with some medically significant conditions, including heart disease, high blood pressure, and prostate cancer.

Baldness can arise for a number of different reasons, including illness, injury, drugs, and heredity. The most common type of baldness seen in men is pattern baldness—technically known as androgenic alopecia—in which hair is lost prematurely from the front and top of the head. More than 95% of hair loss in men is pattern baldness. Although pattern baldness is also seen in women, it is usually expressed weakly as mild thinning of the hair. The trait is stimulated by male sex hormones (androgens), as evidenced by the observation that males castrated at an early age rarely become bald (though this is not recommended as a preventive treatment for baldness).

A strong hereditary influence on pattern baldness has long been recognized, but the exact mode of inheritance has been controversial. An early study suggested that it was autosomal dominant in males and recessive in females, which would make pattern baldness a sex-influenced trait (see Chapter 4). Other evidence and common folklore suggested that a man inherits baldness from his mother's side of the family, exhibiting X-linked inheritance.

Pattern baldness is a hereditary trait. Recent research demonstrated that a gene for pattern baldness is linked to genetic markers located on the X chromosome, leading to the discovery that pattern baldness is influenced by variation in the androgen receptor gene. [© Jose Luis Pelaez Inc/AGE Fotostock.]

In 2005, geneticist Axel Hillmer and his colleagues set out to locate the gene that causes pattern baldness. They suspected that the gene might be located on the X chromosome, but they had no idea where on the X chromosome it might reside. To identify the location of the gene, they conducted a linkage analysis study, in which they looked for an association between the inheritance of pattern baldness and the inheritance of genetic variants known to be located on the X chromosome. The genetic variants used in the study were single-nucleotide polymorphisms (SNPs, pronounced "snips"), which are positions in the genome where individuals in a population differ in a single nucleotide. The geneticists studied the inheritance of pattern baldness and SNPs in 95 families in which at least two brothers developed pattern baldness at an early age.

Hillmer and his colleagues found that pattern baldness and SNPs from the X chromosome were not inherited independently, as predicted by Mendel's principle of independent assortment. Instead, they tended to be inherited together, which occurs when genes are physically linked on the same chromosome and segregate together in meiosis.

As we will learn in this chapter, linkage between genes is broken down by a process called recombination, or crossing over, and the amount of recombination between genes is usually related to the distance between them. In 1911, Thomas Hunt Morgan and his student Alfred Sturtevant demonstrated in fruit flies that genes can be mapped by determining the rates of recombination between them. Using this method for families with pattern baldness, Hillmer and his colleagues demonstrated that the gene for pattern baldness is closely linked to SNPs located at position p12–22 on the X chromosome. This region includes the androgen receptor gene, which encodes a protein that binds male sex hormones. Given the clear involvement of male hormones in the development of pattern baldness, the androgen receptor gene seemed a likely candidate for causing pattern baldness. Further analysis revealed that certain alleles of the androgen receptor gene were closely associated with the inheritance of pattern baldness, and that variation in the androgen receptor gene is almost certainly responsible for many of the differences in pattern baldness seen in the families examined. Additional studies conducted in 2008 found that genes on chromosomes 3 and 20 also appear to contribute to the expression of pattern baldness.

THINK-PAIR-SHARE Question 1

This chapter explores the inheritance of genes located on the same chromosome. These linked genes do not strictly obey Mendel's principle of independent assortment; rather, they tend to be inherited together. This tendency requires a new approach to understanding their inheritance and predicting the types of offspring that will be produced. A critical piece of information necessary for predicting the results of these crosses is the arrangement of the genes on the chromosomes; thus, it will be necessary to think about the relation between genes and chromosomes. A key to understanding the inheritance of linked genes is to make the conceptual connection between the genotypes in a cross and the behavior of chromosomes in meiosis.

We will begin our exploration of linkage by first comparing the inheritance of two linked genes with the inheritance of two genes that assort independently. We will then examine how recombination breaks up linked genes. This knowledge of linkage and recombination will be used for predicting the results of genetic crosses in which genes are linked and for mapping genes. Later in the chapter, we will focus on mapping with molecular markers and genome-wide association studies.

5.1 Linked Genes Do Not Assort Independently

Chapter 3 introduced Mendel's principles of segregation and independent assortment. Let's take a moment to review these two important concepts. The principle of segregation states that each diploid organism possesses two alleles at a locus that separate in meiosis, with one allele going into each gamete. The principle of independent assortment provides additional information about the process of segregation: it tells us that, in the process of separation, the two alleles at a locus act independently of alleles at other loci.

The independent separation of alleles results in *recombination*, the sorting of alleles into new combinations. Consider a cross between individuals that are homozygous for two different pairs of alleles: *AA BB* × *aa bb*. The first parent, *AA BB*, produces gametes with the alleles *A B*, and the second parent, *aa bb*, produces gametes with the alleles *a b*, resulting in F_1 progeny with genotype *Aa Bb* (**Figure 5.1**). Recombination

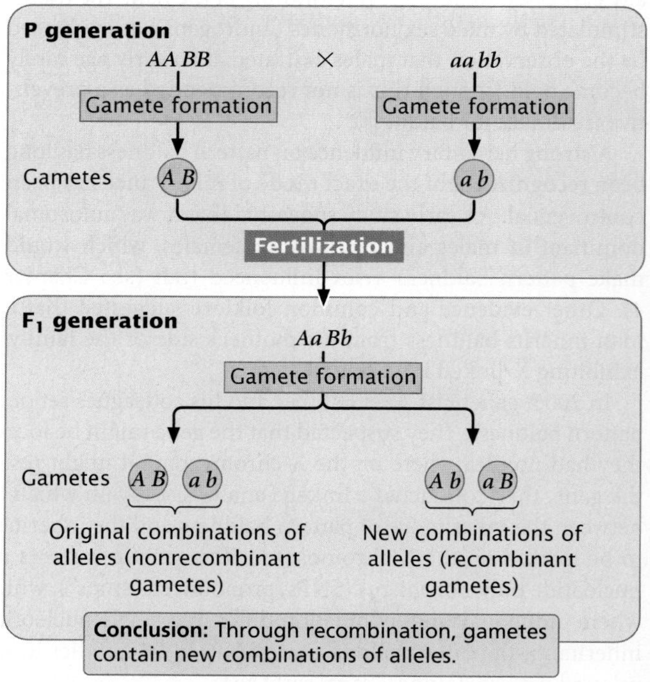

5.1 Recombination is the sorting of alleles into new combinations.

means that when one of the F_1 progeny reproduces, the combination of alleles in its gametes may differ from the combinations in the gametes from its parents. In other words, the F_1 may produce gametes with new allele combinations *A b* or *a B* in addition to parental gametes *A B* or *a b*.

Mendel derived his principles of segregation and independent assortment by observing the progeny of genetic crosses, but he had no idea what biological processes produced these phenomena. In 1903, Walter Sutton proposed a biological basis for Mendel's principles, called the chromosome theory of heredity, which holds that genes are found on chromosomes (see Chapter 3). Let's restate Mendel's two principles in relation to the chromosome theory of heredity. The principle of segregation states that a diploid organism possesses two alleles for a characteristic, each of which is located at the same position, or locus, on each of the two homologous chromosomes. These chromosomes segregate in meiosis, and each gamete receives one homolog. The principle of independent assortment states that, in meiosis, each pair of homologous chromosomes assorts independently of other homologous pairs. With this new perspective, it is easy to see that the number of chromosomes in most organisms is limited and that there are certain to be more genes than chromosomes, so some genes must be present on the same chromosome and should not assort independently. Genes located close together on the same chromosome are called **linked genes** and belong to the same **linkage group**. Linked genes travel together in meiosis, eventually arriving at the same destination (the same gamete), and are not expected to assort independently.

All of the characteristics examined by Mendel in peas did display independent assortment, and the first genetic characteristics studied in other organisms also seemed to assort independently. How could genes be carried on a limited number of chromosomes and yet assort independently?

This apparent inconsistency between the principle of independent assortment and the chromosome theory of heredity soon disappeared as biologists began finding genetic characteristics that did not assort independently. One of the first cases was reported in sweet peas by William Bateson, Edith Rebecca Saunders, and Reginald C. Punnett in 1905. They crossed a homozygous strain of peas that had purple flowers and long pollen grains with a homozygous strain that had red flowers and round pollen grains. All the F_1 had purple flowers and long pollen grains, indicating that purple was dominant over red and long was dominant over round. When they intercrossed the F_1, the resulting F_2 progeny did not appear in the 9 : 3 : 3 : 1 ratio expected with independent assortment (**Figure 5.2**). An excess of F_2 plants had purple flowers and long pollen or red flowers and round pollen (the parental phenotypes). Although Bateson, Saunders, and Punnett were unable to explain these results, we now know that the two loci that they examined lie close together on the same chromosome and therefore do not assort independently.

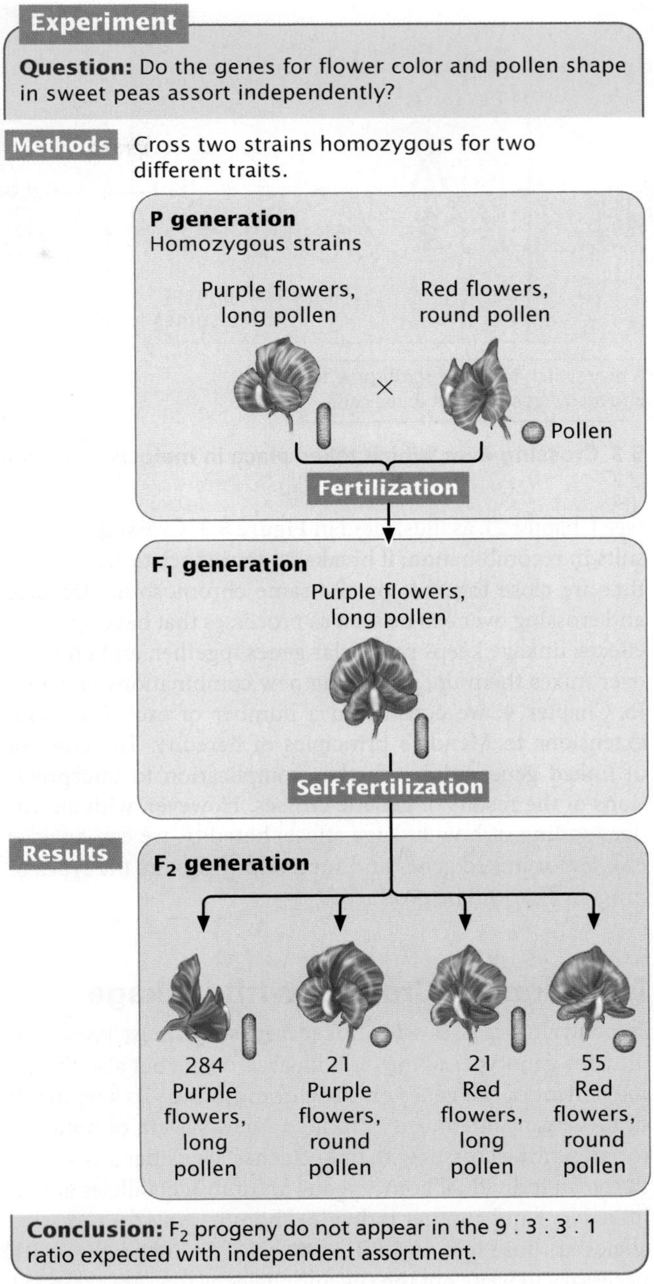

5.2 Nonindependent assortment of flower color and pollen shape in sweet peas.

5.2 Linked Genes Segregate Together, While Crossing Over Produces Recombination Between Them

Genes that are close together on the same chromosome usually segregate as a unit and are therefore inherited together. However, genes occasionally switch from one homologous chromosome to the other through the process of crossing over

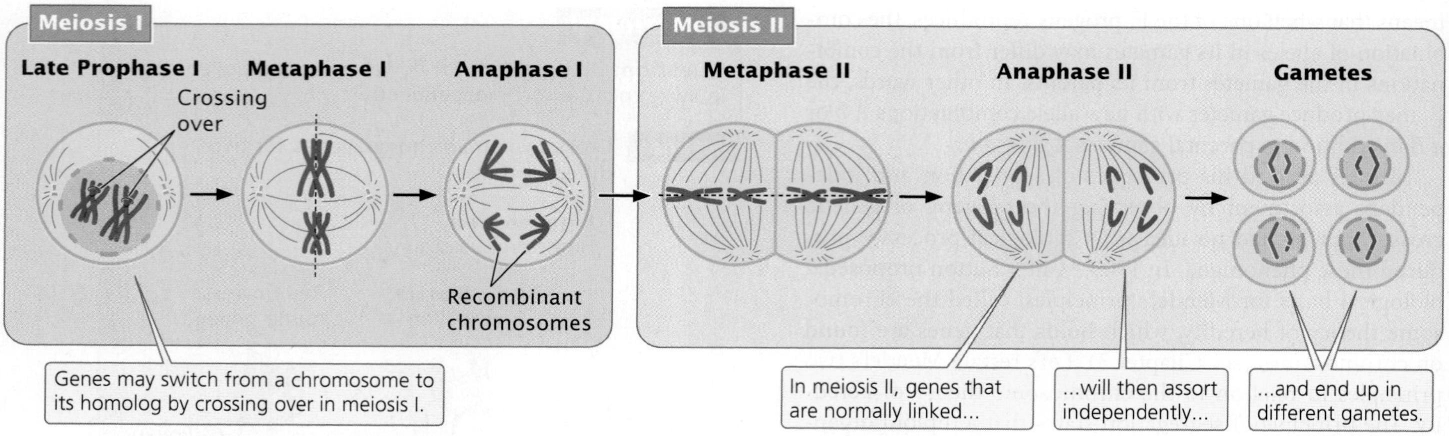

5.3 Crossing over, which takes place in meiosis, is responsible for recombination.

(see Chapter 2), as illustrated in **Figure 5.3**. Crossing over results in recombination; it breaks up the associations of genes that are close together on the same chromosome. Linkage and crossing over can be seen as processes that have opposite effects: linkage keeps particular genes together, and crossing over mixes them up, producing new combinations of genes. In Chapter 4, we considered a number of exceptions and extensions to Mendel's principles of heredity. The concept of linked genes adds a further complication to interpretations of the results of genetic crosses. However, with an understanding of how linkage affects heredity, we can analyze crosses for linked genes and successfully predict the types of progeny that will be produced.

Notation for Crosses with Linkage

In analyzing crosses with linked genes, we must know not only the genotypes of the individuals crossed, but also the arrangement of the genes on the chromosomes. To keep track of this arrangement, we introduce a new system of notation for presenting crosses with linked genes. Consider a cross between an individual homozygous for dominant alleles at two linked loci and another individual homozygous for recessive alleles at those loci ($AA\,BB \times aa\,bb$). For linked genes, it's necessary to write out the specific alleles as they are arranged on each of the homologous chromosomes:

$$\frac{A \qquad B}{A \qquad B} \times \frac{a \qquad b}{a \qquad b}$$

In this notation, each horizontal line represents one of the two homologous chromosomes. Inheriting one chromosome from each parent, the F_1 progeny will have the following genotype:

$$\frac{A \qquad B}{a \qquad b}$$

Here, the importance of designating the alleles on each chromosome is clear. One chromosome has the two dominant alleles *A* and *B*, whereas the homologous chromosome has the two recessive alleles *a* and *b*. The notation can be simplified by drawing only a single line, with the understanding that genes located on the same side of the line lie on the same chromosome:

$$\frac{A \qquad B}{a \qquad b}$$

This notation can be simplified further by separating the alleles on each chromosome with a slash: *AB*/*ab*.

Remember that the two alleles at a locus are always located on different homologous chromosomes and therefore must lie on opposite sides of the line. Consequently, we would *never* write the genotypes as

$$\frac{A \qquad a}{B \qquad b}$$

because the alleles *A* and *a* can *never* be on the same chromosome.

It is also important to always keep the same order of the genes on both sides of the line; thus, we should *never* write

$$\frac{A \qquad B}{b \qquad a}$$

because it would imply that alleles *A* and *b* are allelic (at the same locus).

Complete Linkage Compared with Independent Assortment

We will first consider what happens to genes that exhibit complete linkage, meaning that they are located very close together on the same chromosome and do not exhibit crossing over. Genes are rarely completely linked, but by assuming that no crossing over takes place, we can see the effect of linkage more clearly. We will then consider what happens when genes assort independently. Finally, we will consider the results obtained if the genes are linked but exhibit some crossing over.

A testcross reveals the effects of linkage. For example, if a heterozygous individual is test-crossed with a homozygous recessive individual ($Aa\,Bb \times aa\,bb$), the alleles that are present in the gametes contributed by the heterozygous parent will be expressed in the phenotype of the offspring because the homozygous parent cannot contribute dominant alleles that might mask

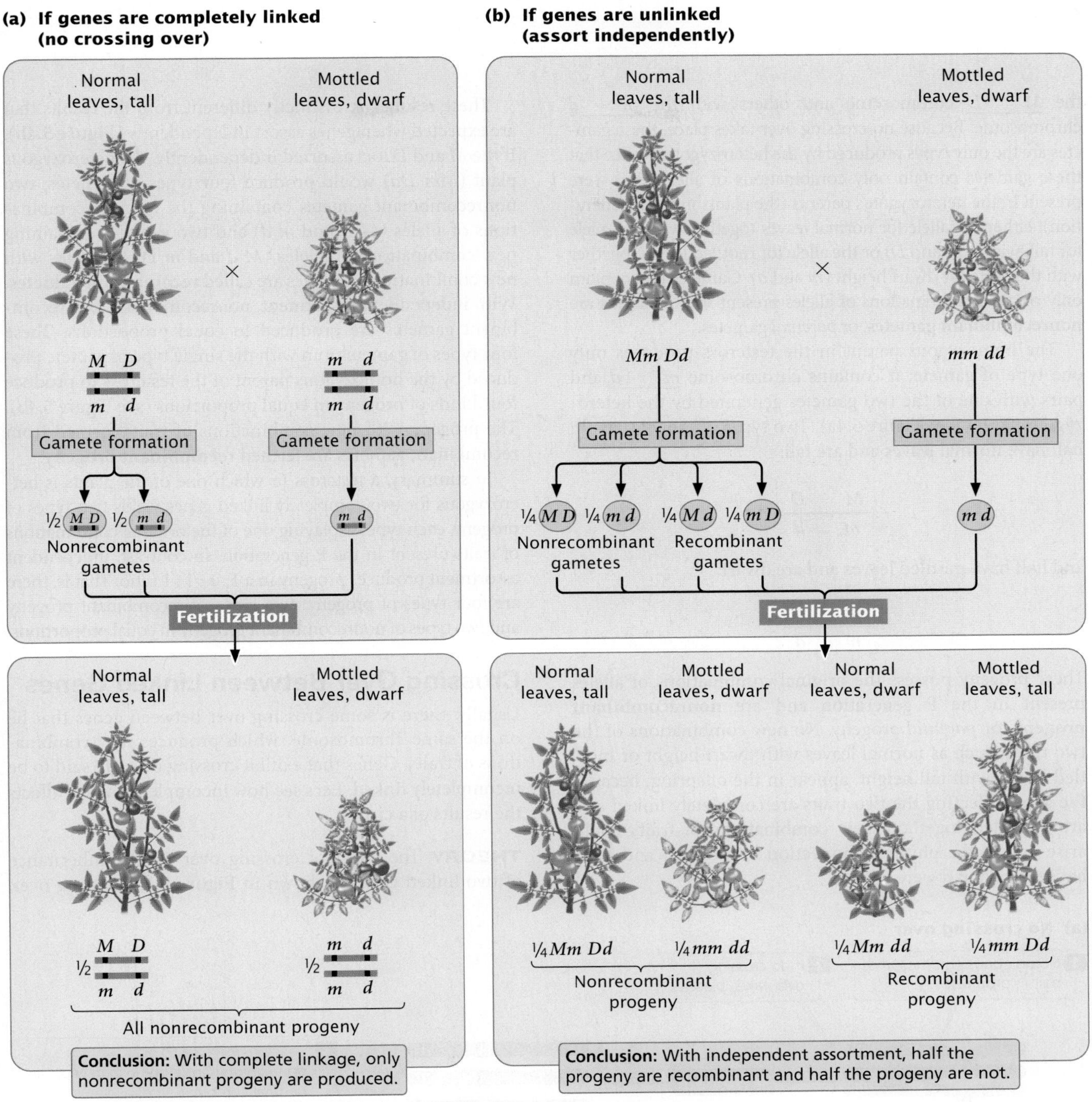

5.4 A testcross reveals the effects of linkage. Results of a testcross for two loci in tomatoes that determine leaf type and plant height.

them. Consequently, traits that appear in the progeny reveal which alleles were transmitted by the heterozygous parent.

Consider a pair of linked genes in tomato plants. One of the genes affects the type of leaf: an allele for mottled leaves (m) is recessive to an allele that produces normal leaves (M). Nearby on the same chromosome, the other gene determines the height of the plant: an allele for dwarf (d) is recessive to an allele for tall (D).

Testing for linkage can be done with a testcross, which requires a plant that is heterozygous for both characteristics. A geneticist might produce this heterozygous plant by crossing a variety of tomato that is homozygous for normal leaves and tall height with a variety that is homozygous for mottled leaves and dwarf height:

$$\text{P} \quad \frac{M \quad D}{M \quad D} \times \frac{m \quad d}{m \quad d}$$

$$\downarrow$$

$$\text{F}_1 \quad \frac{M \quad D}{m \quad d}$$

The geneticist would then use this F_1 heterozygote in a testcross, crossing it with a plant that is homozygous for mottled leaves and dwarf height:

$$\frac{M \quad D}{m \quad d} \times \frac{m \quad d}{m \quad d}$$

The results of this testcross are diagrammed in **Figure 5.4a.** The heterozygote produces two types of gametes: some with the $\underline{M \quad D}$ chromosome and others with the $\underline{m \quad d}$ chromosome. Because no crossing over takes place, these gametes are the only types produced by the heterozygote. Notice that these gametes contain only combinations of alleles that were present in the heterozygote's parents (the plants in the P generation): either the allele for normal leaves together with the allele for tall height (*M* and *D*) or the allele for mottled leaves together with the allele for dwarf height (*m* and *d*). Gametes that contain only original combinations of alleles present in the parents are **nonrecombinant gametes**, or *parental* gametes.

The homozygous parent in the testcross produces only one type of gamete; it contains chromosome $\underline{m \quad d}$ and pairs with one of the two gametes generated by the heterozygous parent (see Figure 5.4a). Two types of progeny result: half have normal leaves and are tall:

$$\frac{M \quad D}{m \quad d}$$

and half have mottled leaves and are dwarf:

$$\frac{m \quad d}{m \quad d}$$

These progeny possess the original combinations of alleles present in the P generation and are **nonrecombinant progeny**, or *parental* progeny. No new combinations of the two traits, such as normal leaves with dwarf height or mottled leaves with tall height, appear in the offspring, because the genes affecting the two traits are completely linked and are inherited together. New combinations of traits could arise only if the physical connection between *M* and *D* or between *m* and *d* were broken.

These results are distinctly different from the results that are expected when genes assort independently (**Figure 5.4b**). If the *M* and *D* loci assorted independently, the heterozygous plant (*Mm Dd*) would produce four types of gametes: two nonrecombinant gametes containing the original combinations of alleles (*M D* and *m d*) and two gametes containing new combinations of alleles (*M d* and *m D*). Gametes with new combinations of alleles are called **recombinant gametes**. With independent assortment, nonrecombinant and recombinant gametes are produced in equal proportions. These four types of gametes join with the single type of gamete produced by the homozygous parent of the testcross to produce four kinds of progeny in equal proportions (see Figure 5.4b). The progeny with new combinations of traits formed from recombinant gametes are termed **recombinant progeny**.

In summary, a testcross in which one of the plants is heterozygous for two completely linked genes yields two types of progeny, each type displaying one of the original combinations of traits present in the P generation. In contrast, independent assortment produces progeny in a 1 : 1 : 1 : 1 ratio. That is, there are four types of progeny: two types of recombinant progeny and two types of nonrecombinant progeny in equal proportions.

Crossing Over Between Linked Genes

Usually, there is some crossing over between genes that lie on the same chromosome, which produces new combinations of traits. Genes that exhibit crossing over are said to be incompletely linked. Let's see how incomplete linkage affects the results of a cross.

THEORY The effect of crossing over on the inheritance of two linked genes is shown in **Figure 5.5**. Crossing over,

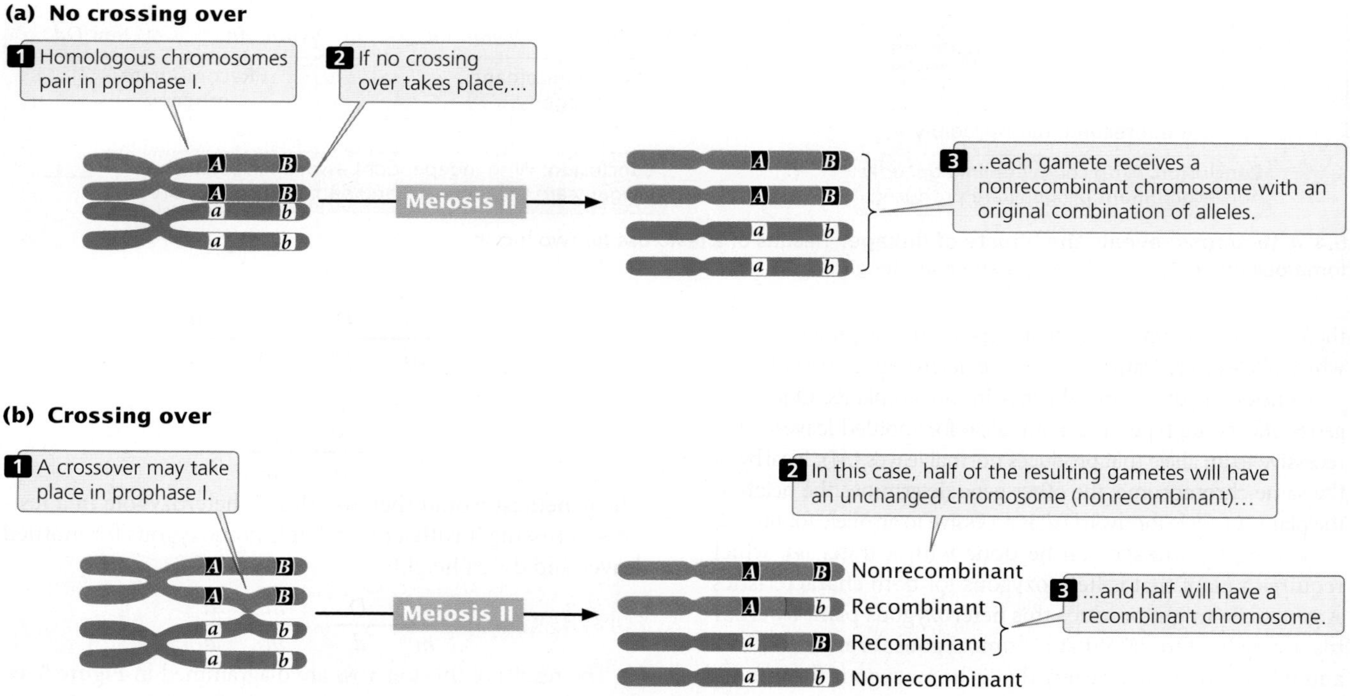

5.5 A single crossover produces half nonrecombinant gametes and half recombinant gametes.

which takes place in prophase I of meiosis, is the exchange of genetic material between nonsister chromatids (see Figures 2.12 and 2.13). After a single crossover has taken place, the two chromatids that did not participate in crossing over are unchanged; gametes that receive these chromatids are nonrecombinants. The other two chromatids, which did participate in crossing over, now contain new combinations of alleles; gametes that receive these chromatids are recombinants. For each meiosis in which a single crossover takes place, two nonrecombinant gametes and two recombinant gametes will be produced. This result is the same as that produced by independent assortment (see Figure 5.4b), so if crossing over between two loci takes place in every meiosis, it is impossible to determine whether the genes are on the same chromosome and crossing over took place or whether the genes are on different chromosomes.

For closely linked genes, crossing over does not take place in every meiosis. In meioses in which there is no crossing over, only nonrecombinant gametes are produced. In meioses in which there is a single crossover, half the gametes are recombinants and half are nonrecombinants (because a single crossover affects only two of the four chromatids). Because each crossover leads to half recombinant gametes and half nonrecombinant gametes, the total percentage of recombinant gametes is always half the percentage of meioses in which crossing over takes place. Even if crossing over between two genes takes place in every meiosis, only 50% of the resulting gametes will be recombinants. Thus, the frequency of recombinant gametes is always half the frequency of crossing over, and the maximum proportion of recombinant gametes is 50%.

CONCEPTS

Linkage between genes causes them to be inherited together and reduces recombination; crossing over breaks up the associations of such genes. In a testcross for two linked genes, each crossover produces two recombinant gametes and two nonrecombinants. The frequency of recombinant gametes is half the frequency of crossing over, and the maximum frequency of recombinant gametes is 50%.

✓ CONCEPT CHECK 1

For single crossovers, the frequency of recombinant gametes is half the frequency of crossing over because

a. a testcross between a homozygote and heterozygote produces ½ heterozygous and ½ homozygous progeny.
b. the frequency of recombination is always 50%.
c. each crossover takes place between only two of the four chromatids of a homologous pair.
d. crossovers take place in about 50% of meioses.

APPLICATION Let's apply what we have learned about linkage and recombination to a cross between tomato plants that differ in the genes that encode leaf type and plant height. Assume now that these genes are linked and that some crossing over takes place between them. Suppose that a geneticist carried out the testcross described earlier:

$$\frac{M \quad D}{m \quad d} \times \frac{m \quad d}{m \quad d}$$

When crossing over takes place between the genes for leaf type and height, two of the four gametes produced are recombinants. When there is no crossing over, all four resulting gametes are nonrecombinants. Because crossing over between two linked genes does not take place in all meioses, most of the gametes will be nonrecombinants. These gametes then unite with gametes produced by the homozygous recessive parent, which contain only the recessive alleles, resulting in mostly nonrecombinant progeny and a few recombinant progeny (**Figure 5.6**). In this cross, we see that 55 of the testcross progeny have normal leaves and are tall and that 53 have mottled leaves and are dwarf. These plants are the nonrecombinant progeny, containing the original combinations of traits that were present in the parents. Of the 123 progeny, 15 have new combinations of traits that were not seen in the parents: 8 have normal leaves and are dwarf, and 7 have mottled leaves and are tall. These plants are the recombinant progeny.

The results of a cross such as the one illustrated in Figure 5.6 reveal several things. A testcross for two independently assorting genes is expected to produce a 1 : 1 : 1 : 1 phenotypic ratio in the progeny. The progeny of this cross clearly do not exhibit such a ratio, so we might suspect that the genes are not assorting independently. When linked genes undergo some crossing over, the result is mostly nonrecombinant progeny and fewer recombinant progeny. This result is what we observe among the progeny of the testcross illustrated in Figure 5.6, so we conclude that the two genes show evidence of linkage with some crossing over.

Calculating Recombination Frequency

The percentage of recombinant progeny produced in a cross is called the **recombination frequency** (or rate of recombination), which is calculated as follows:

$$\text{recombination frequency} = \frac{\text{number of recombinant progeny}}{\text{total number of progeny}} \times 100\%$$

In the testcross shown in Figure 5.6, 15 progeny exhibit new combinations of traits, so the recombination frequency is

$$\frac{8 + 7}{55 + 53 + 8 + 7} \times 100\% = \frac{15}{123} \times 100\% = 12.2\%$$

Thus, 12.2% of the progeny exhibit new combinations of traits resulting from crossing over. The recombination frequency can also be expressed as a decimal fraction (0.122).

▶TRY PROBLEM 9

THINK-PAIR-SHARE Question 3

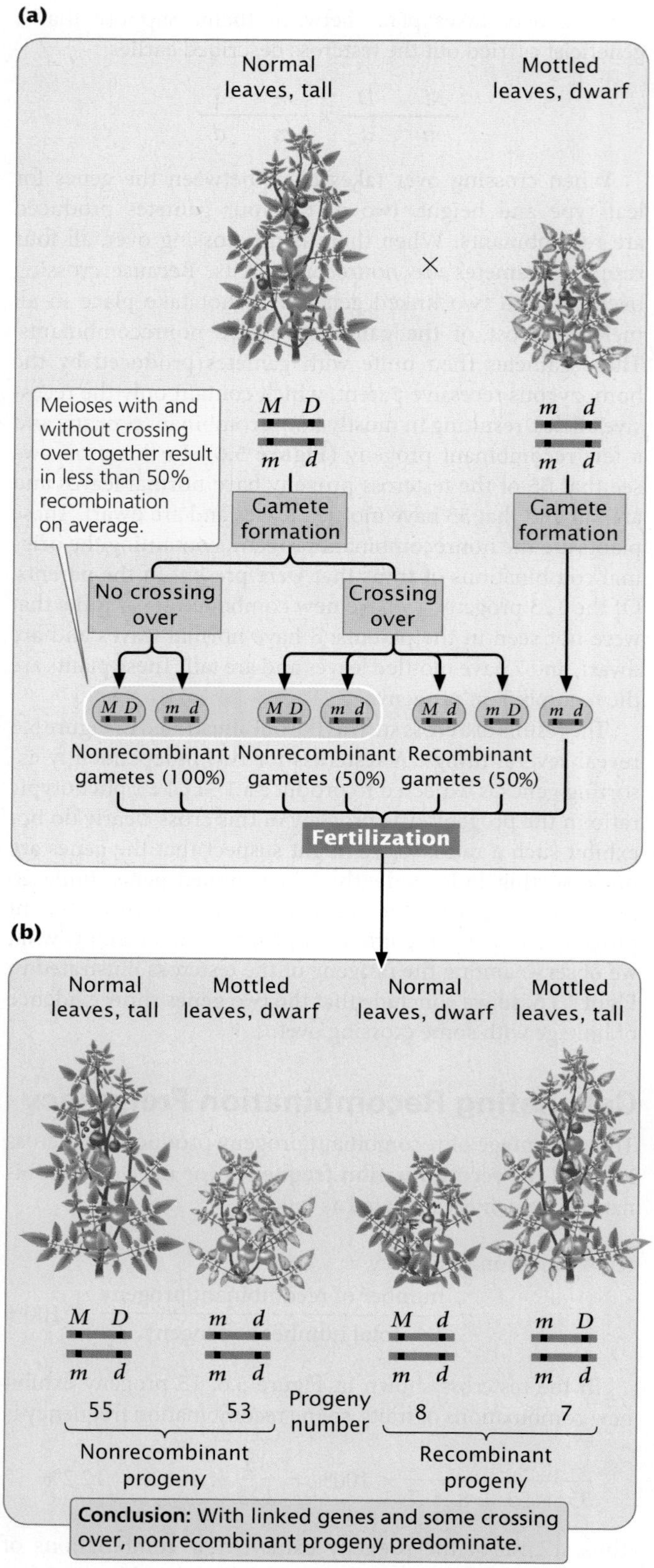

5.6 Crossing over between linked genes produces nonrecombinant and recombinant offspring. In this testcross, genes are linked and there is some crossing over.

Coupling and Repulsion

In crosses for linked genes, the arrangement of alleles on the homologous chromosomes is critical in determining the outcome of the cross. For example, consider the inheritance of two genes in the Australian blowfly, *Lucilia cuprina*. In this species, one locus determines the color of the thorax: a purple thorax (p) is recessive to the normal green thorax (p^+). A second locus determines the color of the puparium: a black puparium (b) is recessive to the normal brown puparium (b^+). The loci for thorax color and puparium color are located close together on the same chromosome. Suppose that we test-cross a fly that is heterozygous at both loci with a fly that is homozygous recessive at both. Because these genes are linked, there are two possible arrangements on the chromosomes of the heterozygous parent. The dominant alleles for green thorax (p^+) and brown puparium (b^+) might reside on one chromosome of the homologous pair, and the recessive alleles for purple thorax (p) and black puparium (b) might reside on the other homologous chromosome:

$$\frac{p^+ \quad b^+}{p \quad b}$$

This arrangement, in which wild-type alleles are found on one chromosome and mutant alleles are found on the other chromosome, is referred to as the **coupling** or **cis configuration**. Alternatively, one chromosome might carry the alleles for green thorax (p^+) and black puparium (b), and the other chromosome might carry the alleles for purple thorax (p) and brown puparium (b^+):

$$\frac{p^+ \quad b}{p \quad b^+}$$

This arrangement, in which each chromosome contains one wild-type and one mutant allele, is called the **repulsion** or **trans configuration**. Whether the alleles in the heterozygous parent are in coupling or repulsion determines which phenotypes will be most common among the progeny of a testcross.

When the alleles are in the coupling configuration, the most numerous progeny types are those with a green thorax and brown puparium and those with a purple thorax and black puparium (**Figure 5.7a**). However, when the alleles of the heterozygous parent are in repulsion, the most numerous progeny types are those with a green thorax and black puparium and those with a purple thorax and brown puparium (**Figure 5.7b**). Notice that the genotypes of the parents in Figure 5.7a and 5.7b are the same ($p^+p\,b^+b \times pp\,bb$) and that the dramatic difference in the phenotypic ratios of the progeny in the two crosses results entirely from the configuration—coupling or repulsion—of the chromosomes. Knowledge of the arrangement of the alleles on the chromosomes is essential to accurately predict the outcome of crosses in which genes are linked.

THINK-PAIR-SHARE Question 4

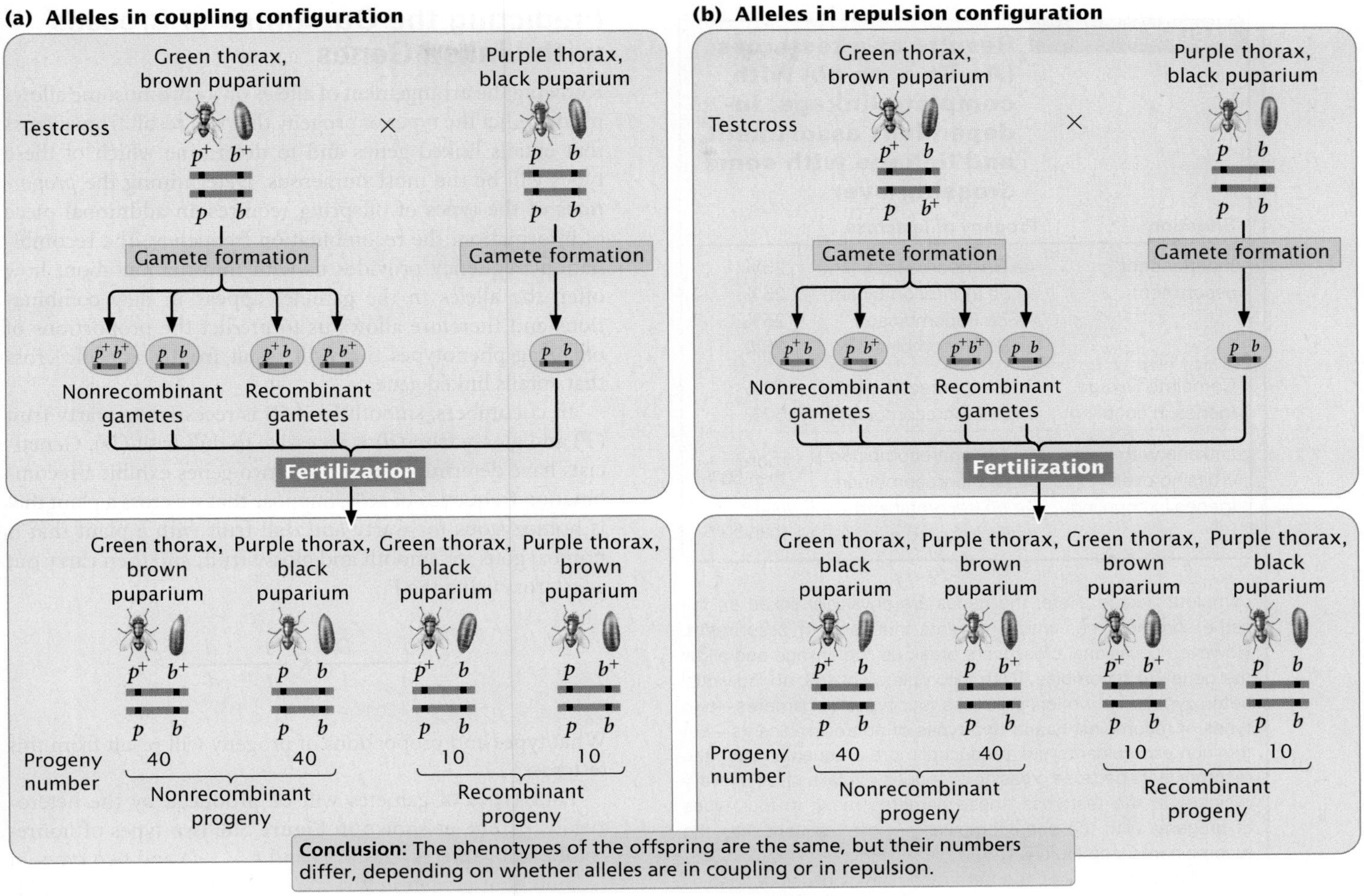

5.7 The arrangement (coupling or repulsion) of linked genes on a chromosome affects the results of a testcross. Linked loci in the Australian blowfly (*Lucilia cuprina*) determine the color of the thorax and that of the puparium.

CONCEPTS

In a cross, the arrangement of linked alleles on the chromosomes is critical for determining the outcome. When two wild-type alleles are on one homologous chromosome and two mutant alleles are on the other, they are in the coupling configuration; when each chromosome contains one wild-type allele and one mutant allele, the alleles are in repulsion.

✓ CONCEPT CHECK 2

The following testcross produces the progeny shown: *Aa Bb* × *aa bb* → 10 *Aa Bb*, 40 *Aa bb*, 40 *aa Bb*, 10 *aa bb*. Were the A and B alleles in the *Aa Bb* parent in coupling or in repulsion?

CONNECTING CONCEPTS

Relating Independent Assortment, Linkage, and Crossing Over

We have now considered three possible situations among genes at different loci. First, the genes may be located on different chromosomes; in this case, they exhibit independent assortment and combine randomly when gametes are formed. An individual heterozygous at two loci (*Aa Bb*) produces four types of gametes (*A B*, *a b*, *A b*, and *a B*) in equal proportions: two types of nonrecombinants and two types of recombinants. In a testcross, these gametes will result in four types of progeny in equal proportions (**Table 5.1**).

Second, the genes may be completely linked—meaning that they are on the same chromosome and lie so close together that crossing over between them is absent. In this case, the genes do not recombine. An individual heterozygous for two completely linked genes in the coupling configuration

$$\frac{A \quad B}{a \quad b}$$

produces only nonrecombinant gametes containing alleles *A B* or *a b*. The alleles do not assort into new combinations such as *A b* or *a B*. In a testcross, completely linked genes will produce only two types of progeny, both nonrecombinants, in equal proportions (see Table 5.1).

The third situation, incomplete linkage, is intermediate between the two extremes of independent assortment and

TABLE 5.1 Results of a testcross (*Aa Bb* × *aa bb*) with complete linkage, independent assortment, and linkage with some crossing over

Situation	Progeny of Testcross	
Independent assortment	*Aa Bb* (nonrecombinant)	25%
	aa bb (nonrecombinant)	25%
	Aa bb (recombinant)	25%
	aa Bb (recombinant)	25%
Complete linkage (genes in coupling)	*Aa Bb* (nonrecombinant)	50%
	aa bb (nonrecombinant)	50%
Linkage with some crossing over (genes in coupling)	*Aa Bb* (nonrecombinant) *aa bb* (nonrecombinant)	more than 50%
	Aa bb (recombinant) *aa Bb* (recombinant)	less than 50%

complete linkage. Here, the genes are physically linked on the same chromosome, which prevents independent assortment. However, occasional crossovers break up the linkage and allow the genes to recombine. With incomplete linkage, an individual heterozygous at two loci produces four types of gametes—two types of recombinants and two types of nonrecombinants—but the nonrecombinants are produced more frequently than the recombinants because crossing over does not take place in every meiosis. In the testcross, these gametes result in four types of progeny, with the nonrecombinants more frequent than the recombinants (see Table 5.1).

Earlier in the chapter, the term recombination was defined as the sorting of alleles into new combinations. We've now considered two types of recombination that differ in their mechanisms. Interchromosomal recombination takes place between genes located on *different* chromosomes. It arises from independent assortment—the random segregation of chromosomes in anaphase I of meiosis—and is the kind of recombination that Mendel discovered while studying dihybrid crosses. A second type of recombination, intrachromosomal recombination, takes place between genes located on the *same* chromosome. This recombination arises from crossing over—the exchange of genetic material in prophase I of meiosis. Both types of recombination produce new allele combinations in the gametes, so they cannot be distinguished by examining the types of gametes produced. Nevertheless, they can often be distinguished by the *frequencies* of types of gametes: interchromosomal recombination produces 50% nonrecombinant gametes and 50% recombinant gametes, whereas intrachromosomal recombination frequently produces more than 50% nonrecombinant gametes and less than 50% recombinant gametes. However, when the genes are very far apart on the same chromosome, crossing over takes place in every meiotic division, leading to 50% recombinant gametes and 50% nonrecombinant gametes. This result is the same as that for the independent assortment of genes located on different chromosomes (interchromosomal recombination). Thus, the intrachromosomal recombination of genes that lie far apart on the same chromosome and interchromosomal recombination are phenotypically indistinguishable.

Predicting the Outcomes of Crosses with Linked Genes

Knowing the arrangement of alleles on a chromosome allows us to predict the types of progeny that will result from a cross that entails linked genes and to determine which of these types will be the most numerous. Determining the *proportions* of the types of offspring requires an additional piece of information: the recombination frequency. The recombination frequency provides us with information about how often the alleles in the gametes appear in new combinations and therefore allows us to predict the proportions of offspring phenotypes that will result from a specific cross that entails linked genes.

In cucumbers, smooth fruit (t) is recessive to warty fruit (T) and glossy fruit (d) is recessive to dull fruit (D). Geneticists have determined that these two genes exhibit a recombination frequency of 16%. Suppose that we cross a plant that is homozygous for warty and dull fruit with a plant that is homozygous for smooth and glossy fruit, and then carry out a testcross using the F_1:

$$\frac{T \quad D}{t \quad d} \times \frac{t \quad d}{t \quad d}$$

What types and proportions of progeny will result from this testcross?

Four types of gametes will be produced by the heterozygous parent, as shown in **Figure 5.8**: two types of nonrecombinant gametes ($\underline{T \quad D}$ and $\underline{t \quad d}$) and two types of recombinant gametes ($\underline{T \quad d}$ and $\underline{t \quad D}$). The recombination frequency tells us that 16% of the gametes produced by the heterozygous parent will be recombinants. Because there are two types of recombinant gametes, each should arise with a frequency of $16\%/2 = 8\%$. This frequency can also be represented as a probability of 0.08. All the other gametes will be nonrecombinants, so they should arise with a frequency of $100\% - 16\% = 84\%$. Because there are two types of nonrecombinant gametes, each should arise with a frequency of $84\%/2 = 42\%$ (or 0.42). The other parent in the testcross is homozygous and therefore produces only a single type of gamete ($\underline{t \quad d}$) with a frequency of 100% (or 1.00).

Four types of progeny result from the testcross (see Figure 5.8). The expected proportion of each progeny type can be determined by using the multiplication rule (see Chapter 3), multiplying together the probability of each gamete. For example, testcross progeny with warty and dull fruit

$$\frac{T \quad D}{t \quad d}$$

appear with a frequency of 0.42 (the probability of inheriting a gamete with chromosome $\underline{T \quad D}$ from the heterozygous

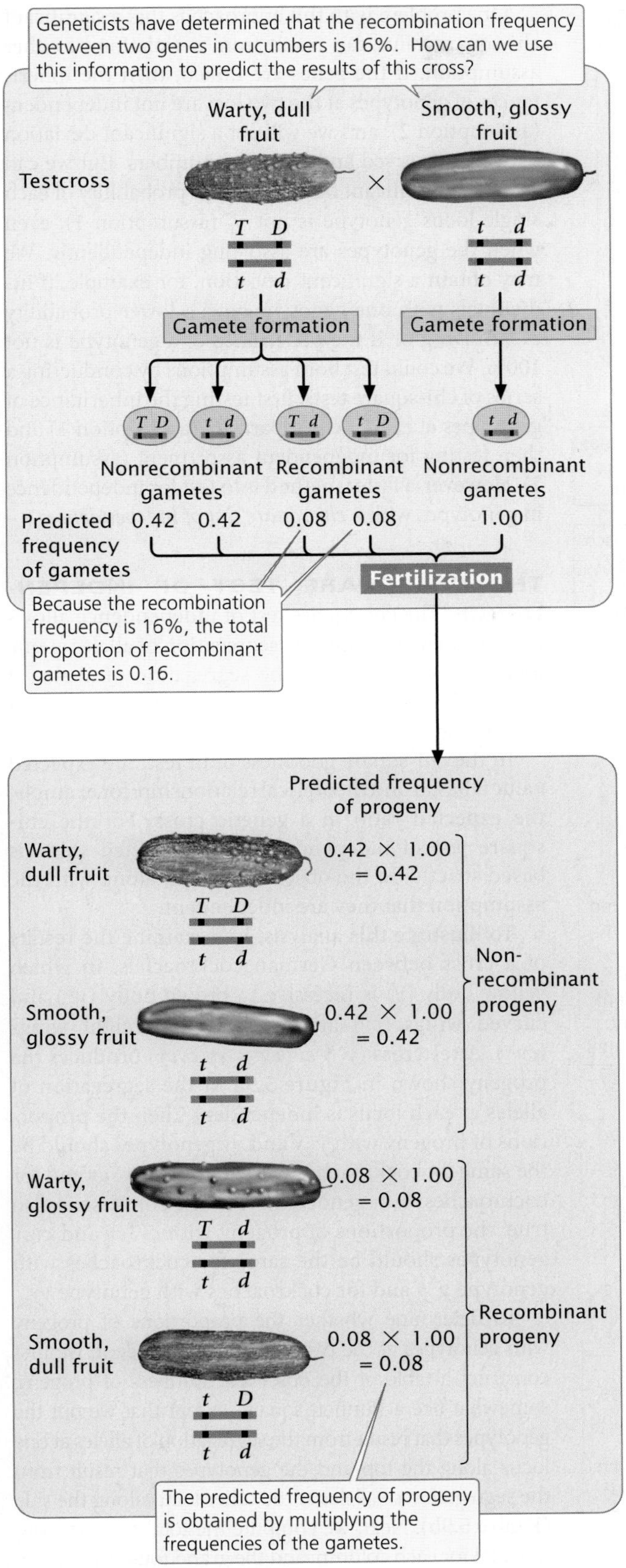

5.8 The recombination frequency allows us to predict the proportions of offspring expected for a cross that entails linked genes.

parent) × 1.00 (the probability of inheriting a gamete with chromosome *t* ——— *d* from the recessive parent) = 0.42. The proportions of the other types of F_2 progeny can be calculated in a similar manner (see Figure 5.8). This method can be used for predicting the outcome of any cross with linked genes for which the recombination frequency is known.

THINK-PAIR-SHARE Question 5

Testing for Independent Assortment

In some crosses, the genes are obviously linked because there are clearly more nonrecombinant progeny than recombinant progeny. In other crosses, the difference between independent assortment and linkage isn't as obvious. For example, suppose that we did a testcross for two pairs of genes, such as *Aa Bb* × *aa bb*, and observed the following numbers of progeny: 54 *Aa Bb*, 56 *aa bb*, 42 *Aa bb*, and 48 *aa Bb*. Is this outcome the 1 : 1 : 1 : 1 ratio that we would expect if *A* and *B* assorted independently? Not exactly, but it's pretty close. Perhaps these genes assorted independently and chance produced the slight deviations between the observed numbers and the expected 1 : 1 : 1 : 1 ratio. Alternatively, the genes might be linked, but considerable crossing over might be taking place between them, so that the number of nonrecombinants is only slightly greater than the number of recombinants. How do we distinguish between the role of chance and the role of linkage in producing deviations from the results expected with independent assortment?

We encountered a similar problem in crosses in which genes were unlinked: the problem of distinguishing between deviations due to chance and those due to other factors. We addressed this problem (in Chapter 3) with the chi-square goodness-of-fit test, which helps us evaluate the likelihood that chance alone is responsible for deviations between the numbers of progeny that we observe and the numbers that we expect according to the principles of inheritance. Here, we are interested in a different question: Is the inheritance of alleles at one locus independent of the inheritance of alleles at a second locus? If the answer to this question is yes, then the genes are assorting independently; if the answer is no, then the genes are probably linked.

A possible way to test for independent assortment is to calculate the expected probability of each progeny type, assuming independent assortment, and then use the chi-square goodness-of-fit test to evaluate whether the observed numbers deviate significantly from the expected numbers. With independent assortment, we expect $^1/_4$ of each phenotype: $^1/_4$ *Aa Bb*, $^1/_4$ *aa bb*, $^1/_4$ *Aa bb*, and $^1/_4$ *aa Bb*. This expected probability of each genotype is based on the multiplication rule of probability (see Chapter 3). For example, if the probability of *Aa* is $^1/_2$ and the probability of *Bb* is $^1/_2$, then the probability of *Aa Bb* is $^1/_2 \times {}^1/_2 = {}^1/_4$. In this calculation, we are making two assumptions: (1) that the probability of each single-locus genotype is $^1/_2$ and (2) that genotypes at the two loci are inherited independently ($^1/_2 \times {}^1/_2 = {}^1/_4$).

(a)

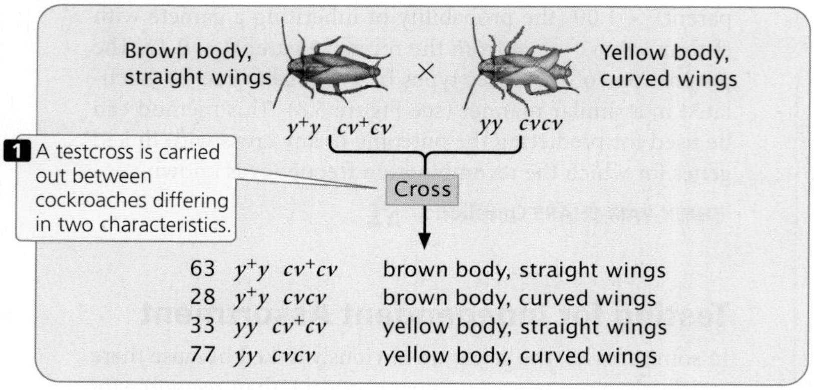

(b) Contingency table

2 To test for independent assortment of alleles encoding the two traits, a table is constructed...

3 ...with genotypes for one locus along the top...

4 ...and genotypes for the other locus along the left side.

5 Numbers of each genotype are placed in the table cells, and the row totals, column totals, and grand total are computed.

Segregation of cv^+ and cv	Segregation of y^+ and y: y^+y	yy	Row totals
cv^+cv	63	33	96
$cv\ cv$	28	77	105
Column totals	91	110	201 (Grand total)

(c)

Genotype	Number observed	Number expected $\left(\frac{\text{row total} \times \text{column total}}{\text{grand total}}\right)$
$y^+y\ cv^+cv$	63	$\frac{96 \times 91}{201} = 43.46$
$y^+y\ cvcv$	28	$\frac{105 \times 91}{201} = 47.54$
$yy\ cv^+cv$	33	$\frac{96 \times 110}{201} = 52.46$
$yy\ cvcv$	77	$\frac{105 \times 110}{201} = 57.46$

6 The expected numbers of progeny, assuming independent assortment, are calculated.

(d)

7 A chi-square value is calculated.

$$\chi^2 = \Sigma \frac{(\text{observed}-\text{expected})^2}{\text{expected}}$$

$$= \frac{(63-43.46)^2}{43.46} + \frac{(28-47.54)^2}{47.54} + \frac{(33-52.54)^2}{52.54} + \frac{(77-57.46)^2}{57.46}$$

$$= 8.79 + 8.03 + 7.27 + 6.64$$

$$= 30.73$$

(e)

df = (number of rows – 1) × (number of columns – 1)

df = (2 – 1) × (2 – 1) = 1 × 1 = 1

$P < 0.005$

8 The probability is less than 0.005, indicating that the *difference* between numbers of observed and expected progeny is probably not due to chance.

Conclusion: The genes for body color and type of wing are not assorting independently and must be linked.

5.9 A chi-square test of independence can be used to determine whether genes at two loci are assorting independently.

One problem with this approach is that a significant chi-square value can result from a violation of either assumption. If the genes are linked, then the inheritances of genotypes at the two loci are not independent (assumption 2), and we will get a significant deviation between observed and expected numbers. But we can also get a significant deviation if the probability of each single-locus genotype is not $^1/_2$ (assumption 1), even when the genotypes are assorting independently. We may obtain a significant deviation, for example, if individuals with one genotype have a lower probability of surviving or if the penetrance of a genotype is not 100%. We could test both assumptions by conducting a series of chi-square tests, first testing the inheritance of genotypes at each locus separately (assumption 1) and then testing for independent assortment (assumption 2). However, a faster method is to test for independence in genotypes with a *chi-square test of independence*.

THE CHI-SQUARE TEST OF INDEPENDENCE The chi-square test of independence allows us to evaluate whether the segregation of alleles at one locus is independent of the segregation of alleles at another locus without making any assumption about the probability of single-locus genotypes.

In the chi-square goodness of fit test, the expected value is based on a theoretical relationship: for example, the expected ratio in a genetic cross. For the chi-square test of independence, the expected value is based strictly on the observed values, along with the assumption that they are independent.

To illustrate this analysis, let's examine the results of a cross between German cockroaches, in which yellow body (y) is recessive to brown body (y^+) and curved wings (cv) are recessive to straight wings (cv^+). A testcross ($y^+y\ cv^+cv \times yy\ cvcv$) produces the progeny shown in **Figure 5.9a**. If the segregation of alleles at each locus is independent, then the proportions of progeny with y^+y and yy genotypes should be the same for cockroaches with genotype cv^+cv and for cockroaches with genotype $cvcv$. The converse is also true: the proportions of progeny with cv^+cv and $cvcv$ genotypes should be the same for cockroaches with genotype y^+y and for cockroaches with genotype yy.

To determine whether the proportions of progeny with genotypes at the two loci are independent, we first construct a table of the observed numbers of progeny, somewhat like a Punnett square, except that we put the genotypes that result from the segregation of alleles at one locus along the top and the genotypes that result from the segregation of alleles at the other locus along the side (**Figure 5.9b**). Next, we compute the total for each row, the total for each column, and the grand total (the sum of all row totals or the sum of all column totals, which should be the same). These totals will be used to compute the expected values for the chi-square test of independence.

Our next step is to compute the expected values for each combination of genotypes (each cell in the table) under the assumption that the segregation of alleles at the *y* locus is independent of the segregation of alleles at the *cv* locus. If the segregation of alleles at each locus is independent, the expected number in each cell can be computed with the following formula:

$$\text{expected number} = \frac{\text{row total} \times \text{column total}}{\text{grand total}}$$

For the cell of the table corresponding to genotype $y^+y\,cv^+cv$ (the upper left-hand cell of the table in Figure 5.9b), the expected number is

$$\frac{96\ (\text{row total}) \times 91\ (\text{column total})}{201\ (\text{grand total})} = \frac{8736}{201} = 43.46$$

With the use of this method, we get the expected numbers for each cell that are given in **Figure 5.9c**.

We now calculate a chi-square value by using the same formula that we used for the chi-square goodness-of-fit test in Chapter 3:

$$\chi^2 = \sum \frac{(\text{observed} - \text{expected})^2}{\text{expected}}$$

Recall that $\sum$ means "sum" and that we are adding together the (observed − expected)^2/expected values for each type of progeny. With the observed and expected numbers of cockroaches from the testcross, the calculated chi-square value is 30.73 (**Figure 5.9d**).

To determine the probability associated with this chi-square value, we need the degrees of freedom. Recall from Chapter 3 that the degrees of freedom are the number of ways in which the observed classes are free to vary from the expected values. In general, for the chi-square test of independence, the degrees of freedom equal the number of rows in the table minus 1 multiplied by the number of columns in the table minus 1 (**Figure 5.9e**), or

$$\text{df} = (\text{number of rows} - 1) \times (\text{number of columns} - 1)$$

In our example, there are two rows and two columns, and so the degrees of freedom are

$$\text{df} = (2 - 1) \times (2 - 1) = 1 \times 1 = 1$$

Therefore, our calculated chi-square value is 30.73, with 1 degree of freedom. We can use Table 3.4 to find the associated probability. Looking at Table 3.4, we find that our calculated chi-square value is larger than the largest chi-square value given for 1 degree of freedom, which has a probability of 0.005. Thus, our calculated chi-square value has a probability less than 0.005. This very small probability indicates that the genotypes are not in the proportions that we would expect if independent assortment were taking place. Our conclusion, then, is that these genes are not assorting independently and must be linked. As is the case for the chi-square goodness-of-fit test, geneticists generally consider that any chi-square value for the test of independence with a probability less than 0.05 is significantly different from the expected values and is therefore evidence that the genes are not assorting independently. ▶TRY PROBLEM 10

Gene Mapping with Recombination Frequencies

Thomas Hunt Morgan and his students developed the idea that physical distances between genes on a chromosome are related to their rates of recombination. They hypothesized that crossover events take place more or less at random up and down the chromosome and that two genes that lie far apart are more likely to undergo a crossover than are two genes that lie close together. They proposed that recombination frequencies could provide a convenient way to determine the order of genes along a chromosome and would give estimates of the relative distances between the genes. Chromosome maps calculated by using the genetic phenomenon of recombination are called **genetic maps**. In contrast, chromosome maps calculated by using physical distances along the chromosome (often expressed as numbers of base pairs) are called **physical maps**.

Distances on genetic maps are measured in **map units** (abbreviated m.u.); one map unit equals a 1% recombination rate. Map units are also called **centiMorgans** (cM), in honor of Thomas Hunt Morgan. Genetic distances measured with recombination rates are approximately additive: if the distance from gene *A* to gene *B* is 5 m.u., the distance from gene *B* to gene *C* is 10 m.u., and the distance from gene *A* to gene *C* is 15 m.u., then gene *B* must be located between genes *A* and *C*. On the basis of the map distances just given, we can draw a simple genetic map for genes *A*, *B*, and *C*, as shown here:

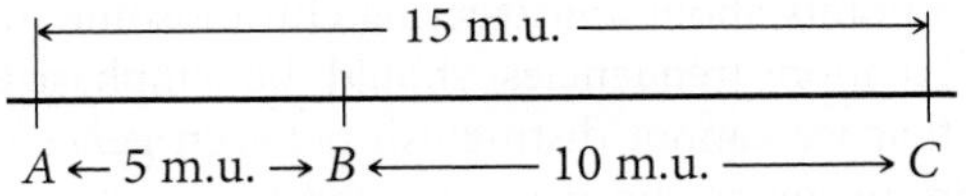

We could just as plausibly draw this map with *C* on the left and *A* on the right:

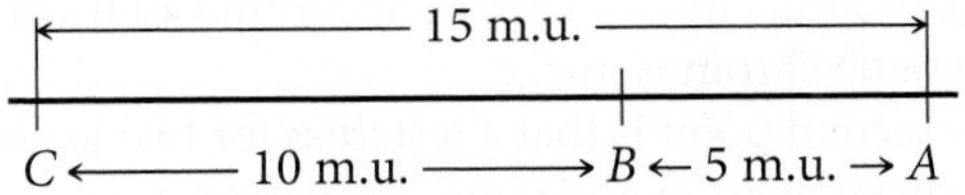

Both maps are correct and equivalent because, with information about the relative positions of only three genes, the most that we can determine is which gene lies in the middle. If we obtained distances to an additional gene, then we could position *A* and *C* relative to that gene. An additional gene *D*, examined through genetic crosses, might yield the following recombination frequencies:

Gene pair	Recombination frequency (%)
A and *D*	8
B and *D*	13
C and *D*	23

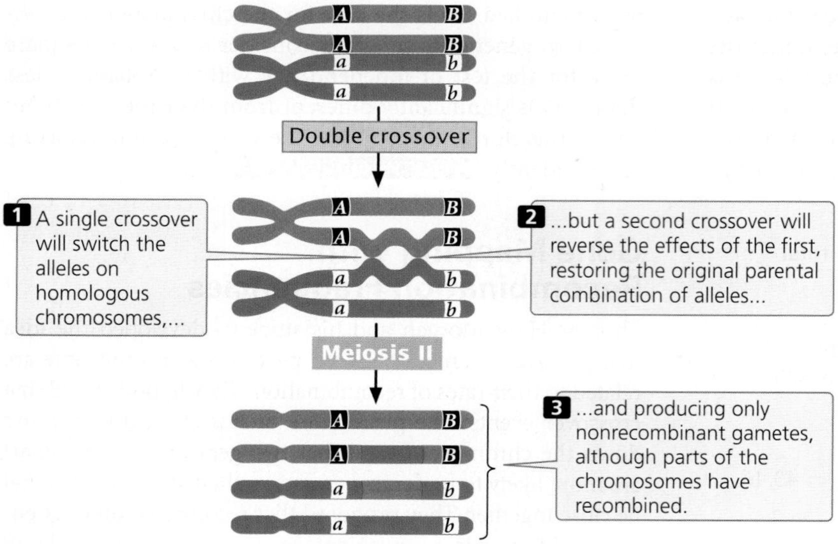

5.10 A two-strand double crossover between two linked genes produces only nonrecombinant gametes.

Notice that *C* and *D* exhibit the highest frequency of recombination; therefore, *C* and *D* must be farthest apart, with genes *A* and *B* between them. Using the recombination frequencies and remembering that 1 m.u. = 1% recombination, we can now add *D* to our map:

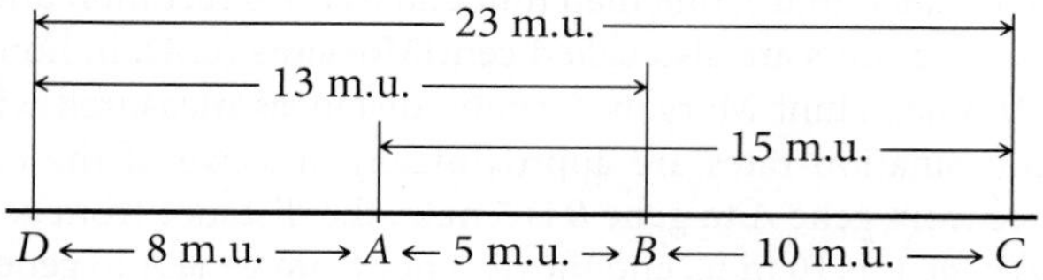

Thus, by doing a series of crosses between pairs of genes, we can construct genetic maps showing the linkage arrangements of a number of genes.

Two points about constructing chromosome maps from recombination frequencies should be emphasized. First, recall that we cannot distinguish between genes on different chromosomes and genes located far apart on the same chromosome. If genes exhibit 50% recombination, the most that can be said about them is that they belong to different linkage groups, either on different chromosomes or far apart on the same chromosome.

The second point is that a testcross for two genes that are relatively far apart on the same chromosome tends to underestimate the true physical distance because the cross does not reveal double crossovers that might take place between the two genes (**Figure 5.10**). A double crossover arises when two separate crossover events take place between two loci. (For now, we will consider only double crossovers that take place between two of the four chromatids of a homologous pair—two-strand double crossovers. Double crossovers that take place among three and four chromatids will be considered later, in the section Effects of Multiple Crossovers.) Whereas a single crossover produces combinations of alleles that were not present on the original parental chromosomes, a second crossover between the same two genes reverses the effects of the first, thus restoring the original parental combination of alleles (see Figure 5.10). We therefore cannot distinguish between the progeny produced by two-strand double crossovers and the progeny produced when there is no crossing over at all. As we will see in the next section, however, we can detect double crossovers if we examine a third gene that lies between the two crossovers. Because double crossovers between two genes go undetected, map distances will be underestimated whenever double crossovers take place. Double crossovers are more frequent between genes that are far apart; therefore, genetic maps based on short distances are usually more accurate than those based on longer distances.

CONCEPTS

A genetic map provides the order of the genes on a chromosome and the approximate distances from one gene to another based on recombination frequencies. In genetic maps, 1% recombination equals 1 map unit, or 1 centiMorgan. Double crossovers between two genes go undetected, so map distances between distant genes tend to underestimate genetic distances.

✓ CONCEPT CHECK 3

How does a genetic map differ from a physical map?

Constructing a Genetic Map with Two-Point Testcrosses

Genetic maps can be constructed by conducting a series of testcrosses. In each testcross, one of the parents is heterozygous for a different pair of genes, and recombination frequencies are calculated between pairs of genes. A testcross between two genes is called a **two-point testcross**, or simply a two-point cross. Suppose that we carried out a series of two-point crosses for four genes, *a*, *b*, *c*, and *d*, and obtained the following recombination frequencies:

Gene loci in testcross	**Recombination frequency (%)**
a and *b*	50
a and *c*	50
a and *d*	50
b and *c*	20
b and *d*	10
c and *d*	28

We can begin constructing a genetic map for these genes by considering the recombination frequencies for each pair of genes. The recombination frequency between *a* and *b* is 50%, which is the recombination frequency expected with independent assortment. Therefore, genes *a* and *b* may either be on different chromosomes or very far apart on the same chromosome; we will place them in different linkage groups with the understanding that they may or may not be on the same chromosome:

Linkage group 1

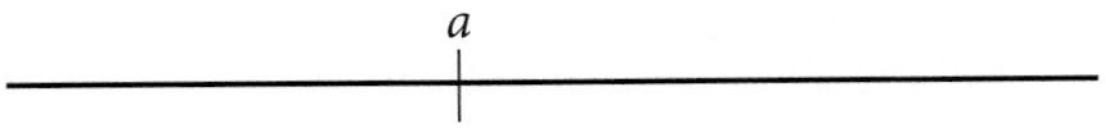

Linkage group 2

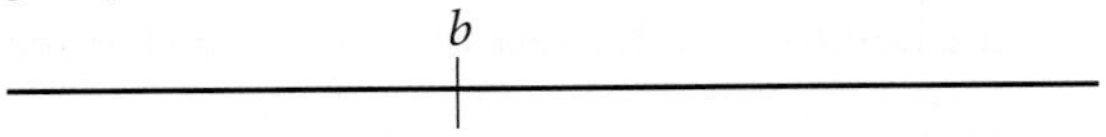

The recombination frequency between *a* and *c* is 50%, indicating that they, too, are in different linkage groups. The recombination frequency between *b* and *c* is 20%, so these genes are linked and separated by 20 map units:

Linkage group 1

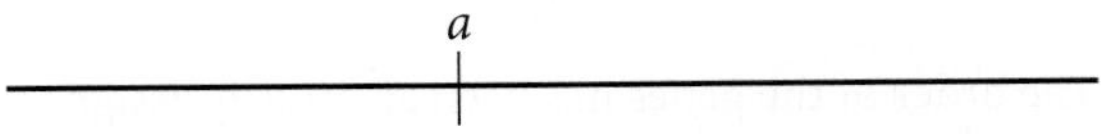

Linkage group 2

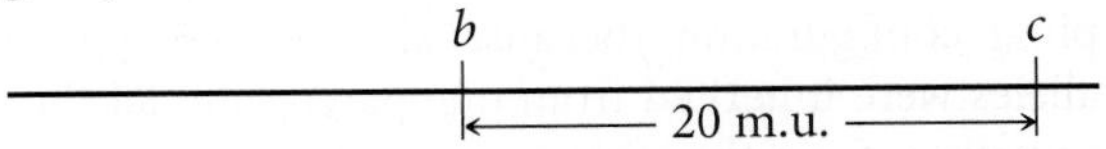

The recombination frequency between *a* and *d* is 50%, indicating that these genes belong to different linkage groups, whereas genes *b* and *d* are linked, with a recombination frequency of 10%. To decide whether gene *d* is 10 m.u. to the left or to the right of gene *b*, we must consult the *c*-to-*d* distance. If gene *d* is 10 m.u. to the left of gene *b*, then the distance between *d* and *c* should be approximately the sum of the distance between *b* and *c* and between *c* and *d*: 20 m.u. + 10 m.u. = 30 m.u. This distance will be only approximate because any double crossovers between the two genes will be missed and the map distance will be underestimated. If, on the other hand, gene *d* lies to the right of gene *b*, then the distance between gene *d* and gene *c* will be much shorter, approximately 20 m.u. − 10 m.u. = 10 m.u. Again, the summed distances will be only approximate because any double crossovers between the two genes will be missed and the map distance will be underestimated.

By examining the recombination frequency between *c* and *d*, we can distinguish between these two possibilities. The recombination frequency between *c* and *d* is 28%, so gene *d* must lie to the left of gene *b*. Notice that the sum of the recombination frequency between *d* and *b* (10%) and between *b* and *c* (20%) is greater than the recombination frequency between *d* and *c* (28%). As already discussed, this discrepancy arises because double crossovers between the two outer genes go undetected, causing an underestimation of the true map distance. The genetic map of these genes is now complete:

Linkage group 1

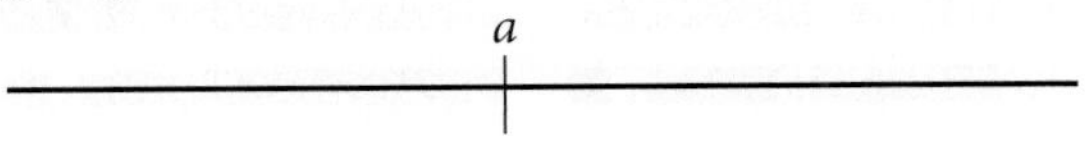

Linkage group 2

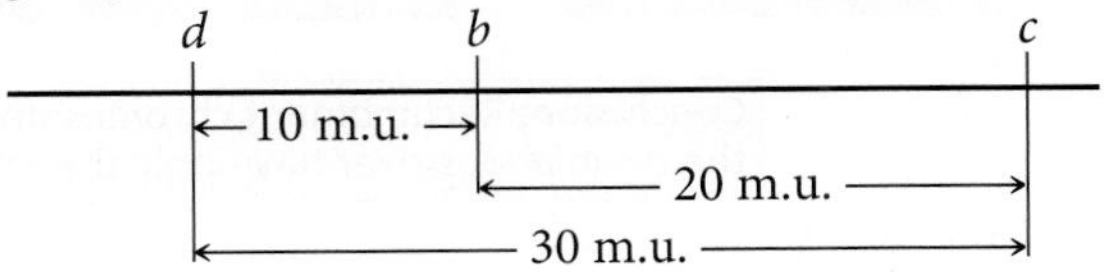

TRY PROBLEM 18

5.3 A Three-Point Testcross Can Be Used To Map Three Linked Genes

While genetic maps can be constructed from a series of testcrosses for pairs of genes, this approach is not particularly efficient because numerous two-point crosses must be carried out to establish the order of the genes and because double crossovers are missed. A more efficient mapping technique is a testcross for three genes—a **three-point testcross**, or three-point cross. With a three-point cross, the order of the three genes can be established in a single set of progeny, and some double crossovers can usually be detected, providing more accurate map distances.

Consider what happens when crossing over takes place among three hypothetical linked genes. **Figure 5.11** illustrates a pair of homologous chromosomes from an individual that is heterozygous at three loci (*Aa Bb Cc*). Notice that the genes are in the coupling configuration: all the dominant alleles are on one chromosome (*A B C*) and all the recessive alleles are on the other chromosome (*a b c*). Three types of crossover events can take place between these three genes: two types of single crossovers (see Figure 5.11a and b) and a double crossover (see Figure 5.11c). In each type of crossover, two of the resulting chromosomes are recombinants and two are nonrecombinants.

Notice that in the recombinant chromosomes resulting from the double crossover, the outer two alleles are the same

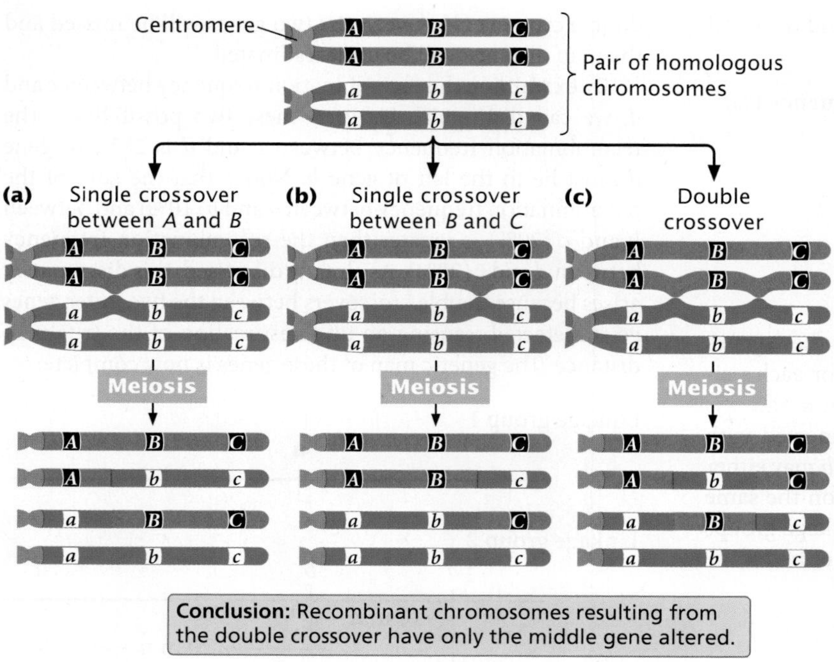

5.11 Three types of crossovers can take place among three linked loci.

as in the nonrecombinants, but the middle allele is different. This result provides us with an important clue about the order of the genes. In progeny that result from a double crossover, only the middle allele should differ from the alleles present in the nonrecombinant progeny.

Constructing a Genetic Map with a Three-Point Testcross

To examine gene mapping with a three-point testcross, we will consider three recessive mutations in the fruit fly *Drosophila melanogaster*. In this species, scarlet eyes (*st*) are recessive to wild-type red eyes (st^+), ebony body color (*e*) is recessive to wild-type gray body color (e^+), and spineless (*ss*)—that is, the presence of small bristles—is recessive to wild-type normal bristles (ss^+). The loci encoding these three characteristics are linked and located on chromosome 3.

We will refer to these three loci as *st*, *e*, and *ss*, but keep in mind that either the recessive alleles (*st*, *e*, and *ss*) or the dominant alleles (st^+, e^+, and ss^+) may be present at each locus. So, when we say that there are 10 m.u. between *st* and *ss*, we mean that there are 10 m.u. between the loci at which mutations *st* and *ss* occur; we could just as easily say that there are 10 m.u. between st^+ and ss^+.

To map these genes, we need to determine their order on the chromosome and the genetic distances between them. First, we must set up a three-point testcross: a cross between a fly heterozygous at all three loci and a fly homozygous for recessive alleles at all three loci. To produce flies heterozygous for all three loci, we might cross a stock of flies that are homozygous for wild-type alleles at all three loci with flies that are homozygous for recessive alleles at all three loci:

$$\text{P} \quad \frac{st^+ \quad e^+ \quad ss^+}{st^+ \quad e^+ \quad ss^+} \times \frac{st \quad e \quad ss}{st \quad e \quad ss}$$

$$\downarrow$$

$$\text{F}_1 \quad \frac{st^+ \quad e^+ \quad ss^+}{st \quad e \quad ss}$$

The order of the genes has been arbitrarily assigned because, at this point, we do not know which one is the middle gene. Additionally, the alleles in these heterozygotes are in coupling configuration (because all the wild-type dominant alleles were inherited from one parent and all the recessive mutations from the other parent), although the testcross can also be done with alleles in repulsion.

In the three-point testcross, we cross the F_1 heterozygotes with flies that are homozygous for all three recessive mutations. In many organisms, it makes no difference whether the heterozygous parent in the testcross is male or female (provided that the genes are autosomal), but in *Drosophila*, no crossing over takes place in males. Because crossing over in the heterozygous parent is essential for determining recombination frequencies, the heterozygous flies in our testcross must be female. So we mate female F_1 flies that are heterozygous for all three traits with male flies that are homozygous for all the recessive traits:

$$\frac{st^+ \quad e^+ \quad ss^+}{st \quad e \quad ss} \text{ Female} \times \frac{st \quad e \quad ss}{st \quad e \quad ss} \text{ Male}$$

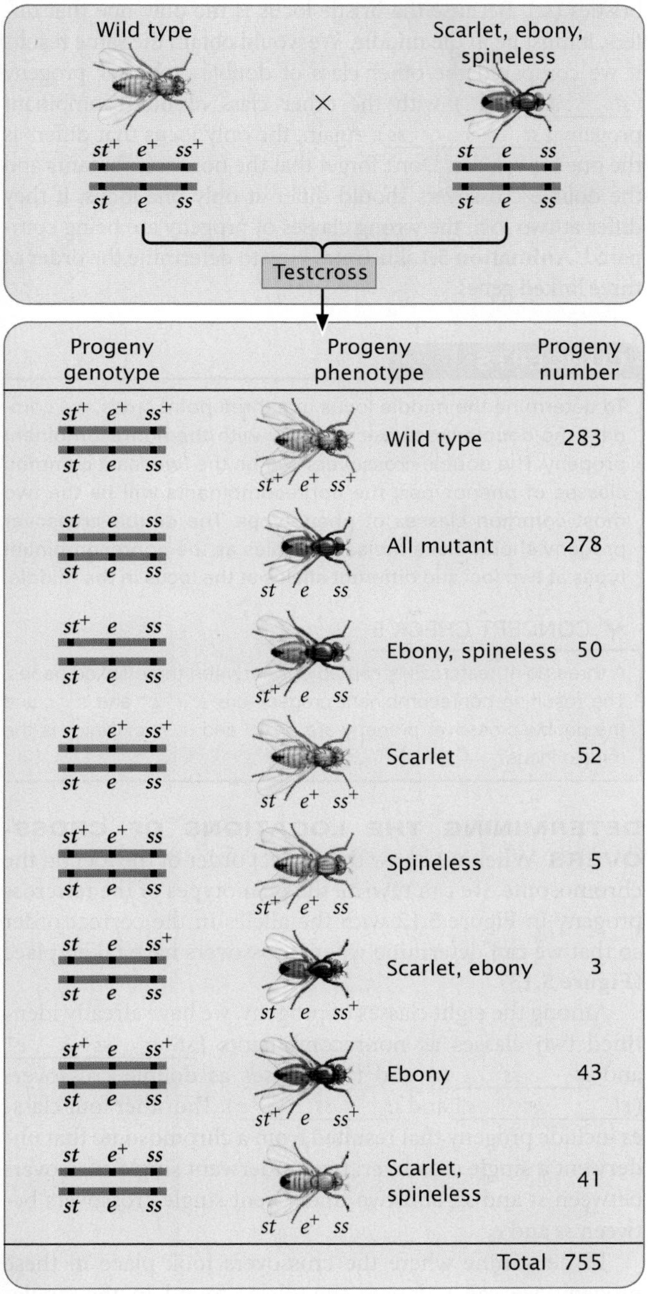

5.12 The results of a three-point testcross can be used to map linked genes. In this three-point testcross of *Drosophila melanogaster*, the recessive mutations that produce scarlet eyes (*st*), ebony body color (*e*), and spineless bristles (*ss*) are at three linked loci. The order of the loci has been assigned arbitrarily. Each phenotypic class includes both male and female flies; the sex of the pictured flies is random.

The progeny produced by this cross are listed in **Figure 5.12**. For each locus, two classes of progeny are produced: progeny that are heterozygous, displaying the dominant trait, and progeny that are homozygous, displaying the recessive trait. With two classes of progeny possible for each of the three loci, there will be $2^3 = 8$ classes of phenotypes possible in the progeny.

In this example, all eight phenotypic classes are present, but in some three-point crosses, one or more of the phenotypes may be missing if the number of progeny is limited. Nevertheless, the absence of a particular class can provide important information about which combination of traits is least frequent and, ultimately, about the order of the genes, as we will see.

To map the genes, we need information about where and how often crossing over has taken place. In the homozygous recessive parent, the two alleles at each locus are the same, and so crossing over will have no effect on the types of gametes produced: with or without crossing over, all gametes from this parent have a chromosome with three recessive alleles ($\underline{st \quad e \quad ss}$). In contrast, the heterozygous parent has different alleles on its two chromosomes, and so crossing over can be detected. The information that we need for mapping, therefore, comes entirely from the gametes produced by the heterozygous parent. Because chromosomes contributed by the homozygous parent carry only recessive alleles, whatever alleles are present on the chromosome contributed by the heterozygous parent will be expressed in the progeny.

As a shortcut, we often do not write out the complete genotypes of the testcross progeny, listing instead only the alleles expressed in the phenotype, which are the alleles inherited from the heterozygous parent. This convention is used in the discussion that follows.

CONCEPTS

To map genes, information about the location and number of crossovers in the gametes that produced the progeny of a cross is needed. An efficient way to obtain this information is to use a three-point testcross, in which an individual heterozygous at three linked loci is crossed with an individual that is homozygous recessive at the same three loci.

✓ CONCEPT CHECK 4

Write the genotypes of all recombinant and nonrecombinant progeny expected from the following three-point cross:

$$\frac{m^+ \quad p^+ \quad s^+}{m \quad\; p \quad\; s} \times \frac{m \quad p \quad s}{m \quad p \quad s}$$

DETERMINING THE GENE ORDER The first task in mapping the genes is to determine their order on the chromosome. In Figure 5.12, we arbitrarily listed the loci in the order *st*, *e*, *ss*, but we had no way of knowing which of the three loci was between the other two. We can now identify the middle locus by examining the double-crossover progeny.

First, determine which progeny are the nonrecombinants; they will be the two most numerous classes of progeny (even if crossing over takes place in every meiosis, the nonrecombinants will constitute at least 50% of the progeny). Among the progeny of the testcross in Figure 5.12, the most numerous are those with all three dominant traits ($\underline{st^+ \quad e^+ \quad ss^+}$) and those with all three recessive traits ($\underline{st \quad e \quad ss}$).

Next, identify the double-crossover progeny. These progeny should always have the two least numerous phenotypes because the probability of a double crossover is always less than the probability of a single crossover. The least common progeny among those listed in Figure 5.12 are progeny with spineless bristles (st^{+} e^{+} ss) and progeny with scarlet eyes and ebony body (st e ss^{+}), so they are the double-crossover progeny.

Three orders of genes on the chromosome are possible: the eye-color locus could be in the middle (e st ss), the body-color locus could be in the middle (st e ss), or the bristle locus could be in the middle (st ss e). To determine which gene is in the middle, we can draw the chromosomes of the heterozygous parent with all three possible gene orders and then see if a double crossover produces the combination of genes observed in the double-crossover progeny. The three possible gene orders and the types of progeny produced by their double crossovers are

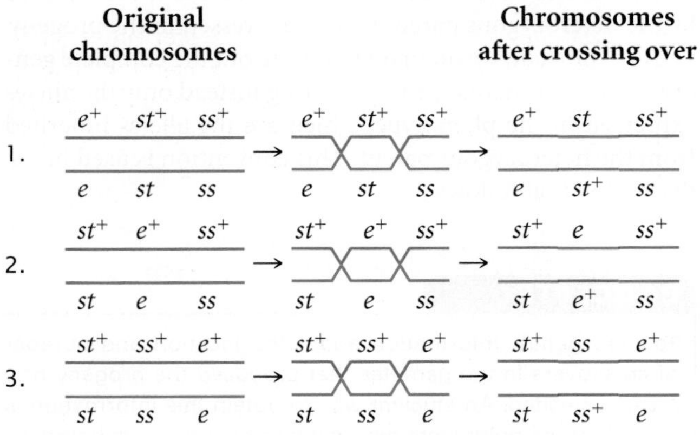

The only gene order that produces chromosomes with the set of alleles observed in the least numerous progeny—the double crossovers (st^{+} e^{+} ss and st e ss^{+} in Figure 5.12)—is the one in which the *ss* locus for bristles lies in the middle (gene order 3). Therefore, this order (st ss e) must be the correct sequence of genes on the chromosome.

With a little practice, we can quickly determine which locus is in the middle without writing out all the gene orders. The phenotypes of the progeny are expressions of the alleles inherited from the heterozygous parent. Recall that when we looked at the results of double crossovers (see Figure 5.11), only the alleles at the middle locus differed from those in the nonrecombinants. If we compare the nonrecombinant progeny with the double-crossover progeny, they should differ only in alleles at the middle locus.

Let's compare the alleles in the double-crossover progeny st^{+} e^{+} ss with those in the nonrecombinant progeny st^{+} e^{+} ss^{+}. We see that both have an allele for red eyes (st^{+}) and both have an allele for gray body (e^{+}), but the nonrecombinants have an allele for normal bristles (ss^{+}), whereas the double crossovers have an allele for spineless bristles (*ss*). Because the bristle locus is the only one that differs, it must lie in the middle. We would obtain the same results if we compared the other class of double-crossover progeny (st e ss^{+}) with the other class of nonrecombinant progeny (st e ss). Again, the only locus that differs is the one for bristles. Don't forget that the nonrecombinants and the double crossovers should differ at only one locus; if they differ at two loci, the wrong classes of progeny are being compared. **Animation 5.1** illustrates how to determine the order of three linked genes.

CONCEPTS

To determine the middle locus in a three-point cross, we compare the double-crossover progeny with the nonrecombinant progeny. The double crossovers will be the two least common classes of phenotypes; the nonrecombinants will be the two most common classes of phenotypes. The double-crossover progeny should have the same alleles as the nonrecombinant types at two loci and different alleles at the locus in the middle.

✓ CONCEPT CHECK 5

A three-point testcross is carried out between three linked genes. The resulting nonrecombinant progeny are s^{+} r^{+} c^{+} and s r c and the double-crossover progeny are s r c^{+} and s^{+} r^{+} c. Which is the middle locus?

DETERMINING THE LOCATIONS OF CROSSOVERS When we know the correct order of the loci on the chromosome, we can rewrite the phenotypes of the testcross progeny in Figure 5.12 with the alleles in the correct order so that we can determine where crossovers have taken place (**Figure 5.13**).

Among the eight classes of progeny, we have already identified two classes as nonrecombinants (st^{+} ss^{+} e^{+} and st ss e) and two classes as double crossovers (st^{+} ss e^{+} and st ss^{+} e). The other four classes include progeny that resulted from a chromosome that underwent a single crossover: two underwent single crossovers between *st* and *ss*, and two underwent single crossovers between *ss* and *e*.

To determine where the crossovers took place in these progeny, we can compare the alleles found in the single-crossover progeny with those found in the nonrecombinants, just as we did for the double crossovers. For example, consider the progeny with chromosome st^{+} ss e. The first allele (st^{+}) came from the nonrecombinant chromosome st^{+} ss^{+} e^{+}, and the other two alleles (*ss* and *e*) must have come from the other nonrecombinant chromosome st ss e through crossing over:

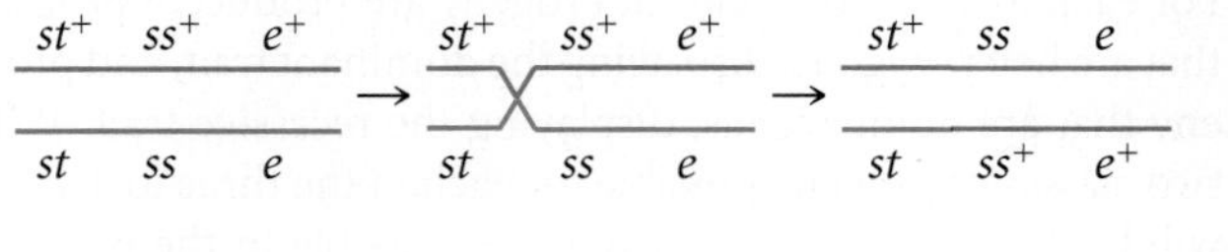

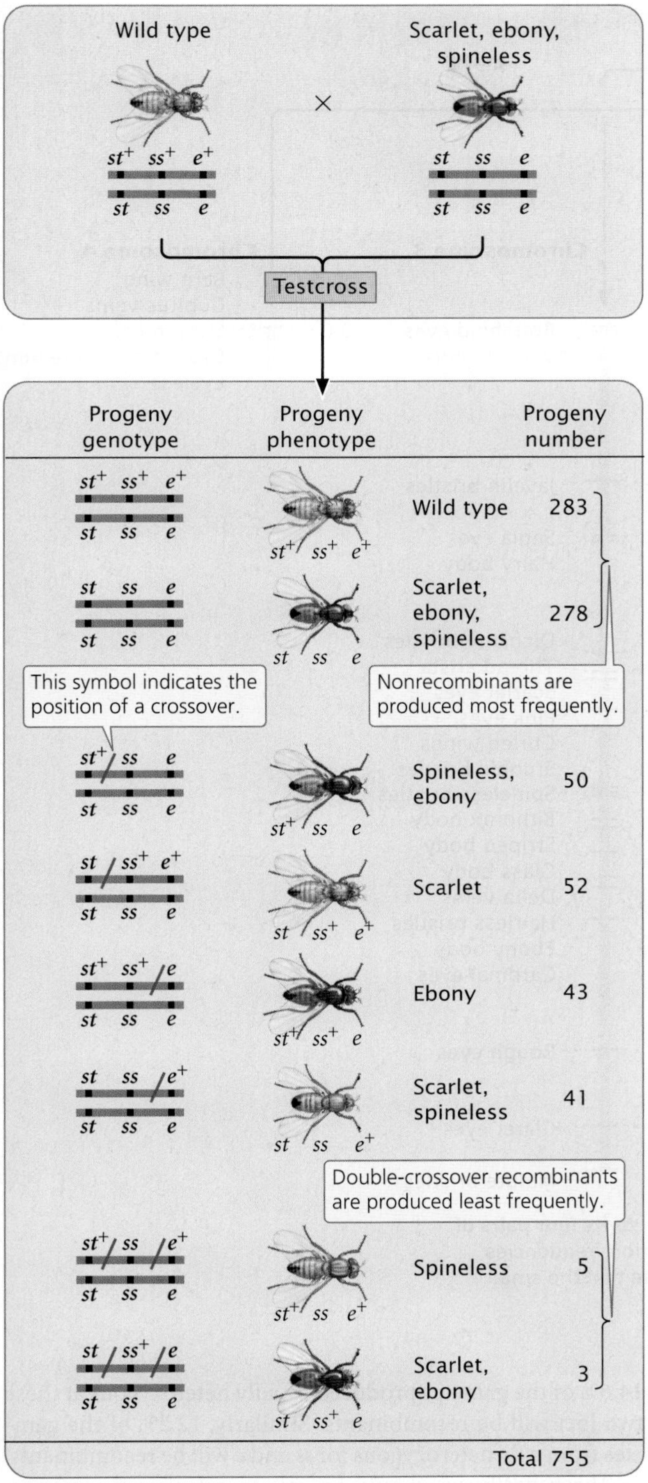

5.13 Writing the results of a three-point testcross with the loci in the correct order allows the locations of crossovers to be determined. These results are from the testcross illustrated in Figure 5.12, with the loci shown in the correct order. The location of a crossover is indicated by a red slash (/). Each phenotypic class includes both male and female flies; the sex of the pictured flies is random.

This same crossover also produces $st \quad ss^+ \quad e^+$ progeny.

This method can also be used to determine the location of crossing over in the other two types of single-crossover progeny. Crossing over between ss and e produces $st^+ \quad ss^+ \quad e$ and $st \quad ss \quad e^+$ chromosomes:

$$\frac{st^+ \quad ss^+ \quad e^+}{st \quad ss \quad e} \rightarrow \frac{st^+ \quad ss^+ \quad e^+}{st \quad ss \quad e} \rightarrow \frac{st^+ \quad ss^+ \quad e}{st \quad ss \quad e^+}$$

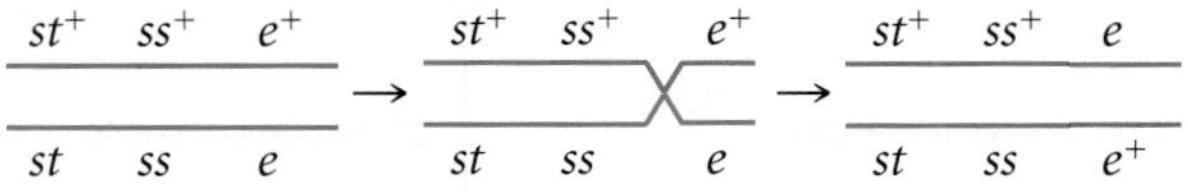

We now know the locations of all the crossovers. Their locations are marked with red slashes in Figure 5.13.

CALCULATING THE RECOMBINATION FREQUENCIES Next, we can determine the frequencies of recombination, which we will use to determine the map distances. We can calculate recombination frequency by adding up all of the recombinant progeny, dividing this number by the total number of progeny from the cross, and multiplying the number obtained by 100%. To determine the map distances accurately, we must include all crossovers (both single and double) that take place between two genes.

Recombinant progeny that possess a chromosome that underwent crossing over between the eye-color locus (st) and the bristle locus (ss) include the single crossovers ($st^+ \quad / \quad ss \quad e$ and $st \quad / \quad ss^+ \quad e^+$) and the two double crossovers ($st^+ \quad / \quad ss \quad / \quad e^+$ and $st \quad / \quad ss^+ \quad / \quad e$) (see Figure 5.13). There are a total of 755 progeny; so the recombination frequency between ss and st is

$st–ss$ recombination frequency =
$$\frac{50 + 52 + 5 + 3}{775} \times 100\% = 14.6\%$$

The distance between the st and ss loci can be expressed as 14.6 m.u.

The map distance between the bristle locus (ss) and the body locus (e) is determined in the same manner. The recombinant progeny that possess a crossover between ss and e are the single crossovers $st^+ \quad ss^+ \quad / \quad e$ and $st \quad ss \quad / \quad e^+$ and the double crossovers $st^+ \quad / \quad ss \quad / \quad e^+$ and $st \quad / \quad ss^+ \quad / \quad e$. The recombination frequency is

$ss–e$ recombination frequency =
$$\frac{43 + 41 + 5 + 3}{755} \times 100\% = 12.2\%$$

Thus, the map distance between ss and e is 12.2 m.u.

Finally, calculate the map distance between the outer two loci, st and e. This map distance can be obtained by summing the map distances between st and ss and between ss and e (14.6 m.u. + 12.2 m.u. = 26.8 m.u.). We can now use the map distances to draw a map of the three genes on the chromosome:

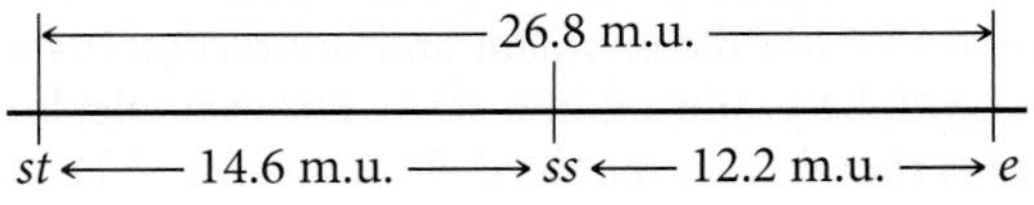

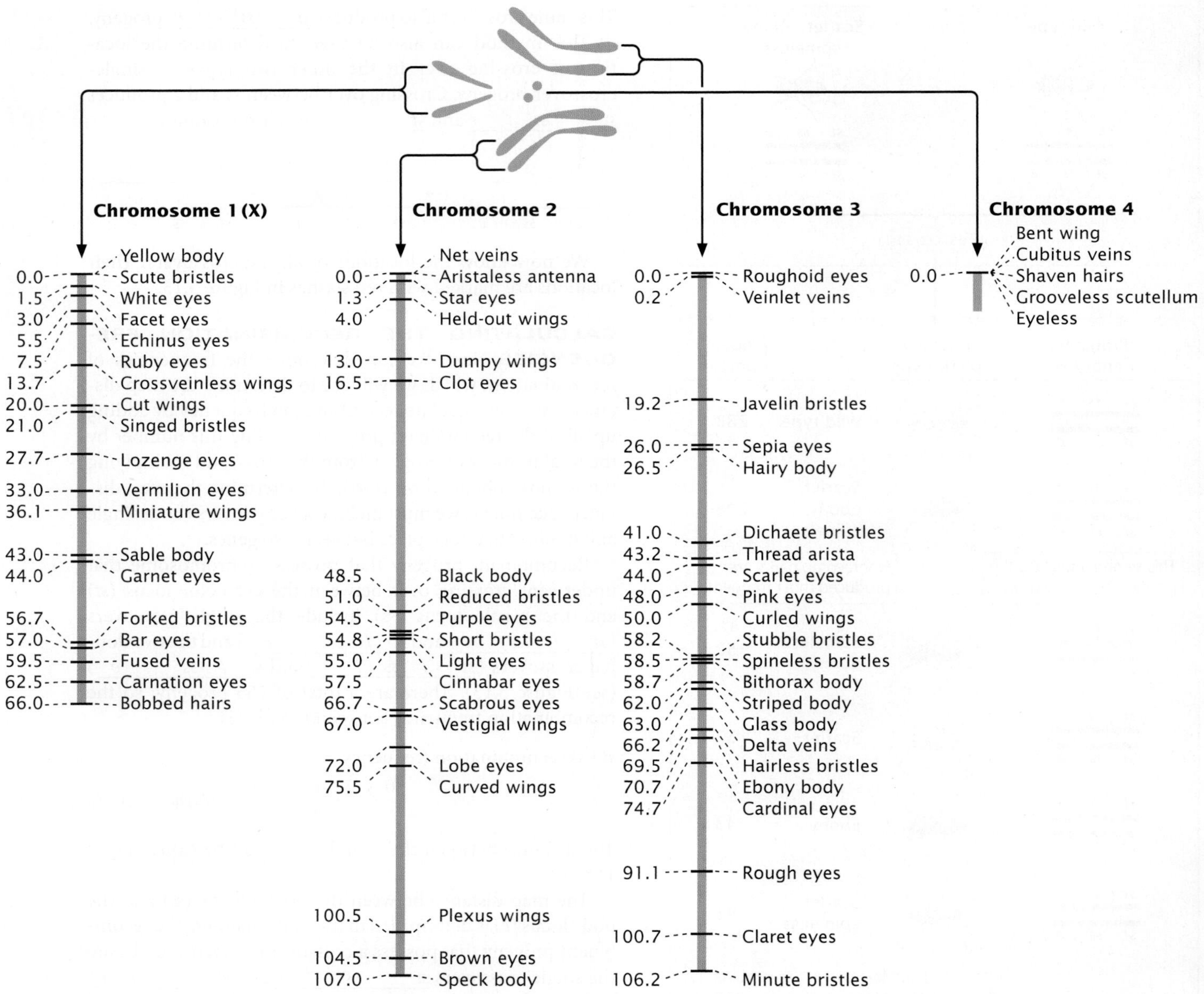

5.14 ***Drosophila melanogaster*** **has four linkage groups corresponding to its four pairs of chromosomes.** These genes were mapped with the use of recombination frequencies. Distances between genes within a linkage group are in map units. Note that the small 4th chromosome never undergoes recombination.

A genetic map of *D. melanogaster* is illustrated in **Figure 5.14**. **Animation 5.2** illustrates how to determine map distances in a three-point cross.

INTERFERENCE AND THE COEFFICIENT OF COINCIDENCE Map distances give us information not only about the distances that separate genes, but also about the proportions of recombinant and nonrecombinant gametes that will be produced in a cross. For example, knowing that genes *st* and *ss* on the third chromosome of *D. melanogaster* are separated by a distance of 14.6 m.u. tells us that 14.6% of the gametes produced by a fly heterozygous at these two loci will be recombinants. Similarly, 12.2% of the gametes from a fly heterozygous for *ss* and *e* will be recombinants.

Theoretically, we should be able to calculate the proportion of double-recombinant gametes by using the multiplication rule of probability (see Chapter 3). Applying this rule, we should find that the proportion (probability) of gametes with double crossovers between *st* and *e* is equal to the probability of recombination between *st* and *ss* multiplied by the probability of recombination between *ss* and *e*, or $0.146 \times 0.122 = 0.0178$. Multiplying this probability by the total number of progeny gives us the *expected* number of double-crossover progeny from the cross: $0.0178 \times 755 = 13.4$. But in our example, only 8 double crossovers—considerably fewer than the 13 expected—were observed in the progeny of the cross (see Figure 5.13).

This phenomenon is common in eukaryotic organisms. The calculation assumes that each crossover event is independent and that the occurrence of one crossover does not influence the occurrence of another. But crossovers are frequently *not* independent events: the occurrence of one crossover tends to inhibit additional crossovers in the same region of the chromosome, so double crossovers are less frequent than expected.

The degree to which one crossover interferes with additional crossovers in the same region is termed the **interference**. To calculate the interference, we first determine the **coefficient of coincidence**, which is the ratio of observed double crossovers to expected double crossovers:

$$\text{coefficient of coincidence} = \frac{\text{number of observed double crossovers}}{\text{number of expected double crossovers}}$$

For the loci that we mapped on the third chromosome of *D. melanogaster* (see Figure 5.13), we find

$$\text{coefficient of coincidence} = \frac{5 + 3}{0.146 \times 0.122 \times 755} = \frac{8}{13.4} = 0.6$$

which indicates that we are actually observing only 60% of the double crossovers that we expected on the basis of the single-crossover frequencies. The interference is calculated as

$$\text{interference} = 1 - \text{coefficient of coincidence}$$

So the interference for our three-point cross is

$$\text{interference} = 1 - 0.6 = 0.4$$

This value of interference tells us that 40% of the double-crossover progeny expected will not be observed because of interference. When interference is complete and no double-crossover progeny are observed, the coefficient of coincidence is 0 and the interference is 1.

Sometimes a crossover *increases* the probability of another crossover taking place nearby and we see *more* double-crossover progeny than expected. In this case, the coefficient of coincidence is greater than 1 and the interference is negative.

Most eukaryotic organisms exhibit interference, which causes crossovers to be more widely spaced along the chromosome than would be expected on a random basis. Interference was first observed in crosses of *Drosophila* in the early 1900s, yet despite years of study, the mechanism by which interference occurs is still not well understood. One proposed model of interference suggests that crossovers occur when stress builds up along the chromosome. Under this model, a crossover releases stress for some distance along the chromosome. Because a crossover relieves the stress that causes crossovers, additional crossovers are less likely to occur in the same area. TRY PROBLEM 23

THINK-PAIR-SHARE Questions 6 and 7

CONCEPTS

The coefficient of coincidence equals the number of double crossovers observed divided by the number of double crossovers expected on the basis of the single-crossover frequencies. The interference equals 1 minus the coefficient of coincidence; it indicates the degree to which one crossover interferes with additional crossovers.

✓ CONCEPT CHECK 6

In analyzing the results of a three-point testcross, a student determines that the interference is −0.23. What does this negative interference value indicate?

a. Fewer double crossovers took place than expected on the basis of single-crossover frequencies.
b. More double crossovers took place than expected on the basis of single-crossover frequencies.
c. Fewer single crossovers took place than expected.
d. A crossover in one region interferes with additional crossovers in the same region.

CONNECTING CONCEPTS

Stepping Through the Three-Point Cross

We have now examined the three-point cross in considerable detail, and we have seen how the information derived from it can be used to map a series of three linked genes. Let's briefly review the steps required to map genes using a three-point cross.

1. **Write out the phenotypes and numbers of progeny produced by the three-point cross.** The progeny phenotypes will be easier to interpret if you use allelic symbols for the traits (such as $st^+\ e^+\ ss$).

2. **Write out the genotypes of the original parents used to produce the triply heterozygous F_1 individual** in the testcross and, if known, the arrangement (coupling or repulsion) of the alleles on their chromosomes.

3. **Determine which phenotypic classes among the progeny of the testcross are the nonrecombinants and which are the double crossovers.** The nonrecombinants will be the two most common phenotypes; the double crossovers will be the two least common phenotypes.

4. **Determine which locus lies in the middle.** Compare the alleles present in the double crossovers with those present in the nonrecombinants; each class of double crossovers should be like one of the nonrecombinants for two loci and should differ for one locus. The locus that differs is the middle one.

5. **Rewrite the phenotypes with the genes in correct order.**

6. **Determine where crossovers must have taken place to give rise to the progeny phenotypes.** To do so, compare each phenotype with the phenotype of the nonrecombinant progeny.

7. Determine the recombination frequencies. Add the numbers of the progeny that possess a chromosome with a single crossover between a pair of loci. Add the double crossovers to this number. Divide this sum by the total number of progeny from the cross, and multiply by 100%; the result is the recombination frequency between the loci, which is the same as the map distance.

8. Draw a map of the three loci. Indicate which locus lies in the middle, and indicate the distances between them.

9. Determine the coefficient of coincidence and the interference. The coefficient of coincidence is the number of observed double-crossover progeny divided by the number of expected double-crossover progeny. The expected number can be obtained by multiplying the product of the two single-recombination probabilities by the total number of progeny in the cross.

WORKED PROBLEM

In *D. melanogaster*, cherub wings (*ch*), black body (*b*), and cinnabar eyes (*cn*) result from recessive alleles that are all located on chromosome 2. A homozygous wild-type fly was mated with a cherub, black, cinnabar fly, and the resulting F_1 females were test-crossed with cherub, black, cinnabar males. The following progeny were produced by the testcross:

ch	b^+	cn	105
ch^+	b^+	cn^+	750
ch^+	b	cn	40
ch^+	b^+	cn	4
ch	b	cn	753
ch	b^+	cn^+	41
ch^+	b	cn^+	102
ch	b	cn^+	5
Total			1800

a. Determine the linear order of the genes on the chromosome (which gene is in the middle).

b. Calculate the map distances between the three loci.

c. Determine the coefficient of coincidence and the interference for these three loci.

Solution Strategy

What information is required in your answer to the problem?

The order of the genes on the chromosome, the map distances among the genes, the coefficient of coincidence, and the interference.

What information is provided to solve the problem?

- A homozygous wild-type fly was mated with a cherub, black, cinnabar fly, and the resulting F_1 females were test-crossed with cherub, black, cinnabar males.
- The numbers of the different types of flies appearing among the progeny of the testcross.

Solution Steps

a. We can represent the crosses in this problem as follows:

$$\text{P} \quad \frac{ch^+ \quad b^+ \quad cn^+}{ch^+ \quad b^+ \quad cn^+} \times \frac{ch \quad b \quad cn}{ch \quad b \quad cn}$$

$$\downarrow$$

$$\text{F}_1 \quad \frac{ch^+ \quad b^+ \quad cn^+}{ch \quad b \quad cn}$$

$$\text{Testcross} \quad \frac{ch^+ \quad b^+ \quad cn^+}{ch \quad b \quad cn} \times \frac{ch \quad b \quad cn}{ch \quad b \quad cn}$$

Note that at this point we do not know the order of the genes; we have arbitrarily put *b* in the middle.

The next step is to determine which of the testcross progeny are nonrecombinants and which are double crossovers. The nonrecombinants should be the most frequent phenotype, so they must be the progeny with phenotypes encoded by $\underline{ch^+ \quad b^+ \quad cn^+}$ and $\underline{ch \quad b \quad cn}$. These genotypes are consistent with the genotypes of the parents, given earlier. The double crossovers are the least frequent phenotypes and are encoded by $\underline{ch^+ \quad b^+ \quad cn}$ and $\underline{ch \quad b \quad cn^+}$.

We can determine the gene order by comparing the alleles present in the double crossovers with those present in the nonrecombinants. The double-crossover progeny should be like one of the nonrecombinants at two loci and unlike it at one locus; the allele that differs should be in the middle. Compare the double-crossover progeny $\underline{ch \quad b \quad cn^+}$ with the nonrecombinant $\underline{ch \quad b \quad cn}$. Both have cherub wings (*ch*) and black body (*b*), but the double-crossover progeny have wild-type eyes (cn^+), whereas the nonrecombinants have cinnabar eyes (*cn*). The locus that determines cinnabar eyes must be in the middle.

b. To calculate the recombination frequencies among the genes, we first write the phenotypes of the progeny with the genes encoding them in the correct order. We have already identified the nonrecombinant and double-crossover progeny, so the other four progeny types must have resulted from single crossovers. To determine *where* single crossovers took place, we compare the alleles found in the single-crossover progeny with those in the nonrecombinants. Crossing over must have taken place where the alleles switch from those found in one nonrecombinant to those found in the other nonrecombinant. The locations of the crossovers are indicated with a slash:

ch	*cn* /	b^+	105	single crossover
ch^+	cn^+	b^+	750	nonrecombinant
ch^+ /	*cn*	*b*	40	single crossover
ch^+ /	*cn* /	b^+	4	double crossover
ch	*cn*	*b*	753	nonrecombinant
ch /	cn^+	b^+	41	single crossover
ch^+	cn^+ /	*b*	102	single crossover
ch /	cn^+ /	*b*	5	double crossover
Total			1800	

Next, we determine the recombination frequencies and draw a genetic map:

$$ch\text{–}cn \text{ recombination frequency} = \frac{40 + 4 + 41 + 5}{1800} \times 100\% = 5\% \text{ (5 m.u.)}$$

$$cn\text{–}b \text{ recombination frequency} = \frac{105 + 4 + 102 + 5}{1800} \times 100\% = 12\% \text{ (12 m.u.)}$$

$$ch\text{–}b \text{ map distance} = 5 \text{ m.u.} + 12 \text{ m.u.} = 17 \text{ m.u.}$$

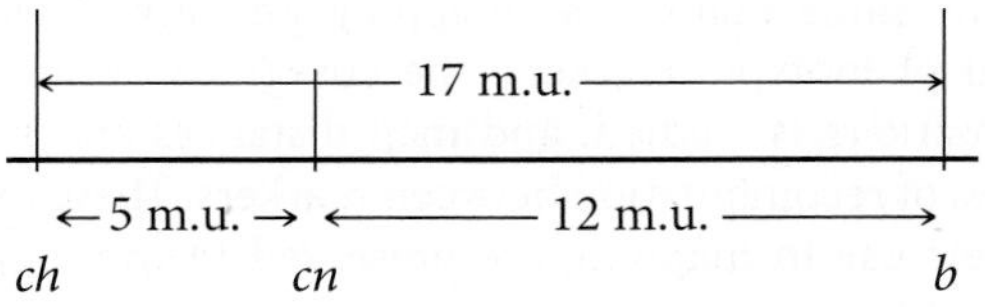

c. The coefficient of coincidence is the number of observed double crossovers divided by the number of expected double crossovers. The number of expected double crossovers is obtained by multiplying the probability of a crossover between *ch* and *cn* (0.05) × the probability of a crossover between *cn* and *b* (0.12) × the total number of progeny in the cross (1800):

$$\text{coefficient of coincidence} = \frac{4 + 5}{0.05 \times 0.12 \times 1800} = 0.83$$

Finally, the interference is equal to 1 – the coefficient of coincidence:

$$\text{interference} = 1 - 0.83 = 0.17$$

▶ To increase your skill with three-point crosses, try working **Problem 20** at the end of this chapter.

Effects of Multiple Crossovers

So far, we have examined the effects of double crossovers that take place between only two of the four chromatids (strands) of a homologous pair. These crossovers are called two-strand crossovers. Double crossovers that include three and even four of the chromatids of a homologous pair may also take place (**Figure 5.15**). If we examine only

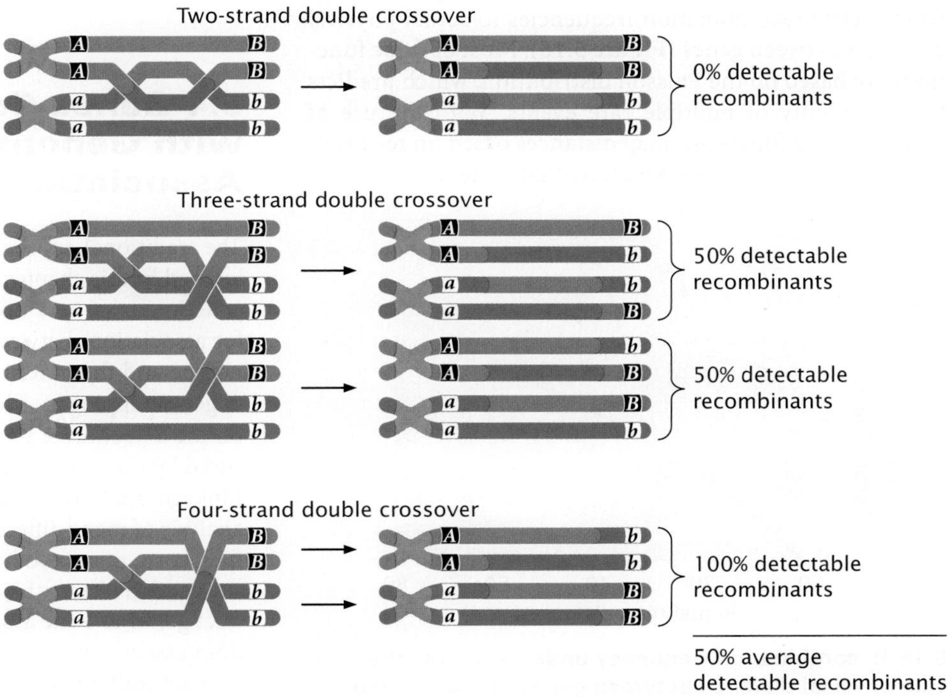

5.15 Results of two-, three-, and four-strand double crossovers on recombination between two genes.

the alleles at loci on either side of both crossover events, two-strand double crossovers result in no new combinations of alleles, and no recombinant gametes are produced (see Figure 5.15). Three-strand double crossovers result in two of the four gametes being recombinant, and four-strand double crossovers result in all four gametes being recombinant. Thus, two-strand double crossovers produce 0% recombination, three-strand double crossovers produce 50% recombination, and four-strand double crossovers produce 100% recombination. The overall result is that all types of double crossovers, taken together, produce an average of 50% recombinant progeny.

As we have seen, two-strand double crossovers cause alleles on either side of the crossovers to remain the same and produce no recombinant progeny. Three-strand and four-strand crossovers produce recombinant progeny, but these progeny are the same types as those produced by single crossovers. Consequently, some multiple crossovers go undetected when the progeny of a genetic cross are observed. Therefore, map distances based on recombination rates will underestimate the true physical distances between genes because some multiple crossovers will not be detected among the progeny of a cross. When genes are very close together, multiple crossovers are unlikely, and genetic distances based on recombination rates accurately correspond to the physical distances on the chromosome. But as the distance between genes increases, more multiple crossovers are likely, and the discrepancy between genetic distances (based on recombination rates) and physical distances increases. To correct for this discrepancy, geneticists have developed mathematical **mapping functions**, which relate recombination frequencies to actual physical distances between genes (**Figure 5.16**). Most of these functions are based on the Poisson distribution, which predicts the probability of multiple rare events. With the use of such mapping functions, map distances based on recombination rates can be more accurately estimated.

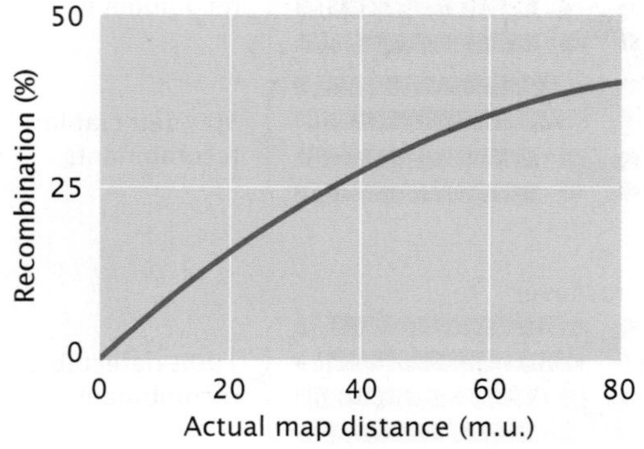

5.16 Recombination frequency underestimates the true physical distance between genes at higher map distances.

Mapping with Molecular Markers

For many years, gene mapping was limited in most organisms by the availability of **genetic markers**: variable genes with easily observable phenotypes whose inheritance could be studied. Traditional genetic markers include genes that encode easily observable characteristics such as flower color, seed shape, blood type, or biochemical differences. The paucity of these types of characteristics in many organisms limited mapping efforts.

In the 1980s, new molecular techniques made it possible to examine variations in DNA itself, providing an almost unlimited number of genetic markers that can be used for creating genetic maps and studying linkage relations. The earliest of these markers consisted of restriction fragment length polymorphisms (RFLPs), which are variations in DNA sequence detected by cutting the DNA with restriction enzymes (see Chapter 14). Later, methods were developed for detecting variable numbers of short DNA sequences repeated in tandem, called microsatellites. Now DNA sequencing allows the direct detection of individual variations in single nucleotides. All of these methods have expanded the availability of genetic markers and greatly facilitated the creation of genetic maps.

Gene mapping with molecular markers is done in essentially the same manner as mapping performed with traditional phenotypic markers: the cosegregation of two or more markers is studied, and map distances are based on the rates of recombination between markers. These methods and their use in mapping are presented in more detail in Chapter 14.

5.4 Genes Can Be Located with Genome-Wide Association Studies

The traditional approach to mapping genes, which we have learned in this chapter, is to examine progeny phenotypes in genetic crosses or among individuals in a pedigree, looking for associations between the inheritance of a particular phenotype and the inheritance of alleles at other loci. This type of gene mapping is called **linkage analysis** because it is based on the detection of physical linkage between genes, as measured by the rate of recombination, in progeny from a cross. Linkage analysis has been a powerful tool in the genetic analysis of many different types of organisms, including fruit flies, corn, mice, and humans.

Another alternative approach to mapping genes is to conduct **genome-wide association studies**, looking for nonrandom associations between the presence of a trait and alleles at many different loci scattered across the genome. Unlike linkage analysis, this approach does not trace the inheritance

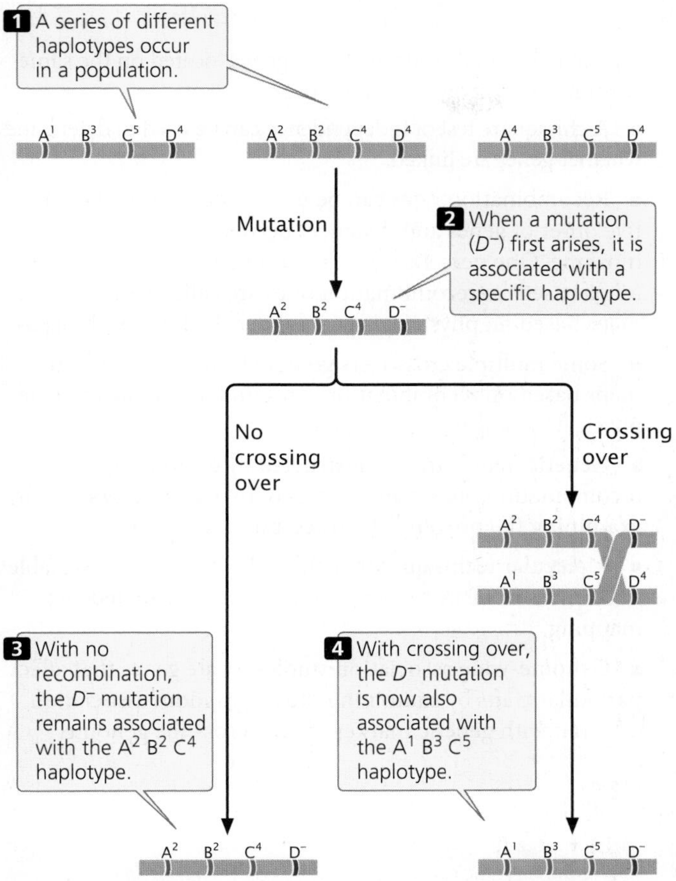

5.17 Genome-wide association studies are based on the nonrandom association of a mutation (D^-) that produces a trait and closely linked genes that constitute the haplotype.

of genetic markers and a trait in a genetic cross or family. Rather, it looks for associations between traits and particular suites of alleles in a *population*.

Imagine that we are interested in finding genes that contribute to bipolar disorder, a psychiatric illness characterized by severe depression and mania. When a mutation that predisposes a person to bipolar disorder first arises in a population, it will occur on a particular chromosome and will be associated with a specific set of alleles on that chromosome. In the example illustrated in **Figure 5.17**, the D^- mutation first arises on a chromosome that has alleles A^2, B^2, and C^4, and therefore the D^- mutation is initially linked to the A^2, B^2, and C^4 alleles. A specific set of linked alleles such as this is called a **haplotype**, and the nonrandom association between alleles in a haplotype is called **linkage disequilibrium**. Because of the physical linkage between the bipolar mutation and the other alleles of the haplotype, bipolar disorder and the haplotype will tend to be inherited together. Crossing over, however, breaks up the association between the alleles of the haplotype (see Figure 5.17), reducing the linkage disequilibrium between them. How long the linkage disequilibrium persists over evolutionary time depends on the amount of recombination between alleles at different loci. When the loci are far apart, linkage disequilibrium breaks down quickly; when the loci are close together, crossing over is less common, and linkage disequilibrium will persist longer. The important point is that linkage disequilibrium provides information about the distance between genes. A strong association between a trait such as bipolar disorder and a set of linked genetic markers indicates that one or more genes contributing to bipolar disorder are likely to be near the genetic markers.

In recent years, geneticists have mapped millions of **single-nucleotide polymorphisms** (SNPs), which are positions in the genome at which people vary in a single nucleotide base (see Chapter 15). Recall that SNPs were used in a linkage analysis that located the gene responsible for pattern baldness, as discussed in the introduction to this chapter. It is now possible to quickly and inexpensively genotype people for hundreds of thousands or millions of SNPs. This genotyping has provided the genetic markers needed for conducting genome-wide association studies, in which SNP haplotypes of people who have a particular disease, such as bipolar disorder, are compared with the haplotypes of healthy people. Nonrandom associations between SNPs and the disease suggest that one or more genes that contribute to the disease are closely linked to the SNPs. Genome-wide association studies do not usually locate specific genes; rather, they associate the inheritance of a trait or disease with a specific chromosomal region. After such an association has been established, geneticists can examine the chromosomal region for genes that might be responsible for the trait. Genome-wide association studies have been instrumental in the discovery of genes or chromosomal regions that influence a number of genetic diseases and important human traits, including bipolar disorder, height, skin pigmentation, eye color, body weight, coronary artery disease, blood lipid concentrations, diabetes, heart attacks, bone density, and glaucoma, among others.

THINK-PAIR-SHARE Question 8

CONCEPTS

The development of molecular techniques for examining variation in DNA sequences has provided a large number of genetic markers that can be used to create genetic maps and study linkage relations. Genome-wide association studies examine the nonrandom association of genetic markers and phenotypes to locate genes that contribute to the expression of traits.

CONCEPTS SUMMARY

- Linked genes do not assort independently. In a testcross for two completely linked genes (no crossing over), only nonrecombinant progeny are produced. When two genes assort independently, recombinant progeny and nonrecombinant progeny are produced in equal proportions. When two genes are linked with some crossing over between them, more nonrecombinant progeny than recombinant progeny are produced.
- Recombination frequency is calculated by summing the number of recombinant progeny, dividing by the total number of progeny produced in the cross, and multiplying by 100%. The recombination frequency is half the frequency of crossing over, and the maximum frequency of recombinant gametes is 50%.
- Coupling and repulsion refer to the arrangement of alleles on a chromosome. Whether genes are in coupling or in repulsion determines which combination of phenotypes will be most frequent in the progeny of a testcross.
- Interchromosomal recombination takes place among genes located on different chromosomes through the random segregation of chromosomes in meiosis. Intrachromosomal recombination takes place among genes located on the same chromosome through crossing over.
- A chi-square test of independence can be used to determine whether genes are linked.
- Recombination rates can be used to determine the relative order of genes and distances between them on a chromosome. One percent recombination equals one map unit. Maps based on recombination rates are called genetic maps; maps based on physical distances are called physical maps.
- Some multiple crossovers go undetected; thus, genetic maps based on recombination rates underestimate the true physical distances between genes.
- Genetic maps can be constructed by examining recombination rates from a series of two-point crosses or by examining the progeny of a three-point testcross.
- Molecular techniques that allow the detection of variable differences in DNA sequence have greatly facilitated gene mapping.
- Genome-wide association studies locate genes that affect particular traits by examining the nonrandom association of a trait with genetic markers from across the genome.

IMPORTANT TERMS

linked genes (p. 119)
linkage group (p. 119)
nonrecombinant (parental) gamete (p. 122)
nonrecombinant (parental) progeny (p. 122)
recombinant gamete (p. 122)
recombinant progeny (p. 122)
recombination frequency (p. 123)
coupling (cis) configuration (p. 124)
repulsion (trans) configuration (p. 124)
genetic map (p. 129)
physical map (p. 129)
map unit (m.u.) (p. 129)
centiMorgan (cM) (p. 129)
two-point testcross (p. 130)
three-point testcross (p. 131)
interference (p. 137)
coefficient of coincidence (p. 137)
mapping function (p. 140)
genetic marker (p. 140)
linkage analysis (p. 140)
genome-wide association studies (p. 140)
haplotype (p. 141)
linkage disequilibrium (p. 141)
single-nucleotide polymorphism (SNP) (p. 141)

ANSWERS TO CONCEPT CHECKS

1. c

2. Repulsion

3. Genetic maps are based on rates of recombination; physical maps are based on physical distances.

4.

$$\frac{m^+\ p^+\ s^+}{m\ \ p\ \ s}\quad \frac{m^+\ p\ s}{m\ \ p\ s}\quad \frac{m\ \ p^+\ s^+}{m\ \ p\ \ s}\quad \frac{m^+\ p^+\ s}{m\ \ p\ \ s}\quad \frac{m\ p\ s^+}{m\ p\ s}\quad \frac{m^+\ p\ s^+}{m\ \ p\ s}\quad \frac{m\ p^+\ s}{m\ p\ \ s}\quad \frac{m\ p\ s}{m\ p\ s}$$

5. The *c* locus.

6. b

WORKED PROBLEMS

Problem 1

A series of two-point crosses were carried out among seven loci (*a*, *b*, *c*, *d*, *e*, *f*, and *g*), producing the following recombination frequencies. Using these recombination frequencies, map the seven loci, showing their linkage groups, the order of the loci in each linkage group, and the distances between the loci of each group.

Loci	Recombination frequency (%)	Loci	Recombination frequency (%)
a and *b*	10	*c* and *d*	50
a and *c*	50	*c* and *e*	8
a and *d*	14	*c* and *f*	50
a and *e*	50	*c* and *g*	12
a and *f*	50	*d* and *e*	50
a and *g*	50	*d* and *f*	50
b and *c*	50	*d* and *g*	50
b and *d*	4	*e* and *f*	50
b and *e*	50	*e* and *g*	18
b and *f*	50	*f* and *g*	50
b and *g*	50		

Solution Strategy

What information is required in your answer to the problem?

The linkage groups for the seven loci, the order of the loci within each linkage group, and the map distances between the loci.

What information is provided to solve the problem?

- Recombination frequencies for each pair of loci.

For help with this problem, review:

Constructing a Genetic Map with Two-Point Testcrosses in Section 5.2.

Solution Steps

To work this problem, remember that 1% recombination equals 1 map unit. The recombination frequency between *a* and *b* is 10%; so these two loci are in the same linkage group, approximately 10 m.u. apart.

A recom-
n frequency
means that
t the two
assorting
dently
d in different
groups).

Linkage group 1

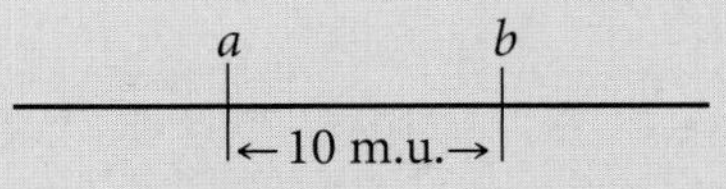

The recombination frequency between *a* and *c* is 50%; so *c* must lie in a second linkage group.

Linkage group 1

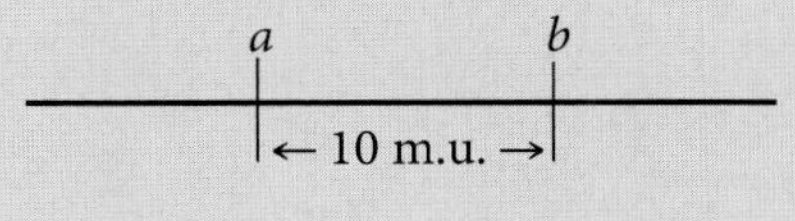

Linkage group 2

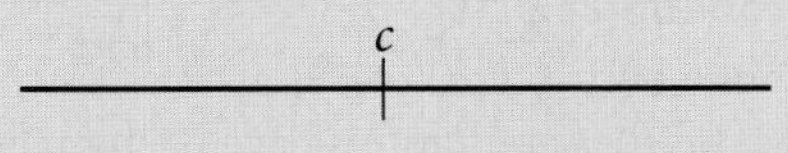

The recombination frequency between *a* and *d* is 14%; so *d* is located in linkage group 1. Is locus *d* 14 m.u. to the right or to the left of gene *a*? If *d* is 14 m.u. to the left of *a*, then the *b*-to-*d* distance should be 10 m.u. + 14 m.u. = 24 m.u. On the other hand, if *d* is to the right of *a*, then the distance between *b* and *d* should be 14 m.u. − 10 m.u. = 4 m.u. The *b*–*d* recombination frequency is 4%; so *d* is 14 m.u. to the right of *a*. The updated map is

Hint: To determine whether locus *d* is to the right or the left of locus *a*, look at the *b*-to-*d* distance.

Linkage group 1

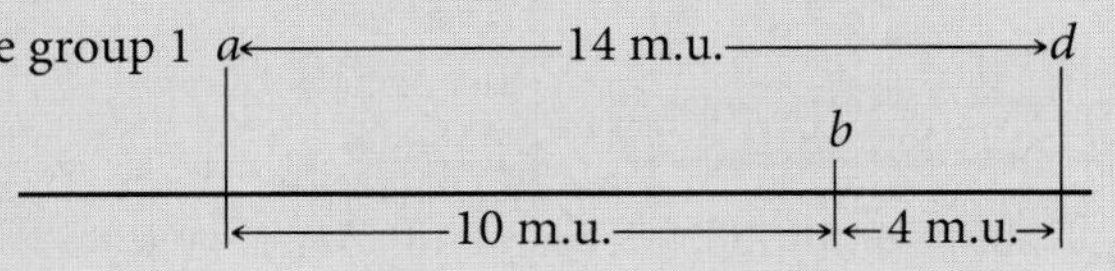

Linkage group 2

The recombination frequencies between each of loci *a*, *b*, and *d* and locus *e* are all 50%; so *e* is not in linkage group 1 with *a*, *b*, and *d*. The recombination frequency between *e* and *c* is 8%; so *e* is in linkage group 2:

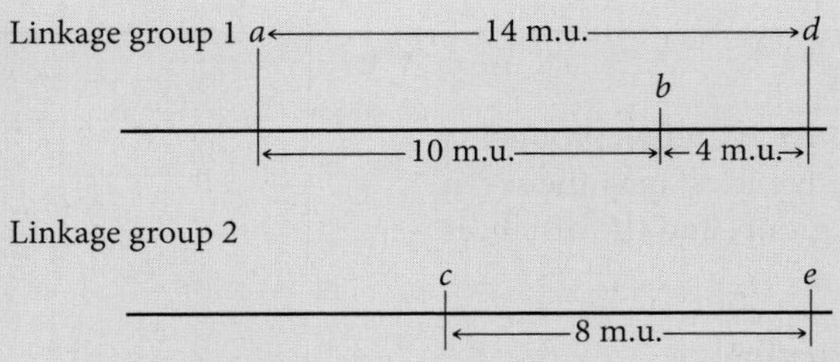

There is 50% recombination between *f* and all the other genes; so *f* must belong to a third linkage group:

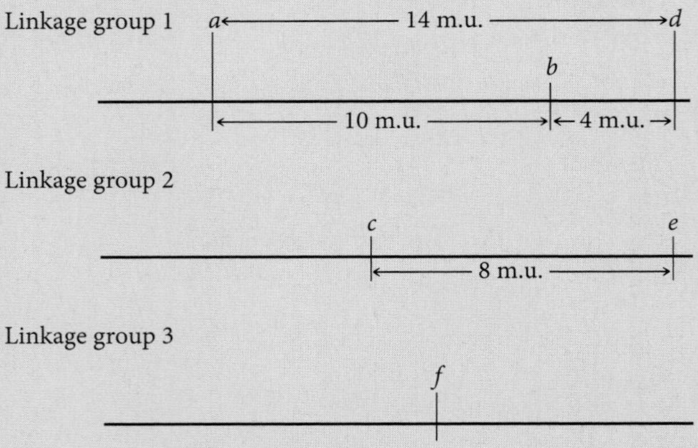

Finally, position locus *g* with respect to the other genes. The recombination frequencies between *g* and loci *a*, *b*, and *d* are all 50%; so *g* is not in linkage group 1. The recombination frequency between *g* and *c* is 12%; so *g* is a part of linkage group 2. To determine whether *g* is 12 m.u. to the right or left of *c*, consult the *g*–*e* recombination frequency. Because this recombination frequency is 18%, *g* must lie to the left of *c*:

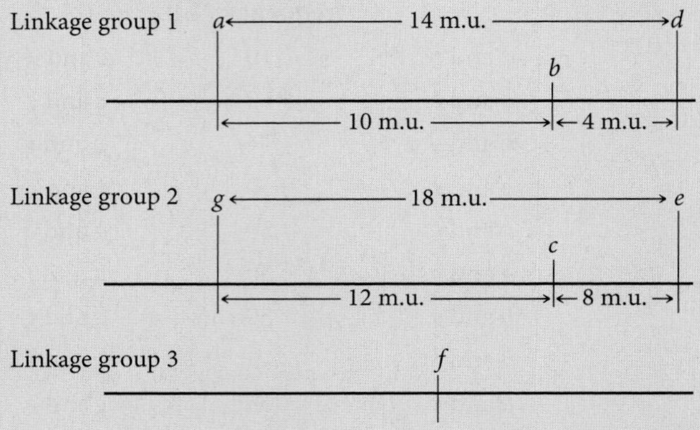

Note that the *g*-to-*e* distance (18 m.u.) is shorter than the sum of the *g*-to-*c* (12 m.u.) and *c*-to-*e* distances (8 m.u.) because of undetectable double crossovers between *g* and *e*.

Recall: Be some double crossovers m undetected, distance betw distant gene as *g* and *e*) n less than the shorter dista as *g* to *c* and

Problem 2

Ebony body color (*e*), rough eyes (*ro*), and brevis bristles (*bv*) are three recessive mutations that occur in fruit flies. The loci for these mutations have been mapped and are separated by the following map distances:

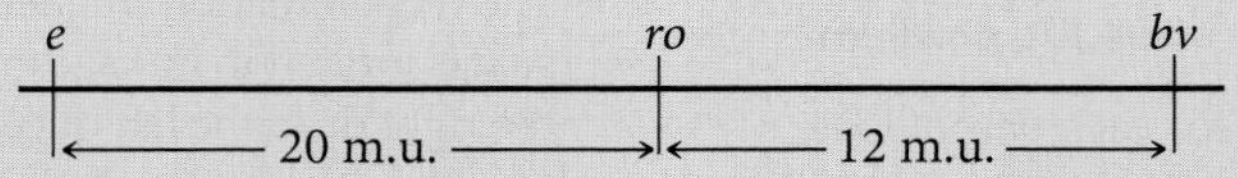

The interference between these genes is 0.4.

A fly with ebony body, rough eyes, and brevis bristles is crossed with a fly that is homozygous for the wild-type traits. The resulting F_1 females are test-crossed with males that have ebony body, rough eyes, and brevis bristles; 1800 progeny are produced. Give the expected numbers of phenotypes in the progeny of the testcross.

Solution Strategy

What information is required in your answer to the problem?

The expected numbers of different phenotypes produced by the testcross.

What information is provided to solve the problem?

- The map distances among the three loci.
- The interference among the loci.
- A testcross is carried out and 1800 progeny are produced.

For help with this problem, review:

Constructing a Genetic Map with a Three-Point Testcross and the Worked Problem in Section 5.3.

Solution Steps

The crosses are

$$\text{P} \quad \frac{e^+ \quad ro^+ \quad bv^+}{e^+ \quad ro^+ \quad bv^+} \times \frac{e \quad ro \quad bv}{e \quad ro \quad bv}$$

$$\downarrow$$

$$\text{F}_1 \quad \frac{e^+ \quad ro^+ \quad bv^+}{e \quad ro \quad bv}$$

$$\downarrow$$

$$\text{Testcross} \quad \frac{e^+ \quad ro^+ \quad bv^+}{e \quad ro \quad bv} \times \frac{e \quad ro \quad bv}{e \quad ro \quad bv}$$

In this case, we know that *ro* is the middle locus because the genes have been mapped. Eight classes of progeny will be produced by this cross:

e^+	ro^+	bv^+	nonrecombinant
e	ro	bv	nonrecombinant
e^+ /	ro	bv	single crossover between *e* and *ro*
e /	ro^+	bv^+	single crossover between *e* and *ro*
e^+	ro^+ /	bv	single crossover between *ro* and *bv*
e	ro /	bv^+	single crossover between *ro* and *bv*
e^+ /	ro /	bv^+	double crossover
e /	ro^+ /	bv	double crossover

[Margin hint, partly cut off:] ...he or-... e genes ...ed by ...tic map.

To determine the numbers of each type, use the map distances, starting with the double crossovers. The expected number of double crossovers is equal to the product of the single-crossover probabilities:

$$\text{expected number of double crossovers} = 0.20 \times 0.12 \times 1800 = 43.2$$

[Margin hint, partly cut off:] ...: The pres-... interference ...hat not all ...d double ...rs will be ...d.

However, there is some interference; so the observed number of double crossovers will be less than the expected. The interference is 1 – coefficient of coincidence; so the coefficient of coincidence is

$$\text{coefficient of coincidence} = 1 - \text{interference}$$

The interference is given as 0.4; so the coefficient of coincidence equals $1 - 0.4 = 0.6$. Recall that the coefficient of coincidence is

$$\text{coefficient of coincidence} = \frac{\text{number of observed double crossovers}}{\text{number of expected double crossovers}}$$

Rearranging this equation, we obtain:

number of observed double crossovers = coefficient of coincidence × number of expected double crossovers

$$\text{number of observed double crossovers} = 0.6 \times 43.2 = 26$$

A total of 26 double crossovers should be observed. Because there are two classes of double crossovers (e^+ ro bv^+ and e ro^+ bv), we expect to observe 13 of each class.

Hint: To obtain the number of single crossover progeny, subtract the number of double crossover progeny from the total number that resulted from recombination.

Next, we determine the number of single-crossover progeny. The genetic map indicates that the distance between *e* and *ro* is 20 m.u.; so 360 progeny (20% of 1800) are expected to have resulted from recombination between these two loci. Some of them will be single-crossover progeny and some will be double-crossover progeny. We have already determined that the number of double-crossover progeny is 26; so the number of progeny resulting from a single crossover between *e* and *ro* is $360 - 26 = 334$, which will be divided equally between the two single-crossover phenotypes (e ro^+ bv^+ and e^+ ro bv).

Hint: Don't forget to subtract the double crossovers from the total number of recombinants.

The distance between *ro* and *bv* is 12 m.u.; so the number of progeny resulting from recombination between these two genes is $0.12 \times 1800 = 216$. Again, some of these recombinants will be single-crossover progeny and some will be double-crossover progeny. To determine the number of progeny resulting from a single crossover, subtract the double crossovers: $216 - 26 = 190$. These single-crossover progeny will be divided between the two single-crossover phenotypes (e^+ ro^+ bv and e ro bv^+); so there will be $^{190}/_2 = 95$ of each of these phenotypes. The remaining progeny will be nonrecombinants, and they can be obtained by subtraction: $1800 - 26 - 334 - 190 = 1250$; there are two nonrecombinants (e^+ ro^+ bv^+ and e ro bv); so there will be $^{1250}/_2 = 625$ of each. The numbers of the various phenotypes are listed here:

Hint: The number of nonrecombinants can be obtained by subtraction.

e^+	ro^+	bv^+	625	nonrecombinant
e	ro	bv	625	nonrecombinant
e^+ /	ro	bv	167	single crossover between *e* and *ro*
e /	ro^+	bv^+	167	single crossover between *e* and *ro*
e^+	ro^+ /	bv	95	single crossover between *ro* and *bv*
e	ro /	bv^+	95	single crossover between *ro* and *bv*
e^+ /	ro /	bv^+	13	double crossover
e /	ro^+ /	bv	13	double crossover
Total			1800	

COMPREHENSION QUESTIONS

Section 5.1

1. What does the term *recombination* mean? What are two causes of recombination?

Section 5.2

2. What effect does crossing over have on linkage?
3. Why is the frequency of recombinant gametes always half the frequency of crossing over?
4. What is the difference between genes in coupling configuration and genes in repulsion? How does the arrangement of linked genes (whether they are in coupling or repulsion) affect the results of a genetic cross?
5. What is the difference between a genetic map and a physical map?

Section 5.3

6. Explain how to determine, using the numbers of progeny from a three-point cross, which of three linked loci is the middle locus.
7. What does the interference tell us about the effect of one crossover on another?

For more questions that test your comprehension of the key chapter concepts, go to **LearningCurve** for this chapter.

APPLICATION QUESTIONS AND PROBLEMS

Section 5.2

8. In the snail *Cepaea nemoralis*, an autosomal allele causing a banded shell (B^B) is recessive to the allele for an unbanded shell (B^O). Genes at a different locus determine the background color of the shell; here, yellow (C^Y) is recessive to brown (C^{Bw}). A banded, yellow snail is crossed with a homozygous brown, unbanded snail. The F_1 are then crossed with banded, yellow snails (a testcross).

   a. What will the results of the testcross be if the loci that control banding and color are linked with no crossing over?

   b. What will the results of the testcross be if the loci assort independently?

   c. What will the results of the testcross be if the loci are linked and 20 m.u. apart?

[Picture Press/Getty Images]

*9. In silkmoths (*Bombyx mori*), red eyes (*re*) and white-banded wings (*wb*) are encoded by two mutant alleles that are recessive to those that produce wild-type traits (re^+ and wb^+); these two genes are on the same chromosome. A moth homozygous for red eyes and white-banded wings is crossed with a moth homozygous for the wild-type traits. The F_1 have normal eyes and normal wings. The F_1 are crossed with moths that have red eyes and white-banded wings in a testcross. The progeny of this testcross are

wild-type eyes, wild-type wings	418
red eyes, wild-type wings	19
wild-type eyes, white-banded wings	16
red eyes, white-banded wings	426

   a. What phenotypic proportions would be expected if the genes for red eyes and for white-banded wings were located on different chromosomes?

   b. What is the rate of recombination between the genes for red eyes and those for white-banded wings?

*10. A geneticist discovers a new mutation in *Drosophila melanogaster* that causes the flies to shake and quiver. She calls this mutation *quiver* (*qu*) and determines that it is due to an autosomal recessive gene. She wants to determine whether the gene encoding *quiver* is linked to the recessive gene for vestigial wings (*vg*). She crosses a fly homozygous for quiver and vestigial traits with a fly homozygous for the wild-type traits and then uses the resulting F_1 females in a testcross. She obtains the following flies from this testcross:

vg^+	qu^+	230
vg	qu	224
vg	qu^+	97
vg^+	qu	99
Total		650

Are the genes that cause vestigial wings and quiver linked? Do a chi-square test of independence to determine whether the genes have assorted independently.

*11. In cucumbers, heart-shaped leaves (*hl*) are recessive to normal leaves (*Hl*) and numerous fruit spines (*ns*) are recessive to few fruit spines (*Ns*). The genes for leaf shape and for number of spines are located on the same chromosome; findings from mapping experiments indicate that they are 32.6 m.u. apart. A cucumber plant having heart-shaped leaves and numerous spines is crossed with a plant that is homozygous for normal leaves and few spines. The F_1 are crossed with plants that have heart-shaped leaves and numerous spines. What phenotypes and phenotypic proportions are expected in the progeny of this cross?

12. In tomatoes, tall (*D*) is dominant over dwarf (*d*), and smooth fruit (*P*) is dominant over pubescent fruit (*p*), which is covered with fine hairs. A farmer has two tall

and smooth tomato plants, which we will call plant A and plant B. The farmer crosses plants A and B with the same dwarf and pubescent plant and obtains the following numbers of progeny:

	Progeny of Plant A	Progeny of Plant B
Dd Pp	122	2
Dd pp	6	82
dd Pp	4	82
dd pp	124	4

a. What are the genotypes of plant A and plant B?

b. Are the loci that determine the height of the plant and pubescence linked? If so, what is the rate of recombination between them?

c. Explain why different proportions of progeny are produced when plant A and plant B are crossed with the same dwarf pubescent plant.

13. Alleles *A* and *a* are at a locus on the same chromosome as is a locus with alleles *B* and *b*. *Aa Bb* is crossed with *aa bb* and the following progeny are produced:

Aa Bb	5
Aa bb	45
aa Bb	45
aa bb	5

What conclusion can be made about the arrangement of the genes on the chromosome in the *Aa Bb* parent?

14. Recombination frequencies between three loci in corn are shown here.

Loci	Recombination frequency (%)
R and W_2	17
R and L_2	35
W_2 and L_2	18

What is the order of the genes on the chromosome?

15. In German cockroaches, bulging eyes (*bu*) are recessive to normal eyes (bu^+), and curved wings (*cv*) are recessive to straight wings (cv^+). Both traits are encoded by autosomal genes that are linked. A cockroach has genotype $bu^+ bu\ cv^+ cv$ and the genes are in repulsion. Which of the following sets of genes will be found in the most common gametes produced by this cockroach?

a. $bu^+\ cv^+$

b. $bu\ cv$

c. $bu^+\ bu$

d. $cv^+\ cv$

e. $bu\ cv^+$

Explain your answer.

***16.** In *Drosophila melanogaster*, ebony body (*e*) and rough eyes (*ro*) are encoded by autosomal recessive genes found on chromosome 3; they are separated by 20 m.u. The gene that encodes forked bristles (*f*) is X-linked recessive and assorts independently of *e* and *ro*. Give the phenotypes of progeny and their expected proportions when a female of each of the following genotypes is testcrossed with a male.

a. $\dfrac{e^+ \quad ro^+}{e \quad ro} \quad \dfrac{f^+}{f}$

b. $\dfrac{e^+ \quad ro}{e \quad ro^+} \quad \dfrac{f^+}{f}$

17. Perform a chi-square test of independence on the data provided in **Figure 5.2** to determine if the genes for flower color and pollen shape in sweet peas are assorting independently. Give the chi-square value, degrees of freedom, and associated probability. What conclusion would you make about the independent assortment of these genes?

***18.** A series of two-point crosses were carried out among seven loci (*a*, *b*, *c*, *d*, *e*, *f*, and *g*), producing the following recombination frequencies. Map the seven loci, showing their linkage groups, the order of the loci in each linkage group, and the distances between the loci of each group.

Loci	Recombination frequency (%)	Loci	Recombination frequency (%)
a and *b*	50	*c* and *d*	50
a and *c*	50	*c* and *e*	26
a and *d*	12	*c* and *f*	50
a and *e*	50	*c* and *g*	50
a and *f*	50	*d* and *e*	50
a and *g*	4	*d* and *f*	50
b and *c*	10	*d* and *g*	8
b and *d*	50	*e* and *f*	50
b and *e*	18	*e* and *g*	50
b and *f*	50	*f* and *g*	50
b and *g*	50		

19. R. W. Allard and W. M. Clement determined recombination rates for a series of genes in lima beans (R. W. Allard and W. M. Clement. 1959. *Journal of Heredity* 50:63–67). The following table lists paired recombination frequencies for eight of the loci (D, Wl, R, S, L_1, Ms, C, and G) that they mapped. On the basis of these data, draw a series of genetic maps for the different linkage groups of the genes, indicating the distances between the genes. Keep in mind that these frequencies are estimates of the true recombination frequencies and that some error is associated with each estimate. An asterisk beside a recombination frequency indicates that the recombination frequency is significantly different from 50%.

Recombination Frequencies (%) among Seven Loci in Lima Beans

	Wl	*R*	*S*	L_1	*Ms*	*C*	*G*
D	2.1*	39.3*	52.4	48.1	53.1	51.4	49.8
Wl		38.0*	47.3	47.7	48.8	50.3	50.4
R			51.9	52.7	54.6	49.3	52.6
S				26.9*	54.9	52.0	48.0
L_1					48.2	45.3	50.4
Ms						14.7*	43.1
C							52.0

*Significantly different from 50%

Section 5.3

*20. Waxy endosperm (*wx*), shrunken endosperm (*sh*), and yellow seedlings (*v*) are encoded by three recessive genes in corn that are linked on chromosome 5. A corn plant homozygous for all three recessive alleles is crossed with a plant homozygous for all the dominant alleles. The resulting F_1 are then crossed with a plant homozygous for the recessive alleles in a three-point testcross. The progeny of the testcross are

wx	*sh*	*V*	87
Wx	*Sh*	*v*	94
Wx	*Sh*	*V*	3,479
wx	*sh*	*v*	3,478
Wx	*sh*	*V*	1,515
wx	*Sh*	*v*	1,531
wx	*Sh*	*V*	292
Wx	*sh*	*v*	280
Total			10,756

a. Determine the order of these genes on the chromosome.

b. Calculate the map distances between the genes.

c. Determine the coefficient of coincidence and the interference among these genes.

21. Priscilla Lane and Margaret Green studied the linkage relations of three genes affecting coat color in mice: mahogany (*mg*), agouti (*a*), and ragged (*Rg*). They carried out a series of three-point crosses, mating mice that were heterozygous at all three loci with mice that were homozygous for the recessive alleles at these loci (P W. Lane and M. C. Green. 1960. *Journal of Heredity* 51:228–230). The following table lists the progeny of the testcrosses:

Phenotype			Number
a	*Rg*	+	1
+	+	*mg*	1
a	+	+	15
+	*Rg*	*mg*	9
+	+	+	16
a	*Rg*	*mg*	36
a	+	*mg*	76
+	*Rg*	+	69
Total			213

Note: + represents a wild-type allele.

a. Determine the order of the loci that encode mahogany, agouti, and ragged on the chromosome, the map distances between them, and the interference and coefficient of coincidence for these genes.

b. Draw a picture of the two chromosomes in the triply heterozygous mice used in the testcrosses, indicating which of the alleles are present on each chromosome.

22. Fine spines (*s*), smooth fruit (*tu*), and uniform fruit color (*u*) are three recessive traits in cucumbers, the genes for which are linked on the same chromosome. A cucumber plant heterozygous for all three traits is used in a testcross, and the following progeny are produced by this testcross:

S	*U*	*Tu*	2
s	*u*	*Tu*	70
S	*u*	*Tu*	21
s	*u*	*tu*	4
S	*U*	*tu*	82
s	*U*	*tu*	21
s	*U*	*Tu*	13
S	*u*	*tu*	17
Total			230

a. Determine the order of these genes on the chromosome.

b. Calculate the map distances between the genes.

c. Determine the coefficient of coincidence and the interference among these genes.

d. List the genes found on each chromosome in the parents used in the testcross.

*23. Raymond Popp studied linkage among genes for pink eye (*p*), shaker-1 (*sh*-1, which causes circling behavior, head tossing, and deafness), and hemoglobin (*Hb*) in mice (R. A. Popp. 1962. *Journal *of Heredity* 53:73–80). He performed a series of testcrosses, in which mice heterozygous for pink eye, shaker-1, and hemoglobin 1 and 2 were crossed with mice that were homozygous for pink eye, shaker-1, and hemoglobin 2.

$$\frac{P\ Sh\text{-}1\ Hb^1}{p\ sh\text{-}1\ Hb^2} \times \frac{p\ sh\text{-}1\ Hb^2}{p\ sh\text{-}1\ Hb^2}$$

The following progeny were produced:

Progeny genotype	**Number**
$\frac{p\ sh\text{-}1\ Hb^2}{p\ sh\text{-}1\ Hb^2}$	274
$\frac{P\ Sh\text{-}1\ Hb^1}{p\ sh\text{-}1\ Hb^2}$	320
$\frac{P\ sh\text{-}1\ Hb^2}{p\ sh\text{-}1\ Hb^2}$	57
$\frac{p\ Sh\text{-}1\ Hb^1}{p\ sh\text{-}1\ Hb^2}$	45
$\frac{P\ Sh\text{-}1\ Hb^2}{p\ sh\text{-}1\ Hb^2}$	6
$\frac{p\ sh\text{-}1\ Hb^1}{p\ sh\text{-}1\ Hb^2}$	5
$\frac{p\ Sh\text{-}1\ Hb^2}{p\ sh\text{-}1\ Hb^2}$	0
$\frac{P\ sh\text{-}1\ Hb^1}{p\ sh\text{-}1\ Hb^2}$	1
Total	708

a. Determine the order of these genes on the chromosome.

b. Calculate the map distances between the genes.

c. Determine the coefficient of coincidence and the interference among these genes.

***24.** In *Drosophila melanogaster*, black body (*b*) is recessive to gray body (b^+), purple eyes (*pr*) are recessive to red eyes (pr^+), and vestigial wings (*vg*) are recessive to normal wings (vg^+). The loci encoding these traits are linked, with the following map distances:

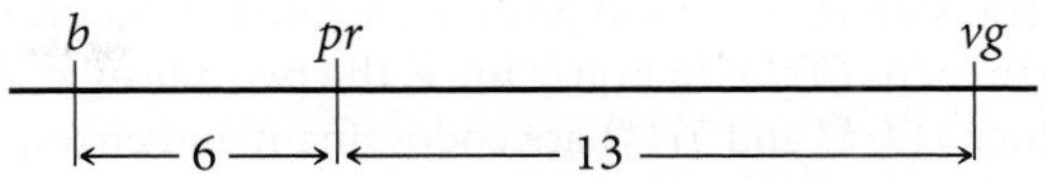

The interference among these genes is 0.5. A fly with a black body, purple eyes, and vestigial wings is crossed with a fly homozygous for a gray body, red eyes, and normal wings. The female progeny are then crossed with males that have a black body, purple eyes, and vestigial wings. If 1000 progeny are produced by this testcross, what will be the phenotypes and proportions of the progeny?

25. *Sepia eyes*, *spineless bristles*, and *striped thorax* are three recessive mutations in *Drosophila* found on chromosome 3. A genetics student crosses a fly homozygous for sepia eyes, spineless bristles, and striped thorax with a fly homozygous for the wild-type traits—red eyes, normal bristles, and solid thorax. The female progeny are then test-crossed with males that have sepia eyes, spineless bristles, and striped thorax. Assume that the interference between these genes is 0.2 and that 400 progeny flies are produced by the testcross. Based on the map distances provided in **Figure 5.14**, predict the phenotypes and proportions of the progeny resulting from the testcross.

Section 5.4

26. Eight DNA sequences from different individuals are given in the diagram below.

	Nucleotide position (1, 5, 10, 15)
Sequence 1	....T C T G G A T C A T C A C A T...
Sequence 2	....A C A G C A T C A T T A C G T...
Sequence 3	....T C A G G A T C A T T A C T A...
Sequence 4	....T C A G G A T C A T T A C A T...
Sequence 5	....A C A G C A T C A T T A C G T...
Sequence 6	....T C T G G A T C A T C A C A T...
Sequence 7	....T C A G G A T C A T T A C A T...
Sequence 8	....A C A G C A T C A T T A C G T...

a. Give the nucleotide positions of all single-nucleotide polymorphisms (SNPs; nucleotide positions where individuals vary in which base is present) in these sequences.

b. How many different haplotypes (sets of linked variants) are found in these eight sequences?

c. Give the haplotype of each sequence by listing the specific bases at each variable position in that particular haplotype. (Hint: See Figure 15.6.)

CHALLENGE QUESTION

Section 5.3

27. Transferrin is a blood protein encoded by the transferrin locus (*Trf*). In house mice, the two alleles at this locus (Trf^a and Trf^b) are codominant and encode three types of transferrin:

DATA ANALYSIS

Genotype	**Phenotype**
Trf^a/Trf^a	Trf-a
Trf^a/Trf^b	Trf-ab
Trf^b/Trf^b	Trf-b

The dilution locus, found on the same chromosome, determines whether the color of a mouse is diluted or full; an allele for dilution (*d*) is recessive to an allele for full color (d^+):

Genotype	**Phenotype**
d^+d^+	d^+ (full color)
d^+d	d^+ (full color)
dd	d (dilution)

Donald Shreffler conducted a series of crosses to determine the map distance between the transferrin locus and the dilution locus (D. C. Shreffler. 1963. *Journal of Heredity* 54:127–129). The following table presents a series of crosses carried out by Shreffler and the progeny resulting from these crosses.

			Progeny phenotypes				
			d⁺	**d⁺**	**d**	**d**	
Cross	**♂**	**♀**	**Trf-ab**	**Trf-b**	**Trf-ab**	**Trf-b**	**Total**
1	$\frac{d^+\ Trf^a}{d\ \ Trf^b}$ ×	$\frac{d\ \ Trf^b}{d\ \ Trf^b}$	32	3	6	21	62
2	$\frac{d\ \ Trf^b}{d\ \ Trf^b}$ ×	$\frac{d^+\ Trf^a}{d\ \ Trf^b}$	16	0	2	20	38
3	$\frac{d^+\ Trf^a}{d\ \ Trf^b}$ ×	$\frac{d\ \ Trf^b}{d\ \ Trf^b}$	35	9	4	30	78
4	$\frac{d\ \ Trf^b}{d\ \ Trf^b}$ ×	$\frac{d^+\ Trf^a}{d\ \ Trf^b}$	21	3	2	19	45
5	$\frac{d^+\ Trf^b}{d\ \ Trf^a}$ ×	$\frac{d\ \ Trf^b}{d\ \ Trf^b}$	8	29	22	5	64
6	$\frac{d\ \ Trf^b}{d\ \ Trf^b}$ ×	$\frac{d^+\ Trf^b}{d\ \ Trf^a}$	4	14	11	0	29

a. Calculate the recombination frequency between the *Trf* and the *d* loci by using the pooled data from all the crosses.

b. Which crosses represent recombination in male gamete formation and which crosses represent recombination in female gamete formation?

c. On the basis of your answer to part *b*, calculate the frequency of recombination among male parents and female parents separately.

d. Are the rates of recombination in males and females the same? If not, what might produce the difference?

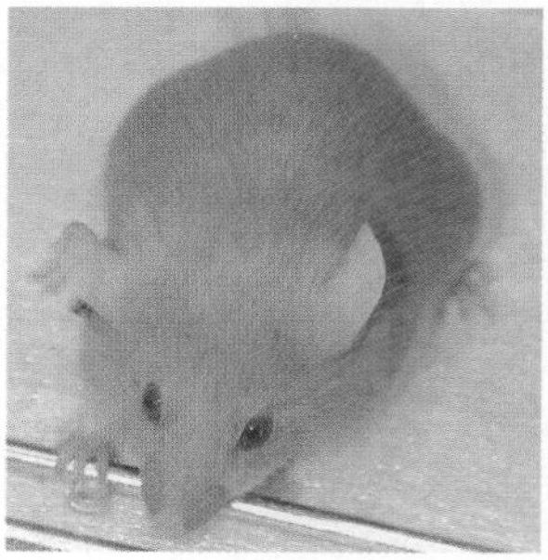

A mouse with the dilution trait. [Montoliu L, Oetting WS, Bennett DC. Color Genes. 3/2010. European Society for Pigment Cell Research. (http://www.espcr.org/micemut). 03/2013.]

THINK-PAIR-SHARE QUESTIONS

Introduction

1. Common folklore says that if a young man wants to know whether he will become bald, he should look at his mother's father. Based on the information provided in the introduction to this chapter, is this folklore scientifically accurate? Why or why not?

Section 5.1

2. Species A has $2n = 10$ chromosomes. Species B has $2n = 40$ chromosomes. On average, will two randomly selected genes from species A be more likely, less likely, or equally likely to assort independently than two randomly selected genes from species B? Explain your reasoning.

Section 5.2

3. The recombination frequency between genes *a* and *b* is 20%. What is the frequency of crossing over between genes *a* and *b*? Explain your reasoning.

4. Is it possible to have a recombination frequency of 80% between two genes? Is it possible for two genes to be separated by 80 map units? Why or why not?

5. The rate of crossing over between two linked genes (*r* and *w*) is 0.44. The following cross is carried out: *Rr Ww* × *rr ww*. What proportion of the gametes of the *Rr Ww* parent were recombinant gametes? Explain your reasoning and state any assumptions you made.

Section 5.3

6. Why is interference important? Why do we calculate it in a three-point cross? Why don't we calculate interference in a two-point cross?

7. Three recessive mutations in *Drosophila melanogaster*, *roughoid* (*ru*, small rough eyes), *javelin* (*jv*, cylindrical bristles), and *sepia eyes* (*se*, dark brown eyes) are linked. A three-point cross was carried out and the following progeny obtained:

jv^+	ru^+	se^+	37
jv^+	ru^+	se	2
jv^+	ru	se	14
jv^+	ru	se^+	146
Jv	ru	se^+	2
Jv	ru	se	35
Jv	ru^+	se	154
Jv	ru^+	se^+	10

   a. Determine the order of the genes on the chromosome.

   b. Determine which progeny contain single crossovers and which contain double crossovers and indicate where among the genes the crossovers occurred.

   c. Calculate the map distances among the genes.

   d. Calculate the coefficient of coincidence and interference among the genes.

Section 5.4

8. Compare and contrast linkage analysis and genome-wide association studies. How are they similar? How are they different?

SaplingPlus Self-study tools that will help you practice what you've learned and reinforce this chapter's concepts are available online. Go to www.macmillanlearning.com/PierceGenetics6e.

6 Chromosome Variation

Building a Better Banana

Many varieties of bananas have multiple sets of chromosomes. [Frankie Angel/Alamy.]

Bananas and plantains (collectively referred to as bananas) are the world's most popular fruit. In many developing countries, they are a critically important source of food, providing starch and calories for hundreds of millions of people. In industrial countries, more bananas are consumed than any other fruit; for example, Americans consume as many pounds of bananas as apples and oranges combined. Over 100 million tons of bananas are produced annually worldwide.

There is no concrete biological distinction between bananas and plantains, but the term "banana" generally refers to the sweeter forms that are eaten uncooked, while the term "plantain" is applied to bananas that are peeled when unripe and cooked before eating. Cultivated bananas differ from their wild relatives by being seedless, which makes them more edible but hinders their reproduction. Farmers propagate bananas vegetatively, by cutting off parts of existing plants and coaxing them to grow into new plants. Because they are propagated by this method, many cultivated bananas are genetically identical.

From a genetic standpoint, bananas are interesting because many varieties have multiple sets of chromosomes. Most eukaryotic organisms in nature are diploid ($2n$), with two sets of chromosomes. Others, such as fungi, are haploid (n), with a single set of chromosomes. Cultivated bananas are often polyploid, with more than two sets of chromosomes ($3n$, $4n$, or higher). Most strains of cultivated bananas were created by crossing plants within and between two diploid species: *Musca acuminata* (genome = AA) and *Musca balbisiana* (genome = BB). Many cultivated bananas are triploid, with three sets of chromosomes, consisting of AAA, AAB, or ABB, and some bananas even have four sets of chromosomes (tetraploid), consisting of AAAA, AAAB, AABB, or ABBB.

In spite of their worldwide importance as a food, modern cultivated bananas are in trouble. The strain most often sold in grocery stores—the Cavendish—is threatened by disease and pests: in recent years, a soil fungus has devastated crops in Asia. The Cavendish's predecessor, called Gros Michel (Big Mike), was the banana of choice until disease in the 1950s and 1960s wiped it out. Because vegetative propagation produces genetically identical plants, cultivated bananas are particularly vulnerable to attack by pathogens and pests.

To help develop a better banana—more disease and pest resistant, as well as more nutritious—geneticists launched an international effort to sequence the genome of the banana, producing a draft sequence in 2012. This research demonstrated that the banana genome consists of over 500 million base pairs (bp) of DNA and contains 36,500 protein-encoding genes. Using this genome sequence, scientists have already identified several genes that play a role in resistance to fungal diseases and are exploring ways to breed and genetically engineer bananas.

THINK-PAIR-SHARE Questions 1 and 2

Most species have a characteristic number of chromosomes, each with a distinct size and structure, and all the tissues of an organism (except for gametes) generally have the same set of chromosomes. Nevertheless, variations in chromosome number—such as the extra sets of chromosomes seen in bananas—do periodically arise. Variations may also arise in chromosome structure: individual chromosomes may lose or gain parts, and the order of genes within a chromosome may become altered. These variations in the number and structure of chromosomes are termed **chromosome mutations**, and they frequently play an important role in agriculture and evolution.

We begin this chapter by briefly reviewing some of the basic concepts of chromosome structure that we learned in Chapter 2. We then consider the different types of chromosome mutations and their features, phenotypic effects, and influences on evolution.

6.1 Chromosome Mutations Include Rearrangements, Aneuploidy, and Polyploidy

Before we consider the different types of chromosome mutations, their effects, and how they arise, it will be helpful to review the basics of chromosome structure.

Chromosome Morphology

Each functional eukaryotic chromosome has a centromere, to which spindle microtubules attach, and two telomeres, which stabilize the chromosome ends (see Figure 2.6). Chromosomes are classified into four basic types (see Figure 2.7):

1. **Metacentric.** The centromere is located approximately in the middle, so the chromosome has two arms of equal length.
2. **Submetacentric.** The centromere is displaced toward one end, creating a long arm and a short arm. (On human chromosomes, the short arm is designated by the letter p and the long arm by the letter q.)
3. **Acrocentric.** The centromere is near one end, producing a long arm and a knob, or satellite, at the other end.
4. **Telocentric.** The centromere is at or very near the end of the chromosome.

The complete set of chromosomes possessed by an organism is called its *karyotype*. An organism's karyotype is usually presented as a picture of metaphase chromosomes lined up in descending order of their size (**Figure 6.1**). Karyotypes are prepared from actively dividing cells, such as white blood cells, bone-marrow cells, or cells from meristematic tissues of plants. After treatment with a chemical (such as colchicine) that prevents them from entering anaphase, the cells are chemically preserved. They are then burst open to release the chromosomes onto a microscope slide, and the chromosomes are stained, and photographed. The photograph is then enlarged, and the individual chromosomes are cut out and arranged in a karyotype. For human chromosomes, karyotypes are routinely prepared by automated machines, which scan a slide using a video camera attached to a microscope, looking for chromosome spreads. When a spread has been located, the camera takes a picture of the chromosomes, the image is digitized, and the chromosomes are sorted and arranged electronically by a computer.

Preparation and staining techniques help to distinguish among chromosomes of similar size and shape. For instance, special preparation and staining of chromosomes with a dye called Giemsa reveals G bands (**Figure 6.2a**), which distinguish areas of DNA that are rich in adenine–thymine (A–T) base pairs (see Chapter 8). Q bands (**Figure 6.2b**) are revealed by staining chromosomes with quinacrine mustard and viewing the chromosomes under ultraviolet light; variation in the brightness of Q bands results from differences in the relative amounts of cytosine–guanine (C–G) and adenine–thymine base pairs. Other techniques

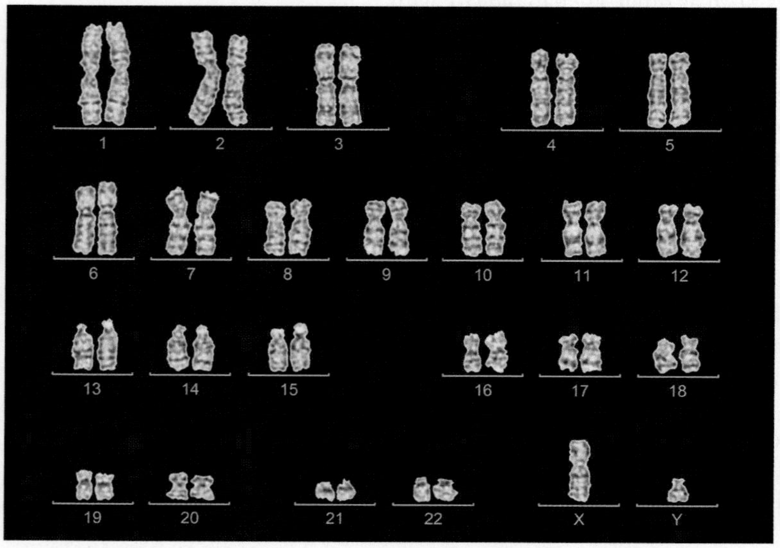

6.1 A human karyotype consists of 46 chromosomes. A karyotype for a male is shown here; a karyotype for a female would have two X chromosomes. [© Dr. Philippe Vago/ISM/Phototake.]

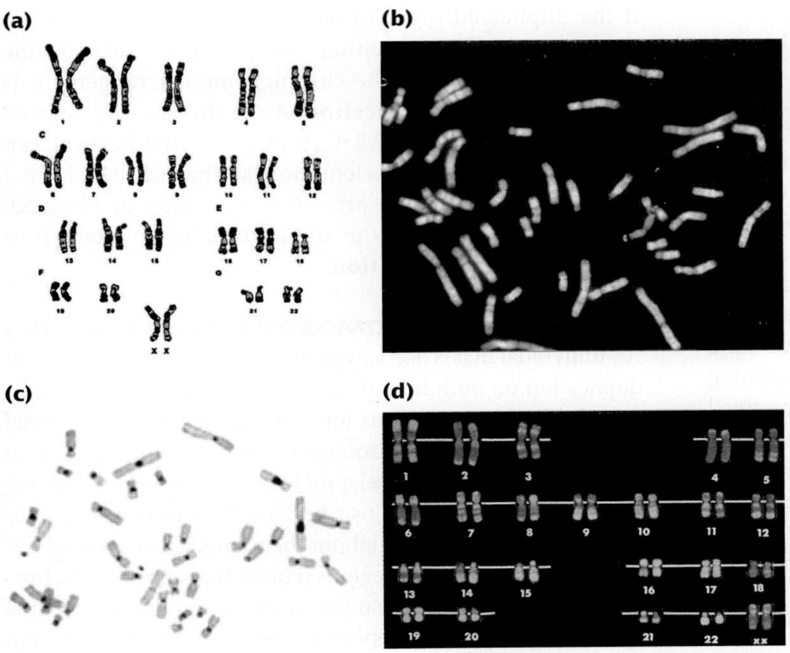

6.2 Chromosome banding is revealed by special staining techniques. (a) G banding. (b) Q banding. (c) C banding. (d) R banding. [Part a: Leonard Lessin/Science Source. Parts b and c: University of Washington Pathology Department. http://pathology.washington.edu. Part d: Dr. Ram Verma/Phototake.]

reveal C bands (**Figure 6.2c**), which are regions of DNA occupied by centromeric heterochromatin, and R bands (**Figure 6.2d**), which are rich in cytosine–guanine base pairs.

Types of Chromosome Mutations

Chromosome mutations can be grouped into three basic categories: chromosome rearrangements, aneuploidy, and polyploidy (**Figure 6.3**). Chromosome rearrangements alter the *structure* of chromosomes; for example, a piece of a chromosome may be duplicated, deleted, or inverted. In aneuploidy, the *number* of chromosomes is altered: one or more individual chromosomes are added or deleted. In polyploidy, one or more complete *sets* of chromosomes are added. A polyploid is any organism that has more than two sets of chromosomes ($3n$, $4n$, $5n$, or more).

THINK-PAIR-SHARE Question 3

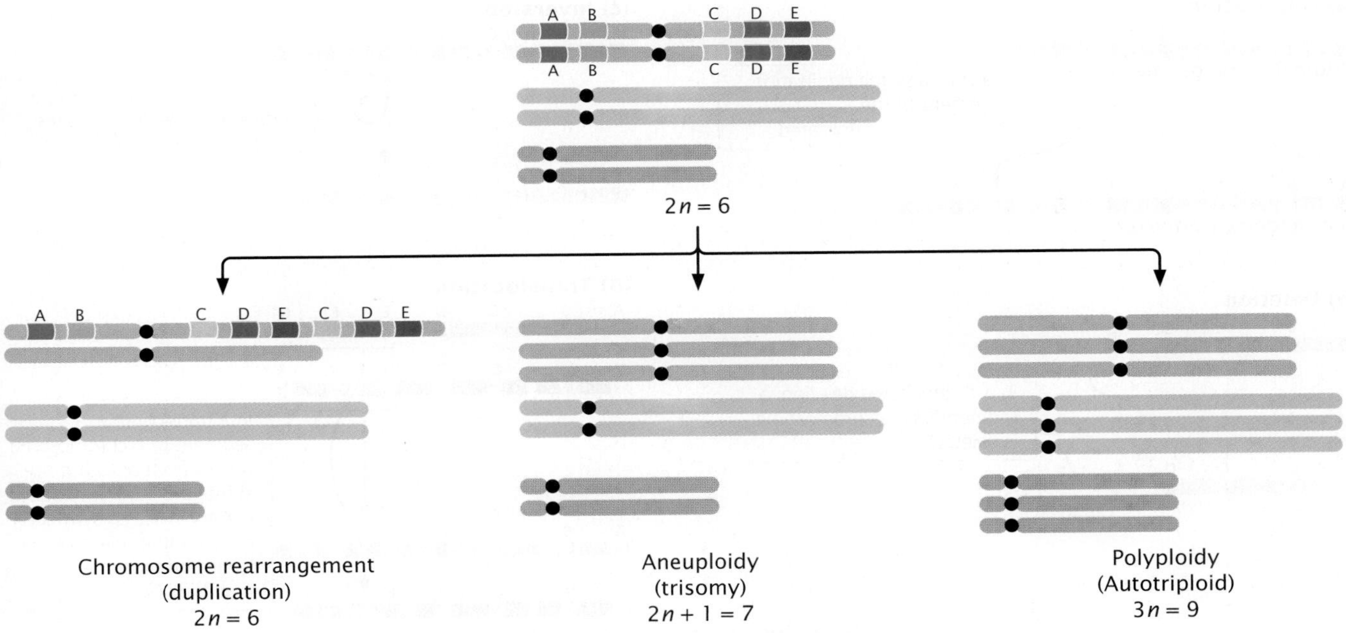

6.3 Chromosome mutations consist of chromosome rearrangements, aneuploidy, and polyploidy. An example of each category of mutation is shown here.

6.2 Chromosome Rearrangements Alter Chromosome Structure

Chromosome rearrangements are mutations that change the structure of individual chromosomes. The four basic types of rearrangements are duplications, deletions, inversions, and translocations (**Figure 6.4**). Many of these chromosome rearrangements originate when double-strand breaks occur in the DNA molecules found within a chromosome. Double-strand breaks in DNA often cause cell death, so organisms have evolved elaborate mechanisms to repair breaks by connecting the broken ends of DNA. If the two broken ends are rejoined correctly, the original chromosome is restored, and no chromosome rearrangement results. Sometimes the wrong ends are connected, however, leading to a chromosome rearrangement. Chromosome rearrangements can also arise through errors in crossing over or when crossing over occurs between repeated DNA sequences.

Duplications

A **chromosome duplication** is a mutation in which part of the chromosome has been doubled (see Figure 6.4a). Consider a chromosome with segments AB•CDEFG, in which • represents the centromere. A duplication might include the EF segments, giving rise to a chromosome with segments AB•CDEFEFG. This type of duplication, in which the duplicated segment is immediately adjacent to the original segment, is called a **tandem duplication**. If the duplicated segment is located some distance from the original segment, either on the same chromosome or on a different one, the chromosome rearrangement is called a **displaced duplication**. An example of a displaced duplication would be AB•CDEFGEF. A duplication can be either in the same orientation as that of the original sequence, as in the two preceding examples, or inverted: AB•CDEFFEG. When the duplication is inverted, it is called a **reverse duplication**.

EFFECTS OF CHROMOSOME DUPLICATIONS An individual that is homozygous for a duplication carries that duplication on both homologous chromosomes, and an individual that is heterozygous for a duplication has one normal chromosome and one chromosome with the duplication. In heterozygotes (**Figure 6.5a**), problems in chromosome pairing arise at prophase I of meiosis because the two chromosomes are not homologous throughout their length. The pairing and synapsis of homologous regions require that one or both chromosomes loop and twist so that these regions are able to line up (**Figure 6.5b**). The appearance of this characteristic loop structure in meiosis is one way to detect duplications.

Duplications may have major effects on the phenotype. Among fruit flies, for example, a fly with the *Bar* mutation has a reduced number of facets in the eye, making the eye smaller and bar shaped instead of oval (**Figure 6.6**). The *Bar* mutation results from a small duplication on the X chromosome that is inherited as an incompletely dominant, X-linked trait: heterozygous female flies have somewhat smaller eyes (the number of facets is reduced; Figure 6.6b), whereas in homozygous female and hemizygous male flies, the number

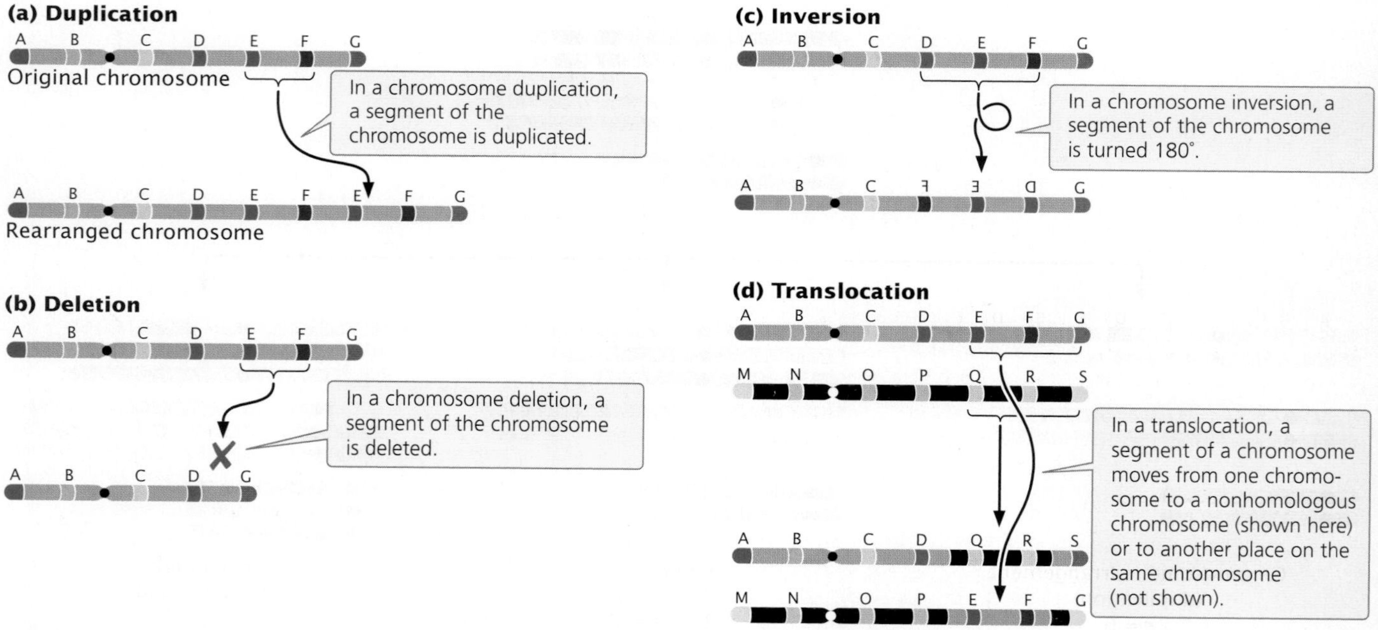

6.4 The four basic types of chromosome rearrangements are duplications, deletions, inversions, and translocations.

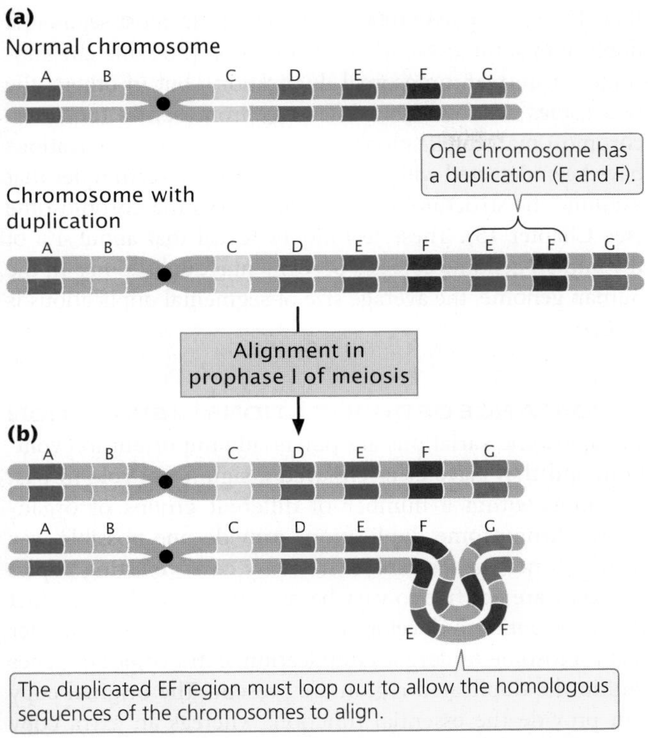

6.5 In an individual heterozygous for a duplication, the chromosome with the duplication loops out during pairing in prophase I.

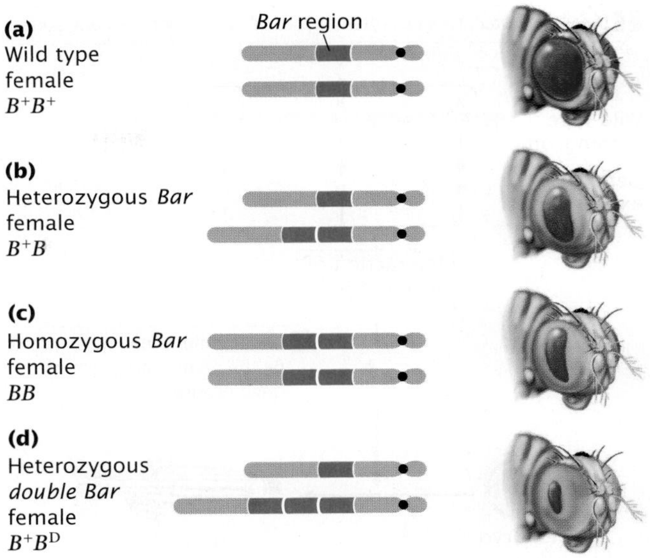

6.6 The *Bar* phenotype in *Drosophila melanogaster* results from an X-linked duplication. (a) Wild-type fruit flies have normal-sized eyes. Flies (b) heterozygous and (c) homozygous for the *Bar* mutation have smaller, bar-shaped eyes. (d) Flies with *double Bar* have three copies of the duplication and much smaller bar-shaped eyes.

of facets is greatly reduced (Figure 6.6c). Occasionally, a fly carries three copies of the *Bar* duplication on its X chromosome; in flies with this mutation, termed *double Bar*, the number of facets is extremely reduced (Figure 6.6d).

Duplications and deletions often arise from unequal crossing over, in which duplicated segments of chromosomes misalign during the process. Unequal crossing over is frequently the cause of red–green color blindness in humans. Perception of color is affected by red and green opsin genes, which are found on the X chromosome and are 98% identical in their DNA sequence. Most people with normal color vision have one red opsin gene and one green opsin gene (although some people have more than one copy of each). Because the red and green opsin genes are similar in sequence, they occasionally pair in prophase I, and unequal crossing over takes place (**Figure 6.7**; see also Figure 13.13). The unequal crossing over produces one chromosome with an extra opsin gene and one chromosome that is missing an opsin gene. When a male inherits the chromosome that is missing one of the opsin genes, red–green color blindness results.

UNBALANCED GENE DOSAGE How does a chromosome duplication alter the phenotype? After all, gene sequences are not usually altered by duplications, and no genetic information is missing; the only change is the presence of additional copies of normal sequences. The answer to this question is not well understood, but the effects are most likely due to imbalances in the amounts of gene products (abnormal gene dosage). The amount of a particular protein synthesized by a cell is often directly related to the number of copies of its corresponding gene: an individual organism with three functional copies of a gene often produces 1.5 times as much of the protein encoded by that gene as an individual with two copies. Because developmental processes require the interaction of many proteins, they often depend

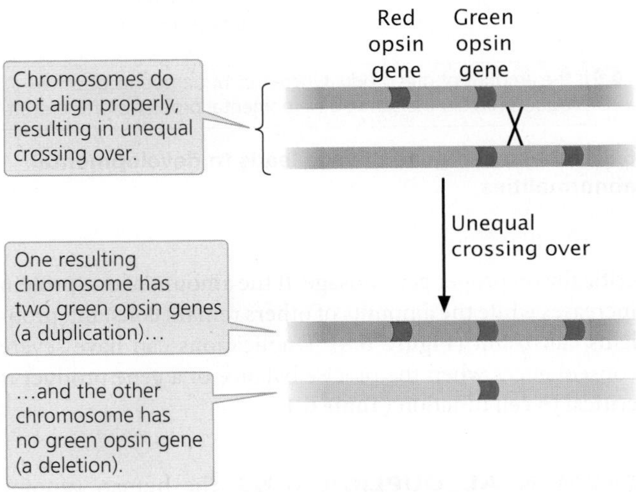

6.7 Unequal crossing over produces duplications and deletions.

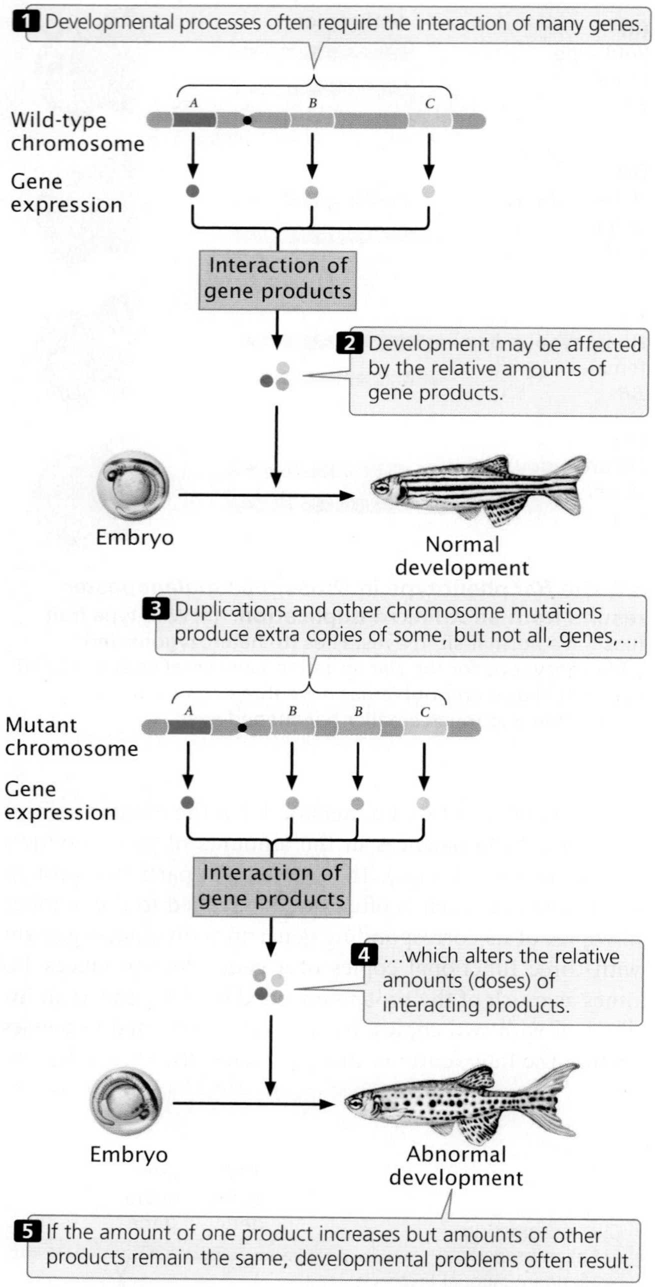

1 Developmental processes often require the interaction of many genes.
A
B
C
Wild-type chromosome
Gene expression
Interaction of gene products
2 Development may be affected by the relative amounts of gene products.
Embryo
Normal development
3 Duplications and other chromosome mutations produce extra copies of some, but not all, genes,...
A
B
B
C
Mutant chromosome
Gene expression
Interaction of gene products
4 ...which alters the relative amounts (doses) of interacting products.
Embryo
Abnormal development
5 If the amount of one product increases but amounts of other products remain the same, developmental problems often result.

6.8 Unbalanced gene dosage leads to developmental abnormalities.

critically on proper gene dosage. If the amount of one protein increases while the amounts of others remain constant, problems can result (**Figure 6.8**). Duplications can have severe consequences when the precise balance of a gene product is critical to cell function (**Table 6.1**).

SEGMENTAL DUPLICATIONS The human genome contains numerous duplicated sequences called **segmental duplications**, which are defined as duplications greater than 1000 base pairs (1000 bp) in length. In most segmental duplications, the two copies are found on the same chromosome (an intrachromosomal duplication), but in others, the two copies are found on different chromosomes (an interchromosomal duplication). Many segmental duplications have been detected with the use of molecular techniques that examine the structure of DNA sequences on a chromosome (see Chapter 15). These techniques reveal that about 4% of the human genome consists of segmental duplications. In the human genome, the average size of segmental duplications is 15,000 bp.

IMPORTANCE OF DUPLICATIONS IN EVOLUTION Chromosome variations are potentially important in evolution, and they have clearly played a significant role in past evolution within a number of different groups of organisms. Chromosome duplications provide one possible way in which new genes evolve. In many cases, existing copies of a gene are not free to vary because they encode a product that is essential to development or function. However, after a chromosome undergoes duplication, extra copies of genes within the duplicated region are present. The original copy can provide the essential function, whereas an extra copy from the duplication is potentially free to undergo mutation and change. Over evolutionary time, the extra copy may acquire enough mutations to assume a new function that benefits the organism. For example, humans have a series of genes that encode different globin chains, which function as oxygen carriers. Some of these globin chains function during adult stages, and others function during embryonic and fetal development. The genes encoding all of these globins arose from an original ancestral gene that underwent a series of duplications.

CONCEPTS

A chromosome duplication is a mutation that doubles part of a chromosome. In individuals heterozygous for a chromosome duplication, the duplicated region of the chromosome loops out when homologous chromosomes pair in prophase I of meiosis. Duplications often have major effects on the phenotype, possibly by altering gene dosage. Segmental duplications are common within the human genome and have played an important role in the evolution of many eukaryotes.

✓ CONCEPT CHECK 1

Chromosome duplications often result in abnormal phenotypes because

a. developmental processes depend on the relative amounts of proteins encoded by different genes.
b. extra copies of the genes within the duplicated region do not pair in meiosis.
c. the chromosome is more likely to break when it loops in meiosis.
d. extra DNA must be replicated, which slows down cell division.

TABLE 6.1	Effects of some human chromosome rearrangements		
Type of rearrangement	**Chromosome**	**Disorder**	**Symptoms**
Duplication	4, short arm	—	Small head, short neck, low hairline, reduced growth, and intellectual disability
Duplication	4, long arm	—	Small head, sloping forehead, hand abnormalities
Duplication	7, long arm	—	Delayed development, asymmetry of the head, fuzzy scalp, small nose, low-set ears
Duplication	9, short arm	—	Characteristic face, variable intellectual disability, high and broad forehead, hand abnormalities
Deletion	5, short arm	*Cri-du-chat* syndrome	Small head, distinctive cry, widely spaced eyes, round face, intellectual disability
Deletion	4, short arm	Wolf–Hirschhorn syndrome	Small head with high forehead, wide nose, cleft lip and palate, severe intellectual disability
Deletion	4, long arm	—	Small head, mild to moderate intellectual disability, cleft lip and palate, hand and foot abnormalities
Deletion	7, long arm	Williams–Beuren syndrome	Facial features, heart defects, mental impairment
Deletion	15, long arm	Prader–Willi syndrome	Feeding difficulty at early age but becoming obese after 1 year of age, mild to moderate intellectual disability
Deletion	18, short arm	—	Round face, large low-set ears, mild to moderate intellectual disability
Deletion	18, long arm	—	Distinctive mouth shape, small hands, small head, intellectual disability

Deletions

A second type of chromosome rearrangement is a **chromosome deletion**, the loss of a chromosome segment (see Figure 6.4b). A chromosome with segments AB•CDEFG that undergoes a deletion of segment EF would generate the mutated chromosome AB•CDG.

A large deletion can be easily detected because the chromosome is noticeably shortened. In individuals heterozygous for deletions, the normal chromosome must loop out during the pairing of homologs in prophase I of meiosis (**Figure 6.9**) to allow the homologous regions of the two chromosomes to align and undergo synapsis. This looping

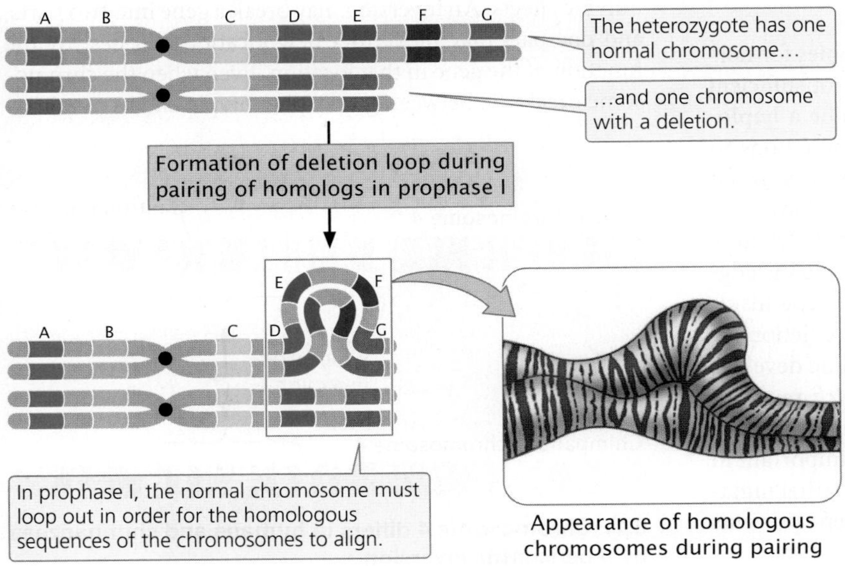

A
B
C
D
E
F
G
The heterozygote has one normal chromosome…
…and one chromosome with a deletion.
Formation of deletion loop during pairing of homologs in prophase I
In prophase I, the normal chromosome must loop out in order for the homologous sequences of the chromosomes to align.
Appearance of homologous chromosomes during pairing

6.9 In an individual heterozygous for a deletion, the normal chromosome loops out during chromosome pairing in prophase I.

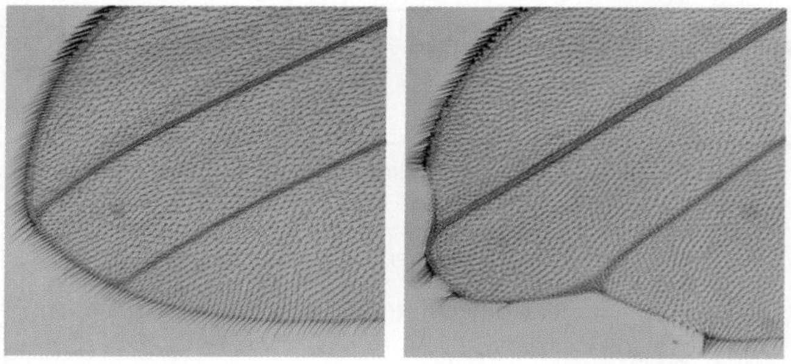

6.10 The Notch phenotype is produced by a chromosome deletion that includes the *Notch* gene. (Left) Normal wing venation. (Right) Wing venation produced by the *Notch* mutation. [Spyros Artavanis-Tsakonas, Kenji Matsuno, and Mark E. Fortini.]

out generates a structure that looks very much like that seen in individuals heterozygous for duplications.

EFFECTS OF DELETIONS The phenotypic consequences of a deletion depend on which genes are located in the deleted region. If the deletion includes the centromere, the chromosome will not segregate in meiosis or mitosis and will usually be lost. Many deletions are lethal in the homozygous state because all copies of any essential genes located in the deleted region are missing.

Even individuals heterozygous for a deletion may have multiple defects, for three reasons. First, the heterozygous condition may produce imbalances in the amounts of gene products, similar to the imbalances produced by extra gene copies. Second, normally recessive mutations on the homologous chromosome lacking the deletion may be expressed when the wild-type allele has been deleted (and is no longer present to mask the recessive allele's expression). The expression of a normally recessive mutation is referred to as **pseudodominance**, and it is an indication that one of the homologous chromosomes has a deletion.

Third, some genes must be present in two copies for normal function. When a single copy of a gene is not sufficient to produce a wild-type phenotype, it is said to be a **haploinsufficient gene**. A series of X-linked wing mutations in *Drosophila* is known as *Notch*. These mutations often result from chromosome deletions. *Notch* deletions behave in a dominant manner: when heterozygous for a *Notch* deletion, a fly has wings that are notched at the tips and along the edges (**Figure 6.10**). The *Notch* mutation is therefore haploinsufficient. Females that are homozygous for a *Notch* deletion (or males that are hemizygous) die early in embryonic development. The *Notch* locus, which is deleted in *Notch* mutants, encodes a receptor that normally transmits signals received from outside the cell to the cell's interior and is important in fly development. The deletion acts as a recessive lethal mutation because the loss of all copies of the *Notch* gene prevents normal development.

CONCEPTS

A chromosome deletion is a mutation in which a part of a chromosome is lost. In individuals heterozygous for a deletion, the normal chromosome loops out during prophase I of meiosis. Deletions cause recessive genes on the homologous chromosome to be expressed and may cause imbalances in gene products.

✓ CONCEPT CHECK 2

What is pseudodominance and how is it produced by a chromosome deletion?

Inversions

A third type of chromosome rearrangement is a **chromosome inversion**, in which a chromosome segment is inverted—turned 180 degrees (see Figure 6.4c). If a chromosome originally had segments AB•CDEFG, then chromosome AB•CFEDG represents an inversion that includes segments DEF. For an inversion to take place, the chromosome must break in two places. Inversions that do not include the centromere, such as AB•CFEDG, are termed **paracentric inversions** (*para* meaning "next to"), whereas inversions that include the centromere, such as ADC•BEFG, are termed **pericentric inversions** (*peri* meaning "around").

Inversion heterozygotes are common in many organisms, including a number of plants, some species of *Drosophila*, mosquitoes, and grasshoppers. Inversions may have played an important role in human evolution: G-banding patterns reveal that several human chromosomes differ from those of chimpanzees by only a pericentric inversion (**Figure 6.11**).

EFFECTS OF INVERSIONS Individual organisms with inversions have neither lost nor gained any genetic material; only the order of the chromosome segment has been altered. Nevertheless, these mutations often have pronounced phenotypic effects. An inversion may break a gene into two parts, and one part may move to a new location and destroy the function of the gene in that location. Even when the chromosome breaks lie between genes, phenotypic effects may arise

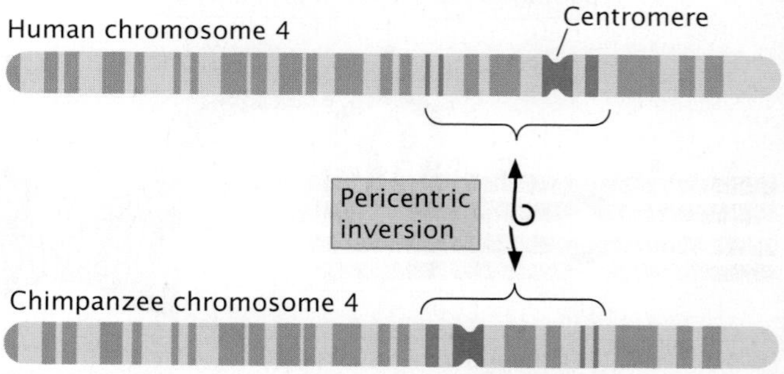

6.11 Chromosome 4 differs in humans and chimpanzees by a pericentric inversion.

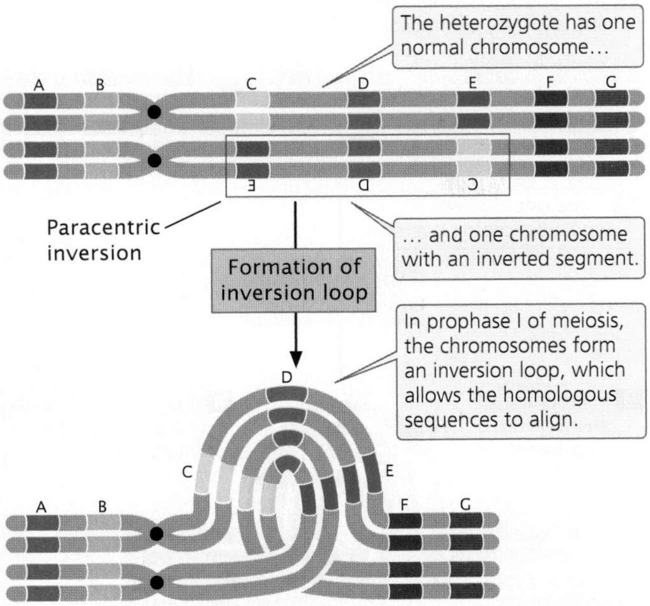

6.12 In an individual heterozygous for a paracentric inversion, the chromosomes form an inversion loop during pairing in prophase I.

from the inverted gene order. Many genes are regulated in a position-dependent manner; if their positions are altered by an inversion, their expression may be altered, an outcome referred to as a **position effect**. For example, when an inversion moves a wild-type allele (that normally encodes red eyes) at the *white* locus in *Drosophila* to a chromosome region that contains highly condensed and inactive chromatin, the wild-type allele is not expressed in some cells, resulting in an eye consisting of red and white spots.

INVERSIONS IN MEIOSIS When an individual is homozygous for a particular inversion, no special problems arise in meiosis, and the two homologous chromosomes can pair and separate normally. However, when an individual is heterozygous for an inversion, the gene order of the two homologs differs, and the homologous sequences can align and pair only if the two chromosomes form an inversion loop (**Figure 6.12**).

Individuals heterozygous for inversions also exhibit reduced recombination among genes located in the inverted region. The frequency of crossing over within the inversion is not actually diminished, but when crossing over does take place, the result is abnormal gametes that do not give rise to viable offspring, and thus no recombinant progeny are observed. Let's see why this happens.

Figure 6.13 illustrates the results of crossing over within a paracentric inversion. The individual is heterozygous for an inversion (see Figure 6.13a), with one wild-type, nonmutated chromosome (AB•CDEFG) and one inverted chromosome (AB•EDCFG). In prophase I of meiosis, an inversion

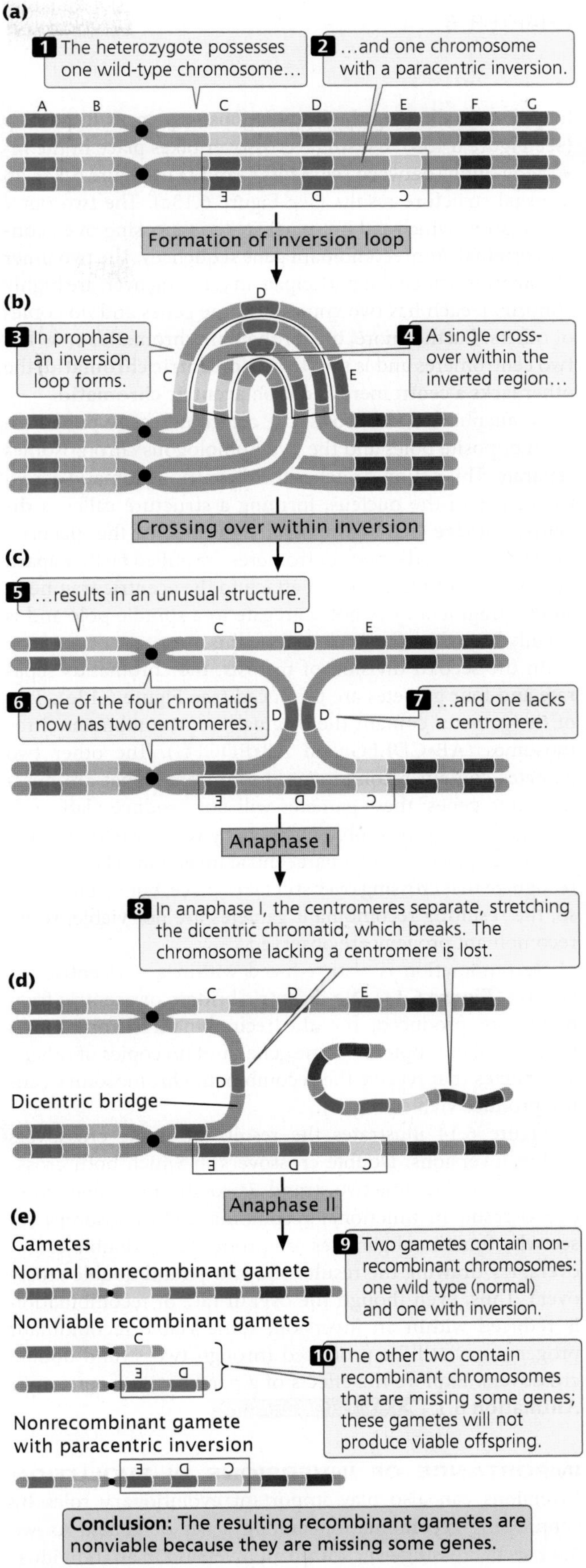

6.13 In a heterozygous individual, a single crossover within a paracentric inversion leads to abnormal gametes.

loop forms, allowing the homologous sequences to pair up (see Figure 6.13b). If a single crossover takes place in the inverted region (between segments C and D in Figure 6.13), an unusual structure results (see Figure 6.13c). The two outer chromatids, which did not participate in crossing over, contain original, nonrecombinant gene sequences. The two inner chromatids, which did participate in crossing over, are highly abnormal: each has two copies of some genes and no copies of others. Furthermore, one of the four chromatids now has two centromeres and is said to be a **dicentric chromatid**; the other lacks a centromere and is an **acentric chromatid**.

In anaphase I of meiosis, the centromeres are pulled toward opposite poles and the two homologous chromosomes separate. This action stretches the dicentric chromatid across the center of the nucleus, forming a structure called a **dicentric bridge** (see Figure 6.13d). Eventually, the dicentric bridge breaks as the two centromeres are pulled farther apart. Spindle microtubules do not attach to the acentric fragment, so this fragment does not segregate to a spindle pole and is usually lost when the nucleus re-forms.

In the second division of meiosis, the chromatids separate and four gametes are produced (see Figure 6.13e). Two of the gametes contain the original, nonrecombinant chromosomes (AB•CDEFG and AB•EDCFG). The other two gametes contain recombinant chromosomes that are missing some genes; these gametes will not produce viable offspring. Thus, no recombinant progeny result when crossing over takes place within a paracentric inversion. The key is to recognize that crossing over still takes place, but when it does so, the resulting recombinant gametes are not viable, so no recombinant progeny are observed.

Recombination is also reduced within a pericentric inversion (**Figure 6.14**). No dicentric bridges or acentric fragments are produced, but the recombinant chromosomes have too many copies of some genes and no copies of others, so gametes that receive the recombinant chromosomes cannot produce viable progeny.

Figure 6.14 illustrates the results of single crossovers within inversions. Double crossovers in which both crossovers are on the same two strands (two-strand double crossovers) result in functional recombinant chromosomes (to see why functional gametes are produced by double crossovers, try drawing the results of a two-strand double crossover). Thus, even though the overall rate of recombination is reduced within an inversion, some viable recombinant progeny may still be produced through two-strand double crossovers. Explore the effects of a paracentric inversion in A **Animation 6.1**. TRY PROBLEM 19

IMPORTANCE OF INVERSIONS IN EVOLUTION Inversions can also play important evolutionary roles by suppressing recombination among a set of genes. As we have seen, crossing over within an inversion in an individual heterozygous for a pericentric or paracentric inversion leads to unbalanced gametes and no recombinant progeny.

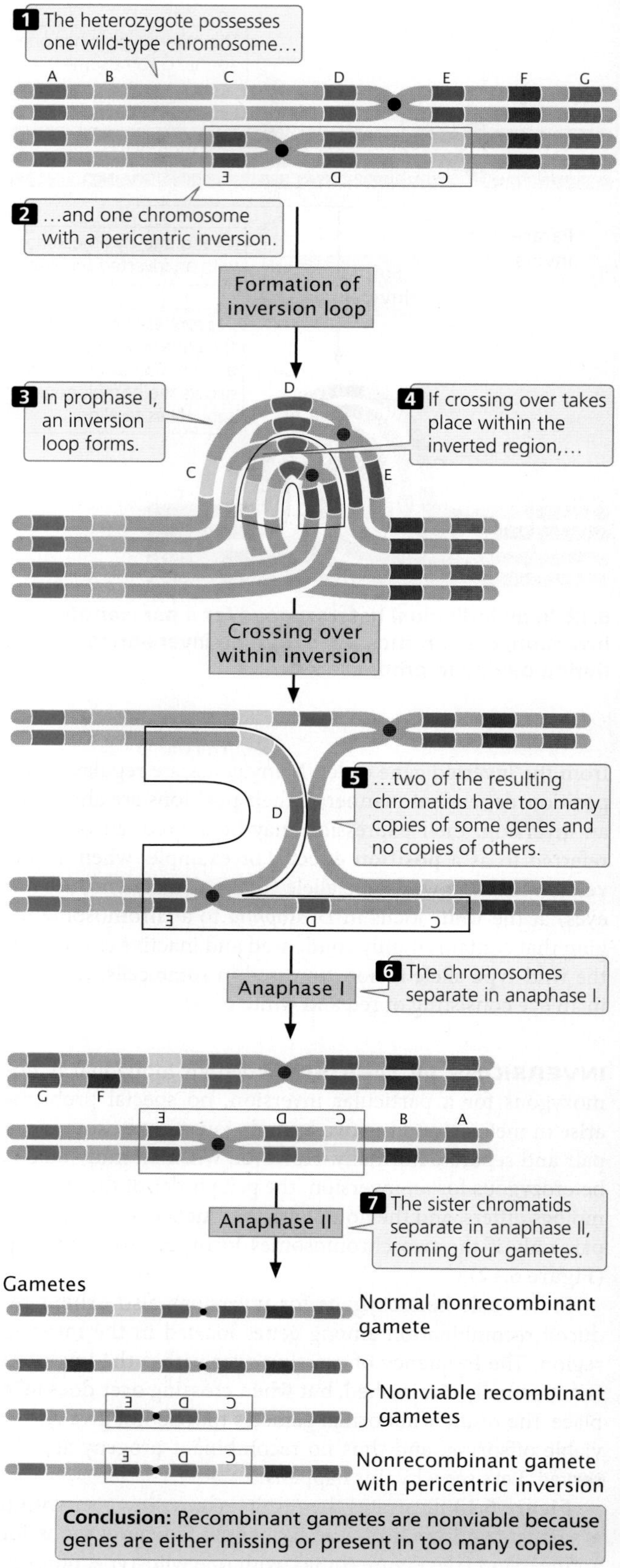

6.14 In a heterozygous individual, a single crossover within a pericentric inversion leads to abnormal gametes.

CONCEPTS

In an inversion, a segment of a chromosome is turned 180 degrees. Inversions cause breaks in some genes and may move others to new locations. In individuals heterozygous for a chromosome inversion, the homologous chromosomes form an inversion loop in prophase I of meiosis. When crossing over takes place within the inverted region, nonviable gametes are usually produced, resulting in a depression in observed recombination frequencies.

✓ CONCEPT CHECK 3

A dicentric chromosome is produced when crossing over takes place in an individual heterozygous for which type of chromosome rearrangement?

a. Duplication
b. Deletion
c. Paracentric inversion
d. Pericentric inversion

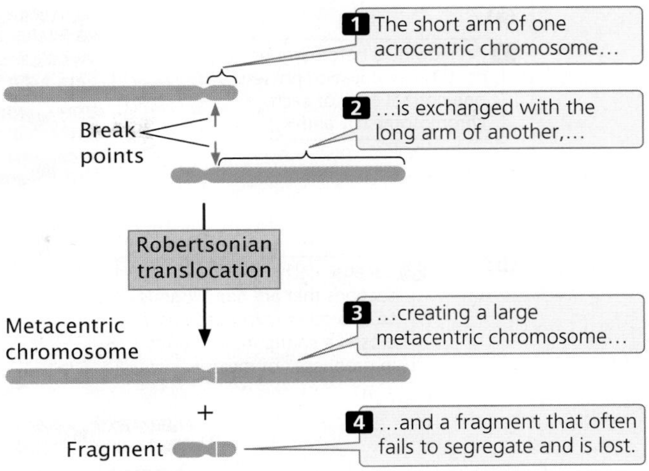

6.15 In a Robertsonian translocation, the short arm of one acrocentric chromosome is exchanged with the long arm of another.

Translocations

A **translocation** entails the movement of genetic material between nonhomologous chromosomes (see Figure 6.4d) or within the same chromosome. Translocation should not be confused with crossing over, in which there is an exchange of genetic material between *homologous* chromosomes.

In a **nonreciprocal translocation**, genetic material moves from one chromosome to another without any reciprocal exchange. Consider the following two nonhomologous chromosomes: AB•CDEFG and MN•OPQRS. If chromosome segment EF moves from the first chromosome to the second without any transfer of segments from the second chromosome to the first, a nonreciprocal translocation has taken place, producing chromosomes AB•CDG and MN•OP<u>EF</u>QRS. More commonly, there is a two-way exchange of segments between the chromosomes, resulting in a **reciprocal translocation**. A reciprocal translocation between chromosomes AB•CDEFG and MN•OPQRS might give rise to chromosomes AB•CD<u>QRS</u> and MN•OP<u>EFG</u>.

EFFECTS OF TRANSLOCATIONS Translocations can affect a phenotype in several ways. First, they can physically link genes that were formerly located on different chromosomes. These new linkage relations may affect gene expression (a position effect): genes translocated to new locations may come under the control of different regulatory sequences or other genes that affect their expression.

Second, the chromosome breaks that bring about translocations may take place within a gene and disrupt its function. Molecular geneticists have used these types of effects to map human genes. Neurofibromatosis is a genetic disease characterized by numerous fibrous tumors of the skin and nervous tissue; it results from an autosomal dominant mutation. Linkage studies first placed the locus that, when mutated, causes neurofibromatosis on chromosome 17, but its precise location was unknown. Geneticists later narrowed down the location when they identified two patients with neurofibromatosis who possessed a translocation affecting chromosome 17. These patients were assumed to have developed neurofibromatosis because one of the chromosome breaks that occurred in the translocation disrupted a particular gene. DNA from the regions around the breaks was sequenced, eventually leading to the identification of the gene responsible for neurofibromatosis.

Deletions frequently accompany translocations. In a **Robertsonian translocation**, for example, the long arms of two acrocentric chromosomes become joined to a common centromere through a translocation, generating a metacentric chromosome with two long arms and another chromosome with two very short arms (**Figure 6.15**). The smaller chromosome is often lost because very small chromosomes do not have enough mass to segregate properly during mitosis and meiosis. The result is an overall reduction in chromosome number. As we will see, Robertsonian translocations are the cause of some cases of Down syndrome, a disorder discussed later in this chapter.

TRANSLOCATIONS IN MEIOSIS The effects of a translocation on chromosome segregation in meiosis depend on the nature of the translocation. Let's consider what happens in an individual heterozygous for a reciprocal translocation. Suppose that the original chromosomes were AB•CDEFG and M•NOPQRST (designated N1 and N2, respectively, for normal chromosomes 1 and 2) and that a reciprocal translocation takes place, producing chromosomes AB•CD<u>QRST</u> and M•NOP<u>EFG</u> (designated T1 and T2, respectively, for translocated chromosomes 1 and 2). An individual heterozygous for this translocation would possess one normal copy of each chromosome and one translocated copy (**Figure 6.16a**). Each of these

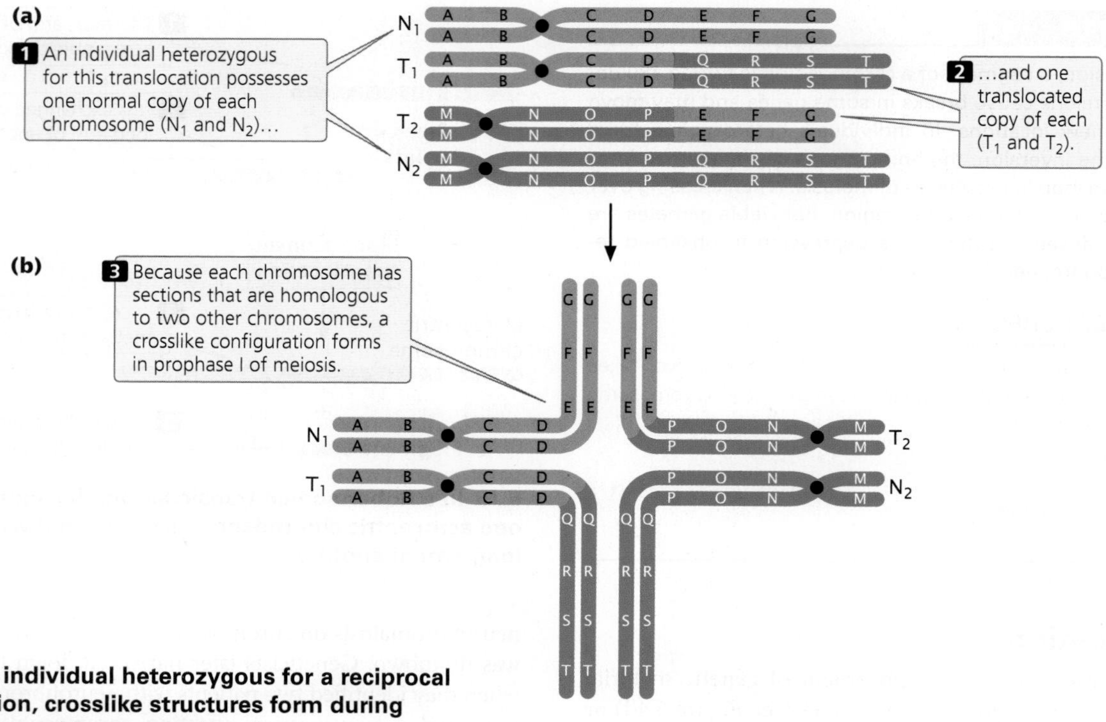

6.16 In an individual heterozygous for a reciprocal translocation, crosslike structures form during homologous pairing in prophase I.

chromosomes contains segments that are homologous to segments of two other chromosomes. Thus, when the homologous sequences pair in prophase I of meiosis, crosslike configurations consisting of all four chromosomes form (**Figure 6.16b**). Whether viable or nonviable gametes are produced depends on how the chromosomes in these crosslike configurations separate. Only about half of the gametes from an individual heterozygous for a reciprocal translocation are expected to be functional, so these individuals frequently exhibit reduced fertility. See the effects of a reciprocal translocation in **Animation 6.2**.

THINK-PAIR-SHARE Question 4

CONCEPTS

In translocations, parts of chromosomes move to other nonhomologous chromosomes or to other regions of the same chromosome. Translocations can affect the phenotype by causing genes to move to new locations, where they come under the influence of new regulatory sequences, or by breaking genes and disrupting their function.

✓ CONCEPT CHECK 4

What is the outcome of a Robertsonian translocation?

a. Two acrocentric chromosomes
b. One large chromosome and one very small chromosome with two very short arms
c. One large metacentric and one large acrocentric chromosome
d. Two large metacentric chromosomes

Fragile Sites

Chromosomes of cells grown in culture sometimes develop constrictions or gaps at particular locations, called **fragile sites** (**Figure 6.17**) because they are prone to breakage under certain conditions. More than 100 fragile sites have been identified on human chromosomes. One of the most intensively studied is a fragile site on the human X chromosome that is associated with **fragile-X syndrome**, a disorder that includes intellectual disability. Fragile-X syndrome, which exhibits X-linked inheritance and arises with a frequency of about 1 in 5000 male births, has been shown to result from an increase in the number of repeats of a CGG trinucleotide (see Chapter 13). Other common fragile sites do not consist of trinucleotide repeats, however, and their nature is still not completely understood.

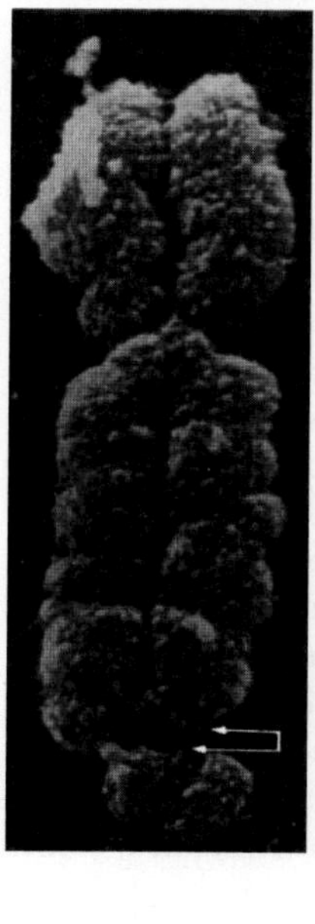

6.17 Fragile sites are chromosome regions susceptible to breakage under certain conditions. Shown here is a fragile site on human chromosome X. [Courtesy of Dr. Christine Harrison.]

Copy-Number Variations

Chromosome rearrangements have traditionally been detected by examination of the chromosomes with a microscope. Visual examination identifies chromosome rearrangements on the basis of changes in the overall size of a chromosome, alteration of banding patterns revealed by chromosome staining, or changes in the behavior of chromosomes in meiosis. Microscopy, however, can detect only large chromosome rearrangements, typically those that are at least 5 million base pairs in length.

With the completion of the Human Genome Project (see Chapter 15), detailed information about DNA sequences found on individual chromosomes became available. Using this information, geneticists can now examine the number of copies of specific DNA sequences present in a cell and detect duplications, deletions, and other chromosome rearrangements that cannot be observed with microscopy alone. This work has been greatly facilitated by the availability of microarrays (see Chapter 15), which allow the simultaneous detection of hundreds of thousands of specific DNA sequences from across the genome. Because these methods measure the number of copies of particular DNA sequences, the variations that they detect are called **copy-number variations** (CNVs). Copy-number variations include duplications and deletions that range in length from thousands of base pairs to several million base pairs. Many of these variants encompass at least one gene and may encompass several genes.

Recent studies of copy-number variations have revealed that submicroscopic chromosome duplications and deletions are quite common: research suggests that each person may possess as many as 1000 copy-number variations. Many probably have no observable phenotypic effects, but some copy-number variations have been implicated in a number of diseases and disorders. For example, Janine Wagenstaller and her colleagues studied copy-number variation in 67 children with unexplained intellectual disability and found that 11 (16%) of them had duplications or deletions. Copy-number variations have also been associated with osteoporosis, autism, schizophrenia, and a number of other diseases and disorders. ▶TRY PROBLEM 14

CONCEPTS

Fragile sites are constrictions or gaps in chromosomes that are prone to breakage under certain conditions. Variations in the number of copies of particular DNA sequences (copy-number variations) are surprisingly common in the human genome.

6.3 Aneuploidy Is an Increase or Decrease in the Number of Individual Chromosomes

In addition to chromosome rearrangements, chromosome mutations include changes in the number of chromosomes. Variations in chromosome number can be classified into two basic types: **aneuploidy**, which is a change in the number of individual chromosomes, and **polyploidy**, which is an increase in the number of chromosome sets.

Aneuploidy can arise in several ways. First, a chromosome may be lost in the course of mitosis or meiosis if, for example, its centromere is deleted. Loss of the centromere prevents the spindle microtubules from attaching, so the chromosome fails to move to the spindle pole and does not become incorporated into a nucleus after cell division. Second, the small chromosome generated by a Robertsonian translocation may be lost in mitosis or meiosis. Third, aneuploidy may arise through **nondisjunction**, the failure of homologous chromosomes or sister chromatids to separate in meiosis or mitosis. Nondisjunction leads to some gametes or cells that contain an extra chromosome and other gametes or cells that are missing a chromosome (**Figure 6.18**). ▶TRY PROBLEM 21

Types of Aneuploidy

We will consider four types of common aneuploid conditions in diploid individuals: nullisomy, monosomy, trisomy, and tetrasomy.

1. **Nullisomy** is the loss of both members of a homologous pair of chromosomes. It is represented as $2n - 2$, where n refers to the haploid number of chromosomes. Thus, among humans, who normally possess $2n = 46$ chromosomes, a nullisomic zygote has 44 chromosomes.
2. **Monosomy** is the loss of a single chromosome, represented as $2n - 1$. A human monosomic zygote has 45 chromosomes.
3. **Trisomy** is the gain of a single chromosome, represented as $2n + 1$. A human trisomic zygote has 47 chromosomes. The gain of a chromosome means that there are three homologous copies of one chromosome. Most cases of Down syndrome, discussed later in this chapter, result from trisomy of chromosome 21.
4. **Tetrasomy** is the gain of two homologous chromosomes, represented as $2n + 2$. A human tetrasomic zygote has 48 chromosomes. Tetrasomy is not the gain of *any* two extra chromosomes, but rather the gain of two homologous chromosomes, so that there are four homologous copies of a particular chromosome.

More than one aneuploid mutation may occur in the same individual organism. An individual that has an extra copy of each of two different (nonhomologous) chromosomes is referred to as being double trisomic and is represented as $2n + 1 + 1$. Similarly, a double monosomic individual has two fewer nonhomologous chromosomes ($2n - 1 - 1$), and a double tetrasomic individual has two extra pairs of homologous chromosomes ($2n + 2 + 2$).

THINK-PAIR-SHARE Question 5

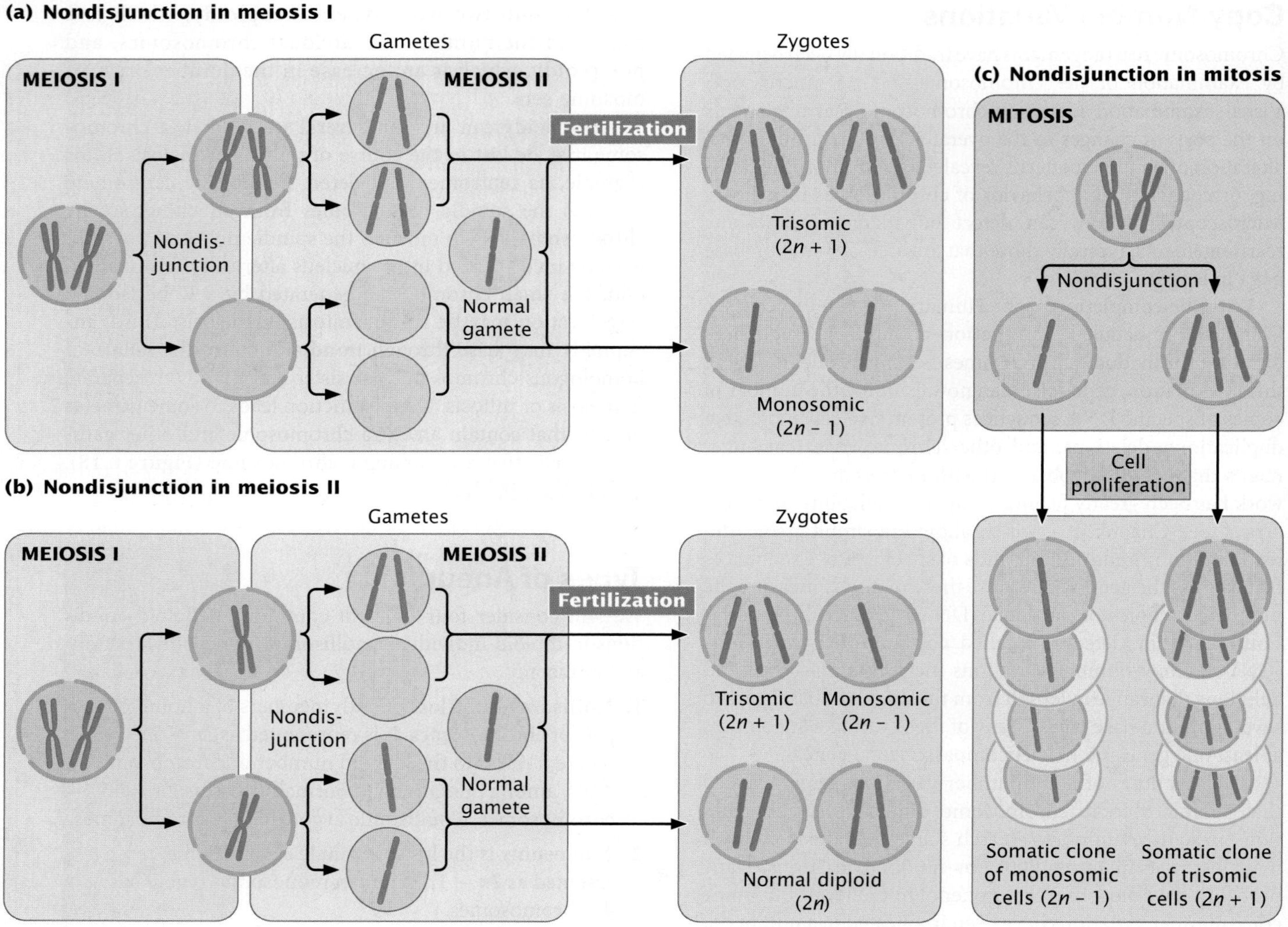

6.18 Aneuploids can be produced through nondisjunction in meiosis I, meiosis II, or mitosis. The gametes that result from meioses with nondisjunction combine with a gamete (with blue chromosome) that results from normal meiosis to produce the zygotes shown.

Effects of Aneuploidy

Aneuploidy usually alters the phenotype drastically. In most animals and many plants, aneuploid mutations are lethal. Because aneuploidy affects the number of gene copies, but not their nucleotide sequences, the effects of aneuploidy are most likely due to abnormal gene dosage. Aneuploidy alters the dosage for some, but not all, genes, disrupting the relative concentrations of gene products and often interfering with normal development.

A major exception to the relation between gene number and gene dosage pertains to genes on the mammalian X chromosome. In mammals, X-chromosome inactivation ensures that males (who have a single X chromosome) and females (who have two X chromosomes) receive the same functional dosage for X-linked genes (see pp. 86–87 in Chapter 4 for further discussion of X-chromosome inactivation). Additional X chromosomes in mammals are inactivated, so we might expect that aneuploidy of the sex chromosomes would be less detrimental in these animals. Indeed, this is the case for mice and humans, for whom aneuploidy of the sex chromosomes is the most common form of aneuploidy seen in living individuals. Y-chromosome aneuploidy is probably common because there is so little information in the Y chromosome.

THINK-PAIR-SHARE Question 6

CONCEPTS

Aneuploidy, the loss or gain of one or more individual chromosomes, may arise from the loss of a chromosome subsequent to translocation or from nondisjunction in meiosis or mitosis. It disrupts gene dosage and often has severe phenotypic effects.

✓ CONCEPT CHECK 5

A diploid organism has $2n = 36$ chromosomes. How many chromosomes will be found in a trisomic member of this species?

Aneuploidy in Humans

For unknown reasons, an incredibly high percentage of all human embryos that are conceived possess chromosome abnormalities. Findings from studies of women who are attempting pregnancy suggest that more than 30% of all conceptions are spontaneously aborted (miscarried), usually so early in development that the woman is not even aware of her pregnancy. Chromosome defects are present in at least 50% of spontaneously aborted human fetuses, with aneuploidy accounting for most of them. This rate of chromosome abnormality in humans is higher than in other organisms that have been studied; in mice, for example, aneuploidy is found in no more than 2% of fertilized eggs. Aneuploidy in humans usually produces such serious developmental problems that spontaneous abortion results. Only about 2% of all fetuses with a chromosome defect survive to birth.

SEX-CHROMOSOME ANEUPLOIDIES The most common aneuploidies seen in living humans are those that involve the sex chromosomes. As is true of all mammals, aneuploidy of the human sex chromosomes is better tolerated than aneuploidy of autosomal chromosomes. Both Turner syndrome and Klinefelter syndrome (described in Chapter 4) result from aneuploidy of the sex chromosomes.

AUTOSOMAL ANEUPLOIDIES Autosomal aneuploidies resulting in live births are less common than sex-chromosome aneuploidies in humans, probably because there is no mechanism of dosage compensation for autosomal chromosomes. Most embryos with autosomal aneuploidies are spontaneously aborted, though occasionally fetuses with aneuploidies of some of the small autosomes, such as chromosome 21, complete development. Because these chromosomes are small and carry fewer genes, the presence of extra copies is less detrimental than it is for larger chromosomes.

DOWN SYNDROME In 1866, John Langdon Down, physician and medical superintendent of the Earlswood Asylum in Surrey, England, noticed a remarkable resemblance among a number of his intellectually disabled patients: all of them possessed a broad, flat face, a small nose, and oval-shaped eyes. Their features were so similar, in fact, that he felt that they might easily be mistaken for children from the same family. Down did not understand the cause of their intellectual disability, but his original description faithfully records the physical characteristics of people with this genetic form of intellectual disability. In his honor, the disorder is today known as Down syndrome.

Down syndrome, also known as **trisomy 21**, is the most common autosomal aneuploidy in humans (**Figure 6.19a**). The worldwide incidence is about 1 in 700 human births, although the incidence increases among children born to older mothers. Approximately 92% of those who have Down syndrome

(a) (b)

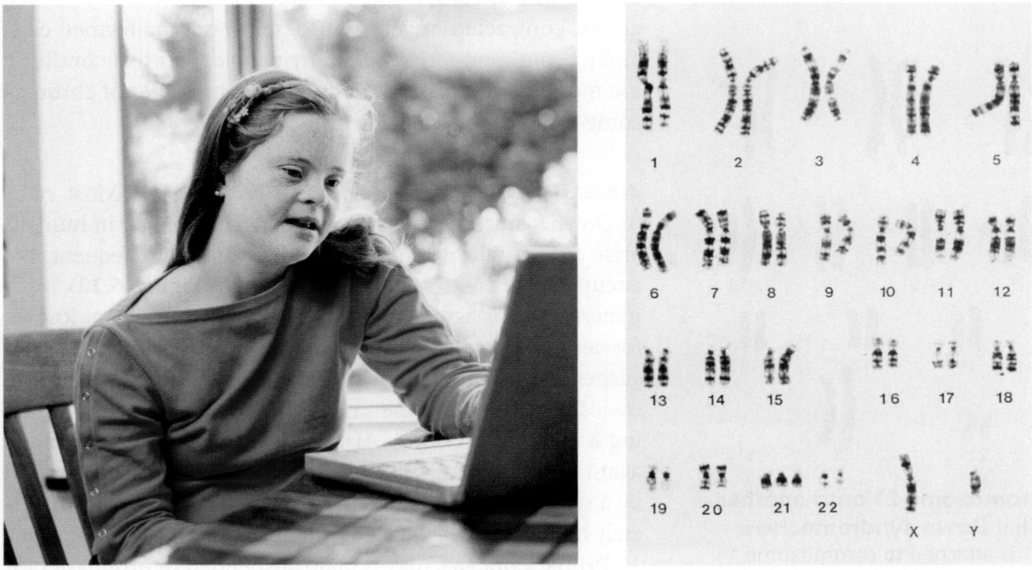

6.19 Down syndrome is caused by trisomy of chromosome 21. [Part a: George Doyle/Getty Images. Part b: L. Wilatt, East Anglian Regional Genetics Service/Science Photo Library/Science Source.]

have three full copies of chromosome 21 (and therefore a total of 47 chromosomes), a condition termed **primary Down syndrome** (**Figure 6.19b**). Primary Down syndrome usually arises from spontaneous nondisjunction in egg formation: about 75% of the nondisjunction events that cause Down syndrome are maternal in origin, most arising in meiosis I. Most children with Down syndrome are born to normal parents, and the failure of the chromosomes to divide has little hereditary tendency. A couple who has conceived one child with primary Down syndrome has only a slightly higher risk of conceiving a second child with Down syndrome (compared with other couples of similar age who have not had any children with Down syndrome). Similarly, the couple's relatives are not more likely to have a child with primary Down syndrome.

THINK-PAIR-SHARE Question 7

About 4% of people with Down syndrome are not trisomic for a complete chromosome 21. Instead, they have 46 chromosomes, but an extra copy of part of chromosome 21 is attached to another chromosome through a translocation. This condition is termed **familial Down syndrome** because it has a tendency to run in families. The phenotypic characteristics of familial Down syndrome are the same as those of primary Down syndrome.

Familial Down syndrome arises in offspring whose parents are carriers of chromosomes that have undergone a Robertsonian translocation, most commonly between chromosome 21 and chromosome 14: the long arm of 21 and the short arm of 14 exchange places (**Figure 6.20**). This exchange produces one chromosome that includes the long arms of chromosomes 14 and 21 and another, very small chromosome that consists of the short arms of chromosomes 21 and 14. The small chromosome is generally lost after several cell divisions. Although exchange between chromosomes 21 and 14 is the most common cause of familial Down syndrome, the condition can also be caused by translocations between 21 and other chromosomes, such as 15.

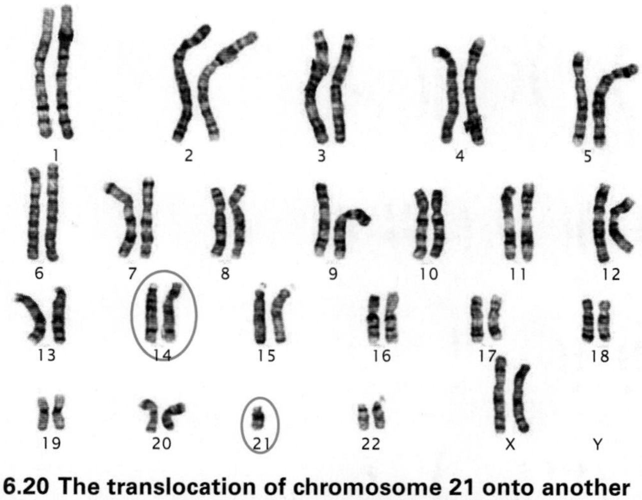

6.20 The translocation of chromosome 21 onto another chromosome results in familial Down syndrome. Here, the long arm of chromosome 21 is attached to chromosome 14. This karyotype is from a translocation carrier, who is phenotypically normal but is at increased risk for producing children with Down syndrome. [© Centre for Genetics Education for and on behalf of the Crown in right of the State of New South Wales.]

People with this type of translocation, called **translocation carriers**, do not have Down syndrome. Although they possess only 45 chromosomes, their phenotypes are normal because they have two copies of the long arms of chromosomes 14 and 21, and apparently the short arms of these chromosomes (which are lost) carry no essential genetic information. Although translocation carriers are completely healthy, they have an increased chance of producing children with Down syndrome (**Figure 6.21**). ▶TRY PROBLEM 22

OTHER HUMAN TRISOMIES Few autosomal aneuploidies in humans besides trisomy 21 result in live births. **Trisomy 18**, also known as **Edward syndrome**, arises with a frequency of approximately 1 in 8000 live births. Babies with Edward syndrome have severe intellectual disability, low-set ears, a short neck, deformed feet, clenched fingers, heart problems, and other disabilities. Few live for more than a year after birth. **Trisomy 13** has a frequency of about 1 in 15,000 live births and produces features that are collectively known as **Patau syndrome**. Characteristics of this condition include severe intellectual disability, a small head, sloping forehead, small eyes, cleft lip and palate, extra fingers and toes, and numerous other problems. About half of children with trisomy 13 die within the first month of life, and 95% die by the age of 3. Rarer still is **trisomy 8**, which arises with a frequency ranging from about 1 in 25,000 to 1 in 50,000 live births. This aneuploidy is characterized by intellectual disability, contracted fingers and toes, low-set malformed ears, and a prominent forehead. Many people with this condition are mosaics, having some cells with three copies of chromosome 8 and other cells with two copies.

ANEUPLOIDY AND MATERNAL AGE Most cases of Down syndrome and other types of aneuploidy in humans arise from maternal nondisjunction, and the frequency of aneuploidy increases with maternal age (**Figure 6.22**). Why maternal age is associated with nondisjunction is not known for certain. Female mammals are born with primary oocytes suspended in prophase I of meiosis. Just before ovulation, meiosis resumes and the first division is completed, producing a secondary oocyte. At this point, meiosis is suspended again and remains so until the secondary oocyte is penetrated by a sperm. The second meiotic division takes place immediately before the nuclei of egg and sperm unite to form a zygote.

Primary oocytes may remain suspended in prophase I for many years before ovulation takes place and meiosis recommences. Components of the spindle and other structures required for chromosome segregation may break down during

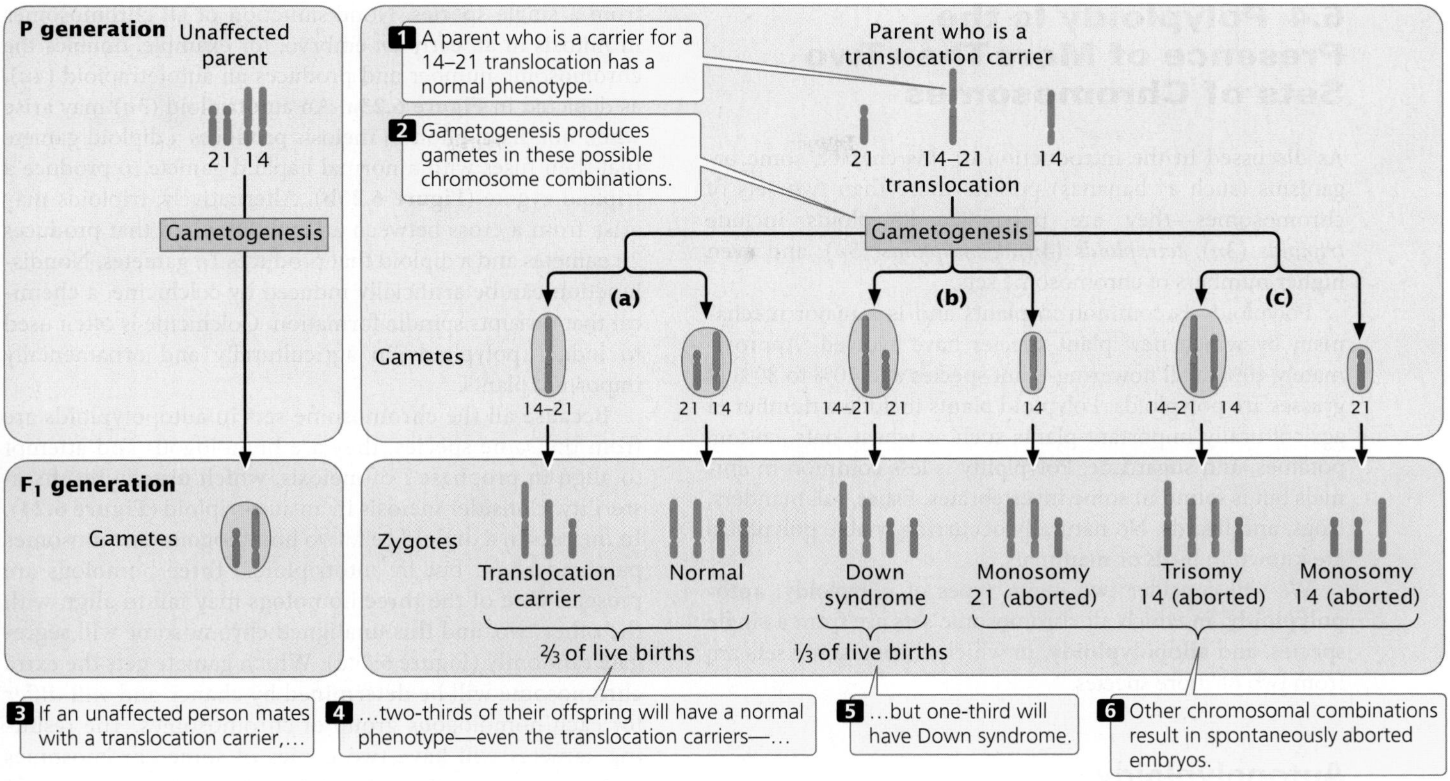

6.21 Translocation carriers are at increased risk for producing children with Down syndrome.

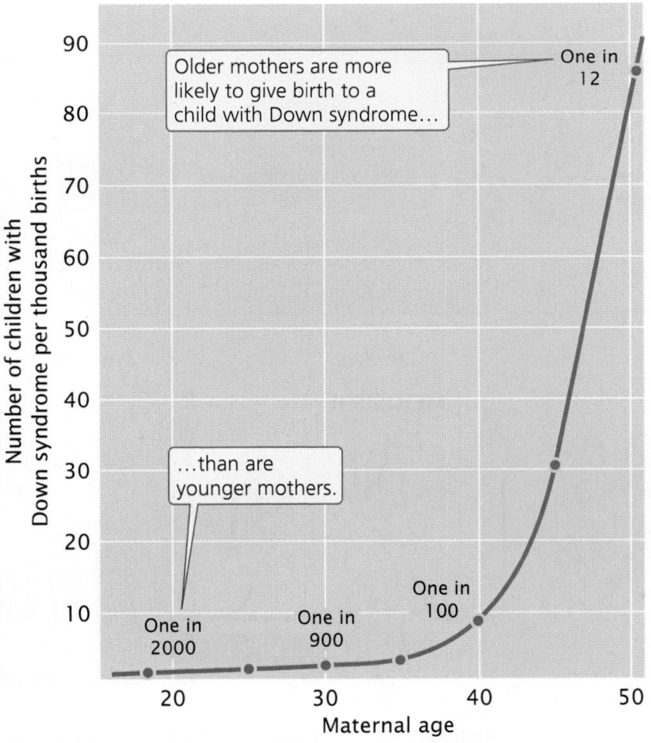

6.22 The incidence of primary Down syndrome and other aneuploidies increases with maternal age.

the long arrest of meiosis, leading to more aneuploidy in children born to older mothers. According to this theory, no age effect is seen in males because sperm are produced continuously after puberty, with no long suspension of the meiotic divisions.

ANEUPLOIDY AND CANCER Many tumor cells have extra chromosomes, missing chromosomes, or both; some types of tumors are consistently associated with specific chromosome mutations, including aneuploidy and chromosome rearrangements. The role of chromosome mutations in cancer will be explored in Chapter 16.

CONCEPTS

In humans, sex-chromosome aneuploidies are more common than autosomal aneuploidies. X-chromosome inactivation prevents problems of gene dosage for X-linked genes. Down syndrome results from three functional copies of chromosome 21, either through trisomy (primary Down syndrome) or a Robertsonian translocation (familial Down syndrome).

✔ CONCEPT CHECK 6

Why are sex-chromosome aneuploidies more common than autosomal aneuploidies in humans and other mammals?

6.4 Polyploidy Is the Presence of More Than Two Sets of Chromosomes

As discussed in the introduction to this chapter, some organisms (such as bananas) possess more than two sets of chromosomes—they are polyploid. Polyploids include *triploids* ($3n$), *tetraploids* ($4n$), *pentaploids* ($5n$), and even higher numbers of chromosome sets.

Polyploidy is common in plants and is a major mechanism by which new plant species have evolved. Approximately 40% of all flowering-plant species and 70% to 80% of grasses are polyploids. Polyploid plants include a number of agriculturally important plants such as wheat, oats, cotton, potatoes, and sugarcane. Polyploidy is less common in animals but is found in some invertebrates, fishes, salamanders, frogs, and lizards. No naturally occurring, viable polyploids are known in birds or mammals.

We will consider two major types of polyploidy: **autopolyploidy**, in which all chromosome sets are from a single species, and **allopolyploidy**, in which chromosome sets are from two or more species.

Autopolyploidy

Autopolyploidy is caused by accidents of mitosis or meiosis that produce extra sets of chromosomes, all derived from a single species. Nondisjunction of all chromosomes in mitosis in an early $2n$ embryo, for example, doubles the chromosome number and produces an autotetraploid ($4n$), as depicted in **Figure 6.23a**. An autotriploid ($3n$) may arise when nondisjunction in meiosis produces a diploid gamete that then fuses with a normal haploid gamete to produce a triploid zygote (**Figure 6.23b**). Alternatively, triploids may arise from a cross between an autotetraploid that produces $2n$ gametes and a diploid that produces $1n$ gametes. Nondisjunction can be artificially induced by colchicine, a chemical that disrupts spindle formation. Colchicine is often used to induce polyploidy in agriculturally and ornamentally important plants.

Because all the chromosome sets in autopolyploids are from the same species, they are homologous and attempt to align in prophase I of meiosis, which usually results in sterility. Consider meiosis in an autotriploid (**Figure 6.24**). In meiosis in a diploid cell, two homologous chromosomes pair and align, but in autotriploids, three homologs are present. One of the three homologs may fail to align with the other two, and this unaligned chromosome will segregate randomly (Figure 6.24a). Which gamete gets the extra chromosome will be determined by chance and will differ for each homologous group of chromosomes. The resulting gametes will have two copies of some chromosomes and one copy of others. Even if all three chromosomes do align, two chromosomes must segregate to one gamete and one chromosome to the other (Figure 6.24b). Occasionally,

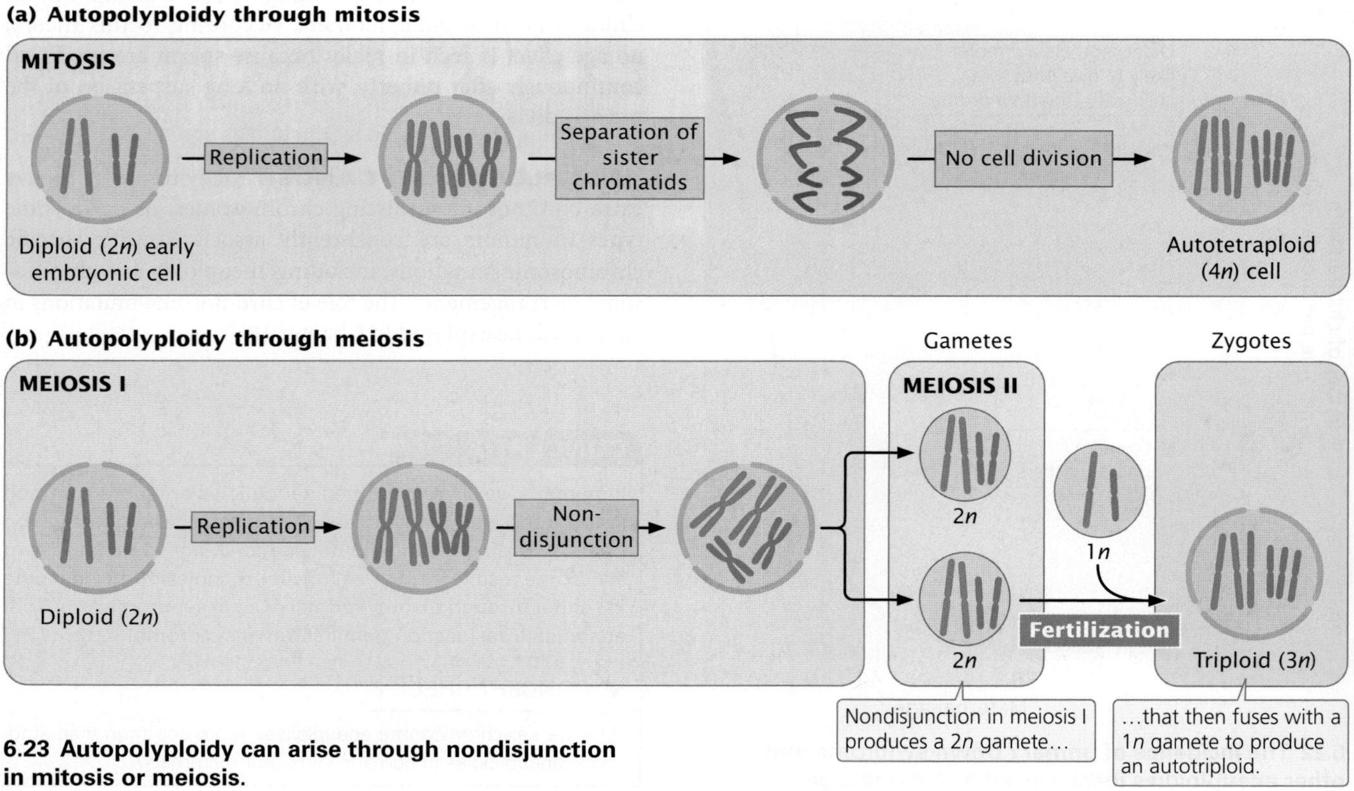

6.23 Autopolyploidy can arise through nondisjunction in mitosis or meiosis.

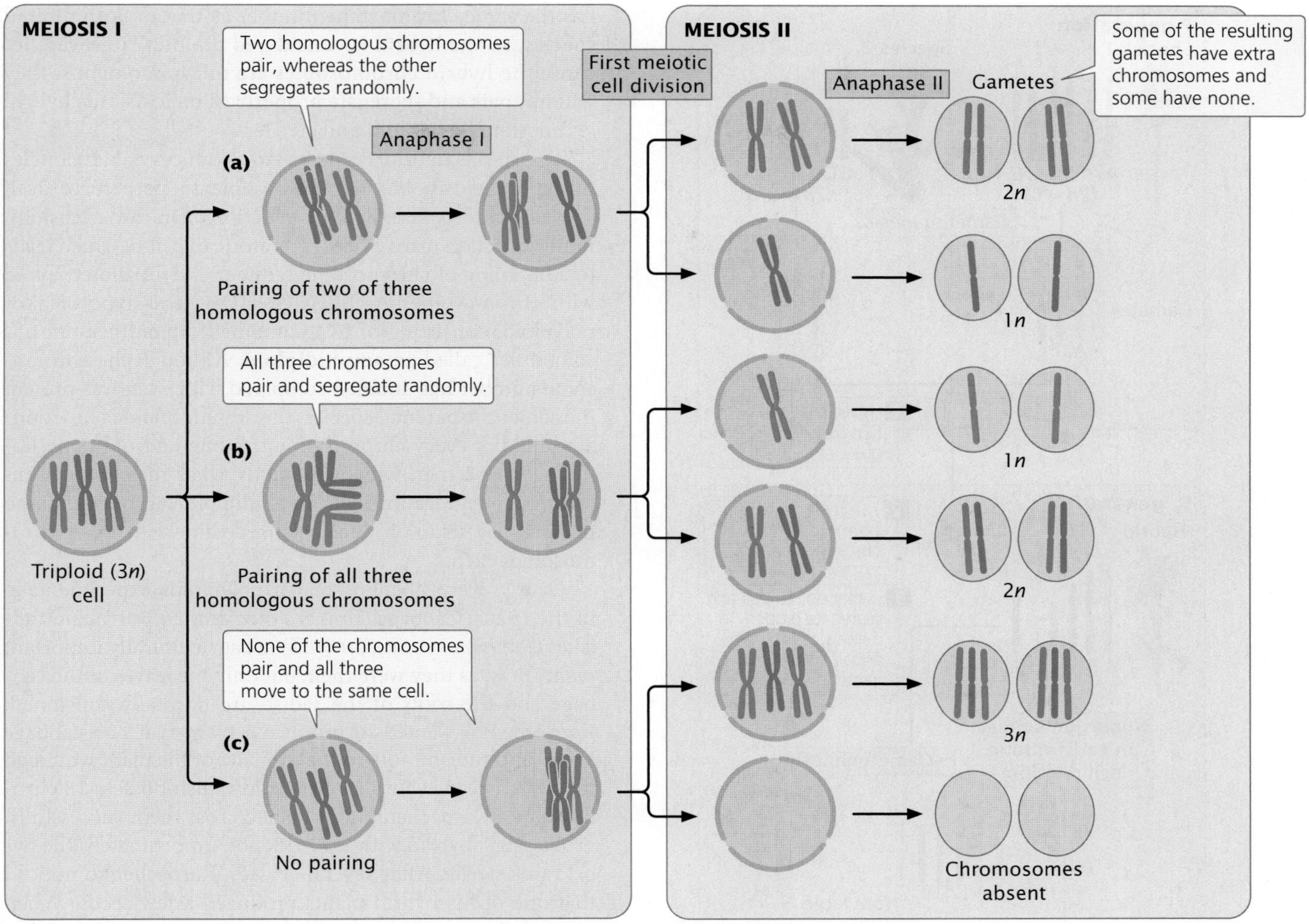

6.24 In meiosis in an autotriploid, homologous chromosomes can pair, or not pair, in three ways. This example illustrates the pairing and segregation of a single set of homologous chromosomes.

the presence of a third chromosome interferes with normal alignment, and all three chromosomes move to the same gamete (Figure 6.24c).

No matter how the three homologous chromosomes align, their random segregation will create **unbalanced gametes**, with various numbers of chromosomes. A gamete produced by meiosis in such an autotriploid might receive, say, two copies of chromosome 1, one copy of chromosome 2, three copies of chromosome 3, and no copies of chromosome 4. When the unbalanced gamete fuses with a normal gamete (or with another unbalanced gamete), the resulting zygote has different numbers of the four types of chromosomes. This difference in number creates unbalanced gene dosage in the zygote, which is often lethal. For this reason, triploids do not usually produce viable offspring.

In even-numbered autopolyploids, such as autotetraploids, the homologous chromosomes can theoretically form pairs and divide equally. However, this event rarely takes place, so these types of autotetraploids also produce unbalanced gametes.

The sterility that usually accompanies autopolyploidy has been exploited in agriculture. As discussed in the introduction to this chapter, triploid bananas ($3n = 33$) are sterile and seedless. Similarly, seedless triploid watermelons have been created and are now widely sold.

Allopolyploidy

Allopolyploidy arises from hybridization between two species; the resulting polyploid carries chromosome sets derived from two or more species. **Figure 6.25** shows how allopolyploidy can arise from two species that are sufficiently related that hybridization takes place between them. Species 1 (AABBCC, $2n = 6$) produces haploid gametes with chromosomes ABC, and species 2

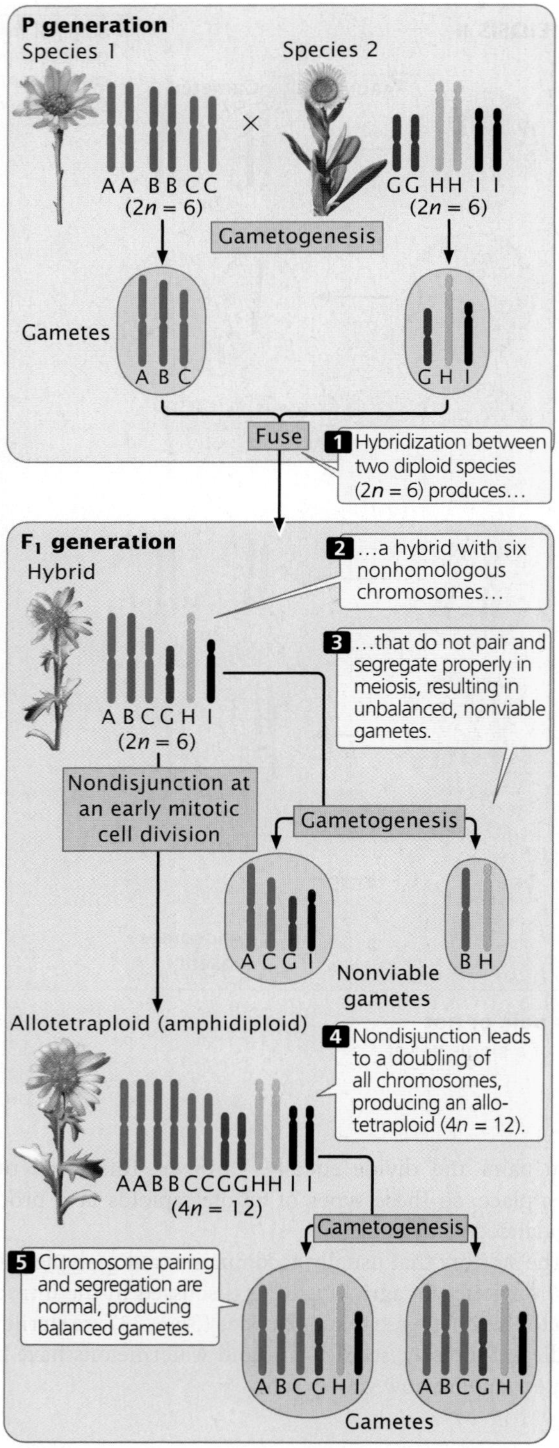

6.25 Most allopolyploids arise from hybridization between two species followed by chromosome doubling.

(GGHHII, $2n = 6$) produces haploid gametes with chromosomes GHI. If gametes from species 1 and 2 fuse, a hybrid with six chromosomes (ABCGHI) is created. The hybrid has the same chromosome number as that of both diploid species, so the hybrid is considered diploid. However, because the hybrid chromosomes are not homologous, they will not pair and segregate properly in meiosis; this hybrid is functionally haploid and sterile.

The sterile hybrid is unable to produce viable gametes through meiosis, but it may be able to perpetuate itself through mitosis (asexual reproduction). On rare occasions, nondisjunction takes place in a mitotic division, which leads to a doubling of chromosome number and an allotetraploid with chromosomes AABBCCGGHHII. This type of allopolyploid, consisting of two combined diploid genomes, is sometimes called an **amphidiploid**. Although the chromosome number has doubled compared with what was present in each of the parental species, the amphidiploid is functionally diploid: every chromosome has one and only one homologous partner, which is exactly what meiosis requires for proper segregation. The amphidiploid can now undergo normal meiosis to produce balanced gametes with six chromosomes each.

George Karpechenko created polyploids experimentally in the 1920s. Cabbage (*Brassica oleracea*, $2n = 18$) and radishes (*Raphanus sativa*, $2n = 18$) are agriculturally important plants now, as they were then, but only the leaves of the cabbage and the roots of the radish are normally consumed. Karpechenko wanted to produce a plant that had cabbage leaves and radish roots so that no part of the plant would go to waste. Because both cabbage and radish possess 18 chromosomes, Karpechenko was able to cross them successfully, producing a hybrid with $2n = 18$, but, unfortunately, the hybrid was sterile. After several crosses, Karpechenko noticed that one of his hybrid plants produced a few seeds. When planted, these seeds grew into plants that were viable and fertile. Analysis of their chromosomes revealed that the plants were allotetraploids, with $2n = 36$ chromosomes. To Karpechencko's great disappointment, however, the new plants possessed the roots of a cabbage and the leaves of a radish.

TRY PROBLEM 27

The Significance of Polyploidy

In many organisms, cell volume is correlated with nuclear volume, which in turn is determined by genome size. Thus, the increase in chromosome number in polyploidy is often associated with an increase in cell size, and many polyploids are physically larger than diploids. Breeders have used this effect to produce plants with larger leaves, flowers, fruits, and seeds. The hexaploid ($6n = 42$) genome of wheat probably contains chromosomes derived from three different wild species (**Figure 6.26**). As a result, the seeds of modern wheat are larger than those of its ancestors. Many other cultivated plants also are polyploid (**Table 6.2**).

Polyploidy is less common in animals than in plants for several reasons. As discussed, allopolyploids require hybridization between different species, which happens less frequently in animals than in plants. Animal behavior often prevents interbreeding among species, and the complexity of animal development causes most interspecific hybrids to be nonviable. Many of the polyploid animals that do arise are in groups that reproduce through parthenogenesis (a type of reproduction in which the animal develops from an unfertilized egg). Thus, asexual reproduction may facilitate the development of polyploids, perhaps because the perpetuation of hybrid individuals through asexual reproduction provides greater opportunities for nondisjunction than does sexual reproduction. Only a few human polyploid babies have been reported, and most died within a few days of birth. Polyploidy—usually triploidy—is seen in about 10% of all spontaneously aborted human fetuses.

THINK-PAIR-SHARE Question 8

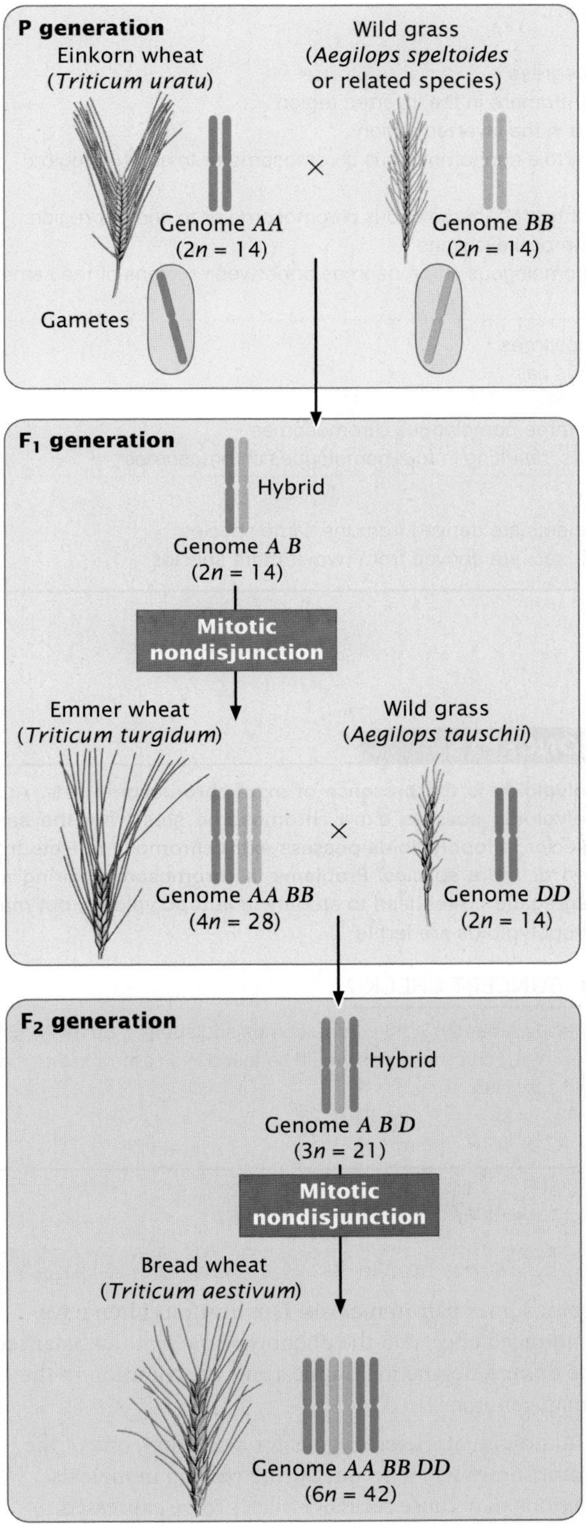

6.26 Modern bread wheat, *Triticum aestivum*, is a hexaploid with genes derived from three different species. Two diploid species, *T. uratu* and probably *Aegilops speltoides* or a related wild grass species, originally crossed to produce a diploid hybrid that underwent chromosome doubling to create *T. turgidum*. A cross between *T. turgidum* and *A. tauschii* produced a triploid hybrid that then underwent chromosome doubling to eventually produce *T. aestivum*, which is a hexaploid.

TABLE 6.2 Examples of polyploid crop plants

Plant	Type of polyploidy	Chromosome sets	Chromosome number
Potato	Autopolyploid	$4n$	48
Banana	Autopolyploid	$3n$	33
Peanut	Autopolyploid	$4n$	40
Sweet potato	Autopolyploid	$6n$	90
Tobacco	Allopolyploid	$4n$	48
Cotton	Allopolyploid	$4n$	52
Wheat	Allopolyploid	$6n$	42
Oats	Allopolyploid	$6n$	42
Sugarcane	Allopolyploid	$8n$	80
Strawberry	Allopolyploid	$8n$	56

Source: After F. C. Elliot, *Plant Breeding and Cytogenetics* (New York: McGraw-Hill, 1958).

TABLE 6.3 Different types of chromosome mutations

Chromosome mutation	Definition
Chromosome rearrangement	Change in chromosome structure
Chromosome duplication	Duplication of a chromosome segment
Chromosome deletion	Deletion of a chromosome segment
Inversion	Chromosome segment inverted 180 degrees
Paracentric inversion	Inversion that does not include the centromere in the inverted region
Pericentric inversion	Inversion that includes the centromere in the inverted region
Translocation	Movement of a chromosome segment to a nonhomologous chromosome or to another region of the same chromosome
Nonreciprocal translocation	Movement of a chromosome segment to a nonhomologous chromosome or to another region of the same chromosome without reciprocal exchange
Reciprocal translocation	Exchange between segments of nonhomologous chromosomes or between regions of the same chromosome
Aneuploidy	Change in number of individual chromosomes
Nullisomy	Loss of both members of a homologous pair
Monosomy	Loss of one member of a homologous pair
Trisomy	Gain of one chromosome, resulting in three homologous chromosomes
Tetrasomy	Gain of two homologous chromosomes, resulting in four homologous chromosomes
Polyploidy	Addition of entire chromosome sets
Autopolyploidy	Polyploidy in which extra chromosome sets are derived from the same species
Allopolyploidy	Polyploidy in which extra chromosome sets are derived from two or more species

The Importance of Polyploidy in Evolution

Polyploidy, particularly allopolyploidy, often gives rise to new species and has been particularly important in the evolution of flowering plants. Occasional genome doubling through polyploidy has been a major contributor to evolutionary success in several groups. For example, *Saccharomyces cerevisiae* (yeast) is a tetraploid, having undergone whole-genome duplication about 100 million years ago. The vertebrate genome has duplicated twice, once in the common ancestor of jawed vertebrates and again in the ancestor of fishes. Certain groups of vertebrates, such as some frogs and some fishes, have undergone additional duplications. Cereal plants have undergone several genome duplication events. The types of chromosome mutations we have discussed in this chapter are summarized in **Table 6.3**.

CONCEPTS

Polyploidy is the presence of extra chromosome sets. Autopolyploids possess extra chromosome sets from the same species; allopolyploids possess extra chromosome sets from two or more species. Problems in chromosome pairing and segregation often lead to sterility in autopolyploids, but many allopolyploids are fertile.

✓ CONCEPT CHECK 7

Species A has $2n = 16$ chromosomes and species B has $2n = 14$. How many chromosomes would be found in an allotriploid of these two species?

a. 21 or 24
b. 42 or 48
c. 22 or 23
d. 45

CONCEPTS SUMMARY

- The three basic types of chromosome mutations are (1) chromosome rearrangements, which are changes in the structure of chromosomes; (2) aneuploidy, which is an increase or decrease in chromosome number; and (3) polyploidy, which is the presence of extra chromosome sets.
- Chromosome rearrangements include duplications, deletions, inversions, and translocations.
- In individuals heterozygous for a duplication, the duplicated region will form a loop when homologous chromosomes pair in meiosis. Duplications often have pronounced effects on the phenotype owing to unbalanced gene dosage. Segmental duplications are common in the human genome.
- In individuals heterozygous for a deletion, one of the chromosomes will loop out during pairing in meiosis. Deletions may cause recessive alleles to be expressed.
- Pericentric inversions include the centromere; paracentric inversions do not. In individuals

heterozygous for an inversion, the homologous chromosomes form inversion loops in meiosis, and reduced recombination takes place within the inverted region.

- In translocation heterozygotes, the chromosomes form crosslike structures in meiosis.
- Fragile sites are constrictions or gaps that appear at particular regions on the chromosomes of cells grown in culture and are prone to breakage under certain conditions.
- Copy-number variations are differences in the number of copies of DNA sequences and include duplications and deletions. These variants are common in the human genome; some are associated with diseases and disorders.
- Nullisomy is the loss of two homologous chromosomes; monosomy is the loss of a single chromosome; trisomy is the addition of a single chromosome; and tetrasomy is the addition of two homologous chromosomes.
- Aneuploidy usually causes drastic phenotypic effects because it leads to unbalanced gene dosage.
- Primary Down syndrome is caused by the presence of three full copies of chromosome 21. Familial Down syndrome is caused by the presence of two normal copies of chromosome 21 and a third copy that is attached to another chromosome through a translocation.
- All the chromosomes in an autopolyploid derive from one species; chromosomes in an allopolyploid come from two or more species.

IMPORTANT TERMS

chromosome mutation (p. 154)
metacentric chromosome (p. 154)
submetacentric chromosome (p. 154)
acrocentric chromosome (p. 154)
telocentric chromosome (p. 154)
chromosome rearrangement (p. 156)
chromosome duplication (p. 156)
tandem duplication (p. 156)
displaced duplication (p. 156)
reverse duplication (p. 156)
segmental duplication (p. 158)
chromosome deletion (p. 159)
pseudodominance (p. 160)
haploinsufficient gene (p. 160)
chromosome inversion (p. 160)
paracentric inversion (p. 160)
pericentric inversion (p. 160)
position effect (p. 161)
dicentric chromatid (p. 162)
acentric chromatid (p. 162)
dicentric bridge (p. 162)
translocation (p. 163)
nonreciprocal translocation (p. 163)
reciprocal translocation (p. 163)
Robertsonian translocation (p. 163)
fragile site (p. 164)
fragile-X syndrome (p. 164)
copy-number variation (CNV) (p. 165)
aneuploidy (p. 165)
polyploidy (p. 165)
nondisjunction (p. 165)
nullisomy (p. 165)
monosomy (p. 165)
trisomy (p. 165)
tetrasomy (p. 165)
Down syndrome (trisomy 21) (p. 167)
primary Down syndrome (p. 168)
familial Down syndrome (p. 168)
translocation carrier (p. 168)
Edward syndrome (trisomy 18) (p. 168)
Patau syndrome (trisomy 13) (p. 168)
trisomy 8 (p. 168)
autopolyploidy (p. 170)
allopolyploidy (p. 170)
unbalanced gametes (p. 171)
amphidiploid (p. 172)

ANSWERS TO CONCEPT CHECKS

1. a

2. Pseudodominance is the expression of a normally recessive mutation that is produced when the dominant wild-type allele in a heterozygous individual is absent due to a deletion on one chromosome.

3. c

4. b

5. 37

6. Dosage compensation prevents the expression of additional copies of X-linked genes in mammals, and there is little information in the Y chromosome, so extra copies of the X and Y chromosomes do not have major effects on development. In contrast, there is no mechanism of dosage compensation for autosomes, and so extra copies of autosomal genes are expressed, upsetting development and causing the spontaneous abortion of aneuploid embryos.

7. c

WORKED PROBLEMS

Problem 1

A chromosome has the following segments, where • represents the centromere:

ABCDE•FG

What types of chromosome mutations are required to change this chromosome into each of the following chromosomes? (In some cases, more than one chromosome mutation may be required.)

a. ABE•FG
b. AEDCB•FG
c. ABABCDE•FG
d. AF•EDCBG
e. ABCDEEDC•FG

Solution Strategy

What information is required in your answer to the problem?

The types of chromosome mutations that would lead to the chromosome shown.

What information is provided to solve the problem?

- The original gene segments found on the chromosome.
- The altered gene segments that occur after the mutations.

For help with this problem, review:

Section 6.2.

Solution Steps

a. The mutated chromosome (ABE•FG) is missing segment CD, so this mutation is a deletion.

b. The mutated chromosome (AEDCB•FG) has one and only one copy of all the gene segments, but segment BCDE has been inverted 180 degrees. Because the centromere has not changed location and is not in the inverted region, this chromosome mutation is a paracentric inversion.

c. The mutated chromosome (ABABCDE•FG) is longer than normal, and we see that segment AB has been duplicated. This mutation is a tandem duplication.

d. The mutated chromosome (AF•EDCBG) is normal length, but the gene order and the location of the centromere have changed; this mutation is therefore a pericentric inversion of region (BCDE•F).

e. The mutated chromosome (ABCDEEDC•FG) contains a duplication (CDE) that is also inverted; so this chromosome has undergone a duplication and a paracentric inversion.

Problem 2

Species I is diploid ($2n = 4$) with chromosomes AABB; related species II is diploid ($2n = 6$) with chromosomes MMNNOO. Give the chromosomes that would be found in individuals with the following chromosome mutations.

a. Autotriploidy in species I
b. Allotetraploidy including species I and II
c. Monosomy in species I
d. Trisomy in species II for chromosome M
e. Tetrasomy in species I for chromosome A
f. Allotriploidy including species I and II
g. Nullisomy in species II for chromosome N

Solution Strategy

What information is required in your answer to the problem?

The chromosomes that will be found in individuals with each type of mutation.

What information is provided to solve the problem?

- Species I is diploid with $2n = 4$.
- Species I has chromosomes AABB.
- Species II is diploid with $2n = 6$.
- Species II has chromosomes MMNNOO.

For help with this problem, review:

Sections 6.3 and 6.4.

First
he the
genome
ment for
ecies. For
, $n = 2$
omosomes
for species
with
somes MNO.

Solution Steps

a. An autotriploid is $3n$, with all the chromosomes coming from a single species; so an autotriploid of species I would have chromosomes AAABBB ($3n = 6$).

b. An allotetraploid is $4n$, with the chromosomes coming from more than one species. An allotetraploid could consist of $2n$ from species I and $2n$ from species II, giving the allotetraploid ($4n = 2 + 2 + 3 + 3 = 10$) chromosomes AABBMMNNOO. An allotetraploid could also possess $3n$ from species I and $1n$ from species II ($4n = 2 + 2 + 2 + 3 = 9$; AAABBBMNO) or $1n$ from species I and $3n$ from species II ($4n = 2 + 3 + 3 + 3 = 11$; ABMMMNNNOOO).

c. A monosomic individual is missing a single chromosome; so monosomy in species I would result in $2n - 1 = 4 - 1 = 3$. The monosomy might include either of the two chromosome pairs, giving chromosomes ABB or AAB.

d. Trisomy requires an extra chromosome; so trisomy in species II for chromosome M would result in $2n + 1 = 6 + 1 = 7$ (MMMNNOO).

e. A tetrasomic individual has two extra homologous chromosomes; so tetrasomy in species I for chromosome A would result in $2n + 2 = 4 + 2 = 6$ (AAAABB).

f. An allotriploid is $3n$ with the chromosomes coming from two different species; so an allotriploid could be $3n = 2 + 2 + 3 = 7$ (AABBMNO) or $3n = 2 + 3 + 3 = 8$ (ABMMNNOO).

g. A nullisomic individual is missing both chromosomes of a homologous pair; so a nullisomy in species II for chromosome N would result in $2n - 2 = 6 - 2 = 4$ (MMOO).

COMPREHENSION QUESTIONS

Section 6.1

1. List the different types of chromosome mutations and define each one.

Section 6.2

2. Why do extra copies of genes sometimes cause drastic phenotypic effects?
3. Draw a pair of chromosomes as they would appear during synapsis in prophase I of meiosis in an individual heterozygous for a chromosome duplication.
4. What is the difference between a paracentric and a pericentric inversion?
5. How can inversions in which no genetic information is lost or gained cause phenotypic effects?
6. Explain why recombination is suppressed in individuals heterozygous for paracentric inversions.
7. How do translocations in which no genetic information is lost or gained produce phenotypic effects?

Section 6.3

8. List four major types of aneuploidy.
9. What is the difference between primary Down syndrome and familial Down syndrome? How does each type arise?

Section 6.4

10. What is the difference between autopolyploidy and allopolyploidy? How does each arise?
11. Explain why autopolyploids are usually sterile, whereas allopolyploids are often fertile.

For more questions that test your comprehension of the key chapter concepts, go to **LearningCurve** for this chapter.

APPLICATION QUESTIONS AND PROBLEMS

Section 6.1

12. Examine the karyotypes shown in **Figures 6.1** and **6.2a**. Are the individuals from whom these karyotypes were made males or females?

*13. Which types of chromosome mutations

a. increase the amount of genetic material in a particular chromosome?

b. increase the amount of genetic material in all chromosomes?

c. decrease the amount of genetic material in a particular chromosome?

d. change the position of DNA sequences in a single chromosome without changing the amount of genetic material?

e. move DNA from one chromosome to a nonhomologous chromosome?

Section 6.2

*14. A chromosome has the following segments, where • represents the centromere:

<u>AB•CDEFG</u>

What types of chromosome mutations are required to change this chromosome into each of the following chromosomes? (In some cases, more than one chromosome mutation may be required.)

a. ABAB•CDEFG
b. AB•CDEABFG
c. AB•CFEDG
d. A•CDEFG
e. AB•CDE
f. AB•EDCFG
g. C•BADEFG
h. AB•CFEDFEDG
i. AB•CDEFCDFEG

15. A chromosome initially has the following segments:

AB•CDEFG

Draw the chromosome, identifying its segments, that would result from each of the following mutations.

a. Tandem duplication of DEF
b. Displaced duplication of DEF
c. Deletion of FG
d. Paracentric inversion that includes DEFG
e. Pericentric inversion of BCDE

16. The following diagram represents two nonhomologous chromosomes:

AB•CDEFG
RS•TUVWX

What type of chromosome mutation would produce each of the following groups of chromosomes?

a. AB•CD
RS•TUVWXEFG
b. AUVB•CDEFG
RS•TWX
c. AB•TUVFG
RS•CDEWX
d. AB•CWG
RS•TUVDEFX

17. The green-nose fly normally has six chromosomes: two metacentric and four acrocentric. A geneticist examines the chromosomes of an odd-looking green-nose fly and discovers that it has only five chromosomes; three of them are metacentric and two are acrocentric. Explain how this change in chromosome number might have taken place.

***18.** A wild-type chromosome has the following segments:

ABC•DEFGHI

Researchers have found individuals that are heterozygous for each of the following chromosome mutations. For each mutation, sketch how the wild-type and mutated chromosomes would pair in prophase I of meiosis, showing all chromosome strands.

a. ABC•DEFDEFGHI
b. ABC•DHI
c. ABC•DGFEHI
d. ABED•CFGHI

***19.** As discussed in this chapter, crossing over within a pericentric inversion produces chromosomes that have extra copies of some genes and no copies of other genes. The fertilization of gametes containing such duplication-containing or deficient chromosomes often results in children with syndromes characterized by developmental delay, intellectual disability, the abnormal development of organ systems, and early death. Maarit Jaarola and colleagues examined individual sperm cells of a male who was heterozygous for a pericentric inversion on chromosome 8 and determined that crossing over took place within the pericentric inversion in 26% of the meiotic divisions (M. Jaarola, R. H. Martin, and T. Ashley. 1998. *American Journal of Human Genetics* 63:218–224).

Assume that you are a genetic counselor and that a couple seeks counseling from you. Both the man and the woman are phenotypically normal, but the woman is heterozygous for a pericentric inversion on chromosome 8. The man is karyotypically normal. What is the probability that this couple will produce a child with a debilitating syndrome as the result of crossing over within the pericentric inversion?

20. An individual heterozygous for a reciprocal translocation possesses the following chromosomes:

AB•CDEFG
AB•CDVWX
RS•TUEFG
RS•TUVWX

Draw the pairing arrangement of these chromosomes in prophase I of meiosis.

Section 6.3

***21.** Red–green color blindness is a human X-linked recessive disorder. A young man with a 47,XXY karyotype (Klinefelter syndrome) is color blind. His 46,XY brother also is color blind. Both parents have normal color vision. Where did the nondisjunction that gave rise to the young man with Klinefelter syndrome take place? Assume that no crossing over took place in prophase I of meiosis.

***22.** Bill and Betty have had two children with Down syndrome. Bill's brother has Down syndrome and his sister has two children with Down syndrome. On the basis of these observations, indicate which of the following statements are most likely correct and which are most likely incorrect. Explain your reasoning.

a. Bill has 47 chromosomes.
b. Betty has 47 chromosomes.
c. Bill and Betty's children each have 47 chromosomes.
d. Bill's sister has 45 chromosomes.
e. Bill has 46 chromosomes.
f. Betty has 45 chromosomes.
g. Bill's brother has 45 chromosomes.

***23.** In mammals, sex-chromosome aneuploids are more common than autosomal aneuploids, but in fishes, sex-chromosome aneuploids and autosomal aneuploids are found with equal frequency. Offer a possible explanation for these differences between mammals and fishes. (Hint: Think about why sex-chromosome aneuploids are more common than autosomal aneuploids in mammals.)

24. Using breeding techniques, Andrei Dyban and V. S. Baranov (*Cytogenetics of Mammalian Embryonic Development*. Oxford: Oxford University Press, Clarendon Press; New York: Oxford University Press, 1987) created mice that were trisomic for each of the different mouse chromosomes. They found that only mice with trisomy 19 completed development. Mice that were trisomic for all other chromosomes died in the course of development. For some of these trisomics, the researchers plotted the length of development (number of days after conception before the embryo died) as a function of the size of the mouse chromosome that was present in three copies (see the adjoining graph). Summarize their findings and provide a possible explanation for the results.

DATA ANALYSIS

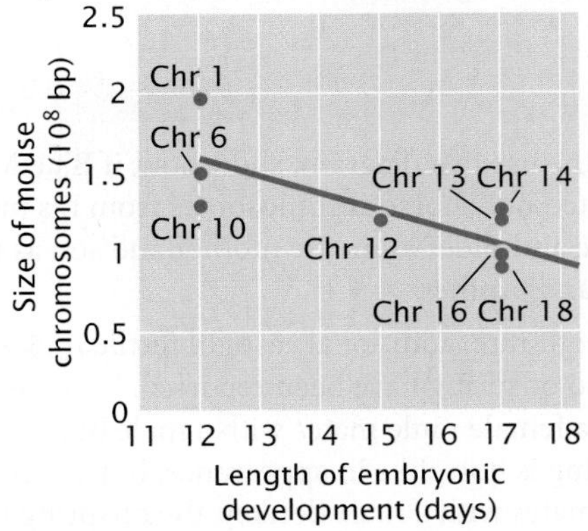

[E. Torres, B. R. Williams, and A. Amon. 2008. *Genetics* 179:737–746, Fig. 2B.]

Section 6.4

25. Species I has $2n = 16$ chromosomes. How many chromosomes will be found per cell in each of the following mutants in this species?

a. Monosomic
b. Autotriploid
c. Autotetraploid
d. Trisomic
e. Double monosomic
f. Nullisomic
g. Autopentaploid
h. Tetrasomic

26. Species I is diploid ($2n = 8$) with chromosomes AABBCCDD; related species II is diploid ($2n = 8$) with chromosomes MMNNOOPP. What types of chromosome mutations do individuals with the following sets of chromosomes have?

a. AAABBCCDD
b. MMNNOOOOPP
c. AABBCDD
d. AAABBBCCCDDD
e. AAABBCCDDD
f. AABBDD
g. AABBCCDDMMNNOOPP
h. AABBCCDDMNOP

***27.** Species I has $2n = 8$ chromosomes and species II has $2n = 14$ chromosomes. What would the expected chromosome numbers be in individuals with the following chromosome mutations? Give all possible answers.

a. Allotriploidy including species I and II
b. Autotetraploidy in species II
c. Trisomy in species I
d. Monosomy in species II
e. Tetrasomy in species I
f. Allotetraploidy including species I and II

28. Suppose that species I in **Figure 6.25** had $2n = 10$ and species II in the figure had $2n = 12$. How many chromosomes would be present in the allotetraploid at the bottom of the figure?

29. Consider a diploid cell that has $2n = 4$ chromosomes: one pair of metacentric chromosomes and one pair of acrocentric chromosomes. Suppose that this cell undergoes nondisjunction, giving rise to an autotriploid cell ($3n$). The triploid cell then undergoes meiosis. Draw the different types of gametes that could result from meiosis in the triploid cell, showing the chromosomes present in each type. To distinguish between the different metacentric and acrocentric chromosomes, use a different color to draw each metacentric chromosome; similarly, use a different color to draw each acrocentric chromosome. (Hint: See **Figure 6.24**).

30. Assume that the autotriploid cell in **Figure 6.24** has $3n = 30$ chromosomes. For each of the gametes produced by this cell, give the chromosome number of the zygote that would result if the gamete fused with a normal haploid gamete.

31. *Nicotiana glutinosa* ($2n = 24$) and *N. tabacum* ($2n = 48$) are two closely related plants that can be intercrossed, but the F_1 hybrid plants that result are usually sterile. In 1925, Roy Clausen and Thomas Goodspeed crossed *N. glutinosa* and *N. tabacum* and obtained one fertile F_1 plant (R. E. Clausen and T. H. Goodspeed. 1925. *Genetics* 10:278–284). They were able to self-pollinate the flowers of this plant to produce an F_2 generation. Surprisingly, the F_2 plants were fully fertile and produced viable seeds. When Clausen and Goodspeed examined the chromosomes of the F_2 plants, they observed 36 pairs of chromosomes in metaphase I and 36 individual chromosomes in metaphase II. Explain the origin of the F_2 plants obtained by Clausen and Goodspeed and the numbers of chromosomes observed.

DATA ANALYSIS

32. What would be the chromosome number of progeny resulting from the following crosses in wheat (see **Figure 6.26**)? What type of polyploid (allotriploid, allotetraploid, etc.) would result from each cross?

a. Einkorn wheat and emmer wheat
b. Bread wheat and emmer wheat
c. Einkorn wheat and bread wheat

33. Karl and Hally Sax crossed *Aegilops cylindrica* ($2n = 28$), a wild grass found in the Mediterranean region, with *Triticum vulgare* ($2n = 42$), a type of wheat (K. Sax and H. J. Sax. 1924. *Genetics* 9:454–464). The resulting F_1 plants from this cross had 35 chromosomes. Examination of metaphase I in the F_1 plants revealed the presence of 7 pairs of chromosomes (bivalents) and 21 unpaired chromosomes (univalents).

DATA ANALYSIS

a. If the unpaired chromosomes segregate randomly, what possible chromosome numbers will appear in the gametes of the F_1 plants?

b. What does the appearance of the bivalents in the F_1 hybrids suggest about the origin of *Triticum vulgare* wheat?

Aegilops cylindrica, jointed goatgrass. [Sam Brinker, MNR-NHIC, 2008/Canadian Food Inspection Agency.]

Triticum vulgare, wheat. [Michael Hieber/123RF .com.]

CHALLENGE QUESTIONS

Section 6.3

34. Red–green color blindness is a human X-linked recessive disorder. Jill has normal color vision, but her father is color blind. Jill marries Tom, who also has normal color vision. Jill and Tom have a daughter who has Turner syndrome and is color blind.

a. How did the daughter inherit color blindness?

b. Did the daughter inherit her X chromosome from Jill or from Tom?

35. Mules result from a cross between a horse ($2n = 64$) and a donkey ($2n = 62$), have 63 chromosomes, and are almost always sterile. However, in the summer of 1985, a female mule named Krause who was pastured with a male donkey gave birth to a foal (O. A. Ryder et al. 1985. *Journal of Heredity* 76:379–381). Blood tests established that the male foal, appropriately named Blue Moon, was the offspring of Krause and that Krause was indeed a mule. Both Blue Moon and Krause were fathered by the same donkey (see the pedigree below). The foal, like his mother, had 63 chromosomes—half of them horse chromosomes and the other half donkey chromosomes. Analyses of genetic markers showed that, remarkably, Blue Moon seemed to have inherited a complete set of horse chromosomes from his mother, instead of the random mixture of horse and donkey chromosomes that would be expected with normal meiosis. Thus, Blue Moon and Krause were not only mother and son, but also brother and sister.

DATA ANALYSIS

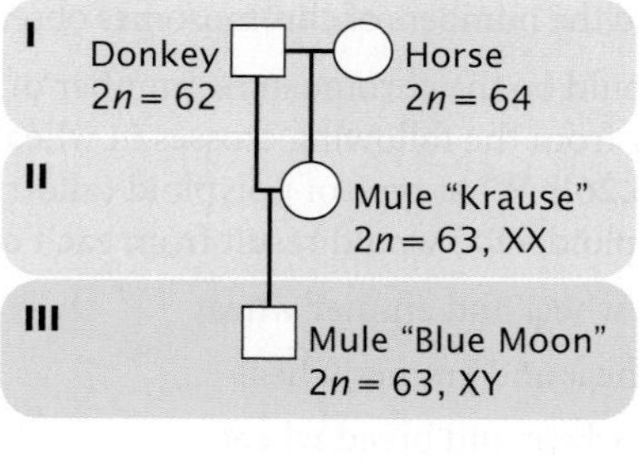

a. With the use of a diagram, show how, if Blue Moon inherited only horse chromosomes from his mother, Krause and Blue Moon are mother and son as well as sister and brother.

b. Although rare, additional cases of fertile mules giving birth to offspring have been reported. In these cases, when a female mule mates with a male horse, the offspring is horselike in appearance, but when a female mule mates with a male donkey, the offspring is mule-like in appearance. Is this observation consistent with the idea that the offspring of fertile female mules inherit only a set of horse chromosomes from their mule mothers? Explain your reasoning.

c. Can you suggest a possible mechanism for how fertile female mules might pass on a complete set of horse chromosomes to their offspring?

Section 6.4

36. Humans and many other complex organisms are diploid, possessing two sets of genes, one inherited from the mother and one from the father. However, a number of eukaryotic organisms spend most of their life cycles in a haploid state. Many of these eukaryotes, such as *Neurospora* and yeast, still undergo meiosis and sexual reproduction, but most of the cells that make up the organism are haploid.

Considering that haploid organisms are fully capable of sexual reproduction and generating genetic variation, why are most complex eukaryotes diploid? In other words, what might be the evolutionary advantage of existing in a diploid state instead of a haploid state? And why might a few organisms, such as *Neurospora* and yeast, exist as haploids?

THINK-PAIR-SHARE QUESTIONS

Introduction

1. What are some possible advantages to producing cultivated bananas that are polyploid? What might be some disadvantages?
2. What is a genetically modified food? Are the currently consumed, polyploid bananas genetically modified? Explain your reasoning.

Section 6.1

3. Why do species usually have a characteristic number of chromosomes? Why don't we see many species in which chromosome number varies within the species, with some individuals, say, having $2n = 20$ and others having $2n = 24$?

Section 6.2

4. An individual is heterozygous for a reciprocal translocation, with the following chromosomes:

   A • B C D E F
   A • B C V W X
   R S T • U D E F
   R S T • U V W X

   a. Draw a picture of these chromosomes pairing in prophase I of meiosis.
   b. Draw the products of alternate, adjacent-1, and adjacent-2 segregations.
   c. Explain why the fertility of this individual is likely to be less than the fertility of an individual without a translocation.

Section 6.3

5. Monozygotic (identical) twins arise when a single egg, fertilized by a single sperm, divides early in development, giving rise to two genetically identical embryos. A recent study examined a pair of monozygotic twins in which one of the twins had trisomy 21 and the other twin had two normal copies of chromosome 21. Explain how two identical twins could differ in their number of chromosomes.
6. The human X chromosome is about the same size as human chromosomes 4 and 5. Trisomy of the X chromosome occurs about once in every thousand human female births, and yet trisomy 4 and trisomy 5 are almost never seen among living humans. Why do these differences in the frequency of trisomic individuals occur, in spite of similar chromosome size?
7. Geneticists have been exploring ways to suppress the expression of the extra chromosome 21 in individuals with Down syndrome in hopes of preventing the medical problems and intellectual disability of individuals with trisomy 21. One approach involves modifying a gene that is already present in human cells and using it to suppress the expression of the extra copy of chromosome 21. What approach, do you think, are they taking, and what may be some of the challenges for using it in patients?

Section 6.4

8. Trisomics often have more developmental problems than triploids. Can you suggest a reason why?

7 Bacterial and Viral Genetic Systems

The Genetics of Medieval Leprosy

Woman with leprosy, a disease caused by the bacterium *Mycobacterium leprae*. The study of *M. leprae* DNA isolated from ancient skeletons of medieval Europeans with leprosy has provided information about the evolution of this bacterium. [Reuters/Rupak de Chowdhuri (India).]

Leprosy, one of the most feared diseases of history, was well known in ancient times, and people with leprosy were frequently ostracized from society. Although leprosy is successfully treated today with antibiotics, it remains a major public health problem: from 2 million to 3 million people worldwide are disabled by leprosy, and over 200,000 new cases are reported each year. In its severest form, leprosy causes paralysis, blindness, and disfigurement. Leprosy is caused by the bacterium *Mycobacterium leprae*, which infects cells of the nervous system—although human genes do play a role in susceptibility to this disease.

In 2013, geneticists isolated DNA of *M. leprae* from five skeletons of medieval Europeans who exhibited signs of leprosy. Their comparisons of the gene sequences of these ancient bacteria with those of modern strains provided insight into the evolution of this organism. Scientists had previously determined the genome sequence of *M. leprae* and found that it contains 3,268,203 base pairs of DNA, 1 million base pairs fewer than the genomes of other mycobacteria. In most bacterial genomes, the vast majority of the DNA encodes proteins—little DNA lies between the protein-encoding genes. In contrast, only 50% of the DNA of *M. leprae* encodes proteins. *M. leprae* also has 2300 fewer genes than its close relative M. tuberculosis. An incredible 27% of *M. leprae*'s genome consists of nonfunctional copies of genes (called pseudogenes) that have been inactivated by mutations. Its reduced DNA content, fewer functional genes, and large number of pseudogenes suggest that, evolutionarily, the genome of *M. leprae* has undergone massive decay over time, losing DNA and acquiring mutations that have inactivated many of its genes. Although the reasons for this decay are not known, it helps account for *M. leprae*'s long generation time—14 days in humans, an incredibly long replication time for a bacterium—and the inability of scientists to culture the bacteria in the laboratory.

Leprosy was common in Europe until the Middle Ages, when it disappeared from the population. Why did it disappear from Europe, despite remaining common in many other parts of the world? To address this question, geneticists extracted DNA from the bones and teeth of five medieval skeletons (dating from the eleventh to the fourteenth centuries) exhumed from cemeteries in Denmark, Sweden, and the United Kingdom. They separated out the DNA of *M. leprae* and determined whole-genome sequences for the bacteria. They then compared the DNA sequences of these medieval strains of *M. leprae* with those of modern strains from India, Thailand, the United States, Brazil, and other locations.

This analysis revealed that the ancient strains of *M. leprae* were remarkably similar to modern strains. Three of the ancient strains were most closely related to modern strains from Iran and Turkey, suggesting a Middle Eastern–European connection to the disease. Some of the ancient strains were closely related to modern *M. leprae* currently found in the United States, suggesting that these North American bacteria originated in Europe. The close similarity between ancient European strains and a modern virulent strain from North America suggests that leprosy did not disappear from Europe because it lost its virulence. More likely, improved social conditions, changes in the immunity of Europeans, or the presence of other infectious diseases brought about the demise of leprosy in Europe. This study of leprosy illustrates the importance of genetic studies of bacteria to human health and shows how modern tools of evolutionary and molecular genetics are being applied to our understanding of bacterial biology.

THINK-PAIR-SHARE Questions 1 and 2

In this chapter, we will examine some of the genetic properties of bacteria and viruses and the mechanisms by which they exchange and recombine their genes. Since the 1940s, the genetic systems of bacteria and viruses have contributed to the discovery of many important concepts in genetics. The study of molecular genetics initially focused almost entirely on their genes. Today, bacteria and viruses are still essential tools for probing the nature of genes in more complex organisms, in part because they possess a number of characteristics that make them suitable for genetic studies (**Table 7.1**).

The genetic systems of bacteria and viruses are also studied because these organisms play important roles in human society. Bacteria are found naturally in the mouth, in the gut, and on the skin, where they are essential to human function and ecology. They have been harnessed to produce a number of economically important substances, and they have immense medical significance, causing many human diseases. In this chapter, we focus on several unique aspects of bacterial and viral genetic systems. Important processes of gene transfer and recombination will be described, and we will see how these processes can be used to map bacterial and viral genes. ▶TRY PROBLEM 12

TABLE 7.1 Advantages of using bacteria and viruses for genetic studies

1. Reproduction is rapid.
2. Many progeny are produced.
3. The haploid genome allows all mutations to be expressed directly.
4. Asexual reproduction simplifies the isolation of genetically pure strains.
5. Growth in the laboratory is easy and requires little space.
6. Genomes are small.
7. Techniques are available for isolating and manipulating bacterial genes.
8. They have medical importance.
9. They can be genetically engineered to produce substances of commercial value.

7.1 The Genetic Analysis of Bacteria Requires Special Methods

Heredity in bacteria is fundamentally similar to heredity in more complex organisms, but the bacterial haploid genome and the small size of bacteria (which makes observation of their phenotypes difficult) necessitate different approaches and methods.

Bacterial Diversity

Prokaryotes are unicellular organisms that lack nuclear membranes and membrane-bounded cell organelles. For many years, biologists considered all prokaryotes to be related, but genome sequence information now provides convincing evidence that prokaryotes are divided into at least two distinct groups: the archaea and the bacteria. The archaea are a group of diverse prokaryotes that are frequently found in extreme environments, such as hot springs and the bottoms of oceans. The bacteria are the remaining prokaryotes, including most of the familiar bacterial species. Although superficially similar in their cell structure, bacteria and archaea are distinct in their genetic makeup, and the differences between them are as great as those between eubacteria and eukaryotes. In fact, the archaea are more similar to eukaryotes than to bacteria in a number of molecular features and genetic processes.

Bacteria are extremely diverse and come in a variety of shapes and sizes. Some are rod-shaped and others are spherical or helical. Most are much smaller than eukaryotic cells, but at least one species isolated from the guts of fish is almost 1 mm long and can be seen with the naked eye. Some bacteria are photosynthetic. Others produce stalks and spores, superficially resembling fungi.

Bacteria have long been considered simple organisms that lack much of the cellular complexity of eukaryotes. However, recent evidence points to a number of similarities and parallels in the structure of bacteria and eukaryotes. For example, a bacterial protein termed FtsZ, which plays an integral part in cell division, is structurally similar to tubulin in eukaryotic cells, a protein that is found in microtubules and helps to segregate chromosomes in mitosis and meiosis. Like eukaryotes, bacteria have proteins that help condense DNA. Other bacterial proteins function much as cytoskeletal proteins do in eukaryotes, helping to give bacterial cells shape and structure.

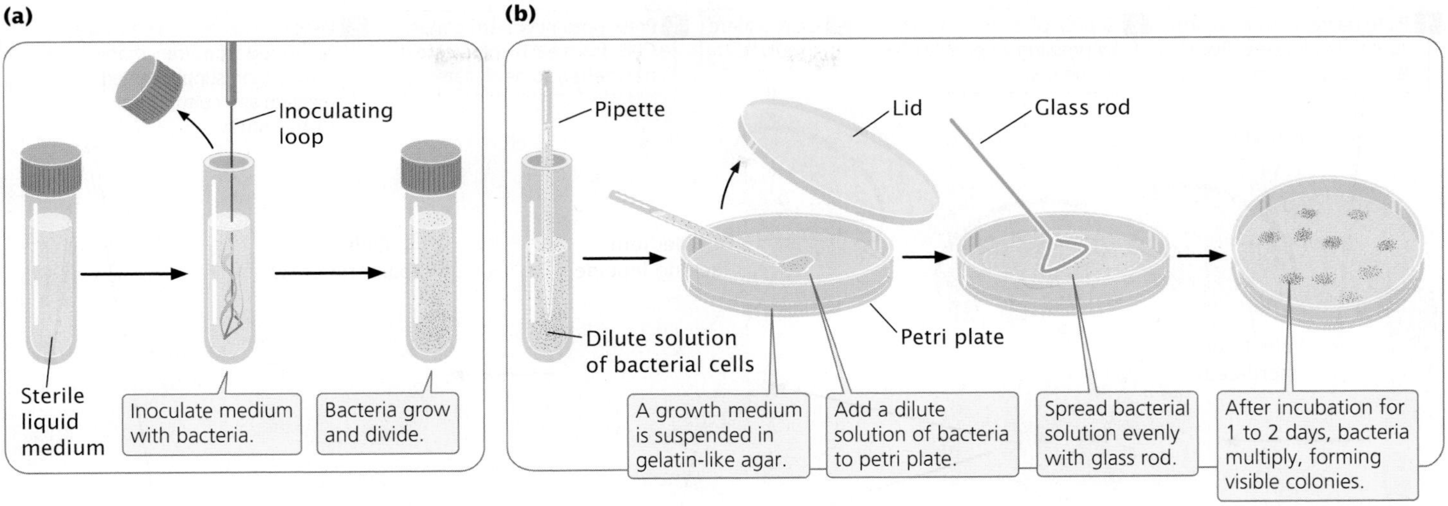

7.1 Bacteria can be grown (a) in liquid medium or (b) on solid medium.

And, though bacteria don't go through mitosis and meiosis, replication of the bacterial chromosome precedes binary fission, and there are bacterial processes that ensure that one copy of the chromosome gets allocated to each daughter cell.

Techniques for Studying Bacteria

The culture and study of bacteria require special techniques. Microbiologists have defined the nutritional needs of a number of bacteria and developed culture media for growing them in the laboratory. These culture media typically contain a carbon source, essential elements such as nitrogen and phosphorus, certain vitamins, and other required ions and nutrients. Wild-type (*prototrophic*) bacteria can use these simple ingredients to synthesize all the compounds that they need for growth and reproduction. A medium that contains only the nutrients required by prototrophic bacteria is termed **minimal medium**. Mutant strains called *auxotrophs* lack one or more enzymes necessary for synthesizing essential molecules and will grow only on medium supplemented with those essential molecules. For example, auxotrophic strains that are unable to synthesize the amino acid leucine will not grow on minimal medium but *will* grow on medium to which leucine has been added. **Complete medium** contains all the substances, such as the amino acid leucine, required by bacteria for growth and reproduction.

Cultures of bacteria are often grown in test tubes that contain sterile liquid medium, or "broth" (**Figure 7.1a**). A few bacteria are added to a broth tube, and they grow and divide until all the nutrients are used up or—more commonly—until the concentration of their waste products becomes toxic. Bacteria may also be grown on agar plates (**Figure 7.1b**), in which melted agar is suspended in growth medium and poured into the bottom half of a petri plate. The agar solidifies when cooled and provides a solid, gel-like base for bacterial growth. In a process called plating, a dilute solution of bacteria is spread over the surface of an agar plate. As each bacterium grows and divides, it gives rise to a visible clump of genetically identical cells (a **colony**). Genetically pure strains of the bacteria can be isolated by collecting bacteria from a single colony and transferring them to a new broth tube or agar plate. The chief advantage of this method is that it allows one to isolate and count bacteria, which individually are too small to see without a microscope.

Microbiologists often study phenotypes that affect the appearance of the colony (**Figure 7.2**) or that can be detected

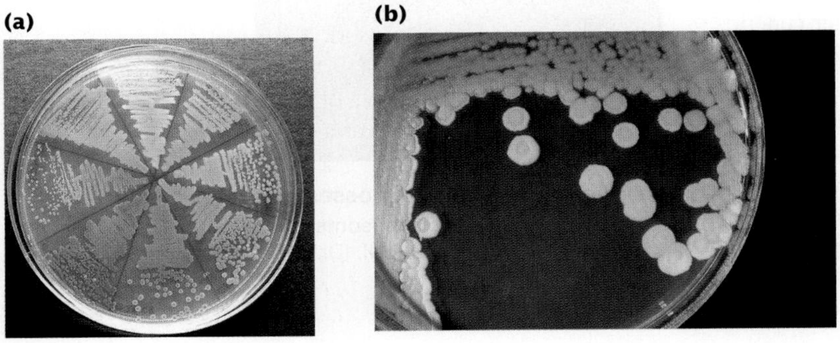

7.2 Bacterial colonies have a variety of phenotypes. (a) *Serratia marcescens* with color variation. (b) *Bacillus cereus*; colony shape varies among different strains of this species. [Part a: Dr. Edward Bottone. Part b: Biophoto Associates/Science Source.]

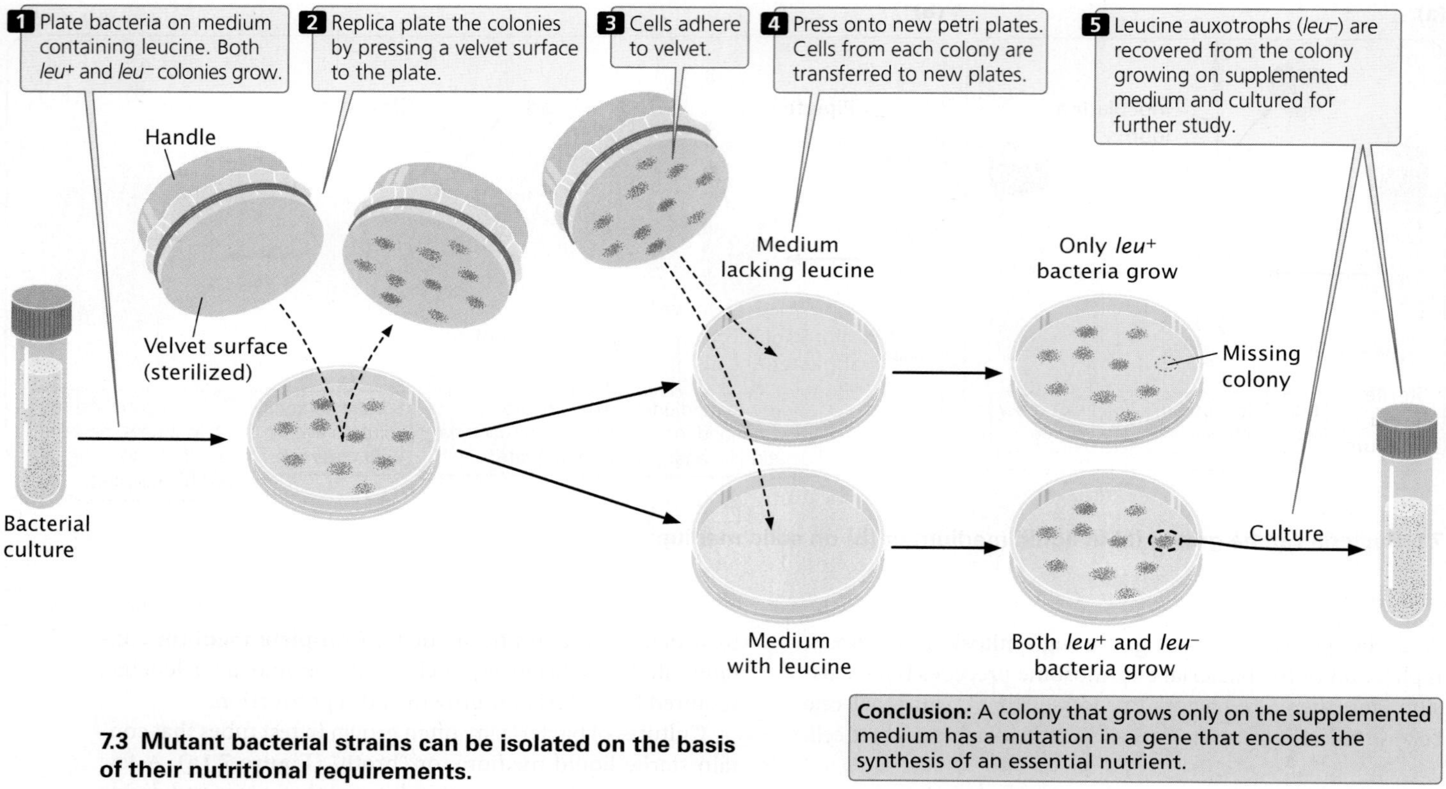

7.3 Mutant bacterial strains can be isolated on the basis of their nutritional requirements.

by simple chemical tests. Auxotrophs are commonly studied phenotypes. Suppose that we want to detect auxotrophs that cannot synthesize leucine (*leu*$^-$ mutants). We first spread the bacteria on a petri plate containing medium that includes leucine; both prototrophs that have the *leu*$^+$ allele and auxotrophs that have the *leu*$^-$ allele will grow on it (**Figure 7.3**). Next, by using a technique called replica plating, we transfer a few cells from each of the colonies on the original plate to two new replica plates: one plate contains medium to which leucine has been added; the other plate contains a medium lacking leucine. A medium that lacks an essential nutrient, such as the medium lacking leucine, is called a *selective* medium. The *leu*$^+$ bacteria will grow on both media, but the *leu*$^-$ mutants will grow only on the medium supplemented with leucine because they cannot synthesize their own leucine. Any colony that grows on medium that contains leucine, but not on medium that lacks leucine, consists of *leu*$^-$ bacteria. The auxotrophs that grow only on the supplemented medium can then be cultured for further study. Auxotrophic mutants are often used to study the results of genetic crosses and other genetic manipulations.

The Bacterial Genome

Bacteria are unicellular organisms that lack a nuclear membrane. Most bacterial genomes that have been studied consist of a circular chromosome that contains a single double-stranded DNA molecule several million base pairs in length (**Figure 7.4**). For example, the genome of *E. coli* has approximately 4.6 million base pairs of DNA. However, some bacteria contain multiple chromosomes. For example, *Vibrio cholerae*, which causes cholera, has two circular chromosomes, and *Rhizobium meliloti* has three chromosomes.

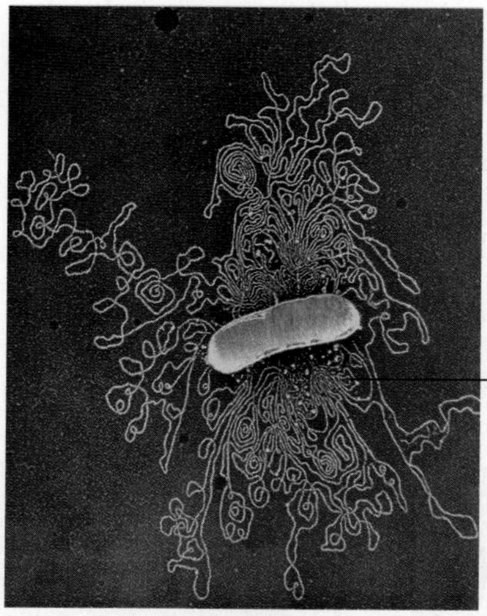

7.4 Most bacterial cells possess a single, circular chromosome. The chromosome shown here is emerging from a ruptured bacterial cell. [Dr. Gopal Murti/Science Source.]

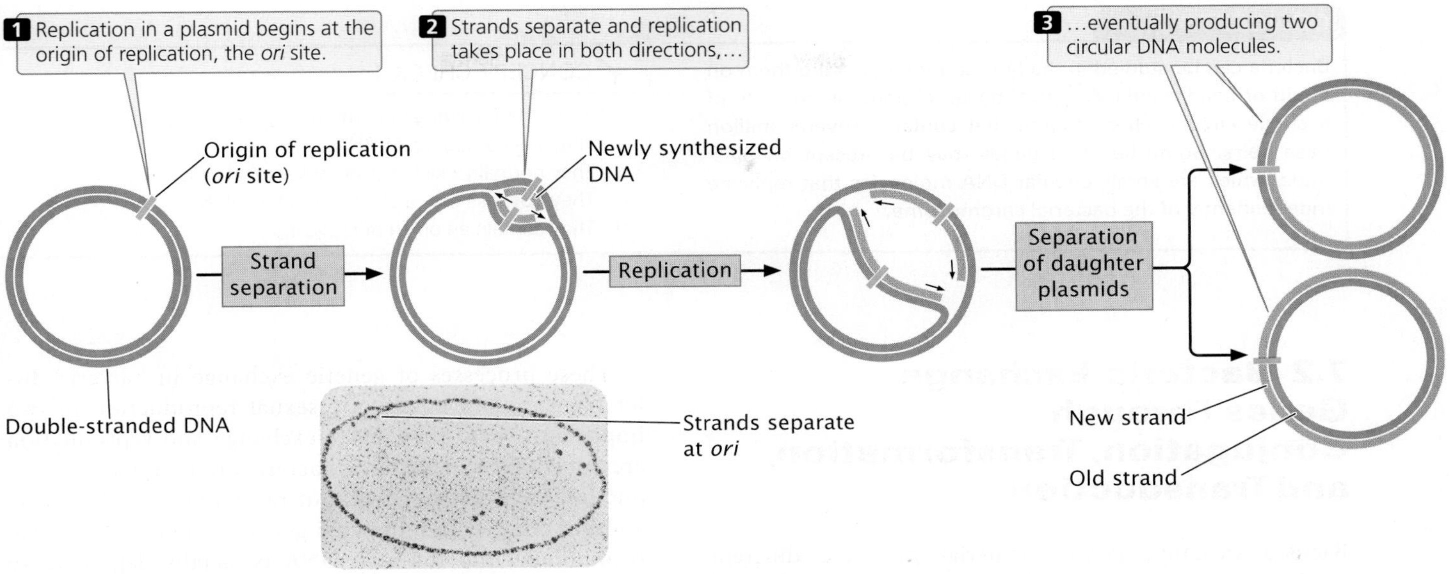

7.5 A plasmid replicates independently of its bacterial chromosome. Replication begins at the origin of replication (*ori*) and continues around the circle. In this diagram, replication is taking place in both directions; in some plasmids, replication is in one direction only. [Photograph: Biology Pics/Science Source.]

There are even a few bacteria that have linear chromosomes. Most bacterial chromosomes consist largely of sequences that encode proteins. For example, more than 90% of the DNA in *E. coli* encodes proteins. In contrast, only about 1% of human DNA encodes proteins.

THINK-PAIR-SHARE Question 3

Plasmids

In addition to having a chromosome, many bacteria possess **plasmids**: small, usually circular DNA molecules that are distinct from the bacterial chromosome. Some plasmids are present in many copies per cell, whereas others are present in only one or two copies. In general, plasmids carry genes that are not essential to bacterial function but may play an important role in the life cycle and growth of their bacterial hosts. There are many different types of plasmids; *E. coli* alone is estimated to have more than 270 different naturally occurring plasmids. Some plasmids promote mating between bacteria; others encode compounds that kill other bacteria. Of importance to human health, plasmids are responsible for the spread of antibiotic resistance among bacteria. Plasmids are also used extensively in genetic engineering (see Chapter 14).

Most plasmids are circular and several thousand base pairs in length, although plasmids consisting of several hundred thousand base pairs have also been found. Each plasmid possesses an origin of replication, a specific DNA sequence where DNA replication is initiated (see Chapter 2). The origin allows a plasmid to replicate independently of the bacterial chromosome (**Figure 7.5**). **Episomes** are plasmids that are capable of replicating freely and are able to integrate into the bacterial chromosomes. The **F** (fertility) **factor** of *E. coli* (**Figure 7.6**) is an episome that controls mating and gene exchange between *E. coli* cells, a process we will discuss shortly.

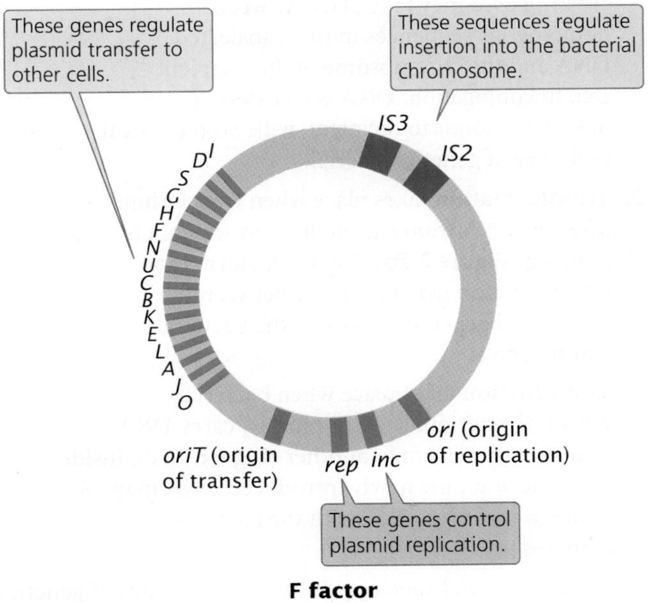

7.6 The F factor, a circular episome of *E. coli*, contains a number of genes that regulate its transfer into a bacterial cell, replication, and insertion into the bacterial chromosome. Replication is initiated at *ori*. Insertion sequences *IS3* and *IS2* control both insertion into the bacterial chromosome and excision from it.

CONCEPTS

Bacteria can be studied in the laboratory by growing them on liquid or solid media. A typical bacterial genome consists of a single circular chromosome that contains several million base pairs. Some bacterial genes may be present on plasmids, which are small, circular DNA molecules that replicate independently of the bacterial chromosome.

✓ CONCEPT CHECK 1

Which of the following statements is true of plasmids?

a. They are composed of RNA.
b. They normally exist outside of bacterial cells.
c. They possess only a single strand of DNA.
d. They contain an origin of replication.

7.2 Bacteria Exchange Genes Through Conjugation, Transformation, and Transduction

Bacteria exchange genetic material by three different mechanisms, all entailing some type of DNA transfer and recombination between the transferred DNA and the bacterial chromosome.

1. **Conjugation** takes place when genetic material passes directly from one bacterium to another (**Figure 7.7a**). In conjugation, two bacteria lie close together and a connection forms between them. A plasmid or a part of the bacterial chromosome passes from one cell (the donor) to the other (the recipient). After conjugation, crossing over may take place between homologous sequences in the transferred DNA and the chromosome of the recipient cell. In conjugation, DNA is transferred only from donor to recipient, with no reciprocal exchange of genetic material.
2. **Transformation** takes place when a bacterium takes up DNA from the medium in which it is growing (**Figure 7.7b**). After transformation, recombination may take place between the introduced genes and those of the bacterial chromosome.
3. **Transduction** takes place when bacterial viruses (bacteriophages, or phages) carry DNA from one bacterium to another (**Figure 7.7c**). Inside the bacterium, the newly introduced DNA may undergo recombination with the bacterial chromosome.

Not all bacterial species exhibit all three types of genetic transfer. Conjugation takes place more frequently in some species than in others. Transformation takes place to a limited extent in many species of bacteria, but laboratory techniques can increase the rate of DNA uptake. Most bacteriophages have a limited host range, so transduction normally takes place between bacteria of the same or closely related species only.

These processes of genetic exchange in bacteria differ from diploid eukaryotic sexual reproduction in two important ways. First, DNA exchange and reproduction are not coupled in bacteria; bacteria often undergo reproduction (cell division) without receiving any DNA from another cell. Second, donated genetic material that is not recombined into the host DNA is usually degraded, so the recipient cell remains haploid. Each type of genetic transfer can be used to map genes, as we will see in the following sections.

CONCEPTS

DNA may be transferred between bacterial cells through conjugation, transformation, or transduction. Each type of genetic transfer consists of a one-way movement of genetic information to the recipient cell, sometimes followed by recombination. These processes are not connected to cellular reproduction in bacteria.

✓ CONCEPT CHECK 2

Which process of DNA transfer in bacteria requires a virus?

a. Conjugation
b. Transduction
c. Transformation
d. All of the above

Conjugation

In 1946, Joshua Lederberg and Edward Tatum demonstrated that bacteria can transfer and recombine genetic information, paving the way for the use of bacteria in genetic studies. In the course of their research, Lederberg and Tatum studied auxotrophic strains of *E. coli*. The Y10 strain required the amino acids threonine (and was genotypically thr^-) and leucine (leu^-) and the vitamin thiamine (thi^-) for growth, but did not require the vitamin biotin (bio^+) or the amino acids phenylalanine (phe^+) and cysteine (cys^+); the genotype of this strain can be written as $thr^-\ leu^-\ thi^-\ bio^+\ phe^+\ cys^+$. The Y24 strain had the opposite set of alleles: it required biotin, phenylalanine, and cysteine in its medium, but it did not require threonine, leucine, or thiamine; its genotype was $thr^+\ leu^+\ thi^+\ bio^-\ phe^-\ cys^-$. In one experiment, Lederberg and Tatum mixed Y10 and Y24 bacteria together and plated

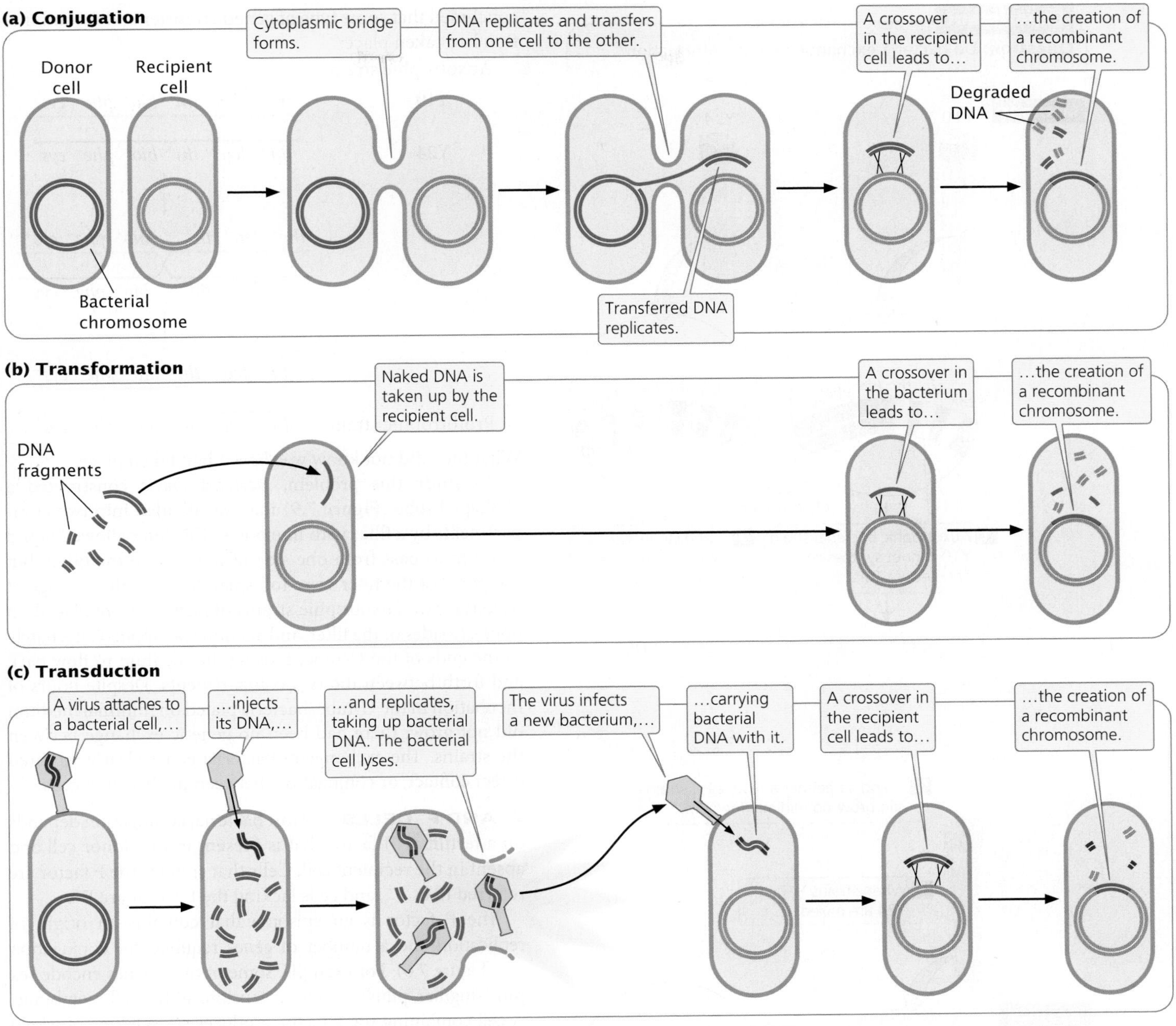

7.7 Conjugation, transformation, and transduction are three processes of gene transfer in bacteria. For the transferred DNA to be stably inherited, all three processes require the transferred DNA to undergo recombination with the bacterial chromosome.

them on minimal medium (**Figure 7.8**). Each strain was also plated separately on minimal medium.

Alone, neither Y10 nor Y24 grew on minimal medium: each strain required nutrients that were absent. Strain Y10 was unable to grow because it required threonine, leucine, and thiamine, which were absent in the minimal medium; strain Y24 was unable to grow because it required biotin, phenylalanine, and cysteine, which also were absent from the minimal medium. When Lederberg and Tatum mixed the two strains, however, a few colonies did grow on the minimal medium. These prototrophic bacteria must have had genotype thr^+ leu^+ thi^+ bio^+ phe^+ cys^+. Where had they come from?

If mutations were responsible for the prototrophic colonies, then some colonies should also have grown on the plates containing Y10 or Y24 alone, but no bacteria grew on those plates. Multiple simultaneous mutations ($thr^- \rightarrow thr^+$, $leu^- \rightarrow leu^+$, and $thi^- \rightarrow thi^+$ in strain Y10 or $bio^- \rightarrow bio^+$, $phe^- \rightarrow phe^+$, and $cys^- \rightarrow cys^+$ in strain Y24) would have been required for either strain to become prototrophic by mutation, which was very improbable. Lederberg and Tatum

Experiment

Question: Do bacteria exchange genetic information?

Methods

Y10 Y24

thr⁻ *leu⁻ thi⁻ bio⁺ phe⁺ cys⁺*

thr⁺ *leu⁺ thi⁺ bio⁻ phe⁻ cys⁻*

Bacterial chromosome

1 Auxotrophic bacterial strain Y10 cannot synthesize Thr, Leu, or Thi…

2 …and strain Y24 cannot synthesize biotin, Phe, or Cys,…

3 …and so neither auxotrophic strain can grow on minimal medium.

4 When strains Y10 and Y24 are mixed,…

Results

thr⁺ *leu⁺ thi⁺ bio⁺ phe⁺ cys⁺*

5 …some colonies grow…

6 …because genetic recombination has taken place and bacteria can synthesize all necessary nutrients.

Conclusion: Yes, genetic exchange and recombination took place between the two mutant strains.

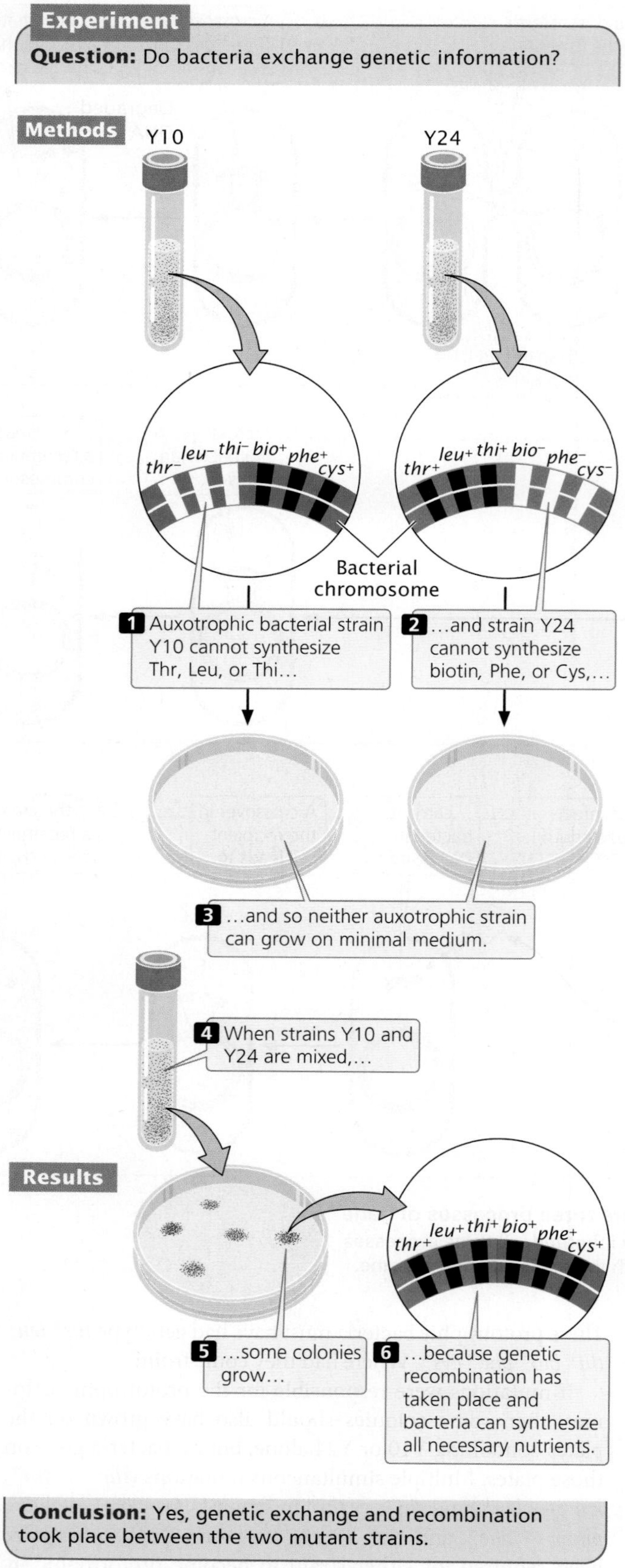

7.8 Lederberg and Tatum's experiment demonstrated that bacteria undergo genetic exchange.

concluded that some type of genetic transfer and recombination had taken place:

Auxotrophic strain

Y10 $thr^- \quad leu^- \quad thi^- \quad bio^+ \quad phe^+ \quad cys^+$

Y24 $thr^+ \quad leu^+ \quad thi^+ \quad bio^- \quad phe^- \quad cys^-$

$thr^- \quad leu^- \quad thi^- \qquad\qquad bio^+ \quad phe^+ \quad cys^+$

$thr^+ \quad leu^+ \quad thi^+ \qquad\qquad bio^- \quad phe^- \quad cys^-$

$thr^- \quad leu^- \quad thi^- \quad bio^- \quad phe^- \quad cys^-$

Prototrophic strain $thr^+ \quad leu^+ \quad thi^+ \quad bio^+ \quad phe^+ \quad cys^+$

What they did not know was *how* it had taken place.

To study this problem, Bernard Davis constructed a U-shaped tube (**Figure 7.9**) that was divided into two compartments by a filter with fine pores. This filter allowed liquid medium to pass from one side of the tube to the other, but the pores of the filter were too small to allow the passage of bacteria. Two auxotrophic strains of bacteria were placed on opposite sides of the filter, and suction was applied alternately to the ends of the U-tube, causing the medium to flow back and forth between the two compartments. Despite hours of incubation in the U-tube, bacteria plated on minimal medium did not grow; there had been no genetic exchange between the strains. The exchange of bacterial genes clearly required direct contact, or conjugation, between the bacterial cells.

F⁺ AND F⁻ CELLS In most bacteria, conjugation depends on a fertility (F) factor that is present in the donor cell and absent in the recipient cell. Cells that contain the F factor are referred to as F⁺, and cells lacking the F factor are F⁻.

The F factor is an episome that contains an origin of replication and a number of genes required for conjugation (see Figure 7.6). For example, some of these genes encode sex **pili** (singular, pilus), slender extensions of the cell membrane. A cell containing the F factor produces sex pili, one of which makes contact with a receptor on an F⁻ cell (**Figure 7.10**) and pulls the two cells together. DNA is then transferred from the F⁺ cell to the F⁻ cell. Conjugation can take place only between a cell that possesses the F factor and a cell that lacks the F factor.

In most cases, the only genes transferred during conjugation between an F⁺ and F⁻ cell are those on the F factor (**Figure 7.11a and b**). Transfer is initiated when one of the DNA strands on the F factor is nicked at an origin of transfer (*oriT*). One end of the nicked DNA separates from the circular F factor and passes into the recipient cell (**Figure 7.11c**). Replication takes place on the nicked strand, proceeding around the F factor in the F⁺ cell and replacing the transferred strand (**Figure 7.11d**). Because the F factor in the donor cell is always nicked at the *oriT* site, this site always enters the recipient cell first, followed by the rest of the

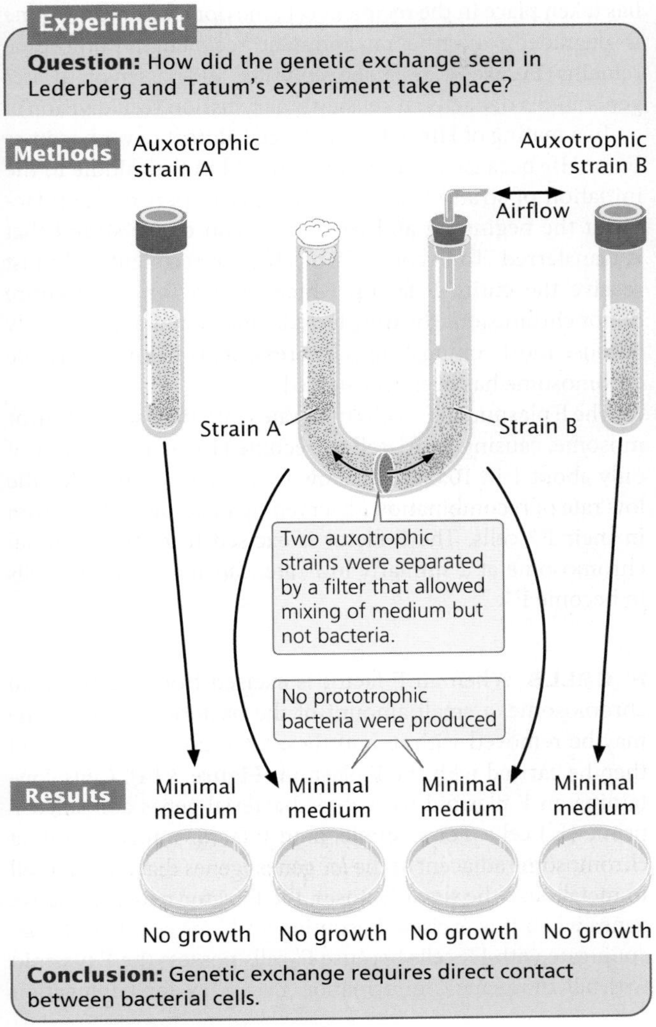

7.9 Davis's U-tube experiment.

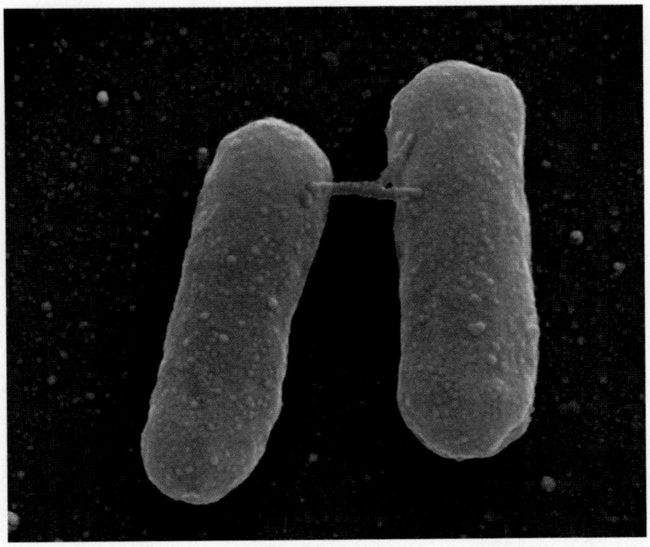

7.10 A sex pilus connects F$^+$ and F$^-$ cells during bacterial conjugation. Here, *E. coli* cells are shown in conjugation. [Eye of Science/Science Source.]

plasmid. Thus, the transfer of genetic material has a defined direction. Inside the recipient cell, the single strand is replicated, producing a circular, double-stranded copy of the F plasmid (**Figure 7.11e**). If the entire F factor is transferred to the recipient F$^-$ cell, that cell becomes an F$^+$ cell.

Hfr CELLS Conjugation transfers genetic material in the F plasmid from F$^+$ to F$^-$ cells, but it does not account for the transfer of chromosomal genes observed by Lederberg and Tatum. In Hfr (high-frequency recombination) bacterial strains, the F factor is integrated into the bacterial chromosome (**Figure 7.12**). Hfr cells behave like F$^+$ cells, forming sex pili and undergoing conjugation with F$^-$ cells.

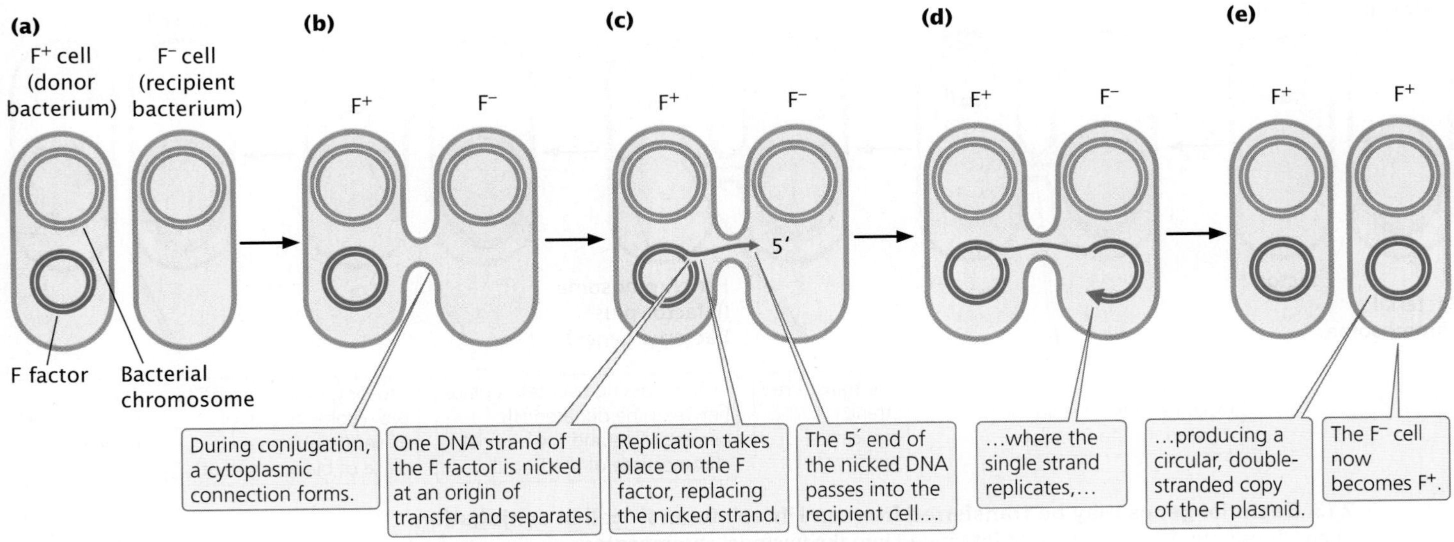

7.11 The F factor is transferred during conjugation between an F$^+$ and an F$^-$ cell.

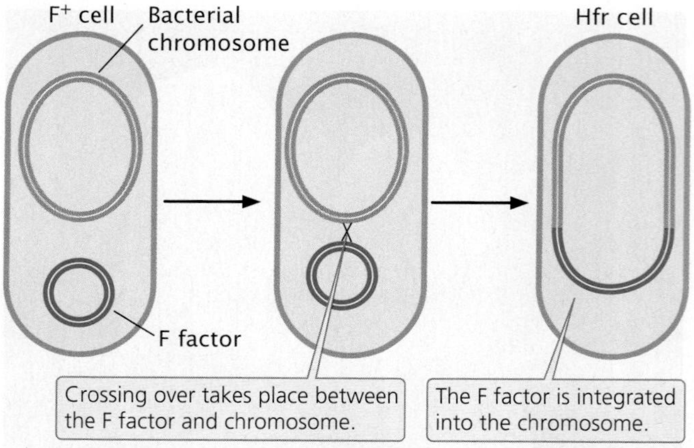

F+ cell
Bacterial chromosome
Hfr cell
F factor
Crossing over takes place between the F factor and chromosome.
The F factor is integrated into the chromosome.

7.12 The F factor is integrated into the bacterial chromosome in an Hfr cell.

In conjugation between Hfr and F$^-$ cells (**Figure 7.13a**), the integrated F factor is nicked, and the end of the nicked strand moves into the F$^-$ cell (**Figure 7.13b**), just as it does in conjugation between F$^+$ and F$^-$ cells. But because, in an Hfr cell, the F factor has been integrated into the bacterial chromosome, the chromosome follows the F factor into the recipient cell. How much of the bacterial chromosome is transferred depends on the length of time that the two cells remain in conjugation.

Inside the recipient cell, the donor DNA strand is replicated (**Figure 7.13c**), and crossing over between it and the original chromosome of the F$^-$ cell (**Figure 7.13d**) may take place. This chromosomal gene transfer between Hfr and F$^-$ cells explains how the recombinant prototrophic cells observed by Lederberg and Tatum were produced. After crossing over has taken place in the recipient cell, the donated chromosome is degraded, and the recombinant recipient chromosome remains (**Figure 7.13e**), to be replicated and passed on to later generations of bacterial cells by binary fission (cell division).

In a mating of Hfr × F$^-$, the F$^-$ cell almost never becomes F$^+$ or Hfr because the F factor is nicked in the middle in the initiation of strand transfer, which places part of the F factor at the beginning and part at the end of the strand that is transferred. To become F$^+$ or Hfr, the recipient cell must receive the entire F factor, which requires that the entire donor chromosome be transferred. This event happens rarely because most conjugating cells break apart before the entire chromosome has been transferred.

The F plasmid in F$^+$ cells integrates into the bacterial chromosome, causing an F$^+$ cell to become Hfr, at a frequency of only about 1 in 10,000. This low frequency accounts for the low rate of recombination observed by Lederberg and Tatum in their F$^+$ cells. The F factor is excised from the bacterial chromosome at a similarly low rate, causing a few Hfr cells to become F$^+$.

F′ CELLS When an F factor is excised from the bacterial chromosome, a small amount of the bacterial chromosome may be removed with it, and these chromosomal genes will then be carried with the F plasmid (**Figure 7.14**). Cells containing an F plasmid with some bacterial genes are called F prime (F′) cells. For example, if an F factor integrates into a chromosome adjacent to the *lac* genes (genes that enable a cell to metabolize the sugar lactose), the F factor may pick up *lac* genes when it is excised, becoming an F′ *lac* cell. F′ cells can conjugate with F$^-$ cells because F′ cells possess the F plasmid with all the genetic information necessary for conjugation

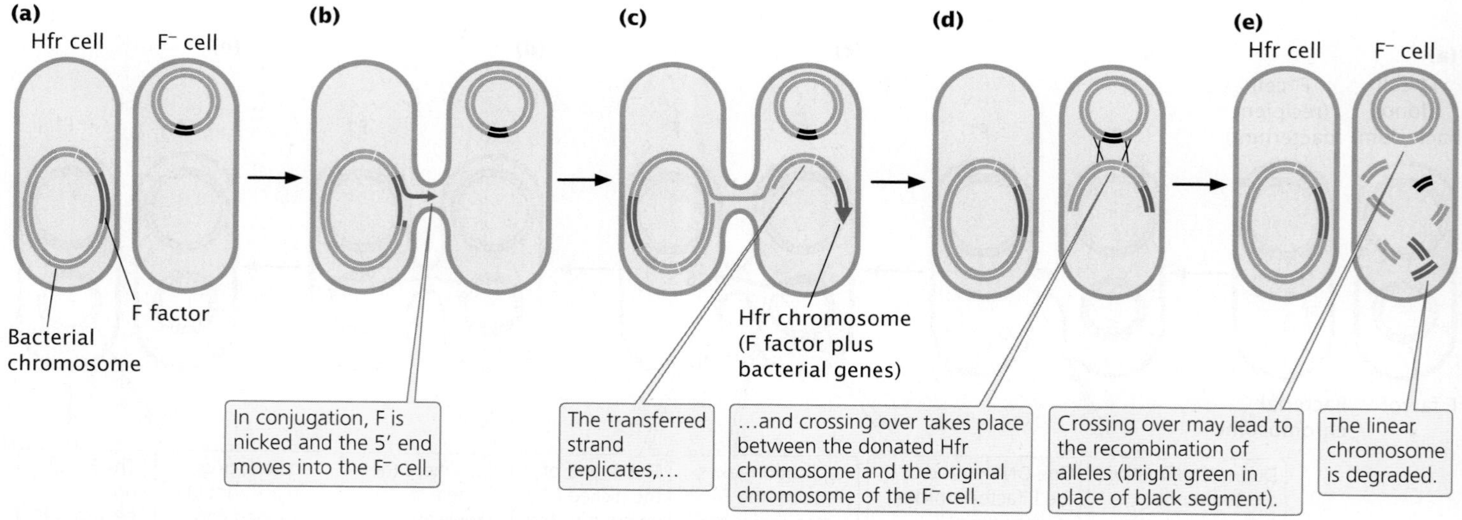

7.13 Bacterial genes may be transferred from an Hfr cell to an F$^-$ cell in conjugation.
In an Hfr cell, the F factor has been integrated into the bacterial chromosome.

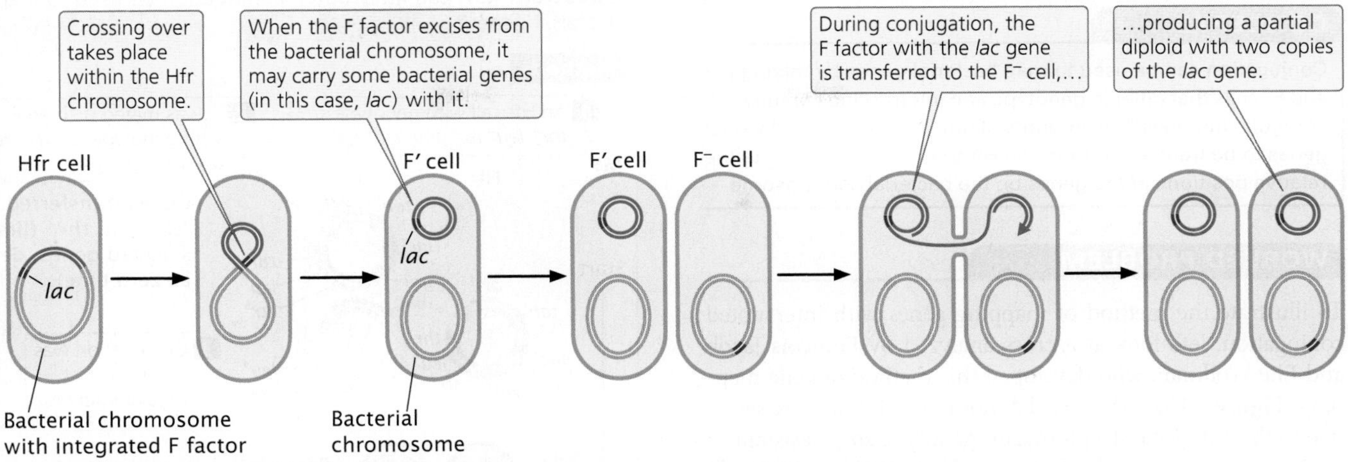

7.14 An Hfr cell may be converted into an F′ cell when the F factor excises from the bacterial chromosome and carries bacterial genes with it. Conjugation produces a partial diploid.

and gene transfer. Characteristics of *E. coli* cells of different F types (called *mating types*) are summarized in **Table 7.2.**

During conjugation between an F′ cell and an F^- cell, the F plasmid is transferred to the F^- cell, which means that any genes on the F plasmid, including those from the bacterial chromosome, may be transferred to F^- recipient cells. This process produces partial diploids, or *merozygotes*, which are cells with two copies of some genes, one on the bacterial chromosome and one on the newly introduced F plasmid. The outcomes of conjugation between different mating types of *E. coli* are summarized in **Table 7.3.**

TABLE 7.2 Characteristics of *E. coli* cells with different types of F factors

Type	F factor characteristics	Role in conjugation
F^+	Present as separate circular plasmid	Donor
F^-	Absent	Recipient
Hfr	Present, integrated into bacterial chromosome	High-frequency donor
F′	Present as separate circular plasmid, carrying some bacterial genes	Donor

TABLE 7.3 Results of conjugation between cells with different types of F factors

Conjugating	Cell types present after conjugation
$F^+ \times F^-$	Two F^+ cells (F^- cell becomes F^+)
Hfr $\times$ F^-	One Hfr cell and one F^- (no change)*
$F' \times F^-$	Two F′ cells (F^- cell becomes F′)

*Rarely, the F^- cell becomes F^+ in an Hfr $\times$ F^- conjugation if the entire chromosome is transferred during conjugation.

CONCEPTS

Conjugation in *E. coli* is controlled by an episome called the F factor. Cells containing the F factor (F^+ cells) are donors of genes; cells lacking the F factor (F^- cells) are recipients. In Hfr cells, the F factor is integrated into the bacterial chromosome; these cells donate DNA to F^- cells. F′ cells contain a copy of the F factor with some chromosomal genes.

✓ CONCEPT CHECK 3

Conjugation between an F^+ and an F^- cell usually results in

a. two F^+ cells.
b. two F^- cells.
c. an F^+ and an F^- cell.
d. an Hfr cell and an F^+ cell.

MAPPING BACTERIAL GENES WITH INTERRUPTED CONJUGATION The transfer of DNA that takes place during conjugation between Hfr and F^- cells allows bacterial genes to be mapped. In conjugation, as we have seen, the chromosome of the Hfr cell is transferred to the F^- cell. Transfer of the entire *E. coli* chromosome requires about 100 minutes; if conjugation is interrupted before 100 minutes have elapsed, only part of the donor chromosome will pass into the F^- cell and have an opportunity to recombine with the recipient chromosome. Chromosome transfer always begins within the integrated F factor and proceeds in a continuous direction, so genes are transferred according to their sequence on the chromosome. This knowledge can be used to map bacterial genes by mixing Hfr and F^- cells that differ in genotype and interrupting conjugation at regular intervals. The times required for individual genes to be transferred indicate their relative positions on the chromosome. In most genetic maps, distances are expressed as recombination frequencies; however, in bacterial gene maps constructed with interrupted conjugation, the basic unit of distance is a minute. View **Animation 7.1** to see how genes are mapped using interrupted conjugation. A

CONCEPTS

Conjugation can be used to map bacterial genes by mixing Hfr and F^- cells that differ in genotype and interrupting conjugation at regular intervals. The amounts of time required for individual genes to be transferred from the Hfr to the F^- cells indicate the relative positions of the genes on the bacterial chromosome.

WORKED PROBLEM

To illustrate the method of mapping genes with interrupted conjugation, let's look at a cross analyzed by François Jacob and Elie Wollman, who developed this method of gene mapping (**Figure 7.15a**). They used donor Hfr cells that were sensitive to the antibiotic streptomycin (genotype str^s), resistant to sodium azide (azi^r) and infection by bacteriophage T1 (ton^r), prototrophic for threonine (thr^+) and leucine (leu^+), and able to break down lactose (lac^+) and galactose (gal^+). They used F^- recipient cells that were resistant to streptomycin (str^r), sensitive to sodium azide (azi^s) and to infection by bacteriophage T1 (ton^s), auxotrophic for threonine (thr^-) and leucine (leu^-), and unable to break down lactose (lac^-) and galactose (gal^-). Thus, the genotypes of the donor and recipient cells were

Donor Hfr cells: str^s leu^+ thr^+ azi^r ton^r lac^+ gal^+

Recipient F^- cells: str^r leu^- thr^- azi^s ton^s lac^- gal^-

The two strains were mixed in nutrient medium and allowed to conjugate. After a few minutes, the medium was diluted to prevent any new pairings. At regular intervals, a sample of cells was removed and agitated vigorously in a kitchen blender to halt all conjugation and DNA transfer. The cells from each sample were plated on a selective medium that contained streptomycin and lacked leucine and threonine. The original donor cells were streptomycin sensitive (str^s) and would not grow on this medium. The F^- recipient cells were auxotrophic for leucine and threonine, and they also failed to grow on this medium. Only recipient cells that underwent conjugation and received at least the leu^+ and thr^+ genes from the Hfr donors could grow on this medium. All str^r leu^+ thr^+ cells were then tested for the presence of other genes that might have been transferred from the donor Hfr strain.

Because Jacob and Wollman used streptomycin to kill all the donor cells, they were not able to examine the transfer of the str^s gene. All of the cells that grew on the selective medium were leu^+ thr^+, so we know that these genes were transferred. In **Figure 7.15b**, the percentages of str^r leu^+ thr^+ cells receiving specific alleles (azi^r, ton^r, leu^+, and gal^+) from the Hfr strain are plotted against the duration of conjugation. What is the order in which the genes are transferred and what are the distances among them?

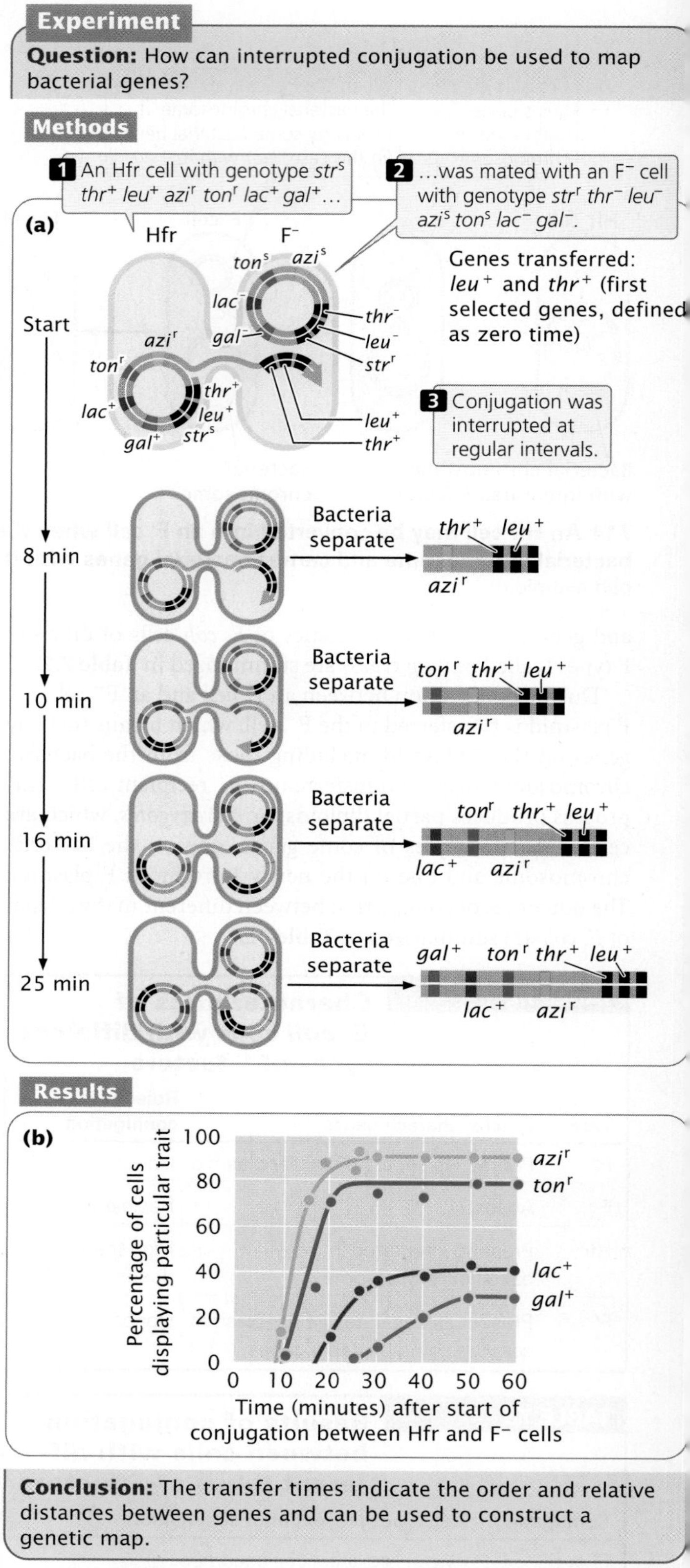

7.15 Jacob and Wollman used interrupted conjugation to map bacterial genes.

Solution Strategy

What information is required in your answer to the problem?

The order of the genes on the bacterial chromosome and the distances among them.

What information is provided to solve the problem?

- The donor cells were str^s leu^+ thr^+ azi^r ton^r lac^+ gal^+ and the recipient cells were str^r leu^- thr^- azi^s ton^s lac^- gal^-.

- The percentages of recipient cells with different traits that appear at various times after the start of conjugation (Figure 7.15b).

Solution Steps

The first donor gene to appear in the recipient cells (at about 9 minutes) was *azi*^{r}. Gene *ton*^{r} appeared next (after about 10 minutes), followed by *lac*$^+$ (at about 18 minutes) and by *gal*$^+$ (after 25 minutes) (see Figure 7.15b). These transfer times indicate the order of gene transfer and the relative distances among the genes.

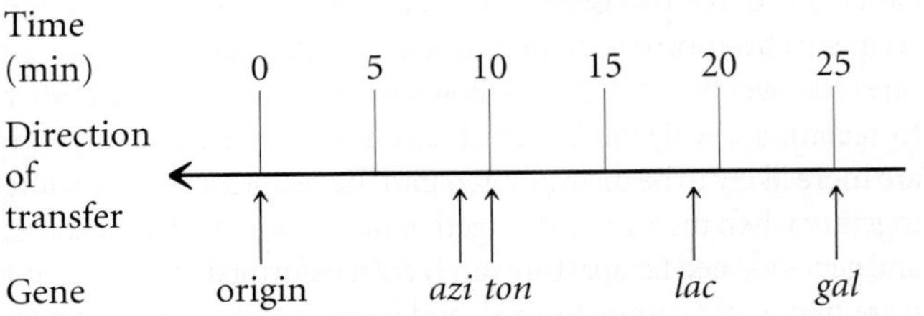

Notice that the frequency of gene transfer from donor to recipient cells decreased with distance from the origin of transfer. For example, about 90% of the recipients received the *azi*^{r} allele, but only about 30% received the *gal*$^+$ allele. The lower percentage for *gal*$^+$ is due to the fact that some conjugating cells spontaneously broke apart before they were disrupted by the blender. The probability of spontaneous disruption increases with time, so fewer cells had an opportunity to receive genes that were transferred later.

THINK-PAIR-SHARE Question 4

▶ For additional practice mapping bacterial genes with interrupted conjugation, try **Problem 14** at the end of the chapter.

Natural Gene Transfer and Antibiotic Resistance

Antibiotics are substances that kill bacteria. Their development and widespread use has greatly reduced the threat of infectious disease and saved countless lives. But many pathogenic bacteria have developed resistance to antibiotics, particularly in environments where antibiotics are routinely used, such as hospitals, livestock operations, and fish farms. In these environments where antibiotics are continually present, the only bacteria that survive are those that possess antibiotic resistance. Released from competition with other bacteria, resistant bacteria multiply quickly and spread. In this way, the presence of antibiotics selects for resistant bacteria, reducing the effectiveness of antibiotic treatment for infections.

Antibiotic resistance in bacteria frequently results from the action of genes located on *R plasmids*, small circular plasmids that can be transferred by conjugation. Some drug-resistant R plasmids convey resistance to several antibiotics simultaneously. Ironic but plausible sources of some of the resistance genes found in R plasmids are the microbes that produce antibiotics in the first place. R plasmids can spread easily throughout the environment, passing between related and unrelated bacteria in a variety of situations.

THINK-PAIR-SHARE Question 5

Transformation

A second way that DNA can be transferred between bacteria is through transformation (see Figure 7.7b). Transformation played an important role in the initial identification of DNA as the genetic material, as we will see in Chapter 8.

Transformation requires both the uptake of DNA from the surrounding medium and its incorporation into a bacterial chromosome or a plasmid. It may occur naturally when dead bacteria break down and release DNA fragments into the environment. In soil and marine environments, transformation may be an important route of genetic exchange for some bacteria. Transformation is also an important technique for transferring genes to bacteria in the laboratory.

MECHANISM OF TRANSFORMATION Cells that take up DNA through their cell membrane are said to be **competent**. Some species of bacteria take up DNA more easily than do others; competence is influenced by growth stage, the concentration of available DNA, and environmental challenges. The DNA taken up by a competent cell need not be bacterial: virtually any type of DNA (bacterial or otherwise) can be taken up by competent cells under the appropriate conditions.

Typically, as a DNA fragment enters a bacterial cell in the course of transformation (**Figure 7.16**), one of the strands

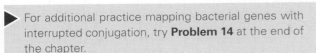

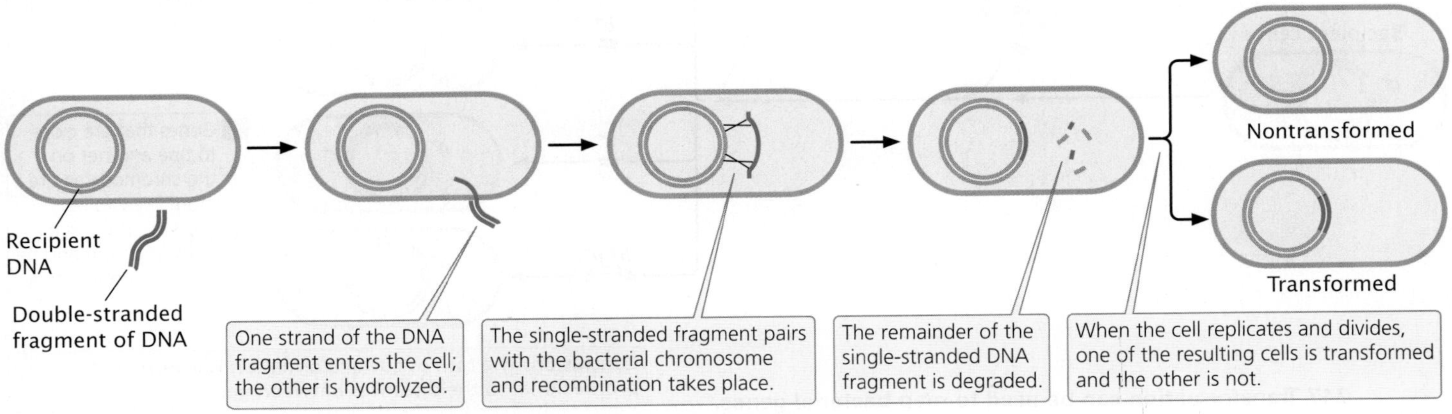

7.16 Genes can be transferred between bacteria through transformation.

is hydrolyzed, whereas the other strand moves across the membrane, where it may pair with a homologous region and become integrated into the bacterial chromosome. This integration requires two crossover events, after which the remaining single-stranded DNA is degraded by bacterial enzymes. In some bacteria, double-stranded DNA moves across the cell membrane and is integrated into the bacterial chromosome. Cells that receive genetic material through transformation are called **transformants**.

Bacterial geneticists have developed techniques to increase the frequency of transformation in the laboratory to introduce particular DNA fragments or whole plasmids into cells. They have also developed strains of bacteria that are more competent than wild-type cells. Treatment with calcium chloride, heat shock, or an electrical field makes bacterial membranes more porous and permeable to DNA. The efficiency of transformation can also be increased by using high concentrations of DNA. These techniques make it possible to transform bacteria such as *E. coli*, which are not naturally competent.

GENE MAPPING WITH TRANSFORMATION Transformation, like conjugation, can be used to map bacterial genes, especially in those species that do not undergo conjugation or transduction (see Figure 7.7a and c). Transformation mapping requires two strains of bacteria that differ in several genetic traits; for example, the recipient strain might be $a^- b^- c^-$ (auxotrophic for three nutrients), and the donor strain might be $a^+ b^+ c^+$ (prototrophic for the same three nutrients). DNA from the donor strain is isolated, purified, and fragmented. The recipient strain is treated to increase competence, and DNA from the donor strain is added to the medium. Fragments of the donor DNA enter the recipient cells and undergo recombination with homologous DNA sequences on the bacterial chromosome.

Genes can be mapped by observing the rate at which two or more genes are transferred together, or **cotransformed**, in transformation. Genes that are physically close on the chromosome are more likely to be present on the same DNA fragment and transferred together, as shown for genes a^+ and b^+ in **Figure 7.17**. Genes that are far apart are unlikely to be present on the same DNA fragment and will rarely be transferred together. Inside the cell, DNA becomes incorporated into the bacterial chromosome through recombination. If two genes are close together on the same fragment, any two crossovers are likely to take place on either side of the two genes, allowing both to become part of the recipient chromosome. If the two genes are far apart, there may be one crossover between them, allowing one gene but not the other to recombine with the bacterial chromosome. Thus, two genes are more likely to be incorporated into the recipient chromosome together when they are close together on the donor chromosome, and genes located far apart are rarely cotransformed. If genes *a* and *b* are frequently cotransformed, and genes *b* and *c* are frequently cotransformed, but genes *a* and *c* are rarely cotransformed, then gene *b* must be between *a* and *c*—the gene order is *a b c*.

CONCEPTS

Bacterial genes can be mapped by taking advantage of transformation—the ability of bacteria to take up DNA from the environment and incorporate it into their chromosomes through crossing over. The relative rate at which pairs of genes are cotransformed indicates the distance between them: the higher the rate of cotransformation, the closer the genes are on the bacterial chromosome.

✓ CONCEPT CHECK 4

A bacterial strain with genotype $his^- \ leu^- \ thr^-$ is transformed with DNA from a strain that is $his^+ \ leu^+ \ thr^+$. A few $leu^+ \ thr^+$ cells and a few $his^+ \ thr^+$ cells are found, but no $his^+ \ leu^+$ cells are observed. Which genes are farthest apart?

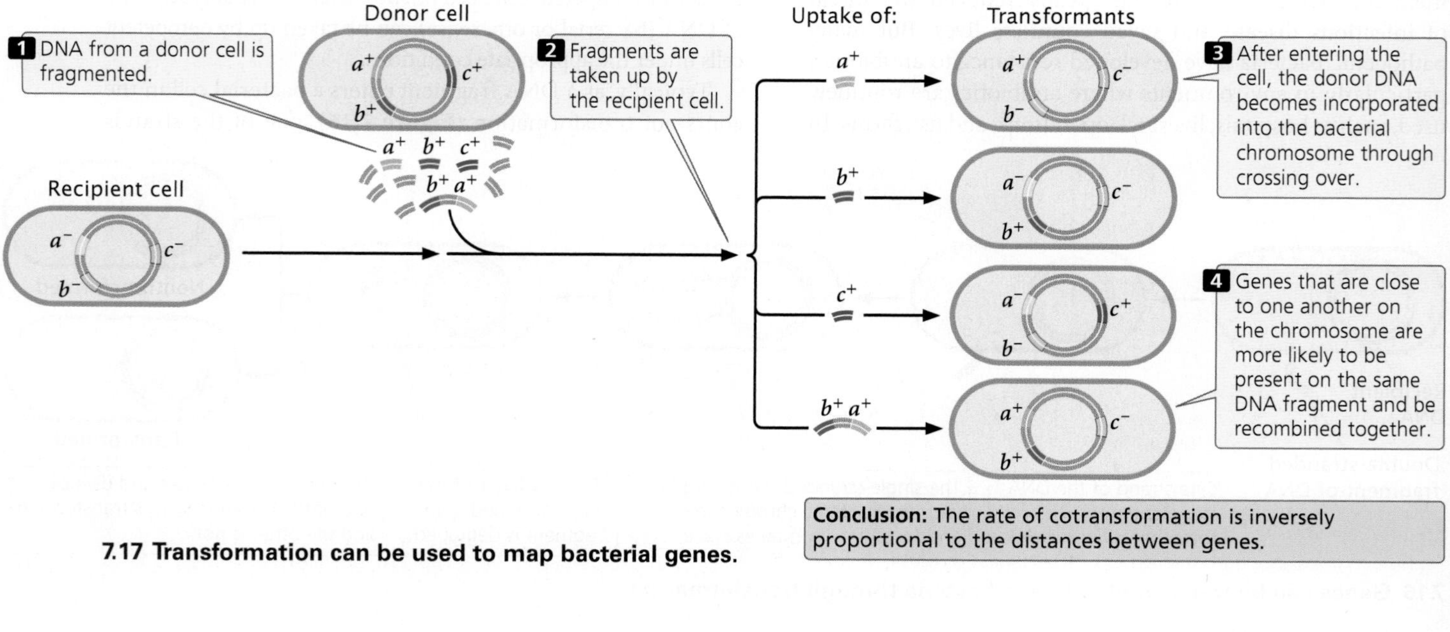

7.17 Transformation can be used to map bacterial genes.

Bacterial Genome Sequences

Genetic maps serve as a foundation for more detailed information provided by DNA sequencing. Geneticists have now determined the complete nucleotide sequences of thousands of bacterial genomes, and many additional microbial sequencing projects are under way. The size and content of bacterial genomes are discussed in Chapter 14.

The availability of genome sequences has provided evidence that many bacteria have acquired genetic information from other species of bacteria—and sometimes even from eukaryotic organisms—in a process called **horizontal gene transfer**. In most eukaryotes, genes are passed only among members of the same species through reproduction (a process called vertical transmission); in horizontal gene transfer, genes can be passed between individual members of different species by nonreproductive mechanisms, such as conjugation, transformation, and transduction. Evidence suggests that horizontal gene transfer has taken place repeatedly among bacteria. For example, as much as 17% of *E. coli*'s genome has been acquired from other bacteria through horizontal gene transfer. Of medical significance, some pathogenic bacteria have acquired the genes necessary for infection, whereas others have acquired genes that confer resistance to antibiotics.

Bacterial Defense Mechanisms

Bacteria and archaea face significant challenges, the most serious of which are viruses, and many have evolved mechanisms to prevent the entry and reproduction of viruses and other invading DNA. Some bacteria turn off the expression of receptors to which viruses attach; others secrete polysaccharide coats that limit the access of viruses to those receptors. Other defense mechanisms block virus replication or use restriction-modification systems (see Chapter 14).

Another recently discovered defense mechanism in bacteria and archaea is CRISPR-Cas systems. These systems are analogous to the immune systems of vertebrates in that they recognize and remember DNA from specific pathogens. After DNA from a virus, plasmid, or other element has invaded a bacterial cell, bacterial proteins cut up the foreign DNA and incorporate short pieces of it into the bacterial chromosome. The foreign pieces are inserted as spacers into sequences called clustered regularly interspaced short palindromic repeats (CRISPRs). The CRISPR sequences are later transcribed and processed into small RNA molecules called CRISPR RNA (crRNA), which form complexes with proteins and sometimes with other RNA molecules. These complexes then recognize and cleave foreign DNA with sequences that are complementary to the crRNAs, destroying the foreign DNA. Because the bacterial cell carries the inserted pieces of foreign DNA in its own chromosome, it can "remember" previously encountered pathogens and quickly destroy their DNA.

CRISPR-Cas systems are widespread in bacteria and archaea: about 85% of all archaea and close to 50% of all bacteria use CRISPR-Cas systems as defenses against pathogens. Geneticists are currently using these systems as powerful tools for genetic engineering (see Chapter 14).

THINK-PAIR-SHARE Question 6

Model Genetic Organism
The Bacterium *Escherichia coli*

The most widely studied prokaryotic organism, and one of the best genetically characterized of all species, is the bacterium *Escherichia coli* (**Figure 7.18**).

Advantages of *E. coli* as a Model Genetic Organism

Escherichia coli is one of the true workhorses of genetics: its twofold advantage is rapid reproduction and small size. Under optimal conditions, this organism can reproduce every 20 minutes; in a mere 7 hours, a single bacterial cell can give rise to more than 2 million descendants. Numerous mutations in *E. coli*, affecting everything from colony appearance to drug resistance, have been isolated and characterized.

Escherichia coli is easy to culture in the laboratory in liquid medium (see Figure 7.1a) or on solid medium in petri plates (see Figure 7.1b). When *E. coli* cells are diluted and spread on the solid medium of a petri plate, individual bacteria reproduce asexually, giving rise to a concentrated clump of 10 million to 100 million genetically identical cells, called a colony. This colony formation makes it easy to isolate genetically pure strains of the bacteria.

The *E. coli* Genome

The *E. coli* genome is on a single chromosome; the common laboratory strain of *E. coli* has 4,638,858 base pairs of DNA in its genome. The *E. coli* genome contains an estimated 4300 genes, many of which have no known function. The haploid genome of *E. coli* makes it easy to isolate mutations because there are no dominant genes at the same locus to suppress and mask recessive mutations.

The *E. coli* Life Cycle

Wild-type *E. coli* is prototrophic and can grow on minimal medium that contains only glucose and some inorganic salts. Mating between bacteria, called conjugation, is controlled by fertility genes normally located on the F plasmid (see p. 187). In conjugation, one bacterium donates genetic material to another bacterium, and genetic recombination integrates that material into the bacterial chromosome. Genetic material can also be exchanged between strains of *E. coli* through transformation and transduction (see Figure 7.7).

Genetic techniques with *E. coli*

Escherichia coli is used in a number of experimental systems in which fundamental genetic processes are studied in detail. One of these applications is genetic engineering (recombinant DNA technology; see Chapter 14). Plasmids have been isolated from *E. coli* and genetically modified to create effective vectors for transferring genes into bacterial

Bacterium
Escherichia coli

ADVANTAGES

- Small size
- Rapid reproduction, dividing every 20 minutes under optimal conditions
- Easy to culture in liquid medium or on petri plates
- Small genome
- Many mutants available
- Numerous methods available for genetic engineering

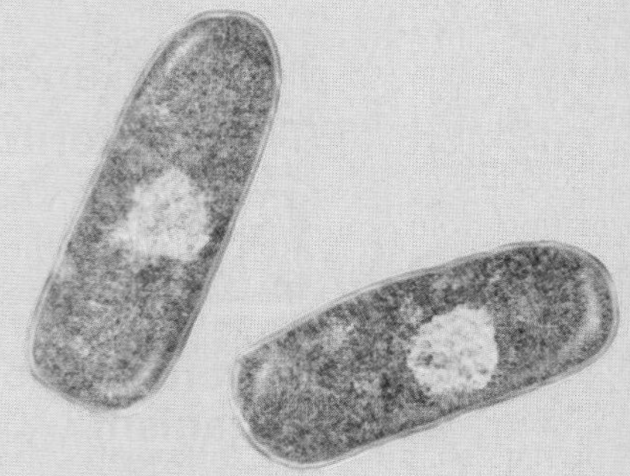

STATS

Taxonomy:	Bacteria
Size:	1–2 μm in length
Anatomy:	Single cell surrounded by cell wall with nucleoid region
Habitat:	Intestinal tract of warm-blooded animals

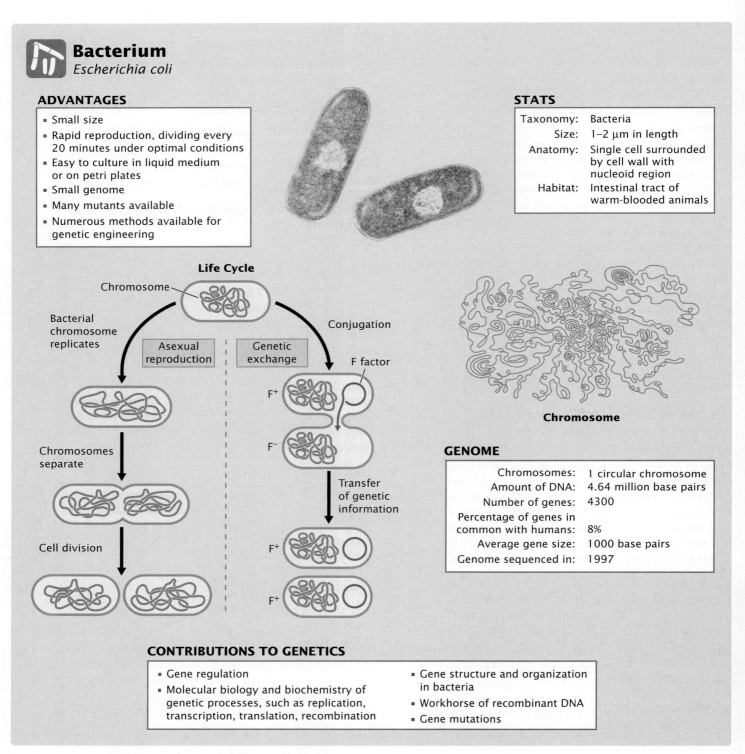

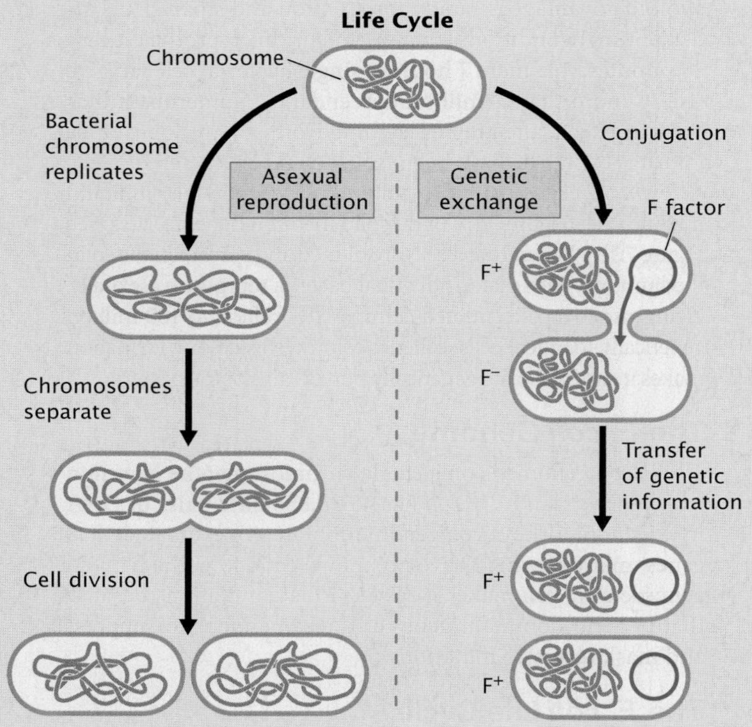

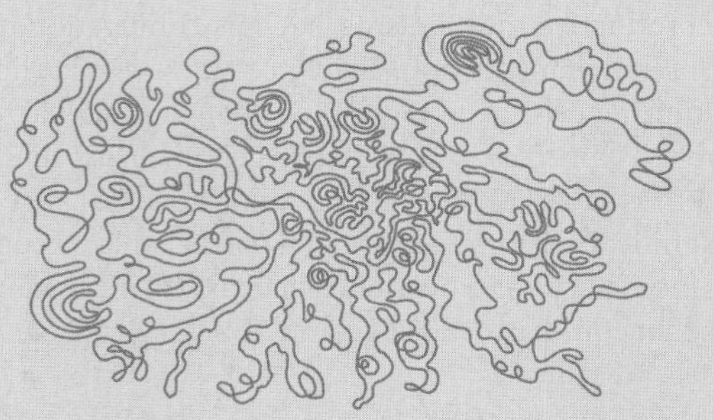

GENOME

Chromosomes:	1 circular chromosome
Amount of DNA:	4.64 million base pairs
Number of genes:	4300
Percentage of genes in common with humans:	8%
Average gene size:	1000 base pairs
Genome sequenced in:	1997

CONTRIBUTIONS TO GENETICS

- Gene regulation
- Molecular biology and biochemistry of genetic processes, such as replication, transcription, translation, recombination
- Gene structure and organization in bacteria
- Workhorse of recombinant DNA
- Gene mutations

7.18 *Escherichia coli* is a model genetic organism.

and eukaryotic cells. Often, new genetic constructs (DNA sequences created in the laboratory) are assembled and cloned in *E. coli* before transfer to other organisms. Methods have been developed to introduce specific mutations into *E. coli* genes, so genetic analysis no longer depends on the isolation of randomly occurring mutations. New DNA sequences produced by recombinant DNA technology can be introduced by transformation into special strains of *E. coli* that are particularly efficient (competent) at taking up DNA.

Because of its powerful advantages as a model genetic organism, *E. coli* has played a leading role in many fundamental discoveries in genetics, including elucidation of the genetic code, probing the nature of replication, and working out the basic mechanisms of gene regulation. ■

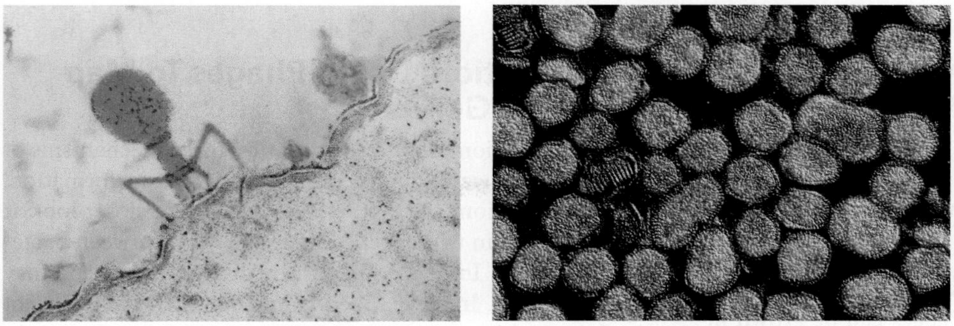

7.19 Viruses have a variety of structures and sizes. (a) T4 bacteriophage (bright orange). (b) Influenza A virus (green structures). [Part a: Biozentrum, University of Basel/Science Source. Part b: Eye of Science/Science Source.]

7.3 Viruses Are Simple Replicating Systems Amenable to Genetic Analysis

All organisms—plants, animals, fungi, and bacteria—are infected by viruses. A **virus** is a simple replicating structure made up of nucleic acid surrounded by a protein coat (see Figure 2.4). Viruses come in a great variety of shapes and sizes (**Figure 7.19**). Enveloped viruses have an outer lipid envelope that is derived from the host's cell membrane. Some viruses have DNA as their genetic material, whereas others have RNA; the nucleic acid may be double stranded or single stranded, linear or circular. Not surprisingly, viruses reproduce in a number of different ways.

Viruses that infect bacteria (bacteriophages, or phages) have played a central role in genetic research since the late 1940s. They are ideal for many types of genetic research because they have small and easily manageable genomes, reproduce rapidly, and produce large numbers of progeny. Bacteriophages have two alternative life cycles: the lytic and the lysogenic cycles. In the lytic cycle, a phage attaches to a receptor on the bacterial cell wall and injects its DNA into the cell (**Figure 7.20**). Inside the host cell, the phage DNA is replicated, transcribed, and translated, producing more phage DNA and phage proteins. New phage particles are assembled from these components. The phages then produce an enzyme that breaks open the host cell, releasing the new phages. **Virulent phages** reproduce strictly through the lytic cycle and kill their host cells.

Temperate phages can undergo either the lytic or the lysogenic cycle. The lysogenic cycle begins like the lytic cycle (see Figure 7.20), but inside the cell, the phage DNA integrates into the bacterial chromosome, where it remains as an inactive **prophage**. The prophage is replicated along with the bacterial DNA and is passed on when the bacterium divides. Certain stimuli, such as ultraviolet light and some chemicals, can cause the prophage

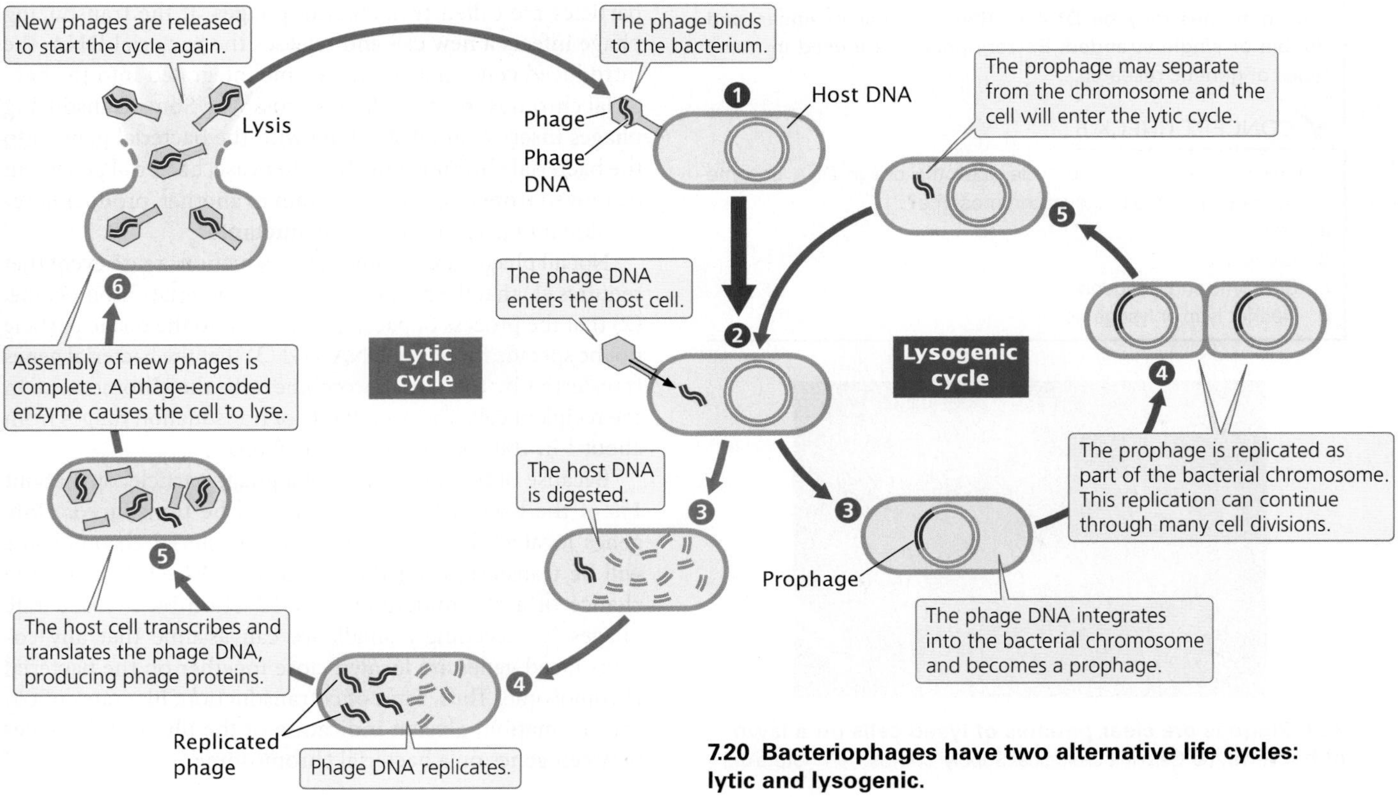

7.20 Bacteriophages have two alternative life cycles: lytic and lysogenic.

to dissociate from the bacterial chromosome and enter the lytic cycle, producing new phage particles and lysing the cell.

THINK-PAIR-SHARE Question 7

Techniques for the Study of Bacteriophages

Viruses reproduce only within host cells, so bacteriophages must be cultured in bacterial cells. Phages can be grown in large liquid cultures of bacteria to generate large numbers of offspring, but to study the characteristics of individual phages, we must isolate them on petri plates. Phages and bacteria are mixed together and plated on solid medium on a petri plate. A high concentration of bacteria is used so that the colonies will grow into one another and produce a continuous layer of bacteria, or "lawn," on the agar. An individual phage infects a single bacterial cell and goes through its lytic cycle. Many new phages are released from the lysed cell and infect additional cells; the cycle is then repeated. Because the bacteria are growing on solid medium, the diffusion of the phages is restricted, so only nearby cells are infected. After several rounds of phage reproduction, a clear patch of lysed cells, or a **plaque**, appears on the plate (**Figure 7.21**). Each plaque represents a single phage that multiplied and lysed many cells. Plating a known volume of a dilute solution of phages on a bacterial lawn and counting the number of plaques that appear can be used to determine the original concentration of phages in the solution.

CONCEPTS

Viral genomes may be DNA or RNA, circular or linear, and double or single stranded. Bacteriophages are used in many types of genetic research.

✓ CONCEPT CHECK 5

In which bacteriophage life cycle does the phage DNA become incorporated into the bacterial chromosome?

a. Lytic
b. Lysogenic
c. Both lytic and lysogenic
d. Neither lytic or lysogenic

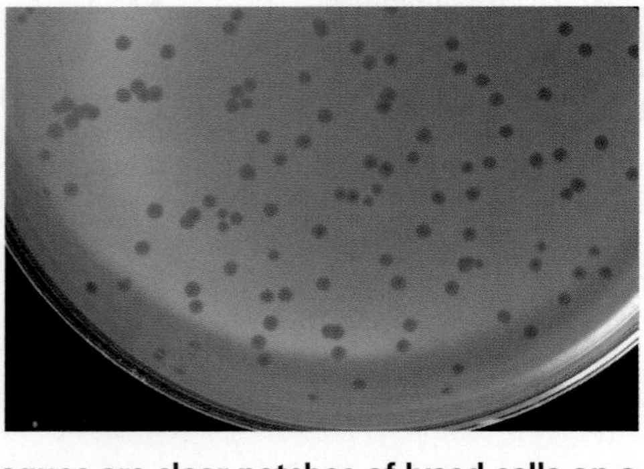

7.21 Plaques are clear patches of lysed cells on a lawn of bacteria. [© Carolina Biological Supply Company / Phototake.]

Transduction: Using Phages To Map Bacterial Genes

In our discussion of bacterial genetics, three mechanisms of gene transfer were identified: conjugation, transformation, and transduction (see Figure 7.7). Let's take a closer look at transduction, in which genes are transferred between bacteria by viruses. In **generalized transduction**, any gene may be transferred. In **specialized transduction**, which will not be discussed in detail here, only a few genes are transferred.

Joshua Lederberg and Norton Zinder discovered generalized transduction in 1952 while trying to produce recombination in the bacterium *Salmonella typhimurium* by conjugation. They mixed a strain of *S. typhimurium* that was phe^+ trp^+ tyr^+ met^- his^- with a strain that was phe^- trp^- tyr^- met^+ his^+ (**Figure 7.22**) and plated the mixture on minimal medium. A few prototrophic recombinants (phe^+ trp^+ tyr^+ met^+ his^+) appeared, suggesting that conjugation had taken place. However, when they tested the two strains in a U-shaped tube similar to the one used by Davis, some phe^+ trp^+ tyr^+ met^+ his^+ prototrophs were obtained on one side of the tube (compare Figure 7.22 with Figure 7.9). In this apparatus, the two strains were separated by a filter with pores too small for the passage of bacteria. How were genes being transferred between bacteria in the absence of conjugation? The results of subsequent studies revealed that the agent of transfer was a bacteriophage.

In the lytic cycle of phage reproduction, the phage degrades the bacterial chromosome into random fragments (**Figure 7.23** on p. 202). In some types of bacteriophage, a piece of the bacterial chromosome, instead of phage DNA, occasionally gets packaged into a phage coat; these phage particles are called **transducing phages**. If the transducing phage infects a new cell and releases the bacterial DNA, the introduced genes may then become integrated into the bacterial chromosome by a double crossover. Some transducing phages insert viral DNA, along with the bacterial gene, into the bacterial chromosome. In either case, bacterial genes can be moved from one bacterial strain to another, producing recombinant bacteria called **transductants**.

Not all phages are capable of transduction, a rare event that requires (1) that the phage degrade the bacterial chromosome, (2) that the process of packaging DNA into the phage particle not be specific for phage DNA, and (3) that the bacterial genes transferred by the virus recombine with the chromosome in the recipient cell. The overall rate of transduction ranges from about 1 in 100,000 to 1 in 1,000,000 phages.

Because of the limited size of a phage particle, only about 1% of the bacterial chromosome can be transduced. Only genes located close together on the bacterial chromosome will be transferred together, or **cotransduced**. Because the chance of a cell undergoing transduction by two separate phages is exceedingly small, we can assume that any cotransduced genes are located close together on the bacterial chromosome. Thus, rates of cotransduction, like rates of cotransformation, give an indication of the physical distances between genes on a bacterial chromosome.

To map genes by using transduction, two bacterial strains with different alleles at several loci are used. The donor strain is infected with phages (**Figure 7.24**), which reproduce within the cells. When the phages have lysed the donor cells, a suspension of the progeny phages is mixed with a recipient strain of bacteria, which is then plated on several different kinds of media to determine the phenotypes of the transducing progeny phages. ▶TRY PROBLEM 20

CONCEPTS

In transduction, bacterial genes become packaged into a viral coat, are transferred to another bacterium by the virus, and become incorporated into the bacterial chromosome by crossing over. Bacterial genes can be mapped with the use of generalized transduction.

✓ CONCEPT CHECK 6

In gene mapping using generalized transduction, bacterial genes that are cotransduced are

a. far apart on the bacterial chromosome.
b. on different bacterial chromosomes.
c. close together on the bacterial chromosome.
d. on a plasmid.

CONNECTING CONCEPTS

Three Methods for Mapping Bacterial Genes

Three methods of mapping bacterial genes have now been outlined: (1) interrupted conjugation mapping, (2) transformation mapping, and (3) transduction mapping. These methods have important similarities and differences.

Mapping with interrupted conjugation is based on the time required for genes to be transferred from one bacterium to another by means of cell-to-cell contact. The key to this technique is that the bacterial chromosome itself is transferred, so the order of genes and the time required for their transfer provide information about the positions of the genes on the chromosome. In contrast with other mapping methods, the distance between genes is measured not in recombination frequencies, but in units of time required for genes to be transferred. Here, the basic unit of mapping is a minute.

In gene mapping with transformation, DNA from the donor strain is isolated, broken up, and mixed with the recipient strain. Some fragments pass into the recipient cells, where the transformed DNA may recombine with the bacterial chromosome. The unit of transfer here is a random fragment of the chromosome. Loci that are close together on the donor chromosome tend to be on the same DNA fragment, so the rates of cotransformation provide information about the relative positions of genes on the chromosome.

Transduction mapping also relies on the transfer of genes between bacteria that differ in two or more traits, but here, the vehicle of gene transfer is a bacteriophage. In a number of respects, transduction mapping is similar to transformation mapping. Small fragments of DNA are carried by the phage from donor to recipient bacteria, and the rates of cotransduction, like the rates of cotransformation, provide information about the relative distances between the genes.

All of the methods use a common strategy for mapping bacterial genes. The movement of genes from donor to recipient is

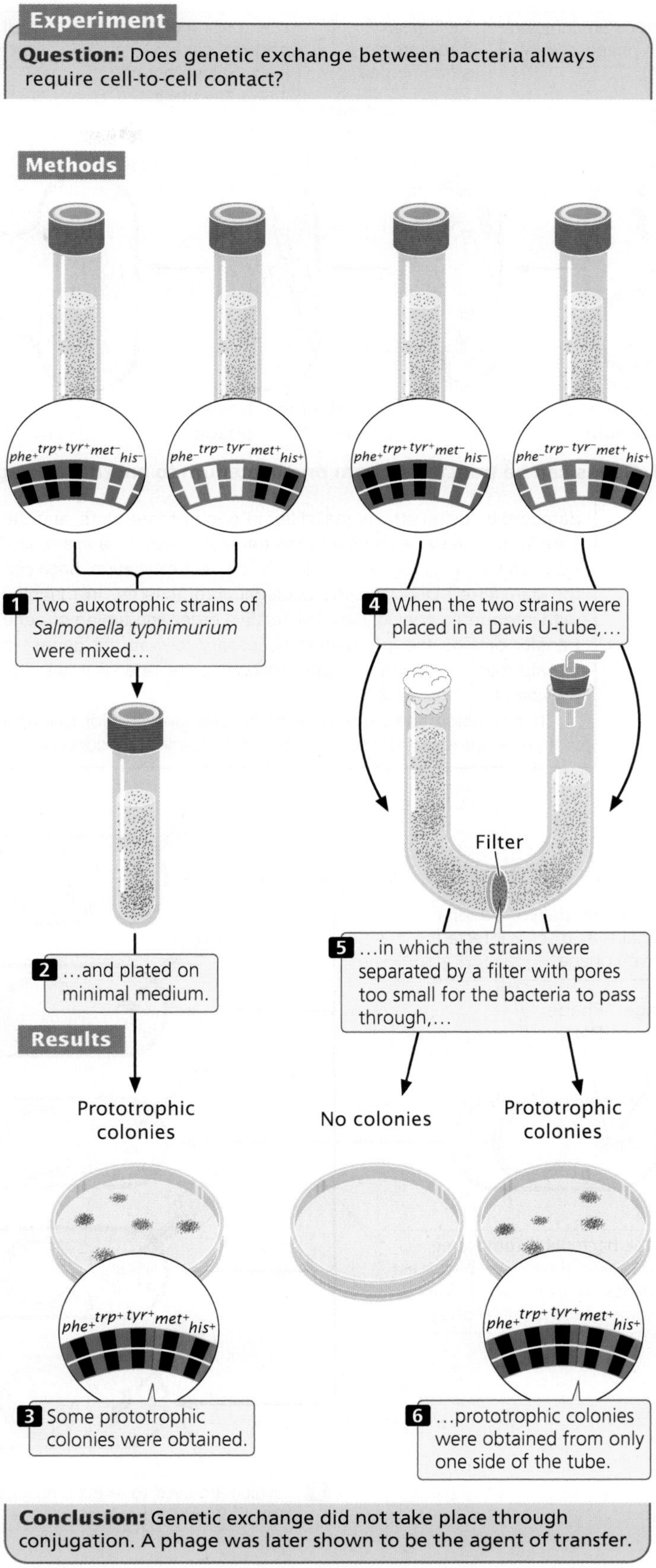

7.22 The Lederberg and Zinder experiment.

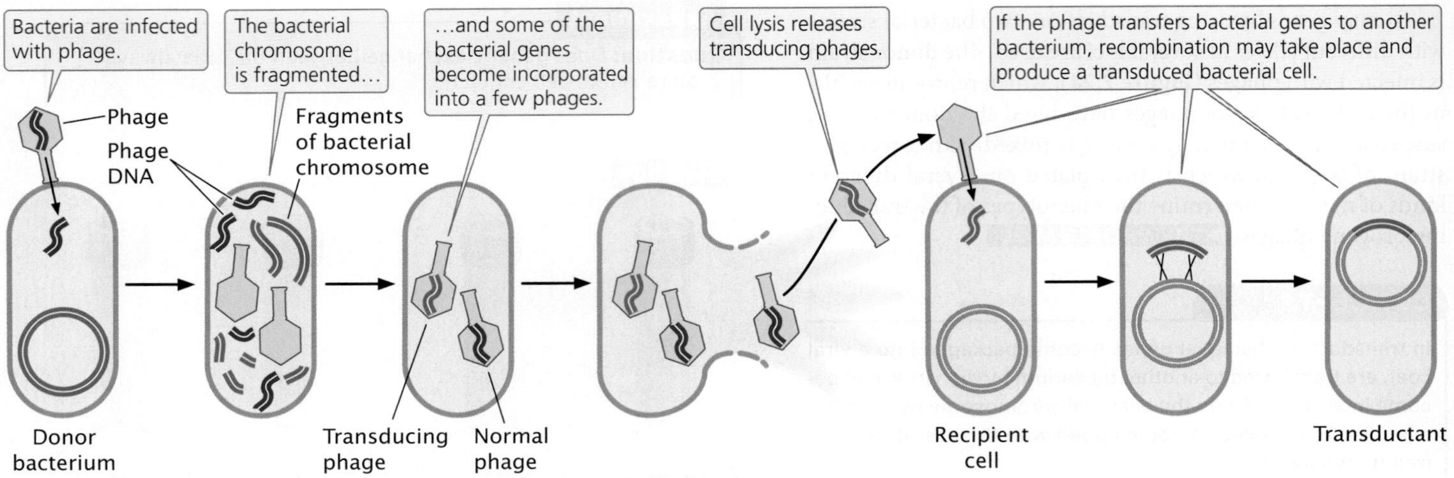

7.23 Genes can be transferred from one bacterium to another through generalized transduction.

detected by using strains that differ in two or more traits, and the transfer of one gene relative to the transfer of others is examined. Additionally, all three methods rely on recombination between the transferred DNA and the bacterial chromosome. In mapping with interrupted conjugation, the relative order and timing of gene transfer provide the information necessary to map the genes; in transformation and transduction mapping, the rate of cotransfer provides this information.

In conclusion, the same basic strategies are used for mapping with interrupted conjugation, transformation, and transduction. The methods differ principally in their mechanisms of transfer: in interrupted conjugation mapping, DNA is transferred though contact between bacteria; in transformation mapping, DNA is transferred as small naked fragments; and, in transduction mapping, DNA is transferred by bacteriophages. Which method is used for mapping often depends on the distances between genes, as the methods have different resolutions. For example, interrupted conjugation has relatively low resolution and can only be used to map genes that are relatively far apart. Transformation and transduction mapping have higher resolutions and are useful when genes are close together.

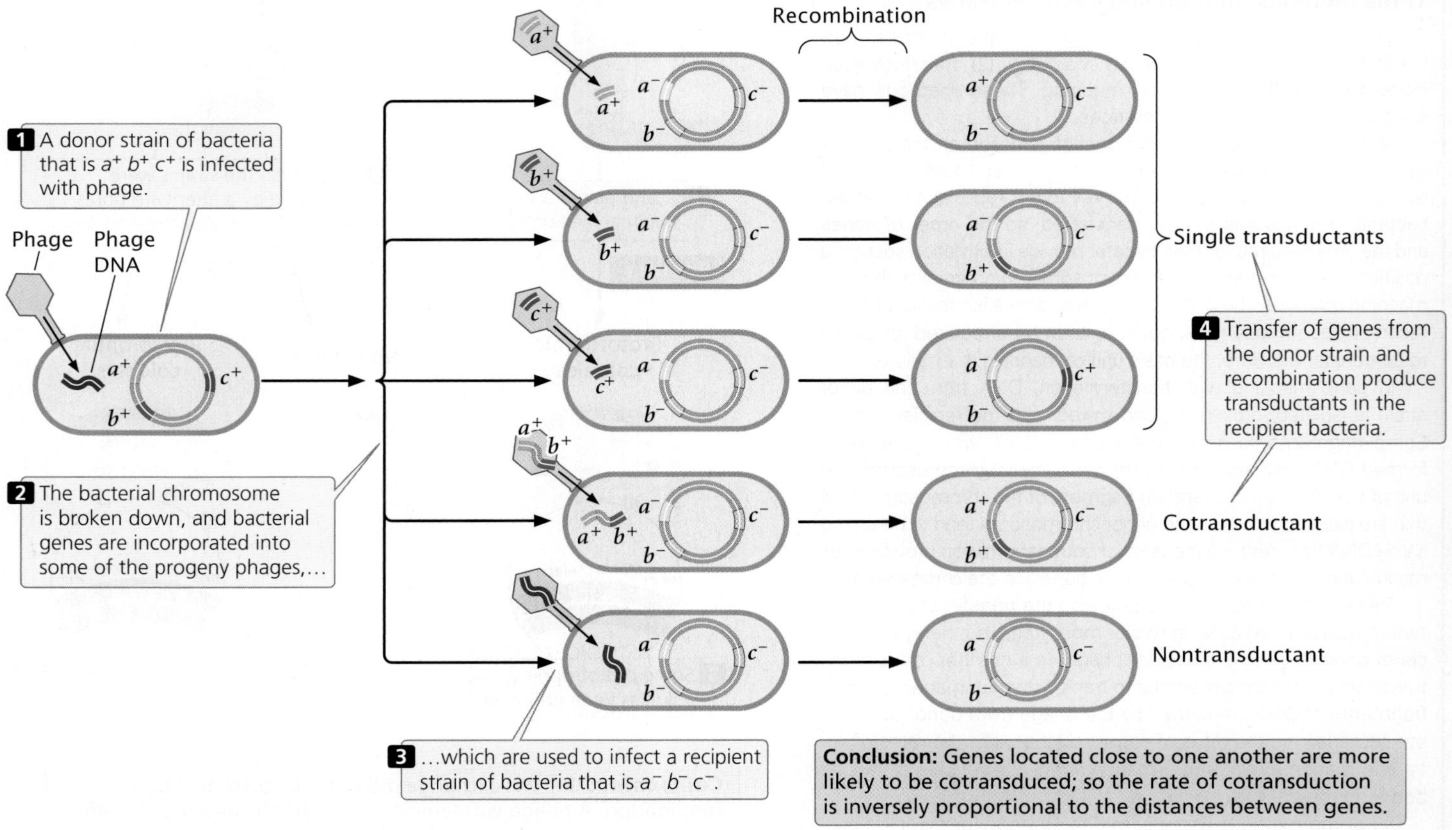

7.24 Generalized transduction can be used to map genes.

Gene Mapping in Phages

The mapping of genes in bacteriophages themselves depends on homologous recombination between phage chromosomes and therefore requires the application of the same principles that are applied to mapping genes in eukaryotic organisms (see Chapter 5). Crosses are made between viruses that differ in two or more genes, and recombinant progeny phages are identified and counted. The proportion of recombinant progeny is then used to estimate the distances between the genes and their linear order on the chromosome.

In 1949, Alfred Hershey and Raquel Rotman examined rates of recombination in the T2 bacteriophage, which has single-stranded DNA. They studied recombination between genes in two strains that differed in plaque appearance and host range (the bacterial strains that the phages could infect). One strain was able to infect and lyse type B *E. coli* cells, but not type B/2 cells (making this strain of phage wild type with a normal host range, h^+), and produced abnormal plaques that were large with distinct borders (r^-). The other strain was able to infect and lyse *both* B *and* B/2 cells (mutant host range, h^-) and produced wild-type plaques that were small with fuzzy borders (r^+).

Hershey and Rotman crossed the $h^+ r^-$ and $h^- r^+$ strains of T2 by infecting type B *E. coli* cells with a mixture of the two strains. They used a high concentration of phages so that most cells could be simultaneously infected by both strains (**Figure 7.25**). Within the bacterial cells, homologous recombination occasionally took place between the chromosomes of the different bacteriophage strains, producing $h^+ r^+$ and $h^- r^-$ chromosomes, which were then packaged into new phage particles. When the cells lysed, the recombinant phages were released, along with the nonrecombinant $h^+ r^-$ phages and $h^- r^+$ phages.

Hershey and Rotman diluted the progeny phages and plated them on a bacterial lawn that consisted of a *mixture* of B and B/2 cells. Phages carrying the h^+ allele (which conferred the ability to infect only B cells) produced a cloudy plaque because the B/2 cells were not lysed. Phages carrying the h^- allele produced a clear plaque because all the cells within the plaque were lysed. The r^+ phages produced small plaques, whereas the r^- phages produced large plaques. The genotypes of these progeny phages could therefore be determined by the appearance of the plaque (see Figure 7.25 and **Table 7.4**).

TABLE 7.4 Progeny phages produced from $h^- r^+ \times h^+ r^-$

Phenotype	Genotype
Clear and small	$h^- r^+$
Cloudy and large	$h^+ r^-$
Cloudy and small	$h^+ r^+$
Clear and large	$h^- r^-$

Experiment

Question: How can we determine the position of a gene on a phage chromosome?

Method

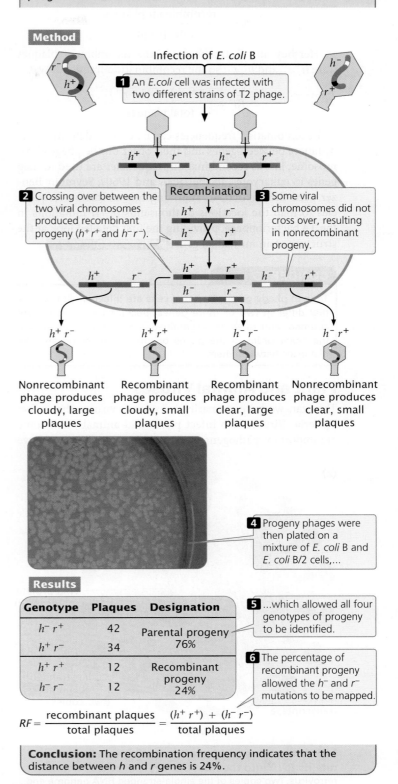

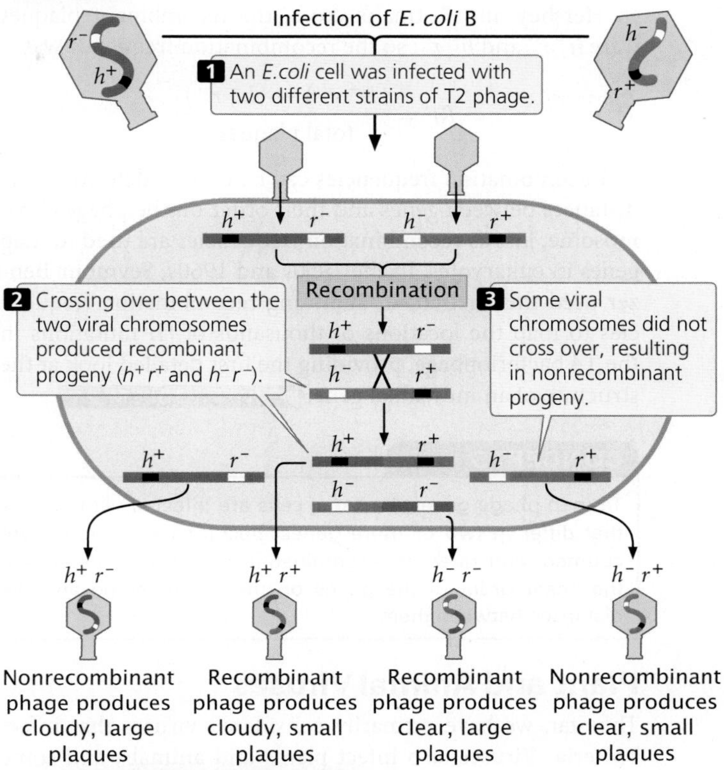

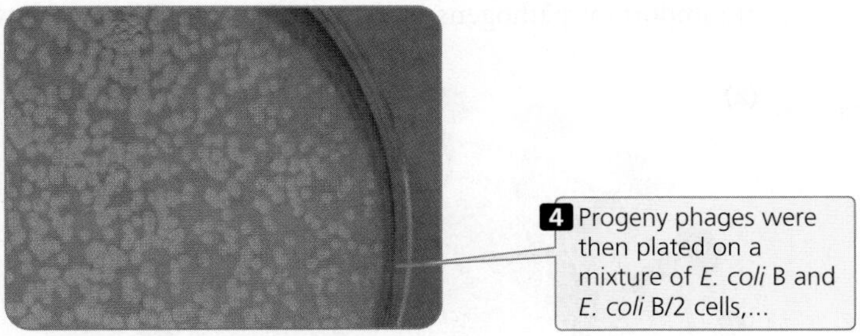

4 Progeny phages were then plated on a mixture of E. coli B and E. coli B/2 cells,...

Results

Genotype	Plaques	Designation
$h^- r^+$	42	Parental progeny 76%
$h^+ r^-$	34	
$h^+ r^+$	12	Recombinant progeny 24%
$h^- r^-$	12	

5 ...which allowed all four genotypes of progeny to be identified.

6 The percentage of recombinant progeny allowed the h^- and r^- mutations to be mapped.

$$RF = \frac{\text{recombinant plaques}}{\text{total plaques}} = \frac{(h^+ r^+) + (h^- r^-)}{\text{total plaques}}$$

Conclusion: The recombination frequency indicates that the distance between *h* and *r* genes is 24%.

7.25 Hershey and Rotman developed a technique for mapping viral genes. [Photograph: Courtesy Steven R. Spilatro.]

In this type of phage cross, the recombination frequency (RF) between the two genes can be calculated by using the following formula:

$$RF = \frac{\text{recombinant plaques}}{\text{total plaques}}$$

In Hershey and Rotman's cross, the recombinant plaques were $h^+ r^+$ and $h^- r^-$; so the recombination frequency was

$$RF = \frac{(h^+ r^+) + (h^- r^-)}{\text{total plaques}}$$

Recombination frequencies can be used to determine the distances between genes and their order on the phage chromosome, just as recombination frequencies are used to map genes in eukaryotes. In the 1950s and 1960s, Seymour Benzer used this method of analyzing recombination frequencies to map the locations of thousands of *rII* mutations in the T4 bacteriophage, providing the first detailed look at the structure of an individual gene. TRY PROBLEM 21

CONCEPTS

To map phage genes, bacterial cells are infected with viruses that differ in two or more genes. Recombinant plaques are counted, and rates of recombination are used to determine the linear order of the genes on the chromosome and the distances between them.

Plant and Animal Viruses

Thus far, we have primarily considered viruses that infect bacteria. Viruses also infect plants and animals, and some are important pathogens in these organisms. What we have

(a)

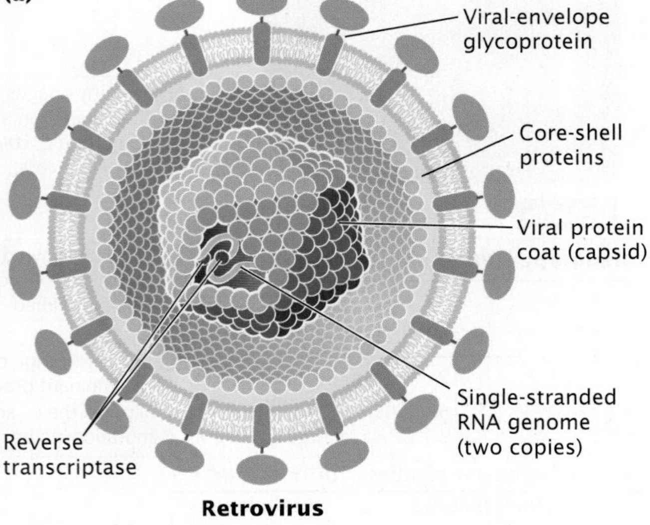

(b)

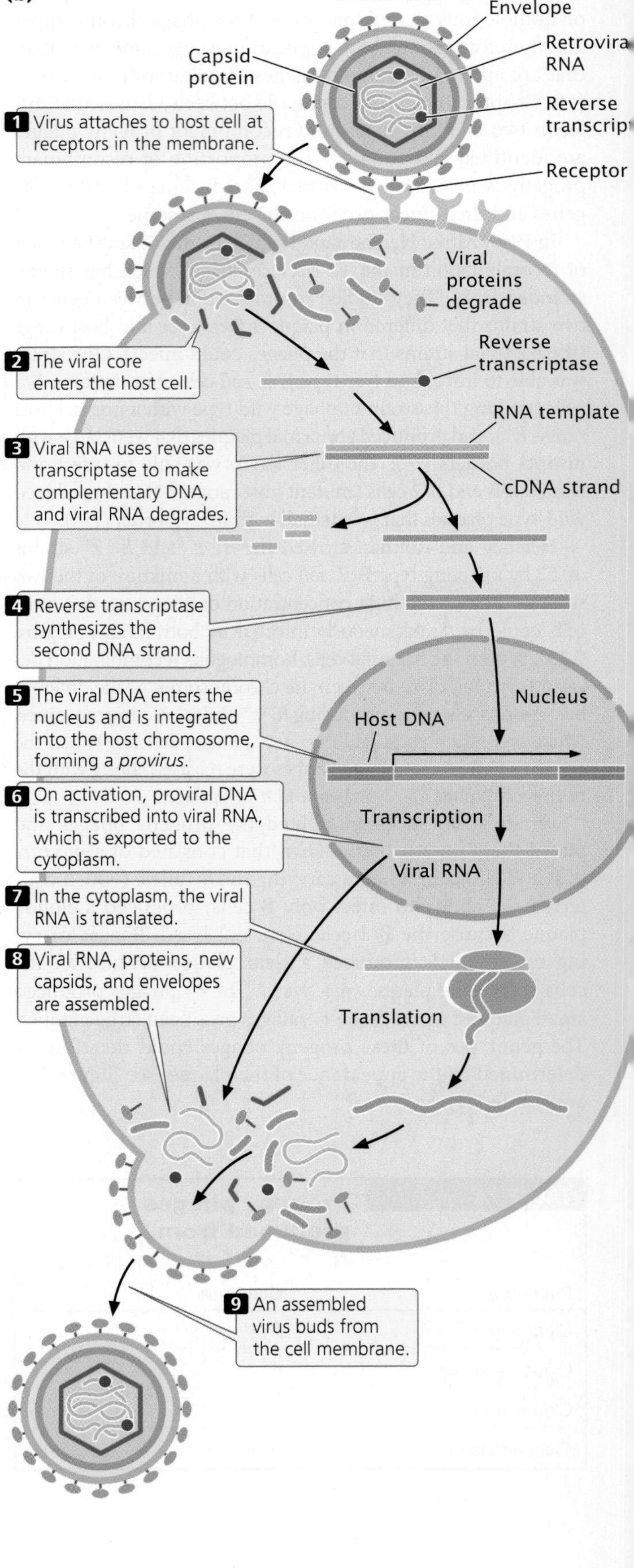

7.26 A retrovirus uses reverse transcription to incorporate its RNA into the host DNA. (a) Structure of a typical retrovirus. Two copies of the single-stranded RNA genome and the reverse transcriptase enzyme are shown enclosed within a protein capsid. The capsid is surrounded by a viral envelope that is studded with viral glycoproteins. (b) The retrovirus life cycle.

learned about bacteriophages has important implications for viruses that infect more complex organisms.

Viral genomes may be encoded in either DNA or RNA and may be double-stranded or single-stranded. Double-stranded DNA viruses include: adenoviruses, which are responsible for gastroenteritis and some types of pneumonia; herpes viruses, which causes genital herpes and cold sores; papilloma viruses, which are associated with some cases of cervical cancer (see chapter 23); and numerous viruses that infect other vertebrate and invertebrate animals. Single-stranded DNA viruses include parvovirus (parvo), which is highly infectious and sometimes lethal in dogs.

Almost all viruses that infect plants have RNA genomes. RNA is also the genetic material of some human viruses, including those that cause colds, influenza, polio, and AIDS. The medical and economic importance of RNA viruses has encouraged their study.

RNA viruses capable of integrating into the genomes of their hosts, much as temperate phages insert themselves into bacterial chromosomes, are called **retroviruses** (**Figure 7.26a**). Because the retroviral genome is RNA, whereas that of the host is DNA, a retrovirus must produce **reverse transcriptase**, an enzyme that synthesizes complementary DNA (cDNA) from either an RNA or a DNA template. A retrovirus uses reverse transcriptase to copy its RNA genome into a single-stranded DNA molecule, and the reverse transcriptase enzyme—or sometimes the host DNA polymerase—copies this single-stranded DNA, creating a double-stranded DNA molecule. The DNA copy of the viral genome then integrates into the host chromosome. A viral genome incorporated into the host chromosome is called a **provirus**. The provirus is replicated by host enzymes when the host chromosome is duplicated (**Figure 7.26b**).

When conditions are appropriate, the provirus undergoes transcription to produce numerous copies of the original RNA genome. This RNA encodes viral proteins and serves as genomic RNA for new viral particles. As these new viruses escape the cell, they collect patches of the cell membrane to use as their envelopes.

All known retroviral genomes have three genes in common: *gag*, *pol*, and *env*, each encoding a precursor protein that is cleaved into two or more functional proteins. The *gag* gene encodes proteins that make up the viral protein coat. The *pol* gene encodes reverse transcriptase and an enzyme, called **integrase**, that inserts the viral DNA into the host chromosome. The *env* gene encodes the glycoproteins that appear on surface of the virus.

Some retroviruses contain **oncogenes** (see Chapter 16) that may stimulate cell division and cause the formation of tumors. The first retrovirus to be isolated, the Rous sarcoma virus, was originally recognized by its ability to produce connective-tissue tumors (sarcomas) in chickens.

THINK-PAIR-SHARE Question 8

Human Immunodeficiency Virus and AIDS

An example of a retrovirus is human immunodeficiency virus (HIV), which causes acquired immune deficiency syndrome (AIDS). AIDS was first recognized in 1982, when a number of homosexual males in the United States began to exhibit symptoms of a new immune system–deficiency disease. In that year, Robert Gallo proposed that AIDS was caused by a retrovirus; eventually, research by Gallo, Luc Montagnier, Francoise Barre-Sinoussi, Jay Levy, and others identified HIV as the causative agent of AIDS. Between 1983 and 1984, as the AIDS epidemic became widespread, the HIV retrovirus was isolated from people with the disease. AIDS is now known to be caused by two different immunodeficiency viruses, HIV-1 and HIV-2, which together have infected more than 60 million people worldwide. Of those infected, 30 million have died. Most cases of AIDS are caused by HIV-1, which now has a global distribution; HIV-2 is primarily found in western Africa.

HIV illustrates the importance of genetic recombination in viral evolution. Studies of the DNA sequences of HIV and other retroviruses reveal that HIV-1 is closely related to the simian immunodeficiency virus found in chimpanzees (SIV_{cpz}). Many wild chimpanzees in Africa are infected with SIV_{cpz}, although it doesn't cause AIDS-like symptoms in these animals. SIV_{cpz} is itself a hybrid that resulted from recombination between a retrovirus found in the red-capped mangabey (a monkey) and a retrovirus found in the greater spot-nosed monkey (**Figure 7.27**). Apparently, one or more chimpanzees became infected with both viruses; recombination between the viruses produced SIV_{cpz}, which was then transmitted to humans through contact with infected chimpanzees. In humans, SIV_{cpz} underwent significant evolution to become HIV-1, which then spread throughout the world to produce the AIDS epidemic. Several independent transfers of SIV_{cpz} to humans gave rise to different strains of HIV-1. In the case of HIV-1 groups O and P, SIV_{cpz} first jumped to gorillas, and then passed from gorillas to humans. HIV-2 evolved from a different retrovirus, SIV_{sm}, found in sooty mangabeys.

HIV is transmitted by sexual contact between humans and through any type of blood-to-blood contact, such as that caused by the sharing of dirty needles by drug users. It can also be transmitted between mother and child during pregnancy and after pregnancy in breast milk. Until screening tests were developed to identify HIV-infected blood, transfusions and clotting factors used to treat hemophilia were sources of infection as well.

HIV principally attacks a class of blood cells called helper T lymphocytes, or simply helper T cells (**Figure 7.28**). HIV enters a helper T cell, undergoes reverse transcription, and integrates into the chromosome. The virus reproduces rapidly, destroying the T cell as new virus particles escape from the cell. Because helper T cells are central to immune function, people with AIDS have a diminished immune response; most

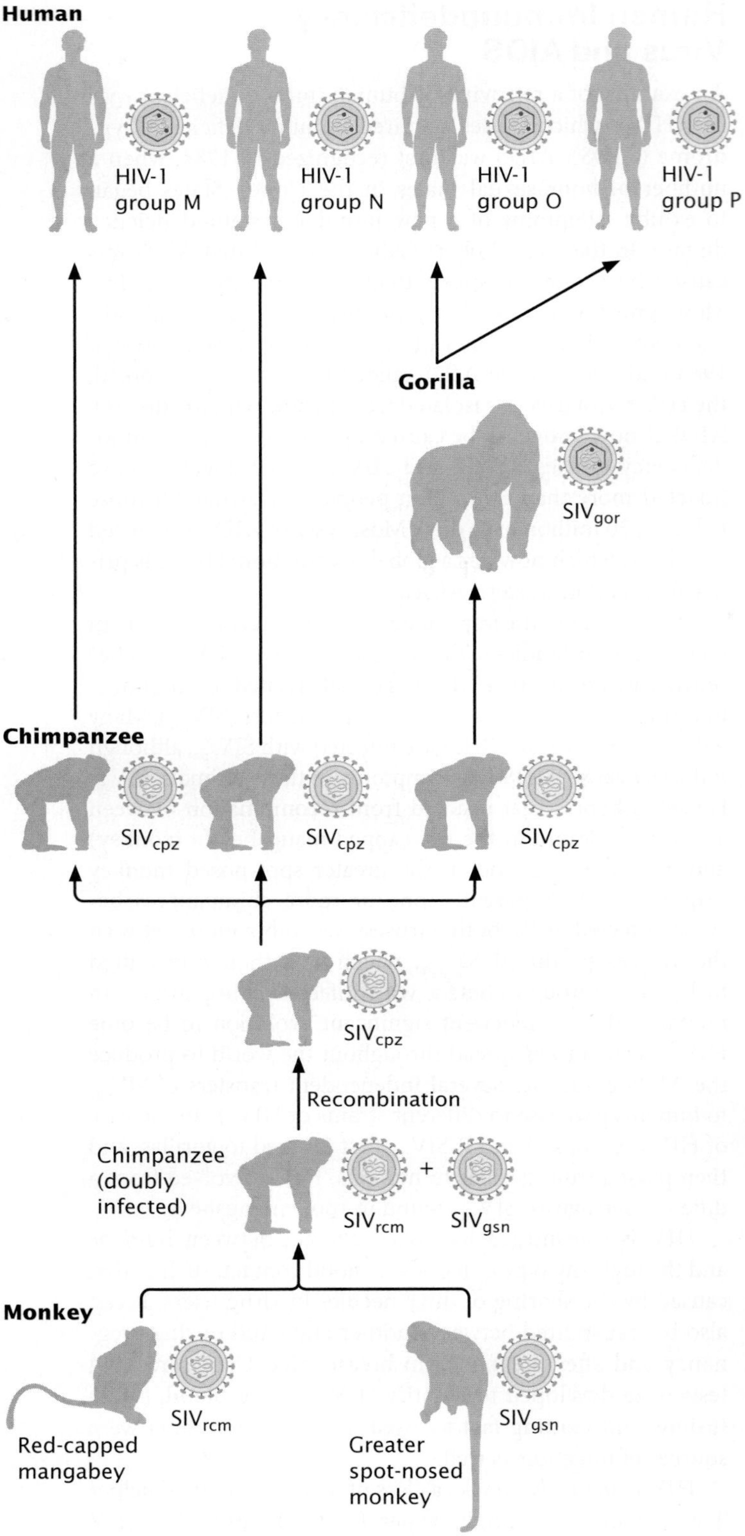

7.27 HIV-1 evolved from a similar virus (SIV_{cpz}) found in chimpanzees and was transmitted to humans. SIV_{cpz} arose from recombination between retroviruses from red-capped mangabeys and from greater spot-nosed monkeys that took place in chimpanzees.

people with the disease die of secondary infections that develop because they have lost the ability to fight off pathogens.

The HIV genome is 9749 nucleotides long and carries *gag*, *pol*, *env*, and six other genes that regulate the life cycle of the virus. HIV's reverse transcriptase is very error prone, giving the virus a high mutation rate and allowing it to evolve rapidly, even within a single host. This rapid evolution makes the development of an effective vaccine against HIV particularly difficult. Genetic variation within the human population also affects the virus. To date, more than 10 loci in humans that affect HIV infection and the progression of AIDS have been identified.

CONCEPTS

A retrovirus is an RNA virus that integrates into its host's chromosome by making a DNA copy of its RNA genome by reverse transcription. Human immunodeficiency virus, the causative agent of AIDS, is a retrovirus. It evolved from related retroviruses found in other primates.

✔ CONCEPT CHECK 7

What enzyme is used by a retrovirus to make a DNA copy of its genome?

Influenza

Influenza demonstrates how rapid changes in a pathogen can arise through the recombination of its genetic material. Influenza, commonly called flu, is a respiratory disease caused by influenza viruses. In the United States, from 5% to 20% of the population is infected with influenza annually, and though most cases are mild, an estimated 36,000 people die from influenza-related causes each year. At certain times, particularly when new strains of influenza virus enter the

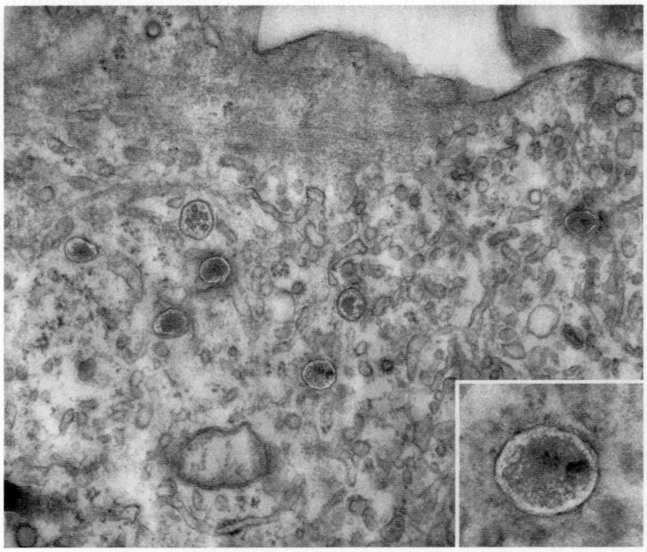

7.28 HIV principally attacks helper T lymphocytes. This electron micrograph shows a T cell (green) infected with HIV (orange). [Thomas Deerinck, NCMIR/Science Source.]

human population, there are worldwide epidemics (called pandemics); for example, in 1918, the Spanish flu virus killed an estimated 20 million to 100 million people worldwide.

Influenza viruses are RNA viruses that infect birds and mammals. The three main types are influenza A, influenza B, and influenza C. Most cases of the common flu are caused by influenza A and B. Influenza A is divided into subtypes based on the types of two proteins, hemagglutinin (HA) and neuraminidase (NA), found on the surface of the virus. The HA and NA proteins affect the ability of the virus to enter host cells and the host organism's immune response to infection. There are 16 types of HA and 9 types of NA, which can exist in a virus in different combinations. For example, common strains of influenza circulating in humans today are H1N1 and H3N2 (**Table 7.5**), along with several strains of influenza B. Most of the different subtypes of influenza A are found in birds.

Although influenza is an RNA virus, it is not a retrovirus: its genome is not copied into DNA and incorporated into the host chromosome like that of a retrovirus. The influenza viral genome consists of seven or eight pieces of RNA that are enclosed in a viral envelope. Each piece of RNA encodes one or two of the virus's proteins. The virus enters a host cell by attaching to specific receptors on the cell membrane. After the viral particle has entered the cell, the viral RNA is released, copied, and translated into viral proteins. Viral RNA molecules and viral proteins are then assembled into new viral particles, which exit the cell and infect additional cells.

One of the dangers of the influenza virus is that it evolves rapidly, so that new strains appear frequently. Influenza evolves in two ways. First, each strain continually changes through mutations arising in the viral RNA. The enzyme that copies the RNA is especially prone to making mistakes, so new mutations are continually introduced into the viral genome. This type of continual change is called **antigenic drift**. Occasionally, major changes in the viral genome take place through **antigenic shift**, in which genetic material from different strains is combined in a process called reassortment. If a host is simultaneously infected with two different strains, the RNAs of both strains may be replicated within the cell and RNA segments from two different stains may be incorporated into the same viral particle, creating a new strain. For example, in 2002, reassortment between the H1N1 and the H3N2 subtypes created a new H1N2 strain that contained the hemagglutinin from H1N1 and the neuraminidase from H3N2. New strains produced by antigenic shift are responsible for most pandemics because no one has immunity to the radically different virus that is produced.

Birds harbor the most different strains of influenza A, but humans are not easily infected with bird influenza. The appearance of new strains in humans is thought to arise most often from viruses that reassort in pigs, which can be infected by viruses from both humans and birds. In 2009, a new strain of H1N1 influenza (called swine flu) emerged in Mexico and quickly spread throughout the world. This virus arose from a series of reassortment events that combined gene sequences from human, bird, and pig influenza viruses to produce the new H1N1 virus (**Figure 7.29**). Farming practices that raise pigs and birds in close proximity may facilitate reassortment among avian, swine, and human strains of influenza.

TABLE 7.5 Strains of influenza virus responsible for major flu pandemics

Year	Influenza pandemic	Strain
1918	Spanish flu	H1N1
1957	Asian flu	H2N2
1968	Hong Kong flu	H3N2
2009	Swine flu	H1N1

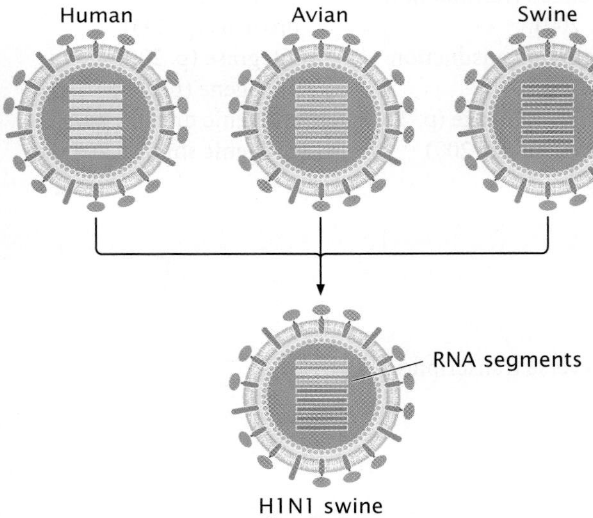

7.29 New strains of influenza virus are created by the reassortment of genetic material from different strains. A new H1N1 virus (swine flu virus) that appeared in 2009 contained material from avian, swine, and human viruses.

CONCEPTS

Influenza is caused by RNA viruses. New strains of influenza appear through antigenic shift, in which new viral genomes are created through the reassortment of RNA molecules of different strains.

CONCEPTS SUMMARY

- Bacteria and viruses are well suited to genetic studies: they are small, have a small haploid genome, undergo rapid reproduction, and produce large numbers of progeny through asexual reproduction.
- The bacterial genome normally consists of a single, circular molecule of double-stranded DNA. Plasmids are small pieces of bacterial DNA that can replicate independently of the bacterial chromosomes.
- DNA may be transferred between bacteria by means of conjugation, transformation, and transduction.
- Conjugation is the transfer of genetic material from one bacterial cell to another. It is controlled by an episome called the F factor. The time it takes for individual genes to be transferred during conjugation provides information about the order of the genes, and the distances between them, on the bacterial chromosome.
- Bacteria take up DNA from the environment through the process of transformation. Frequencies of the cotransformation of genes provide information about the physical distances between chromosomal genes.
- The bacterium *E. coli* is an important model genetic organism that has the advantages of small size, rapid reproduction, and a small genome.
- Viruses are replicating structures with DNA or RNA genomes that may be double stranded or single stranded, linear or circular.
- Bacterial genes become incorporated into phage coats and are transferred to other bacteria by phages through the process of transduction. Rates of cotransduction can be used to map bacterial genes.
- Phage genes can be mapped by infecting bacterial cells with two different phage strains and counting the number of recombinant plaques produced by the progeny phages.
- A number of viruses have RNA genomes. Retroviruses encode a reverse transcriptase enzyme used to make a DNA copy of the viral genome, which then integrates into the host genome as a provirus.
- HIV is a retrovirus that is the causative agent for AIDS.
- Influenza is caused by RNA influenza viruses that evolve both by small changes taking place through mutation (antigenic drift) and by major changes taking place through the reassortment of genetic material from different strains.

IMPORTANT TERMS

minimal medium (p. 185)
complete medium (p. 185)
colony (p. 185)
plasmid (p. 187)
episome (p. 187)
F (fertility) factor (p. 187)
conjugation (p. 188)
transformation (p. 188)
transduction (p. 188)
pili (singular, pilus) (p. 190)
competent (p. 195)
transformant (p. 196)
cotransformation (p. 196)
horizontal gene transfer (p. 197)
virus (p. 199)
virulent phage (p. 199)
temperate phage (p. 199)
prophage (p. 199)
plaque (p. 200)
generalized transduction (p. 200)
specialized transduction (p. 200)
transducing phage (p. 200)
transductant (p. 200)
cotransduction (p. 200)
retrovirus (p. 205)
reverse transcriptase (p. 205)
provirus (p. 205)
integrase (p. 205)
oncogene (p. 205)
antigenic drift (p. 207)
antigenic shift (p. 207)

ANSWERS TO CONCEPT CHECKS

1. d
2. b
3. a
4. *his* and *leu*
5. b
6. c
7. Reverse transcriptase

WORKED PROBLEMS

Problem 1

DNA from a strain of bacteria with genotype $a^+ \ b^+ \ c^+ \ d^+ \ e^+$ was isolated and used to transform a strain of bacteria that was $a^- \ b^- \ c^- \ d^- \ e^-$. The transformed cells were tested for the presence of donated genes. The following genes were cotransformed:

c^+ and d^+ a^+ and d^+ b^+ and e^+ c^+ and e^+

What is the order of genes *a*, *b*, *c*, *d*, and *e* on the bacterial chromosome?

Solution Strategy

What information is required in your answer to the problem?

The order of genes *a*, *b*, *c*, *d*, and *e* on the bacterial chromosome.

What information is provided to solve the problem?

- The donor cells were a^+ b^+ c^+ d^+ e^+ and the recipient cells were a^- b^- c^- d^- e^-.
- The combinations of genes that were cotransformed.

For help with this problem, review:

Transformation in Bacteria in Section 7.2.

Solution Steps

In this transformation experiment, gene c^+ is cotransformed with both gene e^+ and gene d^+, but genes e^+ and d^+ are not cotransformed; therefore, the *c* locus must be between the *d* and *e* loci:

Recall: The rate at which genes are cotransformed is inversely proportional to the distance between them: genes that are close together are frequently cotransformed, whereas genes that are far apart are rarely cotransformed.

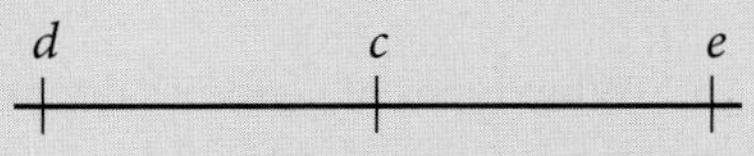

Gene e^+ is also cotransformed with gene b^+, so the *e* and *b* loci must be located close together. Locus *b* could be on either side of locus *e*. To determine whether locus *b* is on the same side of *e* as locus *c*, we look to see whether genes b^+ and c^+ are cotransformed. They are not; so locus *b* must be on the opposite side of *e* from *c*:

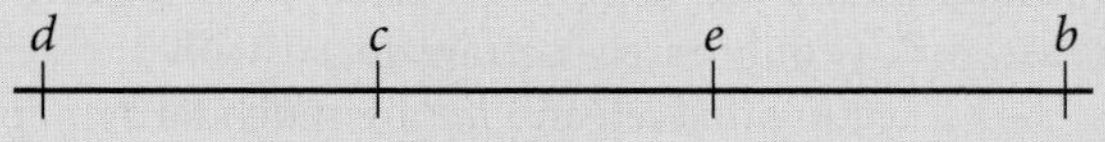

Gene a^+ is cotransformed with gene d^+, so they must be located close together. If locus *a* were located on the same side of *d* as locus *c*, then genes a^+ and c^+ would be cotransformed. Because these genes display no cotransformation, locus *a* must be on the opposite side of locus *d*:

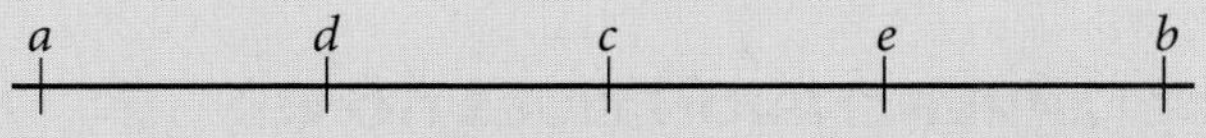

Problem 2

Consider three genes in *E. coli*: thr^+ (the ability to synthesize threonine), ara^+ (the ability to metabolize arabinose), and leu^+ (the ability to synthesize leucine). All three of these genes are close together on the *E. coli* chromosome. Phages are grown in a thr^+ ara^+ leu^+ strain of bacteria (the donor strain). The phage lysate is collected and used to infect a strain of bacteria that is thr^- ara^- leu^-. The recipient bacteria are then tested on selective medium lacking leucine. Bacteria that grow and form colonies on this medium (leu^+ transductants) are then replica plated on medium lacking threonine and on medium lacking arabinose to see which are thr^+ and which are ara^+.

Another group of the recipient bacteria is tested on medium lacking threonine. Bacteria that grow and form colonies on this medium (thr^+ transductants) are then replica plated on medium lacking leucine and on medium lacking arabinose to see which are ara^+ and which are leu^+.
Results from these experiments are as follows:

Selected gene	Cells with cotransduced genes (%)
leu^+	3 thr^+
	76 ara^+
thr^+	3 leu^+
	0 ara^+

How are the loci arranged on the chromosome?

Solution Strategy

What information is required in your answer to the problem?

The order of genes *thr*, *leu*, and *ara* on the bacterial chromosome.

What information is provided to solve the problem?

- The genes are located close together on the *E. coli* chromosome.
- The donor strain is thr^+ ara^+ leu^+ and the recipient strain is thr^- ara^- leu^-.
- The percentage of cells with cotransduced genes.

For help with this problem, review:

Transduction: Using Phages to Map Bacterial Genes in Section 7.3.

Solution Steps

Notice that, when we select for *leu*$^+$ (the top half of the table), most of the selected cells are also *ara*$^+$. This finding indicates that the *leu* and *ara* genes are usually cotransduced, and are therefore located close together. In contrast, *thr*$^+$ is only rarely cotransduced with *leu*$^+$, indicating that *leu* and *thr* are much farther apart. On the basis of these observations, we know that *leu* and *ara* are closer together than are *leu* and *thr*, but we don't yet know the order of the three genes—whether *thr* is on the same side of *ara* as *leu* or on the opposite side, as shown here:

Hint: Genes located close together are more likely to be co-transduced than are genes located far apart.

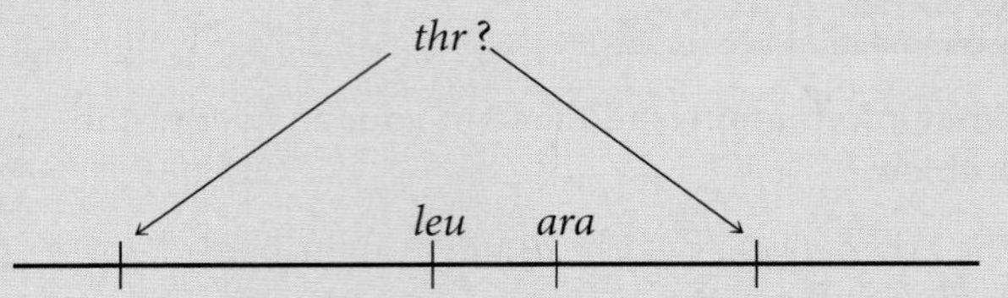

Notice that, although the cotransduction frequency for *thr* and *leu* is 3%, no *thr*$^+$ *ara*$^+$ cotransductants are observed. This finding indicates that *thr* is closer to *leu* than to *ara*, and therefore *thr* must be to the left of *leu*, as shown here:

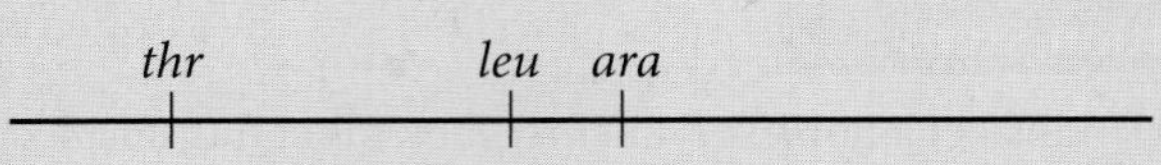

Hint: We can determine the position of *thr* with respect to the other two genes by looking at the cotransduction frequencies when *thr*$^+$ is selected.

COMPREHENSION QUESTIONS

Section 7.1

1. What is the difference between complete medium and minimal medium? How are complete media and minimal media to which one or more nutrients have been added (selective media) used to isolate auxotrophic mutants of bacteria?

Section 7.2

2. Briefly explain the differences between F$^+$, F$^-$, Hfr, and F′ cells.
3. What types of matings are possible between F$^+$, F$^-$, Hfr, and F′ cells? What outcomes do these matings produce? What is the role of the F factor in conjugation?
4. Explain how interrupted conjugation, transformation, and transduction can be used to map bacterial genes. How are these methods similar and how are they different?

Section 7.3

5. List some of the characteristics that make bacteria and viruses ideal organisms for many types of genetic studies.
6. What types of genomes do viruses have?
7. Briefly describe the differences between the lytic cycle of virulent phages and the lysogenic cycle of temperate phages.
8. Briefly explain how genes in phages are mapped.
9. Briefly describe the genetic structure of a typical retrovirus.
10. What are the evolutionary origins of HIV-1 and HIV-2?
11. Most humans are not easily infected by avian influenza. How, then, do DNA sequences from avian influenza become incorporated into human influenza?

For more questions that test your comprehension of the key chapter concepts, go to **LearningCurve** for this chapter.

APPLICATION QUESTIONS AND PROBLEMS

Section 7.2

*12. John Smith is a pig farmer. For the past five years, Smith has been adding vitamins and low doses of antibiotics to his pig food; he says that these supplements enhance the growth of the pigs. Within the past year, however, several of his pigs died from infections of common bacteria, which failed to respond to large doses of antibiotics. Can you explain the increased rate of mortality due to infection in Smith's pigs? What advice might you offer Smith to prevent this problem in the future?

13. In **Figure 7.8**, what do the red and blue parts of the DNA labeled by balloon 6 represent?

*__14.__ Austin Taylor and Edward Adelberg isolated some new strains of Hfr cells that they then used to map several genes in *E. coli* by using interrupted conjugation (A. L. Taylor and E. A. Adelberg. 1960. *Genetics* 45:1233–1243). In one experiment, they mixed cells of Hfr strain AB-312, which were xyl^+ mtl^+ mal^+ met^+ and sensitive to phage T6, with F^- strain AB-531, which was xyl^- mtl^- mal^- met^- and resistant to phage T6. The cells were allowed to undergo conjugation. At regular intervals, the researchers removed a sample of cells and interrupted conjugation by killing the Hfr cells with phage T6. The F^- cells, which were resistant to phage T6, survived and were then tested for the presence of genes transferred from the Hfr strain. The results of this experiment are shown in the accompanying graph. On the basis of these data, give the order of the *xyl*, *mtl*, *mal*, and *met* genes on the bacterial chromosome and indicate the minimum distances between them.

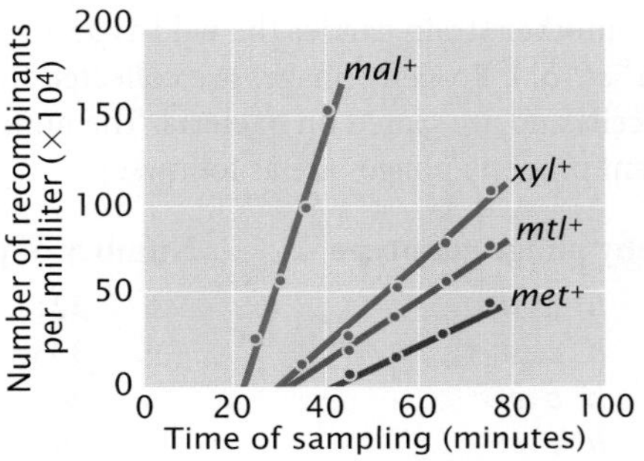

15. DNA from a strain of *Bacillus subtilis* with genotype a^+ b^+ c^+ d^+ e^+ is used to transform a strain with genotype a^- b^- c^- d^- e^-. Pairs of genes are checked for cotransformation, and the following results are obtained:

Pair of genes	Cotransformation	Pair of genes	Cotransformation
a^+ and b^+	No	b^+ and d^+	No
a^+ and c^+	No	b^+ and e^+	Yes
a^+ and d^+	Yes	c^+ and d^+	No
a^+ and e^+	Yes	c^+ and e^+	Yes
b^+ and c^+	Yes	d^+ and e^+	No

On the basis of these results, what is the order of the genes on the bacterial chromosome?

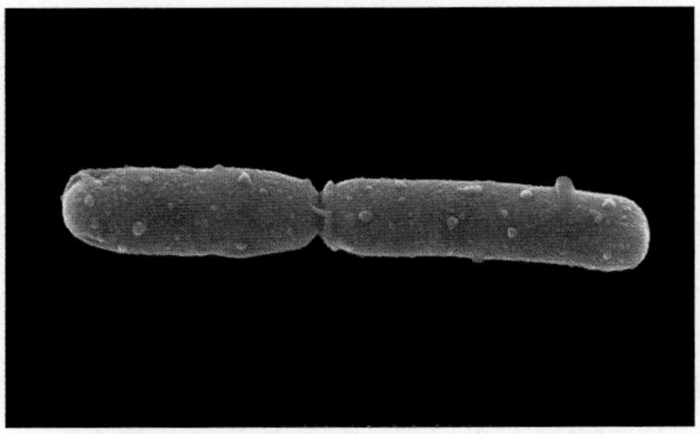

B. subtilis. [Oxford Scientific/Getty Images.]

16. DNA from a bacterial strain that is his^+ leu^+ lac^+ is used to transform a strain that is his^- leu^- lac^-. The following percentages of cells are transformed:

Donor strain	Recipient strain	Genotype of transformed cells	Percentage of all cells
his^+ leu^+ lac^+	his^- leu^- lac^-	his^+ leu^+ lac^+	0.02
		his^+ leu^+ lac^-	0.00
		his^+ leu^- lac^+	2.00
		his^+ leu^- lac^-	4.00
		his^- leu^+ lac^+	0.10
		his^- leu^- lac^+	3.00
		his^- leu^+ lac^-	1.50

a. What conclusions can you draw about the order of these three genes on the chromosome?

b. Which two genes are closest?

*__17.__ Rollin Hotchkiss and Julius Marmur studied transformation in the bacterium *Streptococcus pneumoniae* (R. D. Hotchkiss and J. Marmur. 1954. *Proceedings of the National Academy of Sciences of the United States of America* 40:55–60). They examined four mutations in this bacterium: penicillin resistance (*P*), streptomycin resistance (*S*), sulfanilamide resistance (*F*), and the ability to utilize mannitol (*M*). They extracted DNA from strains of bacteria with different combinations of different mutations and used this DNA to transform wild-type bacterial cells (P^+ S^+ F^+ M^+). The results from one of their transformation experiments are shown below.

Donor DNA	Recipient DNA	Transformants	Percentage of all cells
M S F	M^+ S^+ F^+	M^+ S F^+	4.0
		M^+ S^+ F	4.0
		M S^+ F^+	2.6
		M S F^+	0.41
		M^+ S F	0.22
		M S^+ F	0.0058
		M S F	0.0071

a. Hotchkiss and Marmur noted that the percentage of cotransformation was higher than would be expected on a random basis. For example, the results show that the 2.6% of the cells were transformed into *M* and 4% were transformed into *S*. If the *M* and *S* traits were inherited independently, the expected probability of cotransformation of *M* and *S* (*M S*) would be $0.026 \times 0.04 = 0.001$, or 0.1%. However, they observed 0.41% *M S* cotransformants, four times more than they expected. What accounts for the relatively high frequency of cotransformation of the traits they observed?

b. On the basis of the results, what conclusion can you draw about the order of the *M*, *S*, and *F* genes on the bacterial chromosome?

c. Why is the rate of cotransformation for all three genes (*M S F*) almost the same as that of the cotransformation of *M F* alone?

Section 7.3

***18.** Two mutations that affect plaque morphology in phages (a^- and b^-) have been isolated. Phages carrying both mutations ($a^-\ b^-$) are mixed with wild-type phages ($a^+\ b^+$) and added to a culture of bacterial cells. Subsequent to infection and lysis, samples of the phage lysate are collected and cultured on bacterial cells. The following numbers of plaques are observed:

Plaque phenotype	Number
$a^+\ b^+$	2043
$a^+\ b^-$	320
$a^-\ b^+$	357
$a^-\ b^-$	2134

What is the frequency of recombination between the *a* and *b* genes?

***19.** DATA ANALYSIS T. Miyake and M. Demerec examined proline-requiring mutations in the bacterium *Salmonella typhimurium* (T. Miyake and M. Demerec. 1960. *Genetics* 45:755–762). On the basis of complementation studies, they found four proline auxotrophs: *proA*, *proB*, *proC*, and *proD*. To determine whether *proA*, *proB*, *proC*, and *proD* loci were located close together on the bacterial chromosome, they conducted a transduction experiment. Bacterial strains that were $proC^+$ and had mutations at *proA*, *proB*, or *proD* were used as donors. The donors were infected with bacteriophages, and progeny phages were allowed to infect recipient bacteria with genotype $proC^-\ proA^+\ proB^+\ proD^+$. The recipient bacteria were then plated on a selective medium that allowed only $proC^+$ bacteria to grow. After this, the $proC^+$ transductants were plated on selective media to reveal their genotypes at the other three *pro* loci. The following results were obtained:

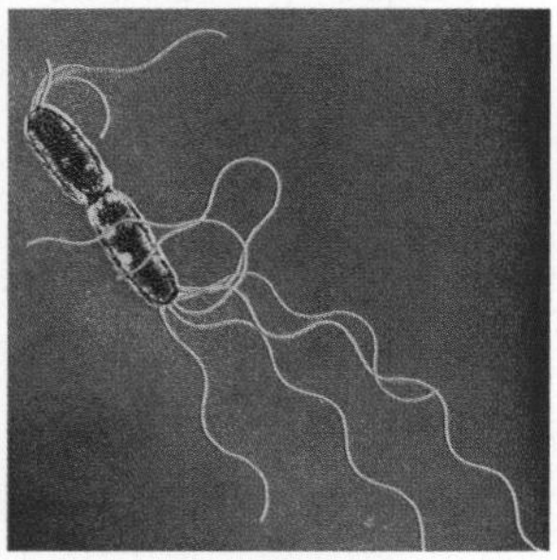

S. typhimurium. [Kwangshin Kim/Science Source.]

Donor genotype	Transductant genotype	Number
$proC^+\ proA^-\ proB^+\ proD^+$	$proC^+\ proA^+\ proB^+\ proD^+$	2765
	$proC^+\ proA^-\ proB^+\ proD^+$	3
$proC^+\ proA^+\ proB^-\ proD^+$	$proC^+\ proA^+\ proB^+\ proD^+$	1838
	$proC^+\ proA^+\ proB^-\ proD^+$	2
$proC^+\ proA^+\ proB^+\ proD^-$	$proC^+\ proA^+\ proB^+\ proD^+$	1166
	$proC^+\ proA^+\ proB^+\ proD^-$	0

a. Why are there no $proC^-$ genotypes among the transductants?

b. Which genotypes represent single transductants and which represent cotransductants?

c. Is there evidence that *proA*, *proB*, and *proD* are located close to *proC*? Explain your answer.

***20.** A geneticist isolates two mutations in a bacteriophage. One mutation causes clear plaques (*c*), and the other produces minute plaques (*m*). Previous mapping experiments have established that the genes responsible for these two mutations are 8 m.u. apart. The geneticist mixes phages with genotype $c^+\ m^+$ and genotype $c^-\ m^-$ and uses the mixture to infect bacterial cells. She collects the progeny phages and cultures a sample of them on plated bacteria. A total of 1000 plaques are observed. What numbers of the different types of plaques ($c^+\ m^+$, $c^-\ m^-$, $c^+\ m^-$, $c^-\ m^+$) should she expect to see?

***21.** *E. coli* cells are simultaneously infected with two strains of phage λ. One strain has a mutant host range, is temperature sensitive, and produces clear plaques (genotype *h st c*); another strain carries the wild-type alleles (genotype $h^+\ st^+\ c^+$). Progeny phages are collected from the lysed cells and are plated on bacteria. The numbers of different progeny phages are as follows:

Progeny phage genotype	Number of plaques
$h^+\ c^+\ st^+$	321
$h\ c\ st$	338
$h^+\ c\ st$	26
$h\ c^+\ st^+$	30
$h^+\ c\ st^+$	106
$h\ c^+\ st$	110
$h^+\ c^+\ st$	5
$h\ c\ st^+$	6

a. Determine the order of the three genes on the phage chromosome.

b. Determine the map distances between the genes.

c. Determine the coefficient of coincidence and the interference (see pp. 134–135 in Chapter 5).

22. A donor strain of bacteria with alleles $a^+\ b^+\ c^+$ is infected with phages to map the donor chromosome using generalized transduction. The phage lysate from the bacterial cells is collected and used to infect a second strain of bacteria that are $a^-\ b^-\ c^-$. Bacteria with the a^+ gene are selected, and the percentages of cells with cotransduced b^+ and c^+ genes are recorded.

Donor	Recipient	Selected gene	Cells with cotransduced gene (%)
$a^+\ b^+\ c^+$	$a^-\ b^-\ c^-$	a^+	25 b^+
		a^+	3 c^+

Is the *b* or *c* gene closer to *a*? Explain your reasoning.

23. For the H1N1 influenza virus shown at the bottom of **Figure 7.29**, viruses from which organism contributed the most RNA to the virus?

CHALLENGE QUESTION

Section 7.2

24. A group of genetics students mix two auxotrophic strains of bacteria: one is *leu*$^+$ *trp*$^+$ *his*$^-$ *met*$^-$ and the other is *leu*$^-$ *trp*$^-$ *his*$^+$ *met*$^+$. After mixing the two strains, they plate the bacteria on minimal medium and observe a few prototrophic colonies (*leu*$^+$ *trp*$^+$ *his*$^+$ *met*$^+$). They assume that some gene transfer has taken place between the two strains. How can they determine whether the transfer of genes is due to conjugation, transduction, or transformation?

THINK-PAIR-SHARE QUESTIONS

Introduction

1. In the ancient world, leprosy was greatly feared, and people with the disease were often ostracized from society. Why was leprosy so feared? Are there modern diseases that evoke similar fears and for which infected people are ostracized? If so, give some examples. What characteristics of a disease produce this response on the part of society? Are any of these modern diseases caused by bacteria or viruses?

2. Genetic analysis of *Mycobacterium leprae,* the bacterium that causes leprosy, reveals that its genome has undergone decay over time, losing DNA and acquiring mutations that make some of its genes nonfunctional. What might be some potential reasons for this evolutionary decay of its genome?

Section 7.1

3. One advantage of using bacteria and viruses for genetic study is the fact that they have haploid genomes. Explain why a haploid genome facilitates genetic analysis.

Section 7.2

4. A series of Hfr strains that have genotype $m^+ n^+ o^+ p^+ q^+ r^+$ are mixed with an F$^-$ strain that has genotype $m^- n^- o^- p^- q^- r^-$. Conjugation is interrupted at regular intervals, and the order of the appearance of genes from the Hfr strain is determined in the recipient cells. The order of gene transfer for each Hfr strain is

Hfr5	$m^+ q^+ p^+ n^+ r^+ o^+$
Hfr4	$n^+ r^+ o^+ m^+ q^+ p^+$
Hfr1	$o^+ m^+ q^+ p^+ n^+ r^+$
Hfr9	$q^+ m^+ o^+ r^+ n^+ p^+$

What is the order of genes on the circular bacterial chromosome? For each Hfr strain, give the location of the F factor in the chromosome and its polarity.

5. Antibiotic resistance genes are often found on R plasmids (see Natural Gene Transfer and Antibiotic Resistance). A likely source of the R plasmids is bacteria that produce the antibiotic. Why would some bacteria produce antibiotics (chemicals that kill bacteria), and why would they carry R plasmids?

6. Bacteria have evolved numerous mechanisms to prevent the invasion of foreign viral DNA (see Bacterial Defense Mechanisms). Yet clearly some bacteria have evolved competence, the ability to take up foreign DNA from the environment. Why do bacteria take up naked DNA from the environment and yet exclude DNA from viruses?

Section 7.3

7. Researchers have recently discovered giant viruses that are 1 μm in length, the same size as some bacterial cells. The genomes of these viruses contain over 2 million base pairs of DNA, which is more DNA than is found in many bacterial genomes, and their genomes contain hundreds—in some cases, thousands—of genes. Given these observations and what you know about viruses, should viruses be considered living or nonliving? Give arguments for and against considering viruses as living organisms.

8. RNA viruses often undergo rapid evolution. What aspects of their biology contribute to their high rate of evolution? What are some consequences of their rapid evolution?

SaplingPlus Self-study tools that will help you practice what you've learned and reinforce this chapter's concepts are available online. Go to www.macmillanlearning.com/PierceGenetics6e.

8 DNA: The Chemical Nature of the Gene

Arctic Treks and Ancient DNA

Greenland is the world's largest island, consisting of over 830,000 square miles (2,200,000 square kilometers), but the vast majority of the land is permanently buried under hundreds of feet of ice. It is one of Earth's most extreme environments. Temperatures along the coast rise a few degrees above freezing during summer days, but then drop to far below zero during much of the winter. The sun moves above the horizon for only a few hours on winter days, and hurricane-force winds, coupled with the extreme cold, create a dangerously inhospitable environment.

In spite of the severe conditions, Arctic peoples have continuously occupied Greenland for almost 5000 years. The earliest inhabitants were the Saqqaq people, who occupied small settlements on Greenland's coast from around 4800 to 2500 years ago, living in small tents and hunting marine mammals and seabirds. The origin of the Saqqaq people had long been a mystery. Did they descend from Native Americans who migrated from Asia into the New World and later moved to Greenland? Or did they descend from the same group that gave rise to the Inuit people, who currently inhabit the New World Arctic? Or, perhaps they originated from yet another group that migrated independently from Asia to Greenland after the ancestors of both the Inuit and Native Americans entered the New World.

Greenland, one of Earth's most extreme environments, was originally settled by the Saqqaq people. The genome from a 4000-year-old male Saqqaq was sequenced in 2010. The remarkable stability of DNA makes analysis of genomes from ancient remains possible. [© Alex Hibbert/age fotostock.]

The mystery of the Saqqaq was solved in 2010, when geneticists determined the entire DNA sequence of a 4000-year-old Saqqaq male—nicknamed Inuk—whose remains were recovered from an archeological site on the western coast of Greenland. Scientists extracted DNA from four hair tufts found in the permafrost. Despite the great age of the sample, they were able to successfully determine Inuk's entire genome sequence, consisting of over 3 billion base pairs of DNA.

By comparing Inuk's DNA with sequences from known populations, the scientists were able to demonstrate that the Saqqaq are most closely related to the Chukchi, a present-day group of indigenous people from Russia. This finding indicates that the Saqqaq originated from hunters who trekked from Siberia eastward across Alaska and Canada to Greenland, arriving in the New World independently of the peoples who gave rise to Native Americans and the Inuit. Further analysis of Inuk's DNA revealed that he was dark-skinned and brown-eyed, had blood type A+, and was probably going bald.

DNA, with its double-stranded spiral, is among the most elegant of all biological molecules. But the double helix is not just a beautiful structure; it also gives DNA incredible stability and permanence, as evidenced by the sequencing of Inuk's 4000-year-old DNA. In an even more remarkable feat, geneticists in 2009 sequenced the entire Neanderthal genome from DNA extracted from 38,000-year-old bones.

THINK-PAIR-SHARE Question 1

This chapter focuses on how DNA was identified as the source of genetic information and how it encodes the genetic instructions for all life. We begin by considering the basic requirements of the genetic material and the history of the study of DNA—how its relation to genes was uncovered and its structure determined. The history of DNA illustrates several important points about the nature of scientific research. As with so many important scientific advances, the structure of DNA and its role as the genetic material were not discovered by any single person, but were gradually revealed over a period of almost 100 years, thanks to the work of many investigators. Our understanding of the relation between DNA and genes was enormously enhanced in 1953, when James Watson and Francis Crick, analyzing data provided by Rosalind Franklin and Maurice Wilkins, proposed a three-dimensional structure for DNA that brilliantly illuminated its role in genetics.

After reviewing the discoveries that led to our current understanding of DNA, we will examine DNA structure. While the structure of DNA is important in its own right, the key genetic concept is the relation between the structure and the function of DNA—how its structure allows it to serve as the genetic material.

8.1 Genetic Material Possesses Several Key Characteristics

Life is characterized by tremendous diversity, but the coding instructions of all living organisms are written in the same genetic language: that of nucleic acids. Surprisingly, the idea that genes are made of nucleic acids was not widely accepted until after 1950. This skepticism was due in part to a lack of knowledge about the structure of deoxyribonucleic acid (DNA). Until the structure of DNA was understood, no one knew how DNA could store and transmit genetic information.

Even before nucleic acids were identified as the genetic material, biologists recognized that, whatever the nature of the genetic material, it must possess four important characteristics:

1. **Genetic material must contain complex information.** First and foremost, genetic material must be capable of storing large amounts of information—instructions for the traits and functions of an organism.
2. **Genetic material must replicate faithfully.** Every organism begins life as a single cell. To produce a complex multicellular creature like you, this single cell must undergo billions of divisions. At each cell division, the genetic instructions must be transmitted to descendant cells with great accuracy. And when organisms reproduce and pass genes on to their progeny, the coding instructions must be copied with fidelity.
3. **Genetic material must encode the phenotype.** Genetic material (the genotype) must have the capacity to be expressed as a phenotype—to code for traits. The product of a gene is often a protein or an RNA molecule, so there must be a mechanism for genetic instructions to be transcribed into RNA and translated into the amino acid sequence of a protein.
4. **Genetic material must have the capacity to vary.** The genetic information must have the ability to vary because different species, and even individual members of a species, differ in their genetic makeup.

THINK-PAIR-SHARE Question 2

CONCEPTS

The genetic material must be capable of carrying large amounts of information, replicating faithfully, and expressing its coding instructions as phenotypes, and it must have the capacity to vary.

✓ CONCEPT CHECK 1

Why was the discovery of the structure of DNA so important for understanding genetics?

8.2 All Genetic Information Is Encoded in the Structure of DNA

Although our understanding of how DNA encodes genetic information is relatively recent, the study of DNA structure stretches back more than 100 years.

Early Studies of DNA

In 1868, Johann Friedrich Miescher graduated from medical school in Switzerland. Influenced by an uncle who believed that the key to understanding disease lay in the chemistry of tissues, Miescher traveled to Tübingen, Germany, to study under Ernst Felix Hoppe-Seyler, an early leader in the emerging field of biochemistry. Under Hoppe-Seyler's direction, Miescher turned his attention to the chemistry of pus, a substance of clear medical importance. Pus contains white blood cells, which have large nuclei. Miescher isolated a novel substance from these nuclei that was slightly acidic and high in phosphorus.

By 1887, several researchers had independently concluded that the physical basis of heredity lies in the nucleus. The material that Miescher had isolated (chromatin) was shown to consist of nucleic acid and proteins, but which of these substances was actually the genetic information was not clear. In the late 1800s, Albrecht Kossel carried out further work on the chemistry of DNA and determined that it contains four nitrogenous bases: adenine, cytosine, guanine, and thymine (abbreviated A, C, G, and T).

Phoebus Aaron Levene later showed that DNA consists of a large number of linked, repeating units called **nucleotides**; each nucleotide contains a sugar, a phosphate, and a base.

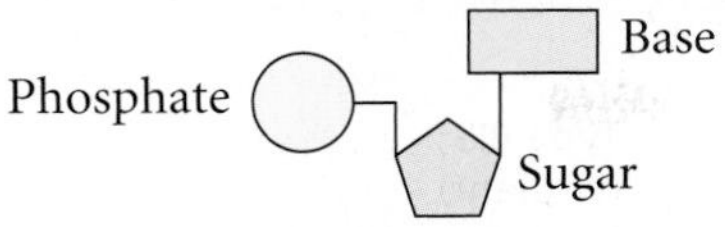

Nucleotide

Levene incorrectly proposed that DNA consists of a series of four-nucleotide units, each containing all four bases—adenine, guanine, cytosine, and thymine—in a fixed sequence. This concept, known as the tetranucleotide hypothesis, implied that the structure of DNA was not variable enough to be the genetic material. The tetranucleotide hypothesis contributed to the idea that protein is the genetic material because the structure of protein, with its 20 different amino acids, could be highly variable.

As additional studies of the chemistry of DNA were completed in the 1940s and 1950s, this notion of DNA as a simple, invariant molecule began to change. Erwin Chargaff and his colleagues carefully measured the amounts of the four bases in DNA from a variety of organisms and found that DNA from different organisms varies greatly in base composition. This finding disproved the tetranucleotide hypothesis. They discovered that, within each species, there is some regularity in the ratios of the bases: the amount of adenine is always equal to the amount of thymine (A = T), and the amount of guanine is always equal to the amount of cytosine (G = C; **Table 8.1**). These findings became known as **Chargaff's rules**. However, the cause of these ratios among the bases wasn't known at the time.

TABLE 8.1 Base composition (percent)*and ratios of bases in DNA from different sources

	Base composition				Ratio		
Source of DNA	A	T	G	C	A/T	G/C	(A + G)/(T + C)
E. coli	26.0	23.9	24.9	25.2	1.09	0.99	1.04
Yeast	31.3	32.9	18.7	17.1	0.95	1.09	1.00
Sea urchin	32.8	32.1	17.7	18.4	1.02	0.96	1.00
Rat	28.6	28.4	21.4	21.5	1.01	1.00	1.00
Human	30.3	30.3	19.5	19.9	1.00	0.98	0.99

*Percentage in moles of nitrogenous constituents per 100 g-atoms of phosphate in hydrolysate corrected for 100% recovery. From E. Chargaff and J. Davidson, Eds. *The Nucleic Acids*, vol. 1. (New York: Academic Press, 1955).

CONCEPTS

Details of the structure of DNA were worked out by a number of scientists. At first, DNA was interpreted as being too regular in structure to carry genetic information, but by the 1940s, DNA from different organisms was shown to vary in its base composition.

DNA as the Source of Genetic Information

While chemists were working out the structure of DNA, biologists were attempting to identify the carrier of genetic information. Mendel identified the basic rules of heredity in 1866, but he had no idea about the physical nature of hereditary information. By the early 1900s, biologists had concluded that genes reside on chromosomes, which were known to contain both DNA and protein. Two sets of experiments, one conducted on bacteria and the other on viruses, provided pivotal evidence that DNA, rather than protein, was the genetic material.

THE DISCOVERY OF THE TRANSFORMING PRINCIPLE An initial step in identifying DNA as the source of genetic information came with the discovery of a phenomenon called *transformation* (see Chapter 7). This phenomenon was first observed in 1928 by Fred Griffith, an English physician whose special interest was the bacterium that causes pneumonia, *Streptococcus pneumoniae*. Griffith had succeeded in isolating several different strains of *S. pneumoniae* (type I, II, III, and so forth). In the virulent (disease-causing) forms of a strain, each bacterium is surrounded by a polysaccharide coat, which makes the bacterial colony appear smooth (S) when grown on an agar plate. Griffith found that these virulent forms occasionally mutated to nonvirulent forms, which lack a polysaccharide coat and produce a rough-appearing colony (R).

Griffith observed that small amounts of living type IIIS bacteria injected into mice caused the mice to develop pneumonia and die; when he examined the dead mice, he found large amounts of type IIIS bacteria in their blood (**Figure 8.1a**). When Griffith injected type IIR bacteria into mice, the mice lived, and no bacteria were recovered from their blood (**Figure 8.1b**). Griffith knew that boiling killed all bacteria and destroyed their virulence; when he injected large amounts of heat-killed type IIIS bacteria into mice, the mice lived, and no type IIIS bacteria were recovered from their blood (**Figure 8.1c**).

The results of these experiments were not unusual. However, Griffith got a surprise when he injected his mice with a small amount of living type IIR bacteria along with a large amount of heat-killed type IIIS bacteria. Because both the type IIR bacteria and the heat-killed type IIIS bacteria were nonvirulent, he expected these mice to live. Surprisingly, 5 days after the injections, the mice became infected with pneumonia and died (**Figure 8.1d**). When Griffith examined blood from the hearts of these mice, he observed live type IIIS bacteria. Furthermore, these bacteria retained their type IIIS characteristics through several generations: the infectivity was heritable.

Griffith concluded that the type IIR bacteria had somehow been transformed, acquiring the genetic virulence of the dead type IIIS bacteria, and that this transformation had produced a permanent genetic change in the bacteria. Although Griffith didn't understand the nature of transformation, he theorized that some substance in the polysaccharide coat of the dead bacteria might be responsible. He called this substance the **transforming principle.** ➤TRY PROBLEM 19

IDENTIFICATION OF THE TRANSFORMING PRINCIPLE At the time of Griffith's report, Oswald Avery was a microbiologist at the Rockefeller Institute. At first, Avery was skeptical of Griffith's results, but after other microbiologists successfully repeated Griffith's experiments with other bacteria, Avery set out to understand the nature of the transforming principle.

After 10 years of research, Avery, Colin MacLeod, and Maclyn McCarty succeeded in isolating and partially purifying the transforming substance. They showed that it had a chemical composition closely matching that of DNA and quite different from that of proteins. Enzymes such as trypsin and chymotrypsin, known to break down proteins, had no effect on the transforming substance. Ribonuclease, an enzyme that destroys RNA, also had no effect. Enzymes capable of destroying DNA, however, eliminated the biological activity of the transforming substance (**Figure 8.2**).

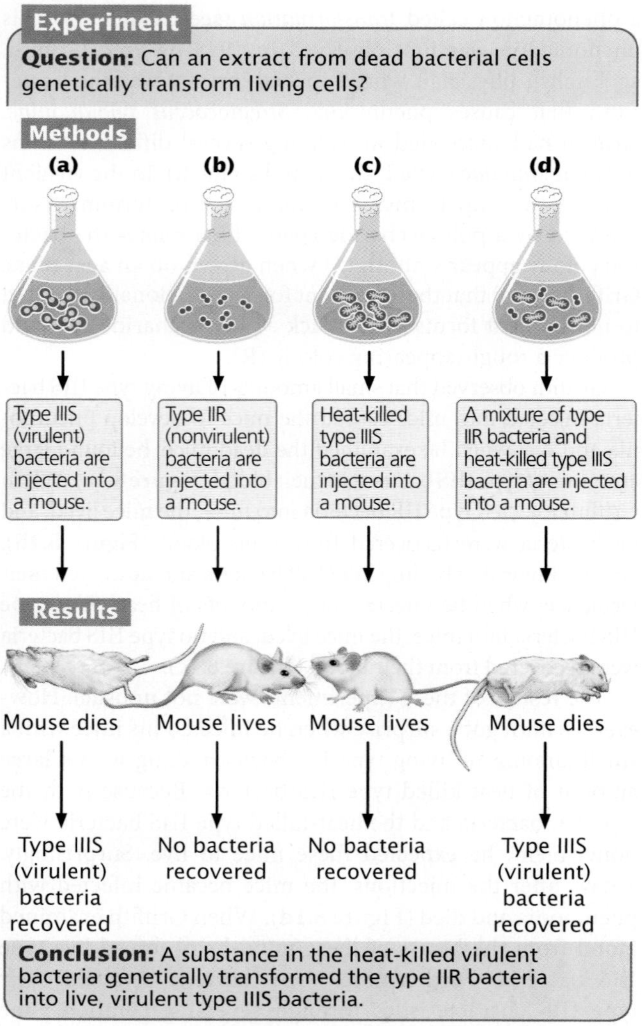

8.1 Griffith's experiments demonstrated transformation in bacteria.

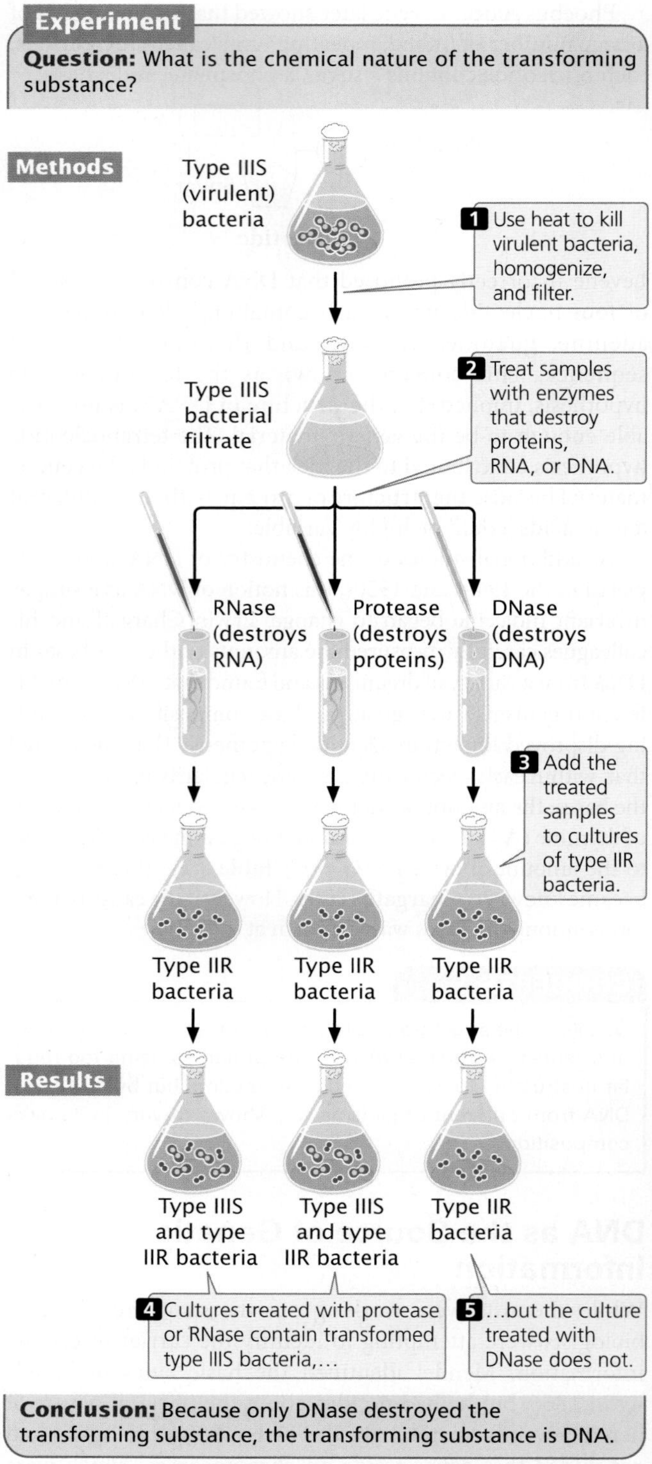

8.2 Avery, MacLeod, and McCarty's experiment revealed the nature of the transforming principle.

Avery, MacLeod, and McCarty showed that the transforming substance precipitated at about the same rate as purified DNA and that it absorbed ultraviolet light at the same wavelengths as DNA. These results, published in 1944, provided compelling evidence that the transforming principle—and therefore genetic information—resides in DNA. However, new theories in science are rarely accepted on the basis of a single experiment, and many biologists continued to prefer the hypothesis that the genetic material is protein.

Phage genome is DNA.

All other parts of the bacteriophage are protein.

CONCEPTS

The process of transformation indicates that some substance—the transforming principle—is capable of genetically altering bacteria. Avery, MacLeod, and McCarty demonstrated that the transforming principle is DNA, providing the first evidence that DNA is the genetic material.

✓ CONCEPT CHECK 2

If Avery, MacLeod, and McCarty had found that samples of heat-killed bacteria treated with RNase and DNase transformed bacteria, but that samples treated with protease did not, what conclusion would they have made?

a. Protease carries out transformation.
b. RNA and DNA are the genetic materials.
c. Protein is the genetic material.
d. RNase and DNase are necessary for transformation.

THE HERSHEY–CHASE EXPERIMENT A second piece of evidence that indicated DNA was the genetic material resulted from a study of the T2 bacteriophage conducted by Alfred Hershey and Martha Chase. The T2 bacteriophage infects the bacterium *Escherichia coli* (**Figure 8.3a**). As we saw in Chapter 7, a phage reproduces by attaching to the outer wall of a bacterial cell and injecting its DNA into the cell, where it replicates and directs the cell to synthesize phage protein. The phage DNA becomes encapsulated within the proteins, producing progeny phages that lyse (break open) the cell and escape (**Figure 8.3b**).

At the time of the Hershey–Chase study (their paper was published in 1952), biologists did not understand exactly how phages reproduce. What they did know was that the T2 phage is approximately 50% protein and 50% DNA, that a phage infects a cell by first attaching to the cell wall, and that progeny phages are ultimately produced within the cell. Because the progeny carry the same traits as the infecting phage, they knew that genetic material from the infecting phage must be transmitted to the progeny, but how this transmission takes place was unknown.

Hershey and Chase designed a series of experiments to determine whether the phage *protein* or the phage *DNA* is transmitted in phage reproduction. To follow the fates of protein and DNA, they used radioactive forms,

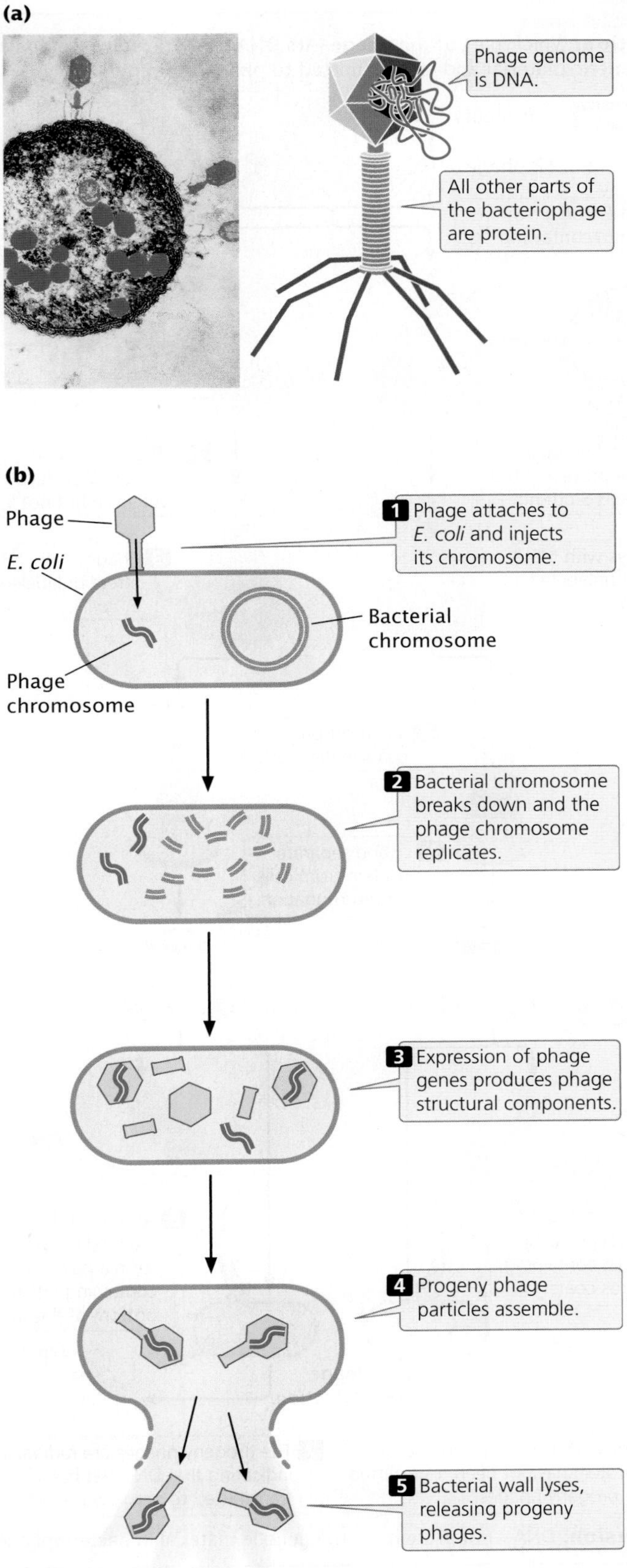

(b)

Phage

E. coli

Phage chromosome

1 Phage attaches to *E. coli* and injects its chromosome.

Bacterial chromosome

2 Bacterial chromosome breaks down and the phage chromosome replicates.

3 Expression of phage genes produces phage structural components.

4 Progeny phage particles assemble.

5 Bacterial wall lyses, releasing progeny phages.

8.3 T2 is a bacteriophage that infects E. coli. (a) T2 phage. (b) Its life cycle. [Micrograph by Lee D. Simon/Science Source.]

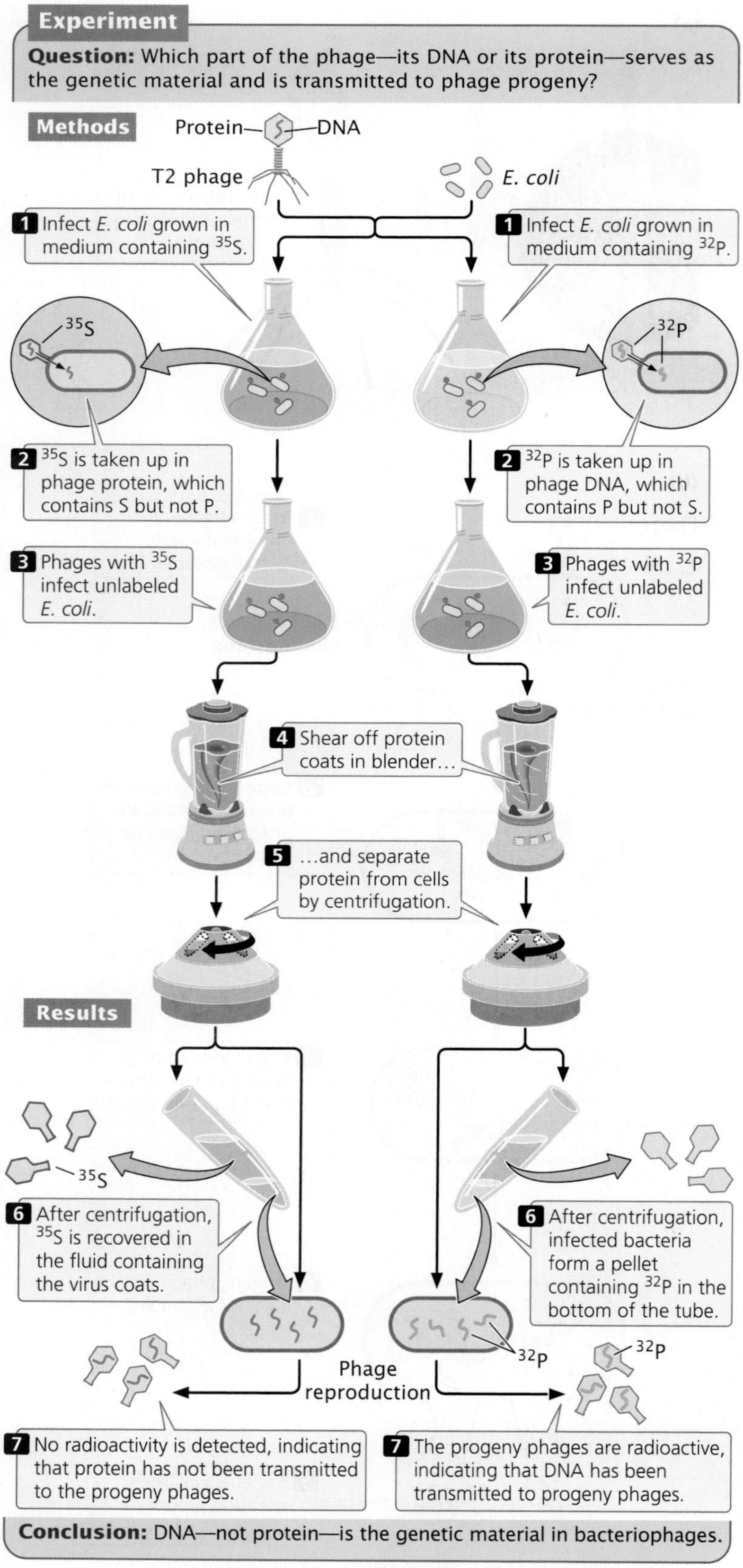

8.4 Hershey and Chase demonstrated that DNA carries the genetic information in bacteriophages.

or **isotopes**, of phosphorus and sulfur. A radioactive isotope can be used as a tracer to identify the location of a specific molecule because any molecule containing the isotope will be radioactive and therefore easily detected. DNA contains phosphorus, but not sulfur. Thus, Hershey and Chase used a radioactive isotope of phosphorus (^{32}P) to follow phage DNA during reproduction. Protein contains sulfur, but not phosphorus, so they used a radioactive isotope of sulfur (^{35}S) to follow the protein.

Hershey and Chase grew one batch of *E. coli* in a medium containing ^{32}P and infected the bacteria with T2 phage so that all the new phages would have DNA labeled with ^{32}P (**Figure 8.4**). They grew a second batch of *E. coli* in a medium containing ^{35}S and infected these bacteria with T2 phage so that all these new phages would have proteins labeled with ^{35}S. Hershey and Chase then infected separate batches of unlabeled *E. coli* with the ^{35}S- and ^{32}P-labeled phages. After allowing time for the phages to infect the cells, they placed the *E. coli* cells in a blender and sheared off the now-empty phage protein coats from the cell walls. They separated out the protein coats and cultured the infected bacterial cells.

When phages labeled with ^{35}S infected the bacteria, most of the radioactivity was detected in the protein coats and little was detected in the cells. Furthermore, when new phages emerged from the cells, they contained almost no ^{35}S (see Figure 8.4). This result indicated that the protein component of a phage does not enter the cell and is not transmitted to progeny phages.

In contrast, when Hershey and Chase infected bacteria with ^{32}P-labeled phages and removed the protein coats, the bacteria were radioactive. Most significantly, after the cells lysed and new progeny phages emerged, many of these phages emitted radioactivity from ^{32}P, demonstrating that DNA from the infecting phages had been passed on to the progeny (see Figure 8.4). These results confirmed that DNA, not protein, is the genetic material of phages.

TRY PROBLEM 36

CONCEPTS

Using radioactive isotopes, Hershey and Chase traced the movement of DNA and protein during phage infection of bacteria. They demonstrated that DNA, not protein, enters the bacterial cell during phage reproduction and that only DNA is passed on to progeny phages.

✓ CONCEPT CHECK 3

Could Hershey and Chase have used a radioactive isotope of carbon instead of ^{32}P? Why or why not?

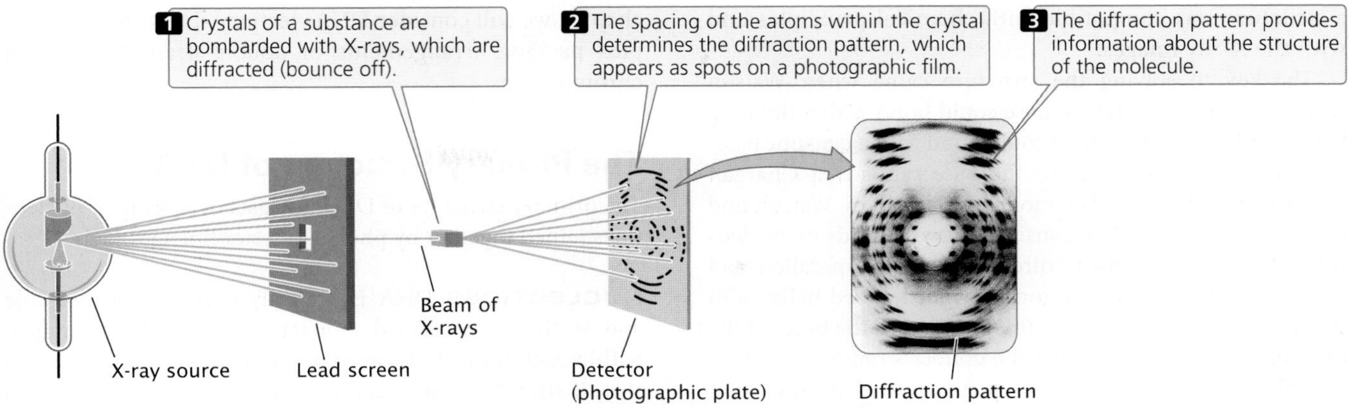

8.5 X-ray diffraction provides information about the structures of molecules. [Photograph: Science Source.]

Watson and Crick's Discovery of the Three-Dimensional Structure of DNA

The experiments on the nature of the genetic material set the stage for one of the most important advances in the history of biology: the discovery of the three-dimensional structure of DNA by James Watson and Francis Crick in 1953. Before Watson and Crick's breakthrough, much of the basic chemistry of DNA had already been determined by Miescher, Kossel, Levene, Chargaff, and others, who had established that DNA consists of nucleotides and that each nucleotide contains a sugar, a base, and a phosphate group. However, how the nucleotides fit together in the three-dimensional structure of the molecule was not at all clear.

In 1947, William Astbury began studying the three-dimensional structure of DNA by using a technique called **X-ray diffraction** (**Figure 8.5**), in which X-rays beamed at a molecule are reflected in specific patterns that reveal aspects of the structure of the molecule. However, his diffraction images did not provide enough resolution to reveal the structure. A research group at King's College in London, led by Maurice Wilkins, also used X-ray diffraction to study DNA. Working in Wilkins's laboratory, Rosalind Franklin obtained strikingly better images of the molecule. However, Wilkins and Franklin's progress in solving the structure was impeded by the personal discord between them.

Watson and Crick investigated the structure of DNA not by collecting new data, but by using all available information about the chemistry of DNA to construct molecular models (**Figure 8.6a**). They used the excellent X-ray diffraction photographs taken by Rosalind Franklin (**Figure 8.6b**), and by applying the laws of structural chemistry, they were able to limit the number of possible structures that DNA could assume. They tested various structures by building models made of wire and metal plates. With their models, they were able to see whether

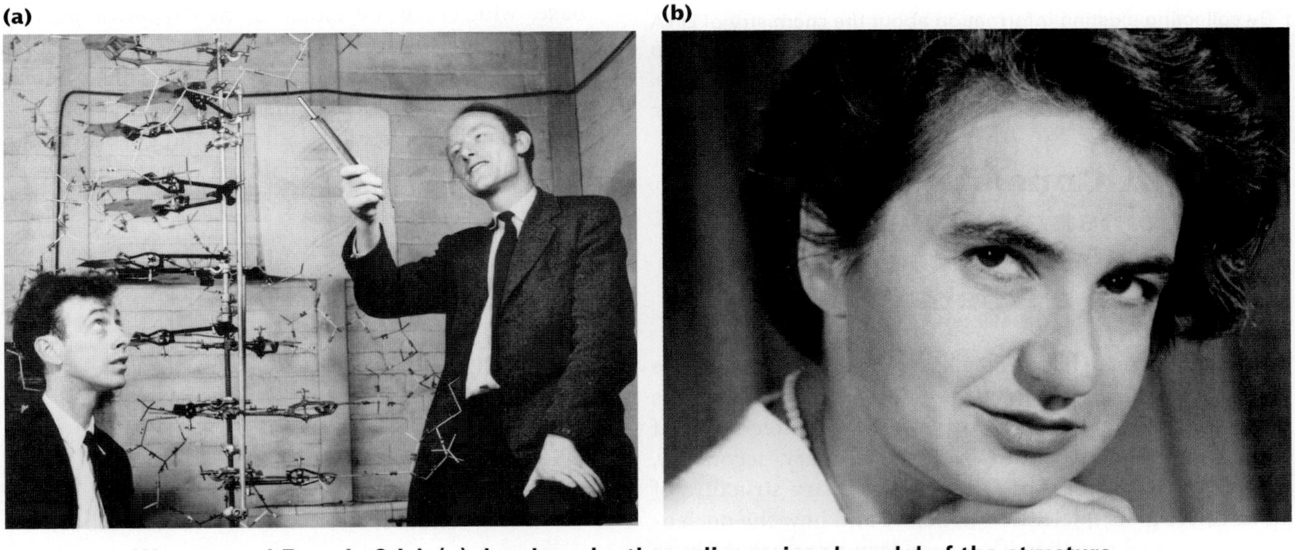

8.6 James Watson and Francis Crick (a) developed a three-dimensional model of the structure of DNA based in part on X-ray diffraction photographs taken by Rosalind Franklin (b). [Part a: A. Barrington/Science Photo Library/Science Source. Part b: National Library of Medicine/Science Source.]

a structure was compatible with chemical principles and with the X-ray images.

The key to solving the structure came when Watson recognized that an adenine base could bond with a thymine base and that a guanine base could bond with a cytosine base; these pairings accounted for the base ratios that Chargaff had discovered earlier. The model developed by Watson and Crick showed that DNA consists of two strands of nucleotides that run in opposite directions (are antiparallel) and wind around each other to form a right-handed helix, with the sugars and phosphates on the outside and the bases in the interior. They recognized that the double-stranded structure of DNA with its specific base pairing provided an elegant means by which the DNA could be replicated. Watson and Crick published an electrifying description of their model in *Nature* in 1953. At the same time, Wilkins and Franklin each published their X-ray diffraction data, which demonstrated that DNA was helical in structure.

Many have called the solving of DNA's structure the most important biological discovery of the twentieth century. For their discovery, Watson and Crick, along with Maurice Wilkins, were awarded the Nobel Prize in Chemistry in 1962. Rosalind Franklin had died of cancer in 1958 and thus could not be considered a candidate for the shared prize, but many scholars and historians believe that she should receive equal credit for solving the structure of DNA.

Following the discovery of DNA's structure, much research focused on how genetic information is encoded within the base sequence and how this information is copied and expressed. Even today, the details of DNA structure and function continue to be the subject of active research.

THINK-PAIR-SHARE Questions 3 and 4

CONCEPTS

By collecting existing information about the chemistry of DNA and building molecular models, Watson and Crick were able to discover the three-dimensional structure of the DNA molecule.

8.3 DNA Consists of Two Complementary and Antiparallel Nucleotide Strands That Form a Double Helix

DNA, though relatively simple in structure, has an elegance and beauty unsurpassed by other large molecules. It is useful to consider the structure of DNA at three levels of increasing complexity, known as the primary, secondary, and tertiary structures of DNA. The primary structure of DNA refers to its nucleotide structure and how the nucleotides are joined together. The secondary structure refers to DNA's stable three-dimensional configuration, the helical structure worked out by Watson and Crick. Later in this chapter, we will consider DNA's tertiary structure, the complex packing arrangements of double-stranded DNA in chromosomes.

The Primary Structure of DNA

The primary structure of DNA consists of a string of nucleotides joined together by phosphodiester linkages.

NUCLEOTIDES DNA is typically a very long molecule and is therefore termed a macromolecule. For example, within each human chromosome is a single DNA molecule that, if stretched out straight, would be several centimeters in length, thousands of times longer than the cell itself. In spite of its large size, DNA has quite a simple structure: it is a polymer—that is, a chain made up of many repeating units linked together. The repeating units of DNA are *nucleotides*, each comprised of three parts: (1) a sugar, (2) a phosphate group, and (3) a nitrogen-containing base.

The sugars of nucleic acids—called pentose sugars—have five carbon atoms, numbered 1′, 2′, 3′, 4′, and 5′ (**Figure 8.7**). The sugars of DNA and RNA are slightly different in structure. RNA's sugar, called **ribose**, has a hydroxyl group (−OH) attached to the 2′-carbon atom, whereas DNA's sugar, or **deoxyribose**, has a hydrogen atom (−H) at this position and therefore contains one oxygen atom fewer overall. This difference gives rise to the names ribonucleic acid (RNA) and *deoxy*ribonucleic acid (DNA). This minor chemical difference is recognized by most of the cellular enzymes that interact with DNA or RNA, thus providing specific functions for each nucleic acid. Furthermore, the additional oxygen atom in the RNA nucleotide makes it more reactive and less chemically stable than DNA. For this reason, DNA is better suited to serve as the long-term carrier of genetic information.

The second component of a nucleotide is its **nitrogenous base**, which may be either of two types: a **purine** or a **pyrimidine** (**Figure 8.8**). Each purine consists of a six-member ring attached to a five-member ring, whereas each pyrimidine consists of a six-member ring only. Both DNA and RNA contain two purines, **adenine** and **guanine** (A and G), which differ in the positions of their double bonds and in the groups attached to the six-member ring. Three pyrimidines

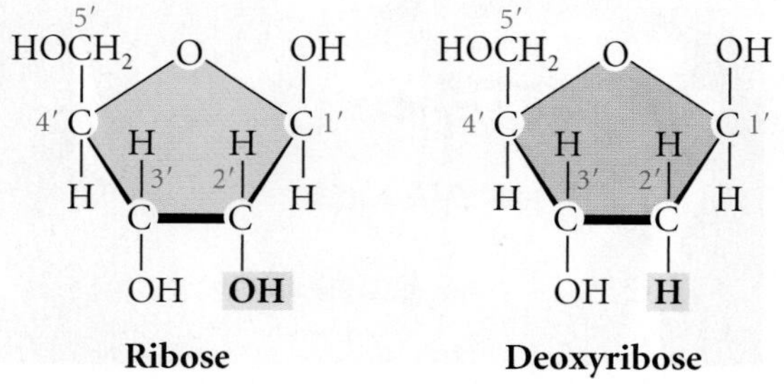

8.7 A nucleotide contains either a ribose sugar (in RNA) or a deoxyribose sugar (in DNA). The carbon atoms are assigned primed numbers.

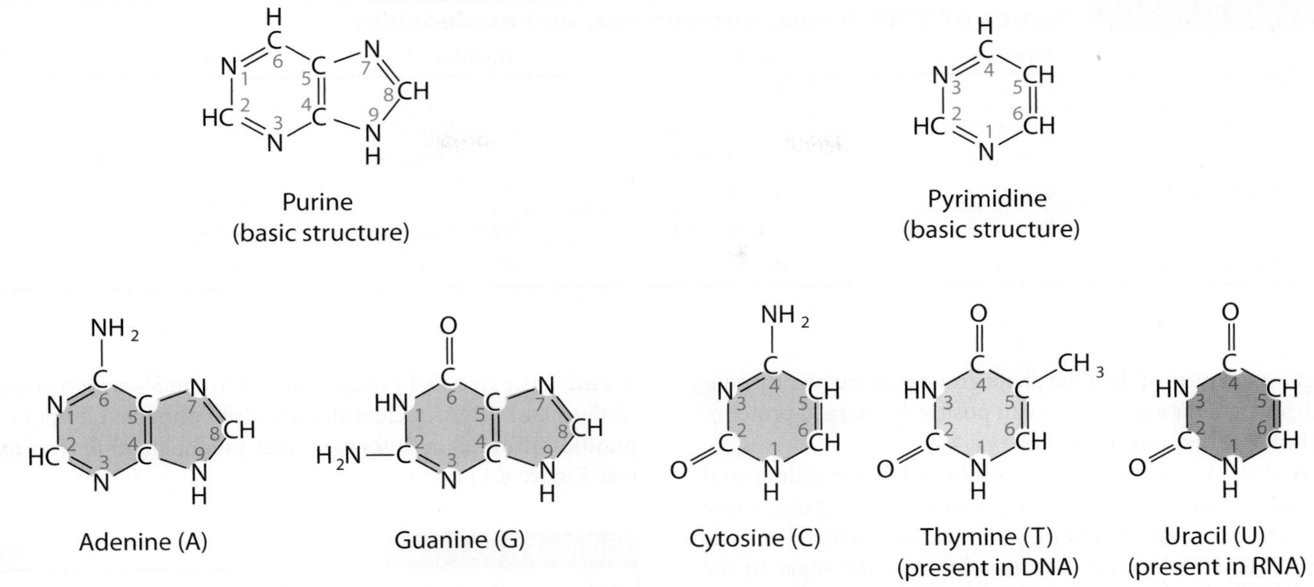

8.8 A nucleotide contains either a purine or a pyrimidine base. The atoms of the rings in the bases are assigned unprimed numbers.

are common in nucleic acids: **cytosine** (C), **thymine** (T), and **uracil** (U). Cytosine is present in both DNA and RNA; however, thymine is restricted to DNA, and uracil is found only in RNA. The three pyrimidines differ in the groups or atoms attached to the carbon atoms of the ring and in the number of double bonds in the ring. In a nucleotide, the nitrogenous base always forms a covalent bond with the 1′-carbon atom of the sugar (see Figure 8.7). A deoxyribose or a ribose sugar and a base together are referred to as a **nucleoside**.

The third component of a nucleotide is the **phosphate group**, which consists of a phosphorus atom bonded to four oxygen atoms (**Figure 8.9**). Phosphate groups are found in every nucleotide and frequently carry a negative charge, which makes DNA acidic. The phosphate group is always bonded to the 5′-carbon atom of the sugar (see Figure 8.7) in a nucleotide.

The DNA nucleotides are properly known as **deoxyribonucleotides**, or deoxyribonucleoside 5′-monophosphates. Because there are four types of bases, there are four different kinds of DNA nucleotides (**Figure 8.10**). The equivalent RNA nucleotides are termed **ribonucleotides**, or ribonucleoside 5′-monophosphates. RNA molecules sometimes contain additional rare bases, which are modified forms of the four common bases. These modified bases will be discussed in more detail when we examine the function of RNA molecules in Chapter 10. The names for DNA bases, nucleotides, and nucleosides are shown in **Table 8.2**.

▶TRY PROBLEM 21

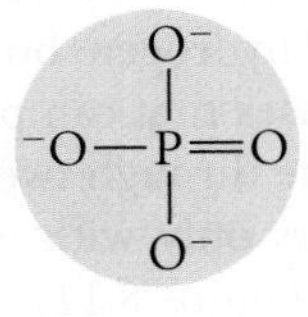

Phosphate

8.9 A nucleotide contains a phosphate group.

CONCEPTS

The primary structure of DNA consists of a string of nucleotides. Each nucleotide consists of a five-carbon sugar, a phosphate group, and a nitrogenous base. There are two types of DNA bases: purines (adenine and guanine) and pyrimidines (thymine and cytosine).

✓ CONCEPT CHECK 4

How do the sugars of RNA and DNA differ?

a. RNA has a six-carbon sugar; DNA has a five-carbon sugar.
b. The sugar of RNA has a hydroxyl group that is not found in the sugar of DNA.
c. RNA contains uracil; DNA contains thymine.
d. DNA's sugar has a phosphorus atom; RNA's sugar does not.

POLYNUCLEOTIDE STRANDS DNA is made up of many nucleotides, which are connected by covalent bonds that link the 5′-phosphate group of one nucleotide to the 3′-hydroxyl group of the next nucleotide (**Figure 8.11**; it should be noted that the structures shown in this illustration are flattened into two dimensions, while the molecule itself is three-dimensional, as shown in Figure 8.12a). These bonds, called **phosphodiester linkages**, are strong covalent bonds; a series of nucleotides linked in this way constitutes a **polynucleotide strand**. The backbone of the polynucleotide strand is composed of alternating sugars and phosphates; the bases project away from the long axis of the strand. The

TABLE 8.2	Names of DNA bases, nucleotides, and nucleosides			
	Adenine	**Guanine**	**Thymine**	**Cytosine**
Base symbol	A	G	T	C
Nucleotide	Deoxyadenosine 5′ monophosphate	Deoxyguanosine 5′ monophosphate	Deoxythymidine 5′ monophosphate	Deoxycytidine 5′ monophosphate
Nucleotide symbol	dAMP	dGMP	dTMP	dCMP
Nucleoside	Deoxyadenosine	Deoxyguanosine	Deoxythymidine	Deoxycytidine
Nucleoside symbol	dA	dG	dT	dC

negative charges of the phosphate groups are frequently neutralized by their association with positively charged proteins, metals, or other molecules.

An important characteristic of the polynucleotide strand is its direction, or polarity. At one end of the strand, a free phosphate group (meaning that it's unattached on one side) is attached to the 5′-carbon atom of the sugar in the nucleotide. This end of the strand is therefore referred to as the **5′ end**. The other end of the strand, referred to as the **3′ end**, has a free OH group attached to the 3′-carbon atom of the sugar. RNA nucleotides are also connected by phosphodiester linkages to form similar polynucleotide strands (see Figure 8.11).

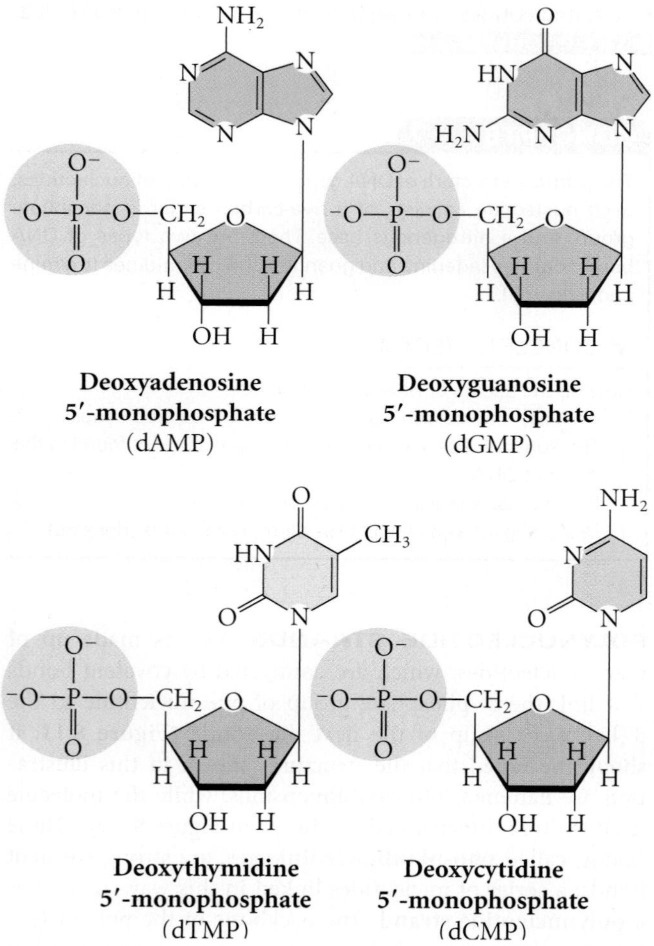

Deoxyadenosine 5′-monophosphate (dAMP) **Deoxyguanosine 5′-monophosphate (dGMP)** **Deoxythymidine 5′-monophosphate (dTMP)** **Deoxycytidine 5′-monophosphate (dCMP)**

8.10 There are four types of DNA nucleotides.

CONCEPTS

The nucleotides of DNA are joined to form polynucleotide strands by phosphodiester linkages that connect the 3′-carbon atom of one nucleotide to the 5′-phosphate group of the next. Each polynucleotide strand has polarity, with a 5′ end and a 3′ end.

Secondary Structures of DNA

The secondary structure of DNA refers to its three-dimensional configuration—its fundamental helical structure. DNA's secondary structure can assume a variety of configurations, depending on its base sequence and the conditions in which it is placed.

THE DOUBLE HELIX A fundamental characteristic of DNA's secondary structure is that it consists of two polynucleotide strands wound around each other: it's a *double* helix. The sugar–phosphate linkages are on the outside of the helix, and the bases are stacked in the interior of the molecule (see Figure 8.11). The two polynucleotide strands run in opposite directions: they are **antiparallel**, which means that the 5′ end of one strand is opposite the 3′ end of the other strand.

The strands are held together by two types of molecular forces. Hydrogen bonds link the bases on opposite strands (see Figure 8.11). These bonds are relatively weak compared with the covalent phosphodiester bonds that connect the sugar and phosphate groups of adjoining nucleotides on the same strand. As we will see, several important functions of DNA require the separation of its two nucleotide strands, and this separation can be readily accomplished because of the relative ease of breaking and reestablishing the hydrogen bonds.

The nature of the hydrogen bond imposes a limitation on the types of bases that can pair. Adenine normally pairs only with thymine through two hydrogen bonds, and cytosine normally pairs only with guanine through three hydrogen bonds (see Figure 8.11). Because three hydrogen bonds form between C and G and only two hydrogen bonds form between A and T, C–G pairing is stronger than

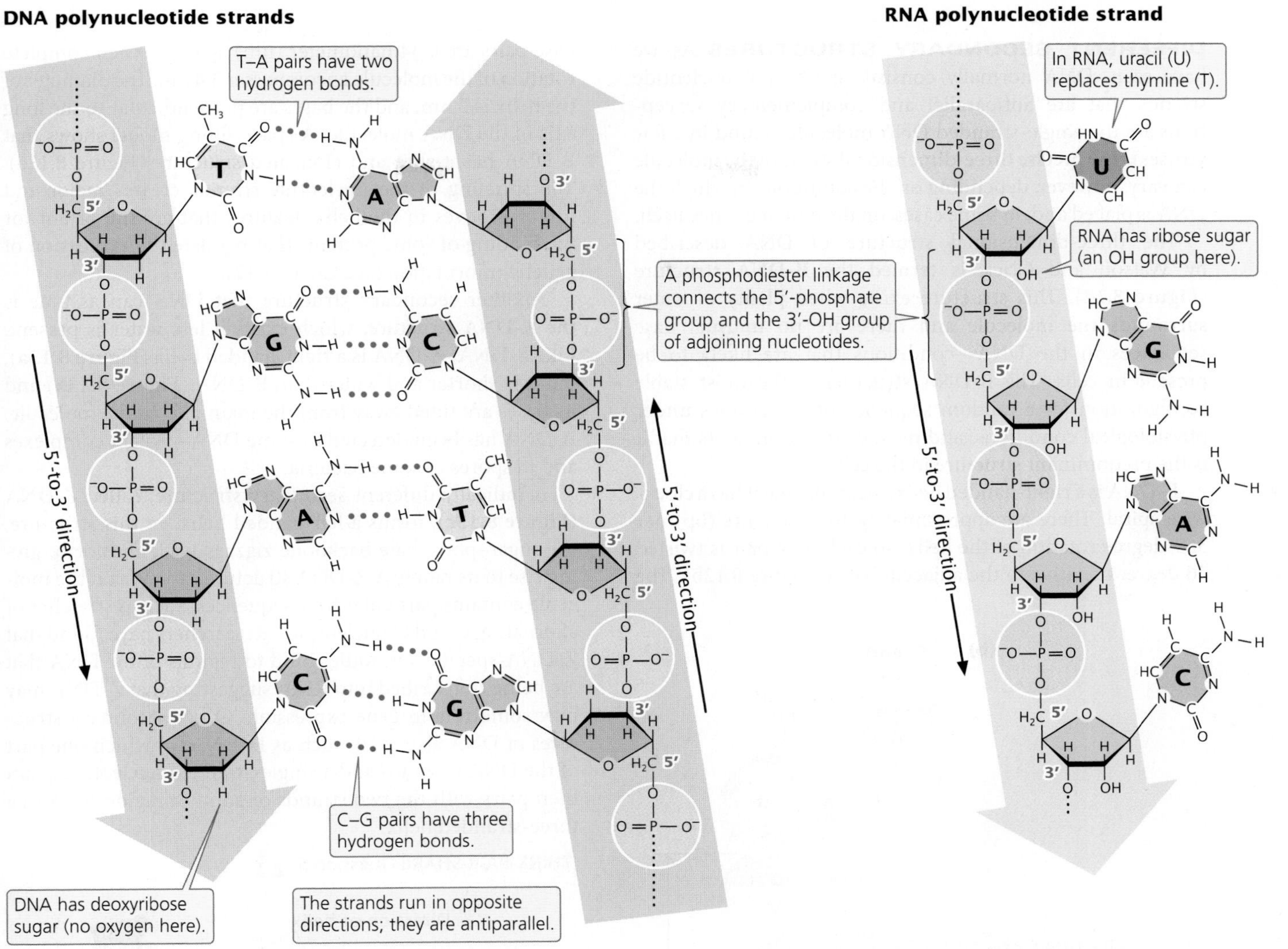

8.11 DNA and RNA are composed of polynucleotide strands. DNA is usually composed of two polynucleotide strands, although single-stranded DNA is found in some viruses.

A–T pairing. The specificity of the base pairing means that wherever there is an A on one strand, there must be a T in the corresponding position on the other strand, and wherever there is a G on one strand, a C must be on the other. The two polynucleotide strands of a DNA molecule are therefore not identical, but rather **complementary DNA strands.** The complementary nature of the two nucleotide strands provides for efficient and accurate DNA replication (as we will see in Chapter 9).

The second force that holds the two DNA strands together is the interaction between the stacked base pairs in the interior of the molecule. Stacking means that adjacent bases are aligned so that their rings are parallel and stack on top of one another. The stacking interactions stabilize the DNA molecule but do not require that any particular base follow another. Thus, the base sequence of the DNA molecule is free to vary, allowing DNA to carry genetic information.

▶ TRY PROBLEMS 25 AND 27

CONCEPTS

DNA consists of two polynucleotide strands. The sugar–phosphate groups of each polynucleotide strand are on the outside of the molecule, and the bases are in the interior. Hydrogen bonding joins the bases of the two strands: guanine pairs with cytosine, and adenine pairs with thymine. The two polynucleotide strands of a DNA molecule are complementary and antiparallel.

✓ CONCEPT CHECK 5

The antiparallel nature of DNA refers to

a. its charged phosphate groups.
b. the pairing of bases on one strand with bases on the other strand.
c. the formation of hydrogen bonds between bases from opposite strands.
d. the opposite direction of the two strands of nucleotides.

DIFFERENT SECONDARY STRUCTURES As we have seen, DNA normally consists of two polynucleotide strands that are antiparallel and complementary (exceptions are the single-stranded DNA molecules found in a few viruses). The precise three-dimensional shape of the molecule can vary, however, depending on the conditions in which the DNA is placed and, in some cases, on the base sequence itself.

The three-dimensional structure of DNA described by Watson and Crick is termed the **B-DNA** structure (**Figure 8.12**). This structure exists when plenty of water surrounds the molecule and there are no unusual base sequences in the DNA—conditions that are likely to be present in cells. The B-DNA structure is the most stable configuration for a random sequence of nucleotides under physiological conditions, and most evidence suggests that it is the predominant structure in the cell.

B-DNA is a right-handed helix, meaning that it has a clockwise spiral. There are approximately 10 base pairs (bp) per 360-degree rotation of the helix, so each base pair is twisted 36 degrees relative to the adjacent bases (Figure 8.12b). The base pairs are 0.34 nanometer (nm) apart; so each complete rotation of the molecule encompasses 3.4 nm. The diameter of the helix is 2 nm, and the bases are perpendicular to the long axis of the DNA molecule. A space-filling model shows that B-DNA has a slim and elongated structure (Figure 8.12a). The spiraling of the nucleotide strands creates major and minor grooves in the helix, features that are important for the binding of some proteins that regulate the expression of genetic information (see Chapter 12).

Another secondary structure that DNA can assume is the **A-DNA** structure, which exists if less water is present. Like B-DNA, A-DNA is a right-handed helix (**Figure 8.13a**), but it is shorter and wider than B-DNA (**Figure 8.13b**) and its bases are tilted away from the main axis of the molecule. A-DNA has been detected in some DNA–protein complexes and in spores of some bacteria.

A radically different secondary structure, called **Z-DNA** (**Figure 8.13c**), forms a left-handed helix. In this structure, the sugar–phosphate backbone zigzags back and forth, giving rise to its name. A Z-DNA structure can result if the molecule contains particular base sequences, such as stretches of alternating C and G nucleotides. Researchers have found that Z-DNA-specific antibodies bind to regions of the DNA that are being transcribed into RNA, suggesting that Z-DNA may play some role in gene expression. Other secondary structures of DNA also exist, such as H-DNA, in which one part of the DNA unwinds and a single-stranded nucleotide chain then pairs with the two strands of another region to form a three-stranded helix.

THINK-PAIR-SHARE Question 5

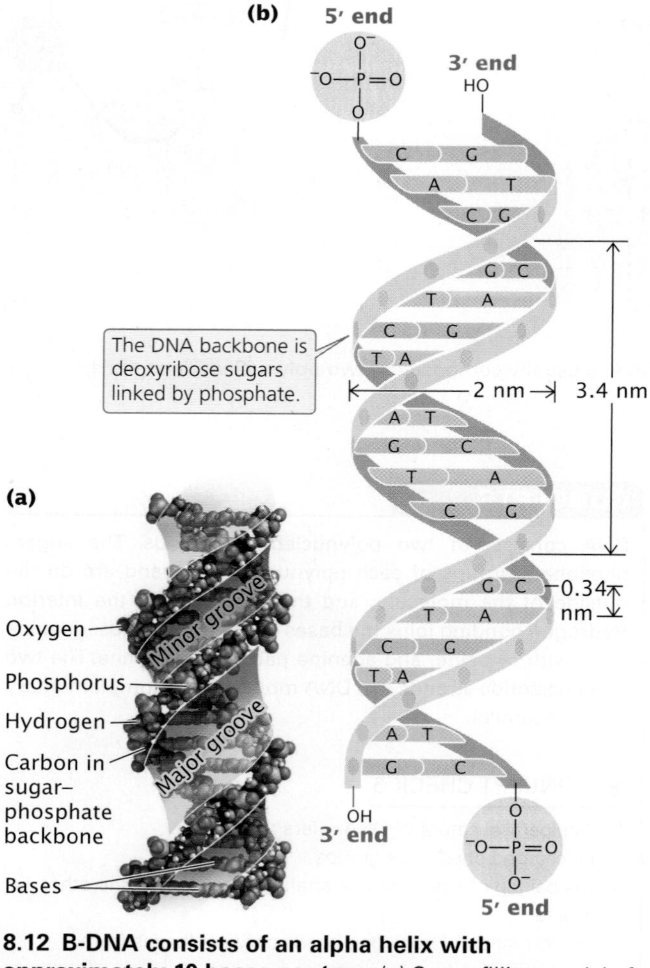

8.12 B-DNA consists of an alpha helix with approximately 10 bases per turn. (a) Space-filling model of B-DNA showing major and minor grooves. (b) Diagrammatic representation.

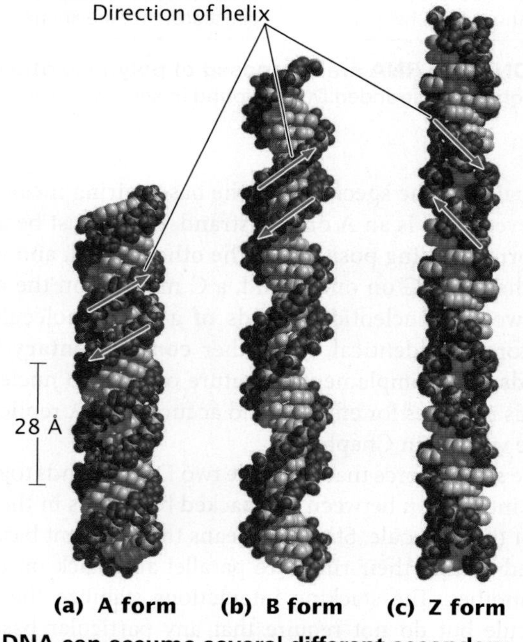

8.13 DNA can assume several different secondary structures. [After J. M. Berg, J. L. Tymoczko, and L. Stryer, *Biochemistry*, 7th ed. (New York: W. H. Freeman and Company, 2012), pp. 115, 116.]

CONCEPTS

DNA can assume different secondary structures, depending on the conditions in which it is placed and on its base sequence. B-DNA is thought to be the most common configuration in the cell.

✓ CONCEPT CHECK 6

How does Z-DNA differ from B-DNA?

Genetic Implications of DNA Structure

Watson and Crick's great contribution was their elucidation of the genotype's chemical structure, making it possible for geneticists to begin to examine genes directly, instead of looking only at the phenotypic consequences of gene action. The determination of the structure of DNA led to the birth of molecular genetics: the study of the chemical and molecular nature of genetic information.

Watson and Crick's structure did more than create the potential for molecular genetic studies, however; it was an immediate source of insight into key genetic processes. At the beginning of this chapter, four fundamental properties of genetic material were identified. First, it must be capable of carrying large amounts of information. Watson and Crick's model suggested that genetic instructions are encoded in the base sequence, the only variable part of the DNA molecule.

A second necessary property of genetic material is its ability to replicate faithfully. The complementary polynucleotide strands of DNA make this replication possible. Watson and Crick proposed that in replication, the two polynucleotide strands unzip, breaking the weak hydrogen bonds between them, and each strand serves as a template on which a new strand is synthesized. The specificity of the base pairing means that only one possible sequence of bases—the complementary sequence—can be synthesized from each template strand. Newly replicated double-stranded DNA molecules are therefore identical with the original double-stranded DNA molecule (see Chapter 9 on DNA replication).

A third essential property of genetic material is the ability to express its instructions as a phenotype. DNA expresses its genetic instructions by first transferring its information to an RNA molecule in a process termed **transcription** (see Chapter 10). The term *transcription* is appropriate because, although the information is transferred from DNA to RNA, the information remains in the language of nucleic acids. In some cases, the RNA molecule then transfers the genetic information to a protein by specifying its amino acid sequence. This process is termed **translation** (see Chapter 11) because the information must be *translated* from the language of nucleotides into the language of amino acids. A fourth property of DNA is that it must be capable of varying. This variation, as we have seen, consists of differences in the sequence of bases found among different individuals.

We can now identify three major pathways of information flow in the cell (**Figure 8.14a**): in **replication**, information passes from one DNA molecule to other DNA molecules; in transcription, information passes from DNA to RNA; and in translation, information passes from RNA to protein. This concept of information flow was formalized by Francis Crick in a concept that he called the **central dogma** of molecular biology. The central dogma states that genetic information passes from DNA to protein in a one-way information pathway. We now realize, however, that the central dogma is an oversimplification. In addition to the three general information pathways of replication, transcription, and translation, other transfers may take place in certain organisms or under special circumstances. Retroviruses (see Chapter 7) and some transposable elements (see Chapter 13) transfer information from RNA to DNA (in **reverse transcription**), and some RNA viruses transfer information from RNA to RNA (in **RNA replication**; **Figure 8.14b**).

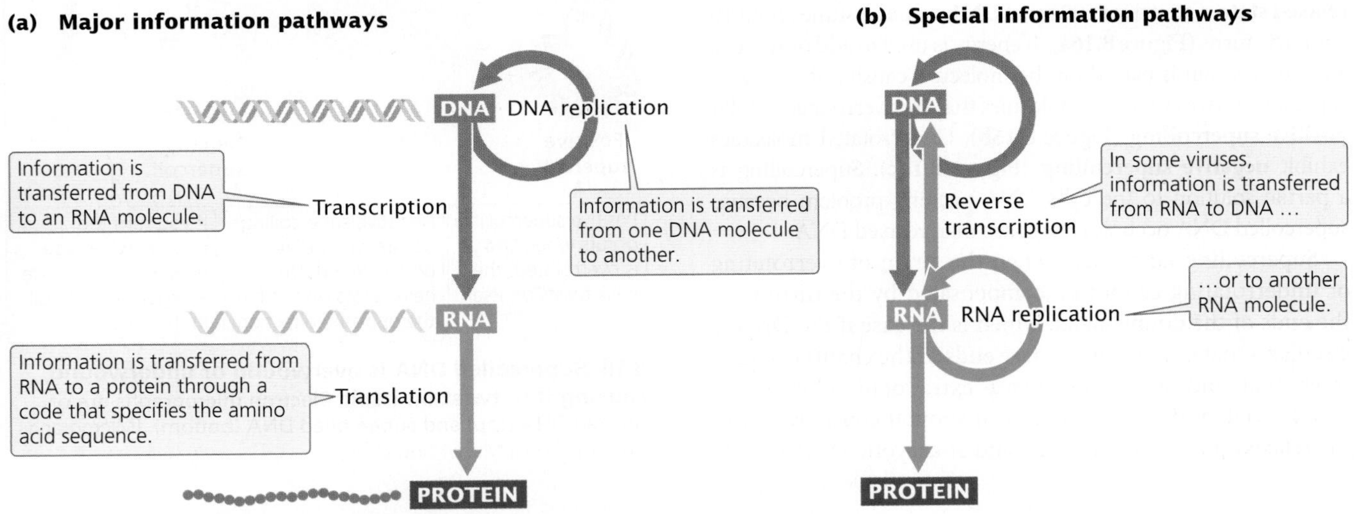

8.14 Pathways of information transfer within the cell.

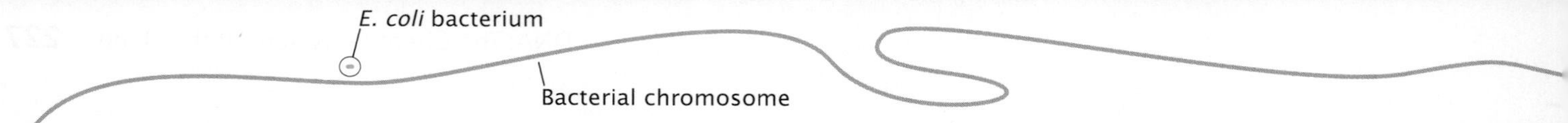

8.15 The DNA in *E. coli* is about 1000 times as long as the cell itself.

8.4 Large Amounts of DNA Are Packed into a Cell

The packaging of tremendous amounts of genetic information into the small space within a cell has been called the ultimate storage problem. Consider the chromosome of the bacterium *E. coli*, a single molecule of DNA with approximately 4.6 million base pairs. Stretched out straight, this DNA would be about 1000 times as long as the cell within which it resides (**Figure 8.15**). Human cells contain more than 6 billion base pairs of DNA, which would measure more than 2 meters (more than 6 feet) stretched end to end. Even the DNA in the smallest human chromosome would stretch 14,000 times the length of the cell nucleus. Clearly, DNA molecules must be tightly packed to fit into such small spaces.

As mentioned earlier, the structure of DNA can be considered at three hierarchical levels: the primary structure of DNA is its nucleotide sequence; the secondary structure is the double-stranded helix; and the tertiary structure is the higher-order folding of DNA. Here, we consider the tertiary structure, which allows DNA to be packed into the confined space of a cell.

CONCEPTS

Chromosomal DNA exists in the form of very long molecules that are tightly packed to fit into the small confines of a cell.

Supercoiling

One type of DNA tertiary structure is **supercoiling**, which takes place when the DNA helix is subjected to strain by being overwound or underwound. B-DNA is in its lowest-energy state when it has approximately 10 bp per turn of its helix. In this **relaxed state**, a stretch of 100 bp of DNA would assume about 10 complete turns (**Figure 8.16a**). If energy is used to add or remove any turns, strain is placed on the molecule, causing the helix to supercoil, or twist on itself. Molecules that are overrotated exhibit **positive supercoiling** (**Figure 8.16b**). Underrotated molecules exhibit **negative supercoiling** (**Figure 8.16c**). Supercoiling is a partial solution to the cell's DNA packing problem because supercoiled DNA occupies less space than relaxed DNA.

Supercoiling takes place when the strain of overrotating or underrotating cannot be compensated by the turning of the ends of the double helix, which is the case if the DNA is circular—that is, there are no free ends. If the chains *can* turn freely, their ends will simply turn as extra rotations are added or removed, and the molecule will spontaneously revert to the relaxed state. Both bacterial and eukaryotic DNA usually

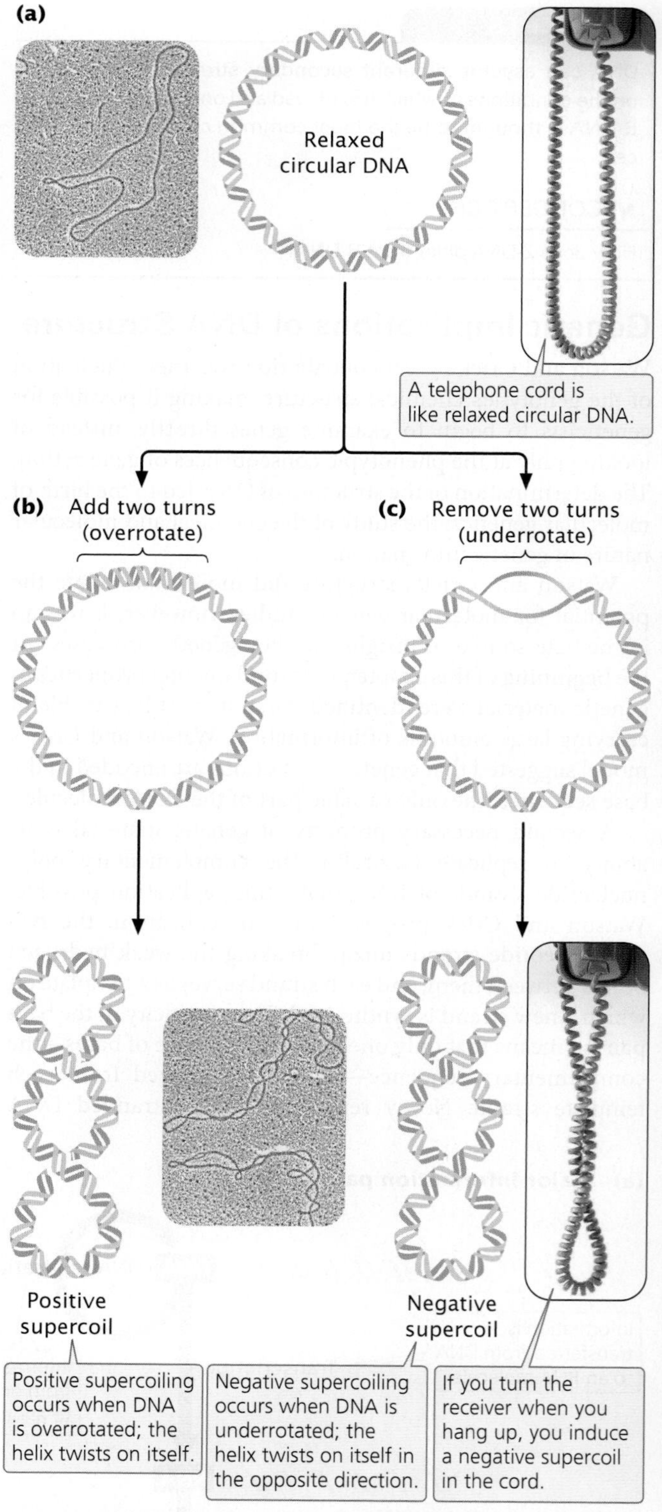

8.16 Supercoiled DNA is overwound or underwound, causing it to twist on itself. Electron micrographs are of relaxed DNA (top) and supercoiled DNA (bottom). [Micrographs from Dr. Gopal Murti/Phototake.]

fold into loops stabilized by proteins (which prevent free rotation of the ends; see Figure 8.17 below), and supercoiling takes place within the loops.

Supercoiling relies on **topoisomerases**, enzymes that add or remove rotations from the DNA helix by temporarily breaking the nucleotide strands, rotating the ends around each other, and then rejoining the broken ends. Thus, topoisomerases can both induce and relieve supercoiling, although not all topoisomerases do both.

Most DNA found in cells is negatively supercoiled. This state has two advantages over relaxed DNA. First, negative supercoiling makes the separation of the two strands of DNA easier during replication and transcription. Negatively supercoiled DNA is underrotated, so separation of the two strands during replication and transcription is more rapid and requires less energy. Second, the supercoiled DNA can be packed into a smaller space than can relaxed DNA.

CONCEPTS

Overrotation or underrotation of a DNA double helix places strain on the molecule, causing it to supercoil. Supercoiling is controlled by topoisomerase enzymes. Most cellular DNA is negatively supercoiled, which eases the separation of nucleotide strands during replication and transcription and allows the DNA to be packed into small spaces.

✓ CONCEPT CHECK 7

A DNA molecule 300 bp long has 20 complete rotations. This DNA molecule is

a. positively supercoiled.
b. negatively supercoiled.
c. relaxed.

The Bacterial Chromosome

Most bacterial genomes consist of a single circular DNA molecule, although linear DNA molecules have been found in a few species. In circular bacterial chromosomes, the DNA does not exist in an open, relaxed circle; the 3 million to 4 million base pairs of DNA found in a typical bacterial genome would be much too large in this state to fit into a bacterial cell (see Figure 8.15). Bacterial DNA is not attached to histone proteins as is eukaryotic DNA (considered later in this chapter), but bacterial DNA is associated with a number of proteins that help to compact it.

When a bacterial cell is viewed with an electron microscope, its DNA frequently appears as a distinct clump, called the **nucleoid**, which is confined to a definite region of the cytoplasm. If a bacterial cell is broken open gently, its DNA spills out in a series of twisted loops (**Figure 8.17a**). The ends of the loops are most likely held in place by proteins (**Figure 8.17b**). Many bacteria contain additional DNA in the form of small circular molecules called plasmids, which replicate independently of the chromosome (see Chapter 7).

CONCEPTS

A typical bacterial chromosome consists of a large circular molecule of DNA that forms a series of twisted loops. Bacterial DNA appears as a distinct clump, the nucleoid, within the bacterial cell.

✓ CONCEPT CHECK 8

How does bacterial DNA differ from eukaryotic DNA?

Eukaryotic Chromosomes

Individual eukaryotic chromosomes contain enormous amounts of DNA. Like a bacterial chromosome, each eukaryotic chromosome consists of a single, extremely long linear molecule of DNA. Fitting all of this DNA into the nucleus requires tremendous packing and folding, the extent of which must change in the course of the cell cycle. The chromosomes are in an elongated, relatively uncondensed state during interphase (see p. 23 in Chapter 2), but the term *relatively* is important here. Although the DNA of interphase

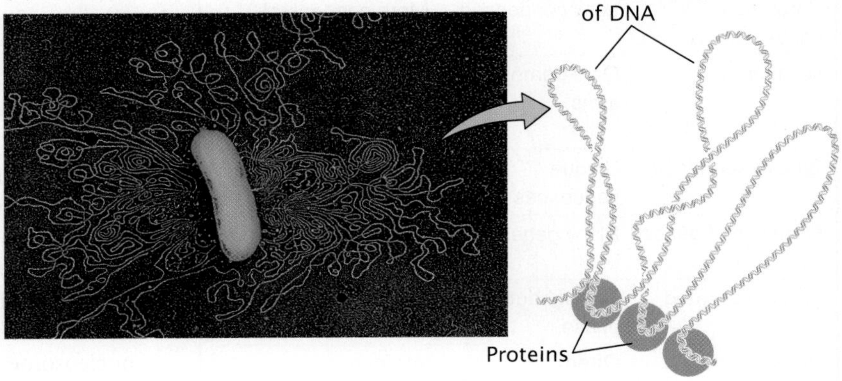

8.17 Bacterial DNA is highly folded into a series of twisted loops. [Part a: G. Murti/ Science Source.]

chromosomes is less tightly packed than the DNA of mitotic chromosomes, it is still highly condensed; it's just *less* condensed. In the course of the cell cycle, the level of DNA packing changes: chromosomes progress from a highly packed state to a state of extreme condensation, which is necessary for chromosome movement in mitosis and meiosis. DNA packing also changes locally during replication and transcription, when the two nucleotide strands must unwind so that particular base sequences are exposed. Thus, the packing of eukaryotic DNA (its tertiary chromosomal structure) is not static; rather, it changes regularly in response to cellular processes.

CHROMATIN STRUCTURE Eukaryotic DNA in the cell is closely associated with proteins. This combination of DNA and protein is called *chromatin*. The two basic types of chromatin are **euchromatin**, which undergoes the normal process of condensation and decondensation in the cell cycle, and **heterochromatin**, which remains in a highly condensed state throughout the cell cycle, even during interphase. Euchromatin constitutes the majority of the chromosomal material and is where most transcription takes place. All chromosomes have constitutive (permanent) heterochromatin at the centromeres and telomeres; the Y chromosome also consists largely of constitutive heterochromatin. Heterochromatin may also occur during certain developmental stages; this material is referred to as facultative heterochromatin. For example, facultative heterochromatin occurs along one entire X chromosome in female mammals when that X becomes inactivated (see pp. 86–87 in Chapter 4). In addition to remaining condensed throughout the cell cycle, heterochromatin is characterized by a general lack of transcription, the absence of crossing over, and replication late in S phase. Differences between euchromatin and heterochromatin are summarized in **Table 8.3**.

TABLE 8.3 Characteristics of euchromatin and heterochromatin

Characteristic	Euchromatin	Heterochromatin
Chromatin condensation	Less condensed	More condensed
Location	On chromosome arms	At centromeres, telomeres, and other specific places
Type of sequences	Unique sequences	Repeated sequences*
Presence of genes	Many genes	Few genes*
When replicated	Throughout S phase	Late S phase
Transcription	Often	Infrequent
Crossing over	Common	Uncommon

*Applies only to constitutive heterochromatin.

The most abundant proteins in chromatin are the *histones*, which are small, positively charged proteins of five major types: H1, H2A, H2B, H3, and H4. All histones have a high percentage of arginine and lysine, positively charged amino acids that give the histones a net positive charge. Their positive charges attract the negative charges on the phosphates of DNA; this attraction holds the DNA in contact with the histones. A heterogeneous assortment of **nonhistone chromosomal proteins** is also found in eukaryotic chromosomes. At times, variant histones, with somewhat different amino acid sequences, are incorporated into chromatin in place of one of the major histone types. These variants alter chromatin structure and influence its function. For example, some specific variant histones are associated with actively transcribed DNA; these variants are assumed to make the chromatin more open and accessible to the enzymes and proteins that carry out transcription. ▶TRY PROBLEM 31

CONCEPTS

Chromatin, which consists of DNA closely associated with proteins, is the material that makes up eukaryotic chromosomes. The most abundant of these proteins are the five types of positively charged histone proteins: H1, H2A, H2B, H3, and H4. Variant histones may at times be incorporated into chromatin in place of the normal histones.

✓ CONCEPT CHECK 9

Neutralizing their positive charges would have which effect on the histone proteins?

a. They would bind the DNA tighter.
b. They would bind less tightly to the DNA.
c. They would no longer be attracted to each other.
d. They would cause supercoiling of the DNA.

THE NUCLEOSOME Chromatin has a highly complex structure with several levels of organization (**Figure 8.18**). The simplest level is the double-stranded helical structure of DNA. At a more complex level, the DNA molecule is closely associated with proteins and is highly folded to produce a chromosome.

When chromatin is isolated from the nucleus of a cell and viewed with an electron microscope, it frequently looks like beads on a string (**Figure 8.19a** on p. 232). If a small amount of nuclease is added to this structure, the enzyme cleaves the "string" between the "beads," leaving individual beads attached to about 200 bp of DNA (**Figure 8.19b**). If more nuclease is added, the enzyme chews up all of the DNA between the beads and leaves a core of proteins attached to a fragment of DNA (**Figure 8.19c**). Such experiments demonstrated that chromatin is not a random association of proteins and DNA; rather, it has a fundamental repeating structure.

The repeating core of protein and DNA produced by digestion with nuclease enzymes is the simplest level of chromatin structure, the **nucleosome** (see Figure 8.18). The nucleosome is a core particle consisting of DNA wrapped about two times around an octamer of eight histone proteins (two copies each of H2A, H2B, H3, and H4), much like thread wound around a spool (**Figure 8.19d**). The DNA in

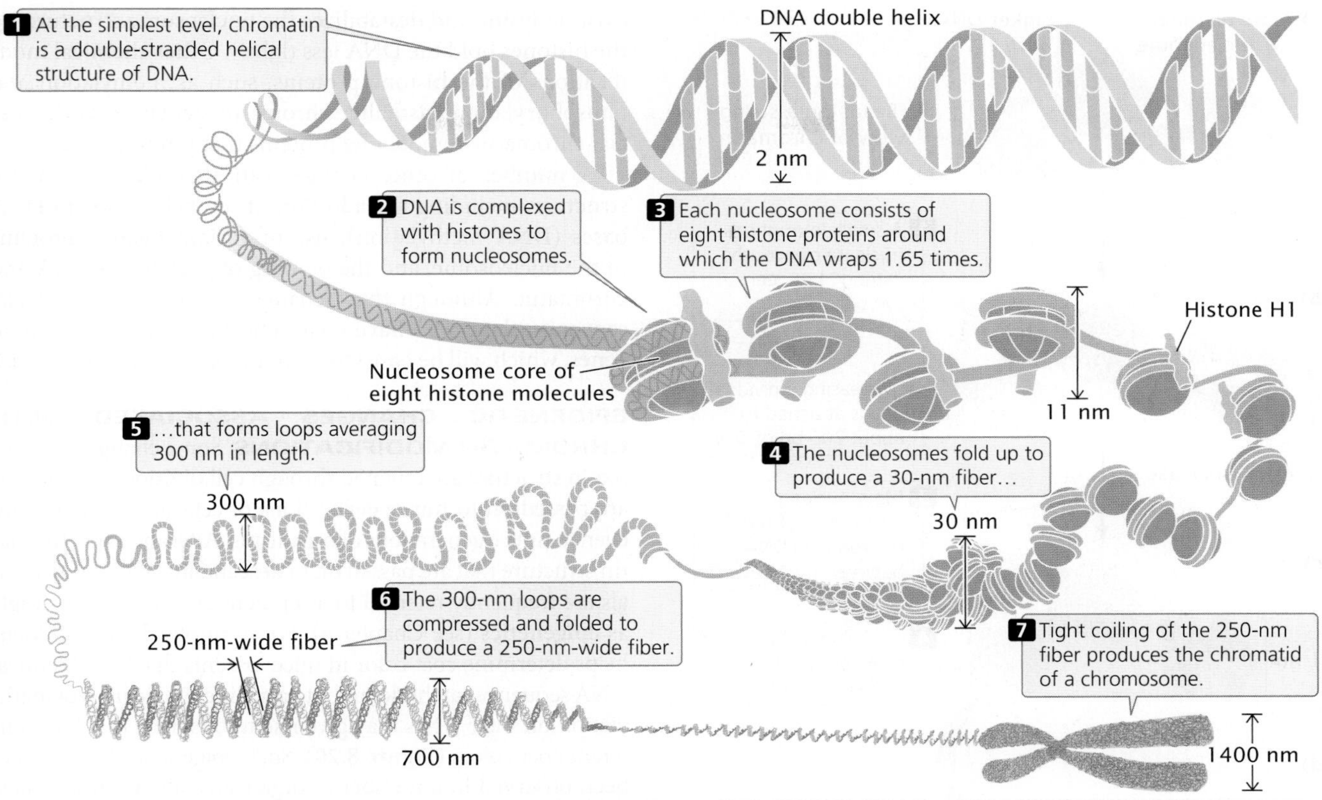

8.18 Chromatin has a highly complex structure with several levels of organization.

direct contact with the histone octamer is between 145 and 147 bp in length.

Each of the histone proteins that make up the nucleosome core particle has a flexible "tail," containing from 11 to 37 amino acids, which extends out from the nucleosome. Positively charged amino acids in the tails of the histones interact with the negative charges of the phosphates on the DNA, keeping the DNA and histones tightly associated (complexed). The tails of one nucleosome may also interact with neighboring nucleosomes, which facilitates compaction of the nucleosomes themselves. Chemical modifications of these histone tails bring about changes in chromatin structure that are necessary for gene expression.

The fifth type of histone, H1, is not a part of the nucleosome core particle, but plays an important role in nucleosome structure. H1 binds to 20–22 bp of DNA where the DNA joins and leaves the histone octamer (see Figure 8.18). It helps to lock the DNA into place, acting as a clamp around the nucleosome.

Each nucleosome encompasses about 167 bp of DNA. Nucleosomes are located at regular intervals along the DNA molecule and are separated from one another by **linker DNA**, which varies in size among cell types; in most cells, linker DNA comprises from about 30 to 40 bp. Nonhistone chromosomal proteins may be associated with this linker DNA, and a few also appear to bind directly to the nucleosome. ▶TRY PROBLEM 32

HIGHER-ORDER CHROMATIN STRUCTURE When chromatin is in a condensed form, adjacent nucleosomes are not separated by space equal to the length of the linker DNA. Rather, the nucleosomes fold on themselves to form a dense, tightly packed structure (see Figure 8.18) that makes up a fiber with a diameter of about 30 nm. The precise molecular structure of the 30-nm fiber remains uncertain.

The next level of chromatin structure is a series of loops of the 30-nm fiber, each anchored at its base by proteins. On average, each loop encompasses some 20,000 to 100,000 bp of DNA and is about 300 nm in length, but the individual loops vary considerably. The 300-nm loops are packed and folded to produce a 250-nm-wide fiber. Tight helical coiling of the 250-nm fiber, in turn, produces the structure that appears in metaphase: an individual chromatid approximately 700 nm in width. You can view the different levels of chromatin structure in **Animation 8.1**. A

CONCEPTS

The nucleosome consists of a core particle of eight histone proteins and the DNA that wraps around them. A single H1 histone associates with each nucleosome. Nucleosomes are separated by linker DNA. Nucleosomes fold to form a 30-nm chromatin fiber, which appears as a series of loops that pack to create a 250-nm fiber. Helical coiling of the 250-nm fiber produces a chromatid.

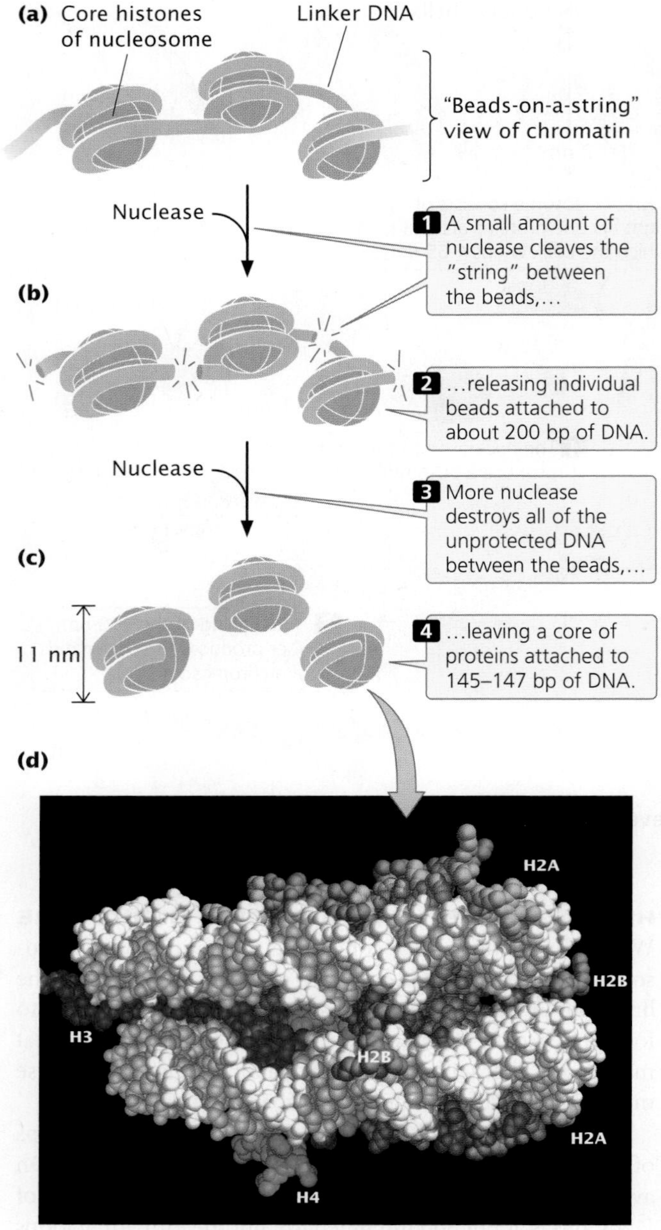

8.19 The nucleosome is the fundamental repeating unit of chromatin. Part d shows a space-filling model of a nucleosome, which consists of two copies each of H2A, H2B, H3, and H4, around which DNA (white) coils. [Part d: Reprinted by permission from Macmillan Publishers Ltd. From K. Luger et al., "Crystal structure of the nucleosome core particle at 2.8 A resolution", Nature 389:251. © 1997. Courtesy of T.H. Richmond, permission conveyed through Copyright Clearance Center, Inc.]

CHANGES IN CHROMATIN STRUCTURE Although eukaryotic DNA must be tightly packed to fit into the cell nucleus, it must also periodically unwind to undergo transcription and replication. One process that alters chromatin structure is acetylation. Enzymes called acetyltransferases attach acetyl groups to lysine amino acids on the histone tails. This modification reduces the positive charges that normally exist on lysine and destabilizes the nucleosome structure, so the histones hold the DNA less tightly. Other chemical modifications of the histone proteins, such as methylation and phosphorylation, also alter chromatin structure, as do special chromatin-remodeling proteins that bind to the DNA.

A number of other changes can also affect chromatin structure, including the addition of methyl groups to DNA bases (DNA methylation), use of variant histone proteins in the nucleosome, and the binding of proteins to DNA and chromatin. Although these changes do not alter the DNA sequence, they often have major effects on the expression of genes, which will be considered in more detail in Chapter 12.

EPIGENETIC CHANGES ASSOCIATED WITH CHROMATIN MODIFICATIONS Some changes in chromatin structure are retained through cell division, so that they are passed on to future generations of cells and, occasionally, even to future generations of organisms. Alterations of chromatin structure that are passed on to descendant cells or individuals are frequently referred to as **epigenetic changes**, or simply as **epigenetics** (see Chapter 4). For example, the *agouti* locus helps determine coat color in mice: parents that have identical DNA sequences at this locus but have different degrees of methylation on their DNA may give rise to offspring that have different coat colors (**Figure 8.20**). Such epigenetic changes have been observed in a number of organisms and are responsible for a variety of phenotypic effects. Unlike mutations, epigenetic changes do not alter the DNA sequence, are capable of being reversed, and are often influenced by environmental factors. One type of epigenetic change is genomic imprinting, in which an allele is differentially expressed depending on whether it is inherited from the maternal or paternal parent (see Chapter 4). Genomic imprinting is caused by differences in DNA methylation of oocytes and sperm. The DNA of sperm and oocytes is differentially methylated at various imprinting control regions in the course of gamete formation, and the different methylation patterns of paternal and maternal alleles are then maintained and passed on to all resulting cells in the zygote.

CONCEPTS

Epigenetic changes are alterations of chromatin or DNA structure that do not include changes in the base sequence, but are stable and are passed on to descendant cells or organisms. Some epigenetic changes result from differences in DNA methylation.

8.5 Eukaryotic Chromosomes Possess Centromeres and Telomeres

As we have seen, chromosomes segregate in mitosis and meiosis and remain stable through many cell divisions. These properties arise in part from special structural features of chromosomes, including centromeres and telomeres.

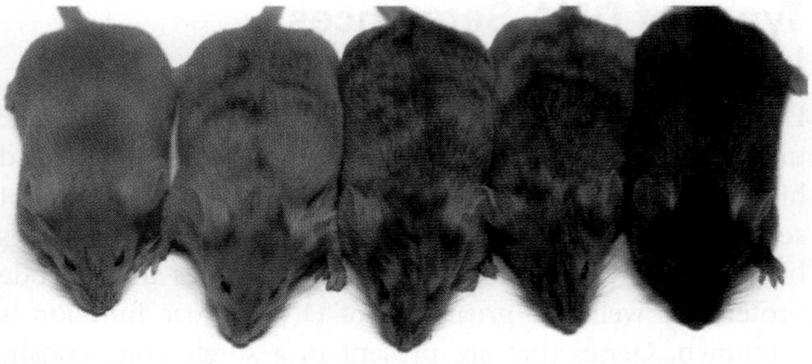

8.20 Variation in DNA methylation at the *agouti* locus produces different coat colors in mice. [Cropley et al. "Germ-line epigenetic modification of the murine Avy allele by nutritional supplementation", PNAS November 14, 2006 vol. 103 no. 46 17308-17312, ©2006 National Academy of Sciences, U.S.A.]

Centromere Structure

The centromere, a constricted region of the chromosome, is essential for proper chromosome movement in mitosis and meiosis (see Figure 2.6). Centromeres are the binding sites for the kinetochore, to which spindle microtubules attach. In *Drosophila*, *Arabidopsis*, and humans, centromeres span hundreds of thousands of base pairs. Most of the centromere is made up of heterochromatin. Surprisingly, there are no specific sequences that are found in all centromeres, which raises the question of what exactly determines where the centromere is. Research suggests that most centromeres are not defined by DNA sequence, but rather by epigenetic changes in chromatin structure. Nucleosomes in the centromeres of most eukaryotes have a variant histone protein called CenH3, which takes the place of the usual H3 histone. The CenH3 histone brings about a change in the nucleosome and chromatin structure, which is believed to promote the formation of the kinetochore and the attachment of spindle microtubules to the chromosome.

CONCEPTS

The centromere is a region of the chromosome to which spindle microtubules attach. Centromeres display considerable variation in their DNA sequences.

Telomere Structure

Telomeres are the natural ends of a chromosome (see Figure 2.6). Pioneering work by Hermann Muller (on fruit flies) and Barbara McClintock (on corn) showed that chromosome breaks produce unstable ends that have a tendency to stick together and enable the chromosome to be degraded. Because attachment and degradation don't happen to the ends of a chromosome that has telomeres, the telomeres must serve as caps that stabilize the chromosome. Telomeres also provide a means of replicating the ends of the chromosome, as we will see in Chapter 9.

Telomeres have now been isolated from protozoans, plants, humans, and other organisms; most are similar in structure. These **telomeric sequences** are usually repeated units of a series of adenine or thymine nucleotides followed by several guanine nucleotides, taking the form $5'$—(A or T)$_m$G$_n$—$3'$, where m ranges from 1 to 4 and n is 2 or more. For example, the repeating unit in human telomeres is $5'$—TTAGGG—$3'$, which may be repeated from hundreds to thousands of times. The sequence is always oriented with the string of Gs and Cs toward the end of the chromosome, as shown here:

toward cetromere ← 5′—TTAGGGTTAGGGTTAGGG—3′ → end of chromosome
3′—AATCCCAATCCCAATCCC—5′

The G-rich strand often protrudes beyond the complementary C-rich strand at the end of the chromosome (**Figure 8.21a**), in which case it is called the $3'$ overhang. The $3'$ overhang in mammalian telomeres is from 50 to 500 nucleotides long. Special proteins bind to the G-rich single-stranded sequence, protecting the telomere from degradation and preventing the ends of chromosomes from sticking together. A multiprotein complex called **shelterin** binds to telomeres and protects the ends of the DNA from being inadvertently repaired as a double-strand break in the DNA. In some cells, the $3'$ overhang may fold over and pair with a short stretch of DNA to form a structure called the t-loop, which also functions in protecting the end of the telomere from degradation (**Figure 8.21b**).

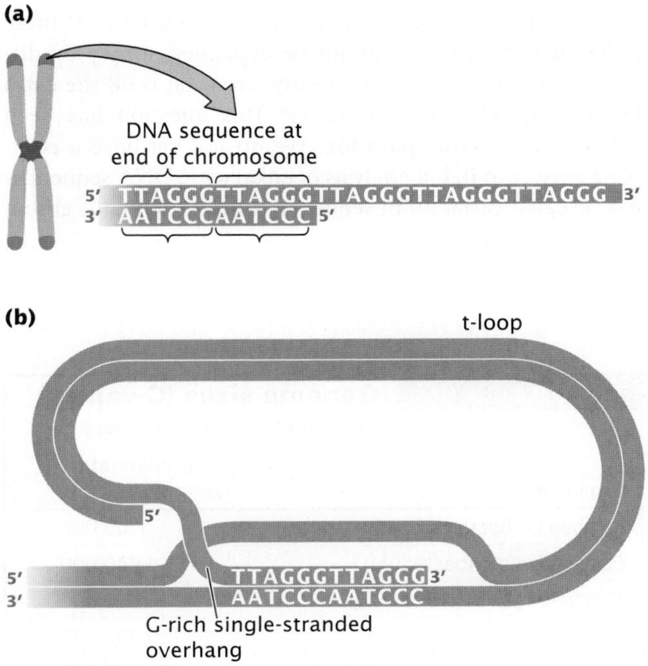

8.21 DNA at the ends of eukaryotic chromosomes consists of telomeric sequences. (a) The G-rich strand at the telomere is longer than the C-rich strand. (b) In some cells, the G-rich strand folds over and pairs with a short stretch of DNA to form a t-loop.

CONCEPTS

A telomere is the stabilizing end of a chromosome. At the end of each telomere are many short telomeric sequences.

✓ CONCEPT CHECK 10

Which of the following is a characteristic of DNA sequences at the telomeres?

a. One strand consists of guanine and adenine or thymine nucleotides.
b. They consist of repeated sequences.
c. One strand protrudes beyond the other, creating some single-stranded DNA at the end.
d. All of the above

8.6 Eukaryotic DNA Contains Several Classes of Sequence Variation

Eukaryotic organisms differ dramatically in the amount of DNA found in each of their cells, a quantity termed an organism's **C-value** (**Table 8.4**). Each cell of a fruit fly, for example, contains 35 times the amount of DNA found in a cell of the bacterium *E. coli*. In general, eukaryotic cells contain more DNA than prokaryotic cells do, but variation among eukaryotes in their C-values is huge. Human cells contain more than 10 times the amount of DNA found in *Drosophila* cells, whereas some salamander cells contain 20 times as much DNA as human cells. Clearly, these differences in C-value cannot be explained simply by differences in organismal complexity. So, what is all the extra DNA in eukaryotic cells doing? This question has been termed the **C-value paradox**. We do not yet have a complete answer to it, but analysis of eukaryotic DNA sequences has revealed variation in sequence complexity that is absent from prokaryotic DNA.

TABLE 8.4 Genome sizes (C-values) of various organisms

Organism	Approximate genome size (bp)
Lambda (λ) bacteriophage	50,000
Escherichia coli (bacterium)	4,640,000
Saccharomyces cerevisiae (yeast)	12,000,000
Arabidopsis thaliana (plant)	125,000,000
Drosophila melanogaster (insect)	170,000,000
Homo sapiens (human)	3,200,000,000
Zea mays (corn)	4,500,000,000
Amphiuma (salamander)	765,000,000,000

Types of DNA Sequences in Eukaryotes

Eukaryotic DNA consists of at least three sequence types: unique-sequence DNA, moderately repetitive DNA, and highly repetitive DNA. **Unique-sequence DNA** consists of sequences that are present only once or, at most, a few times in the genome. This DNA includes sequences that encode proteins as well as a great deal of DNA whose function is unknown. Genes that are present in a single copy constitute roughly 25% to 50% of the protein-encoding genes in most multicellular eukaryotes. Other genes within unique-sequence DNA are present in several similar, but not identical, copies and together are referred to as a **gene family**. Most gene families arose through duplication of an existing gene and include just a few member genes, but some, such as those that encode immunoglobulin proteins in vertebrates, contain hundreds of members. The genes that encode β-like globins are another example of a gene family. In humans, there are seven β-globin genes, clustered together on chromosome 11. The polypeptides encoded by these genes join with α-globin polypeptides to form hemoglobin molecules, which transport oxygen in the blood.

Other sequences, called **repetitive DNA**, exist in many copies. Some eukaryotic organisms have large amounts of repetitive DNA; for example, almost half the human genome consists of repetitive DNA. A major class of repetitive DNA is **moderately repetitive DNA**, which typically consists of sequences from 150 to 300 bp in length (although they may be longer) that are repeated many thousands of times. Some of these sequences perform important functions for the cell; for example, the genes for ribosomal RNAs (rRNAs) and transfer RNAs (tRNAs) make up a part of the moderately repetitive DNA. However, the function of much moderately repetitive DNA is unknown, and indeed, it may have no function.

Moderately repetitive DNA itself encompasses two types of repeats. **Tandem repeat sequences** appear one after another and tend to be clustered at particular locations on the chromosomes. **Interspersed repeat sequences** are scattered throughout the genome. An example of an interspersed repeat is the *Alu* sequence, an approximately 300-bp sequence that is present more than a million times and constitutes 11% of the human genome, although it has no obvious cellular function. Short repeats, such as the *Alu* sequences, are called **SINEs (short interspersed elements)**. Longer interspersed repeats consisting of several thousand base pairs are called **LINEs (long interspersed elements)**. One class of LINE, called LINE1, constitutes about 17% of the human genome. Most interspersed repeats are the remnants of transposable elements, sequences that can multiply and move (see Chapter 13).

The other major class of repetitive DNA is **highly repetitive DNA**. These short sequences, often less than 10 bp in length, are present in hundreds of thousands to millions of copies that are repeated in tandem and clustered in certain regions of the chromosome, especially at centromeres and telomeres. Highly repetitive DNA is rarely transcribed into RNA, and most highly repetitive DNA has no known function.

Organization of Genetic Information in Eukaryotes

The sequencing of eukaryotic genomes also tells us a lot about how genetic information is organized within chromosomes. We now know that the density of genes varies greatly among and within chromosomes. For example, human chromosome 19 has a high density of genes, with about 26 genes per million base pairs. Chromosome 13, on the other hand, has only about 6.5 genes per million base pairs. Gene density can also vary among different regions of the same chromosome: some parts of the long arm of chromosome 13 have only 3 genes per million base pairs, whereas other parts have almost 30 genes per million base pairs. And the short arm of chromosome 13 contains almost no genes, consisting entirely of heterochromatin.

The functional role of DNA sequences that do not encode proteins, including repetitive DNA, has been addressed by the Encyclopedia of DNA Elements (ENCODE) project (see Chapter 15). The purpose of ENCODE was to identify all nucleotides within the human genome that have some function. The project concluded that much of the genome is transcribed and that at least 80% of the sequences are functional. However, some researchers have challenged the conclusion that the vast majority of DNA sequences are functional.

In addition, some DNA in eukaryotes is found associated with cytoplasmic organelles such as mitochondria and chloroplasts. This organelle DNA encodes traits that exhibit cytoplasmic inheritance (see Chapter 4). In contrast with nuclear DNA, organelle DNA is typically circular and is not complexed with histone proteins.

CONCEPTS

Eukaryotic DNA comprises three major classes: unique-sequence DNA, moderately repetitive DNA, and highly repetitive DNA. Unique-sequence DNA consists of sequences that exist in one or a few copies; moderately repetitive DNA consists of sequences that may be several hundred base pairs in length and are present in thousands to hundreds of thousands of copies. Highly repetitive DNA consists of very short sequences repeated in tandem and present in hundreds of thousands to millions of copies. The density of genes varies greatly among and even within chromosomes. DNA in cytoplasmic organelles is usually circular and not complexed with histone proteins.

✔ **CONCEPT CHECK 11**

Most of the genes that encode proteins are found in
a. unique-sequence DNA.
b. moderately repetitive DNA.
c. highly repetitive DNA.
d. All of the above

CONCEPTS SUMMARY

- Genetic material must contain complex information, be replicated accurately, code for the phenotype, and have the capacity to vary.
- Evidence that DNA is the source of genetic information came from the finding by Avery, MacLeod, and McCarty that transformation depended on DNA and from the demonstration by Hershey and Chase that viral DNA is passed on to progeny phages.
- James Watson and Francis Crick, using data provided by Rosalind Franklin and Maurice Wilkins, proposed a model for the three-dimensional structure of DNA in 1953.
- A DNA nucleotide consists of a deoxyribose sugar, a phosphate group, and a nitrogenous base. An RNA nucleotide consists of a ribose sugar, a phosphate group, and a nitrogenous base.
- The bases of a DNA nucleotide are of two types: purines (adenine and guanine) and pyrimidines (cytosine and thymine). RNA contains the pyrimidine uracil instead of thymine.
- Nucleotides are joined together by phosphodiester linkages to form a polynucleotide strand. Each polynucleotide strand has a free phosphate group at its 5′ end and a free hydroxyl group at its 3′ end.
- DNA consists of two nucleotide strands that wind around each other to form a double helix. The sugars and phosphates lie on the outside of the helix, and the bases are stacked in the interior. The two strands are joined together by hydrogen bonding between bases in each strand. The two strands are antiparallel and complementary.
- DNA molecules can form a number of different secondary structures, depending on the conditions in which the DNA is placed and on its base sequence.
- The structure of DNA has several important genetic implications. Genetic information resides in the base sequence of DNA, which ultimately specifies the amino acid sequence of proteins. Complementarity of the bases on DNA's two strands allows genetic information to be replicated.
- The central dogma of molecular biology proposes that information flows in one direction, from DNA to RNA to protein. Exceptions to the central dogma are now known.
- Chromosomes contain very long DNA molecules that are tightly packed. Supercoiling results from strain produced when rotations are added to or removed from a relaxed DNA molecule.

- A bacterial chromosome consists of a single, circular DNA molecule that is bound to proteins and exists as a series of large loops.
- Each eukaryotic chromosome contains a single, long linear DNA molecule that is bound to histone proteins. Euchromatin undergoes the normal cycle of decondensation and condensation in the cell cycle. Heterochromatin remains highly condensed throughout the cell cycle.
- The nucleosome consists of a core of eight histone proteins and the DNA that wraps around the core. Nucleosomes are folded into a 30-nm fiber that forms a series of 300-nm loops; these loops are anchored at their bases by proteins. The 300-nm loops are condensed to form a fiber that is itself tightly coiled to produce a chromatid.
- Epigenetic changes are stable alterations of gene expression that do not require changes in DNA sequences. Epigenetic changes can take place through alterations of chromatin structure.
- Centromeres are chromosomal regions where spindle microtubules attach. Telomeres stabilize the ends of chromosomes.
- Eukaryotic DNA exhibits three classes of sequences. Unique-sequence DNA exists in very few copies. Moderately repetitive DNA consists of moderately long sequences that are repeated from hundreds to thousands of times. Highly repetitive DNA consists of very short sequences that are repeated in tandem from many thousands to millions of times.

IMPORTANT TERMS

nucleotide (p. 217)
Chargaff's rules (p. 217)
transforming principle (p. 218)
isotope (p. 220)
X-ray diffraction (p. 221)
ribose (p. 222)
deoxyribose (p. 222)
nitrogenous base (p. 222)
purine (p. 222)
pyrimidine (p. 222)
adenine (A) (p. 222)
guanine (G) (p. 222)
cytosine (C) (p. 223)
thymine (T) (p. 223)
uracil (U) (p. 223)
nucleoside (p. 223)
phosphate group (p. 223)
deoxyribonucleotide (p. 223)
ribonucleotide (p. 223)
phosphodiester linkage (p. 223)
polynucleotide strand (p. 223)
5′ end (p. 224)
3′ end (p. 224)
antiparallel (p. 224)
complementary DNA strands (p. 225)
B-DNA (p. 226)
A-DNA (p. 226)
Z-DNA (p. 226)
transcription (p. 227)
translation (p. 227)
replication (p. 227)
central dogma (p. 227)
reverse transcription (p. 227)
RNA replication (p. 227)
supercoiling (p. 228)
relaxed state (p. 228)
positive supercoiling (p. 228)
negative supercoiling (p. 228)
topoisomerase (p. 229)
nucleoid (p. 229)
euchromatin (p. 230)
heterochromatin (p. 230)
nonhistone chromosomal protein (p. 230)
nucleosome (p. 230)
linker DNA (p. 231)
epigenetic change (epigenetics) (p. 232)
telomeric sequence (p. 233)
shelterin (p. 233)
C-value (p. 234)
C-value paradox (p. 234)
unique-sequence DNA (p. 234)
gene family (p. 234)
repetitive DNA (p. 234)
moderately repetitive DNA (p. 234)
tandem repeat sequence (p. 234)
interspersed repeat sequence (p. 234)
short interspersed element (SINE) (p. 234)
long interspersed element (LINE) (p. 234)
highly repetitive DNA (p. 234)

ANSWERS TO CONCEPT CHECKS

1. Without knowledge of the structure of DNA, it was impossible to understand how genetic information was encoded or expressed.
2. c
3. No: carbon is found in both protein and nucleic acid.
4. b
5. d
6. Z-DNA has a left-handed helix; B-DNA has a right-handed helix. The sugar–phosphate backbone of Z-DNA zigzags back and forth, whereas the sugar–phosphate backbone of B-DNA forms a smooth continuous ribbon.
7. b
8. Bacterial DNA is not complexed with histone proteins and is circular.
9. b
10. d
11. a

WORKED PROBLEMS

Problem 1

The percentage of cytosine in a double-stranded DNA molecule is 40%. What is the percentage of thymine?

Solution Strategy

What information is required in your answer to the problem?

The percentage of thymine in the DNA molecule.

What information is provided to solve the problem?

- The DNA molecule is double-stranded.
- The percentage of cytosine is 40%.

For help with this problem, review:

The Primary Structure of DNA and Secondary Structures of DNA in Section 8.3.

Solution Steps

If C = 40%, then G also must be 40%. The total percentage of C + G is therefore 40% + 40% = 80%. All the remaining bases must be either A or T; so the total percentage of A + T = 100% − 80% = 20%; because the percentage of A equals the percentage of T, the percentage of T is 20%/2 = 10%.

Recall: In double-stranded DNA, A pairs with T, whereas G pairs with C; so the percentage of A equals the percentage of T, and the percentage of G equals the percentage of C.

Problem 2

Which of the following relations will be true for the percentage of bases in double-stranded DNA?

a. $C + T = A + G$

b. $\frac{C}{A} = \frac{T}{G}$

Solution Strategy

What information is required in your answer to the problem?

Indicate whether $C + T = A + G$ and $\frac{C}{A} = \frac{T}{G}$ are true.

What information is provided to solve the problem?

- The DNA is double-stranded.
- Ratios of different groups of bases.

For help with this problem, review:

The Primary Structure of DNA and Secondary Structures of DNA in Section 8.3.

Solution Steps

An easy way to determine whether the relations are true is to arbitrarily assign percentages to the bases, remembering that, in double-stranded DNA, A = T and G = C. For example, if the percentages of A and T are each 30%, then the percentages of G and C are each 20%. We can substitute these values into the equations to see if the relations are true.

a. $20 + 30 = 30 + 20$. This relation is true.

b. $\frac{20}{30} \neq \frac{30}{20}$. This relation is not true.

Problem 3

A diploid plant cell contains 2 billion base pairs of DNA.

a. How many nucleosomes are present in the cell?

b. Give the number of molecules of each type of histone protein associated with the genomic DNA.

Solution Strategy

What information is required in your answer to the problem?

The number of nucleosomes per cell and the numbers of each type of histone protein associated with the DNA.

What information is provided to solve the problem?

- The cell contains 2 billion base pairs of DNA.

For help with this problem, review:

The Nucleosome in Section 8.4.

Solution Steps

Recall: The repeating unit of the chromosome is a nucleosome, which consists of DNA complexed with histone proteins.

Each nucleosome encompasses about 200 bp of DNA: 145–147 bp of DNA wrapped around the histone core, 20–22 bp of DNA associated with the H1 protein, and another 30–40 bp of linker DNA.

a. To determine how many nucleosomes are present in the cell, we simply divide the total number of base pairs of DNA (2×10^9 bp) by the number of base pairs per nucleosome:

$$\frac{2 \times 10^9 \text{nucleotides}}{2 \times 10^2 \text{nucleotides per nucleosome}} = 1 \times 10^7 \text{nucleosomes}$$

Thus, there are approximately 10 million nucleosomes in the cell.

b. Each nucleosome includes two molecules each of H2A, H2B, H3, and H4 histones. Therefore, there are 2×10^7 molecules each of H2A, H2B, H3, and H4 histones. Each nucleosome has associated with it one copy of the H1 histone, so there are 1×10^7 molecules of H1.

COMPREHENSION QUESTIONS

Section 8.1

1. What four general characteristics must the genetic material possess?

Section 8.2

2. What is transformation? How did Avery and his colleagues demonstrate that the transforming principle is DNA?

3. How did Hershey and Chase show that DNA is passed to new phages in phage reproduction?

Section 8.3

4. Draw and identify the three parts of a DNA nucleotide.

5. How does an RNA nucleotide differ from a DNA nucleotide?

6. Draw a short segment of a single DNA polynucleotide strand, including at least three nucleotides. Indicate the polarity of the strand by identifying the 5′ end and the 3′ end.

7. What are some of the important genetic implications of the DNA structure?

8. What are the three major pathways of information flow within the cell?

Section 8.4

9. How does supercoiling arise? What is the difference between positive and negative supercoiling?

10. What functions does supercoiling serve for the cell?

11. What are some differences between euchromatin and heterochromatin?

12. Describe the composition and structure of the nucleosome.

13. Describe in steps how the double helix of DNA, which is 2 nm wide, gives rise to a chromosome that is 700 nm wide.

14. What are epigenetic changes?

Section 8.5

15. Describe the function and molecular structure of a telomere.

16. Describe the different classes of DNA sequence variation that exist in eukaryotes.

For more questions that test your comprehension of the key chapter concepts, go to **LearningCurve** for this chapter.

APPLICATION QUESTIONS AND PROBLEMS

Introduction

17. The introduction to this chapter, which describes the sequencing of 4000-year-old DNA, emphasizes DNA's extreme stability. What aspects of DNA's structure contribute to the stability of the molecule? Why is RNA less stable than DNA?

18. Match the scientists with the discoveries listed.

a. Kossel
b. Watson and Crick
c. Levene
d. Miescher
e. Hershey and Chase
f. Avery, MacLeod, and McCarty
g. Griffith
h. Franklin and Wilkins
i. Chargaff

___ Took X-ray diffraction pictures used in determining the structure of DNA.
___ Determined that DNA contains nitrogenous bases.
___ Identified DNA as the genetic material in bacteriophage.
___ Discovered regularity in the ratios of different bases in DNA.

___ Determined that DNA is responsible for transformation in bacteria.
___ Worked out the helical structure of DNA by building models.
___ Discovered that DNA consists of repeating nucleotides.
___ Determined that DNA is acidic and high in phosphorus.
___ Demonstrated that heat-killed material from bacteria could genetically transform live bacteria.

Section 8.2

*19. A student mixes some heat-killed type IIS *Streptococcus pneumoniae* bacteria with live type IIR bacteria and injects the mixture into a mouse. The mouse develops pneumonia and dies. The student recovers some type IIS bacteria from the dead mouse. If this is the only experiment conducted by the student, has the student demonstrated that transformation has taken place? What other explanations might explain the presence of the type IIS bacteria in the dead mouse?

20. Predict what would have happened if Griffith had mixed some heat-killed type IIIS bacteria and some heat-killed type IIR bacteria and injected this mixture into a mouse. Would the mouse have contracted pneumonia and died? Explain why or why not.

Section 8.3

*21. DNA molecules of different sizes are often separated with the use of a technique called electrophoresis (see Chapter 14). With this technique, DNA molecules are placed in a gel, an electrical current is applied to the gel, and the DNA molecules migrate toward the positive (+) pole of the current. What aspect of its structure causes a DNA molecule to migrate toward the positive pole?

*22. DATA ANALYSIS Erwin Chargaff collected data on the proportions of nucleotide bases from the DNA of a variety of different organisms and tissues (E. Chargaff, in *The Nucleic Acids: Chemistry and Biology*, vol. 1, E. Chargaff and J. N. Davidson, Eds. New York: Academic Press, 1955). Data from the DNA of several organisms analyzed by Chargaff are shown here.

	Percent			
Organism and tissue	**A**	**G**	**C**	**T**
Sheep thymus	29.3	21.4	21.0	28.3
Pig liver	29.4	20.5	20.5	29.7
Human thymus	30.9	19.9	19.8	29.4
Rat bone marrow	28.6	21.4	20.4	28.4
Hen erythrocytes	28.8	20.5	21.5	29.2
Yeast	31.7	18.3	17.4	32.6
E. coli	26.0	24.9	25.2	23.9
Human sperm	30.9	19.1	18.4	31.6
Salmon sperm	29.7	20.8	20.4	29.1
Herring sperm	27.8	22.1	20.7	27.5

a. For each organism, compute the ratio of (A + G)/(T + C) and the ratio of (A + T)/(C + G).

b. Are these ratios constant or do they vary among the organisms? Explain why.

c. Is the (A + G)/(T + C) ratio different for the sperm samples? Would you expect it to be? Why or why not?

Erwin Chargaff. [Horst Tappe/Getty Images.]

23. DATA ANALYSIS Boris Magasanik collected data on the amounts of the bases of RNA isolated from a number of sources (shown in the next column), expressed relative to a value of 10 for adenine (B. Magasanik, in *The Nucleic Acids: Chemistry and Biology*, vol. 1, E. Chargaff and J. N. Davidson, Eds. New York: Academic Press, 1955).

	Percent			
Organism and tissue	**A**	**G**	**C**	**U**
Rat liver nuclei	10	14.8	14.3	12.9
Rabbit liver nuclei	10	13.6	13.1	14.0
Cat brain	10	14.7	12.0	9.5
Carp muscle	10	21.0	19.0	11.0
Yeast	10	12.0	8.0	9.8

a. For each organism, compute the ratio of (A + G)/(U + C).

b. How do these ratios compare with the (A + G)/(T + C) ratio found in DNA (see Problem 22)? Explain.

24. Which of the following relations or ratios would be true for a double-stranded DNA molecule?

a. $A + T = G + C$

b. $A + T = T + C$

c. $A + C = G + T$

d. $\frac{A + T}{C + G} = 1.0$

e. $\frac{A + G}{C + T} = 1.0$

f. $\frac{A}{C} = \frac{G}{T}$

g. $\frac{A}{G} = \frac{T}{C}$

h. $\frac{A}{T} = \frac{G}{C}$

*25. If a double-stranded DNA molecule is 15% thymine, what are the percentages of all the other bases?

26. DATA ANALYSIS Heinz Shuster collected the following data on the base composition of the ribgrass virus (H. Shuster, in *The Nucleic Acids: Chemistry and Biology*, vol. 3, E. Chargaff and J. N. Davidson, Eds. New York: Academic Press,

1955). On the basis of this information, is the hereditary information of the ribgrass virus RNA or DNA? Is it likely to be single stranded or double stranded?

	Percent				
	A	G	C	T	U
Ribgrass virus	29.3	25.8	18.0	0.0	27.0

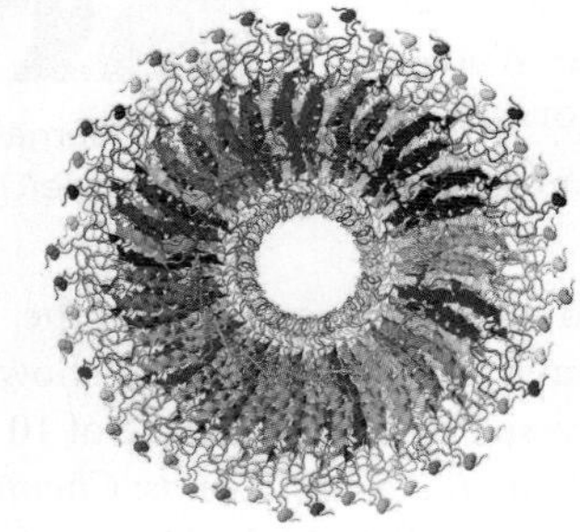

Ribgrass mosaic virus. [Leibniz Institute for Age Research, Fritz Lipmann-Institute.]

*27. For entertainment on a Friday night, a genetics professor proposed that his children diagram a polynucleotide strand of DNA. Having learned about DNA in preschool, his five-year-old daughter Sarah was able to draw a polynucleotide strand, but she made a few mistakes. Sarah's diagram on the right contains at least 10 mistakes.

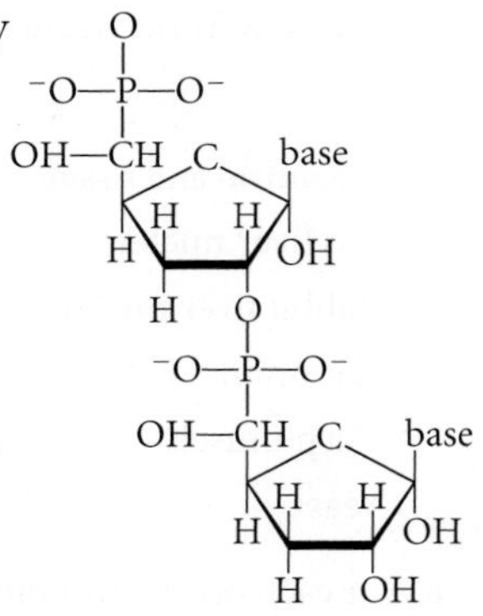

a. Make a list of all the mistakes in the structure of this DNA polynucleotide strand.

b. Draw the correct structure for the polynucleotide strand.

28. One nucleotide strand of a DNA molecule has the base sequence illustrated below.

5′—ATTGCTACGG—3′

Give the base sequence and label the 5′ and 3′ ends of the complementary DNA nucleotide strand.

*29. Chapter 1 considered the theory of the inheritance of acquired characteristics and noted that this theory is no longer accepted. Is the central dogma consistent with the theory of the inheritance of acquired characteristics? Why or why not?

30. Which of the processes of information transfer illustrated in **Figure 8.14** are required for the T2 phage reproduction illustrated in **Figure 8.4**?

Section 8.5

*31. Compare and contrast prokaryotic and eukaryotic chromosomes. How are they alike and how do they differ?

*32. A diploid human cell contains approximately 6.4 billion base pairs of DNA.

a. How many nucleosomes are present in such a cell? (Assume that the linker DNA encompasses 40 bp.)

b. How many histone proteins are complexed with this DNA?

CHALLENGE QUESTIONS

Section 8.1

*33. Suppose that an automated, unmanned probe is sent into deep space to search for extraterrestrial life. After wandering for many light-years among the far reaches of the universe, this probe arrives on a distant planet and detects life. The chemical composition of life on this planet is completely different from that of life on Earth, and its genetic material is not composed of nucleic acids. What predictions can you make about the properties of the genetic material on this planet?

Section 8.2

34. How might ^{32}P and ^{35}S be used to demonstrate that the transforming principle is DNA? Briefly outline an experiment that would show that DNA, rather than protein, is the transforming principle.

Section 8.3

35. Researchers have proposed that early life on Earth used RNA as its source of genetic information and that DNA eventually replaced RNA as the source of genetic information. What aspects of DNA structure might make it better suited than RNA to be the genetic material?

*36. Imagine that you are a student in Alfred Hershey and Martha Chase's laboratory in the late 1940s. You are given five test tubes containing *E. coli* bacteria infected with T2 bacteriophages that have been labeled with either ^{32}P or ^{35}S. Unfortunately, you forget to indicate which tubes are labeled with ^{32}P and which with ^{35}S. You place the contents of each tube in a blender and turn it on for a few seconds to shear off the phage protein coats. You then centrifuge the contents to separate the protein coats and the cells, check for the presence of radioactivity, and obtain the results shown here. Which tubes contained *E. coli* infected with ^{32}P-labeled phage? Explain your answer.

Tube number	Radioactivity present in
1	Cells
2	Protein coats
3	Protein coats
4	Cells
5	Cells

THINK-PAIR-SHARE QUESTIONS

Introduction

1. What do you think might be some of the problems associated with isolating and sequencing DNA from ancient samples, such as that of the 4000-year-old Saqqaq man from Greenland?

Section 8.1

2. Suppose that proteins, instead of nucleic acids, had evolved as the carriers of genetic information. How well would proteins satisfy the four requirements for the genetic material listed in Section 8.1?

Section 8.2

3. Isaac Newton said, "If I have seen further, it is by standing on the shoulders of giants." How does this statement apply to Watson and Crick?

4. Compare and contrast Gregor Mendel's scientific method and approach to science (see Chapter 3) with that of Watson and Crick. How did they differ? Were there any similarities?

Section 8.3

5. How does its structure enable DNA to function effectively as the genetic material? Give some specific examples.

SaplingPlus Self-study tools that will help you practice what you've learned and reinforce this chapter's concepts are available online. Go to www.macmillanlearning.com/PierceGenetics6e.

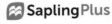

9 DNA Replication and Recombination

Topoisomerase, Replication, and Cancer

The happy tree, *Camptotheca acuminata*, contains camptothecin, a substance used to treat cancer. Camptothecin inhibits cancer by blocking an important component of the replication machinery. [Johnny Pan/Getty Images.]

In 1966, Monroe Wall and Mansukh Wani found a potential cure for cancer in the bark of the happy tree (*Camptotheca acuminata*), a rare plant native to China. Wall and Wani were in the process of screening a large number of natural substances for anticancer activity, hoping to find chemicals that might prove effective in cancer treatment. They discovered that an extract from the happy tree was effective in treating leukemia in mice. Through chemical analysis, they were able to isolate the active compound, which was dubbed camptothecin.

In the 1970s, physicians administered camptothecin to patients with incurable cancers. Although the drug showed some anticancer activity, it had toxic side effects. Eventually, chemists synthesized several analogs of camptothecin that were less toxic and more effective in cancer treatment. Two of those analogs, topotecan and irinotecan, are used today for the treatment of ovarian cancer, small-cell lung cancer, and colon cancer.

For many years, the mechanism by which camptothecin compounds inhibited cancer was unknown. In 1985, almost 20 years after its discovery, scientists at Johns Hopkins University and Smith Kline and French Laboratories (now GlaxoSmithKline) showed that camptothecin worked by inhibiting an important component of the DNA-synthesizing machinery, an enzyme called topoisomerase I.

Cancer chemotherapy is a delicate task because the target cells are the patient's own and the drugs must kill the cancer cells without killing the patient. One of the hallmarks of cancer is proliferation: the division of cancer cells is unregulated, and many cancer cells divide at a rapid rate, giving rise to a tumor with the ability to grow and spread. As we learned in Chapter 2, before a cell can divide, it must successfully replicate its DNA so that each daughter cell receives an exact copy of the genetic material. Checkpoints in the cell cycle ensure that cell division does not proceed if DNA replication is inhibited or faulty, and many cancer treatments focus on interfering with the process of DNA replication.

DNA replication is a complex process that requires a large number of components, the action of which must be intricately coordinated to ensure that DNA is accurately copied. An essential component of replication is topoisomerase. As the DNA unwinds in the replication process, strain builds up ahead of the separation, and the two strands writhe around each other, much as a rope knots up as you pull apart two of its stands. This writhing of the DNA is called supercoiling (see Chapter 8). If the supercoils are not removed, they eventually stop strand separation, and the process of replication comes to a halt. Topoisomerase enzymes remove the supercoils by clamping tightly to the DNA and breaking one or both of its strands. The strands then revolve around each other, removing the supercoiling and strain. After the DNA has relaxed, the topoisomerase reseals the broken ends of the DNA strands.

Camptothecin works by interfering with topoisomerase I, an enzyme used in DNA replication in humans. The drug inserts itself into the gap created by the break in the DNA strand, blocking the topoisomerase from resealing the broken ends. Researchers originally assumed that camptothecin trapped the topoisomerase and blocked the action of other enzymes that were necessary to synthesize the DNA. However, research now indicates that camptothecin poisons the topoisomerase so that it is unable to remove supercoils ahead of replication. Accumulating supercoils halt the replication machinery and prevent the proliferation of the cancer cells. But like many other anticancer drugs, camptothecin also inhibits the replication of normal, noncancerous cells, which is why chemotherapy often makes the patient sick.

THINK-PAIR-SHARE Question 1

This chapter focuses on DNA replication, the process by which a cell doubles its DNA before division. We begin with the basic mechanism of replication that emerged from the structure of DNA discovered by Watson and Crick. We then examine several different modes of replication, the requirements of replication, and the universal direction of DNA synthesis. We examine the enzymes and proteins that participate in this process and, finally, consider the molecular details of recombination, which is closely related to replication and is essential for the segregation of homologous chromosomes, the production of genetic variation, and DNA repair.

9.1 Genetic Information Must Be Accurately Copied Every Time a Cell Divides

In a schoolyard game, a verbal message, such as "John's brown dog ran away from home," is whispered to a child, who runs to a second child and repeats the message. The message is relayed from child to child around the schoolyard until it returns to the original sender. Inevitably, the last child returns with an amazingly transformed message, such as "Joe Brown has a pig living under his porch." The more children playing the game, the more garbled the message becomes. This game illustrates an important principle: errors arise whenever information is copied, and the more times it is copied, the greater the potential number of errors.

A complex multicellular organism faces a problem analogous to that of the children in the schoolyard game: how to faithfully transmit genetic instructions each time that its cells divide. The solution to this problem is central to replication. A single-celled human zygote contains 6.4 billion base pairs of DNA. If a copying error were made only once per million base pairs, 6400 mistakes would be made every time a cell divided—errors that would be compounded at each of the millions of cell divisions that take place in human development.

Not only must the copying of DNA be astoundingly accurate, it must also take place at breakneck speed. The single circular chromosome of *E. coli* contains about 4.6 million base pairs. At a rate of 1000 nucleotides per minute, replication of the entire chromosome would require over 3 days. Yet these bacteria are capable of dividing every 20 minutes. *Escherichia coli* actually replicates its DNA at a rate of 1000 nucleotides per second, with less than one error in a billion nucleotides. How is this extraordinarily accurate and rapid process accomplished?

THINK-PAIR-SHARE Question 2

9.2 All DNA Replication Takes Place in a Semiconservative Manner

When Watson and Crick solved the three-dimensional structure of DNA in 1953 (see Figure 8.12), several important genetic implications were immediately apparent. The complementary nature of the two nucleotide strands in a DNA molecule suggested that, during replication, each strand can serve as a template for the synthesis of a new strand. The specificity of base pairing (adenine with thymine; guanine with cytosine) implied that only one sequence of bases can be specified by each template, and so the two DNA molecules built on the pair of templates will be identical with the original. This process is called **semiconservative replication** because each of the original nucleotide strands remains intact (conserved), despite no longer being combined in the same molecule; the original DNA molecule is half (semi) conserved during replication.

Initially, three models were proposed for DNA replication. In conservative replication (**Figure 9.1a**), the entire double-stranded DNA molecule serves as a template for a whole new molecule of DNA, and the original DNA molecule is *fully* conserved during replication. In dispersive replication (**Figure 9.1b**), both nucleotide strands break down (disperse) into fragments, which serve as templates for the synthesis of new DNA fragments, and then somehow reassemble into two complete DNA molecules. In this model, each resulting DNA molecule contains interspersed fragments of old and new DNA; none of the original molecule is conserved. Semiconservative replication (**Figure 9.1c**) is intermediate between these two models; the two nucleotide strands unwind, and each serves as a template for a new DNA molecule.

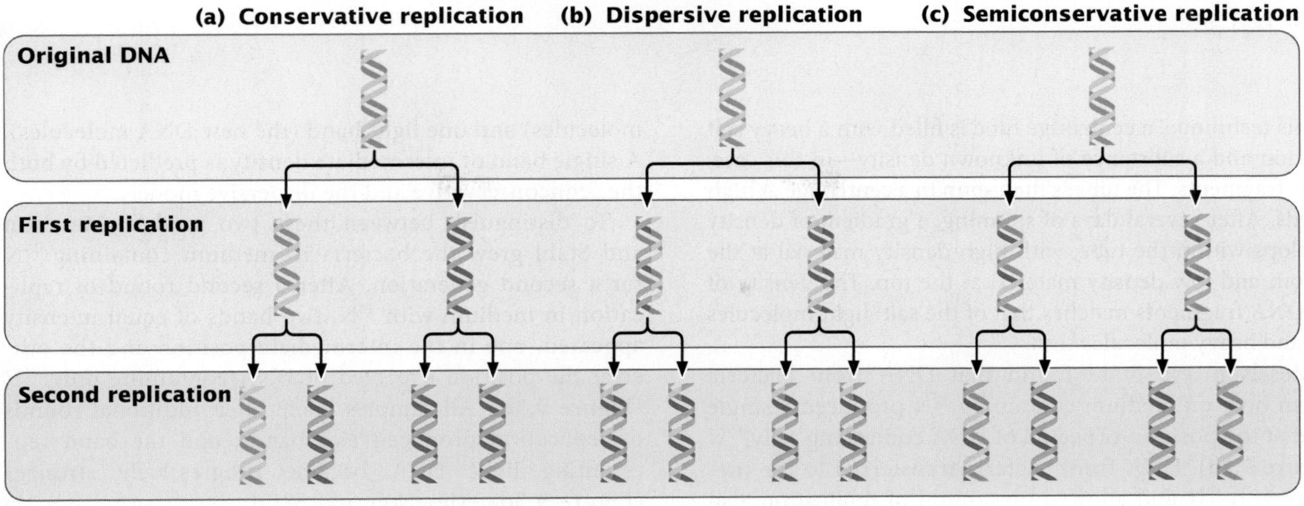

9.1 Three proposed models of replication are conservative replication, dispersive replication, and semiconservative replication.

These three models allow different predictions to be made about the distribution of original DNA and newly synthesized DNA after replication. With conservative replication, after one round of replication, 50% of the molecules would consist entirely of the original DNA and 50% would consist entirely of new DNA. After a second round of replication, 25% of the molecules would consist entirely of the original DNA and 75% would consist entirely of new DNA. With each additional round of replication, the proportion of molecules with new DNA would increase, although the number of molecules with the original DNA would remain constant. Dispersive replication would always produce hybrid molecules, containing some original and some new DNA, but the proportion of new DNA within the molecules would increase with each replication event. In contrast, with semiconservative replication, one round of replication would produce two hybrid molecules, each consisting of half original DNA and half new DNA. After a second round of replication, half the molecules would be hybrid, and the other half would consist of new DNA only. Additional rounds of replication would produce more and more molecules consisting entirely of new DNA, and a few hybrid molecules would persist.

Meselson and Stahl's Experiment

To determine which of the three models of replication applied to *E. coli* cells, Matthew Meselson and Franklin Stahl needed a way to distinguish old and new DNA. They accomplished this by using two isotopes of nitrogen, ^{14}N (the common form) and ^{15}N (a rare, heavy form). Meselson and Stahl grew a culture of *E. coli* in a medium that contained ^{15}N as the sole nitrogen source; after many generations, all the *E. coli* cells had ^{15}N incorporated into all the purine and pyrimidine bases of their DNA (see Figure 8.8). Meselson and Stahl took a sample of these bacteria, switched the rest of the bacteria to a medium that contained only ^{14}N, and then took additional samples of bacteria over the next few cellular generations. In each sample, the bacterial DNA that was synthesized before the change in medium contained ^{15}N and was relatively heavy, whereas any DNA synthesized after the switch contained ^{14}N and was relatively light.

Meselson and Stahl distinguished between the heavy ^{15}N-laden DNA and the light ^{14}N-containing DNA with the use of **equilibrium density gradient centrifugation** (**Figure 9.2**).

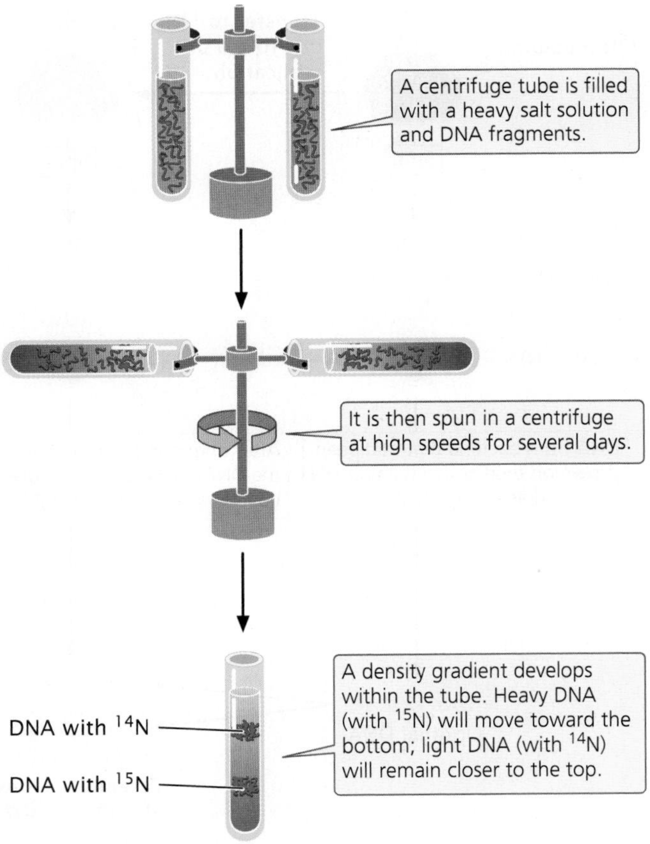

A centrifuge tube is filled with a heavy salt solution and DNA fragments.
It is then spun in a centrifuge at high speeds for several days.
DNA with ¹⁴N
DNA with ¹⁵N
A density gradient develops within the tube. Heavy DNA (with ¹⁵N) will move toward the bottom; light DNA (with ¹⁴N) will remain closer to the top.

9.2 Meselson and Stahl used equilibrium density gradient centrifugation to distinguish between heavy, ^{15}N-laden, DNA and lighter, ^{14}N-laden, DNA.

In this technique, a centrifuge tube is filled with a heavy salt solution and a substance of unknown density—in this case, DNA fragments. The tube is then spun in a centrifuge at high speeds. After several days of spinning, a gradient of density develops within the tube, with high-density material at the bottom and low-density material at the top. The density of the DNA fragments matches that of the salt: light molecules rise and heavy molecules sink.

Meselson and Stahl found that DNA from bacteria grown only on medium containing ^{15}N produced a single band at the position expected of DNA containing only ^{15}N (**Figure 9.3a**). DNA from bacteria transferred to the medium with ^{14}N and allowed one round of replication also produced a single band, but at a position intermediate between that expected of DNA containing only ^{15}N and that expected of DNA containing only ^{14}N (**Figure 9.3b**). This result is inconsistent with the conservative replication model, which predicts one heavy band (the original DNA molecules) and one light band (the new DNA molecules). A single band of intermediate density is predicted by both the semiconservative and the dispersive models.

To distinguish between these two models, Meselson and Stahl grew the bacteria in medium containing ^{14}N for a second generation. After a second round of replication in medium with ^{14}N, two bands of equal intensity appeared, one in the intermediate position and the other at the position expected of DNA containing only ^{14}N (**Figure 9.3c**). All samples taken after additional rounds of replication produced two bands, and the band representing light DNA became progressively stronger (**Figure 9.3d**). Meselson and Stahl's results were exactly as expected for semiconservative replication and were incompatible with those predicated for both conservative and dispersive replication. TRY PROBLEM 18

THINK-PAIR-SHARE Question 3

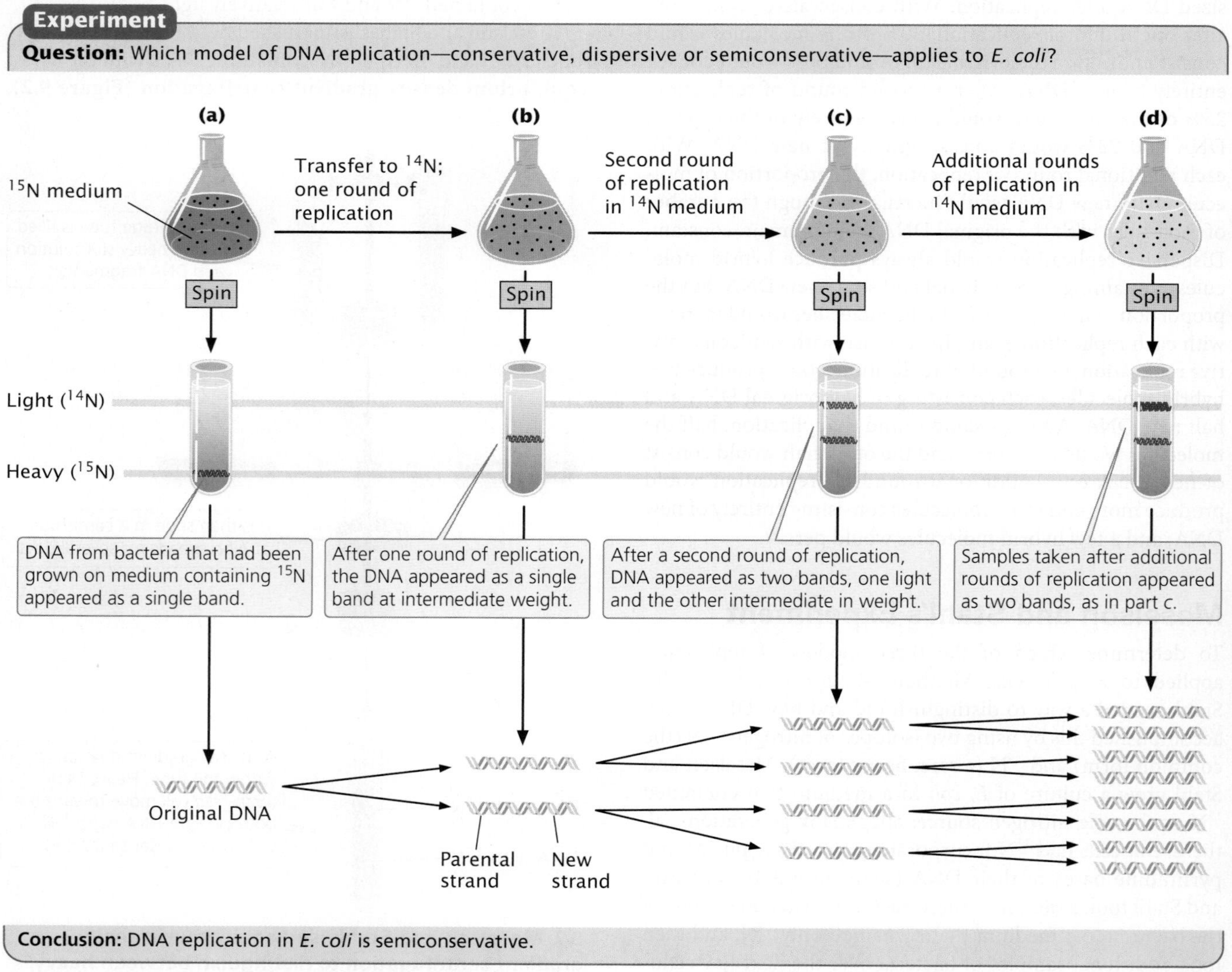

9.3 Meselson and Stahl demonstrated that DNA replication is semiconservative.

CONCEPTS

Meselson and Stahl convincingly demonstrated that replication in *E. coli* is semiconservative: each DNA strand serves as a template for the synthesis of a new DNA molecule.

✓ CONCEPT CHECK 1

How many bands of DNA would be expected in Meselson and Stahl's experiment after two rounds of *conservative* replication?

Modes of Replication

After Meselson and Stahl's work was published, investigators confirmed that other organisms also use semiconservative replication. No evidence was found for conservative or dispersive replication. There are, however, several different ways in which semiconservative replication can take place, differing principally in the nature of the template DNA—whether it is linear or circular.

A segment of DNA that undergoes replication is called a **replicon**; each replicon contains an origin of replication. Replication starts at the origin and continues until the entire replicon has been replicated. Bacterial chromosomes have a single origin of replication, whereas eukaryotic chromosomes contain many.

A common type of replication that takes place in circular DNA, such as that found in *E. coli* and other bacteria, is called **theta replication** (**Figure 9.4**) because it generates an intermediate structure that resembles the Greek letter theta (θ). In Figure 9.4 and all subsequent figures in this chapter, the original (template) strand of DNA is shown in gray and the newly synthesized strand of DNA is shown in red.

In theta replication, double-stranded DNA begins to unwind at the origin of replication, producing single nucleotide strands that then serve as templates on which new DNA can be synthesized. The unwinding of the double helix generates a loop, termed a **replication bubble**. Unwinding occurs at one or both ends of the bubble, making it progressively larger. DNA replication on both of the template strands is simultaneous with unwinding. The point of unwinding, where the two single strands separate from the double-stranded DNA helix, is called a **replication fork**.

If there are two replication forks, one at each end of the replication bubble, the forks proceed outward in both directions in a process called **bidirectional replication**, simultaneously

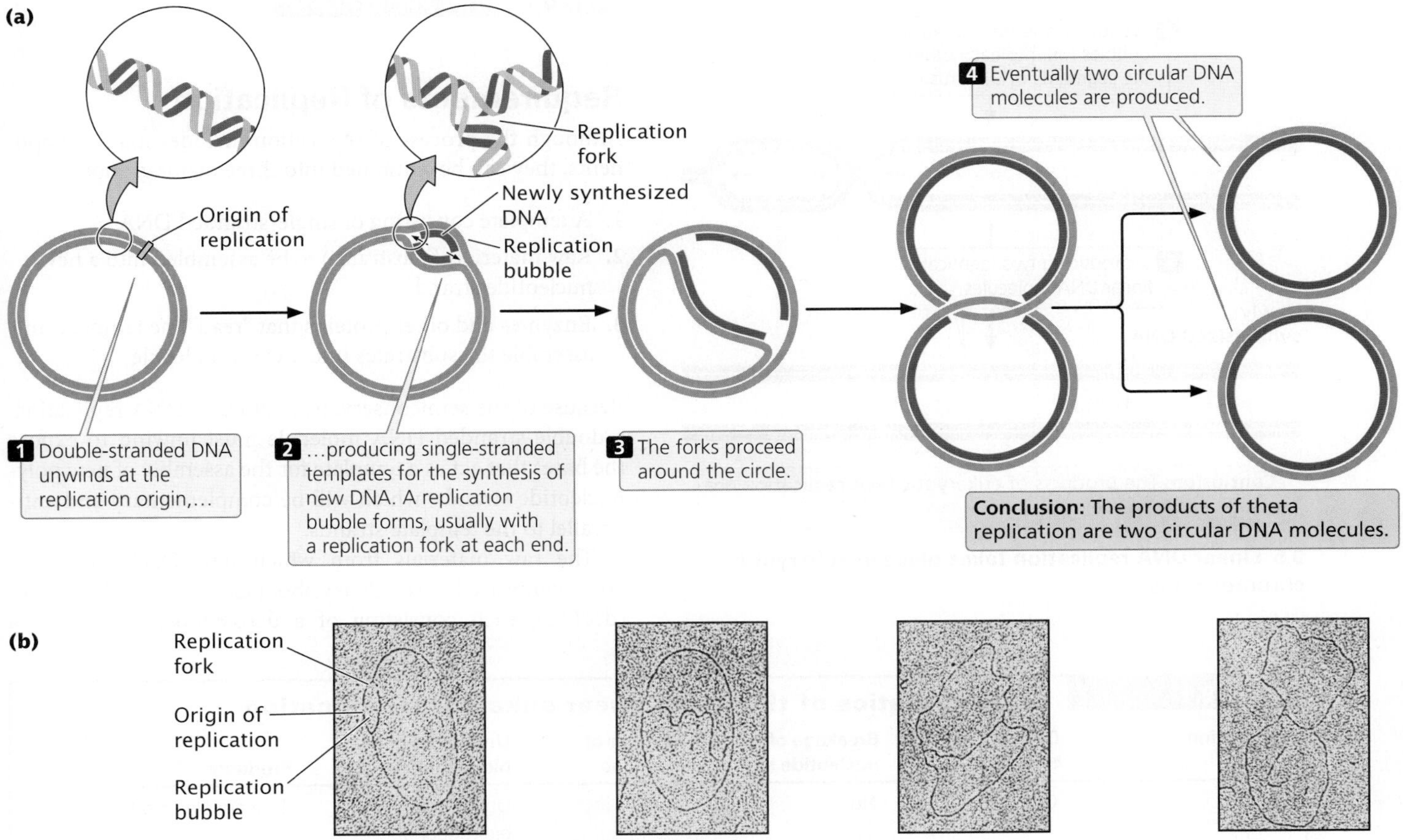

9.4 Theta replication is a type of replication common in *E. coli* and other organisms possessing circular DNA. [Part b: Bernhard Hirt, L'Institut Suisse de Recherche Expérimentale sur le Cancer.]

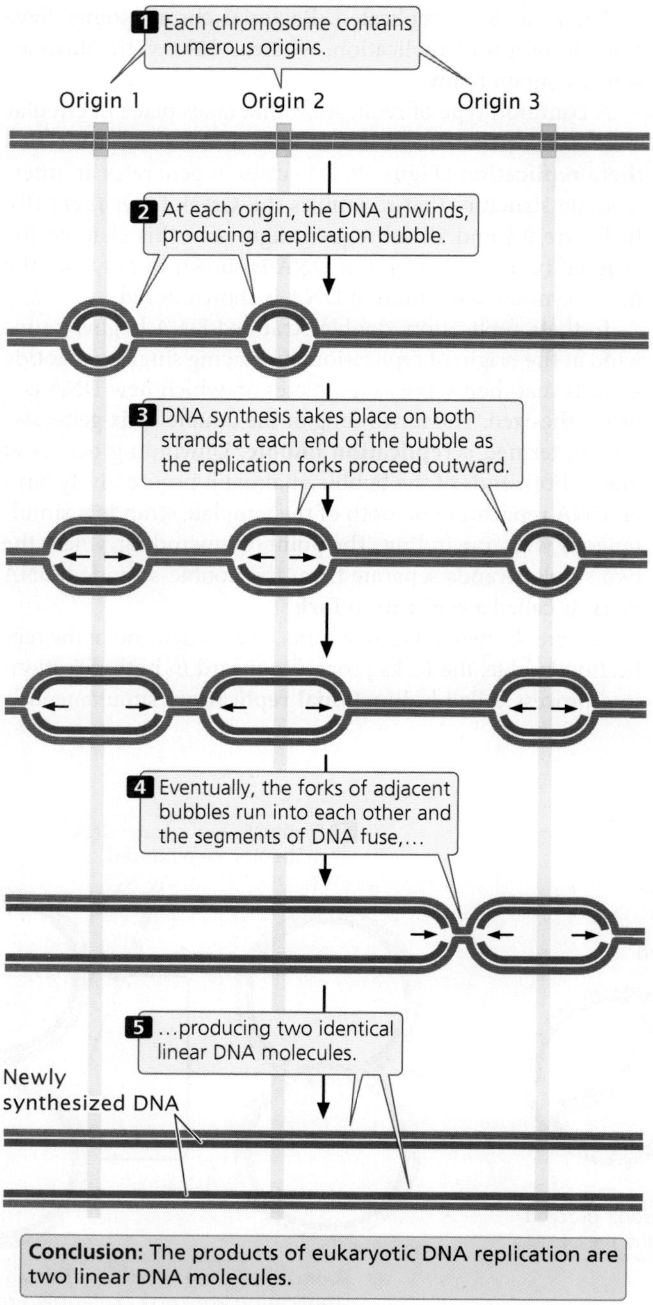

9.5 Linear DNA replication takes place in eukaryotic chromosomes.

unwinding and replicating the DNA until they eventually meet. If unidirectional replication with a single replication fork is present, it proceeds around the entire circle to produce two complete circular DNA molecules, each consisting of one old and one new nucleotide strand.

Circular DNA molecules that undergo theta replication have a single origin of replication. Because of the limited size of these DNA molecules, replication starting from one origin can traverse the entire chromosome in a reasonable amount of time. The large linear chromosomes in eukaryotic cells, however, contain far too much DNA to be replicated speedily from a single origin. The replication of eukaryotic chromosomes is initiated at thousands of origins.

Typical eukaryotic replicons are from 20,000 to 300,000 base pairs in length. At each origin of replication, the DNA unwinds and produces a replication bubble. Replication takes place on both strands at each end of the bubble, with the two replication forks spreading outward. Eventually, the replication forks of adjacent replicons run into each other, and the replicons fuse to form long stretches of newly synthesized DNA (**Figure 9.5**). The replication and fusion of all the replicons lead to two identical DNA molecules. Important features of theta replication and linear eukaryotic replication are summarized in **Table 9.1**. TRY PROBLEM 19

Requirements of Replication

Although the process of replication includes many components, they can be combined into three major groups:

1. A template consisting of single-stranded DNA
2. Raw materials (substrates) to be assembled into a new nucleotide strand
3. Enzymes and other proteins that "read" the template and assemble the substrates into a DNA molecule

Because of the semiconservative nature of DNA replication, a double-stranded DNA molecule must unwind to expose the bases that act as a template for the assembly of new polynucleotide strands, which will be complementary and antiparallel to the template strands.

The raw materials from which new DNA molecules are synthesized are deoxyribonucleoside triphosphates (dNTPs), each consisting of a deoxyribose sugar and a

TABLE 9.1 Characteristics of theta and linear eukaryotic replication

Replication model	DNA template	Breakage of nucleotide strand	Number of replicons	Unidirectional or bidirectional	Products
Theta	Circular	No	1	Unidirectional or bidirectional	Two circular molecules
Linear eukaryotic	Linear	No	Many	Bidirectional	Two linear molecules

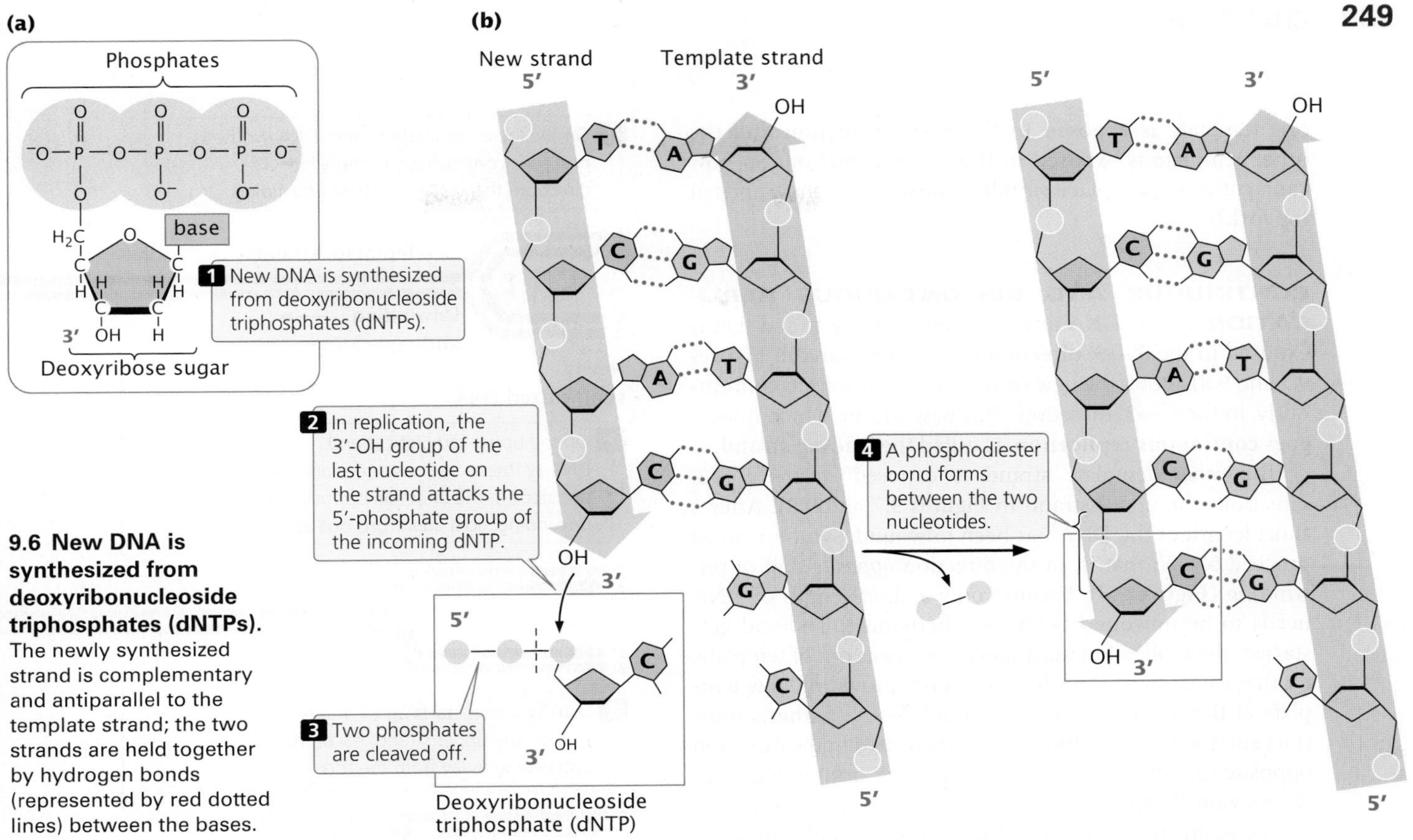

9.6 New DNA is synthesized from deoxyribonucleoside triphosphates (dNTPs). The newly synthesized strand is complementary and antiparallel to the template strand; the two strands are held together by hydrogen bonds (represented by red dotted lines) between the bases.

base (a nucleoside) attached to three phosphate groups (**Figure 9.6a**). In DNA synthesis, nucleotides are added to the 3′-hydroxyl (3′-OH) group of the growing nucleotide strand (**Figure 9.6b**). The 3′-OH group of the last nucleotide on the strand attacks the 5′-phosphate group of the incoming dNTP. Two phosphate groups are cleaved from the incoming dNTP, and a phosphodiester bond is created between the two nucleotides.

DNA synthesis does not happen spontaneously. Rather, it requires a number of enzymes and proteins that function in a coordinated manner. We will examine this complex array of proteins and enzymes in Section 9.3 as we consider the replication process in more detail.

CONCEPTS

DNA replication requires a single-stranded DNA template, deoxyribonucleoside triphosphates, a growing nucleotide strand, and a group of enzymes and proteins.

Direction of Replication

In DNA synthesis, new nucleotides are joined one at a time to the 3′ end of the newly synthesized strand. **DNA polymerases**, the enzymes that synthesize DNA, can add nucleotides only to the 3′ end of the growing strand (not the 5′ end), and so new DNA strands always elongate in the same 5′-to-3′ direction (5′→3′). Because the two single-stranded DNA templates are antiparallel and strand elongation is always 5′→3′, if synthesis on one template proceeds from, say, right to left, then synthesis on the other template must proceed in the opposite direction, from left to right (**Figure 9.7**). As DNA unwinds during replication, the antiparallel nature of the two DNA strands means that

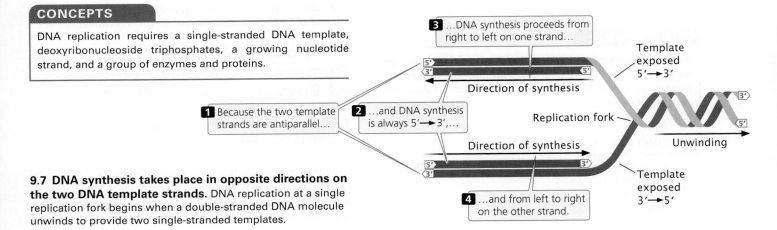

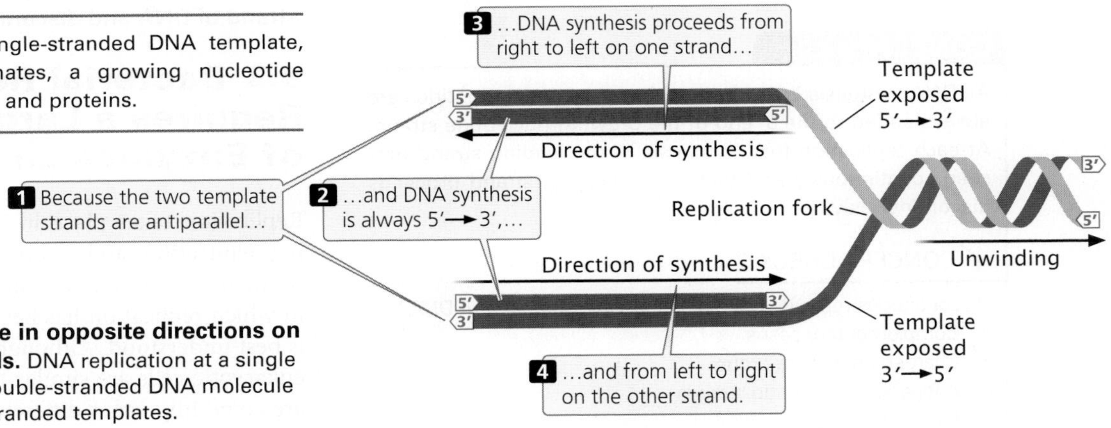

9.7 DNA synthesis takes place in opposite directions on the two DNA template strands. DNA replication at a single replication fork begins when a double-stranded DNA molecule unwinds to provide two single-stranded templates.

one template is exposed in the 5′→3′ direction and the other template is exposed in the 3′→5′ direction. So how can synthesis take place simultaneously on both strands at the fork?

CONTINUOUS AND DISCONTINUOUS REPLICATION As the DNA unwinds, the template strand that is exposed in the 3′→5′ direction (the lower strand in Figures 9.7 and 9.8) allows the new strand to be synthesized continuously, in the 5′→3′ direction. This new strand, which undergoes **continuous replication**, is called the **leading strand**.

The other template strand is exposed in the 5′→3′ direction (the upper strand in Figures 9.7 and 9.8). After a short length of the DNA has been unwound, synthesis must proceed 5′→3′; that is, in the direction *opposite* that of unwinding (**Figure 9.8**). Because only a short length of DNA needs to be unwound before synthesis on this strand gets started, the replication machinery soon runs out of template. By that time, more DNA has unwound, providing new template at the 5′ end of the new strand. DNA synthesis must start anew at the replication fork and proceed in the direction opposite that of the movement of the fork until it runs into the previously replicated segment of DNA. This process is repeated again and again, so synthesis of this strand is in short, discontinuous bursts. The newly made strand that undergoes **discontinuous replication** is called the **lagging strand**.

OKAZAKI FRAGMENTS The short lengths of DNA produced by the discontinuous replication of the lagging strand are called **Okazaki fragments**, after Reiji Okazaki, who discovered them. In bacterial cells, each Okazaki fragment ranges from about 1000 to 2000 nucleotides in length; in eukaryotic cells, they are about 100 to 200 nucleotides long. Okazaki fragments on the lagging strand are linked together to create a continuous new DNA molecule. To see how replication occurs continuously on one strand and discontinuously on the other, view **Animation 9.1**.

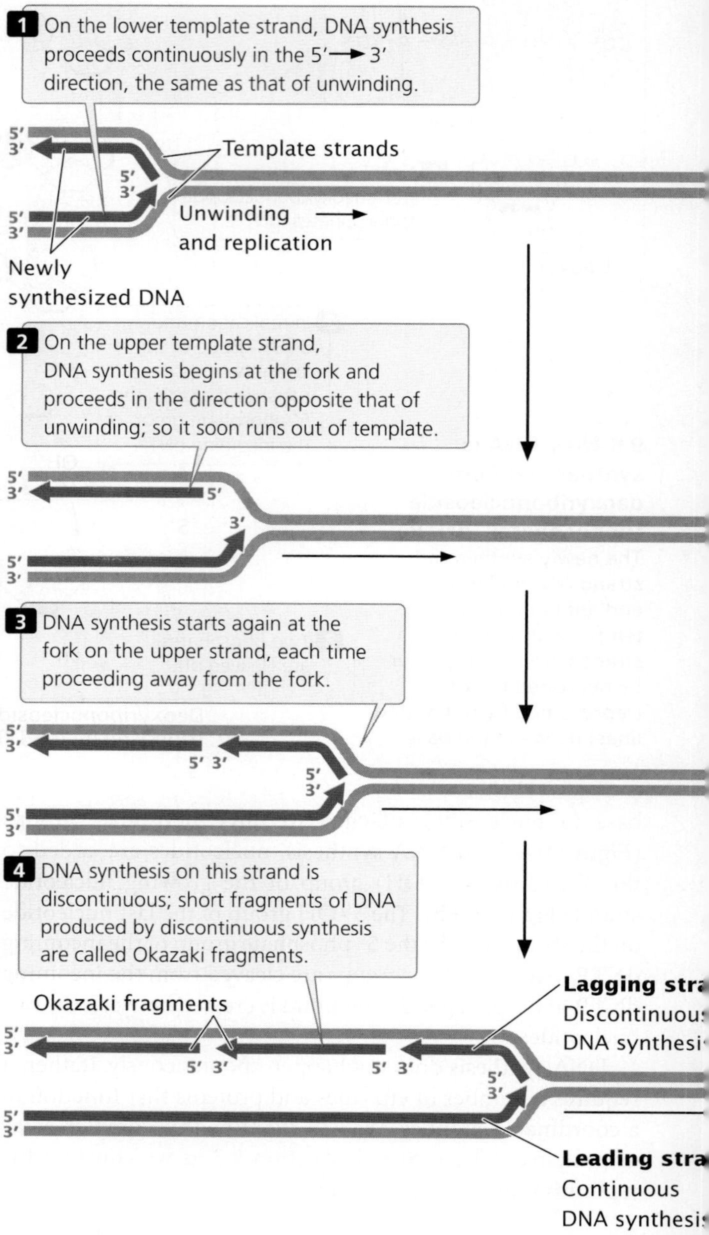

1 On the lower template strand, DNA synthesis proceeds continuously in the 5′→3′ direction, the same as that of unwinding.
Template strands
Unwinding and replication
Newly synthesized DNA
2 On the upper template strand, DNA synthesis begins at the fork and proceeds in the direction opposite that of unwinding; so it soon runs out of template.
3 DNA synthesis starts again at the fork on the upper strand, each time proceeding away from the fork.
4 DNA synthesis on this strand is discontinuous; short fragments of DNA produced by discontinuous synthesis are called Okazaki fragments.
Okazaki fragments

9.8 DNA synthesis is continuous on one template strand of DNA and discontinuous on the other.

CONCEPTS

All DNA synthesis is 5′→3′, meaning that new nucleotides are always added to the 3′ end of the growing nucleotide strand. At each replication fork, synthesis of the leading strand proceeds continuously and that of the lagging strand proceeds discontinuously.

✓ CONCEPT CHECK 2

Discontinuous replication is a result of which property of DNA?

a. Complementary bases
b. Charged phosphate group
c. Antiparallel nucleotide strands
d. Five-carbon sugar

9.3 Bacterial Replication Requires a Large Number of Enzymes and Proteins

Replication takes place in four stages: initiation, unwinding, elongation, and termination. The following discussion of the process of replication will focus on bacterial systems, in which replication has been most thoroughly studied and is best understood. Although many aspects of replication in eukaryotic cells are similar to those in bacterial cells, there are some important differences. We will compare bacterial and eukaryotic replication later in this chapter.

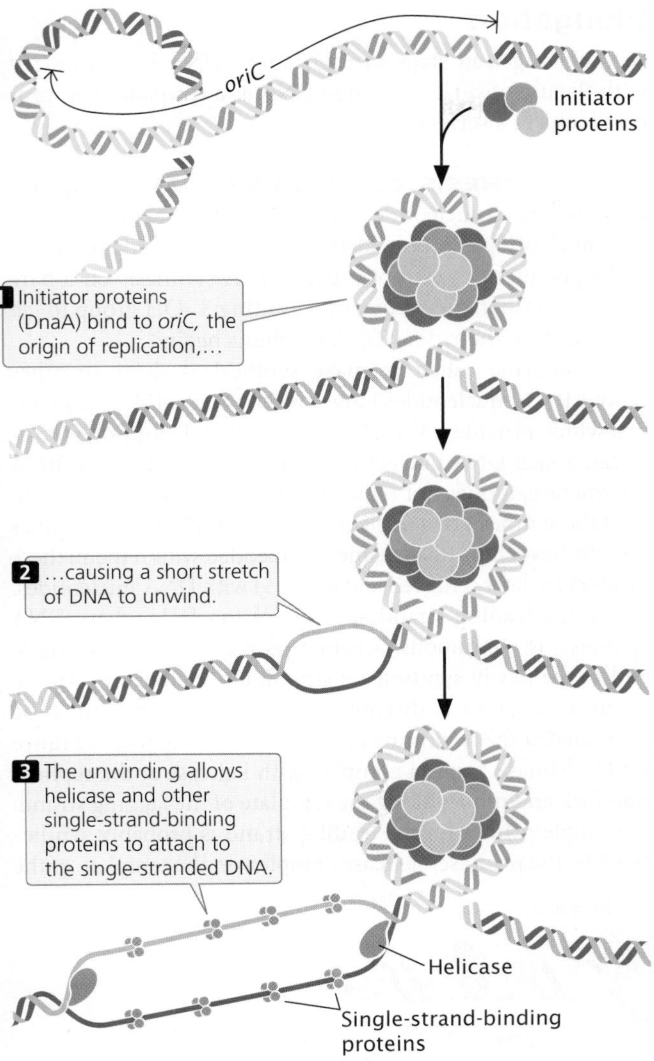

9.9 *E. coli* DNA replication begins when initiator proteins bind to *oriC*, the origin of replication.

Initiation

The circular chromosome of *E. coli* has a single origin of replication (*oriC*). **Initiator proteins** (known as DnaA in *E. coli*) bind to *oriC* and cause a short section of DNA to unwind. This unwinding allows helicase (see next section) and other single-strand-binding proteins to attach to the polynucleotide strand (**Figure 9.9**).

Unwinding

Because DNA synthesis requires a single-stranded template, and therefore double-stranded DNA must be unwound before DNA synthesis can take place, the cell relies on several proteins and enzymes to accomplish the unwinding.

DNA HELICASE A **DNA helicase** breaks the hydrogen bonds that exist between the bases of the two nucleotide strands of a DNA molecule. Helicase cannot *initiate* the unwinding of double-stranded DNA; the initiator proteins first separate DNA strands at the origin, providing a short stretch of single-stranded DNA to which a helicase binds. Helicase binds to the lagging-strand template at each replication fork and moves in the 5′→3′ direction along this strand, thus also moving the replication fork (**Figure 9.10**).

SINGLE-STRAND-BINDING PROTEINS After DNA has been unwound by helicase, **single-strand-binding proteins** (SSBs) attach tightly to the exposed single-stranded DNA (see Figure 9.10). These proteins protect the single-stranded nucleotide chains and prevent the formation of secondary structures that interfere with replication. Unlike many DNA-binding proteins, SSBs are indifferent to base sequence: they will bind to any single-stranded DNA. Single-strand-binding proteins form tetramers (groups of four); each tetramer covers from 35 to 65 nucleotides.

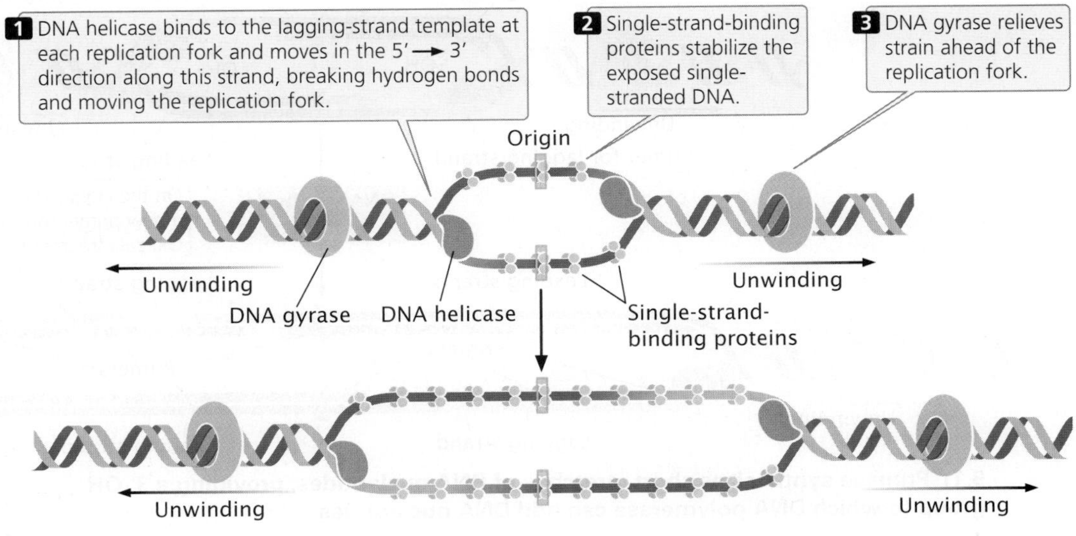

9.10 DNA helicase unwinds DNA by binding to the lagging-strand template at each replication fork and moving in the 5′→3′ direction.

DNA GYRASE Another protein essential for the unwinding process is the enzyme **DNA gyrase**, a topoisomerase. As discussed in Chapter 8 and the introduction to this chapter, topoisomerases control the supercoiling of DNA. They come in two major types: type I topoisomerases alter supercoiling by making single-strand breaks in DNA, while type II topoisomerases create double-strand breaks. DNA gyrase is a type II topoisomerase. In replication, DNA gyrase reduces the torsional strain (torque) that builds up ahead of the replication fork as a result of unwinding (see Figure 9.10). It reduces torque by making a double-strand break in one segment of the DNA helix, passing another segment of the helix through the break, and then resealing the broken ends of the DNA. This action removes a twist in the DNA and reduces the supercoiling.

CONCEPTS

Replication is initiated at an origin of replication, where initiator proteins bind and cause a short stretch of DNA to unwind. DNA helicase breaks hydrogen bonds at a replication fork, and single-strand-binding proteins stabilize the separated strands. DNA gyrase reduces the torsional strain that develops as the two strands of double-helical DNA unwind.

✓ CONCEPT CHECK 3

Place the following components in the order in which they are first used in the course of replication: helicase, single-strand-binding protein, DNA gyrase, initiator proteins.

Elongation

In the elongation stage of replication, DNA is synthesized with the use of single-stranded DNA as a template. This process requires a series of enzymes.

THE SYNTHESIS OF PRIMERS All DNA polymerases require a nucleotide with a 3′-OH group to which a new nucleotide can be added. Because of this requirement, DNA polymerases cannot initiate DNA synthesis on a bare template; rather, they require an existing 3′-OH group to get started. How, then, does DNA synthesis begin?

An enzyme called **primase** synthesizes short stretches (about 10–12 nucleotides long) of RNA nucleotides, or **primers**, which provide a 3′-OH group to which DNA polymerases can attach DNA nucleotides. (Because primase is an RNA polymerase, it does not require a preexisting 3′-OH group to start the synthesis of a nucleotide strand.) All DNA molecules initially have short RNA primers embedded within them; these primers are later removed and replaced with DNA nucleotides.

On the leading strand at a replication fork, where DNA synthesis is continuous, a primer is required only at the 5′ end of the newly synthesized strand. On the lagging strand, where replication is discontinuous, a new primer must be generated at the beginning of each Okazaki fragment (**Figure 9.11**). Primase forms a complex with helicase at the replication fork and moves along the template of the lagging strand. The single primer on the leading strand is probably synthesized by the primase–helicase complex on the template of the

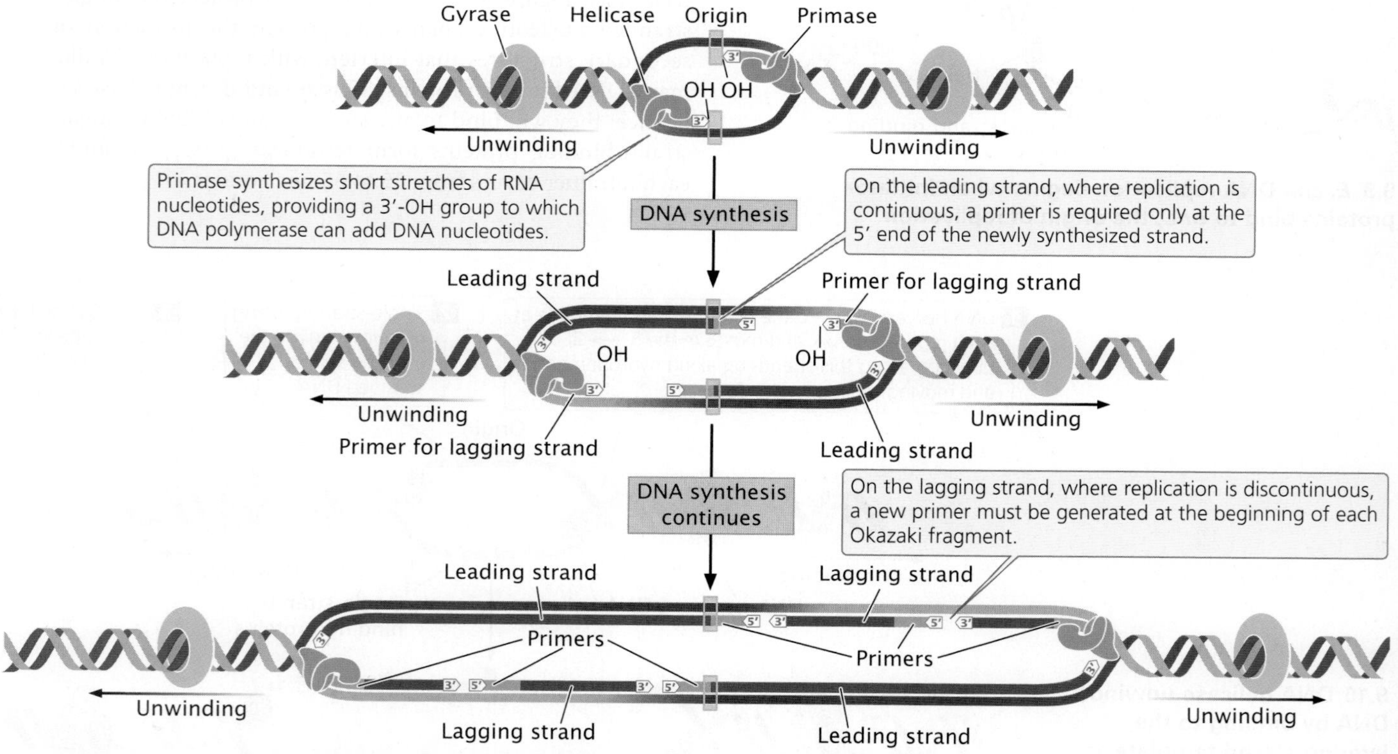

9.11 Primase synthesizes short stretches of RNA nucleotides, providing a 3′-OH group to which DNA polymerase can add DNA nucleotides.

CONCEPTS

Primase synthesizes a short stretch of RNA nucleotides (a primer), which provides a 3′-OH group for the attachment of DNA nucleotides to start DNA synthesis.

✓ CONCEPT CHECK 4

Primers are synthesized where on the lagging strand?

a. Only at the 5′ end of the newly synthesized strand
b. Only at the 3′ end of the newly synthesized strand
c. At the beginning of every Okazaki fragment
d. At multiple places within an Okazaki fragment

lagging strand of the *other* replication fork, at the opposite end of the replication bubble.

DNA SYNTHESIS BY DNA POLYMERASES After DNA has unwound and a primer has been added, DNA polymerases elongate the polynucleotide strand by catalyzing DNA polymerization. The best-studied polymerases are those of *E. coli*, which has at least five different DNA polymerases. Two of them, DNA polymerase I and DNA polymerase III, carry out DNA synthesis in replication (**Table 9.2**); the other three have specialized functions in DNA repair.

DNA polymerase III is a large multiprotein complex that acts as the main workhorse of replication. DNA polymerase III synthesizes nucleotide strands by adding new nucleotides to the 3′ end of a growing DNA strand. This enzyme has two enzymatic activities (see Table 9.2). Its 5′→3′ polymerase activity allows it to add new nucleotides in the 5′→3′ direction. Its 3′→5′ exonuclease activity allows it to remove nucleotides in the 3′→5′ direction, enabling it to correct errors. If a nucleotide with an incorrect base is inserted into the growing DNA strand, DNA polymerase III uses its 3′→5′ exonuclease activity to back up and remove the incorrect nucleotide. It then resumes its 5′→3′ polymerase activity. These two functions together allow DNA polymerase III to efficiently and accurately synthesize new DNA molecules.

The first *E. coli* polymerase to be discovered, **DNA polymerase I**, also has 5′→3′ polymerase and 3′→5′ exonuclease activities (see Table 9.2), which allow the enzyme to synthesize DNA and to correct errors. Unlike DNA polymerase III, however, DNA polymerase I also possesses 5′→3′ exonuclease activity, which is used to remove the primers laid down by primase and replace them with DNA nucleotides by synthesizing in a 5′→3′ direction (see Figure 9.12 on p. 254). The removal and replacement of primers appears to constitute the main function of DNA polymerase I.

Despite their differences, all of *E. coli*'s DNA polymerases

1. synthesize any sequence specified by the template strand.
2. synthesize in the 5′→3′ direction by adding nucleotides to a 3′-OH group.
3. use dNTPs to synthesize new DNA.
4. require a 3′-OH group to initiate synthesis.
5. catalyze the formation of a phosphodiester bond by joining the 5′-phosphate group of the incoming nucleotide to the 3′-OH group of the preceding nucleotide on the growing strand, cleaving off two phosphates in the process.
6. produce newly synthesized strands that are complementary and antiparallel to the template strands.
7. are associated with a number of other proteins.

▶TRY PROBLEM 23

CONCEPTS

DNA polymerases synthesize DNA in the 5′→3′ direction by adding new nucleotides to the 3′ end of a growing nucleotide strand.

PRIMER REPLACEMENT AND DNA LIGASE After DNA polymerase III attaches a DNA nucleotide to the 3′-OH group on the last nucleotide of the RNA primer, each new DNA nucleotide then provides the 3′-OH group needed for the next DNA nucleotide to be added. This process continues as long as template is available (**Figure 9.12a**). DNA polymerase I follows DNA polymerase III and, using its 5′→3′ exonuclease activity, removes the RNA primer. It then uses its 5′→3′ polymerase activity to replace the RNA nucleotides with DNA nucleotides. DNA polymerase I attaches the first nucleotide to the OH group at the 3′ end of the preceding Okazaki fragment and then continues, in the 5′→3′ direction along the nucleotide strand, removing and replacing, one at a time, the RNA nucleotides of the primer (**Figure 9.12b**).

TABLE 9.2 Characteristics of DNA polymerases that function in replication in *E. coli*

DNA polymerase	5′→3′ polymerase activity	3′→5′ exonuclease activity	5′→3′ exonuclease activity	Function
I	Yes	Yes	Yes	Removes and replaces primers
III	Yes	Yes	No	Elongates DNA

Note: DNA polymerases II, IV, and V are involved in DNA repair and translesion synthesis.

After polymerase I has replaced the last nucleotide of the RNA primer with a DNA nucleotide, a break remains in the sugar–phosphate backbone of the new DNA strand. The 3′-OH group of the last nucleotide to have been added by DNA polymerase I is not attached to the 5′-phosphate group of the first nucleotide added by DNA polymerase III (**Figure 9.12c**). This break is sealed by the enzyme **DNA ligase**, which catalyzes the formation of a phosphodiester bond without adding another nucleotide to the strand (**Figure 9.12d**). Some of the major enzymes and proteins required for prokaryotic DNA replication are summarized in **Table 9.3**.

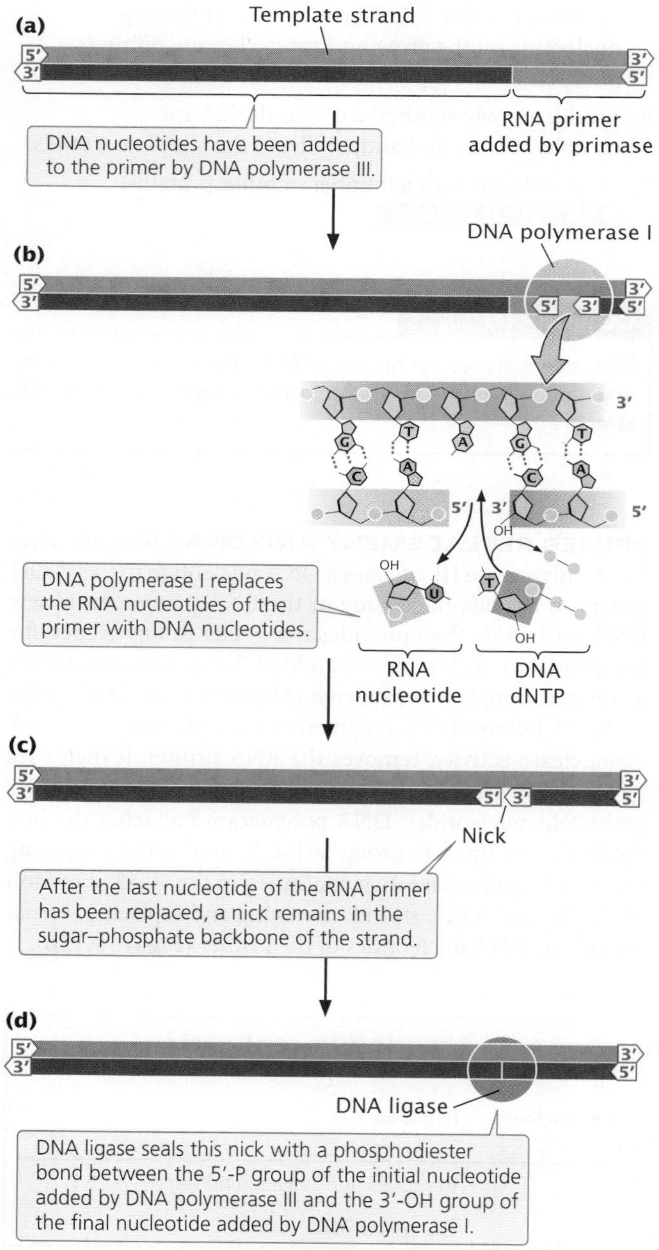

9.12 DNA ligase seals the break left by DNA polymerase I in the sugar–phosphate backbone.

TABLE 9.3 Components required for replication in bacterial cells

Component	Function
Initiator proteins	Bind to origin and separate strands of DNA to initiate replication
DNA helicase	Unwinds DNA at replication fork
Single-strand-binding proteins	Attach to single-stranded DNA and prevent secondary structures from forming
DNA gyrase	Moves ahead of the replication fork, making and resealing breaks in the double-stranded helical DNA to release the torque that builds up as a result of unwinding at the replication fork
DNA primase	Synthesizes a short RNA primer to provide a 3′-OH group for the attachment of DNA nucleotides
DNA polymerase III	Elongates a new nucleotide strand from the 3′-OH group provided by the primer
DNA polymerase I	Removes RNA primers and replaces them with DNA
DNA ligase	Joins Okazaki fragments by sealing breaks in the sugar–phosphate backbone of newly synthesized DNA

CONCEPTS

After primers have been removed and replaced, the break in the sugar–phosphate linkage is sealed by DNA ligase.

✓ CONCEPT CHECK 5

Which bacterial enzyme removes the primers?

a. Primase
b. DNA polymerase I
c. DNA polymerase III
d. Ligase

ELONGATION AT THE REPLICATION FORK Now that the major enzymatic components of elongation—DNA polymerases, helicase, primase, and ligase—have been introduced, let's consider how these components interact at the replication fork. Because the synthesis of both strands takes place simultaneously, two units of DNA polymerase III must be present at the replication fork, one for each strand. These two units of DNA polymerase III are connected (**Figure 9.13**); the lagging-strand template loops around so that it is in position for 5′→3′ replication. In this way, the DNA polymerase III complex is able to carry out 5′→3′ replication simultaneously on both templates, even though they run in opposite directions. After about 1000 bp of new DNA has been synthesized, DNA polymerase III releases the lagging-strand template, and a new loop forms (see Figure 9.13). Primase synthesizes a new primer on the lagging strand, and DNA polymerase III then synthesizes a new Okazaki fragment. See how replication takes place on both strands simultaneously by viewing **Animation 9.2**. A

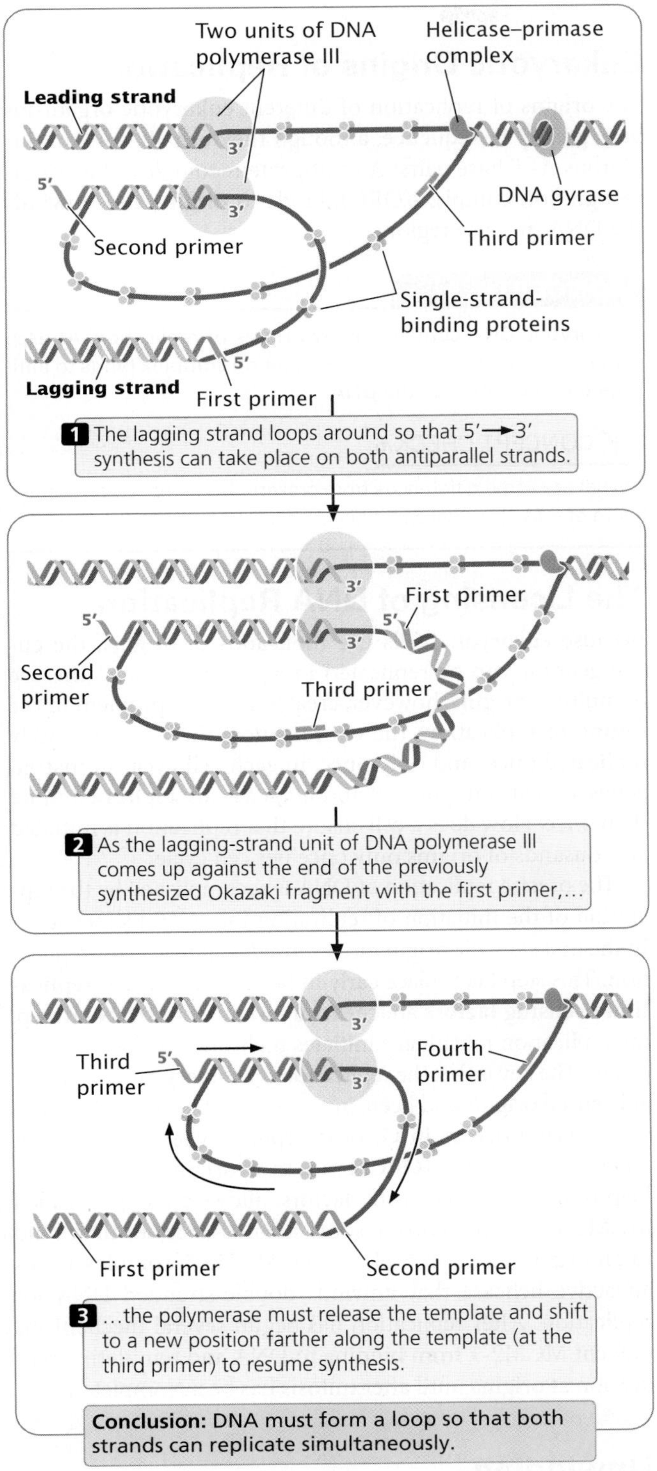

9.13 During DNA replication in *E. coli*, the two units of DNA polymerase III are connected. The lagging-strand template forms a loop so that replication can take place on the two antiparallel DNA strands. Components of the replication machinery at the replication fork are shown at the top.

In summary, each active replication fork requires five basic components:

1. Helicase to unwind the DNA
2. Single-strand-binding proteins to protect the single nucleotide strands and prevent secondary structures
3. The topoisomerase gyrase to remove strain ahead of the replication fork
4. Primase to synthesize primers with a 3′-OH group at the beginning of each DNA fragment
5. DNA polymerase to synthesize the leading and lagging nucleotide strands

You can see how the different components of the replication process work together by viewing **Animations 9.3 and 9.4**. A

Termination

In some DNA molecules, replication is terminated whenever two replication forks meet. In others, specific termination sequences (called *Ter* sites) block further replication. A termination protein, called Tus in *E. coli*, binds to these sequences, creating a Tus-*Ter* complex that blocks the movement of helicase, thus stalling the replication fork and preventing further DNA replication. Each Tus-*Ter* complex blocks a replication fork moving in one direction, but not the other.

The Fidelity of DNA Replication

Overall, the error rate in replication is less than one mistake per billion nucleotides. How is this incredible accuracy achieved?

DNA polymerases are very particular in pairing nucleotides with their complements on the template strand. Errors in nucleotide selection by DNA polymerase arise only about once per 100,000 nucleotides. Most of the errors that do arise in nucleotide selection are corrected in a second process called **proofreading**. When a DNA polymerase inserts an incorrect nucleotide into the growing strand, the 3′-OH group of the mispaired nucleotide is not correctly positioned in the active site of the DNA polymerase for accepting the next nucleotide. The incorrect positioning stalls the polymerization reaction, and the 3′→5′ exonuclease activity of DNA polymerase removes the incorrectly paired nucleotide. DNA polymerase then inserts the correct nucleotide. Together, proofreading and nucleotide selection result in an error rate of only 1 in 10 million nucleotides.

A third process, called **mismatch repair** (considered further in Chapter 13), corrects errors after replication is complete. Any incorrectly paired nucleotides remaining after replication produce a deformity in the secondary structure of the DNA; that deformity is recognized by enzymes that excise the incorrectly paired nucleotide and use the original nucleotide strand as a template to replace the incorrect nucleotide.

THINK-PAIR-SHARE Question 4

CONCEPTS

Replication is extremely accurate, with less than one error per billion nucleotides. The high level of accuracy in DNA replication is produced by nucleotide selection, proofreading, and mismatch repair, each of which catches errors missed by the preceding processes.

CONNECTING CONCEPTS

The Basic Rules of Replication

Bacterial replication requires a number of enzymes (see Table 9.3), proteins, and DNA sequences that function together to synthesize a new DNA molecule. These components are important, but we must not become so immersed in the details of the process that we lose sight of the general principles of replication:

1. Replication is always semiconservative.
2. Replication begins at sequences called origins.
3. DNA synthesis begins with the synthesis of short segments of RNA called primers.
4. The elongation of DNA strands is always in the 5′→3′ direction.
5. New DNA is synthesized from dNTPs; in the polymerization of DNA, two phosphate groups are cleaved from a dNTP and the resulting nucleotide is added to the 3′-OH group of the growing nucleotide strand.
6. Replication is continuous on the leading strand and discontinuous on the lagging strand.
7. New nucleotide strands are complementary and antiparallel to their template strands.
8. Replication takes place at very high rates and is astonishingly accurate, thanks to precise nucleotide selection, proofreading, and mismatch repair.

9.4 Eukaryotic DNA Replication Is Similar to Bacterial Replication but Differs in Several Aspects

Although eukaryotic replication resembles bacterial replication in many respects, replication in eukaryotic cells presents several additional challenges. First, the much greater size of eukaryotic genomes requires that replication be initiated at multiple origins. Second, eukaryotic chromosomes are linear, whereas prokaryotic chromosomes are circular. Third, the DNA template is associated with histone proteins in the form of nucleosomes, and nucleosome assembly must immediately follow DNA replication.

Eukaryotic Origins of Replication

The origins of replication of different eukaryotic organisms vary greatly in sequence, although they usually contain numerous A–T base pairs. A multiprotein complex, the origin-recognition complex (ORC), binds to origins and unwinds the DNA in those regions.

CONCEPTS

Eukaryotic DNA contains many origins of replication. At each origin, a multiprotein origin-recognition complex binds to initiate the unwinding of the DNA.

✓ CONCEPT CHECK 6

What are some differences from prokaryotes in the genome structure of eukaryotic cells that affect how replication takes place?

The Licensing of DNA Replication

Because eukaryotic cells use thousands of origins, the entire genome can be replicated in a timely manner. The use of multiple origins, however, creates a special problem in the timing of replication: the entire genome must be precisely replicated once, and only once, in each cell cycle so that no genes are left unreplicated and no genes are replicated more than once. How does a cell ensure that replication is initiated at thousands of origins only once per cell cycle?

The precise replication of DNA is accomplished by the separation of the initiation of replication into two distinct steps. In the first step, the origins are licensed—approved for replication. This step takes place early in the cell cycle when **replication licensing factors** attache to an origin. In the second step, the replication machinery initiates replication at each *licensed* origin. The key is that the replication machinery functions only at licensed origins and licensing occurs early in the cell cycle.

Licensing occurs in G_1 of interphase when the origin of recognition complex (ORC) binds to an origin. ORC, with the help of additional licensing factors, allows a complex called MCM2-7 (for minichromosome maintenance) to then bind to an origin. Then, in S phase, the MCM2-7 complex forms an active helicase that unwinds double-stranded DNA for replication. After replication has begun several mechanisms prevent MCM2-7 from binding to DNA and reinitiating replication at origins until after mitosis has been completed.

Unwinding

Helicases that separate double-stranded DNA have been isolated from eukaryotic cells, as have single-strand-binding proteins and topoisomerases (which have a function equivalent to the DNA gyrase in bacterial cells). These enzymes and proteins are assumed to function in unwinding eukaryotic DNA in much the same way as their bacterial counterparts do.

Eukaryotic DNA Polymerases

Some significant differences between the processes of bacterial and eukaryotic replication are in the number and functions of DNA polymerases. Eukaryotic cells contain a number of different DNA polymerases that function in replication, recombination, and DNA repair.

Three DNA polymerases carry out most of nuclear DNA synthesis during replication: DNA polymerase α, DNA polymerase δ, and DNA polymerase ε (**Table 9.4**). **DNA polymerase α** has primase activity and initiates nuclear DNA synthesis by synthesizing an RNA primer, followed by a short string of DNA nucleotides. After DNA polymerase α has laid down from 30 to 40 nucleotides, **DNA polymerase δ** completes replication on the lagging strand. Similar in structure and function to DNA polymerase δ, **DNA polymerase ε** replicates the leading strand. **DNA polymerase γ** replicates mitochondrial DNA; a γ-like polymerase also replicates chloroplast DNA. Other specialized DNA polymerases, called translesion polymerases, are able to bypass bulky distortions in DNA structures that block the usual replicating polymerases. Still other DNA polymerases play a role in DNA repair and recombination (see Table 9.4).

CONCEPTS

There are a large number of different DNA polymerases in eukaryotic cells. DNA polymerases α, δ, and ε carry out replication on the leading and lagging strands. Other DNA polymerases replicate organelle DNA, carry out DNA repair, and help bring about recombination.

Replication at the Ends of Chromosomes

A fundamental difference between eukaryotic and bacterial replication arises because eukaryotic chromosomes are linear and thus have ends. The 3′-OH group needed for replication by DNA polymerases is provided at the initiation of replication by RNA primers that are synthesized by primase, as stated earlier. This solution is temporary because, eventually, the primers must be removed and replaced by DNA nucleotides. In a circular DNA molecule, elongation around the circle eventually provides a 3′-OH group immediately in

TABLE 9.4 DNA polymerases in eukaryotic cells

DNA polymerase	5′→3′ polymerase activity	3′→5′ exonuclease activity	Cellular function
α (alpha)	Yes	No	Initiation of nuclear DNA synthesis and DNA repair; has primase activity
δ (delta)	Yes	Yes	Lagging-strand synthesis of nuclear DNA, DNA repair, and translesion DNA synthesis
ε (epsilon)	Yes	Yes	Leading-strand synthesis
γ (gamma)	Yes	Yes	Replication and repair of mitochondrial DNA
ξ (zeta)	Yes	No	Translesion DNA synthesis
η (eta)	Yes	No	Translesion DNA synthesis
θ (theta)	Yes	No	DNA repair
ι (iota)	Yes	No	Translesion DNA synthesis
κ (kappa)	Yes	No	Translesion DNA synthesis
λ (lambda)	Yes	No	DNA repair
μ (mu)	Yes	No	DNA repair
σ (sigma)	Yes	No	Nuclear DNA replication (possibly), DNA repair, and sister-chromatid cohesion
ϕ (phi)	Yes	No	Translesion DNA synthesis
Rev1	Yes	No	DNA repair

Note: The three polymerases listed at the top of the table are those that carry out nuclear DNA replication.

front of the primer (**Figure 9.14a**). After the primer has been removed, the replacement DNA nucleotides can be added to this 3′-OH group. But what happens when a DNA molecule is not circular, but linear?

THE END-REPLICATION PROBLEM In linear chromosomes with multiple origins, the elongation of DNA in adjacent replicons provides a 3′-OH group preceding each primer (**Figure 9.14b**). At the very end of a linear chromosome, however, there is no adjacent stretch of replicated DNA to provide this crucial 3′-OH group. When the terminal primer at the end of the chromosome has been removed, it cannot be replaced by DNA nucleotides, so its removal produces a gap at the end of the chromosome, suggesting that the chromosome should become progressively shorter with each round of replication. This situation has been termed the end-replication problem.

The end-replication problem, as originally proposed, assumed that the terminal primer is located at the very end of the chromosome. Experimental evidence suggests that in some single-celled eukaryotes, such as yeast and some

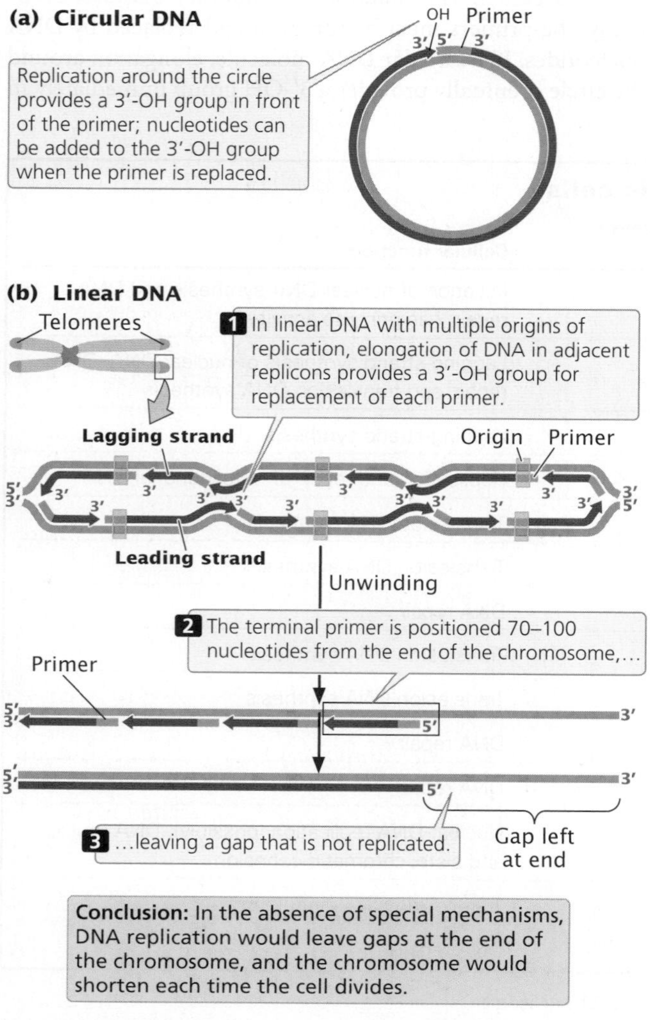

9.14 DNA synthesis at the ends of circular and linear chromosomes must differ.

protozoans, the terminal primer is indeed placed at the very end of the chromosome, but this has not been demonstrated for more complex, multicellular eukaryotes. Furthermore, chromosome ends in humans are known to shorten at a much faster rate than would be expected if only the terminal primer (which is only about 10 nucleotides long) was not replaced. Research has now demonstrated that in replication of human chromosomes the terminal primer is positioned not at the end of the chromosome, but rather some 70 to 100 nucleotides from the end (see Figure 9.14b). This means that 70 to 100 nucleotides of DNA at the end of the chromosome are not replicated during division of somatic cells, and the chromosome shortens by this amount each time the cell divides.

TELOMERES AND TELOMERASE The end-replication problem suggests that chromosomes in eukaryotic cells will shorten with each cell division. However, in single-celled organisms, germ cells, and early embryonic cells, chromosomes do not shorten and self-destruct. So how are the ends of linear chromosomes in these cells replicated? The ends of eukaryotic chromosomes—the telomeres—possess several unique features, one of which is the presence of many copies of a short repeated sequence. In humans, this telomeric repeat is TTAGGG. The strand containing this G-rich repeat typically protrudes beyond the complementary C-rich strand (**Figure 9.15a**; see also the section Telomere Structure in Chapter 8):

toward centromere ←	5′—TTAGGGTTAGGGTTAGGG—3′ 3′—AATCCC—5′	→ end of chromosome

The single-stranded protruding end of the telomere, known as the **G overhang**, can be extended by **telomerase**, an enzyme with both a protein and an RNA component (also known as a ribonucleoprotein). The RNA part of the enzyme contains from 15 to 22 nucleotides that are complementary to the sequence on the G-rich strand. This RNA sequence pairs with the overhanging 3′ end of the DNA (**Figure 9.15b**) and provides a template for the synthesis of additional DNA copies of the repeats. DNA nucleotides are added to the 3′ end of the G overhang one at a time (**Figure 9.15c**); after several nucleotides have been added, the RNA template moves down the DNA and more nucleotides are added to the 3′ end (**Figure 9.15d**). Usually, from 14 to 16 nucleotides are added to the 3′ end of the G-rich strand.

In this way, the telomerase can extend the 3′ end of the chromosome without the use of a complementary DNA template (**Figure 9.15e**). How the complementary C-rich strand is synthesized (**Figure 9.15f**) is not clear. It may be synthesized by conventional replication, with DNA polymerase α synthesizing an RNA primer on the 5′ end of the extended (G-rich) template. The removal of this primer once again leaves a gap at the 5′ end of the chromosome, but this gap does not matter, because the end of the chromosome is extended at each replication by telomerase, so the chromosome does not become shorter overall. View the action of telomerase in **Animation 9.5**. A

Telomerase is present in single-celled eukaryotes, germ cells, early embryonic cells, and certain proliferative

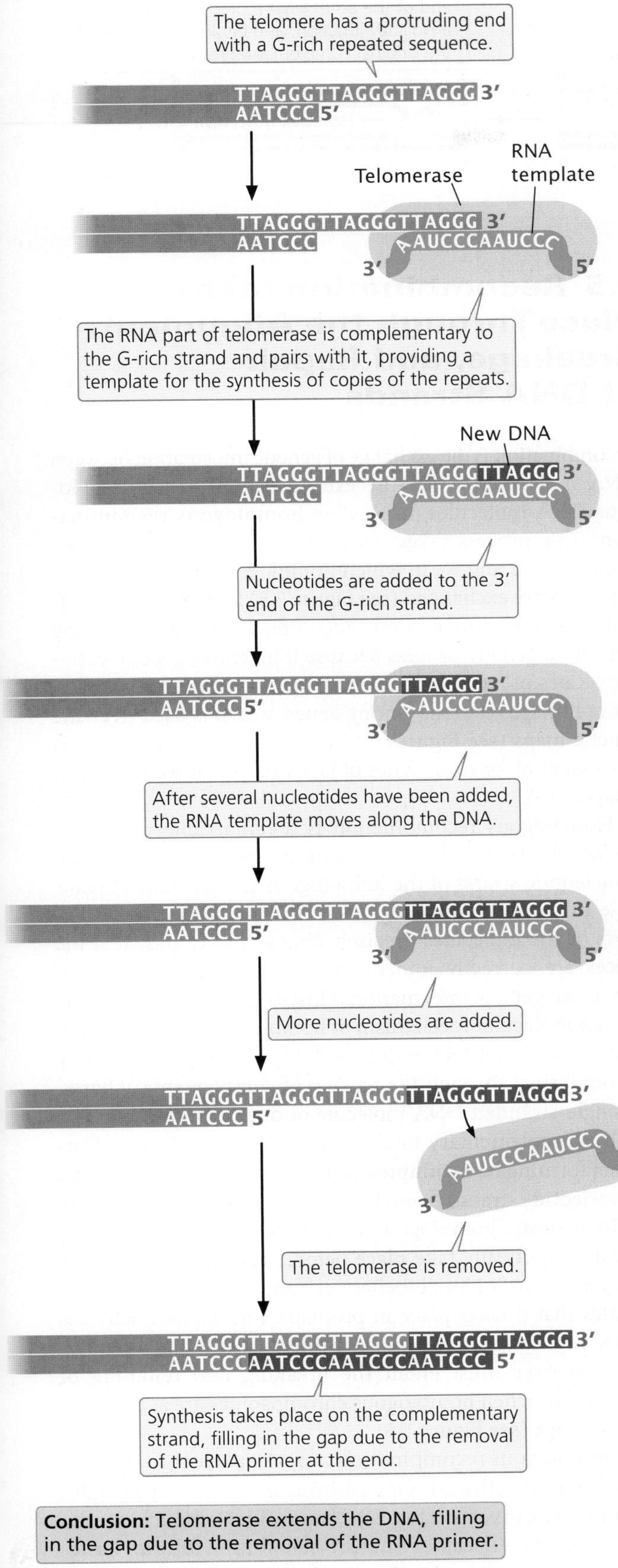

9.15 The enzyme telomerase is responsible for the replication of chromosome ends.

somatic cells (such as bone-marrow cells and cells lining the intestine), all of which must undergo continuous cell division. Most somatic cells have little or no telomerase activity, and chromosomes in these cells progressively shorten with each cell division. These cells are capable of only a limited number of divisions; when the telomeres have shortened beyond a critical point, the chromosomes become unstable, have a tendency to undergo rearrangements, and are degraded. These events lead to cell death.

THINK-PAIR-SHARE Question 5

CONCEPTS

The ends of eukaryotic chromosomes are replicated by an RNA–protein enzyme called telomerase. This enzyme adds extra nucleotides to the G-rich DNA strand of the telomere.

✓ CONCEPT CHECK 7

What would be the result if an organism's telomerase were mutated and nonfunctional?

a. No DNA replication would take place.
b. The DNA polymerase enzyme would stall at the telomere.
c. Chromosomes would shorten each generation.
d. RNA primers could not be removed.

TELOMERASE, AGING, AND DISEASE The shortening of telomeres may contribute to the process of aging. The telomeres of genetically engineered mice that lack a functional telomerase gene (and therefore do not express telomerase in somatic or germ cells) undergo progressive shortening in successive generations. After several generations, these mice show some signs of premature aging, such as graying, hair loss, and delayed wound healing. Through genetic engineering, it is also possible to create somatic cells that express telomerase. In these cells, telomeres do not shorten, cell aging is inhibited, and the cells will divide indefinitely.

Some of the strongest evidence that telomere length is related to aging comes from studies of telomeres in birds. In 2012, scientists in the United Kingdom measured telomere length in red blood cells taken from 99 zebra finches at various times during their lives. The scientists found a strong correlation between telomere length and longevity; birds with longer telomeres lived longer than birds with short telomeres. The strongest predictor of life span was telomere length early in life, at 25 days, which is roughly equivalent to human adolescence. Although these observations suggest that telomere length is associated with aging in some animals, the precise role of telomeres in *human* aging remains uncertain.

Some diseases are associated with abnormalities of telomere replication. People with Werner syndrome, an autosomal recessive disease, show signs of premature aging that begins in adolescence or early adulthood, including wrinkled skin, graying of the hair, baldness, cataracts, and muscle atrophy. They often develop cancer, osteoporosis,

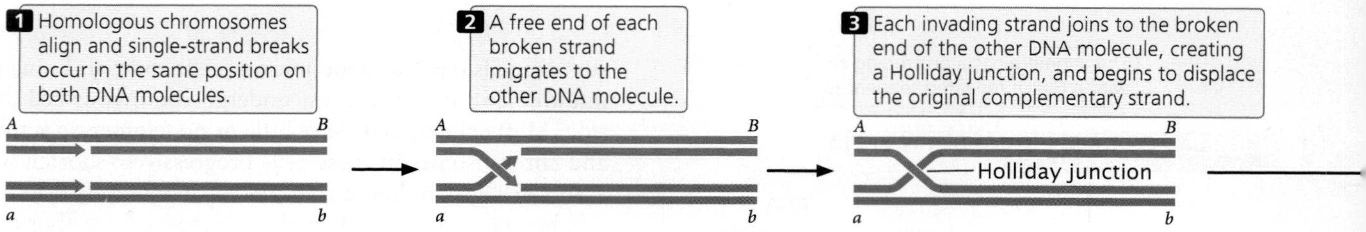

heart and artery disease, and other ailments typically associated with aging. The causative gene, called *WRN*, has been mapped to human chromosome 8 and normally encodes a RecQ helicase enzyme, which is necessary for the efficient replication of telomeres. In people with Werner syndrome, this enzyme is defective, and consequently, the telomeres shorten prematurely.

Telomerase also appears to play a role in cancer. Cancer cells have the capacity to divide indefinitely, and telomerase is expressed in 90% of all cancers. As we will see in Chapter 16, cancer is a complex, multistep process that usually requires mutations in at least several genes. Telomerase activation alone does not lead to cancerous growth in most cells, but it does appear to be required, along with other mutations, for cancer to develop. Some experimental anticancer drugs work by inhibiting the action of telomerase.

One of the difficulties in studying the effect of telomere shortening on the aging process is that the expression of telomerase in somatic cells also promotes cancer. To circumvent this problem, Antonia Tomas-Loba and her colleagues created genetically engineered mice that expressed telomerase and carried genes that made them resistant to cancer. These mice had longer telomeres, lived longer, and exhibited fewer age-related changes, such as skin alterations, decreases in neuromuscular coordination, and degenerative diseases. These results support the idea that telomere shortening contributes to aging. ▶TRY PROBLEM 27

THINK-PAIR-SHARE Question 6

Replication in Archaea

The process of replication in archaea has a number of features in common with replication in eukaryotic cells. Many of the proteins taking part are more similar to those in eukaryotic cells than to those in bacteria. Like bacteria, some archaea have a single origin of replication, but many archaea have more than one origin, similar to the multiple origins seen in eukaryotic genomes. The origins of archaea do not contain the typical sequences recognized by bacterial initiator proteins; instead, they have sequences that are similar to those found in some eukaryotic origins. The initiator proteins of archaea are also more similar to those of eukaryotes than to those of bacteria. These similarities in replication between archaeal and eukaryotic cells reinforce the conclusion that the archaea are more closely related to eukaryotic cells than to the prokaryotic bacteria.

9.5 Recombination Takes Place Through the Alignment, Breakage, and Repair of DNA Strands

Recombination is the exchange of genetic information between DNA molecules; when the exchange is between homologous DNA molecules, it is called **homologous recombination**. This process takes place in crossing over (discussed in Chapters 2 and 5), in which homologous regions of chromosomes are exchanged (see Figure 5.5) and alleles are shuffled into new combinations. Recombination is an extremely important genetic process because it increases genetic variation. Rates of recombination provide important information about linkage relations among genes, which is used to create genetic maps (see Figures 5.12 and 5.13). Recombination is also essential for some types of DNA repair (as we will see in Chapter 13).

Homologous recombination is a remarkable process: a nucleotide strand of one chromosome aligns precisely with a nucleotide strand of the homologous chromosome, breaks arise in corresponding regions of the two DNA molecules, parts of the molecules precisely change place, and then the pieces are correctly joined. In this complicated series of events, no genetic information is lost or gained. Although the precise molecular mechanism of homologous recombination is still not completely understood, the exchange is probably accomplished through the pairing of complementary bases. A single-stranded DNA molecule of one chromosome pairs with a complementary single-stranded DNA molecule of another, forming **heteroduplex DNA**, which is DNA consisting of nucleotide strands from different sources (see Figure 9.16).

In meiosis, homologous recombination (crossing over) could theoretically take place before, during, or after DNA synthesis. Cytological, biochemical, and genetic evidence indicates that it takes place in prophase I of meiosis, whereas DNA replication takes place earlier, in interphase. Thus, crossing over must entail the breaking and rejoining of chromatids when homologous chromosomes are at the four-strand stage (see Figure 5.5).

Homologous recombination can take place through several different pathways. One pathway is initiated by a single-strand break in each of two DNA molecules and includes the formation of a special structure called the **Holliday junction** (**Figure 9.16**). In this mechanism, double-stranded DNA molecules from two homologous chromosomes align

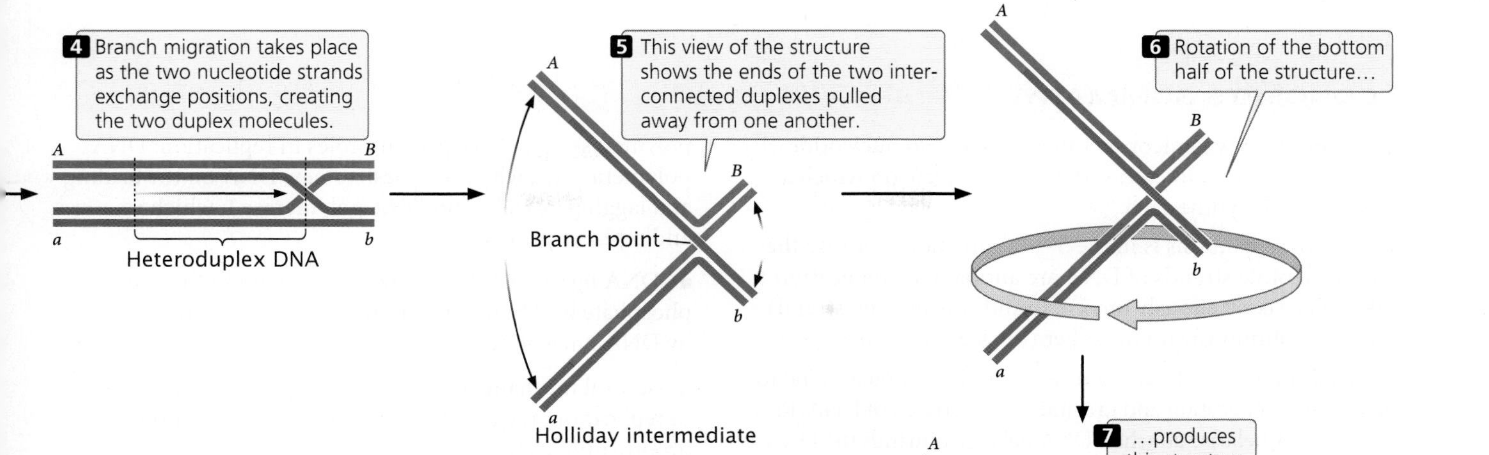

precisely. A single-strand break in one of the DNA molecules provides a free end that invades and joins the free end of the other DNA molecule. Thus, strand invasion and joining take place on both DNA molecules, creating two heteroduplex DNAs, each consisting of one original strand plus one new strand from the other DNA molecule. The point at which nucleotide strands pass from one DNA molecule to the other is the Holliday junction (see Figure 9.16). The junction moves along the molecules in a process called branch migration.

The exchange of nucleotide strands and branch migration produce a structure termed the Holliday intermediate, which can be cleaved in one of two ways. Cleavage in the horizontal plane, followed by rejoining of the strands, produces noncrossover recombinants, in which the genes on either end of the molecules are identical with those originally present (gene *A* with gene *B*, and gene *a* with gene *b*). Cleavage in the vertical plane, followed by rejoining, produces crossover recombinants, in which the genes on either end of the molecules are different from those originally present (gene *A* with gene *b*, and gene *a* with gene *B*). View **Animation 9.6** to see how the Holliday model is resolved.

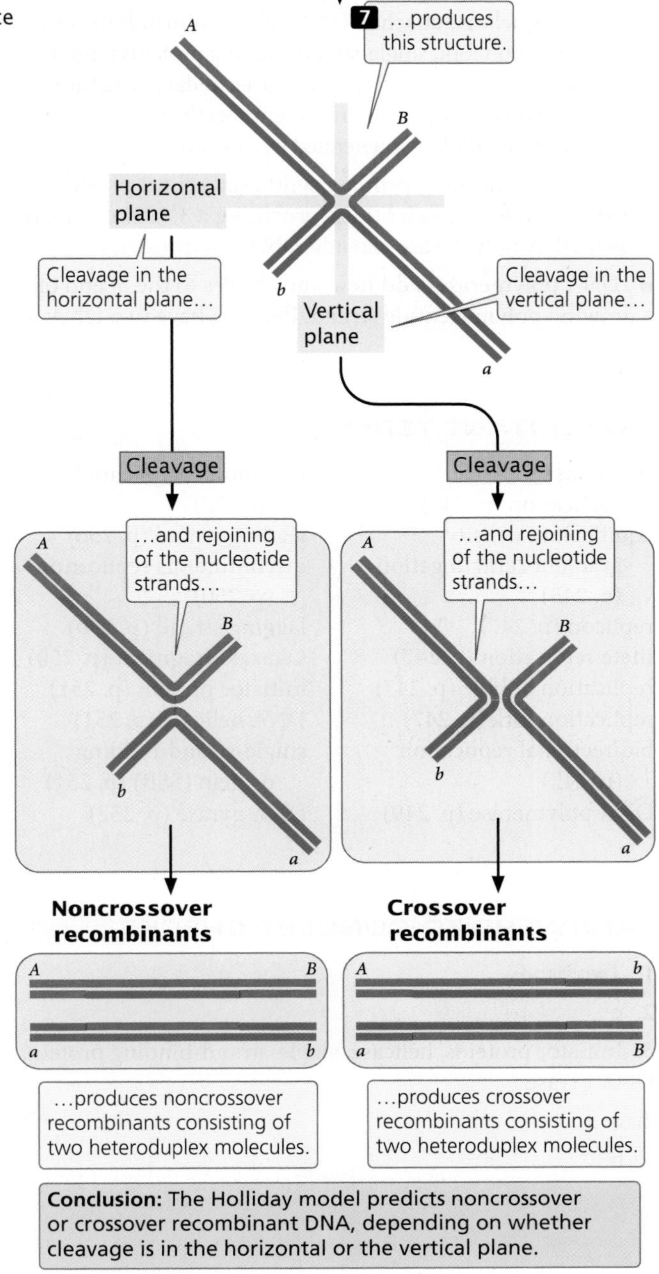

9.16 The Holliday model of homologous recombination. In this model, recombination takes place through single-strand breaks, strand displacement, branch migration, and the resolution of a single Holliday junction.

CONCEPTS

Homologous recombination requires the formation of heteroduplex DNA consisting of one nucleotide strand from each of two homologous chromosomes. In the Holliday model, homologous recombination is accomplished through a single-strand break in each DNA molecule, strand displacement, and branch migration.

✓ CONCEPT CHECK 8

Why is recombination important?

CONCEPTS SUMMARY

- Replication is semiconservative: DNA's two nucleotide strands separate, and each serves as a template on which a new strand is synthesized.
- All DNA synthesis is in the $5' \rightarrow 3'$ direction. Because the two nucleotide strands of DNA are antiparallel, replication takes place continuously on one strand (the leading strand) and discontinuously on the other (the lagging strand).
- Replication in bacteria begins when initiator proteins bind to an origin of replication and unwind a short stretch of DNA, to which DNA helicase attaches. DNA helicase unwinds the DNA at the replication fork, single-strand-binding proteins bind to the single nucleotide strands to prevent secondary structures, and DNA gyrase (a topoisomerase) removes the strain ahead of the replication fork that is generated by unwinding.
- During replication, primase synthesizes short primers consisting of RNA nucleotides, providing a 3′-OH group to which DNA polymerase can add DNA nucleotides.
- DNA polymerases add new nucleotides to the 3′ end of a growing polynucleotide strand. Bacteria have two DNA polymerases that have primary roles in replication: DNA polymerase III, which synthesizes new DNA on the leading and lagging strands, and DNA polymerase I, which removes and replaces primers.
- DNA ligase seals the breaks that remain in the sugar–phosphate backbones when the RNA primers are replaced by DNA nucleotides.
- Several mechanisms ensure the high rate of accuracy in replication, including precise nucleotide selection, proofreading, and mismatch repair.
- Precise replication at multiple origins in eukaryotes is ensured by a licensing factor that must attach to an origin before replication can begin.
- The ends of linear eukaryotic DNA molecules are replicated by the enzyme telomerase.
- Homologous recombination takes place through alignment of homologous DNA segments, breaks in nucleotide strands, and rejoining of the strands.

IMPORTANT TERMS

semiconservative replication (p. 244)
equilibrium density gradient centrifugation (p. 245)
replicon (p. 247)
theta replication (p. 247)
replication bubble (p. 247)
replication fork (p. 247)
bidirectional replication (p. 247)
DNA polymerase (p. 249)
continuous replication (p. 250)
leading strand (p. 250)
discontinuous replication (p. 250)
lagging strand (p. 250)
Okazaki fragment (p. 250)
initiator protein (p. 251)
DNA helicase (p. 251)
single-strand-binding protein (SSB) (p. 251)
DNA gyrase (p. 252)
primase (p. 252)
primer (p. 252)
DNA polymerase III (p. 253)
DNA polymerase I (p. 253)
DNA ligase (p. 254)
proofreading (p. 255)
mismatch repair (p. 256)
replication licensing factor (p. 256)
DNA polymerase α (p. 257)
DNA polymerase δ (p. 257)
DNA polymerase ε (p. 257)
DNA polymerase γ (p. 257)
G overhang (p. 258)
telomerase (p. 258)
homologous recombination (p. 260)
heteroduplex DNA (p. 260)
Holliday junction (p. 260)

ANSWERS TO CONCEPT CHECKS

1. Two bands

2. c

3. Initiator proteins, helicase, single-strand-binding protein, DNA gyrase

4. c

5. b

6. The size of eukaryotic genomes, the linear structure of eukaryotic chromosomes, and the association of DNA with histone proteins

7. c

8. Recombination is important for generating genetic variation.

WORKED PROBLEMS

Problem 1

The following diagram represents the template strands of a replication bubble in a DNA molecule. Draw in the newly synthesized strands and identify the leading and lagging strands.

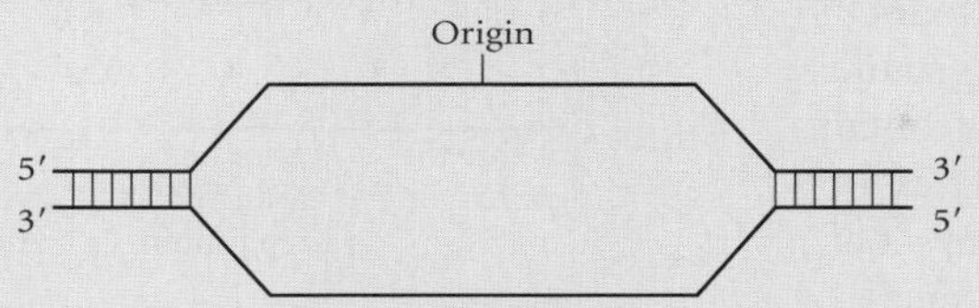

Solution Strategy

What information is required in your answer to the problem?

The diagram above with the newly synthesized strands drawn in and the leading and lagging strands labeled.

What information is provided to solve the problem?

- A diagram of the template DNA with 5′ and 3′ ends labeled.

For help with this problem, review:

Direction of Replication in Section 9.2 and Figure 9.8.

Solution Steps

To determine the leading and lagging strands, first note which end of each template strand is 5′ and which end is 3′. With a pencil, draw in the strands being synthesized on these templates, and identify their 5′ and 3′ ends.

: The two of DNA are llel, so the nthesized hould have osite polarity n) from the e strand.

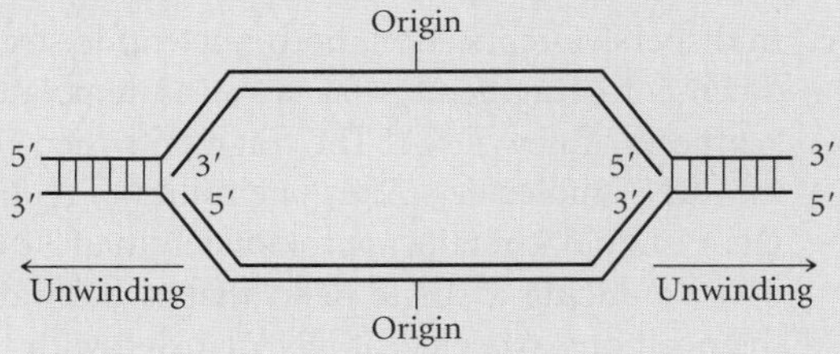

Next, determine the direction of replication for each new strand, which must be 5′→3′. You might draw arrows on the new strands to indicate the direction of replication. After you have established the direction of replication for each strand, look at each fork and determine whether the direction of replication for a strand is the same as the direction of unwinding. The strand on which replication is in the same direction as that of unwinding is the leading strand. The strand on which replication is in the direction opposite that of unwinding is the lagging strand.

Recall: DNA synthesis is always 5′ to 3′.

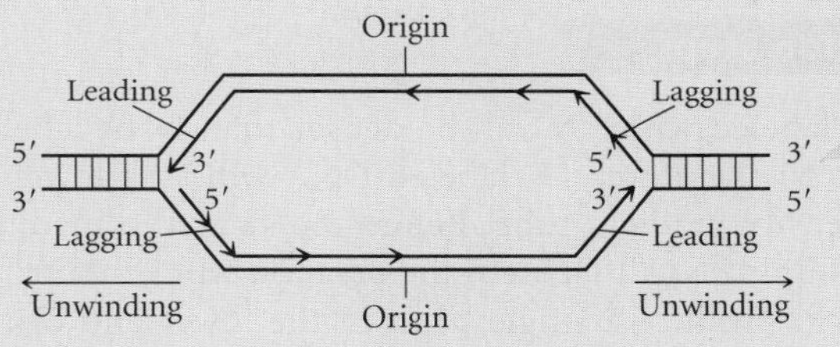

Hint: Each replication fork should have one leading and one lagging strand.

Problem 2

Consider the experiment conducted by Meselson and Stahl in which they used ^{14}N and ^{15}N in cultures of *E. coli* and equilibrium density gradient centrifugation. Draw pictures to represent the bands produced by bacterial DNA in the density-gradient tube before the switch to medium containing ^{14}N and after one, two, and three rounds of replication after the switch to the medium containing ^{14}N. Use a separate set of drawings to show the bands that would appear if replication were (a) semiconservative; (b) conservative; (c) dispersive.

Solution Strategy

What information is required in your answer to the problem?

Drawings that represent the bands produced by bacterial DNA in density-gradient tubes before the switch to medium containing ^{14}N and after one, two, and three rounds of replication following the switch to the medium containing ^{14}N; thus, you should have drawings of four tubes for each model of replication. You will need a separate set of drawings for semiconservative, conservative, and dispersive replication.

What information is provided to solve the problem?

- The bacterial DNA was originally labeled with ^{15}N, and then the bacteria were switched to a medium with ^{14}N (see discussion of experiment on pp. 245–246). Original DNA will have ^{15}N. Newly synthesized DNA will have ^{14}N.
- Equilibrium density gradient centrifugation was performed before switching to ^{14}N and after one, two, and three rounds of replication following the switch.

For help with this problem, review:

Meselson and Stahl's experiment in Section 9.2.

Solution Steps

DNA labeled with ^{15}N will be denser than DNA labeled with ^{14}N; therefore, ^{15}N-labeled DNA will sink lower in the density-gradient tube. Before the switch to medium containing ^{14}N, all DNA in the bacteria will contain ^{15}N and will produce a single band in the lower end of the tube.

Hint: Review the distribution of new and old DNA in semiconservative, conservative, and dispersive replication in Figure 9.1.

a. With semiconservative replication, the two strands separate, and each serves as a template on which a new strand is synthesized. After one round of replication, the original template strand of each molecule will contain ^{15}N and the new strand of each molecule will contain ^{14}N; so a single band will appear in the density-gradient tube halfway between the positions expected of DNA with ^{15}N and of DNA with ^{14}N. In the next round of replication, the two strands again separate and serve as templates for new strands. Each of the new strands contains only ^{14}N; thus, some DNA molecules will contain one strand with the original ^{15}N and one strand with new ^{14}N, whereas the other molecules will contain two strands with ^{14}N. This labeling will produce two bands, one at the intermediate position and one at a higher position in the tube. Additional rounds of replication should produce increasing amounts of DNA that contains only ^{14}N; so the higher band will get darker.

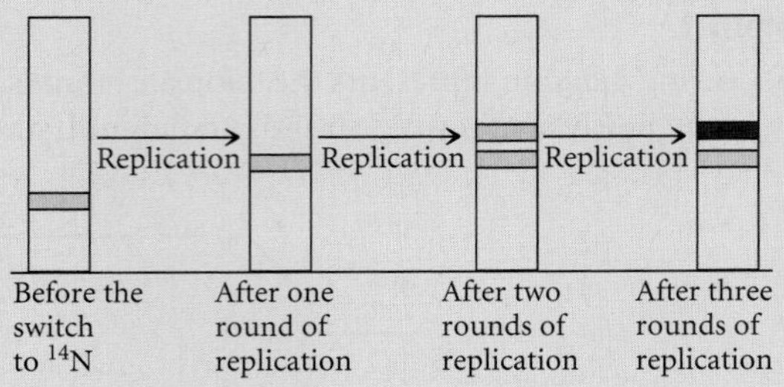

b. With conservative replication, the entire molecule serves as a template. After one round of replication, some molecules will consist entirely of ^{15}N, and others will consist entirely of ^{14}N; therefore, two bands should be present. Subsequent rounds of replication will increase the fraction of DNA consisting entirely of new ^{14}N; thus the upper band should get darker. However, the original DNA with ^{15}N will remain, and so two bands should be present.

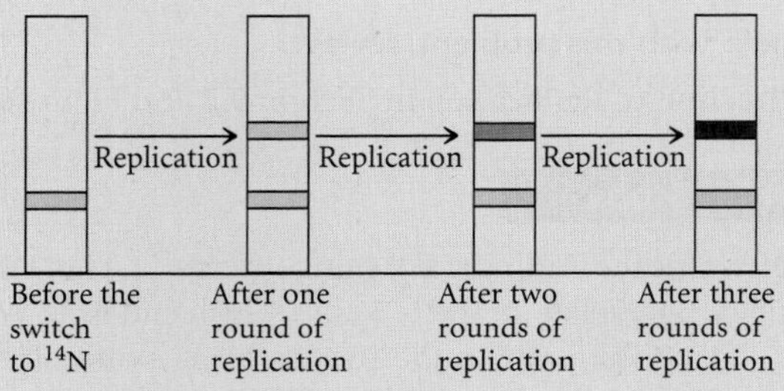

c. In dispersive replication, both nucleotide strands break down into fragments that serve as templates for the synthesis of new DNA. The fragments then reassemble into DNA molecules. After one round of replication, all DNA should contain approximately half ^{15}N and half ^{14}N, producing a single band that is halfway between the positions expected of DNA labeled with ^{15}N and of DNA labeled with ^{14}N. With further rounds of replication, the proportion of ^{14}N in each molecule increases; so a single hybrid band should remain, but its position in the density-gradient tube should move upward. The band should also get darker as the total amount of DNA increases.

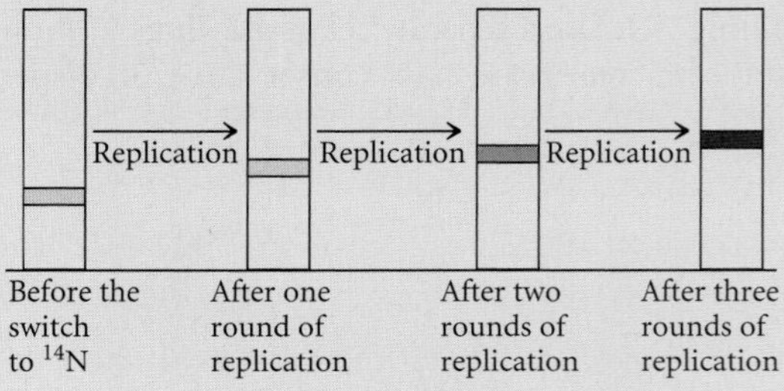

COMPREHENSION QUESTIONS

Section 9.2

1. What is semiconservative replication?
2. How did Meselson and Stahl demonstrate that replication in *E. coli* takes place in a semiconservative manner?
3. Draw a molecule of DNA undergoing theta replication. On your drawing, identify (1) origin, (2) polarity (5′ and 3′ ends) of all template strands and newly synthesized strands, (3) leading and lagging strands, (4) Okazaki fragments, and (5) location of primers.
4. Draw a molecule of DNA undergoing eukaryotic linear replication. On your drawing, identify (1) origin, (2) polarity (5′ and 3′ ends) of all template and newly synthesized strands, (3) leading and lagging strands, (4) Okazaki fragments, and (5) location of primers.
5. What are three major requirements of replication?
6. What substrates are used in the DNA synthesis reaction?

Section 9.3

7. List the different proteins and enzymes taking part in bacterial replication. Give the function of each in the replication process.
8. What similarities and differences exist in the enzymatic activities of DNA polymerases I and III? What is the function of each DNA polymerase in bacterial cells?
9. Why is primase required for replication?
10. Why is DNA gyrase necessary for replication?
11. What three mechanisms ensure the accuracy of replication in bacteria?
12. How does replication licensing ensure that DNA is replicated only once at each origin per eukaryotic cell cycle?
13. In what ways is eukaryotic replication similar to bacterial replication, and in what ways is it different?
14. What is the end-replication problem? Why, in the absence of telomerase, do the ends of linear chromosomes get progressively shorter each time the DNA is replicated?
15. Outline in words and pictures how telomeres at the ends of eukaryotic chromosomes are replicated.

For more questions that test your comprehension of the key chapter concepts, go to **LearningCurve** for this chapter.

APPLICATION QUESTIONS AND PROBLEMS

Section 9.2

16. Suppose that a future scientist explores a distant planet and discovers a novel form of double-stranded nucleic acid. When this nucleic acid is exposed to DNA polymerases from *E. coli*, replication takes place continuously on both strands. What conclusion can you make about the structure of this novel nucleic acid?
17. Phosphorus is required to synthesize the deoxyribonucleoside triphosphates used in DNA replication. A geneticist grows some *E. coli* in a medium containing nonradioactive phosphorus for many generations. A sample of the bacteria is then transferred to a medium that contains a radioactive isotope of phosphorus (^{32}P). Samples of the bacteria are removed immediately after the transfer and after one and two rounds of replication. Assume that newly synthesized DNA contains ^{32}P and that the original DNA contains nonradioactive phosphorus. What will be the distribution of radioactivity in the DNA of the bacteria in each sample? Will radioactivity be detected in neither strand, one strand, or both strands of the DNA?

*18. A line of mouse cells is grown for many generations in a medium with ^{15}N. Cells in G_1 are then switched to a new medium that contains ^{14}N. Draw a pair of homologous chromosomes from these cells at the following stages, showing the two strands of DNA molecules found in the chromosomes. Use different colors to represent strands with ^{14}N and ^{15}N. (See Chapter 2 for a review of the stages of the cell cycle and meiosis.)
   - **a.** Cells in G_1, before switching to medium with ^{14}N
   - **b.** Cells in G_2, after switching to medium with ^{14}N
   - **c.** Cells in anaphase of mitosis, after switching to medium with ^{14}N
   - **d.** Cells in metaphase I of meiosis, after switching to medium with ^{14}N
   - **e.** Cells in anaphase II of meiosis, after switching to medium with ^{14}N

*19. A circular molecule of DNA contains 1 million base pairs. If the rate of DNA synthesis at a replication fork is 100,000 nucleotides per minute, how much time will theta replication require to completely replicate the molecule, assuming that theta replication is bidirectional?

20. A bacterium synthesizes DNA at each replication fork at a rate of 1000 nucleotides per second. If this bacterium completely replicates its circular chromosome by theta replication in 30 minutes, how many base pairs of DNA will its chromosome contain?

Section 9.3

21. In **Figure 9.7**, which is the leading strand and which is the lagging strand?

*__22.__ The following diagram represents a DNA molecule that is undergoing replication. Draw in the strands of newly synthesized DNA and identify (a) through (d):

a. Polarity of newly synthesized strands

b. Leading and lagging strands

c. Okazaki fragments

d. RNA primers

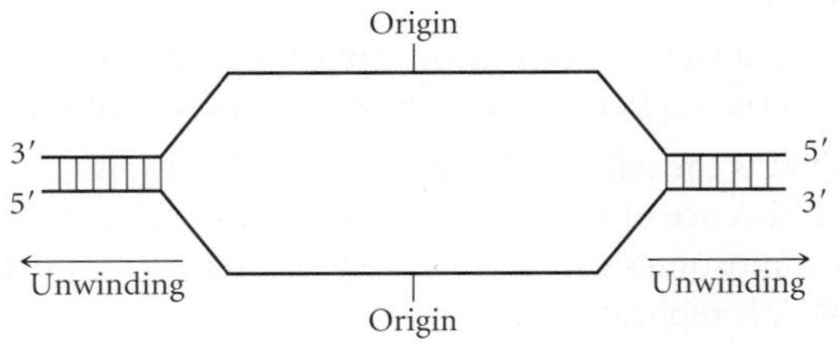

*__23.__ What would be the effect on DNA replication of mutations that destroyed each of the following activities of DNA polymerase I?

a. 3′→5′ exonuclease activity

b. 5′→3′ exonuclease activity

c. 5′→3′ polymerase activity

24. How would DNA replication be affected in a bacterial cell that is lacking DNA gyrase?

25. Arrange the following components of replication in the order in which they first act in the replication process: ligase, DNA polymerase I, helicase, gyrase, primase, single-strand-binding protein, initiator proteins.

Section 9.4

26. A number of scientists who study cancer treatment have become interested in telomerase. Why? How might anticancer therapies that target telomerase work?

*__27.__ The enzyme telomerase is part protein and part RNA. What would be the most likely effect of deleting the gene that encodes the RNA part of telomerase? How would the function of telomerase be affected?

28. DATA ANALYSIS Dyskeratosis congenita (DKC) is a rare genetic disorder characterized by abnormal fingernails and skin pigmentation, the formation of white patches on the tongue and cheek, and progressive failure of the bone marrow. An autosomal dominant form of DKC results from mutations in the gene that encodes the RNA component of telomerase. Tom Vulliamy and his colleagues examined a series of families with autosomal dominant DKC (T. Vulliamy et al. 2004. *Nature Genetics* 36:447–449). They observed that the median age of onset of DKC in parents was 37 years, whereas the median age of onset in the children of affected parents was 14.5 years. Thus, DKC in these families arose at progressively younger ages in successive generations, a phenomenon known as anticipation. The researchers measured the telomere length of members of these families; the measurements are given in accompanying table. Telomeres normally shorten with age, so telomere length was adjusted for age. Note that the age-adjusted telomere length of all members of these families is negative, indicating that their telomeres are shorter than normal. For age-adjusted telomere length, the more negative the number, the shorter the telomere.

a. How does the telomere length of the parents compare with the telomere length of the children? (Hint: Calculate the average telomere length of all parents and the average telomere length of all children.)

b. Explain why the telomeres of people with DKC are shorter than normal.

c. Explain why DKC arises at an earlier age in subsequent generations.

Parent telomere length	Child telomere length
−4.7	−6.1
	−6.6
	−6.0
−3.9	−0.6
−1.4	−2.2
−5.2	−5.4
−2.2	−3.6
−4.4	−2.0
−4.3	−6.8
−5.0	−3.8
−5.3	−6.4
−0.6	−2.5
−1.3	−5.1
	−3.9
−4.2	−5.9

Section 9.5

29. An individual is heterozygous at two loci (*Ee Ff*), and the two genes are in repulsion (see p. 124 in Chapter 5). Assume that single-strand breaks and branch migration occur at the positions shown below. Using different colors to represent the two homologous chromosomes, draw the noncrossover recombinant and crossover recombinant DNA molecules that will result from homologous recombination. (Hint: See **Figure 9.16**.)

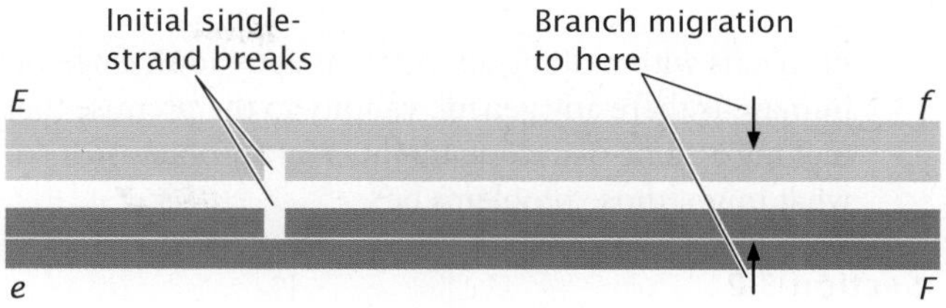

CHALLENGE QUESTIONS

Section 9.3

30. A conditional mutation is one that expresses its mutant phenotype only under certain conditions (the restrictive conditions) and expresses the normal phenotype under other conditions (the permissive conditions). One type of conditional mutation is a temperature-sensitive mutation, which expresses the mutant phenotype only at certain temperatures.

Strains of *E. coli* have been isolated that contain temperature-sensitive mutations in the genes encoding different components of the replication machinery. In each of these strains, the protein produced by the mutated gene is nonfunctional under the restrictive conditions. These strains are grown under permissive conditions and then abruptly switched to the restrictive condition. After one round of replication under the restrictive condition, the DNA from each strain is isolated and analyzed. What characteristics would you expect to see in the DNA isolated from each strain with a temperature-sensitive mutation in its gene that encodes in the following proteins?

a. DNA ligase

b. DNA polymerase I

c. DNA polymerase III

d. Primase

e. Initiator proteins

31. DATA ANALYSIS DNA topoisomerases play important roles in DNA replication and supercoiling (see Chapter 8). These enzymes are also the targets for certain anticancer drugs. Eric Nelson and his colleagues studied m-AMSA, an anticancer compound that acts on topoisomerases. They found that m-AMSA stabilizes an intermediate produced in the course of the topoisomerase's action. The intermediate consists of the topoisomerase bound to the broken ends of the DNA (E. M. Nelson, K. M. Tewey, and L. F. Liu. 1984. *Proceedings of the National Academy of Sciences of the United States of America* 81:1361–1365). Breaks in DNA that are produced by anticancer compounds such as m-AMSA inhibit the replication of the cellular DNA and thus stop cancer cells from proliferating. Explain how m-AMSA and other anticancer agents that target topoisomerase enzymes taking part in replication might lead to DNA breaks and chromosome rearrangements.

THINK-PAIR-SHARE QUESTIONS

Introduction

1. Chemotherapeutic agents such as camptothecin often have unwanted side effects. What are some common negative side effects of chemotherapy, and why do they arise?

Section 9.1

2. In the 1996 movie *Multiplicity*, Doug (played by Michael Keaton) is a construction worker who wants to spend more time with his family. He meets a friendly scientist who has developed a method for cloning humans. Doug decides to make a clone of himself who can take over his work while he spends quality time with his family. The clone, named "Two," seems great at first, but later problems surface in his functioning. Two decides to make a clone of himself, so he won't have to work, and creates Three. A fourth clone is eventually made. Each successive clone seems to have more problems. Ignore, for the present, the technical difficulties and ethical

problems with making an instantaneous adult clone of a human. Is there any genetic validity to the premise that making a clone of a clone might create problems? If so, what might those problems be?

Section 9.2

3. In their experiment, could Meselson and Stahl have used two different isotopes of carbon, instead of ^{14}N and ^{15}N? Why or why not? What about two different isotopes of sulfur?

Section 9.3

4. DNA polymerases cannot act as primers for replication, yet primase and other RNA polymerases can. Some geneticists have speculated that the inability of DNA polymerase to prime replication is a result of its proofreading function. This hypothesis argues that proofreading is essential for the faithful transmission of genetic information and that because DNA polymerases have evolved the ability to proofread, they cannot prime DNA synthesis. Explain why proofreading and priming functions in the same enzyme might be incompatible.

5. HeLa cells are a line of cells grown in laboratory culture that has been used extensively in research. This cell line was originally derived from malignant cervical cancer calls that were removed from a woman named Henrietta Lacks in 1951. They were subsequently grown in culture and shipped to research laboratories around the world, where they have been used in many important experiments. Like HeLa cells, many other cell lines were originally taken from cancerous tissue. Why are cancer cells often used for developing cell lines?

6. For centuries, people have searched for the fabled Fountain of Youth, said to confer the ability to forestall old age and remain young forever. To gain insight into how tackling aging by targeting telomerase might work, researchers looked at mice that lack a telomerase gene. These mice have shorter telomeres than normal mice and age prematurely. But when these mice are engineered to express telomerase in their somatic cells, their telomeres lengthen, and the effects of aging are reversed. This observation suggests that a drug that stimulates the expression of telomerase in somatic cells could prevent telomere shortening and stop the aging process. Do you think such a drug would work well in humans? What might be some potential side effects of such a drug?

SaplingPlus Self-study tools that will help you practice what you've learned and reinforce this chapter's concepts are available online. Go to www.macmillanlearning.com/PierceGenetics6e.

10 From DNA to Proteins: Transcription and RNA Processing

Death Cap Poisoning

On November 8, 2009, 31-year-old Tomasa was hiking the Lodi Lake nature trail east of San Francisco with her husband and cousin when they came across some large white mushrooms that looked very much like edible mushrooms that they enjoyed in their native Mexico. They picked the mushrooms and took them home, cooking and consuming them for dinner. Within hours, Tomasa and her family were sick and went to the hospital. They were later transferred to the critical care unit at California Pacific Medical Center in San Francisco, where Tomasa died of liver failure three weeks later. Her husband eventually recovered after a lengthy hospitalization; her cousin required a liver transplant to survive.

The mushrooms consumed by Tomasa and her family were *Amanita phalloides*, commonly known as the death cap. A single death cap contains enough toxin to kill an adult human. The death rate among those who consume death caps is 22%; among children under the age of 10, it's more than 50%. Death cap mushrooms appear to be spreading in California, leading to a surge in the number of mushroom poisonings.

The death cap mushroom, *Amanita phalloides*, causes death by inhibiting the process of transcription. [© MAP/Jean-Yves Grospas/Age FotoStock America, Inc.]

Death cap poisoning is insidious. Gastrointestinal symptoms—abdominal pain, cramping, vomiting, diarrhea—begin within 6 to 12 hours of consuming the mushrooms, but these symptoms usually subside within a few hours, and the patient seems to recover. Because of this initial remission, the poisoning is often not taken seriously until it's too late to pump the stomach and remove the toxin from the body. After a day or two, serious symptoms begin. Cells in the liver die, causing permanent liver damage and sometimes death within a few days. There is no effective treatment, other than a liver transplant to replace the damaged organ.

How do death caps kill? Their deadly toxin, contained within the fruiting bodies that produce reproductive spores, is the protein α-amanitin, which consists of a short peptide of eight amino acids that forms a circular loop. α-Amanitin is a potent inhibitor of RNA polymerase II, the enzyme that transcribes protein-encoding genes in eukaryotes. RNA polymerase II binds to genes and synthesizes RNA molecules that are complementary to the DNA template. In the process of transcription, the RNA polymerase moves down the DNA template, adding one nucleotide at a time to the growing RNA chain. α-Amanitin binds to RNA polymerase and jams the moving parts of the enzyme, interfering with its ability to move along the DNA template. In the presence of α-amanitin, RNA synthesis slows from its normal rate of several thousand nucleotides per minute to just a few nucleotides per minute. The results are catastrophic. Without transcription, protein synthesis—required for cellular function—ceases, and cells die. The liver, where the toxin accumulates, is irreparably damaged and stops functioning. In severe cases, the patient dies.

THINK-PAIR-SHARE Questions 1 and 2

Death cap poisoning illustrates the extreme importance of transcription and the central role that RNA polymerase plays in the process. This chapter is about the process of transcription—the first step in the central dogma, the pathway of information transfer from DNA (genotype) to protein (phenotype). Transcription is a complex process that requires precursors to RNA nucleotides, a DNA template, and a number of protein components. As we examine the stages of transcription, try to keep all the details in perspective and focus on understanding how they relate to the overall purpose of transcription: the selective synthesis of an RNA molecule.

This chapter begins with a brief review of RNA structure and a discussion of the different classes of RNA. We then consider the major components required for transcription. Finally, we explore the process of transcription. At several points in the text, we'll pause to consider some general principles that emerge.

THINK-PAIR-SHARE Question 3

10.1 RNA, a Single Strand of Ribonucleotides, Participates in a Variety of Cellular Functions

Before we begin our study of transcription, let's review the structure of RNA and consider the different types of RNA molecules.

An Early RNA World

Life requires two basic functions. First, living organisms must be able to store and faithfully transmit genetic information during reproduction. Second, they must have the ability to catalyze the chemical transformations that drive life processes. A long-held belief was that the functions of information storage and chemical transformation are handled by two entirely different types of molecules: genetic information is stored in nucleic acids, whereas the catalysis of chemical transformations is carried out by protein enzymes. This biochemical dichotomy—nucleic acid for information, proteins for catalysis—created a dilemma. Which came first: proteins or nucleic acids? If nucleic acids carry the coding instructions for proteins, how can proteins be generated without them? Nucleic acids are unable to copy themselves, so how can they be generated without proteins? If DNA and proteins each require the other, how can life begin?

This apparent paradox disappeared in 1981 when Thomas Cech and his colleagues discovered that RNA can serve as a biological catalyst. They found that some RNA molecules from the protozoan *Tetrahymena thermophila* can excise 400 nucleotides from its RNA in the absence of any protein. Other examples of catalytic RNAs have now been discovered in different types of cells. Called **ribozymes**, these catalytic RNA molecules can cut out parts of their own sequences, connect some RNA molecules together, replicate others, and even catalyze the formation of peptide bonds between amino acids. The discovery of ribozymes complements other evidence suggesting that the original genetic material was RNA.

Self-replicating ribozymes probably first arose between 3.5 billion and 4 billion years ago and may have begun the evolution of life on Earth. Early life was probably an RNA world, with RNA molecules serving both as carriers of genetic information and as catalysts that drove the chemical reactions needed to sustain and perpetuate life. These catalytic RNAs may have acquired the ability to synthesize protein-based enzymes, which are more efficient catalysts. With enzymes taking over more and more of the catalytic functions, RNA probably became relegated to the role of information storage and transfer. DNA, with its chemical stability and faithful replication, eventually replaced RNA as the primary carrier of genetic information. Nevertheless, today RNA is either produced by or plays a vital role in many biological processes, including transcription, replication, RNA processing, and translation.

CONCEPTS

Early life probably centered on RNA, which served as the original genetic material and as a biological catalyst.

The Structure of RNA

RNA, like DNA, is a polymer consisting of nucleotides joined together by phosphodiester bonds (see Chapter 8 for a discussion of DNA structure). However, there are several important differences in the structures of DNA and RNA. Whereas DNA nucleotides contain deoxyribose sugars, RNA nucleotides have ribose sugars (**Figure 10.1a**). With a free hydroxyl group on the 2′-carbon atom of the ribose sugar, RNA is degraded rapidly under alkaline conditions. The deoxyribose sugar of DNA lacks this free hydroxyl group, so DNA is a more stable molecule. Another important difference is that the pyrimidine uracil is present in RNA instead of thymine, one of the two pyrimidines found in DNA.

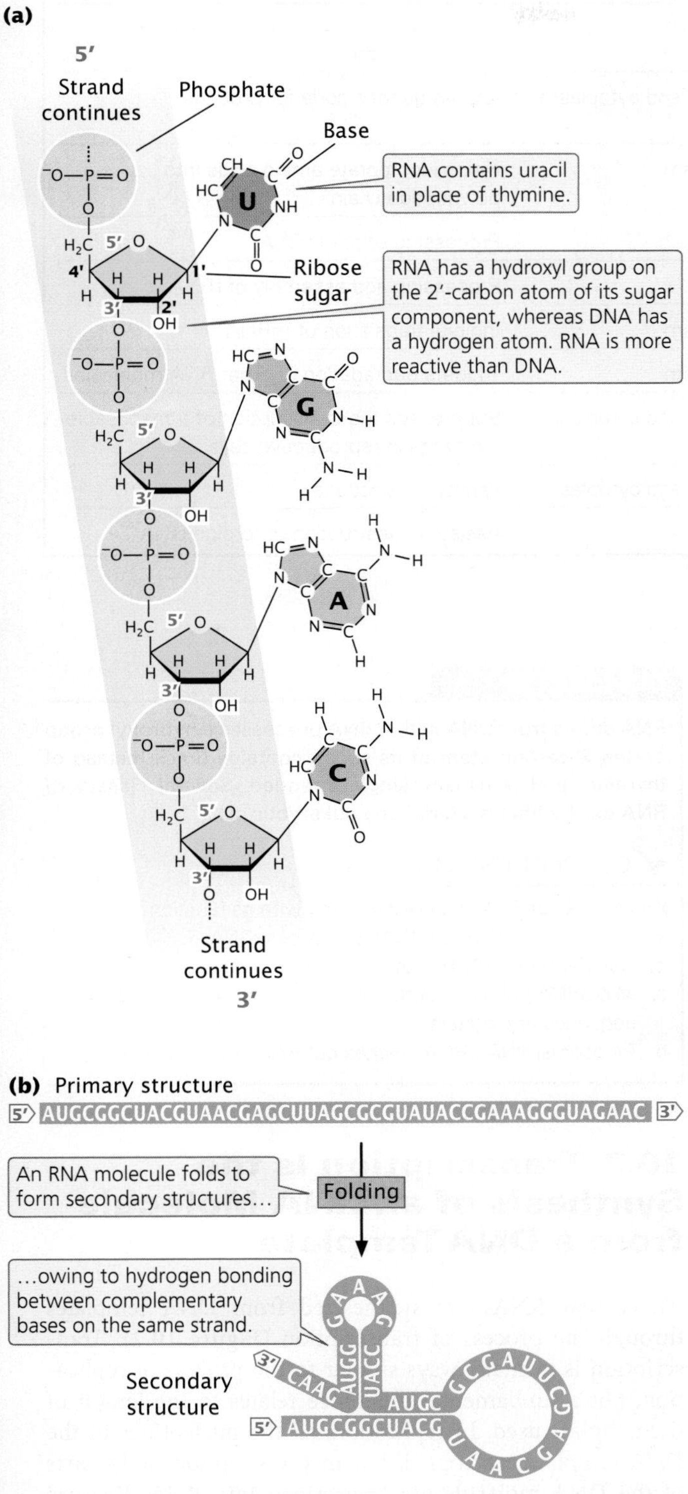

10.1 RNA has a primary and a secondary structure.

A final difference in the structures of DNA and RNA is that RNA usually consists of a single polynucleotide strand, whereas DNA normally consists of two polynucleotide strands joined by hydrogen bonding between complementary bases. Although RNA is usually single stranded, short complementary regions within a nucleotide strand can pair and form secondary structures (**Figure 10.1b**). For example, complementary regions within a transfer RNA molecule fold to form a cloverleaf secondary structure, which is important in the function of the tRNA. Exceptions to the rule that RNA is usually single stranded are found in a few RNA viruses that have double-stranded RNA genomes. Similarities and differences in DNA and RNA structures are summarized in **Table 10.1**. ▶TRY PROBLEM 15

Classes of RNA

RNA molecules perform a variety of functions in the cell. **Ribosomal RNA (rRNA)**, along with ribosomal protein subunits, makes up the ribosome, the site of protein assembly. We'll take a more detailed look at the ribosome later in the chapter. **Messenger RNA (mRNA)** carries the coding instructions for polypeptide chains from DNA to the ribosome. After attaching to a ribosome, an mRNA molecule specifies the sequence of the amino acids in a polypeptide chain and provides a template for the joining of those amino acids. Large precursor molecules, which are termed **pre-messenger RNAs (pre-mRNAs)**, are the immediate products of transcription in eukaryotic cells. Pre-mRNAs (also called primary transcripts) are modified extensively before becoming mRNA and exiting the nucleus for translation into protein. Bacterial cells do not possess pre-mRNA; in these cells, transcription takes place concurrently with translation.

TABLE 10.1 The structures of DNA and RNA compared

Characteristic	DNA	RNA
Composed of nucleotides	Yes	Yes
Type of sugar	Deoxyribose	Ribose
Presence of 2′-OH group	No	Yes
Bases	A, G, C, T	A, G, C, U
Nucleotides joined by phosphodiester bonds	Yes	Yes
Double or single stranded	Usually double	Usually single
Secondary structure	Double helix	Many types
Stability	Stable	Easily degraded

TABLE 10.2 Locations and functions of different classes of RNA molecules

Class of RNA	Cell type	Location of function in eukaryotic cells*	Function
Ribosomal RNA (rRNA)	Prokaryotic and eukaryotic	Cytoplasm	Structural and functional components of the ribosome
Messenger RNA (mRNA)	Prokaryotic and eukaryotic	Nucleus and cytoplasm	Carries genetic code for proteins
Transfer RNA (tRNA)	Prokaryotic and eukaryotic	Cytoplasm	Helps incorporate amino acids into polypeptide chain
Small nuclear RNA (snRNA)	Eukaryotic	Nucleus	Processing of pre-mRNA
Small nucleolar RNA (snoRNA)	Eukaryotic	Nucleus	Processing and assembly of rRNA
MicroRNA (miRNA)	Eukaryotic	Cytoplasm	Inhibits translation of mRNA
Small interfering RNA (siRNA)	Eukaryotic	Cytoplasm	Triggers degradation of other RNA molecules
Piwi-interacting RNA (piRNA)	Eukaryotic	Nucleus and cytoplasm	Suppresses the transcription of transposable elements in reproductive cells
Long noncoding RNA (lncRNA)	Eukaryotic	Nucleus and cytoplasm	Variety of functions
CRISPR RNA (crRNA)	Prokaryotic	—	Assists in destruction of foreign DNA

*All eukaryotic RNAs are synthesized in the nucleus.

Transfer RNA (tRNA) serves as the link between the coding sequence of nucleotides in an mRNA molecule and the amino acid sequence of a polypeptide chain. Each tRNA attaches to one particular type of amino acid and helps to incorporate that amino acid into a polypeptide chain.

Additional classes of RNA molecules are found in the nuclei of eukaryotic cells. **Small nuclear RNAs (snRNAs)** combine with small protein subunits to form **small nuclear ribonucleoproteins** (**snRNPs**, affectionately known as "snurps"). Some snRNAs participate in the processing of RNA, converting pre-mRNA into mRNA. **Small nucleolar RNAs (snoRNAs)** take part in the processing of rRNA. Two types of very small and abundant RNA molecules found in the cytoplasm of eukaryotic cells, termed **microRNAs (miRNAs)** and **small interfering RNAs (siRNAs)**, carry out RNA interference (RNAi), a process in which these small RNA molecules help trigger the degradation of mRNA or inhibit its translation into protein. Another class of small RNA molecules, called **Piwi-interacting RNAs** (**piRNAs**; named after Piwi proteins, with which they interact), has a role in suppressing the expression of transposable elements (see Chapter 13) in reproductive cells. **Long noncoding RNAs (lncRNAs)** are relatively long RNA molecules found in eukaryotes that do not code for proteins. They provide a variety of functions, including regulation of gene expression. An RNA interference–like system has been discovered in prokaryotes, in which small **CRISPR RNAs (crRNAs)** assist in the destruction of foreign DNA molecules. Some of the different classes of RNA molecules are summarized in **Table 10.2**.

CONCEPTS

RNA differs from DNA in that RNA possesses a hydroxyl group on the 2′-carbon atom of its sugar, contains uracil instead of thymine, and is usually single stranded. Several classes of RNA exist within bacterial and eukaryotic cells.

✓ CONCEPT CHECK 1

Which class of RNA is correctly paired with its function?

a. Small nuclear RNA (snRNA): processes rRNA
b. Transfer RNA (tRNA): attaches to an amino acid
c. MicroRNA (miRNA): carries information for the amino acid sequence of a protein
d. Ribosomal RNA (rRNA): carries out RNA interference

10.2 Transcription Is the Synthesis of an RNA Molecule from a DNA Template

All cellular RNAs are synthesized from DNA templates through the process of transcription (**Figure 10.2**). Transcription is in many ways similar to the process of replication, but a fundamental difference relates to the length of the template used. In replication, all the nucleotides in the DNA template are copied, but in transcription, only parts of the DNA molecule are transcribed into RNA. Because not all gene products are needed at the same time or in the same cell, the constant transcription of all of a cell's genes would be highly inefficient. Furthermore, much of the DNA

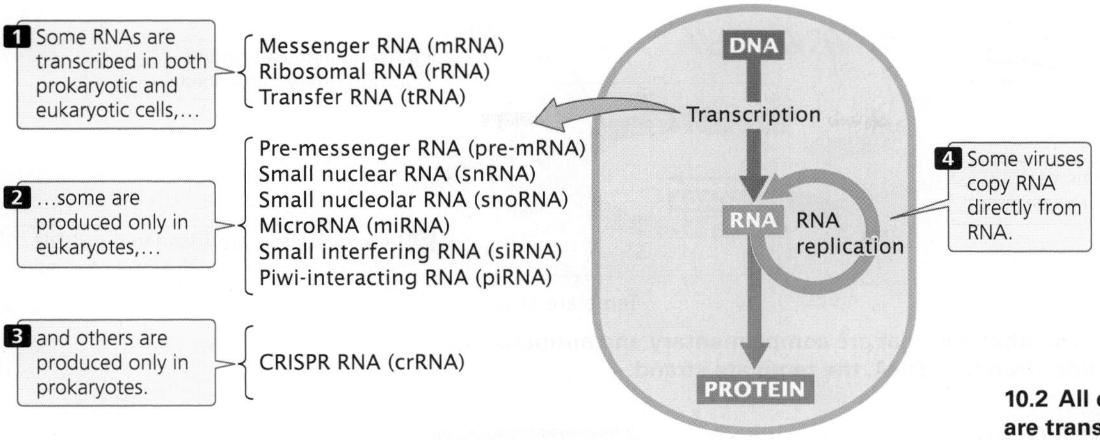

10.2 All cellular types of RNA are transcribed from DNA.

does not encode a functional product, and transcription of such sequences would be pointless. Transcription is, in fact, a highly selective process: individual genes are transcribed only as their products are needed. However, this selectivity imposes a fundamental problem on the cell: how to recognize individual genes and transcribe them at the proper time and place.

Like replication, transcription requires three major components:

1. A DNA template
2. The raw materials (ribonucleotide triphosphates) needed to build a new RNA molecule
3. The transcription apparatus, consisting of the proteins necessary for catalyzing the synthesis of RNA

The Template

In 1970, Oscar Miller, Jr., Barbara Hamkalo, and Charles Thomas used electron microscopy to examine cellular contents and demonstrate that RNA is transcribed from a DNA template. They broke open cells and spread the chromatin onto a fine mesh grid. When viewed with electron microscopy, they observed Christmas-tree-like structures: thin central fibers (the trunk of the tree) to which were attached strings (the branches) with granules (**Figure 10.3a**). The addition of deoxyribonuclease (an enzyme that degrades DNA) caused the central fibers to disappear, indicating that the "tree trunks" were DNA molecules. Ribonuclease (an enzyme that degrades RNA) removed the granular strings, indicating that the branches were RNA. Their conclusion was that each "Christmas tree" represented a gene undergoing transcription (**Figure 10.3b**). The transcription of each gene begins at the top of the tree; there, little of the DNA has been transcribed, and the RNA branches are short. As the transcription apparatus proceeds down the tree, transcribing more of the template, the RNA molecules lengthen, producing the long branches at the bottom.

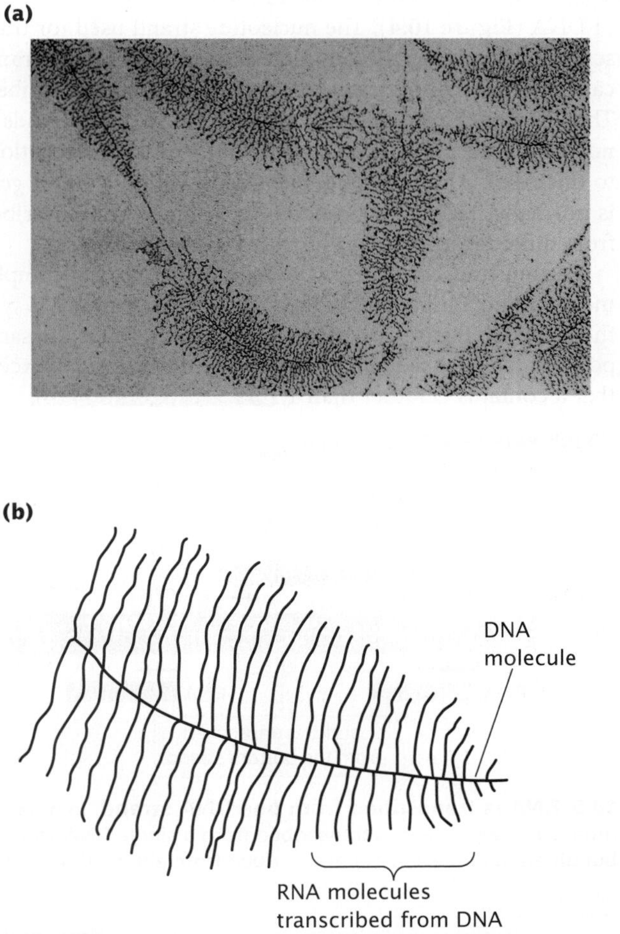

10.3 Under the electron microscope, DNA molecules undergoing transcription exhibit Christmas-tree-like structures. (a) Electronic micrograph of Christmas-tree-like structures. (b) The trunk of each "Christmas tree" (a transcription unit) represents a portion of a DNA molecule; the tree branches (granular strings attached to the DNA) are RNA molecules that have been transcribed from the DNA. As the transcription apparatus proceeds down the DNA, transcribing more of the template, the RNA molecules become longer and longer. [© Phototake, Inc. / Phototake.]

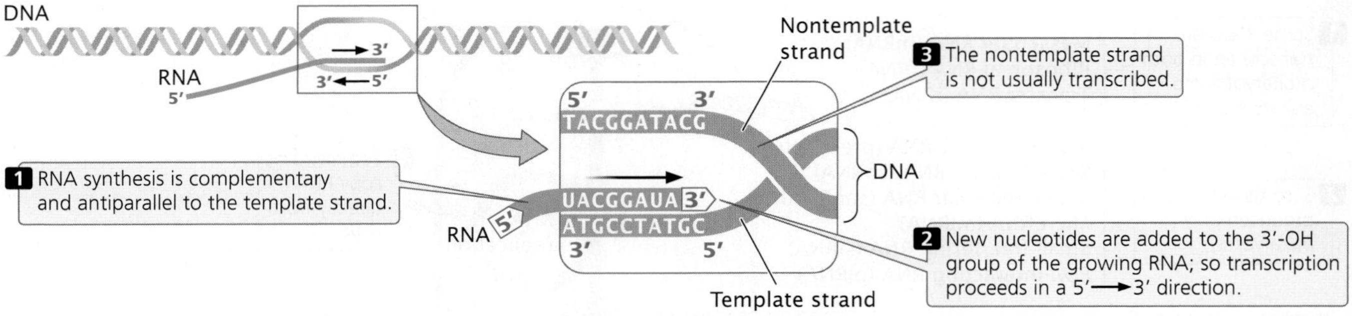

10.4 RNA molecules are synthesized that are complementary and antiparallel to one of the two nucleotide strands of DNA, the template strand.

THE TRANSCRIBED STRAND The template for RNA synthesis, as for DNA synthesis, is a single strand of the DNA double helix. Unlike replication, however, the transcription of a gene takes place on only one of the two nucleotide strands of DNA (**Figure 10.4**). The nucleotide strand used for transcription is termed the **template strand**. The other strand, called the **nontemplate strand**, is not ordinarily transcribed. Thus, within a gene, only one of the nucleotide strands is normally transcribed into RNA (there are some exceptions to this rule). Although only one strand within a single *gene* is normally transcribed, different genes may be transcribed from different strands, as illustrated in **Figure 10.5**.

During transcription, an RNA molecule that is complementary and antiparallel to the DNA template strand is synthesized (see Figure 10.4). The RNA transcript has the same polarity and base sequence as the nontemplate strand, except that it contains U rather than T. ▶TRY PROBLEM 16

THINK-PAIR-SHARE Question 4

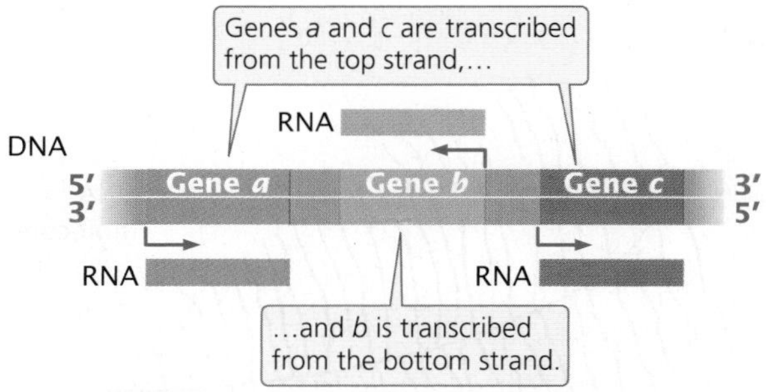

10.5 RNA is transcribed from one DNA strand. In most organisms, each gene is transcribed from a single DNA strand, but different genes may be transcribed from either DNA strand.

CONCEPTS

Within a single gene, only one of the two DNA strands, the template strand, is usually transcribed into RNA.

✔ CONCEPT CHECK 2

What is the difference between the template strand and the nontemplate strand?

THE TRANSCRIPTION UNIT A **transcription unit** is a stretch of DNA that encodes an RNA molecule and the sequences necessary for its transcription. How does the complex of enzymes and proteins that performs transcription—the transcription apparatus—recognize a transcription unit? How does it know which DNA strand to read and where to start and stop? This information is encoded by the DNA sequence.

Included within a transcription unit are three critical regions: a promoter, an RNA-coding region, and a terminator (**Figure 10.6**). The **promoter** is a DNA sequence that the transcription apparatus recognizes and binds. It indicates which of the two DNA strands is to be read as the template and the direction of transcription. The promoter also determines the transcription start site, the first nucleotide that will be transcribed into RNA. In many transcription units, the promoter is located next to the transcription start site but is not itself transcribed.

The second critical region of the transcription unit is the **RNA-coding region**, a sequence of DNA nucleotides that is copied into an RNA molecule. The third component of the

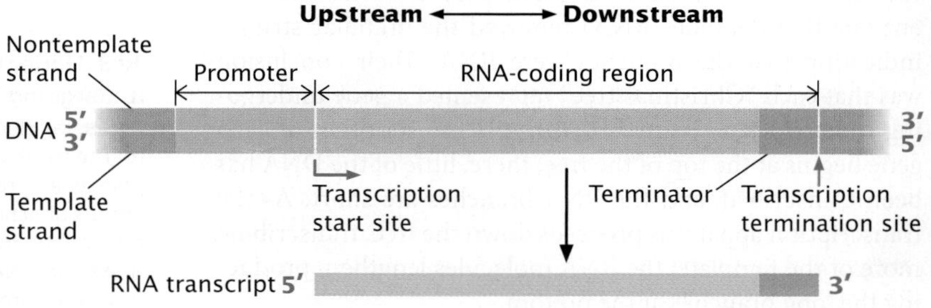

10.6 A transcription unit includes a promoter, a region that encodes RNA, and a terminator.

transcription unit is the **terminator**, a sequence of nucleotides that signals where transcription is to end. Terminators are usually part of the RNA-coding sequence; that is, transcription stops only after the terminator has been copied into RNA.

Molecular biologists often use the terms *upstream* and *downstream* to refer to the direction of transcription and the locations of nucleotide sequences surrounding the RNA-coding sequence. The transcription apparatus is said to move downstream as transcription takes place: it binds to the promoter (which is usually upstream of the start site) and moves toward the terminator (which is downstream of the start site).

When DNA sequences are written out, often the sequence of only one of the two strands is listed. Molecular biologists typically write the sequence of the nontemplate strand because it will be the same as the sequence of the RNA transcribed from the template strand (with the exception that U in RNA replaces T in DNA). By convention, the sequence on the nontemplate strand is written with the 5′ end on the left and the 3′ end on the right. The first nucleotide transcribed (the transcription start site) is numbered +1; nucleotides downstream of the start site are assigned positive numbers, and nucleotides upstream of the start site are assigned negative numbers. So, nucleotide +34 would be 34 nucleotides downstream of the start site, whereas nucleotide −75 would be 75 nucleotides upstream of the start site. There is no nucleotide numbered 0.

CONCEPTS

A transcription unit is a stretch of DNA that encodes an RNA molecule and the sequences necessary for its proper transcription. Each transcription unit includes a promoter, an RNA-coding region, and a terminator.

The Substrate for Transcription

RNA is synthesized from **ribonucleoside triphosphates** (rNTPs; **Figure 10.7**). In synthesis, nucleotides are added one at a time to the 3′-OH group of the growing RNA molecule. Two phosphate groups are cleaved from the incoming ribonucleoside triphosphate; the remaining phosphate group participates in a phosphodiester bond that connects the nucleotide to the growing RNA molecule. The overall chemical reaction for the addition of each nucleotide is

$$RNA_n + rNTP \rightarrow RNA_{n+1} + PP_i$$

in which PP_i represents pyrophosphate. Nucleotides are always added to the 3′ end of the RNA molecule, and the direction of transcription is therefore 5′ → 3′ (**Figure 10.8**), the same as the direction of DNA synthesis in replication. Thus, the newly synthesized RNA is complementary and antiparallel to the template strand. Unlike DNA synthesis, RNA synthesis does not require a primer.

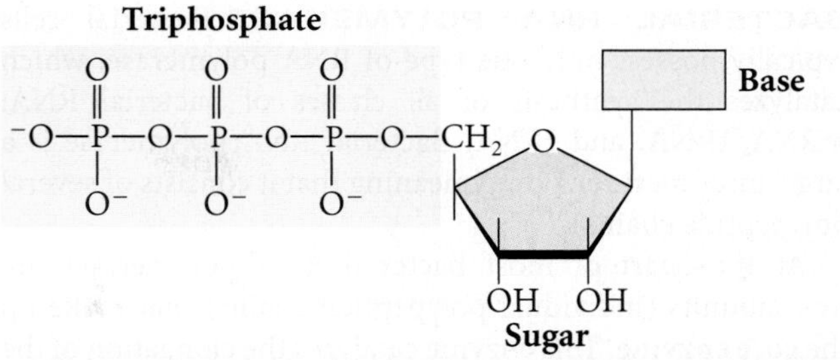

10.7 Ribonucleoside triphosphates are substrates used in RNA synthesis.

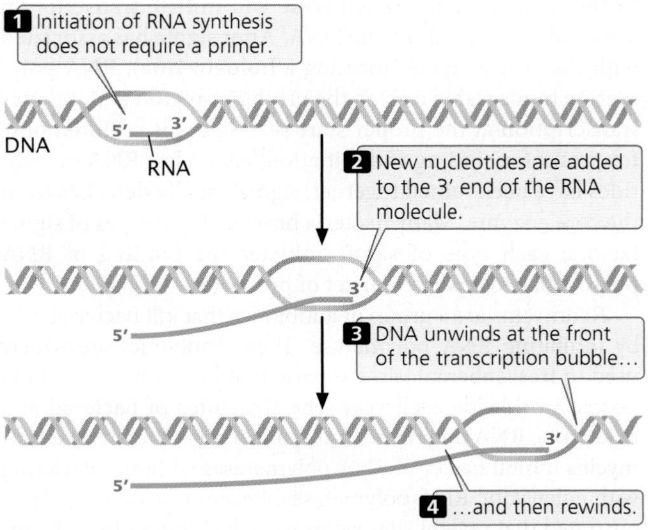

10.8 In transcription, nucleotides are always added to the 3′ end of the RNA molecule.

CONCEPTS

RNA is synthesized from ribonucleoside triphosphates. Transcription is 5′ → 3′: each new nucleotide is joined to the 3′-OH group of the last nucleotide added to the growing RNA molecule.

The Transcription Apparatus

Recall that DNA replication requires a number of different enzymes and proteins. Transcription might initially appear to be quite different because a single enzyme—**RNA polymerase**—carries out all the required steps of transcription, but on closer inspection, the processes are actually similar. The action of RNA polymerase is enhanced by a number of accessory proteins that join and leave the polymerase at different stages of the process. Each accessory protein is responsible for providing or regulating a special function. Thus, transcription, like replication, requires an array of proteins.

BACTERIAL RNA POLYMERASE Bacterial cells typically possess only one type of RNA polymerase, which catalyzes the synthesis of all classes of bacterial RNA: mRNA, tRNA, and rRNA. Bacterial RNA polymerase is a large, multimeric enzyme (meaning that it consists of several polypeptide chains).

At the heart of most bacterial RNA polymerases are five subunits (individual polypeptide chains) that make up the **core enzyme**. This enzyme catalyzes the elongation of the RNA molecule by the addition of RNA nucleotides. Other functional subunits join and leave the core enzyme at particular stages of the transcription process. The **sigma** (σ) **factor** controls the binding of RNA polymerase to the promoter. Without sigma, RNA polymerase will initiate transcription at a random point along the DNA. After sigma has associated with the core enzyme (forming a **holoenzyme**), RNA polymerase binds stably only to the promoter region and initiates transcription at the proper start site. Sigma is required only for promoter binding and initiation; after a few RNA nucleotides have been joined together, sigma usually detaches from the core enzyme. Many bacteria have multiple types of sigma factors; each type of sigma initiates the binding of RNA polymerase to a particular set of promoters.

Rifamycins are a group of antibiotics that kill bacterial cells by inhibiting RNA polymerase. These antibiotics are widely used to treat tuberculosis, a disease that kills almost 2 million people worldwide each year. The structures of bacterial and eukaryotic RNA polymerases are sufficiently different that rifamycins inhibit bacterial RNA polymerases without interfering with eukaryotic RNA polymerases. Recent research has demonstrated that several rifamycins work by binding to and jamming the part of the bacterial RNA polymerase that clamps onto DNA, thus preventing the RNA polymerase from interacting with the promoter on the DNA.

EUKARYOTIC RNA POLYMERASES Most eukaryotic cells possess three distinct types of RNA polymerase, each of which is responsible for transcribing a different class of RNA: **RNA polymerase I** transcribes rRNA; **RNA polymerase II** transcribes pre-mRNAs, snoRNAs, some miRNAs, and some snRNAs; and **RNA polymerase III** transcribes other small RNA molecules—specifically, tRNAs, small rRNAs, some miRNAs, and some snRNAs (**Table 10.3**). RNA polymerases I, II, and III are found in all eukaryotes. Two additional RNA polymerases, named **RNA polymerase IV** and **RNA polymerase V**, have been found in plants. These RNA polymerases transcribe RNAs that play a role in DNA methylation and chromatin structure.

All eukaryotic polymerases are large, multimeric enzymes, typically consisting of more than a dozen subunits. Some subunits are common to all RNA polymerases, whereas others are limited to one type of polymerase. As in bacterial cells, a number of accessory proteins bind to the core enzyme and affect its function.

TABLE 10.3 Eukaryotic RNA polymerases

Type	Present in	Transcribes
RNA polymerase I	All eukaryotes	Large rRNAs
RNA polymerase II	All eukaryotes	Pre-mRNA, some snRNAs, snoRNAs, some miRNAs
RNA polymerase III	All eukaryotes	tRNAs, small rRNAs, some snRNAs, some miRNAs
RNA polymerase IV	Plants	Some siRNAs
RNA polymerase V	Plants	RNA molecules taking part in heterochromatin formation

CONCEPTS

Bacterial cells possess a single type of RNA polymerase, consisting of a core enzyme and other subunits that participate in various stages of transcription. Eukaryotic cells possess several distinct types of RNA polymerase that transcribe different kinds of RNA molecules.

✓ CONCEPT CHECK 3

What is the function of the sigma factor?

10.3 Bacterial Transcription Consists of Initiation, Elongation, and Termination

Now that we've considered some of the major components of transcription, we're ready to take a detailed look at the process. Transcription can be conveniently divided into three stages:

1. Initiation, in which the transcription apparatus assembles on the promoter and begins the synthesis of RNA
2. Elongation, in which DNA is threaded through RNA polymerase, and the polymerase unwinds the DNA and adds new nucleotides, one at a time, to the 3′ end of the growing RNA strand
3. Termination, the recognition of the end of the transcription unit and the separation of the RNA molecule from the DNA template

We will examine each of these steps in bacterial cells, in which the process is best understood.

Initiation

Initiation comprises all the steps necessary to begin RNA synthesis, including (1) promoter recognition, (2) formation of a transcription bubble, (3) creation of the first bonds between rNTPs, and (4) escape of the transcription apparatus from the promoter.

Transcription initiation requires that the transcription apparatus recognize and bind to the promoter. At this step, the selectivity of transcription is enforced: the binding of RNA polymerase to the promoter determines which parts of the DNA template are to be transcribed, and how often. Different genes are transcribed with different frequencies, and promoter binding is important in determining the frequency of transcription for a particular gene. Promoters also have different affinities for RNA polymerase. Even within a single promoter, the affinity can vary with the passage of time, depending on the promoter's interaction with RNA polymerase and a number of other factors.

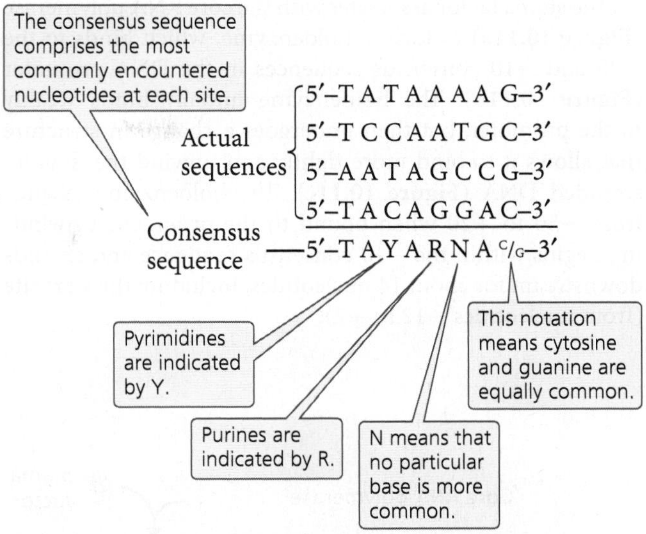

10.9 A consensus sequence consists of the most commonly encountered bases at each position in a group of related sequences.

BACTERIAL PROMOTERS Promoters, as we have seen, are DNA sequences that are recognized by the transcription apparatus and are required for transcription to take place. Essential information for the transcription apparatus—where it will start transcribing, which strand is to be read, and in what direction the RNA polymerase will move—is embedded in the nucleotide sequence of the promoter. In bacterial cells, promoters are usually adjacent to an RNA-coding sequence.

An examination of many promoters in *E. coli* and other bacteria reveals a general feature: although most promoters vary in sequence, short stretches of nucleotides are common to many. Furthermore, the spacing and location of these nucleotides relative to the transcription start site are similar in most promoters. These short stretches of common nucleotides are called consensus sequences. A consensus sequence is the set of most commonly encountered nucleotides among sequences that possess considerably similarity, or *consensus* (**Figure 10.9**). The presence of consensus in a set of nucleotides usually implies that the sequence is associated with an important function. ▶TRY PROBLEM 25

The most commonly encountered consensus sequence, found in almost all bacterial promoters, is centered about 10 bp upstream of the start site. Called the −10 **consensus sequence**, or sometimes the Pribnow box, this consensus sequence,

5′ — TATAAT — 3′
3′ — ATATTA — 5′

is often written simply as TATAAT (**Figure 10.10**). Remember that TATAAT is just the *consensus* sequence—representing the most commonly encountered nucleotides at each of these positions. In most prokaryotic promoters, the actual sequence is not TATAAT.

Another consensus sequence common to most bacterial promoters is TTGACA, which lies approximately 35 nucleotides upstream of the start site and is termed the **−35 consensus sequence** (see Figure 10.10). The nucleotides on either side of the −10 and −35 consensus sequences and those between them vary greatly from promoter to promoter, suggesting that they are not very important in promoter recognition.

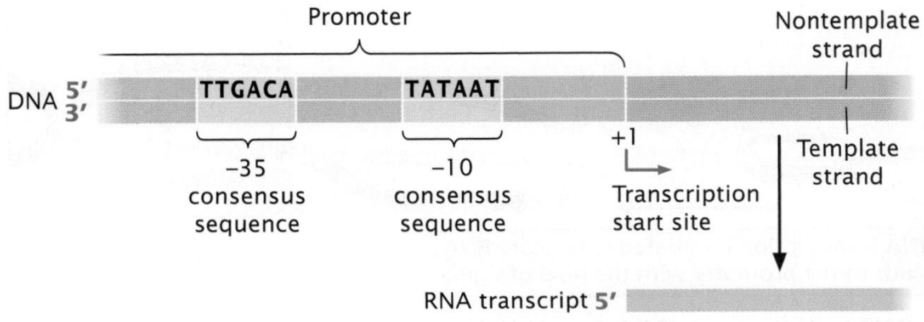

10.10 In bacterial promoters, consensus sequences are found upstream of the start site, approximately at positions −10 and −35.

The sigma factor associates with the core RNA polymerase (**Figure 10.11a**) to form a holoenzyme, which binds to the −35 and −10 consensus sequences in the DNA promoter (**Figure 10.11b**). The holoenzyme initially binds weakly to the promoter, but then undergoes a change in structure that allows it to bind more tightly and unwind the double-stranded DNA (**Figure 10.11c**). The holoenzyme extends from −50 to +20 when bound to the promoter. Unwinding begins within the −10 consensus sequence and extends downstream for about 14 nucleotides, including the start site (from nucleotides −12 to +2).

CONCEPTS

A promoter is a DNA sequence, usually adjacent to a gene, that is required for gene transcription. Promoters contain short consensus sequences that are important in the initiation of transcription.

INITIAL RNA SYNTHESIS After the holoenzyme has attached to the promoter, RNA polymerase is positioned over the start site for transcription (at position +1) and has unwound the DNA to produce a single-stranded template. The orientation and spacing of consensus sequences on a

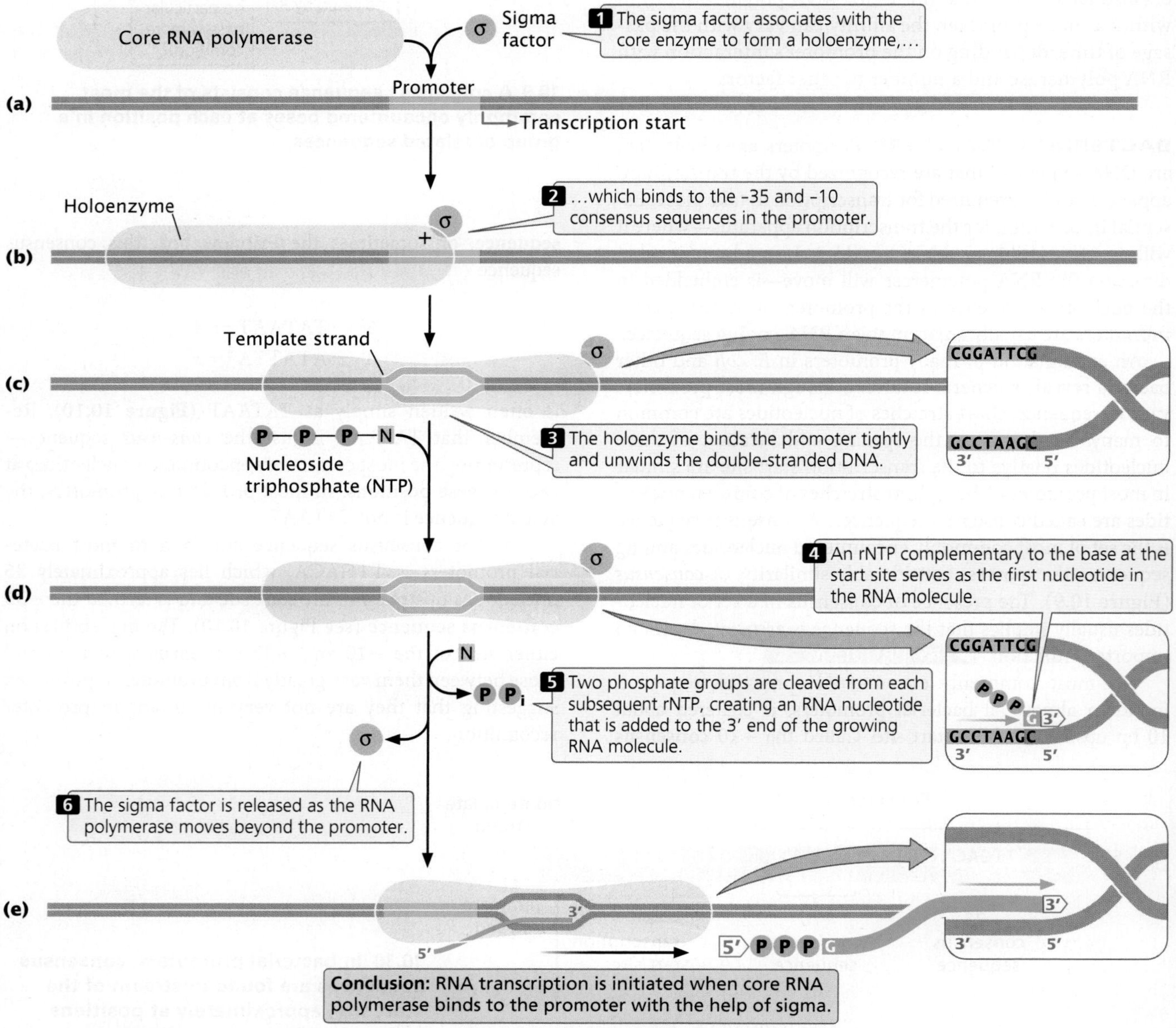

10.11 Transcription in bacteria is catalyzed by RNA polymerase, which must bind to the sigma factor to initiate transcription.

DNA strand determine which strand will be the template for transcription and thereby determine the direction of transcription.

The position of the start site is determined not by the sequences located there, but by the location of the consensus sequences, which position RNA polymerase so that the enzyme's active site is aligned for the initiation of transcription at +1. If the consensus sequences are artificially moved upstream or downstream, the location of the starting point of transcription correspondingly changes.

To begin the synthesis of an RNA molecule, RNA polymerase pairs the base on a ribonucleoside triphosphate with its complementary base at the start site on the DNA template strand (**Figure 10.11d**). No primer is required to initiate the synthesis of the 5′ end of the RNA molecule. Two of the three phosphate groups are cleaved from the ribonucleoside triphosphate as the nucleotide is added to the 3′ end of the growing RNA molecule. However, because the 5′ end of the first ribonucleoside triphosphate does not take part in the formation of a phosphodiester bond, all three of its phosphate groups remain. An RNA molecule therefore possesses, at least initially, three phosphate groups at its 5′ end (**Figure 10.11e**).

Elongation

At the end of initiation, RNA polymerase undergoes a change in conformation (shape) and is thereafter no longer able to bind to the consensus sequences in the promoter. This change allows the polymerase to escape from the promoter and begin transcribing downstream. The sigma factor is usually released after initiation, although some RNA polymerases may retain sigma throughout elongation.

Transcription takes place within a short stretch of about 18 nucleotides of unwound DNA called the transcription bubble. As it moves downstream along the template, RNA polymerase progressively unwinds the DNA at the leading (downstream) edge of the transcription bubble, joining nucleotides to the RNA molecule according to the sequence on the template, and rewinds the DNA at the trailing (upstream) edge of the bubble.

CONCEPTS

Transcription is initiated at the start site, which, in bacterial cells, is determined by the binding of RNA polymerase to the consensus sequences of the promoter. No primer is required. Transcription takes place within the transcription bubble. DNA is unwound ahead of the bubble and rewound behind it.

Termination

RNA polymerase adds nucleotides to the 3′ end of the growing RNA molecule until it transcribes a terminator. Most terminators are found upstream of the site at which termination actually takes place. Transcription therefore does not suddenly stop when polymerase reaches a terminator, like a car stopping at a stop sign. Rather, transcription stops after the terminator has been transcribed, like a car that stops only after running over a speed bump. At the terminator, several overlapping events are needed to bring an end to transcription: RNA polymerase must stop synthesizing RNA, the RNA molecule must be released from RNA polymerase, the newly made RNA molecule must dissociate fully from the DNA, and RNA polymerase must detach from the DNA template.

Bacterial cells possess two major types of terminators. **Rho-dependent terminators** are able to cause the termination of transcription only in the presence of an ancillary protein called the **rho** (ρ) **factor**. **Rho-independent terminators** (also known as intrinsic terminators) are able to cause the end of transcription in the absence of rho.

RHO-DEPENDENT TERMINATORS Genes with rho-dependent terminators have a stretch of RNA upstream of the terminator that is devoid of any secondary structures. This unstructured RNA, called the rho utilization (rut) site, serves as binding site for the rho protein, which binds the RNA and moves toward its 3′ end, following the RNA polymerase (**Figure 10.12**). When RNA polymerase encounters the terminator, it pauses, allowing rho to catch up. The rho

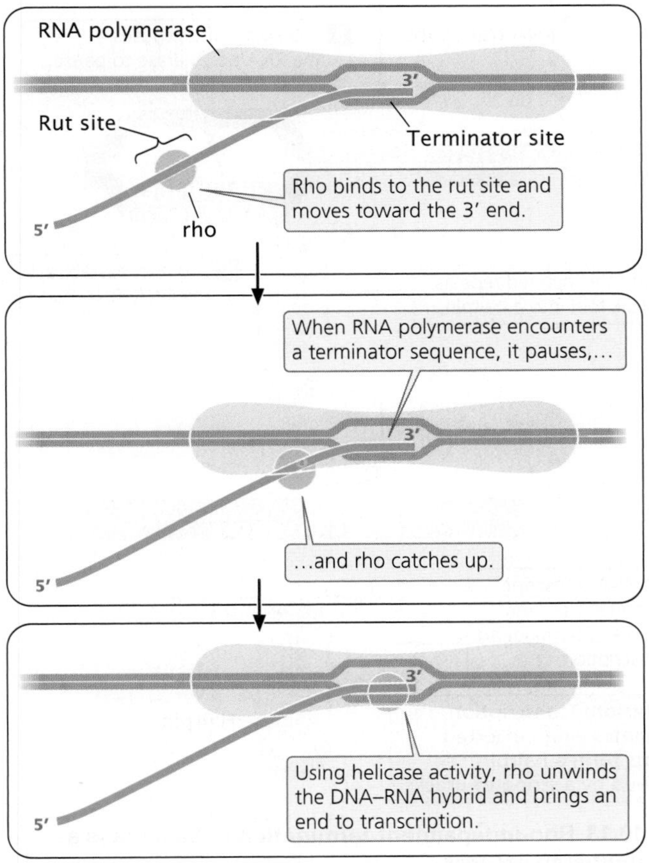

10.12 The termination of transcription in some bacterial genes requires the presence of the rho factor.

protein has helicase activity, which it uses to unwind the DNA–RNA hybrid in the transcription bubble, bringing transcription to an end.

RHO-INDEPENDENT TERMINATORS Rho-independent terminators, which make up about 50% of all terminators in prokaryotes, have two common features. First, they contain inverted repeats (sequences of nucleotides on the same strand that are inverted and complementary). When these inverted repeats are transcribed into RNA and bind to each other, a secondary structure called a **hairpin** forms (**Figure 10.13**). Second, in rho-independent terminators, a string of seven to nine adenine nucleotides follows the inverted repeat in the template DNA. Their transcription produces a string of uracil nucleotides after the hairpin in the transcribed RNA.

The string of uracil nucleotides in the RNA molecule causes the RNA polymerase to pause, allowing time for the hairpin structure to form. Evidence suggests that the formation of the hairpin destabilizes the DNA–RNA pairing, causing the RNA molecule to separate from its DNA template and transcription to terminate.

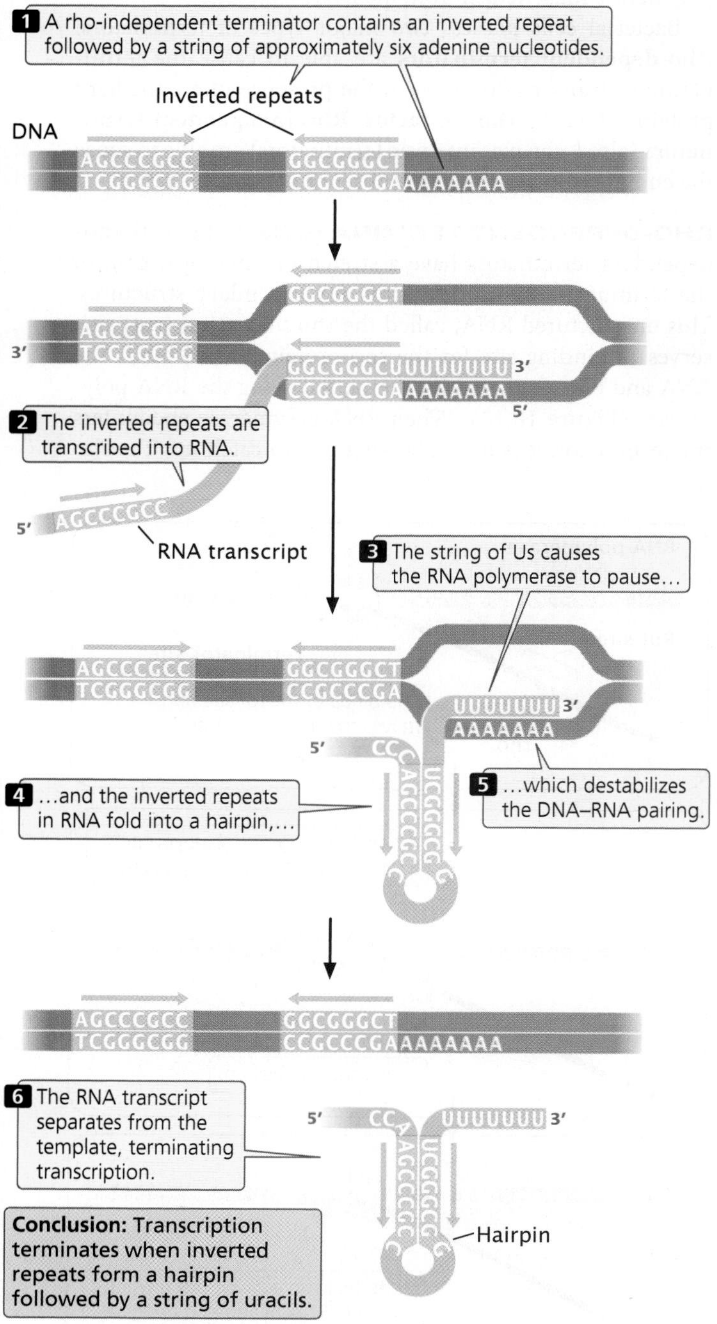

10.13 Rho-independent termination in bacteria is a multistep process.

POLYCISTRONIC RNA In bacteria, a group of genes is often transcribed into a single RNA molecule, which is termed **polycistronic RNA**. Thus, polycistronic RNA is produced when a single terminator is present at the end of a group of several genes that are transcribed together, instead of each gene having its own terminator. Polycistronic mRNA does occur in some eukaryotes such as *Caenorhabditis elegans*, but it is uncommon.

You can view the process of transcription, including initiation, elongation, and termination, in **Animation 10.1**. The animation shows how the different parts of the transcription unit interact to bring about the complete synthesis of an RNA molecule.

THINK-PAIR-SHARE Question 5

CONCEPTS

Transcription ends after RNA polymerase transcribes a terminator. Bacterial cells possess two types of terminators: a rho-independent terminator, which RNA polymerase can recognize by itself; and a rho-dependent terminator, which RNA polymerase can recognize only with the help of the rho protein.

CONNECTING CONCEPTS

The Basic Rules of Transcription

Before we examine gene structure and how RNA molecules are modified after transcription, let's summarize some of the general principles of bacterial transcription.

1. Transcription is a selective process; only certain parts of the DNA are transcribed at any one time.
2. RNA is transcribed from a single strand of DNA. Within a gene, only one of the two DNA strands—the template strand—is usually copied into RNA.
3. Ribonucleoside triphosphates are used as the substrates in RNA synthesis. Two phosphate groups are cleaved from a ribonucleoside triphosphate, and the resulting nucleotide is joined to the 3′-OH group of the growing RNA strand.

4. RNA molecules are antiparallel and complementary to the DNA template strand. Transcription is always in the 5′→3′ direction, meaning that the RNA molecule grows at the 3′ end.
5. Transcription depends on RNA polymerase—a complex, multimeric enzyme. RNA polymerase consists of a core enzyme, which is capable of synthesizing RNA, and other subunits that may join transiently to perform additional functions. A sigma factor enables the core enzyme of RNA polymerase to bind to a promoter and initiate transcription.
6. Promoters contain short sequences crucial to the binding of RNA polymerase to DNA.
7. RNA polymerase binds to DNA at a promoter, begins transcribing at the start site of the gene, and ends transcription after a terminator has been transcribed.

THINK-PAIR-SHARE Question 6

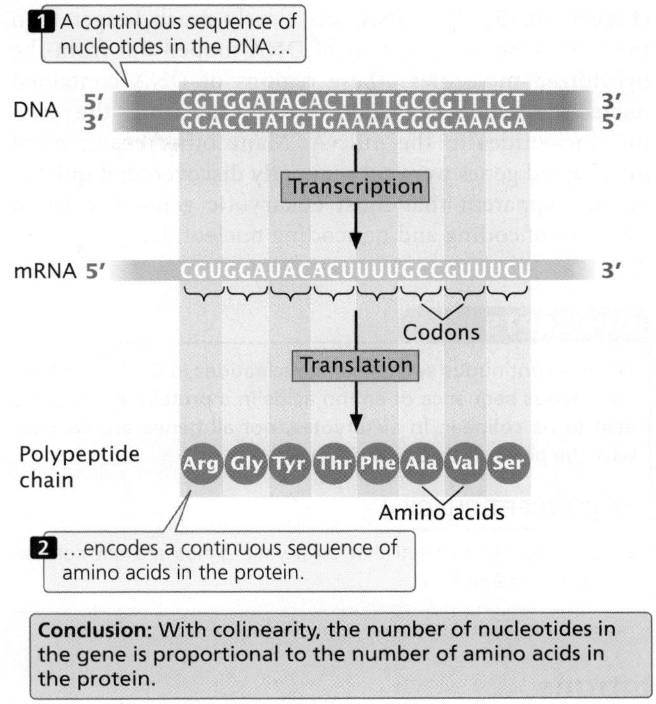

10.14 The concept of colinearity suggests that a continuous sequence of nucleotides in DNA encodes a continuous sequence of amino acids in a protein. As illustrated here, a codon specifies each amino acid.

10.4 Many Genes Have Complex Structures

Having now considered the chemical structure of DNA and how DNA is transcribed into RNA, we can ask: What is a gene? As noted in Chapter 3, the definition of a gene often changes as we explore different aspects of heredity. A gene was defined in Chapter 3 as an inherited factor that determines a characteristic. This definition may have seemed vague because it says only what a gene does rather than what a gene is. Nevertheless, this definition was appropriate at the time because our focus was on how genes influence the inheritance of traits. We did not have to consider the physical nature of the gene in learning the rules of inheritance.

At this point, we can be more precise about what a gene is. Chapter 8 described how genetic information is encoded in the base sequence of DNA; a gene consists of a set of DNA nucleotides. But how many nucleotides constitute a gene, and how is the information in those nucleotides organized? In 1902, Archibald Garrod correctly suggested that genes encode proteins. Proteins are made of amino acids, so a gene contains the nucleotides that specify the amino acids of a protein. Therefore, for many years, the working definition of a gene was a set of nucleotides that specifies the amino acid sequence of a protein. As geneticists learned more about the structure of genes, however, it became clear that this concept of a gene was an oversimplification.

Gene Organization

Early work on gene structure was carried out largely through the examination of mutations in bacteria and viruses. This research led Francis Crick in 1958 to propose that genes and proteins are **colinear**—that there is a direct correspondence between the nucleotide sequence of DNA and the amino acid sequence of a protein (**Figure 10.14**). The concept of colinearity suggests that the number of nucleotides in a gene should be proportional to the number of amino acids in the protein encoded by that gene. In a general sense, this concept is true for genes found in bacterial cells and in many viruses, although those genes are slightly longer than would be expected if colinearity were strictly applied because the mRNAs encoded by the genes contain sequences at their ends that do not specify amino acids. At first, eukaryotic genes and proteins were also assumed to be colinear, but there were hints that eukaryotic gene structure is fundamentally different. Eukaryotic cells were found to contain far more DNA than is required to encode proteins. Furthermore, many large RNA molecules observed in the nucleus were absent from the cytoplasm, suggesting that nuclear RNAs undergo some type of change before they are exported to the cytoplasm.

Most geneticists were nevertheless surprised by the announcement in the 1970s that not all genes are continuous. Researchers observed four coding sequences in a gene from a eukaryotic virus that were interrupted by nucleotides that did not specify amino acids. This discovery was made when the viral DNA was hybridized (paired) with the mRNA transcribed from it and the hybridized structure was examined using an electron microscope

(**Figure 10.15**). The DNA was clearly much longer than the mRNA because regions of DNA looped out from the hybridized molecules. These regions of DNA contained nucleotide sequences that were absent from the coding nucleotides in the mRNA. Many other examples of interrupted genes were subsequently discovered; it quickly became apparent that most eukaryotic genes consist of stretches of coding and noncoding nucleotides.

CONCEPTS

When a continuous sequence of nucleotides in DNA encodes a continuous sequence of amino acids in a protein, the two are said to be colinear. In eukaryotes, not all genes are colinear with the proteins that they encode.

✓ CONCEPT CHECK 4

What evidence indicated that eukaryotic genes are not colinear with their proteins?

Introns

Many eukaryotic genes contain coding regions called **exons** and noncoding regions called intervening sequences, or **introns**. For example, the gene encoding the protein ovalbumin has eight exons and seven introns; the gene for cytochrome *b* has five exons and four introns (**Figure 10.16**). The average human gene contains from eight to nine introns. All the introns and exons are initially transcribed into RNA, but during or after transcription, the introns are removed and the exons are joined to yield the mature RNA.

Introns are common in eukaryotic genes but are rare in bacterial genes. All classes of eukaryotic genes—those that encode rRNA, tRNA, and proteins—may contain introns. The numbers and sizes of introns vary widely: some eukaryotic genes have no introns, whereas others may have more than 60; intron length varies from fewer than 200 nucleotides to more than 50,000. Introns tend to be longer than exons, and most eukaryotic genes contain more noncoding

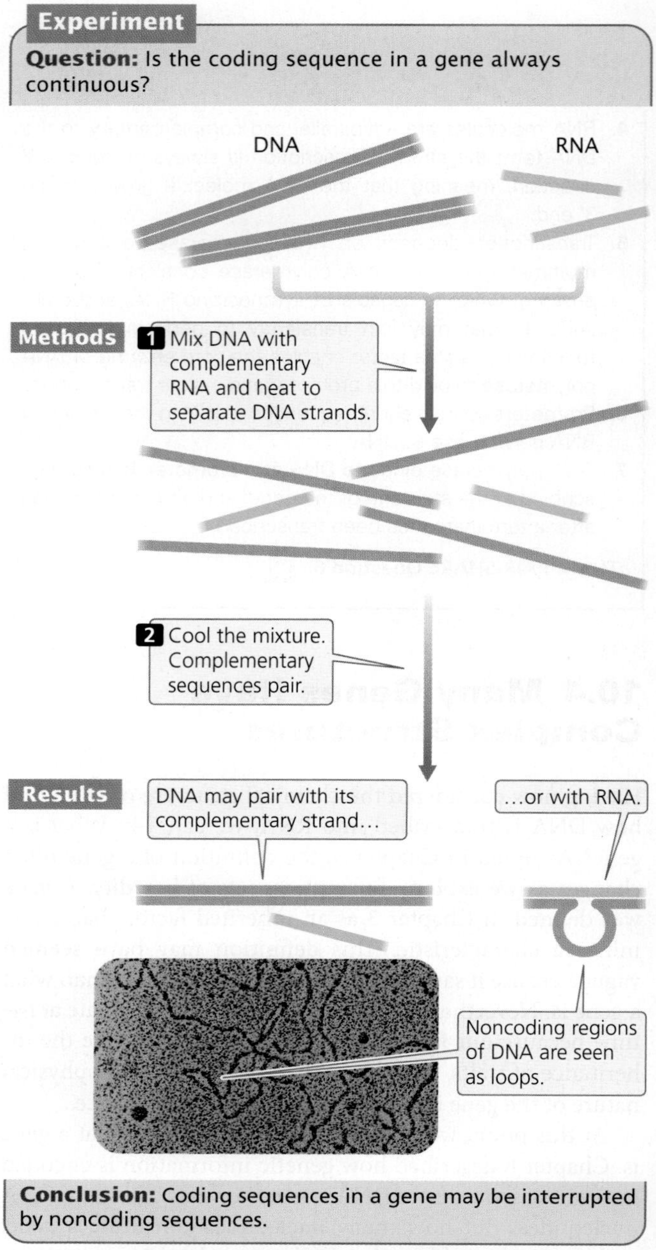

Experiment
Question: Is the coding sequence in a gene always continuous?
DNA
RNA
Methods
1 Mix DNA with complementary RNA and heat to separate DNA strands.
2 Cool the mixture. Complementary sequences pair.
Results
DNA may pair with its complementary strand…
…or with RNA.
Noncoding regions of DNA are seen as loops.
Conclusion: Coding sequences in a gene may be interrupted by noncoding sequences.

10.15 The noncolinearity of eukaryotic genes was discovered by hybridizing DNA and mRNA. [From Susan M. Berget, Claire Moore, and Phillip A. Sharp, Proc. Nati. Acad. Sci. USA Vol. 74, No. 8, pp. 3171–3175, Fig. 4g August 1977 Biochemistry.]

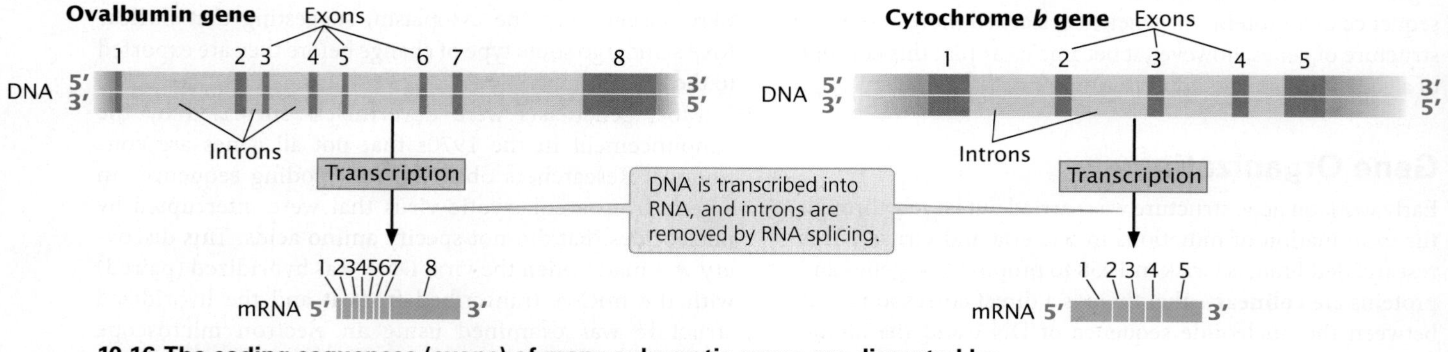

10.16 The coding sequences (exons) of many eukaryotic genes are disrupted by noncoding sequences (introns).

nucleotides than coding nucleotides. Finally, most introns do not encode proteins: an intron of one gene is not usually an exon for a different gene. ▶TRY PROBLEM 27

THINK-PAIR-SHARE Question 7

CONCEPTS

Many eukaryotic genes contain exons and introns, both of which are transcribed into RNA, but introns are later removed by RNA processing. The numbers and sizes of introns vary from gene to gene. Introns are common in eukaryotic genes but uncommon in bacterial genes.

The Concept of the Gene Revisited

How does the presence of introns affect our concept of a gene? To define a gene as a sequence of nucleotides that encodes amino acids in a protein no longer seems appropriate because this definition excludes introns, which do not specify amino acids. This definition also excludes the nucleotides that encode the 5′ and 3′ ends of an mRNA molecule, which, as we will see, are required for translation but do not encode amino acids. And defining a gene in these terms excludes sequences that specify rRNA, tRNA, and other RNAs that do not encode proteins. Given our current understanding of DNA structure and function, we need a more precise definition of a gene.

Many geneticists have broadened the concept of a gene to include all sequences in DNA that are transcribed into a single RNA molecule. Defined this way, a gene includes all exons, introns, and those sequences at the beginning and end of the RNA that are not translated into a protein. This definition also includes DNA sequences that encode rRNAs, tRNAs, and other types of nonmessenger RNA. Some geneticists have expanded the definition of a gene even further to include the entire transcription unit—the promoter, the RNA-coding sequence, and the terminator. However, new evidence now calls into question even this definition. Recent research suggests that much of the genome is transcribed into RNA, although it is unclear what, if anything, much of this RNA does. What is certain is that the process of transcription is more complex than formerly thought, and defining a gene as a sequence that is transcribed into an RNA molecule is not as straightforward as formerly thought. It seems that the more we learn about the nature of genetic information, the more elusive the definition of a gene becomes.

THINK-PAIR-SHARE Question 8

CONCEPTS

The discovery of introns forced a reevaluation of the definition of the gene. Today, a gene is often defined as a DNA sequence that encodes an RNA molecule or the entire DNA sequence required to transcribe and encode an RNA molecule.

10.5 Many RNA Molecules Are Modified after Transcription in Eukaryotes

Many eukaryotic RNAs undergo extensive processing after transcription. In this section, we examine different classes of RNA and the changes that they undergo after having been transcribed from the DNA template.

Messenger RNA Processing

Messenger RNA functions as the template for protein synthesis; it carries genetic information from DNA to a ribosome and helps to assemble amino acids in their correct order. In bacteria, mRNA is transcribed directly from DNA, but in eukaryotes, a pre-mRNA (the primary transcript) is first transcribed from DNA and then processed to yield the mature mRNA. We will reserve the term "mRNA" for RNA molecules that have been completely processed and are ready to undergo translation.

THE STRUCTURE OF mRNA In mRNA, each amino acid in a protein is specified by a set of three nucleotides called a **codon**. Both prokaryotic and eukaryotic mRNAs contain three primary regions (**Figure 10.17**). The **5′ untranslated region** (5′ UTR; sometimes called the leader), a sequence of nucleotides at the 5′ end of the mRNA, does not encode any of the amino acids of a protein. In bacterial mRNA, this region contains a consensus sequence (UAAGGAGGU) called the **Shine–Dalgarno sequence**, which serves as the ribosome-binding site during translation; it is found approximately seven nucleotides upstream of the first codon translated into an amino acid (called the start codon). The Shine–Dalgarno sequence is complementary to sequences found in one of the RNA molecules that make up the ribosome and pairs with those sequences during translation. In eukaryotic cells, ribosomes bind to a modified 5′ end of mRNA, as discussed later in this section.

The next section of mRNA is the **protein-coding region**, which comprises the codons that specify the amino acid sequence of the protein. The protein-coding region begins with a start codon and ends with a stop codon.

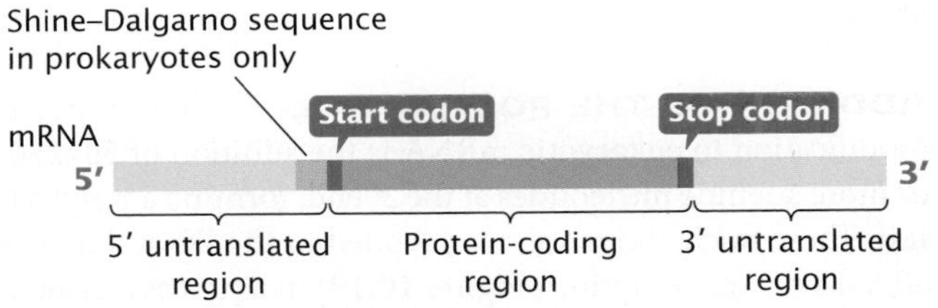

10.17 The three primary regions of mature mRNA are the 5′ untranslated region, the protein-coding region, and the 3′ untranslated region.

The last region of mRNA is the 3′ **untranslated region** (3′ UTR; sometimes called a trailer), a sequence of nucleotides that is at the 3′ end of the mRNA and is not translated into protein. The 3′ untranslated region affects the stability of mRNA and the translation of the mRNA protein-coding sequence. View **Animation 10.2** to see how mutations in different regions of a gene affect the flow of information from genotype to phenotype.

CONCEPTS

Messenger RNA molecules contain three main regions: a 5′ untranslated region, a protein-coding region, and a 3′ untranslated region. The 5′ and 3′ untranslated regions do not encode any amino acids of a protein, but contain information that is important in translation and RNA stability.

In bacterial cells, transcription and translation take place simultaneously: while the 3′ end of an mRNA is undergoing transcription, ribosomes attach to the Shine–Dalgarno sequence near the 5′ end and begin translation. Because transcription and translation are coupled, there is little opportunity for the bacterial mRNA to be modified before protein synthesis. In contrast, transcription and translation are separated in both time and space in eukaryotic cells. Transcription takes place in the nucleus, whereas translation takes place in the cytoplasm; this separation provides an opportunity for eukaryotic RNA to be modified before it is translated. Indeed, eukaryotic mRNA is extensively altered after transcription. Changes are made to the 5′ end, the 3′ end, and the protein-coding section of the RNA molecule.

ADDITION OF THE 5′ CAP One type of modification of eukaryotic pre-mRNA is the addition of a structure called a **5′ cap**. The cap consists of an extra modified nucleotide at the 5′ end of the mRNA as well as methyl groups (CH_3) on the 2′-OH group of the sugar of one or more nucleotides at the 5′ end (**Figure 10.18**). The addition of the cap takes place rapidly after the initiation of transcription. The cap functions in the initiation of translation, as we'll see in Chapter 11. Cap-binding proteins recognize the cap and attach to it; a ribosome then binds to these proteins and moves downstream along the mRNA until the start codon is reached and translation begins. The presence of a 5′ cap also increases the stability of mRNA and influences the removal of introns.

ADDITION OF THE POLY(A) TAIL A second type of modification to eukaryotic mRNA is the addition of 50–250 or more adenine nucleotides at the 3′ end, forming a **poly(A) tail**. These nucleotides are not encoded in the DNA, but are added after transcription (**Figure 10.19**) in a process termed polyadenylation. Many eukaryotic genes are transcribed well beyond the end of the coding sequence; most of the extra material at the 3′ end is then cleaved, and the poly(A) tail is added. For some pre-mRNA molecules, more than 1000 nucleotides may be removed from the 3′ end before polyadenylation.

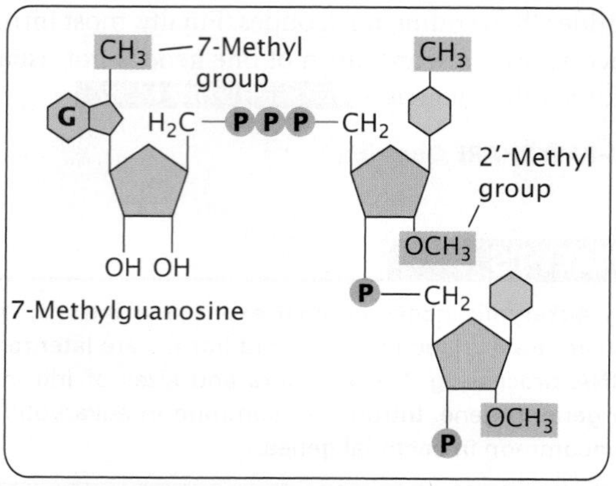

10.18 Most eukaryotic mRNAs have a 5′ cap. The cap consists of a nucleotide with 7-methylguanine that is attached to the 5′ end of pre-mRNA by a unique 5′—5′ bond, as well as methyl groups added to the 2′ position of the sugars in the second and third nucleotides of the pre-mRNA, and sometimes a methyl group added to the base (N) on the initial nucleotide.

Processing of the 3′ end of pre-mRNA requires sequences, termed the polyadenylation signal, located upstream and downstream of the site where cleavage occurs. The consensus sequence AAUAAA is usually from 11 to 30 nucleotides upstream of the cleavage site (see Figure 10.19) and determines the point at which cleavage will take place. A sequence rich in uracil nucleotides (or guanine and uracil nucleotides) is typically downstream of the cleavage site. A large number of proteins take part in finding the cleavage site and removing the 3′ end. After cleavage has been completed, adenine nucleotides are added without a template to the new 3′ end, creating the poly(A) tail.

The poly(A) tail confers stability on many mRNAs, increasing the time during which the mRNA remains intact and available for translation before it is degraded by cellular enzymes. The stability conferred by the poly(A) tail depends on proteins that attach to the tail and on its length. The poly(A) tail also facilitates the attachment of the ribosome to the mRNA and plays a role in export of the mRNA into the cytoplasm.

CONCEPTS

Eukaryotic pre-mRNAs are processed at their 5′ and 3′ ends. A 5′ cap, consisting of a modified nucleotide and several methyl groups, is added to the 5′ end. The cap facilitates the binding of a ribosome, increases the stability of the mRNA, and may affect the removal of introns. Processing at the 3′ end includes cleavage downstream of an AAUAAA consensus sequence and the addition of a poly(A) tail.

RNA SPLICING The other major type of modification of eukaryotic pre-mRNA is the removal of introns by **RNA splicing**.

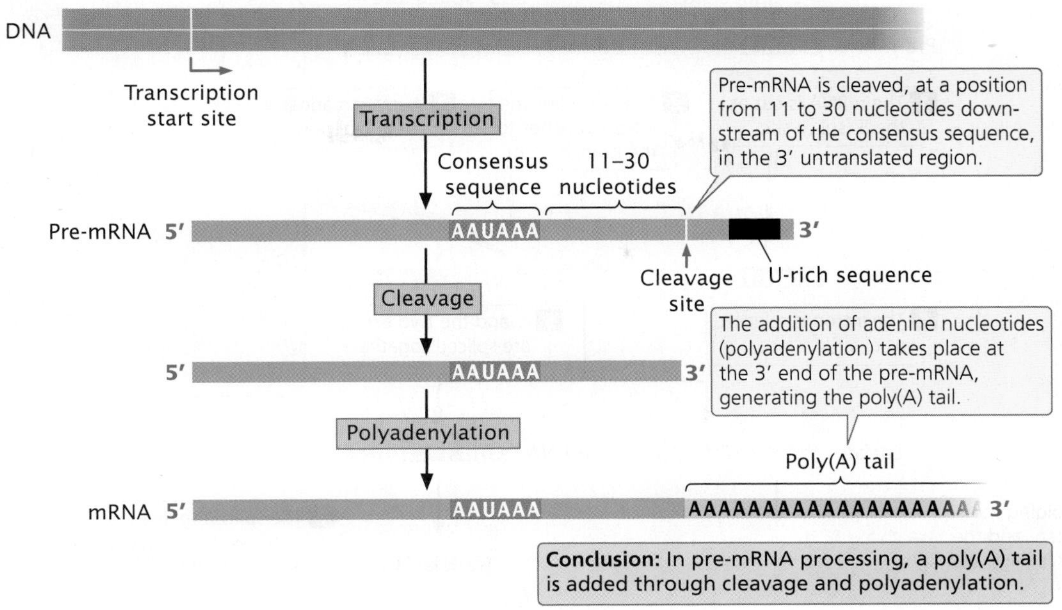

10.19 Most eukaryotic mRNAs have a 3′ poly(A) tail.

This modification takes place in the nucleus, before the RNA moves to the cytoplasm. Splicing requires the presence of three sequences in the intron. One end of the intron is referred to as the **5′ splice site**, and the other end is the **3′ splice site** (**Figure 10.20**); these splice sites possess short consensus sequences. Most introns in pre-mRNAs begin with GU and end with AG, indicating that these sequences play a crucial role in splicing. Indeed, changing a single nucleotide at either of these sites prevents splicing.

The third sequence important for splicing is at the **branch point**, which is an adenine nucleotide that lies from 18 to 40 nucleotides upstream of the 3′ splice site (see Figure 10.20). The sequence surrounding the branch point is a weak consensus sequence. Deletion or mutation of the adenine nucleotide at the branch point prevents splicing.

Splicing takes place within a large structure called the **spliceosome**, which is one of the largest and most complex of all molecular structures. The spliceosome consists of five RNA molecules and almost three hundred proteins. The RNA components are small nuclear RNAs; these snRNAs associate with proteins to form small nuclear ribonucleoprotein particles (snRNPs). Each snRNP contains a single snRNA molecule and multiple proteins. The spliceosome is composed of five snRNPs (U1, U2, U4, U5, and U6) and some proteins not associated with an snRNA.

CONCEPTS

Introns in pre-mRNAs contain three consensus sequences critical to splicing: a 5′ splice site, a 3′ splice site, and a branch point. The splicing of pre-mRNA takes place within a large complex called the spliceosome, which consists of snRNAs and proteins.

✓ CONCEPT CHECK 5

If a splice site were mutated so that splicing did not take place, what would the effect be on the mRNA?

a. It would be shorter than normal.
b. It would be longer than normal.
c. It would be the same length but would encode a different protein.

Before splicing takes place, an intron lies between an upstream exon (exon 1) and a downstream exon (exon 2), as shown in **Figure 10.21**. Pre-mRNA is spliced in two distinct steps. In the first step of splicing, the pre-mRNA is cut at the 5′ splice site. This cut frees exon 1 from the intron, and the 5′ end of the intron attaches to the branch point; that is, the intron folds back on itself, forming a structure called a **lariat**. In this reaction, the guanine nucleotide in the consensus sequence at the 5′ splice site bonds with the adenine nucleotide at the branch point through a transesterification reaction. As a result,

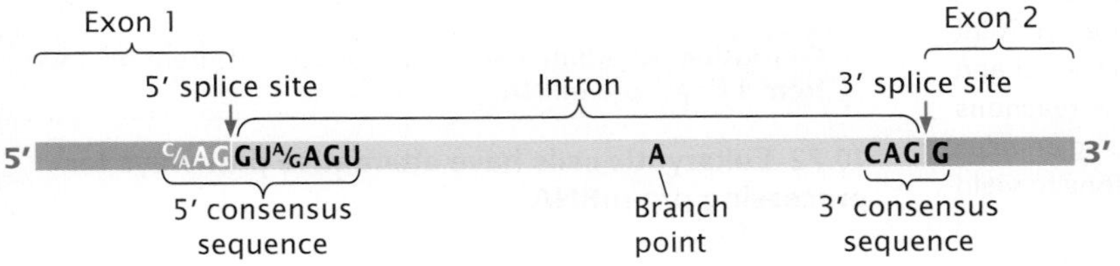

10.20 The splicing of pre-mRNA requires consensus sequences. Critical consensus sequences are present at the 5′ splice site and the 3′ splice site. A weak consensus sequence (not shown) exists at the branch point.

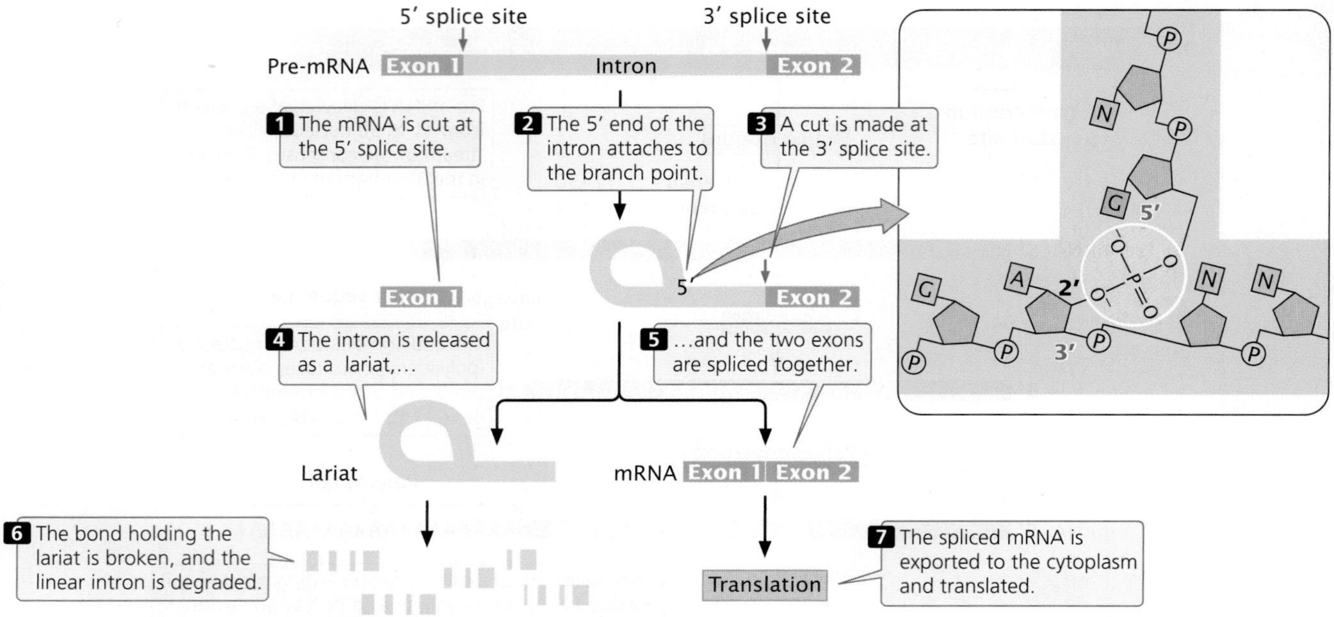

10.21 The splicing of pre-mRNA introns requires a two-step process.

the 5′ phosphate group of the guanine nucleotide becomes attached to the 2′-OH group of the adenine nucleotide at the branch point (see Figure 10.21).

In the second step of RNA splicing, a cut is made at the 3′ splice site, and simultaneously, the 3′ end of exon 1 becomes covalently attached (spliced) to the 5′ end of exon 2. The intron is released as a lariat. Eventually, a lariat debranching enzyme breaks the bond at the branch point, producing a linear intron that is rapidly degraded by nuclear enzymes. The mature mRNA, consisting of the exons spliced together, is exported to the cytoplasm, where it is translated. These splicing reactions take place within the spliceosome, which carries out the splicing reactions.

THINK-PAIR-SHARE Question 9

Many eukaryotic mRNAs undergo **alternative processing**, in which a single pre-mRNA is processed in different ways to produce alternative types of mRNA, resulting in the production of different proteins from the same DNA sequence. One type of alternative processing is **alternative splicing**, in which the same pre-mRNA can be spliced in more than one way to yield multiple mRNAs that are translated into different amino acid sequences and thus different proteins (**Figure 10.22**). Alternative processing is an important source of protein diversity in vertebrates; an estimated 60% of all human genes are alternatively spliced.

CONCEPTS

Intron splicing in pre-mRNAs is a two-step process: (1) the 5′ end of an intron is cleaved and attached to the branch point to form a lariat, and (2) the 3′ end of the intron is cleaved and the two ends of the exon are spliced together. These reactions take place within the spliceosome. Alternative splicing enables exons to be spliced together in different combinations to yield mRNAs that encode different proteins.

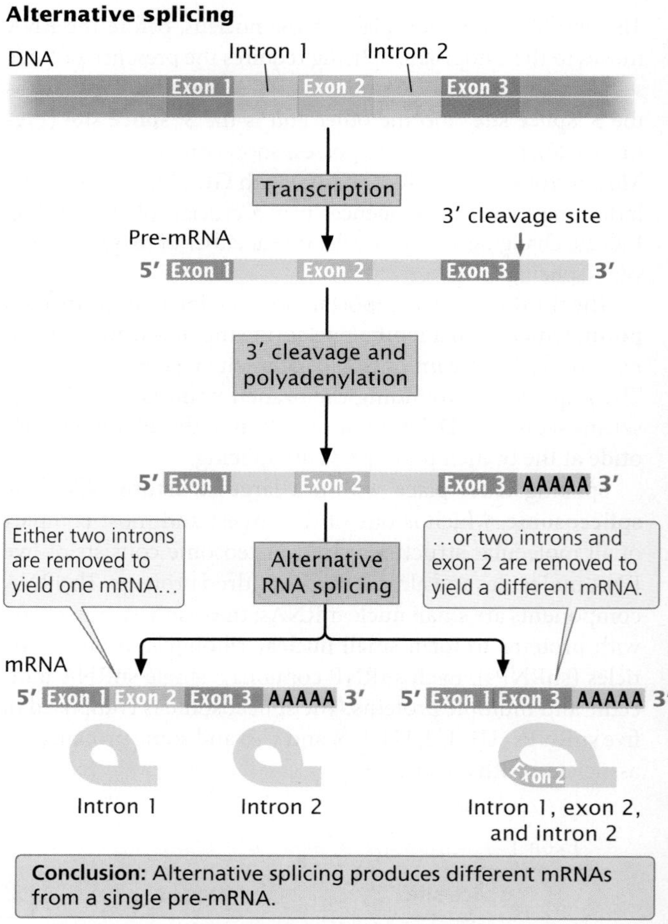

10.22 Eukaryotic cells have alternative pathways for processing pre-mRNA.

CONNECTING CONCEPTS

Eukaryotic Gene Structure and Pre-mRNA Processing

This chapter has introduced a number of different components of genes and RNA molecules, including promoters, 5′ untranslated regions, coding sequences, exons, introns, 3′ untranslated regions, poly(A) tails, and caps. Let's see how some of these components are combined to create a typical eukaryotic gene and how a mature mRNA is produced from them.

The promoter, which typically lies upstream of the transcription start site, is necessary for transcription to take place, but is itself not usually transcribed when protein-encoding genes are transcribed by RNA polymerase II (**Figure 10.23a**). Farther upstream or downstream of the start site there may be sequences called enhancers, DNA sequences that also regulate transcription. In transcription, all the nucleotides between the transcription start site and the termination site are transcribed into pre-mRNA, including exons, introns, and a long 3′ end that is later cleaved from the transcript (**Figure 10.23b**). Notice that the 5′ end of the first exon contains the sequence that encodes the 5′ untranslated region, and that the 3′ end of the last exon contains the sequence that encodes the 3′ untranslated region.

The pre-mRNA is then processed to yield a mature mRNA. The first step in this processing is the addition of a cap to the 5′ end of the pre-mRNA (**Figure 10.23c**). Next, the 3′ end is cleaved at a site downstream of the AAUAAA consensus sequence in the last exon (**Figure 10.23d**). Immediately after cleavage, a poly(A) tail is added to the 3′ end (**Figure 10.23e**). Finally, the introns are removed to yield the mature mRNA (**Figure 10.23f**). The mRNA now contains 5′ and 3′ untranslated regions, which are not translated into amino acids, and the nucleotides that carry the protein-coding sequences. You can explore the consequences of failed RNA processing by viewing and interacting with **Animation 10.3.** A

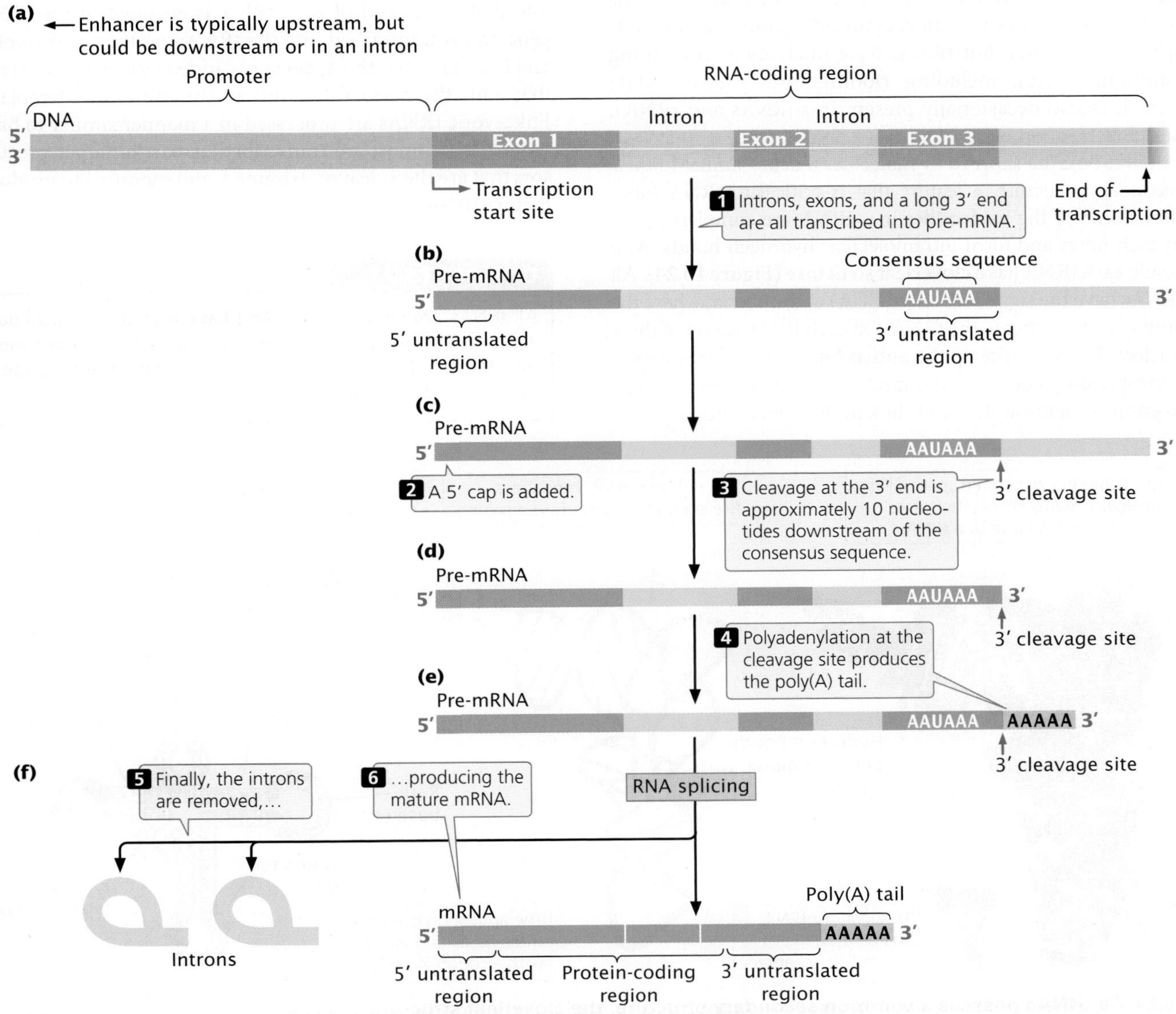

10.23 Mature eukaryotic mRNA is produced when pre-mRNA is transcribed and undergoes several types of processing.

The Structure and Processing of Transfer RNA

Transfer RNA serves as a link between the genetic code in mRNA and the amino acids that make up a protein. Each tRNA attaches to a particular amino acid and carries it to the ribosome, where the tRNA adds its amino acid to the growing polypeptide chain at the position specified by the genetic instructions in the mRNA.

THE STRUCTURE OF TRANSFER RNA Each tRNA is capable of attaching to only one type of amino acid. The complex of tRNA plus its amino acid can be written in abbreviated form by adding a three-letter superscript representing the amino acid to the term "tRNA." For example, a tRNA that attaches to the amino acid alanine is written as $tRNA^{Ala}$.

A unique feature of tRNA is the presence of rare **modified bases**. All classes of RNAs have nucleotides containing the four standard bases (adenine, cytosine, guanine, and uracil) specified by DNA, but tRNAs have nucleotides containing additional bases, including ribothymine, pseudouridine (which is also occasionally present in snRNAs and rRNA), and dozens of others.

All tRNAs are short RNA molecules that are similar in their secondary structure, a feature that is critical to tRNA function. Some of the nucleotides in a tRNA are complementary to each other and form intramolecular hydrogen bonds. As a result, each tRNA has a **cloverleaf structure** (**Figure 10.24**). All tRNAs have the same sequence (CCA) at the 3′ end, where the amino acid attaches to the tRNA. On each tRNA is a set of three nucleotides that make up the **anticodon**, which pairs with the corresponding codon on the mRNA during protein synthesis to ensure that the amino acids link in the correct order.

Although each tRNA molecule folds into a cloverleaf owing to the complementary pairing of bases, the cloverleaf is not the three-dimensional (tertiary) structure of tRNAs found in the cell. The results of X-ray crystallographic studies have shown that the cloverleaf folds on itself to form an L-shaped structure, as illustrated by the space-filling and ribbon models in Figure 10.24.

TRANSFER RNA PROCESSING Both bacterial and eukaryotic tRNAs are extensively modified after transcription. In *E. coli*, several tRNAs are usually transcribed together as one large precursor tRNA, which is then cut up into pieces, each containing a single tRNA. Additional nucleotides may then be removed one at a time from the 5′ and 3′ ends of the tRNA in a process known as trimming. Base-modifying enzymes may then change some of the standard bases into modified bases. In some prokaryotes, the CCA sequence found at the 3′ end of each tRNA is encoded in the tRNA gene and is transcribed into the tRNA; in other prokaryotes and in eukaryotes, this sequence is added by a special enzyme that adds the nucleotides without the use of any template. Eukaryotic tRNAs are processed in a manner similar to bacterial tRNAs: most are transcribed as part of larger precursors that are then cleaved, trimmed, and modified to produce mature tRNAs.

CONCEPTS

All tRNAs are similar in size and have a common secondary structure known as the cloverleaf. Transfer RNAs contain modified bases and are extensively processed after transcription in both bacterial and eukaryotic cells.

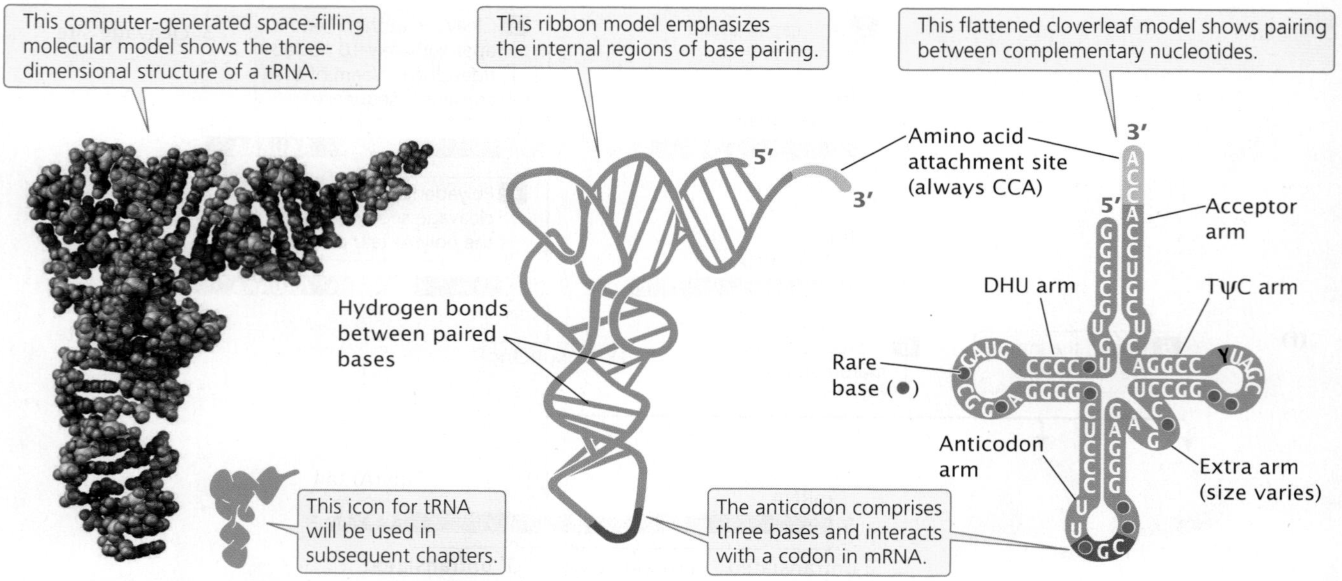

10.24 All tRNAs possess a common secondary structure, the cloverleaf structure. The base sequence in the flattened model is for $tRNA^{Ala}$. The DHU and TψC arms are named for rare bases in nucleotides of the arms.

The Structure and Processing of Ribosomal RNA

Within ribosomes, the genetic instructions contained in mRNA are translated into the amino acid sequences of polypeptides. Ribosomes are complex structures, each consisting of more than 50 different proteins and RNA molecules (**Table 10.4**). A functional ribosome consists of two subunits, a **large ribosomal subunit** and a **small ribosomal subunit**, each of which consists of one or more RNA molecules and a number of proteins. The sizes of the ribosomes and their RNA components are given in Svedberg (S) units (a measure of how rapidly an object sediments in a centrifugal field).

Ribosomal RNA is processed after transcription in both bacterial and eukaryotic cells. A precursor RNA molecule is methylated in several places, then cleaved and trimmed to produce the mature rRNAs that make up the ribosome. In eukaryotes, small nucleolar RNAs (snoRNAs) help to cleave and modify rRNAs and assemble them into a mature ribosome.

CONCEPTS

A ribosome is a complex structure consisting of several rRNA molecules and many proteins. Each functional ribosome consists of a large and a small subunit. Ribosomal RNAs in both bacterial and eukaryotic cells are modified after transcription.

Small RNA Molecules and RNA Interference

Numerous small RNA molecules (most of them 20–30 nucleotides long) have been discovered that greatly influence many basic biological processes, including the formation of chromatin structure, transcription, and translation. These small RNA molecules play important roles in gene expression, development, cancer, and defense against foreign DNA. They are also being harnessed by researchers to study gene function and treat genetic diseases (see Chapter 14).

In 1998, Andrew Fire, Craig Mello, and their colleagues found that potent gene silencing was triggered when double-stranded RNA was injected into the animals. This finding was puzzling because no mechanism by which double-stranded RNA could inhibit translation was known. Several other, previously described types of gene silencing were also found to be triggered by double-stranded RNA.

Subsequent research revealed an astonishing array of small RNA molecules with important cellular functions in eukaryotes, which now include at least three major classes: small interfering RNAs (siRNAs), microRNAs (miRNAs), and Piwi-interacting RNAs (piRNAs). These small RNAs are found in many eukaryotes and are responsible for a variety of different functions (see Table 10.2). An analogous group of small RNAs with silencing functions—CRISPR RNAs (crRNAs)—have been detected in prokaryotes. For their discovery of RNA interference, Fire and Mello were awarded the Nobel Prize in Physiology or Medicine in 2006.

RNA interference (RNAi) is a powerful and precise mechanism used by eukaryotic cells to limit the invasion of foreign genes (from viruses and transposons) and to censor the expression of their own genes. RNA interference is triggered by double-stranded RNA molecules, which may arise in several ways (**Figure 10.25**): by the transcription of inverted repeats into an RNA molecule that then base pairs with itself to form double-stranded RNA; by the simultaneous transcription of two different RNA molecules that are complementary to one another and that pair, forming double-stranded RNA; or by infection by viruses that make double-stranded RNA. These double-stranded RNA molecules are chopped up by an enzyme appropriately called Dicer, resulting in tiny RNA molecules that are unwound to produce siRNAs and miRNAs (see Figure 10.25).

SMALL INTERFERING AND microRNAs Two abundant classes of RNA molecules that function in RNA interference in eukaryotes are small interfering RNAs and microRNAs. Although these two types of RNA differ in how they originate (**Table 10.5**; see also Figure 10.25), they have a number of features in common, and their functions overlap considerably.

Both siRNA and miRNA molecules combine with proteins to form an **RNA-induced silencing complex (RISC)** (see Figure 10.25). The RISC pairs with an mRNA molecule that possesses a sequence complementary to the RISC's siRNA or miRNA component, and then either cleaves the

TABLE 10.4 Composition of ribosomes in bacterial and eukaryotic cells

Cell type	Ribosome size	Subunit	rRNA component	Proteins
Bacterial	70S	Large (50S)	23S (2900 nucleotides), 5S (120 nucleotides)	31
		Small (30S)	16S (1500 nucleotides)	21
Eukaryotic	80S	Large (60S)	28S (4700 nucleotides), 5.8S (160 nucleotides), 5S (120 nucleotides)	49
		Small (40S)	18S (1900 nucleotides)	33

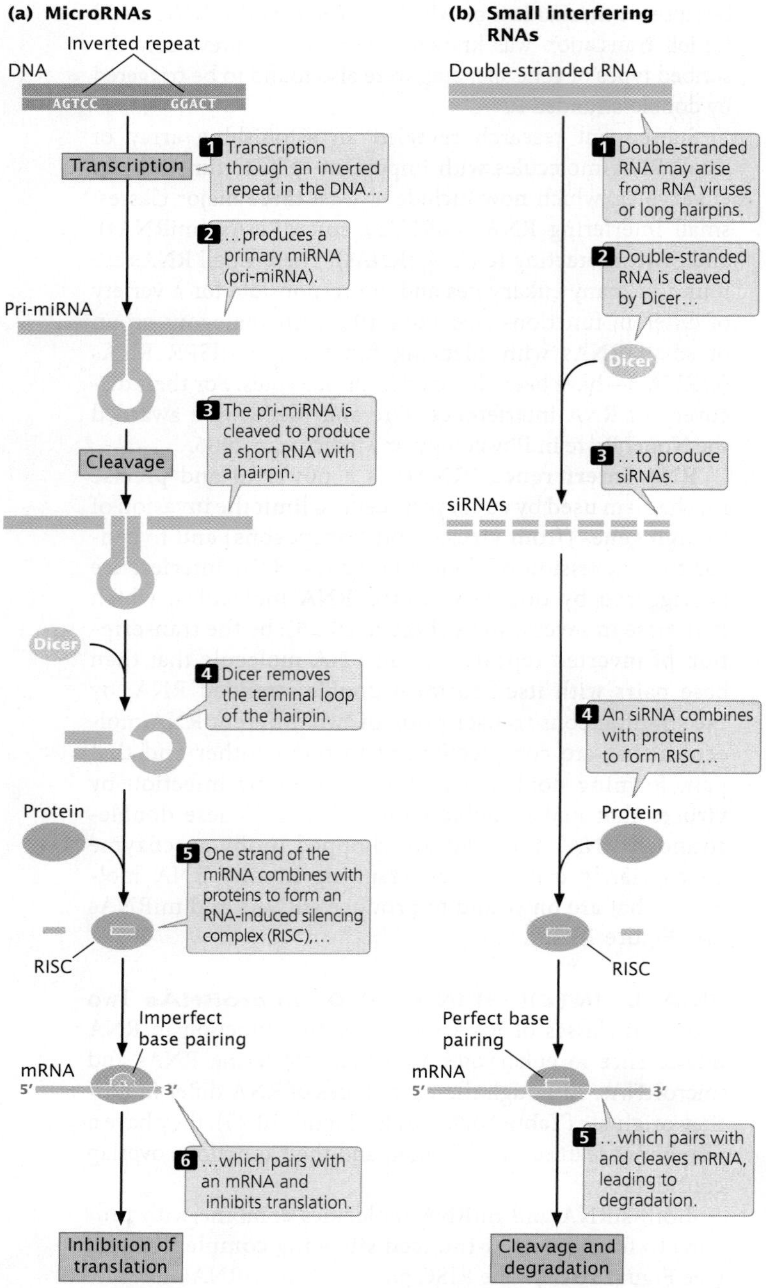

(a) MicroRNAs

Inverted repeat

DNA

AGTCC GGACT

Transcription

1 Transcription through an inverted repeat in the DNA...

2 ...produces a primary miRNA (pri-miRNA).

Pri-miRNA

Cleavage

3 The pri-miRNA is cleaved to produce a short RNA with a hairpin.

Dicer

4 Dicer removes the terminal loop of the hairpin.

Protein

5 One strand of the miRNA combines with proteins to form an RNA-induced silencing complex (RISC),...

RISC

Imperfect base pairing

mRNA
5' 3'

6 ...which pairs with an mRNA and inhibits translation.

1 Double-stranded RNA may arise from RNA viruses or long hairpins.

2 Double-stranded RNA is cleaved by Dicer...

Dicer

3 ...to produce siRNAs.

siRNAs

4 An siRNA combines with proteins to form RISC...

Protein

RISC

Perfect base pairing

mRNA
5' 3'

5 ...which pairs with and cleaves mRNA, leading to degradation.

10.25 Small interfering RNAs and microRNAs are produced from double-stranded RNAs. Although small interfering RNAs and microRNAs differ in how they originate, they share a number of features and their functions overlap.

mRNA, leading to degradation of the mRNA, or represses translation of the mRNA. Some siRNAs also serve as guides for the methylation of complementary sequences in DNA, whereas others alter chromatin structure, both of which affect transcription.

MicroRNAs have been found in all eukaryotic organisms examined to date as well as in viruses: To see how small interfering RNAs and microRNAs affect gene expression, see **Animation 10.4**.

THINK-PAIR-SHARE Question 10

The genes that encode miRNA are transcribed into longer precursors, called primary microRNA (pri-miRNA), that are then cleaved and processed prior to their incorporation into a RISC (Figure 10.25a). Small interfering RNAs are processed in a similar way (see Figure 10.25b).

CRISPR RNA

CRISPR RNAs (crRNAs) are encoded by DNA sequences found in bacterial and archaeal genomes termed *c*lustered *r*egularly *i*nterspaced *s*hort *p*alindromic *r*epeats. Palindromic sequences are sequences that read the same forward and backward on two complementary DNA strands. A CRISPR array consists of a series of such palindromic sequences, separated by spacers that are derived from the DNA of foreign bacteriophage or plasmid genomes.

CRISPR RNAs combine with Cas (CRISPR-associated) proteins to provide defense against the invasion of specific foreign DNA molecules, such as those from bacteriophages and plasmids (see Chapter 7). Because they target specific DNA molecules, CRISPR-Cas systems have been compared to the immune systems of vertebrates. CRISPR-Cas systems are widespread in prokaryotes, occurring in 50% of bacterial species and 90% of archaea.

Foreign DNA from a bacteriophage or plasmid that enters the cell is identified, processed, and inserted into the CRISPR array as a new spacer between the palindromic repeats (**Figure 10.26a**). The spacer DNA then serves as a memory of this foreign DNA. The CRISPR array is transcribed into a long CRISPR precursor RNA (**Figure 10.26b**), which is then cleaved by Cas proteins and processed into crRNAs, each of which contains one spacer sequence that is homologous to the foreign DNA. The crRNA combines with a Cas protein to form an effector complex. When foreign DNA

TABLE 10.5 Differences between siRNAs and miRNAs

Feature	siRNA	miRNA
Origin	mRNA, transposon, or virus	RNA transcribed from distinct gene
Cleavage of	RNA duplex or single-stranded RNA that forms long hairpins	Single-stranded RNA that forms short hairpins of double-stranded RNA
Action	Degradation of mRNA, inhibition of transcription, chromatin modification	Degradation of mRNA, inhibition of translation, chromatin modification
Target	Genes from which they were transcribed	Genes other than those from which they were transcribed

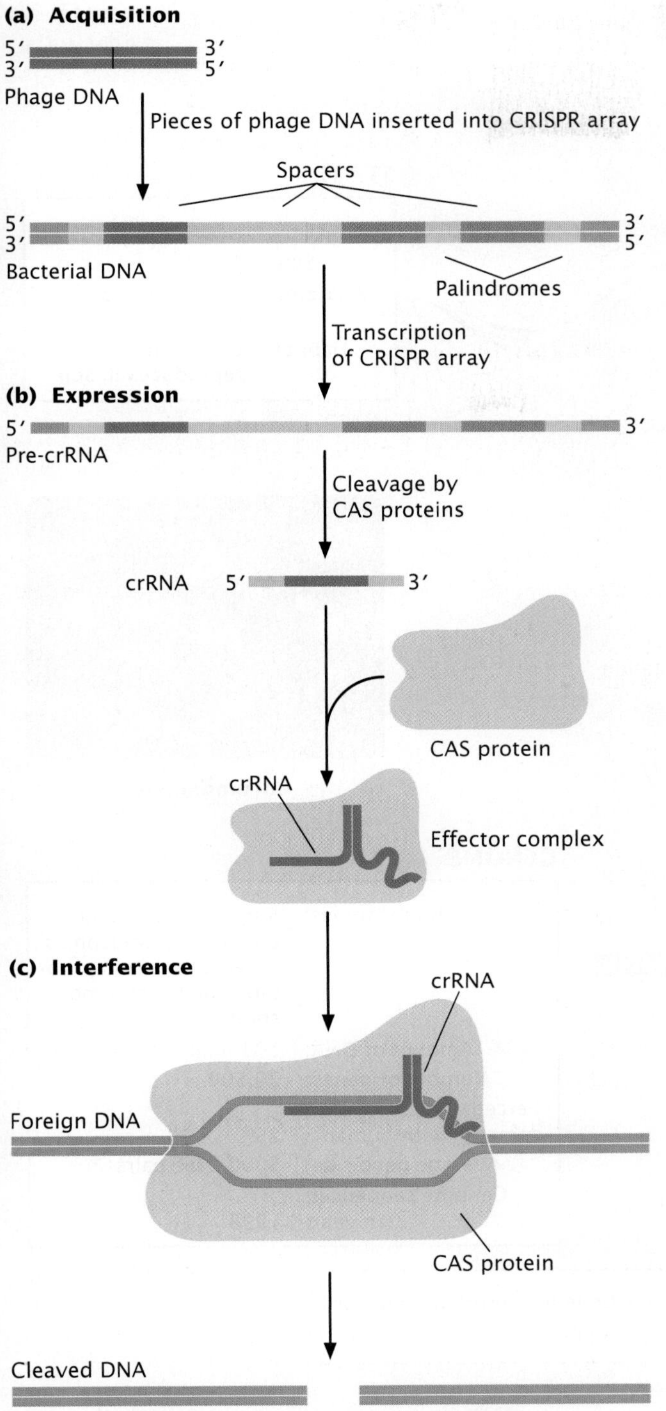

10.26 CRISPR RNAs defend prokaryotic cells against invasion by foreign DNA, such as DNA from bacteriophages and plasmids.

from the same bacteriophage or plasmid enters the cell again, the effector complex binds to the foreign DNA through base paring between the crRNA and the foreign DNA sequence. The Cas protein then cleaves the foreign DNA, rendering it nonfunctional **(Figure 10.26c).** In this way, crRNAs serve as an adaptive RNA defense system against foreign invaders.

CONCEPTS

Small interfering RNAs and microRNAs are tiny RNAs produced when larger double-stranded RNA molecules are cleaved and processed. These RNAs participate in a variety of processes, including mRNA degradation, the inhibition of translation, the methylation of DNA, and chromatin remodeling. CRISPR RNAs are small RNAs found in prokaryotes, where they function in defense against foreign DNA.

Long Noncoding RNAs Regulate Gene Expression

Transcription of eukaryotic genomes produces many long RNA molecules that do not encode proteins. Called long noncoding RNAs (lncRNAs), these RNAs are typically over 200 nucleotides in length and can be as much as 100,000 nucleotides long. Thousands of lncRNAs have been discovered in the last five years. The DNA that encodes them, along with other DNA of unknown function, has been called "the dark matter of the genome."

Although the function of many lncRNAs is still unclear, there is increasing evidence that at least some play a role in controlling gene expression by affecting transcription. One of the best-studied is *Xist* RNA, which plays a central role in dosage compensation in mammalian cells (described in Chapter 4).

Model Genetic Organism The Nematode Worm *Caenorhabditis elegans*

As we have seen, RNA interference was first demonstrated in the nematode *Caenorhabditis elegans* when geneticists discovered that they could silence specific genes in this species by injecting the animals with double-stranded RNA that was complementary to the genes. The geneticists were studying gene expression in *C. elegans* because this species had proved to be an excellent model genetic organism, particularly for studies of how genes influence development. For reasons that are not completely understood, RNAi is particularly effective in this species.

You may be asking what a nematode is and why it is a model genetic organism. Although rarely seen, nematodes are among the most abundant organisms on Earth, inhabiting soils throughout the world. Most of these worms are free living and cause no harm, but a few are important parasites of plants and animals, including humans. Although *C. elegans* has no economic importance, it has become widely used in genetic studies because of its simple body plan, ease of culture, and high reproductive capacity (**Figure 10.27**). First introduced to the study of genetics by Sydney Brenner, who formulated plans in 1962 to use *C. elegans* for the genetic dissection of behavior, this species has made important contributions to the study of development, cell death, aging, and behavior.

The Nematode Worm
Caenorhabditis elegans

ADVANTAGES

- Small size
- Short generation time of 3 days
- Each female can produce 200–1000 eggs
- Easy to culture in laboratory
- Simple body plan
- Transparent
- Capable of self-fertilization or crossing

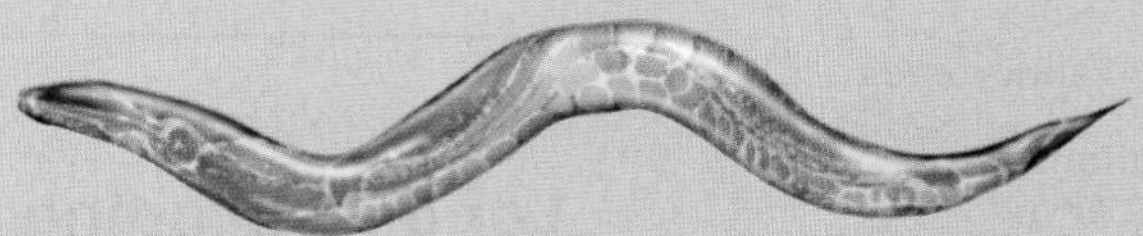

STATS

Taxonomy:	Nematode
Size:	1 mm
Anatomy:	Unsegmented, elongated body
Habitat:	Lives and reproduces in soil

Life Cycle

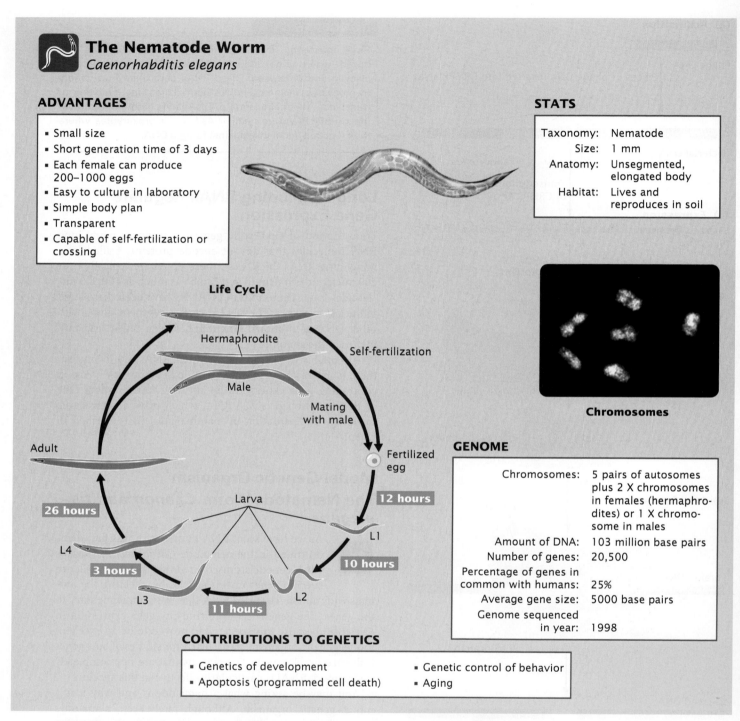

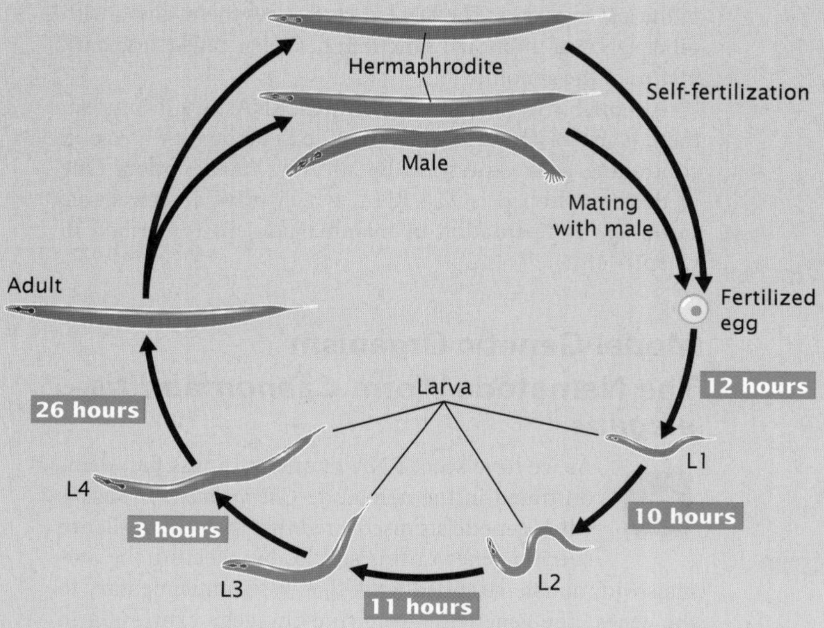

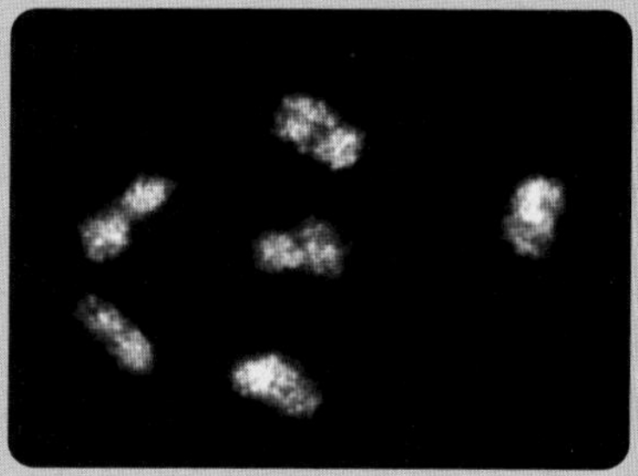

Chromosomes

GENOME

Chromosomes:	5 pairs of autosomes plus 2 X chromosomes in females (hermaphrodites) or 1 X chromosome in males
Amount of DNA:	103 million base pairs
Number of genes:	20,500
Percentage of genes in common with humans:	25%
Average gene size:	5000 base pairs
Genome sequenced in year:	1998

CONTRIBUTIONS TO GENETICS

- Genetics of development
- Apoptosis (programmed cell death)
- Genetic control of behavior
- Aging

10.27 The nematode worm *Caenorhabditis elegans* is a model genetic organism. [Micrograph courtesy of William Goodyer and Monique Zetka.]

Advantages of *Caenorhabditis elegans* as a model genetic organism

An ideal genetic organism, *C. elegans* is small, easy to culture, and produces large numbers of offspring. The adult *C. elegans* is about 1 mm in length. Most investigators grow *C. elegans* on agar-filled petri plates that are covered with a lawn of bacteria, which the nematodes devour. Thousands of worms can be easily cultured in a single laboratory. Compared with most multicellular animals, they have a very short generation time, about 3 days at room temperature. And they are prolific reproducers: a single female produces 250–1000 fertilized eggs in 3 to 4 days.

Another advantage of *C. elegans*, particularly for developmental studies, is that the worm is transparent, allowing easy observation of internal development at all stages. It has a simple body structure, with a small, invariant number of somatic cells: 959 cells in a mature hermaphroditic female and 1031 cells in a mature male.

Life cycle

Most mature adults are hermaphrodites, with the ability to produce both eggs and sperm and undergo self-fertilization. A few are males, which produce only sperm and mate with hermaphrodites. The hermaphrodites have two sex chromosomes (XX); the males possess a single sex chromosome (XO). Thus, hermaphrodites that self-fertilize produce only females (with the exception of a few males that result from nondisjunction of the X chromosomes). When hermaphrodites mate with males, half the progeny are XX hermaphrodites and half are XO males.

Eggs are fertilized internally, either by sperm produced by the hermaphrodite or by sperm contributed by a male (see Figure 10.27). The eggs are then laid, and development is completed externally. Approximately 14 hours after fertilization, a larva hatches from the egg and goes through four larval stages—termed L1, L2, L3, and L4—that are separated by molts. The L4 larva undergoes a final molt to produce the adult worm. Under normal laboratory conditions, worms will live for 2 to 3 weeks.

Genetic techniques

Geneticists began developing plans in 1989 to sequence the genome of *C. elegans*, and the complete genome sequence was obtained in 1998. Compared with the genomes of most multicellular animals, that of *C. elegans*, at 103 million base pairs of DNA, is small, which facilitates genomic analysis. The availability of the complete genome sequence provides a great deal of information about gene structure, function, and organization in this species.

Chemical mutagens are routinely used to generate mutations in *C. elegans*—mutations that are easy to identify and isolate. The ability of hermaphrodites to self-fertilize means that progeny homozygous for recessive mutations can be obtained in a single generation; the existence of males means that genetic crosses can be carried out.

Developmental studies are facilitated by the transparent body of the worm. As mentioned earlier, *C. elegans* has a small and exact number of somatic cells. Researchers studying the development of *C. elegans* have meticulously mapped the entire cell lineage of the species, and so the developmental fate of every cell in the adult body can be traced back to the original single-celled fertilized egg. Developmental biologists often use lasers to destroy (ablate) specific cells in a developing worm and then study the effects on physiology, development, and behavior.

RNA interference has proved to be an effective tool for turning off genes in *C. elegans*. Geneticists inject double-stranded copies of RNA that is complementary to specific genes; the double-stranded RNA then silences the expression of these genes through the RNAi process. The worms can even be fed bacteria that have been genetically engineered to express the double-stranded RNA, thus avoiding the difficulties of microinjection.

CONCEPTS SUMMARY

- Early life used RNA both as the carrier of genetic information and as a biological catalyst.
- RNA is a polymer consisting of nucleotides joined together by phosphodiester bonds. Each RNA nucleotide consists of a ribose sugar, a phosphate, and a base. RNA contains the base uracil and is usually single stranded.
- Cells possess a number of different classes of RNA. Ribosomal RNA is a component of the ribosome, messenger RNA carries coding instructions for proteins, and transfer RNA helps incorporate the amino acids into a polypeptide chain.
- The template for RNA synthesis is single-stranded DNA. In transcription, RNA synthesis is complementary and antiparallel to the DNA template strand. A transcription unit consists of a promoter, an RNA-coding region, and a terminator.
- The substrates for RNA synthesis are ribonucleoside triphosphates.
- RNA polymerase in bacterial cells consists of a core enzyme, which catalyzes the addition of nucleotides to an RNA molecule, and other subunits. The sigma factor controls the binding of the core enzyme to the promoter. Eukaryotic cells contain multiple types of RNA polymerase.
- Transcription begins at the start site, which is determined by consensus sequences. RNA is synthesized from a single template strand of DNA. RNA synthesis ceases after a terminator sequence has been transcribed.
- Introns—noncoding sequences that interrupt the coding sequences (exons) of genes—are common in eukaryotic genes but rare in bacterial genes.
- An mRNA molecule has three primary parts: a 5′ untranslated region, a protein-coding sequence, and a 3′ untranslated region.
- The pre-mRNA of a eukaryotic protein-encoding gene is extensively processed: a modified nucleotide and methyl groups, collectively termed the 5′ cap, are added to the 5′ end of pre-mRNA; the 3′ end is cleaved and a poly(A) tail is added; and introns are removed.
- Transfer RNAs, which attach to amino acids, are short molecules that assume a common secondary structure and contain modified bases.
- Ribosomes, the sites of protein synthesis, are composed of several ribosomal RNA molecules and numerous proteins.
- Small interfering RNAs, microRNAs, Piwi-interacting RNAs, and CRISPR RNAs play important roles in gene silencing and in a number of other biological processes.
- Long noncoding RNAs are RNA molecules that do not encode proteins. Evidence increasingly suggests that many of these molecules function in the control of gene expression.
- *Caenorhabditis elegans* is a nematode that is widely used as a model genetic organism.

IMPORTANT TERMS

ribozyme (p. 270)
ribosomal RNA (rRNA) (p. 271)
messenger RNA (mRNA) (p. 271)
pre-messenger RNA (pre-mRNA) (p. 271)
transfer RNA (tRNA) (p. 272)
small nuclear RNA (snRNA) (p. 272)
small nuclear ribonucleoprotein (snRNP) (p. 272)
small nucleolar RNA (snoRNA) (p. 272)
microRNA (miRNA) (p. 272)
small interfering RNA (siRNA) (p. 272)
Piwi-interacting RNA (pi-RNA) (p. 272)
long noncoding RNA (lncRNA) (p. 272)
CRISPR RNA (crRNA) (p. 272)
template strand (p. 274)
nontemplate strand (p. 274)
transcription unit (p. 274)
promoter (p. 274)
RNA-coding region (p. 274)
terminator (p. 275)
ribonucleoside triphosphate (rNTP) (p. 275)
RNA polymerase (p. 275)
core enzyme (p. 276)
sigma (σ) factor (p. 276)
holoenzyme (p. 276)
RNA polymerase I (p. 276)
RNA polymerase II (p. 276)
RNA polymerase III (p. 276)
RNA polymerase IV (p. 276)
RNA polymerase V (p. 276)
−10 consensus sequence (Pribnow box) (p. 277)
−35 consensus sequence (p. 277)
rho-dependent terminator (p. 279)
rho (ρ) factor (p. 279)
rho-independent terminator (p. 279)
hairpin (p. 280)
polycistronic RNA (p. 280)
colinearity (p. 281)
exon (p. 282)
intron (p. 282)
codon (p. 283)
5′ untranslated region (p. 283)
Shine–Dalgarno sequence (p. 283)
protein-coding region (p. 283)
3′ untranslated region (p. 284)
5′ cap (p. 284)
poly(A) tail (p. 284)
RNA splicing (p. 284)
5′ splice site (p. 285)
3′ splice site (p. 285)
branch point (p. 285)
spliceosome (p. 285)
lariat (p. 285)
alternative processing (p. 286)
alternative splicing (p. 286)
modified base (p. 288)
cloverleaf structure (p. 288)
anticodon (p. 288)
large ribosomal subunit (p. 289)
small ribosomal subunit (p. 289)
RNA interference (RNAi) (p. 289)
RNA-induced silencing complex (RISC) (p. 289)

ANSWERS TO CONCEPT CHECKS

1. b

2. The template strand is the DNA strand that is transcribed into an RNA molecule, whereas the nontemplate strand is not transcribed.

3. The sigma factor controls the binding of RNA polymerase to the promoter.

4. When DNA was hybridized to the mRNA transcribed from it, regions of DNA that did not correspond to RNA looped out.

5. b

WORKED PROBLEMS

Problem 1

DNA from a eukaryotic gene was isolated, the two strands separated, and then hybridized to the mRNA transcribed from the gene. When the hybridized structure was observed with an electron microscope, it looked like this:

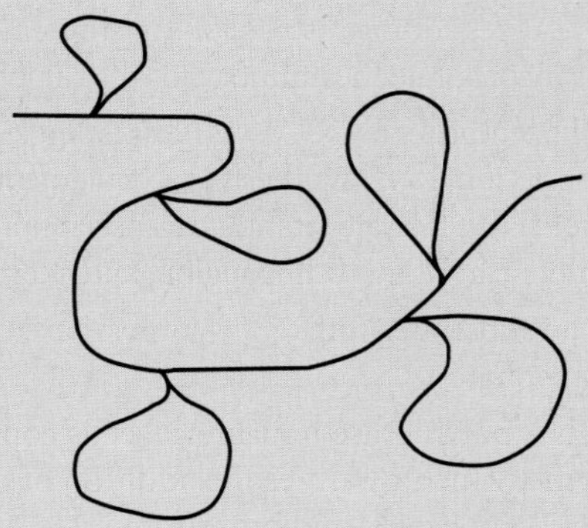

a. How many introns and exons are there in this gene? Explain your answer.

b. Identify the exons and introns in this hybridized structure.

Solution Strategy

What information is required in your answer to the problem?

a. The number of introns and exons and how you arrived at your answer.

b. The locations of the introns and exons labeled on the figure.

What information is provided to solve the problem?

- The DNA and mRNA are from a eukaryote.
- The DNA was denatured and hybridized to the mRNA.
- A picture of the hybridized structure.

For help with this problem, review:

Introns in Section 10.4 and Figure 10.15.

Solution Steps

a. Each of the loops represents a region in which sequences in the DNA do not have corresponding sequences in the RNA; these regions are introns. There are five loops in the hybridized structure; so there must be five introns in the DNA and six exons.

Recall: Introns are noncoding sequences found within eukaryotic genes.

Hint: The number of introns will be one less than the number of exons.

b.

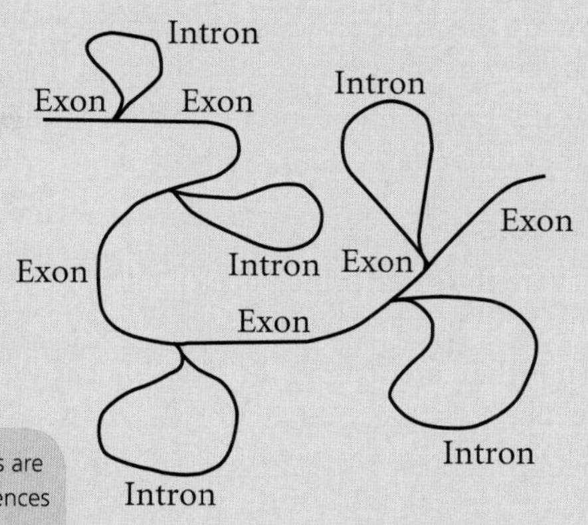

Problem 2

Draw a typical bacterial mRNA and the gene from which it was transcribed. Identify the 5′ and 3′ ends of the RNA and DNA molecules, as well as the following regions or sequences:

a. Promoter
b. 5′ untranslated region
c. 3′ untranslated region
d. Protein-coding sequence
e. Transcription start and termination sites
f. Terminator
g. Shine–Dalgarno sequence
h. Start and stop codons

Solution Strategy

What information is required in your answer to the problem?

A drawing of the mRNA and the gene from which it is transcribed. The 5′ and 3′ ends of the mRNA and DNA molecules. Locations of the listed structures on the drawing.

What information is provided to solve the problem?

- The gene is from a bacterium.
- Different parts of the DNA and RNA that are to be labeled.

For help with this problem, review:

The Template in Section 10.2 and The Structure of mRNA in Section 10.5.

Solution Steps

Hint: Review the structure of a transcription unit in Figure 10.6 and the structure of mRNA in Figure 10.17.

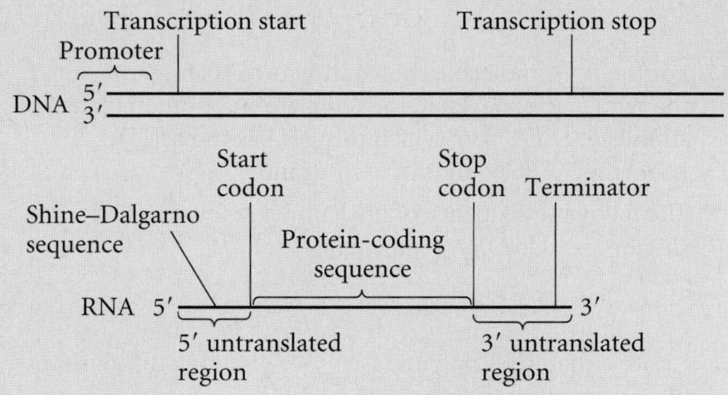

COMPREHENSION QUESTIONS

Section 10.1

1. Draw an RNA nucleotide and a DNA nucleotide, highlighting the differences. How is the structure of RNA similar to that of DNA? How is it different?
2. What are the major classes of cellular RNA?

Section 10.2

3. What parts of DNA make up a transcription unit? Draw a typical bacterial transcription unit and identify its parts.
4. What is the substrate for RNA synthesis? How is this substrate modified and joined together to produce an RNA molecule?

Section 10.3

5. What are the three basic stages of transcription? Describe what takes place at each stage.
6. How are transcription and replication similar and how are they different?

Section 10.5

7. What are the three principal elements in mRNA sequences in bacterial cells?
8. What is the function of the Shine–Dalgarno consensus sequence?
9. What is the 5′ cap?
10. What is the function of the spliceosome?
11. What is alternative splicing? How does it lead to the production of multiple proteins from a single gene?
12. Summarize the different types of processing that can take place in pre-mRNA.
13. Briefly describe the structure of tRNAs.
14. What is the origin of small interfering RNAs and microRNAs? What do these RNA molecules do in the cell?

For more questions that test your comprehension of the key chapter concepts, go to **LearningCurve** for this chapter.

APPLICATION QUESTIONS AND PROBLEMS

Section 10.1

*15. An RNA molecule has the following percentages of bases: A = 23%, U = 42%, C = 21%, and G = 14%.

a. Is this RNA single stranded or double stranded? How can you tell?

b. What would be the percentages of bases in the template strand of the DNA that contains the gene for this RNA?

Section 10.2

*16. The following diagram represents DNA that is part of the RNA-coding sequence of a transcription unit. The bottom strand is the template strand. Give the sequence found on the RNA molecule transcribed from this DNA and identify the 5′ and 3′ ends of the RNA.

5′ — ATAGGCGATGCCA — 3′
3′ — TATCCGCTACGGT — 5′ ← Template strand

17. For the RNA molecule shown in **Figure 10.1a**, write out the sequence of bases on the template and nontemplate strands of DNA from which this RNA is transcribed. Label the 5′ and 3′ ends of each strand.

18. The following sequence of nucleotides is found in a single-stranded DNA template:

ATTGCCAGATCATCCCAATAGAT

Assume that RNA polymerase proceeds along this template from left to right.

a. Which end of the DNA template is 5′ and which end is 3′?

b. Give the sequence and identify the 5′ and 3′ ends of the RNA copied from this template.

19. List at least five properties that DNA polymerases and RNA polymerases have in common. List at least three differences.

20. Most RNA molecules have three phosphate groups at the 5′ end, but DNA molecules never do. Explain this difference.

21. A strain of bacteria possesses a temperature-sensitive mutation in the gene that encodes the sigma factor. The mutant bacteria produce a sigma factor that is unable to bind to RNA polymerase at elevated temperatures. What effect will this mutation have on the process of transcription when the bacteria are raised at elevated temperatures?

22. On **Figure 10.5**, indicate the locations of the promoters and terminators for genes *a*, *b*, and *c*.

23. The following diagram represents a transcription unit on a DNA molecule.

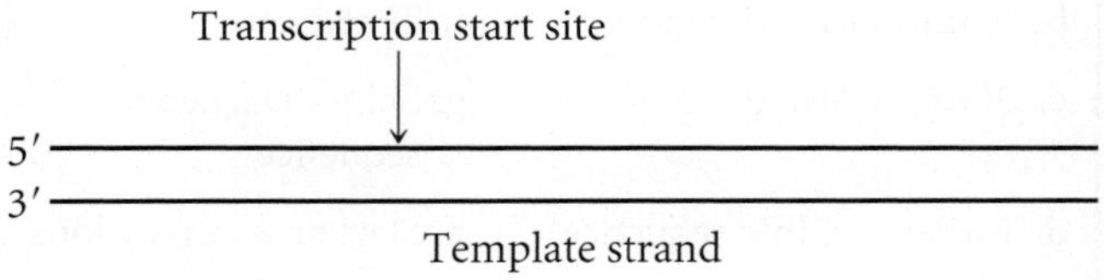

Assume that this DNA molecule is from a bacterial cell. Label the approximate locations of the promoter and terminator for this transcription unit.

Section 10.3

24. Provide the consensus sequence for the *first three* actual sequences shown in **Figure 10.9**.

*25. Write the consensus sequence for the following set of nucleotide sequences.

AGGAGTT
AGCTATT
TGCAATA
ACGAAAA
TCCTAAT
TGCAATT

*26. What would be the most likely effect of a mutation at the following locations in an *E. coli* gene?

a. −8
b. −35
c. −20
d. Start site of transcription

Section 10.4

*27. Duchenne muscular dystrophy is caused by a mutation in a gene that encompasses more than 2 million nucleotides and specifies a protein called dystrophin. However, less than 1% of the gene actually encodes the amino acids in the dystrophin protein. On the basis of what you now know about gene structure and RNA processing in eukaryotic cells, provide a possible explanation for the large size of the dystrophin gene.

28. What would be the most likely effect of moving the AAUAAA consensus sequence shown in **Figure 10.19** ten nucleotides upstream?

29. Suppose that a mutation occurs in the middle of a large intron of a gene encoding a protein. What will the most likely effect of the mutation be on the amino acid sequence of that protein? Explain your answer.

30. A geneticist isolates a gene that contains eight exons. He then isolates the mature mRNA produced by this gene. After making the DNA single stranded, he mixes the single-stranded DNA and RNA. Some of the single-stranded DNA hybridizes (pairs) with the complementary mRNA. Draw a picture of what the DNA–RNA hybrids will look like under the electron microscope.

Section 10.5

*31. Draw a typical eukaryotic gene and the pre-mRNA and mRNA derived from it. Assume that the gene contains three exons. Identify the following items and, for each item, give a brief description of its function:

a. 5′ untranslated region
b. Promoter
c. AAUAAA consensus sequence
d. Transcription start site
e. 3′ untranslated region
f. Introns
g. Exons
h. Poly(A) tail
i. 5′ cap

32. How would the deletion of the Shine–Dalgarno sequence affect a bacterial mRNA?

33. A geneticist discovers that two different proteins are encoded by the same gene. One protein has 56 amino acids, and the other has 82 amino acids. Provide a possible explanation for how the same gene can encode both of these proteins.

34. In the early 1990s, Carolyn Napoli and her colleagues were attempting to genetically engineer a variety of petunias with dark purple petals by introducing numerous copies of a gene that encodes purple petals (C. Napoli, C. Lemieux, and R. Jorgensen. 1990. *Plant Cell* 2:279–289). Their thinking was that extra copies of the gene would cause more purple pigment to be produced and would result in a petunia with an even darker hue of purple. However, much to their surprise, many of the plants carrying extra copies of the purple gene were completely white or had only patches of color. Molecular analysis revealed that the level of the mRNA produced by the purple gene was reduced 50-fold in the engineered plants compared with levels of mRNA in wild-type plants. Somehow, the introduction of extra copies of the purple gene silenced both the introduced copies and the plant's own purple genes. Provide a possible explanation for how the introduction of numerous copies of the purple gene silenced all copies of the purple gene.

Petunia. [© Roger Ashford/Alamy.]

CHALLENGE QUESTIONS

Section 10.3

35. Many genes in both bacteria and eukaryotes contain numerous sequences that potentially cause pauses in or premature terminations of transcription. Nevertheless, the transcription of these genes within a cell normally produces multiple RNA molecules thousands of nucleotides long without pausing or terminating prematurely. However, when a single round of transcription takes place on such templates in a test tube, RNA synthesis is frequently interrupted by pauses and premature terminations, which reduce the rate at which transcription takes place and frequently shorten the length of the mRNA molecules produced. Most pauses and premature terminations appear when RNA polymerase temporarily backtracks (i.e., backs up) for one or two nucleotides along the DNA. Experimental findings have demonstrated that most transcriptional delays and premature terminations disappear if several RNA polymerases are simultaneously transcribing the DNA molecule. Propose an explanation for faster transcription and longer mRNA when the template DNA is being transcribed by multiple RNA polymerases.

Section 10.5

36. Alternative splicing takes place in more than 90% of the human genes that encode proteins. Researchers have found that how a pre-mRNA is spliced is affected by the pre-mRNA's promoter sequence (D. Auboeuf et al. 2002. *Science* 298:416–419). In addition, factors that affect the rate of elongation of the RNA polymerase during transcription affect the type of splicing that takes place. These findings suggest that the process of transcription affects splicing.

Propose one or more mechanisms that would explain how transcription might affect alternative splicing.

37. Duchenne muscular dystrophy (DMD) is an X-linked recessive genetic disease caused by mutations in the gene that encodes dystrophin, a large protein that plays an important role in the development of normal muscle fibers. The gene that encodes dystrophin is immense, spanning 2.5 million base pairs, and includes 79 exons and 78 introns. Many of the mutations that cause DMD produce premature stop codons, which bring protein synthesis to a halt, resulting in a greatly shortened and nonfunctional form of dystrophin. Some geneticists have proposed treating DMD patients by causing the spliceosome to skip the exon containing the stop codon. Exon skipping will produce a protein that is somewhat shortened (because an exon is skipped and some amino acids are missing) but may still result in a protein that has some function (A. Goyenvalle et al. 2004. *Science* 306:1796–1799). Propose a mechanism to bring about exon skipping for the treatment of DMD.

DATA ANALYSIS

THINK-PAIR-SHARE QUESTIONS

Introduction

1. Why would mushrooms produce a substance like α-amanitin? What function might this peptide have in mushrooms?

2. Some cancer researchers have proposed using α-amanitin in cancer therapy. How might α-amanitin be used in the treatment of cancer?

3. RNA polymerase I is insensitive to α-amanitin, RNA polymerase II is highly sensitive, and RNA polymerase III is moderately sensitive. These three RNA polymerases have different functions (see Table 10.3). How might geneticists use α-amanitin as a tool in research? Propose a specific question that could be addressed using α-amanitin.

Section 10.2

4. Usually, within a gene, only one of the DNA strands is transcribed into RNA. Why haven't both strands in a DNA molecule evolved to carry genetic information? In other words, why might we not expect both strands to carry genetic information and be transcribed into RNA?

Section 10.3

5. The following diagram represents one of the Christmas-tree-like structures shown in Figure 10.3. On the diagram, identify parts a through i.

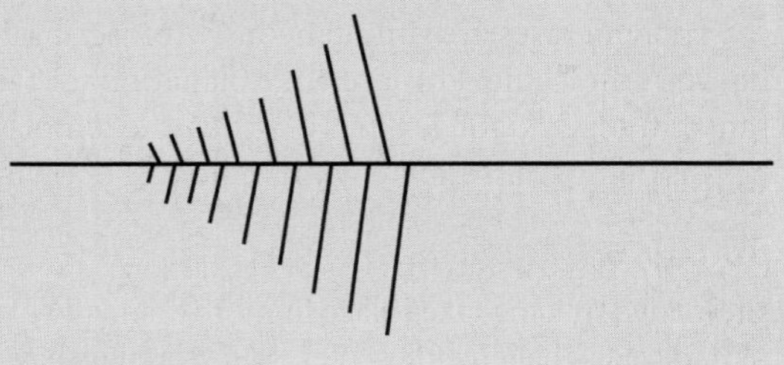

   a. DNA molecule
   b. 5′ and 3′ ends of the template strand of DNA
   c. At least one RNA molecule
   d. 5′ and 3′ ends of at least one RNA molecule
   e. Direction of movement of the transcription apparatus on the DNA molecule
   f. Approximate location of the promoter
   g. Possible location of a terminator
   h. Upstream and downstream directions
   i. Molecules of RNA polymerase (use dots to represent these molecules)

6. Make a table listing similarities and differences between transcription and replication. List as many as you can think of.

Section 10.4

7. Eukaryotic genes are often interrupted by noncoding introns. What might be some possible reasons that organisms have evolved introns? And why might other organisms lose introns?

8. Suppose that you are at a party on Friday night, relaxing after your big genetics exam. Someone comes up to you and, hearing that you just finished your genetics exam, says, "What exactly is a gene?" How would you respond? What are the strengths and weaknesses of your definition of a gene?

Section 10.5

9. Many human genetic diseases are caused by mutations that occur at splice sites. Propose some ways that mutations at the 5′ splice site, 3′ splice site, and branch point might disrupt splicing and alter the phenotype.

10. Small RNA molecules are involved in numerous genetic processes, including replication, translation, mRNA processing and degradation, inhibition of translation, chromatin modification, and protection against viruses and transposable elements. However, small DNA molecules have little or no role in these functions. Why has RNA and not DNA evolved to carry out these functions?

SaplingPlus Self-study tools that will help you practice what you've learned and reinforce this chapter's concepts are available online. Go to www.macmillanlearning.com/PierceGenetics6e.

11 From DNA to Proteins: Translation

A Child Without a Spleen

The spleen is an often underappreciated organ. Brownish in color and weighing about a third of a pound, it sits in the left upper part of your abdomen, storing blood and filtering out bacteria and old blood cells. The spleen is underappreciated because it's widely believed that you can live without a spleen. Indeed, many people who lose their spleen to automobile accidents and other trauma do survive, although they are at increased risk of infection. But a young child without a spleen is in serious trouble.

A small group of children are born without spleens; these kids are highly susceptible to life-threatening bacterial infections, and many die in childhood. This rare disorder, known as isolated congenital asplenia (ICA), is inherited as an autosomal dominant trait. Except for the absence of a spleen, children with ICA are unaffected. But their immune function is severely compromised. When infected with bacteria that the immune system normally eliminates, these children develop raging infections that quickly spread throughout the body. Even when treated with modern antibiotics, they often die.

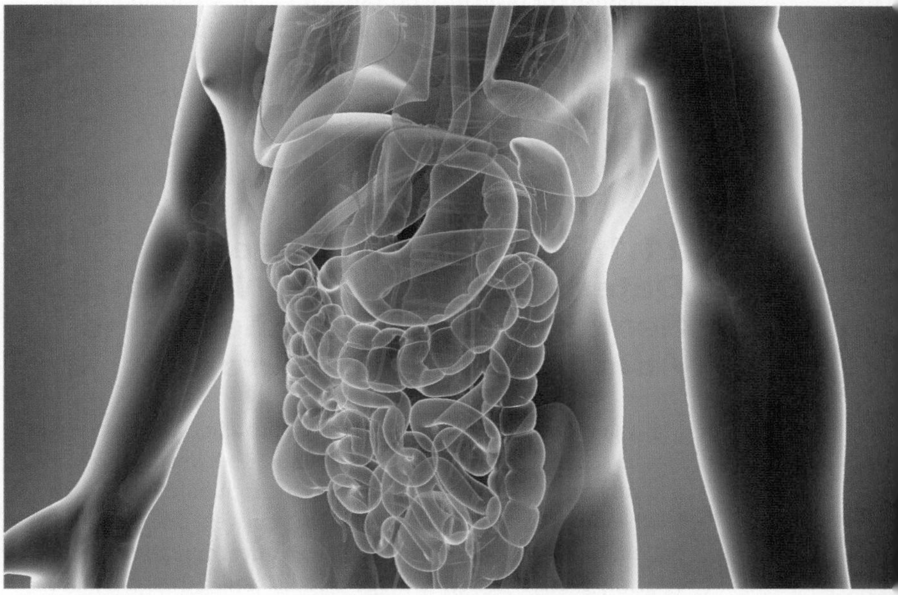

The spleen, an organ found in the upper abdomen, plays an important role in defense against infection. Isolated congenital asplenia is an autosomal dominant condition in which children are born without a spleen. [Sebastian Kaulitzki/Shutterstock.]

In 2013, an international team led by scientists from Rockefeller University discovered the genetic cause of ICA. Using the power of DNA sequencing, they examined all the coding DNA of 23 individuals with ICA and compared their DNA sequences with those of 508 individuals with normal spleens. Statistical analysis pointed to differences in one particular gene that was associated with ICA, a gene encoding ribosomal protein SA (*RPSA*). The RPSA protein is one of the 33 proteins that make up the small subunit of the ribosome, the organelle responsible for protein synthesis. How a defect in the *RPSA* gene results in the absence of a spleen is not known. Diseases such as ICA, which result from defective ribosomes, are referred to as ribosomopathies.

Many, but not all, individuals with ICA have mutations in *RPSA*, indicating that other genes may also be involved in the disorder. The researchers found several different types of mutations in *RPSA* associated with ICA: some caused premature stop codons, halting translation before a functional protein could be made; one was a frameshift mutation, a change that alters the way the mRNA sequence is read during translation; and others changed the amino acid sequence of the RPSA protein.

One interesting but unanswered question is why a defect in *RPSA* affects only the spleen. Inherited mutations in *RPSA* occur in every cell of the body, and protein synthesis—carried out by ribosomes—is essential for numerous life processes, yet these mutations affect only the development of the spleen. Why aren't other organs altered? Why aren't numerous physiological functions affected? Scientists are still studying these important questions.

THINK-PAIR-SHARE Questions 1 and 2

Isolated congenital asplenia illustrates the extreme importance of translation—the process of protein synthesis—which is the focus of this chapter. We begin this chapter by examining the genetic code—the instructions that specify the amino acid sequence of a protein—and then examine the mechanism of protein synthesis. Our primary focus will be protein synthesis in bacterial cells, but we will also examine some of the differences between bacterial and eukaryotic cells. At the end of the chapter, we will look at some additional aspects of protein synthesis.

11.1 The Genetic Code Determines How the Nucleotide Sequence Specifies the Amino Acid Sequence of a Protein

Many genes specify traits by encoding proteins. The first person to suggest the existence of a relation between genotype and proteins was the English physician Archibald Garrod. In 1908, Garrod correctly proposed that genes encode enzymes, but, unfortunately, his proposal made little impression on his contemporaries. Later, George Beadle and Edward Tatum examined the genetic basis of biochemical pathways in the bread mold *Neurospora* and developed the **one-gene, one-enzyme hypothesis**, which suggested that genes function by encoding enzymes and that each gene encodes a separate enzyme. Later research findings showed that some proteins are composed of more than one polypeptide chain and that different polypeptide chains are encoded by separate genes, so this model was modified to become the **one-gene, one-polypeptide hypothesis**.

THINK-PAIR-SHARE Questions 3 and 4

CONCEPTS

Many genes specify traits by encoding proteins. The one-gene, one-enzyme hypothesis proposed that each gene encodes a separate enzyme. This hypothesis was later modified to the one-gene, one-polypeptide hypothesis.

The Structure and Function of Proteins

Proteins are central to all living processes (**Figure 11.1**). Many proteins are enzymes, the biological catalysts that drive the chemical reactions of the cell; others are structural components, providing scaffolding and support for membranes, filaments, bone, and hair. Some proteins help transport substances; others have a regulatory, communication, or defense function.

AMINO ACIDS All proteins are polymers composed of **amino acids**, linked end to end. Twenty common amino acids are found in proteins (**Table 11.1**). All of these amino acids are similar in structure: each consists of a central carbon atom bonded to an amino group, a hydrogen atom, a carboxyl group, and an R (radical) group that differs for each amino acid (**Figure 11.2a**). The amino acids in proteins are joined together by **peptide bonds** (**Figure 11.2b**) to form **polypeptide** chains; a protein consists of one or more polypeptide

11.1 Proteins serve a number of biological functions. (a) The light produced by fireflies is the result of a light-producing reaction between luciferin and ATP catalyzed by the enzyme luciferase. (b) The protein fibroin is the major structural component of spider webs. (c) Castor beans contain a highly toxic protein called ricin. [Part a: Darwin Dale/Science Source. Part b: Rosemary Calvert/Imagestate/Media Bakery. Part c: Paroli Galperti /© Cuboimages/Photoshot.]

TABLE 11.1 The twenty common amino acids found in most proteins

Amino acid	Three-letter abbreviation	One-letter abbreviation	Amino acid	Three-letter abbreviation	One-letter abbreviation
Alanine	Ala	A	Leucine	Leu	L
Arginine	Arg	R	Lysine	Lys	K
Asparagine	Asn	N	Methionine	Met	M
Aspartate	Asp	D	Phenylalanine	Phe	F
Cysteine	Cys	C	Proline	Pro	P
Glutamate	Glu	E	Serine	Ser	S
Glutamine	Gln	Q	Threonine	Thr	T
Glycine	Gly	G	Tyrosine	Tyr	Y
Histidine	His	H	Tryptophan	Trp	W
Isoleucine	Ile	I	Valine	Val	V

chains. Like nucleic acids, polypeptides have polarity at physiological pH: one end has a free amino group (NH_3^+) and the other end has a free carboxyl group (COO^-). Some polypeptide chains consist of only a few amino acids, whereas others may have thousands.

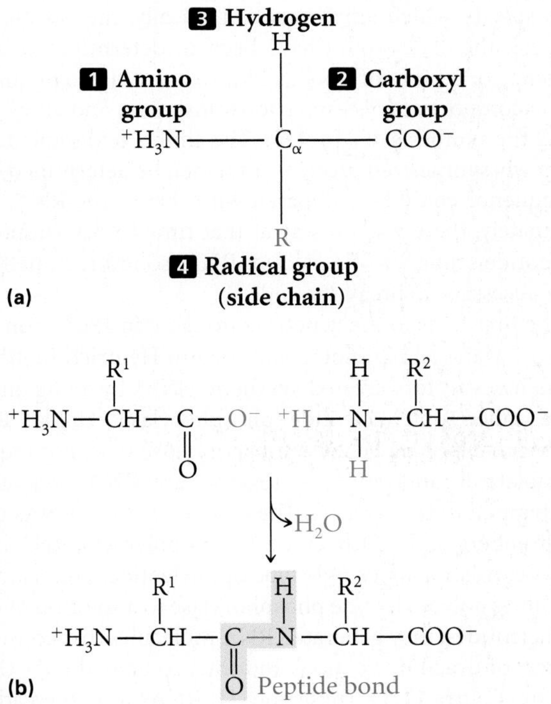

11.2 The common amino acids that make up proteins have similar structures. (a) Each amino acid consists of a central carbon atom (C_α) attached to (1) an amino group (NH_3^+); (2) a carboxyl group (COO^-); (3) a hydrogen atom (H); and (4) a radical group, designated R. (b) Amino acids are joined together by peptide bonds. In a peptide bond (pink shading), the carboxyl group of one amino acid is covalently attached to the amino group of another amino acid.

PROTEIN STRUCTURE Like that of nucleic acids, the molecular structure of proteins has several levels of organization. The primary structure of a protein is its sequence of amino acids (**Figure 11.3a**). Through interactions between neighboring amino acids, a polypeptide chain folds and twists into a secondary structure (**Figure 11.3b**). Two common secondary structures found in proteins are the beta (β) pleated sheet and the alpha (α) helix. Secondary structures interact and fold further to form a tertiary structure (**Figure 11.3c**), which is the overall, three-dimensional shape of the protein. The secondary and tertiary structures of a protein are ultimately determined by the primary structure—the amino acid sequence—of the protein. Finally, some proteins consist of two or more polypeptide chains that associate to produce a quaternary structure (**Figure 11.3d**).

THINK-PAIR-SHARE Question 5

CONCEPTS

Proteins are polymers consisting of amino acids linked by peptide bonds. The amino acid sequence of a protein is its primary structure. This structure folds to create the secondary and tertiary structures; two or more polypeptide chains may associate to create a quaternary structure.

✓ **CONCEPT CHECK 1**

What determines the secondary and tertiary structures of a protein?

Breaking the Genetic Code

In 1953, James Watson and Francis Crick solved the structure of DNA and identified its base sequence as the carrier of genetic information (as we saw in Chapter 8). However, the way in which the base sequence of DNA specifies the amino acid sequences of proteins (the genetic code) remained elusive for another 10 years.

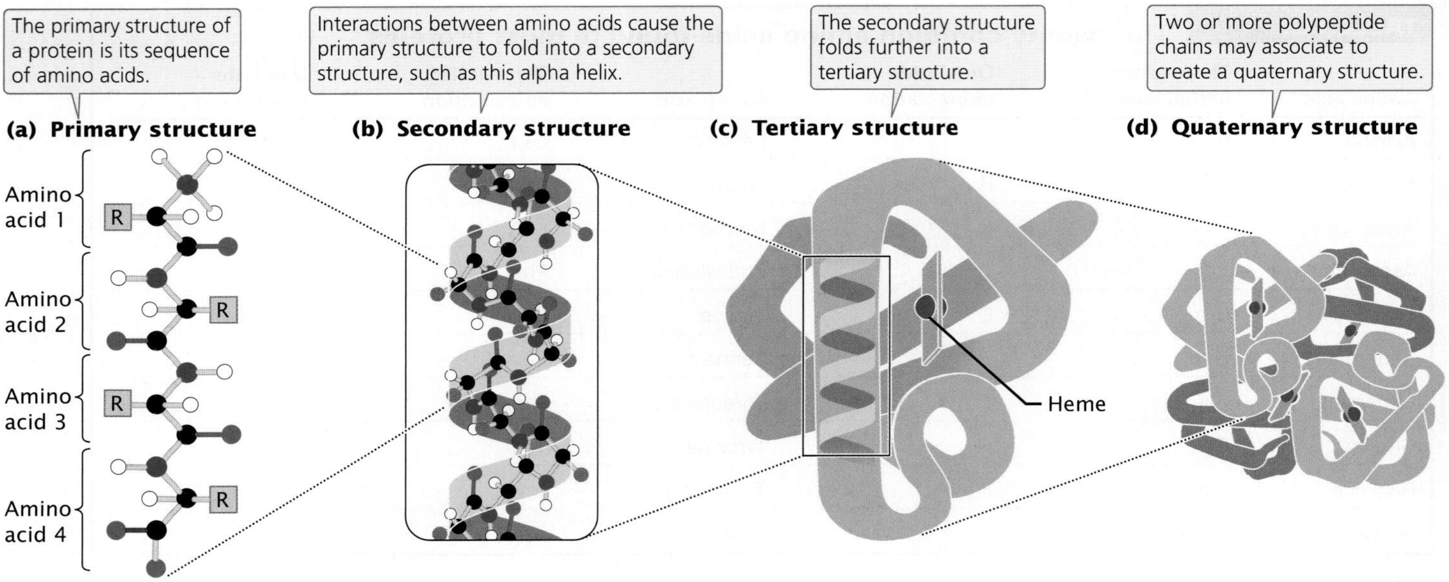

11.3 Proteins have several levels of structural organization. Atoms are represented in color as follows: blue, nitrogen; white, hydrogen; black, carbon; red, oxygen.

One of the first questions about the genetic code to be addressed was how many nucleotides are necessary to specify a single amino acid. The set of nucleotides that encodes a single amino acid—the basic unit of the genetic code—is called a codon (see Chapter 10). Many early investigators recognized that codons must contain a minimum of three nucleotides. Each nucleotide position in mRNA can be occupied by one of four bases: A, G, C, or U. If a codon consisted of a single nucleotide, only four different codons (A, G, C, and U) would be possible, which is not enough to encode the 20 different amino acids commonly found in proteins. If codons were made up of two nucleotides each (i.e., GU, AC, etc.), there would be $4 \times 4 = 16$ possible codons—still not enough to encode all 20 amino acids. With three nucleotides per codon, there are $4 \times 4 \times 4 = 64$ possible codons, which is more than enough to specify 20 different amino acids. Therefore, a *triplet code* requiring three nucleotides per codon would be the most efficient way to encode all 20 amino acids. Using mutations in bacteriophage, Francis Crick and his colleagues confirmed, in 1961, that the genetic code is indeed a triplet code.

TRY PROBLEMS 12 AND 13

THINK-PAIR-SHARE Question 6

CONCEPTS

The genetic code is a triplet code, in which three nucleotides encode each amino acid in a protein.

✓ CONCEPT CHECK 2

A codon is

a. one of three nucleotides that encode an amino acid.
b. three nucleotides that encode an amino acid.
c. three amino acids that encode a nucleotide.
d. one of four bases in DNA.

Once it had been firmly established that the genetic code consists of codons that are three nucleotides in length, the next step was to determine which groups of three nucleotides specify which amino acids. Logically, the easiest way to break the code would have been to determine the base sequence of a piece of RNA, add it to a test tube containing all the components necessary for translation, and allow it to direct the synthesis of a protein. The amino acid sequence of the newly synthesized protein could then be determined, and its sequence could be compared with that of the RNA. Unfortunately, there was no way at that time to determine the nucleotide sequence of a piece of RNA, so indirect methods were necessary to break the code.

The first clues to the genetic code came in 1961, from the work of Marshall Nirenberg and Johann Heinrich Matthaei. These investigators created synthetic RNAs by using an enzyme called polynucleotide phosphorylase. Unlike RNA polymerase, polynucleotide phosphorylase does not require a template; it randomly links together any RNA nucleotides that happen to be available. The first synthetic mRNAs used by Nirenberg and Matthaei were homopolymers, RNA molecules consisting of a single type of nucleotide. For example, by adding polynucleotide phosphorylase to a solution of uracil nucleotides, they generated RNA molecules that consisted entirely of uracil nucleotides and thus contained only UUU codons (**Figure 11.4**). These poly(U) RNAs were then added to 20 test tubes, each containing the components necessary for translation and all 20 amino acids. A different amino acid was radioactively labeled in each of the 20 tubes. Radioactive protein appeared in only one of the tubes: the one containing labeled phenylalanine (see Figure 11.4). This result showed that the codon UUU specifies the amino acid phenylalanine. The results of similar experiments using poly(C) and poly(A)

RNA demonstrated that CCC encodes proline and AAA encodes lysine; for technical reasons, the results from poly(G) were uninterpretable. Other experiments provided additional information about the genetic code, and it was fully understood by 1968.

Characteristics of the Genetic Code

The genetic code is so important to modern biology that Francis Crick compared its place to that of the periodic table of the elements in chemistry. We will now examine a number of features of the genetic code.

THE DEGENERACY OF THE CODE One amino acid is encoded by three consecutive nucleotides in mRNA, and each nucleotide can have one of four possible bases (A, G, C, and U), so there are $4^3 = 64$ possible codons (**Figure 11.5**). Three of these codons are stop codons, which specify the end of translation, as we'll see shortly. Thus, 61 codons, called **sense codons**, encode amino acids. Because there are 61 sense codons and only 20 different amino acids commonly found in proteins, the code contains more information than is needed to specify the amino acids and is said to be **degenerate**. This expression does not mean that the genetic code is depraved; *degenerate* is a term that Francis Crick borrowed from quantum physics, where it describes multiple physical states that have equivalent meaning. The degeneracy of the genetic code means that the code is redundant: amino acids may be specified by more than one codon. Only tryptophan and methionine are encoded by a single codon (see Figure 11.5). Other amino acids are specified by two or more codons, and some, such as leucine, are specified by six different codons. Codons that specify the same amino acid are said to be **synonymous codons**, just as synonymous words are different words that have the same meaning.

As we learned in Chapter 10, tRNAs serve as adapter molecules that bind particular amino acids and deliver them to a ribosome, where the amino acids are then assembled into polypeptide chains. Each type of tRNA attaches to a single type of amino acid. The cells of most organisms possess from about 30 to 50 different tRNAs, and yet there are only 20 different amino acids commonly found in proteins. Thus, some amino acids are carried by more than one tRNA. Different tRNAs that accept the same amino acid but have different anticodons are called **isoaccepting tRNAs**.

Even though some amino acids can pair with multiple (isoaccepting) tRNAs, there are still more codons than anticodons. One anticodon can sometimes pair with different codons through flexibility in base pairing at the third position of the codon. Examination of Figure 11.5 reveals that many synonymous codons differ only in the third position. For example, serine is encoded by the codons

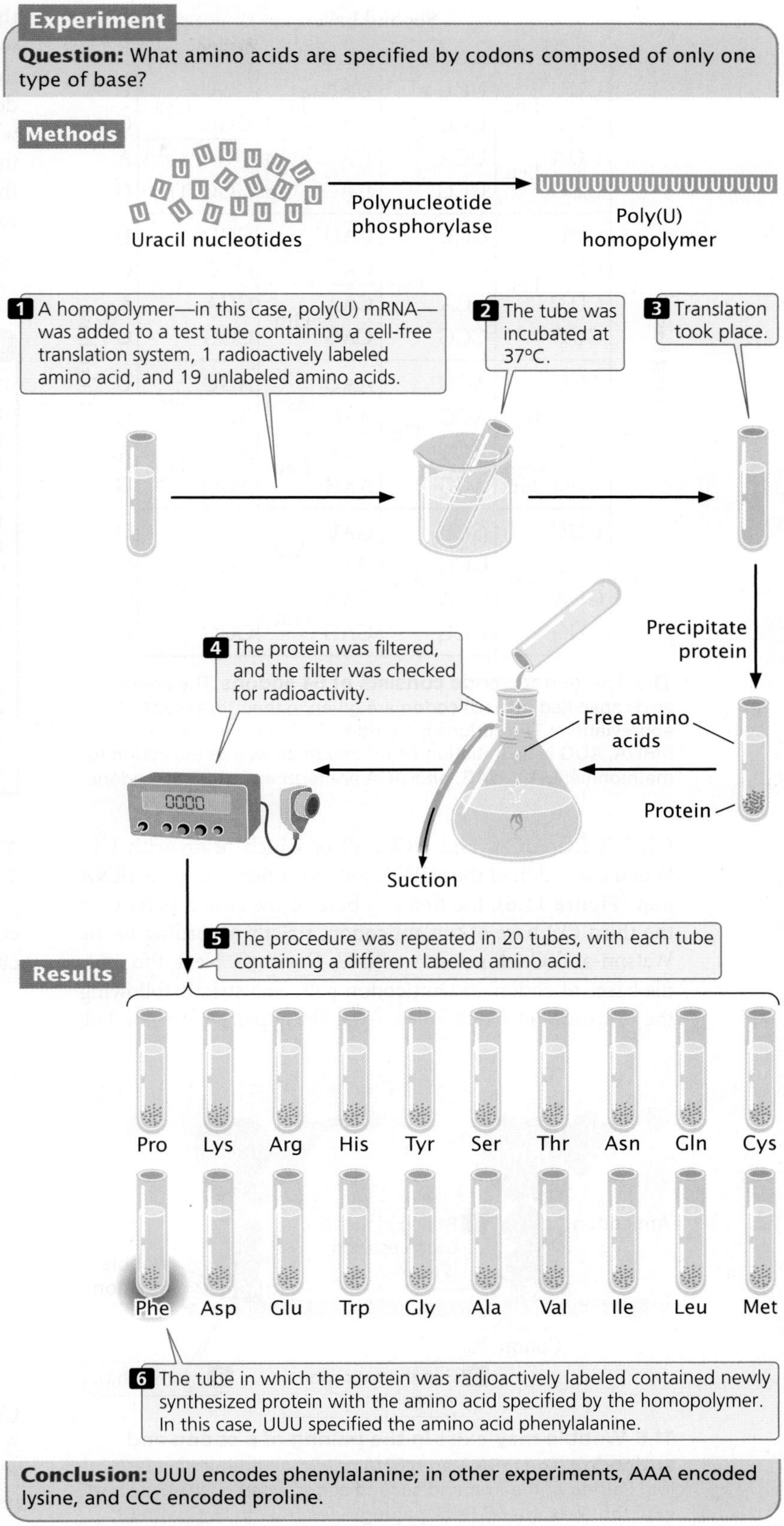

11.4 Nirenberg and Matthaei developed a method for identifying the amino acid specified by a homopolymer.

First base	Second base: U		C		A		G		Third base
U	UUU	Phe	UCU	Ser	UAU	Tyr	UGU	Cys	U
	UUC		UCC		UAC		UGC		C
	UUA	Leu	UCA		UAA	Stop	UGA	Stop	A
	UUG		UCG		UAG	Stop	UGG	Trp	G
C	CUU	Leu	CCU	Pro	CAU	His	CGU	Arg	U
	CUC		CCC		CAC		CGC		C
	CUA		CCA		CAA	Gln	CGA		A
	CUG		CCG		CAG		CGG		G
A	AUU	Ile	ACU	Thr	AAU	Asn	AGU	Ser	U
	AUC		ACC		AAC		AGC		C
	AUA		ACA		AAA	Lys	AGA	Arg	A
	AUG	Met	ACG		AAG		AGG		G
G	GUU	Val	GCU	Ala	GAU	Asp	GGU	Gly	U
	GUC		GCC		GAC		GGC		C
	GUA		GCA		GAA	Glu	GGA		A
	GUG		GCG		GAG		GGG		G

11.5 The genetic code consists of 64 codons. The amino acids specified by each codon are given in their three-letter abbreviations. The codons are written 5'→3', as they appear in the mRNA. AUG is an initiation (start) codon as well as the codon for methionine; UAA, UAG, and UGA are termination (stop) codons.

UCU, UCC, UCA, and UCG, all of which begin with UC. When the codon of the mRNA and the anticodon of the tRNA join (**Figure 11.6**), the first (5') base of the codon pairs with the third (3') base of the anticodon, strictly according to the Watson-and-Crick rules: A with U; C with G. Next, the middle bases of codon and anticodon pair, also strictly following the Watson-and-Crick rules. After these pairs have bonded,

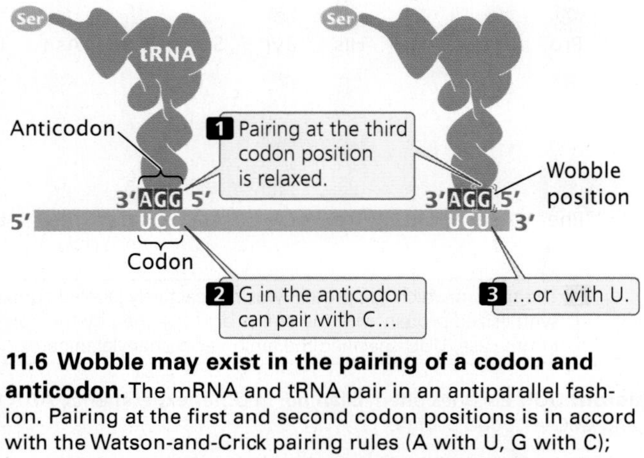

11.6 Wobble may exist in the pairing of a codon and anticodon. The mRNA and tRNA pair in an antiparallel fashion. Pairing at the first and second codon positions is in accord with the Watson-and-Crick pairing rules (A with U, G with C); however, pairing rules are relaxed at the third position of the codon, and G on the anticodon can pair with either U or C on the codon in this example.

the third bases pair weakly, and there may be flexibility, or **wobble**, in their pairing; for example, a G in the anticodon may pair with either a C or a U in the third position of the codon. In 1966, Francis Crick developed the wobble hypothesis, which proposed that there could be some nonstandard pairings of bases at the third position of a codon. The important thing to remember about wobble is that it allows some tRNAs to pair with more than one codon on an mRNA.

THINK-PAIR-SHARE Question 7

CONCEPTS

The genetic code consists of 61 sense codons that specify the 20 common amino acids. The code is degenerate, meaning that some amino acids are encoded by more than one codon. Isoaccepting tRNAs are tRNAs with different anticodons that specify the same amino acid. Wobble at the third position of the codon allows different codons to specify the same amino acid.

✓ CONCEPT CHECK 3

Through wobble, a single __________ can pair with more than one __________.

a. codon, anticodon
b. group of three nucleotides in DNA, codon in mRNA
c. tRNA, amino acid
d. anticodon, codon

THE READING FRAME AND INITIATION CODONS Findings from early studies indicated that the genetic code is generally **nonoverlapping**. An overlapping code would be one in which a single nucleotide might be included in more than one codon, as follows:

Nucleotide sequence: A U A C G A G U C

Nonoverlapping code: A U A | C G A | G U C → Ile, Arg, Val

Overlapping code: A U A C G A G U
A U A → Ile
U A C → Tyr
A C G → Thr

Usually, however, each nucleotide is part of a single codon. A few overlapping genes are found in viruses, but codons within the same gene do not overlap, and the genetic code is generally considered to be nonoverlapping.

For any sequence of nucleotides, there are three potential sets of codons—three ways in which the sequence can be read in groups of three. Each different way of reading the sequence is called a **reading frame**, and any sequence of nucleotides has three potential reading frames. The three reading frames have completely different sets of codons and will therefore specify proteins with entirely different amino acid sequences. Thus, it is essential for the translational machinery to use the correct reading frame. How is the correct reading frame established? The reading frame is set by the **initiation codon** (or **start codon**), which is the first codon of the mRNA to specify an amino acid. After the initiation codon, the other codons are read as successive groups of three nucleotides. No bases are skipped between the codons, so there are no punctuation marks to separate the codons.

The initiation codon is usually AUG, although GUG and UUG are used on rare occasions. The initiation codon is not just a sequence that marks the beginning of translation; it also specifies an amino acid. In bacterial cells, the first AUG encodes a modified type of methionine, *N*-formylmethionine; all proteins in bacteria initially begin with this amino acid, but its formyl group (or, in some cases, the entire amino acid) may be removed after the protein has been synthesized. When the codon AUG is at an internal position in a gene, it encodes unformylated methionine. In archaeal and eukaryotic cells, AUG specifies unformylated methionine both at the initiation position and at internal positions. In both bacteria and eukaryotes, there are different tRNAs for the initiator methionine and internal methionine.

TERMINATION CODONS Three codons—UAA, UAG, and UGA—do not encode amino acids. These codons, which signal the end of the protein in both bacterial and eukaryotic cells, are called **stop codons**, **termination codons**, or **nonsense codons**. No tRNAs have anticodons that pair with termination codons.

THINK-PAIR-SHARE Question 8

THE UNIVERSALITY OF THE CODE For many years, the genetic code was assumed to be **universal**, meaning that each codon specifies the same amino acid in all organisms. We now know that the genetic code is almost, but not completely, universal; a few exceptions have been found. Many of these exceptions are termination codons, but there are some cases in which one sense codon substitutes for another. Most exceptions are found in mitochondrial genes; a few nonuniversal codons have also been detected in the nuclear genes of protozoans and in bacterial DNA. ▶ TRY PROBLEM 15

THINK-PAIR-SHARE Question 9

CONCEPTS

Each sequence of nucleotides possesses three potential reading frames. The correct reading frame is set by the initiation codon. The end of a protein-coding sequence is marked by a termination codon. With a few exceptions, all organisms use the same genetic code.

CONNECTING CONCEPTS

Characteristics of the Genetic Code

We have now considered a number of characteristics of the genetic code. Let's take a moment to review these characteristics.

1. The genetic code consists of a sequence of nucleotides in DNA or RNA. There are four letters in the code, corresponding to the four bases—A, G, C, and U (T in DNA).
2. The genetic code is a triplet code. Each amino acid is encoded by a sequence of three consecutive nucleotides, called a codon.
3. The genetic code is degenerate; that is, of 64 codons, 61 codons encode only 20 amino acids in proteins (3 codons are termination codons). Some codons are synonymous, specifying the same amino acid.
4. Isoaccepting tRNAs are tRNAs with different anticodons that accept the same amino acid. Wobble allows the anticodon on one type of tRNA to pair with more than one codon on mRNA.
5. The genetic code is generally nonoverlapping; each nucleotide in an mRNA sequence belongs to a single reading frame.
6. The reading frame is set by an initiation codon, which is usually AUG.
7. When a reading frame has been set, codons are read as successive groups of three nucleotides.
8. Any one of three termination codons (UAA, UAG, or UGA) can signal the end of a protein; no amino acids are encoded by the termination codons.
9. The genetic code is almost universal.

11.2 Amino Acids Are Assembled into a Protein Through Translation

Now that we are familiar with the genetic code, we can study how amino acids are assembled into proteins. Because more is known about translation in bacteria than in eukaryotes, we will focus primarily on bacterial translation. In most respects, eukaryotic translation is similar, although some significant differences will be noted.

Remember that only mRNAs are translated into proteins. Translation takes place on ribosomes; indeed, ribosomes can be thought of as moving protein-synthesizing machines. Through a variety of techniques, a detailed view of the structure of the ribosome has been produced in recent years, which has greatly improved our understanding of translation. A ribosome attaches near the 5′ end of an mRNA strand and moves toward the 3′ end, translating the codons as it goes (**Figure 11.7**). Synthesis begins at the amino end of the protein, and the protein is elongated by the addition of new amino acids to the carboxyl end.

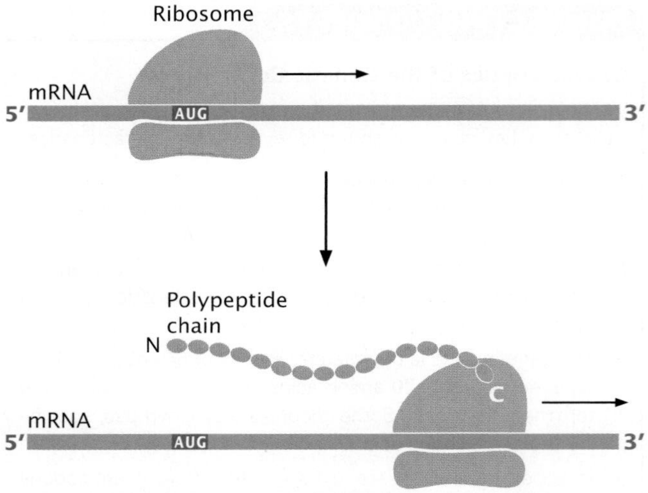

11.7 The translation of an mRNA molecule takes place on a ribosome. The letter N represents the amino end of the protein; C represents the carboxyl end.

Protein synthesis can be conveniently divided into four stages: (1) tRNA charging, which entails the binding of amino acids to tRNAs; (2) initiation, in which the components necessary for translation are assembled at the ribosome; (3) elongation, in which amino acids are joined, one at a time, to the growing polypeptide chain; and (4) termination, in which protein synthesis halts at the termination codon and the translation components are released from the ribosome.

The Binding of Amino Acids To Transfer RNAs

The first stage of translation is the binding of tRNA molecules to their appropriate amino acids. As we have seen, although there may be several different tRNAs for a particular amino acid, each tRNA is specific for only one amino acid. The key to specificity between an amino acid and its tRNA is a set of enzymes called **aminoacyl-tRNA synthetases**. A cell has 20 different aminoacyl-tRNA synthetases, one for each of the 20 amino acids. Each synthetase recognizes a particular amino acid as well as all the tRNAs that accept that amino acid.

The attachment of a tRNA to its appropriate amino acid, termed **tRNA charging**, requires energy, which is supplied by adenosine triphosphate (ATP):

$$\text{amino acid} + \text{tRNA} + \text{ATP} \rightarrow \text{aminoacyl-tRNA} + \text{AMP} + \text{PP}_i$$

The carboxyl group (COO^-) of the amino acid is attached to the adenine nucleotide at the 3′ end of the tRNA (**Figure 11.8**). To identify the resulting aminoacylated tRNA, we write the three-letter abbreviation for the amino acid in front of the tRNA; for example, the amino acid alanine (Ala) attaches to its tRNA (tRNA^{Ala}), giving rise to its aminoacyl-tRNA ($\text{Ala-tRNA}^{\text{Ala}}$).

CONCEPTS

Amino acids are attached to specific tRNAs by aminoacyl-tRNA synthetases in a reaction that requires ATP.

✓ CONCEPT CHECK 4

Amino acids bind to which part of the tRNA?

a. anticodon
b. codon
c. 3′ end
d. 5′ end

The Initiation of Translation

The second stage in the process of protein synthesis is initiation. At this stage, all the components necessary for protein synthesis assemble: (1) mRNA; (2) the small and large subunits of the ribosome; (3) a set of three proteins called initiation factors; (4) initiator tRNA with *N*-formylmethionine attached ($\text{fMet-tRNA}^{\text{fMet}}$); and (5) guanosine triphosphate (GTP). Initiation comprises three major steps. First, mRNA binds to the small subunit of the ribosome. Second, initiator tRNA binds to the mRNA through base pairing between the codon and the anticodon. Third, the large ribosome joins the initiation complex. Let's look at each of these steps more closely.

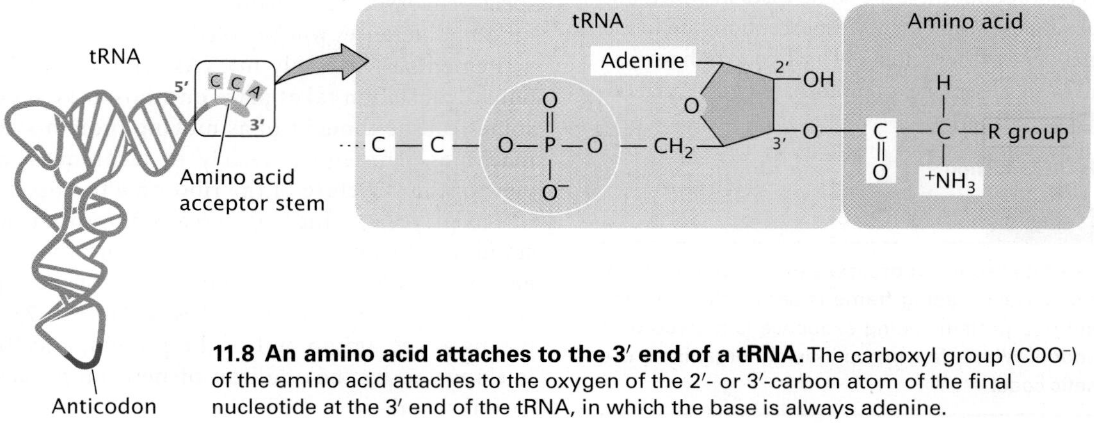

11.8 An amino acid attaches to the 3′ end of a tRNA. The carboxyl group (COO^-) of the amino acid attaches to the oxygen of the 2′- or 3′-carbon atom of the final nucleotide at the 3′ end of the tRNA, in which the base is always adenine.

INITIATION IN BACTERIA The functional ribosome of bacteria exists as two subunits, the small 30S subunit and the large 50S subunit (**Figure 11.9a**). An mRNA molecule can bind to the small ribosome subunit only when the subunits are separate. **Initiation factor 3 (IF-3)** binds to the small subunit of the ribosome and prevents the large subunit from binding during initiation (**Figure 11.9b**). Another factor, **initiation factor 1 (IF-1)**, enhances the disassociation of the large and small ribosomal subunits.

The initiator tRNA, fMet-tRNA^{fMet}, then attaches to the initiation codon (**Figure 11.9c**). This attachment requires **initiation factor 2 (IF-2)**, which forms a complex with GTP.

At this point, the initiation complex consists of (1) the small subunit of the ribosome; (2) the mRNA; (3) the initiator tRNA with its amino acid (fMet-tRNA^{fMet}); (4) one molecule of GTP; and (5) several initiation factors. These components are collectively known as the **30S initiation complex** (see Figure 11.9c). In the final step of initiation, the initiation factors disassociate from the small subunit, allowing the large subunit of the ribosome to join the initiation complex (**Figure 11.9d**). When the large subunit has joined the initiation complex, the complex is called the **70S initiation complex**.

INITIATION IN EUKARYOTES Similar events take place in the initiation of translation in eukaryotic cells, but there are some important differences. In bacterial cells, sequences in 16S rRNA of the small subunit of the ribosome bind to the Shine–Dalgarno sequence in mRNA. No analogous consensus sequence exists in eukaryotic mRNA. Instead, the cap at the 5′ end of eukaryotic mRNA plays a critical role in the initiation of translation. The small subunit of the eukaryotic ribosome, initiation factors, and the initiator tRNA with its amino acid (Met-tRNA^{Met}) form an initiation complex that recognizes the cap and binds there. The initiation complex then moves along (scans) the mRNA until it locates the first AUG codon. The identification of the start codon is facilitated by the presence of a consensus sequence (called the Kozak sequence) that surrounds the start codon:

Kozak sequence

5′–ACCAUGG–3′

Start codon

Another difference is that eukaryotic initiation requires at least 12 initiation factors. The poly(A) tail at the 3′ end of eukaryotic mRNA also plays a role in the initiation of translation. During initiation, proteins that attach to the poly(A) tail interact with proteins that bind to the 5′ cap, enhancing the binding of the small subunit of the ribosome to the 5′ end of the mRNA. This interaction indicates that the 3′ end of the mRNA bends over and associates with the 5′ cap during the initiation of translation, forming a circular structure known as the closed loop (**Figure 11.10**).

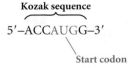

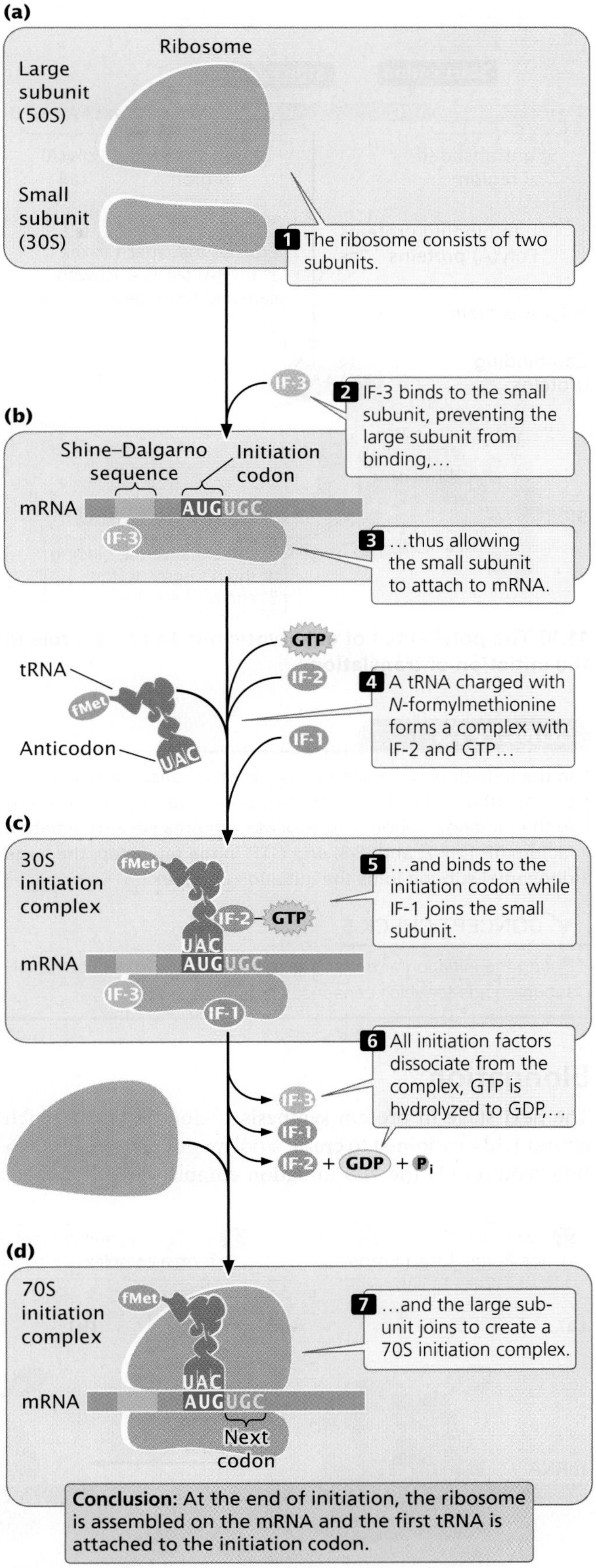

11.9 The initiation of translation requires several initiation factors and GTP.

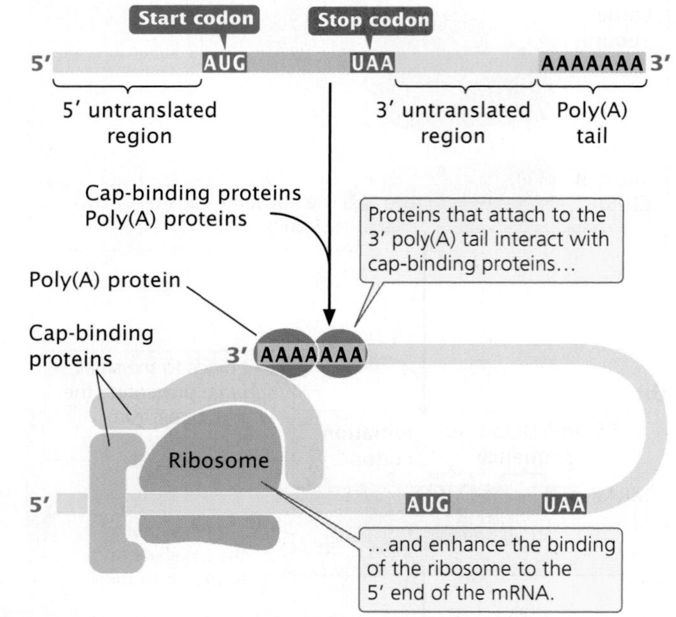

11.10 The poly(A) tail of eukaryotic mRNA plays a role in the initiation of translation.

CONCEPTS

In the initiation of translation in bacterial cells, the small ribosomal subunit attaches to mRNA, and initiator tRNA attaches to the initiation codon. This process requires several initiation factors (IF-1, IF-2, and IF-3) and GTP. In the final step, the large ribosomal subunit joins the initiation complex.

✓ CONCEPT CHECK 5

During the initiation of translation in bacteria, the small ribosomal subunit binds to which consensus sequence?

Elongation

The next stage in protein synthesis is elongation, in which amino acids are joined to create a polypeptide chain. Elongation requires (1) the 70S initiation complex just described; (2) tRNAs charged with their amino acids; (3) several elongation factors; and (4) GTP.

A ribosome has three sites that can be occupied by tRNAs: the **aminoacyl (A) site**, the **peptidyl (P) site**, and the **exit (E) site** (**Figure 11.11a**). The initiator tRNA immediately occupies the P site (the only site to which the fMet-tRNA^{fMet} is able to bind), but all other tRNAs first enter the A site. At the end of initiation, the ribosome is attached to the mRNA, and fMet-tRNA^{fMet} is positioned over the AUG start codon in the P site; the adjacent A site is unoccupied (see Figure 11.11a).

Elongation takes place in three steps. In the first step (**Figure 11.11b**), a charged tRNA binds to the A site. This binding takes place when **elongation factor Tu (EF-Tu)** joins with GTP and then with a charged tRNA to form a three-part complex. This complex enters the A site of the ribosome, where the anticodon on the tRNA pairs with the codon on the mRNA. Once the charged tRNA is in the A site, GTP is cleaved to form GDP, and the EF-Tu–GDP complex is released (**Figure 11.11c**). **Elongation factor Ts (EF-Ts)** regenerates EF-Tu–GTP from EF-Tu–GDP. In eukaryotic cells, a similar set of reactions delivers a charged tRNA to the A site.

The second step of elongation is the formation of a peptide bond between the amino acids that are attached to tRNAs in the P and A sites (**Figure 11.11d**). The formation of this peptide bond releases the amino acid in the P site from its tRNA. Peptide-bond formation occurs within the large subunit of the ribosome. The catalytic activity that creates the peptide bond is a property of ribosomal RNA in the large subunit (the 23S rRNA in bacteria, the 28S RNA in eukaryotes); this rRNA acts as a ribozyme (see p. 270 in Chapter 10).

The third step in elongation is **translocation** (**Figure 11.11e**), the movement of the ribosome down the mRNA in the 5′ → 3′ direction. This step positions the ribosome over the next codon and requires **elongation factor G (EF-G)** and the hydrolysis of GTP to GDP. Because the tRNAs in the P and A sites are still attached to the mRNA by codon–anticodon pairing, they do not move with the ribosome as it translocates. Consequently, the ribosome shifts so that the tRNA that previously occupied the P site now occupies the E site, from which it moves into the cytoplasm, where it can

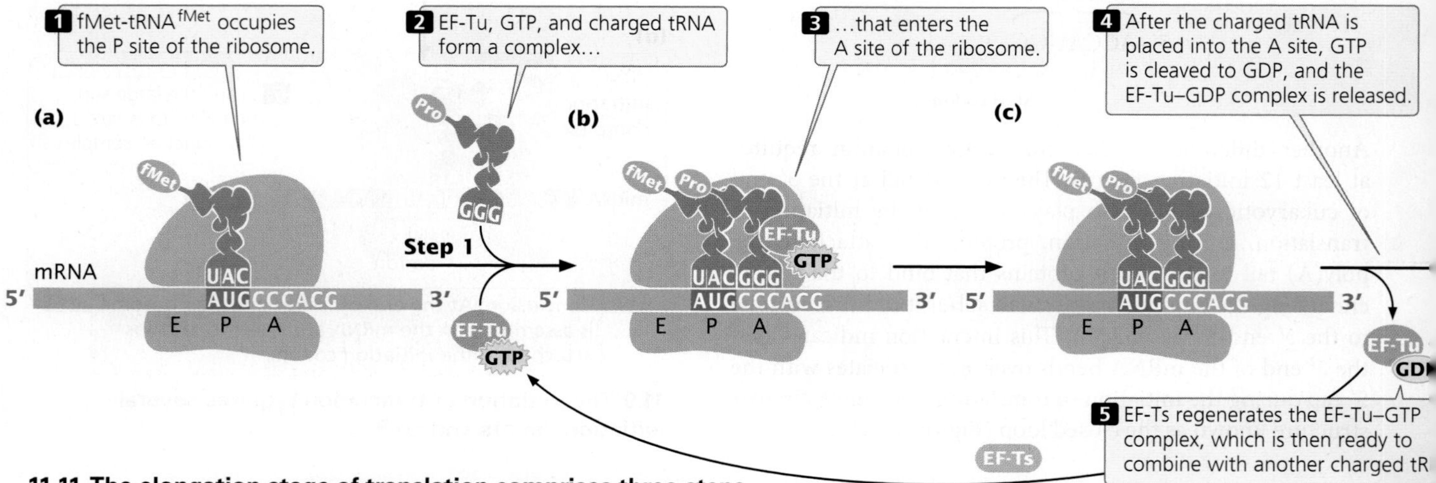

11.11 The elongation stage of translation comprises three steps.

be recharged with another amino acid. Translocation also causes the tRNA that occupied the A site (which is attached to the growing polypeptide chain) to be in the P site, leaving the A site open. Thus, the progress of each tRNA through the ribosome in the course of elongation can be summarized as follows: cytoplasm → A site → P site → E site → cytoplasm. As stated earlier, the initiator tRNA is an exception: it attaches directly to the P site and never occupies the A site.

After translocation, the A site of the ribosome is empty and ready to receive the tRNA specified by the next codon. The elongation cycle (see Figure 11.11b through e) repeats itself: a charged tRNA and its amino acid occupy the A site, a peptide bond is formed between the amino acids in the A and P sites, and the ribosome translocates to the next codon. Throughout the cycle, the polypeptide chain remains attached to the tRNA in the P site.

Elongation in eukaryotic cells takes place in a similar manner. Eukaryotes possess at least three elongation factors, one of which also acts in initiation and termination. Another of these elongation factors, called eukaryotic elongation factor 2 (eEF2), is the target of a toxin produced by the bacteria that cause diphtheria, a disease that until recently was a leading killer of children. The diphtheria toxin inhibits eEF2, preventing the translocation of the ribosome along the mRNA, and protein synthesis ceases.

CONCEPTS

Elongation consists of three steps: (1) a charged tRNA enters the A site, (2) a peptide bond is created between amino acids in the A and P sites, and (3) the ribosome translocates to the next codon. Elongation requires several elongation factors and GTP.

✓ CONCEPT CHECK 6

In elongation, the creation of peptide bonds between amino acids is catalyzed by

a. rRNA.
b. protein in the small subunit.
c. protein in the large subunit.
d. tRNA.

Termination

Protein synthesis ends when the ribosome translocates to a termination codon. Because there are no tRNAs with anticodons complementary to the termination codons, no tRNA enters the A site of the ribosome (**Figure 11.12a**). Instead, proteins called **release factors** bind to the ribosome (**Figure 11.12b**). *Escherichia coli* has three release factors: RF-1, RF-2, and RF-3. Release factor 1 binds to the termination codons UAA and UAG, and RF-2 binds to UGA and UAA. The binding of RF-1 or RF-2 to the A site of the ribosome promotes the cleavage of the tRNA in the P site from the polypeptide chain and the release of the polypeptide chain. Release factor 3 binds to the ribosome and forms a complex with GTP. This complex brings about a conformational change in the ribosome, releasing RF-1 or RF-2 from the A site and causing the tRNA in the P site to move to the E site; in the process, GTP is hydrolyzed to GDP. The tRNA is released from the E site, mRNA is released from the ribosome, and the ribosome disassociates (**Figure 11.12c**).

It is important to note that the termination codon is not located at the 3′ end of the mRNA; rather, the termination codon is followed by a number of nucleotides that constitute the 3′ untranslated region (UTR) of the mRNA. The 3′ UTR often contains sequences that affect the stability of the mRNA and influence whether translation takes place (see Chapter 12).

THINK-PAIR-SHARE Question 10

Recent research shows that some bacterial ribosomes engage in a type of proofreading, similar to the proofreading that DNA polymerases perform during replication. After translocation, the ribosome checks the interaction between the mRNA and the tRNA in the P site. If the wrong tRNA was added, the alignment between the mRNA and tRNA will be incorrect, which triggers premature termination of translation. Evidence suggests that an important function of RF-3 is to help bring about termination of translation when the wrong tRNA has been used. You can explore the

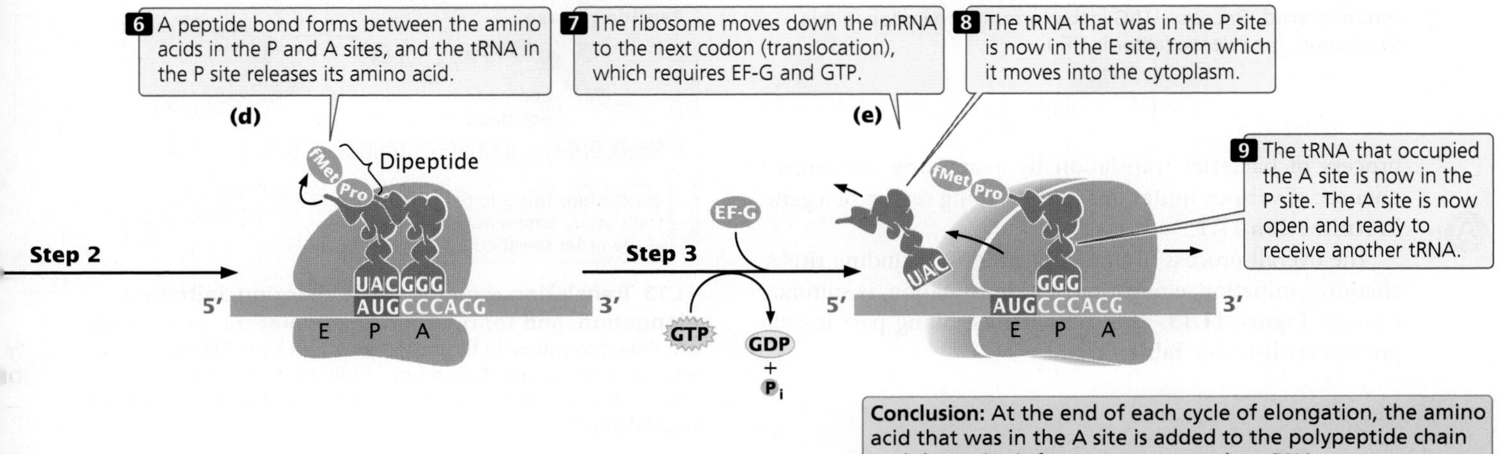

6 A peptide bond forms between the amino acids in the P and A sites, and the tRNA in the P site releases its amino acid.
7 The ribosome moves down the mRNA to the next codon (translocation), which requires EF-G and GTP.
8 The tRNA that was in the P site is now in the E site, from which it moves into the cytoplasm.
(d)
(e)
Dipeptide
fMet
Pro
Step 2
Step 3
EF-G
UAC GGG
AUGCCCACG
5′
3′
E P A
GTP
GDP
+
P_i
UAC
GGG
9 The tRNA that occupied the A site is now in the P site. The A site is now open and ready to receive another tRNA.

Conclusion: At the end of each cycle of elongation, the amino acid that was in the A site is added to the polypeptide chain and the A site is free to accept another tRNA.

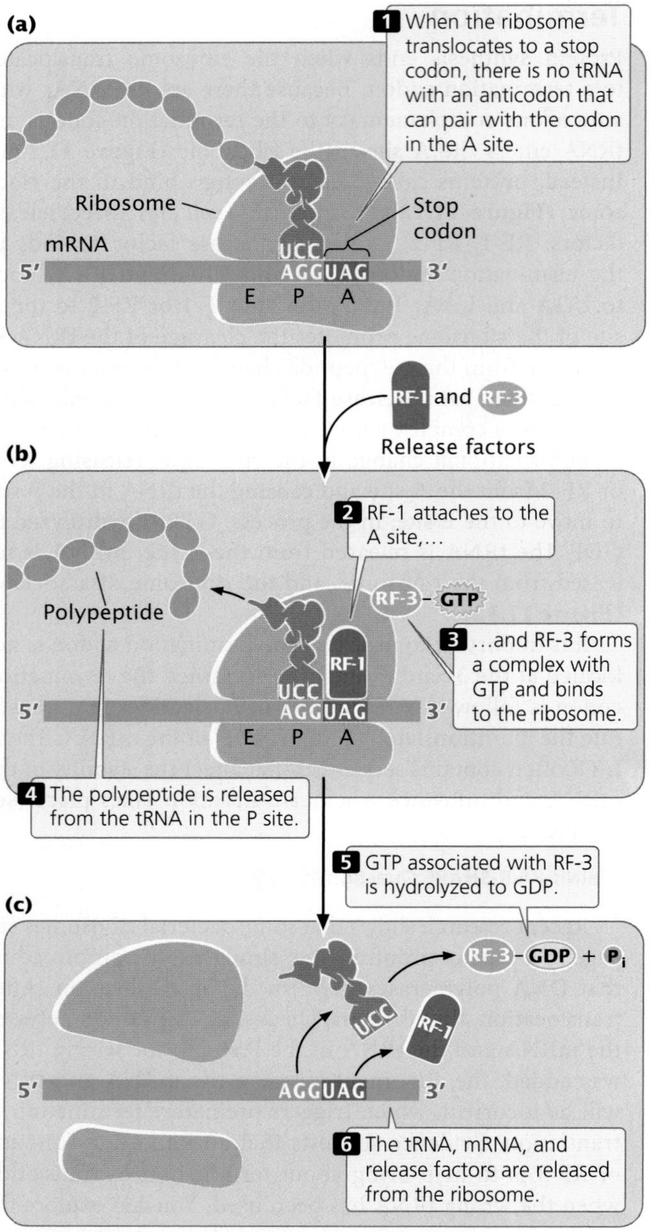

(a)

1 When the ribosome translocates to a stop codon, there is no tRNA with an anticodon that can pair with the codon in the A site.

Ribosome

Stop codon

mRNA

5′ UCC AGGUAG 3′

E P A

RF-1 and RF-3

Release factors

(b)

2 RF-1 attaches to the A site,…

RF-3 – GTP

3 …and RF-3 forms a complex with GTP and binds to the ribosome.

Polypeptide

RF-1

5′ UCC AGGUAG 3′

E P A

4 The polypeptide is released from the tRNA in the P site.

5 GTP associated with RF-3 is hydrolyzed to GDP.

(c)

RF-3 – GDP + P_i

UCC RF-1

5′ AGGUAG 3′

6 The tRNA, mRNA, and release factors are released from the ribosome.

11.12 Translation ends when a stop codon is encountered. Because UAG is the termination codon in this illustration, the release factor is RF-1.

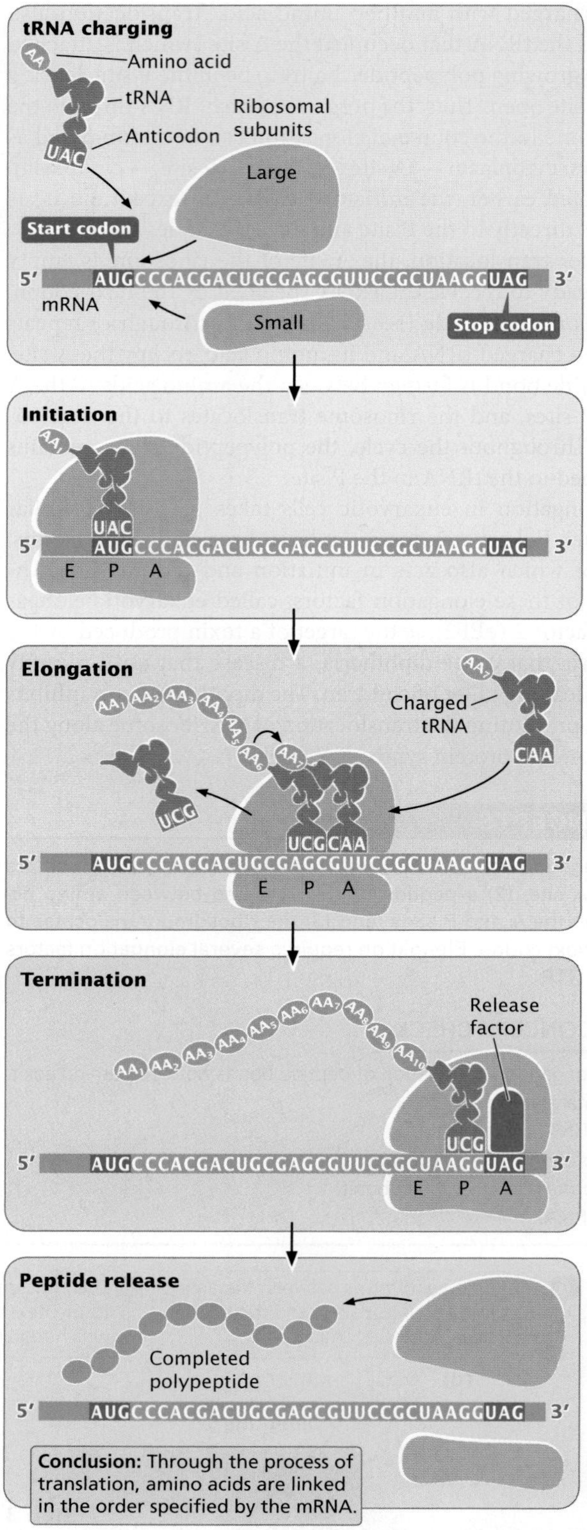

AA — Amino acid

— tRNA

Ribosomal subunits

— Anticodon

UAC

Large

Start codon

5′ AUGCCCACGACUGCGAGCGUUCCGCUAAGGUAG 3′

mRNA

Small

Stop codon

Initiation

AA₁

UAC

5′ AUGCCCACGACUGCGAGCGUUCCGCUAAGGUAG 3′

E P A

Elongation

AA₁ AA₂ AA₃ AA₄ AA₅ AA₆ AA₇

Charged tRNA

UCG

UCGCAA

CAA

5′ AUGCCCACGACUGCGAGCGUUCCGCUAAGGUAG 3′

E P A

Termination

AA₁ AA₂ AA₃ AA₄ AA₅ AA₆ AA₇ AA₈ AA₉ AA₁₀

Release factor

UCG

5′ AUGCCCACGACUGCGAGCGUUCCGCUAAGGUAG 3′

E P A

Peptide release

Completed polypeptide

5′ AUGCCCACGACUGCGAGCGUUCCGCUAAGGUAG 3′

Conclusion: Through the process of translation, amino acids are linked in the order specified by the mRNA.

11.13 Translation requires tRNA charging, initiation, elongation, and termination. In this process, amino acids are linked together in the order specified by mRNA to create a polypeptide chain. A number of initiation, elongation, and release factors take part in the process, and energy is supplied by ATP and GTP.

process of bacterial translation by examining the consequences of various mutations in the coding region of a gene in **Animation 11.1**.

The overall process of protein synthesis, including tRNA charging, initiation, elongation, and termination, is summarized in **Figure 11.13**. The components taking part in this process are listed in **Table 11.2**.

TABLE 11.2 Components required for protein synthesis in bacterial cells

Stage	Component	Function
tRNA charging	Amino acids	Building blocks of proteins
	tRNAs	Deliver amino acids to ribosomes
	Aminoacyl-tRNA synthetases	Attach amino acids to tRNAs
	ATP	Provides energy for binding amino acids to tRNAs
Initiation	mRNA	Carries coding instructions
	fMet-$tRNA^{fMet}$	Provides first amino acid in peptide
	30S ribosomal subunit	Attaches to mRNA
	50S ribosomal subunit	Stabilizes tRNAs and amino acids
	Initiation factor 1	Enhances dissociation of large and small subunits of ribosome
	Initiation factor 2	Binds GTP; delivers fMet-$tRNA^{fMet}$ to initiation codon
	Initiation factor 3	Binds to 30S subunit and prevents association with 50S subunit
Elongation	70S initiation complex	Functional ribosome with A, P, and E sites where protein synthesis takes place
	Charged tRNAs	Bring amino acids to ribosome and help assemble them in order specified by mRNA
	Elongation factor Tu	Binds GTP and charged tRNA; delivers charged tRNA to A site
	Elongation factor Ts	Regenerates active elongation factor Tu
	Elongation factor G	Stimulates movement of ribosome to next codon
	GTP	Provides energy
	rRNA in 50S ribosomal subunit	Creates peptide bond between amino acids in A site and P site
Termination	Release factors 1, 2, and 3	Bind to ribosome when stop codon is reached and terminate translation

CONCEPTS

Translation ends when the ribosome reaches a termination codon. Release factors bind to the termination codon, causing the release of the polypeptide from the last tRNA, of the tRNA from the ribosome, and of the mRNA from the ribosome.

CONNECTING CONCEPTS

A Comparison of Bacterial and Eukaryotic Translation

We have now considered the process of translation in bacterial cells and noted some distinctive differences that exist in eukaryotic cells. Let's reflect on some of the important similarities and differences between protein synthesis in bacterial and in eukaryotic cells.

First, we should emphasize that the genetic code of bacterial and eukaryotic cells is virtually identical; the only difference is in the amino acid specified by the initiation codon. In bacterial cells, AUG encodes a modified type of methionine, *N*-formylmethionine, whereas in eukaryotic cells, AUG encodes unformylated methionine. One consequence of the fact that bacteria and eukaryotes use the same code is that eukaryotic genes can be translated in bacterial systems, and vice versa; this feature makes genetic engineering possible, as we will see in Chapter 14.

Another difference is that transcription and translation take place simultaneously in bacterial cells, but the nuclear envelope separates these processes in eukaryotic cells. The physical separation of transcription and translation has important implications for the control of gene expression, which we will consider in Chapter 12, and it allows for extensive modification of eukaryotic mRNAs, as discussed in Chapter 10.

Yet another difference is that mRNA in bacterial cells is short-lived, typically lasting only a few minutes, but mRNA in eukaryotic cells can last for hours or days. The 5′ cap and 3′ poly(A) tail found on eukaryotic mRNAs add to their stability (see Chapter 10).

In both bacterial and eukaryotic cells, aminoacyl-tRNA synthetases attach amino acids to their appropriate tRNAs by the same chemical process. There are significant differences, however, in the sizes and compositions of bacterial and eukaryotic ribosomal subunits. For example, the large subunit of the eukaryotic ribosome contains three rRNAs, whereas that of the bacterial ribosome contains only two (see Table 10.4). These differences allow antibiotics and other substances to inhibit bacterial translation while having no effect on the translation of eukaryotic nuclear genes, as we will see later in this chapter.

Other fundamental differences lie in the process of initiation. In bacterial cells, the small subunit of the ribosome attaches directly to the region surrounding the start codon through hydrogen bonding between the Shine–Dalgarno consensus sequence in the 5′ untranslated region of the mRNA and a sequence at the 3′ end of the 16S rRNA. In contrast, the small subunit of a eukaryotic ribosome first binds to proteins attached to the 5′ cap on mRNA and then migrates down the mRNA, scanning the sequence until it encounters the first AUG initiation codon. Additionally, a larger number of initiation factors take part in eukaryotic initiation than in bacterial initiation.

Elongation and termination are similar in bacterial and eukaryotic cells, although different elongation and termination factors are used. In both types of organisms, mRNAs are translated multiple times and are simultaneously attached to several ribosomes, as we will see next.

11.3 Additional Properties of Translation and Proteins

Now that we have considered the process of translation in some detail, let's examine some additional aspects of protein synthesis.

Polyribosomes

In both prokaryotic and eukaryotic cells, mRNA molecules are translated simultaneously by multiple ribosomes (**Figure 11.14**). The resulting structure—an mRNA with several ribosomes attached—is called a **polyribosome** (or often just a polysome). Each ribosome successively attaches to the ribosome-binding site at the 5′ end of the mRNA and moves toward the 3′ end; the polypeptide associated with each ribosome becomes progressively longer as the ribosome moves along the mRNA. In prokaryotic cells, transcription and translation are simultaneous; multiple ribosomes may be attached to the 5′ end of the mRNA while transcription is still taking place at the 3′ end, as shown in Figure 11.14. In eukaryotes, transcription and translation are separated in time and space: transcription takes place in the nucleus and translation takes place in the cytoplasm.

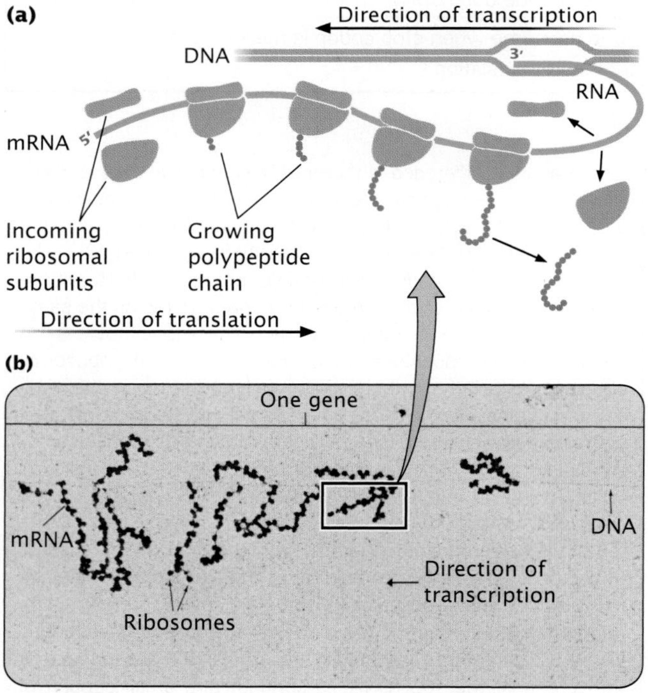

11.14 An mRNA molecule may be translated simultaneously by several ribosomes. (a) Four ribosomes are translating an mRNA molecule; the ribosomes are depicted as moving from the 5′ end to the 3′ end of the mRNA. (b) In this electron micrograph, the long horizontal filament is DNA, the dark-staining spheres are polyribosomes, and the thin filaments connecting the ribosomes are mRNAs. Transcription of the DNA is proceeding from right to left; the mRNAs on the right are shorter than those on the left. Each mRNA is being translated by multiple ribosomes. [Part b: From O.L. Miller, Jr., B.A. Hamkalo, and C.A. Thomas, Jr. "Visualization of Bacterial Genes in Action", Science 169(1970):392. Reprinted with permission from AAAS, permission conveyed through Copyright Clearance Center, Inc.]

CONCEPTS

In both prokaryotic and eukaryotic cells, multiple ribosomes may be attached to a single mRNA, generating a structure called a polyribosome.

✓ CONCEPT CHECK 7

In a polyribosome, the polypeptides associated with which ribosomes will be the longest?

a. Those at the 5′ end of mRNA
b. Those at the 3′ end of mRNA
c. Those in the middle of mRNA
d. All polypeptides will be the same length.

Folding and Posttranslational Modifications of Proteins

The functions of many proteins critically depend on the proper folding of the polypeptide chain. Some proteins spontaneously fold into their correct shapes, but for others, correct folding may initially require the participation of other molecules called **molecular chaperones**.

After translation, proteins in both prokaryotic and eukaryotic cells may undergo alterations termed posttranslational modifications. A number of different types of modifications are possible. Some proteins are synthesized as larger precursor proteins and must be cleaved and trimmed by enzymes before the proteins can become functional. For others, the attachment of carbohydrates may be required for activation. Amino acids within a protein may be modified: phosphates, carboxyl groups, and methyl groups are added to some amino acids. In eukaryotic cells, the amino end of a protein is often acetylated after translation.

THINK-PAIR-SHARE Question 11

CONCEPTS

Many proteins undergo posttranslational modifications after their synthesis.

Translation and Antibiotics

Antibiotics are drugs that kill bacteria. To make an effective antibiotic—not just any poison will do—the trick is to kill the bacteria without harming the patient.

Translation is frequently the target of antibiotics because it is essential to all living organisms and differs significantly between bacterial and eukaryotic cells. A number of antibiotics bind selectively to bacterial ribosomes and inhibit specific steps in translation, but they do not affect eukaryotic ribosomes. Tetracyclines, for instance, are a class of antibiotics that bind to the A site of a bacterial ribosome and block the entry of charged tRNAs, yet they have no effect on eukaryotic ribosomes. Chloramphenicol binds to the large subunit of the ribosome and blocks peptide-bond formation. Streptomycin binds to the small subunit of the ribosome and inhibits initiation, and erythromycin blocks translocation. Although chloramphenicol and streptomycin are potent inhibitors of translation in bacteria, they do not inhibit translation in archaea. Because different antibiotics block different steps in protein synthesis, antibiotics are frequently used to study protein synthesis.

CONCEPTS SUMMARY

- Amino acids in a protein are linked together by peptide bonds. Chains of amino acids fold and associate to produce the secondary, tertiary, and quaternary structures of proteins.
- The genetic code is a triplet code: three nucleotides specify a single amino acid. It is also degenerate (meaning that more than one codon may specify an amino acid), nonoverlapping, and almost universal.
- The reading frame is set by the initiation codon. The end of the protein-coding sequence of an mRNA is marked by one of three termination codons.
- Protein synthesis comprises four steps: (1) the binding of amino acids to the appropriate tRNAs, (2) initiation, (3) elongation, and (4) termination.
- The binding of an amino acid to tRNA requires the presence of a specific aminoacyl-tRNA synthetase and ATP.
- In bacterial translation initiation, the small subunit of the ribosome attaches to the mRNA and is positioned over the initiation codon. It is joined by the first tRNA and its associated amino acid (*N*-formylmethionine in bacterial cells) and, later, by the large subunit of the ribosome. Initiation requires several initiation factors and GTP.
- In elongation, a charged tRNA enters the A site of a ribosome, a peptide bond is formed between amino acids in the A and P sites, and the ribosome moves (translocates) along the mRNA to the next codon. Elongation requires several elongation factors and GTP.
- Translation is terminated when the ribosome encounters one of the three termination codons. Release factors and GTP are required to bring about termination.
- Each mRNA may be simultaneously translated by several ribosomes, producing a structure called a polyribosome.
- Many proteins undergo posttranslational modification.

IMPORTANT TERMS

one-gene, one-enzyme hypothesis (p. 300)
one-gene, one-polypeptide hypothesis (p. 300)
amino acid (p. 300)
peptide bond (p. 300)
polypeptide (p. 300)
sense codon (p. 303)
degenerate genetic code (p. 303)
synonymous codon (p. 303)
isoaccepting tRNA (p. 303)
wobble (p. 304)
nonoverlapping genetic code (p. 304)
reading frame (p. 305)
initiation (start) codon (p. 305)
stop (termination or nonsense) codon (p. 305)
universal genetic code (p. 305)
aminoacyl-tRNA synthetase (p. 306)
tRNA charging (p. 306)
initiation factor 1 (IF-1) (p. 307)
initiation factor 2 (IF-2) (p. 307)
initiation factor 3 (IF-3) (p. 307)
30S initiation complex (p. 307)
70S initiation complex (p. 307)
aminoacyl (A) site (p. 308)
peptidyl (P) site (p. 308)
exit (E) site (p. 308)
elongation factor Tu (EF-Tu) (p. 308)
elongation factor Ts (EF-Ts) (p. 308)
translocation (p. 308)
elongation factor G (EF-G) (p. 308)
release factor (p. 309)
polyribosome (p. 312)
molecular chaperone (p. 312)

ANSWERS TO CONCEPT CHECKS

1. The amino acid sequence (primary structure) of the protein

2. b

3. d

4. c

5. The Shine–Dalgarno sequence

6. a

7. b

WORKED PROBLEMS

Problem 1

If there were five different types of bases in mRNA instead of four, what would be the minimum codon size (number of nucleotides) required to specify the following numbers of different amino acid types: (a) 4, (b) 20, (c) 30?

Solution Strategy

What information is required in your answer to the problem?

The minimum codon size if there were five bases instead of four.

What information is provided to solve the problem?

- There are five different types of bases in mRNA.
- The number of different amino acids that need to be specified by the code.

For help with this problem, review:

Section 11.2.

Solution Steps

To answer this question, we must determine the number of combinations (codons) possible when there are different numbers of bases and different codon lengths. In general, the number of different codons possible will be equal to

$$b^{lg}$$

where *b* equals the number of different types of bases and *lg* equals the number of nucleotides in each codon (codon length). If there are five different types of bases, then

Hint: The n… of possible co… must be grea… than or equa… number of a… acids specifie…

$5^1 = 5$ possible codons

$5^2 = 25$ possible codons

$5^3 = 125$ possible codons

A codon length of one nucleotide could specify 4 different amino acids, a codon length of two nucleotides could specify 20 different amino acids, and a codon length of three nucleotides could specify 30 different amino acids: (a) one, (b) two, (c) three.

Problem 2

A template strand in bacterial DNA has the following base sequence

5′—AGGTTTAACGTGCAT—3′

What amino acids are encoded by this sequence?

Solution Strategy

What information is required in your answer to the problem?

The list of amino acids encoded by the given sequence.

What information is provided to solve the problem?

- The DNA sequence of the template strand.
- The 5′ and 3′ ends of the template sequence.
- The amino acids encoded by different codons (**Figure 11.5**).

For help with this problem, review:

The Degeneracy of the Code in Section 11.2.

Solution Steps

To answer this question, we must first work out the mRNA sequence that will be transcribed from this DNA sequence.

Recall: Th… is antiparallel… complementa… DNA templat…

DNA template strand: 5′—AGGTTTAACGTGCAT—3′
mRNA copied from DNA: 3′—UCCAAAUUGCACGUA—5′

An mRNA is translated 5′→3′; so it will be helpful if we turn the RNA molecule around with the 5′ end on the left:

mRNA copied from DNA: 5′—AUGCACGUUAAACCU—3′

The codons consist of groups of three nucleotides that are read successively after the first AUG codon; by referring to **Figure 11.5**, we can determine that the amino acids are

5′–AUG—CAC—GUU—AAA—CCU–3′
↓ ↓ ↓ ↓ ↓
fMet—His—Val—Lys—Pro

COMPREHENSION QUESTIONS

Section 11.1

1. What is the one-gene, one-enzyme hypothesis?
2. What are isoaccepting tRNAs?
3. What is the significance of the fact that many synonymous codons differ only in the third nucleotide position?
4. Define the following terms as they apply to the genetic code:
   a. Reading frame
   b. Overlapping code
   c. Nonoverlapping code
   d. Initiation codon
   e. Termination codon
   f. Sense codon
   g. Nonsense codon
   h. Universal code
   i. Nonuniversal codons
5. How is the reading frame of a nucleotide sequence set?

Section 11.2

6. How are tRNAs linked to their corresponding amino acids?
7. What role do the initiation factors play in protein synthesis?
8. What events bring about the termination of translation?
9. Compare and contrast the process of protein synthesis in bacterial and eukaryotic cells, giving similarities and differences in the process of translation in these two types of cells.

Section 11.3

10. What are some types of posttranslational modification of proteins?
11. Explain how some antibiotics work by affecting the process of protein synthesis.

For more questions that test your comprehension of the key chapter concepts, go to **LearningCurve** for this chapter.

APPLICATION QUESTIONS AND PROBLEMS

Section 11.1

*12. Assume that the number of different types of bases in RNA is four. What would be the minimum codon size (number of nucleotides) required to specify all amino acids if the number of different types of amino acids in proteins were (a) 2, (b) 8, (c) 17, (d) 45, (e) 75?

*13. How many codons would be possible in a triplet code if only three bases (A, C, and U) were used?

14. Using the genetic code presented in **Figure 11.5**, indicate which amino acid is encoded by each of the following mRNA codons.
   a. 5′—CCC—3′
   b. 5′—UUG—3′
   c. 5′—CUG—3′
   d. 5′—AGA—5′
   e. 5′—UAA—3′

*15. Referring to the genetic code presented in **Figure 11.5**, give the amino acids specified by the following bacterial mRNA sequences, and indicate the amino and carboxyl ends of the polypeptide produced. Hint: Remember that AUG is the initiation codon.
   a. 5′—AUGUUUAAAUUUAAAUUUUGA—3′
   b. 5′—AGGGAAAUCAGAUGUAUAUAUAUAUAUGA—3′
   c. 5′—UUUGGAUUGAGUGAAACGAUGGAUGAAAGAUUUCUCGCUUGA—3′
   d. 5′—GUACUAAGGAGGUUGUAUGGGUUAGGGGACAUCAUUUUGA—3′

16. A nontemplate strand of bacterial DNA has the following base sequence. What amino acid sequence will be encoded by this sequence?

5′—ATGATACTAAGGCCC—3′

*17. The following amino acid sequence is found in a tripeptide: Met-Trp-His. Give all possible nucleotide sequences on the mRNA, on the template strand of DNA, and on the nontemplate strand of DNA that can encode this tripeptide.

18. How many different mRNA sequences can encode a polypeptide chain with the amino acid sequence Met-Leu-Arg? (Be sure to include the stop codon.)

19. The following anticodons are found in a series of tRNAs. Refer to the genetic code in **Figure 11.5** and give the amino acid carried by each of these tRNAs.
   a. 5′—GUA—3′
   b. 5′—AUU—3′
   c. 5′—GGU—3′
   d. 5′—CCU—3′

20. Which of the following amino acid changes could result from a mutation that changed a single base? For each change that could result from the alteration of a single base, determine which position of the codon (first, second, or third nucleotide) in the mRNA must be altered for the change to result.
   a. Leu → Gln
   b. Phe → Ser
   c. Phe → Ile
   d. Pro → Ala
   e. Asn → Lys
   f. Ile → Asn

Section 11.2

*21. Arrange the following components of translation in the approximate order in which they would appear or be used in prokaryotic protein synthesis:

70S initiation complex
30S initiation complex
Release factor 1
Elongation factor Tu
Initiation factor 3
Elongation factor G
fMet-tRNA^{fMet}

22. The following diagram illustrates a step in the process of translation. Identify the following elements on the diagram.

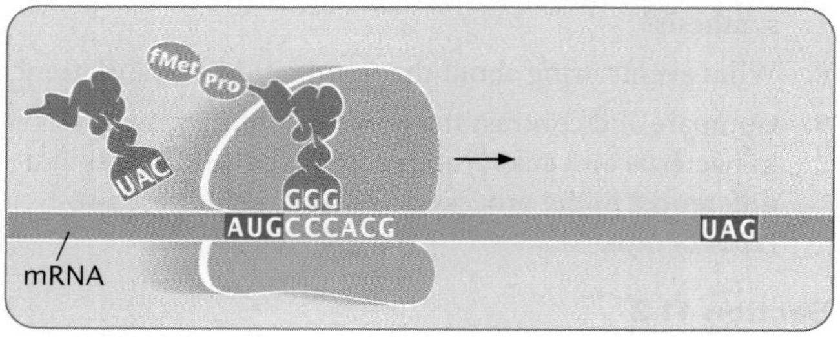

- **a.** 5′ and 3′ ends of the mRNA
- **b.** A, P, and E sites
- **c.** Start codon
- **d.** Stop codon
- **e.** Amino and carboxyl ends of the newly synthesized polypeptide chain
- **f.** Approximate location of the next peptide bond that will be formed
- **g.** Place on the ribosome where release factor 1 will bind

23. Refer to the diagram in Problem 22 to answer the following questions.

- **a.** What will be the anticodon of the next tRNA added to the A site of the ribosome?
- **b.** What will be the next amino acid added to the growing polypeptide chain?

***24.** A synthetic mRNA added to a cell-free protein-synthesizing system produces a peptide with the following amino acid sequence: Met-Pro-Ile-Ser-Ala. What would be the effect on translation if the following components were omitted from the cell-free protein-synthesizing system? What, if any, type of protein would be produced? Explain your reasoning.

- **a.** Initiation factor 3
- **b.** Initiation factor 2
- **c.** Elongation factor Tu
- **d.** Elongation factor G
- **e.** Release factors RF-1, RF-2, and RF-3
- **f.** ATP
- **g.** GTP

25. For each of the sequences in the following table, place a check mark in the appropriate space to indicate the process *most immediately* affected by deleting the sequence. Choose only one process for each sequence (i.e., one check mark per sequence).

Sequence deleted	Process most immediately affected by deletion: Replication	Transcription	RNA processing	Translation
a. *ori* site	____	____	____	____
b. 3′ splice-site consensus	____	____	____	____
c. Poly(A) tail	____	____	____	____
d. Terminator	____	____	____	____
e. Start codon	____	____	____	____
f. −10 consensus	____	____	____	____
g. Shine–Dalgarno	____	____	____	____

26. Give the amino acid sequence of the protein encoded by the mRNA in **Figure 11.13**.

CHALLENGE QUESTIONS

Section 11.1

27. The redundancy of the genetic code means that some amino acids are specified by more than one codon. For example, the amino acid leucine is encoded by six different codons. Within a genome, synonymous codons are not present in equal numbers; some synonymous codons appear much more frequently than others, and the preferred codons differ among different species. For example, in one species, the codon UUA might be used most often to encode leucine, whereas, in another species, the codon CUU might be used most often. Speculate on a reason for this bias in codon usage and why the preferred codons are not the same in all organisms.

Section 11.2

***28.** Several experiments were conducted to obtain information about how the eukaryotic ribosome recognizes the AUG start codon. In one experiment, the gene that encodes methionine initiator tRNA ($tRNA_i^{Met}$) was located and changed. The nucleotides that specify the anticodon on $tRNA_i^{Met}$ were mutated so that the anticodon in the tRNA was 5′—CCA—3′ instead of 5′—CAU—3′. When this mutated gene was placed in a eukaryotic cell, protein synthesis took place, but the proteins produced were abnormal. Some of the proteins produced contained extra amino acids, and others contained fewer amino acids than normal.

- **a.** What do these results indicate about how the ribosome recognizes the starting point for translation in eukaryotic cells? Explain your reasoning.
- **b.** If the same experiment had been conducted on bacterial cells, what results would you expect?
- **c.** Explain why some proteins contained extra amino acids while others contained fewer amino acids than normal.

THINK-PAIR-SHARE QUESTIONS

Introduction

1. Propose some possible reasons why mutations in the *RPSA* gene affect only the spleen and not other tissues where ribosomes carry out translation.
2. What are some possible reasons that researchers might be interested in identifying the gene that causes a genetic disease such as ICA? In other words, what benefits might result from this research?

Section 11.1

3. Archibald Garrod was an English physician who first proposed that genes encode enzymes. As with the work of Gregor Mendel (see Chapter 3), Garrod's discovery had little impact on his contemporaries and was not widely accepted until many years later. Why are important discoveries in science sometimes not accepted immediately? Why does it often take years before they are generally accepted by other scientists?
4. Do you think the one gene, one polypeptide hypothesis is correct today? Why or why not?
5. Compare and contrast the structure and function of proteins and nucleic acids. How are they similar and how are they different? How do their different structures contribute to their different functions?
6. A triplet code with three nucleotides per codon is the most efficient way to encode the 20 different amino acids. Why would cells be expected to use the most efficient code? In other words, what might be the advantages of using an efficient code?
7. Some evidence suggests that synonymous codons are not really synonymous. Sometimes mutations that change a codon to a synonymous codon produce an effect on the phenotype, even though the amino acid sequence of the protein is the same. Propose some ways in which synonymous codons might have different phenotypic effects.
8. There are three termination codons (UAA, UAG, UGA) but usually only one initiation codon (AUG) is used. Propose some possible reasons for why there are more termination codons than initiation codons.
9. Exceptions to the universality of the genetic code were once thought to be rare, but more and more exceptions are being found. For example, one study found that a significant proportion of bacteria in the environment use nonuniversal codons. Can you think of any advantages for an organism using nonuniversal codons?

Section 11.2

10. Structural analysis of bacterial release factor 1 (RF-1) and release factor 2 (RF-2) reveals that these proteins are similar in size and shape to a tRNA molecule. This similarity has sometimes been called molecular mimicry. Why might RF-1 and RF-2 have evolved to mimic tRNAs?

Section 11.3

11. In humans, there may be three times as many proteins as genes. If each gene encodes a protein, how can there be more proteins than genes?

12 Control of Gene Expression

Operons and the Noisy Cell

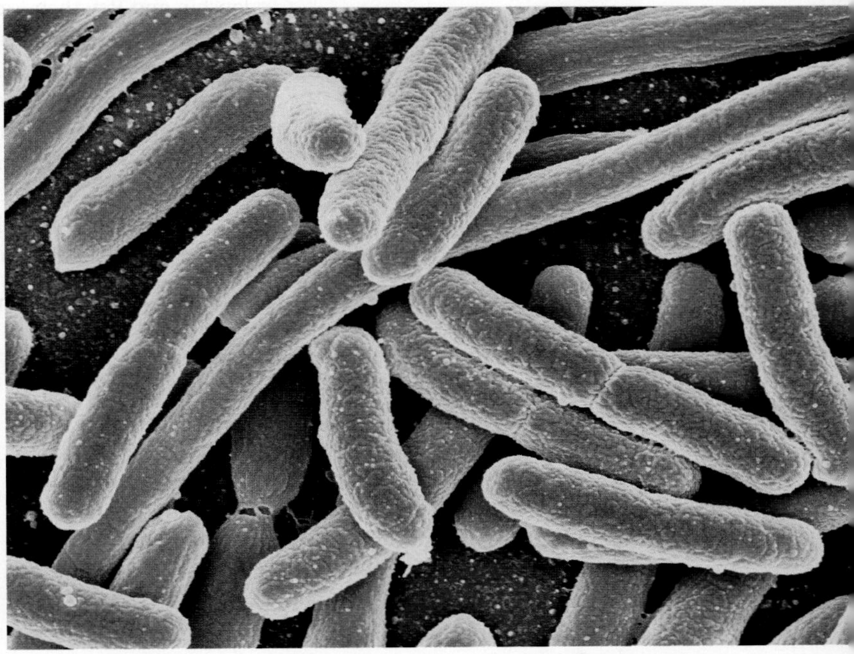

The expression of genes in bacteria is often regulated through operons, groups of genes that are transcribed as a unit. Shown here is *Escherichia coli*, a common bacterium found in the intestinal tracts of mammals. [Pasieka/Science Source.]

In 2011, geneticists from around the world celebrated the 50th anniversary of the operon. An operon is a group of genes that share a common promoter and are transcribed as a unit, producing a single mRNA molecule that encodes several proteins. Often, these proteins are functionally related in some way; for example, the *trp* operon in the bacterium *Escherichia coli* encodes components of three enzymes that work together to synthesize the amino acid tryptophan. Operons control the expression of genes such as these by regulating their transcription. Genes in bacteria and archaea are often organized into operons, but operons are much less common in eukaryotes.

The operon was discovered through the elegant research of François Jacob and Jacques Monod, who worked at opposite ends of the attic floor of the Pasteur Institute in Paris. Jacob was studying bacteriophage λ, a virus that infects *E. coli*; Monod was analyzing the properties of β-galactosidase, an enzyme *E. coli* uses to metabolize the sugar lactose. One summer evening in 1958, Jacob had a flash of inspiration: he saw a connection between the research taking place at the two ends of the attic. Jacob recognized that the genes that induce phage reproduction were controlled in the same way as the genes that control production of β-galactosidase in *E. coli*. This realization led to an important collaboration between Jacob and Monod, who eventually uncovered the structure and function of the *lac* operon in 1961. In 1965, they were awarded the Nobel Prize in Physiology or Medicine along with their collaborator Andre Lwoff.

Following Jacob and Monod's discovery of the *lac* operon, other operons were discovered in bacteria and a great deal of research was focused on the mechanism of operon function. Despite extensive research on *how* operons work, much less is known about *why* operons exist: Why do prokaryotes have them while eukaryotes don't? Why are some genes included in operons and others are not? These questions intrigued Oleg Igoshin, at Rice University, and Christian Ray, at the University of Texas MD Anderson Cancer Center. Igoshin and Ray are computational biologists, a new breed of scientists who use complex mathematics to study fundamental problems of biology. They took what might seem like an unlikely approach to the question of why operons exist. Instead of growing bacteria, inducing mutations, and examining DNA, they developed a series of mathematical models of gene networks that could be run on a computer. Using these models, they looked at how genes functioned when grouped into operons and when regulated separately.

Igoshin and Ray knew that random fluctuations in the levels of transcription and translation occur naturally. Because of these fluctuations—noise in the system—the amounts of different proteins can vary widely; the amount of a protein produced might be more or less than is optimal for cell growth and survival. Igoshin and Ray hypothesized that coordinating the transcription of several genes through an operon structure might reduce noise in the system and permit more finely tuned control over gene expression.

To test their hypothesis, Igoshin and Ray ran computer models for six different types of interactions between the products of the kinds of genes that are potentially found in operons. The models showed that for some types of protein interactions, grouping genes together in an operon decreases the biochemical noise. For other types of protein interactions, grouping the genes in operons actually increases the noise. Thus, operon structure has the potential to increase or decrease noise, depending on which genes are grouped together.

Igoshin and Ray then examined genes that are actually found in operons in *E. coli* and discovered that operons containing genes whose interactions decrease noise were more common than expected on a random basis. Conversely, operons whose gene interactions increase noise were less common than expected. They concluded that operons have evolved as a way for the cell to couple transcription of genes so as to reduce biochemical noise in the cell, allowing the cell to more finely tune the relative proportions of the proteins encoded by the operon. Igoshin and Ray speculated that operons are less common in eukaryotes because the larger volume of eukaryotic cells reduces the effect of random fluctuations and, perhaps, because eukaryotes have other mechanisms (such as changes in chromatin structure) to couple the transcription of different genes.

THINK-PAIR-SHARE Questions 1, 2, and 3

Gene regulation, which encompasses the mechanisms and systems that control the expression of genes, is critical for the control of numerous life processes in all organisms and is the focus of this chapter. We begin by considering the levels at which gene expression is controlled and the difference between genes and regulatory elements. We then examine gene regulation in bacteria, including the structure and function of operons, as well as gene regulation in eukaryotes. Next we consider some of the similarities and differences in gene regulation of bacteria and eukaryotes. Finally, we explore the role of epigenetic effects in the control of gene expression.

12.1 The Regulation of Gene Expression Is Critical for All Organisms

A major theme of molecular genetics is the central dogma, which states that genetic information flows from DNA to RNA to proteins (see Figure 8.14). Although the central dogma provided a molecular basis for the connection between genotype and phenotype, it failed to address a critical question: How is the flow of information along the molecular pathway *regulated*?

Consider *E. coli*, a bacterium that resides in your large intestine. Your eating habits completely determine the nutrients available to this bacterium: it can neither seek out nourishment when nutrients are scarce nor move away when confronted with an unfavorable environment. But *E. coli* makes up for its inability to alter the external environment by being internally flexible. For example, if glucose is present, *E. coli* uses it to generate ATP; if there's no glucose, it uses lactose, arabinose, maltose, xylose, or any of a number of other sugars. When amino acids are available, *E. coli* uses them to synthesize proteins; if a particular amino acid is absent, *E. coli* produces the enzymes needed to synthesize that amino acid. Thus, *E. coli* responds to environmental changes by rapidly altering its biochemistry. This biochemical flexibility, however, has a high price. Constantly producing all the enzymes necessary for every environmental condition would be energetically expensive. So how does *E. coli* maintain biochemical flexibility while optimizing energy efficiency?

The answer is gene regulation. Bacteria carry the genetic information for synthesizing many proteins, but only a subset of that genetic information is expressed at any time. When the environment changes, new genes are expressed, and proteins appropriate for the new environment are synthesized. For example, if a carbon source appears in the environment, genes encoding enzymes that take up and metabolize that carbon source are quickly transcribed and translated. When that carbon source disappears, the genes that encode the enzymes are shut off.

Multicellular eukaryotic organisms face a further challenge. Individual cells in a multicellular organism are specialized for particular tasks. The proteins produced by a nerve cell, for example, are quite different from those produced by a white blood cell. But although they differ in shape and function, a nerve cell and a blood cell still carry the same genetic instructions. A multicellular organism's challenge is to bring about the specialization of different cell types that have a common set of genetic instructions (the process of development). This challenge is met through gene regulation: all of an organism's cells carry the same genetic information, but only a subset of genes are expressed in each cell type. Genes needed for other cell types are not expressed. Gene regulation is therefore the key to both unicellular flexibility and multicellular specialization, and it is critical to the success of all living organisms.

CONCEPTS

In bacteria, gene regulation maintains internal flexibility, turning genes on and off in response to environmental changes. In multicellular eukaryotic organisms, gene regulation also brings about cellular differentiation.

The mechanisms of gene regulation were first investigated in bacterial cells, in which the availability of mutants and the ease of laboratory manipulation made it possible to unravel those mechanisms. When the study of gene regulation mechanisms in eukaryotic cells began, eukaryotic gene regulation seemed very different from bacterial gene regulation. However, as more and more information has accumulated about gene regulation, a number of common themes have emerged. Today, many aspects of gene regulation in bacterial and eukaryotic cells are recognized to be similar. Before examining specific elements of bacterial and eukaryotic gene regulation, we will briefly consider some themes of gene regulation common to all organisms.

Genes and Regulatory Elements

In considering gene regulation in both bacteria and eukaryotes, we must distinguish between the DNA sequences that are transcribed and the DNA sequences that regulate the expression of other sequences. **Structural genes** encode proteins that are used in metabolism or biosynthesis or that play a structural role in the cell. The products of **regulatory genes**, either RNA or proteins, interact with other DNA sequences and affect the transcription or translation of those sequences. In many cases, the products of regulatory genes are DNA-binding proteins or RNA molecules that affect gene expression. Bacteria and eukaryotes use regulatory genes to control the expression of many of their structural genes. However, a few structural genes, particularly those that encode essential cellular functions, are expressed continually and are therefore said to be **constitutive**. Constitutive genes are not regulated.

We will also encounter DNA sequences that are not transcribed at all, but still play a role in regulating genes and other nucleotide sequences. These **regulatory elements** affect the expression of DNA sequences to which they are physically linked. Regulatory elements are common in both bacterial and eukaryotic cells, and much of gene regulation in both types of organisms takes place through the action of proteins produced by regulatory genes that recognize and bind to regulatory elements.

The regulation of gene expression can occur through processes that stimulate gene expression, termed *positive control*, or through processes that inhibit gene expression, termed *negative control*. Bacteria and eukaryotes use both positive and negative control mechanisms to regulate their genes.

THINK-PAIR-SHARE Question 4

CONCEPTS

Regulatory elements are DNA sequences that are not transcribed, but affect the expression of genes. Positive control mechanisms stimulate gene expression, whereas negative control mechanisms inhibit gene expression.

✓ CONCEPT CHECK 1

What is a constitutive gene?

Levels of Gene Regulation

In both bacteria and eukaryotes, genes can be regulated at a number of levels along the pathway of information flow from genotype to phenotype (**Figure 12.1**).

First, genes may be regulated through the alteration of DNA or chromatin structure; this type of gene regulation takes place primarily in eukaryotes. Modifications to DNA

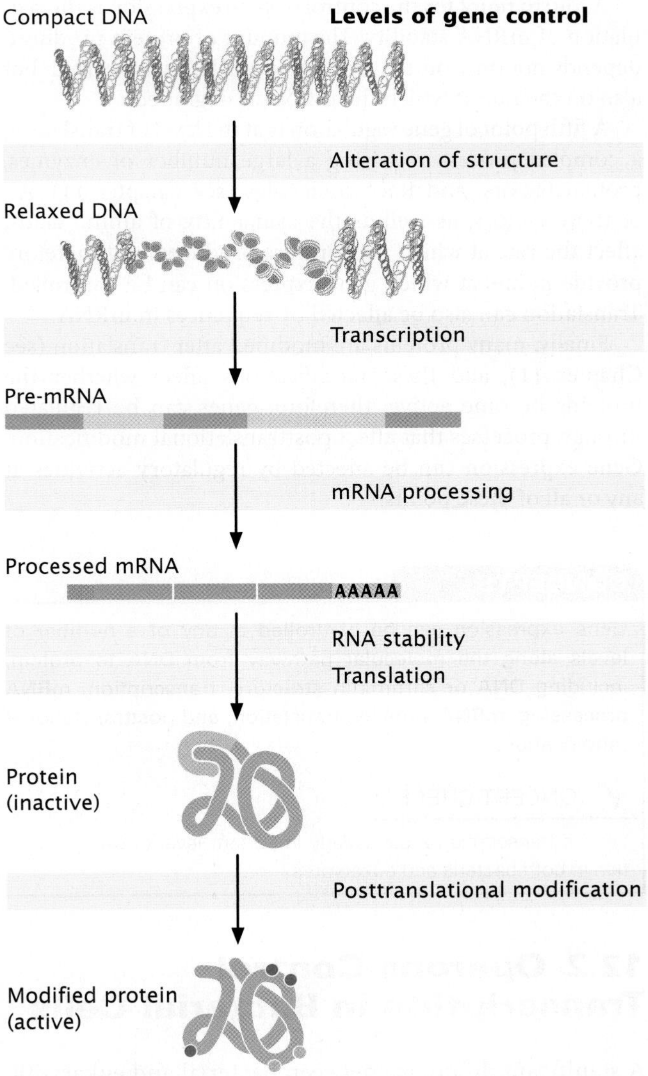

12.1 Gene expression can be controlled at multiple levels.

or its packaging may help to determine which sequences are available for transcription or the rate at which sequences are transcribed. DNA methylation and changes in chromatin are two processes that play a pivotal role in gene regulation.

A second point at which a gene can be regulated is at the level of transcription. For the sake of cellular economy, limiting the production of a protein early in the process makes sense, and transcription is an important point of gene regulation in both bacterial and eukaryotic cells.

A third potential point of gene regulation is mRNA processing. Eukaryotic mRNA is extensively modified before it is translated: a 5′ cap is added, the 3′ end is cleaved and polyadenylated, and introns are removed (see Chapter 10). These modifications determine the stability of the mRNA, the movement of the mRNA into the cytoplasm, whether the mRNA can be translated, the rate of translation, and the amino acid sequence of the protein produced. There is growing evidence that a number of regulatory mechanisms in eukaryotic cells operate at the level of mRNA processing.

A fourth point for the control of gene expression is the regulation of mRNA stability. The amount of protein produced depends not only on the amount of mRNA synthesized, but also on the rate at which the mRNA is degraded.

A fifth point of gene regulation is at the level of translation, a complex process requiring a large number of enzymes, protein factors, and RNA molecules (see Chapter 11). All of these factors, as well as the availability of amino acids, affect the rate at which proteins are produced and therefore provide points at which gene expression can be controlled. Translation can also be affected by sequences in mRNA.

Finally, many proteins are modified after translation (see Chapter 11), and these modifications affect whether the proteins become active; therefore, genes can be regulated through processes that affect posttranslational modification. Gene expression can be affected by regulatory activities at any or all of these points.

CONCEPTS

Gene expression can be controlled at any of a number of levels along the molecular pathway from DNA to protein, including DNA or chromatin structure, transcription, mRNA processing, mRNA stability, translation, and posttranslational modification.

✓ CONCEPT CHECK 2

Why is transcription a particularly important level of gene regulation in both bacteria and eukaryotes?

12.2 Operons Control Transcription in Bacterial Cells

A significant difference between bacterial and eukaryotic gene control lies in the organization of functionally related genes. As we saw in the introduction to this chapter, many bacterial genes that have related functions are clustered together and are under the control of a single promoter. These genes are often transcribed together into a single mRNA molecule. A group of bacterial structural genes that are transcribed together, along with their promoter and additional sequences that control their transcription, is called an **operon**. The operon regulates the expression of the structural genes by controlling transcription, which, in bacteria, is usually the most important level of gene regulation.

Operon Structure

The organization of a typical operon is illustrated in **Figure 12.2**. At one end of the operon is a set of structural genes, shown in Figure 12.2 as gene *a*, gene *b*, and gene *c*. These structural genes are transcribed into a single mRNA, which is translated to produce enzymes A, B, and C. These enzymes carry out a series of biochemical reactions that convert precursor molecule X into product Y. The transcription of structural genes *a*, *b*, and *c* is under the control of a promoter, which lies upstream of the first structural gene. RNA polymerase binds to the promoter and then moves downstream, transcribing the structural genes.

A **regulator gene** helps to control the expression of the structural genes of the operon by increasing or decreasing their transcription. Although it affects operon function, the regulator gene is not considered part of the operon. The regulator gene has its own promoter and is transcribed into a short mRNA, which is translated into a small protein. This **regulator protein** can bind to a region of the operon called the **operator** and affect whether transcription can take place. The operator usually overlaps the 3′ end of the promoter and sometimes the 5′ end of the first structural gene (see Figure 12.2).

CONCEPTS

Functionally related genes in bacterial cells are frequently clustered together in a single transcriptional unit termed an operon. A typical operon includes several structural genes, a promoter for those structural genes, and an operator to which the product of a regulator gene binds.

✓ CONCEPT CHECK 3

What is the difference between a structural gene and a regulator gene?

a. Structural genes are transcribed into mRNA, but regulator genes are not.
b. Structural genes have complex structures; regulator genes have simple structures.
c. Structural genes encode proteins that function in the structure of the cell; regulator genes carry out metabolic reactions.
d. Structural genes encode proteins; regulator genes control the transcription of structural genes.

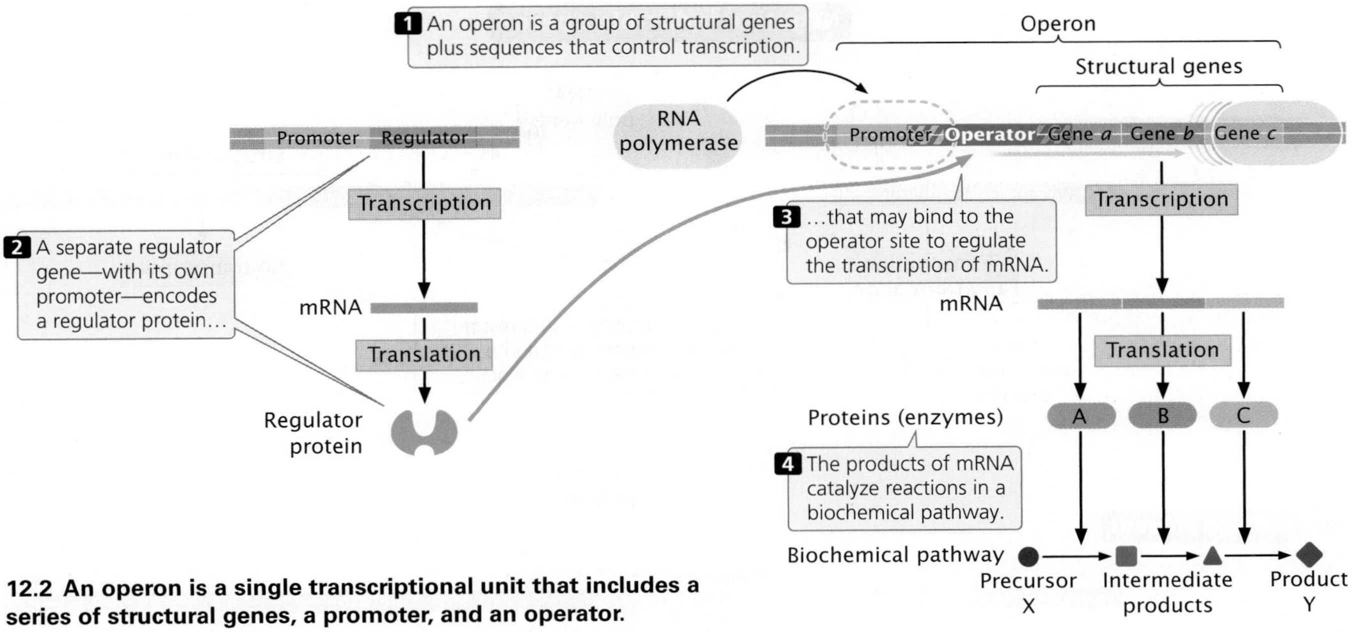

12.2 An operon is a single transcriptional unit that includes a series of structural genes, a promoter, and an operator.

Negative and Positive Control: Inducible and Repressible Operons

There are two types of transcriptional control: **negative control**, in which a regulatory protein is a repressor, binding to DNA and inhibiting transcription; and **positive control**, in which a regulatory protein is an activator, stimulating transcription. Operons can also be either inducible or repressible. **Inducible operons** are those in which transcription is normally off (not taking place); something must happen to induce transcription, or turn it on. **Repressible operons** are those in which transcription is normally on (taking place); something must happen to repress transcription, or turn it off. In the next sections, we will consider several varieties of these basic control mechanisms.

NEGATIVE INDUCIBLE OPERONS The regulator gene for a negative inducible operon encodes an active *repressor* protein that readily binds to the operator (**Figure 12.3a**). Because the operator overlaps the promoter, the binding of this protein to the operator physically blocks the binding of RNA polymerase to the promoter and prevents transcription. For transcription to take place, something must happen to prevent the binding of the repressor at the operator. This type of system is said to be *inducible* because transcription is normally off (inhibited) and must be turned on (induced).

Transcription is turned on when a small molecule called an **inducer** binds to the repressor (**Figure 12.3b**). Regulatory proteins frequently have two binding sites: one that binds to DNA and another that binds to a small molecule such as an inducer. The binding of the inducer (precursor V in Figure 12.3b) alters the shape of the repressor, preventing it from binding to DNA. Proteins of this type, which change shape upon binding to another molecule, are called **allosteric proteins**.

When the inducer is absent, the repressor binds to the operator, the structural genes are not transcribed, and enzymes D, E, and F (which metabolize precursor V) are not synthesized (see Figure 12.3a). This mechanism is an adaptive one: because no precursor V is available, synthesis of the enzymes would be wasteful because they would have no substrate to metabolize. As soon as precursor V becomes available, some of it binds to the repressor, rendering the repressor inactive and unable to bind to the operator. RNA polymerase can now bind to the promoter and transcribe the structural genes. The resulting mRNA is then translated into enzymes D, E, and F, which convert substrate V into product W (see Figure 12.3b). So, an operon with negative inducible control regulates the synthesis of the enzymes economically: the enzymes are synthesized only when their substrate (V) is available.

Inducible operons usually control proteins that carry out degradative processes—proteins that break down molecules. For these types of proteins, inducible control makes sense because the proteins are not needed unless the substrate (which is broken down by the proteins) is present.

NEGATIVE REPRESSIBLE OPERONS Some operons with negative control are *repressible*, meaning that transcription *normally* takes place and must be turned off, or repressed. The regulator protein that acts on this type of operon is also a repressor, but it is synthesized in an *inactive* form that cannot by itself bind to the operator. Because no repressor is bound to the operator, RNA polymerase readily

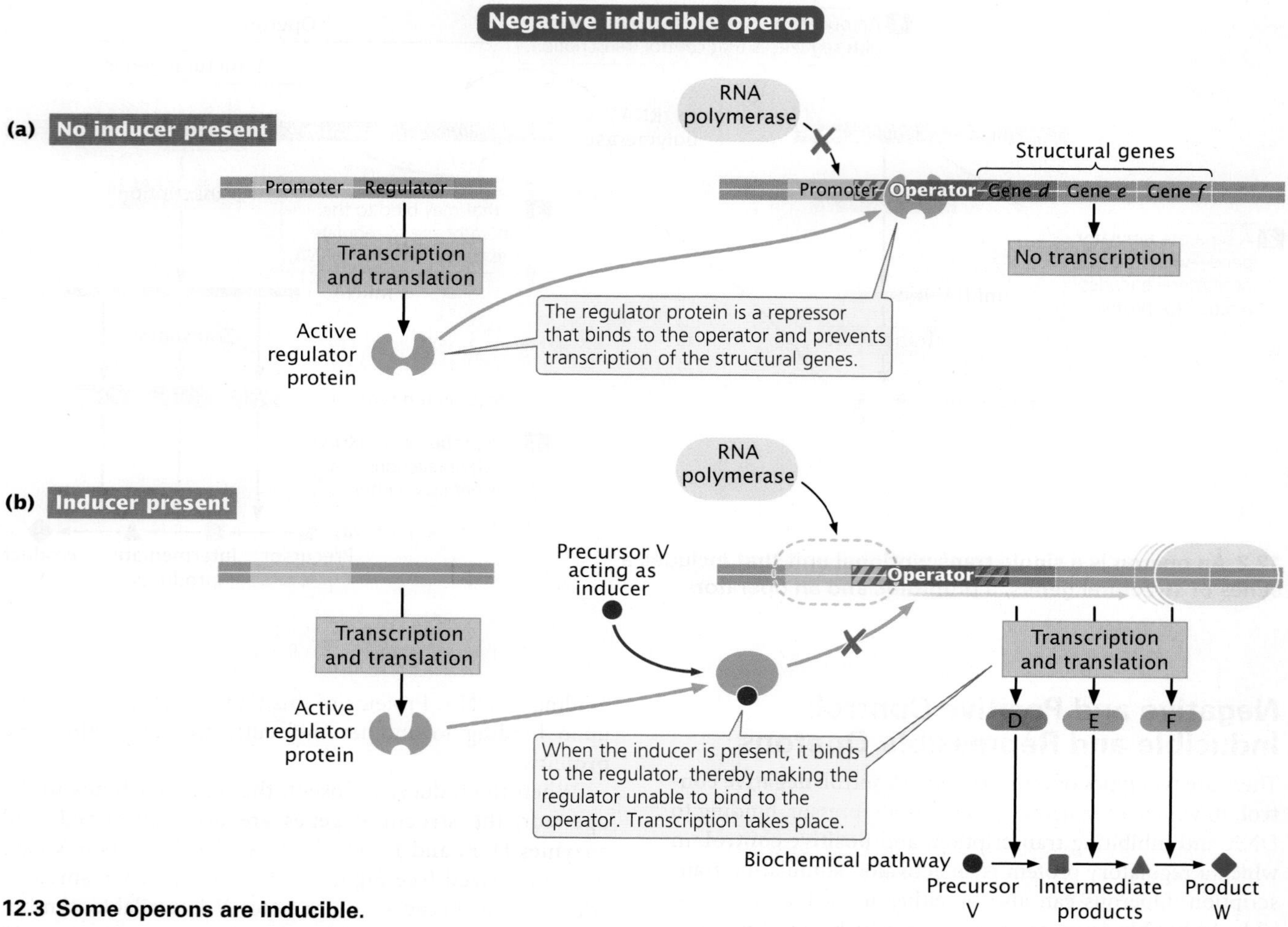

12.3 Some operons are inducible.

binds to the promoter and transcription of the structural genes takes place (**Figure 12.4a**).

To turn transcription off, something must happen to make the repressor active. A small molecule called a **corepressor** binds to the repressor and makes it capable of binding to the operator. In the example illustrated (see Figure 12.4a), the product (U) of the metabolic reaction controlled by the operon is the corepressor. As long as the level of product U is high, it is available to bind to the repressor and activate it, preventing transcription (**Figure 12.4b**). With the operon repressed, enzymes G, H, and I are not synthesized, and no more product U is produced from precursor T. However, when all of product U is used up, the repressor is no longer activated by product U and cannot bind to the operator. The inactivation of the repressor allows the transcription of the structural genes and the synthesis of enzymes G, H, and I, resulting in the conversion of precursor T into product U. Like inducible operons, repressible operons are economical: the proteins they encode are synthesized only as needed.

Repressible operons usually control enzymes that carry out the biosynthesis of molecules needed in the cell, such as amino acids. For these types of operons, repressible control makes sense because the product produced by the enzymes is always needed by the cell. Thus, these operons are normally turned on and are turned off only when there are adequate amounts of the product already present.

Note that both the inducible and the repressible operons that we have considered are forms of negative control, in which the regulator protein is a repressor. We will now consider positive control, in which a regulator protein stimulates transcription.

POSITIVE CONTROL With positive control, a regulatory protein is an activator: it binds to DNA (usually at a site other than the operator) and stimulates transcription. In the *lac* operon, for example, an activator protein binds to the promoter and increases the efficiency with which RNA polymerase binds to the promoter and transcribes the structural genes, as we'll see below. Positive control can be inducible or repressible.

In a positive *inducible* operon, transcription is normally turned off because the regulatory protein (an activator) is produced in an inactive form. Transcription takes place when an inducer becomes attached to and activates the regulatory protein. Logically, the inducer should be the precursor of the reaction controlled by the operon so that the necessary

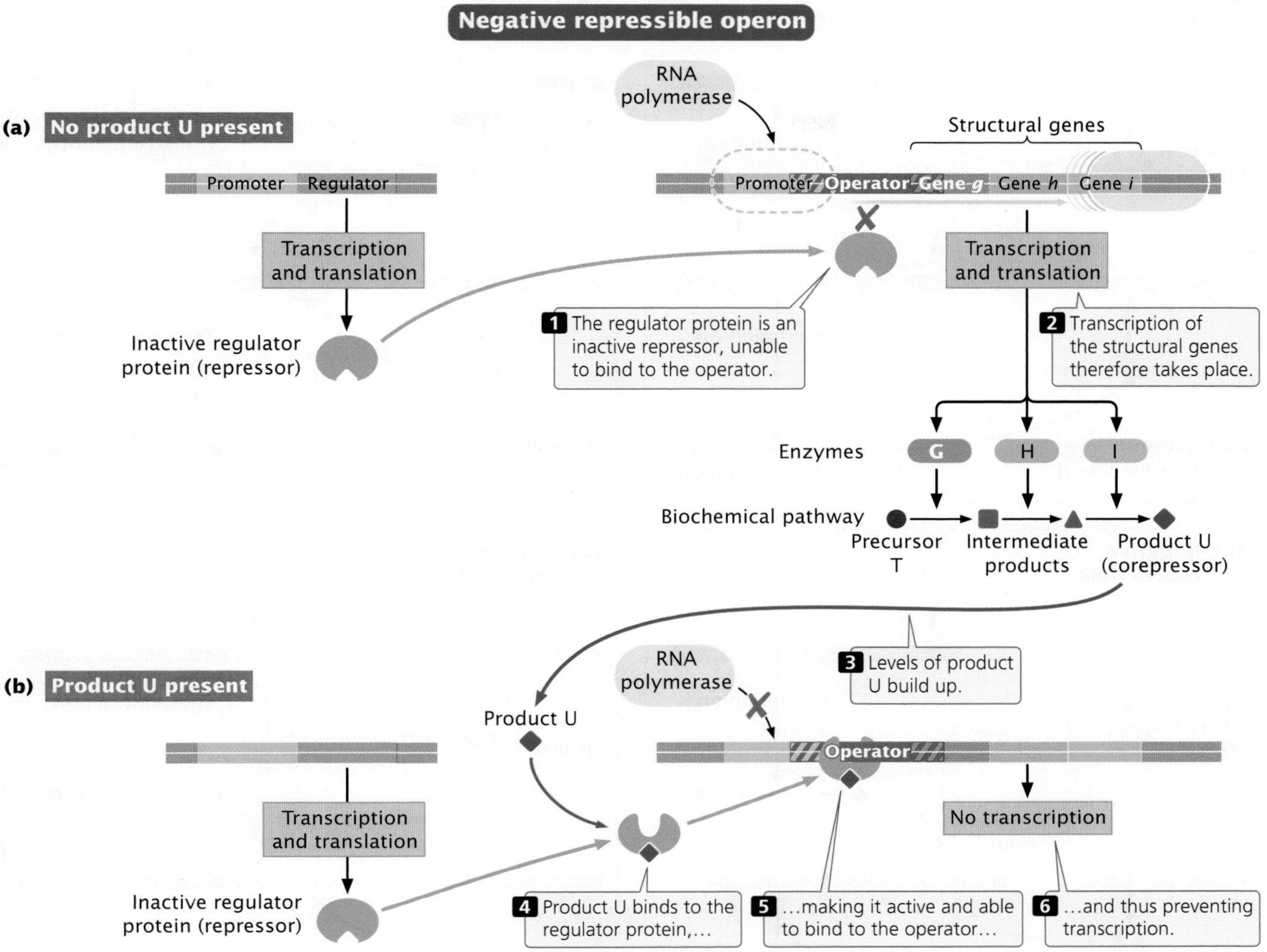

12.4 Some operons are repressible.

enzymes will be synthesized only when the substrate for their reaction is present.

A positive operon can also be repressible, in which case the regulatory protein is produced in a form that readily binds to DNA, meaning that transcription normally takes place and has to be repressed. In this case, the regulatory protein is produced in a form that readily binds to DNA and stimulates transcription. Transcription is inhibited when a substance becomes attached to the activator and renders it unable to bind to the DNA, so that transcription is no longer stimulated. Here, the product of the reaction controlled by the operon should logically be the repressing substance because prevention of the transcription of genes that allow the synthesis of that product when plenty of that product is already available would be economical for the cell. The characteristics of positive and negative control in inducible and repressible operons are summarized in **Figure 12.5**.

▶TRY PROBLEM 17

THINK-PAIR-SHARE Question 5

CONCEPTS

There are two basic types of transcriptional control: negative and positive. In negative control, a regulator protein is a repressor; when it binds to DNA, transcription is inhibited. In positive control, a regulatory protein is an activator; when it binds to DNA, transcription is stimulated. Some operons are inducible: their transcription is normally off and must be turned on. Other operons are repressible: their transcription is normally on and must be turned off.

✓ CONCEPT CHECK 4

In a negative repressible operon, the regulator protein is synthesized as

a. an active activator.
b. an inactive activator.
c. an active repressor.
d. an inactive repressor.

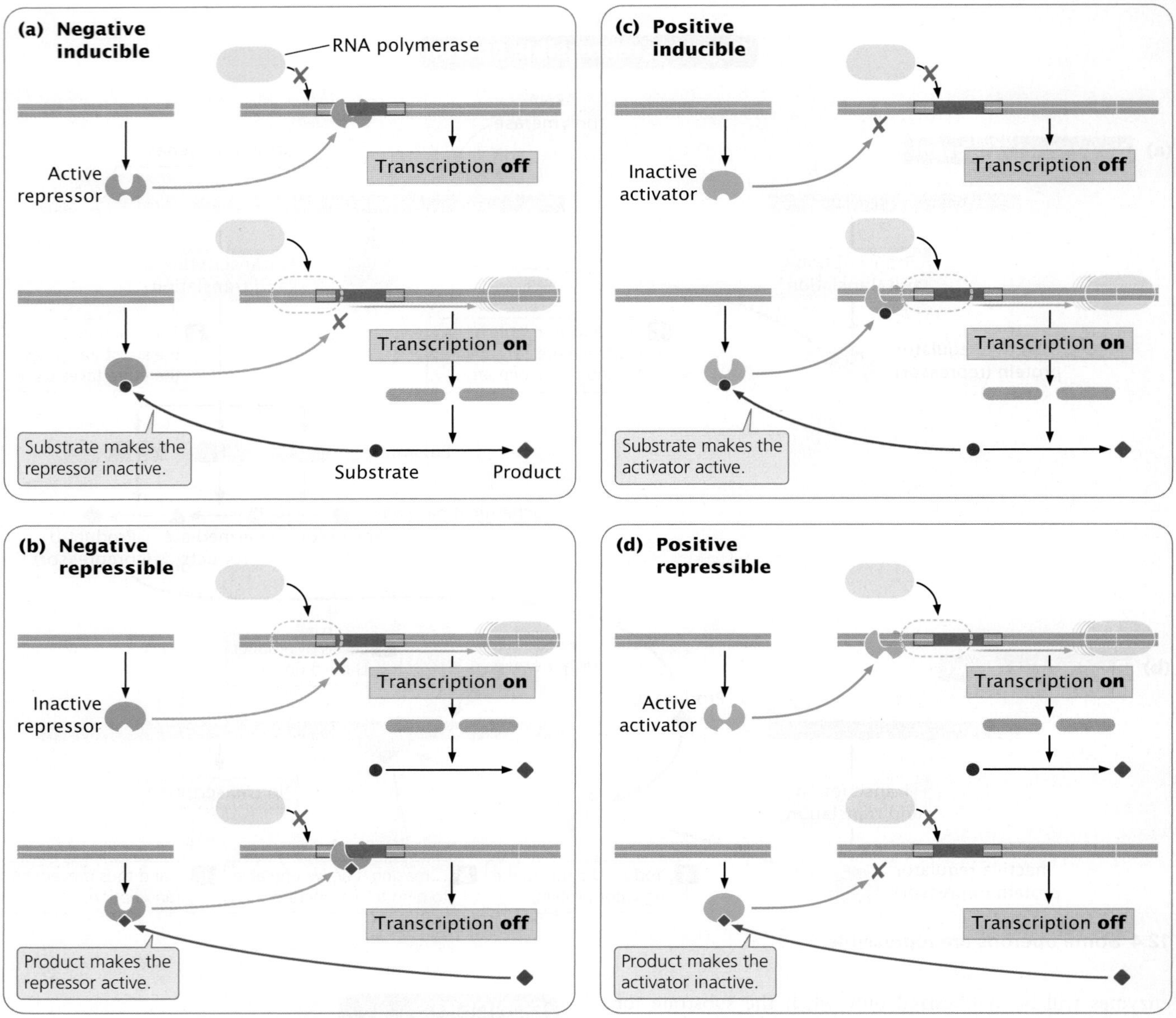

12.5 A summary of the characteristics of positive and negative control in inducible and repressible operons.

The *lac* Operon of *Escherichia coli*

As we saw at the start of this chapter, François Jacob and Jacques Monod first described the "operon model" of the genetic control of lactose metabolism in *E. coli*. Their work and subsequent research on the genetics of lactose metabolism established the operon as the basic unit of transcriptional control in bacteria. At the time, no methods were available for determining nucleotide sequences; Jacob and Monod deduced the structure of the operon *genetically* by analyzing the interactions of mutations that interfered with the normal regulation of lactose metabolism. We will examine the effects of some of these mutations after seeing how the *lac* operon regulates lactose metabolism.

LACTOSE METABOLISM Lactose is one of the major carbohydrates found in milk; it can be metabolized by *E. coli* bacteria that reside in the mammalian gut. Lactose does not easily diffuse across the *E. coli* cell membrane and must be actively transported into the cell by the protein permease (**Figure 12.6**). To use lactose as an energy source, *E. coli* must first break it into glucose and galactose, a reaction catalyzed by the enzyme β-galactosidase. This enzyme can also convert lactose into allolactose, a compound that plays an important role in regulating lactose metabolism. A third enzyme, thiogalactoside transacetylase, is also produced by the *lac* operon, but its function in lactose metabolism is not yet clear. One possible function is detoxification, preventing the accumulation of thiogalactosides that are transported into the cell along with lactose by lactose permease.

REGULATION OF THE *lac* OPERON The *lac* operon of *E. coli* is an example of a negative inducible operon.

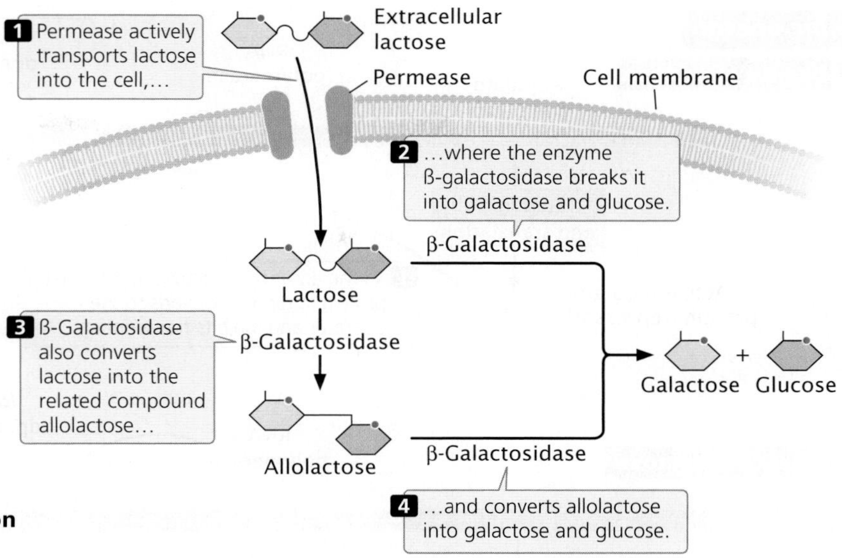

1 Permease actively transports lactose into the cell,...
Extracellular lactose
Permease
Cell membrane
2 ...where the enzyme β-galactosidase breaks it into galactose and glucose.
β-Galactosidase
Lactose
3 β-Galactosidase also converts lactose into the related compound allolactose...
β-Galactosidase
Galactose
Glucose
Allolactose
β-Galactosidase
4 ...and converts allolactose into galactose and glucose.

12.6 Lactose, a major carbohydrate found in milk, consists of two six-carbon sugars linked together.

The proteins β-galactosidase, permease, and transacetylase are encoded by adjacent structural genes in the *lac* operon (**Figure 12.7a**) and have a common promoter (*lacP* in Figure 12.7a). β-Galactosidase is encoded by the *lacZ* gene, permease by the *lacY* gene, and transacetylase by the *lacA* gene. When lactose is absent from the medium in which *E. coli* grows, few molecules of each protein are produced. If lactose is added to the medium and glucose is absent, the rate of synthesis of all three proteins simultaneously increases about a thousandfold within 2 to 3 minutes. This boost in protein synthesis, which results from the simultaneous transcription of *lacZ*, *lacY*, and *lacA*, exemplifies **coordinate induction**, stimulation of the simultaneous synthesis of several proteins by a specific molecule, the inducer (**Figure 12.7b**).

Although lactose might appear to be the inducer in this case, allolactose is actually responsible for induction. Upstream of *lacP* is a regulator gene, *lacI*, which has its own promoter (P_I). The *lacI* gene is transcribed into a short mRNA that is translated into a repressor. The repressor consists of four identical polypeptides and has two types of binding sites: one type of site binds to allolactose and the other binds to DNA. In the absence of lactose (and, therefore, allolactose), the repressor binds to the *lac* operator *lacO* (see Figure 12.7a). The location of the operator relative to the promoter and the *lacZ* gene is shown in **Figure 12.8**.

RNA polymerase binds to the promoter and moves down the DNA molecule, transcribing the structural genes. When the repressor is bound to the operator, the binding of RNA polymerase is blocked, and transcription is prevented. When lactose is present, some of it is converted into allolactose, which binds to the repressor and causes the repressor to be released from the DNA. In the presence of lactose, then, the repressor is inactivated, the binding of RNA polymerase is no longer blocked, the transcription of *lacZ*, *lacY*, and *lacA* takes place, and the *lac* operon proteins are produced.

Have you spotted the flaw in the explanation just given for the induction of the *lac* operon proteins? You might recall that permease is required to transport lactose into the cell. If the *lac* operon is repressed and no permease is being produced, how does lactose get into the cell to inactivate the repressor and turn on transcription? Furthermore, the inducer is actually allolactose, which must be produced from lactose by β-galactosidase. If β-galactosidase production is repressed, how can lactose metabolism be induced?

The answer is that repression never *completely* shuts down transcription of the *lac* operon. Even with active repressor bound to the operator, there is a low level of transcription, and a few molecules of β-galactosidase, permease, and transacetylase are synthesized. When lactose appears in the medium, the permease that is present transports a small amount of lactose into the cell. There, the few molecules of β-galactosidase that are present convert some of the lactose into allolactose, which then induces transcription.

THINK-PAIR-SHARE Question 6

CONCEPTS

The *lac* operon of *E. coli* controls the transcription of three genes needed in lactose metabolism: the *lacZ* gene, which encodes β-galactosidase; the *lacY* gene, which encodes permease; and the *lacA* gene, which encodes thiogalactoside transacetylase. The *lac* operon is a negative inducible operon: a regulator gene produces a repressor that binds to the operator and prevents the transcription of the structural genes. The presence of allolactose inactivates the repressor and allows the transcription of the *lac* operon.

✓ CONCEPT CHECK 5

In the presence of allolactose, the *lac* operon repressor

a. binds to the operator.
b. binds to the promoter.
c. cannot bind to the operator.
d. binds to the regulator gene.

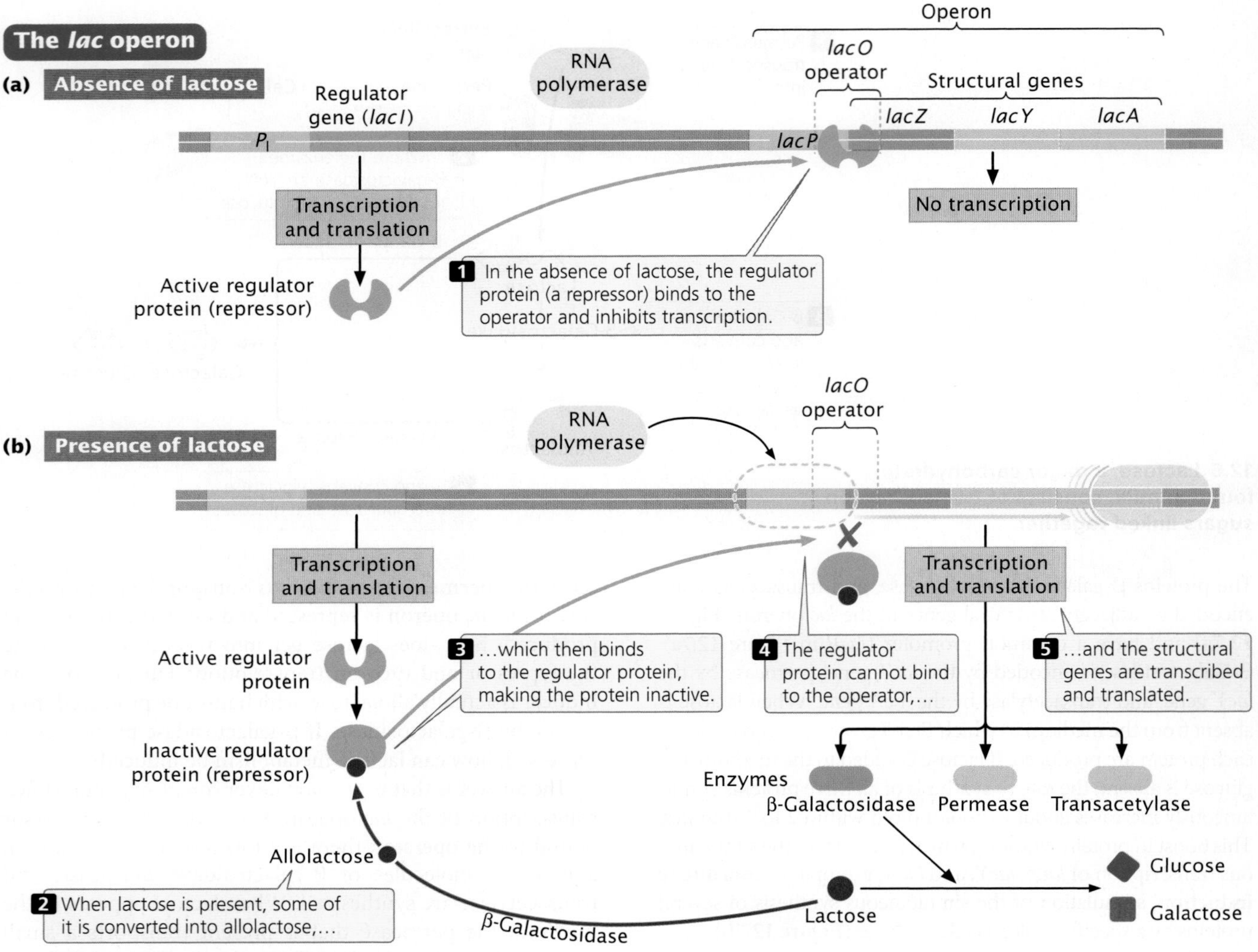

12.7 The *lac* operon regulates lactose metabolism.

Mutations Affecting the *lac* Operon

Jacob and Monod worked out the structure and function of the *lac* operon by analyzing mutations that affected lactose metabolism. To help define the roles of the different components of the operon, they used **partial diploid** strains of *E. coli*. The cells of these strains possessed two different DNA molecules: the full bacterial chromosome and an extra piece of DNA. Jacob and Monod created these strains by allowing conjugation to take place between two bacteria (see Chapter 7). In conjugation, a small circular piece of DNA (the F plasmid, see Chapter 7) is transferred from one bacterium to another. The F plasmid used by Jacob and Monod contained the *lac* operon, so the recipient bacterium became partly diploid, possessing two copies of the *lac* operon. By using different combinations of mutations on the bacterial and plasmid DNA, Jacob and Monod determined that some parts of the *lac* operon are cis acting (able to control

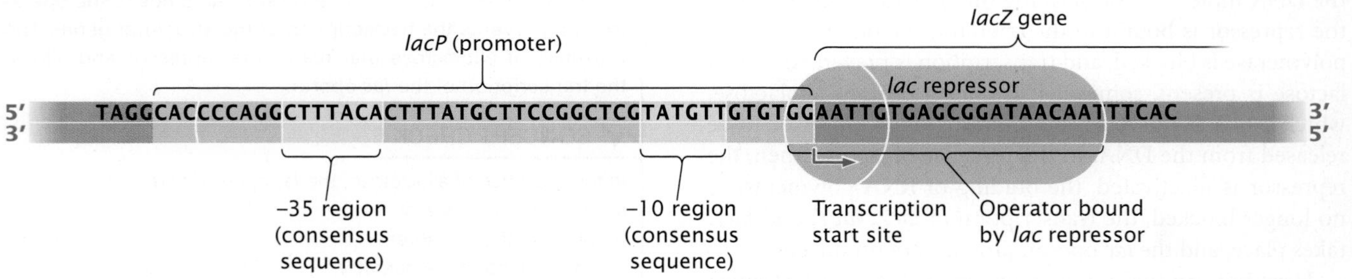

12.8 In the *lac* operon, the operator overlaps the promoter and the 5′ end of the first structural gene.

the expression of genes only when on the same piece of DNA), whereas other parts are trans acting (able to control the expression of genes on other DNA molecules).

STRUCTURAL-GENE MUTATIONS Jacob and Monod first discovered some mutant strains that had lost the ability to synthesize either β-galactosidase or permease. (They did not study the effects of mutations on the transacetylase enzyme in detail, so transacetylase will not be considered here.) The mutations in those mutant strains mapped to the *lacZ* or *lacY* structural genes and altered the amino acid sequences of the proteins encoded by these genes. These mutations clearly affected the *structure* of the proteins, but not the regulation of their synthesis.

Through the use of partial diploids, Jacob and Monod were able to establish that mutations at the *lacZ* and *lacY* genes were independent and usually affected only the product of the gene in which the mutation occurred. Partial diploids with $lacZ^+$ $lacY^-$ on the bacterial chromosome and $lacZ^-$ $lacY^+$ on the plasmid functioned normally, producing β-galactosidase and permease in the presence of lactose. (The genotype of a partial diploid is written by separating the genes on each DNA molecule with a slash: $lacZ^+$ $lacY^-/lacZ^-$ $lacY^+$.) In this partial diploid, a single functional β-galactosidase gene ($lacZ^+$) is sufficient to produce β-galactosidase; whether the functional β-galactosidase gene is coupled to a functional ($lacY^+$) or a defective ($lacY^-$) permease gene makes no difference. The same is true of the $lacY^+$ gene.

REGULATOR-GENE MUTATIONS Jacob and Monod also isolated mutations that affected the *regulation* of protein production. Mutations in the *lacI* gene affected the production of both β-galactosidase and permease because the genes for both proteins are in the same operon and are regulated coordinately by the *lac* repressor protein.

Some of these mutations were **constitutive mutations**, causing the *lac* proteins to be produced all the time, whether lactose was present or not. Such mutations in the regulator gene were designated $lacI^-$. The construction of partial diploids demonstrated that a $lacI^+$ gene is dominant over a $lacI^-$ gene; a single copy of $lacI^+$ (genotype $lacI^+/lacI^-$) was sufficient to bring about normal regulation of protein production. Furthermore, $lacI^+$ restored normal control to an operon even if the operon was located on a different DNA molecule, showing that $lacI^+$ can be trans acting. A partial diploid with genotype $lacI^+$ $lacZ^-/lacI^-$ $lacZ^+$ functioned normally, synthesizing β-galactosidase only when lactose was present (**Figure 12.9**). In this strain, the $lacI^+$ gene on the bacterial chromosome was functional, but the $lacZ^-$ gene was defective; on the plasmid, the $lacI^-$ gene was defective, but the $lacZ^+$ gene was functional. The fact that a $lacI^+$ gene could regulate a $lacZ^+$ gene located on a different DNA molecule indicated to Jacob and Monod that the $lacI^+$ gene product was able to operate on either the plasmid or the chromosome.

Some *lacI* mutations isolated by Jacob and Monod prevented transcription from taking place even in the presence of lactose. These mutations were referred to as superrepressors ($lacI^s$), because they produced defective repressors that could not be inactivated by an inducer. The $lacI^s$ mutations produced a repressor with an altered inducer-binding site, which made the inducer unable to bind to the repressor; consequently, the repressor was always able to attach to the operator and prevent transcription of the *lac* genes. Superrepressor mutations were dominant over $lacI^+$: partial diploids with genotype $lacI^s$ $lacZ^+/lacI^+$ $lacZ^+$ were unable to synthesize either β-galactosidase or permease, whether or not lactose was present (**Figure 12.10**).

OPERATOR MUTATIONS Jacob and Monod mapped a second set of constitutive mutations to a site adjacent to *lacZ*. These mutations occurred at the operator and were referred to as $lacO^c$ ("*O*" stands for operator and "c" for constitutive). The $lacO^c$ mutations altered the sequence of DNA at the operator so that the repressor protein was no longer able to bind to it. A partial diploid with genotype $lacI^+$ $lacO^c$ $lacZ^+/lacI^+$ $lacO^+$ $lacZ^+$ exhibited constitutive synthesis of β-galactosidase, indicating that $lacO^c$ is dominant over $lacO^+$.

Analyses of other partial diploids showed that the *lacO* gene is cis acting, affecting only genes on the same DNA molecule. For example, a partial diploid with genotype $lacI^+$ $lacO^+$ $lacZ^-/lacI^+$ $lacO^c$ $lacZ^+$ was constitutive, producing β-galactosidase in the presence or absence of lactose (**Figure 12.11a** on p. 331), but a partial diploid with $lacI^+$ $lacO^+$ $lacZ^+/lacI^+$ $lacO^c$ $lacZ^-$ produced β-galactosidase only in the presence of lactose (**Figure 12.11b**). In the constitutive partial diploid ($lacI^+$ $lacO^+$ $lacZ^-/lacI^+$ $lacO^c$ $lacZ^+$; see Figure 12.11a), the $lacO^c$ mutation and the functional $lacZ^+$ gene were present on the same DNA molecule, but in $lacI^+$ $lacO^+$ $lacZ^+/lacI^+$ $lacO^c$ $lacZ^-$ (see Figure 12.11b), the $lacO^c$ mutation and the functional $lacZ^+$ gene were on different molecules. Thus, the $lacO^c$ mutation affects only genes to which it is physically connected, as is true of all operator mutations. Such mutations prevent the binding of a repressor protein to the operator and thereby allow RNA polymerase to transcribe genes on the same DNA molecule. However, they cannot prevent a repressor from binding to normal operators on other DNA molecules. Watch **Animation 12.1** to observe the effects of different combinations of *lacI* and *lacO* mutations on the expression of the *lac* operon. ▶TRY PROBLEM 26

PROMOTER MUTATIONS Mutations affecting lactose metabolism have also been isolated at the promoter; these mutations, designated $lacP^-$, interfere with the binding of RNA polymerase to the promoter. Because this binding is essential for the transcription of the structural genes, *E. coli* strains with $lacP^-$ mutations don't produce *lac* proteins either in the presence or in the absence of lactose. Like

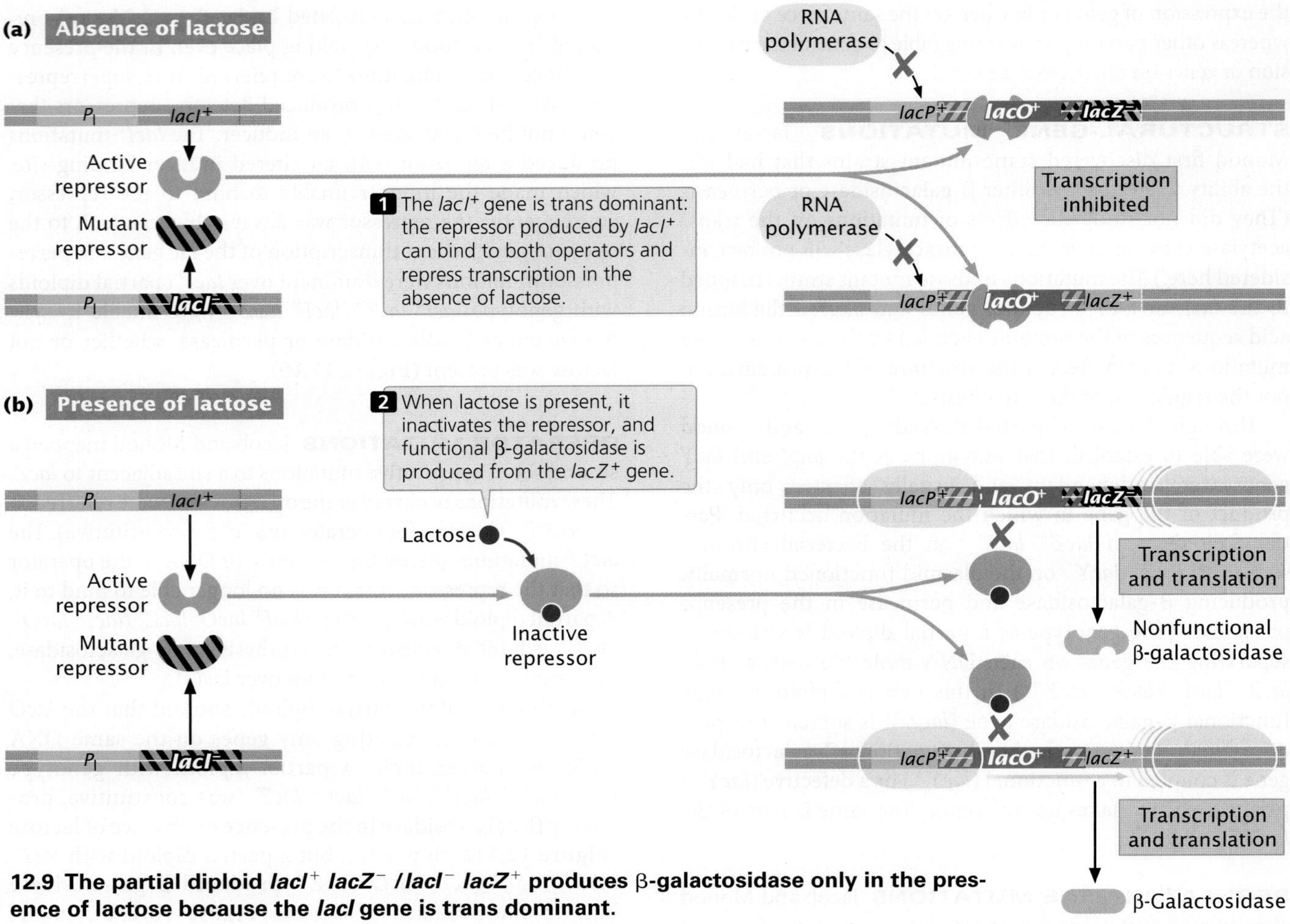

12.9 The partial diploid $lacI^+$ $lacZ^-$ / $lacI^-$ $lacZ^+$ produces β-galactosidase only in the presence of lactose because the *lacI* gene is trans dominant.

operator mutations, $lacP^-$ mutations are cis acting and thus affect only genes on the same DNA molecule. The partial diploid $lacI^+$ $lacP^+$ $lacZ^+$/$lacI^+$ $lacP^-$ $lacZ^+$ exhibits normal synthesis of β-galactosidase, whereas $lacI^+$ $lacP^-$ $lacZ^+$/$lacI^+$ $lacP^+$ $lacZ^-$ fails to produce β-galactosidase whether or not lactose is present.

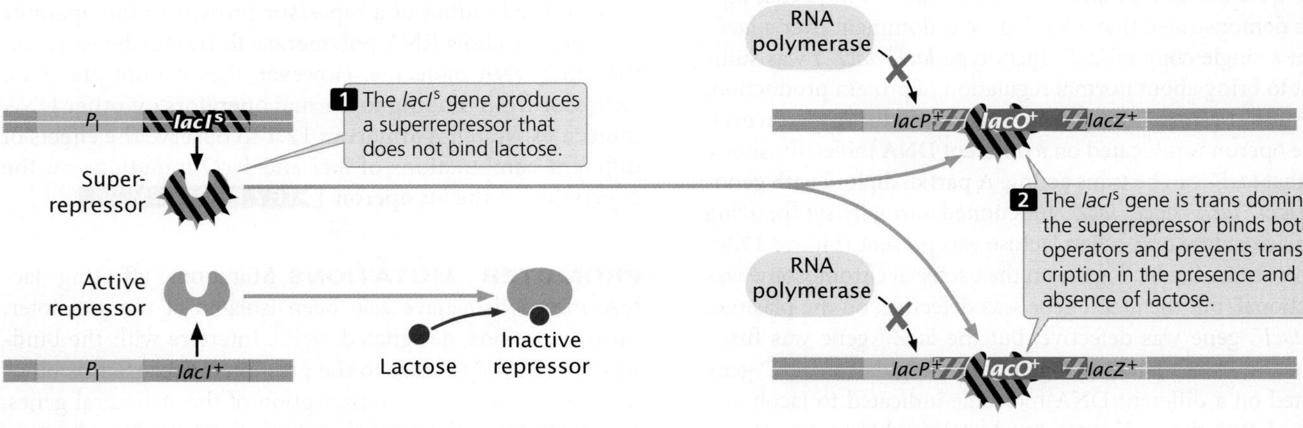

12.10 The partial diploid $lacI^s$ $lacZ^+$/$lacI^+$ $lacZ^+$ fails to produce β-galactosidase in the presence and absence of lactose because the $lacI^s$ gene encodes a superrepressor.

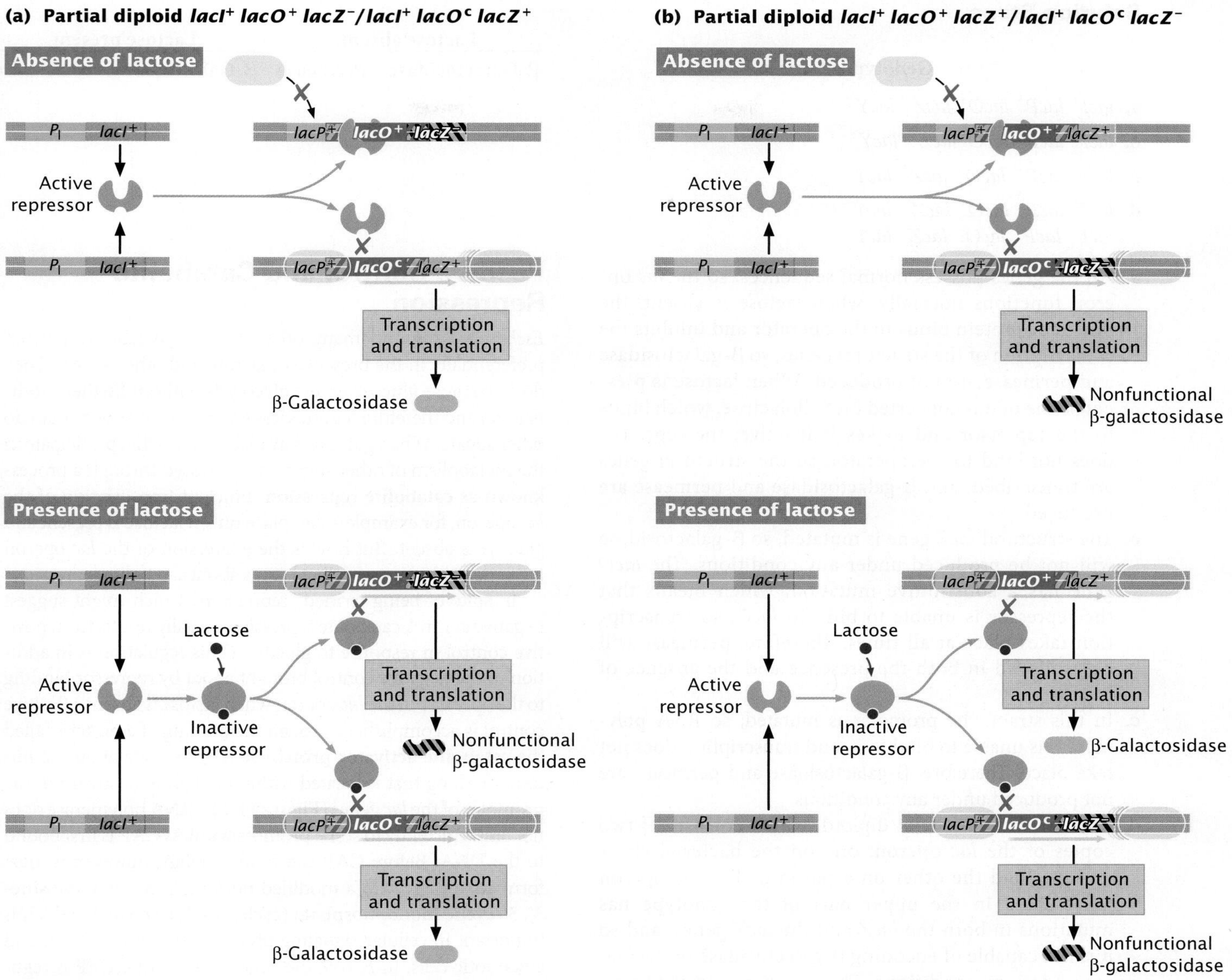

12.11 Mutations in *lacO* are constitutive and cis acting. (a) The partial diploid $lacI^+$ $lacO^+$ $lacZ^-$/$lacI^+$ $lacO^c$ $lacZ^+$ is constitutive, producing β-galactosidase in the presence and absence of lactose. (b) The partial diploid $lacI^+$ $lacO^+$ $lacZ^+$/$lacI^+$ $lacO^c$ $lacZ^-$ is inducible (produces β-galactosidase only when lactose is present), demonstrating that the *lacO* gene is cis acting.

WORKED PROBLEM

For *E. coli* strains with the following *lac* genotypes, make a table and use a plus sign (+) to indicate the synthesis of β-galactosidase and permease and a minus sign (−) to indicate no synthesis of the proteins when lactose is absent and when it is present.

Genotype of strain

a. $lacI^+$ $lacP^+$ $lacO^+$ $lacZ^+$ $lacY^+$
b. $lacI^+$ $lacP^+$ $lacO^c$ $lacZ^-$ $lacY^+$
c. $lacI^+$ $lacP^-$ $lacO^+$ $lacZ^+$ $lacY^-$
d. $lacI^+$ $lacP^+$ $lacO^+$ $lacZ^-$ $lacY^-$/
$lacI^-$ $lacP^+$ $lacO^+$ $lacZ^+$ $lacY^+$

Solution Strategy

What information is required in your answer to the problem?

An indication of whether or not β-galactosidase and permease are produced by each genotype when lactose is present and when lactose is absent by placing a plus sign (+) or minus sign (−) for each enzyme and condition in the table.

What information is provided to help solve the problem?

- The genotype of each strain.

Solution Steps

Genotype of strain	Lactose absent		Lactose present	
	β-Galactosidase	Permease	β-Galactosidase	Permease
a. $lacI^{+}\ lacP^{+}\ lacO^{+}\ lacZ^{+}\ lacY^{+}$	−	−	+	+
b. $lacI^{+}\ lacP^{+}\ lacO^{c}\ lacZ^{-}\ lacY^{+}$	−	+	−	+
c. $lacI^{+}\ lacP^{-}\ lacO^{+}\ lacZ^{+}\ lacY^{-}$	−	−	−	−
d. $lacI^{+}\ lacP^{+}\ lacO^{+}\ lacZ^{-}\ lacY^{-}$/ $lacI^{-}\ lacP^{+}\ lacO^{+}\ lacZ^{+}\ lacY^{+}$	−	−	+	+

a. All the genes possess normal sequences, so the *lac* operon functions normally: when lactose is absent, the regulator protein binds to the operator and inhibits the transcription of the structural genes, so β-galactosidase and permease are not produced. When lactose is present, some of it is converted into allolactose, which binds to the repressor and makes it inactive; the repressor does not bind to the operator, so the structural genes are transcribed, and β-galactosidase and permease are produced.

b. The structural *lacZ* gene is mutated, so β-galactosidase will not be produced under any conditions. The *lacO* gene has a constitutive mutation, which means that the repressor is unable to bind to *lacO*, so transcription takes place at all times. Therefore, permease will be produced in both the presence and the absence of lactose.

c. In this strain, the promoter is mutated, so RNA polymerase is unable to bind to it, and transcription does not take place. Therefore, β-galactosidase and permease are not produced under any conditions.

d. This strain is a partial diploid, which consists of two copies of the *lac* operon: one on the bacterial chromosome and the other on a plasmid. The *lac* operon represented in the upper part of the genotype has mutations in both the *lacZ* and the *lacY* genes, and so it is not capable of encoding β-galactosidase or permease under any conditions. The *lac* operon in the lower part of the genotype has a defective regulator gene, but the normal regulator gene in the upper operon produces a repressor that is capable of diffusing to other molecules (trans acting), so that it binds to the lower operon in the absence of lactose and inhibits transcription. Therefore, no β-galactosidase or permease is produced when lactose is absent. In the presence of lactose, the repressor cannot bind to the operator, and so the lower operon is transcribed and β-galactosidase and permease are produced.

▶ Now try your own hand at predicting the outcome of different *lac* mutations by working **Problem 24** at the end of the chapter.

Positive Control and Catabolite Repression

Escherichia coli and many other bacteria metabolize glucose preferentially in the presence of lactose and other sugars. They do so because glucose enters glycolysis without further modification and therefore requires less energy to metabolize than do other sugars. When glucose is available, genes that participate in the metabolism of other sugars are repressed through a process known as **catabolite repression.** Efficient transcription of the *lac* operon, for example, takes place only if lactose is present and glucose is absent. But how is the expression of the *lac* operon influenced by glucose? What brings about catabolite repression?

In spite of being termed "repression", which might suggest negative control, catabolite repression actually results from positive control in response to glucose. (This regulation is in addition to the negative control brought about by repressor binding to the operator of the *lac* operon when lactose is absent.) Positive control is accomplished through the binding of a protein called the **catabolite activator protein (CAP)** to a site about 22 nucleotides long that is located within or slightly upstream of the promoter of the *lac* genes (**Figure 12.12**). RNA polymerase does not bind efficiently to some promoters unless CAP is first bound to the DNA. Before CAP can bind to DNA, however, it must form a complex with a modified nucleotide called **adenosine-3′, 5′-cyclic monophosphate** (cyclic AMP, or **cAMP**), which is important in cellular signaling processes in both bacterial and eukaryotic cells. In *E. coli*, the concentration of cAMP is regulated so that it is inversely proportional to the level of available glucose. High concentrations of glucose within the cell lower the amount of cAMP, so little cAMP–CAP complex is available to bind to the DNA. Consequently, RNA polymerase has poor affinity for the *lac* promoter, and little transcription of the *lac* operon takes place. Low concentrations of glucose stimulate high levels of cAMP, resulting in increased cAMP–CAP binding to DNA. This increase enhances the binding of RNA polymerase to the promoter and increases transcription of the *lac* genes by approximately 50-fold. ▶TRY PROBLEM 20

THINK-PAIR-SHARE Question 7

CONCEPTS

In spite of its name, catabolite repression is a type of positive control of the *lac* operon. The catabolite activator protein (CAP), complexed with cAMP, binds to a site near the promoter and stimulates the binding of RNA polymerase. Cellular levels of cAMP in the cell are controlled by glucose; a low glucose level increases the abundance of cAMP and enhances the transcription of the *lac* structural genes.

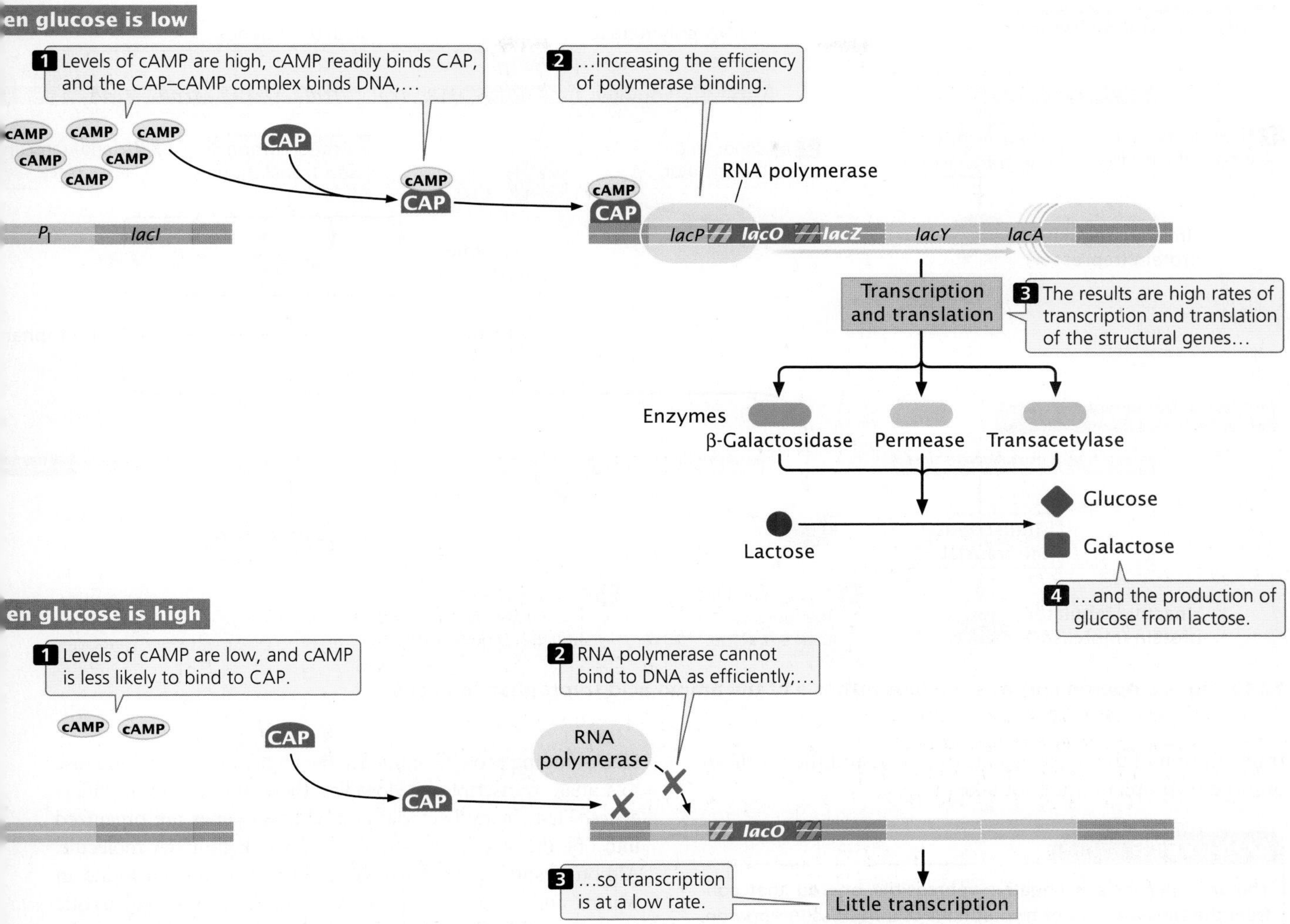

12.12 The catabolite activator protein (CAP) binds to the promoter of the *lac* operon and stimulates transcription. CAP must complex with adenosine-3′,5′-cyclic monophosphate (cAMP) before binding to the promoter of the *lac* operon. The binding of cAMP–CAP to the promoter activates transcription by facilitating the binding of RNA polymerase. Levels of cAMP are inversely related to glucose: low glucose stimulates high cAMP; high glucose stimulates low cAMP.

The *trp* Operon of *E. coli*

The *lac* operon just discussed is an inducible operon, one in which transcription does not normally take place and must be turned on. Other operons are repressible; transcription in these operons is normally turned on and must be repressed. The tryptophan (*trp*) operon in *E. coli*, which controls the biosynthesis of the amino acid tryptophan, is an example of a negative repressible operon.

The *trp* operon contains five structural genes (*trpE*, *trpD*, *trpC*, *trpB*, and *trpA*) that produce the components of three enzymes (two of the enzymes consist of two polypeptide chains). These enzymes convert chorismate into tryptophan (**Figure 12.13**). Upstream of the structural genes is the *trp* promoter. When tryptophan levels are low, RNA polymerase binds to the promoter and transcribes the five structural genes into a single mRNA, which is then translated into the enzymes that convert chorismate into tryptophan.

Some distance from the *trp* operon is a regulator gene, *trpR*, which encodes a repressor that is normally inactive (see Figure 12.13). Like the *lac* repressor, the tryptophan repressor has two binding sites, one that binds to DNA at the operator and another that binds to tryptophan (the corepressor). Binding to tryptophan causes a conformational change in the repressor that makes it capable of binding to the operator, which overlaps the promoter. When the operator is occupied by the repressor, RNA polymerase cannot bind to the promoter, and the structural genes cannot be transcribed. Thus, when cellular levels of tryptophan are low, transcription of the *trp* operon takes place and more tryptophan is synthesized; when cellular levels of tryptophan are high,

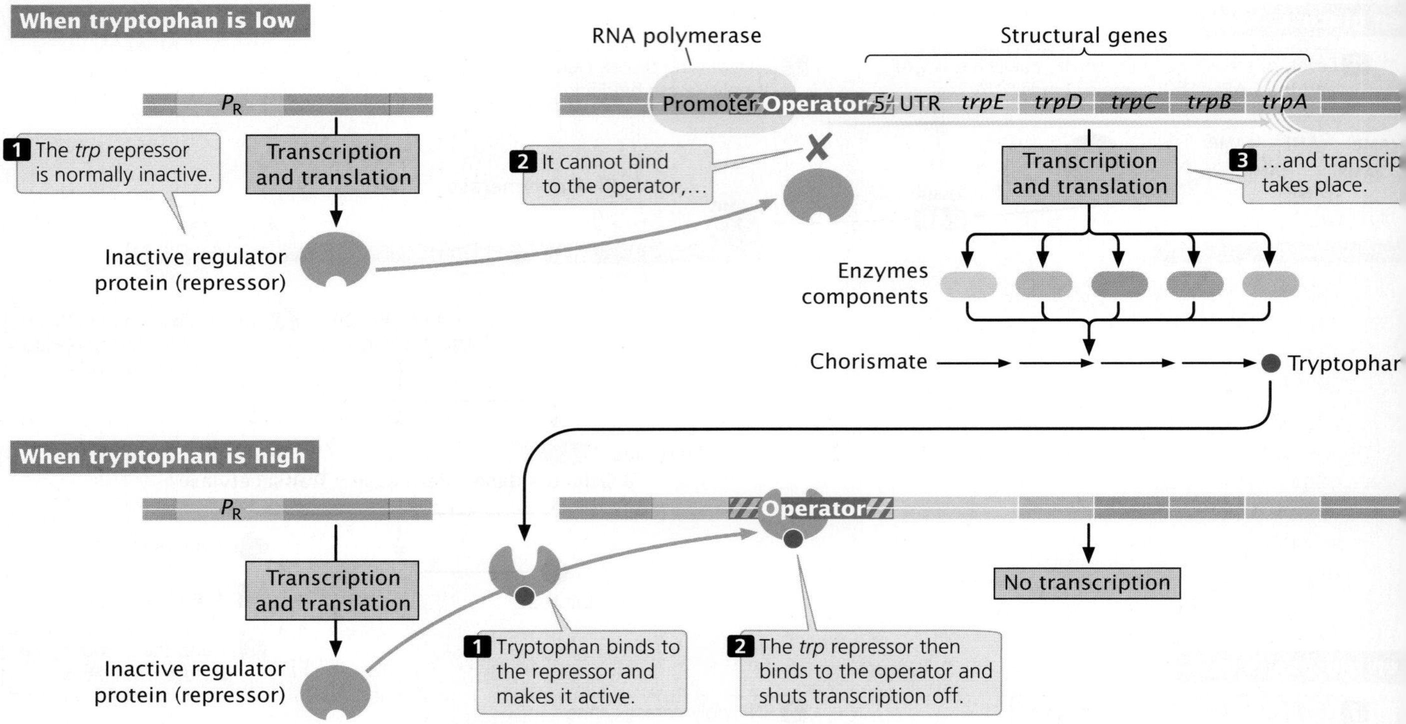

12.13 The *trp* operon controls the biosynthesis of the amino acid tryptophan in *E. coli*.

transcription of the *trp* operon is inhibited and the synthesis of more tryptophan does not take place.

CONCEPTS

The *trp* operon is a negative repressible operon that controls the biosynthesis of tryptophan. In a repressible operon, transcription is normally turned on and must be repressed: this is accomplished through the binding of tryptophan to the repressor, which renders the repressor active. The active repressor binds to the operator and prevents RNA polymerase from transcribing the structural genes.

✓ CONCEPT CHECK 6

In the *trp* operon, what happens to the *trp* repressor in the absence of tryptophan?

a. It binds to the operator and represses transcription.
b. It cannot bind to the operator and transcription takes place.
c. It binds to the regulator gene and represses transcription.
d. It cannot bind to the regulator gene and transcription takes place.

12.3 Gene Regulation in Eukaryotic Cells Takes Place at Multiple Levels

Many features of gene regulation are common to both bacterial and eukaryotic cells. For example, in both types of cells, DNA-binding proteins influence the ability of RNA polymerase to initiate transcription. However, there are also some differences. First, many bacterial and archaeal genes are organized into operons and are transcribed into a single RNA molecule. Although some operon-like gene clusters have been found in worms and even some primitive chordates, most eukaryotic genes have their own promoters and are transcribed separately. Second, chromatin structure affects gene expression in eukaryotic cells: DNA must partly unwind from the histone proteins before transcription can take place. Finally, the presence of the nuclear membrane in eukaryotic cells separates transcription and translation in time and space. Therefore, the regulation of gene expression in eukaryotic cells is characterized by a greater diversity of mechanisms that act at different points in the transfer of information from DNA to protein.

Eukaryotic gene regulation is less well understood than bacterial gene regulation, partly owing to the larger genomes of eukaryotes, their greater sequence complexity, and the difficulty of isolating and manipulating mutations that can be used in the study of gene regulation. Nevertheless, great advances in our understanding of the regulation of eukaryotic genes have been made in recent years.

Changes in Chromatin Structure

One type of gene regulation in eukaryotic cells is accomplished through the modification of chromatin structure. In the nucleus, histone proteins associate to form octamers, around which helical DNA tightly coils to create chromatin (see Figure 8.18). In a general sense, this chromatin structure represses gene expression. For a gene to be transcribed, proteins called transcription factors (see Transcription Factors and Transcriptional Regulator Proteins below) must bind to the DNA. Other regulator proteins and RNA polymerase must also bind to the DNA for transcription to take place. How can these events take place with DNA wrapped tightly around histone proteins? The answer is that, before transcription, chromatin structure changes so that the DNA becomes more accessible to the transcription apparatus.

CHROMATIN REMODELING Some transcription factors and other regulatory proteins alter chromatin structure without altering the chemical structure of the histones directly. These proteins are called **chromatin-remodeling complexes**. They bind directly to particular sites on DNA and reposition the nucleosomes, allowing transcription factors and RNA polymerase to bind to promoters and initiate transcription.

One of the best-studied examples of a chromatin-remodeling complex is SWI–SNF, which is found in yeast, humans, *Drosophila*, and other eukaryotes. This complex uses energy derived from the hydrolysis of ATP to reposition nucleosomes, exposing promoters in the DNA to the action of other regulatory proteins and RNA polymerase.

THINK-PAIR-SHARE Question 8

HISTONE MODIFICATION The histones in the octamer core of a nucleosome have two domains: **(1)** a globular domain that associates with other histones and the DNA and **(2)** a positively charged tail domain that interacts with the negatively charged phosphate groups on the backbone of DNA. The tails of histone proteins are often modified by the addition or removal of phosphate groups, methyl groups, or acetyl groups. Another modification of histones is ubiquitination, in which small molecules called ubiquitin are added or removed from the histones. All of these modifications have sometimes been collectively called the **histone code** because they encode information that affects how genes are expressed. The histone code affects gene expression by altering chromatin structure directly or, in some cases, by providing recognition sites for proteins that bind to DNA and then regulate transcription.

METHYLATION OF HISTONES One type of histone modification is the addition of methyl groups to the tails of histone proteins. These modifications can bring about either the activation or the repression of transcription, depending on which histone is modified, which particular amino acids in the tails are methylated, and how many methyl groups are added. Enzymes called histone methyltransferases add methyl groups to specific amino acids (usually lysine or arginine) of histones. Other enzymes, called histone demethylases, remove methyl groups from histones. Many of the enzymes and proteins that modify histones, such as methyltransferases and demethylases, do not bind to specific DNA sequences and must be recruited to specific chromatin sites. Sequence-specific DNA-binding proteins, preexisting histone modifications, and RNA molecules serve to recruit histone-modifying enzymes to specific sites.

A common modification is the addition of three methyl groups to lysine 4 in the tail of the H3 histone protein, abbreviated H3K4me3 (K is the abbreviation for lysine). Histones containing the H3K4me3 modification are frequently found near promoters of transcriptionally active genes. Studies have identified proteins that recognize and bind to H3K4me3, including nucleosome-remodeling factor (NURF). NURF and other proteins that recognize H3K4me3 have a common protein-binding domain that binds to the H3 histone tail and then alters chromatin packing, allowing transcription to take place. Some transcription factors also bind to H3K4me3.

ACETYLATION OF HISTONES Another type of histone modification that affects chromatin structure is acetylation, the addition of acetyl groups (CH_3CO) to histone. The acetylation of histones usually stimulates transcription. For example, the addition of a single acetyl group to lysine 16 in the tail of the H4 histone prevents the formation of the 30-nm chromatin fiber, causing the chromatin to be in an open configuration and available for transcription (see Figure 8.18). In general, acetyl groups destabilize chromatin structure, allowing transcription to take place (**Figure 12.14**). Acetyl groups are added to histone proteins by acetyltransferase enzymes; other enzymes called deacetylases strip acetyl groups from histones and restore chromatin structure, which represses transcription. Certain proteins that regulate transcription

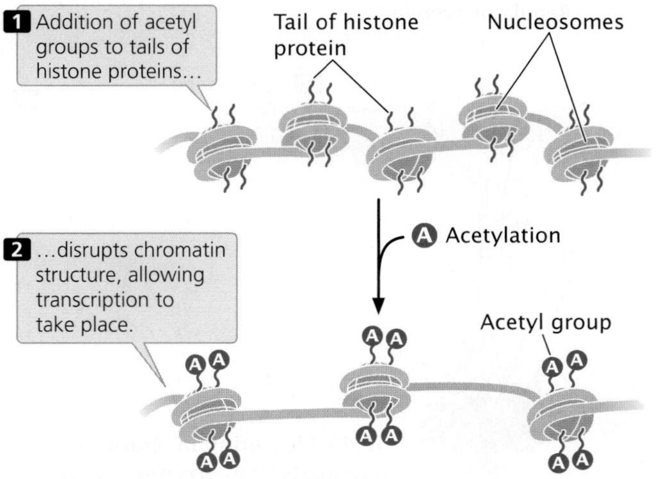

12.14 The acetylation of histone proteins alters chromatin structure and permits some transcription factors to bind to DNA.

either have acetyltransferase activity themselves or attract acetyltransferases to the DNA.

THE ACETYLATION OF HISTONES CONTROLS FLOWERING IN *ARABIDOPSIS* The importance of histone acetylation in gene regulation is demonstrated by the control of flowering in *Arabidopsis thaliana*, a plant with a number of characteristics that make it an excellent genetic model for plant systems. The time at which flowering takes place is critical to the life of a plant: if flowering is initiated at the wrong time of year, pollinators may not be available to fertilize the flowers or environmental conditions may be unsuitable for the survival and germination of the seeds. Consequently, flowering time in most plants is carefully regulated in response to multiple internal and external cues such as plant size, photoperiod, and temperature.

Among the many genes that control flowering in *Arabidopsis* is *flowering locus C* (*FLC*), which plays an important role in suppressing flowering until after an extended period of cold (a process called vernalization). The *FLC* gene encodes a regulatory protein that represses the activity of other genes that affect flowering (**Figure 12.15**). As long as *FLC* is active, flowering remains suppressed.

The activity of *FLC* is controlled by another locus called *flowering locus D* (*FLD*), the key role of which is to stimulate flowering by repressing the action of *FLC*. In essence, flowering is stimulated because *FLD* represses the repressor. How does *FLD* repress *FLC*? *FLD* encodes a deacetylase enzyme, which removes acetyl groups from histone proteins in the chromatin surrounding *FLC* (see Figure 12.15). The removal of these acetyl groups alters the chromatin structure and inhibits transcription. The inhibition of transcription prevents *FLC* from being transcribed and removes its repression on flowering. In short, *FLD* stimulates flowering in *Arabidopsis* by deacetylating the chromatin that surrounds *FLC*, thereby removing its inhibitory effect on flowering.

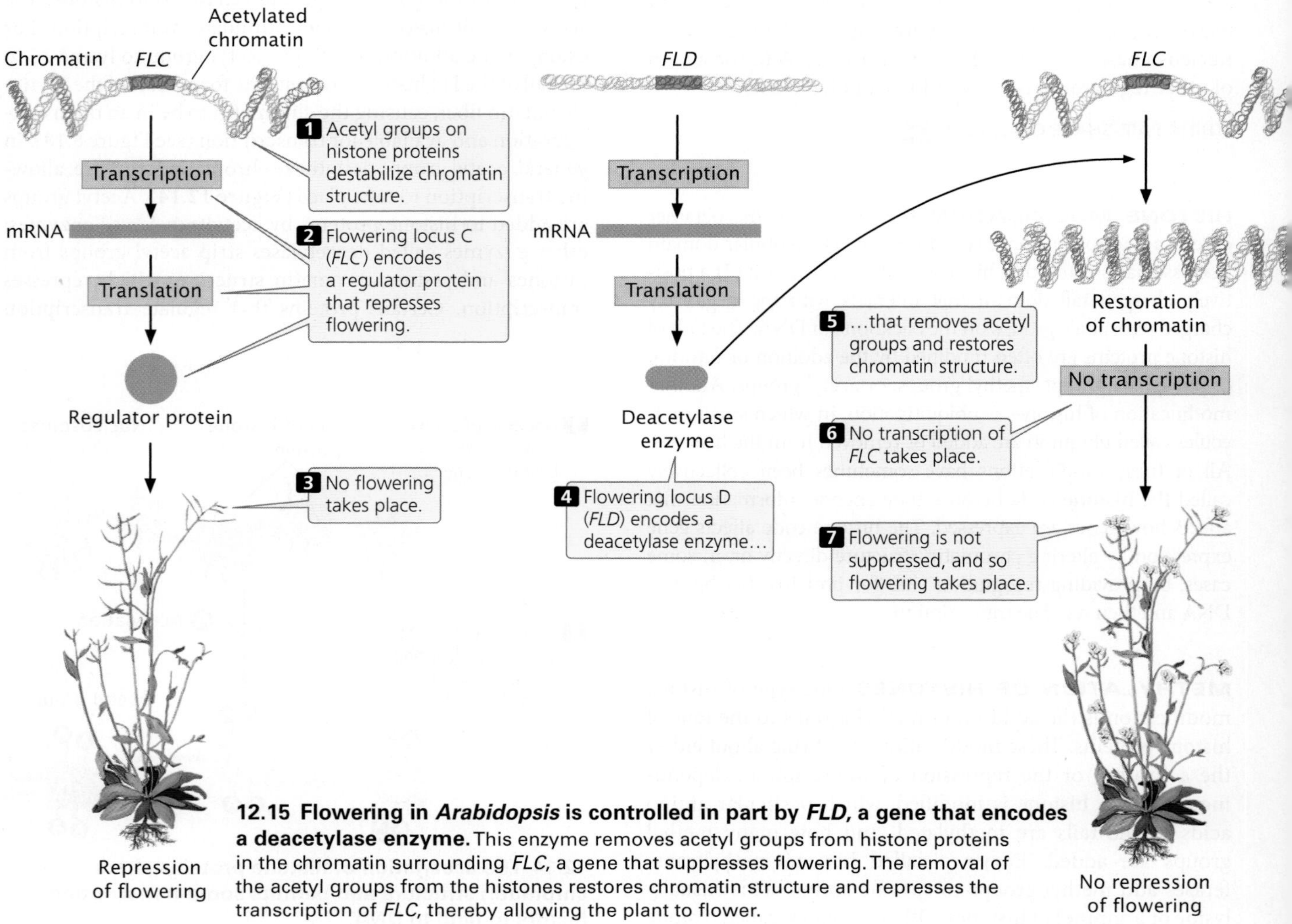

12.15 Flowering in *Arabidopsis* is controlled in part by *FLD*, a gene that encodes a deacetylase enzyme. This enzyme removes acetyl groups from histone proteins in the chromatin surrounding *FLC*, a gene that suppresses flowering. The removal of the acetyl groups from the histones restores chromatin structure and represses the transcription of *FLC*, thereby allowing the plant to flower.

DNA METHYLATION Another change in chromatin structure associated with transcription is the methylation of cytosine bases, which yields 5-methylcytosine. The methylation of cytosine in DNA is distinct from the methylation of histone proteins mentioned earlier. Heavily methylated DNA is associated with the repression of transcription in vertebrates and plants, whereas transcriptionally active DNA is usually unmethylated in these organisms. Abnormal patterns of methylation are also associated with some types of cancer.

Evidence indicates that an association exists between DNA methylation and the deacetylation of histones, both of which repress transcription. Certain proteins that bind tightly to methylated sequences form complexes with other proteins that act as histone deacetylases. In other words, methylation appears to attract deacetylases, which remove acetyl groups from the histone tails, stabilizing the nucleosome structure and repressing transcription. The demethylation of DNA allows acetyltransferases to add acetyl groups, disrupting nucleosome structure and permitting transcription. The role of methylation in chromatin structure and epigenetics is discussed further in Section 12.4.

CONCEPTS

Chromatin structure can be altered by chromatin-remodeling complexes that reposition nucleosomes, by modifications of histone proteins, and by the methylation of DNA.

Transcription Factors and Transcriptional Regulator Proteins

Transcription is an important level of control in eukaryotic cells, and this control requires a number of different types of proteins and regulatory elements. Eukaryotic promoters for genes that encode proteins consist of two parts: a core promoter, which is located immediately upstream of the gene, and a regulatory promoter, which lies upstream of the core promoter. Proteins called **general transcription factors** and RNA polymerase assemble into a basal transcription apparatus, which binds to a core promoter. The basal transcription apparatus is capable of minimal levels of transcription, but **transcriptional regulator proteins** are required to bring about normal levels of transcription. These proteins bind to the regulatory promoter and affect the levels of transcription that take place (**Figure 12.16**). Transcriptional regulator proteins also bind to **enhancers**, which are regulatory elements that may be located some distance from the gene. The exact position and orientation of an enhancer in relation to the promoter it affects can vary. Some transcriptional regulator proteins are activators; others are repressors.

TRANSCRIPTIONAL ACTIVATORS AND REPRESSORS **Transcriptional activator proteins** stimulate and stabilize the basal transcription apparatus at the core promoter. The activators may interact directly with the basal transcription apparatus or indirectly through protein **coactivators**. Some activators and coactivators,

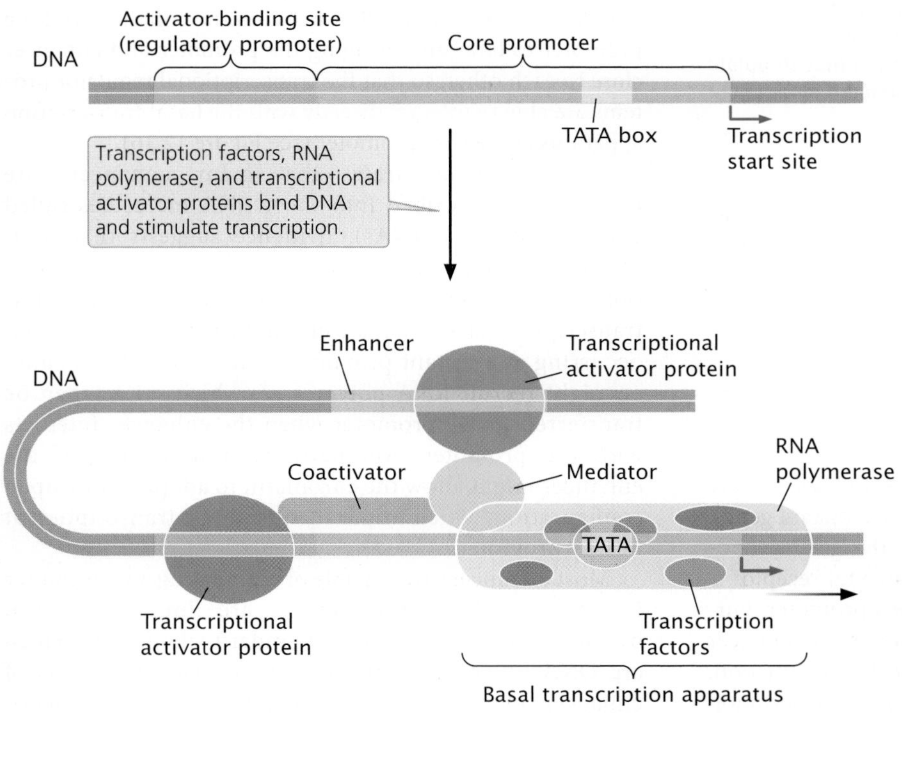

12.16 Transcriptional activator proteins bind to sites on DNA and stimulate transcription. Most act by stimulating or stabilizing the assembly of the basal transcription apparatus.

as well as general transcription factors, also have acetyltransferase activity and so further stimulate transcription by altering chromatin structure. Some regulatory proteins in eukaryotic cells act as repressors, inhibiting transcription. These repressors bind to sequences in the regulatory promoter or to **silencers**. Silencers, like enhancers, affect transcription at distant promoters and are position and orientation independent. Unlike repressors in bacteria, most eukaryotic repressors do not directly block RNA polymerase. These repressors may compete with transcriptional activators for DNA binding sites: when a site is occupied by an activator, transcription is activated, but if a repressor occupies that site, there is no activation. Alternatively, a repressor may bind to sites near an activator-binding site and prevent the activator from contacting the basal transcription apparatus. A third mechanism of repressor action is direct interference with the assembly of the basal transcription apparatus, which blocks the initiation of transcription.

Within the regulatory promoter are typically several different consensus sequences to which different transcriptional activators and repressors can bind. Among different promoters, these binding sites are mixed and matched in different combinations (**Figure 12.17**), so each promoter is regulated by a unique combination of transcriptional activator and repressor proteins.

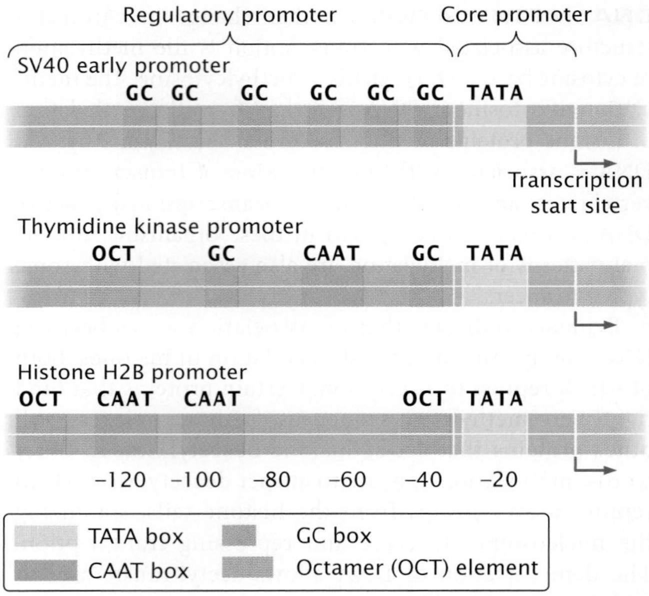

12.17 The consensus sequences in the promoters of three eukaryotic genes illustrate the principle that these sequences can be mixed and matched in different combinations in different promoters. A different transcriptional activator protein binds to each consensus sequence, and so each promoter responds to a unique combination of activator proteins. SV40 is a virus found in monkeys and humans.

THINK-PAIR-SHARE Question 9

CONCEPTS

Transcriptional regulator proteins in eukaryotic cells can influence the initiation of transcription by affecting the stability or assembly of the basal transcription apparatus. Some transcriptional regulator proteins are activators that stimulate transcription; others are repressors that inhibit the initiation of transcription.

✓ CONCEPT CHECK 7

Most transcriptional activator proteins affect transcription by interacting with

a. introns.
b. the basal transcription apparatus.
c. DNA polymerase.
d. nucleosomes.

ENHANCERS AND INSULATORS Enhancers are regulatory elements that affect transcription at distant genes. For example, an enhancer that regulates the gene encoding the alpha chain of the T-cell (T-lymphocyte) receptor is located 69,000 bp downstream of the gene's promoter. Furthermore, the exact position and orientation of an enhancer relative to that of the promoter is not critical to its function. How can an enhancer affect the initiation of transcription at a promoter that is tens of thousands of base pairs away? In many cases, transcriptional regulator proteins bind to the enhancer and cause the DNA between the enhancer and the promoter to loop out, bringing the promoter and enhancer close to each other, so that the transcriptional regulator proteins are able to interact directly with the basal transcription apparatus at the core promoter (see Figure 12.16).

Research demonstrates that many enhancers are themselves transcribed into short RNA molecules called enhancer RNAs (eRNAs). Evidence suggests that transcription of enhancers is often associated with transcription at the promoters that the enhancers affect. How transcription at the enhancer might affect transcription occurring at a distant promoter is not clear. The enhancer might recruit RNA polymerase, which might then be transferred to the promoter when the enhancer interacts with the promoter. Alternatively, transcription of the enhancer might allow the chromatin to adopt a more open configuration, which would then facilitate transcription at nearby promoters.

Most enhancers are capable of stimulating any promoter in their vicinities. Their effects are limited, however, by **insulators** (also called boundary elements), which are DNA sequences that block or insulate the effects of enhancers in a position-dependent manner. If the insulator

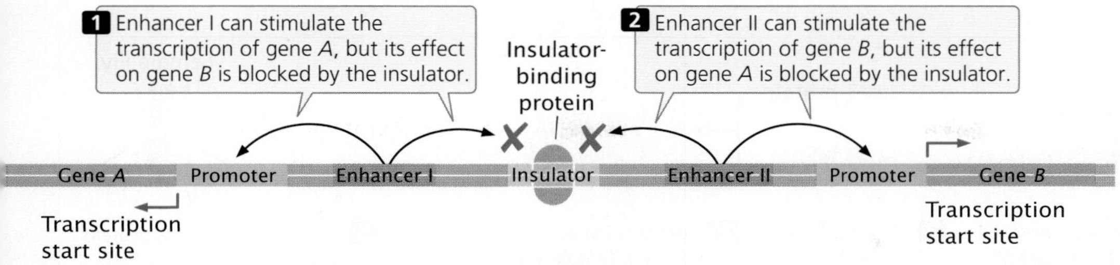

12.18 An insulator blocks the action of an enhancer on a promoter when the insulator lies between the enhancer and the promoter.

lies between the enhancer and the promoter, it blocks the action of the enhancer, but if the insulator lies outside the region between the two, it has no effect (**Figure 12.18**). Specific proteins bind to insulators and play a role in their blocking activity. Some insulators also limit the spread of changes in chromatin structure that affect transcription. Some enhancer-like elements are found in prokaryotes. ▶TRY PROBLEM 34

THINK-PAIR-SHARE Question 10

CONCEPTS

Some regulatory proteins bind to enhancers, which are regulatory elements that are distant from the gene for which they stimulate transcription. Insulators are DNA sequences that block the action of enhancers.

✓ CONCEPT CHECK 8

How does the binding of regulatory proteins to enhancers affect transcription at genes that are thousands of base pairs away?

COORDINATED GENE REGULATION Although most eukaryotic cells do not possess operons, several eukaryotic genes may be activated by the same stimulus. For example, many eukaryotic cells respond to extreme heat and other stresses by producing **heat-shock proteins**, which help to prevent damage from such stressing agents. Heat-shock proteins are produced by a large number of different genes. During times of environmental stress, transcription of all the heat-shock genes is greatly elevated. Groups of bacterial genes are often coordinately expressed (turned on and off together) because they are physically clustered as an operon and have the same promoter, but coordinately expressed genes in eukaryotic cells are not clustered. How, then, is the transcription of eukaryotic genes coordinated if they are not organized into an operon?

Genes that are coordinately expressed in eukaryotic cells are able to respond to the same stimulus because they share short regulatory sequences in their promoters or enhancers. For example, different eukaryotic heat-shock genes possess a common regulatory element upstream of their start sites. Such DNA regulatory sequences are called **response elements**; they typically contain the same short consensus sequences at varying distances from the gene being regulated. The response elements are binding sites for transcriptional activators, which bind to the response elements and elevate transcription.

Gene Regulation by RNA Processing and Degradation

In bacteria, transcription and translation take place simultaneously. In eukaryotes, transcription takes place in the nucleus and the pre-mRNAs are then processed before moving to the cytoplasm for translation, allowing opportunities for gene control after transcription. Consequently, posttranscriptional gene regulation assumes an important role in eukaryotic cells. A common level of gene regulation in eukaryotes is RNA processing and degradation.

GENE REGULATION THROUGH RNA PROCESSING Alternative splicing allows a pre-mRNA to be spliced in multiple ways, generating different proteins in different tissues or at different times in development (see Chapter 10). Many eukaryotic genes undergo alternative splicing, and the regulation of splicing is an important means of controlling gene expression in eukaryotic cells.

An example of regulation by alternative mRNA splicing is sex determination in fruit flies (see Sex Determination in *Drosophila melanogaster* in Section 4.1). Sex differentiation in *Drosophila* arises from a cascade of gene regulation. In XX fly embryos, a female-specific promoter is activated early in development and stimulates the transcription of the *sex-lethal* (*Sxl*) gene (**Figure 12.19**). The protein encoded by *Sxl* regulates the splicing of the pre-mRNA transcribed from another gene called *transformer* (*tra*). The splicing of *tra* pre-mRNA results in the production of Tra protein (see Figure 12.19). Together with another protein (Tra-2), Tra stimulates the female-specific splicing of pre-mRNA from yet another gene called *doublesex* (*dsx*). This event produces a female-specific Dsx protein, which causes the embryo to develop female characteristics.

In XY fly embryos, the promoter that transcribes the *Sxl* gene in females is inactive, so no Sxl protein is produced. In the absence of Sxl protein, *tra* pre-mRNA is spliced at a different 3′ splice site to produce a nonfunctional form of

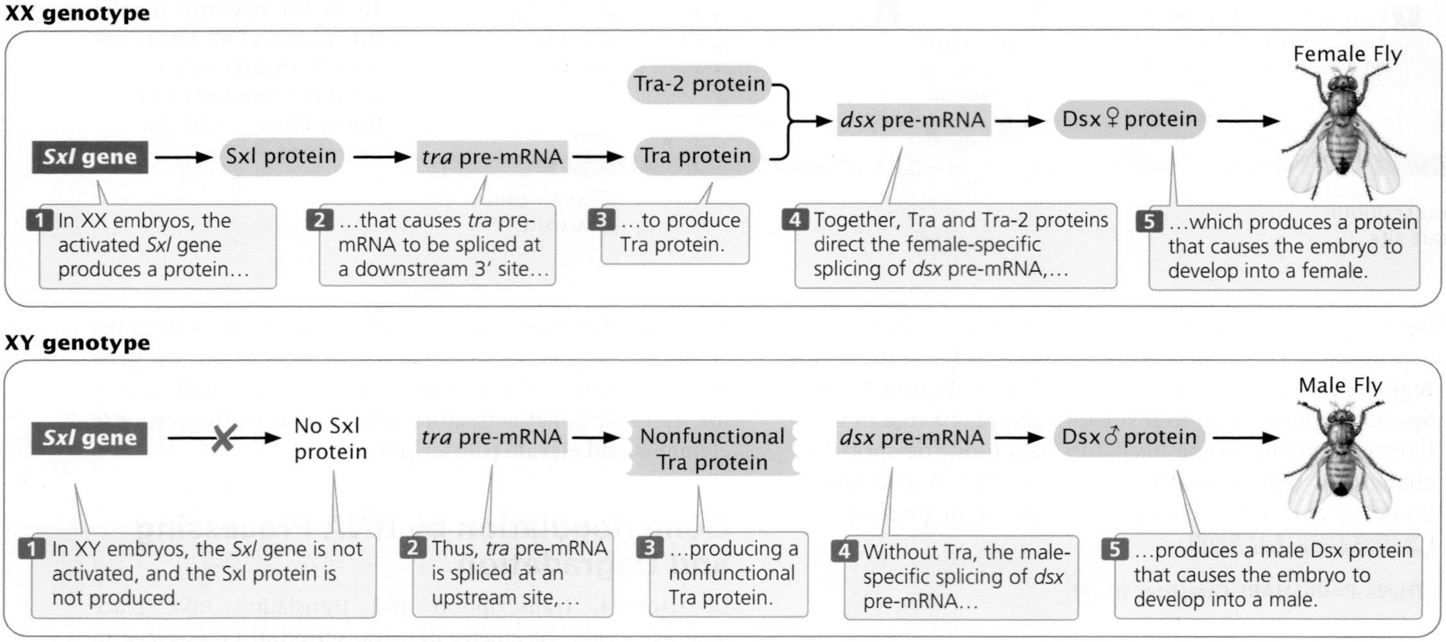

12.19 Alternative splicing controls sex determination in *Drosophila*.

Tra protein (**Figure 12.20**). In turn, the presence of this nonfunctional Tra in males causes *dsx* pre-mRNAs to be spliced differently from that in females, and a male-specific Dsx protein is produced (see Figure 12.19). This event causes the development of male-specific traits.

THINK-PAIR-SHARE Question 11

CONCEPTS

Eukaryotic genes can be regulated through the control of mRNA processing. The selection of alternative splice sites leads to the production of different proteins.

THE DEGRADATION OF RNA The amount of a protein that is synthesized depends on the amount of corresponding mRNA available for translation. The amount of available mRNA, in turn, depends on both the rate of mRNA synthesis and the rate of mRNA degradation. Eukaryotic mRNAs are generally more stable than bacterial mRNAs, which typically last only a few minutes before being degraded. Nonetheless, there is great variability in the stability of eukaryotic mRNAs: some mRNAs persist for only a few minutes, whereas others last for hours, days, or even months. These variations can produce large differences in the amount of protein that is synthesized.

Various factors, including the 5′ cap and the poly(A) tail (see Chapter 10), affect the stability of eukaryotic mRNA. Poly(A)-binding proteins (PABPs) normally bind to the poly(A) tail and contribute to its stability-enhancing effect. The presence of these proteins at the 3′ end of the mRNA protects the 5′ cap. When the poly(A) tail has been shortened below a critical limit, the 5′ cap is removed, and the mRNA is degraded by the removal of nucleotides from the 5′ end. These observations suggest that the 5′ cap and the 3′ poly(A) tail of eukaryotic mRNA physically interact with each other, most likely by the poly(A) tail bending around so that the PABPs make contact with the 5′ cap.

Much of RNA degradation in eukaryotes takes place in specialized complexes called P bodies. P bodies help control the expression of genes by regulating which RNA molecules are degraded and which are sequestered for later release. RNA degradation facilitated by small interfering RNAs (see next section) can also take place within P bodies.

Other parts of eukaryotic mRNA, including sequences in the 5′ untranslated region (5′ UTR), the coding region, and the 3′ UTR, also affect mRNA stability. Some short-lived eukaryotic mRNAs have one or more copies of the consensus sequence 5′—AUUUAUAA—3′, referred to as the AU-rich element, in the 3′ UTR. The mRNAs containing AU-rich elements are degraded by a mechanism in which microRNAs (see next section) take part.

CONCEPTS

The stability of mRNA influences gene expression by affecting the amount of mRNA available to be translated. The stability of mRNA is affected by the 5′ cap, the poly(A) tail, the 5′ UTR, the coding region, and sequences in the 3′ UTR.

✓ CONCEPT CHECK 9

How does the poly(A) tail affect mRNA stability?

RNA Interference and Gene Regulation

The expression of a number of eukaryotic genes is controlled through RNA interference, also known as RNA silencing and posttranscriptional gene silencing (see Chapter 10). Research suggests that as many as 30% of human genes are regulated by RNA interference. RNA interference is widespread in

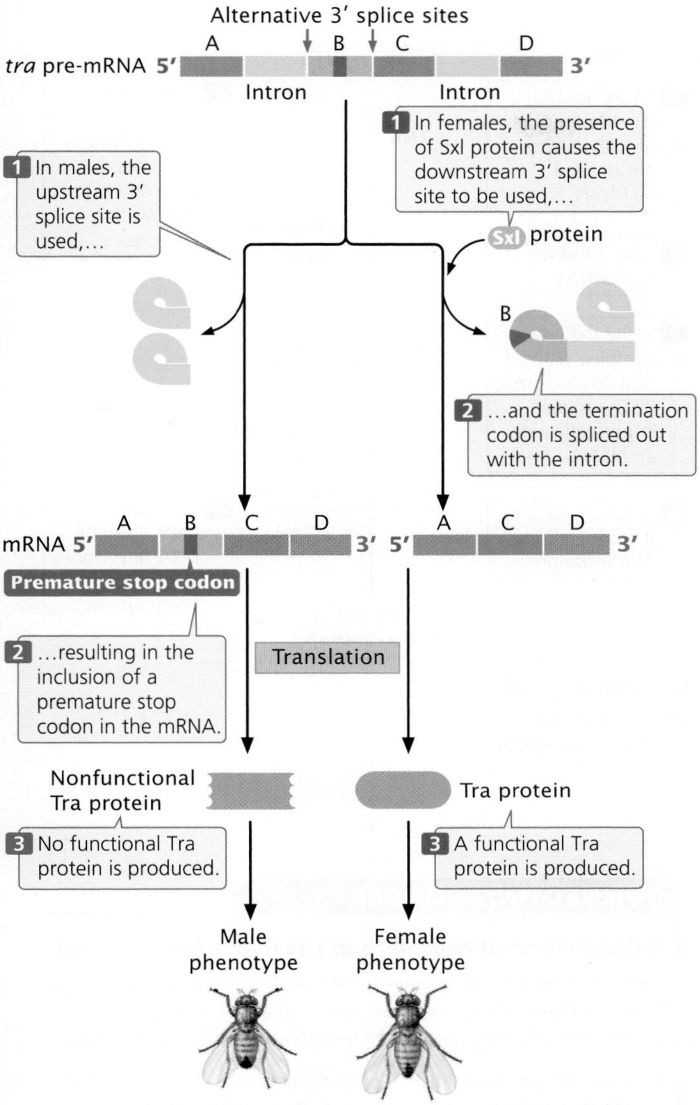

12.20 Alternative splicing of *tra* pre-mRNA. Two alternative 3′ splice sites are present.

eukaryotes, existing in fungi, plants, and animals. A similar group of small RNA molecules with silencing and activating functions have been detected in prokaryotes. The mechanism of RNA silencing is also widely used as a powerful technique for artificially regulating gene expression in genetically engineered organisms (see Chapter 14).

RNA interference is triggered by small RNA molecules know as microRNAs (miRNAs) and small interfering RNAs (siRNAs), depending on their origin and mode of action (see Chapter 10). An enzyme called Dicer cleaves and processes double-stranded RNA to produce single-stranded siRNAs or miRNAs 21–25 nucleotides in length, which pair with proteins to form an RNA-induced silencing complex (RISC). The RNA component of the RISC then pairs with complementary base sequences in specific mRNA molecules, most often with sequences in the 3′ UTR of the mRNA. Small interfering RNAs and microRNAs regulate gene expression through at least three distinct mechanisms: (1) cleavage of mRNA, (2) inhibition of translation, or (3) inhibition of transcription (**Figure 12.21**).

RNA CLEAVAGE RISCs that contain an siRNA (and some that contain an miRNA) pair with mRNA molecules and cleave the mRNA near the middle of the bound siRNA (Figure 12.21a). This cleavage is carried out by a protein that is sometimes referred to as Slicer. After cleavage, the mRNA is further degraded. Thus, the presence of siRNAs and miRNAs increases the rate at which mRNAs are broken down and decreases the amount of protein produced.

INHIBITION OF TRANSLATION Some miRNAs regulate genes by inhibiting the translation of their complementary mRNAs (Figure 12.21b). For example, an important gene in flower development in *Arabidopsis thaliana* is *APETALA2*. The expression of this gene is regulated by an miRNA that base pairs with nucleotides in the coding region of *APETALA2* mRNA and inhibits its translation.

TRANSCRIPTIONAL SILENCING Other siRNAs silence transcription by altering chromatin structure. These siRNAs combine with proteins to form a complex called RITS (for RNA-induced transcriptional silencing), which is analogous to RISC (Figure 12.21c). The siRNA component of a RITS then binds to its complementary sequence in DNA or an RNA molecule in the process of being transcribed, where it represses transcription by attracting enzymes that methylate the tails of histone proteins. The addition of these methyl groups to the histones causes them to bind DNA more tightly, restricting the access of proteins and enzymes necessary to carry out transcription (see the sections on histone modification earlier in this chapter).

CONCEPTS

RNA interference is initiated by double-stranded RNA molecules that are cleaved and processed. The resulting siRNAs or miRNAs combine with proteins to form complexes that bind to complementary sequences in mRNA or DNA. The siRNAs and miRNAs affect gene expression by cleaving mRNA, inhibiting translation, or altering chromatin structure.

✓ CONCEPT CHECK 10

In RNA silencing, siRNAs and miRNAs usually bind to which part of the mRNA molecules that they control?

a. 5′ UTR
b. 5′ cap
c. 3′ poly(A) tail
d. 3′ UTR

Gene Regulation in the Course of Translation and Afterward

Ribosomes, aminoacyl tRNAs, initiation factors, and elongation factors are all required for the translation of mRNA

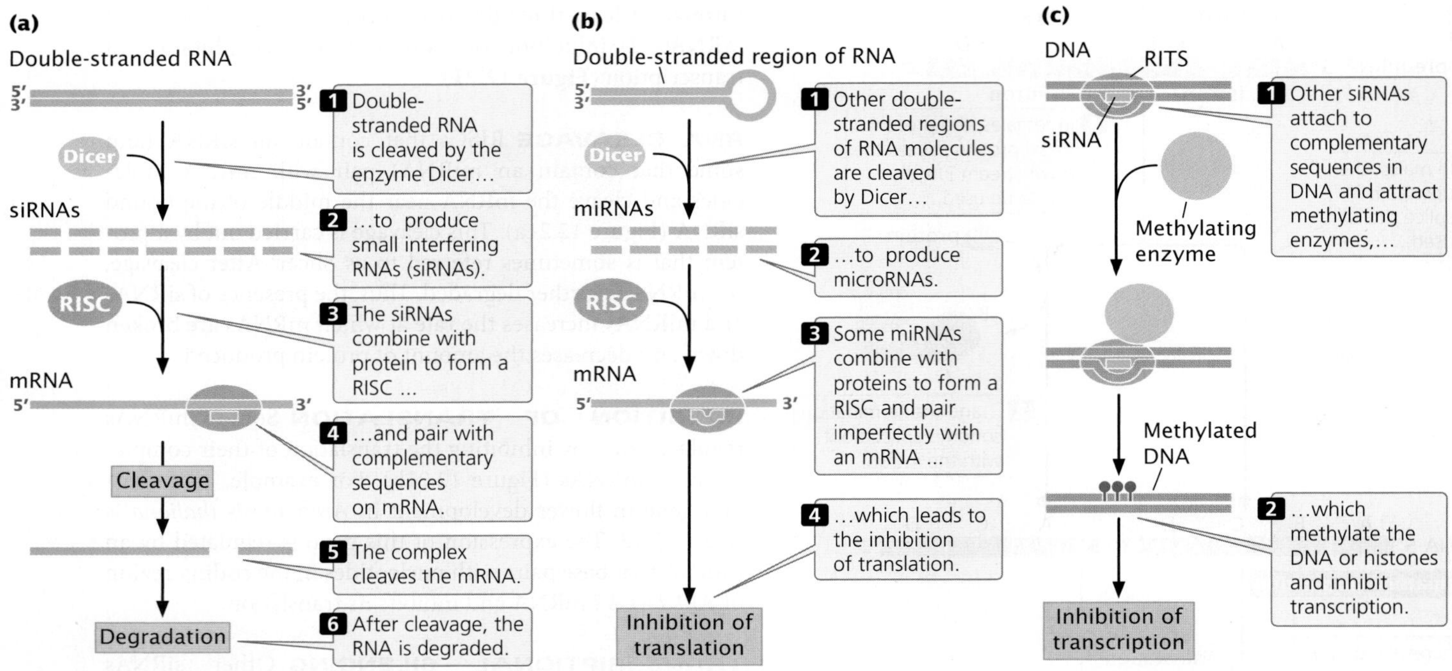

12.21 RNA silencing leads to the degradation of mRNA or to the inhibition of translation or transcription. (a) Small interfering RNAs (siRNAs) degrade mRNA by cleavage. (b) MicroRNAs (miRNAs) lead to the inhibition of translation. (c) Some siRNAs cause the methylation of histone proteins or DNA, inhibiting transcription.

molecules, as we saw in Chapter 11. The availability of these components affects the rate of translation and therefore influences gene expression.

Mechanisms also exist for the regulation of translation of specific mRNAs. The initiation of translation in some mRNAs is regulated by proteins that bind to an mRNA's 5′ UTR and inhibit the binding of ribosomes, in a manner similar to the repressor proteins that bind to operators and prevent the transcription of structural genes in prokaryotes. The translation of some mRNAs is affected by the binding of proteins to sequences in the 3′ UTR.

Many eukaryotic proteins are extensively modified after translation by the selective cleavage and trimming of amino acids from the ends, by acetylation, or by the addition of phosphate groups, carboxyl groups, methyl groups, carbohydrates, or ubiquitin (a small protein). These modifications affect the transport, function, stability, and activity of proteins and therefore have the ability to affect gene expression.

CONCEPTS

The initiation of translation may be affected by proteins that bind to specific sequences near the 5′ end of mRNA. The availability of ribosomes, tRNAs, initiation and elongation factors, and other components of the translational apparatus may affect the rate of translation.

CONNECTING CONCEPTS

A Comparison of Bacterial and Eukaryotic Gene Control

Now that we have considered the major types of gene regulation in bacteria and eukaryotes, let's consider some of the similarities and differences in bacterial and eukaryotic gene control.

1. Much of gene regulation in bacterial cells is at the level of transcription (although it does exist at other levels). Gene regulation in eukaryotic cells takes place at multiple levels, including chromatin structure, transcription, mRNA processing, mRNA stability, translation, and posttranslational control.
2. Complex biochemical and developmental events in bacterial and eukaryotic cells may require a cascade of gene regulation, in which the activation of one set of genes stimulates the activation of another set.
3. Much of gene regulation in both bacterial and eukaryotic cells is accomplished through proteins that bind to specific sequences in DNA.
4. Chromatin structure plays a role in eukaryotic (but not bacterial) gene regulation. In general, condensed chromatin represses gene expression; chromatin structure must be altered before transcription can take place. Chromatin structure is altered by chromatin-remodeling complexes, the modification of histone proteins, and DNA methylation.
5. In bacterial cells, genes are often clustered in operons and are coordinately expressed by transcription into a single mRNA molecule. In contrast, many eukaryotic genes have their own promoters and are transcribed independently. Coordinated regulation in eukaryotic cells often takes

place through common response elements located in the promoters and enhancers of the genes. Different genes that share the same response element are influenced by the same regulatory protein.

6. Regulatory proteins that affect transcription exhibit two basic types of control: *repressors* inhibit transcription (negative control) and *activators* stimulate transcription (positive control). Both negative control and positive control are found in bacterial and eukaryotic cells.
7. The initiation of transcription is a relatively simple process in bacterial cells, and regulatory proteins function by blocking or stimulating the binding of RNA polymerase to DNA. In contrast, eukaryotic transcription requires complex machinery that includes RNA polymerase, general transcription factors, and transcriptional activators and repressors, allowing, transcription to be influenced by multiple factors.
8. Some eukaryotic transcriptional regulator proteins function at a distance from the gene by binding to enhancers, causing the formation of a loop in the DNA, which brings the promoter and enhancer into close proximity. Some distant-acting sequences analogous to enhancers have been described in bacterial cells, but they appear to be less common.
9. The greater time lag between transcription and translation in eukaryotic cells than in bacterial cells allows mRNA stability and mRNA processing to play larger roles in eukaryotic gene regulation.
10. Regulation by small RNAs occurs in both eukaryotes and bacteria.

These similarities and differences in gene regulation in bacteria and eukaryotes are summarized in **Table 12.1**.

TABLE 12.1 Comparison of gene expression in bacteria and eukaryotes

Characteristic	Bacterial gene control	Eukaryotic gene control
Levels of regulation	Primarily transcription	Many levels
Cascades of gene regulation	Present	Present
DNA-binding proteins	Important	Important
Role of chromatin structure	Absent	Important
Presence of operons	Common	Uncommon
Negative and positive control	Present	Present
Initiation of transcription	Relatively simple	Relatively complex
Enhancers	Less common	More common
Transcription and translation	Occur simultaneously	Occur separately
Regulation by small RNAs	Common	Common

Model Genetic Organism
The Plant *Arabidopsis thaliana*

Much of the early work in genetics was carried out on plants, including Mendel's seminal discoveries in pea plants as well as the discoveries of important aspects of heredity, gene mapping, chromosome genetics, and quantitative inheritance in corn, wheat, beans, and other plants. However, by the mid-twentieth century, many geneticists had turned to bacteria, viruses, yeast, *Drosophila*, and mouse genetic models. Because a good genetic plant model did not exist, plants were relatively neglected, particularly for the study of molecular genetic processes.

This neglect of plants ended in the last part of the twentieth century with the widespread introduction of a new genetic model organism: the thale cress, *Arabidopsis thaliana* (**Figure 12.22**). *Arabidopsis thaliana* was identified in the sixteenth century, and the first mutant was reported in 1873, but this species was not commonly studied until the first detailed genetic maps appeared in the early 1980s. Today, *Arabidopsis* figures prominently in the study of genome structure, gene regulation, development, and evolution in plants, and it provides important basic information about plant genetics that is applied to other economically important plant species.

Advantages of *Arabidopsis* as a model genetic organism

Arabidopsis thaliana is a member of the mustard (Brassicaceae) family and grows as a weed in many parts of the world. Except in its role as a model genetic organism, *Arabidopsis* has no economic importance, but it has a number of characteristics that make it suitable for the study of genetics. As an angiosperm, it has features in common with other flowering plants, some of which play critical roles in the ecosystem or are important sources of food, fiber, building materials, and pharmaceuticals. *Arabidopsis*' chief advantages are its small

The Plant
Arabidopsis thaliana

ADVANTAGES

- Small size
- Short generation time of 6 weeks
- Each plant can produce from 10,000 to 40,000 seeds
- Ability to grow in laboratory
- Small genome for a plant
- Many variants available
- Self-fertilizes and outcrosses

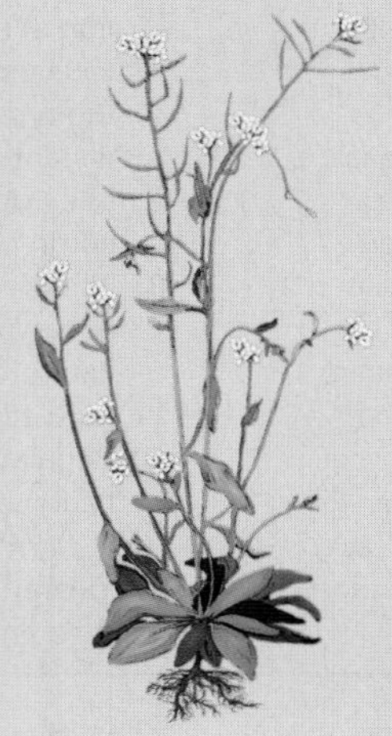

STATS

Taxonomy:	Flowering plant
Size:	10–20 cm
Anatomy:	Roots, one primary shoot, simple leaves, flowers
Habitat:	Meadows

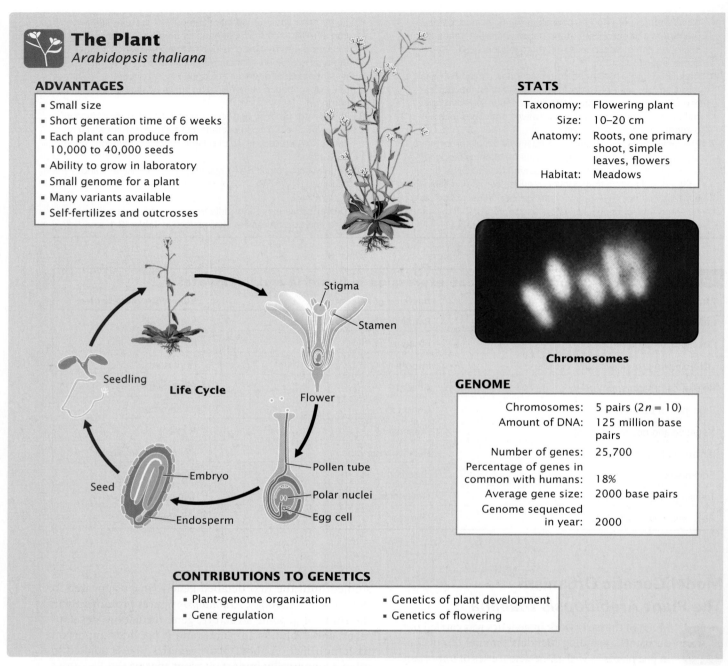

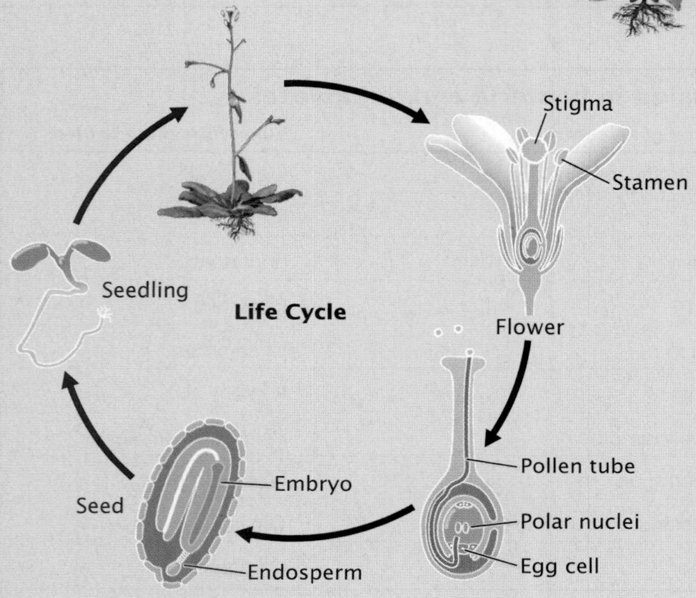

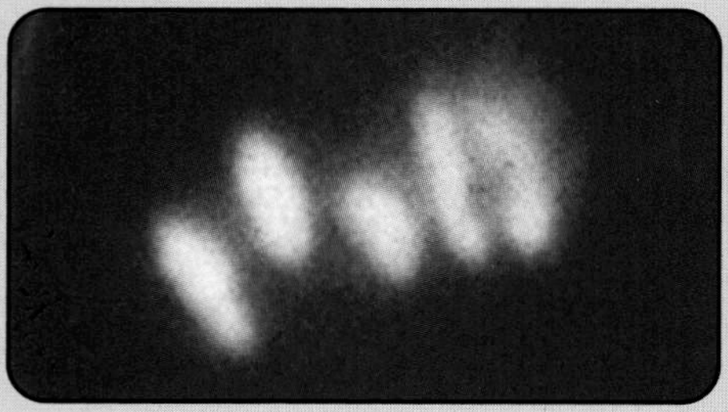

Chromosomes

GENOME

Chromosomes:	5 pairs ($2n = 10$)
Amount of DNA:	125 million base pairs
Number of genes:	25,700
Percentage of genes in common with humans:	18%
Average gene size:	2000 base pairs
Genome sequenced in year:	2000

CONTRIBUTIONS TO GENETICS

- Plant-genome organization
- Gene regulation
- Genetics of plant development
- Genetics of flowering

12.22 *Arabidopsis thaliana* is a model genetic organism that serves as an important subject for research on genetic processes in plants. [Photograph courtesy of Anand P. Tyagi and Luca Comai.]

size (maximum height of 10 to 20 cm), prolific reproduction, and small genome (see Figure 12.22).

Arabidopsis thaliana completes development—from seed germination to seed production—in about 6 weeks. Its small size and ability to grow under low illumination make it ideal for laboratory culture. Each plant is capable of producing from 10,000 to 40,000 seeds, and the seeds typically have a high rate of germination, so large numbers of progeny can be obtained from single genetic crosses.

Another key advantage for molecular studies is *Arabidopsis*' small genome, which consists of only 125 million base pairs of DNA on five pairs of chromosomes (compared with 2.5 billion base pairs in the corn genome and 16 billion in the wheat genome). The genome of *A. thaliana* was completely sequenced in 2000, providing detailed information about gene structure and organization in this species. A number of variants of *A. thaliana*—called ecotypes—that vary in shape, size, physiological characteristics, and DNA sequence are available for study.

Life cycle of *Arabidopsis*

The *Arabidopsis* life cycle is fairly typical of flowering plants (see Figures 2.19 and 12.22). The main, vegetative part of the plant is diploid; haploid gametes are produced in the pollen and ovaries. When a pollen grain lands on the stigma of a flower, a pollen tube grows into the pistil and ovary. Two haploid sperm nuclei contained in each pollen grain travel down the pollen tube and enter the embryo sac. There, one of the haploid sperm cells fertilizes the haploid egg cell to produce a diploid zygote. The other haploid sperm cell fuses with two haploid nuclei to form the $3n$ endosperm, which provides tissue that will nourish the growing embryonic plant. The zygotes develop within the seeds, which are produced in a long pod.

Under appropriate conditions, the embryo germinates and begins to grow into a plant. The shoot grows upward and the roots downward. A compact rosette of leaves is produced, and under the right conditions, the shoot enlarges and differentiates into flower structures. At maturity, *A. thaliana* is a low-growing plant with roots, a main shoot with branches that bear mature leaves, and small white flowers at the tips of the branches.

Genetic techniques with *Arabidopsis*

A number of traditional and modern molecular techniques that are commonly used with *Arabidopsis* provide it with special advantages for genetic studies. *Arabidopsis* can self-fertilize, which means that any recessive mutation appearing in the germ line can be recovered in the immediate progeny. Cross-fertilization is also possible by removing the anther from one plant and dusting pollen on the stigma of another plant—essentially the same technique used by Gregor Mendel with pea plants (see Figure 3.4).

As already mentioned, many naturally occurring variants of *Arabidopsis* are available for study, and new mutations can be produced by exposing its seeds to chemical mutagens, radiation, or transposable elements that randomly insert into genes. The large number of offspring produced by *Arabidopsis* facilitates screening for rare mutations.

Genes from other organisms can be transferred to *Arabidopsis* by means of the Ti plasmid from the bacterium *Agrobacterium tumefaciens*, which naturally infects plants and transfers the Ti plasmid to plant cells. Subsequent to the transfer, the Ti plasmid randomly inserts into the DNA of the plant that it infects, thereby generating mutations in the plant DNA in a process called *insertional mutagenesis*. Geneticists have modified the Ti plasmid to carry a *GUS* gene, which has no promoter of its own. The *GUS* gene encodes an enzyme that converts a colorless compound (X-Glu) into a blue dye. Because the *GUS* gene has no promoter, it is expressed only when inserted into the coding sequence of a plant gene. When that happens, the enzyme encoded by *GUS* is synthesized and converts X-Glu into a blue dye that stains the cell. This dye provides a means to visually determine the expression pattern of a gene that has been interrupted by Ti DNA, producing information about the expression of genes that are mutated by insertional mutagenesis. ■

12.4 Epigenetic Effects Influence Gene Expression

In eukaryotic cells, changes in chromatin structure that affect gene expression (discussed in Section 12.3) are at least partly responsible for the phenomena of epigenetics, which we discussed in Chapters 4 and 8.

The term *epigenetics* comes from the Greek root *epi*, which means "over" or "above"; the term has come to represent the inheritance of variation above and beyond differences in DNA sequence. Epigenetics usually refers to phenotypes and processes that are transmitted to other cells and sometimes to future generations, but are not the result of differences in the DNA base sequence. Many epigenetic effects are caused by changes in gene expression that result from alterations to chromatin structure or other aspects of DNA structure, such as DNA methylation. Some have broadened the definition of epigenetics to refer to any alteration of chromatin or DNA structure that affects gene expression. Here, we will use the term *epigenetics* to refer to changes in gene expression or phenotype that are potentially heritable without alteration of the underlying DNA base sequence.

Many epigenetic changes are stable, persisting across cell divisions or even generations. However, epigenetic alterations can also be influenced by environmental factors. The fact that epigenetic traits may be induced by environmental effects and transmitted to future generations has been interpreted by some to mean that through epigenetics, genes have memory—that environmental factors acting on individuals can have effects that are transmitted to future generations. Epigenetics has been called "inheritance, but not as we know it." ▶TRY PROBLEM 37

THINK-PAIR-SHARE Question 12

CONCEPTS

Epigenetic effects are phenotypes and processes that are passed to other cells and sometimes to future generations, but are not the result of differences in the DNA base sequence.

Molecular Mechanisms of Epigenetic Changes

Most evidence suggests that epigenetic effects are brought about by physical changes to chromatin structure. We will consider three types of molecular mechanisms that alter chromatin structure and underlie many epigenetic phenotypes: (1) changes in patterns of DNA methylation;

(2) chemical modifications of histone proteins; and (3) RNA molecules that affect chromatin structure and gene expression.

DNA METHYLATION The best-understood mechanism of epigenetic change is methylation of DNA. As we have seen, DNA methylation refers to the addition of methyl groups to the nucleotide bases. In eukaryotes, the predominant type of DNA methylation is the methylation of cytosine to produce 5-methylcytosine, which is often associated with repression of transcription. DNA methylation often occurs on cytosine nucleotides that are immediately adjacent to guanine nucleotides, together referred to as CpG dinucleotides (p represents the phosphate group that connects the C and G nucleotides). DNA regions that have many CpG dinucleotides are referred to as CpG islands. In mammalian cells, CpG islands are often located in or near the promoters of genes. These CpG islands are usually not methylated when genes are being actively transcribed. However, methylation of CpG islands near a gene leads to repression of transcription.

The fact that epigenetic changes are passed to other cells and (sometimes) to future generations means that the changes in chromatin structure associated with epigenetic phenotypes must be faithfully maintained when chromosomes replicate. How are epigenetic changes retained and replicated through the process of cell division?

Methylation of CpG dinucleotides means that two methylated cytosine bases sit diagonally across from each other on opposite strands. Before replication, cytosine bases on both strands are methylated (**Figure 12.23**). Immediately after semiconservative replication, the cytosine base on the template strand has a methyl group, but the cytosine base on the newly replicated strand does not. Special methyltransferase enzymes recognize the hemimethylated state of CpG dinucleotides and add methyl groups to the unmethylated cytosine bases, resulting in two new DNA molecules that are fully methylated. In this way, the methylation pattern of DNA is maintained across cell division.

HISTONE MODIFICATIONS Epigenetic changes can also occur through modification of histone proteins. As we saw in Section 12.3, these modifications can alter chromatin structure and affect the transcription of genes. These types of modifications have been called **epigenetic marks**. For example, the addition of three methyl groups to lysine 4 in the H3 histone (H3K4me3) is often found near transcriptionally active genes. Methylation of lysine 36 in the H3 histone (H3K36me3) is also associated with increased transcription. On the other hand, the addition of three methyl groups to lysine 9 in H3 (H3K9me3) and to lysine 20 in histone 4 (H4K20me3) is associated with repression of transcription. Many additional histone modifications have been shown to be associated with the level of transcription.

Several models have been proposed to explain how histone modifications are faithfully transmitted to daughter cells. During the process of DNA replication, nucleosomes are disrupted and the original histone proteins are distributed randomly between the two new DNA molecules. Newly synthesized histones are then added to complete the formation of new nucleosomes (see Chapter 8). Most models assume that after replication, the epigenetic marks remain on the original histones; these

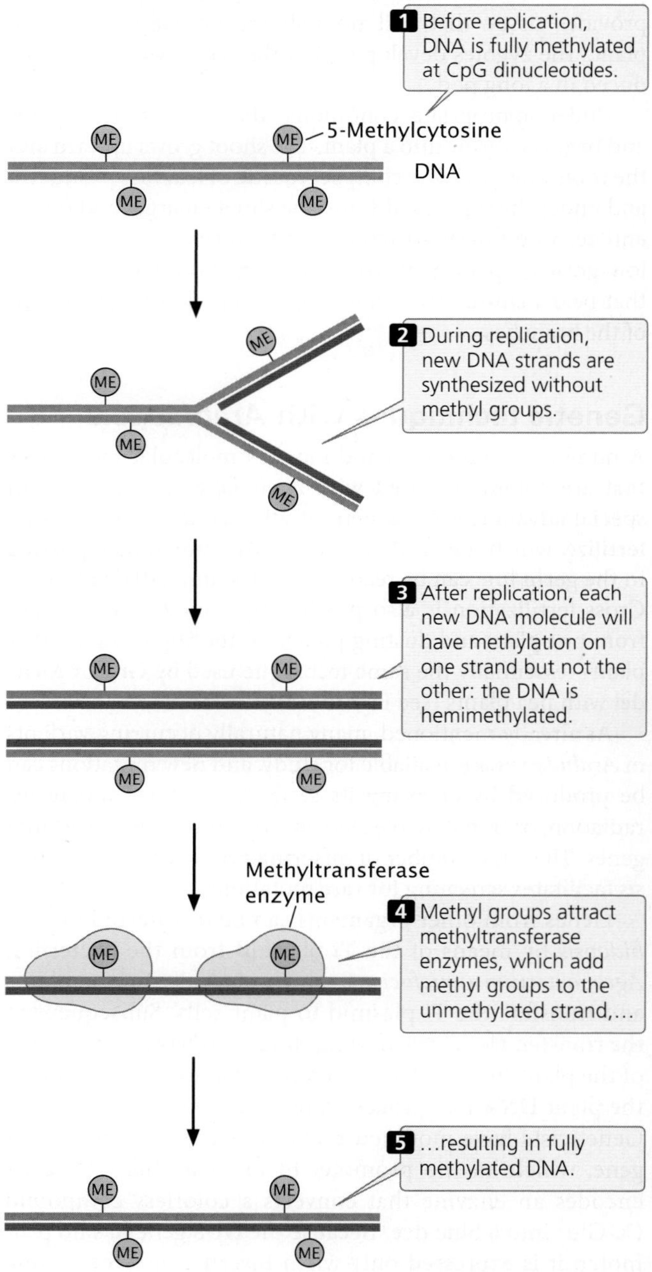

12.23 DNA methylation is stably maintained through DNA replication.

marks then recruit enzymes that make similar changes to the new histones, maintaining the histone modifications across cell division.

EPIGENETIC EFFECTS OF RNA MOLECULES Evidence increasingly demonstrates that RNA molecules play an important role in bringing about epigenetic effects. The first discovered and still best-understood example of RNA mediation of epigenetic change is X inactivation, in which a long noncoding RNA called *Xist* suppresses transcription on one of the X chromosomes in female mammals. The *Xist* lncRNA coats one X chromosome and then attracts a protein called polycomb repressive complex 2, which deposits methyl groups on lysine 27 of histone H3, creating a H3K27me3 epigenetic mark that alters chromatin structure and represses transcription.

CONCEPTS

DNA methylation, histone modifications, and RNA molecules bring about alterations of chromatin structure. Some of these modifications are passed to daughter cells during cell division and to future generations, producing epigenetic effects.

Epigenetic Effects

A number of examples of epigenetic effects have been documented. In some cases, the molecular basis of these effects is at least partly understood.

PARAMUTATION Paramutation is defined as an interaction between alleles that leads to a heritable change in expression of one of the alleles. Surprisingly, paramutation produces these differences in phenotype without any alteration in the DNA base sequence of the converted allele. The phenomenon of paramutation has several important features. First, the newly established expression pattern of the converted allele is transmitted to future generations, even when the allele that brought about the alteration is no longer present with it. Second, the altered allele is now able to convert other alleles to the new phenotype. And third, there are no associated DNA sequence differences in the altered alleles. A number of examples of paramutation have now been discovered in different organisms, and geneticists have begun to unravel the molecular mechanisms of this curious phenomenon.

Paramutation occurs in alleles at the *b1* locus in corn, which is involved in determining pigmentation. The *b1* locus helps to determine the amount of purple anthocyanin that a corn plant produces. The locus actually encodes a transcription factor that regulates genes involved in pigment production. Plants homozygous for the *B-I* allele (*B-I B-I*) show high expression of the *b1* locus and are dark purple (**Figure 12.24**). Plants homozygous for the *B′* allele (*B′B′*) have lower expression of *b1* and are lightly pigmented.

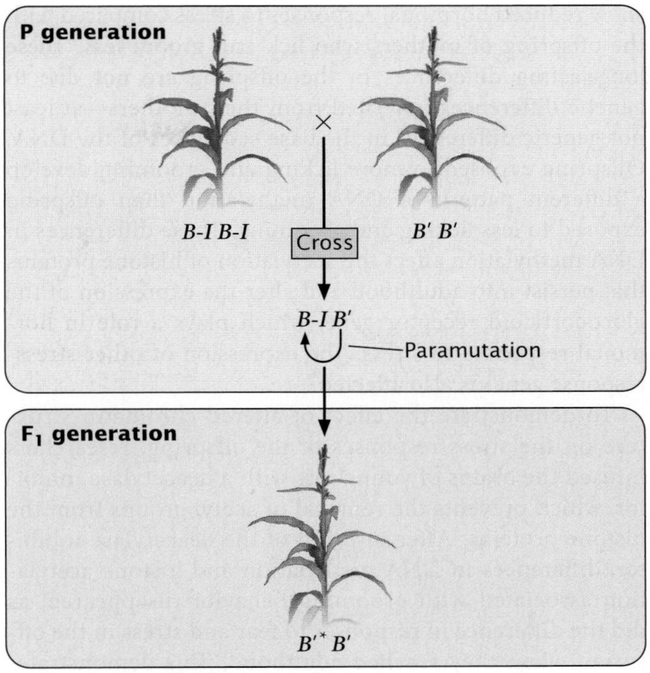

12.24 In paramutation at the *b1* locus in corn, a copy of the *B′* allele converts the *B-I* allele to *B′, which has the same phenotype as *B′*.** The *B-I B-I* genotype produces a pigmented plant, while *B′B′* and *B′B′** genotypes are lightly pigmented.

However, the DNA sequences of the *B-I* and *B′* alleles are identical. Genetically identical alleles such as these, which produce heritable differences in phenotypes through epigenetic processes, are referred to as **epialleles**.

THINK-PAIR-SHARE Question 13

In plants that are heterozygous *B-I B′*, the *B-I* allele is converted to *B′*, with the result that the heterozygous plants are lightly pigmented (see Figure 12.24), just like the *B′B′* homozygotes. The newly converted allele is usually designated *B′**. Importantly, there is no functional difference between *B′* and *B′**; the *B′** allele is now fully capable of converting other *B-I* alleles into *B′** alleles in subsequent generations. Current evidence suggests that siRNA molecules are involved in the conversion of *B-I* to *B′**, and that this conversion probably involves a change in the chromatin states of the alleles.

THINK-PAIR-SHARE Question 14

EPIGENETIC CHANGES INDUCED BY MATERNAL BEHAVIOR A fascinating example of epigenetics is seen in the long-lasting effects of maternal behavior in rats. A mother rat licks and grooms her offspring, usually while she arches her back and nurses them. The offspring of mothers who display more licking and grooming behavior are less fearful as adults and show reduced hormonal responses to stress compared with the offspring of mothers who lick and groom less. These long-lasting differences in the offspring are not due to genetic differences inherited from their mothers—at least not genetic differences in the base sequences of the DNA. Offspring exposed to more licking and grooming develop a different pattern of DNA methylation than offspring exposed to less licking and grooming. These differences in DNA methylation affect the acetylation of histone proteins that persist into adulthood and alter the expression of the glucocorticoid receptor gene, which plays a role in hormonal responses to stress. The expression of other stress-response genes is also affected.

To demonstrate the effect of altered chromatin structure on the stress responses of the offspring, researchers infused the brains of young rats with a deacetylase inhibitor, which prevents the removal of acetyl groups from the histone proteins. After infusion of the deacetylase inhibitor, differences in DNA methylation and histone acetylation associated with grooming behavior disappeared, as did the difference in responses to fear and stress in the offspring when they reached adulthood. This demonstrates that the mother rat's licking and grooming behavior brings about epigenetic changes in the offspring's chromatin, which causes long-lasting differences in their behavior. ▶TRY PROBLEM 38

EPIGENETIC EFFECTS OF ENVIRONMENTAL CHEMICALS Because some chemicals are capable of modifying chromatin structure, researchers have looked for long-term effects of environmental toxicants on chromatin structure and epigenetic traits. There has been much recent interest in chemicals called endocrine disruptors, which mimic or interfere with natural hormones. Endocrine disruptors are capable of interfering with processes regulated by natural hormones, such as sexual development and reproduction. One endocrine disruptor is vinclozolin, a common fungicide used to control fungal diseases in fruits—particularly wine grapes—and vegetables and to treat turf on golf courses.

In one study, researchers found that the exposure of female rats to vinclozolin led to reduced sperm production not only in their male offspring (when they reached puberty), but also in several subsequent generations. Increased DNA methylation was seen in sperm of the males that were exposed to vinclozolin, and these patterns of methylation were inherited. This study and others have raised concerns that, through epigenetic changes, environmental exposure to some chemicals might have effects on the health of future generations.

THINK-PAIR-SHARE Question 15

EPIGENETIC EFFECTS IN MONOZYGOTIC TWINS Monozygotic (identical) twins develop from a single egg fertilized by a single sperm that divides and gives rise to two zygotes. Monozygotic twins are genetically identical, in the sense that they possess identical DNA sequences, but they often differ somewhat in appearance, health, and behavior. The nature of these differences in the phenotypes of identical twins is not well understood, but recent evidence suggests that at least some of these differences may be due to epigenetic changes. In one study, Mario Fraga at the Spanish National Cancer Center and his colleagues examined 80 pairs of identical twins and compared the degree and location of their DNA methylation and histone acetylation. They found that DNA methylation and histone acetylation in pairs of identical twins were similar early in life, but that older twin pairs had remarkable differences in their overall content and distribution of DNA methylation and histone acetylation. Furthermore, these differences affected gene expression in the twins. This research suggests that identical twins do differ epigenetically and that phenotypic differences between them may be caused by differential gene expression.

THINK-PAIR-SHARE Question 16

The Epigenome

As we have seen, epigenetic information is contained within chromatin structure, and that information is heritable and affects how the DNA base sequence is expressed. The overall pattern of chromatin modifications in a genome has been termed the **epigenome**.

Over the past few years, a number of techniques have become available for detecting and describing epigenetic modifications across the genome. Using these techniques, geneticists have compared the epigenomes of different types of cells. For example, researchers have compared the epigenomes of cancer cells and normal cells and have observed distinct epigenetic marks associated with cancer. Similarly, researchers have mapped the genomic locations of histone modifications in different cell types. These studies have detected specific histone modifications associated with promoters and enhancers of active genes. In one study, researchers mapped nine different epigenetic marks in nine different types of human cells. They were able to determine how the chromatin marks varied across cell types and to compare the epigenetic marks associated with active and repressed genes. Because specific epigenetic marks are often associated with regulatory elements such as promoters and enhancers, researchers have used the presence of these marks to map the locations of these regulatory elements throughout the genome.

CONCEPTS

The epigenome is the complete set of chromatin modifications possessed by an individual organism.

CONCEPTS SUMMARY

- Gene expression can be controlled at different levels along the molecular pathway from DNA to protein, including the alteration of gene structure, transcription, mRNA processing, mRNA stability, translation, and posttranslational modification. Much of gene regulation takes place through the action of regulatory gene products that recognize and bind to regulatory elements.
- Genes in bacterial cells are typically clustered into operons—groups of functionally related structural genes and the sequences that control their transcription. Structural genes in an operon are transcribed together as a single mRNA molecule.
- In negative control, a repressor protein binds to DNA and inhibits transcription. In positive control, an activator protein binds to DNA and stimulates transcription. In inducible operons, transcription is normally off and must be turned on; in repressible operons, transcription is normally on and must be turned off.
- The *lac* operon of *E. coli* is a negative inducible operon. In the absence of lactose, a repressor binds to the operator and prevents the transcription of genes that encode β-galactosidase, permease, and transacetylase. When lactose is present, some of it is converted into allolactose, which binds to the repressor and makes it inactive, allowing the structural genes to be transcribed.
- Positive control in the *lac* and other operons occurs through catabolite repression. When complexed with cAMP, the catabolite activator protein binds to a site in or near the promoter and stimulates the transcription of the structural genes. Levels of cAMP are inversely correlated with glucose, so low levels of glucose stimulate transcription and high levels inhibit transcription.
- The *trp* operon of *E. coli* is a negative repressible operon that controls the biosynthesis of tryptophan.
- Eukaryotic cells differ from bacteria in several ways that affect gene regulation, including, in eukaryotes, the absence of operons, the presence of chromatin, and the presence of a nuclear membrane.
- In eukaryotic cells, chromatin structure represses gene expression. In transcription, chromatin structure may be altered by chromatin-remodeling complexes that reposition nucleosomes and by modifications of histone proteins, including acetylation, phosphorylation, and methylation. The methylation of DNA also affects transcription.
- The initiation of eukaryotic transcription is controlled by general transcription factors that assemble into the basal transcription apparatus and by transcriptional regulator proteins that stimulate or repress normal levels of transcription by binding to regulatory promoters and enhancers.
- Enhancers affect the transcription of distant genes. Regulatory proteins bind to enhancers and interact with the basal transcription apparatus by causing the intervening DNA to loop out. Insulators limit the action of enhancers by blocking their action in a position-dependent manner.
- Coordinately controlled genes in eukaryotic cells respond to the same factors because they have common response elements that are stimulated by the same transcriptional activator.
- Gene expression in eukaryotic cells can be influenced by RNA processing and by changes in mRNA stability. The 5′ cap, the poly(A) tail, the 5′ UTR, the coding region, and sequences in the 3′ UTR are important in controlling the stability of eukaryotic mRNAs.
- RNA interference plays an important role in eukaryotic gene regulation. Small RNA molecules (siRNAs and miRNAs) combine with proteins and bind to sequences on mRNA or DNA. These complexes cleave RNA, inhibit translation, affect RNA degradation, and silence transcription.
- Posttranslational modification of proteins may play a role in the regulation of gene expression.
- *Arabidopsis thaliana* possesses a number of characteristics that make it an ideal model genetic organism.
- Epigenetic effects—inherited changes in gene expression not due to changes in the DNA base sequence—are frequently caused by DNA methylation and alterations in chromatin structure. Epigenetic changes are stable but can be affected by environmental factors.
- Many epigenetic phenotypes result from changes to chromatin structure. Epigenetic effects occur through DNA methylation, histone modifications, and RNA molecules.
- Paramutation is a heritable alteration of one allele by another allele, without any change in DNA sequence.
- Early life experiences can produce epigenetic changes that have long-lasting effects on behavior. Environmental chemicals may produce epigenetic effects that are passed to later generations. Phenotypic differences between genetically identical monozygotic twins may result from epigenetic effects.
- The epigenome is the complete set of chromatin modifications possessed by an individual organism.

IMPORTANT TERMS

gene regulation (p. 320)
structural gene (p. 321)
regulatory gene (p. 321)
constitutive gene (p. 321)
regulatory element (p. 321)
operon (p. 322)
regulator gene (p. 322)
regulator protein (p. 322)
operator (p. 322)
negative control (p. 323)
positive control (p. 323)
inducible operon (p. 323)
repressible operon (p. 323)
inducer (p. 323)
allosteric protein (p. 323)
corepressor (p. 324)
coordinate induction (p. 327)
partial diploid (p. 328)
constitutive mutation (p. 329)
catabolite repression (p. 332)
catabolite activator protein (CAP) (p. 332)
adenosine-3′,5′-cyclic monophosphate (cAMP) (p. 332)
chromatin-remodeling complex (p. 335)
histone code (p. 335)
general transcription factor (p. 337)
transcriptional regulator protein (p. 337)
enhancer (p. 337)
transcriptional activator protein (p. 337)
coactivator (p. 337)
silencer (p. 338)
insulator (p. 338)
heat-shock protein (p. 339)
response element (p. 339)
epigenetic marks (p. 346)
paramutation (p. 347)
epialleles (p. 347)
epigenome (p. 348)

ANSWERS TO CONCEPT CHECKS

1. A constitutive gene is not regulated and is expressed continually.

2. Transcription is the first step in the process of information transfer from DNA to protein. For cellular efficiency, gene expression is often regulated early in the process of protein production.

3. d

4. d

5. c

6. b

7. b

8. The DNA between the enhancer and the promoter loops out, so that transcription activators bound to the enhancer are able to interact directly with the basal transcription apparatus.

9. The poly(A) tail stabilizes the 5′ cap, which must be removed before the mRNA molecule can be degraded from the 5′ end.

10. d

WORKED PROBLEMS

Problem 1

The *fox* operon, which has sequences *A*, *B*, *C*, and *D* (which may represent either structural genes or regulatory sequences), encodes enzymes 1 and 2. Mutations in sequences *A*, *B*, *C*, and *D* have the following effects, where a plus sign (+) indicates that the enzyme is synthesized and a minus sign (−) indicates that the enzyme is not synthesized.

	Fox absent		Fox present	
Mutation in sequence	**Enzyme 1**	**Enzyme 2**	**Enzyme 1**	**Enzyme 2**
No mutation	−	−	+	+
A	−	−	−	+
B	−	−	−	−
C	−	−	+	−
D	+	+	+	+

a. Is the *fox* operon inducible or repressible?

b. Indicate which sequence (*A*, *B*, *C*, or *D*) is part of the following components of the operon. Each sequence should be used only once.

Regulator gene ______________ Structural gene for enzyme 1 ______________

Promoter ______________ Structural gene for enzyme 2 ______________

Solution Strategy

What information is required in your answer to the problem?

a. Whether the *fox* operon is inducible or repressible.

b. Which sequence represents each part of the operon.

What information is provided to solve the problem?

For each mutation, whether enzyme 1 and enzyme 2 are produced in the presence and absence of Fox.

eview Figure
a summary
n structure.

For help with this problem, review:

Section 12.2.

Solution Steps

a. When no mutations are present, enzymes 1 and 2 are produced in the presence of Fox but not in its absence, indicating that the operon is inducible and that Fox is the inducer.

b. The mutation in *A* allows the production of enzyme 2 in the presence of Fox, but enzyme 1 is not produced in the presence or absence of Fox, so *A* must have a mutation in the structural gene for enzyme 1. With the mutation in *B*, neither enzyme is produced under any conditions, so this mutation most likely occurs in the promoter and prevents RNA polymerase from binding. The mutation in *C* affects only enzyme 2, which is not produced in the presence or absence of Fox; enzyme 1 is produced normally (only in the presence of Fox), so the mutation in *C* most likely occurs in the structural gene for enzyme 2. The mutation in *D* is constitutive, allowing the production of enzymes 1 and 2 whether or not Fox is present. This mutation most likely occurs in the regulator gene, producing a defective repressor that is unable to bind to the operator under any conditions.

Regulator gene	*D*
Promoter	*B*
Structural gene for enzyme 1	*A*
Structural gene for enzyme 2	*C*

Problem 2

What would be the effect of a mutation that caused poly(A)-binding proteins to be nonfunctional?

Solution Strategy

What information is required in your answer to the problem?

The effect of a mutation that eliminates the function of the poly(A)-binding protein.

What information is provided to help solve the problem?

- A mutation occurs in the gene that encodes poly(A) binding protein.
- The mutation causes the poly(A) protein to be nonfunctional.

For help with this problem, review:

- The Degradation of RNA in Section 12.3.
- Addition of the Poly(A) Tail in Section 10.5.

Solution Steps

Recall: The poly(A) tail affects the stability of mRNA.

Degradation of mRNA from the 5′ end requires the removal of the 5′ cap and is usually preceded by the shortening of the poly(A) tail. Poly(A)-binding proteins bind to the poly(A) tail and prevent it from being shortened. Thus, the presence of these proteins on the poly(A) tail protects the 5′ cap, which prevents RNA degradation. If the gene for poly(A)-binding protein were mutated in such a way that nonfunctional poly(A)-binding protein was produced, the protein would not bind to the poly(A) tail. The tail would be shortened prematurely, the 5′ cap removed, and mRNA degraded more easily. The end result would be less mRNA and thus less protein synthesis.

COMPREHENSION QUESTIONS

Section 12.1

1. Why is gene regulation important for bacterial cells?
2. Name six different levels at which gene expression might be controlled.

Section 12.2

3. Draw a picture illustrating the general structure of an operon and identify its parts.
4. What is the difference between positive and negative control? What is the difference between inducible and repressible operons?
5. Briefly describe the *lac* operon and how it controls the metabolism of lactose.
6. What is catabolite repression? How does it allow a bacterial cell to use glucose in preference to other sugars?

Section 12.3

7. What changes take place in chromatin structure and what role do these changes play in eukaryotic gene regulation?
8. What is the histone code?
9. Briefly explain how transcriptional activator and repressor proteins affect the level of transcription of eukaryotic genes.
10. What is an enhancer? How does it affect the transcription of distant genes?
11. What role does mRNA stability play in gene regulation? What controls mRNA stability in eukaryotic cells?
12. Briefly list some of the ways in which siRNAs and miRNAs regulate genes.
13. How does bacterial gene regulation differ from eukaryotic gene regulation? How are they similar?

Section 12.4

14. What are epigenetic effects? How do they differ from other genetic traits?
15. What types of changes are thought to be responsible for epigenetic traits?
16. How are patterns of DNA methylation maintained across cell divisions?

For more questions that test your comprehension of the key chapter concepts, go to **LearningCurve** for this chapter.

APPLICATION QUESTIONS AND PROBLEMS

Section 12.2

*17. For each of the following types of transcriptional control, indicate whether the protein produced by the regulator gene will be synthesized initially as an active repressor, inactive repressor, active activator, or inactive activator.

  a. Negative control in a repressible operon

  b. Positive control in a repressible operon

  c. Negative control in an inducible operon

  d. Positive control in an inducible operon

*18. A mutation at the operator prevents the regulator protein from binding. What effect will this mutation have in the following types of operons?

  a. Regulator protein is a repressor in a repressible operon.

  b. Regulator protein is a repressor in an inducible operon.

19. The *blob* operon produces enzymes that convert compound A into compound B. The operon is controlled by a regulatory gene *S*. Normally, the enzymes are synthesized only in the absence of compound B. If gene *S* is mutated, the enzymes are synthesized in the presence *and* in the absence of compound B. Does gene *S* produce a regulatory protein that exhibits positive or negative control? Is this operon inducible or repressible?

*20. A mutation prevents the catabolite activator protein (CAP) from binding to the promoter in the *lac* operon. What will the effect of this mutation be on the transcription of the operon?

21. Under which of the following conditions would a *lac* operon produce the greatest amount of β-galactosidase? The least? Explain your reasoning.

	Lactose present	Glucose present
Condition 1	Yes	No
Condition 2	No	Yes
Condition 3	Yes	Yes
Condition 4	No	No

22. A mutant strain of *E. coli* produces β-galactosidase in the presence *and* in the absence of lactose. Where in the operon might the mutation in this strain be located?

23. Examine **Figure 12.7**. What would be the effect of a drug that altered the structure of allolactose so that it was unable to bind to the regulator protein?

*24. For *E. coli* strains with the *lac* genotypes given on the next page, use a plus sign (+) to indicate the synthesis of β-galactosidase and permease and a minus sign (−) to indicate no synthesis of the proteins.

Genotype of strain	Lactose absent		Lactose present	
	β-Galactosidase	Permease	β-Galactosidase	Permease
$lacI^+\ lacP^+\ lacO^+\ lacZ^+\ lacY^+$				
$lacI^-\ lacP^+\ lacO^+\ lacZ^+\ lacY^+$				
$lacI^+\ lacP^+\ lacO^c\ lacZ^+\ lacY^+$				
$lacI^-\ lacP^+\ lacO^+\ lacZ^+\ lacY^-$				
$lacI^-\ lacP^-\ lacO^+\ lacZ^+\ lacY^+$				
$lacI^+\ lacP^+\ lacO^+\ lacZ^-\ lacY^+/lacI^-\ lacP^+\ lacO^+\ lacZ^+\ lacY^-$				
$lacI^-\ lacP^+\ lacO^c\ lacZ^+\ lacY^+/lacI^+\ lacP^+\ lacO^+\ lacZ^-\ lacY^-$				
$lacI^-\ lacP^+\ lacO^+\ lacZ^+\ lacY^-/lacI^+\ lacP^-\ lacO^+\ lacZ^-\ lacY^+$				
$lacI^+\ lacP^-\ lacO^c\ lacZ^-\ lacY^+/lacI^-\ lacP^+\ lacO^+\ lacZ^+\ lacY^-$				
$lacI^+\ lacP^+\ lacO^+\ lacZ^+\ lacY^+/lacI^+\ lacP^+\ lacO^+\ lacZ^+\ lacY^+$				
$lacI^s\ lacP^+\ lacO^+\ lacZ^+\ lacY^-/lacI^+\ lacP^+\ lacO^+\ lacZ^-\ lacY^+$				
$lacI^s\ lacP^-\ lacO^+\ lacZ^-\ lacY^+/lacI^+\ lacP^+\ lacO^+\ lacZ^+\ lacY^+$				

25. Give all possible genotypes of a *lac* operon that produces β-galactosidase and permease under the following conditions. Do not give partial-diploid genotypes.

	Lactose absent		Lactose present	
	β-Galactosidase	Permease	β-Galactosidase	Permease
a.	−	−	+	+
b.	−	−	−	+
c.	−	−	+	−
d.	+	+	+	+
e.	−	−	−	−
f.	+	−	+	−
g.	−	+	−	+

***26.** Explain why mutations in the *lacI* gene are trans in their effects, but mutations in the *lacO* gene are cis in their effects.

27. The *mmm* operon, which has sequences *A*, *B*, *C*, and *D* (which may be structural genes or regulatory sequences), encodes enzymes 1 and 2. Mutations in sequences *A*, *B*, *C*, and *D* have the following effects, where a plus sign (+) indicates that the enzyme is synthesized and a minus sign (−) indicates that the enzyme is not synthesized.

Mutation in sequence	Mmm absent		Mmm present	
	Enzyme 1	Enzyme 2	Enzyme 1	Enzyme 2
No mutation	+	+	−	−
A	−	+	−	−
B	+	+	+	+
C	+	−	−	−
D	−	−	−	−

a. Is the *mmm* operon inducible or repressible?

b. Indicate which sequence (*A*, *B*, *C*, or *D*) is part of the following components of the operon:

Regulator gene ____________

Promoter ____________

Structural gene for enzyme 1 ____________

Structural gene for enzyme 2 ____________

28. DATA ANALYSIS Ellis Engelsberg and his coworkers examined the regulation of genes taking part in the metabolism of arabinose, a sugar (E. Engelsberg et al. 1965. *Journal of Bacteriology* 90:946–957). Four structural genes encode enzymes that help metabolize arabinose (genes *A*, *B*, *D*, and *E*). An additional gene *C* is linked to genes *A*, *B*, and *D*. These genes are in the order *D-A-B-C*. Gene *E* is distant from the other genes. Engelsberg and his colleagues isolated mutations at the *C* gene that affected the expression of structural genes *A*, *B*, *D*, and *E*. In one set of experiments, they created various genotypes at the *A* and *C* loci and determined whether arabinose isomerase (the enzyme encoded by gene *A*) was produced in the presence or absence of arabinose (the substrate of arabinose isomerase). Results from this experiment are shown in the following table, where a plus sign (+) indicates that the arabinose isomerase was synthesized and a minus sign (−) indicates that the enzyme was not synthesized.

Genotype	Arabinose absent	Arabinose present
1. $C^+\ A^+$	−	+
2. $C^-\ A^+$	−	−
3. $C^-\ A^+/C^+\ A^-$	−	+
4. $C^c\ A^-/C^-\ A^+$	+	+

a. On the basis of the results of these experiments, is the *C* gene an operator or a regulator gene? Explain your reasoning.

b. Do these experiments suggest that the arabinose operon is negatively or positively controlled? Explain your reasoning.

c. What type of mutation is C^c?

29. In *E. coli*, three structural genes (*A*, *D*, and *E*) encode enzymes A, D, and E, respectively. Gene *O* is an operator. The genes are in the order *O-A-D-E* on the chromosome. These enzymes catalyze the biosynthesis of valine. Mutations were isolated at the *A*, *D*, *E*, and *O* genes to study the production of enzymes A, D, and E (T. Ramakrishnan and E. A. Adelberg. 1965. *Journal of Bacteriology* 89:654–660). Levels of the enzymes produced by partial-diploid *E. coli* with various combinations of mutations are shown in the following table.

	Amount of enzyme produced		
Genotype	E	D	A
1. $E^+ D^+ A^+ O^+ / E^+ D^+ A^+ O^+$	2.40	2.00	3.50
2. $E^+ D^+ A^+ O^- / E^+ D^+ A^+ O^+$	35.80	38.60	46.80
3. $E^+ D^- A^+ O^- / E^+ D^+ A^- O^+$	1.80	1.00	47.00
4. $E^+ D^+ A^- O^- / E^+ D^- A^+ O^+$	35.30	38.00	1.70
5. $E^- D^+ A^+ O^- / E^+ D^- A^+ O^+$	2.38	38.00	46.70

a. Is the regulator protein that binds to the operator of this operon a repressor (negative control) or an activator (positive control)? Explain your reasoning.

b. Are genes *A*, *D*, and *E* all under the control of operator *O*? Explain your reasoning.

c. Propose an explanation for the low level of enzyme E produced in genotype 3.

Section 12.3

30. A geneticist is trying to determine how many genes are found in a 300,000-bp region of DNA. Analysis shows that four different areas within the 300,000-bp region have H3K4me3 modifications. What might their presence suggest about the number of genes located there?

31. In a line of human cells grown in culture, a geneticist isolates a temperature-sensitive mutation at a locus that encodes an acetyltransferase enzyme; at temperatures above 38°C, the mutant cells produce a nonfunctional form of the enzyme. What would be the most likely effect of this mutation when the cells are grown at 40°C?

32. X31b is an experimental compound that is taken up by rapidly dividing cells. Research has shown that X31b stimulates the methylation of DNA. Some cancer researchers are interested in testing X31b as a possible drug for treating prostate cancer. Offer a possible explanation for why X31b might be an effective anticancer drug.

33. What would be the effect of moving the insulator shown in **Figure 12.18** to a position between enhancer II and the promoter for gene B?

*****34.** An enhancer is surrounded by four genes (*A*, *B*, *C*, and *D*), as shown in the diagram below. An insulator lies between gene *C* and gene *D*. On the basis of the positions of the genes, the enhancer, and the insulator, the transcription of which genes is most likely to be stimulated by the enhancer? Explain your reasoning.

35. Some eukaryotic mRNAs have an AU-rich element in the 3′ untranslated region. What would be the effect on gene expression if this element were mutated or deleted?

36. A strain of *Arabidopsis thaliana* possesses a mutation in the *APETALA2* gene, in which much of the 3′ untranslated region of mRNA transcribed from the gene is deleted. What is the most likely effect of this mutation on the expression of the *APETALA2* gene?

Section 12.4

*****37.** How do epigenetic traits differ from traditional genetic traits, such as the differences in color and shape of peas that Mendel studied?

*****38.** A scientist does an experiment in which she removes the offspring of rats from their mother at birth and has her genetics students feed and rear the offspring. Assuming that the students do not lick and groom the baby rats like their mothers normally do, what long-term behavioral and epigenetic effects would you expect to see in the rats when they grow up?

39. Pregnant female rats were exposed to a daily dose of 100 or 200 mg/kg of vinclozolin, a fungicide commonly used in the wine industry (M. D. Anway et al. 2005, *Science* 308:1466–1469). The F_1 offspring of the exposed females were interbred, producing F_2, F_3, and F_4 rats. None of the F_2, F_3, or F_4 rats were exposed to vinclozolin. Testes from the F_1–F_4 male rats were examined and compared with those of control rats descended from females that had not been exposed to vinclozolin. There were higher percentages of apoptotic cells (cells that underwent controlled cell death) in the testes of F_1–F_4 male descendants of females who were exposed to vinclozolin than in descendants of control females (graph a). Furthermore, sperm numbers (graph b) and motilities (graph c) were lower in the F_1–F_4 descendants of vinclozolin-exposed females than in those of control females. In addition, 8% of the F_1–F_4 males descended from vinclozolin-exposed females developed complete infertility, compared with 0% of the F_1–F_4 males descended from control females. Molecular analysis of the testes demonstrated that DNA methylation patterns differed between descendants of vinclozolin-exposed females and descendants of control females. Explain the transgenerational effects of vinclozolin on male fertility.

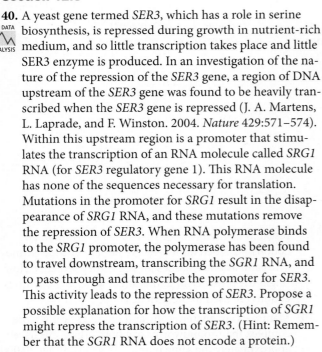

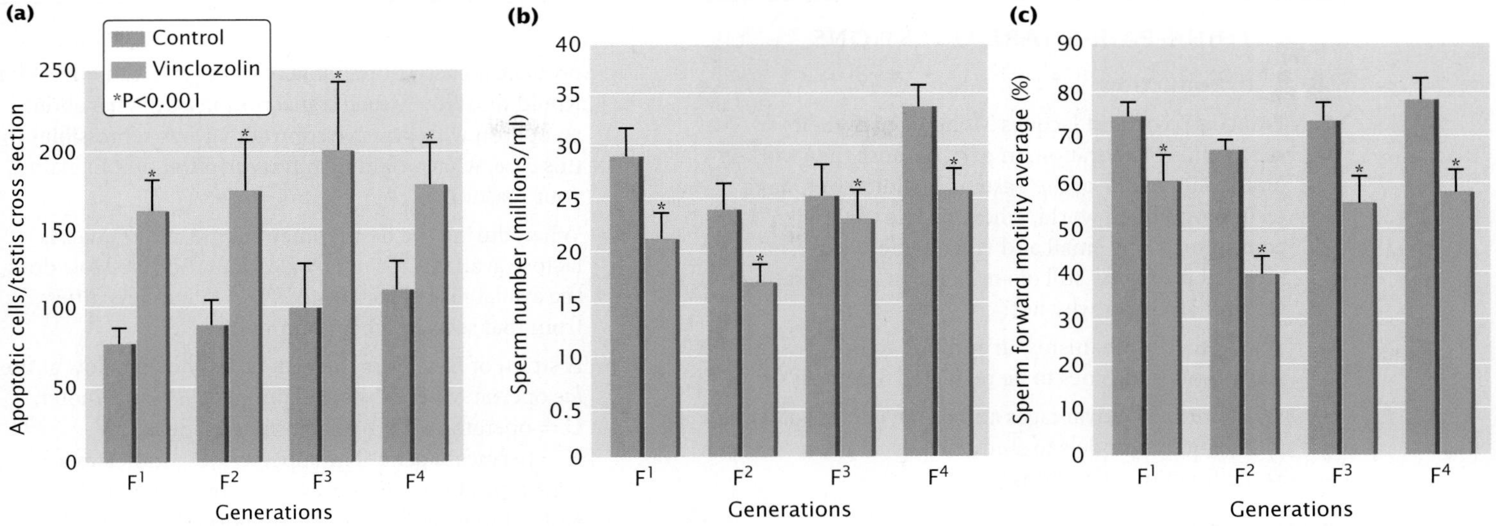

[Graphs after M. D. Anway et al. 2005. Science 308:1466–1469.]

CHALLENGE QUESTIONS

Section 12.3

40. A yeast gene termed *SER3*, which has a role in serine biosynthesis, is repressed during growth in nutrient-rich medium, and so little transcription takes place and little SER3 enzyme is produced. In an investigation of the nature of the repression of the *SER3* gene, a region of DNA upstream of the *SER3* gene was found to be heavily transcribed when the *SER3* gene is repressed (J. A. Martens, L. Laprade, and F. Winston. 2004. *Nature* 429:571–574). Within this upstream region is a promoter that stimulates the transcription of an RNA molecule called *SRG1* RNA (for *SER3* regulatory gene 1). This RNA molecule has none of the sequences necessary for translation. Mutations in the promoter for *SRG1* result in the disappearance of *SRG1* RNA, and these mutations remove the repression of *SER3*. When RNA polymerase binds to the *SRG1* promoter, the polymerase has been found to travel downstream, transcribing the *SGR1* RNA, and to pass through and transcribe the promoter for *SER3*. This activity leads to the repression of *SER3*. Propose a possible explanation for how the transcription of *SGR1* might repress the transcription of *SER3*. (Hint: Remember that the *SGR1* RNA does not encode a protein.)

41. A common feature of many eukaryotic mRNAs is the presence of a rather long 3′ UTR, which often contains consensus sequences. Creatine kinase B (CK-B) is an enzyme important in cellular metabolism. Certain cells—termed U937D cells—have lots of CK-B mRNA, but no CK-B enzyme is present. In these cells, the 5′ end of the CK-B mRNA is bound to ribosomes, but the mRNA is apparently not translated. Something inhibits the translation of the CK-B mRNA in these cells.

Researchers introduced numerous short segments of RNA containing only 3′ UTR sequences into U937D cells. As a result, the U937D cells began to synthesize the CK-B enzyme, but the total amount of CK-B mRNA did not increase. The introduction of short segments of other RNA sequences did not stimulate the synthesis of CK-B; only the 3′ UTR sequences turned on the translation of the enzyme.

Based on these results, propose a mechanism for how CK-B translation is inhibited in U937D cells. Explain how the introduction of short segments of RNA containing the 3′ UTR sequences might remove the inhibition.

Section 12.4

42. Techniques have been developed to clone mammals through a process called nuclear transfer, in which the nucleus of a somatic cell is transferred to an egg cell from which the nuclear material has been removed. Research has demonstrated that when a nucleus from a *differentiated somatic cell* is transferred to an egg cell, only a small percentage of the resulting embryos complete development and many of those that do die shortly after birth. In contrast, when a nucleus from an *undifferentiated embryonic stem cell* is transferred into an egg cell, the percentage of embryos that complete development is significantly higher (W. M. Rideout, K. Eggan, and R. Jaenisch. 2001. *Science* 293:1095–1098). Why might the successful development of cloned embryos be higher when the nucleus transferred comes from an undifferentiated embryonic stem cell?

THINK-PAIR-SHARE QUESTIONS

Introduction

1. François Jacob and Jacques Monod began their successful collaboration, in part, because they worked on the same floor of the Pasteur Institute. In today's technology-filled world, where people can easily communicate by email and videoconferencing, is physical proximity still of any value in collaborating in research? Why or why not?
2. What other mechanisms, in addition to operons, might allow groups of genes to be regulated together?
3. What are some advantages and disadvantages of using mathematical models to study biological processes such as the function of operons?

Section 12.1

4. What types of genes would you expect to be constitutive?

Section 12.2

5. Suppose that an operon that exhibited positive control was inducible. Would the regulator gene in this case produce an active or an inactive activator protein? What would most likely turn transcription on? What about an operon with positive control that was repressible? In this case, what would turn transcription off? Explain your reasoning.
6. When during the development of most mammals is lactose available for use by *E. coli* in the gut? How does the availability of lactose to *E. coli* in humans differ from that seen in other mammals?
7. A strain of *E. coli* has the genotypes shown below at the *lac* operon, where I = regulator gene, P = promoter, O = operator, Z = β-galactosidase gene, and Y = permease gene. The superscript + indicates a wild-type allele, *c* indicates a constitutive mutation, and − indicates a defective mutation. For each genotype, indicate whether the enzyme will be synthesized or not synthesized when lactose is present or absent by placing a + for *synthesis occurring* and a − for *synthesis not occurring* in the appropriate blank. Explain your reasoning for each answer.

	Inducer (lactose absent)		Inducer (lactose present)	
Genotype	***β*-Galactosidase**	**Permease**	***β*-Galactosidase**	**Permease**
a. $lacI^{+}\ lacP^{+}\ lacO^{+}\ lacZ^{+}\ lacY^{+}$				
b. $lacI^{+}\ lacP^{+}\ lacO^{c}\ lacZ^{+}\ lacY^{+}$				
c. $lacI^{-}\ lacP^{+}\ lacO^{+}\ lacZ^{-}\ lacY^{+}$				
d. $lacI^{-}\ lacP^{+}\ lacO^{+}\ lacZ^{+}\ lacY^{+}/$ $lacI^{+}\ lacP^{-}\ lacO^{+}\ lacZ^{+}\ lacY^{-}$				
e. $lacI^{-}\ lacP^{+}\ lacO^{c}\ lacZ^{-}\ lacY^{+}/$ $lacI^{+}\ lacP^{+}\ lacO^{+}\ lacZ^{+}\ lacY^{+}$				
f. $lacI^{+}\ lacP^{-}\ lacO^{c}\ lacZ^{+}\ lacY^{-}/$ $lacI^{-}\ lacP^{+}\ lacO^{+}\ lacZ^{-}\ lacY^{+}$				

Section 12.3

8. Mutations in genes encoding chromatin-remodeling complexes have been identified in high frequencies in some human cancers. For example, mutations in components of the human SWI–SNF remodeling complex were found in 19% of tumors in a group of cancers. How might a mutation in a chromatin-remodeling complex contribute to cancer?
9. Some DNA nucleotides are located within the coding regions of genes, where they specify the amino acid sequence of a protein. Other nucleotides are found in regulatory elements, where they serve as binding sites for regulatory proteins. It has long been assumed that sequences within coding regions and sequences within regulatory elements are independent, but recent research has determined that about 15% of human codons—dubbed duons—serve both to encode amino acids and as binding sites for regulatory proteins. How might the presence of duons affect which synonymous codons are used in the genetic code?
10. Mammals are anatomically and physiologically more complex than roundworms, yet both organisms have approximately the same number of genes, about 20,000. Some biologists have argued that mammals and other vertebrates have evolved increased complexity by means of pleiotropy—in which each gene encodes multiple characteristics—and that pleiotropy was made possible by the evolution of additional enhancers. Propose an explanation for how additional enhancers might produce increased pleiotropy

11. Research has demonstrated that the composition of a gene's promoter can affect alternative splicing of its RNA transcript. For example, promoters that are activated by certain transcriptional activator proteins cause one form of alternative splicing; when these activators are not present, a different form of splicing takes place. Propose some possible mechanisms by which the promoter could affect alternative splicing.

Section 12.4

12. Epigenetics has been described as "inheritance, but not as we know it." Do you think this is a good definition? Why or why not?

13. How are epialleles different from genetic alleles, such as those that encode differences in the shape of peas or blood types?

14. How is paramutation similar to normal gene mutation? How does it differ? Make a list of similarities and differences.

15. Some people have said that epigenetics provides genes with a memory. Explain how epigenetics allows genes to remember.

16. People often say that monozygotic twins are genetically identical. Do you think that is a correct statement? Present arguments for and against this statement.

SaplingPlus Self-study tools that will help you practice what you've learned and reinforce this chapter's concepts are available online. Go to www.macmillanlearning.com/PierceGenetics6e.

13 Gene Mutations, Transposable Elements, and DNA Repair

Lou Gehrig and Expanding Nucleotide Repeats

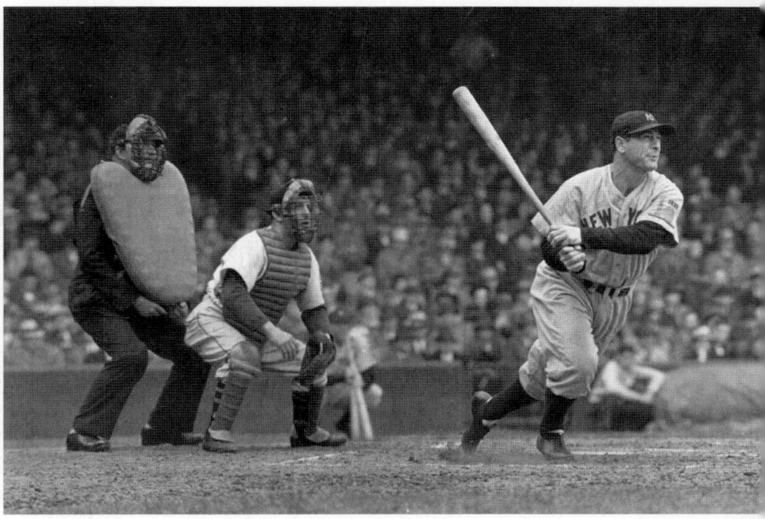

Lou Gehrig at bat. Gehrig, who played baseball for the New York Yankees from 1923 to 1939, was diagnosed with amyotrophic lateral sclerosis, a disease that in some people is caused by expanding nucleotide repeats. [Transcendental Graphics/Getty Images.]

Lou Gehrig was the finest first baseman ever to play major league baseball. A left-handed power hitter, Gehrig played for the New York Yankees from 1923 to 1939. He compiled a lifetime batting average of .340 and drove in more than 100 runs every season for 13 years. During his career, he hit a total of 23 grand slams (home runs with bases loaded). But Gehrig's greatest baseball record, which stood for more than 50 years, was his record of playing 2130 consecutive games.

In 1938, Gehrig fell into a strange slump. His batting average dropped below .300, and in the World Series that year, he managed only four hits—all singles. Nevertheless, he finished the season convinced that he was undergoing a temporary slump. When the 1939 season began, however, it was clear to everyone that something was wrong. Gehrig had no power in his swing; he was awkward and clumsy at first base. His condition worsened, and on May 2 he voluntarily removed himself from the lineup. On June 20, his medical diagnosis was made public: Lou Gehrig was suffering from a rare, progressive disease known as amyotrophic lateral sclerosis (ALS). Since then, ALS has commonly been known as Lou Gehrig disease.

Gehrig experienced symptoms typical of ALS: progressive weakness and wasting of skeletal muscles. Most cases of ALS are sporadic, but about 10% of cases run in families, and in these cases, the disease is inherited as an autosomal dominant trait. Mutations in several genes can cause familial cases of ALS, the most common of which occur in a gene called *chromosome 9 open reading frame 72* (*C9orf72*). The alterations of *C9orf72* that are associated with ALS belong to an unusual group of mutations called expanding nucleotide repeats, in which the number of copies of a set of nucleotides is increased. Most people have somewhere between 2 and 23 repeats of the nucleotide sequence GGGGCC in their *C9orf72* gene, but this number is massively expanded in some people with ALS, who typically possess 700 to 1600 repeats of the sequence.

How the expansion of the repeat in *C9orf72* leads to symptoms of ALS is unknown, but recent research demonstrates that the repeats are translated into one or more proteins that are toxic to nerve cells. The repeats are translated in an unusual and intriguing way: the GGGGCC sequences on both the template and nontemplate strands of the gene are transcribed into RNAs that are translated without a start codon. Because there is no start codon to set the reading frame, all three reading frames on both mRNAs are translated into proteins, resulting in five proteins, each with a different series of repeating dipeptides: glycine-alanine, glycine-proline, proline-alanine, glycine-arginine, and proline-arginine. Research suggests that proteins with the glycine-arginine and proline-arginine dipeptides are responsible for the neurodegeneration that occurs in ALS. The toxicity of these proteins may result from the fact that they mimic RNA-binding proteins and interfere with splicing of pre-mRNA and the processing of rRNA.

THINK-PAIR-SHARE Questions 1 and 2

The story of ALS and expanding nucleotide repeats illustrates the central importance of mutations: the analysis of individuals with mutations is often a source of key insights into diseases and important biological processes. This chapter focuses on gene mutations. We begin with a brief examination of the different types of mutations, including their phenotypic effects, how they can be suppressed, and their rates of occurrence. The next section explores how mutations spontaneously arise, as well as how chemicals and radiation can induce them. We then turn to transposable elements, which often generate mutations. Finally, we look at DNA repair.

13.1 Mutations Are Inherited Alterations in the DNA Sequence

DNA is a highly stable molecule that is replicated with amazing accuracy (as we saw in Chapters 8 and 9), but changes in DNA structure and errors of replication do take place. These changes may alter the genetic information of the DNA. A **mutation** is defined as an inherited change in genetic information; the descendants may be cells or organisms.

The Importance of Mutations

Mutations are both the sustainer of life and the cause of great suffering. On the one hand, mutation is the source of all genetic variation, the raw material of evolution. The ability of organisms to adapt to environmental change depends critically on the presence of genetic variation in natural populations, and genetic variation is produced—at least partly—by mutation. On the other hand, many mutations have detrimental effects, and mutation is the source of many diseases and disorders.

Much of the study of genetics focuses on how genetic variations produced by mutation are inherited; genetic crosses are meaningless if all individual members of a species are identically homozygous for the same alleles. Much of Gregor Mendel's success in unraveling the principles of inheritance can be traced to his use of carefully selected variants of the garden pea. Similarly, Thomas Hunt Morgan and his students discovered many basic principles of genetics by analyzing mutant fruit flies.

Mutations are also useful for examining fundamental biological processes. Finding or creating mutations that affect different components of a biological system and studying their effects can often lead to an understanding of the system. This method, referred to as genetic dissection, is analogous to figuring out how an automobile works by breaking different parts of a car and observing the effects; for example, smash the radiator and the engine overheats, revealing that the radiator cools the engine. The use of mutations to disrupt function can likewise be a source of insight into biological processes. For example, geneticists have begun to unravel the molecular details of development by studying mutations that interrupt various embryonic stages in *Drosophila*. Scientists have also used the analysis of mutations to reveal the different parts of the *lac* operon (discussed in Chapter 12) and how they function in gene regulation. Although breaking "parts" to determine their function might seem like a crude approach to understanding a system, it is actually very powerful and has been used extensively in biochemistry, developmental biology, physiology, and behavioral science. But this method is *not* recommended for learning how your car works!

THINK-PAIR-SHARE Question 3

CONCEPTS

Mutations are heritable changes in DNA. They are essential to the study of genetics and are useful in many other biological fields.

Categories of Mutations

In multicellular organisms, we can distinguish between two broad categories of mutations: somatic mutations and germ-line mutations. **Somatic mutations** arise in somatic tissues, which do not produce gametes (**Figure 13.1**). When

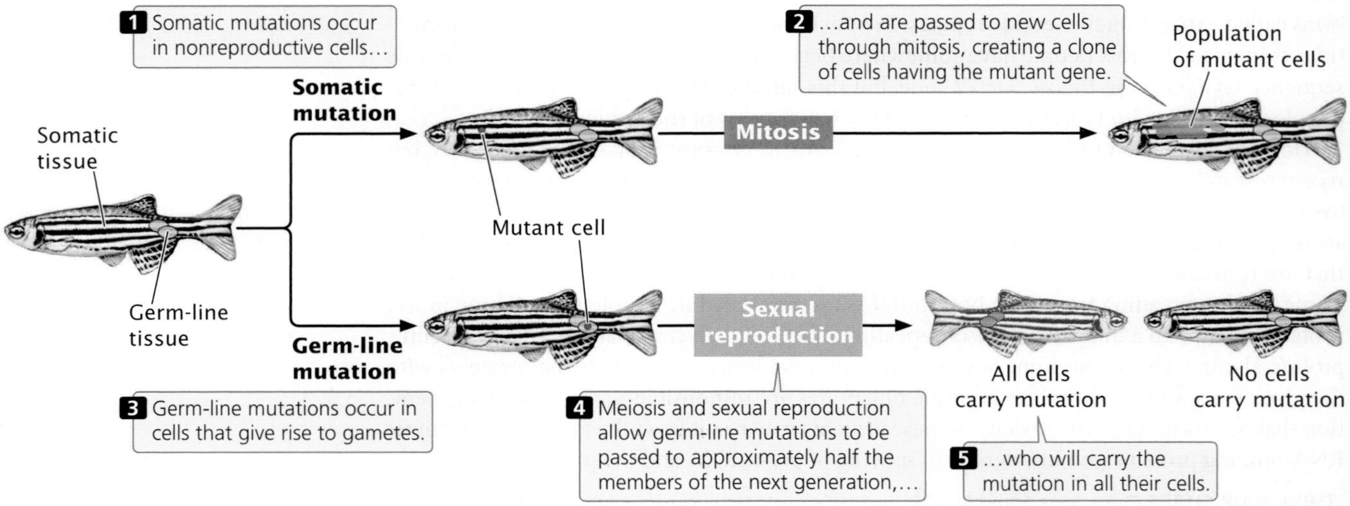

13.1 The two basic classes of mutations are somatic mutations and germ-line mutations.

a somatic cell with a mutation divides (by mitosis), the mutation is passed on to the daughter cells, leading to a population of genetically identical cells (a clone). The earlier in development that a somatic mutation takes place, the larger the clone of cells will be that contain the mutation.

Because of the huge number of cells present in a typical eukaryotic organism, somatic mutations are numerous. For example, there are about 10^{14} cells in the human body. Typically, a mutation arises once in every million cell divisions, so hundreds of millions of somatic mutations must arise in each person. Many somatic mutations have no obvious effect on the phenotype of the organism because the function of the mutant cell is taken over by a normal cell, or the mutant cell dies and is replaced by normal cells. However, cells with a somatic mutation that stimulates cell division can increase in number and spread; this type of mutation can give rise to cells with a selective advantage and is the basis for many cancers (see Chapter 16).

Germ-line mutations arise in cells that ultimately produce gametes. A germ-line mutation can be passed to future generations, producing individuals that carry the mutation in all their somatic and germ-line cells (see Figure 13.1). When we speak of mutations in multicellular organisms, we're usually talking about germ-line mutations.

Historically, mutations have been partitioned into those that affect a single gene, called *gene mutations*, and those that affect the number or structure of chromosomes, called *chromosome mutations*. This distinction arose because chromosome mutations could be observed directly, by looking at chromosomes with a microscope, whereas gene mutations could be detected only by observing their phenotypic effects. Now, with the development of DNA sequencing, gene mutations and chromosome mutations are distinguished somewhat arbitrarily on the basis of the size of the DNA lesion. Nevertheless, it is practical to use *chromosome mutation* for a large-scale genetic alteration that affects chromosome structure or the number of chromosomes and to use **gene mutation** for a relatively small DNA lesion that affects a single gene. This chapter focuses on gene mutations; chromosome mutations were discussed in Chapter 6.

Types of Gene Mutations

There are a number of ways to classify gene mutations. Some classification schemes are based on the nature of the phenotypic effect, others are based on the causative agent of the mutation, and still others focus on the molecular nature of the defect. Here, we will categorize mutations primarily on the basis of their molecular nature, but we will also encounter some terms that relate to the causes and the phenotypic effects of mutations.

BASE SUBSTITUTIONS The simplest type of gene mutation is a **base substitution**, the alteration of a single nucleotide in the DNA (**Figure 13.2a**). There are two types of base substitutions. In a **transition**, a purine is replaced by a different purine or, alternatively, a pyrimidine is replaced by a different pyrimidine (**Figure 13.3**). In a **transversion**, a purine is replaced by a pyrimidine or a pyrimidine is replaced by a purine. The number of possible transversions (see Figure 13.3) is twice the number of possible transitions, but transitions arise more frequently because transforming a purine into different purine or a pyrimidine into different pyrimidine is easier than transforming a purine into pyrimidine, or vice versa. ▶TRY PROBLEM 13

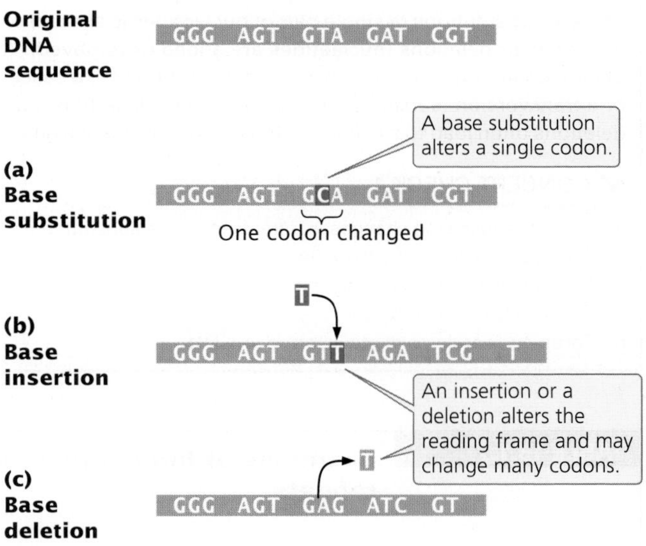

13.2 Three basic types of gene mutations are base substitutions, insertions, and deletions.

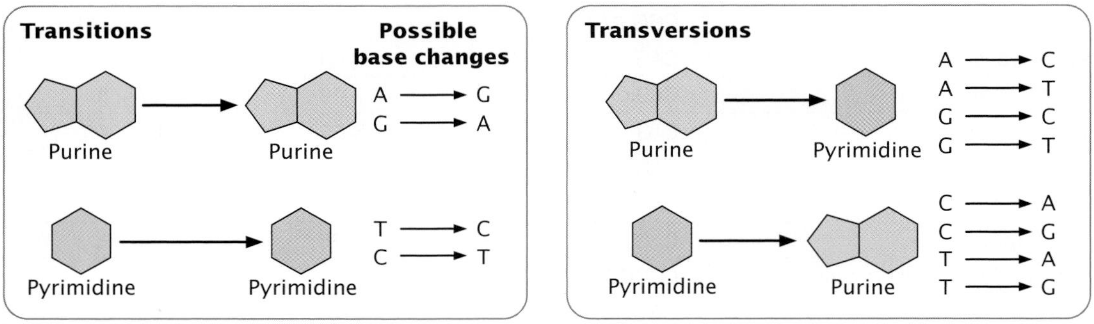

13.3 A transition is the substitution of a purine for a purine or of a pyrimidine for a pyrimidine; a transversion is the substitution of a pyrimidine for a purine or of a purine for a pyrimidine.

INSERTIONS AND DELETIONS Another class of gene mutations contains **insertions** and **deletions** (collectively called indels): the addition or removal, respectively, of one or more nucleotide pairs (**Figure 13.2b and c**). Although base substitutions are often assumed to be the most common type of mutation, molecular analysis has revealed that insertions and deletions are frequently more common. Insertions and deletions within sequences that encode proteins may lead to **frameshift mutations**: changes in the reading frame (see pp. 304–305 in Chapter 11) of the gene. Frameshift mutations usually alter all amino acids encoded by the nucleotides following the mutation, so they generally have drastic effects on the phenotype. Some frameshifts also introduce premature stop codons, terminating protein synthesis early and resulting in a shortened (truncated) protein. Not all insertions and deletions lead to frameshifts, however; insertions and deletions consisting of any multiple of three nucleotides leave the reading frame intact, although the addition or removal of one or more amino acids may still affect the phenotype. Indels that do not affect the reading frame are called **in-frame insertions** and **in-frame deletions**.

CONCEPTS

Gene mutations consist of changes in a single gene and can be base substitutions (a single pair of nucleotides is altered) or insertions or deletions (nucleotides are added or removed). A base substitution can be a transition (substitution of like bases) or a transversion (substitution of unlike bases). Insertions and deletions often lead to a change in the reading frame of a gene.

✓ CONCEPT CHECK 1

Which of the following changes is a transition base substitution?

a. Adenine is replaced by thymine.
b. Cytosine is replaced by adenine.
c. Guanine is replaced by adenine.
d. Three nucleotide pairs are inserted into DNA.

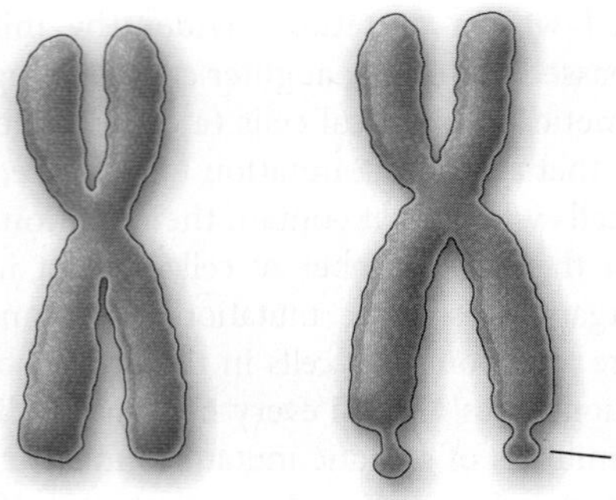

13.4 Fragile-X syndrome is associated with a characteristic constriction (fragile site) on the long arm of the X chromosome. [© CMSP/Custom Medical Stock Photo.]

EXPANDING NUCLEOTIDE REPEATS Mutations in which the number of copies of a set of nucleotides increases are called **expanding nucleotide repeats**. This type of mutation was first observed in 1991 in a gene called *FMR-1*, which causes fragile-X syndrome, the most common hereditary cause of intellectual disability. The disorder is so named because, in specially treated cells from persons having the condition, the tip of each long arm of the X chromosome is attached by only a slender-appearing part of the chromosome (**Figure 13.4**). The normal *FMR-1* allele (not containing the mutation) has 60 or fewer copies of CGG, but in persons with fragile-X syndrome, the allele may harbor hundreds or even thousands of copies.

Expanding nucleotide repeats have been found in a number of other human genetic diseases, several of which are listed in **Table 13.1**. Most of these diseases are caused by the expansion of a set of three nucleotides (called a trinucleotide), most often CNG, where N can be any nucleotide. However, some diseases are caused by repeats of four, five, and even twelve nucleotides. The number of copies of the nucleotide repeat often correlates with the severity

TABLE 13.1 Examples of human genetic diseases caused by expanding nucleotide repeats

		Number of copies of repeat	
Disease	**Repeated sequence**	**Normal range**	**Disease range**
Spinal and bulbar muscular atrophy	CAG	11–33	40–62
Fragile-X syndrome	CGG	6–54	50–1500
Jacobsen syndrome	CGG	11	100–1000
Spinocerebellar ataxia (several types)	CAG	4–44	21–130
Autosomal dominant cerebellar ataxia	CAG	7–19	37–220
Myotonic dystrophy	CTG	5–37	44–3000
Huntington disease	CAG	9–37	37–121
Friedreich ataxia	GAA	6–29	200–900
Dentatorubral-pallidoluysian atrophy	CAG	7–25	49–75
Myoclonus epilepsy of the Unverricht–Lundborg type	CCCCGCCCCGCG	2–3	12–13
Amyotrophic lateral sclerosis	GGGGCC	2–23	700–1600

or age of onset of the disease. The number of copies of the repeat also corresponds to its instability: when more repeats are present, the probability of expansion to even more repeats increases.

Increases in the number of nucleotide repeats can produce disease symptoms in different ways. In several diseases (e.g., Huntington disease), the nucleotide expansion occurs within the coding part of a gene, producing a toxic protein that has extra glutamine residues (the amino acid encoded by CAG). In other diseases, the repeat is outside the coding region of a gene and affects its expression. In fragile-X syndrome, the additional copies of the nucleotide repeat cause the DNA to become methylated, which turns off the transcription of an essential gene.

A possible source of nucleotide repeat expansion is the formation of hairpins and other special DNA structures, which can cause nucleotides in the template strand to be replicated twice, thus increasing the number of repeats on the newly synthesized strand (**Figure 13.5**). Watching **Animation 13.1** will help you understand how the nucleotide repeats increase in number.

CONCEPTS

Expanding nucleotide repeats are regions of DNA that consist of repeated copies of sets of nucleotides. Increased numbers of nucleotide repeats are associated with several genetic diseases.

Functional Effects of Mutations

Another way that mutations are classified is on the basis of their functional effects. At the most general level, we can distinguish a mutation on the basis of its phenotype compared with the wild-type phenotype. A mutation that alters the wild-type allele is called a **forward mutation**, whereas a **reverse mutation** (a *reversion*) changes a mutant allele back into the wild-type allele.

Geneticists use other terms to describe the effects of mutations on protein structure. A base substitution that results in a different amino acid in the protein is referred to as a **missense mutation** (**Figure 13.6a**). A **nonsense mutation** changes a sense codon (one that specifies an amino acid) into a nonsense codon (one that terminates translation), as shown in **Figure 13.6b**. If a nonsense mutation occurs early in the mRNA sequence, the protein will be truncated and usually nonfunctional.

Because of the redundancy of the genetic code, some different codons specify the same amino acid (see Figure 11.5). A **silent mutation** changes a codon to a synonymous codon that specifies the same amino acid (**Figure 13.6c**), altering the DNA sequence without changing the amino acid sequence of the protein. Not all silent mutations, however, are truly silent: some do have phenotypic effects. For example,

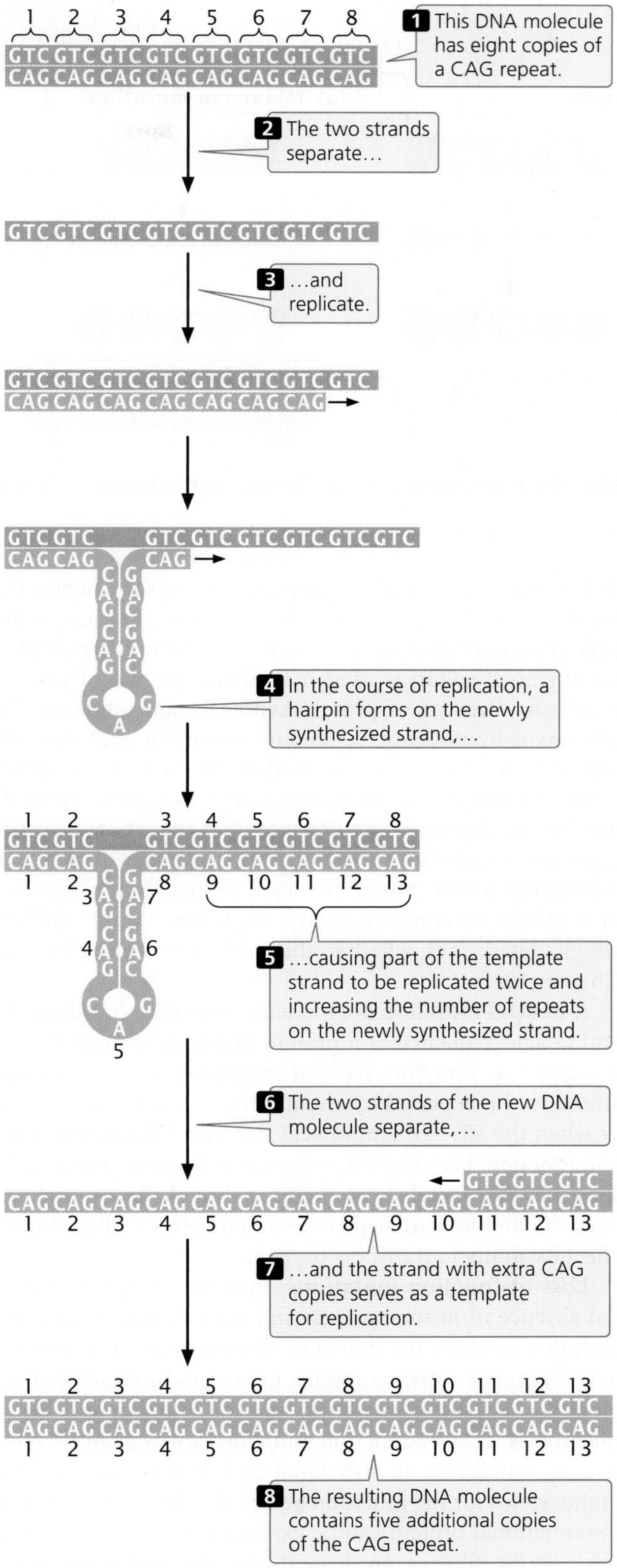

13.5 A model for how the number of copies of a nucleotide repeat can increase in replication.

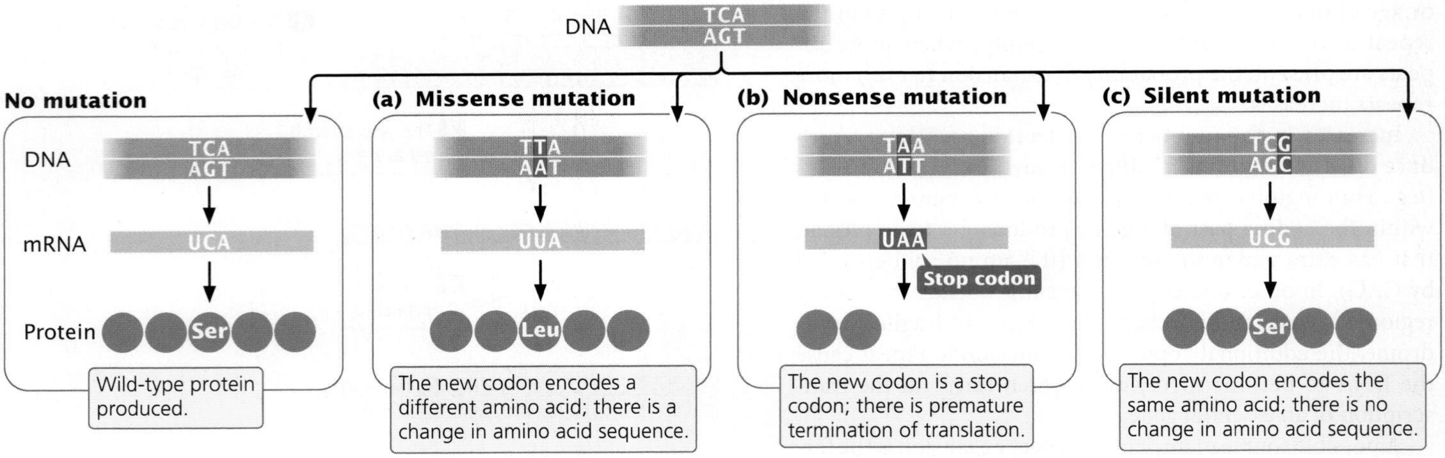

13.6 Base substitutions can cause (a) missense, (b) nonsense, or (c) silent mutations.

different tRNAs (called isoaccepting tRNAs, see Chapter 11) are used for different synonymous codons. Because some isoaccepting tRNAs are more abundant than others, which synonymous codon is used may affect the rate of protein synthesis. The rate of protein synthesis can influence the phenotype by affecting the amount of protein present in the cell and, in a few cases, the folding of the protein. Other silent mutations may alter nucleotides that serve as binding sites for regulatory proteins or alter sequences near the exon–intron junctions that affect splicing (see Chapter 10). Additional silent mutations can influence the binding of miRNAs to complementary sequences in the mRNA, which determines whether the mRNA is translated (see Chapter 10).

A **neutral mutation** is a missense mutation that alters the amino acid sequence of a protein but does not significantly change its function. Neutral mutations occur when one amino acid is replaced by another that is chemically similar or when the affected amino acid has little influence on protein function. For example, some neutral mutations occur in the genes that encode hemoglobin; although these mutations alter the amino acid sequence of hemoglobin, they do not affect its ability to transport oxygen.

Loss-of-function mutations cause the complete or partial absence of normal protein function. A loss-of-function mutation so alters the structure of the protein that the protein no longer works correctly, or occurs in regulatory regions that affect the transcription, translation, or splicing of the protein. Loss-of-function mutations are frequently recessive, in which case an individual diploid organism must be homozygous for the mutation before the effects of the loss of the functional protein can be exhibited. The mutations that cause cystic fibrosis are loss-of-function mutations: these mutations produce a nonfunctional form of the cystic fibrosis transmembrane conductance regulator protein, which normally regulates the movement of chloride ions into and out of the cell (see Chapter 4).

In contrast, a **gain-of-function mutation** causes the cell to produce a protein or gene product whose function is not normally present. The result could be an entirely new gene product or one produced in an inappropriate tissue or at an inappropriate time in development. For example, a mutation in a gene that encodes a receptor for a growth factor might cause the mutated receptor to stimulate growth all the time, even in the absence of the growth factor. Gain-of-function mutations are frequently dominant in their expression because a single copy of the mutation leads to the presence of a new gene product.

Still other types of mutations are **conditional mutations**, which are expressed only under certain conditions. For example, some conditional mutations affect the phenotype only at elevated temperatures. Another type of mutation is a **lethal mutation**, one that causes premature death (Chapter 4). TRY PROBLEM 17

THINK-PAIR-SHARE Question 4

Suppressor Mutations

A **suppressor mutation** is a genetic change that hides or suppresses the effect of another mutation. This type of mutation is different from a reverse mutation, in which the mutated site changes back into the original wild-type sequence (**Figure 13.7**). A suppressor mutation occurs at a site that is distinct from the site of the original mutation; thus, an individual with a suppressor mutation is a double mutant, possessing both the original mutation and the suppressor mutation but exhibiting the phenotype of a nonmutated wild type. Geneticists distinguish between two classes of suppressor mutations: intragenic and intergenic.

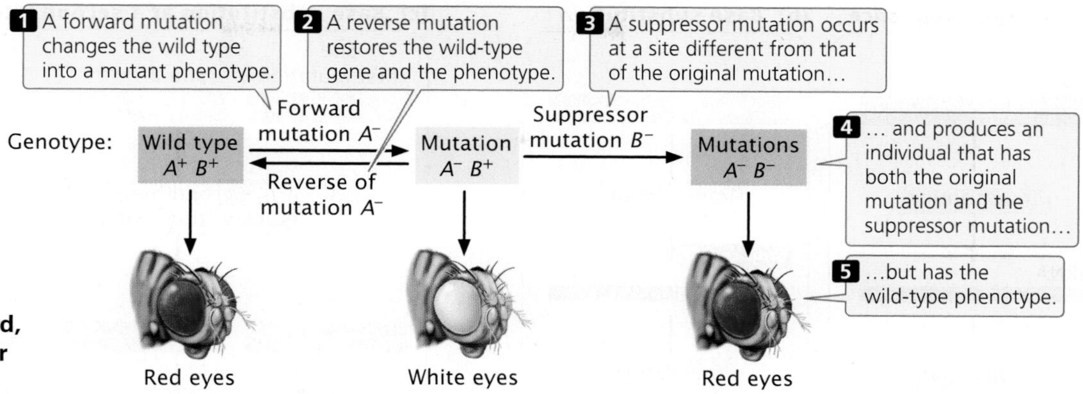

13.7 Relation of forward, reverse, and suppressor mutations.

INTRAGENIC SUPPRESSOR MUTATIONS An **intragenic suppressor mutation** takes place in the same gene that contains the mutation being suppressed. It may work in any of several ways. The suppressor may change a second nucleotide in the same codon altered by the original mutation, producing a codon that specifies the same amino acid that was specified by the original, nonmutated codon (**Figure 13.8**).

Intragenic suppressors may also work by suppressing a frameshift mutation. If the original mutation, for example, is a one-base deletion, then the addition of a single base elsewhere in the gene will restore the former reading frame. Consider the following nucleotide sequence in DNA and the amino acids that it encodes:

DNA	3′—AAA	TCA	CTT	GGC	GTA	CAA—5′
mRNA	5′—UUU	AGU	GAA	CCG	CAU	GUU—3′
Amino acids	Phe	Ser	Glu	Pro	His	Val

Suppose that a one-base deletion occurs in the first nucleotide of the second codon. This deletion shifts the reading frame by one nucleotide and alters all the amino acids that follow the mutation.

One-nucleotide deletion

DNA	3′—AAA	~~T~~CAC	TTG	GCG	TAC	AA—5′
mRNA	5′—UUU	GUG	AAC	CGC	AUG	UU—3′
Amino acids	Phe	Val	Asn	Arg	Met	

If a single nucleotide is added to the third codon (the suppressor mutation), the reading frame is restored, although two of the amino acids differ from those specified by the original sequence.

One-nucleotide insertion

DNA	3′—AAA	CAC	TTT	GGC	GTA	CAA—5′
mRNA	5′—UUU	GUG	AAA	CCG	CAU	GUU—3′
Amino acids	Phe	Val	Lys	Pro	His	Val

Similarly, a mutation due to an insertion may be suppressed by a subsequent deletion in the same gene.

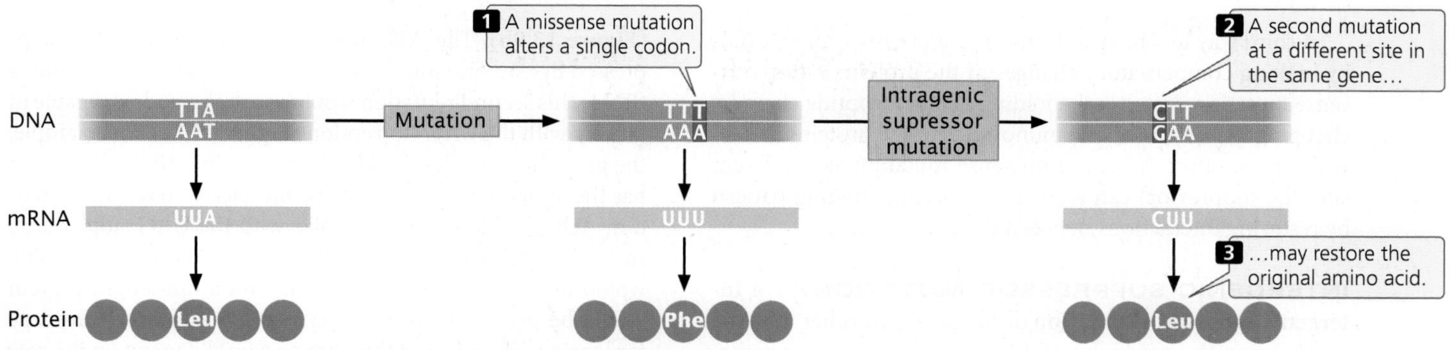

13.8 An intragenic suppressor mutation occurs in the gene containing the mutation being suppressed.

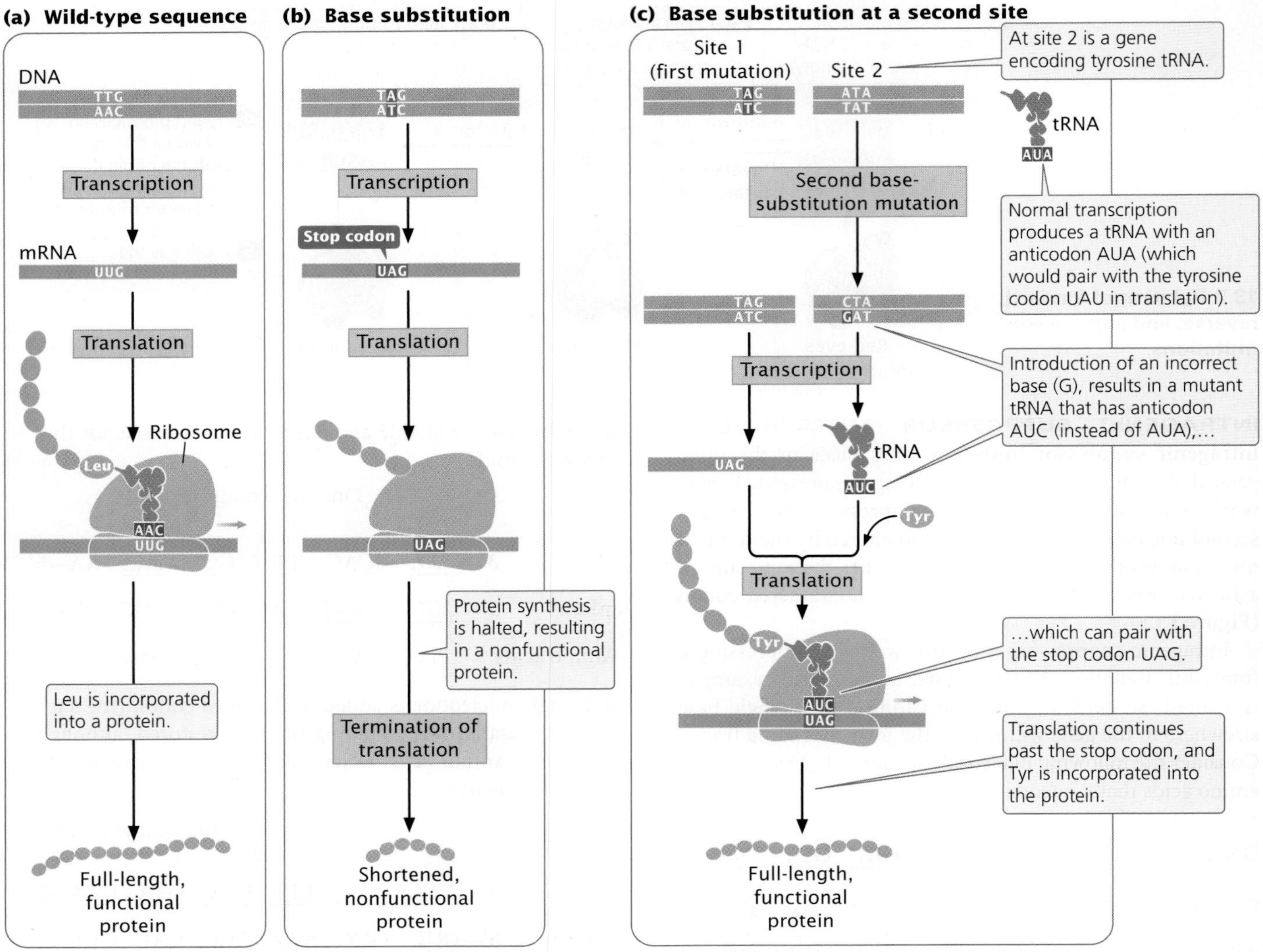

13.9 An intergenic suppressor mutation occurs in a gene other than the one bearing the original mutation. (a) The wild-type sequence produces a full-length, functional protein. (b) A base substitution at a site in the same gene produces a premature stop codon, resulting in a truncated, nonfunctional protein. (c) A base substitution at a site in another gene, which in this case encodes tRNA, alters the anticodon of $tRNA^{Tyr}$ so that $tRNA^{Tyr}$ can pair with the stop codon produced by the original mutation, allowing tyrosine to be incorporated into the protein and translation to continue.

A third way in which an intragenic suppressor may work is by making compensatory changes in the protein. A first missense mutation can alter the folding of a polypeptide chain by changing the way in which amino acids in the protein interact with one another. A second missense mutation at a different site (the suppressor) can recreate the original folding pattern by restoring interactions between the amino acids.

INTERGENIC SUPPRESSOR MUTATIONS An **intergenic suppressor mutation** occurs in a gene other than the one bearing the original mutation. These suppressors sometimes work by changing the way that the mRNA is translated. In the example illustrated in **Figure 13.9a**, the original DNA sequence is AAC (UUG in the mRNA) and specifies leucine. This sequence mutates to AT̲C (UA̲G in mRNA), a stop codon (**Figure 13.9b**). The ATC nonsense mutation could be suppressed by a second mutation in a different gene that encodes a tRNA; this second mutation would result in a codon capable of pairing with the UAG stop codon (**Figure 13.9c**). For example, the gene that encodes the tRNA for tyrosine ($tRNA^{Tyr}$), which has the anticodon AUA, might be mutated to have the anticodon AUC̲, which would then pair with the UAG stop codon. Instead of translation terminating at the UAG codon, tyrosine would be inserted into the protein, and a full-length protein would be produced, although tyrosine would now substitute for leucine. The effect of this change would depend on the role of this amino acid in the overall structure of the protein, but the effect of the suppressor mutation would probably be less detrimental than the effect of the nonsense mutation, which would halt translation prematurely.

TABLE 13.2 **Characteristics of different types of mutations**

Type of mutation	Definition
Base substitution	Changes a single DNA nucleotide
Transition	Base substitution in which a purine replaces a purine or a pyrimidine replaces a pyrimidine
Transversion	Base substitution in which a purine replaces a pyrimidine or a pyrimidine replaces a purine
Insertion	Addition of one or more nucleotides
Deletion	Deletion of one or more nucleotides
Frameshift mutation	Insertion or deletion that alters the reading frame of a gene
In-frame deletion or insertion	Deletion or insertion of a multiple of three nucleotides that does not alter the reading frame
Expanding nucleotide repeats	Increases the number of copies of a set of nucleotides
Forward mutation	Changes the wild-type phenotype to a mutant phenotype
Reverse mutation	Changes a mutant phenotype back to the wild-type phenotype
Missense mutation	Changes a sense codon into a different sense codon, resulting in the incorporation of a different amino acid in the protein
Nonsense mutation	Changes a sense codon into a nonsense (stop) codon, causing premature termination of translation
Silent mutation	Changes a sense codon into a synonymous codon, leaving the amino acid sequence of the protein unchanged
Neutral mutation	Changes the amino acid sequence of a protein without altering its ability to function
Loss-of-function mutation	Causes a complete or partial loss of function
Gain-of-function mutation	Causes the appearance of a new trait or function or causes the appearance of a trait in inappropriate tissue or at an inappropriate time
Lethal mutation	Causes premature death
Suppressor mutation	Suppresses the effect of an earlier mutation at a different site
Intragenic suppressor mutation	Suppresses the effect of an earlier mutation within the same gene
Intergenic suppressor mutation	Suppresses the effect of an earlier mutation in another gene

Because cells in many organisms have multiple copies of tRNA genes, other nonmutated copies of $tRNA^{Tyr}$ would remain available to recognize tyrosine codons in the transcripts of the mutant gene in question and in other genes being expressed concurrently. We might expect that the tRNAs that have undergone the suppressor mutation just described would also suppress the normal stop codons at the ends of other coding sequences, resulting in the production of longer-than-normal proteins, but this event does not usually take place.

Characteristics of some of the different types of mutations are summarized in **Table 13.2**.

CONCEPTS

A suppressor mutation overrides the effect of an earlier mutation at a different site. An intragenic suppressor mutation occurs within the *same* gene as that containing the original mutation; an intergenic suppressor mutation occurs in a *different* gene.

✓ CONCEPT CHECK 2

How does a suppressor mutation differ from a reverse mutation?

Mutation Rates

The frequency with which a wild-type allele at a locus changes into a mutant allele is referred to as the **mutation rate**. It is generally expressed as the number of mutations per biological unit, which may be mutations per cell division, per gamete, or per round of replication. For example, achondroplasia is a type of hereditary dwarfism in humans that results from a dominant mutation. On average, about four achondroplasia mutations arise in every 100,000 gametes, and so the mutation rate is $^4/_{100,000}$, or 0.00004, mutations per gamete. The mutation rate provides information about how often a mutation arises.

Mutation rates vary among genes and species (**Table 13.3**), but we can draw several general conclusions about mutation rates. First, spontaneous mutation rates are low for all organisms studied. Typical mutation rates for bacterial genes range from about 1 to 100 mutations per 10 billion cells (from 1×10^{-8} to 1×10^{-10}). The mutation rates for most eukaryotic genes are a bit higher, from about 1 to 10 mutations per million gametes (from 1×10^{-5} to 1×10^{-6}). These higher values in eukaryotes may be due to the fact that the rates are calculated *per gamete*, and that several cell divisions are required to produce a gamete, whereas mutation rates in prokaryotic cells are calculated *per cell division*.

TABLE 13.3 Mutation rates of different genes in different organisms

Organism	Mutation	Rate	Unit
Bacteriophage T2	Lysis inhibition	1×10^{-8}	Per replication
	Host range	3×10^{-9}	
Escherichia coli	Lactose fermentation	2×10^{-7}	Per cell division
	Histidine requirement	2×10^{-8}	
Neurospora crassa	Inositol requirement	8×10^{-8}	Per asexual spore
	Adenine requirement	4×10^{-8}	
Corn	Kernel color	2.2×10^{-6}	Per gamete
Drosophila	Eye color	4×10^{-5}	Per gamete
	Allozymes	5.14×10^{-6}	
Mouse	Albino coat color	4.5×10^{-5}	Per gamete
	Dilution coat color	3×10^{-5}	
Human	Huntington disease	1×10^{-6}	Per gamete
	Achondroplasia	1×10^{-5}	
	Neurofibromatosis (Michigan)	1×10^{-4}	
	Hemophilia A (Finland)	3.2×10^{-5}	
	Duchenne muscular dystrophy (Wisconsin)	9.2×10^{-5}	

The differences in mutation rates among species may be due to differing abilities to repair mutations, unequal exposures to mutagens, or biological differences in rates of spontaneously arising mutations. Even within a single species, spontaneous rates of mutation vary among genes. The reason for this variation is not entirely understood, but some regions of DNA are known hotspots for mutations.

Recent research suggests that fewer mutations occur in DNA sequences that are associated with nucleosomes (see Chapter 8). Reduced mutation rates may occur in these sequences because DNA associated with nucleosomes is less exposed to mutagens, but they could also be explained by the effect of nucleosomes on DNA repair, recombination, or replication, all of which influence the rate of mutation.

THINK-PAIR-SHARE Question 5

CONCEPTS

Mutation rate is the frequency with which a wild-type allele at a locus changes into a mutant allele. Rates of mutations are generally low and are affected by environmental and genetic factors.

13.2 Mutations May Be Caused by a Number of Different Factors

Mutations result from both internal and external factors. Those that occur under normal conditions are termed **spontaneous mutations**, whereas those that result from changes caused by environmental chemicals or radiation are **induced mutations**.

Spontaneous Replication Errors

Replication is amazingly accurate: less than one error in a billion nucleotides arises in the course of DNA synthesis (see Chapter 9). However, spontaneous replication errors do occasionally occur.

THINK-PAIR-SHARE Question 6

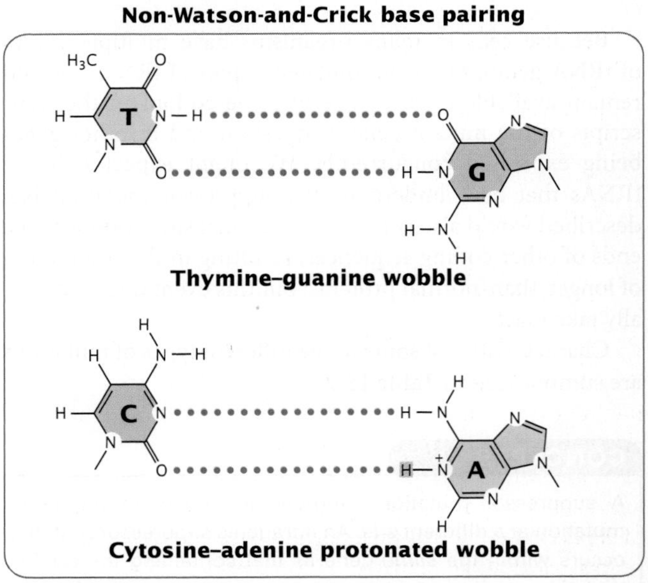

13.10 Nonstandard base pairings can occur as a result of the flexibility in DNA structure. Thymine and guanine can pair through wobble between normal bases. Cytosine and adenine can pair through wobble when adenine is protonated (has an extra hydrogen).

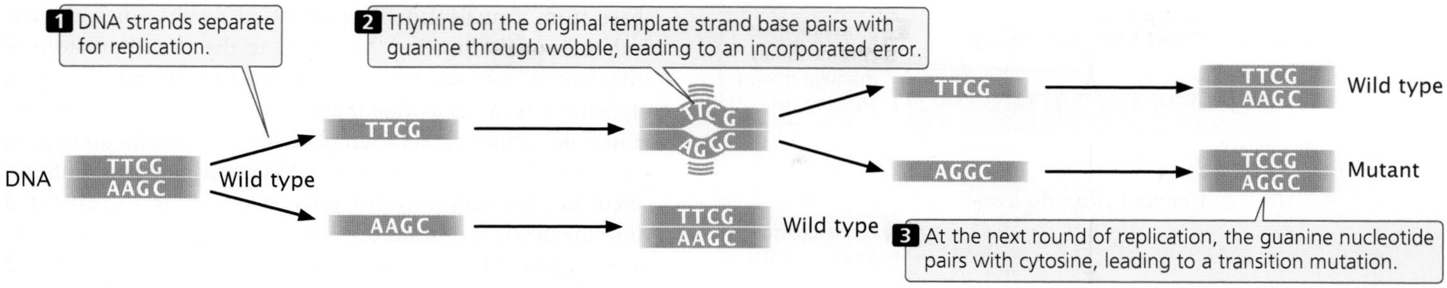

13.11 Wobble base pairing may lead to a replicated error.

TAUTOMERIC SHIFTS The primary cause of spontaneous replication errors was at one time thought to be tautomeric shifts, in which the positions of protons in the DNA bases change. Each of the four bases exists in different chemical forms called tautomers. The two tautomeric forms of each base are in dynamic equilibrium, although one form is much more common than the other. The standard Watson-and-Crick base pairings—adenine with thymine, and cytosine with guanine—occur between the common forms of the bases, but if the bases are in their rare tautomeric forms, other base pairings are possible. For example, the common form of cytosine pairs with guanine, but the rare tautomer of cytosine pairs with adenine.

Watson and Crick proposed that tautomeric shifts might produce mutations, and for many years their proposal was the accepted model for spontaneous replication errors. However, there has never been convincing evidence that the rare tautomers are the cause of spontaneous mutations. Furthermore, research now shows little evidence of tautomers in DNA.

MISPAIRING DUE TO OTHER STRUCTURES Mispairings often arise through wobble (see Chapter 11), in which normal, protonated, and other forms of the bases are able to pair because of flexibility in the DNA helical structure (**Figure 13.10**). These structures have been detected in DNA molecules and are now thought to be responsible for many of the mispairings in replication.

INCORPORATED ERRORS AND REPLICATED ERRORS When a mispaired base has been incorporated into a newly synthesized nucleotide chain, an **incorporated error** is said to have occurred. Suppose that, in replication, thymine (which normally pairs with adenine) mispairs with guanine through wobble (**Figure 13.11**). In the next round of replication, the two mismatched bases separate, and each serves as template for the synthesis of a new nucleotide strand. This time, thymine pairs with adenine, producing another copy of the original DNA sequence. On the other strand, however, the incorrectly incorporated guanine serves as the template and pairs with cytosine, producing a new DNA molecule that has an error: a C • G pair in place of the original T • A pair (a T • A→C • G base substitution). The original incorporated error leads to a **replicated error** (the C • G base pair instead of the original T • A base pair), which creates a permanent mutation because all the base pairings are correct and there is no mechanism for repair systems to detect the error.

CAUSES OF DELETIONS AND INSERTIONS Small insertions and deletions can arise spontaneously in replication and crossing over. **Strand slippage** can occur when one nucleotide strand forms a small loop (**Figure 13.12**). If the looped-out nucleotides are on the newly synthesized strand, an insertion results. At the next round of replication, the insertion will be replicated and both strands will contain the insertion. If the looped-out nucleotides are on the template strand, then the newly replicated strand will have a deletion, and this deletion will be perpetuated in subsequent rounds of replication.

Another process that produces insertions and deletions is unequal crossing over. In normal crossing over, the homologous sequences of the two DNA molecules align, and crossing over produces no net change in the number of nucleotides in either molecule. Misaligned pairing can

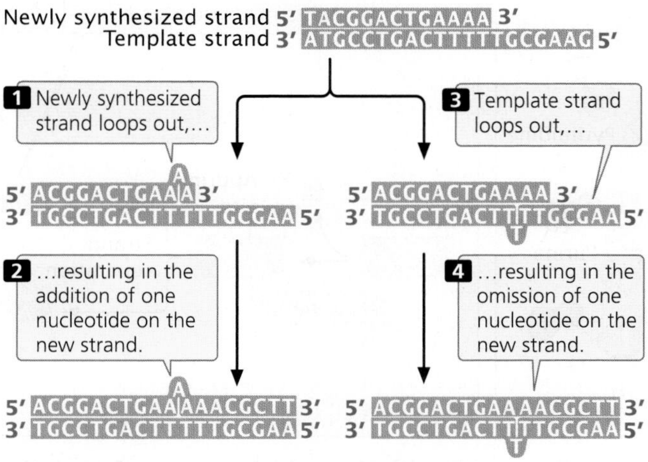

13.12 Insertions and deletions can result from strand slippage.

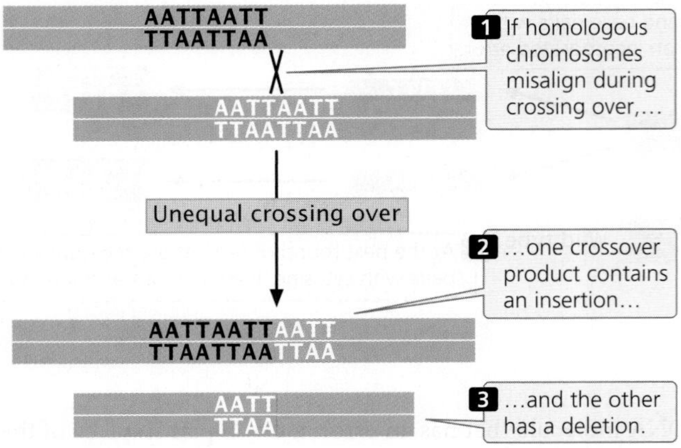

13.13 Unequal crossing over produces insertions and deletions.

cause **unequal crossing over**, which results in one DNA molecule with an insertion and the other with a deletion (**Figure 13.13**).

> **CONCEPTS**
>
> Spontaneous replication errors arise from altered base structures and from wobble. Small insertions and deletions can occur through strand slippage in replication and through unequal crossing over.

Spontaneous Chemical Changes

In addition to spontaneous mutations that arise in replication, mutations also result from spontaneous chemical changes in DNA. One such change is **depurination**, the loss of a purine base from a nucleotide. Depurination results when the covalent bond connecting the purine to the 1′-carbon atom of the deoxyribose sugar breaks (**Figure 13.14a**), producing an apurinic site, a nucleotide that lacks its purine base. An apurinic site cannot act as a template for a complementary base in replication. In the absence of base-pairing constraints, an incorrect nucleotide (most often adenine) is incorporated into the newly synthesized DNA strand opposite the apurinic site (**Figure 13.14b**), frequently leading to an incorporated error. The incorporated error is then transformed into a replication error at the next round of replication. Depurination is a common cause of spontaneous mutation; a mammalian cell in culture loses approximately 10,000 purines every day. Loss of pyrimidine bases also occurs, but at a much lower rate than depurination.

Another spontaneously occurring chemical change that takes place in DNA is **deamination**, the loss of an amino group (NH_2) from a base. Deamination may be spontaneous or may be induced by mutagenic chemicals.

Deamination can alter the pairing properties of a base: the deamination of cytosine, for example, produces uracil (**Figure 13.15a**), which pairs with adenine in replication. After another round of replication, the adenine will pair with thymine, creating a T • A pair in place of the original C • G pair (C • G → U • A → T • A); this chemical change is a transition mutation. This type of mutation is usually prevented by enzymes that remove uracil whenever it is found in DNA. Their ability to recognize the product of cytosine deamination may explain why thymine, not uracil, is found in DNA. In mammals, including humans, some cytosine bases in DNA are naturally methylated and exist in the form of 5-methylcytosine (5mC). When deaminated, 5mC becomes thymine (**Figure 13.15b**). Because thymine pairs with adenine in

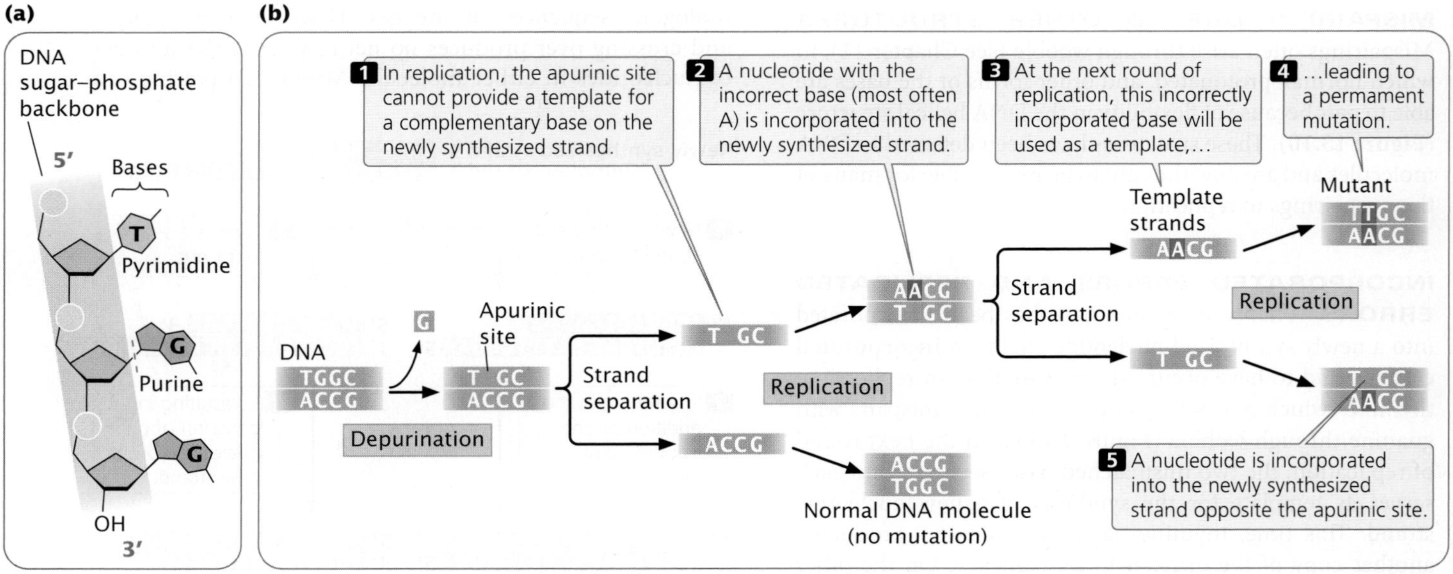

13.14 Depurination (the loss of a purine base from a nucleotide) may lead to a base substitution. (a) Depurination occurs when the covalent bond connecting a purine to the 1′ carbon is broken (indicated by dotted red line). (b) Replication of a template strand with an apurinic site may lead to an incorporation error.

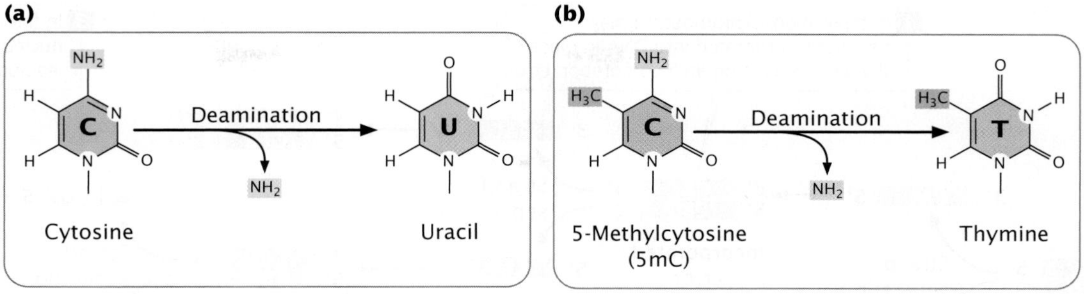

13.15 Deamination alters DNA bases.

replication, the deamination of 5-methylcytosine changes an original C • G pair to T • A (C • G → 5mC • G → T • G → T • A). Consequently, C • G → T • A transitions are frequent in mammalian cells, and 5mC sites are mutation hotspots in humans. ▶TRY PROBLEM 20

CONCEPTS

Some mutations arise from spontaneous alterations in DNA structure, such as depurination and deamination, which can alter the pairing properties of the bases and cause errors in subsequent rounds of replication.

Chemically Induced Mutations

Although many mutations arise spontaneously, a number of environmental agents, including certain chemicals and radiation, are capable of damaging DNA. Any environmental agent that significantly increases the rate of mutation above the spontaneous rate is called a **mutagen**.

BASE ANALOGS One class of chemical mutagens consists of **base analogs**, chemicals with structures similar to those of any of the four standard bases of DNA. DNA polymerases cannot distinguish these analogs from the standard bases, so if base analogs are present during replication, they may be incorporated into newly synthesized DNA molecules. For example, 5-bromouracil (5BU) is an analog of thymine; it has the same structure as thymine except that it has a bromine (Br) atom on the 5-carbon atom instead of a methyl group (**Figure 13.16a**). Normally, 5-bromouracil pairs with adenine just as thymine does, but it occasionally mispairs with guanine (**Figure 13.16b**), leading to a transition (T • A → 5BU • A → 5BU • G → C • G), as shown in **Figure 13.17**. Through mispairing, 5-bromouracil can also be incorporated into a newly synthesized DNA strand opposite guanine. In the next round of replication, 5-bromouracil pairs with adenine, leading to another transition (G • C → G • 5BU → A • 5BU → A • T). In the laboratory, mutations caused by base analogs can be reversed by treatment with the same analog or by treatment with a different analog.

ALKYLATING AGENTS Alkylating agents are chemicals that donate alkyl groups, such as methyl (CH_3) and ethyl (CH_3–CH_2) groups, to nucleotide bases. For example, ethylmethylsulfonate (EMS) adds an ethyl group to guanine, producing O^6-ethylguanine, which pairs with thymine (**Figure 13.18a**). Thus, EMS produces C • G → T • A transitions. Ethylmethylsulfonate is also capable of adding an ethyl group to thymine, producing 4-ethylthymine, which then pairs with guanine, leading to a T • A → C • G transition. Because EMS produces both C • G → T • A and T • A → C • G transitions, mutations produced by EMS can be reversed by additional treatment with EMS.

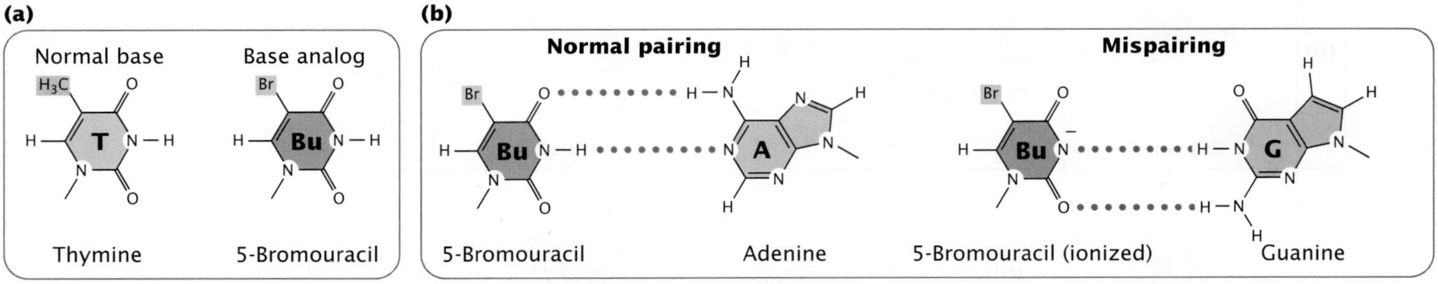

13.16 5-Bromouracil (a base analog) resembles thymine, except that it has a bromine atom in place of a methyl group on the 5-carbon atom. Because of the similarity in their structures, 5-bromouracil may be incorporated into DNA in place of thymine. Like thymine, 5-bromouracil normally pairs with adenine, but when ionized, it may pair with guanine through wobble.

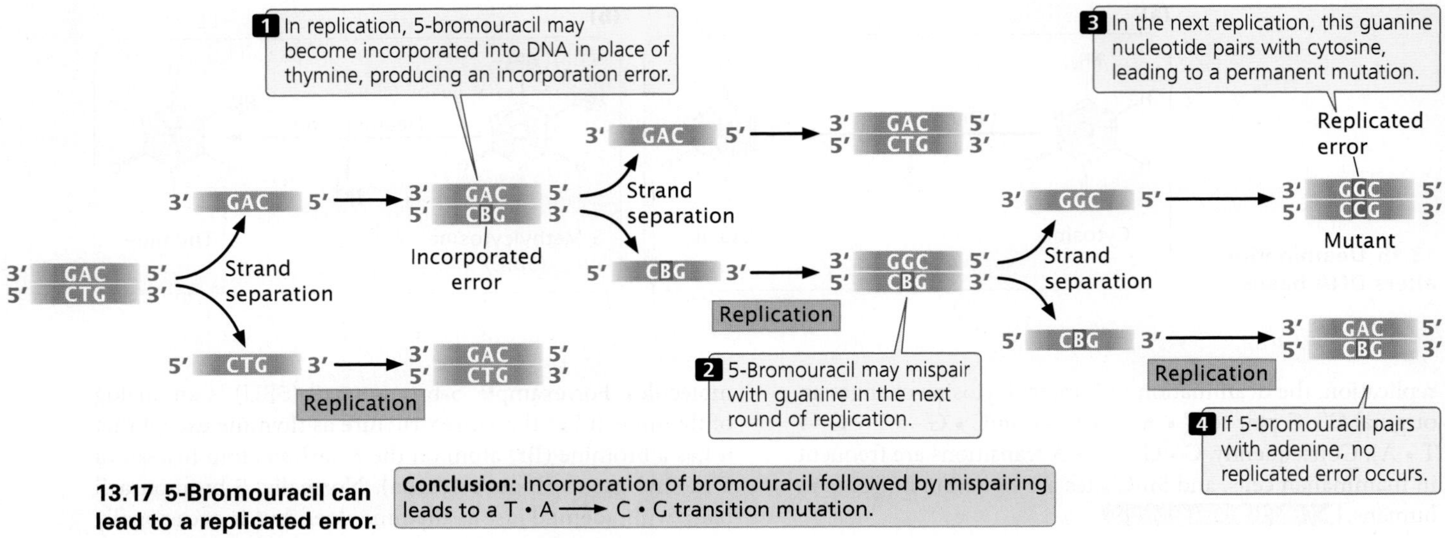

13.17 5-Bromouracil can lead to a replicated error.

DEAMINATION In addition to its spontaneous occurrence (see Figure 13.15), deamination can be induced by some chemicals. For instance, nitrous acid deaminates cytosine, creating uracil, which in the next round of replication pairs with adenine (**Figure 13.18b**), producing a C • G → T • A transition mutation. Nitrous acid also changes adenine into hypoxanthine, which pairs with cytosine, leading to a T • A → C • G transition. In addition, nitrous acid deaminates guanine, producing xanthine, which pairs with cytosine just as guanine does; however, xanthine can also pair with thymine, leading to a C • G → T • A transition. Nitrous acid produces exclusively transition mutations, and because both C • G → T • A and T • A → C • G transitions are produced, these mutations can be reversed with nitrous acid.

HYDROXYLAMINE Hydroxylamine is a very specific base-modifying mutagen that adds a hydroxyl group to cytosine, converting it into hydroxylaminocytosine (**Figure 13.18c**). This conversion increases the frequency of a rare tautomer that pairs with adenine instead of guanine and

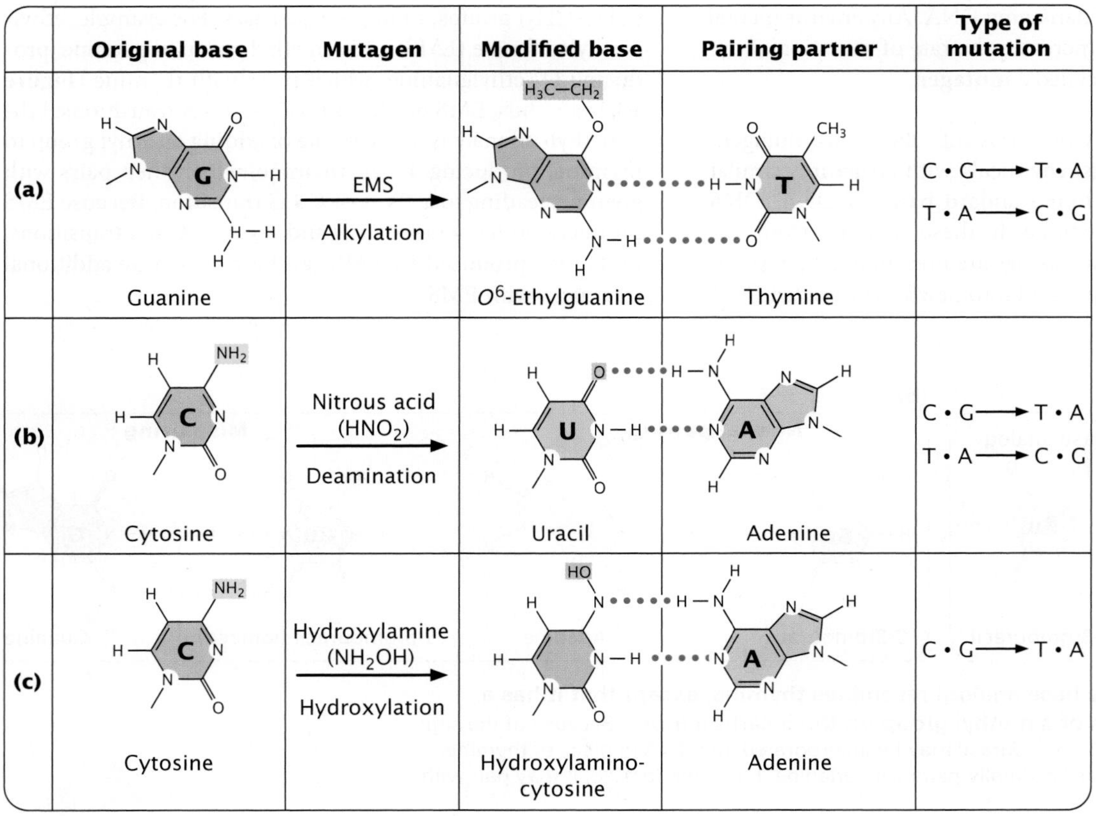

Original base
Mutagen
Modified base
Pairing partner
Type of mutation
(a)
Guanine
EMS
Alkylation
O^6-Ethylguanine
Thymine
C • G ⟶ T • A
T • A ⟶ C • G
(b)
Cytosine
Nitrous acid (HNO_2)
Deamination
Uracil
Adenine
C • G ⟶ T • A
T • A ⟶ C • G
(c)
Cytosine
Hydroxylamine (NH_2OH)
Hydroxylation
Hydroxylamino-cytosine
Adenine
C • G ⟶ T • A

13.18 Chemicals can alter DNA bases. Shown here are a few examples of mutations produced by chemical agents.

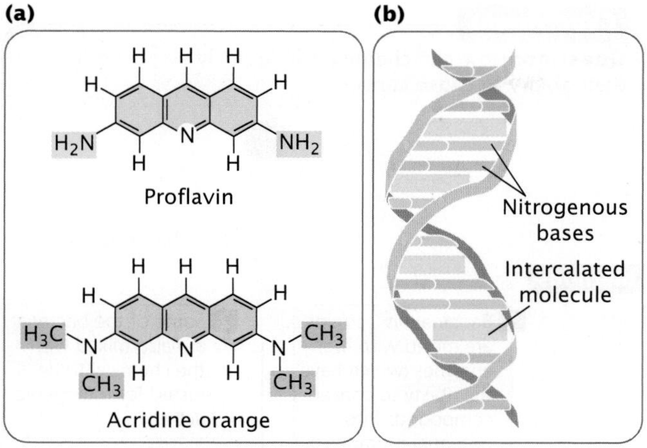

13.19 Intercalating agents. Intercalating agents, such as proflavin and acridine orange (a), insert themselves between adjacent bases in DNA, distorting the three-dimensional structure of the helix (b).

leads to C • G → T • A transitions. Because hydroxylamine acts only on cytosine, it will not generate T • A → C • G transitions; thus, hydroxylamine will not reverse the mutations that it produces.

INTERCALATING AGENTS Proflavin, acridine orange, ethidium bromide, and dioxin are **intercalating agents** (**Figure 13.19a**), which produce mutations by sandwiching themselves (intercalating) between adjacent bases in DNA, distorting the three-dimensional structure of the helix and causing single-nucleotide insertions and deletions in replication (**Figure 13.19b**). These insertions and deletions frequently produce frameshift mutations, so the mutagenic effects of intercalating agents are often severe. Because intercalating agents generate both additions and deletions, they can reverse the mutations they produce.

CONCEPTS

Chemicals can produce mutations by a number of mechanisms. Base analogs are incorporated into DNA and frequently pair with the wrong base. Alkylating agents, deaminating chemicals, hydroxylamine, and other chemicals change the structure of DNA bases, thereby altering their pairing properties. Intercalating agents wedge between the bases and cause single-base insertions and deletions in replication.

✓ CONCEPT CHECK 3

Base analogs are mutagenic because of which characteristic?

a. They produce changes in DNA polymerase that cause it to malfunction.
b. They distort the structure of DNA.
c. They are similar in structure to the normal bases.
d. They chemically modify the normal bases.

Radiation

In 1927, Hermann Muller demonstrated that mutations in fruit flies could be induced by X-rays. The results of subsequent studies showed that X-rays greatly increase mutation rates in all organisms. Because of their high energies, X-rays, gamma rays, and cosmic rays are all capable of penetrating tissues and damaging DNA. These forms of radiation, called ionizing radiation, dislodge electrons from the atoms that they encounter, changing stable molecules into free radicals and reactive ions, which then alter the structures of bases and break phosphodiester bonds in DNA. Ionizing radiation also frequently results in double-strand breaks in DNA. Attempts to repair these breaks can produce chromosome mutations (discussed in Chapter 6).

Ultraviolet (UV) light has less energy than ionizing radiation and does not eject electrons, but is nevertheless highly mutagenic. Pyrimidine bases readily absorb UV light, resulting in the formation of chemical bonds between adjacent pyrimidine molecules on the same strand of DNA, which create **pyrimidine dimers** (**Figure 13.20a**). Pyrimidine dimers consisting of two thymine bases (called thymine dimers) are most frequent, but cytosine dimers and thymine–cytosine dimers can also form. These dimers are bulky lesions that distort the configuration of DNA (**Figure 13.20b**) and often block replication. Most pyrimidine dimers are immediately repaired by mechanisms discussed in Section 13.4, but some escape repair and inhibit replication and transcription.

When pyrimidine dimers block replication, cell division is inhibited and the cell usually dies; for this reason, UV light kills bacteria and is an effective sterilizing agent. For a mutation to occur, the replication block must be overcome. Bacteria can sometimes circumvent replication blocks produced by pyrimidine dimers and other types of DNA damage by means of the **SOS system**. This system allows replication blocks to be overcome, but in the process, it makes numerous mistakes and greatly increases the rate of mutation. Indeed, the very reason that replication can proceed in the presence of a block is that the enzymes in the SOS system do not strictly adhere to the base-pairing rules. The trade-off is that replication can continue and the cell survives, but only by sacrificing the normal accuracy of DNA synthesis.

THINK-PAIR-SHARE Question 7

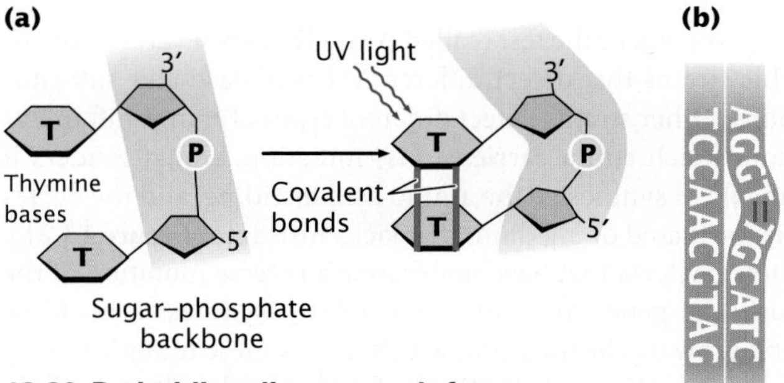

13.20 Pyrimidine dimers result from ultraviolet light. (a) Formation of thymine dimer. (b) Distorted DNA.

CONCEPTS

Ionizing radiation such as X-rays and gamma rays damages DNA by dislodging electrons from atoms; these electrons then break phosphodiester bonds and alter the structure of bases. Ultraviolet light causes mutations primarily by producing pyrimidine dimers that disrupt replication and transcription. The SOS system enables bacteria to overcome replication blocks but introduces mistakes in replication.

Detecting Mutations with the Ames Test

People in industrial societies are surrounded by a multitude of artificially produced chemicals: more than 50,000 different chemicals are in commercial and industrial use today, and from 500 to 1000 new chemicals are introduced each year. Some of these chemicals are potential carcinogens, and some natural products are also potentially carcinogenic. One method for testing the cancer-causing potential of substances is to administer them to laboratory animals (rats or mice) and compare the incidence of cancer in the treated animals with that in control animals. Unfortunately, these tests are time-consuming and expensive. Furthermore, the ability of a substance to cause cancer in rodents is not always indicative of its effect on humans. After all, we aren't rats!

In 1974, Bruce Ames developed a simple test for evaluating the potential of chemicals to cause cancer. The **Ames test** is based on the principle that both cancer and mutations result from damage to DNA, and the results of experiments have demonstrated that 90% of known carcinogens are also mutagens. Ames proposed that mutagenesis in bacteria could serve as an indicator of carcinogenesis in humans.

The Ames test uses auxotrophic strains of the bacterium *Salmonella typhimurium* that have defects in the lipopolysaccharide coat, which normally protects the bacteria from chemicals in the environment. Furthermore, the DNA-repair system in these strains has been inactivated, enhancing their susceptibility to mutagens. Some compounds are not active carcinogens but can be converted into cancer-causing compounds in the body. To make the Ames test sensitive to such *potential* carcinogens, a compound to be tested is first incubated in mammalian liver extract that contains metabolic enzymes.

A version of the test (called Ames II) uses several auxotrophic strains that detect different types of base-pair substitutions. Other strains detect different types of frameshift mutations. Each strain carries a his^- mutation, which renders it unable to synthesize the amino acid histidine, and the bacteria are plated on medium that lacks histidine (**Figure 13.21**). Only bacteria that have undergone a reverse mutation of the histidine gene ($his^- \rightarrow his^+$) are able to synthesize histidine and grow on the medium, which makes these mutations easy to detect. Different dilutions of a chemical to be tested are added to plates inoculated with the bacteria, and the number of mutated bacterial colonies that appear on each plate is compared with the number that appear on control plates with no chemical (i.e., that arose through spontaneous mutation). Any chemical that significantly increases the number of colonies appearing on a treated plate is mutagenic and probably also carcinogenic.

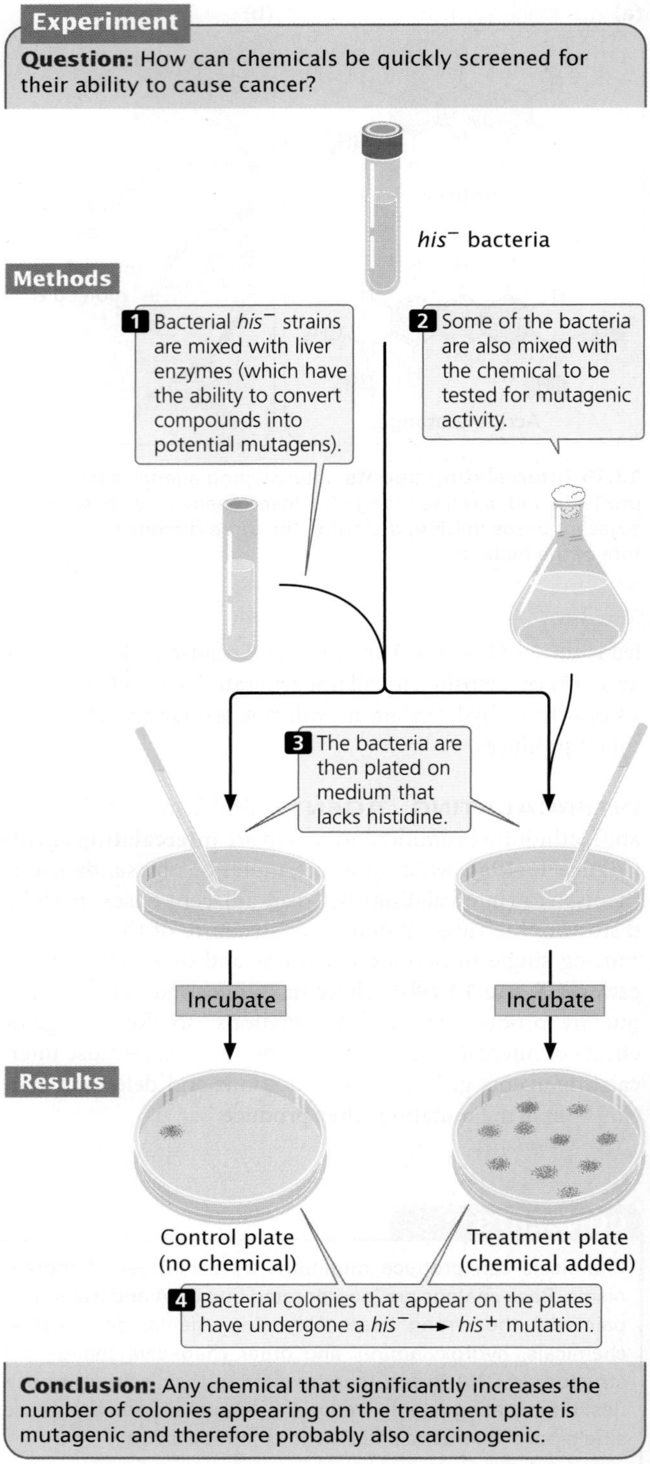

Experiment
Question: How can chemicals be quickly screened for their ability to cause cancer?
his⁻ bacteria
Methods
1 Bacterial his⁻ strains are mixed with liver enzymes (which have the ability to convert compounds into potential mutagens).
2 Some of the bacteria are also mixed with the chemical to be tested for mutagenic activity.
3 The bacteria are then plated on medium that lacks histidine.
Incubate
Incubate
Results
Control plate (no chemical)
Treatment plate (chemical added)
4 Bacterial colonies that appear on the plates have undergone a his⁻ → his⁺ mutation.
Conclusion: Any chemical that significantly increases the number of colonies appearing on the treatment plate is mutagenic and therefore probably also carcinogenic.

13.21 The Ames test is used to identify chemical mutagens.

CONCEPTS

The Ames test uses *his*⁻ strains of bacteria to test chemicals for their ability to produce $his^- \rightarrow his^+$ mutations. Because mutagenic activity and carcinogenic potential are closely correlated, the Ames test is widely used to screen chemicals for their cancer-causing potential.

13.3 Transposable Elements Are Mobile DNA Sequences Capable of Inducing Mutations

Transposable elements—DNA sequences that can move about in the genome—are often a cause of mutations. They are found in the genomes of all organisms and are abundant in many: for example, they make up at least 45% of human DNA. Most transposable elements are able to insert themselves at many different locations in the genome, relying on mechanisms that are distinct from homologous recombination. They often cause mutations, either by inserting into a gene and disrupting it or by promoting DNA rearrangements such as deletions, duplications, and inversions (see Chapter 6).

General Characteristics of Transposable Elements

There are many different types of transposable elements: some have simple structures, encompassing only those sequences necessary for their own transposition (movement), whereas others have complex structures and encode a number of functions not directly related to transposition. Despite this variation, many transposable elements have certain features in common.

Short **flanking direct repeats** from 3 to 12 bp long are present on both sides of most transposable elements. The sequences of these repeats vary, but their length is constant for each type of transposable element. These repeats are not a part of the transposable element and do not travel with it. Rather, they are generated in the process of transposition at the point of insertion. The presence of flanking direct repeats indicates that staggered cuts are made in the target DNA when a transposable element inserts itself, as shown in **Figure 13.22**. The staggered cuts leave short single-stranded pieces of DNA on either side of the transposable element. Replication of the single-stranded DNA then creates the flanking direct repeats.

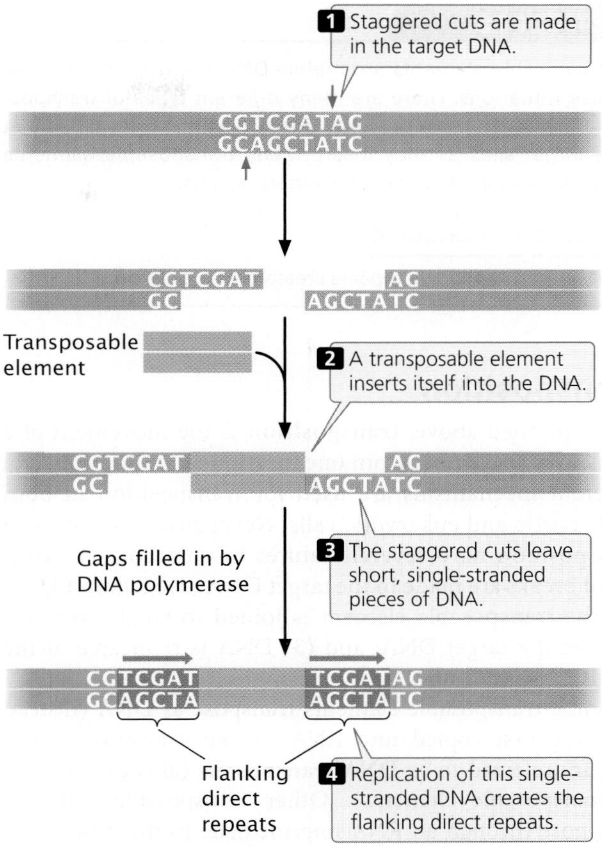

13.22 Flanking direct repeats are generated when a transposable element inserts into DNA.

At the ends of many, but not all, transposable elements are **terminal inverted repeats**, which are sequences from 9 to 40 bp in length that are inverted complements of one another. For example, the following sequences are inverted repeats:

5′—ACAGTTCAG . . . CTGAACTGT—3′
3′—TGTCAAGTC . . . GACTTGACA—5′

On the same strand, the two sequences are not simple inversions, as their name might imply; rather, they are inverted complements of one another. (Notice that the sequence from left to right in the top strand is the same as the sequence from right to left in the bottom strand.) Terminal inverted repeats are recognized by enzymes that catalyze transposition and are required for transposition to take place. **Figure 13.23** summarizes the general characteristics of transposable elements. TRY PROBLEM 23

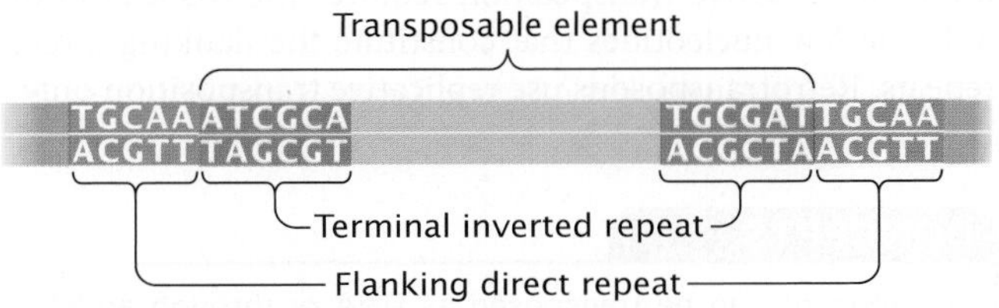

13.23 Many transposable elements have common characteristics. Most transposable elements generate flanking direct repeats on each side of the point of insertion into target DNA. Many transposable elements also possess terminal inverted repeats.

CONCEPTS

Transposable elements are mobile DNA sequences that often cause mutations. There are many different types of transposable elements; most generate short flanking direct repeats at the target sites as they insert. Many transposable elements also possess short terminal inverted repeats.

✓ CONCEPT CHECK 4

How are flanking direct repeats created in transposition?

Transposition

As mentioned above, **transposition** is the movement of a transposable element from one location to another. Several different mechanisms are used for transposition in both prokaryotic and eukaryotic cells. Nevertheless, all types of transposition have several features in common: **(1)** staggered breaks are made in the target DNA (see Figure 13.22); **(2)** the transposable element is joined to single-stranded ends of the target DNA; and **(3)** DNA is replicated at the single-stranded gaps.

Some transposable elements transpose as DNA (instead of being first copied into RNA, as retrotransposons are) and are referred to as **DNA transposons** (also called Class II transposable elements). Other transposable elements transpose through an RNA intermediate. In this case, RNA is transcribed from the transposable element (DNA) and is then copied back into DNA by a special enzyme called reverse transcriptase. Elements that transpose through an RNA intermediate are called **retrotransposons** (also called Class I transposons). Most transposable elements found in bacteria are DNA transposons. Both DNA transposons and retrotransposons are found in eukaryotes, although retrotransposons are more common.

Among DNA transposons, transposition may be replicative or nonreplicative. In **replicative transposition**, a new copy of the transposable element is introduced at a new site while the old copy remains behind at the original site, so the number of copies of the transposable element increases as a result of transposition. In **nonreplicative transposition**, the transposable element excises from the old site and inserts at a new site without any increase in the number of its copies. Nonreplicative transposition requires the replication of only the few nucleotides that constitute the flanking direct repeats. Retrotransposons use replicative transposition only.

CONCEPTS

Transposons can be transposed as DNA or through an RNA intermediate. In replicative transposition, a new copy of the transposable element inserts in a new location and the old copy stays behind; in nonreplicative transposition, the old copy excises from the old site and moves to a new site.

The Mutagenic Effects of Transposition

Because transposable elements can insert into genes and disrupt their function, transposition is generally mutagenic. In fact, more than half of all spontaneously occurring mutations in *Drosophila* result from the insertion of a transposable element in or near a functional gene.

A number of cases of human genetic disease have been traced to the insertion of a transposable element into a vital gene. For example, insertion of the L1 transposable element into the gene for blood clotting factor VIII has caused hemophilia. Although most mutations resulting from transposition are detrimental, transposition can occasionally activate a gene or change the phenotype of the cell in a beneficial way. For instance, bacterial transposable elements sometimes carry genes that encode antibiotic resistance, and several transposable elements have created mutations that confer insecticide resistance in insects.

A dramatic example of the mutagenic effect of transposable elements is seen in the color of grapes, which come in black, red, and white varieties (**Figure 13.24**). Black and red grapes result from the production of red pigments, called anthocyanins, in the skin, which are lacking in white grapes.

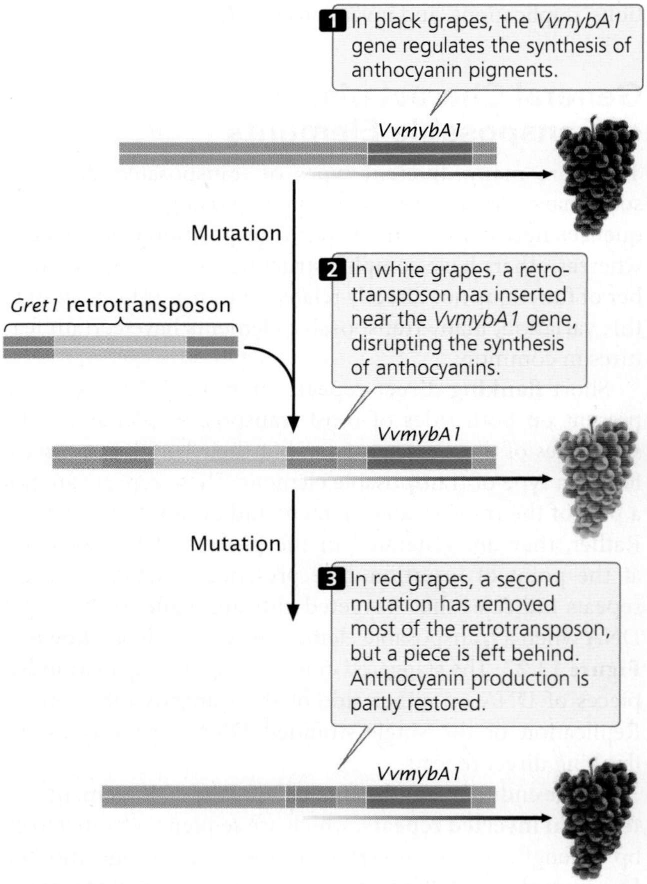

13.24 White and red color in grapes resulted from the insertion and deletion of a retrotransposon, respectively.

A mutation in black grapes that turned off the production of anthocyanins produced white grapes. This mutation consisted of the insertion of a 10,422-bp retrotransposon called *Gret1* near a gene that promotes the production of anthocyanins. The *Gret1* retrotransposon apparently disrupted sequences that regulate the gene, effectively shutting down anthocyanin production and producing a white grape with no anthocyanins. Interestingly, red grapes resulted from a second mutation that occurred in white grapes. This mutation (probably due to faulty recombination) removed most, but not all, of the retrotransposon, switching anthocyanin production back on, though not as intensely as in the original black grapes.

Because transposition entails the exchange of DNA sequences and recombination, it often leads to DNA rearrangements. Homologous recombination between multiple copies of transposons also leads to duplications, deletions, and inversions, as shown in **Figure 13.25**.

TRANSPOSABLE ELEMENTS IN HUMANS As we have seen, about 45% of the human genome consists of sequences derived from transposable elements, although most of these elements are now inactive and no longer capable of transposing. A comparison of human and chimpanzee genomes suggests that almost 11,000 transposition events have taken place since these two species diverged approximately 6 million years ago.

One of the most common transposable elements in the human genome is *Alu*. Every human cell contains more than 1 million related, but not identical, copies of *Alu* in its chromosomes. *Alu* sequences are similar to the gene that encodes the 7S RNA molecule, which transports newly synthesized proteins across the endoplasmic reticulum. *Alu* sequences create short flanking direct repeats when they insert into DNA and have characteristics that suggest that they have transposed through an RNA intermediate.

Evolutionary Significance of Transposable Elements

Transposable elements have clearly played an important role in shaping the genomes of many organisms. The large size of many eukaryotic genomes is due primarily to the abundance of transposable elements, particularly retrotransposons. Homologous recombination between copies of transposable elements has been an important force in producing gene duplications and other DNA rearrangements. Furthermore, some transposable elements may carry extra DNA with them when they transpose to a new site, providing the potential to move DNA sequences that regulate genes to new sites, where they may alter the expression of genes.

CONCEPTS

Increases in copy numbers of transposable elements have contributed to the large size of many eukaryotic genomes.

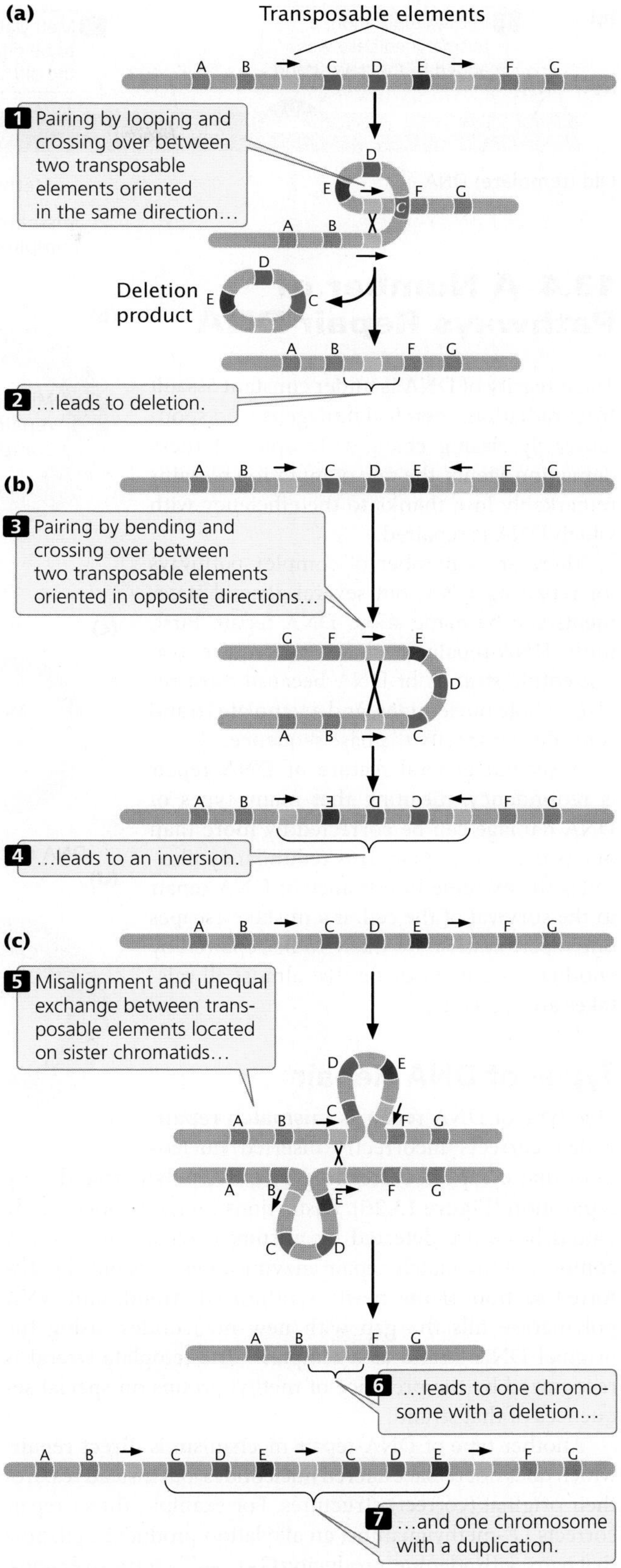

13.25 Many chromosomal rearrangements are generated by transposition.

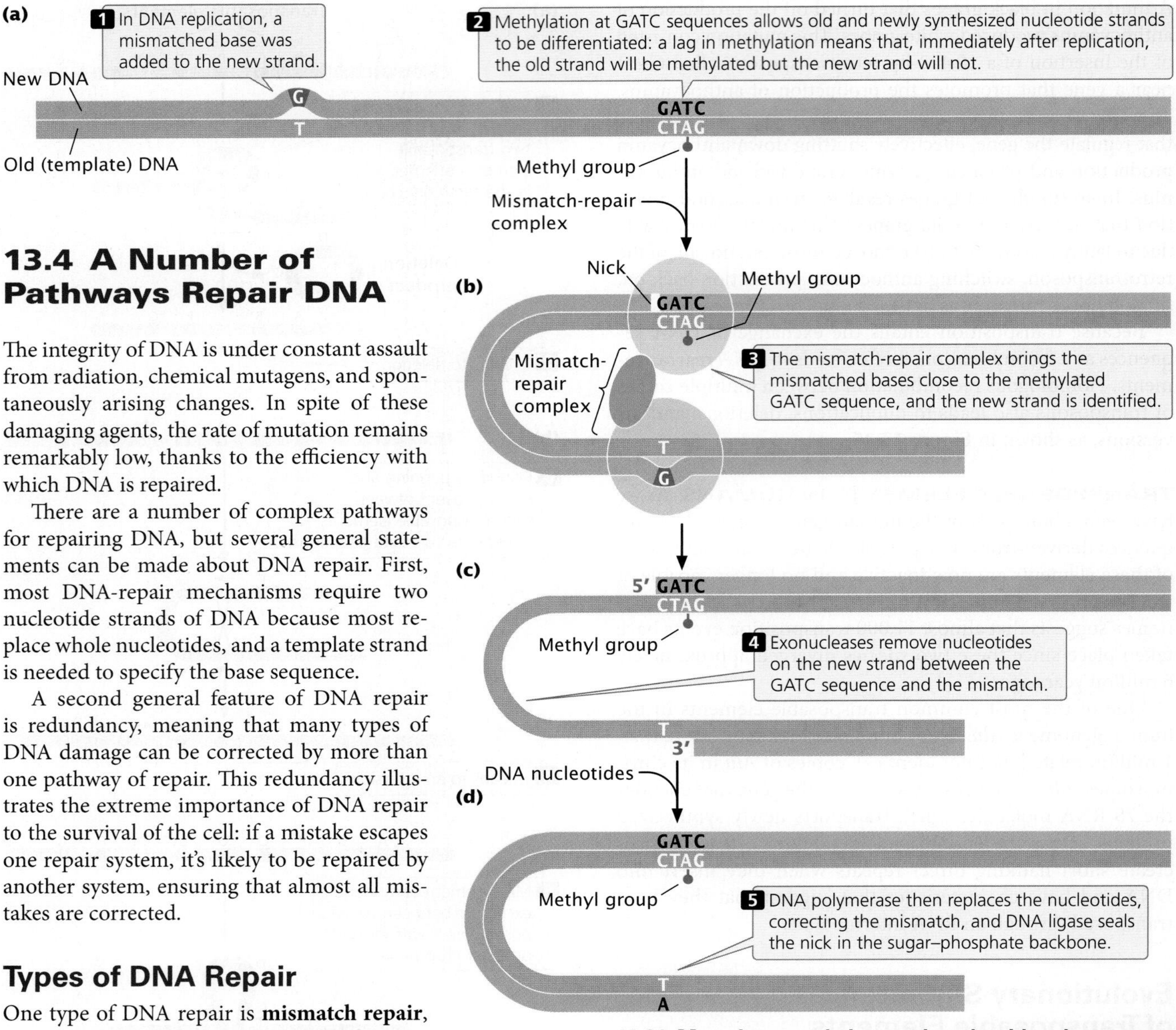

13.26 Many incorrectly inserted nucleotides that escape proofreading are corrected by mismatch repair.

13.4 A Number of Pathways Repair DNA

The integrity of DNA is under constant assault from radiation, chemical mutagens, and spontaneously arising changes. In spite of these damaging agents, the rate of mutation remains remarkably low, thanks to the efficiency with which DNA is repaired.

There are a number of complex pathways for repairing DNA, but several general statements can be made about DNA repair. First, most DNA-repair mechanisms require two nucleotide strands of DNA because most replace whole nucleotides, and a template strand is needed to specify the base sequence.

A second general feature of DNA repair is redundancy, meaning that many types of DNA damage can be corrected by more than one pathway of repair. This redundancy illustrates the extreme importance of DNA repair to the survival of the cell: if a mistake escapes one repair system, it's likely to be repaired by another system, ensuring that almost all mistakes are corrected.

Types of DNA Repair

One type of DNA repair is **mismatch repair**, which corrects incorrectly inserted nucleotides that escape proofreading by DNA polymerase during replication (**Figure 13.26**). Distortions caused by incorrectly paired bases are detected by mismatch-repair enzymes. A complex of mismatch-repair enzymes then cuts out the distorted section of the newly synthesized strand, and DNA polymerase fills the gap with new nucleotides, using the original DNA strand as a template. The template strand is recognized by the presence of methyl groups on special sequences of that strand.

Another type of DNA-repair mechanism is **direct repair**, which does not replace altered nucleotides, but instead restores their original (correct) structures. For example, direct repair corrects O^6-methylguanine, an alkylation product of guanine that pairs with adenine, producing G • C → T • A transversions. An enzyme called O^6-methylguanine-DNA methyltransferase removes the methyl group from O^6-methylguanine, restoring the base to guanine (**Figure 13.27**).

1 In DNA replication, a mismatched base was added to the new strand.

2 Methylation at GATC sequences allows old and newly synthesized nucleotide strands to be differentiated: a lag in methylation means that, immediately after replication, the old strand will be methylated but the new strand will not.

3 The mismatch-repair complex brings the mismatched bases close to the methylated GATC sequence, and the new strand is identified.

4 Exonucleases remove nucleotides on the new strand between the GATC sequence and the mismatch.

5 DNA polymerase then replaces the nucleotides, correcting the mismatch, and DNA ligase seals the nick in the sugar–phosphate backbone.

In **base-excision repair**, a modified base is first excised and then the entire nucleotide is replaced. The excision of modified bases is catalyzed by a set of enzymes called DNA glycosylases, each of which recognizes and removes a specific type of modified base. Uracil glycosylase, for example, recognizes and removes uracil produced by the deamination of cytosine.

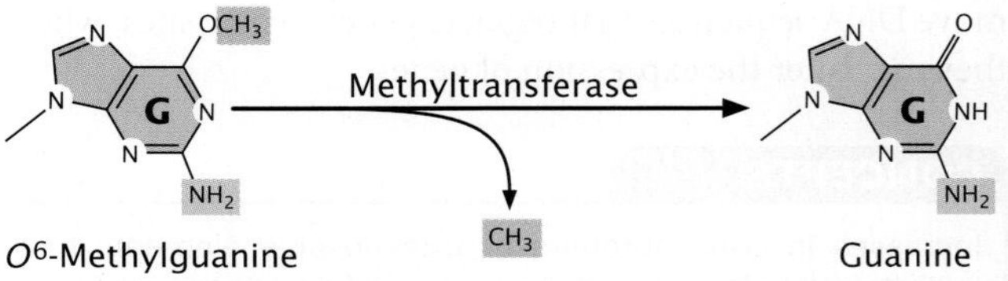

13.27 Direct repair restores the original structures of nucleotides.

Other glycosylases recognize hypoxanthine, 3-methyladenine, 7-methylguanine, and other modified bases.

A final repair pathway that we'll consider is **nucleotide-excision repair**, which removes bulky DNA lesions (such as pyrimidine dimers) that distort the double helix. In nucleotide-excision repair, the two strands of DNA are separated and a section of the DNA containing the distortion is removed; the resulting gap is filled in by DNA polymerase, and DNA ligase seals the gap in the sugar–phosphate linkage. Nucleotide-excision repair can repair many different types of DNA damage. It is found in cells of all organisms from bacteria to humans and is among the most important of all repair mechanisms.

CONCEPTS

A number of pathways exist for the repair of DNA. Most require two nucleotide strands because a template strand is needed to specify the correct base sequence.

Genetic Diseases and Faulty DNA Repair

Several human diseases are connected to defects in DNA repair. These diseases are often associated with high incidences of specific cancers because defects in DNA repair lead to increased rates of mutation. This phenomenon is discussed further in Chapter 16.

Among the best-studied of the human DNA-repair diseases is xeroderma pigmentosum (**Figure 13.28**), a rare autosomal recessive condition that includes abnormal skin pigmentation and acute sensitivity to sunlight. Persons who have this disease also have a strong predisposition to skin cancer, with an incidence ranging from 1000 to 2000 times that found in unaffected people. Sunlight includes a strong UV component, so exposure to sunlight produces pyrimidine dimers in the DNA of skin cells. Most pyrimidine dimers in humans can be corrected by nucleotide-excision repair. However, the cells of most people with xeroderma pigmentosum are defective in nucleotide-excision repair, and many of their pyrimidine dimers remain uncorrected and may lead to cancer.

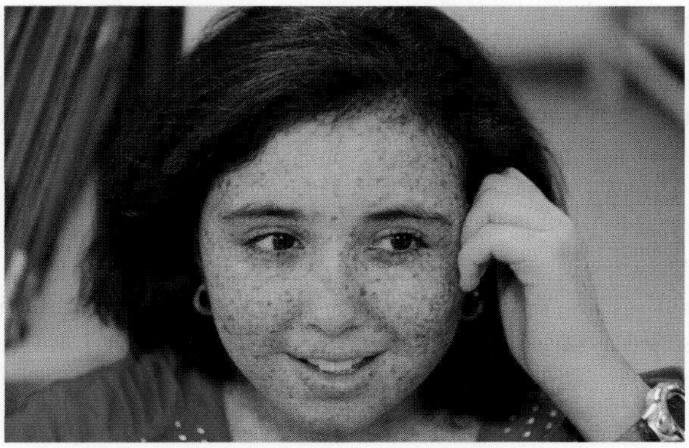

13.28 Xeroderma pigmentosum results from defects in DNA repair. The disease is characterized by freckle-like spots on the skin (shown here) and a predisposition to skin cancer. [Stephane AUDRAS/REA/Redux.]

Another genetic disease caused by faulty DNA repair is an inherited form of colon cancer called hereditary nonpolyposis colon cancer (HNPCC). It is one of the most common hereditary cancers, accounting for about 15% of colon cancers. Research findings indicate that HNPCC arises from mutations in the proteins that carry out mismatch repair. Some genetic diseases associated with defective DNA repair are summarized in **Table 13.4**. TRY PROBLEM 27

THINK-PAIR-SHARE Question 8

CONCEPTS

Defects in DNA repair are the underlying cause of several genetic diseases. Many of these diseases are characterized by a predisposition to cancer.

✓ CONCEPT CHECK 5

Why are defects in DNA repair often associated with increases in cancer?

TABLE 13.4 Genetic diseases associated with defects in DNA-repair systems

Disease	Symptoms	Genetic defect
Xeroderma pigmentosum	Freckle-like spots on skin, sensitivity to sunlight, predisposition to skin cancer	Defects in nucleotide-excision repair
Cockayne syndrome	Dwarfism, sensitivity to sunlight, premature aging, deafness, intellectual disability	Defects in nucleotide-excision repair
Trichothiodystrophy	Brittle hair, skin abnormalities, short stature, immature sexual development, characteristic facial features	Defects in nucleotide-excision repair
Hereditary nonpolyposis colon cancer	Predisposition to colon cancer	Defects in mismatch repair
Fanconi anemia	Increased skin pigmentation; abnormalities of skeleton, heart, and kidneys; predisposition to leukemia	Possibly defects in the repair of inter-strand cross-links
Li–Fraumeni syndrome	Predisposition to cancer in many different tissues	Defects in DNA damage response
Werner syndrome	Premature aging, predisposition to cancer	Defect in homologous recombination

CONCEPTS SUMMARY

- Mutations are heritable changes in genetic information. Somatic mutations occur in somatic cells; germ-line mutations occur in cells that give rise to gametes.
- The simplest type of mutation is a base substitution, a change in a single base pair of DNA. Transitions are base substitutions in which purines are replaced by purines, or pyrimidines are replaced by pyrimidines. Transversions are base substitutions in which a purine replaces a pyrimidine, or a pyrimidine replaces a purine.
- Insertions are additions of nucleotides, and deletions are removals of nucleotides; these mutations often change the reading frame of the gene.
- Expanding nucleotide repeats are mutations in which the number of copies of a set of nucleotides increases with the passage of time; they are responsible for several human genetic diseases.
- A missense mutation alters the coding sequence so that one amino acid substitutes for another. A nonsense mutation changes a codon that specifies an amino acid into a stop codon. A silent mutation produces a synonymous codon that specifies the same amino acid as does the original sequence, whereas a neutral mutation alters the amino acid sequence but does not change the functioning of the protein.
- A suppressor mutation reverses the effect of a mutation at a different site and may be intragenic (within the same gene as the original mutation) or intergenic (within a different gene).
- The mutation rate is the frequency with which a wild-type allele at a locus changes into a mutant allele. Mutation rates are influenced by both genetic and environmental factors.
- Some mutations occur spontaneously. These mutations include the mispairing of bases in replication and spontaneous depurination and deamination.
- Insertions and deletions can arise from strand slippage in replication or from unequal crossing over.
- Base analogs can become incorporated into DNA in the course of replication and pair with the wrong base in subsequent replication events. Alkylating agents, deaminating chemicals, and hydroxylamine lead to mutations by modifying the chemical structure of bases. Intercalating agents insert themselves into the DNA molecule and cause single-nucleotide additions and deletions.
- Ionizing radiation is mutagenic, altering base structures and breaking phosphodiester bonds. Ultraviolet light produces pyrimidine dimers, which block replication.
- The Ames test uses bacteria to assess the mutagenic potential of chemical substances.
- Transposable elements are mobile DNA sequences that insert into many locations within a genome and often cause mutations and DNA rearrangements.
- Most transposable elements have two common characteristics: terminal inverted repeats and the generation of short direct repeats in DNA at the point of insertion.
- A transposon can be transposed as a DNA molecule or through the production of an RNA molecule that is then reverse transcribed into DNA. Transposition may be replicative, in which the transposable element is copied and the copy moves to a new site, or nonreplicative, in which the transposable element excises from the old site and moves to a new site.
- Transposons are mutagenic and have played an important role in genome evolution.
- Damage to DNA is often corrected by DNA-repair mechanisms. Most repair pathways require two strands of DNA. The various pathways exhibit some redundancy in the types of damage repaired.
- Defects in DNA repair are the underlying cause of several genetic diseases.

IMPORTANT TERMS

mutation (p. 360)
somatic mutation (p. 360)
germ-line mutation (p. 361)
gene mutation (p. 361)
base substitution (p. 361)
transition (p. 361)
transversion (p. 361)
insertion (p. 362)
deletion (p. 362)
frameshift mutation (p. 362)
in-frame insertion (p. 362)
in-frame deletion (p. 362)
expanding nucleotide repeat (p. 362)
forward mutation (p. 363)
reverse mutation (reversion) (p. 363)
missense mutation (p. 363)
nonsense mutation (p. 363)
silent mutation (p. 363)
neutral mutation (p. 364)
loss-of-function mutation (p. 364)
gain-of-function mutation (p. 364)
conditional mutation (p. 364)
lethal mutation (p. 364)
suppressor mutation (p. 364)
intragenic suppressor mutation (p. 365)
intergenic suppressor mutation (p. 366)
mutation rate (p. 367)
spontaneous mutation (p. 368)
induced mutation (p. 368)
incorporated error (p. 369)
replicated error (p. 369)
strand slippage (p. 369)
unequal crossing over (p. 370)

depurination (p. 370)
deamination (p. 370)
mutagen (p. 371)
base analog (p. 371)
intercalating agent (p. 373)
pyrimidine dimer (p. 373)
SOS system (p. 373)
Ames test (p. 374)
transposable element (p. 375)
flanking direct repeat (p. 375)
terminal inverted repeat (p. 375)
transposition (p. 376)
DNA transposon (p. 376)
retrotransposon (p. 376)
replicative transposition (p. 376)
nonreplicative transposition (p. 376)
mismatch repair (p. 378)
direct repair (p. 378)
base-excision repair (p. 378)
nucleotide-excision repair (p. 379)

ANSWERS TO CONCEPT CHECKS

1. c

2. A reverse mutation restores the original phenotype by changing the DNA sequence back to the wild type. A suppressor mutation restores the phenotype by causing an additional change in the DNA at a site that is different from that of the original mutation.

3. c

4. In transposition, staggered cuts are made in DNA and the transposable element inserts into the cut. Later, replication of the single-stranded pieces of DNA creates short repeats on either side of the inserted transposable element.

5. Changes in DNA structure may not undergo repair in people with defects in DNA-repair mechanisms. Consequently, increased numbers of mutations occur at all genes, including those that predispose to cancer. This observation indicates that cancer arises from mutations in DNA.

WORKED PROBLEM

A codon that specifies the amino acid Asp undergoes a single-base substitution that yields a codon that specifies Ala. Give all possible DNA sequences for the original and the mutated codon. Is the mutation a transition or a transversion?

Solution Strategy

What information is required in your answer to the problem?

A list of all possible original codons and mutated codons that would cause a change from Asp to Ala.

What information is provided to solve the problem?

- A single-base substitution occurred.
- The amino acid Asp changed to Ala.

For help with this problem, review:

Types of Mutations in Section 13.1 and the genetic code in Figure 11.5.

Solution Steps

Hint: See Figure 11.5 for a list of codons and the amino acids they specify.

There are two possible RNA codons for Asp: GAU and GAC. The DNA sequences that encode these codons will be complementary to the RNA codons: CTA and CTG. There are four possible RNA codons for Ala: GCU, GCC, GCA, and GCG, which correspond to DNA sequences CGA, CGG, CGT, and CGC. If we organize the original and the mutated sequences as shown in the following table, the types of mutations that may have occurred can be easily seen:

Possible original sequence for Asp	Possible mutated sequence for Ala
CTA	CGA
CTG	CGG
	CGT
	CGC

If the mutation is confined to a single-base substitution, then the only mutations possible are that CTA mutated to CGA or that CTG mutated to CGG. In both, there is a T $\rightarrow$ G transversion in the middle nucleotide of the codon.

COMPREHENSION QUESTIONS

Section 13.1

1. What is the difference between a transition and a transversion? Which type of base substitution is more common?
2. Briefly describe expanding nucleotide repeats.
3. What is the difference between a missense mutation and a nonsense mutation? A silent mutation and a neutral mutation?
4. Briefly describe two different ways in which intragenic suppressors can reverse the effects of mutations.

Section 13.2

5. How do insertions and deletions arise?
6. How do base analogs lead to mutations?
7. What is the purpose of the Ames test? How are *his*⁻ bacteria used in this test?

Section 13.3

8. What general characteristics are found in many transposable elements?
9. Describe the differences between replicative and nonreplicative transposition.
10. What is a retrotransposon and how does it move?

Section 13.4

11. List at least three different types of DNA repair and briefly explain how each is carried out.

For more questions that test your comprehension of the key chapter concepts, go to **LearningCurve** for this chapter.

APPLICATION QUESTIONS AND PROBLEMS

Section 13.1

12. A codon that specifies the amino acid Gly undergoes a single-base substitution to become a nonsense mutation. In accord with the genetic code given in **Figure 11.5**, is this mutation a transition or a transversion? At which position of the codon does the mutation occur?

*13. Refer to the genetic code in **Figure 11.5** to answer the following questions:

a. If a single transition occurs in a codon that specifies Phe, what amino acids can be specified by the mutated sequence?

b. If a single transversion occurs in a codon that specifies Phe, what amino acids can be specified by the mutated sequence?

c. If a single transition occurs in a codon that specifies Leu, what amino acids can be specified by the mutated sequence?

d. If a single transversion occurs in a codon that specifies Leu, what amino acids can be specified by the mutated sequence?

14. Hemoglobin is a complex protein that contains four polypeptide chains. The normal hemoglobin found in adults—called adult hemoglobin—consists of two alpha and two beta polypeptide chains, which are encoded by different loci. Sickle-cell hemoglobin, which causes sickle-cell anemia, arises from a mutation in the beta chain of adult hemoglobin. Adult hemoglobin and sickle-cell hemoglobin differ in a single amino acid: the sixth amino acid from one end in adult hemoglobin is glutamic acid, whereas sickle-cell hemoglobin has valine at this position. After consulting the genetic code provided in **Figure 11.5**, indicate the type and location of the mutation that gave rise to sickle-cell anemia.

*15. The following nucleotide sequence is found on the template strand of DNA. First, determine the amino acids of the protein encoded by this sequence by using the genetic code provided in **Figure 11.5**. Then give the altered amino acid sequence of the protein that will be found in each of the following mutations:

Sequence of DNA template → 3′—TAC TGG CCG TTA GTT GAT ATA ACT—5′

Nucleotide number → 1 ... 24

a.	Mutant 1:	A transition at nucleotide 11
b.	Mutant 2:	A transition at nucleotide 13
c.	Mutant 3:	A one-nucleotide deletion at nucleotide 7
d.	Mutant 4:	A T → A transversion at nucleotide 15
e.	Mutant 5:	An addition of TGG after nucleotide 6
f.	Mutant 6:	A transition at nucleotide 9

16. Draw a hairpin structure like that shown in **Figure 13.5** for the repeated sequence found in fragile-X syndrome (see **Table 13.1**).

***17.** A polypeptide has the following amino acid sequence.

Met-Ser-Pro-Arg-Leu-Glu-Gly

The amino acid sequence of this polypeptide was determined in a series of mutants listed in parts *a* through *e*. For each mutant, indicate the type of mutation that occurred in the DNA (single-base substitution, insertion, deletion) and the phenotypic effect of the mutation (nonsense mutation, missense mutation, frameshift, etc.).

a. Mutant 1: Met-Ser-Ser-Arg-Leu-Glu-Gly

b. Mutant 2: Met-Ser-Pro

c. Mutant 3: Met-Ser-Pro-Asp-Trp-Arg-Asp-Lys

d. Mutant 4: Met-Ser-Pro-Glu-Gly

e. Mutant 5: Met-Ser-Pro-Arg-Leu-Leu-Glu-Gly

18. A gene encodes a protein with the following amino acid sequence:

Met-Trp-His-Arg-Ala-Ser-Phe

A mutation occurs in the gene. The mutant protein has the following amino acid sequence:

Met-Trp-His-Ser-Ala-Ser-Phe

An intragenic suppressor restores the amino acid sequence to that of the original protein:

Met-Trp-His-Arg-Ala-Ser-Phe

Give at least one example of base changes that could produce the original mutation and the intragenic suppressor. (Consult the genetic code in **Figure 11.5**).

Section 13.2

19. The following nucleotide sequence is found in a short stretch of DNA:

5′—ATGT—3′
3′—TACA—5′

If this sequence is treated with hydroxylamine, what sequences will result after replication?

***20.** The following nucleotide sequence is found in a short stretch of DNA:

5′—AG—3′
3′—TC—5′

a. Give all the mutant sequences that can result from spontaneous depurination in this stretch of DNA.

b. Give all the mutant sequences that can result from spontaneous deamination in this stretch of DNA.

21. DATA ANALYSIS Mary Alexander studied the effects of radiation on mutation rates in the sperm of *Drosophila melanogaster.* She irradiated *Drosophila* larvae with either 3000 roentgens (r) or 3975 r, collected the adult males that developed from irradiated larvae, mated them with unirradiated females that were homozygous for recessive alleles at eight loci. She then counted the number of F_1 flies that carried a new mutation at each locus. All mutant flies that appeared were used in subsequent crosses to determine if their mutant phenotypes were genetic. For the roughoid locus, she obtained the following results (M. L. Alexander. 1954. *Genetics* 39:409–428):

Group	Number of offspring	Offspring with a mutation at the *roughoid* locus
Control (0 r)	45,504	0
Irradiated (3000 r)	49,512	5
Irradiated (3975 r)	50,159	16

a. Calculate the mutation rates at the *roughoid* locus of the control group and the two groups of irradiated flies.

b. On the basis of these data, do you think radiation has any effect on mutation? Explain your answer.

22. What conclusion would you draw if the number of bacterial colonies in **Figure 13.21** were the same on the control plate and the treatment plate? Explain your reasoning.

Section 13.3

***23.** A particular transposable element generates flanking direct repeats that are 4 bp long. Give the sequence that will be found on both sides of the transposable element if this transposable element inserts at the position indicated on each of the following sequences.

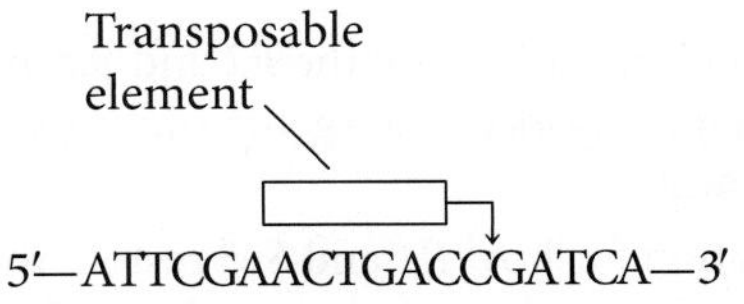

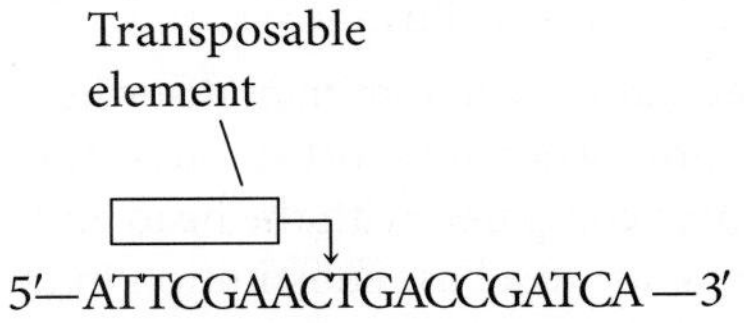

24. What factor determines the length of the flanking direct repeats that are produced in a transposition?

25. Zidovudine (AZT) is a drug used to treat patients with AIDS. AZT works by blocking the reverse-transcriptase enzyme used by the human immunodeficiency virus (HIV), the causative agent of AIDS. Do you expect that AZT would have any effect on transposable elements? If so, what type of transposable elements would be affected, and what would be the most likely effect?

26. A transposable element is found to encode a reverse-transcriptase enzyme. On the basis of this information, what conclusions can you make about the likely method of transposition of this element?

Section 13.4

***27.** A plant breeder wants to isolate mutants in tomatoes that are defective in DNA repair. However, this breeder does not have the expertise or equipment to study enzymes in DNA-repair systems. How can the breeder identify tomato plants that are deficient in DNA repair? What are the traits to look for?

CHALLENGE QUESTIONS

Section 13.1

28. DATA ANALYSIS Robert Bost and Richard Cribbs studied a strain of *E. coli* (*araB14*) that possessed a nonsense mutation in the structural gene that encodes L-ribulokinase, an enzyme that allows the bacteria to metabolize the sugar arabinose (R. Bost and R. Cribbs. 1969. *Genetics* 62:1–8). From the *araB14* strain, they isolated some bacteria that possessed mutations that caused the bacteria to revert back to wild type. Genetic analysis of these revertants showed that they possessed two different suppressor mutations. One suppressor mutation (*R1*) was linked to the original mutation in the L-ribulokinase and probably occurred at the same locus. By itself, this mutation allowed the production of L-ribulokinase, but the enzyme was not as effective in metabolizing arabinose as the enzyme encoded by the wild-type allele. The second suppressor mutation (Su^B) was not linked to the original mutation. In conjunction with the *R1* mutation, Su^B allowed the production of L-ribulokinase, but Su^B by itself was not able to suppress the original mutation.

a. On the basis of this information, are the *R1* and Su^B mutations intragenic suppressors or intergenic suppressors? Explain your reasoning.

b. Propose an explanation for how *R1* and Su^B restore the ability of *araB14* to metabolize arabinose and why Su^B is able to more fully restore this ability.

29. Achondroplasia is an autosomal dominant disorder characterized by disproportionate short stature—the legs and arms are short compared with the head and trunk. The disorder is due to a base substitution in the gene, located on the short arm of chromosome 4, for fibroblast-growth-factor receptor 3 (FGFR3).

Although achondroplasia is clearly inherited as an autosomal dominant trait, more than 80% of the people who have achondroplasia are born to parents with normal stature. This high percentage indicates that most cases are caused by newly arising mutations; these cases (not inherited from an affected parent) are referred to as sporadic. Studies have demonstrated that sporadic cases of achondroplasia are almost always caused by mutations inherited from the father (paternal mutations). In addition, the occurrence of achondroplasia is higher among the children of older fathers; approximately 50% of children with achondroplasia are born to fathers older than 35 years of age. There is no association with maternal age. The mutation rate for achondroplasia (about 4×10^{-5} mutations per gamete) is high compared with those for other genetic disorders. Explain why most spontaneous mutations for achondroplasia are paternal in origin and why the occurrence of achondroplasia is higher among older fathers.

A family of three who have achondroplasia. [AP Photo/ Gail Burton.]

30. *Ochre* and *amber* are two types of nonsense mutations. Before the genetic code was worked out, Sydney Brenner, Anthony O. Stretton, and Samuel Kaplan applied different types of mutagens to bacteriophages in an attempt to determine the bases present in the codons responsible for *amber* and *ochre* mutations. They knew that the *ochre* and *amber* mutations were suppressed by different types of mutations, which demonstrated that each is a different stop codon. They obtained the following results:

1. A single-base substitution could convert an *ochre* mutation into an *amber* mutation.
2. Hydroxylamine induced both *ochre* and *amber* mutations in wild-type phages.
3. 2-Aminopurine caused *ochre* to mutate to *amber*.
4. Hydroxylamine did not cause *ochre* to mutate to *amber*.

These data do not allow the complete nucleotide sequence of the *amber* and *ochre* codons to be worked out, but they do provide some information about the bases found in the nonsense mutations.

a. What conclusions about the bases found in the codons of *amber* and *ochre* mutations can be made from these observations?

b. Of the three nonsense codons (UAA, UAG, UGA), which represents the *ochre* mutation?

THINK-PAIR-SHARE QUESTIONS

Introduction

1. Propose some ways that the new information provided by research on the role of the GGGGCC repeat in ALS might be used to design potential treatments for the disease.
2. Using the genetic code illustrated in Figure 11.5, show how translation of the GGGGCC repeat without a start codon results in the production of five proteins with different dipeptide repeats. (Hint: Consider all reading frames of the two RNAs copied from this sequence.)

Section 13.1

3. Are mutations good or bad? Explain your response to this question.
4. Explain why loss-of-function mutations are frequently recessive, whereas gain-of-function mutations are frequently dominant.
5. The overall rate of mutations in humans is estimated to be about 1×10^{-8} mutations per base pair per generation. How many new mutations would you expect each person to carry, on average, based on this mutation rate? Other studies have estimated that each person carries about 100 new loss-of-function mutations. How does this number compare with your estimate of the number of mutations based on the mutation rate? What might account for any differences?

Section 13.2

6. What accounts for the amazingly accurate replication of DNA, which keeps the mutation rate low?
7. To determine whether radiation associated with the atomic bombings of Hiroshima and Nagasaki produced recessive germ-line mutations, scientists examined the sex ratio of the children of the survivors of the blasts. Can you explain why an increase in germ-line mutations might be expected to alter the sex ratio?

Section 13.4

8. Research has shown that more mutations accumulate in regions of a chromosome that consist of compact chromatin, such as heterochromatin. Offer an explanation for why mutation rates would be higher where chromatin is more compact.

14 Molecular Genetic Analysis and Biotechnology

Editing the Genome with CRISPR-Cas9

For over 40 years, entertainer Jerry Lewis hosted the Jerry Lewis MDA Labor Day Telethon, raising almost $2.5 billion to support research, education, and medical services for neuromuscular diseases, including muscular dystrophy. CRISPR/Cas9 genome editing techniques are being developed for possible use in treating Duchenne muscular dystrophy. [Chris Farina/Getty Images]

One of the most remarkable of all DNA sequences is the human gene for dystrophin. Located on the X chromosome, the gene is enormous, encompassing 2.2 million base pairs, including 79 exons and 78 introns. Its product, dystrophin, connects the cytoskeleton of a muscle cell to its extracellular matrix and is critical for muscle contraction.

Mutations in the dystrophin gene are the cause of Duchenne muscular dystrophy, a fatal muscle disorder that strikes nearly 1 in 3500 boys. At present, there is no cure for the disease. Duchenne muscular dystrophy was first fully described in 1861 by Benjamin A. Duchenne, a Paris physician. In spite of early recognition of its X-linked pattern of inheritance, the molecular basis of Duchenne muscular dystrophy remained a mystery for many years. In 1985, Louis Kunkel and his colleagues at Harvard Medical School pinpointed the location on the X chromosome of a gene that, when mutated, caused Duchenne muscular dystrophy. They went on to locate and clone the piece of DNA responsible for the disease and to discover the dystrophin gene.

Ever since the discovery of the dystrophin gene, researchers and physicians have dreamed of treating Duchenne muscular dystrophy with gene therapy—restoring dystrophin production by inserting a corrected copy of the dystrophin gene into the patient's cells—but the size of the dystrophin gene has made this approach challenging. Recently, however, geneticists have developed a novel and interesting approach to gene therapy that makes use of a new technique for editing the genome. Many of the mutations that cause Duchenne muscular dystrophy consist of duplications or deletions that cause frameshift mutations, leading to premature stop codons that produce a truncated, nonfunctional dystrophin. The huge dystrophin protein consists of several domains, some of which are not essential for muscle function. Researchers proposed that if the exons where mutations occurred could be removed, the premature stop codons would be eliminated. The resulting protein would be missing some amino acids, but a dystrophin that provided some muscle function might still be produced.

The trick to making this strategy work is the ability to selectively edit DNA sequences that control the splicing of pre-mRNA. Within the past few years, a powerful new tool, called CRISPR-Cas9, has been developed that allows precision editing of the genome. In recent experiments, CRISPR-Cas9 was used to engineer changes in the genomes of mice that bring about exon skipping in the dystrophin gene, demonstrating the potential feasibility of this approach.

The geneticists used mice that possess a premature stop codon in exon 23 of the dystrophin gene and exhibit many of the symptoms of the human disorder. They loaded viruses with genetic material encoding CRISPR-Cas9 components, then injected the mice with the viruses. The results were remarkable. Delivered by the virus, CRISPR-Cas9 entered muscle cells and produced double-stranded cuts on either side of exon 23. Exon 23, along with the disease-causing premature stop codon it contained, was removed from the genomes of some cells. The result was synthesis of a partially functional dystrophin protein that enhanced muscle function in the mice. The mice weren't completely cured, and much work needs to be done before these methods can be applied to humans, but the results are encouraging and suggest that CRISPR-Cas9 might be used in a similar way to help restore muscle function to boys with Duchenne muscular dystrophy.

THINK-PAIR-SHARE Questions 1 and 2

The possibility of using CRISPR-Cas9 to genetically engineer a treatment for muscular dystrophy illustrates the power of molecular techniques for manipulating DNA sequences. This chapter introduces some of the techniques used in molecular genetic analysis. We begin by considering the challenges of working at the molecular level. We then examine a number of methods used to analyze and alter DNA, placing emphasis on several transformative techniques that are used to cut, clone, amplify, and sequence DNA. Finally, we explore some of the applications of molecular genetic analysis.

14.1 Genetics Has Been Transformed by the Development of Molecular Techniques

A vast array of molecular methods is now available for probing the nature of hereditary information and revealing the details of genetic processes. These molecular techniques have drastically altered the way in which genes are studied. Previously, information about the structure and organization of genes was gained by examining their phenotypic effects, but molecular genetic analysis now allows the nucleotide sequences themselves to be read. This analysis has provided new information about the structure and function of genes and has altered many fundamental concepts of genetics. Our detailed understanding of genetic processes such as replication, transcription, translation, RNA processing, and gene regulation has been obtained through the use of molecular genetic techniques. These techniques are used in many other fields as well, including biochemistry, microbiology, developmental biology, neurobiology, evolution, and ecology. Molecular genetic techniques are also being used to create a number of commercial products, including drugs, hormones, enzymes, and crops. The **biotechnology** industry has grown up around the use of these techniques to develop new products. In medicine, molecular genetic analysis is being used to probe the nature of cancer, diagnose genetic and infectious diseases, produce drugs, and treat hereditary disorders.

THINK-PAIR-SHARE Question 3

CONCEPTS

Techniques of molecular genetics are used to locate, analyze, alter, sequence, study, and recombine DNA sequences. These techniques are used to probe the structure and function of genes, address questions in many areas of biology, create commercial products, and diagnose and treat diseases.

Working at the Molecular Level

The manipulation of genes at the molecular level presents a serious challenge, often requiring strategies that may not, at first, seem obvious. The basic problems are that genes are minute and that every cell contains thousands of them. Individual nucleotides cannot be seen, and no physical features mark the beginning or the end of a gene.

Let's consider a typical situation faced by a molecular geneticist. Suppose we want to use bacteria to produce large quantities of a human protein. The first and most formidable problem is to find the gene that encodes the desired protein. A haploid human genome consists of 3.2 billion base pairs of DNA. Let's assume that the gene that we want to isolate is 3000 bp long; our target gene occupies only one-millionth of the genome, so searching for our gene in the huge expanse of genomic DNA is more difficult than looking for the proverbial needle in a haystack. But even if we are able to locate the gene, how do we separate it from the rest of the DNA?

If we succeed in locating and isolating the desired gene, we next need to insert it into a bacterial cell. Linear fragments of DNA are quickly degraded by bacteria, so the gene must be inserted in a stable form. It must also be able to successfully replicate, or it will not be passed on when the cell divides. If we succeed in transferring our gene to bacteria in a stable form, we must still ensure that the gene is properly transcribed and translated.

Finally, the methods used to isolate and transfer genes are inefficient: of a million cells that are subjected to these procedures, only *one* cell might successfully take up and express the human gene. So we must search through many bacterial cells to find the one containing the recombinant DNA. We are back to the problem of the needle in a haystack.

Although these problems might seem insurmountable, molecular techniques have been developed to overcome them, and human genes are now routinely transferred to bacterial cells in which the genes are expressed.

CONCEPTS

Molecular genetic analyses require special techniques because individual genes make up a tiny fraction of the cellular DNA and cannot be seen.

14.2 Molecular Techniques Are Used to Cut and Visualize DNA Sequences

Techniques for accurately and efficiently cleaving DNA helped usher in the first major revolution in molecular genetics: the development of **recombinant DNA technology**, a set of molecular techniques for locating, isolating, altering, and studying DNA segments. The term *recombinant* is used because frequently the goal is to combine DNA from two distinct sources. Genes from two different bacteria might be joined, for example, or a human gene might be inserted into a viral chromosome. Commonly called **genetic engineering**, recombinant DNA technology now encompasses many molecular techniques that can be used to analyze, alter, and recombine virtually any DNA sequences from any number of sources.

Restriction Enzymes

A key event in the development of recombinant DNA technology was the discovery in the late 1960s of **restriction enzymes** (also called **restriction endonucleases**), which recognize specific nucleotide sequences in DNA and make double-stranded cuts at those sequences (called *restriction sites*). Restriction enzymes are produced naturally by bacteria and are used in defense against viruses. A bacterium protects its own DNA from a restriction enzyme by modifying the recognition sequence, usually by adding methyl groups to its DNA.

More than 800 different restriction enzymes that recognize and cut DNA at more than 100 different sequences have been isolated from bacteria. Many of these enzymes are commercially available; examples of some commonly used restriction enzymes are given in **Table 14.1**. The name of each restriction enzyme begins with an abbreviation that signifies its bacterial origin.

The sequences recognized by restriction enzymes are usually from 4 to 8 bp long; most of these enzymes recognize a

TABLE 14.1 Characteristics of some common restriction enzymes used in recombinant DNA technology

Enzyme	Microorganism from which enzyme is produced	Recognition sequence	Type of fragment end produced
*Bam*HI	*Bacillus amyloliquefaciens*	↓ 5′—GGATCC—3′ 3′—CCTAGG—3′ ↑	Cohesive
*Cof*I	*Clostridium formicoaceticum*	↓ 5′—GCGC—3′ 3′—CGCG—5′ ↑	Cohesive
*Eco*RI	*Escherichia coli*	↓ 5′—GAATTC—3′ 3′—CTTAAG—5′ ↑	Cohesive
*Eco*RII	*Escherichia coli*	↓ 5′—CCAGG—3′ 3′—GGTCC—5′ ↑	Cohesive
*Hae*III	*Haemophilus aegyptius*	↓ 5′—GGCC—3′ 3′—CCGG—5′ ↑	Blunt
*Hind*III	*Haemophilus influenzae*	↓ 5′—AAGCTT—3′ 3′—TTCGAA—5′ ↑	Cohesive
*Pvu*II	*Proteus vulgaris*	↓ 5′—CAGCTG—3′ 3′—GTCGAC—5′ ↑	Blunt

Note: The first three letters of the abbreviation for each restriction enzyme refer to the bacterial species from which the enzyme was isolated (e.g., *Eco* refers to *E. coli*). A fourth letter may refer to the strain of bacteria from which the enzyme was isolated (the "R" in *Eco*RI indicates that this enzyme was isolated from the RY13 strain of *E. coli*). Roman numerals that follow the letters identify different enzymes from the same species.

sequence of 4 or 6 bp. Most recognition sequences are palindromic—sequences that read the same (5′ to 3′) on the two complementary DNA strands.

Some restriction enzymes make staggered cuts in the DNA. For example, *Hind*III recognizes the following sequence:

↓
5′—AAGCTT—3′
3′—TTCGAA—5′
↑

*Hind*III cuts the sugar–phosphate backbone of each strand at the point indicated by the arrow, generating fragments with short, single-stranded overhanging ends:

5′—A   AGCTT—3′
3′—TTCGA   A—5′

Such ends are called **cohesive ends**, or *sticky ends*, because they are complementary to each other and can spontaneously pair to connect the fragments. Thus DNA fragments with sticky ends can be "glued" together: any two such fragments cleaved by the same enzyme will have complementary ends and will pair (**Figure 14.1**). When their cohesive ends have paired, the two DNA fragments can be joined together permanently by DNA ligase, which seals nicks between the sugar–phosphate groups of the fragments.

Not all restriction enzymes produce staggered cuts and sticky ends (see Figure 14.1a). *Pvu*II cuts in the middle of its recognition sequence, and the cuts on the two strands are directly opposite each other, producing blunt-ended fragments that must be joined together in other ways:

↓
5′—CAGCTG—3′
3′—GTCGAC—5′
↑

↓

5′—CAG   CTG—3′
3′—GTC   GAC—5′

The sequences recognized by a restriction enzyme are located randomly within the genome. Accordingly, there is a relation between the length of the recognition sequence and the number of times it is present in a genome: there will be fewer longer recognition sequences than shorter ones, because the probability of the occurrence of a particular sequence consisting of, say, six specific bases is less than the probability of the occurrence of a particular sequence of four specific bases. Therefore, restriction enzymes that recognize longer sequences will cut a given piece of DNA into fewer and longer fragments than will restriction enzymes that recognize shorter sequences.

Restriction enzymes are used whenever DNA fragments must be cut or joined. In a typical restriction reaction, a concentrated solution of purified DNA is placed in a small tube with a buffer solution and a small amount of restriction enzyme. The reaction mixture is then heated at the optimal temperature for the enzyme, usually 37°C. Often within an hour, the enzyme cuts the appropriate restriction sites in all the DNA molecules, producing a set of DNA fragments.

▶TRY PROBLEM 19

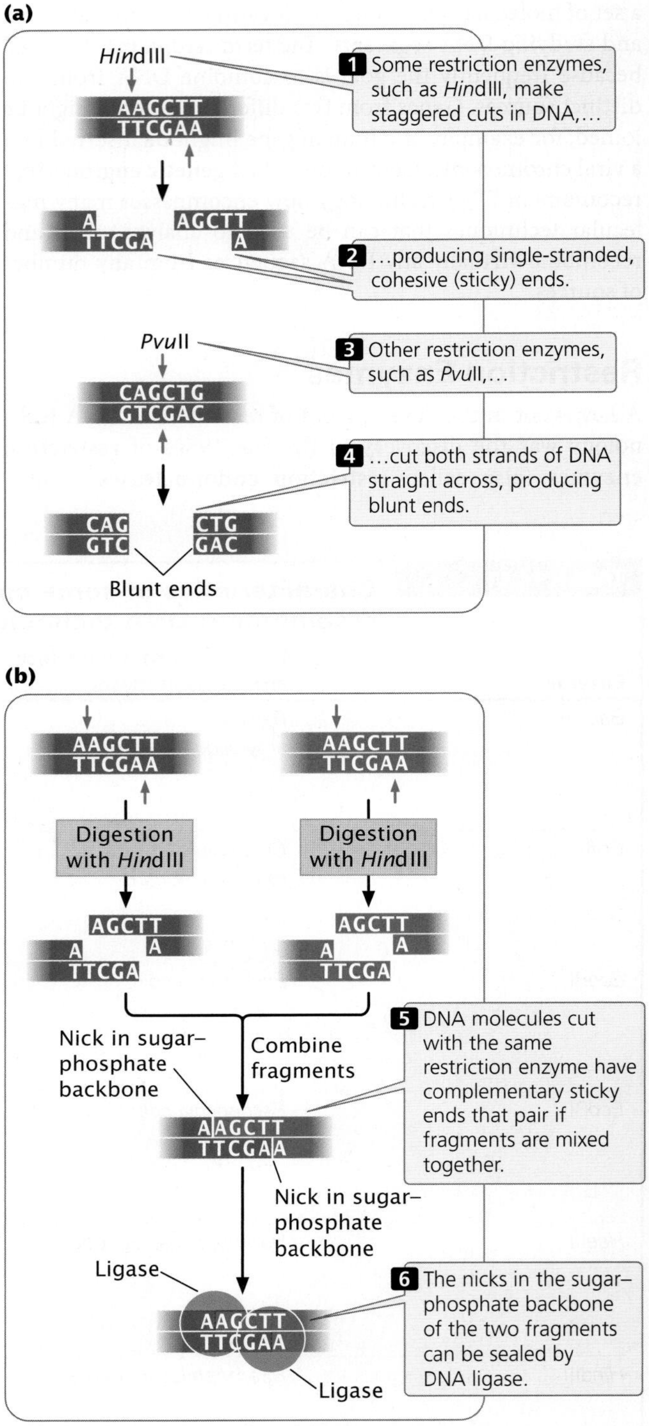

14.1 Restriction enzymes make double-stranded cuts in DNA, producing cohesive, or sticky, ends.

CONCEPTS

Restriction enzymes cut DNA at specific base sequences that are palindromic. Some restriction enzymes make staggered cuts, producing DNA fragments with cohesive ends; others cut both strands straight across, producing blunt-ended fragments. There are fewer long recognition sequences in DNA than short recognition sequences.

✓ CONCEPT CHECK 1

Where do restriction enzymes come from?

Engineered Nucleases

One limitation of restriction enzymes is that because their recognition sequences are short, typically 4–8 bp in length, they occur at random many times within a genome. For example, consider the 6-bp recognition sequence for *Bam*HI:

5′—GGATCC—3′

3′—CCTAGG—5′

There are over 47,000 *Bam*HI restriction sites in the human genome. Cutting human DNA with *Bam*HI or another restriction enzyme results in thousands of fragments. Thus, it is impossible to precisely cut genomic DNA at a single location with restriction enzymes.

To overcome this problem of numerous cuts, geneticists have sought nucleases that recognize longer DNA sequences. In recent years, geneticists have designed complex enzymes, termed **engineered nucleases**, that are capable of making unique double-stranded cuts in DNA at predetermined sequences. Engineered nucleases consist of the part of a restriction enzyme that cleaves DNA nonspecifically, coupled with another protein that recognizes and binds to a specific DNA sequence; the particular sequence to which the protein binds is determined by the protein's amino acid sequence. By altering the amino acid sequence of the binding protein, geneticists can custom-design a nuclease to bind to and cut any particular DNA sequence. Engineered nucleases include **zinc-finger nucleases (ZFNs)**, which use a DNA-binding domain called a zinc finger attached to a restriction enzyme and **transcription activator–like effector nuclease (TALEN)**, in which a protein that normally binds to sequences in promoters is attached to a restriction enzyme.

CRISPR-Cas Genome Editing

Another molecular tool for precisely cutting DNA is the **CRISPR-Cas system** that has been developed in recent years. This technique has revolutionized the field of genetics, providing a powerful way of editing the genome that has now been applied to DNA sequences in bacteria, yeast, nematodes, plants, fruit flies, zebrafish, mice, rats, monkeys, humans, and many more organisms.

CRISPR-CAS IMMUNITY IN BACTERIA AND ARCHAEA CRISPR-Cas systems occur naturally in bacteria and archaea and are used to protect these organisms against bacteriophages, plasmids, and other invading DNA elements (see Chapters 7 and 10). CRISPR RNAs (crRNAs) are encoded by DNA sequences called Clustered Regularly Interspaced Short Palindromic Repeats (CRISPR). A CRISPR array consists of a series of such palindromic sequences (sequences that read the same forward and backward on two complementary DNA strands) separated by unique spacers, which consist of sequences derived from bacteriophages or foreign plasmids (see Figure 10.26). When bacteriophage or plasmid DNA enters a prokaryotic cell, proteins cut up the foreign DNA and insert bits of it into a CRISPR array, which then serves as a memory of the invader.

The CRISPR array is transcribed into a long precursor CRISPR RNA (pre-crRNA), which is cleaved into short crRNAs. The crRNAs combine with proteins called CRISPR-associated (Cas) proteins to form effector complexes. The Cas proteins have nuclease activity—the ability to cut DNA. If the same foreign DNA enters the cell in the future, a CRISPR-Cas complex recognizes and attaches to it. The crRNA in the complex binds to its complementary sequence in the foreign DNA, and the Cas protein cleaves the foreign DNA, rendering it nonfunctional. In this way, CRISPR-Cas serves as an adaptive RNA defense system that remembers and destroys foreign invaders.

GENOME EDITING WITH CRISPR-CAS A variety of CRISPR-Cas systems have been found in different bacterial and archaeal species. Geneticists have engineered some of these systems to serve as molecular editing tools. The most widely used system is CRISPR-Cas9. This system naturally requires two RNA molecules, crRNA and another RNA molecule termed tracrRNA. These two RNAs pair and then combine with Cas9 to form an effector complex. To facilitate the use of this system in genome editing, researchers have engineered the crRNA and the tracrRNA into a single guide RNA (sgRNA) (**Figure 14.2**). A 20-nucleotide region of the sgRNA pairs with DNA, although the nucleotides within an 8–12-nucleotide "seed" sequence are most important in pairing. By altering the sequence of the sgRNA, it is possible to direct the action of the effector complex to any specific DNA sequence desired. This relatively long recognition sequence makes CRISPR-Cas9 much more specific than restriction enzymes, meaning that it can be directed to unique sites within the genome.

An important feature of the CRISPR-Cas9 system is the required presence of a sequence in the target DNA (the DNA to be cleaved) called a protospacer-adjacent motif (PAM; see Figure 14.2). The CRISPR-Cas9 effector complex first associates with a PAM, and then Cas9 unwinds the DNA nearby. This unwinding allows the sgRNA to pair with its complementary sequence in the DNA. Once the target DNA and sgRNA are paired, Cas9 makes double-stranded cuts in the DNA.

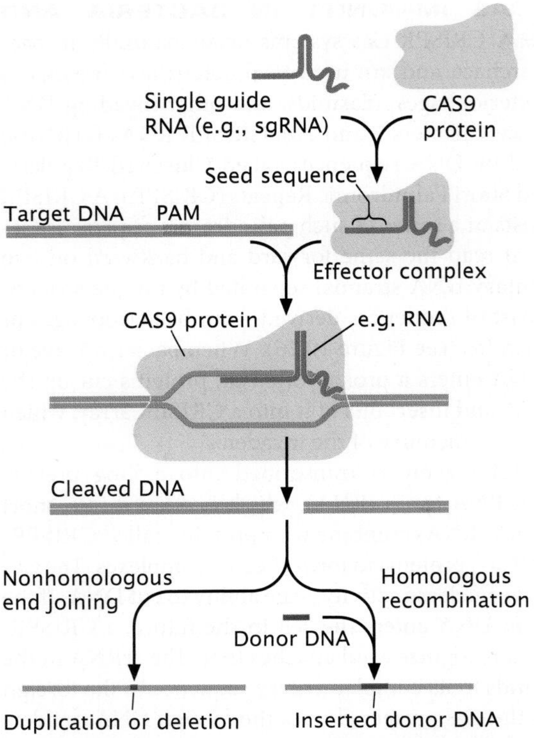

14.2 CRISPR-Cas9 is a technique for precisely editing the genome.

REPAIR OF BREAKS PRODUCED BY CRISPR-CAS9 Once the DNA of a cell has been cleaved by CRISPR-Cas9, the cell immediately uses its DNA-repair mechanism to try to repair the break. This feature provides a mechanism for editing the target sequence. There are two major pathways by which double-strand breaks are repaired within cells. One, called nonhomologous end joining, joins together the two ends of DNA without using any template. This process tends to introduce small insertions and deletions when the two ends are joined, a side effect that allows geneticists to disable a gene. The CRISPR-Cas9 system can be targeted to a specific gene; when that gene is cleaved by Cas9 and then repaired by nonhomologous end joining, the introduction of insertions or deletions at the break site often produces frameshift mutations that disrupt the coding sequence and disable the gene (see Figure 14.2).

The other mechanism used by cells to repair double-strand breaks is homologous recombination. This mechanism functions when a DNA template is provided for repairing the break. A researcher can provide a donor piece of double-stranded DNA that has ends complementary to the sequences at the ends of the break made by Cas9; homologous recombination may insert the donor DNA sequence into the break. In this way, researchers can selectively insert a desired sequence into a genome (see Figure 14.2). Unfortunately, homologous recombination is not highly efficient, and often the two ends are connected without insertion of the donor DNA.

An advantage of CRISPR-Cas9 over restriction enzymes is the relatively long nucleotide sequence recognized by the sgRNA, which allows researchers to produce unique cuts within genomic DNA. By changing the sequence of the sgRNA, precise edits can be made almost anywhere in the genome. CRISPR-Cas has great potential for genetic engineering and biotechnology. It can be applied to many different species, including species in which other methods of DNA manipulation have not worked well. It can be used to introduce new DNA sequences into whole animals and humans. A limitation in the use of CRISPR-Cas9 for genome editing is the potential for off-target cleavage. Some mismatches between the sgRNA and the complementary DNA sequence are tolerated by the Cas9 protein, so the system may cleave DNA at sites other than the desired target sequence. One concern about CRISPR-Cas9 technology is that its ease of use and potential to alter almost any DNA sequence might mean that it could be used to genetically modify humans in ethically questionable ways. View an explanation of CRISPR-Cas genome editing in **Animation 14.1.**

THINK-PAIR-SHARE Questions 4, 5, and 6

CONCEPTS

The development of CRISPR-Cas9 technology provides a powerful means of cutting and editing the genome. CRISPR-Cas9 combines a single guide RNA with a nuclease, which together attach to specific DNA sequences and make double-stranded cuts at specific locations. Repair of these cuts by nonhomologous end joining or homologous recombination provides the means to introduce alterations to the genome.

Separating and Viewing DNA Fragments

We have now discussed several techniques for precisely cutting a DNA molecule. After cutting, it is often necessary to separate the resulting fragments and to verify that the DNA was altered in the expected fashion. Gel electrophoresis provides a way to separate and visualize DNA fragments.

Electrophoresis is a standard technique for separating molecules on the basis of their size and electrical charge. There are a number of different types of electrophoresis; to separate DNA molecules, **gel electrophoresis** is used. A porous gel is often made from agarose (a polysaccharide isolated from seaweed), which is melted in a buffer solution and poured into a plastic mold. As it cools, the agarose solidifies, making a gel that looks something like stiff gelatin.

Small wells are made at one end of the gel to hold solutions of DNA fragments (**Figure 14.3a),** and an electrical current is passed through the gel. Because the phosphate group on each DNA nucleotide carries a negative charge, the DNA fragments migrate toward the positive end of the gel. During this migration, the porous gel acts as a sieve, separating the DNA fragments by size. Small DNA fragments

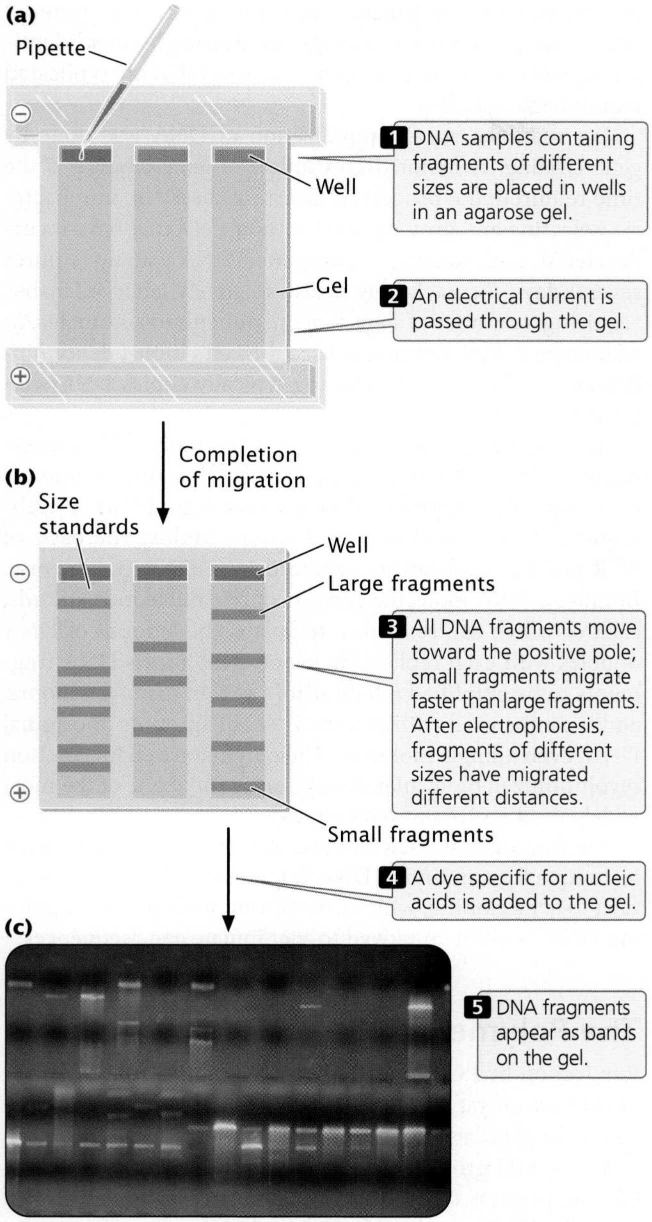

14.3 Gel electrophoresis can be used to separate DNA molecules on the basis of their size. [Klaus Guldbrandsen/Science Source.]

migrate more rapidly than large ones do, so over time, the fragments separate on the basis of their size. Typically, DNA fragments of known length (a marker sample) are placed in one of the wells. By comparing the migration distance of the unknown fragments with the distance traveled by the marker sample, one can determine the approximate size of the unknown fragments (**Figure 14.3b**).

The DNA fragments are still too small to see, so the problem of visualizing the DNA needs to be addressed. Visualization can be accomplished in several ways. The simplest procedure is to stain the gel with a dye specific for nucleic acids, such as ethidium bromide, which wedges itself tightly (intercalates) between the bases of DNA and fluoresces orange when exposed to UV light, producing brilliant orange bands on the gel (**Figure 14.3c**).

Alternatively, DNA fragments can be visualized by adding a label to the DNA before it is placed in the gel. For example, chemical labels can be detected by adding antibodies or other substances that carry a dye and will attach to the relevant DNA, which can then be visualized directly.

TRY PROBLEM 20

CONCEPTS

DNA fragments can be separated, and their sizes determined, with the use of gel electrophoresis. The fragments can be viewed by using a dye that is specific for nucleic acids or by labeling the fragments with a chemical tag.

CONCEPT CHECK 2

DNA fragments that are 500 bp, 1000 bp, and 2000 bp in length are separated by gel electrophoresis. Which fragment will migrate farthest in the gel?

a. The 2000-bp fragment.
b. The 1000-bp fragment.
c. The 500-bp fragment.
d. All will migrate equal distances.

Locating DNA Fragments with Probes

If a small piece of DNA, such as a plasmid, is cut by a restriction enzyme, the few fragments produced can be seen as distinct bands on an electrophoretic gel. In contrast, if genomic DNA from a cell is cut by a restriction enzyme, a large number of fragments of different sizes are produced. A restriction enzyme that recognizes a four-base sequence would theoretically cut about once every 256 bp. The human genome, with 3.2 billion base pairs, would generate more than 12 million fragments when cut by this restriction enzyme. When separated by electrophoresis and visualized, this large set of fragments would appear as a continuous smear on the gel because of the presence of so many fragments of differing sizes. Usually, researchers are interested in only a few of these fragments, perhaps those carrying a specific gene. How do they locate the desired fragments in such a large pool of DNA?

One approach is to use a **probe**, which is a DNA or RNA molecule with a base sequence complementary to a sequence in the gene of interest. The bases on the probe will pair only with the bases on a complementary sequence, so if suitably labeled, the probe can be used to locate a specific gene or other DNA sequence. To use a probe, a researcher first cuts the DNA into fragments by using one or more restriction enzymes and then separates the fragments with gel electrophoresis. Next, the separated fragments must be denatured (the two strands separated) and transferred to a permanent solid medium, such as a nitrocellulose or nylon membrane. **Southern blotting** (named after Edwin M. Southern) is one technique used to transfer the denatured, single-stranded fragments from a gel to a membrane.

After the single-stranded DNA fragments have been transferred, the membrane is placed in a hybridization solution containing a labeled probe. The probe binds to (hybridizes with) any DNA fragments on the membrane that bear complementary sequences. Often, a probe binds to only a part of the DNA fragment, so the DNA fragment may contain sequences not found in the probe. The membrane is then washed to remove any unbound probe; a biochemical method then reveals the presence of the bound probe. Thus, Southern blotting can reveal the presence of a specific DNA fragment in a genome or sample of DNA.

RNA can be transferred from a gel to a solid medium by a related procedure called **Northern blotting** (not named after anyone, but capitalized to match Southern blotting). Hybridization with a probe can reveal the size of a particular mRNA molecule, its relative abundance, or the tissues in which the mRNA is transcribed. **Western blotting** is the transfer of protein from a gel to a membrane. Here, the probe is usually an antibody, used to determine the size of a particular protein and the pattern of the protein's expression.

CONCEPTS

Labeled probes, which are sequences of RNA or DNA that are complementary to the sequence of interest, can be used to locate individual genes or DNA sequences. Southern blotting can be used to transfer DNA fragments from a gel to a membrane.

14.3 Specific DNA Fragments Can Be Amplified

Many of the methods used to manipulate and analyze DNA sequences cannot be carried out on single molecules, requiring instead numerous copies of a specific DNA fragment. A major problem in working at the molecular level is that each gene is a tiny fraction of the total cellular DNA. Because each gene is so rare, it must be isolated and amplified before it can be studied. There are two basic approaches to amplifying a specific DNA fragment: replicating the DNA within cells (in vivo) or replicating the DNA enzymatically outside of cells (in vitro).

In the in vivo approach, a DNA fragment is inserted into a bacterial cell and the cell is allowed to replicate the DNA. Each time the cell divides, one or more copies of the DNA fragment are passed on to each daughter cell. Most bacterial cells divide rapidly, so within a short time (usually a few days), a large number of genetically identical cells are produced, each carrying one or more copies of the DNA fragment. The cells are then lysed to release their DNA, and the desired fragment is isolated from the rest of the bacterial DNA. This procedure is termed **gene cloning** because identical copies (clones) of the original piece of DNA are replicated within bacterial cells.

For many years, all amplification of DNA was done by gene cloning. A major disadvantage of gene cloning is the time required: the process of inserting the DNA into bacteria, selecting and growing the bacterial cells that have incorporated it, and isolating the amplified DNA usually requires several days. Gene cloning is also relatively labor-intensive, requiring a number of steps that are difficult to automate. An advantage is that, because it uses the cell's high-fidelity replication machinery, gene cloning typically copies DNA with great accuracy.

The second, in vitro, approach is to amplify DNA enzymatically in a test tube outside of cells. This amplification is done with the **polymerase chain reaction (PCR)**, a technique first developed in 1983 by Kary Mullis. The basis of PCR is DNA replication catalyzed by a DNA polymerase. Because a DNA molecule consists of two nucleotide strands, each of which can serve as a template, the amount of DNA doubles with each replication event. PCR allows DNA fragments to be amplified a billion-fold within just a few hours, and it can be used with extremely small amounts of original DNA, even a single molecule. The polymerase chain reaction revolutionized molecular biology and is now one of the most widely used molecular techniques.

We first discuss PCR because it is the most widely used technique for amplifying DNA fragments today. We then return to gene cloning, which, in addition to its use in amplifying DNA, is often employed to manipulate gene sequences.

The Polymerase Chain Reaction

Replication by PCR requires a DNA template (the fragment of interest, or target DNA) from which a new DNA strand can be copied, and a pair of single-stranded primers, each with a 3′-OH group to which new nucleotides can be added. The primers used in PCR are short fragments of DNA, typically 17–25 nucleotides long, that are complementary to known sequences on the template. To carry out PCR, researchers begin with a solution that includes the target DNA, DNA polymerase, all four deoxyribonucleoside triphosphates (dNTPs—the substrates for DNA polymerase), primers, and magnesium ions and other salts that are necessary for the reaction to proceed.

A typical polymerase chain reaction includes three steps (**Figure 14.4**). In step 1, a starting solution of DNA is heated to 90°C–100°C to break the hydrogen bonds between the strands and thus produce the necessary single-stranded templates. The reaction mixture is held at this temperature for only a minute or two. In step 2, the DNA solution is cooled quickly to 30°C–65°C and held at this temperature for a minute or less. During this short interval, the DNA strands do not have a chance to reanneal, but the primers

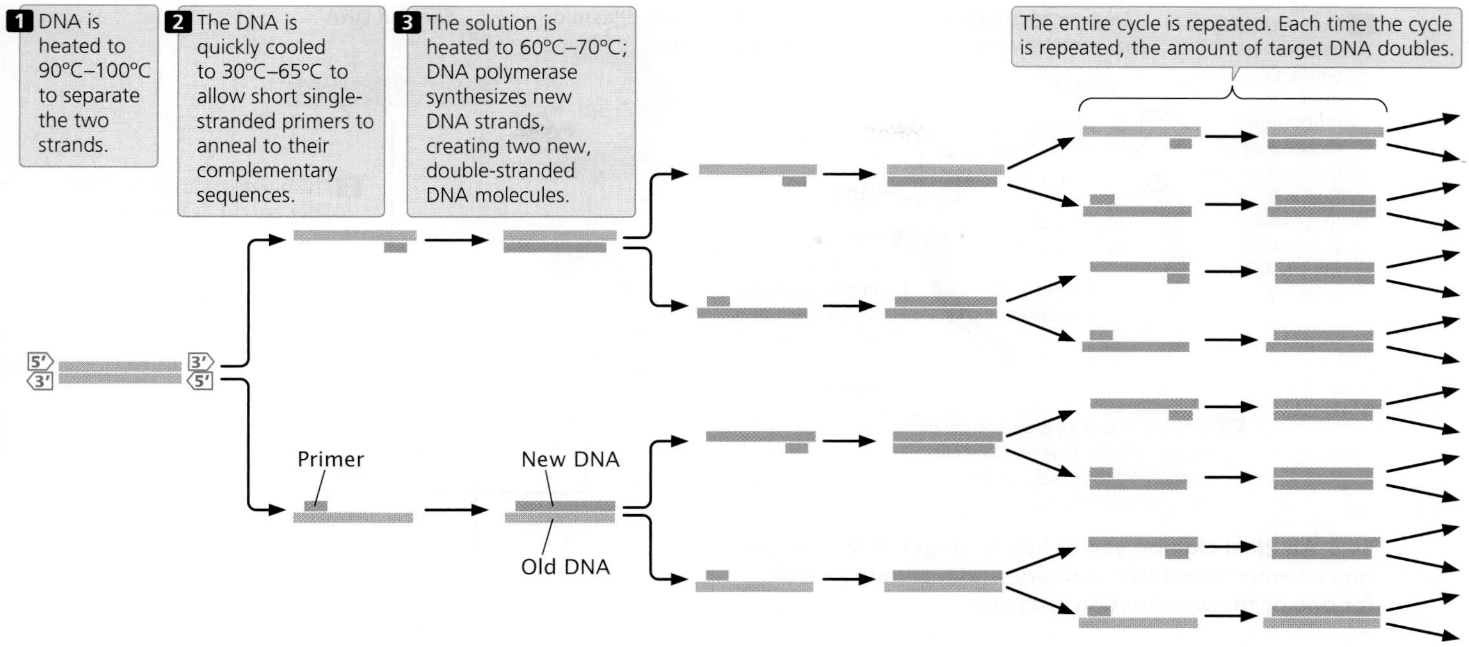

14.4 The polymerase chain reaction can be used to amplify even very small samples of DNA.

are able to attach to the template strands. In step 3, the solution is heated for a minute or less to 72°C, the temperature at which DNA polymerase can synthesize new DNA strands. At the end of the cycle, two new double-stranded DNA molecules are produced for each original molecule of DNA.

The whole cycle is then repeated. With each cycle, the amount of target DNA doubles, so the amount of target DNA increases geometrically. One molecule of DNA increases to more than 1000 molecules in 10 PCR cycles, to more than 1 million molecules in 20 cycles, and to more than 1 billion molecules in 30 cycles. Each cycle is completed within a few minutes, so a large amplification of DNA can be achieved within a few hours. To see how the polymerase chain reaction quickly increases the number of copies of a DNA fragment, view **Animation 14.2**.

A key innovation that facilitated the use of PCR in the laboratory was the discovery of a DNA polymerase that is stable at the high temperatures used in the first step of PCR. The DNA polymerase from *E. coli* that was originally used in PCR denatures at 90°C, so fresh enzyme had to be added to the reaction mixture in *each* cycle. This obstacle was overcome when DNA polymerase was isolated from the bacterium *Thermus aquaticus*, which lives in the boiling springs of Yellowstone National Park. This enzyme, dubbed ***Taq* polymerase**, is remarkably stable at high temperatures and is not denatured in the strand separation step of PCR; it can be added to the reaction mixture at the beginning of the PCR process and continues to function through many cycles.

CONCEPTS

The polymerase chain reaction is an enzymatic, in vitro (in a test tube) method for rapidly amplifying DNA. In this process, DNA is heated to separate the two strands, short primers attach to the target DNA, and DNA polymerase synthesizes new DNA strands from the primers. Each cycle of PCR doubles the amount of DNA.

Cloning Genes

Despite the widespread use of PCR, some DNA sequences are still amplified by gene cloning. In addition, gene cloning provides a powerful means of altering and manipulating DNA sequences. It is often used to alter cells so that they have desired properties or produce substances of commercial value. Here, we outline some of the methods used to manipulate DNA sequences and introduce them into bacterial cells.

Many cloning techniques require a **cloning vector**, a stable, replicating DNA molecule to which a foreign DNA fragment can be attached for introduction into a cell. An effective cloning vector has three important characteristics (**Figure 14.5**): (1) an origin of replication, which ensures that the vector is replicated within the cell; (2) selectable markers, which enable any cells containing the vector to be selected or identified; and (3) one or more unique restriction sites into which a DNA fragment can be inserted. The vector must have only one (unique) restriction site for each restriction enzyme used; if a vector is cut at multiple restriction sites, several pieces of DNA are generated, and getting these pieces back together in the correct order is possible, but extremely difficult.

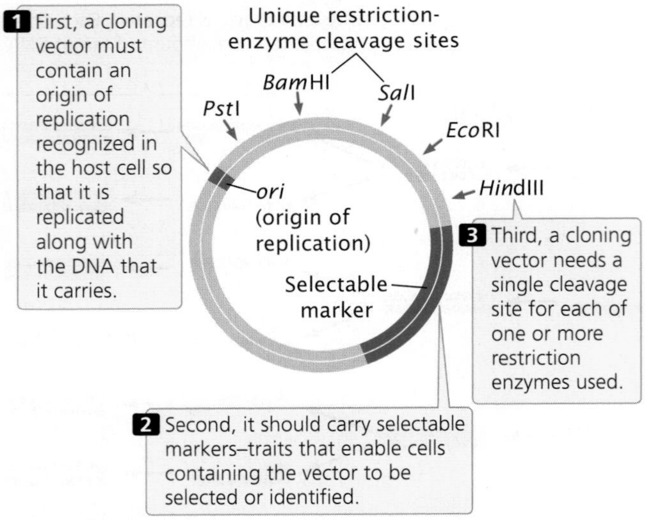

14.5 An ideal cloning vector has an origin of replication, one or more selectable markers, and recognition sites for one or more restriction enzymes.

PLASMID VECTORS Plasmids, circular DNA molecules that exist naturally in bacteria (see Chapter 7), are commonly used as vectors for cloning DNA fragments in bacteria. They contain origins of replication and are therefore able to replicate independently of the bacterial chromosome. The plasmids typically used in cloning have been artificially constructed from larger, naturally occurring bacterial plasmids and have recognition sequences for multiple enzymes, an origin of replication, and selectable markers.

The easiest method for inserting a particular DNA sequence into a plasmid vector is to cut the foreign DNA (containing a DNA fragment of interest) and the plasmid with the same restriction enzyme (**Figure 14.6**). If the restriction enzyme makes staggered cuts in the DNA, complementary sticky ends are produced on the foreign DNA and the plasmid. The DNA and plasmid are then mixed together; some of the foreign DNA fragments will pair with the cut ends of the plasmid. DNA ligase is used to seal the nicks in the sugar–phosphate backbone, creating a recombinant plasmid that contains the foreign DNA fragment. You can learn more about plasmid cloning by viewing **Animation 14.3**.

TRANSFORMATION Once a DNA fragment of interest has been placed inside a plasmid, the plasmid must be introduced into bacterial cells. This task is usually accomplished by *transformation*, the mechanism by which bacterial cells take up DNA from the external environment (see Chapter 7). Some types of cells undergo transformation naturally; others must be treated chemically or physically before they will undergo transformation. Inside the cells, the plasmids replicate and multiply, and the cells themselves multiply, producing many copies of the introduced gene.

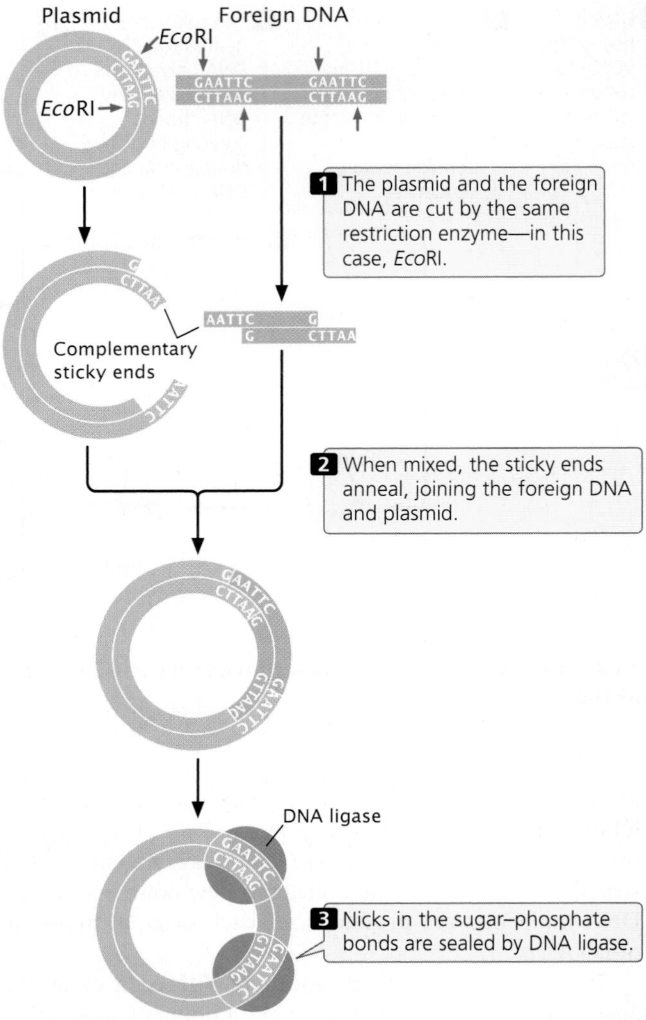

14.6 A foreign DNA fragment can be inserted into a plasmid with the use of restriction enzymes.

SCREENING CELLS FOR RECOMBINANT PLASMIDS Cells bearing recombinant plasmids can be detected by using the selectable markers on the plasmid. Genes that confer resistance to an antibiotic are commonly used as selectable markers; any cell that contains such a plasmid will be able to live in the presence of the antibiotic, which normally kills bacterial cells.

A common way to screen cells for the presence of a recombinant plasmid is to use a plasmid with a copy of the *lacZ* gene as a vector (**Figure 14.7**). The *lacZ* gene contains a series of unique restriction sites at which a piece of DNA can be inserted. In the absence of an inserted fragment, the *lacZ* gene is active and produces β-galactosidase (see Section 12.2). When foreign DNA is inserted into the restriction site, it disrupts the *lacZ* gene, and β-galactosidase is not produced. The plasmid also usually contains a second selectable marker, which may be a gene that confers resistance to an antibiotic such as ampicillin.

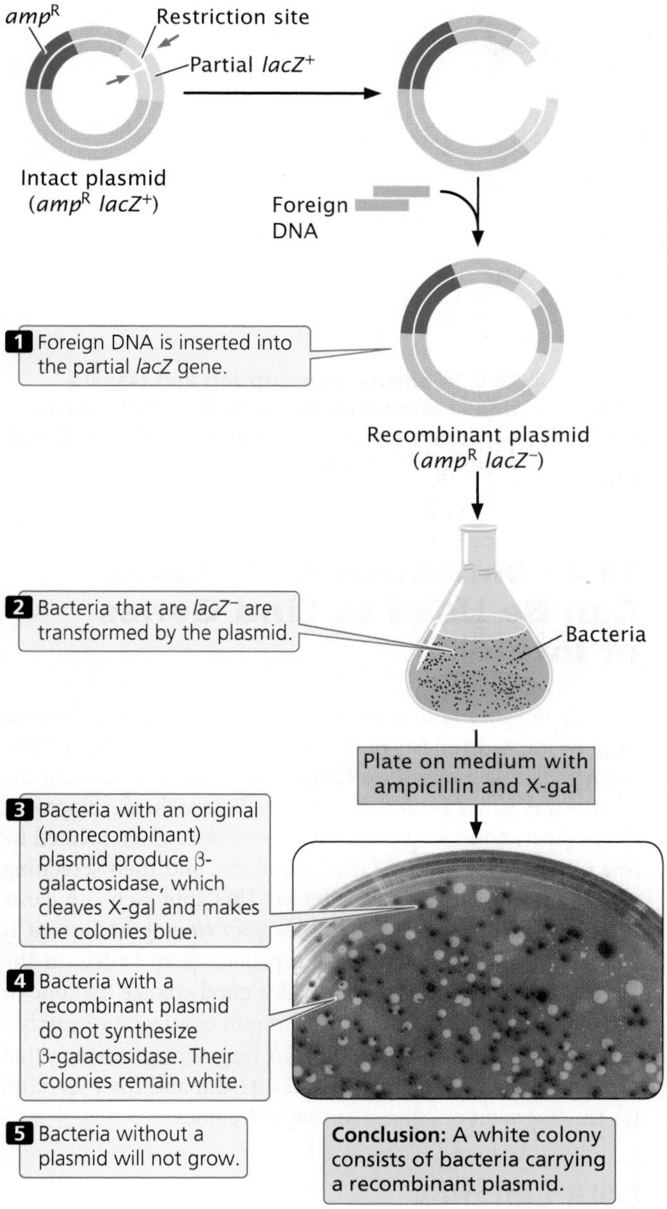

ampR
Restriction site
Partial lacZ+
Intact plasmid (ampR lacZ+)
Foreign DNA
1 Foreign DNA is inserted into the partial lacZ gene.
Recombinant plasmid (ampR lacZ−)
2 Bacteria that are lacZ− are transformed by the plasmid.
Bacteria
Plate on medium with ampicillin and X-gal
3 Bacteria with an original (nonrecombinant) plasmid produce β-galactosidase, which cleaves X-gal and makes the colonies blue.
4 Bacteria with a recombinant plasmid do not synthesize β-galactosidase. Their colonies remain white.
5 Bacteria without a plasmid will not grow.
Conclusion: A white colony consists of bacteria carrying a recombinant plasmid.

14.7 The *lacZ* gene can be used to screen for bacteria containing recombinant plasmids. A special plasmid carries a fragment of the *lacZ* gene and an ampicillin-resistance gene. [Courtesy of Edovotek.]

Bacteria that are *lacZ*⁻ are exposed to the plasmids and then plated on a medium that contains ampicillin. Only cells that have been successfully transformed and thus contain a plasmid with the ampicillin-resistance gene will survive and grow. Some of these cells will contain an intact plasmid, whereas others will possess a recombinant plasmid. The medium also contains the chemical X-gal, which produces a blue substance when cleaved. Bacterial cells with an intact original plasmid—without an inserted DNA fragment—have a functional *lacZ* gene (the front end of which is provided by the plasmid and the back end by the bacterium). These bacteria can synthesize β-galactosidase, which cleaves X-gal and turns the bacteria blue. Bacterial cells with a recombinant plasmid, however, have the front end of the β-galactosidase gene disrupted by the inserted DNA; they do not synthesize β-galactosidase and remain white. (In these experiments, the bacterium's own *lacZ* gene has been inactivated, and so only bacteria with the plasmid turn blue.) Thus, the color of a bacterial colony allows quick determination of whether a recombinant or intact plasmid is present in the cells. Once cells with the recombinant plasmid have been identified, they can be grown in large numbers to replicate the inserted fragment of DNA.

Plasmids make ideal cloning vectors, but can hold only DNA fragments less than about 15 kb in size. When large DNA fragments are inserted into a plasmid vector, the plasmid becomes unstable and tends to spontaneously lose DNA.

OTHER CLONING VECTORS A number of other vectors have been developed for cloning larger pieces of DNA in bacteria. For example, bacteriophage λ (lambda), which infects *E. coli*, can be used to clone up to 23,000 bp of foreign DNA; it transfers DNA into bacterial cells with high efficiency. **Cosmids** are plasmids that are packaged into empty viral protein coats and transferred to bacteria by viral infection. They can carry more than twice as much foreign DNA as can a phage vector. **Bacterial artificial chromosomes** (BACs) are vectors originally constructed from the F plasmid (a special plasmid that controls mating and the transfer of genetic material in some bacteria; see Chapter 7) and can hold very large fragments of DNA that can be as long as 300,000 bp. **Table 14.2** compares the properties of plasmids, phage λ, cosmids, and BACs.

TABLE 14.2 Comparison of plasmids, phage λ, cosmids, and bacterial artificial chromosomes

Cloning vector	Size of DNA that can be cloned	Method of propagation	Introduction to bacteria
Plasmid	As large as 15 kb	Plasmid replication	Transformation
Phage λ	As large as 23 kb	Phage reproduction	Phage infection
Cosmid	As large as 44 kb	Plasmid reproduction	Phage infection
Bacterial artificial chromosome	As large as 300 kb	Plasmid reproduction	Electroporation

Note: 1 kb = 1000 bp. Electroporation is the use of electrical pulses to increase the permeability of a membrane.

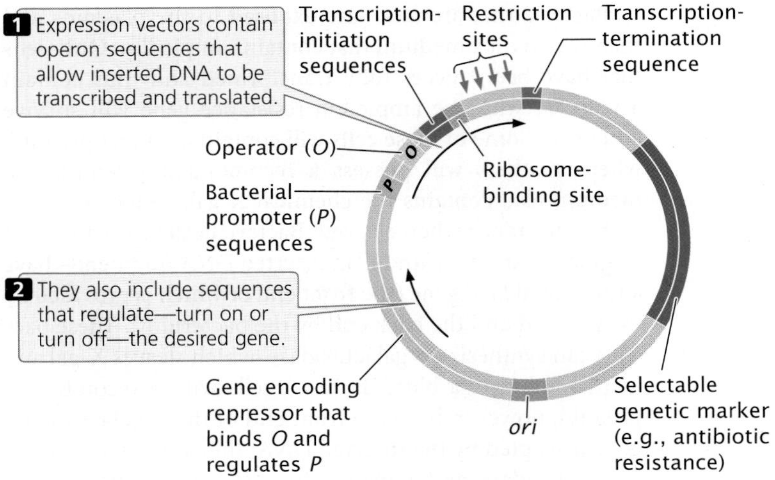

14.8 To ensure transcription and translation, a foreign gene can be inserted into an expression vector—in this example, an *E. coli* expression vector.

Sometimes the goal in gene cloning is not just to replicate the gene, but also to produce the protein that it encodes. To ensure transcription and translation, a foreign gene is usually inserted into an **expression vector**, which, in addition to the usual origin of replication, unique restriction sites, and selectable markers, contains sequences required for transcription and translation in bacterial cells (**Figure 14.8**).

Although manipulating genes in bacteria is simple and efficient, the goal may be to transfer a gene into eukaryotic cells. For example, it might be desirable to transfer a gene conferring herbicide resistance into a crop plant or to transfer a gene for clotting factor into a person suffering from hemophilia. Many eukaryotic proteins are modified after translation (e.g., sugar groups may be added). Such modifications are essential for proper function, but bacteria do not have the ability to carry them out; thus, a functional protein can be produced only in a eukaryotic cell. A number of cloning vectors have been developed that allow the insertion of genes into eukaryotic cells. For example, the Ti plasmid, a large plasmid from the soil bacterium *Agrobacterium tumefaciens*, has been genetically engineered to transfer genes to plants, including genes that confer economically significant attributes such as resistance to herbicides, plant viruses, and insect pests. TRY PROBLEM 21

THINK-PAIR-SHARE Question 7

CONCEPTS

DNA fragments can be inserted into cloning vectors, stable pieces of DNA that will replicate within a cell. A cloning vector must have an origin of replication, one or more unique restriction sites, and selectable markers. An expression vector contains additional sequences that allow a cloned gene to be transcribed and translated. Special cloning vectors have been developed for introducing genes into eukaryotic cells.

✓ CONCEPT CHECK 3

How is a gene inserted into a plasmid cloning vector?

14.4 Molecular Techniques Can Be Used to Find Genes of Interest

To analyze a gene or to transfer it to another organism, the gene must be located and isolated. Today, most genes are located by sequencing a genome (see Chapter 15) and determining the locations of genes from the sequence. Before the development of low-cost sequencing methods, genes were often located by first creating libraries of DNA sequences and then screening those libraries for genes of interest. This approach—to clone first and search later—is called *shotgun cloning* because it is like hunting with a shotgun: the pellets spray widely in the general direction of the target, with a good chance that one or more of the pellets will hit it. In shotgun cloning, a researcher first clones a large number of DNA fragments, knowing that one or more contains the DNA of interest, and then searches for the fragment of interest among the clones.

DNA Libraries

A collection of clones containing all the DNA fragments from one source is called a **DNA library**. For example, we might isolate genomic DNA from human cells, break it into fragments, insert the fragments into vectors, and clone them in bacterial cells. The set of bacterial colonies or phages containing these fragments is a human **genomic library**, containing all the DNA sequences found in the human genome. In contrast, a **cDNA library** contains only DNA sequences that are transcribed into mRNA; a cDNA library is created from mRNA that is first converted into DNA and then cloned in bacteria.

CREATING A GENOMIC LIBRARY To create a genomic library, cells are disrupted, which causes them to release their DNA and other cellular contents into an aqueous solution, and the DNA is extracted from the solution. After the DNA has been isolated, it is incubated with a restriction enzyme for a limited amount of time so that only

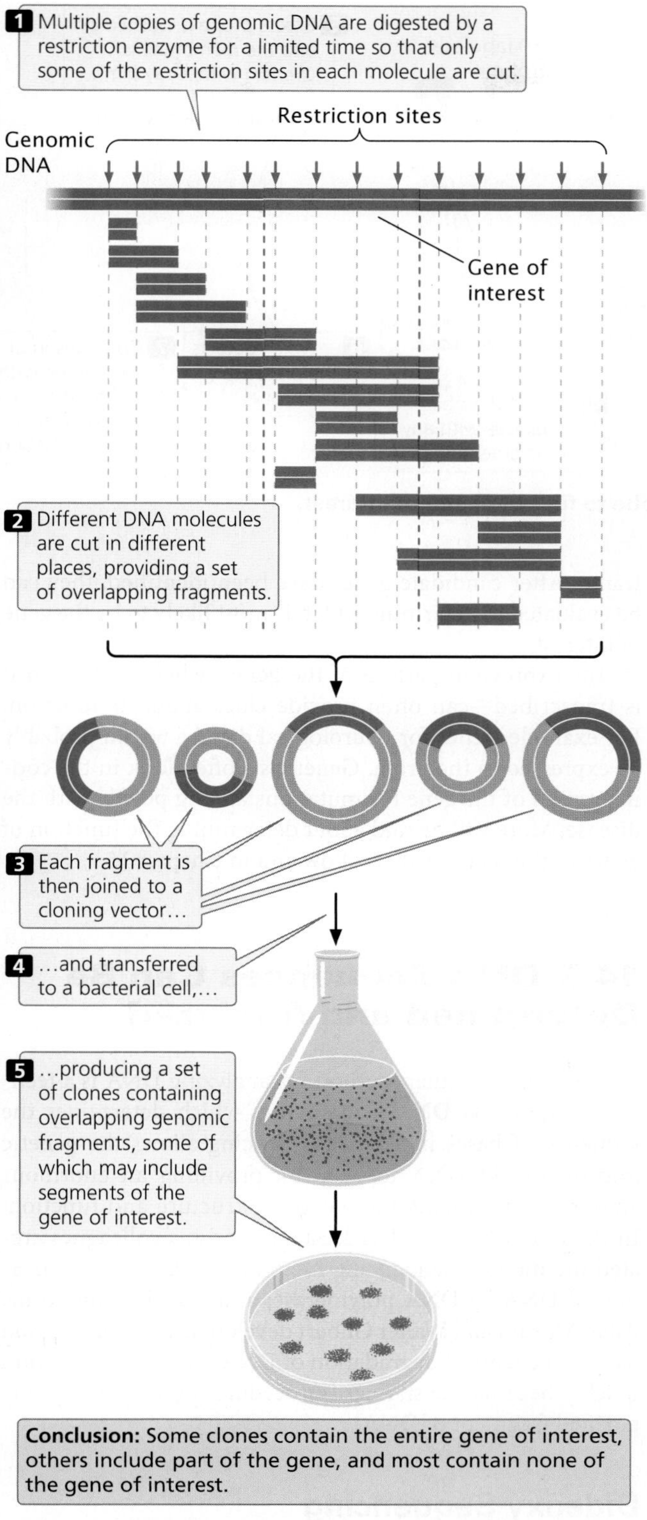

14.9 A genomic library contains all the DNA sequences found in an organism's genome.

some of the restriction sites in each DNA molecule are cut (a partial digestion). Because which sites are cut is random, different DNA molecules will be cut in different places, and a set of overlapping fragments will be produced (**Figure 14.9**). The fragments are then joined to vectors, which can be transferred to bacteria. A few of the clones contain the entire gene of interest (if the gene is not too large), and a few contain parts of the gene, but most contain fragments that have no part of the gene of interest.

A genomic library must contain a large number of clones to ensure that all DNA sequences in the genome are represented in the library. A library of the human genome formed by using cosmids, each carrying a random DNA fragment from 35,000 to 44,000 bp long, would require about 350,000 cosmid clones to provide a 99% chance that every sequence is included in the library.

CONCEPTS

One method of finding a gene begins with the creation of a DNA library. A genomic library is created by cutting genomic DNA into overlapping fragments and cloning those fragments in bacteria. A cDNA library is created from mRNA that is converted into cDNA and cloned in bacteria.

SCREENING DNA LIBRARIES Creating a genomic or cDNA library is relatively easy compared with screening the library to find clones that contain the gene of interest. The screening procedure used depends on what is known about the gene.

The first step in screening is to plate the clones of the library. If a plasmid or cosmid vector was used to construct the library, the bacterial cells are diluted and plated so that each bacterium grows into a distinct colony. If a phage vector was used, the phages are allowed to infect a lawn of bacteria on a petri plate. Each resulting bacterial colony or plaque contains a single cloned DNA fragment that must be screened for the gene of interest.

A common way to screen libraries is with probes. To use a probe, replicas of the plated colonies or plaques must first be made. **Figure 14.10** illustrates this procedure for a cosmid library.

How is a probe obtained when the gene of interest has not yet been isolated? One option is to use a similar gene from another organism as the probe. For example, if we wanted to screen a human genomic library for the growth-hormone gene and a growth-hormone gene had already been isolated from rats, we could use a purified rat-gene sequence as the probe to find the human gene for growth hormone. Successful hybridization does not require perfect complementarity between the probe and the target sequence, so a related sequence can often be used as a probe.

Another method of screening a library is to look for the protein product of a gene. The clones can be tested for the presence of the protein by using an antibody that recognizes the protein or by using a chemical test for the protein. This method depends on the existence of an antibody or test for the protein produced by the gene. DNA libraries can also be screened using PCR or by sequencing.

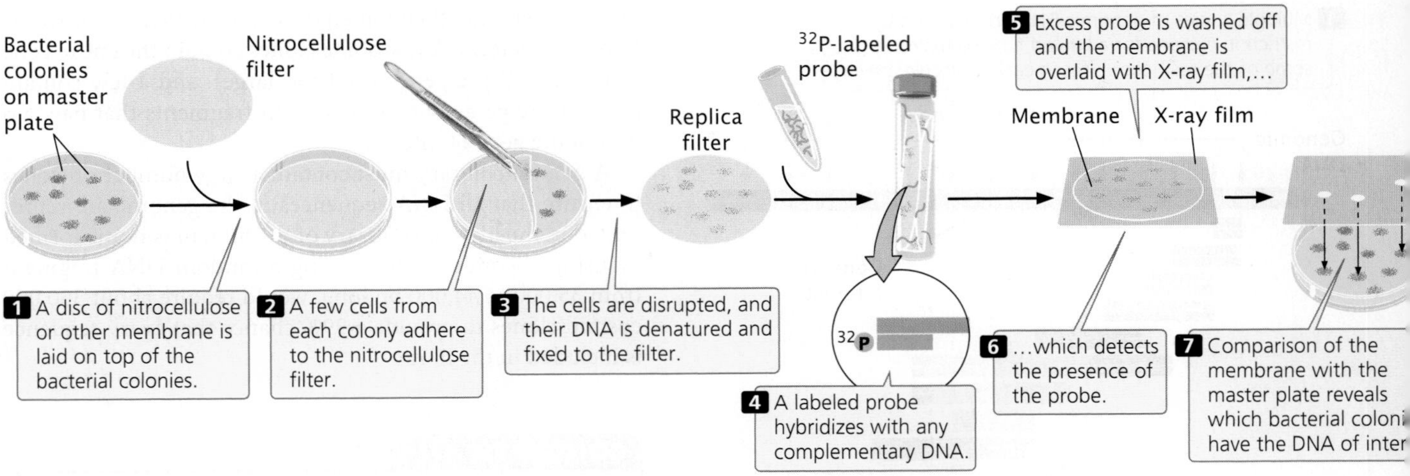

14.10 Genomic and cDNA libraries can be screened with a probe to find the gene of interest.

CONCEPTS

A DNA library can be screened for a specific gene by using probes that hybridize to the gene. Alternatively, clones in a DNA library can be examined for the protein product of the gene.

Positional Cloning

For many genes with important functions, no associated protein product is yet known. The biochemical bases of many human genetic diseases, for example, are still unknown. How can these genes be isolated? One approach is to first determine the general location of the gene on the chromosome by using recombination frequencies derived from crosses or pedigrees (see Chapter 5). After the chromosomal region where the gene is found has been pinpointed, genes in that region can be cloned and identified. Then other techniques can be used to determine which of the "candidate" genes might be the one that causes the disease. This approach—isolation of genes on the basis of their position on a gene map—is called **positional cloning**.

In the first step of positional cloning, geneticists use mapping studies (see Chapter 5) to establish linkage between molecular markers and the phenotype of interest. Demonstration of linkage between the phenotype and one or more molecular markers provides information about which chromosome carries the locus that codes for the phenotype and its general location on that chromosome. The next step is to narrow down the location of the locus by using additional molecular markers clustered in the chromosomal region where the locus resides.

After the locus has been mapped, clones that cover the region can be isolated from a genomic library and all genes located within the region can be identified. Genes can be distinguished from other sequences by the presence of characteristic features, such as consensus sequences in the promoter and start and stop codons within the same reading frame. After candidate genes have been identified, they can be evaluated to determine which is most likely to be the gene of interest.

The expression pattern of the gene—where and when it is transcribed—can often provide clues about its function. For example, genes for neurological disease would probably be expressed in the brain. Geneticists often look in the coding region of the gene for mutations among people with the disease. More will be said about determining the function of genes in the sections that follow and in Chapter 15.

14.5 DNA Sequences Can Be Determined and Analyzed

A powerful molecular method for analyzing DNA is a technique known as **DNA sequencing**, which determines the sequence of bases in DNA. Sequencing allows the genetic information in DNA to be read, providing an enormous amount of information about gene structure and function. In the mid-1970s, Frederick Sanger and his colleagues created the dideoxy sequencing method, based on the elongation of DNA by DNA polymerase; at about the same time, Allan Maxam and Walter Gilbert developed a second method based on chemical degradation of DNA. The Sanger method quickly became the standard procedure for sequencing any purified fragment of DNA.

Dideoxy Sequencing

The dideoxy (or Sanger) method of DNA sequencing is based on replication. The fragment to be sequenced is used as a template to make a series of new DNA molecules. In the process, replication is sometimes terminated when a specific base is encountered, producing DNA strands of different lengths, each of which ends in the same base.

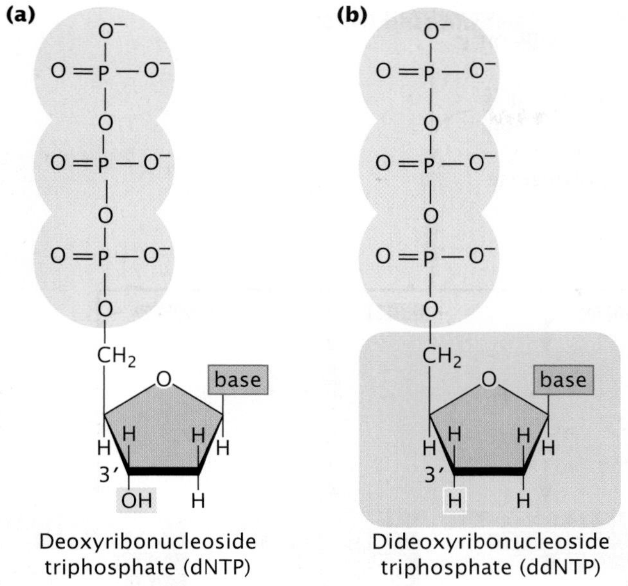

14.11 The dideoxy sequencing reaction requires a special substrate for DNA synthesis. (a) Structure of deoxyribonucleoside triphosphate, the normal substrate for DNA synthesis. (b) Structure of dideoxyribonucleoside triphosphate, which lacks an OH group on the 3′-carbon atom.

The method relies on the use of a special substrate for DNA synthesis. Normally, DNA is synthesized from deoxyribonucleoside triphosphates (dNTPs), which have an OH group on the 3′-carbon atom (**Figure 14.11a**). In the Sanger method, a special nucleotide, called a **dideoxyribonucleoside triphosphate** (ddNTP; **Figure 14.11b**), is also used as a substrate. The ddNTPs are identical to dNTPs, except that they lack a 3′-OH group. In the course of DNA synthesis, ddNTPs are incorporated into a growing DNA strand. However, after a ddNTP has been incorporated into the DNA strand, no more nucleotides can be added, because there is no 3′-OH group to form a phosphodiester bond with an incoming nucleotide. Thus, ddNTPs terminate DNA synthesis.

Although the sequencing of a single DNA molecule is technically possible, most sequencing procedures in use today require a considerable amount of DNA; any DNA fragment to be sequenced must first be amplified by PCR or by cloning in bacteria. Copies of the target DNA are then isolated and split into four samples (**Figure 14.12**). Each sample is placed in a different test tube to which the following ingredients are added:

1. Many copies of a primer that is complementary to one end of the target DNA strand
2. All four types of deoxyribonucleoside triphosphates, the normal precursors of DNA synthesis
3. A small amount of one of the four types of dideoxyribonucleoside triphosphates (ddATP, ddCTP, ddGTP, or ddTTP), which will terminate DNA synthesis as soon as it is incorporated into any growing chain (each of the four tubes receives a different ddNTP)
4. DNA polymerase

Either the primer or the dNTPs are radioactively or chemically labeled so that newly produced DNA can be detected.

Within each of the four tubes, the DNA polymerase synthesizes DNA. Let's consider the reaction in one of the four tubes: the one that received ddATP. Within this tube, each of the single strands of target DNA serves as a template for DNA synthesis. The primer pairs with its complementary sequence at one end of each template strand, providing a 3′-OH group for the initiation of DNA synthesis. DNA polymerase elongates a new strand of DNA from this primer. Wherever DNA polymerase encounters a T on the template strand, it uses, at random, either a dATP or a ddATP to introduce an A into the newly synthesized strand. Because there is more dATP than ddATP in the reaction mixture, dATP is incorporated most often, allowing DNA synthesis to continue. Occasionally, however, ddATP is incorporated into the strand, and synthesis terminates. The incorporation of ddATP into the new strand takes place randomly at different positions in different copies, producing a set of DNA fragments of different lengths (12, 7, and 2 nucleotides long in the example illustrated in Figure 14.12), each ending in a nucleotide that contains adenine.

Equivalent reactions take place in the other three tubes, except that synthesis is terminated at nucleotides with a different base in each tube. After the completion of the polymerization reactions, all of the DNA in the tubes is denatured, and the single-stranded products of each reaction are separated by gel electrophoresis.

The contents of the four tubes are separated side by side on an acrylamide gel (which allows finer separation than agarose) so that DNA strands differing in length by only a single nucleotide can be distinguished. After electrophoresis, the locations, and therefore the sizes, of the DNA strands in the gel are revealed by the radioactive or chemical labels.

Reading the DNA sequence is the simplest and shortest part of the procedure. In Figure 14.12, you can see that the band closest to the bottom of the gel is from the tube that contained the ddGTP reaction, which means that the first nucleotide added was guanine (G). The next band up is from the tube that contained ddATP; so the next nucleotide in the sequence is adenine (A), and so forth. In this way, the sequence is read from the bottom to the top of the gel, with the nucleotides near the bottom corresponding to the 5′ end of the newly synthesized DNA strand and those near the top corresponding to the 3′ end. Keep in mind that the sequence obtained is not that of the target DNA, but that of its *complement*. To see dideoxy sequencing in action, view **Animation 14.4.** TRY PROBLEM 25 A

For many years, DNA sequencing was done largely by hand and was laborious and expensive. Today, sequencing is usually carried out by automated machines that use fluorescent dyes and laser scanners to sequence thousands of

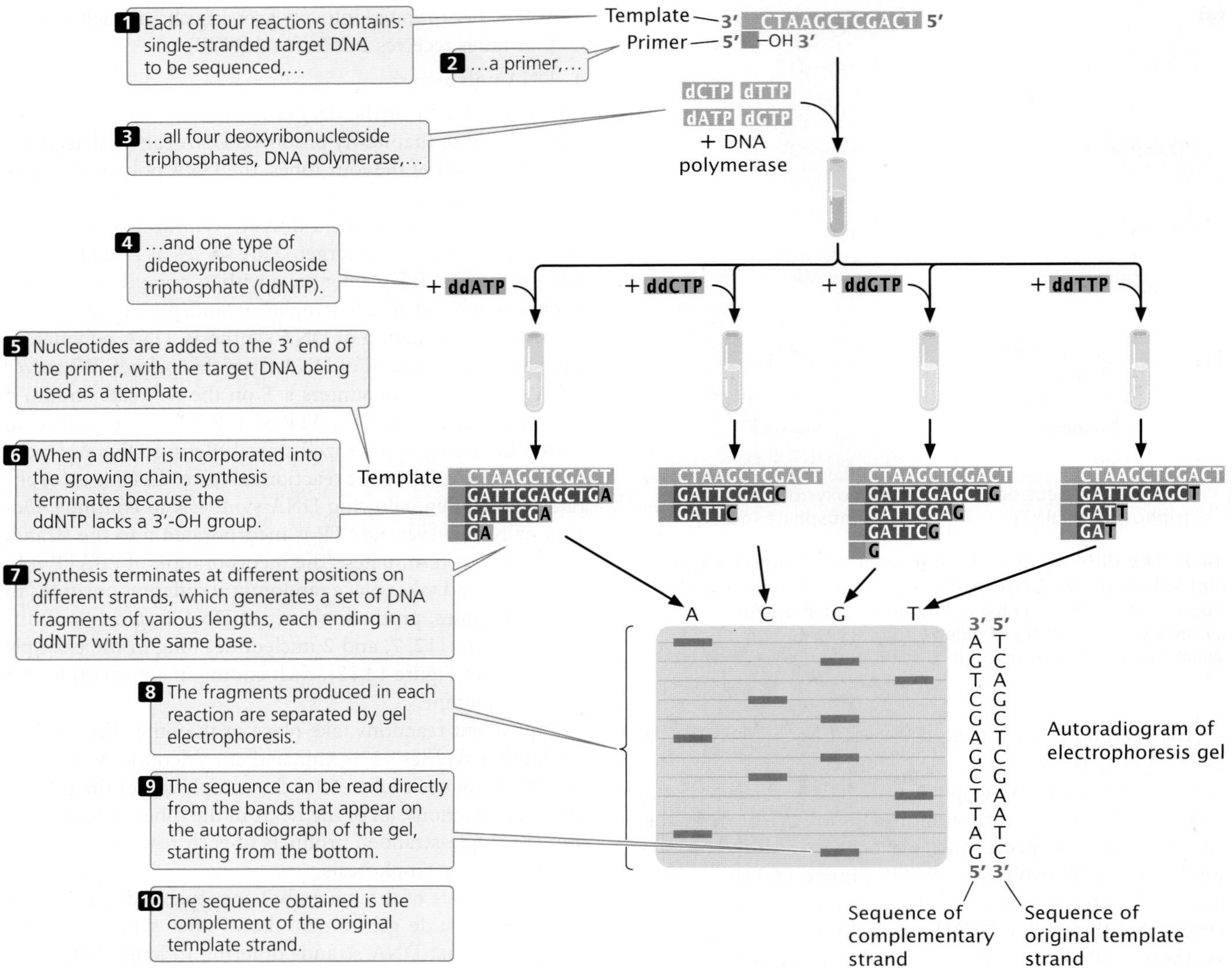

14.12 The dideoxy method of DNA sequencing is based on the termination of DNA synthesis.

base pairs in a few hours (**Figure 14.13**). The dideoxy reaction is still used, but the ddNTPs used in the reaction are labeled with a fluorescent dye, and a different-colored dye is used for each type of ddNTP. In this case, the four sequencing reactions can take place in the same test tube and can be placed in the same well for electrophoresis. Sequencing machines carry out electrophoresis in gel-containing capillary tubes. The different-sized fragments produced by the sequencing reaction separate within a tube and migrate past a laser beam and detector. As the fragments pass the laser, their fluorescent dyes are activated, and the resulting fluorescence is detected by an optical scanner. Each colored dye emits fluorescence of a characteristic wavelength, which is read by the optical scanner. The information is fed into a computer for interpretation, and the results are printed out as a set of peaks on a graph (see Figure 14.13). Automated sequencing machines may contain up to 96 or more capillary tubes, allowing from 50,000 to 60,000 bp of sequence to be read in a few hours.

CONCEPTS

The dideoxy sequencing method uses ddNTPs, which terminate DNA synthesis at specific bases.

✓ CONCEPT CHECK 4

In the dideoxy sequencing reaction, what terminates DNA synthesis at a particular base?

a. The absence of a base on the ddNTP halts the DNA polymerase.
b. The ddNTP causes a break in the sugar–phosphate backbone.
c. DNA polymerase will not incorporate a ddNTP into the growing DNA strand.
d. The absence of a 3′-OH group on the ddNTP prevents the addition of another nucleotide.

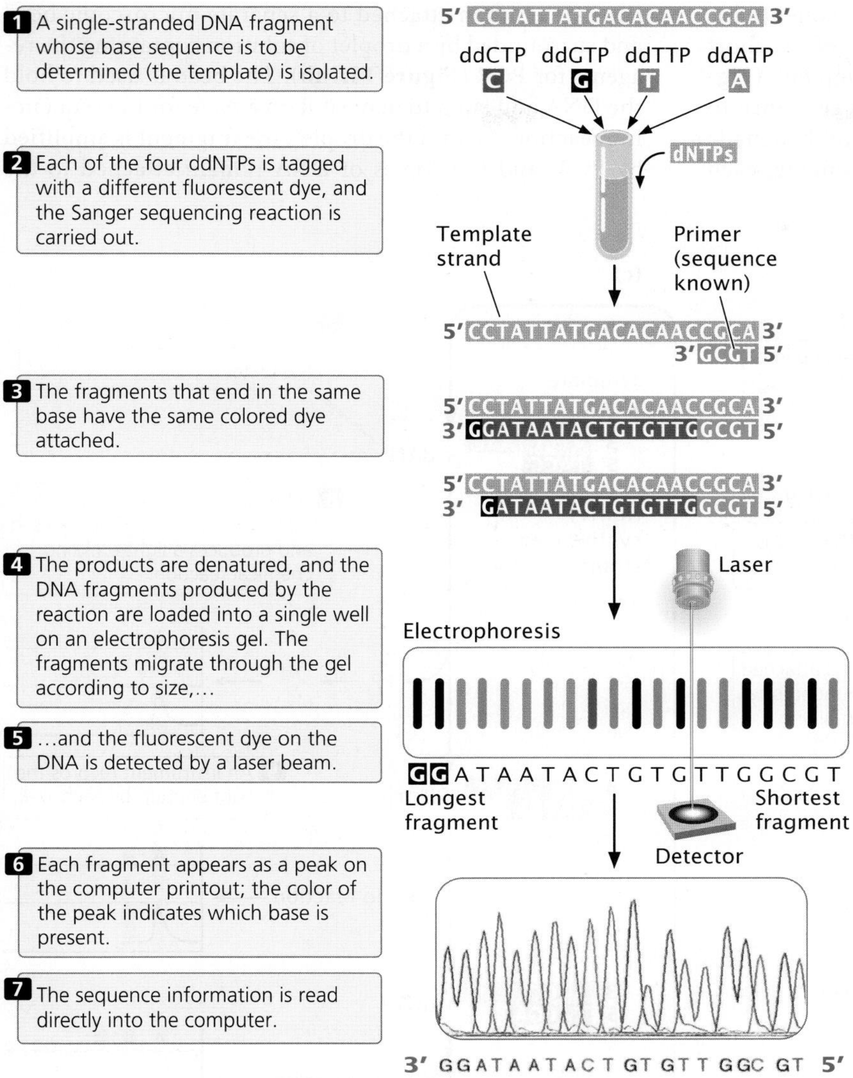

14.13 The dideoxy sequencing method has been automated.

Next-Generation Sequencing Technologies

Newer methods, called **next-generation sequencing technologies**, have made sequencing hundreds of times faster and less expensive that the traditional Sanger sequencing method. Most next-generation sequencing technologies do sequencing in parallel, which means that hundreds of thousands or even millions of DNA fragments can be sequenced simultaneously, allowing, for example, a human genome to be sequenced in days instead of years.

ILLUMINA SEQUENCING Illumina sequencing employs a technology similar to that of the Sanger dideoxy method. Special nucleotides are used that have a fluorescent tag attached, with a different-colored tag for each type of nucleotide. Each nucleotide also has a chemical group (a terminator) that, once incorporated into the growing DNA chain, prevents the incorporation of any additional nucleotides, as ddNTPs do in Sanger sequencing. In this case, however, the terminator is reversible—it can be chemically removed. To carry out sequencing, the DNA is first fragmented into millions of short overlapping fragments. The fragments are attached to a slide and then amplified, creating clusters of up to a thousand copies of each fragment in close proximity on the slide. The fragments are then denatured, and a solution of primers, DNA polymerase, and the special nucleotides is added. The primer attaches to each DNA template, and the first nucleotide is incorporated into the newly synthesized strand. The solution is washed away, and the tag on the incorporated nucleotide is excited with a laser, which causes it to fluoresce. As mentioned previously, each type of nucleotide (A, T, G, or C) has a different-colored fluorescent tag, so the color of the light produced reveals the type of the nucleotide just added. The terminator and the fluorescent tag are then chemically removed, and the process is repeated. As the nucleotides are added one at a time, the sequence is read as a series of flashes of colored light from each cluster of DNA. Hundreds of thousands of DNA clusters, each consisting of copies of a different DNA fragment, are sequenced simultaneously, allowing large amounts of DNA to be sequenced in a short time.

Most next-generation sequencing techniques read shorter DNA fragments than the Sanger sequencing method, but because hundreds of thousands or millions of fragments are sequenced simultaneously, these methods are much faster than traditional Sanger sequencing technology. But these techniques only determine the sequences of the short DNA fragments; they do not, by themselves, allow the sequences of these fragments to be reassembled into the sequence of the entire original piece of DNA. How sequences of small fragments are reassembled into a continuous sequence o DNA is explained in Chapter 15.

PYROSEQUENCING Several other forms of next-generation sequencing have been developed. One type, called pyrosequencing, is based on DNA synthesis: nucleotides are added one at a time in the order specified by template DNA. The addition of a particular nucleotide is detected with a flash of light, which is generated as the nucleotide is added.

To carry out pyrosequencing, DNA to be sequenced is first fragmented. An adapter, consisting of a short string of nucleotides, is added to each fragment (**Figure 14.14a**). The adapter provides a known sequence to prime a PCR reaction. The DNA fragments are then made single stranded. In one version of pyrosequencing, each fragment is then attached to a separate microscopic bead and surrounded by a droplet of solution containing the reagents for PCR (**Figure 14.14b**). The bead is used to hold the DNA and later, to deposit it on a plate for the sequencing reaction. Within the droplet, the fragment is amplified by PCR, and the copies of DNA remain attached to the

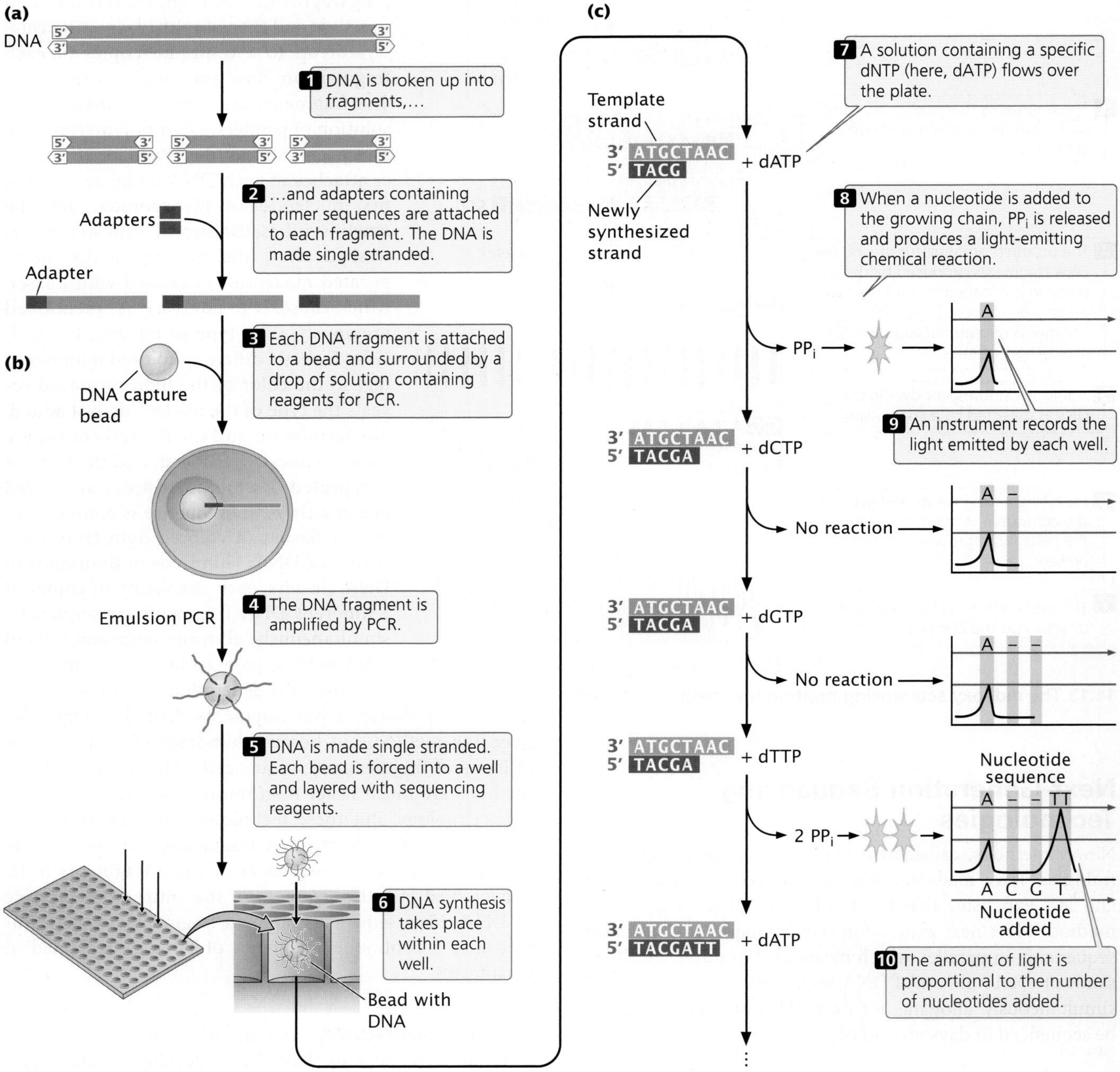

14.14 Next-generation sequencing methods are able to simultaneously determine the sequences of hundreds of thousands or millions of DNA fragments. Pyrosequencing is illustrated here.

bead. After amplification, the beads are mixed with DNA polymerase and deposited on plates containing more than a million wells (small holes), with each bead going into a separate well.

The sequencing reaction that takes place in each well is based on DNA synthesis. Recall from Chapter 9 that deoxyribonucleoside triphosphate—the substrate for DNA synthesis—consists of a deoxyribose sugar attached to a base and three phosphate groups. In the process of DNA synthesis, two phosphate groups (pyrophosphate, or PP_i) are cleaved off, and the resulting nucleotide is attached to the 3′ end of the growing DNA chain. A solution containing one particular type of dNTP—say, dATP—is passed across the wells (**Figure 14.14c**). If the template within a particular well specifies an adenine nucleotide in the next position of the growing chain, then pyrophosphate is cleaved from the dATP, and the adenine nucleotide is added. A chemical reaction uses the pyrophosphate produced in the reaction to generate a flash of light, which is measured by an optical detector. The amount of light emitted in each well is proportional to the number of nucleotides added: if the template in a well specifies three successive adenine nucleotides, then three nucleotides are added, and three times more light is emitted than if a single A is added. If the position in the template specifies a base other than adenine, no nucleotide is added, no pyrophosphate is produced, and no light is emitted.

As mentioned, the first solution passed over the plate contains adenosine triphosphate and allows adenine nucleotides to be added to the template. Each well with a template that specifies adenine in the next position will generate a flash of light. Then a solution with a different type of dNTP—say, dGTP—is passed across the wells. Any fragment that specifies a G in the next position of its growing chain will add a guanine nucleotide and emit a flash of light. The nucleotide triphosphates are passed across the wells in a predetermined order, and the light emitted by each well is measured. In this way, hundreds of thousands or millions of fragments of DNA are sequenced simultaneously on the basis of the order in which nucleotides are added to the 3′ end of the growing chain.

THIRD-GENERATION SEQUENCING TECHNOLOGY

Even more advanced and rapid sequencing methods, typically called third-generation sequencing methods, are being developed. Nanopore sequencing, for example, can determine the sequence of a single molecule of DNA. In this method, a single strand of DNA is passed through a tiny hole—a nanopore—in a membrane. As the molecule passes, one nucleotide at a time, through the nanopore, it disrupts an electrical current in the membrane, and the nature of the disruption is affected by the shape of the nucleotide passing through the nanopore. Each of the four types of nucleotides of DNA causes a characteristic disruption, so the sequence of DNA can be read by analyzing the membrane current as the strand passes through the nanopore. Hundreds of thousands of nanopores can be created on a single chip so that many DNA fragments can be read simultaneously.

> **CONCEPTS**
>
> New next- and third-generation sequencing methods can sequence many DNA fragments simultaneously, providing much faster and less expensive determination of DNA base sequences.

DNA Fingerprinting

The use of DNA sequences to identify individual persons is called **DNA fingerprinting** or DNA profiling. Because some parts of the genome are highly variable, each person's DNA sequence is unique and, like a traditional fingerprint, provides a distinctive characteristic that allows identification.

Today, most DNA fingerprinting uses **microsatellites**, or **short tandem repeats** (**STRs**), which are very short DNA sequences repeated in tandem (see Section 8.6). These repeated sequences are found at many loci throughout the human genome. People vary in the number of copies of repeat sequences they possess at each of these loci. The STRs are typically detected with PCR, using primers flanking the repeats so that a DNA fragment containing the repeated sequences is amplified (**Figure 14.15**). The length of the amplified segment depends on the number of repeats; DNA from a person with more repeats will produce a longer amplified segment than will DNA from a person with fewer repeats.

In DNA fingerprinting, the primers used in the PCR reaction are tagged with fluorescent labels so that the resulting DNA fragments can be detected with a laser. Primers for different STR loci are given different-colored labels so that similar-sized products of different loci can be differentiated. After PCR, the fragments are separated on a gel or by a capillary electrophoresis machine, in which the presence of each fragment is detected as it migrates past a laser, and a computer then calculates the size of each fragment based on its rate of migration. The fragments are represented as peaks on a graph; the distance on the horizontal axis represents the size of the fragment, while the height of the peak represents the amount of DNA (**Figure 14.16**). Homozygotes for an STR allele have a single tall peak; heterozygotes have two shorter peaks. When several different microsatellite loci are examined, the probability that two people will have the same set of patterns becomes vanishingly small, unless they are identical twins.

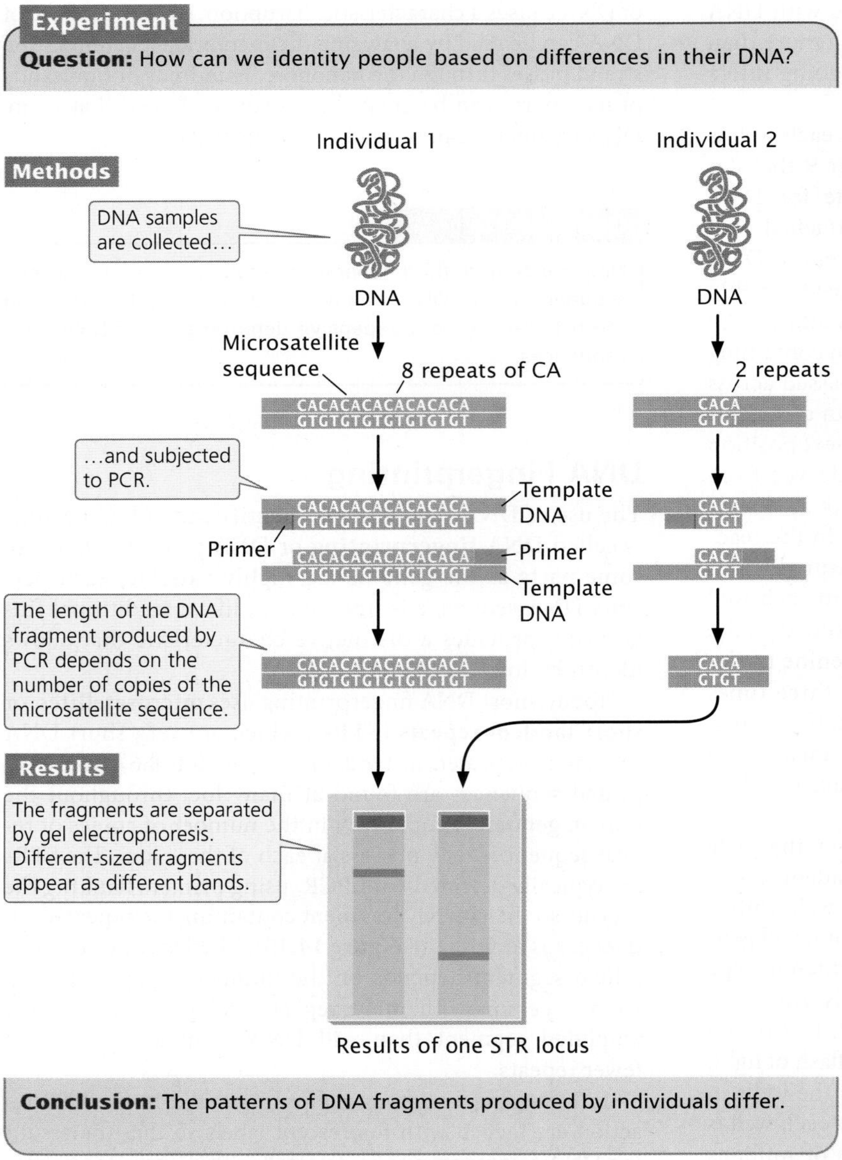

14.15 DNA fingerprinting can be used to identify a person.

The Federal Bureau of Investigation has developed a system of 13 STR loci (**Table 14.3**) that are commonly used for identifying people and solving crimes. These loci make up the Combined DNA Index System (CODIS). Each STR locus in CODIS has a large number of alleles and is located on a different human chromosome, so variation at each locus assorts independently. When all 13 CODIS loci are used together, the probability of two randomly selected people having the same DNA profile is less than 1 in 10 billion.

In a typical application, DNA fingerprinting might be used to confirm that a suspect was present at the scene of a crime. A sample of DNA from blood, semen, hair, or other body tissue is collected from the crime scene. If the sample is very small, PCR can be used to amplify it so that enough DNA is available for testing. Additional DNA samples are collected from one or more suspects. The pattern of DNA fragments produced by DNA fingerprinting of the sample is then compared with the patterns produced by DNA fingerprinting of the suspects. A match between the sample and a suspect can provide evidence that the suspect was present at the scene of the crime (**Figure 14.17**).

Since its introduction in the 1980s, DNA fingerprinting has helped convict a number of suspects in murder and rape cases. Suspects in other cases have been proved innocent when their DNA failed to match that from the crime scenes. Initially, calculations of the odds of a match (the probability that two people could have the same pattern) were controversial, and there were concerns about quality control (such as the accidental contamination of samples and the reproducibility of results) in laboratories where DNA analysis was done. Today, DNA fingerprinting has become widely accepted as an important tool in forensic investigations. In addition to its application in solving crimes, DNA fingerprinting is used to assess paternity, study genetic relationships among individual organisms in natural populations, identify specific strains of pathogenic bacteria, and identify human remains. For example, DNA fingerprinting was used to determine that several samples of anthrax mailed to different people in 2001 were all from the same source.

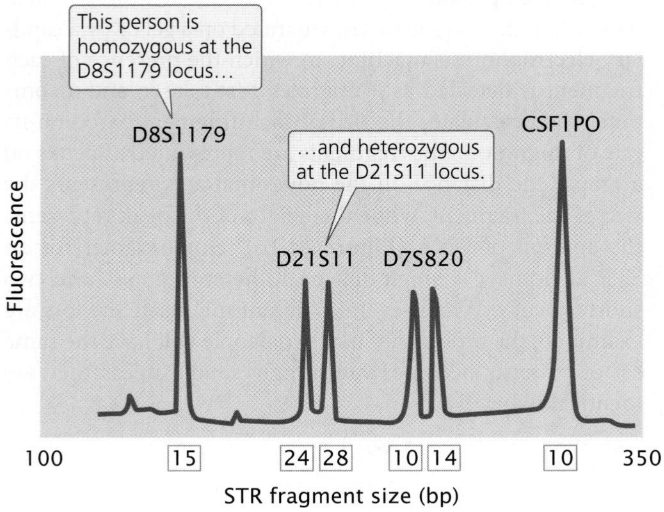

14.16 A DNA profile represents the pattern of DNA fragments produced by PCR of the STR loci. This profile shows the results from four STR loci (D8S1179, D21S11, D7S820, and CSF1P0). The number below each peak represents the number of STRs in that DNA fragment.

TABLE 14.3 Characteristics of the 13 STR loci used in CODIS for DNA fingerprinting

Locus name	Chromosome	Number of repeats	Number of alleles*
CSF1PO	5	5–17	10
FGA	4	12–51	23
TH01	11	3–14	8
TPOX	2	4–16	8
VWA	12	10–25	10
D3S1358	3	6–26	10
D5S818	5	4–29	9
D7S820	7	5–16	11
D8S1179	8	6–20	11
D13S317	13	5–17	8
D16S539	16	4–17	7
D18S51	18	5–40	19
D21S11	21	12–43	22

Source: Data from J. M. Butler and C. R. Hill, *Forensic Science Review* 24:15–26, 2012.
*U.S. population.

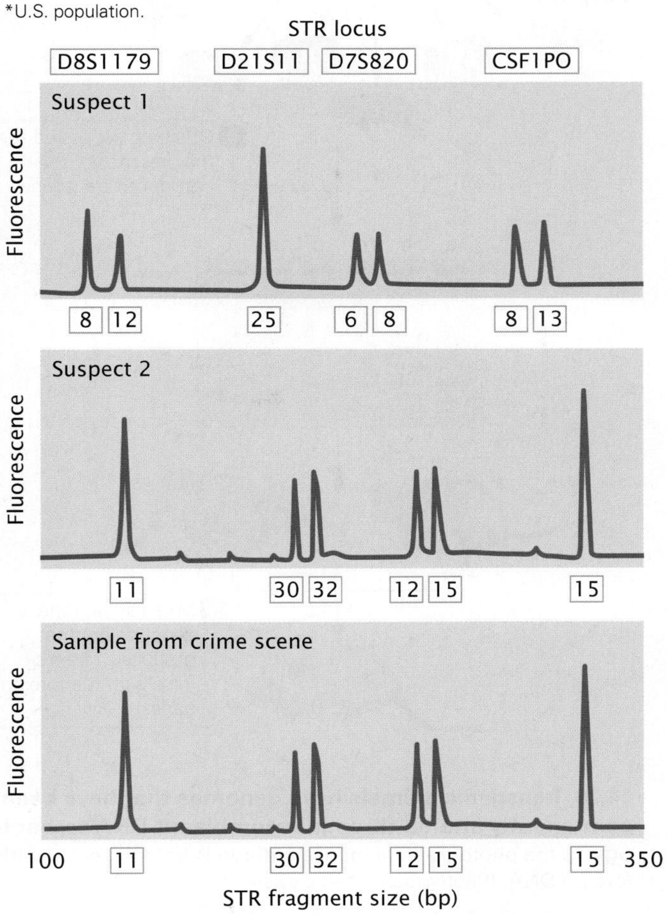

14.17 DNA fingerprinting can be used to determine the presence of a suspect at a crime scene. The DNA profile of suspect 2 matches that of DNA evidence collected at the crime scene. Shown here are results from 4 STR loci.

CONCEPTS

DNA fingerprinting detects genetic differences among people by analyzing highly variable regions of chromosomes.

✓ CONCEPT CHECK 5

How are microsatellites detected?

14.6 Molecular Techniques Are Increasingly Used to Analyze Gene Function

In the preceding sections, we learned that powerful molecular techniques are available for isolating, recombining, and analyzing DNA sequences. Although these methods provide a great deal of information about the organization and nature of gene sequences, the ultimate goal of many molecular studies is to better understand the function of these sequences. In this section, we will explore some advanced molecular techniques that are frequently used to determine gene function and to better understand the genetic processes that these sequences undergo.

Forward and Reverse Genetics

The traditional approach to the study of gene function begins with the identification of mutant organisms. For example, suppose that a geneticist were interested in genes that affect cardiac function in mammals. A first step would be to find individuals—perhaps mice—that have hereditary defects in heart function. The mutations causing the cardiac problems in the mice could then be mapped, and the implicated genes could be isolated and sequenced. Next, the proteins produced by the genes could be predicted from the gene sequences and isolated. Finally, the biochemistry of the proteins could be studied and their role in heart function discerned. This approach, which begins with a phenotype (a mutant individual) and proceeds to a gene that encodes the phenotype, is called **forward genetics**.

An alternative approach is to begin with a genotype—a DNA sequence—and proceed to the phenotype by altering the sequence or inhibiting its expression. A geneticist might begin with a gene of unknown function, induce mutations in it, and then look to see what effects these mutations have on the phenotype of the organism. This approach is called **reverse genetics**. Today, both forward- and reverse-genetics approaches are widely used in the analysis of gene function.

CONCEPTS

Forward genetics begins with a phenotype and detects and analyzes the genotype that causes that phenotype. Reverse genetics begins with a gene sequence and determines the phenotype it encodes.

✓ CONCEPT CHECK 6

A geneticist interested in immune function induces random mutations in a number of specific genes in mice and then determines which of the resulting mutant mice have impaired immune function. This is an example of

a. forward genetics.
b. reverse genetics.
c. both forward and reverse genetics.
d. neither forward nor reverse genetics.

Transgenic Techniques

Another way in which gene function can be analyzed is by adding DNA sequences of interest to the genome of an organism that normally lacks such sequences and then observing the effect of the introduced sequence on the organism's phenotype. This method is a form of reverse genetics. An organism that has been permanently altered by the addition of a DNA sequence to its genome is said to be *transgenic*, and the foreign DNA that it carries is called a **transgene** (**Figure 14.18**). Here, we consider techniques for the creation of transgenic mice, which are often used in the study of the function of human genes because they can be genetically manipulated in ways that are impossible with humans, and as mammals, they are more similar to humans than are fruit flies, fish, and other model genetic organisms.

The oocytes of mice and other mammals are large enough that DNA can be injected into them directly. Immediately after penetration by a sperm, a fertilized mouse egg contains two pronuclei, one from the sperm and one from the egg; these pronuclei later fuse to form the nucleus of the embryo. Mechanical devices can manipulate extremely fine, hollow glass needles to inject DNA directly into one of the pronuclei of a fertilized egg (**Figure 14.19**). Typically, a few hundred

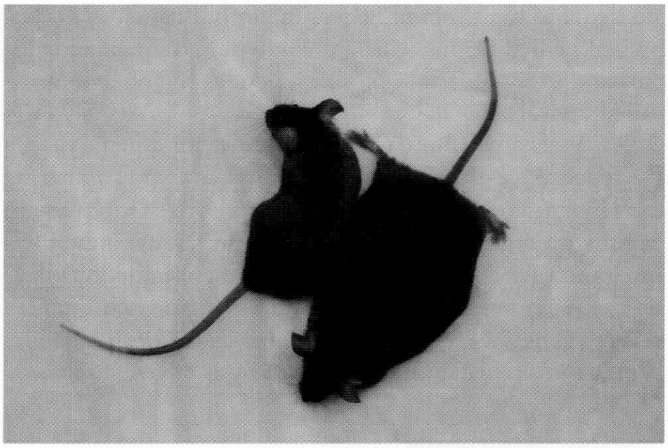

14.18 The genome of a transgenic organism has been permanently altered by genetic engineering. The mouse on the right is transgenic for a gene that causes obesity. [Dr. Liangyou Rui, Professor, Molecular & Integrative Physiology, University of Michigan.]

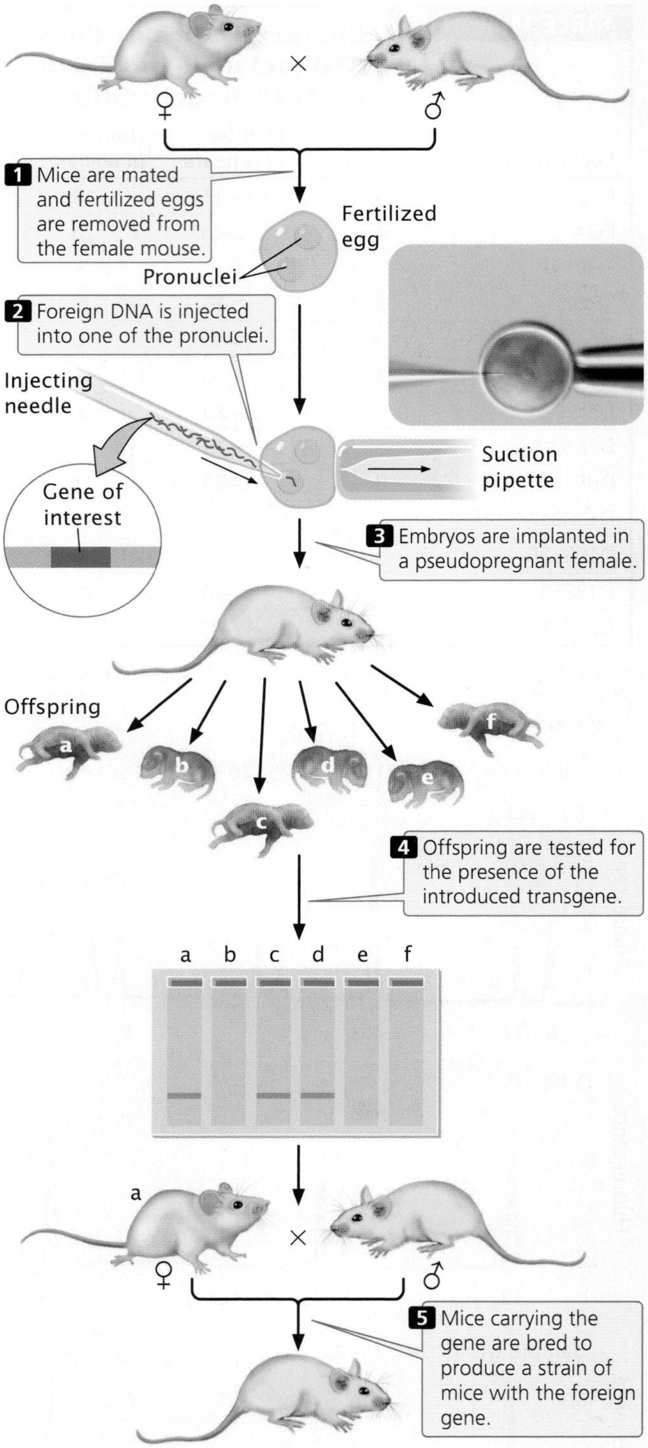

14.19 Transgenic animals have genomes that have been permanently altered through recombinant DNA technology. In the photograph, a mouse embryo is being injected with foreign DNA. [Photograph: Chad Davis/PhotoDisc.]

copies of cloned linear DNA are injected into a pronucleus. In a few of the injected eggs, copies of the cloned DNA integrate randomly into one of the chromosomes through a process called nonhomologous recombination. After injection, the embryos are implanted in a pseudopregnant female, a surrogate mother that has been physiologically prepared for pregnancy by mating with a vasectomized male.

Only about 10% to 30% of the injected embryos survive, and of those that do survive, only a few have a copy of the cloned DNA stably integrated into a chromosome. Nevertheless, if several hundred embryos are injected and implanted, there is a good chance that one or more mice whose chromosomes contain the foreign DNA will be born. Moreover, because the DNA was injected at the one-cell stage of the embryo, these mice usually carry the cloned DNA in every cell of their bodies, including their reproductive cells, and will therefore pass the foreign DNA on to their progeny. Through interbreeding, a strain of mice that carry the foreign gene can be created.

Transgenic mice have proved useful in the study of gene function. For example, proof that the *SRY* gene (see Chapter 4) is the male-determining gene in mice was obtained by injecting a copy of the *SRY* gene into XX embryos and observing that these mice developed as males. In addition, researchers have created a number of transgenic mouse strains that serve as experimental models for human genetic diseases.

Knockout Mice

A useful variant of the transgenic approach is to produce mice in which a normal gene has been not just mutated, but fully disabled. These animals, called **knockout mice**, are particularly helpful in determining the function of a gene: the phenotype of the knockout mouse often gives a good indication of the function of the missing gene.

A variant of the knockout procedure is to insert a particular DNA sequence into a known chromosome location. For example, researchers might insert the sequence of a human disease-causing allele into the same locus in mice, creating a precise mouse model of the human disease. Mice that carry inserted sequences at specific locations are called **knock-in mice**.

CONCEPTS

A transgenic mouse is produced by injecting cloned DNA into the pronucleus of a fertilized egg, and then implanting the egg in a female mouse. In knockout mice, the injected DNA contains a mutation that disables a gene.

Model Genetic Organism The Mouse *Mus musculus*

The ability to create transgenic, knockout, and knock-in mice has greatly facilitated the study of human genetics, and these techniques illustrate the power of the mouse as a model genetic organism. The common house mouse, *Mus musculus*, is among the oldest and most valuable subjects for genetic study: it's small, prolific, and easy to keep, with a short generation time (**Figure 14.20**).

Advantages of the mouse as a model genetic organism

Foremost among the many advantages that *Mus musculus* has as a model genetic organism is its close evolutionary relationship to humans. Being a mammal, the mouse is genetically, behaviorally, and physiologically more similar to humans than are other organisms used in genetic studies, making the mouse the model of choice for many studies of human and medical genetics. Other advantages include a short generation time compared with those of most other mammals. *Mus musculus* is well adapted to life in the laboratory and can be easily raised and bred in cages that require little space; thus, several thousand mice can be raised within the confines of a small laboratory room. Mice have large litters (8–10 pups) and are docile and easy to handle. Finally, a large number of mutations have been isolated and studied in captive-bred mice, providing an important source of variation for genetic analysis.

Life cycle of the mouse

The production of gametes and reproduction in the mouse are very similar to those processes in humans (see Figure 14.20). Male mice begin producing sperm at puberty and continue sperm production throughout the remainder of their lives. Starting at puberty, female mice go through an estrous cycle about every 4 days. After mating and fertilization, the diploid embryo implants in the uterus. Gestation typically takes about 21 days. Mice reach puberty in about 5 to 6 weeks and will live for about two years. A complete generation can be completed in about 8 weeks.

Genetic techniques with the mouse

The mouse genome, which contains about 2.7 billion base pairs of DNA, is similar in size to the human genome. For most human genes, there are homologous genes in the mouse, and the linkage relations of many mouse genes are similar to those in humans. The mouse genome is distributed across 19 pairs of autosomes and one pair of sex chromosomes (see Figure 14.20).

We have already considered three powerful techniques that have been developed for use in the mouse: (1) the creation of transgenic mice by the injection of DNA into a mouse embryo, (2) the ability to disrupt specific genes to create knockout mice, and (3) the ability to insert specific sequences into specific loci. These techniques are made possible by the ability to manipulate the mouse reproductive cycle, including the ability to induce ovulation using hormones, to isolate unfertilized oocytes from the ovary, and to implant fertilized embryos in the uterus of a surrogate mother.

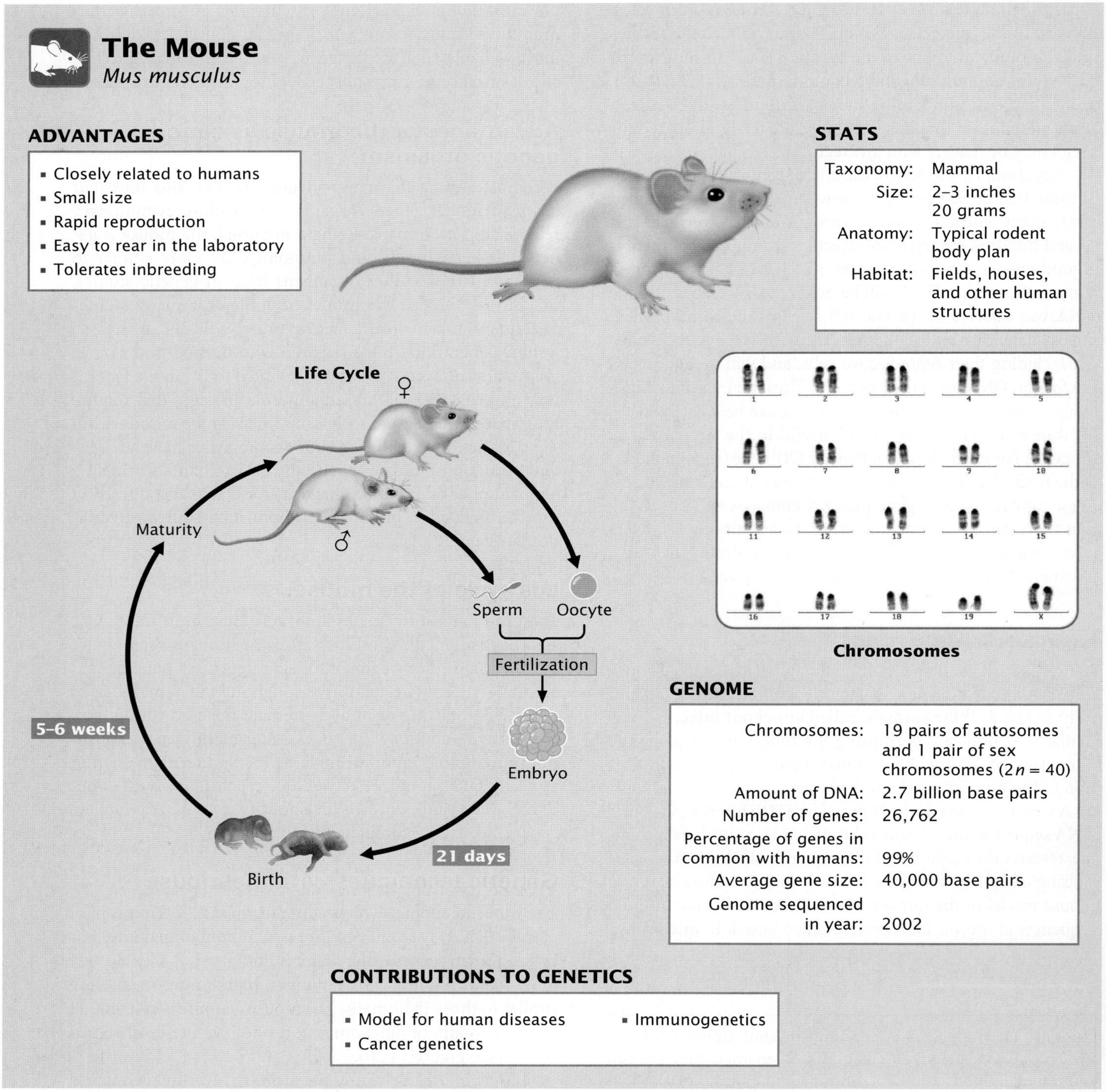

The Mouse
Mus musculus
ADVANTAGES
▪ Closely related to humans
▪ Small size
▪ Rapid reproduction
▪ Easy to rear in the laboratory
▪ Tolerates inbreeding
STATS
Taxonomy: Mammal
Size: 2–3 inches 20 grams
Anatomy: Typical rodent body plan
Habitat: Fields, houses, and other human structures
Life Cycle
♀
♂
Maturity
Sperm
Oocyte
Fertilization
Embryo
5–6 weeks
21 days
Birth
Chromosomes
GENOME
Chromosomes: 19 pairs of autosomes and 1 pair of sex chromosomes (2n = 40)
Amount of DNA: 2.7 billion base pairs
Number of genes: 26,762
Percentage of genes in common with humans: 99%
Average gene size: 40,000 base pairs
Genome sequenced in year: 2002
CONTRIBUTIONS TO GENETICS
▪ Model for human diseases
▪ Cancer genetics
▪ Immunogenetics

14.20 The mouse *Mus musculus* is a model genetic organism. [Chromosome photograph courtesy of Ellen C. Akeson and Muriel T. Davisson, The Jackson Laboratory, Bar Harbor, Maine.]

A large number of mouse models of specific human diseases have been created—in some cases, by isolating and inbreeding mice with naturally occurring mutations, and in other cases, by using knockout and knock-in techniques to disable and modify specific genes. Mice tolerate inbreeding well, and inbred strains of mice are easily created by brother–sister mating. ■

Silencing Genes by Using RNA Interference

In earlier sections, we saw that introducing mutations or new DNA sequences into the genome and observing the resulting phenotypes could provide information about the function of the altered or introduced DNA. We could also analyze gene function by temporarily turning a gene off and observing the effect of the absence of the gene product on the phenotype. For many years, there was no such method for selectively affecting gene expression. However, the discovery of siRNAs (small interfering RNAs) and miRNAs (microRNAs) (see Chapters 10 and 12) provided powerful tools for controlling the expression of individual genes.

Recall that siRNAs and miRNAs are small RNA molecules that combine with proteins to form a RNA-induced silencing complex (RISC). In a process called RNA interference, or RNAi, the RISC pairs with complementary sequences on mRNA and either cleaves the mRNA or prevents it from being translated. Molecular geneticists have exploited this natural machinery for turning off the expression of specific genes. Silencing a gene with siRNAs or miRNAs and observing the effect can often be a source of insight into the gene's function.

14.7 Biotechnology Harnesses the Power of Molecular Genetics

In addition to providing valuable new information about the nature and function of genes, molecular genetic techniques have many practical applications, including the production of pharmaceutical products and other chemicals, specialized bacteria, agriculturally important plants, and genetically engineered farm animals. These techniques are also used extensively in medical testing and, in a few cases, are even being used to correct human genetic defects. Hundreds of firms now specialize in developing products through genetic engineering, and many large multinational corporations have invested enormous sums of money in molecular genetics research. As discussed earlier, DNA analysis is also used in criminal investigations and for the identification of human remains.

Pharmaceutical Products

The first commercial products to be developed with the use of genetic engineering were pharmaceutical products used in the treatment of human diseases and disorders. In 1979, Eli Lilly and Company began selling human insulin produced with the use of recombinant DNA technology. The gene for human insulin was inserted into plasmids and transferred to bacteria, which then produced human insulin. Previously, insulin was isolated from pig and cow pancreases; a few diabetics developed allergic reactions to this foreign protein. Recombinant insulin has the advantage of being the same as that produced in the human body. Other pharmaceutical products produced through recombinant DNA technology include human growth hormone (for children with growth deficiencies), clotting factors (for hemophiliacs), and tissue plasminogen activator (used to dissolve blood clots in heart attack patients).

Specialized Bacteria

Bacteria play important roles in many industrial processes, including the production of ethanol from plant material, the leaching of minerals from ore, and the treatment of sewage and other wastes. The bacteria used in these processes are modified by genetic engineering so that they work more efficiently. New strains of technologically useful bacteria are being developed that will break down toxic chemicals and pollutants, enhance oil recovery, increase nitrogen uptake by plants, and inhibit the growth of pathogenic bacteria and fungi.

Agricultural Products

Recombinant DNA technology has had a major effect on agriculture, where it is now used to create crop plants and domestic animals with valuable traits. For many years, plant pathologists had recognized that plants infected with mild strains of viruses are resistant to infection by virulent strains. Using this knowledge, geneticists created resistance to viruses in plants by transferring genes for viral proteins to the plant cells. A genetically engineered squash, called Freedom II, carries genes from the watermelon mosaic virus 2 and the zucchini yellow mosaic virus, which protect the squash against viral infections.

Another objective has been to genetically engineer pest resistance into plants to reduce dependence on chemical pesticides. A gene from the bacterium *Bacillus thuringiensis* that produces an insecticidal toxin has been transferred into corn, tomatoes, potatoes, cotton, and other plants. These Bt crops are now grown worldwide. Other genes that confer resistance to viruses and herbicides have been introduced into a number of crop plants; resistance to herbicides allows growers to use herbicides for weed control without harming the crops. During 2016, 18 million farmers worldwide planted 185 million hectares of genetically engineered crops. In the United States, 92% of all corn, 96% of all cotton, and 94% of all soybeans grown in 2017 were genetically engineered.

In the future, CRISPR-Cas systems will increasingly be used to precisely edit the genomes of domestic plants and animals. For example, CRISPR-Cas9 has already been used to create transgenic wheat that is resistant to powdery mildew, a fungus that infects wheat. Because of its ease of use and low cost, CRISPR-Cas could be applied to less widely used, specialty crops and animals, potentially widening the use of genetically modified organisms. The ability to make very precise genetic changes with CRISPR-Cas, however, makes these alterations difficult to detect and track once they are made, raising questions about the labeling of products as genetically modified.

The genetic engineering of agricultural products is controversial. One area of concern focuses on the potential effects of releasing novel organisms produced by genetic engineering into the environment. There is also concern that transgenic organisms may hybridize with native organisms and transfer their genetically engineered traits. Other concerns focus on health-safety matters associated with the presence of engineered products in natural foods. Some critics have advocated the labeling of all genetically engineered foods that contain transgenic DNA or protein. Such labeling is required in countries of the European Union but not in the United States.

On the other hand, the use of genetically engineered crops and domestic animals has potential benefits. Genetically engineered crops that are pest resistant have the potential to reduce the use of environmentally harmful chemicals, and research findings indicate that lower amounts of pesticides are being used in the United States as a result of the adoption of transgenic plants. Studies conducted in China show that when Bt crops are used, farmers spray less chemical insecticide, allowing more predatory insects to survive and provide natural pest control. Transgenic crops also increase yields, providing more food per acre, which reduces the amount of land that must be used for agriculture. Genetically engineered plants offer the potential for the greater yields that may be necessary to feed the world's future population.

THINK-PAIR-SHARE Question 8

CONCEPTS

Recombinant DNA technology is used to create a wide range of commercial products, including pharmaceutical products, specialized bacteria, genetically engineered crops, and transgenic domestic animals.

Genetic Testing

The identification and cloning of many important disease-causing human genes have allowed the development of tests for detecting disease-causing mutations. Prenatal testing is already available for many genetic disorders. Additionally, presymptomatic genetic tests for adults and children are available for an increasing number of disorders. Some genetic tests are offered directly to consumers, without requiring the participation of a health care provider. These direct-to-consumer genetic tests are potentially available for many genetic conditions. The U.S. Food and Drug Administration has limited the sale of direct-to-consumer genetic tests for medical conditions, and today, only a few such medical tests are available in the United States.

Gene Therapy

Perhaps the ultimate application of recombinant DNA technology is **gene therapy**, the direct transfer of genes into humans to treat disease. Although it is still experimental, thousands of patients have received gene therapy, and many clinical trials are under way. Gene therapy is being used to treat genetic diseases, cancer, heart disease, and even some infectious diseases such as AIDS.

In spite of the growing number of clinical trials for gene therapy, significant problems remain in transferring foreign genes into human cells, getting them expressed, and limiting immune responses to the gene products and the vectors used to transfer the genes. There are also heightened concerns about the safety of gene therapy. In 1999, a patient participating in a gene-therapy trial had a fatal immune reaction after he was injected with a viral vector carrying a gene to treat his metabolic disorder. And five children who had undergone gene therapy for severe combined immunodeficiency disease developed leukemia that appeared to be directly related to the insertion of the retroviral gene vectors into cancer-causing genes. Despite these setbacks, gene-therapy research has moved on. Unequivocal results demonstrating positive benefits from gene therapy for several different diseases have now been published

Gene therapy conducted to date has targeted only somatic (nonreproductive) cells. Correcting a genetic defect in these cells (termed *somatic gene therapy*) may provide positive benefits to patients, but will not affect the genes of future generations. Gene therapy that alters germ-line (reproductive) cells (termed *germ-line gene therapy*) is technically possible, but raises a number of significant ethical issues because it has the ability to alter the gene pool of future generations.

THINK-PAIR-SHARE Question 9

CONCEPTS

Gene therapy is the direct transfer of genes into humans to treat disease. Gene therapy is being used to treat genetic diseases, cancer, and infectious diseases.

CONCEPTS SUMMARY

- Techniques of molecular genetics are used to locate, analyze, alter, sequence, study, and recombine DNA sequences.
- Restriction endonucleases are enzymes that make double-stranded cuts in DNA at specific base sequences.
- Engineered nucleases are proteins that are capable of making double-stranded cuts at specific DNA sequences. They include zinc-finger nucleases (ZFNs) and transcription activator–like effector nucleases (TALENs).
- The CRISPR-Cas9 system can be used to cut and edit the genome. It combines a single guide RNA with a nuclease, which together attach to DNA sequences and make double-stranded cuts at specific locations.
- DNA fragments can be separated with the use of gel electrophoresis and visualized by staining the gel with a dye that is specific for nucleic acids or by labeling the fragments with a radioactive or chemical tag.

- The polymerase chain reaction is a method for amplifying DNA without cloning.
- In gene cloning, a gene or a DNA fragment is inserted into a bacterial cell, where it will multiply as the cell divides.
- Plasmids, small circular pieces of DNA, are often used as vectors to ensure that a cloned gene is stable and replicated within the recipient cells. Expression vectors contain sequences necessary for foreign DNA to be transcribed and translated.
- Genes can be isolated by creating a DNA library: a set of bacterial colonies or viral plaques that each contain a different cloned fragment of DNA. A genomic library contains the entire genome of an organism; a cDNA library is created from mRNA that is converted into cDNA.
- Positional cloning uses linkage relations to determine the location of genes without any knowledge of their products.
- The Sanger (dideoxy) method of DNA sequencing uses special substrates for DNA synthesis (dideoxyribonucleoside triphosphates) that terminate synthesis after they have been incorporated into the newly made DNA. Next-generation and third-generation sequencing methods sequence many DNA fragments simultaneously, providing a much faster and less expensive determination of DNA base sequences.
- Short tandem repeats, or microsatellites, are used to identify people by their DNA sequences (DNA fingerprinting).
- Forward genetics begins with a phenotype and conducts analyses to locate the responsible genes. Reverse genetics starts with a DNA sequence and conducts analyses to determine its phenotypic effect.
- Transgenic animals, produced by injecting DNA into fertilized eggs, contain foreign DNA that is integrated into a chromosome. Knockout mice are transgenic mice in which a normal gene is disabled. Knock-in mice are transgenic mice in which a particular DNA sequence is inserted in a known location.
- The mouse (*Mus musculus*) is an excellent model genetic organism because of its similarity to humans, small size, and short generation time.
- RNA interference is used to silence the expression of specific genes.
- Techniques of molecular genetics are being used to create products of commercial importance, to develop diagnostic tests, and to treat diseases.

IMPORTANT TERMS

biotechnology (p. 388)
recombinant DNA technology (p. 388)
genetic engineering (p. 389)
restriction enzyme (p. 389)
restriction endonuclease (p. 389)
cohesive end (p. 390)
engineered nuclease (p. 391)
zinc-finger nuclease (ZFN) (p. 391)
transcription activator–like effector nuclease (TALEN) (p. 391)
CRISPR-Cas system (p. 391)
gel electrophoresis (p. 392)
probe (p. 393)
Southern blotting (p. 394)
Northern blotting (p. 394)
Western blotting (p. 394)
gene cloning (p. 394)
polymerase chain reaction (PCR) (p. 394)
Taq polymerase (p. 395)
cloning vector (p. 395)
cosmid (p. 397)
bacterial artificial chromosome (BAC) (p. 397)
expression vector (p. 398)
DNA library (p. 398)
genomic library (p. 398)
cDNA library (p. 398)
positional cloning (p. 400)
DNA sequencing (p. 400)
dideoxyribonucleoside triphosphate (ddNTP) (p. 401)
next-generation sequencing technologies (p. 403)
DNA fingerprinting (p. 405)
microsatellite (p. 405)
short tandem repeat (STR) (p. 405)
forward genetics (p. 407)
reverse genetics (p. 407)
transgene (p. 408)
knockout mice (p. 409)
knock-in mice (p. 409)
gene therapy (p. 412)

ANSWERS TO CONCEPT CHECKS

1. Restriction enzymes exist naturally in bacteria, which use them in defense against viruses.

2. c

3. The gene and plasmid are cut with the same restriction enzyme and mixed together. DNA ligase is used to seal nicks in the sugar–phosphate backbone.

4. d

5. Microsatellites are detected by using PCR with primers that flank the region containing the repeats.

6. b

WORKED PROBLEMS

Problem 1

A molecule of double-stranded DNA that is 5 million base pairs long has a base composition that is 62% G + C. How many times, on average, are the following restriction sites likely to be present in this DNA molecule?

a. *Bam*HI (recognition sequence is GGATCC)

b. *Hind*III (recognition sequence is AAGCTT)

c. *Hpa*II (recognition sequence is CCGG)

Solution Strategy

What information is required in your answer to the problem?

The number of restriction sites likely to be present in the DNA molecule for each of the specified restriction enzymes.

What information is provided to solve the problem?

- The size of the DNA molecule.
- The G + C base composition of the DNA molecule.
- The recognition sequences for each restriction enzyme.

For help with this problem, review:

Restriction Enzymes in Section 14.2.

Solution Steps

Hint: If you know the percentage of any base in the DNA, you can determine the percentages of all the other bases, because G = C and A = T.

The percentages of G and C are equal in double-stranded DNA; so, if G + C = 62%, then %G = %C = 62%/2 = 31%. The percentage of A + T = (100% − G − C) = 38%, and %A = %T = 38%/2 = 19%. To determine the probability of finding a particular base sequence, we use the multiplication rule (see Chapter 3), multiplying together the probability of finding each base at a particular site.

Remember: The multiplication rule states that the probability of two or more independent events is calculated by multiplying their independent probabilities.

a. The probability of finding the sequence GGATCC = $0.31 \times 0.31 \times 0.19 \times 0.19 \times 0.31 \times 0.31 = 0.0003333$. To determine the average number of recognition sequences in a 5 million-base-pair piece of DNA, we multiply 5,000,000 bp × 0.00033 = 1666.5 recognition sequences.

b. The number of AAGCTT recognition sequences is $0.19 \times 0.19 \times 0.31 \times 0.31 \times 0.19 \times 0.19 \times 5{,}000{,}000 = 626$ recognition sequences.

c. The number of CCGG recognition sequences is $0.31 \times 0.31 \times 0.31 \times 0.31 \times 5{,}000{,}000 =$ 46,176 recognition sequences.

Problem 2

You are given the following DNA fragment to sequence: 5′—GCTTAGCATC—3′. You first clone the fragment in bacterial cells to produce sufficient DNA for sequencing. You isolate the DNA from the bacterial cells and carry out the dideoxy sequencing method. You then separate the products of the polymerization reactions by gel electrophoresis. Draw the bands that should appear on the gel from the four sequencing reactions.

Solution Strategy

What information is required in your answer to the problem?

The positions of the bands on the sequencing gel.

What information is provided to solve the problem?

- The base sequence of the DNA fragment to be sequenced.

For help with this problem, review:

Section 14.5, DNA Sequences Can Be Determined and Analyzed.

Solution Steps

Remember: In dideoxy sequencing, a new DNA strand is synthesized, and that strand is what is sequenced. Thus, the bands that appear on the gel represent the complement of the original sequence.

The first task is to write out the sequence of the newly synthesized fragment, which will be complementary and antiparallel to the original fragment. The original sequence is 5′—GCTTAGCATC—3′, so the sequence of the newly synthesized strand will be

3′—CGAATCGTAG—5′

Written 5′→3′, the sequence is 5′—GATGCTAAGC—3′. Bands representing this sequence will appear on the gel, with the bands representing nucleotides near the 5′ end of the molecule at the bottom of the gel.

Hint: Small fragments, those nearer the 5′ end of the newly synthesized strand, will migrate faster and will appear near the bottom of the gel.

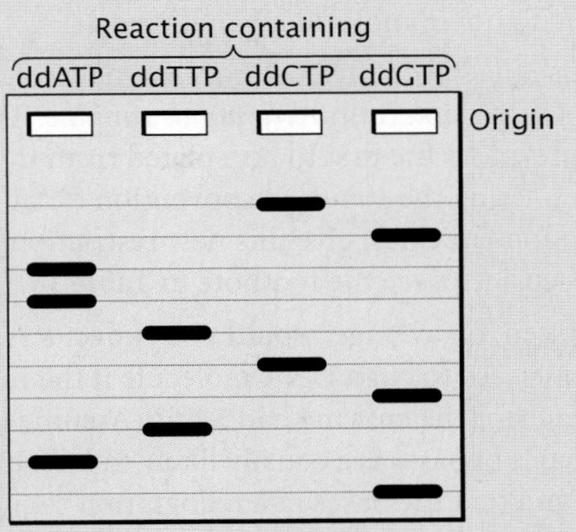

COMPREHENSION QUESTIONS

Section 14.2

1. What role do restriction enzymes play in bacteria? How do bacteria protect their own DNA from the action of restriction enzymes?
2. Explain how gel electrophoresis is used to separate DNA fragments of different lengths.
3. Give three important characteristics of cloning vectors.

Section 14.3

4. Briefly explain how the polymerase chain reaction is used to amplify a specific DNA sequence.
5. Briefly explain how an antibiotic-resistance gene and the *lacZ* gene can be used as markers to determine which cells contain a particular plasmid.

Section 14.4

6. How does a genomic library differ from a cDNA library?
7. Briefly explain how a gene can be isolated through positional cloning.

Section 14.5

8. What is the purpose of the dideoxyribonucleoside triphosphates in the dideoxy sequencing reaction?
9. What is DNA fingerprinting? What types of sequences are examined in DNA fingerprinting?

Section 14.6

10. How does a reverse-genetics approach differ from a forward-genetics approach?
11. What are knockout mice and for what are they used?
12. How is RNA interference used in the analysis of gene function?

Section 14.7

13. What is gene therapy?

For more questions that test your comprehension of the key chapter concepts, go to **LearningCurve** for this chapter.

APPLICATION QUESTIONS AND PROBLEMS

Section 14.2

14. CRISPR-Cas9 was first developed as a molecular tool in 2012; during the next few years, its use in molecular biology exploded, as scientists around the world began applying it to many different research problems, and hundreds of research papers describing its application were published. Explain why CRISPR-Cas is such a powerful tool in molecular genetics.

***15.** Suppose that a geneticist discovers a new restriction enzyme in the bacterium *Aeromonas ranidae*. This restriction enzyme is the first to be isolated from this bacterial species. Using the standard convention for abbreviating restriction enzymes, give this new restriction enzyme a name (for help, see the footnote to **Table 14.1**).

16. How often, on average, would you expect a restriction endonuclease to cut a DNA molecule if the recognition sequence for the enzyme had 5 bp? (Assume that the four types of bases are equally likely to be found in the DNA and that the bases in a recognition sequence are independent.) How often would the endonuclease cut the DNA if the recognition sequence had 8 bp?

***17.** A microbiologist discovers a new restriction endonuclease. When DNA is digested by this enzyme, fragments that average 1,048,500 bp in length are produced. What is the most likely number of base pairs in the recognition sequence of this enzyme?

18. Will restriction sites for an enzyme that has 4 bp in its recognition sequence be closer together, farther apart, or similarly spaced, on average, compared with those of an enzyme that has 6 bp in its recognition sequence? Explain your reasoning.

***19.** About 60% of the base pairs in a human DNA molecule are AT. If the human genome has 3.2 billion base pairs of DNA, about how many times will the following restriction sites be present?

a. *Bam*HI (recognition sequence is 5′—GGATCC—3′)

b. *Eco*RI (recognition sequence is 5′—GAATTC—3′)

c. *Hae*III (recognition sequence is 5′—GGCC—3′)

***20.** A linear piece of DNA has the following *Eco*RI restriction sites:

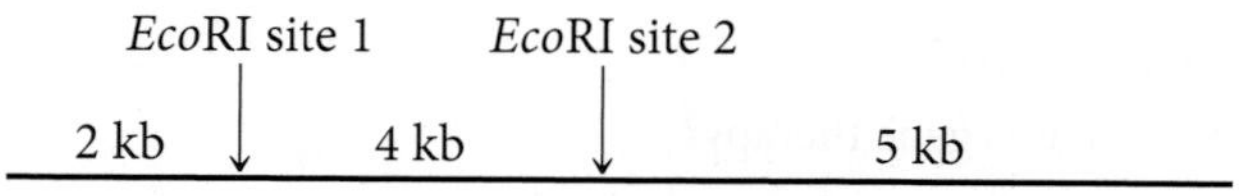

a. This piece of DNA is cut by *Eco*RI, the resulting fragments are separated by gel electrophoresis, and the gel is stained with ethidium bromide. Draw a picture of the bands that will appear on the gel.

b. If a mutation that alters *Eco*RI site 1 occurs in this piece of DNA, how will the banding pattern on the gel differ from the one that you drew in part *a*?

c. If mutations that alter *Eco*RI sites 1 and 2 occur in this piece of DNA, how will the banding pattern on the gel differ from the one that you drew in part *a*?

d. If 1000 bp of DNA were inserted between the two restriction sites, how would the banding pattern on the gel differ from the one that you drew in part *a*?

e. If 500 bp of DNA between the two restriction sites were deleted, how would the banding pattern on the gel differ from the one that you drew in part *a*?

Section 14.3

***21.** Which vectors (plasmid, phage λ, cosmid, bacterial artificial chromosome) can be used to clone a continuous fragment of DNA with the following lengths?

a. 4 kb

b. 20 kb

c. 35 kb

d. 100 kb

22. A geneticist uses a plasmid for cloning that has the *lacZ* gene and a gene that confers resistance to penicillin. The geneticist inserts a piece of foreign DNA into a restriction site that is located within the *lacZ* gene and uses the plasmid to transform bacteria. Explain how the geneticist can identify bacteria that contain a copy of a plasmid with the foreign DNA.

Section 14.4

23. Suppose that you have just graduated from college and have started working at a biotechnology firm. Your first assignment is to clone the pig gene for the hormone prolactin. Assume that the pig gene for prolactin has not yet been isolated, sequenced, or mapped; however, the mouse gene for prolactin has been cloned and the amino acid sequence of mouse prolactin is known. Briefly explain two different strategies that you might use to find and clone the pig gene for prolactin.

Section 14.5

25. Suppose that you want to sequence the following DNA fragment:

5′—TCCCGGGAAA-primer site—3′

You first use PCR to amplify the fragment, so that there is sufficient DNA for sequencing. You carry out dideoxy sequencing of the fragment. You then separate the products of the polymerization reactions by gel electrophoresis. Draw the bands that should appear on the gel from the four sequencing reactions.

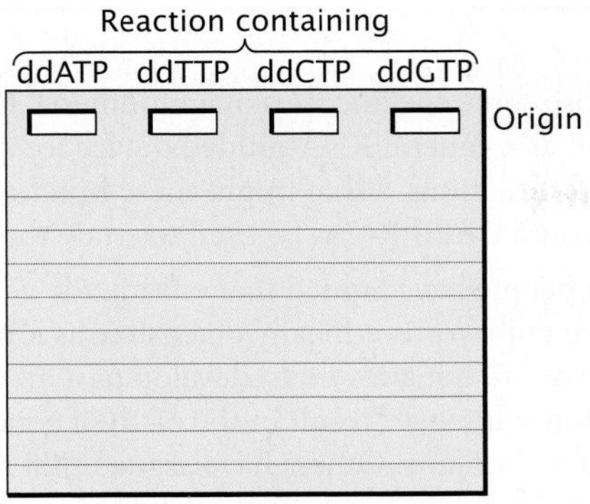

*26. Suppose that you are given a short fragment of DNA to sequence. You amplify the fragment with PCR and set up a series of four dideoxy reactions. You then separate the products of the reactions by gel electrophoresis and obtain the following banding pattern:

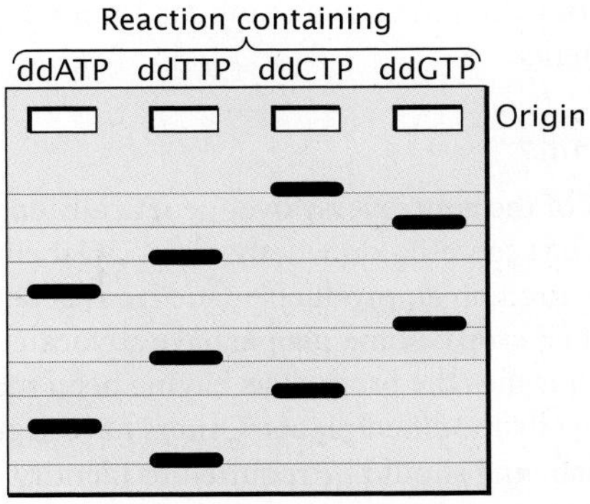

Write out the base sequence of the original fragment that you were given.

Original sequence: 5′— ____________________ —3′

27. The following picture is a sequencing gel from the original study that first sequenced the cystic fibrosis gene (J. R. Riordan, et al. 1989. *Science* 245:1066–1073). From the picture, determine the sequence of the normal copy of the gene and the sequence of the mutated copy of the gene. Identify the location of the mutation that causes cystic fibrosis (CF). Hint: The CF mutation is a 3 bp deletion.

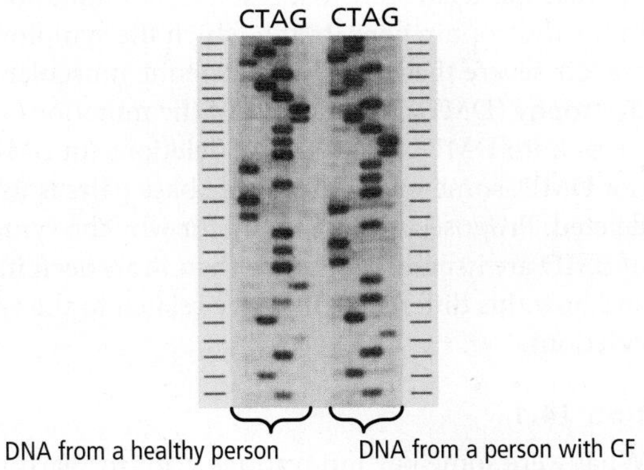

[From Riordan, J. et al. "Identification of the cystic fibrosis gene: cloning and characterization of complementary DNA", Science 245: 1066-1073, © 1989. Reprinted with permission from AAAS, permission conveyed through Copyright Clearance Center, Inc.]

Section 14.6

28. You have discovered a gene in mice that is similar to a gene in yeast. How might you determine whether this gene is essential for development in mice?

CHALLENGE QUESTIONS

Section 14.6

29. Suppose that you are hired by a biotechnology firm to produce a strain of giant fruit flies by using recombinant DNA technology so that genetics students will not be forced to strain their eyes when looking at tiny flies. You go to the library and learn that growth in fruit flies is normally inhibited by a hormone called shorty substance P (SSP). You decide that you can produce giant fruit flies if you can somehow turn off the production of SSP. Shorty substance P is synthesized from a compound called XSP in a single-step reaction catalyzed by the enzyme runtase:

$$\text{XSP} \xrightarrow[\text{Runtase}]{} \text{SSP}$$

A researcher has already isolated cDNA for runtase and has sequenced it, but the location of the runtase gene in the *Drosophila* genome is unknown. In attempting to devise a strategy for turning off the production of SSP and producing giant flies by using standard recombinant DNA techniques, you discover that deleting, inactivating, or otherwise mutating this DNA sequence in *Drosophila* turns out to be extremely difficult. Therefore, you must restrict your genetic engineering to gene augmentation (adding new genes to cells). Describe the methods that you will use to turn off SSP and produce giant flies by using recombinant DNA technology.

THINK-PAIR-SHARE QUESTIONS

Introduction

1. Why does Duchenne muscular dystrophy affect only boys? Is it possible for it to occur in girls? If so, how might this take place?
2. Becker muscular dystrophy (BMD) is another type of muscular dystrophy that results from mutations in the dystrophin gene, but in which the symptoms are less severe than those of Duchenne muscular dystrophy (DMD). About 60% of the mutations that cause both DMD and BMD are deletions. In BMD, but not DMD, some multiple of three base pairs is usually deleted. Propose an explanation for why the symptoms of BMD are usually less severe than those seen in DMD and how this difference might be related to the types of deletions.

Section 14.1

3. What were some key innovations prior to Watson, Crick, Franklin, and Wilkins's discovery of the three-dimensional structure of DNA that were important in shaping the study of genetics?

Section 14.2

4. Now that you understand how the CRISPR-Cas9 system works, think back to the experiments discussed in the introduction to this chapter, in which researchers used CRISPR-Cas9 genome editing to treat mice with Duchenne muscular dystrophy. Why did the researchers choose to cut out the entire exon 23 in the mice with the disorder? Why not replace the specific mutation using a donor piece of DNA and homologous recombination? Propose some possible explanations.
5. Propose some specific uses of a modified CRISPR-Cas system as a general RNA-guided device for altering cellular functions. What might these functions be, and how could CRISPR-Cas be used to study them?
6. Some people have argued that editing the genomes of human embryos is ethically defensible as long as the embryos are not allowed to develop past an early stage and thus will never result in the birth of genetically modified humans. Others have argued that no genome editing should be carried out on human embryos. Present arguments for and against using genome editing on human embryos.

Section 14.3

7. Make a table comparing the advantages and disadvantages of PCR and gene cloning for amplifying DNA fragments.

Section 14.7

8. Much of the controversy over genetically engineered foods has centered on whether special labeling should be required on all products made from genetically modified crops. Some people have advocated labeling that identifies the product as having been made from genetically modified plants. Others have argued that food labeling should be required to identify only the ingredients, not the process by which they were produced. Choose a side in this issue and justify your stand.
9. Why is it that somatic gene therapy is allowed in many countries and yet germ-line gene therapy is often restricted? What are some arguments for and against germ-line gene therapy?

15 Genomics and Proteomics

Building a Chromosome for Class

In the spring of 2014, an international group of scientists reported the first synthesis of a completely artificial eukaryotic chromosome. When substituted for chromosome III in *Saccharomyces cerevisiae*, the synthetic chromosome functioned well. Remarkably, much of the work in constructing this artificial chromosome was carried out by undergraduate students at Johns Hopkins University, who completed the task as a part of their Build a Genome class.

Saccharomyces cerevisiae, commonly known as baker's yeast, has been used for thousands of years for making bread and brewing beer, and today it plays an important role in the food and chemical industries. It's also an important model genetic organism. Baker's yeast, a single-celled eukaryote, is easy to cross and grow in the laboratory. It was the first eukaryote to have its genome completely sequenced, and a number of molecular techniques have been developed to manipulate its DNA. Scientists are now working to create a completely synthetic genome for *S. cerevisiae*, a flexible genome that can be altered at will to endow yeast with selected genes and special properties.

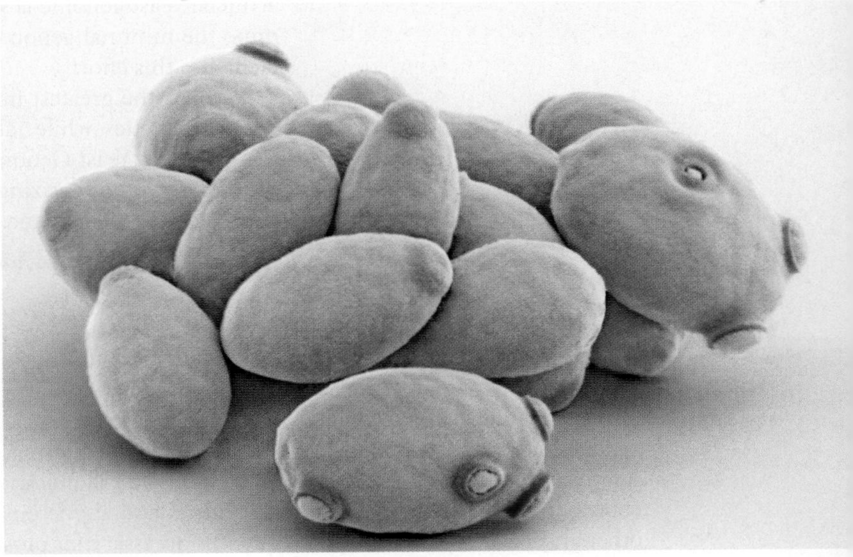

Researchers and students synthesized an artificial chromosome for baker's yeast, Saccharomyces cerevisiae, shown here. [Thomas Deerinck, NCMIR/Science Source.]

The first step in creating a synthetic genome for yeast was synthesis of an artificial chromosome III. Scientists began by designing the new chromosome on the computer, using the basic sequence of natural chromosome III but deleting nonessential DNA, including introns, repeated sequences near the telomeres, transposable elements, and transfer RNA genes. They designed the artificial chromosome with special sequences that allow easy deletion of genes and rearrangement of the genome. They also changed all the TAG stop codons to TAA, freeing up the TAG codon for future use in encoding an additional, artificial amino acid.

Once the sequence of the chromosome was digitally designed, actual construction began. As a first step, students in the Build a Genome class stitched together small pieces of DNA into 750-bp building blocks. Each student started with a series of overlapping 60–79-bp DNA fragments called oligonucleotides, which were synthesized by machine and provided to the class by a commercial firm. The students assembled these fragments into building blocks using methods based on the polymerase chain reaction. After verifying that the correct sequences were joined together, the students cloned the building blocks in bacteria to make many copies. Scientists then assembled the building blocks into 127 larger, overlapping fragments of approximately 7000 bp, called minichunks. Finally, they systematically replaced the native sequences of yeast chromosome III with the minichunks of artificial DNA until a completely synthetic chromosome was achieved.

The resulting artificial chromosome, dubbed synIII, consisted of 272,871 bp of DNA, replacing the original 316,617 bp of the natural chromosome. When inserted into yeast cells in place of the natural chromosome III, synIII functioned normally, replicating and expressing its genetic instructions. Cells with synIII had the same morphology as natural yeast cells and grew well. There appeared to be no fitness differences between synIII cells and natural yeast, and the genomes of the synIII cells were stable over 125 mitotic divisions.

In the design of synIII, geneticists inserted special sequences, called loxPsym sites, on either side of many of its genes, providing an easy way to create variations of the chromosome by deleting and rearranging genes. The loxPsym sites are recognized by an enzyme called Cre, which, when activated by a chemical added by the researcher, brings about recombination between loxPsym sites, resulting in deletions and rearrangements of the genes. This method of altering the genome is called Synthetic Chromosome Rearrangement and Modification by LoxP-Mediated Evolution, or SCRaMbLE for short. One reason scientists wanted to create an artificial yeast genome is so that they could systematically take out groups of genes to determine the minimal genome necessary for a cell to function. The SCRaMbLE system greatly facilitates this effort.

Perhaps the greatest impact of the project was on the students, who helped advance the field of genetics while taking an undergraduate course. In a now-expanded effort called the Synthetic Yeast Genome Project, additional students and scientists from schools around the world are synthesizing the other 15 chromosomes of yeast, with the goal of creating an entirely synthetic genome.

THINK-PAIR-SHARE Questions 1 and 2

Genomics is the field of genetics that attempts to understand the content, organization, function, and evolution of the genetic information contained in whole genomes. The field of genomics is at the cutting edge of modern biology; information resulting from research in this field has made significant contributions to human health, agriculture, and numerous other areas. It has provided gene sequences necessary for producing medically important proteins through recombinant DNA technology, and comparisons of genome sequences from different organisms are leading to a better understanding of evolution and the history of life.

We begin this chapter by examining genetic and physical maps and methods for sequencing entire genomes. Next, we explore functional genomics—how genes are identified in genome sequences and how their functions are defined. The sequence of a genome, by itself, is of limited use, and now that the sequencing of genomes has become routine, much of genomics is focused on deciphering the function of the sequences that have been obtained. We also discuss how genome sequence and organization varies among different organisms and how genomes evolve. In the final section of the chapter, we consider proteomics, the study of the complete set of proteins found in a cell.

15.1 Structural Genomics Determines the DNA Sequences of Entire Genomes

Structural genomics is the study of the organization and sequence of the genetic information contained within a genome. An early step in characterizing a genome is to prepare genetic and physical maps of its chromosomes. These maps provide information about the relative locations of genes, molecular markers, and chromosome segments, which is often essential for positioning chromosome segments and aligning stretches of sequenced DNA into a whole-genome sequence. These maps are also the foundation for positional cloning (see Chapter 14), in which genes that influence specific traits are located and sequenced.

Genetic Maps

Everyone has used a map at one time or another. Maps are indispensable for finding a new friend's house, the way to an unfamiliar city in your state, or the location of a country. Each of these examples requires a map with a different scale. To find a friend's house, you would probably use a city street map; to find your way to an unknown city, you might pick up a state highway map; to find a country such as Kazakhstan, you would need a world atlas. Similarly, navigating a genome requires maps of different types and scales.

Genetic maps (also called linkage maps) provide a rough approximation of the locations of genes relative to the locations of other known genes (**Figure 15.1**). These maps are based on the genetic process of recombination (hence the name *genetic map*). The basic principles of constructing genetic maps are discussed in detail in Chapter 5. In short, individual organisms of known genotype are crossed, and the frequency of recombination between loci is determined by examining the progeny. For linked genes, the rate of recombination is proportional to the physical distance between the loci. Distances on genetic maps are measured in recombination frequencies (centiMorgans, cM), or map units.

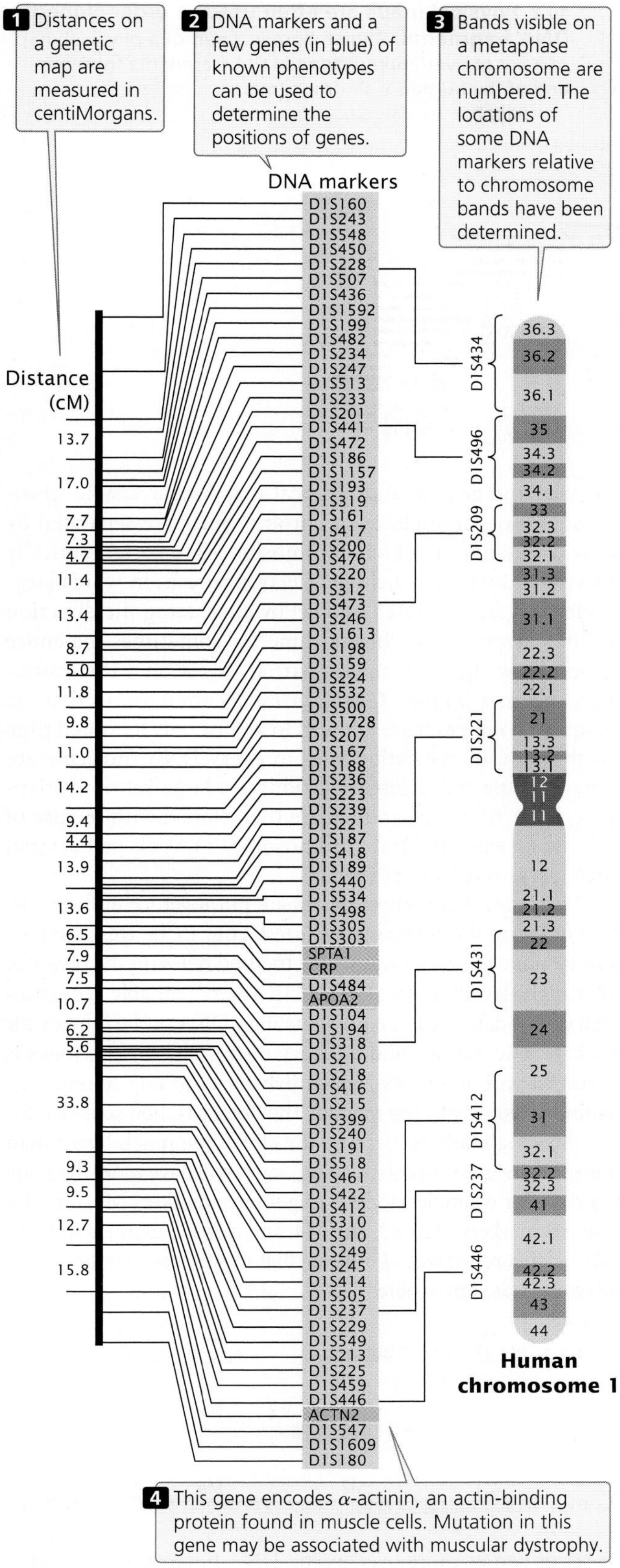

15.1 Genetic maps are based on rates of recombination. Shown here is a genetic map of human chromosome 1.

Physical Maps

Physical maps are based on the direct analysis of DNA, and they place genes in relation to distances measured in number of base pairs, kilobases, or megabases (**Figure 15.2**). A common type of physical map is one that connects isolated pieces of genomic DNA that have been cloned in bacteria or yeast. Physical maps generally have higher resolution and are more accurate than genetic maps. A physical map is analogous to a neighborhood map that shows the location of every house along a street, whereas a genetic map is analogous to a highway map that shows the general locations of major towns and cities.

One of the techniques that has been used for creating physical maps is restriction mapping, which determines the positions of restriction sites on DNA. When a piece of DNA is cut with a restriction enzyme and the fragments are separated by gel electrophoresis, the number of restriction sites in the DNA and the distances between them can be determined by the number and positions of bands on the gel, but this information does not tell us the order or the precise location of the restriction sites. To map restriction sites, a sample of the DNA is cut with one restriction enzyme, and another sample is cut with a different restriction enzyme. A third sample is cut with both restriction enzymes together (a double digest). The DNA fragments produced by these restriction digests are then separated by gel electrophoresis, and their sizes are compared. Overlap in the sizes of the fragments produced by the digests can be used to position the restriction sites on the original DNA molecule. Although restriction mapping has been an important tool for creating physical maps in the past, the rapid, low-cost methods of DNA sequencing that are available today (see Chapter 14) have largely replaced restriction mapping.

CONCEPTS

Both genetic and physical maps provide information about the relative positions of and distances between genes, molecular markers, and chromosome segments. Genetic maps are based on rates of recombination and are measured in recombination frequencies. Physical maps are based on physical distances and are measured in base pairs.

Sequencing an Entire Genome

The ultimate goal of structural genomics is to determine the ordered nucleotide sequences of entire genomes of organisms. In Chapter 14, we considered some of the methods used to sequence small fragments of DNA. The main obstacle to sequencing a whole genome is the immense size of most genomes. Bacterial genomes are usually at least several million base pairs long; many eukaryotic genomes are billions of base pairs long and are distributed among dozens of chromosomes. Furthermore, for technical reasons, sequencing cannot begin at one end of a chromosome and continue straight through to the other end; only small fragments of DNA—usually no more than 500 to 700 nucleotides—are sequenced at one time. Therefore, determining the sequence for an entire genome requires that

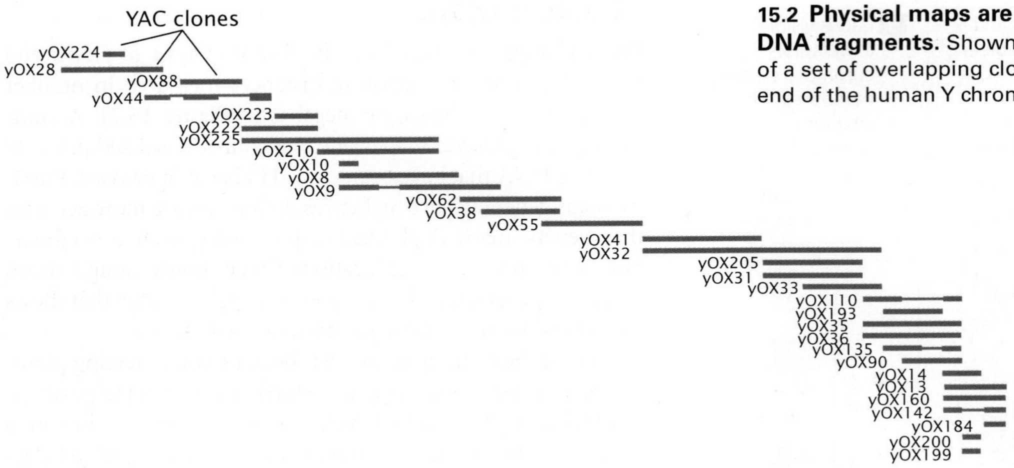

15.2 Physical maps are often used to order cloned DNA fragments. Shown here is a part of a physical map of a set of overlapping cloned DNA fragments from one end of the human Y chromosome.

the DNA be broken into thousands or millions of small overlapping fragments that can then be sequenced. The difficulty lies in putting those short sequences back together in the correct order. Two different approaches have been used to assemble the short sequenced fragments into a complete genome: map-based sequencing and whole-genome shotgun sequencing. We will consider these two approaches in the context of the Human Genome Project, where both approaches were used.

The Human Genome Project

By 1980, methods for mapping and sequencing DNA fragments had been sufficiently developed that geneticists began seriously proposing that the entire human genome be sequenced. An international collaboration was planned to undertake the Human Genome Project; initial estimates suggested that 15 years and $3 billion would be required to accomplish the task.

The Human Genome Project officially began in October 1990. Initial efforts focused on developing new and automated methods for cloning and sequencing DNA and on generating detailed physical and genetic maps of the human genome. By 1993, large-scale physical maps were completed for all 23 pairs of human chromosomes. At the same time, automated sequencing techniques had been developed that made large-scale sequencing feasible.

The initial effort to sequence the genome was a public project consisting of an international collaboration among 20 research groups and hundreds of individual researchers who formed the International Human Genome Sequencing Consortium. This group used a map-based strategy for sequencing the human genome.

MAP-BASED SEQUENCING In **map-based sequencing**, short sequenced fragments are assembled into a whole-genome sequence by first creating detailed genetic and physical maps of the genome, which provide known locations of genetic markers (restriction sites, known genes, or known DNA sequences) at regularly spaced intervals along each chromosome. These markers are later used to help align the short sequenced fragments into their correct order.

After the genetic and physical maps are available, chromosomes or large pieces of chromosomes are separated by flow cytometry, in which chromosomes are sorted optically by size, or by pulsed-field gel electrophoresis, in which large molecules are separated in a gel by alternating the direction of the current. Each chromosome (or sometimes the entire genome) is then cut up by partial digestion with restriction enzymes (**Figure 15.3**). *Partial digestion* means that the restriction enzymes are allowed to act for only a limited time so that not all restriction sites in every DNA molecule are cut. Thus, partial digestion produces a large set of overlapping DNA fragments, which are then cloned with the use of cosmids, yeast artificial chromosomes (YACs), or bacterial artificial chromosomes (BACs).

Next, these large-insert clones are put together in their correct order on the chromosome (see Figure 15.3). This assembly can be done in several ways. One method relies on the presence of a high-density map of genetic markers (variable sequences that can be detected). A complementary DNA probe is made for each genetic marker, and a library of the large-insert clones is screened with the probe, which hybridizes to any colony containing a clone with the marker. The library is then screened for neighboring markers. Because the clones are much larger than the markers used as probes, some clones will have more than one marker. For example, clone A might have markers M1 and M2, clone B markers M2, M3, and M4, and clone C markers M4 and M5. Such a result would indicate that these clones contain areas of overlap, as shown here:

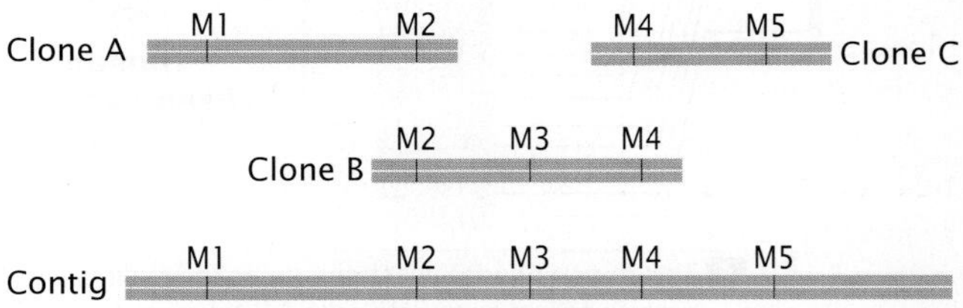

A set of two or more overlapping DNA fragments that form a contiguous stretch of DNA is called a **contig**. This approach was used in 1993 to create a contig consisting of 196 overlapping clones (see Figure 15.2) of the human Y chromosome.

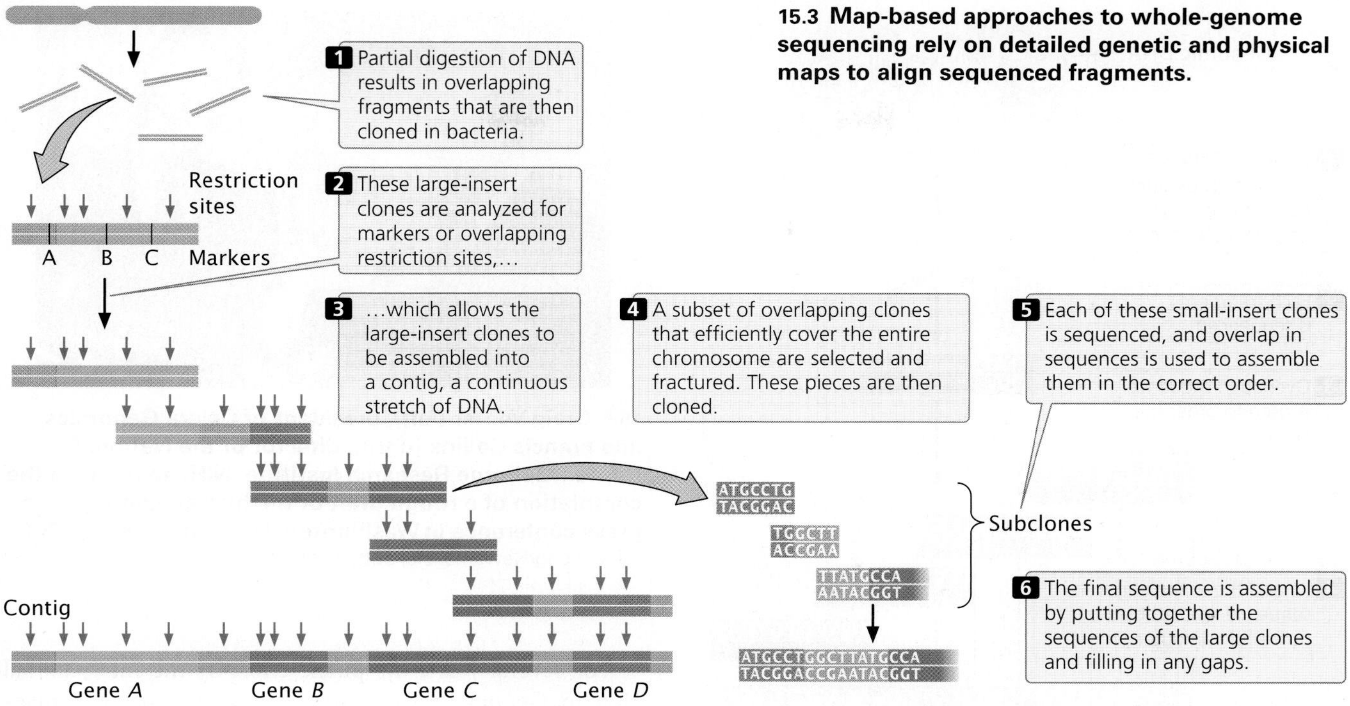

15.3 Map-based approaches to whole-genome sequencing rely on detailed genetic and physical maps to align sequenced fragments.

The order of clones can also be determined without the use of preexisting genetic maps. For example, each clone can be cut with a series of restriction enzymes and the resulting fragments then separated by gel electrophoresis. This method generates a unique set of restriction fragments, called a fingerprint, for each clone. The restriction patterns for the clones are stored in a database. A computer program is then used to examine the restriction patterns of all the clones and look for areas of overlap. The overlap is then used to arrange the clones in order, as shown here:

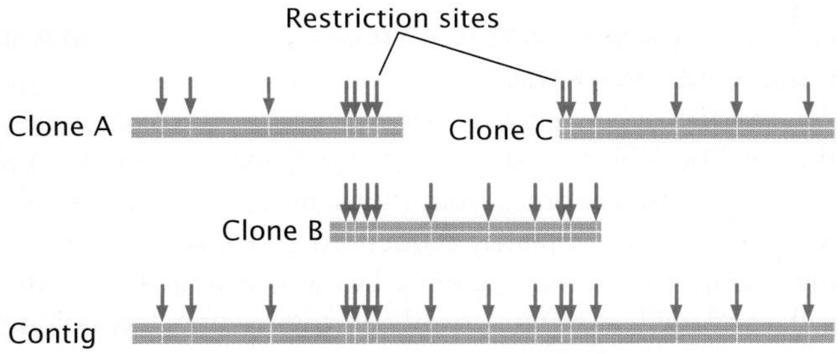

Other genetic markers can be used to help position contigs along the chromosome.

When the large-insert clones have been assembled into the correct order on the chromosome, a subset of overlapping clones that efficiently cover the entire chromosome can be chosen for sequencing; the goal is to select the minimum number of clones that is necessary to represent the chromosome. Each of the selected large-insert clones is fractured into smaller overlapping fragments, which are themselves cloned (see Figure 15.3). These smaller clones (called small-insert clones or subclones) are then sequenced. The sequences of the small-insert clones are examined for overlap, which allows them to be correctly assembled to give the sequence of the large-insert clones. Enough overlapping small-insert clones are usually sequenced to ensure that the entire genome is sequenced several times. Finally, the whole genome is assembled by putting together the sequences of all overlapping contigs. Often, gaps in the genome sequence still exist and must be filled in by using other methods.

The International Human Genome Sequencing Consortium used a map-based approach to sequencing the human genome. Many copies of the human genome were cut up into fragments of about 150,000 bp each, which were inserted into bacterial artificial chromosomes. Restriction fingerprints and other genetic markers were used to assemble the BAC clones into contigs, which were positioned on the chromosomes with the use of genetic markers and probes. The individual BAC clones were sheared into smaller overlapping fragments and sequenced, and the whole genome was assembled by putting together the sequence of the BAC clones.

WHOLE-GENOME SHOTGUN SEQUENCING In 1998, Craig Venter announced that he would lead a company called Celera Genomics in a private effort to sequence the human genome. He proposed using a shotgun sequencing approach, which he suggested would be quicker than the map-based approach employed by the International Human Genome Sequencing Consortium. In **whole-genome shotgun sequencing** (**Figure 15.4**), the entire genome is assembled based on sequence overlap. Small-insert clones are prepared directly from genomic DNA and sequenced. Powerful computer programs then assemble the entire genome by examining overlap in the nucleotide sequences among the small-insert clones. One advantage of shotgun sequencing is that the small-insert clones can be placed in plasmids, which are simple and easy to manipulate.

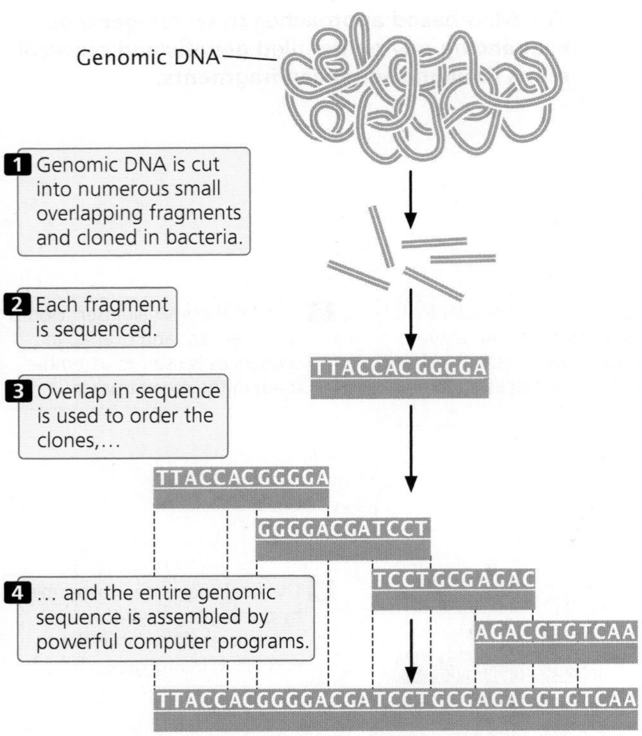

Genomic DNA
1 Genomic DNA is cut into numerous small overlapping fragments and cloned in bacteria.
2 Each fragment is sequenced.
TTACCACGGGGA
3 Overlap in sequence is used to order the clones,...
TTACCACGGGGA
GGGGACGATCCT
TCCTGCGAGAC
AGACGTGTCAA
4 ... and the entire genomic sequence is assembled by powerful computer programs.
TTACCACGGGGACGATCCTGCGAGACGTGTCAA

15.4 Whole-genome shotgun sequencing uses sequence overlap to align sequenced fragments.

The requirement for overlap means that most of the genome must be sequenced multiple times. The average number of times a nucleotide in the genome is sequenced is called the sequencing coverage. For example, 10× coverage means that an average nucleotide in the genome has been sequenced 10 times.

Shotgun sequencing was initially used for assembling small genomes such as those of bacteria. When Venter proposed the use of this approach for sequencing the human genome, it was not at all clear that the approach could successfully assemble a complex genome consisting of billions of base pairs. Today, virtually all genomes are sequenced using the whole-genome shotgun approach. ▶TRY PROBLEM 15

CONCEPTS

Sequencing a genome requires breaking it up into small overlapping fragments whose DNA sequences can be determined in a sequencing reaction. In map-based sequencing, sequenced fragments are ordered into the final genome sequence with the use of genetic and physical maps. In whole-genome shotgun sequencing, the genome is assembled by means of overlap in the sequences of small fragments.

✔ CONCEPT CHECK 1

A contig is

a. a set of molecular markers used in genetic mapping.
b. a set of overlapping fragments that form a continuous stretch of DNA.
c. a set of fragments generated by a restriction enzyme.
d. a small DNA fragment used in sequencing.

15.5 Craig Venter (left), president of Celera Genomics, and Francis Collins (right), director of the National Human Genome Research Institute, NIH, announced the completion of a rough draft of the human genome at a press conference in Washington, D.C., on June 26, 2000. [Alex Wong/Newsmakers/Getty Images.]

For several years, the public effort by the International Human Genome Sequencing Consortium, using a map-based approach, and the private Celera effort, using shotgun sequencing, moved forward simultaneously. In the summer of 2000, both public and private sequencing projects announced the completion of a rough draft that included most of the sequence of the human genome, five years ahead of schedule (**Figure 15.5**). An analysis of this sequence was published 6 months later. The human genome sequence was declared completed in the spring of 2003, although some gaps still remain. For most chromosomes, the finished sequence is 99.999% accurate, with less than one error per 100,000 bp, an accuracy rate 10 times that of the initial goal.

RESULTS AND IMPLICATIONS OF THE HUMAN GENOME PROJECT With the first human genome determined, sequencing individual genomes is much easier. It is now possible to sequence an entire human genome in a few hours, and genomes from thousands of different people have now been completely sequenced. The cost of sequencing a complete human genome has also dropped dramatically and will continue to fall as sequencing technology improves.

The availability of the complete sequence of the human genome is proving to be of great benefit. The sequence has provided tools for detecting and mapping genetic variants across the human genome, greatly facilitating gene mapping in humans. For example, several million sites at which people differ in a single nucleotide (called single-nucleotide polymorphisms; see Section 15.2) have now been identified, and these sites are being used in genome-wide association studies to locate genes that affect diseases and traits in humans (see Chapter 5 and p. 426). The sequence is also providing important information about development and many basic cellular processes.

Next-generation sequencing techniques that allow rapid and inexpensive sequencing of genomic DNA (see Chapter 14) are being used to address fundamental questions in many areas. For example, the genomes of a number of cancer cells have now been completely sequenced and compared with the sequence of healthy cells from the same person, allowing complete determination of all the mutations that lead to tumor formation and cancer progression. The complete genome of an unborn baby has been sequenced from fetal DNA isolated from its mother's blood. The 1000 Genomes Project, completed in 2015, sequenced and and compared the genomes of 2,504 individuals from 26 populations in Africa, East Asia, Europe, South Asia, and the Americas, with the goal of detecting most of the common variations that exist in the human species. Sequencing of the complete genomes of parents and their children has allowed a direct estimate of mutation rates.

DNA has been extracted from the bones of ancient humans, including Neanderthals and Denisovans (a little-known group of humans that appear to be closely related to Neanderthals), and completely sequenced. Comparisons of the modern human genome with the sequences of these and other species are adding to our understanding of human evolution as well as our knowledge of the evolution and the history of all life.

In spite of these benefits and successes, some people have been disappointed by the lack of tangible results from the Human Genome Project. At the time it was proposed, there was speculation that the sequencing of the human genome would immediately revolutionize the practice of medicine, leading to new insights for treating common diseases and resulting in the development of powerful new drugs. Although genome sequence data have produced numerous new and exciting research findings and have led to a better understanding of many diseases, the data are still seldom used by practicing physicians in the treatment of patients.

Along with the many potential benefits of complete genome sequence information, there are concerns about its misuse. With the knowledge gained from genome sequencing, many more genes for diseases, disorders, and behavioral and physical traits have been identified, increasing the number of genetic tests that can be performed to make predictions about the future phenotype and health of a person. There is concern that information from genetic testing might be used to discriminate against people who are carriers of disease-causing genes or who might be at risk for some future disease. This problem has been addressed to some extent in the United States with the passage of the Genetic Information Nondiscrimination Act, which prohibits health insurers and employers from using genetic information to make decisions about health insurance coverage and employment (although it does not apply to life, disability, or long-term care insurance). Questions also arise about who should have access to a person's genome sequence. What about relatives, who have similar genomes and might also be at risk for some of the same diseases? There are also questions about the use of this information to select for specific traits in future offspring. All of these concerns are legitimate and must be addressed if we are to use the information from genome sequencing responsibly.

THINK-PAIR-SHARE Question 3

CONCEPTS

The Human Genome Project was an effort to sequence the entire human genome. Rough drafts were completed by two competing teams, in 2000. The entire sequence was completed in 2003. The ability to rapidly sequence human genomes raises a number of ethical questions.

Single-Nucleotide Polymorphisms

Genome sequencing has identified a number of sites where humans differ in their genomic sequences. Imagine that you are riding in an elevator with a random stranger. How much of your genome do you have in common with this person? Studies of variation in the human genome indicate that you and the stranger will be identical at about 99.9% of your DNA sequences. This difference is very small in *relative* terms, but because the human genome is so large (3.2 billion base pairs), you and the stranger will actually be different at more than 3 million base pairs of your genomic DNA. These differences are what make each of us unique, and they greatly affect our physical features, our health, and possibly even our intelligence and personality.

A site in the genome at which individual members of a species differ in a single base pair is called a **single-nucleotide polymorphism** (**SNP**, pronounced "snip"). Arising through mutation, SNPs are inherited as allelic variants (just as alleles that produce phenotypic differences, such as blood types, are, although SNPs do not usually produce phenotypic differences). Single-nucleotide polymorphisms are numerous and are present throughout genomes. In a comparison of the same chromosome from two different people, a SNP can be found approximately every 1000 bp.

THINK-PAIR-SHARE Question 4

Most SNPs present within a population arose once from a single mutation that occurred on a particular chromosome and subsequently spread throughout the population. Thus, each SNP is initially associated with other SNPs (as well as with other types of genetic variants or alleles) that were present on the particular chromosome on which the mutation arose. The specific set of SNPs and other genetic variants observed on a single chromosome or part of a chromosome is called a **haplotype** (**Figure 15.6**). Single-nucleotide polymorphisms within a haplotype are physically linked and therefore tend to be inherited together. New haplotypes can arise through mutation or crossing over, which breaks up the particular set of SNPs in a haplotype.

Because of their variability and widespread occurrence throughout the genome, SNPs are valuable as markers in

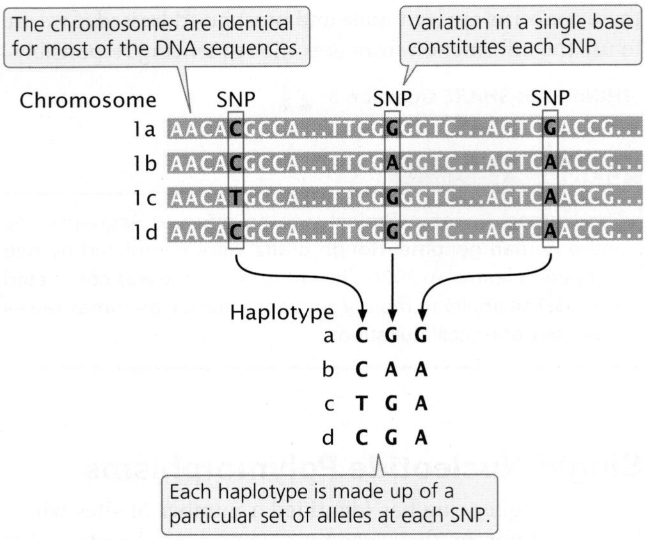

15.6 A haplotype is a specific set of single-nucleotide polymorphisms (SNPs) and other genetic variants observed on a single chromosome or part of a chromosome. Chromosomes 1a, 1b, 1c, and 1d represent different copies of a chromosome that might be found in a population.

linkage studies. When a SNP is physically close to a disease-causing locus, it will tend to be inherited along with the disease-causing allele. Therefore, people with the disease will tend to have different SNPs than healthy people. A comparison of SNP haplotypes in people with the disease and in healthy people can reveal the presence of genes that affect the disease; because the disease gene and the SNP are closely linked, the location of the disease-causing gene can be determined from the locations of associated SNPs. The large number of SNPs identified has provided a dense set of variable markers covering the entire genome that can be used effectively in mapping. Study of SNPs has also provided important information about the function and evolution of the human genome.

GENOME-WIDE ASSOCIATION STUDIES Many common diseases are caused by complex interactions among multiple genes; the availability of SNPs has greatly facilitated the search for these genes. **Genome-wide association studies** use numerous SNPs scattered across the genome to find genes of interest. In one of the most successful applications of SNPs for finding disease associations, researchers genotyped 17,000 people in the United Kingdom for 500,000 SNPs in 2007. They detected strong associations between 24 genes or chromosome segments and the incidence of seven common diseases, including coronary artery disease, Crohn disease, rheumatoid arthritis, bipolar disorder, hypertension, and two types of diabetes. The importance of this study is its demonstration that genome-wide association studies using SNPs can successfully locate genes that contribute to complex diseases caused by multiple genetic and environmental factors.

Within the past few years, SNPs have been used in genome-wide association studies to successfully locate genes that influence many additional traits, such as the age of puberty and menopause in women, variation in facial features, skin pigmentation, eye color, body mass index, bone density, glaucoma, and even susceptibility to infectious diseases such as meningococcal disease and tuberculosis. Unfortunately, the genes identified often explain only a modest proportion of the genetic influence on the trait. For example, one huge genome-wide association study combined data from over 100,000 human subjects in an attempt to locate genes encoding blood lipids involved in cardiovascular disease. Although the study identified 95 different loci associated with lipid traits, those genes corresponded to only 25%–30% of the total genetic variation in these traits. DNA sequences that encode the majority of the missing genetic variation in such traits—sometimes called the "dark matter of the genome"—have thus far remained largely undetected. The low percentage of variation explained by most current genome-wide association studies means that the genes identified are not, by themselves, useful predictors of the risk of inheriting the disease or trait. Nevertheless, the identification of specific genes that influence a disease or trait can lead to a better understanding of the biological processes that produce the phenotype.

Bioinformatics

Complete genome sequences have now been determined for numerous organisms, and many additional projects are under way. These studies are producing tremendous quantities of sequence data. Cataloging, storing, retrieving, and analyzing this huge data set are major challenges of modern genetics. **Bioinformatics** is an interdisciplinary field that combines molecular biology and computer science. It centers on developing databases, computer-search algorithms, gene-prediction software, and other analytical tools that are used to make sense of DNA-, RNA-, and protein-sequence data. Bioinformatics develops and applies these tools to "mine the data," extracting the useful information from sequencing projects. The development and use of algorithms and computer software for analyzing DNA-, RNA-, and protein-sequence data have helped to make molecular biology a more quantitative field. Sequence data in publicly available online databases enable scientists and students throughout the world to access this tremendous resource.

CONCEPTS

Sequencing projects are collecting databases of single-nucleotide polymorphisms (SNPs), which are nucleotides that vary among individual organisms. Bioinformatics is an interdisciplinary field that combines molecular biology and computer science. It develops databases of DNA, RNA, and protein sequences and tools for analyzing those sequences.

Metagenomics

Advances in sequencing technology, which have made sequencing faster and less expensive, now provide the possibility of sequencing not just the genomes of individual species, but the genomes of entire communities of organisms. **Metagenomics** is an emerging field in which the genome sequences of an entire group of organisms that inhabit a common environment are sampled and determined.

Thus far, metagenomics has been applied largely to microbial communities. Traditionally, bacteria have been studied by growing and analyzing them in the laboratory. However, many bacteria cannot be cultured with the use of laboratory techniques. Metagenomics analyzes microbial communities by extracting DNA from the environment, determining its sequences, and reconstructing community composition and function based on those sequences. This technique allows the identification and genetic analysis of species that cannot be grown in the laboratory and have never been studied by traditional microbiological methods. The entire genomes of some dominant species have been reconstructed from environmental samples, providing scientists with a great deal of information on the biology of these microbes.

An early metagenomic study analyzed the microbial community found in acid drainage from a mine and determined that this community consisted of only a few dominant bacterial species. Another study, called the Global Ocean Sampling Expedition, followed the route of Darwin's voyage on H.M.S. *Beagle* in the 1800s. Scientists collected water samples and used metagenomic methods to determine their microbial communities. In this study, scientists cataloged sequences for more than 6 million proteins, including more than 1700 new protein families.

Other metagenomic studies have examined the genes of bacteria that inhabit the human intestinal tract. These bacteria, along with those that inhabit the skin and other parts of the human body, are termed the human **microbiome**. The microbiome of a typical person includes over 100 trillion cells—more than 10 times the number of human cells—and contains 100 times as many genes as the human genome.

Research is demonstrating that the human microbiome plays an important role in human health. One study examined the gut microflora of obese and lean people. Two groups of bacteria are common in the human gut: Bacteroidetes and Firmicutes. Researchers discovered that obese people have relatively more Firmicutes than do lean people and that the proportion of Firmicutes decreases in obese people who lose weight on a low-calorie diet. These same results were observed in obese and lean mice. In an elegant experiment, researchers transferred bacteria from obese to lean mice. Lean mice that received bacteria from obese mice extracted more calories from their food and stored more fat, suggesting that gut microflora might play some role in obesity.

Ecological communities have traditionally been characterized on the basis of the species they contain. Metagenomics provides the possibility of characterizing communities on the basis of their component genes, an approach that has been termed gene-centric. A gene-centric approach leads to new questions: Are certain types of genes more common in some communities than in others? Are some genes essential for energy flow and nutrient recycling within a community? Because of the larger sizes of their genomes, eukaryotic communities have not yet been the focus of these approaches, but many researchers predict that they will be in the future.

CONCEPTS

Metagenomic studies examine the genomes of communities of organisms that inhabit a common environment. This approach has been applied to microbial communities and allows the composition and genetic makeup of a community to be determined without having to cultivate and isolate individual species. Metagenomic studies are sources of important new insights into microbial communities.

Synthetic Biology

The ability to sequence and study whole genomes, coupled with an increased understanding of what genetic information is required for basic biological processes, now provides the possibility of creating—entirely from scratch—novel organisms that have never before existed. **Synthetic biology** is a new field that seeks to design organisms that might provide useful functions, such as microbes that provide clean energy or break down toxic wastes.

Synthetic biologists have already mixed and matched parts from different organisms to synthesize microbes. In 2002, geneticists recreated the poliovirus by joining together pieces of DNA that were synthesized in the laboratory. Even more impressively, in 2010, Daniel Gibson and his colleagues synthesized from scratch the complete 1.08 million-base-pair genome of the bacterium *Mycoplasma mycoides*. They started with a thousand pieces of DNA that were synthesized in the laboratory and joined them together in successively larger pieces until they had assembled a complete copy of the genome. Within their synthetic genome, they included a set of DNA sequences that spelled out—in code—an e-mail address, the names of the researchers who participated in the project, and several well-known quotations. Finally, the researchers transplanted the artificial genome into a cell of a different bacterial species, *M. capricolum*, whose original genome had been removed. The new cell then began expressing the traits specified by the synthetic genome.

Synthetic biology has also been extended to eukaryotic cells, as we saw in the introduction to this chapter. In 2014, geneticists created a synthetic chromosome that successfully replaced a natural chromosome in yeast cells, and efforts are currently underway to engineer a completely synthetic genome in yeast.

These types of experiments have raised a number of concerns. The ability to make novel genomes and to mix and match parts from different organisms creates the potential to synthesize dangerous microbes, which might create ecological havoc if they escaped from the laboratory or might be used in biological warfare or bioterrorism. Ongoing discussions among geneticists, ethicists, security experts, and politicians are addressing these concerns and whether synthetic genomes can be safely made and used.

THINK-PAIR-SHARE Question 5

15.2 Functional Genomics Determines the Function of Genes by Using Genome-Based Approaches

A genome sequence is, by itself, of limited use. Merely knowing the sequence would be like having a huge set of encyclopedias without being able to read: you could recognize the different letters, but the text would be meaningless. **Functional genomics** characterizes what sequences do—their function. The goals of functional genomics include the identification of all the RNA molecules transcribed from a genome, called the **transcriptome** of that genome, and all the proteins encoded by the genome, called the **proteome**. Functional genomics uses both bioinformatics and laboratory-based experimental approaches in its effort to define the function of DNA sequences.

Predicting Function from Sequence

The nucleotide sequence of a gene can be used to predict the amino acid sequence of the protein it encodes. The protein can then be synthesized or isolated and its properties studied to determine its function. However, this biochemical approach to understanding gene function is both time-consuming and expensive. A major goal of functional genomics has been to develop computational methods that allow gene function to be identified from DNA sequence alone, bypassing the laborious process of isolating and characterizing individual proteins.

One computational method (often the first employed) for determining gene function is to conduct a homology search, which relies on comparisons of DNA and protein sequences from the same organism and from different organisms. Genes that are evolutionarily related, which are referred to as **homologous genes**, are likely to have similar sequences. Databases containing genes and proteins found in a wide array of organisms are available for homology searches. Powerful computer programs, such as BLAST (Basic Local Alignment Search Tool), have been developed for scanning these databases to look for particular sequences. Suppose that a geneticist sequences a genome and locates a gene that encodes a protein of unknown function. A homology search conducted on databases containing the DNA or protein sequences of other organisms may identify one or more homologous sequences. If a function is known for a protein encoded by one of those sequences, that function may provide information about the function of the newly discovered protein.

CONCEPTS

The function of an unknown gene can sometimes be determined by finding genes with similar sequences whose function is known.

Gene Expression and Microarrays

Many important clues about gene function come from knowing when and where the genes are expressed. The development of microarrays has allowed the expression of thousands of genes to be monitored simultaneously.

Microarrays rely on nucleic acid hybridization, in which a known DNA fragment is used as a probe to find complementary sequences (**Figure 15.7**). Numerous known DNA fragments are fixed to a solid support in an orderly pattern or array, usually as a series of dots. These DNA fragments (the probes) usually correspond to known genes from a particular organism. An array containing tens of thousands of probes can be applied to a glass slide or silicon chip just a few square centimeters in size.

After the microarray has been constructed, mRNA, DNA, or cDNA isolated from experimental cells is labeled with fluorescent nucleotides and applied to the array. Any of the DNA or RNA molecules that are complementary to probes on the array will hybridize with them and emit fluorescence, which can be detected by an automated scanner.

Used with cDNA, microarrays can provide information about the expression of thousands of genes, enabling scientists to determine which genes are active in particular tissues. They can also be used to investigate how gene expression changes in the course of biological processes such as development or disease progression. In one study, researchers used microarrays to examine the expression patterns of 25,000 genes from primary tumors of 78 young women who had breast cancer (see Figure 15.7). Messenger RNA from cancer and noncancer cells was converted into cDNA and labeled with red and green fluorescent nucleotides, respectively. The labeled cDNAs were mixed and hybridized to a DNA chip, which contained DNA probes from numerous genes. Hybridization of the red (cancer) and green (noncancer) cDNA was proportional to the relative amounts of mRNA in the samples. The fluorescence of each spot was assessed by microscopic scanning and appeared as a single color. Red indicated the overexpression of a gene in the cancer cells relative to that in the noncancer cells (more red-labeled cDNA hybridized), whereas green indicated the underexpression of a gene in the cancer cells relative to that in the noncancer cells (more green-labeled cDNA hybridized). Yellow indicated

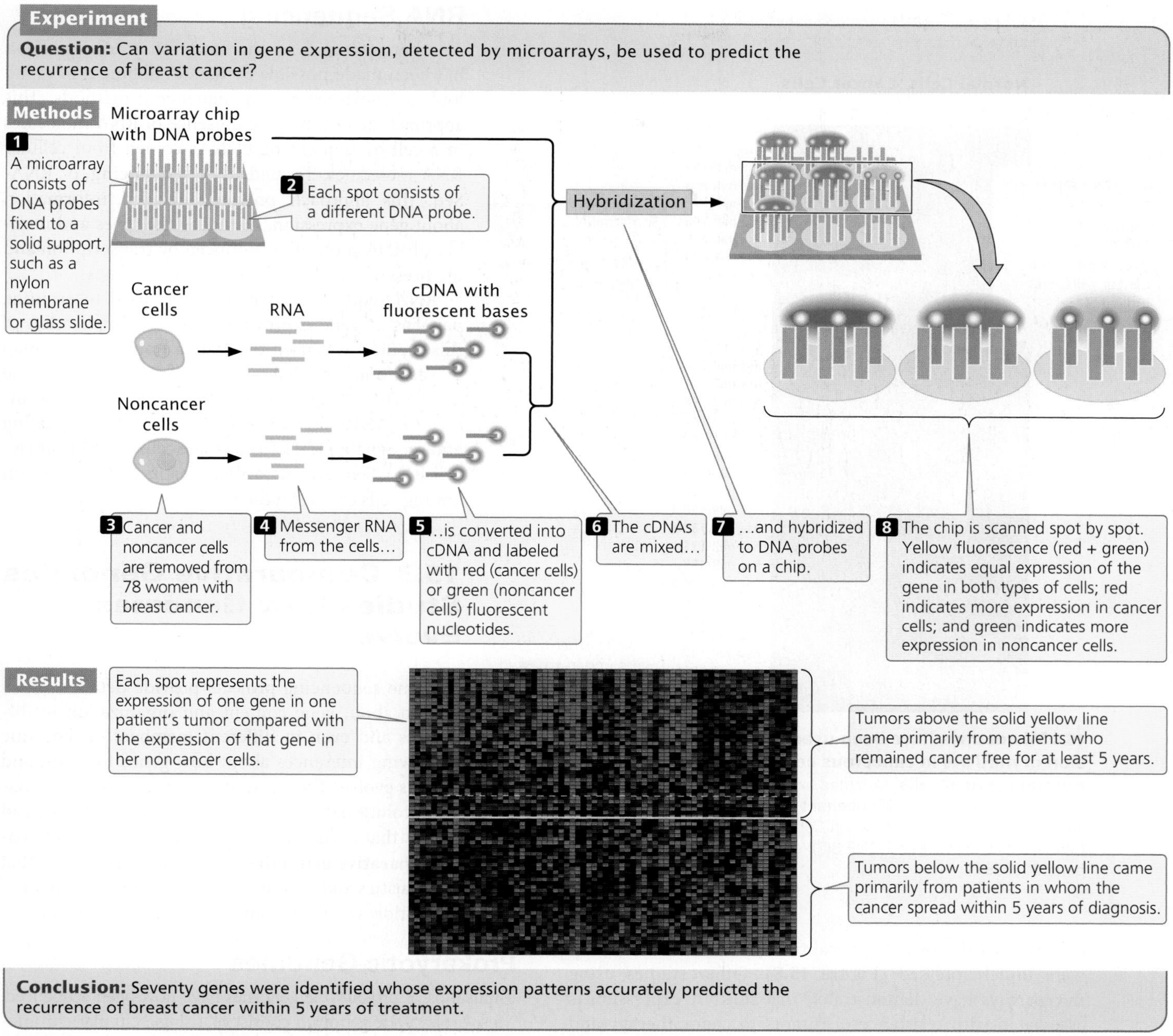

Experiment
Question: Can variation in gene expression, detected by microarrays, be used to predict the recurrence of breast cancer?
Methods
Microarray chip with DNA probes
1 A microarray consists of DNA probes fixed to a solid support, such as a nylon membrane or glass slide.
2 Each spot consists of a different DNA probe.
Hybridization
Cancer cells
RNA
cDNA with fluorescent bases
Noncancer cells
3 Cancer and noncancer cells are removed from 78 women with breast cancer.
4 Messenger RNA from the cells…
5 …is converted into cDNA and labeled with red (cancer cells) or green (noncancer cells) fluorescent nucleotides.
6 The cDNAs are mixed…
7 …and hybridized to DNA probes on a chip.
8 The chip is scanned spot by spot. Yellow fluorescence (red + green) indicates equal expression of the gene in both types of cells; red indicates more expression in cancer cells; and green indicates more expression in noncancer cells.
Results
Each spot represents the expression of one gene in one patient's tumor compared with the expression of that gene in her noncancer cells.
Tumors above the solid yellow line came primarily from patients who remained cancer free for at least 5 years.
Tumors below the solid yellow line came primarily from patients in whom the cancer spread within 5 years of diagnosis.
Conclusion: Seventy genes were identified whose expression patterns accurately predicted the recurrence of breast cancer within 5 years of treatment.

15.7 Microarrays can be used to examine gene expression associated with disease progression. Each row in the microarray represents a tumor from one patient. [Reprinted by permission from Macmillan Publishers Ltd. Van'T Veer, Laura J., et al., "Gene expression profiling predicts clinical outcome of breast cancer," Nature 415:532. © 2002, permission conveyed through Copyright Clearance Center, Inc]

equal expression in both types of cells (equal hybridization of red- and green-labeled cDNAs), and no color indicated no expression in either type of cell. In 34 of the 78 women, the cancer later spread to other sites; the other 44 women remained free of breast cancer for five years after their initial diagnoses. The researchers identified a subset of 70 genes whose expression patterns in the initial tumors accurately predicted whether the cancer would later spread (see Figure 15.7). This degree of prediction was much higher than that of traditional predictive measures, which are based on the size and appearance of the tumor.

Researchers have also used microarrays to examine the expression of microRNAs (miRNAs) in human cancers. Recent research indicates that miRNAs are frequently expressed abnormally in cancerous tissue and may contribute to the progression of cancer (see Chapter 16). For example, one study using microarrays found that several miRNAs were overexpressed in cancerous cervical tissue

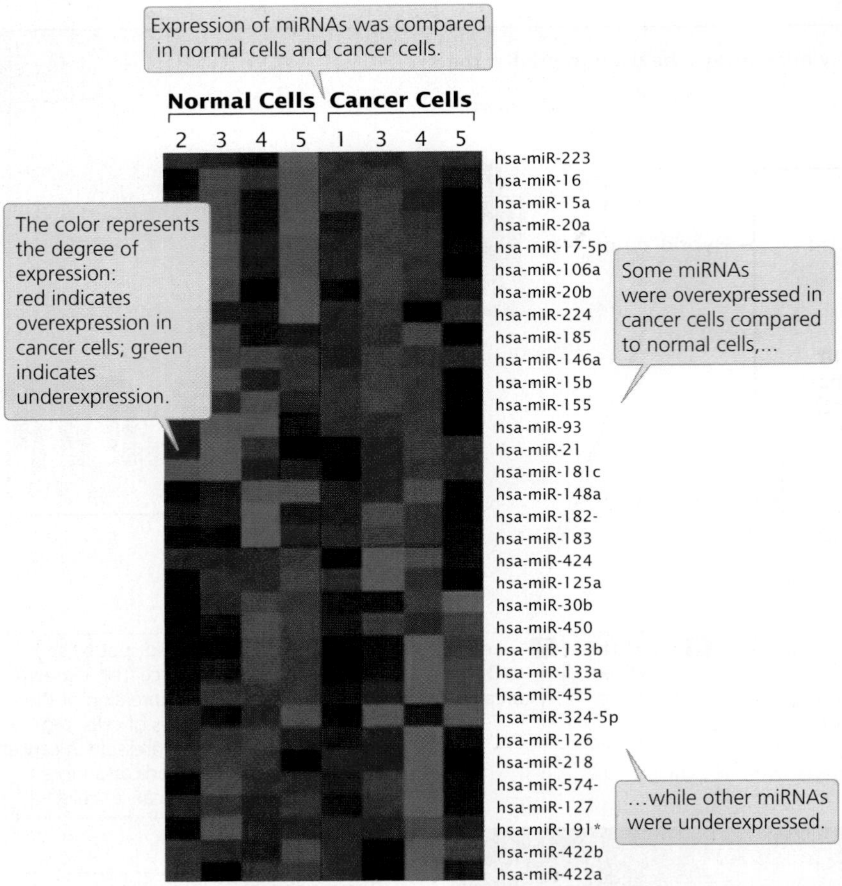

15.8 Microarrays have been used to compare the expression of miRNAs in cancerous cervical cells compared to normal cervical cells. [Adapted from Wang X, Tang S, Le S-Y, Lu R, Rader JS, et al. (2008) Aberrant Expression of Oncogenic and Tumor-Suppressive MicroRNAs in Cervical Cancer Is Required for Cancer Cell Growth. PLoS ONE 3(7): e2557. doi:10.1371/journal.pone.0002557.]

compared with normal cervical tissue, while other miRNAs were underexpressed (**Figure 15.8**). Other studies using microarrays have demonstrated that miRNA expression is associated with resistance of tumors to chemotherapy and radiation, and that miRNA expression can be used to predict the responses of some tumors to cancer treatment. Results such as these suggest that gene-expression data obtained from microarrays can be a powerful tool in cancer research and treatment, and the products of genes that show differences in expression are being examined as possible targets for drug therapy. ▶TRY PROBLEM 18

CONCEPTS

Microarrays, consisting of DNA probes attached to a solid support, can be used to determine which RNA molecules are being synthesized and can thus be used to examine changes in gene expression.

RNA Sequencing

Another approach to the study of gene expression has been made possible by the development of rapid, low-cost next-generation sequencing methods. This approach determines the presence of RNA molecules in a cell by sequencing cDNAs copied from cellular RNA molecules. Termed **RNA sequencing**, or RNA-Seq, this approach provides detailed information about gene expression, including the types and number of RNA molecules produced by transcription and the presence of alternatively processed RNA.

RNA sequencing is now widely used to examine patterns of gene expression. An early application of this method was used to create a transcriptome map of yeast and revealed, surprisingly, that 75% of the nonrepetitive sequences in the yeast genome are transcribed. Methods have been developed for isolating and sequencing RNA from single cells (termed single-cell RNA-Seq) so that differences in gene expression among cells can be studied.

15.3 Comparative Genomics Studies How Genomes Evolve

Genome sequencing projects provide detailed information about gene content and organization in different species and even in different members of the same species, allowing inferences about how genes function and how genomes evolve. They also provide important information about evolutionary relationships among organisms and about factors that influence the speed and direction of evolution. **Comparative genomics** is the field of genomics that studies similarities and differences in gene content, function, and organization among genomes of different organisms.

Prokaryotic Genomes

Thousands of prokaryotic genomes have now been sequenced. Most prokaryotic genomes consist of a single circular chromosome. However, there are exceptions, such as *Vibrio cholerae*, the bacterium that causes cholera, which has two circular chromosomes, and *Borrelia burgdorferi*, which has 1 large linear chromosome and 21 smaller chromosomes.

GENOME SIZE AND NUMBER OF GENES The total amount of DNA in prokaryotic genomes ranges from 490,885 bp in *Nanoarchaeum equitans*, an archaean that lives entirely within another archaean, to 9,105,828 bp in *Bradyrhizobium japonicum*, a soil bacterium (**Table 15.1**). Although this range in genome size might seem extensive, it is much less than the enormous range of genome sizes seen in eukaryotes, which can vary from a few million base pairs to hundreds of billions of base pairs. Most prokaryotic genomes

TABLE 15.1 Characteristics of representative prokaryotic genomes that have been completely sequenced

Species	Size (millions of base pairs)	Number of predicted genes
Archaea		
Archaeoglobus fulgidus	2.18	2407
Methanobacterium thermoautotrophicum	1.75	1869
Nanoarchaeum equitans	0.490	536
Bacteria		
Bacillus subtilis	4.21	4100
Bradyrhizobium japonicum	9.11	8317
Escherichia coli	4.64	4289
Haemophilus influenzae	1.83	1709
Mycobacterium tuberculosis	4.41	3918
Mycoplasma genitalium	0.58	480
Staphylococcus aureus	2.88	2697
Vibrio cholerae	4.03	3828

Source: Data from the Genome Atlas of the Center for Biological Sequence Analysis, http://www.cbs.dtu.dk/services/GenomeAtlas/.

are several million base pairs in size. *Escherichia coli*, the bacterium most widely used for genetic studies, has a fairly typical genome size at 4.6 million base pairs. Archaea and bacteria are similar in their ranges of genome size. Surprisingly, genome size shows extensive variation within some species; for example, different strains of *E. coli* vary in genome size by more than 1 million base pairs.

Among prokaryotes, the number of genes typically varies from 1000 to 2000, but some species have as many as 6700 and others as few as 480. Interestingly, the density of genes is rather constant across all species, with an average of about 1 gene per 1000 base pairs. Thus, prokaryotes with larger genomes have more genes, in contrast to eukaryotes, for which there is little association between genome size and number of genes. The evolutionary factors that determine the sizes of genomes in prokaryotes (as well as in eukaryotes) are still largely unknown.

Only about half of the genes identified in prokaryotic genomes can be assigned a function (**Figure 15.9**). Almost a quarter of those genes have no significant sequence similarity to any known genes in other bacteria.

CONCEPTS

Comparative genomics compares the content and organization of whole-genome sequences from different organisms. Prokaryotic genomes are small, usually ranging from 1 million to 3 million base pairs of DNA and containing several thousand genes.

✓ **CONCEPT CHECK 2**

What is the relation between genome size and gene number in prokaryotes?

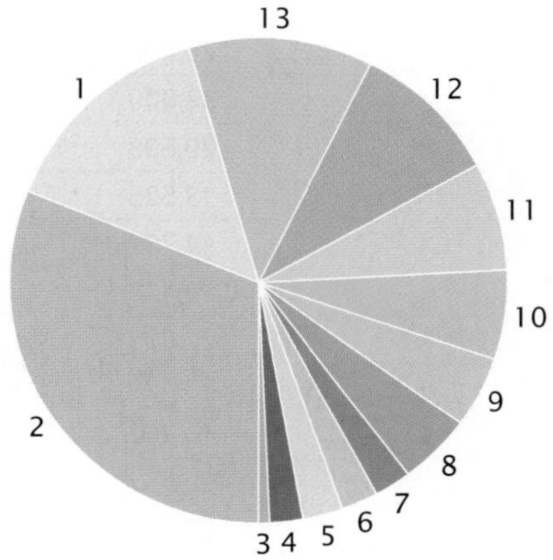

1. Metabolism
2. Unknown
3. Ionic homeostasis
4. Protein synthesis
5. Energy
6. Transport facilitation
7. Cellular biogenesis
8. Intracellular transport
9. Protein destination
10. Cellular communication and signal transduction
11. Cell rescue, defense, cell death, and aging
12. Cell growth, cell division, and DNA synthesis
13. Transcription

15.9 The functions of many genes in prokaryotes cannot be determined by comparison with genes in other prokaryotes. The proportion of the circle occupied by each color represents the proportion of genes affecting various known and unknown functions in *E. coli.*

Eukaryotic Genomes

The genomes of many eukaryotic species have been completely sequenced, including genomes of numerous fungi and protozoan species, cucumbers, papayas, corn, rice, sorghum, grapevine, strawberries, silkworms, a number of fruit flies, aphids, mosquitoes, anemones, tunicates, ctenophores, mice, rats, dogs, pigs, sheep, cows, horses, minke whales, over 200 species of birds, orangutans, gorillas, chimpanzees, humans, and many more species. Even the genomes of some extinct organisms have now been sequenced, including those of the woolly mammoth and Neanderthals. It is important to note that, even though the genomes of these organisms have been "completely sequenced," many of the final assembled sequences contain gaps, and regions of heterochromatin may not have been sequenced at all. Thus, the sizes of eukaryotic genomes are often estimates, and the number of base pairs given for the genome size of a particular species may vary. Predicting the number of genes that are present in a genome also is difficult and may vary, depending on the assumptions made and the particular gene-finding software used.

GENOME SIZE AND NUMBER OF GENES The genomes of eukaryotic organisms (**Table 15.2**) are larger than those of prokaryotes, and in general, multicellular eukaryotes have more DNA than do simple, single-celled eukaryotes such as yeast. However, there is no close relation between genome size and complexity among the multicellular eukaryotes. For example, the mosquito (*Anopheles gambiae*) and fruit fly (*Drosophila melanogaster*) are both insects with similar structural complexity, yet the mosquito has 60% more DNA than the fruit fly.

In general, eukaryotic genomes also contain more genes than do the genomes of prokaryotes (although some large bacteria have more genes than single-celled yeast), and the genomes of multicellular eukaryotes have more genes than do the genomes of single-celled eukaryotes. In contrast to prokaryotes, there is no correlation between genome size and the number of genes in eukaryotes. Nor is the number of genes among multicellular eukaryotes obviously related to phenotypic complexity: humans have more genes than do invertebrates, but only twice as many as fruit flies and fewer than the plant *A. thaliana*. The nematode *C. elegans* has more genes than *D. melanogaster*, but is less complex. The pufferfish has only about one-tenth the amount of DNA present in humans and mice, but about as many genes.

Eukaryotic genomes contain multiple copies of many genes, indicating that gene duplication has been an important process in genome evolution. Many genes in eukaryotes are interrupted by introns. In the more complex eukaryotes, both the number and the length of the introns are greater.

NONCODING DNA Most eukaryotic organisms contain vast amounts of DNA that do not encode proteins. For example, only about 1.5% of the human genome consists of DNA that directly specifies the amino acids of proteins. The function of the remaining DNA sequences, called noncoding DNA, has long been in question. Some research has suggested that much of the genome is "junk DNA" with no function. For example, Marcelo Nóbrega and his colleagues genetically engineered mice that were missing a large chromosomal region with no protein-encoding genes (called a **gene desert**). In one experiment, they created mice that were missing a 1,500,000-base-pair gene desert from

TABLE 15.2 Characteristics of representative eukaryotic genomes that have been completely sequenced

Species	Genome size (millions of base pairs)	Number of predicted genes
Saccharomyces cerevisiae (yeast)	12	6,144
Physcomitrella patens (moss)	480	38,354
Arabidopsis thaliana (plant)	125	25,706
Zea mays (corn)	2,400	32,000
Caenorhabditis elegans (nematode)	103	20,598
Drosophila melanogaster (fruit fly)	170	13,525
Anopheles gambiae (mosquito)	278	14,707
Danio rerio (zebrafish)	1,465	22,409
Takifugu rubripes (tiger pufferfish)	329	22,089
Xenopus tropicalis (clawed frog)	1,510	18,429
Anolis carolinensis (anole lizard)	1,780	17,792
Mus musculus (mouse)	2,627	26,762
Pan troglodytes (chimpanzee)	2,733	22,524
Homo sapiens (human)	3,223	20,000

Source: Data from Ensembl website: http://useast.ensembl.org/index.html and plants.ensembl.org/index.html.

mouse chromosome 3; in another, they created mice missing an 845,000-base-pair gene desert from chromosome 19. Remarkably, these mice appeared healthy and were indistinguishable from control mice. The researchers concluded that large regions of the mammalian genome can be deleted without major phenotypic effects and may, in fact, be superfluous.

Other research, however, has suggested that gene deserts may contain sequences that have a functional role. For example, genome-wide association studies demonstrated that DNA sequences contained within a gene desert on human chromosome 9 are associated with coronary artery disease, and subsequent studies have demonstrated the presence of 33 enhancers in this gene desert.

In 2002, the Encyclopedia of DNA Elements (ENCODE) project was undertaken to determine whether noncoding DNA had any function. Researchers cataloged all nucleotides within the genome that provide some function, including sequences that encode proteins and RNA molecules and those that serve as control sites for gene expression. This 10-year project was carried out by a team of over 400 scientists from around the world. In a series of papers published in 2012, the ENCODE team concluded that at least 80% of the human genome is involved in some type of gene function. Many of the functional sequences consisted of sites where proteins bind and influence the expression of genes. The ENCODE study suggests that there is little nonfunctional DNA in the human genome, but other researchers have questioned this conclusion. TRY PROBLEM 21

TRANSPOSABLE ELEMENTS A substantial part of the genomes of most multicellular organisms consists of moderately and highly repetitive sequences (see Chapter 8), and the percentage of repetitive sequences is usually higher in those species with larger genomes (**Table 15.3**). Most of these repetitive sequences appear to have arisen through transposition. In the human genome, 45% of the DNA is derived from transposable elements, many of which are defective and no longer able to move. In corn, 85% of the genome is derived from transposable elements.

TABLE 15.3 Percentage of genome consisting of interspersed repeats derived from transposable elements

Organism	Percentage of genome
Arabidopsis thaliana (plant)	10.5
Zea mays (corn)	85.0
Caenorhabditis elegans (nematode)	6.5
Drosophila melanogaster (fruit fly)	3.1
Takifugu rubripes (tiger pufferfish)	2.7
Homo sapiens (human)	44.4

PROTEIN DIVERSITY In spite of only a modest increase in gene number, vertebrates have considerably more protein diversity than do invertebrates. One way to measure protein diversity is by counting the number of protein domains, which are characteristic parts of proteins that are often associated with a function. Vertebrate genomes do not encode more protein domains than do invertebrate genomes; for example, there are 1262 domains in humans, compared with 1035 in fruit flies. However, the existing domains in humans are assembled into more combinations, leading to many more types of proteins.

CONCEPTS

Genome size varies greatly among eukaryotic species. For multicellular eukaryotic organisms, there is no clear relation between organismal complexity and amount of DNA or gene number. A substantial part of the genome in eukaryotic organisms consists of repetitive DNA, much of which is derived from transposable elements.

The Human Genome

The human genome, which is fairly typical of mammalian genomes, has been extensively studied and analyzed because of its importance to human health and evolution. It is 3.2 billion base pairs in length, but only about 1.5% of it encodes proteins. Active genes are often separated by vast regions of noncoding DNA, much of which consists of repeated sequences derived from transposable elements.

The average gene in the human genome is approximately 27,000 bp in length, with about 9 exons. (One exceptional gene has 234 exons.) The introns of human genes are much longer, and there are more of them, than in other genomes (**Figure 15.10**). The human genome does not encode substantially more protein domains, but its domains are combined in more ways to produce a relatively diverse proteome. Gene functions encoded by the human genome are presented in **Figure 15.11**. As in bacteria, the functions of

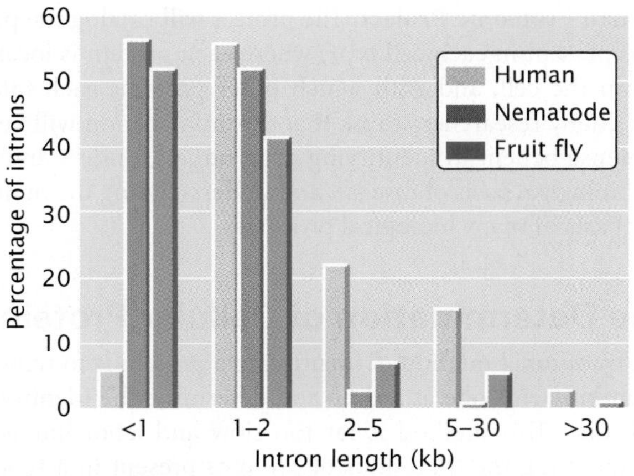

15.10 The introns of genes in humans are generally longer than the introns of genes in nematodes and fruit flies.

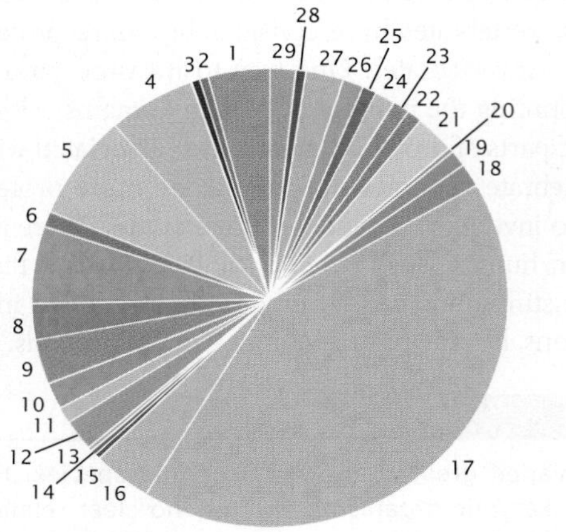

15.11 Functions for many human genes have yet to be determined. The proportion of the circle occupied by each color represents the proportion of genes affecting various known and unknown functions.

many genes in the human genome are still unknown. A single gene often encodes multiple proteins through alternative splicing; each gene encodes, on average, two or three different mRNAs.

15.4 Proteomics Analyzes the Complete Set of Proteins Found in a Cell

DNA sequence data offer tremendous insight into the biology of an organism, but they are not the whole story. In recent years, molecular biologists have turned their attention to the analysis of the protein content of cells. Their ultimate goal is to determine the proteome: the complete set of proteins found in a given cell. The study of the proteome is termed **proteomics**.

Plans are under way to identify and characterize all proteins in the human body, an effort that has been called the **Human Proteome Project**. The project will catalog the proteins present in each cell type, where each protein is located within the cell, and with which other proteins each interacts. Many researchers think that this information will be of immense benefit in identifying drug targets, understanding the biological basis of disease, and understanding the molecular basis of many biological processes.

The Determination of Cellular Proteins

The traditional method of identifying a protein is to remove its amino acids one at a time and determine the identity of each one. This method is far too slow and labor-intensive for analyzing the thousands of proteins present in a typical cell. Today, researchers use **mass spectrometry**, which is a method for precisely determining the molecular mass of a molecule. In mass spectrometry, a molecule is ionized and its migration rate in an electrical field is determined. Because small molecules migrate more rapidly than do larger molecules, the migration rate can accurately determine the mass of the molecule.

To analyze a protein with mass spectrometry, researchers break it into small peptide fragments using a protein-digesting enzyme (**Figure 15.12a**). Mass spectrometry is then used to separate those peptides on the basis of their mass-to-charge (*m*/*z*) ratio (**Figure 15.12b**). This separation produces a profile of peaks, in which each peak corresponds to the mass-to-charge ratio of one peptide (**Figure 15.12c**). A computer program then searches through a database of proteins to find a match between the profile generated and the profile of a known protein (**Figure 15.12d**). Using bioinformatics, the computer creates "virtual digests" and predicts the profiles of all proteins found in a genome, given the DNA sequences of the protein-encoding genes.

Mass spectrometric methods can also be used to measure the amount of each protein identified. In this procedure, a complex mixture of proteins (such as those from a tissue sample) is digested and analyzed with mass spectrometry. The computer program then sorts out the proteins present in the sample by analyzing the peptide profiles.

Mary Lipton and her colleagues used this approach to study the proteome of *Deinococcus radiodurans*, an exceptional bacterium that is able to withstand high doses of ionizing radiation that are lethal to all other organisms. The genome of *D. radiodurans* had already been sequenced. Lipton and her colleagues extracted proteins from the bacterium, broke them up into small peptide fragments, separated the fragments, and then used mass spectrometry to determine the proteins from the peptide fragments. They were able to identify 1910 proteins, which represented more than 60% of the proteins predicted on the basis of the genome sequence.

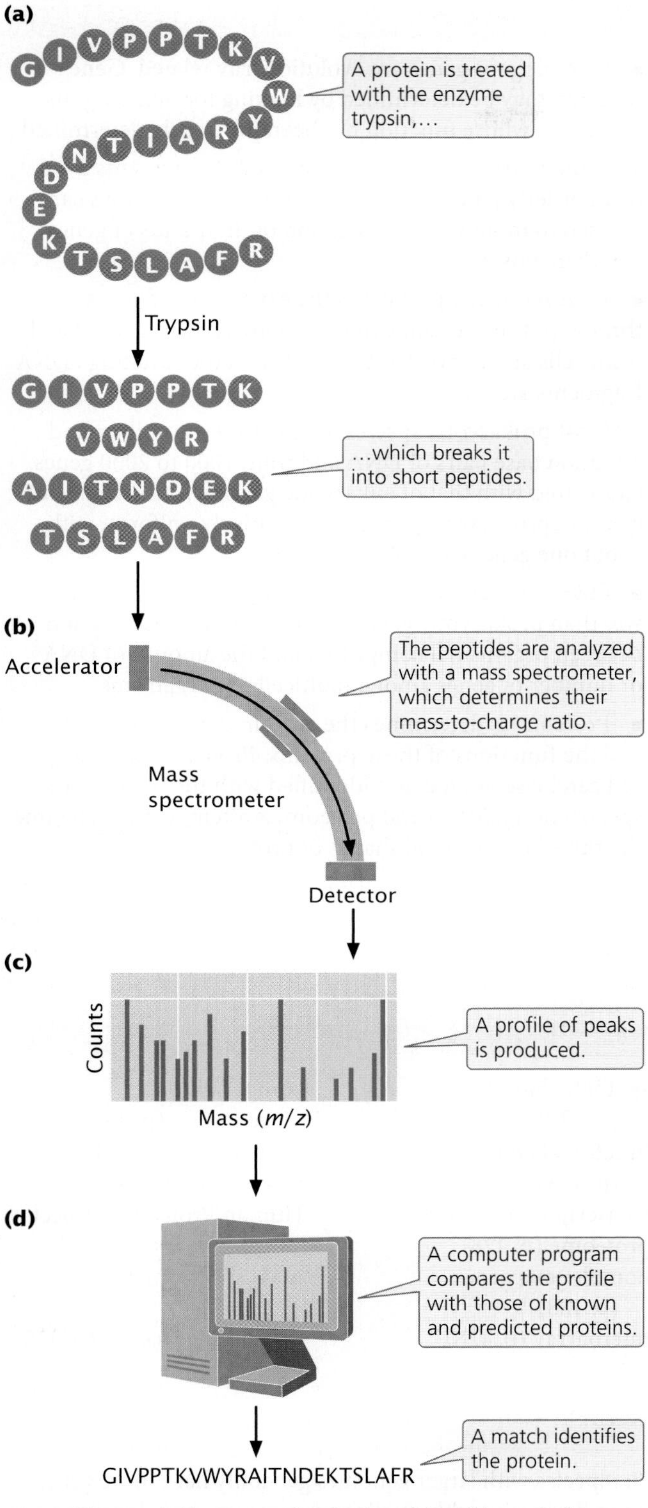

15.12 Mass spectrometry is used to identify proteins.

Protein Microarrays

Protein–protein interactions can be analyzed with the use of **protein microarrays**, which are similar to the microarrays used for examining gene expression. With this technique, a large number of different proteins are applied to a solid support as an orderly array of spots, with each spot containing a different protein. In one application, each spot contains an antibody for a different protein and is labeled with a tag that fluoresces when bound. An extract of tissue is applied to the protein microarray. A spot of fluorescence appears when a protein in the extract binds to an antibody, indicating the presence of that particular protein in the tissue.

Structural Proteomics

The high-resolution structure of a protein provides a great deal of useful information. It is often a source of insight into the function of an unknown protein; it may also suggest the locations of active sites and provide information about other molecules that interact with the protein. Furthermore, knowledge of a protein's structure often suggests targets for potential drugs that might interact with the protein. Therefore, one goal of proteomics is to determine the structure of every protein found in a cell.

Two procedures are currently used to solve the structures of complex proteins: (1) X-ray crystallography, in which crystals of the protein are bombarded with X-rays and the diffraction patterns of the X-rays are used to determine the structure (see Chapter 8); and (2) nuclear magnetic resonance (NMR), which provides information on the positions of specific atoms within a molecule by using the magnetic properties of their nuclei.

Both X-ray crystallography and NMR require human intervention at many stages and are too slow for determining the structures of the thousands of proteins that may exist within a cell. Because the structures of hundreds of thousands of proteins are required for studies of the proteome, researchers ultimately hope to be able to predict the structure of a protein from its amino acid sequence. That is not possible at the present time, but the hope is that, if enough high-resolution protein structures are solved, it will be possible in the future to model the structures from the amino acid sequence alone. As scientists work on automated methods that will speed the structural determination of proteins, bioinformaticists are developing better computer programs for predicting protein structure from sequence. TRY PROBLEM 22

CONCEPTS

The proteome is the complete set of proteins found in a cell. Techniques of protein separation and mass spectrometry are used to identify the proteins present within a cell. Microarrays are used to identify sets of interacting proteins. Structural proteomics attempts to determine the structures of all proteins.

CONCEPTS SUMMARY

- Genomics is the field of genetics that attempts to understand the content, organization, and function of the genetic information contained in whole genomes.
- Genetic maps position genes relative to other genes by determining rates of recombination and are measured in recombination frequencies. Physical maps are based on the physical distances between genes and are measured in base pairs.
- The Human Genome Project was an effort to determine the entire sequence of the human genome.
- Sequencing a whole genome requires breaking it into small overlapping fragments whose DNA sequences can be determined in sequencing reactions. The individual sequences can be ordered into a whole-genome sequence using a map-based approach, in which fragments are assembled in order using previously created genetic and physical maps, or using a whole-genome shotgun approach, in which overlap between fragments is used to assemble them into a whole-genome sequence. Today, almost all genomes are sequenced using whole-genome shotgun sequencing.
- Single-nucleotide polymorphisms are single-base differences in DNA between individual organisms and are valuable as markers in linkage studies.
- Bioinformatics is a synthesis of molecular biology and computer science that develops tools to store, retrieve, and analyze DNA-, RNA-, and protein-sequence data.
- Metagenomics studies the genomes of entire groups of organisms. Synthetic biology is developing techniques for creating novel genomes and organisms.
- Homologous genes are evolutionarily related. Gene function may be determined by looking for homologous sequences whose function has been previously determined.
- A microarray consists of known DNA fragments fixed in an orderly pattern to a solid support. Microarrays can be used to monitor the expression of thousands of genes simultaneously.
- RNA sequencing analyzes the expression of genes throughout the genome. In this approach, RNA is isolated from cells and converted to cDNA, and the resulting cDNA fragments are sequenced.
- Most prokaryotic species have between 1 million and 3 million base pairs of DNA and from 1000 to 2000 genes. Compared with that of eukaryotic genomes, the density of genes in prokaryotic genomes is relatively uniform, with about one gene per 1000 base pairs.
- Eukaryotic genomes are larger and more variable in size than prokaryotic genomes. There is no clear relation between organismal complexity and the amount of DNA or number of genes among multicellular organisms.
- Proteomics determines the protein content of a cell and the functions of those proteins. Proteins within a cell can be separated and identified with the use of mass spectrometry. Structural proteomics attempts to determine the three-dimensional shapes of proteins.

IMPORTANT TERMS

genomics (p. 420)
structural genomics (p. 420)
genetic map (p. 420)
physical map (p. 421)
map-based sequencing (p. 422)
contig (p. 422)
whole-genome shotgun sequencing (p. 423)
single-nucleotide polymorphism (SNP) (p. 425)
haplotype (p. 425)
genome-wide association study (p. 426)
bioinformatics (p. 426)
metagenomics (p. 427)
microbiome (p. 427)
synthetic biology (p. 427)
functional genomics (p. 428)
transcriptome (p. 428)
proteome (p. 428)
homologous genes (p. 428)
microarray (p. 428)
comparative genomics (p. 430)
gene desert (p. 432)
proteomics (p. 434)
Human Proteome Project (p. 434)
mass spectrometry (p. 434)
protein microarray (p. 435)

ANSWERS TO CONCEPT CHECKS

1. b

2. Species with larger genomes generally have more genes than do species with smaller genomes, so gene density is relatively constant.

WORKED PROBLEM

A linear piece of DNA that is 30 kb long is first cut with *Bam*HI, then with *Hpa*II, and finally with both *Bam*HI and *Hpa*II together. Fragments of the following sizes were obtained from these reactions:

*Bam*HI: 20-kb, 6-kb, and 4-kb fragments

*Hpa*II: 21-kb and 9-kb fragments

*Bam*HI and *Hpa*II: 20-kb, 5-kb, 4-kb, and 1-kb fragments

Draw a restriction map of the 30-kb piece of DNA, indicating the locations of the *Bam*HI and *Hpa*II restriction sites.

Solution Strategy

What information is required in your answer to the problem?

A map that includes the number and relative locations of restriction sites for *Bam*HI and *Hpa*II and the distances in base pairs between the sites.

What information is provided to solve the problem?

- The piece of DNA is 30 kb long.
- The sizes of the fragments produced when the DNA is cut with *Bam*HI, with *Hpa*II, and with both enzymes together.

For help with this problem, review:

Physical Maps in Section 15.1.

Solution Steps

e: This problem e solved correctly gh a variety roaches; this on applies one le approach.

t: For linear the number triction sites e less than the er of fragments uced.

t: Look for frag- ts in the double st that sum to length of a frag- t present in the le digest.

When cut by *Bam*HI alone, the linear piece of DNA is cleaved into three fragments; so there must be two *Bam*HI restriction sites. When cut with *Hpa*II alone, a clone of the same piece of DNA is cleaved into only two fragments; so there is a single *Hpa*II site.

Let's begin to determine the locations of these sites by examining the *Hpa*II fragments. Notice that the 21-kb fragment produced when the DNA is cut by *Hpa*II is not present in the fragments produced when the DNA is cut by *Bam*HI and *Hpa*II together (the double digest); this result indicates that the 21-kb *Hpa*II fragment has within it a *Bam*HI site. If we examine the fragments produced by the double digest, we see that the 20-kb and 1-kb fragments sum to 21 kb; so a *Bam*HI site must be 20 kb from one end of the fragment and 1 kb from the other end.

Similarly, we see that the 9-kb *Hpa*II fragment does not appear in the double digest and that the 5-kb and 4-kb fragments in the double digest add up to 9 kb; so another *Bam*HI site must be 5 kb from one end of this fragment and 4 kb from the other end.

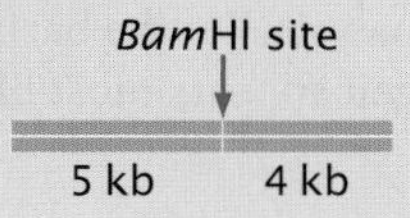

Now, let's examine the fragments produced when the DNA is cut by *Bam*HI alone. The 20-kb and 4-kb fragments are also present in the double digest; so neither of these fragments contains an *Hpa*II site. The 6-kb fragment, however, is not present in the double digest, and the 5-kb and 1-kb fragments in the double digest sum to 6 kb; so this fragment contains an *Hpa*II site that is 5 kb from one end and 1 kb from the other end.

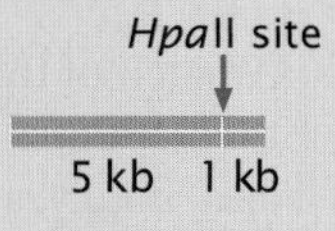

We have accounted for all the restriction sites, but we must still determine the order of the sites on the original 30-kb fragment. Notice that the 5-kb fragment must be adjacent to both the 1-kb and the 4-kb fragments; so it must be in between those two fragments.

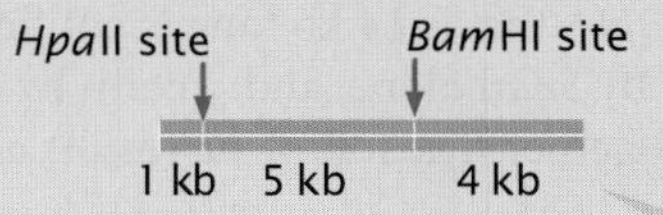

Hint: Two fragments in the double digest that were produced by cutting a fragment in the single digest must be adjacent to one another.

We have also established that the 1-kb and 20-kb fragments are adjacent; because the 5-kb fragment is on one side, the 20-kb fragment must be on the other, completing the restriction map:

COMPREHENSION QUESTIONS

Section 15.1

1. What is the difference between a genetic map and a physical map? Which generally has higher resolution and accuracy and why?
2. What is the difference between a map-based approach to sequencing a whole genome and a whole-genome shotgun approach?
3. What is a single-nucleotide polymorphism (SNP)? How are SNPs used in genomic studies?
4. What is a haplotype?
5. How is a genome-wide association study carried out?
6. Give some examples of important findings from metagenomic studies.

Section 15.2

7. What is a microarray? How can it be used to obtain information about gene function?

Section 15.3

8. What is the relation between genome size and gene number in prokaryotes?
9. DNA content varies considerably among different multicellular organisms. Is this variation closely related to the number of genes and the complexity of the organism? If not, what accounts for the variation?
10. What was the focus of the ENCODE project?

Section 15.4

11. How does proteomics differ from genomics?
12. How is mass spectrometry used to identify proteins in a cell?

For more questions that test your comprehension of the key chapter concepts, go to **LearningCurve** for this chapter.

APPLICATION QUESTIONS AND PROBLEMS

Section 15.1

13. A 22-kb piece of DNA has the following restriction sites:

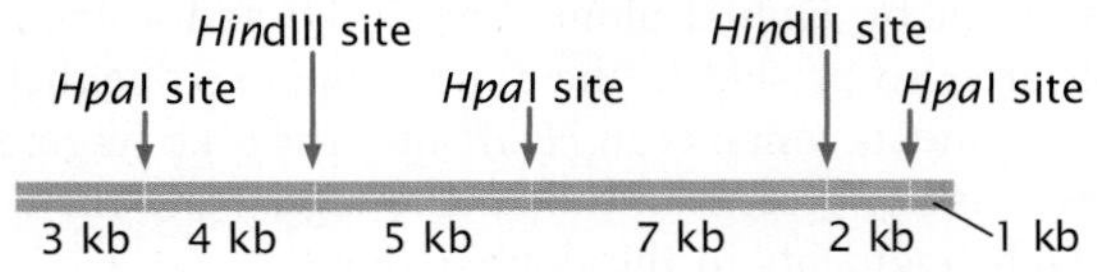

A batch of this DNA is first fully digested by *Hpa*I alone, then another batch is fully digested by *Hin*dIII alone, and, finally, a third batch is fully digested by both *Hpa*I and *Hin*dIII together. The fragments resulting from each of the three digestions are placed in separate wells of an agarose gel, separated by gel electrophoresis, and stained by ethidium bromide. Draw the bands as they would appear on the gel.

*14. A linear piece of DNA that is 14 kb long is cut first by *Eco*RI alone, then by *Sma*I alone, and, finally, by both *Eco*RI and *Sma*I together. The following results are obtained:

Digestion by *Eco*RI alone	Digestion by *Sma*I alone	Digestion by both *Eco*RI and *Sma*I
3-kb fragment	7-kb fragment	2-kb fragment
5-kb fragment	7-kb fragment	3-kb fragment
6-kb fragment		4-kb fragment
		5-kb fragment

Draw a map of the *Eco*RI and *Sma*I restriction sites on this 14-kb piece of DNA, indicating the relative positions of the restriction sites and the distances between them.

*15. A linear piece of DNA was broken into random, overlapping fragments and each fragment was sequenced. The sequence of each fragment is shown below.

Fragment 1: 5′—TAGTTAAAAC—3′
Fragment 2: 5′—ACCGCAATACCCTAGTTAAA—3′
Fragment 3: 5′—CCCTAGTTAAAAC—3′
Fragment 4: 5′—ACCGCAATACCCTAGTT—3′
Fragment 5: 5′—ACCGCAATACCCTAGTTAAA—3′
Fragment 6: 5′—ATTTACCGCAAT—3′

On the basis of overlap in sequence, assemble the fragments into a contig.

16. DATA ANALYSIS In recent years, honeybee colonies throughout North America have been decimated by colony collapse disorder (CCD), which results in the rapid death of worker bees. First noticed by beekeepers in 2004, the disorder has been responsible for the loss of 50% to 90% of beekeeping operations in the United States. Evidence suggests that CCD is caused by a pathogen. Diana Cox-Foster and her colleagues (D. Cox-Foster et al. 2007. *Science* 318:283–287) used a metagenomic approach to try to identify the causative agent of CCD by isolating DNA from normal honeybee hives and from hives that had experienced CCD. A number of different bacteria, fungi, and viruses were identified in the metagenomic analysis. The following table gives the percentages of CCD hives and non-CCD hives that tested positive for four potential pathogens identified in the metagenomic analysis. On the basis of these data, which potential pathogen appears most likely to be responsible

for CCD? Explain your reasoning. Do these data prove that this pathogen is the cause of CCD? Explain.

Potential pathogen	CCD hives infected ($n = 30$)	Non-CCD hives infected ($n = 21$)
Israeli acute paralysis virus	83.3%	4.8%
Kashmir bee virus	100%	76.2%
Nosema apis	90%	47.6%
Nosema cernae	100%	80.8%

17. James Noonan and his colleagues (J. Noonan et al. 2005. *Science* 309:597–599) set out to study the genome sequence of an extinct species of cave bear. They extracted DNA from 40,000-year-old bones from a cave bear and used a metagenomic approach to isolate, identify, and sequence the cave-bear DNA. Why did they use a metagenomic approach when their objective was to sequence the genome of one species (the cave bear)?

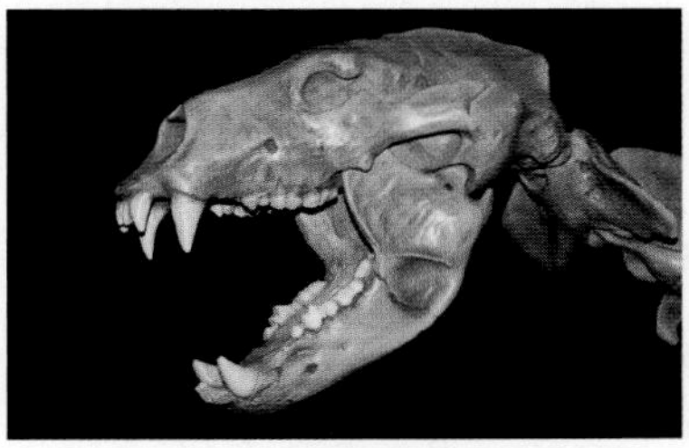

[Larry Miller/Science Source.]

Section 15.2

***18.** Microarrays can be used to determine relative levels of gene expression. In one type of microarray, hybridization of red (experimental) and green (control) cDNAs is proportional to the relative amounts of mRNA in the samples. Red indicates the overexpression of a gene and green indicates the underexpression of a gene in the experimental cells relative to the control cells, yellow indicates equal expression in experimental and control cells, and no color indicates no expression in either experimental or control cells.

In one experiment, mRNA from a strain of antibiotic-resistant bacteria (experimental cells) is converted into cDNA and labeled with red fluorescent nucleotides; mRNA from a nonresistant strain of the same bacteria (control cells) is converted into cDNA and labeled with green fluorescent nucleotides. The cDNAs from the resistant and nonresistant cells are mixed and hybridized to a chip containing spots of DNA from genes 1 through 25. The results are shown in the adjoining illustration. What conclusions can you draw about which genes might be implicated in antibiotic resistance in these bacteria? How might this information be used to design new antibiotics that are less vulnerable to resistance?

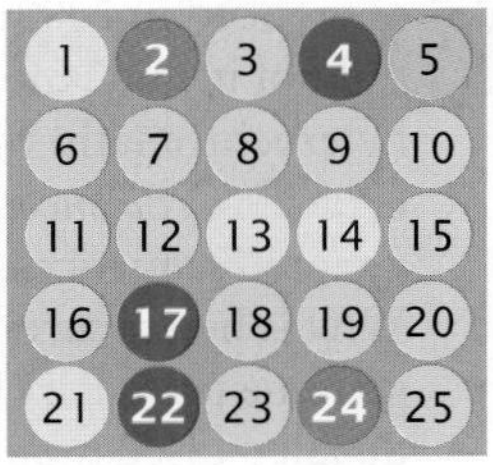

19. Of the genes illustrated in the microarray shown in the lower part of **Figure 15.7**, are most overexpressed or underexpressed in tumors from patients that remained cancer free for at least five years? Explain your reasoning.

Section 15.3

20. *Dictyostelium discoideum* is a soil-dwelling social amoeba: much of the time, the organism consists of single, solitary cells, but during times of starvation, amoebae come together to form aggregates that have many characteristics of multicellular organisms. Biologists have long debated whether *D. discoideum* is a unicellular or a multicellular organism. In 2005, its genome was completely sequenced. The table below lists some genomic characteristics of *D. discoideum* and other eukaryotes (L. Eichinger et al. 2005. *Nature* 435:43–57).

a. On the basis of the organisms other than *D. discoideum* listed in the table, what are some differences in genome characteristics between unicellular and multicellular organisms?

b. On the basis of these data, do you think that the genome of *D. discoideum* is more like those of unicellular eukaryotes or more like those of multicellular eukaryotes? Explain your answer.

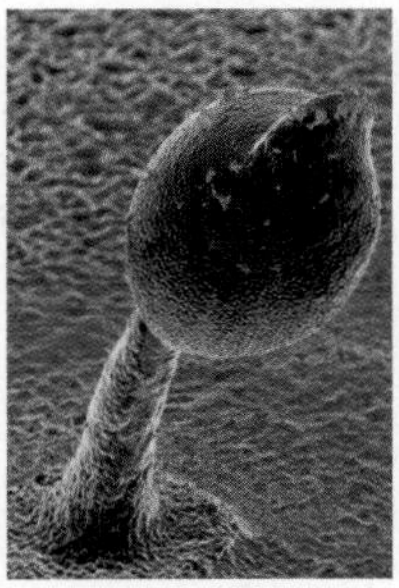

Dictyostelium discoideum. [David Scharf/Science Source.]

***21.** A group of 250 scientists sequenced and analyzed the genomes of 12 species of *Drosophila* (*Drosophila* 12 Genomes Consortium. 2007. *Nature* 450:203–218). Data on genome sizes and numbers of protein-encoding genes from this study are given in the accompanying table. Plot the number of protein-encoding genes as a function of genome size for the 12 species of *Drosophila*. Is there a relation between genome size and number of genes in fruit flies? How does this result compare with the relation between genome size and number of genes across all eukaryotes?

Table for Problem 20: Genomic characteristics of *D. discoideum* and other eukaryotes

Feature	*D. discoideum*	*P. falciparum*	*S. cerevisiae*	*A. thaliana*	*D. melanogaster*	*C. elegans*	*H. sapiens*
Organism	Amoeba	Malaria parasite	Yeast	Plant	Fruit fly	Nematode	Human
Cellularity	?	Uni	Uni	Multi	Multi	Multi	Multi
Genome size (millions of base pairs)	34	23	13	125	180	103	2,851
Number of genes	12,500	5,268	5,538	25,498	13,676	19,893	22,287
Average gene length (bp)	1,756	2,534	1,428	2,036	1,997	2,991	27,000
Genes with introns (%)	69	54	5	79	38	5	85
Mean number of introns	1.9	2.6	1.0	5.4	4.0	5.0	8.1
Mean intron size (bp)	146	179	nd*	170	nd*	270	3,365
Mean G + C (exons)	27%	24%	28%	28%	55%	42%	45%

*nd = Not determined.

Table for Problem 21: Characteristics of 12 *Drosophila* species genomes

Species	Genome size (millions of base pairs)	Number of protein-encoding genes
D. melanogaster	200	13,733
D. simulans	162	15,983
D. sechellia	171	16,884
D. yakuba	190	16,423
D. erecta	135	15,324
D. ananassae	217	15,276
D. pseudoobscura	193	16,363
D. persimilis	193	17,325
D. willistoni	222	15,816
D. virilis	364	14,680
D. mojavensis	130	14,849
D. grimshawi	231	15,270

Section 15.4

*22. A scientist determines the complete genomes and proteomes of a liver cell and a muscle cell from the same person. Would you expect bigger differences in the genomes or in the proteomes of these two cell types? Explain your answer.

CHALLENGE QUESTIONS

Section 15.1

23. Some synthetic biologists have proposed the creation of an entirely new, free-living organism with a minimal genome—the smallest set of genes that allows for replication of the organism in a particular environment. This genome could be used to design and create, from "scratch," novel organisms that might perform specific tasks such as the breakdown of toxic materials in the environment.

   a. How might the minimal genome required for life be determined?

   b. What, if any, social and ethical concerns might be associated with the construction of an entirely new organism with a minimal genome?

24. DATA ANALYSIS The genome of the fruit fly *Drosophila melanogaster* was sequenced in 2000. However, this "completed" sequence did not include most heterochromatin regions. The heterochromatin was not sequenced until 2007 (Hoskins et al. 2007. *Science* 316:1625–1628). Most completed genome sequences do not include heterochromatin. Why is heterochromatin usually not sequenced in genome projects? (Hint: See Chapter 8 for a more detailed discussion of heterochromatin.)

THINK-PAIR-SHARE QUESTIONS

Introduction

1. What are some possible research questions and practical applications that could be addressed by creating organisms with artificial chromosomes and synthetic genomes? What might be some potential safety, environmental, social, and ethical concerns about creating organisms with synthetic genomes?

2. What do you think about taking a course that focuses entirely on building an artificial chromosome? Would this be an effective way to learn genetics? Propose some arguments for and against learning genetics in this way.

Section 15.1

3. Increasingly, whole-genome sequencing of individuals is being done to help identify and treat medical conditions. Genome sequencing invariably identifies a number of variations, some common and some rare, that might be clinically relevant. For example, suppose a person had their genome sequenced to help determine their risk for cardiovascular disease and, just by chance, the sequence revealed that they carry one or more variants that predispose them to cancer or Alzheimer disease. Does the sequencing laboratory or physician have an obligation to report this finding, which was not the purpose of the sequencing and which the patient did not request? What about reporting variants for which no or limited information can be provided about their clinical significance? Does the answer to this question differ for sequencing done on children?

4. As pointed out in the text, you and a complete stranger are 99.9% identical in DNA sequence. But you also differ at more than 3 million base pairs. Is this a large or a small difference? What are some of the consequences of these similarities and differences?

5. Researchers systematically replaced 414 essential genes in yeast with similar genes from humans. Almost half of these transplants (47%) were successful: cells with the humanized gene were able to function and grow. What does this observation tell us about differences between yeast and humans? How might this information be used?

16 Cancer Genetics

Palladin and the Spread of Cancer

Villa designed by Renaissance architect Andrea Palladio, for whom the *palladin* gene is named. The *palladin* gene encodes an essential component of a cell's cytoskeleton; when mutated, *palladin* contributes to the spread of pancreatic cancer. [Gianni Dagli Orti/The Art Archive at Art Resource, NY.]

Pancreatic cancer is among the most serious of all cancers. With about 53,670 new cases each year in the United States, it is only the twelfth most common form of the disease, but it is the third leading cause of death due to cancer, killing more than 43,000 people each year. Most people with pancreatic cancer survive less than 6 months after they are diagnosed; only 7% survive more than five years. A primary reason for pancreatic cancer's lethality is its propensity to spread rapidly to the lymph nodes and other organs. Most symptoms don't appear until the cancer is advanced and has invaded other organs. So what makes it so likely to spread?

In 2006, researchers identified a key gene that contributes to the development of pancreatic cancer—an important source of insight into the disease's aggressive nature. Geneticists at the University of Washington in Seattle found a unique family in which nine members over three generations had been diagnosed with pancreatic cancer (**Figure 16.1**). Nine additional family members had precancerous growths that were likely to develop into pancreatic cancer. In this family, pancreatic cancer was inherited as an autosomal dominant trait.

By using gene-mapping techniques, geneticists determined that the gene causing pancreatic cancer in the family was located within a region on the long arm of chromosome 4. Unfortunately, this region encompasses 16 million base pairs and includes 250 genes. To determine which of the 250 genes in the delineated region might be responsible for cancer in the family, researchers designed a unique microarray (see Chapter 15) that contained sequences from the region. They used this microarray to examine gene expression in pancreatic tumors and precancerous growths in family members, as well as in sporadic (nonhereditary) pancreatic tumors in other people and in normal pancreatic tissue from unaffected people. The researchers reasoned that the cancer gene might be overexpressed or underexpressed in the tumors relative to normal tissue. Data from the microarray revealed that the most overexpressed gene in the pancreatic tumors and precancerous growths was a gene encoding a critical component of the cytoskeleton, referred to as the *palladin* gene. Sequencing demonstrated that those members of the family with pancreatic cancer all had an identical mutation in exon 2 of the *palladin* gene.

The *palladin* gene is named for Renaissance architect Andrea Palladio because it plays a central role in the architecture of the cell. The Palladin protein functions as a scaffold for the binding of the other cytoskeletal proteins that are necessary for maintaining cell shape, movement, and differentiation. The ability of a cancer cell to spread is directly related to its

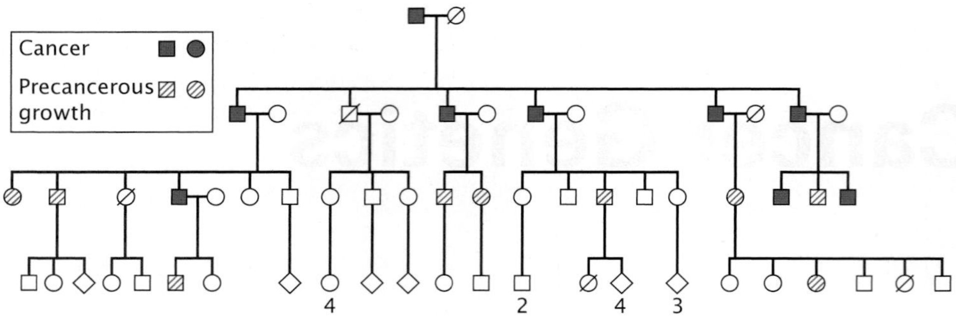

16.1 Pancreatic cancer is inherited as an autosomal dominant trait in a family that possesses a mutated *palladin* gene. [After K. L. Pogue et al., *PLoS Medicine* 3:2216–2228, 2006.]

cytoskeleton: cells that spread typically have poor cytoskeletal architecture, which enables them to detach easily from a primary tumor mass and migrate through other tissues. To determine whether mutations in the *palladin* gene affect cell mobility, researchers genetically engineered cells with a mutated copy of the *palladin* gene and tested the ability of these cells to migrate. The cells with a mutated copy of the *palladin* gene were 33% more efficient at migrating than were cells with the normal gene, demonstrating that the *palladin* gene contributes to the ability of pancreatic cancer cells to spread.

THINK-PAIR-SHARE Questions 1, 2, and 3

The discovery of *palladin*'s link to pancreatic cancer illustrates the power of modern molecular genetics for unraveling the biological nature of cancer. In this chapter, we examine the nature of cancer, a disease that is fundamentally genetic but is often not inherited. We begin by considering the nature of cancer and how multiple genetic alterations are required to transform a normal cell into a cancerous one. We then consider some of the types of genes that contribute to cancer, including oncogenes and tumor-suppressor genes, genes that control the cell cycle, genes encoding DNA-repair systems and telomerase, and genes that, like *palladin*, contribute to the spread of cancer. To illustrate the processes involved, we take a detailed look at how specific genes contribute to the progression of colon cancer. Next, we discuss chromosome mutations associated with cancer and genomic instability. Finally, we examine the role of viruses in some cancers.

16.1 Cancer Is a Group of Diseases Characterized by Cell Proliferation

About one of every five women and one of every four men in the United States will die from cancer, and cancer treatments cost billions of dollars per year. Cancer is not a single disease; rather, it is a heterogeneous group of disorders characterized by the presence of cells that do not respond to the normal controls on division. Cancer cells divide rapidly and continuously, creating tumors that crowd out normal cells and eventually rob healthy tissues of nutrients (**Figure 16.2**). The cells of an advanced tumor can separate from the tumor and travel to distant sites in the body, where they may take up residence and develop into new tumors. The most common cancers in the United States are those of the breast, prostate gland, lung, colon and rectum, and blood (**Table 16.1**).

THINK-PAIR-SHARE Question 4

Tumor Formation

Normal cells grow, divide, mature, and die in response to a complex set of internal and external signals. A normal cell receives both stimulatory and inhibitory signals, and its growth and division are regulated by a delicate balance between these opposing forces. In a cancer cell, one or more of these signals has been disrupted, which causes the cell

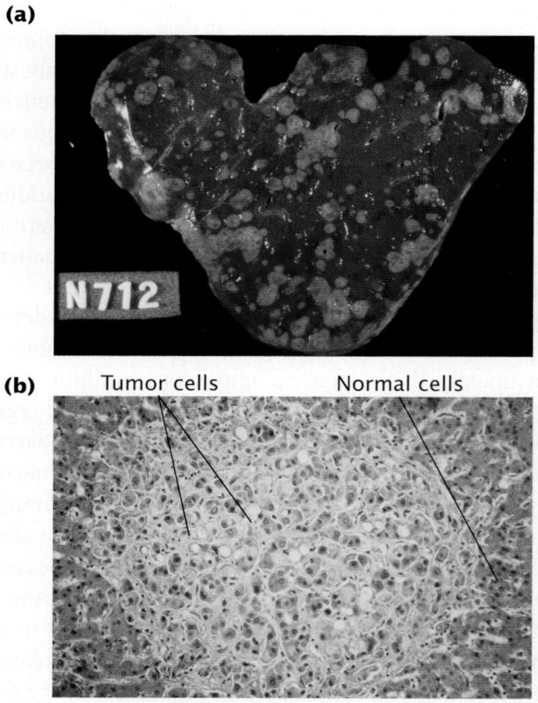

16.2 Abnormal proliferation of cancer cells produces a tumor that crowds out normal cells. (a) Metastatic breast cancer masses (white protrusions) growing in a human liver. (b) A light micrograph of a liver section with tumors. The cancer cells are the light, pale-stained cells; the darker cells are healthy liver cells. [CNRI/Science Source.]

TABLE 16.1 Estimated incidences of various cancers and cancer mortality in the United States in 2017

Type of cancer	New cases per year	Deaths per year
Breast	255,180	41,070
Lung and bronchus	222,500	155,870
Prostate	161,360	26,730
Colon and rectum	135,430	50,260
Melanoma	87,110	9,730
Lymphoma	80,500	21,210
Bladder	79,030	16,870
Thyroid	56,870	2,010
Kidney	63,990	14,400
Leukemia	62,130	24,500
Uterus	61,380	10,920
Pancreas	53,670	43,090
Oral cavity and pharynx	49,670	9,700
Liver	40,710	28,920
Myeloma	30,280	12,590
Stomach	28,000	10,960
Brain and nervous system	23,800	16,700
Ovary	22,440	14,080
Esophagus	16,940	15,690
Larynx	13,360	3,660
Uterine cervix	12,820	4,210
Cancers of soft tissues including heart	12,390	4,990
All cancers	1,688,780	600,920

Source: American Cancer Society, *Cancer Facts and Figures, 2017* (Atlanta: American Cancer Society, 2017), p. 4.

to proliferate at an abnormally high rate. As they lose their response to the normal controls, cancer cells gradually lose their regular shape and boundaries, eventually forming a distinct mass of abnormal cells—a tumor. If tumor cells remain localized, the tumor is said to be benign; if the cells invade other tissues, the tumor is said to be **malignant**. Cells that travel to other sites in the body, where they establish secondary tumors, have undergone **metastasis**.

Cancer as a Genetic Disease

Cancer arises as a result of fundamental defects in the regulation of cell division, and its study therefore has significance not only for public health, but also for our basic understanding of cell biology. Through the years, many ideas have been put forth to explain cancer, but we now recognize that most, if not all, cancers arise from defects in DNA.

EVIDENCE FOR THE GENETIC THEORY OF CANCER Early observations suggested that cancer might result from genetic damage. First, many agents that cause mutations, such as ionizing radiation and chemicals, also cause cancer (are carcinogens; see Chapter 13). Second, some cancers are consistently associated with particular chromosome abnormalities. About 90% of people with chronic myeloid leukemia, for example, have a reciprocal translocation between chromosome 22 and chromosome 9. Third, some specific types of cancers tend to run in families. Retinoblastoma, a rare childhood cancer of the retina, appears with high frequency in a few families, in which it is inherited as an autosomal dominant trait, suggesting that a single gene is responsible for these cases of the disease.

Although these observations hinted that genes play some role in cancer, the theory of cancer as a genetic disease had several significant problems. If cancer is inherited, every cell in the body should receive the cancer-causing gene, and therefore every cell should become cancerous. In the types of cancer that run in families, however, tumors typically appear only in certain tissues and often only when the person reaches an advanced age. Finally, many cancers do not run in families at all, and even in those cancers that generally do, isolated cases crop up in families with no history of the disease.

KNUDSON'S MULTISTEP MODEL OF CANCER In 1971, Alfred Knudson proposed a model to explain the genetic basis of cancer. Knudson was studying retinoblastoma, which usually develops in only one eye but occasionally appears in both. Knudson found that when retinoblastoma appears in both eyes, it presents itself at an early age, and that many children with bilateral retinoblastoma have close relatives who are also affected.

Knudson proposed that retinoblastoma results from two separate genetic defects, both of which are necessary for cancer to develop (**Figure 16.3**). He suggested that in the cases in which the disease affects just one eye, a single cell in one eye undergoes two successive mutations. Because the chance of these two mutations occurring in a single cell is remote, retinoblastoma is rare and typically develops in only one eye. Knudson proposed that children with bilateral retinoblastoma have inherited one of the two mutations required for the cancer, and so every cell contains this initial mutation. In these cases, all that is required for cancer to develop is for one eye cell to undergo the second mutation. Because each eye possesses millions of cells, the probability that the second mutation will occur in at least one cell of each eye is high, so tumors may occur in both eyes at an early age.

Knudson's proposal suggests that cancer is the result of a multistep process that requires several mutations. If one or more of the required mutations are inherited, fewer additional mutations are required to produce cancer, and the cancer will tend to run in families. Knudson's idea has been called the "two-hit hypothesis" because, in retinoblastoma, only two mutations are necessary to cause a tumor. In most

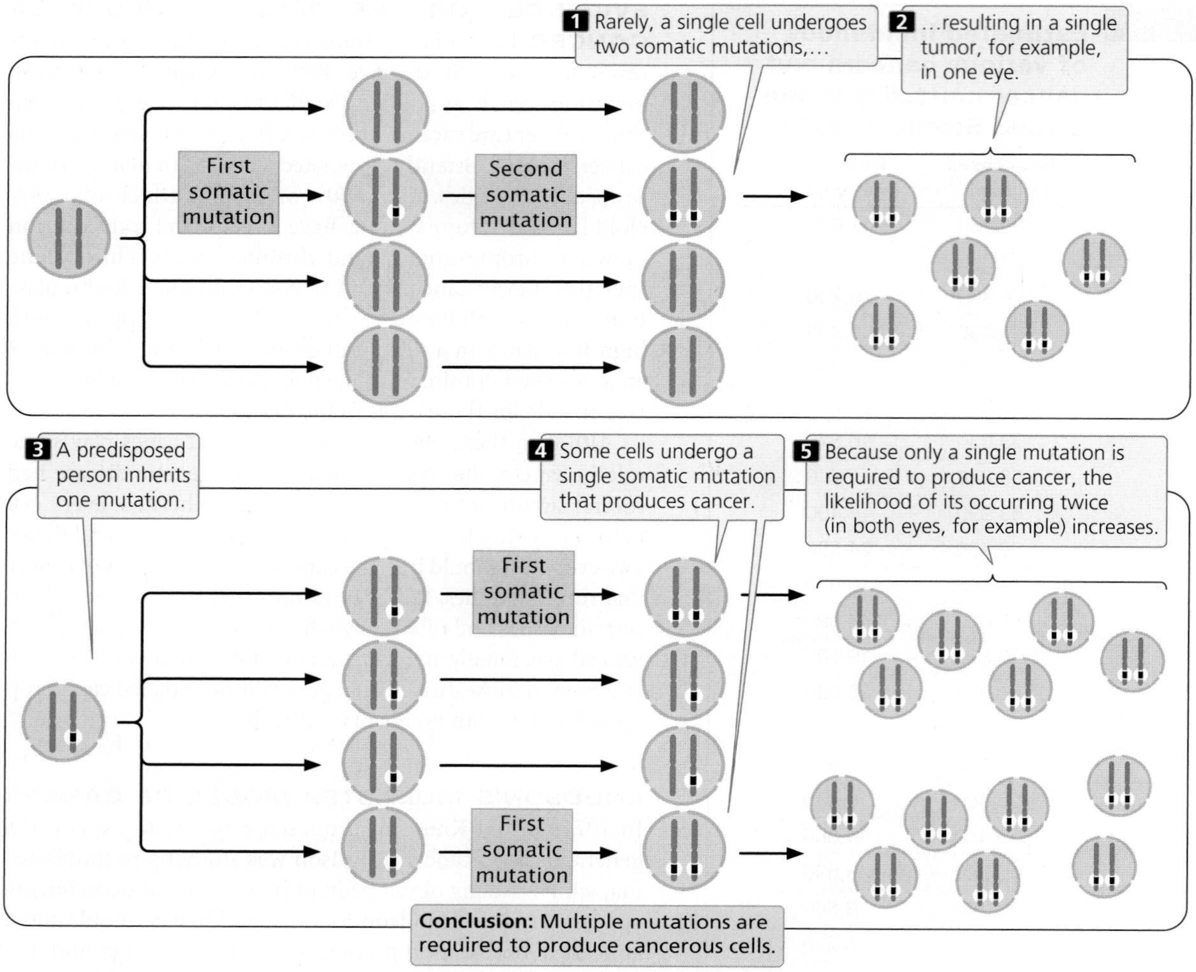

1 Rarely, a single cell undergoes two somatic mutations,...
2 ...resulting in a single tumor, for example, in one eye.
First somatic mutation
Second somatic mutation
3 A predisposed person inherits one mutation.
4 Some cells undergo a single somatic mutation that produces cancer.
5 Because only a single mutation is required to produce cancer, the likelihood of its occurring twice (in both eyes, for example) increases.
First somatic mutation
First somatic mutation
Conclusion: Multiple mutations are required to produce cancerous cells.

16.3 Alfred Knudson proposed that retinoblastoma results from two separate genetic defects, both of which are necessary for cancer to develop.

cancers, however, more than two mutations are involved in the transformation of normal cells into cancer cells. In the case of retinoblastoma, the two required mutations occur at the same locus (both alleles become mutated), but mutations at different loci are required for the development of many other cancers. The idea that cancer results from multiple mutations turns out to be correct for most cancers.

Knudson's genetic theory of cancer has been confirmed by the identification of genes that, when mutated, cause cancer. Today, we recognize that cancer is fundamentally a genetic disease, although few cancers are actually inherited. Most tumors arise from somatic mutations that accumulate over a person's life span, either through spontaneous mutation or in response to environmental mutagens.

THINK-PAIR-SHARE Question 5

THE CLONAL EVOLUTION OF TUMORS Cancer begins when a single somatic cell undergoes a mutation that causes the cell to divide at an abnormally rapid rate. The cell proliferates, giving rise to a clone of cells, each of which carries the same mutation. Because the cells of the clone divide more rapidly than normal, they soon outgrow other cells. An additional somatic mutation that arises in some of the clone's cells may further enhance the ability of those cells to proliferate, and cells carrying both mutations soon become the most common cells in the clone. Eventually, they may be overtaken by cells that contain yet more mutations that enhance proliferation. In this process, called **clonal evolution**, the tumor cells acquire more somatic mutations that allow them to become increasingly more aggressive in their proliferative properties (**Figure 16.4**).

The rate of clonal evolution depends on the frequency with which new mutations arise. Any genetic defect that allows more mutations to arise will accelerate cancer progression. Genes that regulate DNA repair are often found to have been mutated in the cells of advanced cancers, and inherited disorders of DNA repair are usually characterized by increased incidences of cancer. Because DNA-repair mechanisms normally eliminate many of the mutations that arise, cells with defective DNA-repair systems are more likely to retain mutations, including mutations in genes that regulate

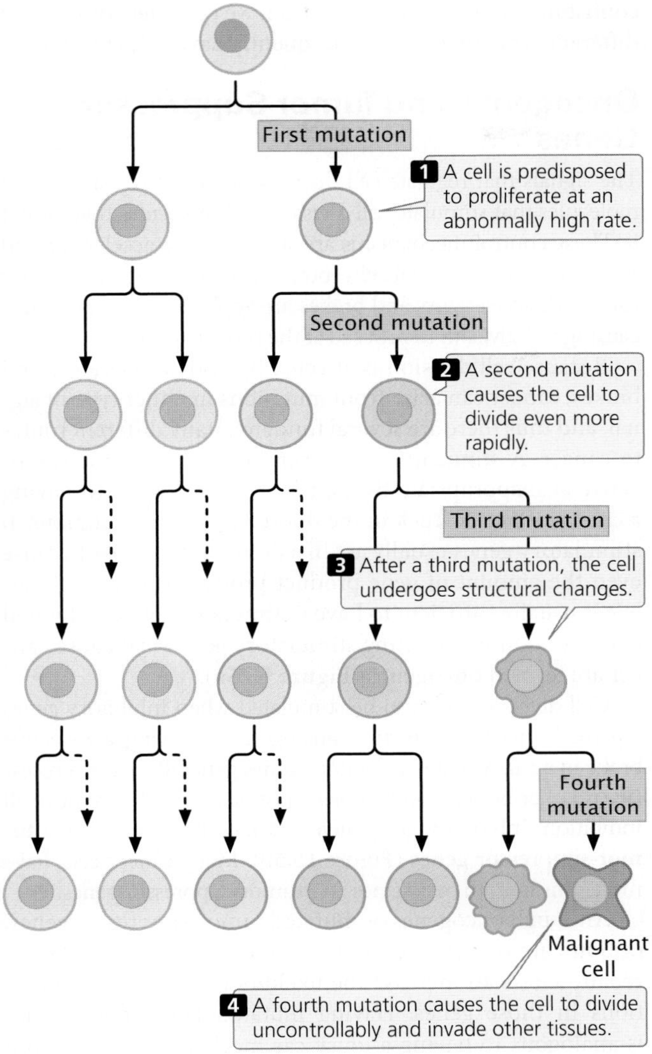

16.4 Through clonal evolution, tumor cells acquire multiple mutations that allow them to become increasingly aggressive and proliferative. To conserve space, a dashed arrow is used to represent a second cell of the same type in each case.

cell division, than are normal cells. Xeroderma pigmentosum, for example, is a rare disorder caused by a defect in DNA repair (see Chapter 13). People with this condition have elevated rates of skin cancer when exposed to sunlight (which induces mutation). Similarly, breast cancer can be caused by mutations in *BRCA1* and *BRCA2*, two genes that function in DNA repair.

Mutations in genes that affect chromosome segregation may also contribute to the clonal evolution of tumors. Many cancer cells are aneuploid (contain extra or missing copies of individual chromosomes; see Chapter 6), and clearly, chromosome mutations contribute to cancer progression by duplicating some genes (those on extra chromosomes) and eliminating others (those on deleted chromosomes). Cellular defects that interfere with chromosome separation increase aneuploidy and may therefore accelerate cancer progression.

CONCEPTS

Cancer is fundamentally a genetic disease. Mutations in several genes are usually required to produce cancer. If one of those mutations is inherited, fewer somatic mutations are necessary for cancer to develop, and the person may have a predisposition to cancer. Clonal evolution is the accumulation of mutations in a clone of cells.

✓ CONCEPT CHECK 1

How does the multistep model of cancer explain the observation that sporadic cases of retinoblastoma usually appear in only one eye, whereas inherited forms of the cancer appear in both eyes?

The Role of Environmental Factors in Cancer

Although cancer is a genetic disease, most cancers are not inherited, and many are influenced by environmental factors. The role of environmental factors in cancer is suggested by differences in the incidence of specific cancers throughout the world (**Table 16.2**). The results of studies show that migrant populations typically take on the cancer incidence of their host country. For example, the overall rates of colon cancer are considerably lower in Japan than in Hawaii. However, within a single generation after migration to Hawaii, Japanese people develop colon cancer at rates similar to or exceeding those of Caucasian Hawaiians. The increased

TABLE 16.2 Examples of geographic variation in the incidence of cancer

Type of cancer	Location	Incidence rate*
Lip	Canada (Newfoundland)	15.1
	Brazil (Fortaleza)	1.2
Nasopharynx	Hong Kong	30.0
	United States (Utah)	0.5
Colon	United States (Iowa)	30.1
	India (Mumbai)	3.4
Lung	United States (New Orleans, African Americans)	110.0
	Costa Rica	17.8
Prostate	United States (Utah)	70.2
	China (Shanghai)	1.8
Bladder	United States (Connecticut, Whites)	25.2
	Philippines (Rizal)	2.8
All cancer	Switzerland (Basel)	383.3
	Kuwait	76.3

Source: C. Muir et al., *Cancer Incidence in Five Continents*, vol. 5 (Lyon: International Agency for Research on Cancer, 1987), Table 12-2.
*The incidence rate is the age-standardized rate in males per 100,000 population.

TABLE 16.3 Percentage of cancer cases in the United Kingdom attributed to environmental factors

Factor	Percentage of cancer cases
Tobacco	19.4
Diet	9.2
Overweight and obesity	5.5
Alcohol	4.0
Occupation	3.7
Radiation (UV)	3.5
Infections	3.1
Radiation (ionizing)	1.8
All environmental factors	42.7

Source: Data from D. M. Parkin, L. Boyd, and L. C. Walker, Fraction of cancer attributable to lifestyle and environmental factors in the UK in 2010, *British Journal of Cancer* 105:S77–S81, 2011.

cancer among the migrants is due to the fact that they are exposed to the same environmental factors as the natives are.

A number of environmental factors contribute to cancer, but those that have the greatest effects include tobacco use, diet, obesity, alcohol, and UV radiation (**Table 16.3**). Other environmental factors that induce cancer are certain types of chemicals, such as benzene (used as an industrial solvent), benzo[a]pyrene (found in cigarette smoke), and polychlorinated biphenyls (PCBs; used in industrial transformers and capacitors). Most environmental factors associated with cancer cause somatic mutations that stimulate cell division or affect the process of cancer.

THINK-PAIR-SHARE Question 6

Environmental factors may interact with genetic predispositions to cancer. Lung cancer, for example, is clearly associated with smoking, an environmental factor. Genome-wide association studies (see Chapter 15) revealed that variation at several genes predisposes some people to smoking-induced lung cancer. Variants at some of these genes cause people to be more likely to become addicted to smoking. Other predisposing genes encode receptors that bind potential carcinogens in cigarette smoke. ▶TRY PROBLEM 17

16.2 Mutations in Several Types of Genes Contribute to Cancer

As we have seen, cancer is a disease caused by alterations in DNA. More than 350 different human genes have been identified that contribute to the development of cancer, and the actual number is probably much higher. Research with mice suggests that more than 2000 genes can, when mutated, contribute to cancer. In this section, we consider some of the different types of genes that frequently have roles in cancer.

Oncogenes and Tumor-Suppressor Genes

The signals that regulate cell division fall into two basic types: molecules that stimulate cell division and molecules that inhibit it. These control mechanisms are similar to the accelerator and brake of a car. In normal cells (but, one would hope, not in your car), both accelerators and brakes are applied at the same time, causing cell division to proceed at the proper speed.

Because cell division is affected by both accelerators and brakes, cancer can arise from mutations in either type of signal, and thus there are several fundamentally different routes to cancer. A stimulatory gene can be made hyperactive, or active at inappropriate times, which is analogous to having a car's accelerator stuck in the floored position. Mutations in stimulatory genes usually act in a dominant manner because even the amount of gene product produced by a single allele is usually sufficient to have a stimulatory effect. Mutated copies of dominant-acting stimulatory genes that cause cancer are termed **oncogenes** (**Figure 16.5a**).

Cell division may also be stimulated when inhibitory genes are made inactive, which is analogous to having a defective brake in a car. Mutated inhibitory genes generally act in a recessive manner because both copies must be mutated to remove all inhibition. Inhibitory genes involved in cancer are termed **tumor-suppressor genes** (**Figure 16.5b**). Many cancer cells have mutations in both oncogenes and tumor-suppressor genes.

Although oncogenes or mutated tumor-suppressor genes, or both, are required to produce cancer, mutations in DNA-repair genes can increase the likelihood of acquiring mutations in those genes. Having mutated DNA-repair genes is analogous to having a lousy car mechanic who does not make the necessary repairs on a broken accelerator or brake.

ONCOGENES Oncogenes were the first type of cancer-causing genes to be identified. In 1909, a farmer brought physician Peyton Rous a hen with a large connective-tissue tumor (sarcoma) growing on its breast. When Rous injected pieces of this tumor into other hens, they also developed sarcomas. Rous conducted experiments that demonstrated that the tumors were being transmitted by a retrovirus, which became known as the Rous sarcoma virus, as mentioned in Chapter 7. A number of other cancer-causing viruses were subsequently isolated from various animal tissues. These viruses were generally assumed to carry a cancer-causing gene that was transferred to the host cell. The first oncogene, called *src*, was isolated from the Rous sarcoma virus in 1970.

In 1975, Michael Bishop, Harold Varmus, and their colleagues began to use probes for viral oncogenes to search for related sequences in normal cells. They discovered that the genomes of all normal cells carry DNA sequences that are closely related to oncogenes. These normal cellular genes are called **proto-oncogenes**. They are responsible for

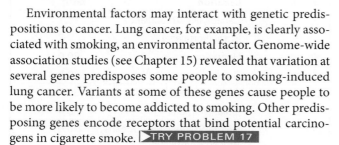

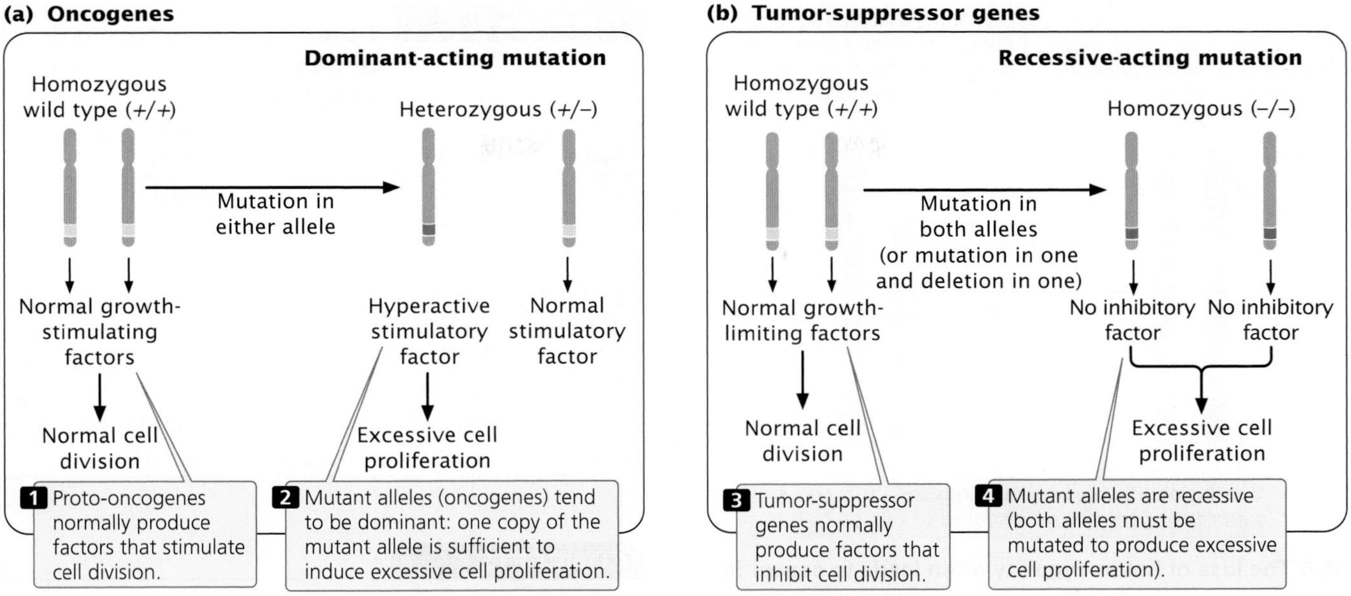

16.5 Both oncogenes (a) and tumor-suppressor genes (b) contribute to cancer, but differ in their modes of action and dominance.

basic cellular functions in normal cells, but when mutated, they become oncogenes that contribute to the development of cancer. When a virus infects a cell, a proto-oncogene may become incorporated into the viral genome through recombination. Within the viral genome, the proto-oncogene may mutate to an oncogene that, when inserted into a host cell, causes rapid cell division and cancer. Because the proto-oncogenes are more likely to undergo mutation or recombination within a virus than within the host cell, viral infection is often associated with some cancers.

Proto-oncogenes can be converted into oncogenes in viruses in several different ways. The sequence of the proto-oncogene may be altered or truncated as it is incorporated into the viral genome. This mutated copy of the gene may then produce an altered protein in the host cell that causes uncontrolled cell proliferation. Alternatively, through recombination, a proto-oncogene may end up next to a viral promoter or enhancer, which causes the gene to be overexpressed in the host cell. Finally, the function of a proto-oncogene in the host cell may be altered when a virus inserts its own DNA into the gene, disrupting its normal function. While viruses are capable of converting proto-oncogenes into oncogenes, most proto-oncogenes are mutated to form oncogenes without the involvement of a virus.

Many oncogenes have been identified by experiments in which selected fragments of DNA are added to cells in culture. Some of the cells take up the DNA, and if these cells become cancerous, then the DNA fragment that was added to the culture must contain an oncogene. The fragments can then be sequenced and the oncogene identified. A large number of oncogenes have now been discovered (**Table 16.4**). About 90% of all cancer genes are thought to be dominant oncogenes.

TABLE 16.4 Some oncogenes and functions of their corresponding proto-oncogenes

Gene	Normal function	Cancer in which gene is mutated
erbB	Part of growth factor receptor	Many types of cancer
fos	Transcription factor	Osteosarcoma and endometrial carcinoma
jun	Transcription factor, cell cycle control	Lung cancer, breast cancer
myc	Transcription factor	Lymphomas, leukemias, neuroblastoma
ras	GTP binding and GTPase	Many types of cancer
sis	Growth factor	Glioblastomas and other cancers
src	Protein tyrosine kinase	Many types of cancer

TUMOR-SUPPRESSOR GENES Tumor-suppressor genes are more difficult to identify than oncogenes because they *inhibit* cancer and are recessive; both alleles must be mutated before the inhibition of cell division is removed. Because it is the *failure* of their function that promotes cell proliferation, tumor-suppressor genes cannot be identified by adding them to cells and looking for cancer. About 10% of cancer-causing genes are thought to be tumor-suppressor genes.

Defects in both copies of a tumor-suppressor gene are usually required to cause cancer. An organism can inherit one

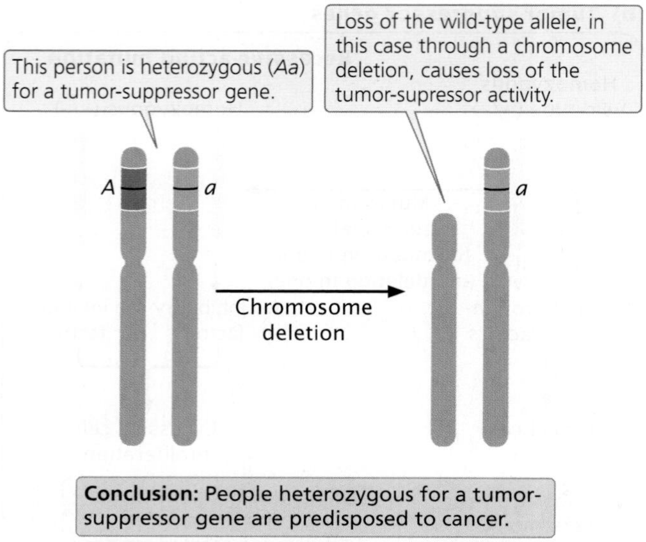

16.6 The loss of heterozygosity often leads to cancer in a person heterozygous for a tumor-suppressor gene.

defective copy of the tumor-suppressor gene (be heterozygous for the cancer-causing mutation) and not have cancer because the remaining normal allele produces the tumor-suppressing product. However, these heterozygotes are often predisposed to cancer because the inactivation or loss of that one remaining allele is all that is required to completely eliminate the tumor-suppressor product. Inactivation of the remaining wild-type allele in heterozygotes is referred to as the **loss of heterozygosity**. A common mechanism for the loss of heterozygosity is a deletion on the chromosome that carried the normal copy of the tumor-suppressor gene (**Figure 16.6**).

Among the first tumor-suppressor genes to be identified was the gene that causes retinoblastoma. In 1985, Raymond White and Webster Cavenne showed that large segments of chromosome 13 were missing in cells from retinoblastoma tumors, and later, the tumor-suppressor gene was isolated from those segments. Another example of a tumor-suppressor gene is *BRCA1*, mutations of which are associated with increased risk of breast and ovarian cancer. *BRCA1* produces a protein that normally helps in the repair of double-strand breaks in DNA by homologous recombination. That protein also acts as a transcription factor and interacts with histone deacetylase enzymes, which affect transcription. A number of tumor-suppressor genes have now been discovered (**Table 16.5**).

Sometimes the mutation or loss of a single allele of a recessive tumor-suppressor gene is sufficient to cause cancer. This effect—the appearance of the trait in an individual cell or organism that is heterozygous for a normally recessive trait—is called **haploinsufficiency**. This phenomenon is thought to be due to dosage effects: the heterozygote produces only half as much of the product encoded by the tumor-suppressing gene as the homozygote does. Normally, this amount is sufficient for the cellular processes that prevent tumor formation, but it is less than the optimal amount, and other factors may sometimes combine with the lowered amount of tumor-suppressor product to cause cancer. TRY PROBLEM 18

TABLE 16.5 Some tumor-suppressor genes and their normal functions

Gene	Normal function	Cancer in which gene is mutated
APC	Scaffold protein, interacts with microtubules	Colorectal
CDKN2A	Regulates cell division	Melanoma
BRCA1	DNA repair, transcription factor	Breast and ovarian
NF1	GTPase activator	Neurofibromatosis
p53	Regulates cell division	Many types of cancer
RB	Regulates cell division	Retinoblastoma

CONCEPTS

Proto-oncogenes are genes that control normal cellular functions; when mutated, they become oncogenes that stimulate cell proliferation. They tend to be dominant in their action. Tumor-suppressor genes normally inhibit cell proliferation; when mutated, they allow cells to proliferate. Tumor-suppressor genes tend to be recessive in their action. Individual organisms that are heterozygous for tumor-suppressor genes are often predisposed to cancer.

✓ CONCEPT CHECK 2

Why are oncogenes usually dominant in their action, whereas tumor-suppressor genes are recessive?

Genes That Control the Cell Cycle

The cell cycle is the normal process by which cells undergo growth and division. Normally, progression through the cell cycle is tightly regulated so that cells divide only when additional cells are needed, when all the components necessary for division are present, and when the DNA has been replicated without damage. Sometimes, however, errors arise in one or more of the components that regulate the cell cycle. These errors often cause cells to divide at inappropriate times or rates, leading to cancer. Indeed, many proto-oncogenes and tumor-suppressor genes function normally by helping to control the cell cycle. Before considering how errors in this system contribute to cancer, we must first understand how the cell cycle is usually regulated.

CONTROL OF THE CELL CYCLE As discussed in Chapter 2, the cell cycle consists of the period from one cell division to the next. Cells that are actively dividing pass through the G_1, S, and G_2 phases of interphase and then move directly into the M phase, in which cell division takes place. Nondividing cells pass from G_1 into the G_0 phase, in which they are functional but not actively growing or dividing. Progression from one phase of the cell cycle to another is influenced by a number of internal and external signals and is regulated at key points in the cycle, called checkpoints.

Key events of the cell cycle are controlled by **cyclin-dependent kinases** (**CDKs**), which are enzymes that phosphorylate (add phosphate groups to) other proteins. In some cases, phosphorylation activates the other protein, and in others, it inactivates the other protein. As their name implies, CDKs are functional only when associated with another type of protein, called a **cyclin**. The levels of cyclins oscillate over the course of the cell cycle; when bound to a CDK, a cyclin specifies which proteins the CDK will phosphorylate. Each cyclin appears at a specific point in the cell cycle, usually because its synthesis and destruction are regulated by another cyclin. Cyclins and CDKs are called by different names in different organisms; here, we will use the terms applied to these molecules in mammals.

Let's look at the G_1-to-S transition. As stated in Chapter 2, progression through the cell cycle is regulated at several checkpoints, which ensure that all cellular components are present and in good working order before the cell proceeds to the next phase. The G_1/S checkpoint is at the end of G_1, just before the cell enters the S phase and replicates its DNA. The cell is prevented from passing through the G_1/S checkpoint by a molecule called the retinoblastoma (RB) protein (**Figure 16.7**), which binds to another molecule called E2F and keeps it inactive. During G_1, cyclin D and cyclin E continuously increase in concentration and combine with their associated CDKs. Cyclin-D–CDK and cyclin-E–CDK both phosphorylate molecules of RB. By late in G_1, phosphorylation of RB is completed, which inactivates RB. Without the inhibitory effects of RB, E2F is released. The E2F protein stimulates the transcription of genes that produce enzymes necessary for replication of the DNA, and the cell moves into the S phase of the cell cycle.

Another important checkpoint controls the G_2-to-M transition. This checkpoint is also regulated by CDKs and cyclins. Other checkpoints control the assembly of the mitotic spindle apparatus and the cell's exit from mitosis.

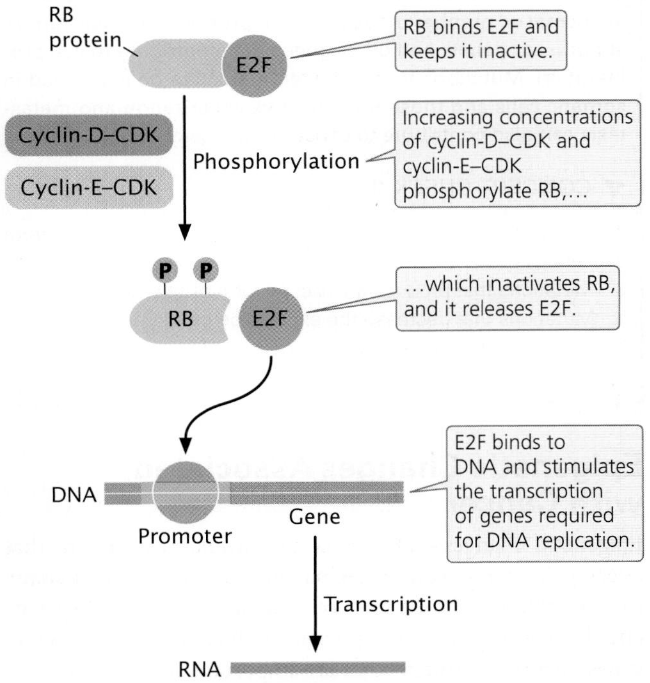

16.7 The RB protein helps control the progression through the G_1/S checkpoint by binding transcription factor E2F.

MUTATIONS IN CELL-CYCLE CONTROLS AND CANCER Many cancers are caused by defects in the cell cycle's regulatory machinery. For example, mutations in the gene that encodes the RB protein—which normally holds the cell in G_1 until the DNA is ready to be replicated—are associated with many cancers, including retinoblastoma. When the *RB* gene is mutated, cells pass through the G_1/S checkpoint without the normal controls that prevent cell proliferation. Overexpression of the gene that encodes cyclin D (which stimulates the passage of cells through the G_1/S checkpoint) takes place in about 50% of all breast cancers as well as in some cases of esophageal and skin cancer. Likewise, the tumor-suppressor gene *p53*, which is mutated in about 75% of all colon cancers, regulates a potent inhibitor of CDK activity.

Some proto-oncogenes and tumor-suppressor genes have roles in **apoptosis**, a process of programmed cell death in which a cell's DNA is degraded, its nucleus and cytoplasm shrink, and the cell undergoes phagocytosis by other cells without the leakage of its contents. Cells have the ability to assess themselves, and if they are abnormal or damaged, they normally undergo apoptosis. Many cancer cells have chromosome mutations, DNA damage, or other cellular anomalies that would normally stimulate apoptosis and prevent their proliferation. Often, these cells have mutations in genes that regulate apoptosis, and therefore they do not undergo programmed cell death. The ability of a cell to initiate apoptosis in response to DNA damage, for example, depends on *p53*, which is inactive in many human cancers.

CONCEPTS

Progression through the cell cycle is controlled at checkpoints, which are regulated by interactions between cyclins and cyclin-dependent kinases. Genes that control the cell cycle are frequently mutated in cancer cells.

DNA-Repair Genes

As we have seen, cancer arises from the accumulation of multiple mutations in a single cell. The rate at which mutations occur is affected not only by the rate at which they arise, but also by the efficiency with which errors are corrected by DNA-repair systems (see Chapter 13). Defects in genes that encode components of these repair systems have been consistently associated with a number of cancers. People with xeroderma pigmentosum, for example, are defective in nucleotide-excision repair, an important cellular repair system that normally corrects DNA damage caused by a number of mutagens, including ultraviolet light. Likewise, about 13% of colorectal, endometrial, and stomach cancers have cells that are defective in mismatch repair, another major repair system in the cell.

Genes That Regulate Telomerase

Another factor that can contribute to the progression of cancer is the inappropriate activation of the enzyme telomerase. Recall from Chapter 9 that in most somatic cells, the ends of chromosomes cannot be replicated, and the telomeres become shorter with each cell division. This shortening eventually leads to the destruction of the chromosomes and cell death, so somatic cells are capable of only a limited number of cell divisions.

In germ cells and stem cells, telomerase replicates the chromosome ends (see pp. 257–260 in Chapter 9), thereby maintaining the telomeres, but this enzyme is not normally expressed in somatic cells. In many tumor cells, however, sequences that regulate the expression of the telomerase gene are mutated, allowing the enzyme to be expressed and the cell to divide indefinitely. Although the expression of telomerase appears to contribute to the development of many cancers, its precise role in tumor progression is unknown and is under investigation.

Genes That Promote Vascularization and the Spread of Tumors

Another important group of factors that contribute to the progression of cancer includes genes that affect the growth and spread of tumors. Oxygen and nutrients, which are essential to the survival and growth of tumors, are supplied by blood vessels, and the growth of new blood vessels (angiogenesis) is important to tumor progression. Angiogenesis is stimulated by growth factors and other proteins encoded by genes whose expression is carefully regulated in normal cells. In tumor cells, genes encoding these proteins are often overexpressed compared with normal cells, and inhibitors of angiogenesis-promoting factors may be inactivated or underexpressed. At least one inherited cancer—von Hippel–Lindau disease, in which people develop multiple types of tumors—is caused by the mutation of a gene that affects angiogenesis.

In the development of many cancers, the primary tumor gives rise to cells that spread to distant sites, producing secondary tumors. This process of metastasis, which is the cause of death in 90% of human cancer deaths, is influenced by cellular changes induced by somatic mutation. As discussed in the introduction to this chapter, the *palladin* gene, when mutated, contributes to the metastasis of pancreatic tumors. By using microarrays to measure levels of gene expression in tumors, researchers have identified other genes that are transcribed at a significantly higher rate in metastatic cells than in nonmetastatic cells. For example, one study detected a set of 95 genes that were overexpressed or underexpressed in a population of metastatic breast-cancer cells that were strongly metastatic to the lung, compared with a population of cells that were only weakly metastatic to the lung. Genes that contribute to metastasis often encode components of the extracellular matrix and the cytoskeleton. Others encode adhesion proteins, which help hold cells together.

Advances in sequencing technology have now made it possible to completely sequence the DNA of tumor cells to see how their genomes differ from those of normal cells. In one experiment, researchers sequenced the entire genome of cells from a metastasized breast-cancer tumor and compared it with the genome of noncancer cells from the same person. They also compared the genome of the metastasized tumor with the genome of the primary tumor (from which the metastasis originated), which had been removed from the patient nine years earlier. The researchers found 32 different somatic mutations in the coding regions of genes from the metastasized tumor cells, 19 of which were not detected in the primary tumor. This finding suggests that the metastasized tumor underwent considerable genetic changes during its nine-year evolution from the primary tumor. In contrast, another study of a breast-cancer metastasis found only two mutations that were not present in the primary tumor, but in this case, the metastasis had evolved in only one year.

CONCEPTS

Mutations in genes that encode components of DNA-repair systems are often associated with cancer; these mutations increase the rate at which mutations are retained and result in an increased number of mutations in proto-oncogenes, tumor-suppressor genes, and other genes that contribute to cell proliferation. Mutations that allow telomerase to be expressed in somatic cells and those that affect vascularization and metastasis can also contribute to cancer progression.

✓ **CONCEPT CHECK 3**

Which type of mutation in telomerase can be associated with cancer cells?

a. Mutations that produce an inactive form of telomerase
b. Mutations that decrease the expression of telomerase
c. Mutations that increase the expression of telomerase
d. All of the above

Epigenetic Changes Associated with Cancer

Epigenetic changes—alterations to chromatin structure that affect gene expression (see Chapter 12)—are seen in many cancer cells. Two broad lines of evidence suggest that epigenetic changes play an important role in cancer progression. First, genes encoding proteins that are important regulators of epigenetic changes are often mutated in some types of cancer. For example, almost 90% of cases of follicular lymphoma exhibit mutations in the *MLL2* gene, which encodes a histone methyltransferase; this enzyme adds methyl groups to DNA, a type of epigenetic modification that alters chromatin structure and affects transcription. Similarly, the *UTX* gene, which encodes a histone demethylase (an enzyme that removes methyl groups from DNA), is mutated in a number of different types of cancer.

A second line of evidence suggesting that epigenetic alterations are important in cancer comes from recent genomic studies that have compared the chromatin structure of cancer cells with that of normal cells from the same individual. These studies often find that the cancer cells have significant alterations to DNA methylation patterns and histone structure.

One type of epigenetic alteration often observed in cancer cells is an overall lower level of DNA methylation (hypomethylation). As discussed in Chapter 12, DNA methylation is often associated with repression of transcription. It is assumed that hypomethylation leads to transcription of oncogenes, which then stimulate cancer. Some evidence also suggests that hypomethylation causes chromosome instability, a hallmark of many tumors. Tumor cells from mice that have been genetically engineered to have reduced DNA methylation show increased gains and losses of chromosomes, but how hypomethylation might cause chromosome instability is unclear.

Research has also demonstrated that the histone proteins in nucleosomes, the fundamental units of chromatin, are often abnormally modified in cancer cells. Modification of histone proteins, including methylation and acetylation, alters chromatin structure and affects whether transcription occurs (see Chapter 12). Genome-wide patterns of histone acetylation are often altered in cancer cells. Epigenetic processes are receiving increasing attention from cancer researchers because they may be amenable to drug therapy.

CONCEPTS

Epigenetic changes, including DNA methylation and histone modification, are often associated with cancer.

16.3 Colorectal Cancer Arises Through the Sequential Mutation of a Number of Genes

Most cancers arise from mutations in several genes, often a combination of oncogenes and tumor suppressor genes. Colorectal cancer is an excellent example of how the accumulation of successive genetic defects can lead to cancer.

Colorectal cancers arise in the cells lining the colon and rectum. More than 135,000 new cases of colorectal cancer are diagnosed each year in the United States, where this cancer is responsible for more than 50,000 deaths annually. If detected early, colorectal cancer can be treated successfully; consequently, there has been much interest in identifying the molecular events responsible for the initial stages of colorectal cancer.

Colorectal cancer is thought to originate as benign tumors called adenomatous polyps (**Figure 16.8**). Initially, these

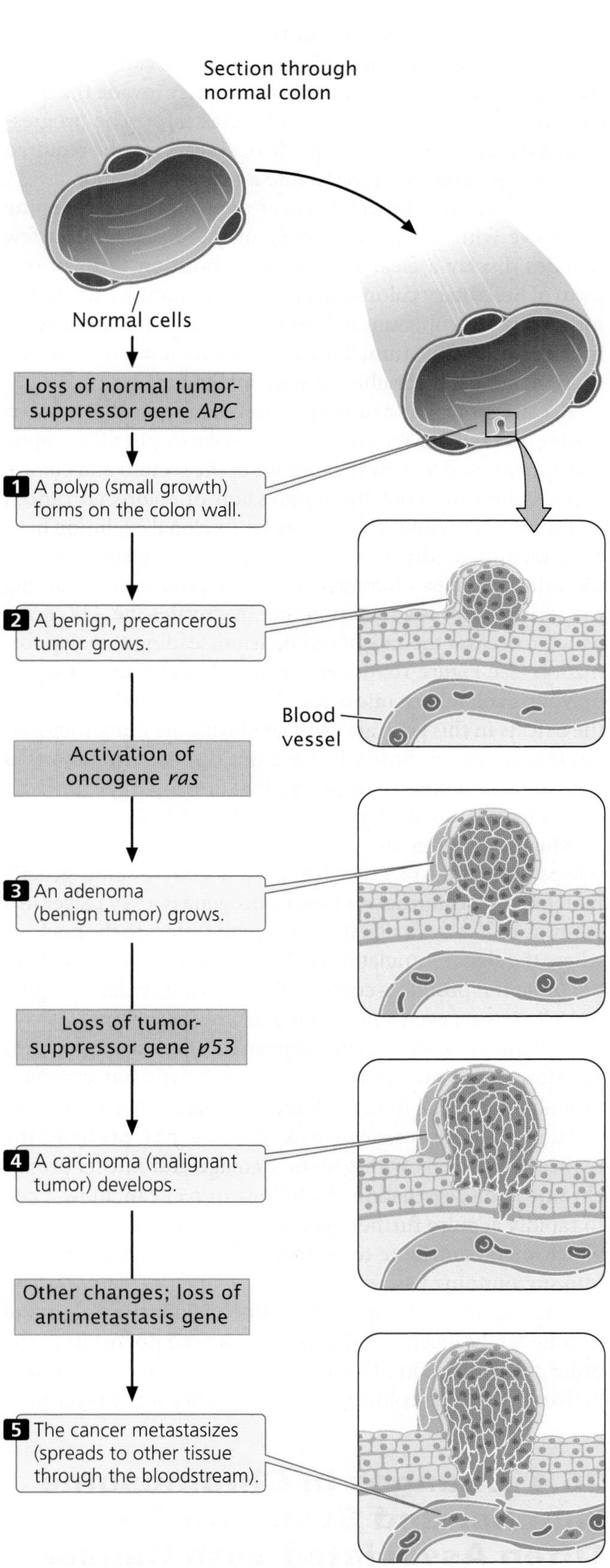

16.8 Mutations in multiple genes contribute to the progression of colorectal cancer.

polyps are microscopic, but in time they enlarge, and their cells acquire the abnormal characteristics of cancer cells. In the later stages of the disease, the tumor may invade the muscle layer surrounding the gut and metastasize. The progression of the disease is slow: from 10 to 35 years may be required for a benign tumor to develop into a malignant tumor.

Most cases of colorectal cancer are sporadic, developing in people with no family history of the disease, but a few families display a clear genetic predisposition to it. In one form of hereditary colon cancer, known as familial adenomatous polyposis coli, hundreds or thousands of polyps develop in the colon and rectum; if these polyps are not removed, one or more almost invariably become malignant.

Because polyps and tumors of the colon and rectum can be easily observed and removed with a colonoscope (a fiber-optic instrument used to view the interior of the rectum and colon), much is known about the progression of colorectal cancer, and some of the genes responsible for its clonal evolution have been identified. Mutations in these genes are responsible for the different steps of colorectal-cancer progression. Among the earliest steps is a mutation that inactivates the *APC* gene, which increases the rate of cell division, leading to polyp formation (see Figure 16.8). Persons with familial adenomatous polyposis coli inherit one defective copy of the *APC* gene, and the defects in this gene are associated with the numerous polyps that appear in those who have this disorder. Mutations in *APC* are also found in the polyps that develop in people who do not have familial adenomatous polyposis coli.

Mutations of the *ras* oncogene usually occur later, in larger polyps consisting of cells that have acquired some genetic mutations. The normal *ras* proto-oncogene is a key player in a pathway that relays a signal from growth factors to the nucleus, where the signal stimulates cell division. When *ras* is mutated, the protein it produces continually relays a stimulatory signal for cell division even when growth factors are absent.

Mutations in other genes appear still later in tumor progression; these mutations are rare in polyps but common in malignant cells. Many colorectal cancers have mutations in tumor-suppressor gene *p53*. Because *p53* prevents the replication of cells with genetic damage and affects proper chromosome segregation, mutations in *p53* can allow a cell to rapidly acquire further gene and chromosome mutations, which then contribute to further proliferation and invasion into surrounding tissues.

The sequence of steps just outlined is not the only route to colorectal cancer, and the mutations need not occur in the order presented here. However, this sequence is a common pathway by which colon and rectal cells become cancerous.

16.4 Changes in Chromosome Number and Structure Are Often Associated with Cancer

Most tumors contain cells with chromosome mutations. For many years, geneticists argued about whether these chromosome mutations were the cause or the result of cancer. Some types of tumors are consistently associated with *specific* chromosome mutations; for example, most cases of chronic myelogenous leukemia are associated with a specific reciprocal translocation. These types of associations suggest that chromosome mutations contribute to the cause of the cancer. Yet many cancers are not associated with specific types of chromosome abnormalities, and individual *gene* mutations are now known to contribute to many types of cancer. Nevertheless, as we have noted, chromosome instability is a general feature of cancer cells. That instability causes them to accumulate chromosome mutations, which then affect individual genes that may contribute to the cancer process. Thus, chromosome mutations appear to be both a *cause* and a *result* of cancer.

At least three types of chromosome rearrangements—deletions, inversions, and translocations—are associated with certain types of cancer. Deletions can result in the loss of one or more tumor-suppressor genes. Inversions and translocations contribute to cancer in several ways. First, the chromosome breaks that accompany these mutations can lie within tumor-suppressor genes, disrupting their function and leading to cell proliferation.

Second, translocations and inversions can bring together sequences from two different genes, generating a fusion protein that stimulates some aspect of the cancer process. Fusion proteins are seen in most cases of chronic myelogenous leukemia, which affects bone-marrow cells. Most people with chronic myelogenous leukemia have a reciprocal translocation between the long arm of chromosome 22 and the tip of the long arm of chromosome 9 (**Figure 16.9**). This translocation produces a shortened chromosome 22, called the Philadelphia chromosome because it was first discovered in Philadelphia. At the end of a normal chromosome 9 is a

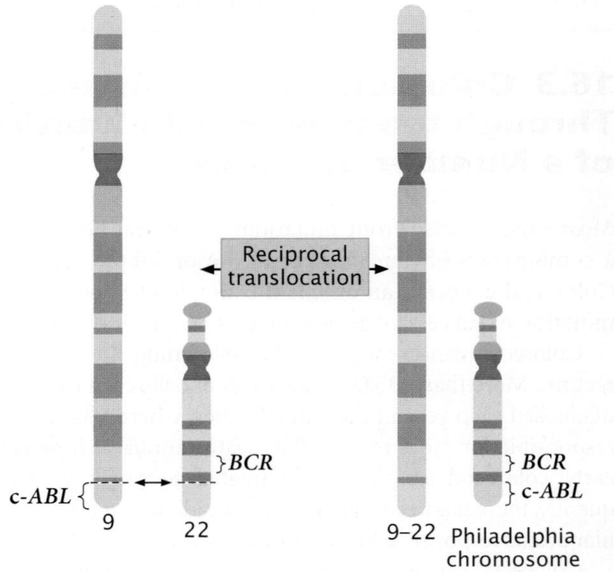

16.9 A reciprocal translocation between chromosomes 9 and 22 causes chronic myelogenous leukemia.

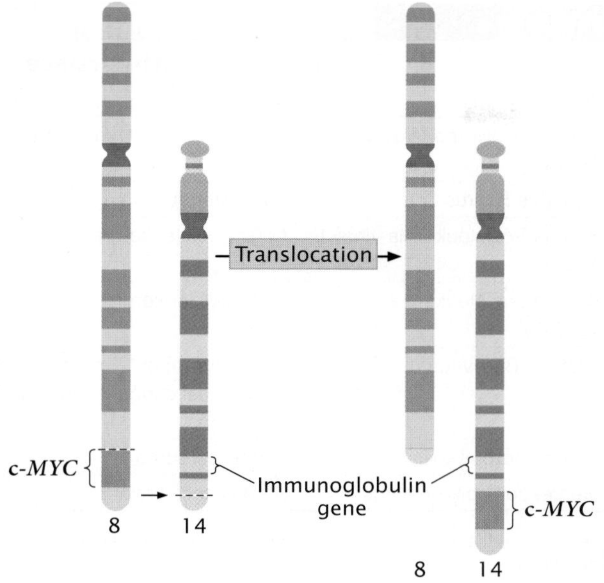

16.10 A reciprocal translocation between chromosomes 8 and 14 causes Burkitt lymphoma.

potential cancer-causing gene called *c-ABL*. As a result of the translocation, part of the *c-ABL* gene is fused with the *BCR* gene from chromosome 22. The protein produced by this *BCR–c-ABL* fusion gene is much more active than the protein produced by the normal *c-ABL* gene; the fusion protein stimulates increased, unregulated cell division and eventually leads to leukemia.

A third mechanism by which chromosome rearrangements can produce cancer is the transfer of a potential cancer-causing gene to a new location, where it is activated by different regulatory sequences. Burkitt lymphoma is a cancer of the B cells, the lymphocytes that produce antibodies. Many people with Burkitt lymphoma possess a reciprocal translocation between chromosome 8 and chromosome 2, 14, or 22 (**Figure 16.10**). This translocation relocates a gene called *c-MYC* from the tip of chromosome 8 to a position on chromosome 2, 14, or 22 that is next to a gene that encodes an immunoglobulin protein. At this new location, *c-MYC*, a cancer-causing gene, comes under the control of regulatory sequences that normally activate the production of immunoglobulins, and *c-MYC* is therefore expressed in B cells. The c-MYC protein stimulates the division of the B cells and leads to Burkitt lymphoma.

CONCEPTS

Many tumors contain a variety of types of chromosome mutations. Some types of tumors are associated with specific deletions, inversions, and translocations. Deletions can eliminate or inactivate genes that control the cell cycle; inversions and translocations can cause breaks in genes that suppress tumors, fuse genes to produce cancer-causing proteins, or move genes to new locations, where they are under the influence of different regulatory sequences.

✓ CONCEPT CHECK 4

Chronic myelogenous leukemia is usually associated with which type of chromosome rearrangement?

a. Duplication
b. Deletion
c. Inversion
d. Translocation

Most advanced tumors contain cells that exhibit a dramatic variety of chromosome anomalies, including extra chromosomes, missing chromosomes, and chromosome rearrangements (**Figure 16.11**). Some cancer researchers believe that cancer is initiated when genetic changes take place that cause the genome to become unstable, generating numerous chromosome abnormalities that then alter the expression of oncogenes and tumor-suppressor genes.

A number of genes that contribute to genomic instability and lead to missing or extra chromosomes (aneuploidy) have now been identified. For example, mutations in genes that encode parts of the mitotic spindle apparatus may contribute to abnormal chromosome segregation and lead to chromosome abnormalities. The *APC* gene, as we have seen, is a tumor-suppressor gene that is often mutated in colon-cancer cells. This gene has several functions, one of which is to interact with the ends of the microtubules that associate with the kinetochore. Dividing mouse cells that have defective copies of the *APC* gene give rise to cells with many chromosome defects.

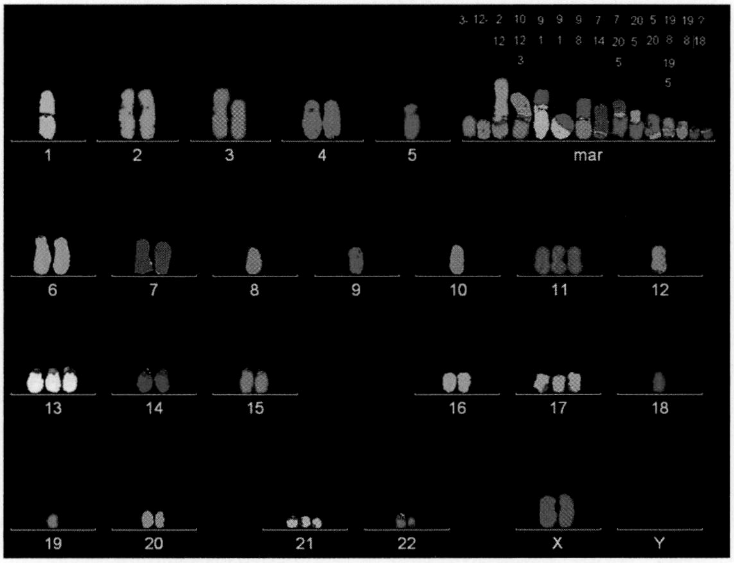

16.11 Cancer cells often possess chromosome abnormalities, including extra chromosomes, missing chromosomes, and chromosome rearrangements. Shown here are chromosomes from a colon-cancer cell, which has numerous chromosome abnormalities. For comparison, see a normal karyotype in Figure 2.5. [Courtesy Dr. Peter Duesberg, UC Berkeley.]

16.5 Viruses Are Associated with Some Cancers

As mentioned earlier in this chapter, viruses are responsible for a number of cancers in animals, and there is evidence that viruses contribute to at least a few cancers in humans (**Table 16.6**).

TABLE 16.6 Some human cancers associated with viruses

Virus	Cancer
Human papilloma viruses (HPVs)	Cervical, penile, and vulvar cancers
Hepatitis B virus	Liver cancer
Human T-cell leukemia virus 1 (HTLV-1)	Adult T-cell leukemia
Human T-cell leukemia virus 2 (HTLV-2)	Hairy-cell leukemia
Epstein–Barr virus	Burkitt lymphoma, nasopharyngeal cancer, Hodgkin lymphoma
Human herpes virus	Kaposi sarcoma
Merkel cell polyomavirus	Merkel cell carcinoma

Note: Some of these associations between cancer and viruses exist only in certain populations and geographic areas.

For example, about 95% of all women with cervical cancer are infected with **human papilloma viruses** (HPVs). Similarly, infection with the virus that causes hepatitis B increases the risk of liver cancer in some people. The Epstein–Barr virus, which is responsible for mononucleosis, has been linked to several types of cancer that are prevalent in parts of Africa, including Burkitt lymphoma.

THINK-PAIR-SHARE Question 7

Many of the viruses that cause cancer in animals are retroviruses (see Chapter 7). Earlier in this chapter, we learned how studies of the Rous sarcoma retrovirus in chickens led to the identification of oncogenes in humans. Retroviruses sometimes cause cancer by mutating and rearranging host genes, converting proto-oncogenes into oncogenes (**Figure 16.12a**).

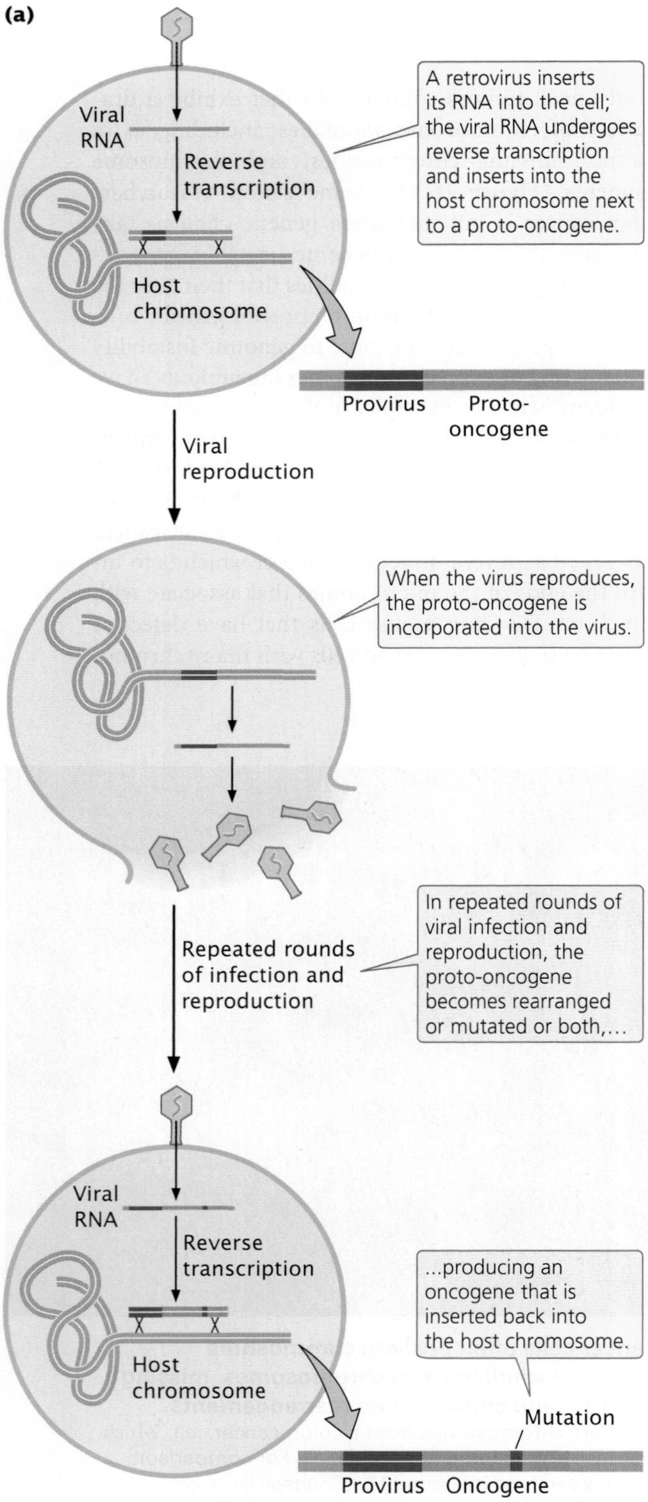

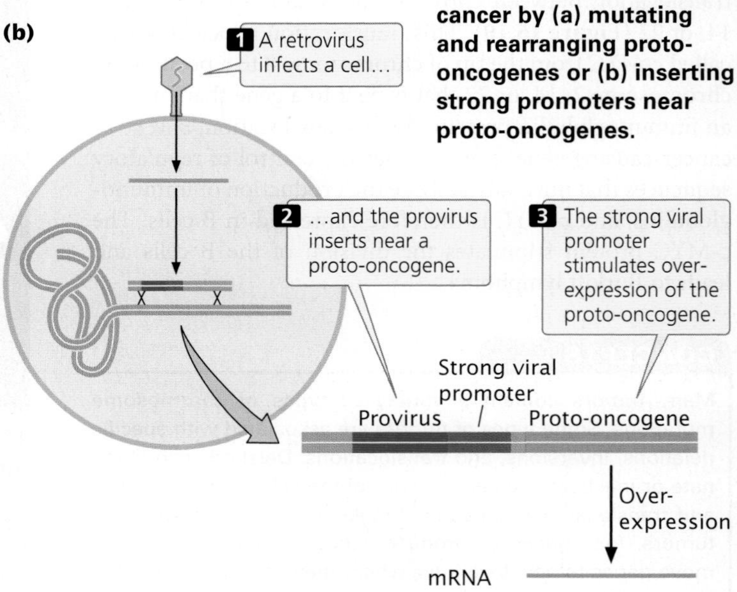

16.12 Retroviruses cause cancer by (a) mutating and rearranging proto-oncogenes or (b) inserting strong promoters near proto-oncogenes.

Another way in which viruses can contribute to cancer is by altering the expression of host genes (**Figure 16.12b**). Retroviruses often contain strong promoters to ensure that their own genetic material is transcribed by the host cell. If the provirus inserts near a proto-oncogene, viral promoters can stimulate high levels of expression of the proto-oncogene, leading to cell proliferation.

CONCEPTS

Viruses contribute to a few cancers in humans by mutating and rearranging host genes that then contribute to cell proliferation, or by altering the expression of host genes.

CONCEPTS SUMMARY

- Cancer is fundamentally a genetic disorder, arising from somatic mutations in multiple genes that affect cell division and proliferation. If one or more mutations are inherited, then fewer additional mutations are required for cancer to develop.
- A mutation that allows a cell to divide rapidly provides the cell with a growth advantage; that cell gives rise to a clone of cells having the same mutation. Within this clone, other mutations occur that provide additional growth advantages, and cells with these additional mutations become dominant in the clone. In this way, the clone evolves.
- Environmental factors play an important role in the development of many cancers by increasing the rate of somatic mutations.
- Oncogenes are dominant mutated copies of normal genes (proto-oncogenes) that stimulate cell division. Tumor-suppressor genes normally inhibit cell division; recessive mutations in these genes may contribute to cancer. Sometimes, the mutation of a single allele of a tumor-suppressor gene is sufficient to cause cancer, a phenomenon known as haploinsufficiency.
- The cell cycle is controlled by cyclins and cyclin-dependent kinases. Mutations in genes that control the cell cycle are often associated with cancer.
- Defects in DNA-repair genes often increase the overall mutation rate of other genes, leading to defects in proto-oncogenes and tumor-suppressor genes that can contribute to cancer progression.
- Mutations in sequences that regulate telomerase allow cells to divide indefinitely, contributing to cancer progression. Tumor progression is also affected by mutations in genes that promote vascularization and the spread of tumors.
- Epigenetic changes are frequently associated with cancer.
- Colorectal cancer offers a model system for understanding tumor progression in humans. Initial mutations stimulate cell division, leading to a small benign polyp. Additional mutations allow the polyp to enlarge, invade the muscle layer of the gut, and eventually spread to other sites. Mutations in particular genes affect different stages of this progression.
- Some cancers are associated with specific chromosome mutations, including chromosome deletions, inversions, and translocations. Mutations in some genes cause or allow the missegregation of chromosomes, leading to aneuploidy that can contribute to cancer.
- Viruses are associated with some cancers; they contribute to cell proliferation by mutating and rearranging host genes or by altering the expression of host genes.

IMPORTANT TERMS

malignant (p. 445)
metastasis (p. 445)
clonal evolution (p. 446)
oncogene (p. 448)
tumor-suppressor gene (p. 448)
proto-oncogene (p. 448)
loss of heterozygosity (p. 450)
haploinsufficiency (p. 450)
cyclin-dependent kinase (CDK) (p. 451)
cyclin (p. 451)
apoptosis (p. 451)
human papilloma virus (HPV) (p. 456)

ANSWERS TO CONCEPT CHECKS

1. Retinoblastoma results from at least two separate genetic defects, both of which are necessary for cancer to develop. In sporadic cases, two successive mutations must occur in a single cell, which is unlikely, and therefore the cancer typically occurs in only one eye. In people who have inherited one of the two required mutations, every cell contains this mutation, meaning only a single additional mutation is required for cancer to develop. Given the millions of cells in each eye, there is a high probability that the second mutation will occur in at least one cell of each eye, producing tumors in both eyes and the inheritance of this type of retinoblastoma.

2. Oncogenes stimulate cell proliferation. Mutations in oncogenes are usually dominant because a mutation in a single copy of the gene is usually sufficient to produce a stimulatory effect. Tumor-suppressor genes inhibit cell proliferation. Mutations in tumor-suppressor genes are generally recessive because both copies must be mutated to remove all inhibition.

3. c

4. d

WORKED PROBLEM

In a sample of cancer cells, a specific gene has become duplicated many times. Is this gene likely to be an oncogene or a tumor-suppressor gene? Explain your reasoning.

Solution Strategy

What information is required in your answer to the problem?

Whether the gene is likely to be an oncogene or a tumor-suppressor gene, and why.

What information is provided to solve the problem?

- In cancer cells, the gene has been amplified many times.

For help with this problem, review:

Oncogenes and Tumor-Suppressor Genes in Section 16.2.

Solution Steps

The gene is likely to be an oncogene. Oncogenes stimulate cell proliferation and act in a dominant manner. Therefore, extra copies of an oncogene will result in cell proliferation and cancer. Tumor-suppressor genes, on the other hand, suppress cell proliferation and act in a recessive manner; a single copy of a tumor-suppressor gene is sufficient to prevent cell proliferation. Therefore, extra copies of the tumor-suppressor gene will not lead to cancer.

Recall: An oncogene is an ac erator of cell divis while a tumor-suppressor gene is brake.

COMPREHENSION QUESTIONS

Section 16.1

1. What types of evidence indicate that cancer arises from genetic changes?
2. How can it be true that many types of cancer are genetic and yet not inherited?
3. Outline Knudson's two-hit hypothesis of retinoblastoma and describe how it helps to explain unilateral and bilateral cases of retinoblastoma.
4. Briefly explain how cancer arises through clonal evolution.

Section 16.2

5. What is the difference between an oncogene and a tumor-suppressor gene? Give some examples of the functions of proto-oncogenes and tumor-suppressor genes in normal cells.
6. How do cyclins and CDKs differ? How do they interact in controlling the cell cycle?
7. Briefly outline the events that control the progression of cells through the G_1/S checkpoint in the cell cycle.
8. Why do mutations in genes that encode DNA-repair enzymes often produce a predisposition to cancer?
9. What role do telomeres and telomerase play in cancer progression?

Section 16.3

10. Briefly outline some of the genetic changes that are commonly associated with the progression of colorectal cancer.

Section 16.4

11. Explain how chromosome deletions, inversions, and translocations can cause cancer.
12. Briefly outline how the Philadelphia chromosome leads to chronic myelogenous leukemia.
13. What is genomic instability? Give some ways in which genomic instability can arise.

Section 16.5

14. How do viruses contribute to cancer?

For more questions that test your comprehension of the key chapter concepts, go to LearningCurve for this chapter.

APPLICATION QUESTIONS AND PROBLEMS

Introduction

15. What characteristics of the pedigree shown in **Figure 16.1** suggest that pancreatic cancer in this family is inherited as an autosomal dominant trait?

Section 16.1

16. If cancer is fundamentally a genetic disease, how might an environmental factor such as smoking cause cancer?

*17. Both genes and environmental factors contribute to cancer. Prostate cancer is 30 times more common among people in Utah than among people in Shanghai (see **Table 16.2**). Briefly outline how you might determine whether these differences in the incidence of prostate cancer are due to differences in the genetic makeup of the two populations or to differences in their environments.

Section 16.2

*18. The *palladin* gene, which plays a role in pancreatic cancer (see the introduction to this chapter), is said to be an oncogene. Which of its characteristics suggest that it is an oncogene rather than a tumor-suppressor gene?

19. Mutations in the *RB* gene are often associated with cancer. Explain how a mutation that results in a nonfunctional RB protein contributes to cancer.

20. Cells in a tumor contain mutated copies of a particular gene that promotes tumor growth. Gene therapy can be used to introduce a normal copy of this gene into the tumor cells. Would you expect this therapy to be effective if the mutated gene were an oncogene? A tumor-suppressor gene? Explain your reasoning.

21. Some cancers have been treated with drugs that demethylate DNA. Explain how these drugs might work. Do you think the cancer-causing genes that respond to the demethylation are likely to be oncogenes or tumor-suppressor genes? Explain your reasoning.

Section 16.4

22. Some cancers are consistently associated with the deletion of a particular part of a chromosome. Does the deleted region contain an oncogene or a tumor-suppressor gene? Explain.

CHALLENGE QUESTIONS

Section 16.2

23. Many cancer cells are immortal (will divide indefinitely) because they have mutations that allow telomerase to be expressed. How might this knowledge be used to design anticancer drugs?

24. Bloom syndrome is an autosomal recessive disease that exhibits haploinsufficiency. A recent survey showed that people heterozygous for mutations at the *BLM* locus are at increased risk of colon cancer. Suppose that you are a genetic counselor. A young woman whose mother has Bloom syndrome is referred to you. The young woman's father has no family history of Bloom syndrome, and she asks whether she is likely to experience any other health problems associated with her family history of Bloom syndrome. What advice would you give her?

25. Radiation is known to cause cancer, yet radiation is often used as treatment for some types of cancer. How can radiation be a contributor to both the cause and the treatment of cancer?

26. Imagine that you discover a large family in which bladder cancer is inherited as an autosomal dominant trait. Briefly outline a series of studies that you might conduct to identify the gene that causes bladder cancer in this family.

THINK-PAIR-SHARE QUESTIONS

Introduction

1. Pancreatic cancer is clearly inherited as an autosomal dominant trait in the family illustrated in **Figure 16.1.** Yet most cases of pancreatic cancer are sporadic, appearing as isolated cases in families with no obvious inheritance. How can a trait be strongly inherited in one family and not inherited in another?

2. Is it correct to say that the *paladin* gene causes cancer? Everyone has a *paladin* gene, but not everyone gets pancreatic cancer. What might be a more accurate way to talk about the link between the *paladin* gene and cancer?

3. The mutation associated with pancreatic cancer in the family in **Figure 16.1** was located in an exon of the *paladin* gene. In general, would you expect to find more cancer-causing mutations in exons or in introns? Explain your answer.

Section 16.1

4. The chapter points out that about one of every five women and one of every four men in the United States

will die from cancer. Why are rates of death from cancer different in men and women? Provide some possible explanations.

5. A couple has one child with bilateral retinoblastoma. The mother is free from cancer, but the father has unilateral retinoblastoma and he has a brother who has bilateral retinoblastoma.
   a. If the couple has another child, what is the probability that this next child will have retinoblastoma?
   b. If the next child has retinoblastoma, is it likely to be bilateral or unilateral?
   c. Explain why the father's case of retinoblastoma is unilateral, whereas his son's and brother's cases are bilateral.

6. **Table 16.3** lists occupation as an environmental factor associated with cancer. What occupations do you think might be associated with higher rates of cancer? What about these occupations might create higher cancer rates?

Section 16.5

7. In 2007, then–Texas governor Rick Perry became the first governor in the United States to mandate by executive order the vaccination of all Texas girls age 11–12 for HPV, arguing that the vaccine prevented cancer. Some conservatives criticized Perry, arguing that the vaccine would encourage sexual promiscuity among children and young adults. The Texas legislature overturned the executive order, and Perry later reversed himself, saying the order was a mistake. Do you think states should require boys and girls to be vaccinated for HPV? Discuss reasons for and against such a requirement.

SaplingPlus Self-study tools that will help you practice what you've learned and reinforce this chapter's concepts are available online. Go to www.macmillanlearning.com/PierceGenetics6e.

17 Quantitative Genetics

Corn Oil and Quantitative Genetics

Methods of quantitative genetics coupled with molecular techniques have been used to identify a gene that determines oil content in corn. [Jim Craigmyle/Corbis.]

In 2017, the world's population surpassed 7.5 billion. The United Nations projects that by 2050 it will increase by another 2.2 billion. Feeding those additional people will be a major challenge for agriculture in the next few decades. Crop plants will have to provide most of the calories and nutrients required for the world's future population. Because of dwindling petroleum supplies and concerns about global warming, plants are also increasingly being used as sources of biofuels, placing additional demands on crop production.

To help meet the need for increased crop yields, plant breeders are using the latest genetic techniques in their quest to develop higher-yielding, more efficient crop plants. The power of this approach is demonstrated by research aimed at increasing the oil content of corn. The oil content of corn is an inherited characteristic, but its inheritance is more complex than that of the characteristics that we have studied so far; it is not a simple single-gene characteristic, like seed shape in peas. Numerous genes and environmental factors contribute to the oil content of corn: it results from the interaction of several loci, and their expression is affected by environmental factors.

Can the inheritance of a complex characteristic such as oil content be studied? Is it possible to predict the oil content of a plant on the basis of its breeding? The answers are yes—at least in part—but these questions cannot be addressed with the methods that we used for simple genetic characteristics. Instead, we must use statistical procedures that have been developed for analyzing complex characteristics.

In 2008, geneticists used a combination of quantitative genetic and molecular techniques to identify a gene that plays a key role in controlling the oil content of corn. First, they conducted crosses between high-oil and low-oil corn plants to identify chromosome regions that play an important role in determining oil production. Chromosome regions containing genes that influence a quantitative trait are termed quantitative trait loci (QTLs). Through these crosses, the geneticists located several QTLs that affected oil content; one of them was on corn chromosome 6.

Fine-scale genetic mapping further narrowed the QTL down to a small region of 4.2 map units (centiMorgans) on chromosome 6. Researchers sequenced DNA from the region and found that it contained five genes, one of which was *DGAT1-2*, a gene known to encode an enzyme that catalyzes the final step in a pathway for triacylglycerol biosynthesis. DNA from the *DGAT1-2* gene in a high-oil-producing strain of corn contained an insertion of a codon that added phenylalanine to the enzyme, an insertion that was missing from a low-oil-producing strain. The researchers confirmed the effect of the additional phenylalanine codon on oil production by producing transgenic corn that contained the extra codon; these transgenic strains produced more oil than transgenic strains without the extra codon. Interestingly, the extra phenylalanine codon is present in wild relatives of corn, suggesting that the codon was lost in the process of domestication or subsequent breeding of modern varieties.

This research suggests that the oil content of corn and other plants might be increased by genetically modifying their *DGAT1-2* genes to contain the extra codon for phenylalanine. Other studies that similarly combine quantitative and molecular analyses have led to the identification of genes that increase the vitamin A content of rice and sugar production in tomatoes.

THINK-PAIR-SHARE Questions 1 and 2

The genetic analysis of complex characteristics such as the oil content of corn is known as **quantitative genetics**. We begin this chapter by considering the differences between quantitative and qualitative characteristics and why the expression of some characteristics varies continuously. We'll see that quantitative characteristics are often influenced by many genes, each of which has a small effect on the phenotype. Next, we'll examine statistical procedures for describing and analyzing quantitative characteristics. We'll consider the question of how much of phenotypic variation can be attributed to genetic and environmental influences. Finally, we'll look at the effects of selection on quantitative characteristics.

17.1 Many Quantitative Characteristics Are Influenced by Alleles at Multiple Loci

Qualitative, or discontinuous, characteristics possess only a few distinct phenotypes (**Figure 17.1a**); these characteristics are the types studied by Mendel (e.g., round and wrinkled peas) and have been the focus of our attention thus far. However, many characteristics vary continuously along a scale of measurement, with many overlapping phenotypes (**Figure 17.1b**). They are referred to as *continuous characteristics*, and they are also called *quantitative characteristics* because any individual's phenotype must be described by a quantitative measurement. Examples of quantitative characteristics include height, weight, and blood pressure in humans, growth rate in mice, seed weight in plants, and milk production in cattle.

Quantitative characteristics arise from two phenomena. First, many are polygenic: they are influenced by genes at many loci. If many loci take part, many genotypes are possible, each producing a slightly different phenotype. Second, quantitative characteristics often arise when environmental factors affect the phenotype because environmental differences result in a single genotype producing a range of phenotypes. Most continuously varying characteristics are *both* polygenic *and* influenced by environmental factors, and these characteristics are said to be multifactorial.

The Relation Between Genotype and Phenotype

For some discontinuous characteristics, the relation between genotype and phenotype is straightforward: each genotype produces a single phenotype, and most phenotypes are encoded by a single genotype. Dominance and epistasis may allow two or three different genotypes to produce the same

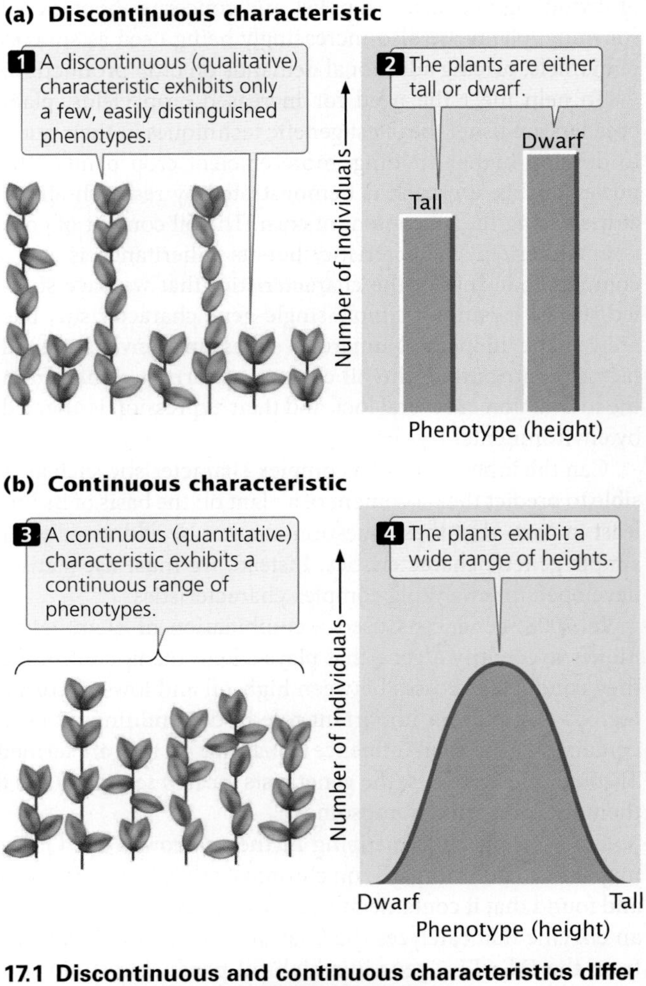

17.1 Discontinuous and continuous characteristics differ in the number of phenotypes exhibited.

phenotype, but the relation remains simple. This simple relation between genotype and phenotype allowed Mendel to decipher the basic rules of inheritance from his crosses with pea plants; it also permits us to predict the outcome of genetic crosses and to assign genotypes to individuals.

For quantitative characteristics, the relation between genotype and phenotype is usually more complex. If the characteristic is polygenic, many different genotypes are possible, several of which may produce the same phenotype. For instance, consider a plant whose height is determined by three loci (*A*, *B*, and *C*), each of which has two alleles. Assume that one allele at each locus (A^+, B^+, and C^+) encodes a plant hormone that causes the plant to grow 1 cm above its baseline height of 10 cm. The other allele at each locus (A^-, B^-, and C^-) does not encode a plant hormone and thus does not contribute to additional height. If we consider only the two alleles at a single locus, 3 genotypes are possible (A^+A^+, A^+A^-, and A^-A^-). If all three loci are taken into account, there are a total of $3^3 = 27$ possible multilocus genotypes ($A^+A^+\ B^+B^+\ C^+C^+$, $A^+A^-\ B^+B^+\ C^+C^+$, etc.). Although there are 27 genotypes, they produce only seven phenotypes (10 cm, 11 cm, 12 cm, 13 cm, 14 cm, 15 cm, and 16 cm in height) because some of the genotypes produce the same phenotype (**Table 17.1**). For example, genotypes $A^+A^-\ B^-B^-\ C^-C^-$, $A^-A^-\ B^+B^-\ C^-C^-$, and $A^-A^-\ B^-B^-\ C^+C^-$ each have one gene that encodes a plant hormone. Each of these genotypes produces one dose of the hormone and results in a plant that is 11 cm tall. Even in this simple example with only three loci, the relation between genotype and phenotype is quite complex. The more loci encoding a characteristic, the greater the complexity. As the number of loci encoding a characteristic increases, the number of potential phenotypes increases, and differences between individual phenotypes become more difficult to distinguish.

The influence of environment on a characteristic can also complicate the relation between genotype and phenotype. Because of environmental effects, the same genotype can produce a range of potential phenotypes. Furthermore, the phenotypic ranges of different genotypes can overlap, making it difficult to know whether individuals differ in phenotype because of genetic or environmental differences (**Figure 17.2**).

In summary, the simple relation between genotype and phenotype that exists for many qualitative (discontinuous) characteristics is absent in quantitative characteristics, and it is impossible to assign a genotype to an individual on the basis of its phenotype alone. The methods used for analyzing qualitative characteristics (examining the phenotypic ratios of progeny from a genetic cross) will not work with quantitative characteristics. Our goal remains the same: we wish to make predictions about the phenotypes of offspring produced in a genetic cross. We may also want to know how much of the variation in a characteristic results from genetic differences and how much results from environmental differences. To answer these questions, we must turn to statistical methods that allow us to make predictions about the inheritance of phenotypes in the absence of information about the underlying genotypes. ▶TRY PROBLEM 11

TABLE 17.1 Hypothetical example of plant height determined by pairs of alleles at each of three loci

Plant genotype	Doses of hormone	Height (cm)
$A^-A^-\ B^-B^-\ C^-C^-$	0	10
$A^+A^-\ B^-B^-\ C^-C^-$	1	11
$A^-A^-\ B^+B^-\ C^-C^-$		
$A^-A^-\ B^-B^-\ C^-C^+$		
$A^+A^+\ B^-B^-\ C^-C^-$	2	12
$A^-A^-\ B^+B^+\ C^-C^-$		
$A^-A^-\ B^-B^-\ C^+C^+$		
$A^+A^-\ B^+B^-\ C^-C^-$		
$A^+A^-\ B^-B^-\ C^+C^-$		
$A^-A^-\ B^+B^-\ C^+C^-$		
$A^+A^+\ B^+B^-\ C^-C^-$	3	13
$A^+A^+\ B^-B^-\ C^+C^-$		
$A^+A^-\ B^+B^+\ C^-C^-$		
$A^-A^-\ B^+B^+\ C^+C^-$		
$A^+A^-\ B^-B^-\ C^+C^+$		
$A^-A^-\ B^+B^-\ C^+C^+$		
$A^+A^-\ B^+B^-\ C^+C^-$		
$A^+A^+\ B^+B^+\ C^-C^-$	4	14
$A^+A^+\ B^+B^-\ C^+C^-$		
$A^+A^-\ B^+B^+\ C^+C^-$		
$A^+A^+\ B^-B^-\ C^+C^+$		
$A^+A^-\ B^+B^-\ C^+C^+$		
$A^-A^-\ B^+B^+\ C^+C^+$		
$A^+A^+\ B^+B^+\ C^+C^-$	5	15
$A^+A^-\ B^+B^+\ C^+C^+$		
$A^+A^+\ B^+B^-\ C^+C^+$		
$A^+A^+\ B^+B^+\ C^+C^+$	6	16

Note: Each + allele contributes 1 cm in height above a baseline of 10 cm.

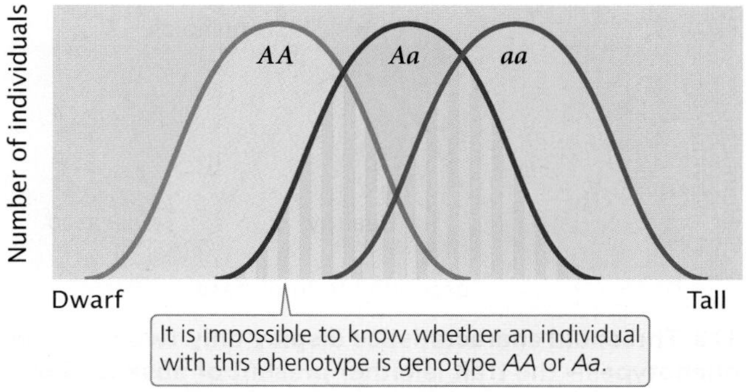

17.2 For a quantitative characteristic influenced by environmental factors, each genotype can produce a range of possible phenotypes. In this hypothetical example, the phenotypes produced by genotypes *AA*, *Aa*, and *aa* overlap.

Types of Quantitative Characteristics

Before we look more closely at polygenic characteristics and relevant statistical methods, we need to more clearly define what is meant by a quantitative characteristic. Thus far, we have considered only quantitative characteristics that vary continuously in a population. A *continuous characteristic* can theoretically assume any value between two extremes; the number of phenotypes is limited only by our ability to precisely measure the phenotype. Human height is a continuous characteristic because, within certain limits, people can theoretically have any height. Although the number of phenotypes possible with a continuous characteristic is infinite, we often group similar phenotypes together for convenience; we may say that two people are both 5 feet 11 inches tall, but careful measurement may show that one is slightly taller than the other.

Some characteristics are not continuous but are nevertheless considered quantitative because they are determined by multiple genetic and environmental factors. **Meristic characteristics**, for instance, are measured in whole numbers. An example is litter size: a female mouse can have 4, 5, or 6 pups, but not 4.13 pups. A meristic characteristic has a limited number of distinct phenotypes, but the underlying determination of the characteristic can still be quantitative. These characteristics must therefore be analyzed with the same techniques that we use to study continuous quantitative characteristics.

Another type of quantitative characteristic is a **threshold characteristic**, which is simply present or absent. Although threshold characteristics exhibit only two phenotypes, they are considered quantitative because they, too, are determined by multiple genetic and environmental factors. The expression of the characteristic depends on an underlying susceptibility (usually referred to as liability or risk) that varies continuously. When the susceptibility is larger than a threshold value, a specific trait is expressed (**Figure 17.3**). Diseases are often threshold characteristics because many factors, both genetic and environmental, contribute to disease susceptibility. If enough of the susceptibility factors are present, the disease develops; otherwise, it is absent. Although we focus on the genetics of continuous characteristics in this chapter, the same principles apply to many meristic and threshold characteristics.

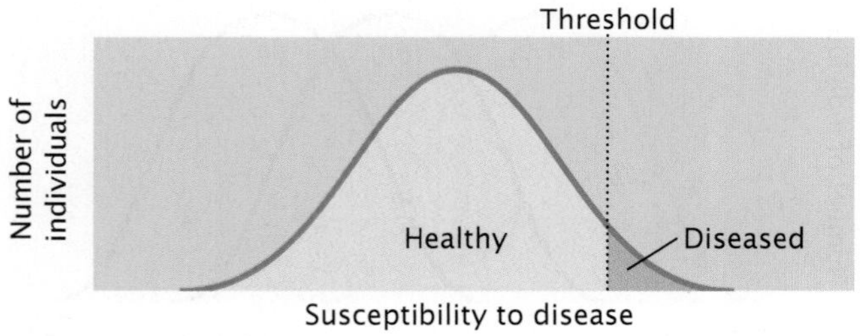

17.3 Threshold characteristics display only two possible phenotypes—the trait is either present or absent—but they are quantitative because the underlying susceptibility to the characteristic varies continuously. When the susceptibility exceeds a threshold value, the characteristic is expressed.

It is important to point out that just because a characteristic can be measured on a continuous scale does not mean that it exhibits quantitative variation. One of the characteristics studied by Mendel was height of the pea plant, which can be described by measuring the length of the plant's stem. However, Mendel's particular plants exhibited only two distinct phenotypes (some were tall and others short), and these differences were determined by alleles at a single locus. The differences that Mendel studied were therefore discontinuous in nature.

CONCEPTS

Characteristics for which the phenotypes vary continuously are quantitative characteristics. For most quantitative characteristics, the relation between genotype and phenotype is complex. Some characteristics for which the phenotypes do not vary continuously are also considered quantitative because the phenotypes are influenced by multiple genes and environmental factors.

Polygenic Inheritance

After the rediscovery of Mendel's work in 1900, questions soon arose about the inheritance of continuously varying characteristics. These characteristics had already been the focus of a group of biologists and statisticians, led by Francis Galton, who used statistical procedures to examine the inheritance of quantitative characteristics such as human height and intelligence. The results of these studies showed that quantitative characteristics were at least partly inherited, although the mechanism of inheritance was not yet known. Some biometricians argued that the inheritance of quantitative characteristics could not be explained by Mendelian principles, whereas others believed that Mendel's principles acting on numerous genes (polygenes) could adequately account for the inheritance of quantitative characteristics.

This conflict began to be resolved through independent work by Wilhelm Johannsen, George Udny Yule, and Herman Nilsson-Ehle, each of whom studied continuous variation in plants. The argument was finally laid to rest in 1918, when Ronald Fisher demonstrated that the inheritance of quantitative characteristics could indeed be explained by the cumulative effects of many genes, each following Mendel's rules.

Kernel Color in Wheat

To illustrate how multiple genes acting on a characteristic can produce a continuous range of phenotypes, let us examine one of the first demonstrations of polygenic inheritance. Nilsson-Ehle studied kernel color in wheat and found that the intensity of red pigmentation was determined by three unlinked loci, each of which had two alleles.

NILSSON-EHLE'S CROSS Nilsson-Ehle obtained several homozygous varieties of wheat that differed in color. Like Mendel, he performed crosses between these homozygous varieties and studied the ratios of phenotypes in the progeny. In one experiment, he crossed a variety of wheat that possessed white kernels with a variety that possessed purple kernels and obtained the following results:

P: Plants with white kernels × Plants with purple kernels

↓

F_1: Plants with red kernels

↓

F_2:
$^1/_{16}$ plants with purple kernels
$^4/_{16}$ plants with dark-red kernels
$^6/_{16}$ plants with red kernels
$^4/_{16}$ plants with light-red kernels
$^1/_{16}$ plants with white kernels

INTERPRETATION OF THE CROSS Nilsson-Ehle interpreted this phenotypic ratio as the result of the segregation of alleles at two loci (although he found alleles at three loci that affect kernel color, the two varieties used in this cross differed at only two of the loci). He proposed that there were two alleles at each locus: one that produced red pigment and another that produced no pigment. We'll designate the alleles that encoded pigment A^+ and B^+ and the alleles that encoded no pigment A^- and B^-. Nilsson-Ehle recognized that the effects of the genes were additive. Each gene seemed to contribute equally to color; so the overall phenotype could be determined by adding the effects of all the alleles, as shown in the following table:

Genotype	Doses of pigment	Phenotype
$A^+A^+\ B^+B^+$	4	Purple
$A^+A^+\ B^+B^-$ $A^+A^-\ B^+B^+$	3	Dark red
$A^+A^+\ B^-B^-$ $A^-A^-\ B^+B^+$ $A^+A^-\ B^+B^-$	2	Red
$A^+A^-\ B^-B^-$ $A^-A^-\ B^+B^-$	1	Light red
$A^-A^-\ B^-B^-$	0	White

Notice that the purple and white phenotypes are each encoded by a single genotype, but the other phenotypes may result from several different genotypes.

From these results, we see that five phenotypes are possible when alleles at two loci influence the phenotype and the effects of the genes are additive. If alleles at more than two loci influenced the phenotype, more phenotypes would be possible, and the color would appear to vary continuously between white and purple. If environmental factors had influenced the characteristic, individuals of the same genotype would vary somewhat in color, making it even more difficult to distinguish between discrete phenotypic classes. Luckily, environment played little role in determining kernel color in Nilsson-Ehle's crosses, and only a few loci encoded color, so Nilsson-Ehle was able to distinguish among the different phenotypic classes. This ability allowed him to see the Mendelian nature of the characteristic.

Let's now see how Mendel's principles explain the ratio obtained by Nilsson-Ehle in his F_2 progeny. Remember that Nilsson-Ehle crossed the homozygous purple variety ($A^+A^+\ B^+B^+$) with the homozygous white variety ($A^-A^-\ B^-B^-$), producing F_1 progeny that were heterozygous at both loci ($A^+A^-\ B^+B^-$). This is a dihybrid cross, like those that we worked in Chapter 3, except that both loci encode the same trait. Each of the F_1 plants possessed two pigment-producing alleles, so each plant received two doses of pigment, which produced red kernels. The types and proportions of progeny expected in the F_2 can be found by applying Mendel's principles of segregation and independent assortment.

Let's first examine the effects of each locus separately. At the first locus, two heterozygous F_1s are crossed ($A^+A^- \times A^+A^-$). As we learned in Chapter 3, when two heterozygotes are crossed, we expect progeny in the proportions $^1/_4\ A^+A^+$, $^1/_2\ A^+A^-$, and $^1/_4\ A^-A^-$. At the second locus, two heterozygotes also are crossed, and again, we expect progeny in the proportions $^1/_4\ B^+B^+$, $^1/_2\ B^+B^-$, and $^1/_4\ B^-B^-$.

To obtain the probability of combinations of genes at both loci, we must use the multiplication rule (see Chapter 3), the use of which assumes Mendel's principle of independent assortment. The expected proportion of F_2 progeny with genotype $A^+A^+\ B^+B^+$ is the product of the probability of obtaining genotype A^+A^+ ($^1/_4$) and the probability of obtaining genotype B^+B^+ ($^1/_4$), or $^1/_4 \times {}^1/_4 = {}^1/_{16}$ (**Figure 17.4**). The probabilities of each of the phenotypes can then be obtained by adding the probabilities of all the genotypes that produce that phenotype. For example, the red phenotype is produced by three genotypes:

Genotype	Phenotype	Probability
$A^+A^+\ B^-B^-$	Red	$^1/_4 \times {}^1/_4 = {}^1/_{16}$
$A^-A^-\ B^+B^+$	Red	$^1/_4 \times {}^1/_4 = {}^1/_{16}$
$A^+A^-\ B^+B^-$	Red	$^1/_2 \times {}^1/_2 = {}^1/_4$

Thus, the overall probability of obtaining red kernels in the F_2 progeny is $^1/_{16} + {}^1/_{16} + {}^1/_4 = {}^6/_{16}$. Figure 17.4 shows that the phenotypic ratio expected in the F_2 is $^1/_{16}$ purple,

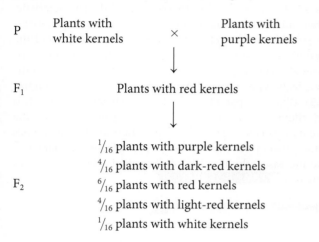

Experiment

Question: How is a continous trait, such as kernel color in wheat, inherited?

Methods Cross wheat with white kernels and wheat with purple kernels. Intercross the F_1 to produce F_2.

P generation

$A^+A^+B^+B^+$ × $A^-A^-B^-B^-$
Purple White

Results

F_1 generation

$A^+A^-B^+B^-$ Red

Break into simple crosses

A^+A^- × A^+A^- B^+B^- × B^+B^-

$\frac{1}{4}A^+A^+$ $\frac{1}{2}A^+A^-$ $\frac{1}{4}A^-A^-$ $\frac{1}{4}B^+B^+$ $\frac{1}{2}B^+B^-$ $\frac{1}{4}B^-B^-$

Combine results

F_2 generation		Number of pigment genes	Phenotype

$\frac{1}{4}A^+A^+$
→ $\frac{1}{4}B^+B^+$ → $\frac{1}{4}×\frac{1}{4}=\frac{1}{16}$ 4 $A^+A^+B^+B^+$ Purple
→ $\frac{1}{2}B^+B^-$ → $\frac{1}{4}×\frac{1}{2}=\frac{2}{16}$ 3 $A^+A^+B^+B^-$ Dark red
→ $\frac{1}{4}B^-B^-$ → $\frac{1}{4}×\frac{1}{4}=\frac{1}{16}$ 2 $A^+A^+B^-B^-$ Red

$\frac{1}{2}A^+A^-$
→ $\frac{1}{4}B^+B^+$ → $\frac{1}{2}×\frac{1}{4}=\frac{2}{16}$ 3 $A^+A^-B^+B^+$ Dark red
→ $\frac{1}{2}B^+B^-$ → $\frac{1}{2}×\frac{1}{2}=\frac{4}{16}$ 2 $A^+A^-B^+B^-$ Red
→ $\frac{1}{4}B^-B^-$ → $\frac{1}{2}×\frac{1}{4}=\frac{2}{16}$ 1 $A^+A^-B^-B^-$ Light red

$\frac{1}{4}A^-A^-$
→ $\frac{1}{4}B^+B^+$ → $\frac{1}{4}×\frac{1}{4}=\frac{1}{16}$ 2 $A^-A^-B^+B^+$ Red
→ $\frac{1}{2}B^+B^-$ → $\frac{1}{4}×\frac{1}{2}=\frac{2}{16}$ 1 $A^-A^-B^+B^-$ Light red
→ $\frac{1}{4}B^-B^-$ → $\frac{1}{4}×\frac{1}{4}=\frac{1}{16}$ 0 $A^-A^-B^-B^-$ White

Combine common phenotypes

F_2 ratio

Frequency	Number of pigment genes	Phenotype
$\frac{1}{16}$	4	Purple
$\frac{4}{16}$	3	Dark red
$\frac{6}{16}$	2	Red
$\frac{4}{16}$	1	Light red
$\frac{1}{16}$	0	White

Conclusion: Kernel color in wheat is inherited according to Mendel's principles acting on alleles at two loci.

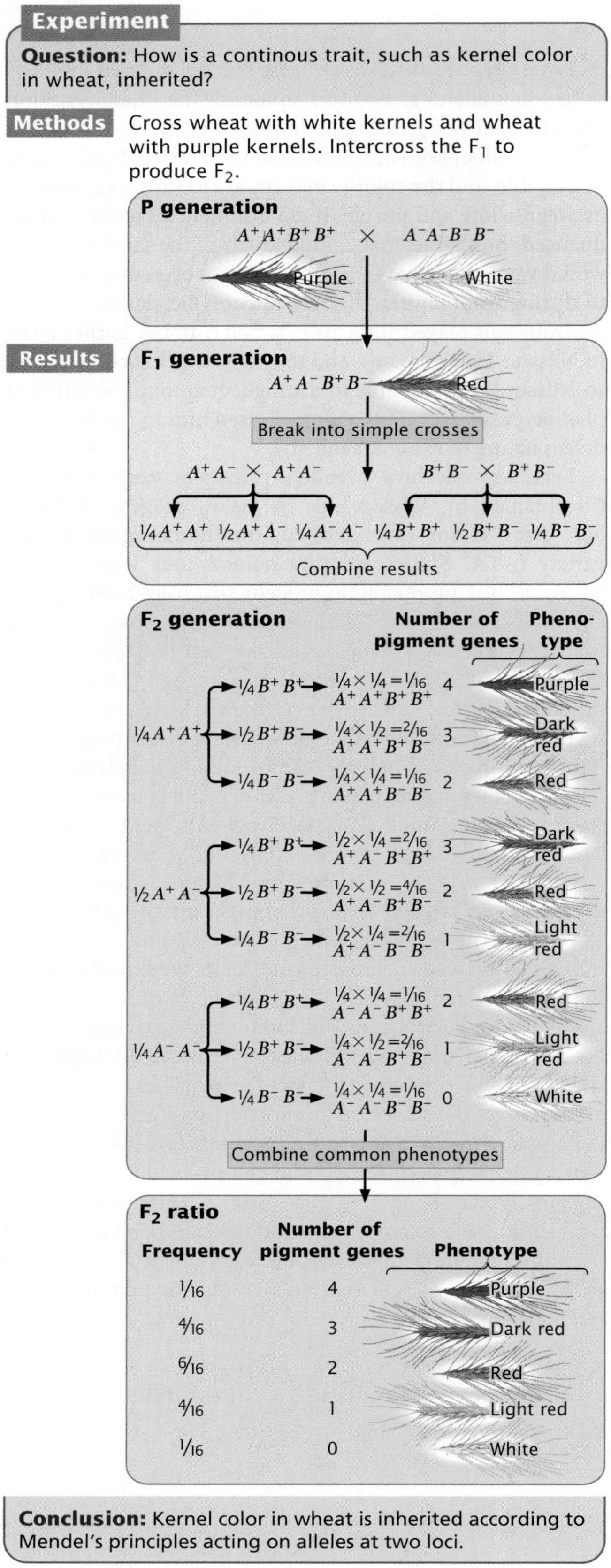

17.4 Nilsson-Ehle demonstrated that kernel color in wheat is inherited according to Mendelian principles.
The ratio of phenotypes in the F_2 can be determined by breaking the dihybrid cross into two simple single-locus crosses and combining the results by using the multiplication rule.

$\frac{4}{16}$ dark red, $\frac{6}{16}$ red, $\frac{4}{16}$ light red, and $\frac{1}{16}$ white. This phenotypic ratio is precisely what Nilsson-Ehle observed in his F_2 progeny, demonstrating that the inheritance of a continuously varying characteristic such as kernel color does indeed occur according to Mendel's basic principles.

CONCLUSIONS AND IMPLICATIONS Nilsson-Ehle's crosses demonstrated that the difference between the inheritance of genes influencing quantitative characteristics and the inheritance of genes influencing discontinuous characteristics is in the *number* of loci that determine the characteristic. When multiple loci affect a characteristic, more genotypes are possible, so the relation between the genotype and the phenotype is less obvious. For example, in a cross of F_1 individuals heterozygous for alleles at a single locus with additive effects, 3 phenotypes appear among the progeny (**Figure 17.5**). When the parents of the cross are heterozygous at two loci, there are 5 phenotypes in the progeny, and when the parents are heterozygous at five loci, there are 11 phenotypes in the progeny. As the number of loci affecting a characteristic increases, the number of phenotypic classes in the F_2 increases. ▶TRY PROBLEM 12

THINK-PAIR-SHARE Question 3

One locus, $Aa × Aa$

Two loci, $Aa\,Bb × Aa\,Bb$

1 As the number of loci affecting the trait increases, . . .

Five loci, $Aa\,Bb\,Cc\,Dd\,Ee × Aa\,Bb\,Cc\,Dd\,Ee$

Many loci

2 . . . the number of phenotypic classes increases.

Relative number of progeny

Phenotypic classes

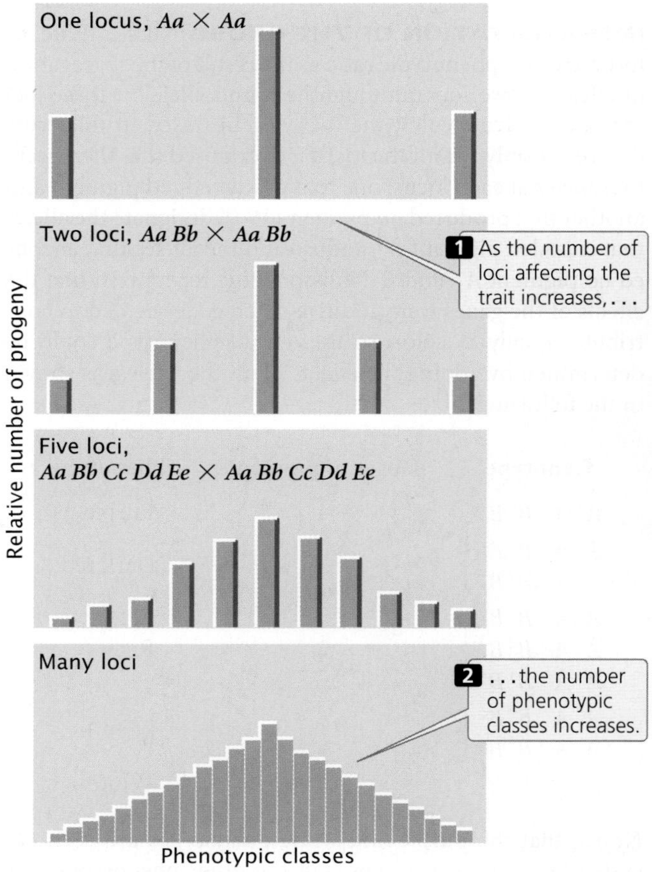

17.5 The results of crossing individuals heterozygous for different numbers of loci affecting a characteristic.

CONCEPTS

The same principles determine the inheritance of quantitative and discontinuous characteristics, but more genes take part in the determination of quantitative characteristics.

17.2 Statistical Methods Are Required for Analyzing Quantitative Characteristics

Because quantitative characteristics are described by a measurement and are influenced by multiple factors, their inheritance must be analyzed statistically.

Distributions

Understanding the genetic basis of any characteristic begins with a description of the numbers and kinds of phenotypes present in a group of individuals. Phenotypic variation in a group can be conveniently represented by a **frequency distribution**, which is a graph of the frequencies (numbers or proportions) of the different phenotypes (**Figure 17.6**). In a typical frequency distribution, the phenotypic classes are plotted on the horizontal (*x*) axis, and the numbers (or proportions) of individuals in each class are plotted on the vertical (*y*) axis. A frequency distribution is a concise method of summarizing all phenotypes of a quantitative characteristic.

Connecting the points of a frequency distribution with a line creates a curve that is characteristic of the distribution. Many quantitative characteristics exhibit a symmetrical (bell-shaped) curve called a **normal distribution** (**Figure 17.7a**). Normal distributions arise when a large number of independent factors contribute to a measurement, as is often the case in quantitative characteristics. Two other common types of distributions (skewed and bimodal) are illustrated in **Figure 17.7b** and **c**.

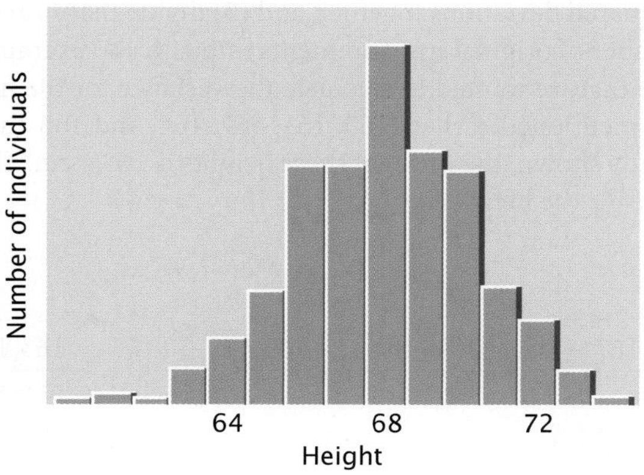

17.6 A frequency distribution is a graph that displays the numbers or proportions of different phenotypes. Phenotypic values (height in inches) are plotted on the horizontal axis, and the numbers (or proportions) of individuals with each phenotype are plotted on the vertical axis.

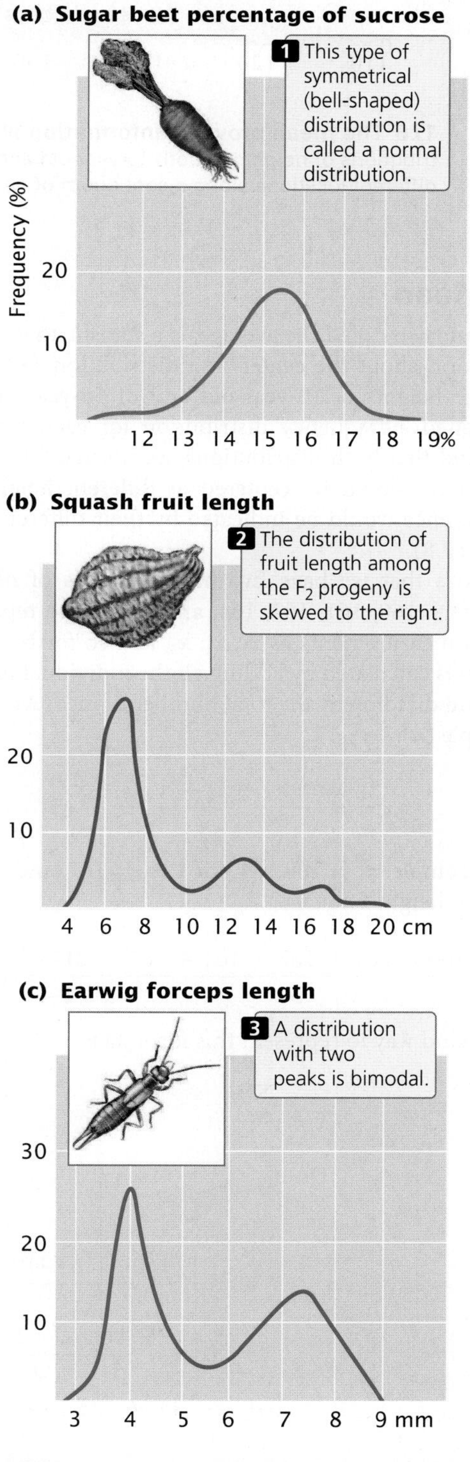

17.7 Distributions of phenotypes can assume several different shapes.

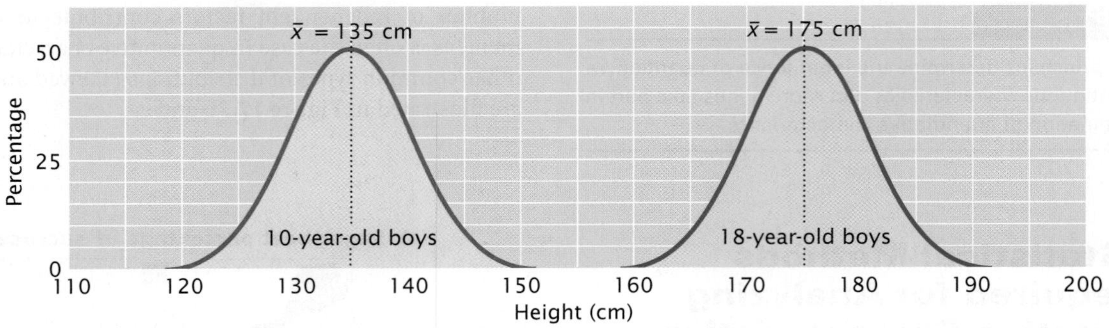

17.8 The mean provides information about the center of a distribution. The distributions of heights of both 10-year-old and 18-year-old boys are normal, but they have different locations along a continuum of height, which makes their means different.

The Mean

The **mean**, also called the average, is a statistic that provides information about the center of a distribution. If we measured the heights of 10-year-old and of 18-year-old boys, then plotted a frequency distribution for each group, we would find that both distributions are normal, but the two distributions would be centered at different heights, and this difference would be indicated by their different means (**Figure 17.8**).

Suppose that we have five measurements of height in centimeters: 160, 161, 167, 164, and 165. If we represent a group of measurements as x_1, x_2, x_3, and so forth, then the mean ($\bar{x}$) is calculated by adding all the individual measurements and dividing by the total number of measurements in the sample (n):

$$\bar{x} = \frac{x_1 + x_2 + x_3 + \ldots + x_n}{n} \tag{17.1}$$

In our example, $x_1 = 160$, $x_2 = 161$, $x_3 = 167$, and so forth. The mean height $\bar{x}$ equals

$$\bar{x} = \frac{160 + 161 + 167 + 164 + 165}{5} = \frac{817}{5} = 163.4$$

A shorthand way to represent this formula is

$$\bar{x} = \frac{\sum x_i}{n}. \tag{17.2}$$

or

$$\bar{x} = \frac{1}{n}\sum x_i \tag{17.3}$$

where the symbol $\sum$ means "the summation of" and x_i represents the individual x values.

The Variance

A statistic that provides key information about a distribution is its **variance**, which indicates the variability of a group of measurements, or how spread out the distribution is. Distributions can have the same mean but different variances (**Figure 17.9**). The larger the variance, the greater the spread of measurements in a distribution around its mean.

The variance (s^2) is defined as the average squared deviation from the mean:

$$s^2 = \frac{\sum (x_i - \bar{x})^2}{n - 1} \tag{17.4}$$

To calculate the variance, we **(1)** subtract the mean from each measurement and square the value obtained, **(2)** add all of these squared deviations together, and **(3)** divide that sum by the number of original measurements minus 1. For example, suppose that we wanted to calculate the variance for the five heights mentioned earlier (160, 161, 167, 164, and 165 cm). As already shown, the mean of these heights is 163.4 cm. The variance for the heights is

$$\begin{aligned} s^2 &= \frac{(160 - 163.4)^2 + (161 - 163.4)^2 + (167 - 163.4)^2 + (164 - 163.4)^2 + (165 - 163.4)^2}{5 - 1} \\ &= \frac{(-3.4)^2 + (-2.4)^2 + (3.6)^2 + (0.6)^2 + (1.6)^2}{4} \\ &= \frac{11.56 + 5.76 + 12.96 + 0.36 + 2.56}{4} \\ &= 8.3 \end{aligned}$$

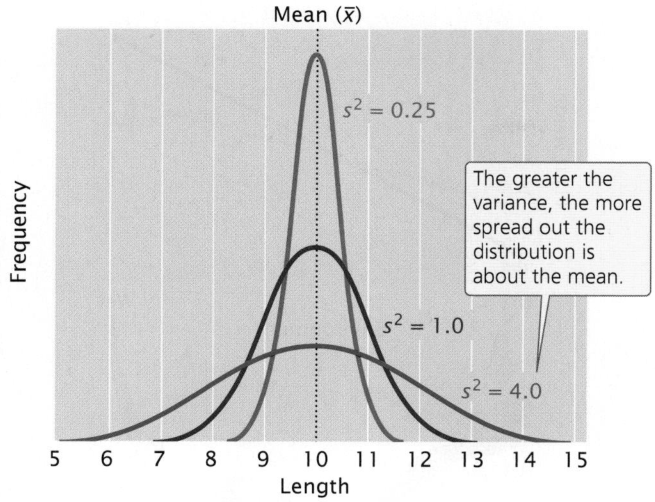

17.9 The variance provides information about the variability of a group of phenotypes. Shown here are three distributions with the same mean but different variances.

CONCEPTS

The mean and the variance are statistics that describe a frequency distribution: the mean provides information about the location of the center of a distribution, and the variance provides information about its variability.

Applying Statistics to the Study of a Polygenic Characteristic

Edward East carried out an early statistical study of polygenic inheritance on the length of flowers in tobacco (*Nicotiana longiflora*). He obtained two varieties of tobacco that differed in flower length: one variety had a mean flower length of 40.5 mm, and the other had a mean flower length of 93.3 mm (**Figure 17.10**). Each of these two varieties had been inbred for many generations and was homozygous at all loci contributing to flower length. Thus, there was no genetic variation in the original parental strains; the small variations in flower length within each strain were due to environmental effects on flower length.

When East crossed the two strains, he found that the mean flower length in the F_1 was about halfway between those of the two parents (see Figure 17.10), as would be expected if the genes determining the differences in the two strains were additive in their effects. The variance of flower length in the F_1 was similar to that seen in the parental strains because all the F_1 had the same genotype, as did each parental strain (the F_1 were all heterozygous at the genes that differed between the two parental varieties).

East then interbred the F_1 to produce F_2 progeny. The mean flower length of the F_2 was similar to that of the F_1, but the variance of the F_2 was much greater (see Figure 17.10). This greater variation indicates that not all of the F_2 progeny had the same genotype.

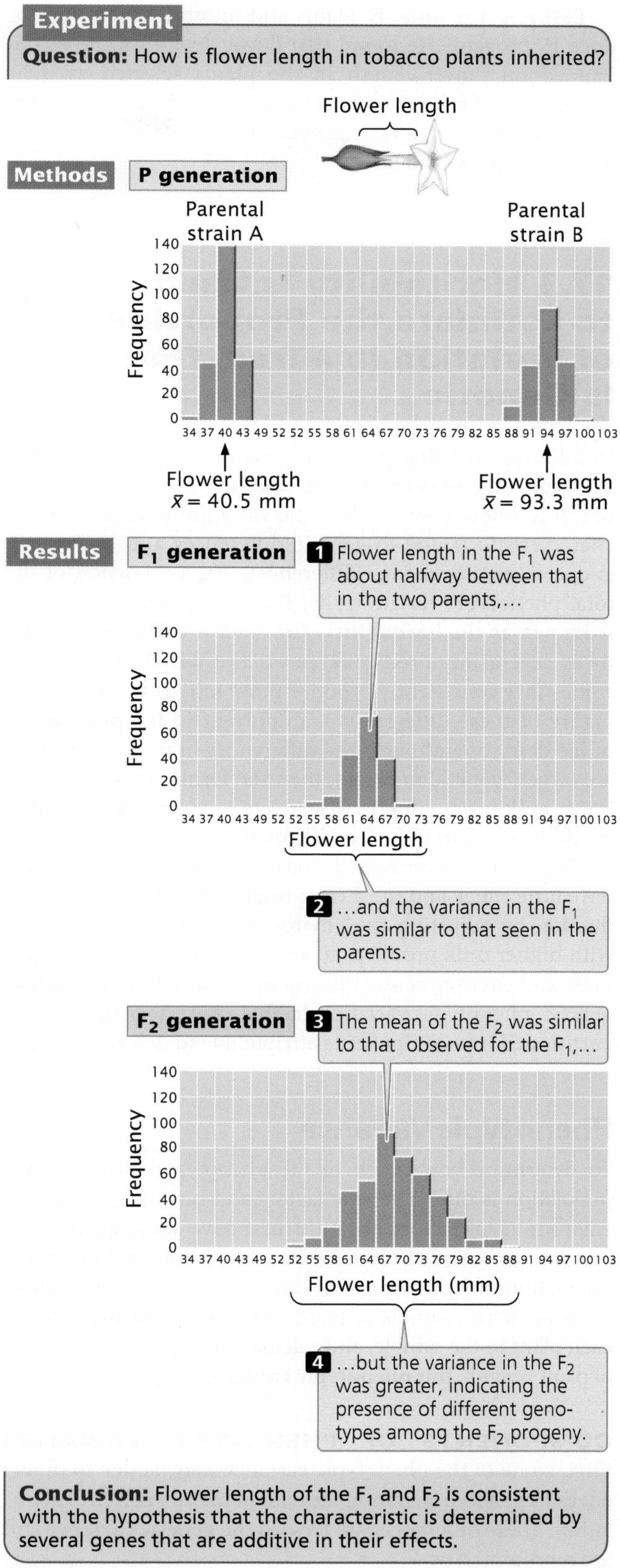

17.10 Edward East conducted an early statistical study of the inheritance of flower length in tobacco.

East selected some F_2 plants and interbred them to produce F_3 progeny. He found that flower length in the F_3 depended on flower length in the plants selected as their parents. This finding demonstrated that flower-length differences in the F_2 were partly genetic and were therefore passed to the next generation.

17.3 Heritability Is Used to Estimate the Proportion of Variation in a Trait That Is Genetic

In addition to being polygenic, quantitative characteristics are frequently influenced by environmental factors. It is often useful to know how much of the variation in a quantitative characteristic is due to genetic differences and how much is due to environmental differences. The proportion of the total phenotypic variation that is due to genetic differences is known as the **heritability**. Consider a dairy farmer who owns several hundred milk cows. The farmer notices that some cows consistently produce more milk than others. The nature of these differences is important to the profitability of his dairy operation. If the differences in milk production are largely genetic in origin, then the farmer may be able to boost milk production by selectively breeding the cows that produce the most milk. On the other hand, if the differences are largely environmental in origin, selective breeding will have little effect, and the farmer might better boost milk production by adjusting the environmental factors associated with higher milk production. To determine the extent of genetic and environmental influences on variation in a characteristic, phenotypic variation in that characteristic must be partitioned into components attributable to different factors.

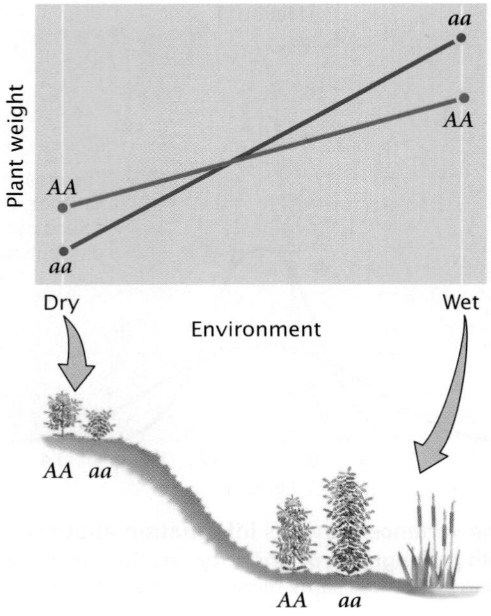

17.11 Genetic–environmental interaction variance is obtained when the effect of a gene depends on the specific environment in which it is found. In this theoretical example, the genotype affects plant weight, but the environmental conditions determine which genotype produces the heavier plant.

Phenotypic Variance

To determine how much of the phenotypic variation in a population is due to genetic factors and how much is due to environmental factors, we must first have some quantitative measure of the phenotype under consideration. Consider a population of wild plants that differ in size. We could collect a representative sample of plants from the population, weigh each plant in the sample, and calculate the mean and variance of plant weight. This **phenotypic variance** is represented by V_P.

COMPONENTS OF PHENOTYPIC VARIANCE

First, some of the phenotypic variance may be due to differences in genotypes among individual members of the population. These differences are termed the **genetic variance** and are represented by V_G.

Second, some of the differences in phenotype may be due to environmental differences among the plants; these differences are termed the **environmental variance**, V_E. Environmental variance includes differences in environmental factors such as the amount of light or water that the plant receives; it also includes random differences in development that cannot be attributed to any specific factor. Any variation in phenotype that is not inherited is, by definition, a part of the environmental variance.

Third, **genetic–environmental interaction variance** (V_{GE}) arises when the effect of a gene depends on the specific environment in which it is found. An example is shown in **Figure 17.11**. In a dry environment, genotype *AA* produces a plant that averages 12 g in weight, and genotype *aa* produces a smaller plant that averages 10 g. In a wet environment, genotype *aa* produces the larger plant, averaging 24 g in weight, whereas genotype *AA* produces a plant that averages 20 g. In this example, there are clearly differences in the two environments: both genotypes produce heavier plants in the wet environment. There are also differences in the weights of the two genotypes, but the relative performances of the two genotypes depend on whether the plants are grown in a wet or a dry environment. In this case, the influences on phenotype cannot be neatly allocated into genetic and environmental components because the expression of the genotype depends on the environment in which the plant grows. The phenotypic variance must therefore include a component that accounts for the way in which genetic and environmental factors interact.

In summary, the total phenotypic variance can be apportioned into three components:

$$V_P = V_G + V_E + V_{GE} \qquad (17.5)$$

COMPONENTS OF GENETIC VARIANCE Genetic variance can be further subdivided into components consisting of different types of genetic effects. First, **additive genetic variance** (V_A) comprises the additive effects of genes on the phenotype, which can be summed to determine the overall effect on the phenotype. For example, suppose that in a plant, allele A^1 contributes 2 g in weight and allele A^2 contributes 4 g. If the alleles are strictly additive, then the genotypes would have the following weights:

$$A^1A^1 = 2 + 2 = 4\,\text{g}$$

$$A^1A^2 = 2 + 4 = 6\,\text{g}$$

$$A^2A^2 = 4 + 4 = 8\,\text{g}$$

The genes that Nilsson-Ehle studied, which affect kernel color in wheat, are additive in this way. It is the additive genetic variance that primarily determines the resemblance between parents and offspring. For example, if all of the phenotypic variance is due to additive genetic variance, then the average phenotype of the offspring will be exactly intermediate between those of the parents.

Second, there is **dominance genetic variance** (V_D), in which the effects of some genes have a dominance component. In this case, the alleles at a locus are not additive; rather, the effect of an allele depends on the identity of the other allele at that locus. For example, with a dominant allele (*T*), genotypes *TT* and *Tt* have the same phenotype. Here, we cannot simply add the effects of the alleles together because the effect of the small *t* allele is masked by the presence of the large *T* allele. Instead, we must add a component (V_D) to the genetic variance to account for the way in which alleles interact.

Third, genes at different loci may interact in the same way that alleles at the same locus interact. When this gene interaction takes place, the effects of the genes are not additive. Coat color in Labrador retrievers, for example, exhibits gene interaction, as described in Chapter 4; genotypes *BB ee* and *bb ee* both produce yellow dogs because the effects of alleles at the *B* locus are masked when two *e* alleles are present at the *E* locus. With gene interaction, we must include a third component, called **gene interaction variance** (V_I), to the genetic variance:

$$V_G = V_A + V_D + V_I \qquad (17.6)$$

SUMMARY EQUATION We can now integrate these components into one equation to represent all the potential contributions to the phenotypic variance:

$$V_P = V_A + V_D + V_I + V_E + V_{GE} \qquad (17.7)$$

Equation 17.7 provides us with a model that describes the potential causes of differences that we observe among individual phenotypes. Note that this model deals strictly with the observable *differences* (variance) in phenotypes among individual members of a population; it says nothing about the absolute value of the characteristic or about the underlying genotypes that produce these differences.

Types of Heritability

The model of phenotypic variance that we've just developed can be used to address the question of how much of the phenotypic variance in a characteristic is due to genetic differences. **Broad-sense heritability** (H^2) represents the proportion of phenotypic variance that is due to genetic variance. It is calculated by dividing the genetic variance by the phenotypic variance:

$$\text{broad-sense heritability} = H^2 = \frac{V_G}{V_P} \qquad (17.8)$$

The symbol H^2 represents broad-sense heritability because it is a measure of variance, which is given in units squared.

Broad-sense heritability can potentially range from 0 to 1. A value of 0 indicates that none of the phenotypic variance results from differences in genotype and all of the differences in phenotype result from environmental variation. A value of 1 indicates that all of the phenotypic variance results from differences in genotype. A heritability value between 0 and 1 indicates that both genetic and environmental factors influence the phenotypic variance.

Often, geneticists are more interested in the proportion of the phenotypic variance that results from the additive genetic variance because, as mentioned earlier, the additive genetic variance primarily determines the resemblance between parents and offspring. **Narrow-sense heritability** (h^2) is equal to the additive genetic variance divided by the phenotypic variance:

$$\text{narrow-sense heritability} = h^2 = \frac{V_A}{V_P} \qquad (17.9)$$

▶TRY PROBLEM 15

Calculating Heritability

Now that we have considered the components that contribute to phenotypic variance and developed a general concept of heritability, we can ask how we might go about estimating these different components and calculating heritability. It is not possible to determine all the components of phenotypic variance from observation of the phenotypes alone, but the heritability of a characteristic can often be measured. The mathematical theory that underlies these calculations of heritability is complex, so here we will focus on developing a general understanding of how heritability is measured.

Most methods for calculating heritability compare the degree of resemblance between related and unrelated individuals or between individuals with different degrees of relatedness. For example, one method compares the phenotypes of parents and offspring with those of unrelated individuals. Another method compares the similarity of identical twins (which have 100% of their genes in common) with the similarity of nonidentical twins (which have 50% of their genes in common). The general idea is that if genes influence phenotypic differences among individuals, then individuals that are more closely related will be more similar in phenotype. Statistical methods called regression and correlation are used to determine the degree to which more closely related individuals are more similar in phenotype.

An example of calculating heritability by comparing the phenotypes of parents and offspring is illustrated in **Figure 17.12**. Here, the mean phenotype of the parents is plotted against the mean phenotype of the offspring. Each data point on the graph represents one family. The line represents the best fit to all the points on the graph (that is, deviations of the points from the line are minimized). A complex mathematical proof (which we will not go into here) demonstrates that the slope of the line equals the narrow-sense heritability.

All estimates of heritability depend on the assumption that the environments of related individuals are not more similar than those of unrelated individuals. This assumption is difficult to meet in human studies because related people are usually reared together. Heritability estimates for humans should therefore be viewed with caution.

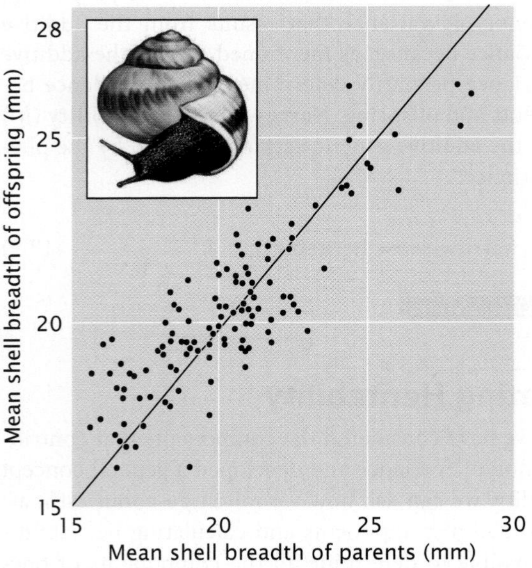

17.12 The heritability of shell breadth in snails can be determined by plotting the mean phenotype of offspring against the mean phenotype of the parents. [From L. M. Cook, *Evolution* 19:86–94, 1965.]

CONCEPTS

Broad-sense heritability is the proportion of phenotypic variance that is due to genetic variance. Narrow-sense heritability is the proportion of phenotypic variance that is due to additive genetic variance. Heritability can be measured by comparing the degree of resemblance between related and unrelated individuals or between individuals having different degrees of relatedness.

✓ CONCEPT CHECK 1

If the environmental variance (V_E) increases and all other variance components remain the same, what will the effect be?

a. Broad-sense heritability will decrease.
b. Broad-sense heritability will increase.
c. Narrow-sense heritability will increase.
d. Broad-sense heritability will increase, but narrow-sense heritability will decrease.

The Limitations of Heritability

Knowledge of heritability has great practical value because it allows us to predict the phenotypes of offspring on the basis of their parents' phenotypes. It also provides useful information about how characteristics will respond to selection (see Section 17.4). In spite of its importance, heritability is frequently misunderstood. It does not provide information about an individual's genes or the environmental factors that control the development of a characteristic, and it says nothing about the nature of differences between groups. This section outlines some limitations and common misconceptions of broad- and narrow-sense heritability.

HERITABILITY DOES NOT INDICATE THE DEGREE TO WHICH A CHARACTERISTIC IS GENETICALLY DETERMINED Heritability is the proportion of the phenotypic variance that is due to genetic variance; it says nothing about the degree to which genes determine a characteristic. Heritability indicates only the degree to which genes determine *variation* in a characteristic. The determination of a characteristic and the determination of variation in a characteristic are two very different things.

Consider polydactyly (the presence of extra digits) in rabbits, which can be caused either by environmental factors or by a dominant gene. Suppose that we have a group of rabbits all homozygous for a gene that produces the usual numbers of digits. None of the rabbits in this group carries a gene for polydactyly, but a few of the rabbits are polydactylous because of environmental factors. Broad-sense heritability for polydactyly in this group is zero because there is no genetic variation for polydactyly; all the variation is due to environmental factors. However, it would be incorrect for us to conclude that genes play no role in determining the number of digits in rabbits. Indeed, we know that there are specific alleles that can produce extra digits (although these alleles are not present in the group of rabbits under consideration). Heritability indicates nothing about whether genes control the development of a characteristic; it provides information only about causes of the variation in a characteristic within a defined group.

AN INDIVIDUAL DOES NOT HAVE HERITABILITY Broad- and narrow-sense heritabilities are statistical values based on the genetic and phenotypic variances found in a *group* of individuals. Heritability cannot be calculated for an individual, and heritability has no meaning for a specific individual. Suppose that we calculate the narrow-sense heritability of adult body weight for the students in a biology class and obtain a value of 0.6. We could conclude that 60% of the variation in adult body weight among the students in this class is determined by additive genetic variance. We should not, however, conclude that 60% of any particular student's body weight is due to additive genes.

THERE IS NO UNIVERSAL HERITABILITY FOR A CHARACTERISTIC The value of heritability for a characteristic is specific for a given population in a given environment. Recall that broad-sense heritability is genetic variance divided by phenotypic variance. Genetic variance depends on which alleles are present, which often differs between populations. In the example of polydactyly in rabbits, there were no alleles for polydactyly in our group of rabbits, so the heritability of the characteristic was zero. A different group of rabbits might contain many alleles for polydactyly, and the heritability of the characteristic might then be high.

Environmental differences can also affect heritability because V_P is composed of both genetic and environmental variance. When the environmental differences that affect a characteristic differ between two groups, the heritabilities for the two groups will often differ as well. Because heritability is specific to a defined population in a given environment, it is important not to extrapolate heritabilities from one population to another.

EVEN WHEN HERITABILITY IS HIGH, ENVIRONMENTAL FACTORS CAN INFLUENCE A CHARACTERISTIC High heritability does not mean that environmental factors cannot influence the expression of a characteristic. High heritability indicates only that the environmental variation to which the population is *currently* exposed is not responsible for variation in the characteristic. Let's look at human height. In most developed countries, the heritability of human height is high, indicating that genetic differences are responsible for most of the variation in height. It would be wrong for us to conclude, however, that human height cannot be changed by the alteration of the environment. Indeed, in several European cities during World War II, height decreased owing to hunger and disease; furthermore, height can be increased dramatically by the administration of growth hormone to children. The absence of environmental variation in a characteristic does not mean that the characteristic will not respond to environmental change.

HERITABILITIES INDICATE NOTHING ABOUT DIFFERENCES AMONG POPULATIONS A common misconception about heritability is that it provides information about the nature of population differences in a characteristic. Heritability is specific for a given population in a given environment, so it cannot be used to draw conclusions about why populations differ in a characteristic.

Suppose that we measured heritability for human height in two groups. One group is from a small town in a developed country, where everyone consumes a high-protein diet. Because there is little variation in the environmental factors that affect human height in this group, and because there is some genetic variation, the heritability of height in this group is high. The second group comprises the inhabitants of a single village in a developing country. These people consume only 25% as much protein as those in the first group, so their average adult height is several centimeters less than that in the group from the developed country. Again, there is little variation in the environmental factors that determine height in this group because everyone in the village eats the same types of food and is exposed to the same diseases. Because there is little environmental variation and there is some genetic variation, the heritability of height in this group also is high. Thus, the heritability of height in both groups is high, and the average height in the two groups is considerably different. We might be tempted to conclude that the difference in height between the two groups is genetically based—that the people in the developed country are genetically taller than the people in the developing country. This conclusion is obviously wrong, however, because these differences in height are due largely to diet—an environmental factor. Heritability provides no information about the causes of differences between populations.

These limitations of heritability have often been ignored, particularly in arguments about the possible social implications of genetic differences between humans. For example, the results of a number of modern studies indicate that human intelligence, as measured by IQ (intelligence quotient) and other intelligence tests, has a moderately high heritability (usually from 0.4 to 0.8). On the basis of this observation, some people have argued that intelligence is innate and that enhanced educational opportunities cannot boost intelligence. This argument is based on the misconception that, when heritability is high, changing the environment will not alter the characteristic. In addition, because heritabilities of intelligence range from 0.4 to 0.8, a considerable amount of the variance in intelligence originates from environmental differences.

TABLE 17.2	Limitations of heritability
1. Heritability does not indicate the degree to which a characteristic is genetically determined.	
2. An individual does not have heritability.	
3. There is no universal heritability for a characteristic.	
4. Even when heritability is high, environmental factors can influence a characteristic.	
5. Heritabilities indicate nothing about the nature of population differences in a characteristic.	

Another argument based on a misconception about heritability is that ethnic differences in measures of intelligence are genetically based. Because the results of some genetic studies show that IQ has moderately high heritability, and because other studies find differences in average IQ among ethnic groups, some people have suggested that ethnic differences in IQ are genetically based. As in the example of the effects of diet on human height, heritability provides no information about causes of differences among groups; it indicates only the degree to which phenotypic variance within a single group is genetically based. High heritability for a characteristic does not mean that phenotypic differences among ethnic groups are genetic. We should also remember that separating genetic and environmental effects in humans is very difficult, so heritability estimates themselves may be unreliable. These limitations of heritability are summarized in **Table 17.2.** ▶TRY PROBLEM 20

THINK-PAIR-SHARE Questions 4 and 5

CONCEPTS

Heritability provides information only about the degree to which *variation* in a characteristic is genetically determined. There is no universal heritability for a characteristic; heritability is specific for a given population in a specific environment. Environmental factors can potentially affect characteristics with high heritability, and heritability says nothing about the nature of population differences in a characteristic.

✓ CONCEPT CHECK 2

Suppose that you just learned that the narrow-sense heritability of blood pressure measured among a group of African Americans in Detroit, Michigan, is 0.4. What does this heritability tell us about genetic and environmental contributions to blood pressure?

Locating Genes That Affect Quantitative Characteristics

The statistical methods we have just described can be used both to make predictions about the average phenotype expected in offspring and to estimate the overall contribution of genes to variation in a characteristic. These methods do not, however, allow us to identify and determine the influence of individual genes that affect quantitative characteristics. As stated in the introduction to this chapter, chromosome regions with genes that control polygenic characteristics are referred to as **quantitative trait loci (QTLs)**. Although quantitative genetics has made important contributions to basic biology and to plant and animal breeding, our inability to identify QTLs and to measure their individual effects severely limited the application of quantitative genetic methods until recently.

MAPPING QTLs In recent years, numerous genetic markers have been identified and mapped with the use of molecular techniques, making it possible to identify QTLs by linkage analysis. The underlying idea is simple: if the inheritance of a genetic marker is associated consistently with the inheritance of a particular phenotype, then that marker must be linked to a QTL that affects that phenotype. Any correlation between the inheritance of a particular marker allele and a quantitative phenotype in a series of crosses suggests that a QTL is physically linked to that marker. If enough markers are used, the detection of all the QTLs affecting a characteristic is theoretically possible.

It is important to recognize that a QTL is not a gene; rather, a QTL is a map location for a chromosome region that is associated with a quantitative trait. After a QTL has been identified, it can be studied for the presence of one or more specific genes or other sequences that influence the trait. The introduction to this chapter considers how this approach was used to identify a major gene that affects the oil content of corn. QTL mapping has been used to detect genes affecting a variety of characteristics in plant and animal species (**Figure 17.13** and **Table 17.3**).

17.13 QTL mapping is used to identify genes that influence many important quantitative traits, including muscle mass in pigs. [USDA.]

TABLE 17.3 Examples of quantitative characteristics for which QTLs have been detected

Organism	Quantitative characteristic
Tomato	Soluble solids
	Fruit mass
	Fruit pH
	Growth
	Leaflet shape
	Height
Corn	Height
	Leaf length
	Tiller number
	Glume hardness
	Grain yield
	Number of ears
	Thermotolerance
Common bean	Number of root nodules
Mung bean	Seed weight
Cow pea	Seed weight
Wheat	Preharvest sprout
Pig	Growth
	Length of small intestine
	Average back fat
	Abdominal fat
Mouse	Epilepsy
Rat	Hypertension

Source: After S. D. Tanksley, Mapping polygenes, *Annual Review of Genetics* 27:218, 1993.

GENOME-WIDE ASSOCIATION STUDIES The traditional method of identifying QTLs is to carry out crosses between varieties that differ in a quantitative trait and then genotype numerous progeny for many markers. Although effective, this method is slow and labor-intensive.

An alternative technique for identifying genes that affect quantitative traits is to conduct genome-wide association studies (introduced in Chapter 5). Unlike traditional linkage analysis, which examines the association of a trait and gene markers among the *progeny of a cross*, genome-wide association studies look for associations between traits and genetic markers in a *biological population*, a group of interbreeding individuals. The presence of an association between genetic markers and a trait indicates that the genetic markers are closely linked to one or more genes that affect variation in the trait. Genome-wide association studies have been facilitated by the identification of single-nucleotide polymorphisms (SNPs), which are positions in the genome at which individual organisms vary in a single base pair (see Chapter 15). It is often possible to quickly and inexpensively genotype individual organisms for numerous SNPs, which provide the genetic markers necessary to conduct genome-wide association studies.

Genome-wide association studies have been widely used to locate genes that affect quantitative traits in humans, including disease susceptibility, obesity, intelligence, and height. A number of quantitative traits in plants have also been studied, including kernel composition, size, color and taste, disease resistance, and starch quality. Genome-wide association studies in domestic animals have identified chromosome segments affecting body weight, body composition, reproductive traits, hormone levels, hair characteristics, and behaviors.

CONCEPTS

The availability of numerous genetic markers revealed by molecular methods makes it possible to map chromosome segments containing genes that contribute to polygenic characteristics. Genome-wide association studies locate genes that affect quantitative traits by detecting associations between genetic markers and a trait within a population of individuals.

17.4 Genetically Variable Traits Change in Response to Selection

Evolution is genetic change that takes place among members of a population over time. Several different forces are potentially capable of bringing about evolution, and we will explore those forces and the process of evolution more fully in Chapter 18. Here, we consider how one of those forces—natural selection—can bring about genetic change in a quantitative characteristic.

Charles Darwin proposed the idea of natural selection in his book *On the Origin of Species* in 1859. **Natural selection** arises through the differential reproduction of individuals with different genotypes. Because of the genes they possess, some individuals produce more offspring than others. The more successful reproducers give rise to more offspring, which inherit the genes that confer a reproductive advantage. Thus, the frequencies of the genes that confer a reproductive advantage increase with the passage of time, and the population evolves. Natural selection is among the most important of the forces that bring about evolutionary change. Through natural selection, organisms become genetically suited to their environments; as environments change, groups of organisms change in ways that make them better able to survive and reproduce.

For thousands of years, humans have practiced a form of selection by promoting the reproduction of organisms with traits perceived as desirable. This form of selection, called **artificial selection**, has produced the domestic plants and animals that make modern agriculture possible.

Predicting the Response to Selection

When a quantitative characteristic is subjected to natural or artificial selection, it is likely to change with the passage of time, provided that there is genetic variation for that characteristic in the population. Suppose that a dairy farmer wants to increase milk production among the cows in his herd. Variation at several loci potentially affects milk production in cows; some alleles at these loci confer high milk production, whereas other alleles confer low milk production. The dairy farmer breeds only those cows in his herd that have the highest milk production. If there is genetic variation in milk production (i.e., there are different alleles at the loci that control milk production), the mean milk production in the daughters of the selected cows should be higher than the mean milk production of the original herd. This increased production is due to the fact that the selected cows possess more alleles for high milk production than does the average cow, and these alleles are passed on to the offspring. Thus, the offspring of the selected cows possess a higher proportion of alleles for high milk production, and therefore produce more milk, than the average cow in the initial herd.

The extent to which a characteristic subjected to selection changes in one generation is termed the **response to selection**. Suppose that the average cow in a dairy herd produces 80 liters of milk per week. A farmer selects for increased milk production by breeding the highest milk producers, and the female progeny of those selected cows produce 100 liters of milk per week on average. The response to selection is calculated by subtracting the mean phenotype of the original population (80 liters) from the mean phenotype of the offspring (100 liters), obtaining a response to selection of 100 − 80 = 20 liters per week.

FACTORS INFLUENCING RESPONSE TO SELECTION The response to selection is determined primarily by two factors. First, it is affected by narrow-sense heritability, which largely determines the degree of resemblance between parents and offspring. When narrow-sense heritability is high, offspring will tend to resemble their parents; conversely, when narrow-sense heritability is low, there will be little resemblance between parents and offspring.

The second factor that determines the response to selection is how much selection there is. If the farmer is very stringent in the choice of parents and breeds only the highest milk producers in the herd (say, the top 3 cows), then all the offspring will receive genes for high milk production. If the farmer is less selective and breeds the top 20 milk producers in the herd, then the offspring will not carry as many genes for high milk production, and on average, they will not produce as much milk as the offspring of the top 3 producers would. The response to selection thus depends on the degree to which the selected parents differ from the rest of the population; this difference is measured by the **selection differential**, defined as the difference between the mean phenotype of the selected parents and the mean phenotype of the original population. If the average milk production of the original herd is 80 liters and the farmer breeds cows with an average milk production of 120 liters, then the selection differential is 120 − 80 = 40 liters.

CALCULATION OF RESPONSE TO SELECTION The response to selection (R) depends on narrow-sense heritability (h^2) and the selection differential (S):

$$R = h^2 \times S \tag{17.10}$$

This equation can be used to predict the magnitude of change in a characteristic when a given selection differential is applied. G. A. Clayton and his colleagues estimated the response to selection that would take place in the abdominal bristle number of *Drosophila melanogaster*. By using several different methods, they first estimated the narrow-sense heritability of abdominal bristle number in one population of fruit flies to be 0.52. The mean number of bristles in the original population was 35.3. They selected individual flies with a mean bristle number of 40.6 and intercrossed them to produce the next generation. The selection differential was 40.6 − 35.3 = 5.3; so they predicted a response to selection to be

$$R = 0.52 \times 5.3 = 2.8$$

The response to selection of 2.8 is the expected increase in the characteristic in the offspring above the mean of the original population. They therefore predicted that the average number of abdominal bristles in the offspring of their selected flies would be 35.3 + 2.8 = 38.1. Indeed, they found an average bristle number of 37.9 in these flies.

ESTIMATING HERITABILITY FROM RESPONSE TO SELECTION Rearranging Equation 17.10 provides another way to calculate narrow-sense heritability:

$$h^2 = \frac{R}{S} \tag{17.11}$$

In this way, h^2 can be calculated by conducting a response-to-selection experiment. First, the selection differential is obtained by subtracting the population mean from the mean of selected parents. The selected parents are then interbred, and the mean phenotype of their offspring is measured. The difference between the mean of the offspring and that of the initial population is the response to selection, which can be used with the selection differential to estimate the heritability. Heritability determined by a response-to-selection experiment is usually termed **realized heritability**. If certain assumptions are met, realized heritability is identical to narrow-sense heritability.

One of the longest-running selection experiments is a study of oil and protein content in corn seeds (**Figure 17.14**). This experiment began at the University of Illinois on 163 ears of corn with an oil content ranging from 4% to 6%. Corn plants with high oil content and corn plants with low

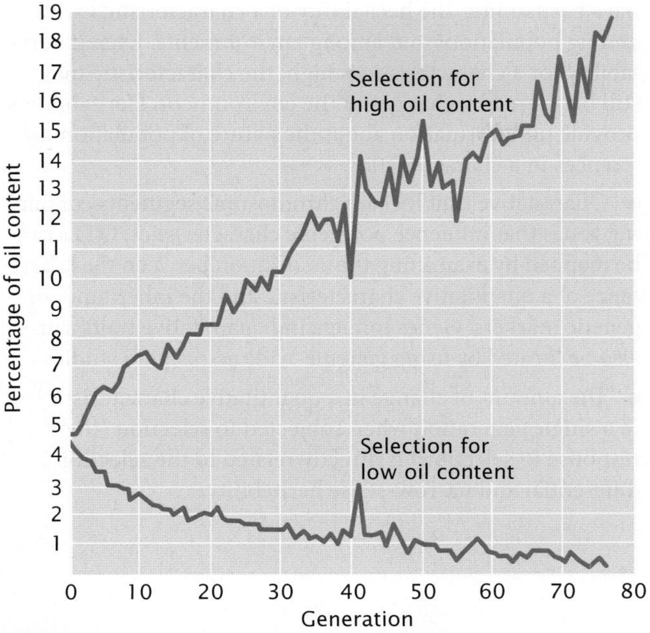

17.14 In a long-term response-to-selection experiment, selection for oil content in corn increased oil content in one line to about 20%, whereas oil content was almost eliminated in another line.

oil content were selected and interbred. Response to selection for increased oil content (the upper line in Figure 17.14) reached about 20%, whereas response to selection for decreased oil content reached a lower limit near zero. Genetic analyses of the high- and low-oil-content strains revealed that at least 20 loci take part in determining oil content, one of which we explored in the introduction to this chapter. ▶TRY PROBLEM 23

THINK-PAIR-SHARE Question 6

CONCEPTS

The response to selection is influenced by narrow-sense heritability and the selection differential.

✓ CONCEPT CHECK 3

The narrow-sense heritability for a trait is 0.4 and the selection differential is 0.5. What is the predicted response to selection?

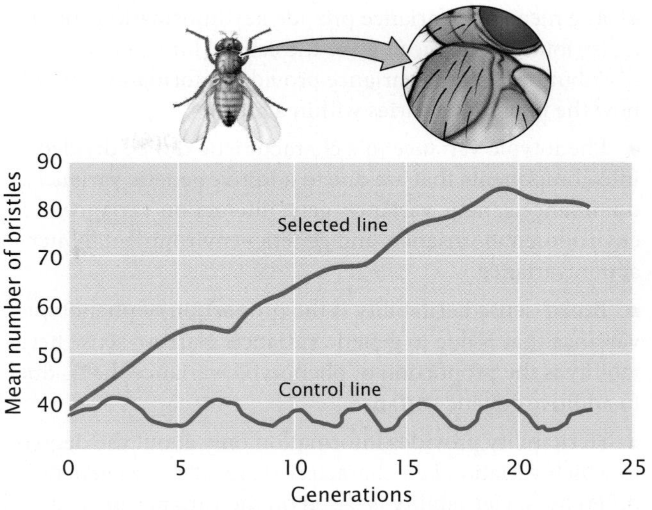

17.15 The response of a population to selection often levels off at some point. In a response-to-selection experiment in which fruit flies were selected for increased numbers of bristles on the abdomens of females, the number of bristles increased steadily for about 20 generations and then leveled off.

Limits to Selection Response

When a characteristic has been selected for many generations, the response may eventually level off, and the characteristic no longer responds to selection (**Figure 17.15**). A potential reason for this leveling off is that the genetic variation in the population may be exhausted; at some point, all individuals in the population may have become homozygous for the alleles that encode the selected trait. When there is no more additive genetic variance, heritability equals zero, and there can be no further response to selection.

Sometimes the response to selection levels off even while some genetic variation remains in the population. This leveling off takes place because natural selection opposes further change in the characteristic. The response to selection for small body size in mice, for example, eventually levels off because the smallest animals are sterile and cannot pass on their genes for small body size. In this case, artificial selection for small size is opposed by natural selection for fertility, and the population can no longer respond to the artificial selection.

THINK-PAIR-SHARE Question 7

CONCEPTS SUMMARY

- Quantitative genetics focuses on the inheritance of complex characteristics whose phenotypes vary continuously. For many quantitative characteristics, the relation between genotype and phenotype is complex because many genes and environmental factors influence a characteristic.
- The individual genes that influence a polygenic characteristic follow the same Mendelian principles that govern discontinuous characteristics, but because many genes participate, the expected ratios of phenotypes are obscured.
- A frequency distribution, in which phenotypes are represented on one axis and the number of individuals possessing each phenotype is represented on the other axis, is a convenient means of summarizing phenotypes found in a group of individuals.

- The mean and variance provide key information about a distribution: the mean gives the central location of the distribution, and the variance provides information about how the phenotype varies within a group.
- Phenotypic variance in a characteristic can be divided into components that are due to additive genetic variance, dominance genetic variance, gene interaction variance, environmental variance, and genetic–environmental interaction variance.
- Broad-sense heritability is the proportion of phenotypic variance that is due to genetic variance. Narrow-sense heritability is the proportion of phenotypic variance that is due to additive genetic variance.
- Heritability provides information only about the degree to which variation in a characteristic results from genetic differences. Heritability is based on the variance present within a group of individuals, and an individual does not have heritability. The heritability of a characteristic varies among populations and among environments. Even if heritability for a characteristic is high, the characteristic may still be altered by changes in the environment. Heritabilities provide no information about the nature of population differences in a characteristic.
- Quantitative trait loci are chromosome segments containing genes that influence polygenic characteristics. QTLs can be mapped by examining the association between the inheritance of a quantitative characteristic and the inheritance of genetic markers. Genes influencing quantitative traits can also be located by using genome-wide association studies.
- The amount of change in a quantitative characteristic in a single generation when subjected to selection (the response to selection) is directly related to the selection differential and narrow-sense heritability.

IMPORTANT TERMS

quantitative genetics (p. 462)
meristic characteristic (p. 464)
threshold characteristic (p. 464)
frequency distribution (p. 467)
normal distribution (p. 471)
mean (p. 468)
variance (p. 468)
heritability (p. 470)
phenotypic variance (p. 470)
genetic variance (p. 470)
environmental variance (p. 470)
genetic–environmental interaction variance (p. 470)
additive genetic variance (p. 471)
dominance genetic variance (p. 471)
gene interaction variance (p. 471)
broad-sense heritability (p. 471)
narrow-sense heritability (p. 471)
quantitative trait locus (QTL) (p. 475)
natural selection (p. 475)
artificial selection (p. 475)
response to selection (p. 476)
selection differential (p. 476)
realized heritability (p. 476)

ANSWERS TO CONCEPT CHECKS

1. a

2. The heritability indicates that about 40% of the differences in blood pressure among African Americans in Detroit are due to additive genetic differences. It neither provides information about the heritability of blood pressure in other groups of people nor indicates anything about the nature of differences in blood pressure between African Americans in Detroit and people in other groups.

3. 0.2

WORKED PROBLEMS

Problem 1

Seed weight in a particular plant species is determined by pairs of alleles at two loci (a^+a^- and b^+b^-) that are additive and equal in their effects. Plants with genotype $a^-a^-\ b^-b^-$ have seeds that average 1 g in weight, whereas plants with genotype $a^+a^+\ b^+b^+$ have seeds that average 3.4 g in weight. A plant with genotype $a^-a^-\ b^-b^-$ is crossed with a plant of genotype $a^+a^+\ b^+b^+$.

a. What is the predicted weight of seeds from the F_1 progeny of this cross?

b. If the F_1 plants are intercrossed, what are the expected seed weights and proportions of the F_2 plants?

Solution Strategy

What information is required in your answer to the problem?

- The predicted weight of seeds from the F_1 progeny.
- The expected seed weights and their proportions among the F_2 progeny.

What information is provided to solve the problem?

- Seed weight is determined by pairs of alleles at two loci (a^+a^- and b^+b^-).
- The alleles are additive and equal in their effects.
- Plants with genotype $a^-a^-\ b^-b^-$ have seeds that average 1 g in weight.
- Plants with genotype $a^+a^+\ b^+b^+$ have seeds that average 3.4 g in weight.
- A plant with genotype $a^-a^-\ b^-b^-$ is crossed with a plant of genotype $a^+a^+\ b^+b^+$.

For help with this problem, review:

Polygenic Inheritance in Section 17.1.

Solution Steps

The difference in average seed weight between the two parental genotypes is 3.4 g − 1 g = 2.4 g. These two genotypes differ in four genes; so, if the genes have equal and additive effects, each gene difference contributes an additional 2.4 g/4 = 0.6 g of weight to the 1-g weight of a plant ($a^-a^-\ b^-b^-$) that has none of the contributing genes.

Hint: Because the alleles are equal and additive, each allele contributes the same amount to seed weight.

The cross between the two homozygous genotypes produces the F_1 and F_2 progeny shown below.

a. The F_1 are heterozygous at both loci ($a^+a^-\ b^+b^-$) and possess two genes that contribute an additional 0.6 g each to the 1-g weight of a plant that has no contributing genes. Therefore, the seeds of the F_1 should average 1 g + 2(0.6 g) = 2.2 g.

b. The F_2 have the following phenotypes and proportions: $\frac{1}{16}$ 1 g; $\frac{4}{16}$ 1.6 g; $\frac{6}{16}$ 2.2 g; $\frac{4}{16}$ 2.8 g; and $\frac{1}{16}$ 3.4 g.

P: $a^-a^-b^-b^-$ (1 g) × $a^+a^+\ b^+b^+$ (3.4 g)

↓

F_1: $a^+a^-\ b^+b^-$ (2.2 g)

↓

Recall: The probability of each two-locus genotype can be determined by multiplying the probabilities of the single-locus genotypes.

F_2:

Genotype	Probability		Number of contributing genes	Average seed weight
$a^-a^-\ b^-b^-$	$\frac{1}{4} \times \frac{1}{4} = \frac{1}{16}$		0	1 g + (0 × 0.6 g) = 1 g
$a^+a^-\ b^-b^-$	$\frac{1}{2} \times \frac{1}{4} = \frac{1}{8}$	$\frac{2}{8} = \frac{4}{16}$	1	1 g + (1 × 0.6 g) = 1.6 g
$a^-a^-\ b^+b^-$	$\frac{1}{4} \times \frac{1}{2} = \frac{1}{8}$			
$a^+a^+\ b^-b^-$	$\frac{1}{4} \times \frac{1}{4} = \frac{1}{16}$	$\frac{1}{16} + \frac{1}{16} + \frac{1}{4} = \frac{6}{16}$	2	1 g + (2 × 0.6 g) = 2.2 g
$a^-a^-\ b^+b^+$	$\frac{1}{4} \times \frac{1}{4} = \frac{1}{16}$			
$a^+a^-\ b^+b^-$	$\frac{1}{2} \times \frac{1}{2} = \frac{1}{4}$			
$a^+a^+\ b^+b^-$	$\frac{1}{4} \times \frac{1}{2} = \frac{1}{8}$	$\frac{2}{8} = \frac{4}{16}$	3	1 g + (3 × 0.6 g) = 2.8 g
$a^+a^-\ b^+b^+$	$\frac{1}{2} \times \frac{1}{4} = \frac{1}{8}$			
$a^+a^+\ b^+b^+$	$\frac{1}{4} \times \frac{1}{4} = \frac{1}{16}$		4	1 g + (4 × 0.6 g) = 3.4 g

Problem 2

A farmer is raising rabbits. The average body weight in his population of rabbits is 3 kg. The farmer selects the 10 largest rabbits in his population, whose average body weight is 4 kg, and interbreeds them. If the heritability of body weight in the rabbit population is 0.7, what is the expected body weight among offspring of the selected rabbits?

Solution Strategy

What information is required in your answer to the problem?

- The expected weight of the offspring of the selected rabbits.

What information is provided to solve the problem?

- The average body weight in the population is 3 kg.
- The average body weight of selected rabbits is 4 kg.
- The heritability of body weight is 0.7.

For help with this problem, review:

Predicting the Response to Selection in Section 17.4.

Solution Steps

The farmer has carried out a response-to-selection experiment. The selection differential equals the difference in average weight between the selected rabbits and the entire population: 4 kg − 3 kg = 1 kg. The narrow-sense heritability is given as 0.7; so the expected response to selection is $R = h^2 \times S = 0.7 \times 1$ kb = 0.7 kg. This value is the increase in weight that is expected in the offspring of the selected parents; so the average weight of the offspring is expected to be 3 kg + 0.7 kg = 3.7 kg.

Recall: The response to selection equals the selection differential multiplied by the narrow-sense heritability.

COMPREHENSION QUESTIONS

Section 17.1

1. How does a quantitative characteristic differ from a discontinuous characteristic?
2. Briefly explain why the relation between genotype and phenotype is frequently complex for quantitative characteristics.
3. Why do polygenic characteristics have many phenotypes?

Section 17.2

4. What information do the mean and variance provide about a distribution?

Section 17.3

5. List all the components that contribute to the phenotypic variance and define each component.
6. How do the broad-sense and narrow-sense heritabilities differ?
7. Briefly describe common misunderstandings or misapplications of the concept of heritability.
8. Briefly explain how genes affecting a polygenic characteristic are located with the use of QTL mapping.

Section 17.4

9. How is the response to selection related to narrow-sense heritability and the selection differential? What information does the response to selection provide?
10. Why does the response to selection often level off after many generations of selection?

For more questions that test your comprehension of the key chapter concepts, go to LearningCurve for this chapter.

APPLICATION QUESTIONS AND PROBLEMS

Section 17.1

*11. For each of the following characteristics, indicate whether it would be considered a discontinuous characteristic or a quantitative characteristic. Briefly justify your answer.

a. Kernel color in a strain of wheat, in which two codominant alleles segregating at a single locus determine the color. Thus, there are three phenotypes present in this strain: white, light red, and medium red.

b. Body weight in a family of Labrador retrievers. An autosomal recessive allele that causes dwarfism is present in this family. Two phenotypes are recognized: dwarf (less than 13 kg) and normal (greater than 23 kg).

c. Presence or absence of leprosy. Susceptibility to leprosy is determined by multiple genes and numerous environmental factors.

d. Number of toes in guinea pigs, which is influenced by genes at many loci.

e. Number of fingers in humans. Extra (more than five) fingers are caused by the presence of an autosomal dominant allele.

***12.** Assume that plant weight is determined by a pair of alleles at each of two independently assorting loci (*A* and *a*, *B* and *b*) that are additive in their effects. Further assume that each allele represented by an uppercase letter contributes 4 g to weight and that each allele represented by a lowercase letter contributes 1 g to weight.

a. If a plant with genotype *AA BB* is crossed with a plant with genotype *aa bb*, what weights are expected in the F_1 progeny?

b. What is the distribution of weight expected in the F_2 progeny?

***13.** Assume that three loci, each with two alleles (*A* and *a*, *B* and *b*, *C* and *c*), determine the difference in height between two homozygous strains of a plant. These genes are additive and equal in their effects on plant height. One strain (*aa bb cc*) is 10 cm in height. The other strain (*AA BB CC*) is 22 cm in height. The two strains are crossed, and the resulting F_1 are interbred to produce F_2 progeny. Give the phenotypes and the expected proportions of the F_2 progeny.

14. Seed size in a plant is a polygenic characteristic. A grower crosses two pure-breeding varieties of the plant and measures seed size in the F_1 progeny. She then backcrosses the F_1 plants to one of the parental varieties and measures seed size in the backcross progeny. The grower finds that seed size in the backcross progeny has a higher variance than does seed size in the F_1 progeny. Explain why the backcross progeny are more variable.

Section 17.3

***15.** Phenotypic variation in the tail length of mice has the following components:

Additive genetic variance (V_A)	= 0.5
Dominance genetic variance (V_D)	= 0.3
Gene interaction variance (V_I)	= 0.1
Environmental variance (V_E)	= 0.4
Genetic–environmental interaction variance (V_{GE})	= 0.0

a. What is the narrow-sense heritability of tail length?

b. What is the broad-sense heritability of tail length?

16. The narrow-sense heritability of ear length in Reno rabbits is 0.4. The phenotypic variance (V_P) is 0.8, and the environmental variance (V_E) is 0.2. What is the additive genetic variance (V_A) for ear length in these rabbits?

17. A characteristic has a narrow-sense heritability of 0.6.

a. If the dominance variance (V_D) increases and all other variance components remain the same, what will happen to the narrow-sense heritability? Will it increase, decrease, or remain the same? Explain.

b. What will happen to the broad-sense heritability? Explain.

c. If the environmental variance (V_E) increases and all other variance components remain the same, what will happen to the narrow-sense heritability? Explain.

d. What will happen to the broad-sense heritability? Explain.

18. What conclusion can you draw from **Figure 17.12** about the proportion of phenotypic variation in shell breadth that is due to genetic differences? Explain your reasoning.

19. Many researchers have estimated the heritability of human traits by comparing the correlation coefficients of identical and nonidentical twins. An assumption in using this method is that two identical twins experience environments that are no more similar to each other than those experienced by two nonidentical twins. How might this assumption be violated? Give some specific examples of how the environments of two identical twins might be more similar than the environments of two nonidentical twins.

***20.** A genetics researcher determines that the broad-sense heritability of height among Southwestern University undergraduate students is 0.90. Which of the following conclusions would be reasonable? Explain your answer.

a. Since Sally is a Southwestern University undergraduate student, 10% of her height is determined by nongenetic factors.

b. Ninety percent of variation in height among all undergraduate students in the United States is due to genetic differences.

c. Ninety percent of the height of Southwestern University undergraduate students is determined by genes.

d. Ten percent of the variation in height among Southwestern University undergraduate students is determined by variation in nongenetic factors.

e. Because the heritability of height among Southwestern University students is so high, any change in the students' environment will have minimal effect on their height.

21. *Drosophila buzzati* is a fruit fly that feeds on the rotting fruits of cacti in Australia. Timothy Prout and Stuart Barker calculated the heritabilities of body size, as measured by thorax length, for a natural population of *D. buzzati* raised in the wild and for a population of *D. buzzati* collected in the wild but raised in the laboratory (T. Prout and J. S. F. Barker. 1989. *Genetics* 123:803–813). They found the following heritabilities.

Population	**Heritability of body size ($\pm$ standard error)**
Wild population	0.0595 ± 0.0123
Laboratory-reared population	0.3770 ± 0.0203

Why do you think that the heritability measured in the laboratory-reared population is higher than that measured in the natural population raised in the wild?

22. Mr. Jones is a pig farmer. For many years, he has fed his pigs the food left over from the local university cafeteria, which is known to be low in protein, deficient in vitamins, and downright untasty. However, the food is free, and his pigs don't complain. One day, a salesman from a feed company visits Mr. Jones. The salesman claims that his company sells a new, high-protein, vitamin-enriched feed that enhances weight gain in pigs. Although the feed is expensive, the salesman claims that the increased weight gain of the pigs will more than pay for the cost of the feed, increasing Mr. Jones's profit. Mr. Jones responds that he took a genetics class at the university and that he has conducted some genetic experiments on his pigs; specifically, he has calculated the narrow-sense heritability of weight gain for his pigs and found it to be 0.98. Mr. Jones says that this heritability value indicates that 98% of the variance in weight gain among his pigs is determined by genetic differences, and therefore the new pig feed can have little effect on the growth of his pigs. He concludes that the feed would be a waste of his money. The salesman doesn't dispute Mr. Jones's heritability estimate, but he still claims that the new feed can significantly increase weight gain in Mr. Jones's pigs. Who is correct and why?

Section 17.4

*23. Joe is breeding cockroaches in his dorm room. He finds that the average wing length in his population of cockroaches is 4 cm. He chooses six cockroaches that have the largest wings; the average wing length among these selected cockroaches is 10 cm. Joe interbreeds these selected cockroaches. From earlier studies, he knows that the narrow-sense heritability for wing length in his population of cockroaches is 0.6.

a. Calculate the selection differential and expected response to selection for wing length in these cockroaches.

b. What should be the average wing length of the progeny of the selected cockroaches?

24. Three characteristics in beef cattle—body weight, fat content, and tenderness—are measured, and the following variance components are estimated:

	Body weight	**Fat content**	**Tenderness**
V_A	22	45	12
V_D	10	25	5
V_I	3	8	2
V_E	42	64	8
V_{GE}	0	0	1

In this population, which characteristic would respond best to selection? Explain your reasoning.

*25. A rancher determines that the average amount of wool produced by a sheep in her flock is 22 kg per year. In an attempt to increase the wool production of her flock, the rancher picks five male and five female sheep with the greatest wool production; the average amount of wool produced per sheep by those selected is 30 kg. She interbreeds these selected sheep and finds that the average wool production among the progeny of the selected sheep is 28 kg. What is the narrow-sense heritability for wool production among the sheep in the rancher's flock?

26. A strawberry farmer determines that the average weight of individual strawberries produced by plants in his garden is 2 g. He selects the 10 plants that produce the largest strawberries; the average weight of strawberries among these selected plants is 6 g. He interbreeds these selected plants. The progeny of these selected plants produce strawberries that weigh 5 g. If the farmer were to select plants that produce an average strawberry weight of 4 g, what would be the predicted weight of strawberries produced by the progeny of these selected plants?

27. Pigs have been domesticated from wild boars. Would you expect to find higher heritability for weight among domestic pigs or wild boars? Explain your answer.

28. Has the response to selection leveled off in the strain of corn selected for high oil content shown in **Figure 17.14**? What does this observation suggest about genetic variation in the strain selected for high oil content?

CHALLENGE QUESTIONS

Section 17.1

29. Bipolar illness is a psychiatric disorder with a strong hereditary basis, but the exact mode of inheritance is not known. Research has shown that siblings of patients with bipolar illness are more likely to develop the disorder than are siblings of unaffected persons. Findings from one study demonstrated that the ratio of bipolar brothers to bipolar sisters is higher when the patient is male than when the patient is female. In other words, relatively more brothers of patients with bipolar illness also have the disease when the patient is male than when the patient is female. What does this observation suggest about the inheritance of bipolar illness?

Section 17.3

30. We have explored some of the difficulties in separating the genetic and environmental components of human behavioral characteristics. Considering these difficulties and what you know about calculating heritability, propose an experimental design for accurately measuring the heritability of musical ability.

Section 17.4

31. DATA ANALYSIS Eugene Eisen selected for increased 12-day litter weight (total weight of a litter of offspring 12 days after birth) in a population of mice (E. J. Eisen. 1972. *Genetics* 72:129–142). The 12-day litter weight of the population steadily increased, but then leveled off after about 17 generations. At generation 17, Eisen took one family of mice from the selected population and reversed the selection procedure: in this group, he selected for *decreased* 12-day litter weight. This group immediately responded to the reversed selection: the 12-day litter weight dropped 4.8 g within 1 generation and dropped 7.3 g after 5 generations. On the basis of the results of the reverse selection, what is the most likely explanation for the leveling off of 12-day litter weight in the original population?

[J & C Sohns/Age Fotostock America, Inc.]

THINK-PAIR-SHARE QUESTIONS

Introduction

1. What are some problems other than food supply associated with the increasing human population, and how is genetics being used to address them?
2. Researchers produced transgenic corn that contained an extra codon for phenylalanine and had a high oil content. Suppose an existing commercial variety of corn was modified in the same way and sold to farmers. Would you consider this variety to be a genetically modified organism (GMO; see Chapter 14)? Should the corn produced by this variety be labeled as a genetically modified food product?

Section 17.1

3. To locate genes associated with quantitative traits, geneticists often use QTL mapping (see the introduction to this chapter). To carry out QTL mapping, a first step is crossing two strains that differ in a quantitative trait, such as a strain of corn with high oil content and a strain with low oil content. The F_1 progeny of this cross are then interbred or backcrossed to produce an F_2 generation. Researchers then look for statistical associations between genetic markers and the value of the quantitative trait (e.g., oil content) in the F_2 generation. Why do the geneticists look for statistical associations in the F_2 plants? Why not use the F_1 progeny?

Section 17.3

4. A student who has just learned about quantitative genetics says, "Heritability estimates are worthless! They don't tell you anything about the genes that affect a characteristic. They don't provide any information about the types of offspring to expect from a cross. Heritability estimates measured in one population can't be used for other populations, so they don't even give you any general information about how much of a characteristic is genetically determined. Heritabilities don't do anything but make undergraduate students sweat during tests." How would you respond to this statement? Is the student correct? What good are heritabilities, and why do geneticists bother to calculate them?
5. Do you think genes influence a person's knowledge of the game of football? Explain your answer.

Section 17.4

6. We have learned that the response to selection is equal to the selection differential times the narrow-sense heritability, and that the narrow-sense heritability

includes only the additive genetic variance. Why aren't the dominance genetic variance and the gene interaction variance included? Why don't they contribute to the genetic variation that is acted on by selection?

7. A geneticist selects for increased body weight in a population of fruit flies that she is raising in her laboratory. She measures body weight in her population and selects the five heaviest males and the five heaviest females and uses them as the parents for the next generation. From the progeny produced by these parents, she selects the five heaviest males and five heaviest females and mates them. She repeats this procedure each generation. The average body weight of flies in the original population was 1.1 mg. The flies respond to selection, and their body weight steadily increases. After 20 generations of selection, the average body weight is 2.3 mg. However, after about 20 generations, the response to selection in subsequent generations levels off, and the average body weight of the flies no longer increases. At this point, the geneticist takes a long vacation; while she is gone, the fruit flies in her population interbreed randomly. When she returns from vacation, she finds that the average body weight of the flies in the population has decreased to 2.0 mg. Provide an explanation for why the response to selection leveled off after 20 generations. Why did the average size of the fruit flies decrease when selection was no longer applied during the geneticist's vacation?

18 Population and Evolutionary Genetics

The Wolves of Isle Royale

Wolves on Isle Royale have been studied for over 50 years. Inbreeding and low genetic variation has led to the population's decline. [PA Images/Alamy]

Rugged, isolated, and pristine, Isle Royale is surrounded by the deep, cold waters of northern Lake Superior. Though small and isolated, the island is famous for its wolves and is the home of the longest-running study of predator–prey dynamics.

Moose first made it to Isle Royale in the early 1900s, most likely by swimming the 24 km of water that separates the island from the mainland. Because these moose had no natural predators, their numbers on Isle Royale ballooned and they decimated the island's trees by browsing. But this situation changed in 1949, when a single breeding pair of wolves trekked across the frozen lake in winter and set up residence on the island. The wolves preyed on the moose and kept their population in check. By 1959, there were 20 wolves and about 500 moose on the island.

Over the next 50 years, the numbers of wolves and moose on Isle Royale waxed and waned. In 1980, the wolf population reached an all-time high of 50 animals. But the population crashed shortly afterward due to a disease epidemic. After the epidemic, wolf numbers climbed from 14 to about 20, but remained low throughout the 1980s and early 1990s. The wolf population suffered from inbreeding and low genetic variation.

Then, in 1997, a remarkable event occurred. A lone male wolf, later dubbed Old Gray Guy, migrated across the winter ice to the island. He joined the Isle Royale wolves and introduced fresh genes that invigorated the population. Old Gray Guy was a prolific breeder. He eventually produced 34 offspring, and his genes came to dominate the population—by 2008, 59% of the wolf genes on the island were his. The population climbed to over 30 wolves, and moose numbers declined. Old Gray Guy died in 2006. Unfortunately, no more new wolves joined the population, and levels of inbreeding reached a new high. There was no reproduction in 2012, and the population dropped further in the following years—by 2016, only a single male and female pair remained. Genetic studies revealed that this pair was highly inbred. Freed from wolf predation, the moose population boomed again.

As the number of wolves on Isle Royale became precariously low, some biologists proposed that additional wolves be transported to the island to counteract the effects of inbreeding. The introduction of new genetic variation into an inbred population, called genetic rescue, often dramatically improves the health of such populations and can better ensure their long-term survival. Others felt that nature should be left to take its course, even if it led to the extinction of the Isle Royale wolves. In 2014, the National Park Service, which traditionally avoids human intervention in wilderness areas, decided against introducing new wolves to rescue the population. As of 2016, the National Park Service is debating the long-term fate of wolves on Isle Royale.

THINK-PAIR-SHARE Questions 1 and 2

The wolves of Isle Royale illustrate several important principles of genetics: small populations lose genetic variation with the passage of time through inbreeding and genetic drift, often with catastrophic consequences for survival and reproduction. Migration, whether naturally or through genetic rescue, introduces new genetic variation that counteracts the effects of genetic drift and inbreeding. These principles have important implications for wildlife management, as well as for how organisms evolve in the natural world.

This chapter introduces population genetics, the branch of genetics that studies the genetic makeup of groups of individuals and how a group's genetic composition changes with time. Population geneticists usually focus their attention on a **Mendelian population**, which is a group of interbreeding, sexually reproducing individuals that have a common set of genes—the **gene pool**. A population evolves through changes in its gene pool; therefore, population genetics is also the study of evolution. Population geneticists study the variation in alleles within and between groups and the evolutionary forces responsible for shaping the patterns of genetic variation found in nature. In this chapter, we will learn how the gene pool of a population is described and what factors are responsible for shaping it. At the end of the chapter, we turn to the evolutionary changes that bring about the appearance of new species and examine patterns of evolutionary change at the molecular level.

18.1 Genotypic and Allelic Frequencies Are Used to Describe the Gene Pool of a Population

An obvious and pervasive feature of life is variation. Students in a typical college class vary in eye color, hair color, skin pigmentation, height, weight, facial features, blood type, and susceptibility to numerous diseases and disorders. No two students in the class are likely to be the same in appearance.

Humans are not unique in their extensive variation (**Figure 18.1a**); almost all organisms exhibit variation in phenotype. For instance, lady beetles are highly variable in their spotting patterns (**Figure 18.1b**), mice vary in color, snails have different numbers of stripes on their shells, and plants vary in their susceptibility to pests. Much of this phenotypic variation is hereditary. Recognition of the extent of phenotypic variation led Charles Darwin to the idea of evolution through natural selection. Genetic variation is the basis of all evolution, and the extent of genetic variation within a population affects its potential to adapt to environmental change.

In fact, even more genetic variation exists in populations than is visible in the phenotype. Much variation exists at the molecular level, owing in part to the redundancy of the genetic code, which allows different codons to specify the same amino acid. Thus, two members of a population can produce the same protein even if their DNA sequences are different. DNA sequences between genes and the introns within genes do not encode proteins; some of the variation in these sequences probably has little effect on the phenotype. But while this variation may not affect the phenotype, it is often useful for determining evolutionary relationships among organisms and understanding the evolutionary forces that have shaped a species.

Before we can explore the evolutionary processes that shape genetic variation, we must be able to describe the

18.1 All organisms exhibit genetic variation. (a) Extensive variation among humans. (b) Variation in the spotting patterns of Asian lady beetles. [Part a: Michael Dwyer/Alamy.]

genetic structure of a population. The usual way of describing the genetic structure of a population is to enumerate the types and frequencies of genotypes and alleles in that population. A frequency is simply a proportion or a percentage, usually expressed as a decimal fraction. For example, if 20% of the alleles at a particular locus in a population are *A*, we would say that the frequency of the *A* allele in the population is 0.20. For large populations, for which it is not practical to determine the genotypes of all individual members, a sample of the population is usually taken and the genotypic and allelic frequencies are calculated for this sample. The genotypic and allelic frequencies of the sample are then used to represent the gene pool of the population.

Calculating Genotypic Frequencies

To calculate a **genotypic frequency**, we simply add up the number of individuals possessing a genotype and divide by the total number of individuals in the sample (*N*). For a locus with three genotypes, *AA*, *Aa*, and *aa*, the frequency (*f*) of each genotype is

$$
\begin{aligned}
f(AA) &= \frac{\text{number of } AA \text{ individuals}}{N} \\
f(Aa) &= \frac{\text{number of } Aa \text{ individuals}}{N} \\
f(aa) &= \frac{\text{number of } aa \text{ individuals}}{N}
\end{aligned}
\qquad (18.1)
$$

The sum of all the genotypic frequencies always equals 1.

Calculating Allelic Frequencies

The gene pool of a population can also be described in terms of **allelic frequencies**. There are always fewer alleles than genotypes, so the gene pool of a population can be described in fewer terms when allelic frequencies are used. In a sexually reproducing population, the genotypes are only temporary assemblages of the alleles. As described by Mendel's principle of segregation, individual alleles, not genotypes, are passed from generation to generation through the gametes, and genotypes reform from alleles with each generation. Thus, the types and numbers of alleles, rather than genotypes, have real continuity from one generation to the next.

Allelic frequencies can be calculated from **(1)** the numbers or **(2)** the frequencies of the genotypes. To calculate the allelic frequency from the numbers of genotypes, we count the number of copies of a particular allele present among the genotypes and divide by the total number of all alleles in the sample:

$$
\text{frequency of an allele} = \frac{\text{number of copies of the allele}}{\text{number of copies of all alleles at the locus}} \qquad (18.2)
$$

For a locus with only two alleles (*A* and *a*), the frequencies of the alleles are usually represented by the symbols *p* and *q*. The frequencies can be calculated as follows:

$$
\begin{aligned}
p = f(A) &= \frac{2n_{AA} + n_{Aa}}{2N} \\
q = f(a) &= \frac{2n_{aa} + n_{Aa}}{2N}
\end{aligned}
\qquad (18.3)
$$

where n_{AA}, n_{Aa}, and n_{aa} represent the numbers of *AA*, *Aa*, and *aa* individuals, and *N* represents the total number of individuals in the sample. To obtain the number of copies of the allele in the numerator of the equation, we add twice the number of homozygotes (because each has two copies of the allele for which the frequency is being calculated) to the number of heterozygotes (because each has a single copy of the allele). We divide by 2*N* because each diploid individual has two alleles at a locus. The sum of the allelic frequencies always equals 1 ($p + q = 1$); so, after *p* has been obtained, *q* can be determined by subtraction: $q = 1 - p$.

Alternatively, allelic frequencies can be calculated from the genotypic frequencies. This calculation is useful if the genotypic frequencies have already been calculated and the numbers of the different genotypes are not available. To calculate allelic frequencies from genotypic frequencies, we add the frequency of the homozygote for each allele to half the frequency of the heterozygote (because half of the heterozygote's alleles are of each type):

$$
\begin{aligned}
p &= f(A) = f(AA) + \tfrac{1}{2}f(Aa) \\
q &= f(a) = f(aa) + \tfrac{1}{2}f(Aa)
\end{aligned}
\qquad (18.4)
$$

We obtain the same values of *p* and *q* whether we calculate the allelic frequencies from the numbers of genotypes (Equation 18.3) or from the genotypic frequencies (Equation 18.4). A sample calculation of allelic frequencies is provided in the next Worked Problem. TRY PROBLEM 23

LOCI WITH MULTIPLE ALLELES We can use the same principles to determine the frequencies of alleles for loci with more than two alleles. To calculate the allelic frequencies from the numbers of genotypes, we count up the number of copies of an allele by adding twice the number of homozygotes to the number of heterozygotes that possess the allele and divide this sum by twice the number of individuals in the sample. For a locus with three alleles (A^1, A^2, and A^3) and six genotypes (A^1A^1, A^1A^2, A^2A^2, A^1A^3, A^2A^3, and A^3A^3), the frequencies (*p*, *q*, and *r*) of the alleles are

$$
\begin{aligned}
p = f(A^1) &= \frac{2n_{A^1A^1} + n_{A^1A^2} + n_{A^1A^3}}{2N} \\
q = f(A^2) &= \frac{2n_{A^2A^2} + n_{A^1A^2} + n_{A^2A^3}}{2N} \\
r = f(A^3) &= \frac{2n_{A^3A^3} + n_{A^1A^3} + n_{A^2A^3}}{2N}
\end{aligned}
\qquad (18.5)
$$

Alternatively, we can calculate the frequencies of multiple alleles from the genotypic frequencies by extending Equation 18.4. Once again, we add the frequency of the homozygote to half the frequency of each heterozygous genotype that possesses the allele:

$$\begin{aligned} p &= f(A^1A^1) + \tfrac{1}{2}f(A^1A^2) + \tfrac{1}{2}f(A^1A^3) \\ q &= f(A^2A^2) + \tfrac{1}{2}f(A^1A^2) + \tfrac{1}{2}f(A^2A^3) \\ r &= f(A^3A^3) + \tfrac{1}{2}f(A^1A^3) + \tfrac{1}{2}f(A^2A^3) \end{aligned} \quad (18.6)$$

CONCEPTS

Population genetics concerns the genetic composition of a population and how it changes with time. The gene pool of a population can be described by the frequencies of genotypes and alleles in the population.

WORKED PROBLEM

The human MN blood-type antigens are determined by two codominant alleles, L^M and L^N (see Section 4.3). The MN blood types and corresponding genotypes of 398 Finns in Karjala are tabulated here.

Phenotype	Genotype	Number
MM	L^ML^M	182
MN	L^ML^N	172
N	L^NL^N	44

Source: W. C. Boyd, *Genetics and the Races of Man* (Boston: Little, Brown, 1950).

Calculate the genotypic and allelic frequencies at the MN locus for the Karjala population.

Solution Strategy

What information is required in your answer to the problem?

The genotypic and allelic frequencies of the population.

What information is provided to solve the problem?

- The numbers of the different MN genotypes in the sample.

Solution Steps

The genotypic frequencies for the population are calculated with the following formula:

$$\text{genotypic frequency} = \frac{\text{number of individuals with genotype}}{\text{total number of individuals in sample } (N)}$$

$$f(L^ML^M) = \frac{\text{number of } L^ML^M \text{ individuals}}{N} = \frac{182}{398} = 0.457$$

$$f(L^ML^N) = \frac{\text{number of } L^ML^N \text{ individuals}}{N} = \frac{172}{398} = 0.432$$

$$f(L^NL^N) = \frac{\text{number of } L^NL^N \text{ individuals}}{N} = \frac{44}{398} = 0.111$$

The allelic frequencies can be calculated from either the numbers or the frequencies of the genotypes. To calculate allelic frequencies from the numbers of genotypes, we add the number of copies of the allele and divide by the number of copies of all alleles at that locus:

$$\text{frequency of an allele} = \frac{\text{number of copies of the allele}}{\text{number of copies of all alleles}}$$

$$p = f(L^M) = \frac{(2n_{L^ML^M}) + (n_{L^ML^N})}{2N} = \frac{2(182) + 172}{2(398)} = \frac{536}{796} = 0.673$$

$$q = f(L^N) = \frac{(2n_{L^NL^N}) + (n_{L^ML^N})}{2N} = \frac{2(44) + 172}{2(398)} = \frac{260}{796} = 0.327$$

To calculate the allelic frequencies from genotypic frequencies, we add the frequency of the homozygote for that genotype to half the frequency of each heterozygote that contains that allele:

$$\begin{aligned} p = f(L^M) &= f(L^ML^M) + \tfrac{1}{2}f(L^ML^N) \\ &= 0.457 + \tfrac{1}{2}(0.432) = 0.673 \\ q = f(L^N) &= f(L^NL^N) + \tfrac{1}{2}f(L^ML^N) \\ &= 0.111 + \tfrac{1}{2}(0.432) = 0.327 \end{aligned}$$

▶ Now try your hand at calculating genotypic and allelic frequencies by working Problem 24 at the end of the chapter.

Models in Population Genetics

An important but frequently misunderstood tool used in population genetics is the mathematical model. Models are used, for example, to explore the influence of various forces on the genetic variation within and among populations. Let's take a moment to consider what a model is and how it can be used. A mathematical model usually describes a process as an equation. Factors that may influence the process are represented by variables in the equation; the equation defines the way in which the variables influence the process. Most models are simplified representations of a process because simultaneous consideration of all the factors involved is impossible; some factors must be ignored in order to examine the effects of others. At first, a model might consider only one or a few factors, but after their effects have been understood, the model can be improved by the addition of more details. Even a simple model, however, can be a source of valuable insight into how a process is influenced by key variables.

18.2 The Hardy–Weinberg Law Describes the Effect of Reproduction on Genotypic and Allelic Frequencies

The primary goal of population genetics is to understand the processes that shape a population's gene pool. First, we must ask what effects reproduction and Mendelian principles have on the genotypic and allelic frequencies: How do the segregation of alleles in gamete formation and the combining of alleles in fertilization influence the gene pool? The answer to this question lies in the **Hardy–Weinberg law**, among the most important principles of population genetics.

The Hardy–Weinberg law was formulated independently by Godfrey H. Hardy and Wilhelm Weinberg in 1908 (similar conclusions were reached by several other geneticists at about the same time). The law is actually a mathematical model that evaluates the effect of reproduction on the genotypic and allelic frequencies of a population. It makes several simplifying assumptions about the population and provides two key predictions if these assumptions are met. For an autosomal locus with two alleles, the Hardy–Weinberg law can be stated as follows:

Assumptions If a population is large, randomly mating, and not affected by mutation, migration, or natural selection, then

Prediction 1 the allelic frequencies of a population do not change; and

Prediction 2 the genotypic frequencies stabilize (will not change) after one generation in the proportions p^2 (the frequency of *AA*), $2pq$ (the frequency of *Aa*), and q^2 (the frequency of *aa*), where p equals the frequency of allele *A* and q equals the frequency of allele *a*.

The Hardy–Weinberg law indicates that when its assumptions are met, reproduction alone does not alter allelic or genotypic frequencies, and the allelic frequencies determine the frequencies of genotypes.

The statement that genotypic frequencies stabilize after one generation means that they may change after the first generation, because one generation of random mating is required to produce Hardy–Weinberg proportions of the genotypes. Afterward, the genotypic frequencies, like the allelic frequencies, do not change as long as the population continues to meet the assumptions of the Hardy–Weinberg law. When genotypes are in the expected proportions of p^2, $2pq$, and q^2, the population is said to be in **Hardy–Weinberg equilibrium**.

CONCEPTS

The Hardy–Weinberg law describes how reproduction and Mendelian principles affect the allelic and genotypic frequencies of a population.

✓ CONCEPT CHECK 1

Which statement is not an assumption of the Hardy–Weinberg law?

a. The allelic frequencies (p and q) are equal.
b. The population is randomly mating.
c. The population is large.
d. Natural selection has no effect.

Genotypic Frequencies at Hardy–Weinberg Equilibrium

How do the conditions of the Hardy–Weinberg law lead to genotypic proportions of p^2, $2pq$, and q^2? Mendel's principle of segregation says that each individual organism possesses two alleles at a locus and that each of those two alleles has an equal probability of passing into a gamete. Thus, the frequencies of alleles in gametes will be the same as the frequencies of alleles in the parents. Suppose that we have a Mendelian population in which the frequencies of alleles *A* and *a* are p and q, respectively. These frequencies will also be those in the gametes. If mating is random (one of the assumptions of the Hardy–Weinberg law), the gametes will come together in random combinations, which can be represented by a Punnett square (**Figure 18.2**).

The multiplication rule (see Chapter 3) can be used to determine the probability of various gametes pairing. For example, the probability of a sperm containing allele *A* is p and the probability of an egg containing allele *A* is p. Applying the multiplication rule, we find that the probability that these two gametes will combine to produce an *AA* homozygote is $p \times p = p^2$. Similarly, the probability of a sperm containing allele *a* combining with an egg containing allele *a* to produce an *aa* homozygote is $q \times q = q^2$. An *Aa* heterozygote can be produced in one of two ways: **(1)** a sperm containing allele *A* may combine with an egg containing allele *a* ($p \times q$)

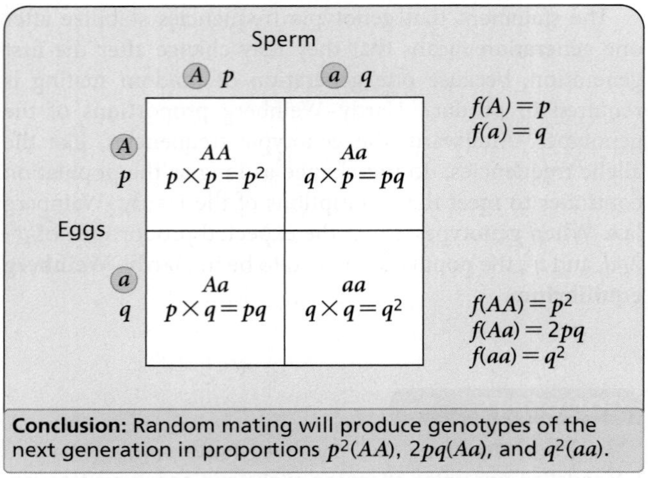

18.2 Random mating produces genotypes in the proportions p^2, $2pq$, and q^2. Note that this Punnett square represents mating in a population, not an individual cross.

or **(2)** an egg containing allele *A* may combine with a sperm containing allele *a* ($p \times q$). Thus, the probability of alleles *A* and *a* combining to produce an *Aa* heterozygote is $2pq$. In summary, whenever the frequencies of alleles in a randomly mating population are p and q, the frequencies of the genotypes in the next generation will be p^2, $2pq$, and q^2. Figure 18.2 demonstrates that only a single generation of random mating is required to produce the Hardy–Weinberg genotypic proportions.

Closer Examination of the Hardy–Weinberg Law

Before we consider the implications of the Hardy–Weinberg law, we need to take a closer look at the three assumptions that it makes about a population. First, it assumes that the population is large. How big is "large"? Theoretically, the Hardy–Weinberg law requires that a population be infinitely large in size, but this requirement is obviously unrealistic. In practice, many large populations have genotypes in the predicted Hardy–Weinberg proportions, and significant deviations arise only when population size is rather small. Later in the chapter, we will examine the effects of small population size on allelic frequencies.

The second assumption of the Hardy–Weinberg law is that members of the population mate randomly, which means that each genotype mates relative to its frequency. For example, suppose that three genotypes are present in a population in the following proportions: $f(AA) = 0.6$, $f(Aa) = 0.3$, and $f(aa) = 0.1$. With random mating, the frequency of mating between two *AA* homozygotes ($AA \times AA$) will be equal to the product of their frequencies: $0.6 \times 0.6 = 0.36$, whereas the frequency of mating between two *aa* homozygotes ($aa \times aa$) will be only $0.1 \times 0.1 = 0.01$.

The third assumption of the Hardy–Weinberg law is that the allelic frequencies of the population are not affected by natural selection, migration, or mutation. Although mutation occurs in every population, its rate is so low that it has little short-term effect on the predictions of the Hardy–Weinberg law (although it may shape allelic frequencies over long periods when no other forces are acting). Although natural selection and migration are significant factors in real populations, we must remember that the purpose of the Hardy–Weinberg law is to examine only the effect of reproduction on the gene pool. When this effect is known, the effects of other factors (such as migration and natural selection) can be examined.

A final point is that the assumptions of the Hardy–Weinberg law apply to a single locus. No real population mates randomly for all traits, and no population is completely free of natural selection for all traits. The Hardy–Weinberg law, however, does not require random mating and the absence of selection, migration, and mutation for all traits; it requires these conditions only for the locus under consideration. A population may be in Hardy–Weinberg equilibrium for one locus but not for others.

Implications of the Hardy–Weinberg Law

The Hardy–Weinberg law has several important implications for the genetic structure of a population. One implication is that a population cannot evolve if it meets the Hardy–Weinberg assumptions because evolution consists of change in the allelic frequencies of a population. Therefore, the Hardy–Weinberg law tells us that reproduction alone will not bring about evolution. Other processes—such as mutation, migration, or natural selection—or chance events are required for populations to evolve.

A second important implication is that when a population is in Hardy–Weinberg equilibrium, the genotypic frequencies are determined by the allelic frequencies. The heterozygote frequency never exceeds 0.5 when a population is in Hardy–Weinberg equilibrium. Furthermore, when the frequency of one allele is low, homozygotes for that allele will be rare, and most of the copies of a rare allele will be present in heterozygotes.

A third implication of the Hardy–Weinberg law is that a single generation of random mating produces the equilibrium frequencies of p^2, $2pq$, and q^2. The fact that genotypes are in Hardy–Weinberg proportions does not prove that the population is free from natural selection, mutation, and migration. It means only that these forces have not acted since the last time random mating took place.

A final implication is that when a population is not in Hardy–Weinberg equilibrium, one of the assumptions of the law has not been met, although without further investigation it will not be apparent which assumption is violated. Finding that a population is not in equilibrium often leads to other studies to determine what evolutionary forces are acting on the population.

Testing for Hardy–Weinberg Proportions

To determine if a population's genotypes are in Hardy–Weinberg equilibrium, the genotypic proportions expected under the Hardy–Weinberg law must be compared with the observed genotypic frequencies. To do so, we first calculate the allelic frequencies, then find the expected genotypic frequencies by using the square of the allelic frequencies, and finally, compare the observed and expected genotypic frequencies by using a chi-square test (see Chapter 3). The next Worked Problem shows how this is done.

WORKED PROBLEM

Jeffrey Mitton and his colleagues found three genotypes (R^2R^2, R^2R^3, and R^3R^3) at a locus encoding the enzyme peroxidase in ponderosa pine trees growing at Glacier Lake, Colorado. The observed numbers of these genotypes were

Genotypes	Number observed
R^2R^2	135
R^2R^3	44
R^3R^3	11

Are the ponderosa pine trees at Glacier Lake in Hardy–Weinberg equilibrium at the peroxidase locus?

Solution Strategy

What information is required in your answer to the problem?

The results of a chi-square test to determine whether the population is in Hardy–Weinberg equilibrium.

What information is provided to solve the problem?

- The numbers of the different genotypes in a sample of the population.

Solution Steps

If the frequency of the R^2 allele equals p and the frequency of the R^3 allele equals q, the frequency of the R^2 allele is

$$p = f(R^2) = \frac{(2n_{R^2R^2}) + (n_{R^2R^3})}{2N} = \frac{2(135) + 44}{2(190)} = 0.826$$

The frequency of the R^3 allele is obtained by subtraction:

$$q = f(R^3) = 1 - p = 0.174$$

The frequencies of the genotypes expected under Hardy–Weinberg equilibrium are then calculated by using p^2, $2pq$, and q^2:

$$R^2R^2 = p^2 = (0.826)^2 = 0.683$$
$$R^2R^3 = 2pq = 2(0.826)(0.174) = 0.287$$
$$R^3R^3 = q^2 = (0.174)^2 = 0.03$$

Multiplying each of these expected genotypic frequencies by the total number of observed genotypes in the sample (190), we obtain the numbers expected for each genotype:

$$R^2R^2 = 0.683 \times 190 = 129.8$$
$$R^2R^3 = 0.287 \times 190 = 54.5$$
$$R^3R^3 = 0.03 \times 190 = 5.7$$

By comparing these expected numbers with the observed numbers of each genotype, we see that there are more R^2R^2 and R^3R^3 homozygotes and fewer R^2R^3 heterozygotes in the population than we expect at equilibrium.

A chi-square goodness-of-fit test is used to determine whether the differences between the observed and the expected numbers of each genotype are due to chance:

$$X^2 = \sum \frac{(\text{observed} - \text{expected})^2}{\text{expected}}$$
$$= \frac{(135 - 129.8)^2}{129.8} + \frac{(44 - 54.5)^2}{54.5} + \frac{(11 - 5.7)^2}{5.7}$$
$$= 0.21 + 2.02 + 4.93 = 7.16$$

The calculated chi-square value is 7.16. To obtain the probability associated with this chi-square value, we determine the appropriate degrees of freedom.

So far, the chi-square test for assessing Hardy–Weinberg equilibrium has been identical with the chi-square tests that we used in Chapter 3 to assess progeny ratios in a genetic cross, where the degrees of freedom were $n - 1$ and n equaled the number of expected genotypes. For the Hardy–Weinberg test, however, we must subtract an additional degree of freedom because the expected numbers are based on the observed allelic frequencies; therefore, the expected numbers are not completely free to vary. In general, the degrees of freedom for a chi-square test of Hardy–Weinberg equilibrium equal the number of expected genotypic classes minus the number of associated alleles. For this particular Hardy–Weinberg test, the degree of freedom is $3 - 2 = 1$.

After we have calculated both the chi-square value and the degrees of freedom, the probability associated with this value can be sought in a chi-square table (see Table 3.4). With 1 degree of freedom, a chi-square value of 7.16 has a probability between 0.01 and 0.001. Thus, the observed values differ significantly from the expected values, and the peroxidase genotypes observed at Glacier Lake are not likely to be in Hardy–Weinberg proportions.

▶ For additional practice, determine whether the genotypic frequencies in **Problem 27** at the end of the chapter are in Hardy–Weinberg equilibrium.

CONCEPTS

The observed number of genotypes in a population can be compared with the expected Hardy–Weinberg proportions by using a chi-square goodness-of-fit test.

Estimating Allelic Frequencies by Using the Hardy–Weinberg Law

A practical use of the Hardy–Weinberg law is that it allows us to calculate allelic frequencies when dominance is present. For example, cystic fibrosis is a life-threatening autosomal recessive disorder characterized by frequent and severe respiratory infections, incomplete digestion, and abnormal sweating (see Section 4.3). Among North American Caucasians, the incidence of the disease is approximately 1 person in 2000. The formula for calculating allelic frequency (see Equation 18.3) requires that we know the numbers of homozygotes and heterozygotes, but cystic fibrosis is a recessive disease, so we cannot easily distinguish between homozygous unaffected persons and heterozygous carriers. Although molecular tests are available for identifying heterozygous carriers of the cystic fibrosis gene, the low frequency of the disease makes widespread screening impractical. In such situations, the Hardy–Weinberg law can be used to estimate the allelic frequencies.

If we assume that a population is in Hardy–Weinberg equilibrium with regard to this locus, then the frequency of the recessive genotype (*aa*) is q^2, and the allelic frequency is the square root of the genotypic frequency:

$$q = \sqrt{f(aa)} \qquad (18.7)$$

If the frequency of cystic fibrosis in North American Caucasians is approximately 1 in 2000, or 0.0005, then $q = \sqrt{0.0005} = 0.02$. Thus, about 2% of the alleles in the Caucasian population encode the defective protein that causes cystic fibrosis. We can calculate the frequency of the normal allele by subtracting: $p = 1 - q = 1 - 0.02 = 0.98$. After we have calculated p and q, we can use the Hardy–Weinberg law to determine the frequencies of homozygous unaffected people and heterozygous carriers of the cystic fibrosis allele:

$$f(AA) = p^2 = (0.98)^2 = 0.960$$

$$f(Aa) = 2pq = 2(0.02)(0.98) = 0.0392$$

Thus, about 4% (1 of 25) of Caucasians are heterozygous carriers of the allele that causes cystic fibrosis. ➤TRY PROBLEM 29

THINK-PAIR-SHARE Question 3

CONCEPTS

Although allelic frequencies cannot be calculated directly for a locus that exhibits dominance, the Hardy–Weinberg law can be used to estimate allelic frequencies if the population is in Hardy–Weinberg equilibrium for that locus. The frequency of the recessive allele will be equal to the square root of the frequency of the recessive trait.

✓ CONCEPT CHECK 2

In cats, all-white color is dominant over colors other than all-white. In a population of 100 cats, 19 are all-white cats. Assuming that the population is in Hardy–Weinberg equilibrium, what is the frequency of the all-white allele in this population?

Nonrandom Mating Alters Genotype Frequencies

An assumption of the Hardy–Weinberg law is that mating is random with respect to genotype. Although it does not alter the frequencies of alleles, nonrandom mating affects the way in which alleles combine to form genotypes and alters the genotypic frequencies of a population.

One form of nonrandom mating is **inbreeding**, which is preferential mating between related individuals. Inbreeding causes a departure from the Hardy–Weinberg equilibrium frequencies of p^2, $2pq$, and q^2. More specifically, it leads to an increase in the proportion of homozygotes and a decrease in the proportion of heterozygotes in a population. Close inbreeding is often harmful because it increases the proportion of homozygotes and thereby boosts the probability that deleterious and lethal recessive alleles will combine to produce homozygotes with a harmful trait. This increased appearance of lethal and deleterious traits with inbreeding is termed **inbreeding depression**.

CONCEPTS

Nonrandom mating alters the frequencies of genotypes but not the frequencies of alleles. Inbreeding is preferential mating between related individuals. With inbreeding, the frequency of homozygotes increases, whereas the frequency of heterozygotes decreases.

18.3 Several Evolutionary Forces Can Cause Changes in Allelic Frequencies

The Hardy–Weinberg law indicates that allelic frequencies do not change as a result of reproduction. Processes that bring about change in allelic frequencies include mutation, migration, genetic drift (random effects due to small population size), and natural selection.

Mutation

Before evolution can take place, genetic variation must exist within a population; consequently, all evolution depends on processes that generate genetic variation. Although new *combinations* of existing genes may arise through recombination in meiosis, all genetic variants ultimately arise through mutation.

THE EFFECT OF MUTATION ON ALLELIC FREQUENCIES Mutation can influence the rate at which one genetic variant increases at the expense of another. Consider a single locus in a population of 25 diploid individuals. Each individual possesses two alleles at the locus under consideration, so the gene pool of the population consists of 50 allele copies. Let's assume that there are two different alleles, designated G^1 and G^2, with frequencies p and q, respectively. If there are 45 copies of G^1 and 5 copies of G^2 in the population, $p = 0.90$ and $q = 0.10$. Now suppose that a mutation changes a G^1 allele into a G^2 allele. After this mutation, there are 44 copies of G^1 and 6 copies of G^2, and the frequency of G^2 has increased from 0.10 to 0.12. Mutation has changed the allelic frequencies.

If copies of G^1 continue to mutate to G^2, the frequency of G^2 will increase and the frequency of G^1 will decrease (**Figure 18.3**). The amount that G^2 will change as a result of mutation depends on **(1)** the rate of G^1-to-G^2 mutation and **(2)** p, the frequency of G^1 in the population. When p is large, many copies of G^1 are available to mutate to G^2 and the amount of change will be relatively large. As more mutations occur and p decreases, fewer copies of G^1 will be available to mutate to G^2.

So far, we have considered only the effects of $G^1 \rightarrow G^2$ forward mutations. Reverse $G^2 \rightarrow G^1$ mutations also occur, but at a rate that will probably differ from the forward mutation rate. Whenever a reverse mutation occurs, the frequency of G^2 decreases and the frequency of G^1 increases (see Figure 18.3).

REACHING EQUILIBRIUM Consider a population that begins with a high frequency of G^1 and a low frequency of G^2 (Figure 18.3a). In this population, many copies of G^1 are initially available to mutate to G^2, and the increase in G^2 due to forward mutation will be relatively large. However, as the frequency of G^2 increases as a result of forward mutations, fewer copies of G^1 are available to mutate, so the number of forward mutations decreases. On the other hand, few copies of G^2 are initially available to undergo a reverse mutation to G^1, but as the frequency of G^2 increases, the number of copies of G^2 available to undergo reverse mutation to G^1 increases; therefore, the number of genes undergoing reverse mutation will increase (Figure 18.3b). Eventually, the number of genes undergoing forward mutation will be counterbalanced by the number of genes undergoing reverse mutation (Figure 18.3c). At this point, the increase in q due to forward mutation will be equal to the decrease in q due to reverse mutation, and there will be no net change in allelic frequency, in spite of the fact that forward and reverse mutations continue to occur. A point at which there is no change in the allelic frequencies of a population is referred to as an **equilibrium** (see Figure 18.3c). At mutational equilibrium, the frequency of G^2 is determined solely by the forward and reverse mutation rates.

SUMMARY OF EFFECTS When the only evolutionary force acting on a population is mutation, allelic frequencies change over time because some alleles mutate into others. Eventually, these allelic frequencies reach equilibrium and are determined only by the forward and reverse mutation rates.

The mutation rates for most genes are low, so change in allelic frequencies due to mutation in one generation is very small, and long periods are required for a population to reach mutational equilibrium. Nevertheless, if mutation is the only force acting on a population for long periods, mutation rates will determine allelic frequencies.

Because most alleles are G^1, there are more forward mutations than reverse mutations.

Forward mutation (○⟶●)

Reverse mutation (○⟵●)

G^1 (p) G^2 (q)

Forward mutations increase the frequency of G^2,...

...which increases the number of alleles undergoing reverse mutation.

G^1 (p) G^2 (q)

Eventually, an equilibrium is reached, where the number of forward mutations equals the number of reverse mutations.

Equilibrium

G^1 (p) G^2 (q)

Conclusion: At equilibrium, the allelic frequencies do not change even though mutation in both directions continues.

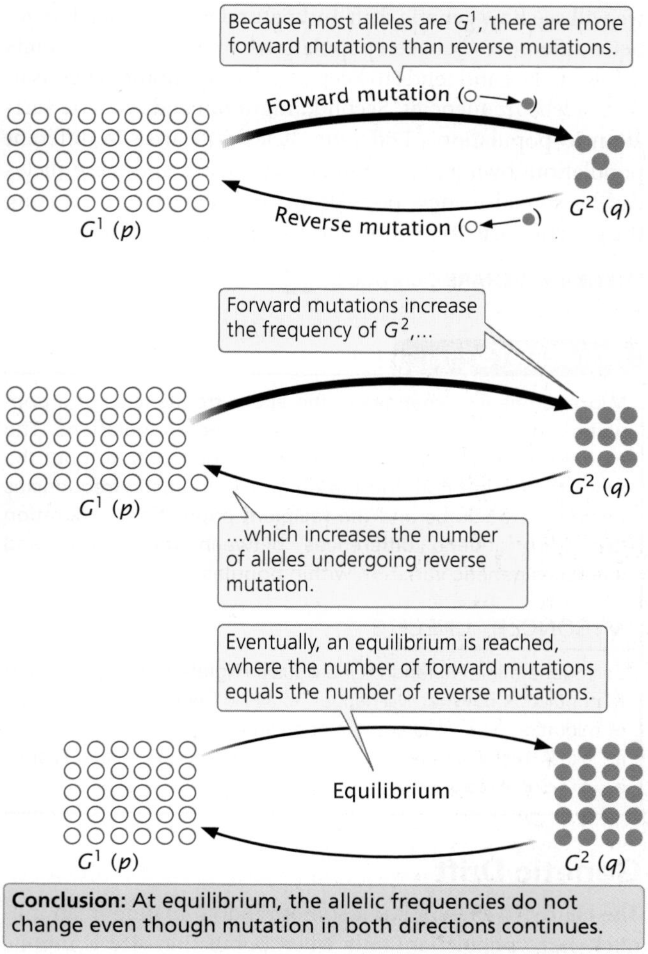

18.3 Recurrent mutation changes allelic frequencies. Forward and reverse mutations eventually lead to equilibrium.

CONCEPTS

Recurrent mutation causes changes in the frequencies of alleles. At equilibrium, allelic frequencies are determined by forward and reverse mutation rates. Because mutation rates are low, the effect of mutation on allelic frequencies per generation is very small.

Migration

Another process that may bring about change in a population's allelic frequencies is the influx of genes from other populations, commonly called **migration** or **gene flow**. One of the assumptions of the Hardy–Weinberg law is that migration does not take place, but many natural populations do experience migration from other populations. The overall effect of migration is twofold: (1) it prevents populations from becoming genetically different from one another, and (2) it increases genetic variation within populations.

THE EFFECT OF MIGRATION ON ALLELIC FREQUENCIES Let's consider the effects of migration by looking at a simple, unidirectional model of migration between two populations that differ in the frequency of an allele *a* (**Figure 18.4**). In each generation, a representative sample of the individuals in population I migrates to population II and reproduces, adding its alleles to population II's gene pool. Migration is only from population I to population II (unidirectional), and all the conditions of the Hardy–Weinberg law (large population size, random mating, etc.) apply except the absence of migration.

After migration, population II consists of two types of individuals: **(1)** migrants with alleles from population I and **(2)** the original residents with alleles from population II. The allelic frequencies in population II after migration depend on the contributions of alleles from the migrants and from the original residents. The amount of change in the frequency of allele *a* in population II is directly proportional to the amount of migration; as the amount of migration increases, the change in allelic frequency increases. The magnitude of change is also affected by the difference in allelic frequencies between the two populations; when the difference is large, the change in allelic frequency will be large.

With each generation of migration, the allelic frequencies of the two populations become more and more similar until eventually, the frequencies of population II equal those of population I. When the allelic frequencies are equal, there will be no further change in the allelic frequencies of population II, in spite of the fact that migration continues. If migration between two populations takes place for a number of generations with no other evolutionary forces present, an equilibrium is reached at which the allelic frequencies of the recipient population equal those of the source population.

The simple model of unidirectional migration between two populations just outlined can be expanded to accommodate multidirectional migration between several populations.

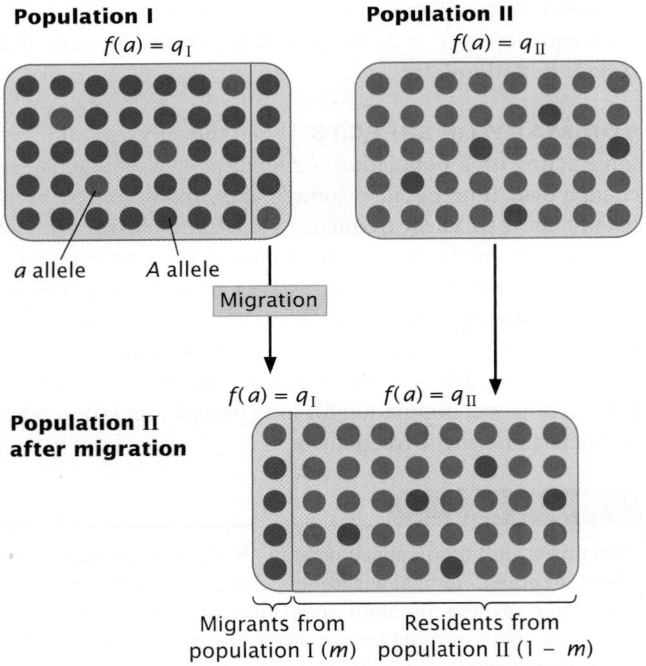

Conclusion: The frequency of allele *a* in population II after migration is $q'_{II} = q_I m + q_{II}(1 - m)$.

18.4 The amount of change in allelic frequencies due to migration between populations depends on the difference between the populations in their allelic frequencies and on the extent of migration. Shown here is a model of the effect of unidirectional migration on allelic frequencies.

THINK-PAIR-SHARE Question 4

THE OVERALL EFFECT OF MIGRATION Migration has two major effects. First, it causes the gene pools of separate populations to become more similar. Later in this section, we will see how genetic drift and natural selection lead to genetic differences between populations; migration counteracts this tendency and tends to keep populations homogeneous in their allelic frequencies. Second, migration adds genetic variation to populations. Different alleles may arise in different populations owing to rare mutational events, and these alleles can be spread to new populations by migration, increasing the genetic variation within the recipient population.

THINK-PAIR-SHARE Question 5

CONCEPTS

Migration causes changes in the allelic frequencies of a population by introducing alleles from other populations. The magnitude of change due to migration depends on both the amount of migration and the difference in allelic frequencies between the source and the recipient populations. Migration decreases genetic differences between populations and increases genetic variation within populations.

✓ CONCEPT CHECK 3

Each generation, 10 random individuals migrate from population A to population B. What will happen to allelic frequency *q* as a result of migration when *q* is equal in populations A and B?

a. *q* in A will decrease.
b. *q* in B will increase.
c. *q* will not change in either A or B.
d. *q* in B will become q^2.

Genetic Drift

The Hardy–Weinberg law assumes random mating in an infinitely large population; only when population size is infinite will the gametes carry genes that perfectly represent the parental gene pool. But no real population is infinitely large, and when population size is limited, the gametes that unite to form individuals of the next generation carry a sample of alleles present in the parental gene pool. Just by chance, the composition of this sample will often deviate from that of the parental gene pool, and this deviation may cause allelic frequencies to change. The smaller the gametic sample, the greater the chance that its composition will deviate from that of the entire gene pool.

The role of chance in altering allelic frequencies is analogous to the flip of a coin. Each time we flip a coin, we have a 50% chance of getting a head and a 50% chance of getting a tail. If we flip a coin 1000 times, the observed ratio of heads to tails will be very close to the expected 50 : 50 ratio. If, however, we flip a coin only 10 times, there is a good chance that we will obtain not exactly five heads and five tails, but perhaps seven heads and three tails, or eight tails and two heads. This kind of deviation from an expected ratio due to limited sample size is referred to as **sampling error**.

Sampling error arises when gametes unite to produce progeny. Many organisms produce a large number of gametes, but when population size is small, a limited number of gametes unite to produce the individuals of the next generation. Chance influences which alleles are present in this limited sample. In this way, sampling error may lead to **genetic drift**, or changes in allelic frequencies. Because the deviations from the expected ratios are random, the direction of change is unpredictable. We can nevertheless predict the magnitude of the change.

THE MAGNITUDE OF GENETIC DRIFT The amount of change resulting from genetic drift is determined largely by the population size (N): genetic drift is higher when the population size is small. For ecological and demographic studies, population size is usually defined as the number of individuals in a group. The evolution of a gene pool depends, however, only on those individuals who contribute genes to the next generation. Population geneticists usually define population size as the equivalent number of breeding adults, referred to as the **effective population size** (N_e).

CONCEPTS

Genetic drift is change in allelic frequencies due to chance. The amount of change in allelic frequencies due to genetic drift is inversely related to the effective population size (the equivalent number of breeding adults in a population).

✓ CONCEPT CHECK 4

Which of the following is an example of genetic drift?

a. Allele *g* for fat production increases in a small population because birds with more body fat have higher survivorship in a harsh winter.
b. Random mutation increases the frequency of allele *A* in one population but not in another.
c. Allele *R* reaches a frequency of 1.0 because individuals with genotype *rr* are sterile.
d. Allele *m* is lost when a virus kills all but a few individuals and just by chance, none of the survivors possess allele *m*.

THE CAUSES OF GENETIC DRIFT All genetic drift arises from sampling error, but sampling error can arise in several different ways. First, a population can be reduced in size for a number of generations because of limitations in space, food, or some other critical resource. Genetic drift in a small population over multiple generations can significantly affect the composition of a population's gene pool.

A second way in which sampling error can arise is through the **founder effect**, which occurs when a population is established by a small number of individuals. The population of wolves on Isle Royale, discussed in the introduction to this chapter, underwent a founder effect. Although the population can increase and become quite large, the genes carried by all its members are derived from the few genes originally present in the founders (assuming no migration or mutation). Chance events affecting which genes were present in the founders have an important influence on the makeup of the entire population.

A third way in which genetic drift arises is through a **genetic bottleneck**, which develops when a population undergoes a drastic reduction in size. An example is seen in northern elephant seals (**Figure 18.5**). Before 1800, thousands of northern elephant seals were found along the California coast, but hunting between 1820 and 1880 devastated the population. By 1884, as few as 20 seals survived on a remote beach of Isla de Guadalupe west of Baja California, Mexico. Restrictions on hunting enacted by the United States and Mexico allowed the seals to recover, and there are now an estimated 100,000 seals. All the seals in the population today are genetically similar, however, because they have only those genes that were carried by the few survivors of the population bottleneck.

THINK-PAIR-SHARE Question 6

THE EFFECTS OF GENETIC DRIFT Genetic drift has several important effects on the genetic composition of a population. First, it produces change in allelic frequencies within a population. Because drift is random, the frequency

18.5 Northern elephant seals underwent a severe genetic bottleneck between 1820 and 1880. Today, these seals have low levels of genetic variation. [PhotoDisc/Getty Images.]

of any allele is just as likely to increase as it is to decrease and will wander with the passage of time (hence the name *genetic drift*). **Figure 18.6** illustrates a computer simulation of genetic drift in five populations over 30 generations, starting with $q = 0.5$ and maintaining a constant population size of 10 males and 10 females. These allelic frequencies change randomly from generation to generation.

A second effect of genetic drift is reduction of genetic variation within populations. Through random change, an allele may eventually reach a frequency of either 1 or 0, at which point all individuals in the population are homozygous for one allele. When an allele has reached a frequency of 1, we say that it has reached **fixation**. Other alleles are lost (reach a frequency of 0) and can be restored only by migration from another population or by mutation. Fixation, then, leads to a loss of genetic variation within a population. This loss can be seen in the northern elephant seals just described. Today, these seals have low levels of genetic variation; a study of 24 protein-encoding genes found no individual or population differences in these genes. A subsequent study of sequence variation in mitochondrial DNA also revealed low levels of genetic variation. In contrast, southern elephant seals have much higher levels of mitochondrial DNA variation. Southern elephant seals, which are found in Antarctic and sub-Antarctic waters, were also hunted, but their population size never dropped below 1000; therefore, unlike the northern elephant seals, they did not experience a genetic bottleneck.

Given enough time, all small populations will become fixed for one allele or another. Which allele becomes fixed is random but is influenced by the initial frequency of the allele. If the population begins with two alleles, each with a frequency of 0.5, both alleles have an equal probability of fixation. However, if one allele is initially more common, it is more likely to become fixed.

A third effect of genetic drift is that different populations diverge genetically with time. In Figure 18.6, all five populations begin with the same allelic frequency ($q = 0.5$) but, because drift is random, the frequencies in different populations do not change in the same way, and so populations gradually acquire genetic differences. Eventually, all the populations will reach fixation; some will become fixed for one allele, and others will become fixed for the alternative allele.

The effect of genetic drift on variation among populations is illustrated by a study conducted by Luca Cavalli-Sforza and his colleagues. They studied variation in blood types among villagers in the Parma Valley of Italy, where the amount of migration between villages was limited. They found that variation in allelic frequencies was greatest between small isolated villages in the upper valley, but decreased between larger villages and towns farther down the valley. This result is exactly what we expect with genetic drift: there should be more genetic drift, and thus more variation among villages, when population sizes are small.

The three results of genetic drift (allelic frequency change, loss of genetic variation within populations, and genetic divergence between populations) take place simultaneously, and all result from sampling error. The first two results take place *within* populations, whereas the third takes place *between* populations. ▶TRY PROBLEM 32

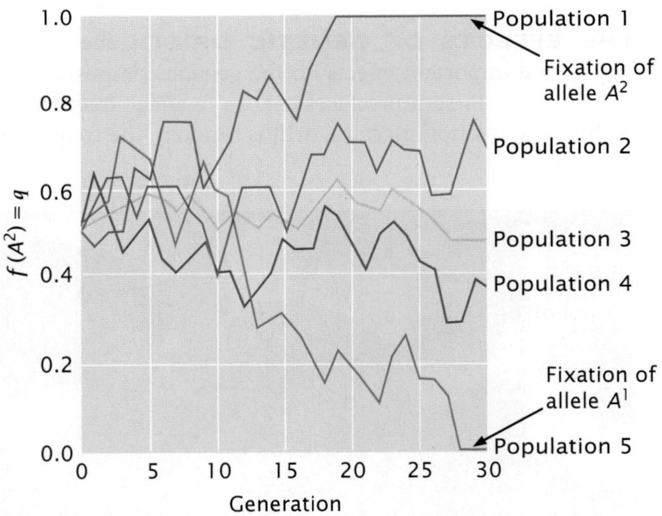

18.6 Genetic drift changes allelic frequencies within populations, leading to a reduction in genetic variation through fixation and genetic divergence among populations. Shown here is a computer simulation of changes in the frequency of allele A^2 (q) in five different populations due to random genetic drift. Each population consists of 10 males and 10 females and begins with $q = 0.5$.

CONCEPTS

Genetic drift results from continuous small population size, the founder effect (when a population is established by a few founders), and the genetic bottleneck effect (when population size is drastically reduced). Genetic drift causes change in allelic frequencies within a population, a loss of genetic variation through the fixation of alleles, and genetic divergence between populations.

Natural Selection

A final process that brings about changes in allelic frequencies is *natural selection*, the differential reproduction of genotypes (see Section 17.4). Natural selection takes place when individuals with adaptive traits produce a greater number of offspring than that produced by others in the population. If the adaptive traits have a genetic basis, they are inherited by the offspring and appear with greater frequency in the next generation. A trait that provides a reproductive advantage thereby increases over time, enabling populations to become better suited—that is, better adapted—to their environments. Natural selection is unique among evolutionary forces in that it promotes adaptation (**Figure 18.7**).

18.7 Natural selection leads to adaptations, such as those seen in the polar bears that inhabit the extreme Arctic environment. These bears blend into the snowy background, which helps them in hunting seals. The hairs of their fur stay erect even when wet, and thick layers of blubber provide insulation, which protects against subzero temperatures. Their digestive tracts are adapted to a seal-based carnivorous diet. [DigitalVision/Getty Images.]

FITNESS AND THE SELECTION COEFFICIENT The effect of natural selection on the gene pool of a population depends on the fitness values of the genotypes in the population. **Fitness** is defined as the relative reproductive success of a genotype. Here, the term *relative* is critical: fitness is the reproductive success of one genotype compared with the reproductive successes of other genotypes in the population.

Fitness values range from 0 to 1. Suppose that the average number of viable offspring produced by three genotypes is

Genotypes:	A^1A^1	A^1A^2	A^2A^2
Mean number of offspring produced:	10	5	2

To calculate fitness (W) for each genotype, we take the mean number of offspring produced by a genotype and divide it by the mean number of offspring produced by the most prolific genotype:

$$\begin{array}{ccc} A^1A^1 & A^1A^2 & A^2A^2 \\ W_{11} = \frac{10}{10} = 1.0 & W_{12} = \frac{5}{10} = 0.5 & W_{22} = \frac{2}{10} = 0.2 \end{array} \quad (18.8)$$

The fitness of genotype A^1A^1 is designated W_{11}, that of A^1A^2 is W_{12}, and that of A^2A^2 is W_{22}.

A related variable is the **selection coefficient** (s), which is the relative intensity of selection against a genotype. We usually speak of selection for a particular genotype, but keep in mind that, when selection is *for* one genotype, selection is automatically *against* at least one other genotype. The selection coefficient is equal to $1 - W$; so the selection coefficients for the preceding three genotypes are

$$\begin{array}{ccc} A^1A^1 & A^1A^2 & A^2A^2 \\ s_{11} = 0 & s_{12} = 0.5 & s_{22} = 0.8 \end{array}$$

CONCEPTS

Natural selection is the differential reproduction of genotypes. It is measured as fitness, which is the reproductive success of a genotype compared with that of other genotypes in a population.

✓ **CONCEPT CHECK 5**

The average numbers of offspring produced by three genotypes are $GG = 6$; $Gg = 3$; $gg = 2$. What is the fitness of Gg?

a. 3 b. 0.5 c. 0.3 d. 0.27

THE RESULTS OF SELECTION The results of selection depend on the fitnesses of the genotypes in a population. In a population with three genotypes (A^1A^1, A^1A^2, and A^2A^2) with fitnesses W_{11}, W_{12}, and W_{22}, we can identify six different types of natural selection (**Table 18.1**).

In type 1 selection, a dominant allele A^1 confers a fitness advantage; in this case, the fitnesses of genotypes A^1A^1

TABLE 18.1 Types of natural selection

Type	Fitness relation	Form of selection	Result
1	$W_{11} = W_{12} > W_{22}$	Directional selection against recessive allele A^2	A^1 increases, A^2 decreases
2	$W_{11} = W_{12} < W_{22}$	Directional selection against dominant allele A^1	A^2 increases, A^1 decreases
3	$W_{11} > W_{12} > W_{22}$	Directional selection against incompletely dominant allele A^2	A^1 increases, A^2 decreases
4	$W_{11} < W_{12} < W_{22}$	Directional selection against incompletely dominant allele A^1	A^2 increases, A^1 decreases
5	$W_{11} < W_{12} > W_{22}$	Overdominance	Stable equilibrium, both alleles maintained
6	$W_{11} > W_{12} < W_{22}$	Underdominance	Unstable equilibrium

Note: W_{11}, W_{12}, and W_{22} represent the fitnesses of genotypes A^1A^1, A^1A^2, and A^2A^2, respectively.

and A^1A^2 are equal and higher than the fitness of A^2A^2 ($W_{11} = W_{12} > W_{22}$). Because both the heterozygote and the A^1A^1 homozygote have copies of the A^1 allele and produce more offspring than the A^2A^2 homozygote does, the frequency of the A^1 allele will increase with time, and the frequency of the A^2 allele will decrease. This form of selection, in which one allele or trait is favored over another, is termed **directional selection**.

Type 2 selection is directional selection against a dominant allele A^1 ($W_{11} = W_{12} < W_{22}$). Type 3 and type 4 selection are also directional selection, but in these cases, there is incomplete dominance, and the heterozygote has a fitness that is intermediate between the two homozygotes ($W_{11} > W_{12} > W_{22}$ for type 3; $W_{11} < W_{12} < W_{22}$ for type 4). Eventually, all four types of directional selection lead to fixation of the favored allele and elimination of the other allele, as long as no other evolutionary forces act on the population.

The last two types of selection (types 5 and 6) are special situations that lead to equilibrium, where there is no further change in allelic frequency. Type 5 selection is referred to as **overdominance** or *heterozygote advantage*. Here, the heterozygote has higher fitness than either homozygote ($W_{11} < W_{12} > W_{22}$). With overdominance, both alleles are favored in the heterozygote, and neither allele is eliminated from the population. The allelic frequencies change with overdominant selection until a stable equilibrium is reached, at which point there is no further change. The allelic frequency at equilibrium ($\hat{q}$) depends on the fitnesses (usually expressed as selection coefficients) of the two homozygotes:

$$\hat{q} = f(A^2) = \frac{s_{11}}{s_{11} + s_{22}} \quad (18.9)$$

where s_{11} represents the selection coefficient of the A^1A^1 homozygote and s_{22} represents the selection coefficient of the A^2A^2 homozygote.

An example of overdominance is sickle-cell anemia in humans, a disease that results from a mutation in one of the genes that encodes hemoglobin. People who are homozygous for the sickle-cell mutation produce only sickle-cell hemoglobin, have severe anemia, and often have tissue damage. People who are heterozygous—with one normal copy and one mutated copy of the gene—produce both normal and sickle-cell hemoglobin, but their red blood cells contain enough normal hemoglobin to prevent sickle-cell anemia. However, heterozygotes are resistant to malaria and have higher fitness than do homozygotes for normal hemoglobin or homozygotes for sickle-cell anemia.

The last type of selection (type 6) is **underdominance**, in which the heterozygote has lower fitness than either homozygote ($W_{11} > W_{12} < W_{22}$). Underdominance leads to an unstable equilibrium; here, allelic frequencies do not change as long as they are at equilibrium, but if they are disturbed from the equilibrium point by some other evolutionary force, they will move away from equilibrium until one allele eventually becomes fixed.

CONCEPTS

Natural selection changes allelic frequencies; the direction and magnitude of change depend on the intensity of selection, the dominance relations of the alleles, and the allelic frequencies. Directional selection favors one allele over another and eventually leads to fixation of the favored allele. Overdominance leads to a stable equilibrium with maintenance of both alleles in the population. Underdominance produces an unstable equilibrium because the heterozygote has lower fitness than either homozygote.

CONNECTING CONCEPTS

The General Effects of Forces That Change Allelic Frequencies

You now know that four processes can bring about change in the allelic frequencies of a population: mutation, migration, genetic drift, and natural selection. Their short- and long-term effects on allelic frequencies are summarized in **Table 18.2**. In some cases, the change continues until one allele is eliminated and the other becomes fixed in the population. Genetic drift and directional selection eventually result in fixation, provided that these forces are the only ones acting on a population. With the other evolutionary forces, allelic frequencies change until an equilibrium point is reached, after which there is no further change in allelic frequencies. Mutation, migration, and some forms of natural selection can lead to stable equilibria.

These evolutionary forces affect both genetic variation within populations and genetic divergence between populations. Evolutionary forces that maintain or increase genetic variation within populations are listed in the upper left quadrant of **Figure 18.8**. These forces include some types of natural selection, such as overdominance, in which both alleles are favored. Mutation and

TABLE 18.2 Effects of different evolutionary forces on allelic frequencies within populations

Force	Short-term effect	Long-term effect
Mutation	Change in allelic frequencies	Equilibrium reached between forward and reverse mutations
Migration	Change in allelic frequencies	Equilibrium reached when allelic frequencies of source and recipient population are equal
Genetic drift	Change in allelic frequencies	Fixation of one allele
Natural selection	Change in allelic frequencies	Directional selection: fixation of one allele Overdominant selection: equilibrium reached Underdominant selection: unstable equilibrium

	Within populations	Between populations
Increase genetic variation	**Mutation** **Migration** **Some types of natural selection**	**Mutation** **Genetic drift** **Some types of natural selection**
Decrease genetic variation	**Genetic drift** **Some types of natural selection**	**Migration** **Some types of natural selection**

18.8 Mutation, migration, genetic drift, and natural selection have different effects on genetic variation within populations and on genetic divergence between populations.

migration also increase genetic variation within populations because they introduce new alleles to a population. Evolutionary forces that decrease genetic variation within populations are listed in the lower left quadrant of Figure 18.8. These forces include genetic drift, which decreases variation through the fixation of alleles, and some forms of natural selection, such as directional selection.

These same evolutionary forces also affect genetic divergence between populations. Natural selection increases divergence between populations if different alleles are favored in different populations, but it can also decrease divergence between populations by favoring the same allele in different populations. Mutation almost always increases divergence between populations because different mutations arise in each population. Genetic drift also increases divergence between populations because changes in allelic frequencies due to drift are random and are likely to occur in different directions in separate populations. Migration, on the other hand, decreases divergence between populations because it makes populations more similar in their genetic composition.

Migration and genetic drift act in opposite directions: migration increases genetic variation within populations and decreases divergence between populations, whereas genetic drift decreases genetic variation within populations and increases divergence between populations. Mutation increases both variation within populations and divergence between populations. Natural selection can either increase or decrease variation within populations, and it can increase or decrease divergence between populations.

An important point to keep in mind is that real populations are simultaneously affected by many evolutionary forces. We have examined the effects of mutation, migration, genetic drift, and natural selection in isolation so that the effect of each process will be clear. In the real world, however, populations are commonly affected by several evolutionary forces at the same time, and evolution results from the complex interplay of numerous processes.

18.4 Evolution Occurs Through Genetic Change Within Populations

The concept of evolution is one of the foundational principles of all of biology. Theodosius Dobzhansky, an important early leader in the field of evolutionary genetics, once remarked, "Nothing in biology makes sense except in the light of evolution." Indeed, evolution is an all-encompassing theory that helps to make sense of much of the natural world, from the sequences of DNA found in our cells to the types of organisms that surround us. The evidence for evolution is overwhelming. Evolution has been directly observed numerous times; for example, hundreds of different insect species evolved resistance to common pesticides introduced after World War II. The theory of evolution is supported by the fossil record, comparative anatomy, embryology, the distributions of plants and animals (biogeography), and molecular genetics.

THINK-PAIR-SHARE Question 7

Biological Evolution

In spite of its vast importance to all fields of biology, evolution is often misunderstood and misinterpreted. In our society, the term *evolution* frequently refers to any type of change. However, **evolution**, in the biological sense, refers only to a specific type of change: genetic change taking place in a group of organisms. Two aspects of this definition should be emphasized. First, biological evolution includes genetic change only. Many nongenetic changes take place in living organisms, such as the development of a complex intelligent person from a single-celled zygote. Although remarkable, this change is not evolution because it does not include genetic changes. Second, biological evolution takes place in *groups* of organisms. An individual organism does not evolve; what evolves is the gene pool common to a group of organisms.

Evolution as a Two-Step Process

Evolution can be thought of as a two-step process. First, genetic variation arises. Genetic variation has its origin in the processes of mutation, which produces new alleles, and recombination, which shuffles alleles into new combinations. Both of these processes are random and produce genetic variation continually, regardless of evolution's requirement for it. The second step in the process of evolution is change in the frequencies of genetic variants. The various evolutionary forces discussed in the previous section cause some alleles in the gene pool to increase in frequency and other alleles to decrease in frequency. This shift in the composition of the gene pool constitutes evolutionary change.

THINK-PAIR-SHARE Question 8

Types of Evolution

We can differentiate between two types of evolution that take place within a group of organisms connected by reproduction. **Anagenesis** refers to evolution taking place in a single lineage (a group of organisms connected by ancestry) over time (**Figure 18.9**). Another type of evolution is **cladogenesis**, the splitting of one lineage into two. When a

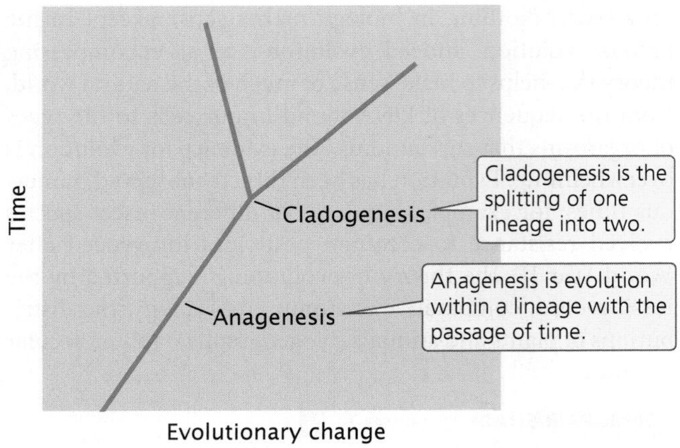

18.9 Anagenesis and cladogenesis are two different types of evolutionary change. Anagenesis is change within an evolutionary lineage; cladogenesis is the splitting of lineages.

lineage splits, the two branches no longer have a common gene pool and evolve independently of each other. New species arise through cladogenesis. ▶TRY PROBLEM 36

CONCEPTS

Biological evolution is genetic change that takes place within a group of organisms. Anagenesis is evolution that takes place within a single lineage; cladogenesis is the splitting of one lineage into two.

18.5 New Species Arise Through the Evolution of Reproductive Isolation

The term *species* literally means "kind" or "appearance"; **species** are different kinds or types of living organisms. In many cases, species differences are easy to recognize: a horse is clearly a different species than a chicken. Sometimes, however, species differences are not so clear. Some species of *Plethodon* salamanders are so similar in appearance that they can be distinguished only by looking at their proteins or genes.

The concept of a species has two primary uses in biology. First, a *species* is a particular type of organism to which a unique name has been given. For effective communication, biologists must use a standard set of names for the organisms that they study, and species names serve that purpose. When a geneticist talks about conducting crosses with *Drosophila melanogaster*, other biologists immediately understand which organism was used. The second use of the term *species* is in an evolutionary context: a species is considered an evolutionarily independent group of organisms.

The Biological Species Concept

What kinds of differences are required to consider two organisms different species? A widely used definition of species is the **biological species concept**, first fully developed by evolutionary biologist Ernst Mayr in 1942. Mayr was primarily interested in the biological characteristics that are responsible for separating organisms into independently evolving units. He defined a species as a group of organisms whose members are capable of interbreeding with one another but are reproductively isolated from the members of other species. In other words, members of the same species have the biological potential to exchange genes, and members of different species cannot exchange genes. Because different species do not exchange genes, each species evolves independently.

Not all biologists adhere to the biological species concept, and there are several problems associated with it. For example, reproductive isolation, on which the biological species concept is based, cannot be determined from fossils, and in practice, it is often difficult to determine whether even living species are biologically capable of exchanging genes. Furthermore, the biological species concept cannot be applied to asexually reproducing organisms, such as bacteria. In practice, many species are distinguished on the basis of phenotypic (usually anatomical) differences. Biologists often assume that phenotypic differences represent underlying genetic differences; if the phenotypes of two organisms are quite different, then they probably cannot and do not interbreed in nature. Because of these problems, some biologists have proposed alternative definitions for a species. For example, the *morphospecies concept* defines a species based entirely on phenotypic (morphological) similarities and differences. The *phylogenetic species concept* defines a species as the smallest recognizable group that has a unique evolutionary history. Here, we will use the biological species concept because it is widely used and is based on reproductive differences.

Reproductive Isolating Mechanisms

The key to species differences under the biological species concept is reproductive isolation—biological characteristics that prevent genes from being exchanged between different species. Any biological characteristic or mechanism that prevents genes from being exchanged with members of other species is termed a **reproductive isolating mechanism.**

Some species are separated by **prezygotic reproductive isolating mechanisms,** which prevent gametes from two different species from fusing and forming a hybrid zygote. This type of reproductive isolation can arise in a number of different ways (**Table 18.3**). Other species are separated by **postzygotic reproductive isolating mechanisms,** in which gametes of two species may fuse and form a zygote, but there is no gene flow between the two species, either because the resulting hybrids are inviable or sterile or because reproduction breaks down in subsequent generations (see Table 18.3).

THINK-PAIR-SHARE Questions 9 and 10

TABLE 18.3 Types of reproductive isolating mechanisms

Type	Characteristic
Prezygotic	**Mechanisms that act before a zygote has formed**
Ecological	Differences in habitat; individuals do not meet
Temporal	Reproduction takes place at different times
Mechanical	Anatomical differences prevent copulation
Behavioral	Differences in mating behavior prevent mating
Gametic	Gametes are incompatible or not attracted to each other
Postzygotic	**Mechanisms that act after a zygote has formed**
Hybrid inviability	Hybrid zygote does not survive to reproduction
Hybrid sterility	Hybrid is sterile
Hybrid breakdown	F_1 hybrids are viable and fertile, but F_2 are inviable or sterile

CONCEPTS

The biological species concept defines a species as a group of potentially interbreeding organisms that are reproductively isolated from the members of other species. Under this concept, species are separated by prezygotic reproductive isolating mechanisms or postzygotic reproductive isolating mechanisms.

✓ CONCEPT CHECK 6

Which of the following is an example of postzygotic reproductive isolation?

a. Sperm of species A dies in the oviduct of species B before fertilization can take place.
b. Zygotes that are hybrids between species A and B are spontaneously aborted early in development.
c. The mating seasons of species A and B do not overlap.
d. Males of species A are not attracted to the pheromones produced by the females of species B.

Modes of Speciation

Speciation is the process by which new species arise. In the terms of the biological species concept, speciation comes about through the evolution of reproductive isolating mechanisms.

New species arise in two principal ways. **Allopatric speciation** occurs when a geographic barrier splits a population into two groups and blocks the exchange of genes between them. The interruption of gene flow then leads to the evolution of genetic differences that result in reproductive isolation. **Sympatric speciation** is speciation that arises in the absence of any external barrier to gene flow; reproductive isolating mechanisms evolve within a single population. We will take a more detailed look at each of these mechanisms next.

ALLOPATRIC SPECIATION Allopatric speciation is initiated when a geographic barrier splits a population into two or more groups and prevents gene flow between those groups (**Figure 18.10a**). Geographic barriers can take a number of forms. Uplifting of a mountain range may split a population of lowland plants into separate groups on each side of the mountains. Oceans serve as effective barriers for many types of terrestrial organisms, separating individuals on different islands from one another and from those on the mainland. Rivers often separate populations of fish located in separate drainages. The erosion of mountains may leave populations of alpine plants isolated on separate mountain peaks.

After two populations have been separated by a geographic barrier that prevents gene flow between them, they evolve independently (**Figure 18.10b**). The genetic isolation allows each population to accumulate genetic differences that are not found in the other population; genetic differences arise

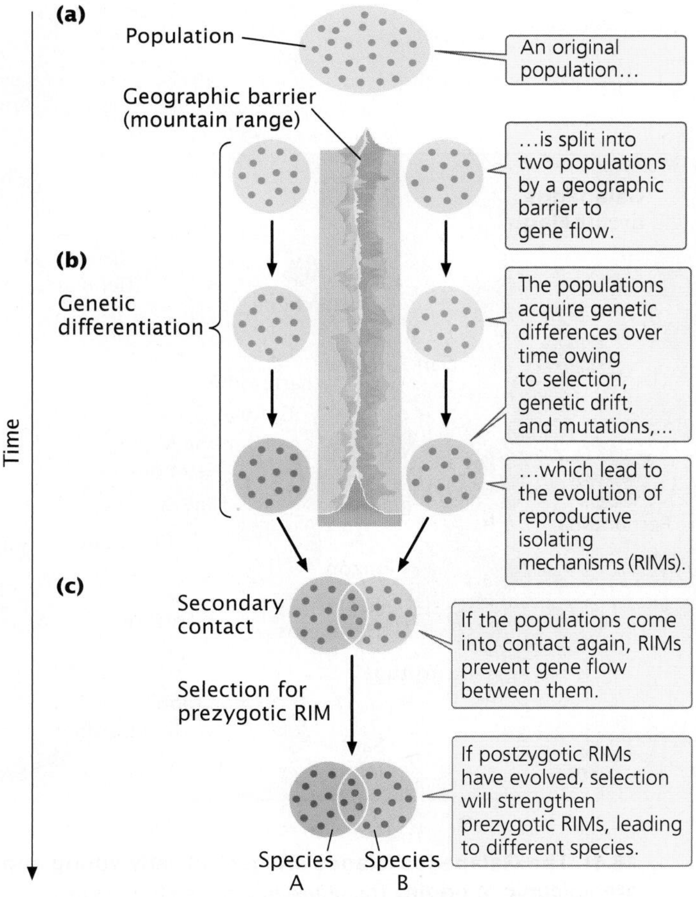

18.10 Allopatric speciation is initiated by a geographic barrier to gene flow between two populations.

through natural selection, unique mutations, and genetic drift (if the populations are small). Genetic differentiation may eventually lead to prezygotic and postzygotic reproductive isolation. It is important to note that both types of reproductive isolation arise simply as a consequence of genetic divergence.

If the populations later come into secondary contact (**Figure 18.10c**), several outcomes are possible. If limited genetic differentiation has taken place during the separation of the populations, reproductive isolating mechanisms may not have evolved or may be incomplete. In this case, the populations will remain a single species. A second possible outcome is that genetic differentiation during separation has led to prezygotic reproductive isolating mechanisms; in this case, the two populations are different species. A third possible outcome is that, during their time apart, some genetic differentiation took place between the populations, leading to postzygotic reproductive isolation. If postzygotic isolating mechanisms have evolved, any mating between individuals from the different populations will produce hybrid offspring that are inviable or sterile. Individuals that mate only with members of the same population will have higher fitness than individuals that mate with members of the other population; therefore, natural selection will increase the frequency of any trait that prevents interbreeding between members of the different populations. Over time, prezygotic reproductive isolating mechanisms will evolve. In short, if some postzygotic reproductive isolation exists, natural selection will favor the evolution of prezygotic reproductive isolating mechanisms to prevent wasted reproductive effort by individuals mating with members of the other population. This process of postzygotic reproductive isolation leading to the evolution of prezygotic isolating mechanisms is termed *reinforcement*.

An excellent example of allopatric speciation can be found in Darwin's finches, a group of birds that Charles Darwin discovered on the Galápagos Islands during his voyage aboard the *Beagle*. The Galápagos Islands form an archipelago located approximately 900 kilometers off the coast of South America (**Figure 18.11**). Consisting of more than a dozen large islands and many smaller ones, the Galápagos formed from volcanoes that erupted over a geological hot spot that has remained stationary while the geological plate over it moved eastward in the past 3 million years. The movement of the geological plate pulled newly formed islands eastward, so that the islands to the east (San Cristóbal and Española) are older than those to the west (Isabela and Fernandina). Over time, the number of islands in the archipelago increased as new volcanoes arose.

Darwin's finches consist of 14 species found on various islands in the Galápagos archipelago. The birds vary in the shapes and sizes of their beaks, which are adapted for eating different types of food. Recent studies of the development of finch embryos have helped to reveal some of the molecular details of how differences in beak shapes have evolved. Genetic studies have demonstrated that all the birds are closely related and that they evolved from a single ancestral species that migrated to the islands from the coast of South America some 2 million to 3 million years ago. The evolutionary relationships among the 14 species, based on studies of microsatellite data (see Chapter 14), are depicted in **Figure 18.12**. Most of the species are separated by a behavioral isolating mechanism (song in particular), but some of the species can and do hybridize in nature.

The first finches to arrive in the Galápagos probably colonized one of the larger eastern islands. A breeding population became established and increased with time. At some point, a few birds dispersed to another island, where they were effectively isolated from the original population, and established a new population. The new population underwent genetic differentiation

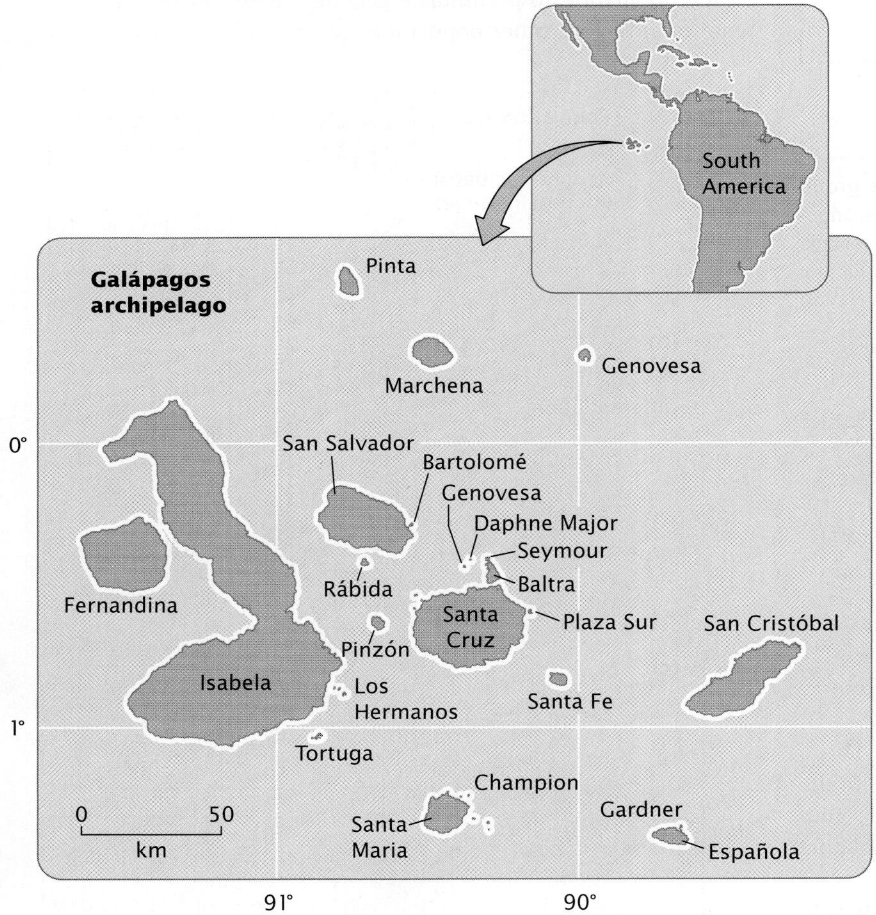

18.11 The Galápagos Islands are geologically young and are volcanic in origin. The oldest islands are to the east. [After *Philosophical Transactions of the Royal Society of London Series B* 351:756–772, 1996.]

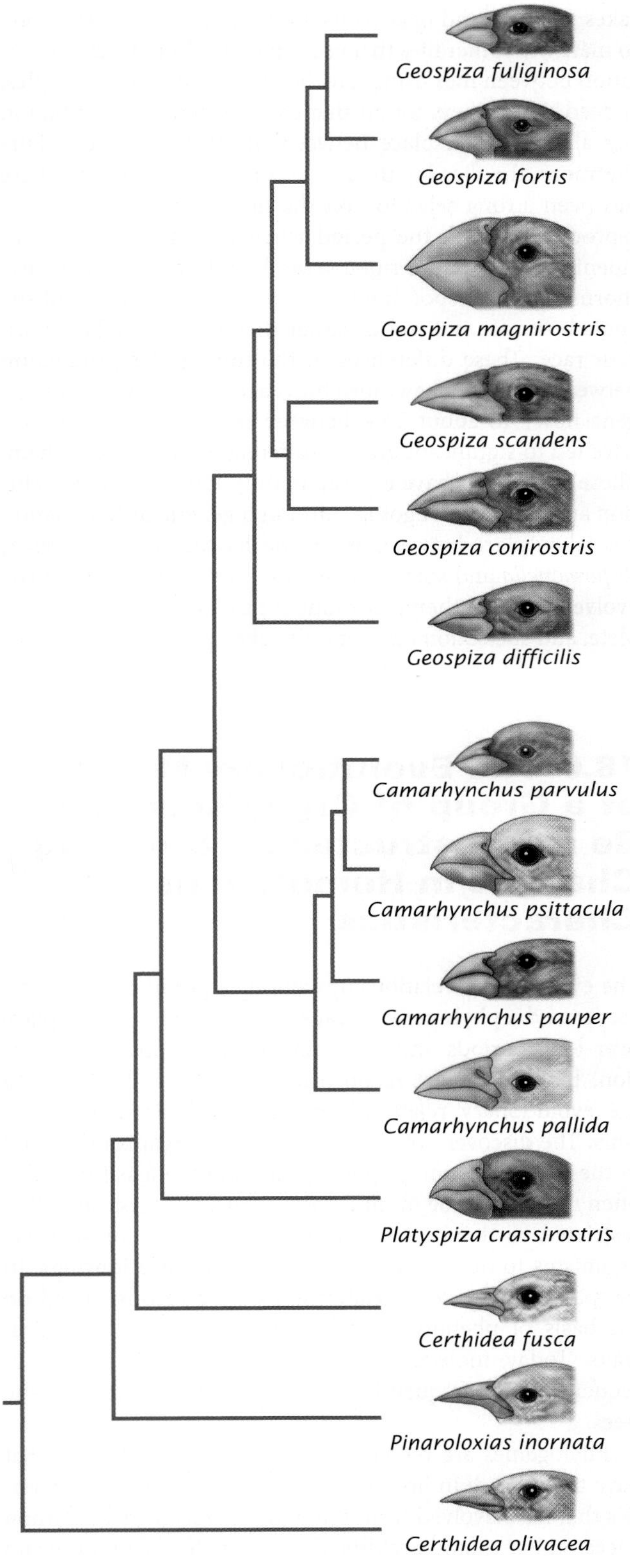

18.12 Darwin's finches consist of 14 species that evolved from a single ancestral species that migrated to the Galápagos Islands and underwent repeated allopatric speciation. [After B. R. Grant and P. R. Grant, *BioScience* 53:965–975, 2003.]

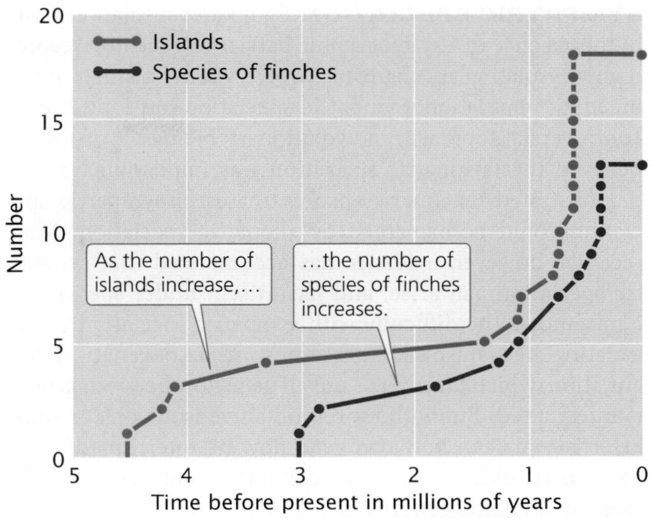

18.13 The number of species of Darwin's finches present at various times in the past corresponds to the number of islands in the Galápagos archipelago. [Data from P. R. Grant, B. R. Grant, and J. C. Deutsch. Speciation and hybridization in island birds. *Philosophical Transactions of the Royal Society of London Series B* 351:765–772, 1996.]

owing to genetic drift and adaptation to the local conditions of the island, and it eventually became reproductively isolated from the original population. Individual birds from the new population then dispersed to other islands and gave rise to additional species. This process was repeated many times. Occasionally, newly evolved species dispersed to an island where another species was already present, giving rise to secondary contact between the species. Today, many of the islands have more than one resident finch species.

The ages of the 14 species have been estimated with data from mitochondrial DNA. As **Figure 18.13** shows, there is a strong correspondence between the number of bird species present at various times in the past and the number of islands in the archipelago. This correspondence is one of the most compelling pieces of evidence for the theory that the different species of finches arose through allopatric speciation.

THINK-PAIR-SHARE Question 11

CONCEPTS

Allopatric speciation is initiated when a geographic barrier to gene flow splits a single population into two or more populations. Over time, the populations evolve genetic differences that bring about reproductive isolation. After postzygotic reproductive isolating mechanisms have evolved, selection favors the evolution of prezygotic reproductive isolating mechanisms.

✓ CONCEPT CHECK 7

What role does genetic drift play in allopatric speciation?

SYMPATRIC SPECIATION Sympatric speciation arises in the absence of any geographic barrier to gene flow; reproductive isolating mechanisms evolve within a single interbreeding population. Sympatric speciation has long been a controversial topic within evolutionary biology. Ernst Mayr believed that sympatric speciation was impossible, and he demonstrated that many apparent cases of sympatric speciation could be explained by allopatric speciation. More recently, however, evidence has accumulated that sympatric speciation can arise, and has arisen, under special circumstances. The difficulty with sympatric speciation is that isolating mechanisms arise as a *consequence* of genetic differentiation, which takes place only if gene flow between groups is interrupted. But without reproductive isolation (or some external barrier), how can gene flow be interrupted? How can genetic differentiation arise within a single group that is freely exchanging genes?

Most models of sympatric speciation assume that genetic differentiation is initiated by strong selection favoring different phenotypes within a single population. An example of how sympatric speciation might arise is seen in the apple maggot flies (*Rhagoletis pomonella*) (**Figure 18.14**). The flies of this species feed on the fruits of a specific host tree. Mating takes place near the fruits, and the flies lay their eggs on the ripened fruits, where their larvae grow and develop. *Rhagoletis pomonella* originally lived only on the fruits of hawthorn trees, which are native to North America. One hundred fifty years ago, *R. pomonella* was first observed on cultivated apples, which are related to hawthorns but a different species. Infestations of apples by this new apple host race of *R. pomonella* quickly spread, and today, many apple trees throughout North America are infested with the flies.

The apple host race of *R. pomonella* probably originated when a few flies acquired a mutation that allowed them to feed on apples instead of hawthorn fruits. Because mating takes place on and near fruits, flies that use apples are likely to mate with other flies that use apples, leading to genetic isolation between flies using hawthorns and those using apples. Indeed, researchers found that some genetic differentiation has already taken place between the two host races. Furthermore, the flies lay their eggs on ripening fruit, so there has been strong selection on the flies to synchronize their reproduction with the period when their host species has ripening fruit. Apples ripen several weeks earlier than hawthorns do. Correspondingly, the peak mating period of the apple host race is 3 weeks earlier than that of the hawthorn host race. These differences in the timing of reproduction between apple and hawthorn host races have further reduced gene flow—to about 2%—between the two host races and have led to significant genetic differentiation between them. These differences have evolved in only 150 years, and evolution appears to be ongoing. Although genetic differentiation has taken place between apple and hawthorn host races of *R. pomonella* and some degree of reproductive isolation has evolved between them, reproductive isolation is not yet complete, and speciation has not fully taken place.

18.14 Host races of the apple maggot fly (*Rhagoletis pomonella*) have evolved some reproductive isolation without any geographic barrier to gene flow. [Joseph Berger, Bugwood.org.]

18.6 The Evolutionary History of a Group of Organisms Can Be Reconstructed by Studying Changes in Homologous Characteristics

The evolutionary relationships among a group of organisms are termed a **phylogeny**. Because most evolution takes place over long periods and is not amenable to direct observation, biologists must reconstruct phylogenies by inferring the evolutionary relationships among present-day organisms. The discovery of fossils of ancestral organisms can aid in the reconstruction of phylogenies, but the fossil record is often too poor to be of much help. Thus, biologists are often restricted to the analysis of characteristics in present-day organisms to determine their evolutionary relationships. In the past, phylogenetic relationships were reconstructed on the basis of phenotypic characteristics—often, anatomical traits. Today, molecular data, including protein and DNA sequences, are frequently used to construct phylogenetic trees.

Phylogenies are reconstructed by inferring changes that have taken place in homologous characteristics: characteristics that have evolved from the same character in a common ancestor. For example, although the front leg of a mouse and the wing of a bat look different and have different functions, close examination of their structure and development reveals that they are indeed homologous; both evolved from the forelimb of an early mammal that was an ancestor to both mouse and bat. And, because mouse and bat have these homologous features and others in common, we know that they are both mammals. Similarly, DNA sequences are homologous if two present-day sequences evolved from a single sequence found in an ancestor. For example, all eukaryotic organisms have a gene for cytochrome *c*, an enzyme that helps carry out oxidative respiration. This gene is assumed to have arisen in a single organism in the distant past and was then passed down to descendants of that early ancestor. Today, all copies of the gene for cytochrome *c* are homologous because they all evolved from the same original copy in the distant ancestor of all organisms that possess this gene.

A graphical representation of a phylogeny is called a **phylogenetic tree**. As shown in **Figure 18.15**, a phylogenetic tree depicts the evolutionary relationships among different organisms, similarly to the way in which a pedigree represents the genealogical relationships among family members. A phylogenetic tree consists of branches and nodes. The **branches** are the evolutionary connections between organisms. In some phylogenetic trees, the lengths of the branches represent the amount of evolutionary divergence that has taken place between organisms. The **nodes** are the points where the branches split; they represent common ancestors that existed before divergence between organisms took place. In most cases, the nodes represent past ancestors that are inferred from the analysis. When one node represents a common ancestor to all other nodes on the tree, the tree is said to be **rooted**. Trees are often rooted by including in the analysis one or more organisms that are distantly related to all the others; this distantly related organism is referred to as an *outgroup*.

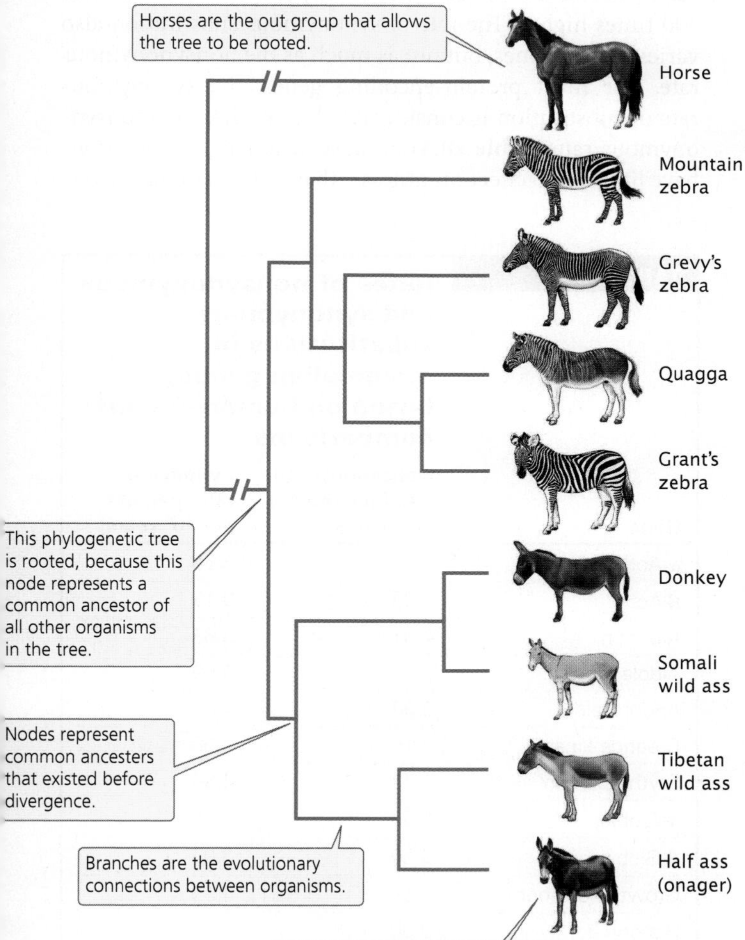

18.15 A phylogenetic tree is a graphical representation of the evolutionary relationships among a group of organisms. This phylogeny of zebras and asses is based on DNA sequences from 20,374 protein-encoding genes obtained by complete sequencing of the genomes of the species. Domestic horses are used as an outgroup to root the tree.

CONCEPTS

A phylogeny represents the evolutionary relationships among a group of organisms. It is often depicted graphically by a phylogenetic tree, which consists of nodes representing the organisms and branches representing their evolutionary connections.

The Construction of Phylogenetic Trees

Consider a simple phylogeny that depicts the evolutionary relationships among three organisms: humans, chimpanzees, and gorillas. Charles Darwin originally proposed that chimpanzees and gorillas were closely related to humans, and modern research supports a close relationship between these three species. There are three possible phylogenetic trees for humans, chimpanzees, and gorillas (**Figure 18.16**). The goal of the evolutionary biologist is to determine which of the trees is correct. Molecular data applied to this question strongly suggest a close relationship between humans and chimpanzees (as shown in Figure 18.16b).

There are different approaches to inferring evolutionary relationships and constructing phylogenetic trees. In one approach, termed the *distance approach*, evolutionary relationships are inferred on the basis of the overall degree of similarity between organisms. Typically, a number of different phenotypic characteristics or gene sequences are examined, and the organisms are grouped on the basis of their overall similarity, considering all the examined characteristics and sequences. Another approach, called the *parsimony approach*, infers phylogenetic relationships on the basis of the minimum number of evolutionary changes that must have taken place since the organisms last had an ancestor in common. With both the distance and the parsimony approaches, a number of different numerical methods are

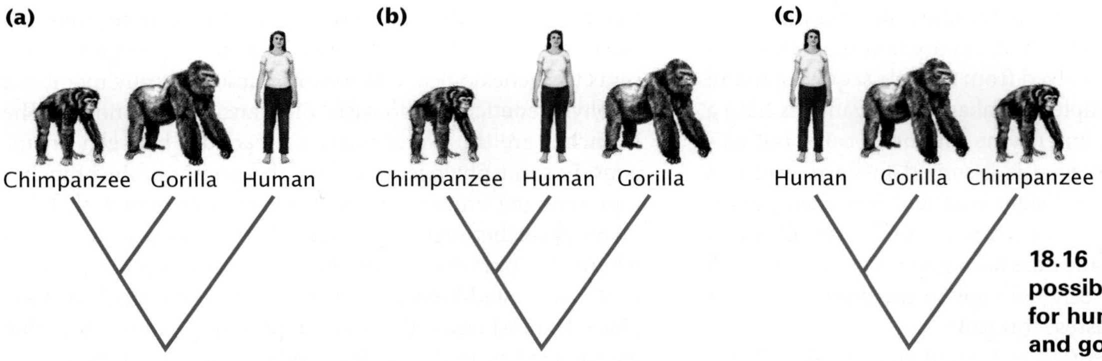

18.16 There are three possible phylogenetic trees for humans, chimpanzees, and gorillas.

available for the construction of phylogenetic trees. All include certain assumptions that help limit the number of different trees that must be considered; most rely on computer programs that compare phenotypic characteristics or sequence data to sequentially group organisms in the construction of the tree. Other methods for constructing phylogenetic trees have been developed that incorporate models of how evolution occurs.

CONCEPTS

Molecular data can be used to infer phylogenies of groups of living organisms. Two approaches to reconstructing a phylogeny are the distance approach, which uses the overall degree of similarity between organisms, and the parsimony approach, which uses the minimum number of evolutionary steps required to connect the organisms.

18.7 Patterns of Evolution Are Revealed by Changes at the Molecular Level

The ability to analyze genetic variation at the molecular level has revealed a number of evolutionary processes and features that were formerly unsuspected. This section considers several aspects of evolution at the molecular level.

Rates of Molecular Evolution

Findings from molecular studies of numerous genes have demonstrated that different genes, and even different parts of the same gene, may evolve at different rates. Rates of evolutionary change in nucleotide sequences are usually measured as the rate of nucleotide substitution, which is the number of substitutions taking place per nucleotide site per year.

NONSYNONYMOUS AND SYNONYMOUS RATES OF SUBSTITUTION Nucleotide changes in a gene that alter the amino acid sequence of a protein are referred to as nonsynonymous substitutions. Nucleotide changes, particularly those at the third position of a codon, that do not alter the amino acid sequence are called synonymous substitutions. The rate of nonsynonymous substitution varies widely among mammalian genes. The rate for the α-actin protein is only 0.01×10^{-9} substitutions per site per year, whereas the rate for interferon γ is 2.79×10^{-9}, almost 300 times higher. The rate of synonymous substitution also varies among genes, but not as much as the nonsynonymous rate. For most protein-encoding genes, the synonymous rate of substitution is considerably higher than the nonsynonymous rate (**Table 18.4**) because synonymous mutations have little or no effect on fitness—that is, they are selectively

TABLE 18.4 Rates of nonsynonymous and synonymous substitutions in mammalian genes based on human–rodent comparisons

Gene	Nonsynonymous rate (per site per 10^9 years)	Synonymous rate (per site per 10^9 years)
α-Actin	0.01	3.68
β-Actin	0.03	3.13
Albumin	0.91	6.63
Aldolase A	0.07	3.59
Apoprotein E	0.98	4.04
Creatine kinase	0.15	3.08
Erythropoietin	0.72	4.34
α-Globin	0.55	5.14
β-Globin	0.80	3.05
Growth hormone	1.23	4.95
Histone 3	0.00	6.38
Immunoglobulin heavy chain (variable region)	1.07	5.66
Insulin	0.13	4.02
Interferon α 1	1.41	3.53
Interferon γ	2.79	8.59
Luteinizing hormone	1.02	3.29
Somatostatin-28	0.00	3.97

Source: After W. Li and D. Graur, *Fundamentals of Molecular Evolution* (Sunderland, Mass.: Sinauer Associates, 1991), p. 69.

neutral. Nonsynonymous mutations, on the other hand, alter the amino acid sequence of the protein and in many cases are detrimental to the fitness of the organism; most of these mutations are eliminated by natural selection.

THINK-PAIR-SHARE Question 12

SUBSTITUTION RATES FOR DIFFERENT PARTS OF A GENE Different parts of a gene also evolve at different rates. The highest rates of substitution occur in those regions of the gene that have the least effect on function, such as the third position of a codon, flanking regions, and introns (**Figure 18.17**). The 5′ and 3′ flanking regions of genes are not transcribed into RNA; therefore, substitutions in these regions do not alter the amino acid sequence of the protein, although they may affect gene expression (see Chapter 12).

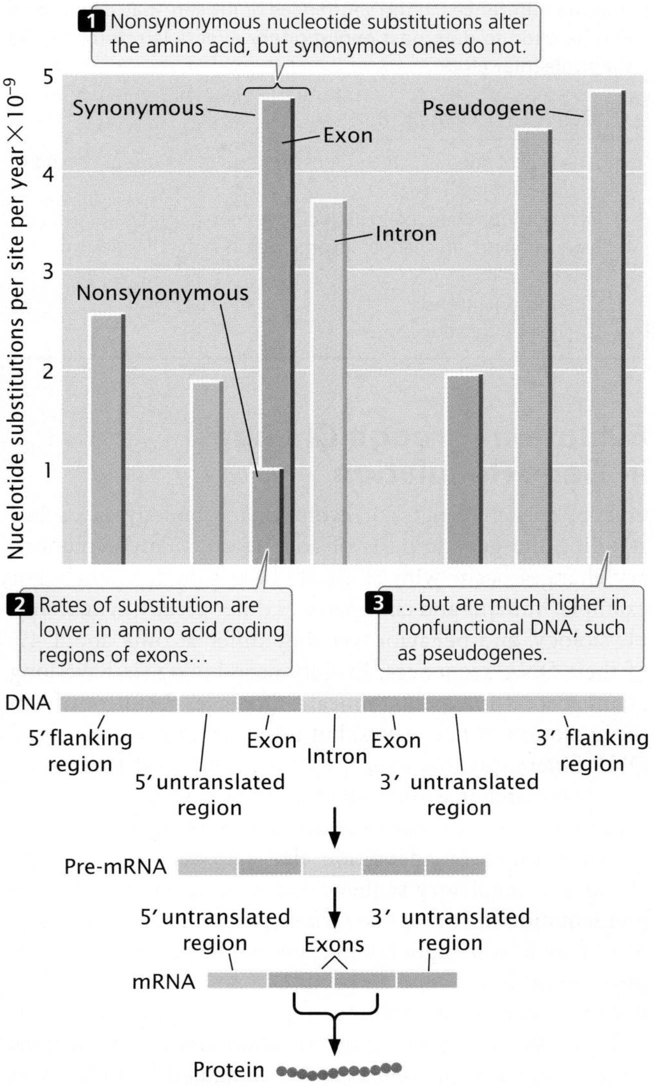

18.17 Different parts of genes evolve at different rates. The highest rates of nucleotide substitution are in sequences that have the least effect on protein function.

Rates of substitution in introns are nearly as high. Although these nucleotides do not encode amino acids, introns must be spliced out of the pre-mRNA for a functional protein to be produced, and particular sequences are required at the 5′ splice site, 3′ splice site, and branch point for correct splicing (see Chapter 10).

Substitution rates are somewhat lower in the 5′ and 3′ untranslated regions of a gene. As we know from Chapters 10 and 11, these regions are transcribed into RNA but do not encode amino acids. The 5′ untranslated region contains the ribosome-binding site, which is essential for translation, and the 3′ untranslated region contains sequences that can function in regulating mRNA stability and translation, so substitutions in these regions can have deleterious effects on organismal fitness and may not be tolerated.

The lowest rates of substitution are seen for nonsynonymous changes in the coding region because these substitutions always alter the amino acid sequence of the protein and are often deleterious. High rates of substitution take place in *pseudogenes*, most of which are duplicate copies of genes that have been rendered nonfunctional by mutations. Such genes no longer produce a functional product, so mutations in pseudogenes have little effect on the fitness of the organism.

In summary, there is a relation between the function of a sequence and its rate of evolution. The highest rates of change are found where the changes have the least effect on function.

The Molecular Clock

The neutral-mutation hypothesis is the idea that evolutionary change at the molecular level takes place primarily through the fixation of neutral mutations by genetic drift. The rate at which one neutral mutation replaces another depends only on the mutation rate, which should be fairly constant for any particular gene and the protein it encodes. If the rate at which a protein evolves is roughly constant over time, the amount of molecular change that a protein has undergone can be used as a **molecular clock** to date evolutionary events.

For example, we could examine the enzyme cytochrome *c* in two organisms known from fossil evidence to have had a common ancestor 400 million years ago. By determining the number of differences between the two organisms in their cytochrome *c* amino acid sequences, we could calculate the number of substitutions that have occurred per amino acid site. If we knew when these organisms last shared a common ancestor, we could determine the rate at which substitutions are occurring. Knowing how fast the molecular clock ticks then allows us to use the number of molecular differences in cytochrome *c* to date other evolutionary events.

The molecular clock was proposed by Emile Zuckerkandl and Linus Pauling in 1965 as a possible means of dating evolutionary events on the basis of molecules in present-day organisms. A number of studies have examined the rate of evolutionary change in proteins (**Figure 18.18**) and in genes, and the molecular clock has been widely used to

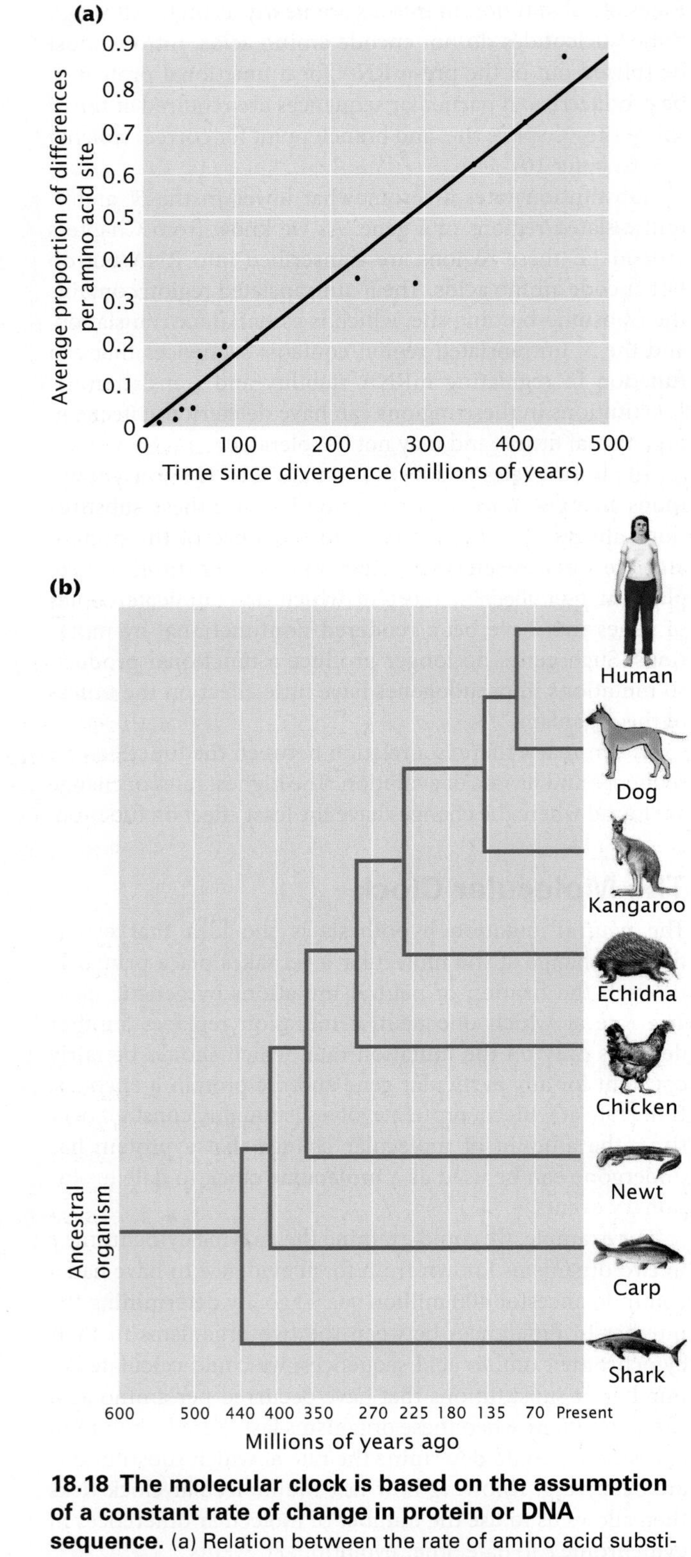

18.18 The molecular clock is based on the assumption of a constant rate of change in protein or DNA sequence. (a) Relation between the rate of amino acid substitution and time since divergence, based in part on amino acid sequences of α-globin from the eight species shown in part *b*. The rate of evolution in protein and DNA sequences has been used as a molecular clock to date past evolutionary events. (b) Phylogeny of eight of the species plotted in part *a* and their approximate times of divergence, based on the fossil record.

date evolutionary events when the fossil record is absent or ambiguous. For example, researchers used a molecular clock to estimate when Darwin's finches diverged from a common ancestor that originally colonized the Galápagos Islands. This clock was based on DNA sequence differences in the cytochrome *b* gene. The researchers concluded that the ancestor of Darwin's finches arrived in the Galápagos and begin diverging some 2 million–3 million years ago. The results of several studies have shown, however, that the molecular clock does not always tick at a constant rate, particularly over shorter time periods, and this method remains controversial.

CONCEPTS

Different genes and different parts of the same gene evolve at different rates. Those parts of genes that have the least effect on function tend to evolve at the highest rates. The idea that individual proteins and genes evolve at a constant rate and that the differences in the sequences of present-day organisms can be used to date past evolutionary events is referred to as the molecular clock.

✓ CONCEPT CHECK 8

In general, changes in which types of sequences are expected to exhibit the slowest rates?

a. Synonymous changes in amino acid–coding regions of exons
b. Nonsynonymous changes in amino acid–coding regions of exons
c. Changes in introns
d. Changes in pseudogenes

Evolution Through Changes in Gene Regulation

One of the challenges of evolutionary biology is understanding the genetic basis of adaptation. Many evolutionary changes occur with relatively little genetic change. For example, humans and chimpanzees differ greatly in anatomy, physiology, and behavior, yet they differ at only about 4% of their DNA sequences. Evolutionary biologists have long assumed that many anatomical differences result not from the evolution of new genes, but rather from relatively small DNA differences that alter the expression of existing genes. Recent research in evolutionary genetics has focused on how evolution occurs through alteration of gene expression.

An example of adaptation that has occurred through changes in regulatory sequences is seen in the evolution of pigmentation in *Drosophila melanogaster* fruit flies in Africa. Most fruit flies are light tan in color, but flies in some African populations have much darker abdomens. These darker flies usually occur in mountainous regions at higher elevations. Indeed, 59% of pigmentation variation among populations within sub-Saharan Africa can be explained by differences in elevation (**Figure 18.19**). Researchers have demonstrated that these differences are genetically determined and that

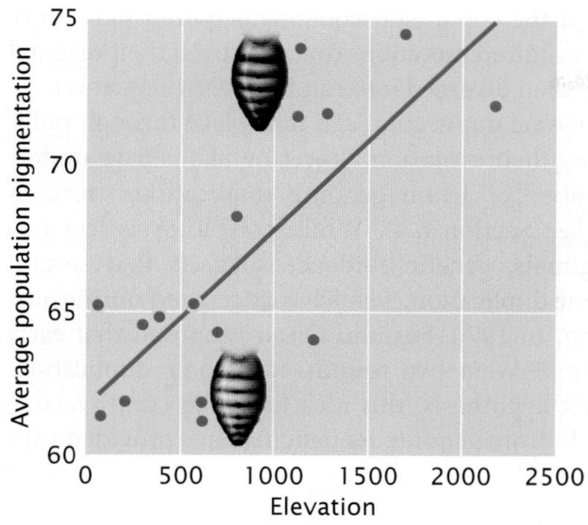

18.19 Sub-Saharan African populations of the fruit fly *Drosophila melanogaster* exhibit a positive association between pigmentation and elevation.

natural selection has favored darker pigmentation at high elevations. High-elevation populations are exposed to lower temperatures, and the darker pigmentation is assumed to help flies absorb more solar radiation and better regulate their body temperature in these environments.

How did flies at high elevations evolve darker color? Genetic studies indicate that the dark abdominal pigmentation seen in flies from these populations results from variation at or near a locus called *ebony*. The *ebony* locus encodes a multifunctional enzyme that produces a yellow exoskeleton; the absence of this enzyme produces a dark phenotype. Sequencing of the *ebony* locus of flies from light and dark populations found no differences in the coding region of the *ebony* gene. However, molecular analysis revealed a marked reduction in the amount of *ebony* mRNA in darker flies, suggesting that the difference in pigmentation is due not to mutations in the *ebony* gene itself, but rather to differences in its expression.

Further investigation detected genetic differences within an enhancer that is about 3600 bp upstream of the *ebony* gene. Dark and light flies differed in over 120 nucleotides scattered over 2400 bp of the enhancer. By experimentally creating enhancers with different combinations of these mutations, researchers determined that five of the mutations are responsible for the majority of the difference in pigmentation.

These studies suggested that over time, high-elevation populations accumulated multiple mutations in the enhancer, which reduced the expression of the *ebony* locus and caused darker pigmentation. Further analysis suggested that these mutations were added sequentially. Some of the mutations are widespread throughout Africa; it is assumed that these existing mutations were favored by natural selection in high-elevation populations and increased in frequency because they helped the flies thermoregulate in colder environments. Other mutations that are seen only in the high-elevation populations probably arose as new mutations within those populations and were quickly favored by natural selection.

THINK-PAIR-SHARE Question 13

Genome Evolution

The vast store of sequence data now available in DNA databases has been a source of insight into evolutionary processes. Whole-genome sequences are also providing new information about how genomes evolve and the processes that shape the size, complexity, and organization of genomes.

GENE DUPLICATION New genes have evolved through the duplication of whole genes and their subsequent divergence. As we saw in Chapter 8, this process creates gene families: sets of genes that are similar in sequence but encode different products. For example, humans possess 13 different genes that encode globin molecules, which take part in oxygen transport (**Figure 18.20**). All of these genes

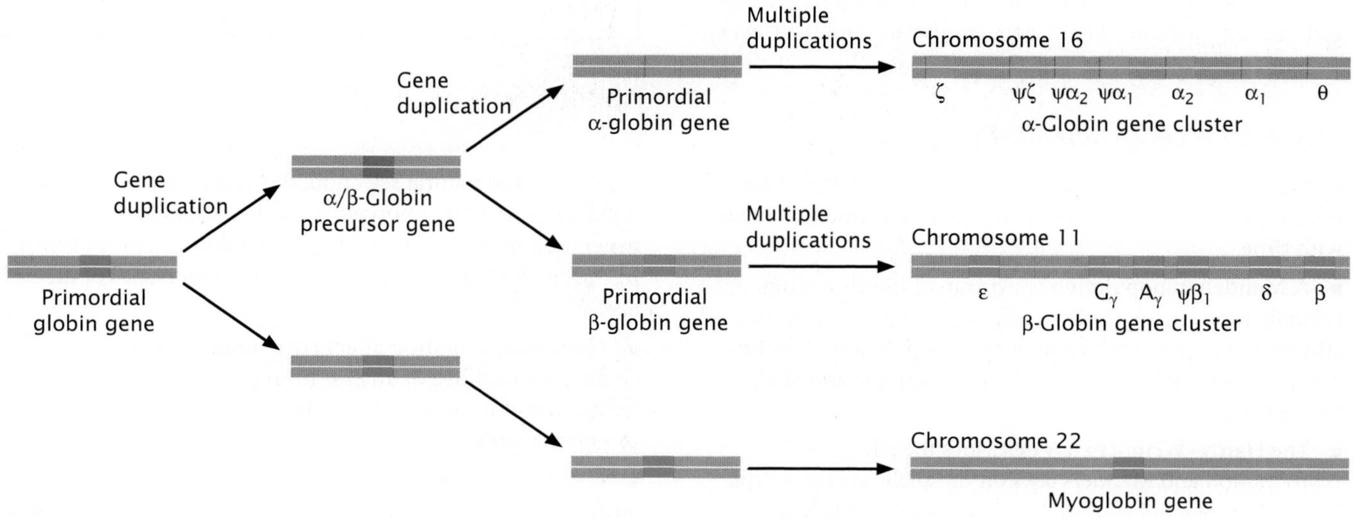

18.20 The human globin gene family has evolved through successive gene duplications.

have a similar structure, with three exons separated by two introns. They are assumed to have evolved through repeated duplication and divergence from a single globin gene in a distant ancestor. This ancestral gene is thought to have been most similar to the present-day myoglobin gene, and it was probably first duplicated to produce an α/β-globin precursor gene and the myoglobin gene. The α/β-globin gene then underwent another duplication to give rise to a primordial α-globin gene and a primordial β-globin gene. Subsequent duplications led to multiple α-globin and β-globin genes. Similarly, vertebrates contain four clusters of *Hox* genes, each cluster comprising from 9 to 11 genes. *Hox* genes play an important role in development.

Some gene families include genes that are arrayed in tandem on the same chromosome; others are dispersed among different chromosomes. Gene duplication is a common occurrence in eukaryotic genomes; for example, about 5% of the human genome consists of duplicated segments.

Gene duplication provides a mechanism for the addition of new genes with novel functions. Once a gene has been duplicated, there are two copies of that gene, one of which is then free to change and potentially take on a new function. The extra copy of the gene may, for example, become active at a different time in development, or be expressed in a different tissue, or even diverge and encode a protein containing different amino acids. The most common fate of duplicate genes, however, is that one copy acquires a mutation that renders it nonfunctional and becomes a pseudogene. Pseudogenes are common in the genomes of complex eukaryotes; the human genome is estimated to contain as many as 20,000 pseudogenes.

WHOLE-GENOME DUPLICATION In addition to the duplication of individual genes, whole genomes of some organisms have been duplicated in the past. For example, a comparison of the genome of the yeast *Saccharomyces cerevisiae* with the genomes of other fungi reveals that *S. cerevisiae*, or one of its immediate ancestors, underwent a whole-genome duplication, which generated two copies of every gene. Many of the copies subsequently acquired new functions; others acquired mutations that destroyed their original function and then diverged into random DNA sequences.

Whole-genome duplication can take place through polyploidy. During their evolution, flowering plants have undergone a number of whole-genome duplications through polyploidy (see Section 6.4). While polyploidy is less frequent in animals, genetic evidence suggests that several whole-genome duplication events have occurred during animal evolution. In 1970, Susumu Ohno proposed that early vertebrates underwent two rounds of genome duplication. Called the 2R hypothesis, this idea has been controversial, but recent data from genome sequencing have provided support for it.

HORIZONTAL GENE TRANSFER Traditionally, scientists assumed that organisms acquire their genomes through vertical transmission—transfer through reproduction of genetic information from parents to offspring—and most phylogenetic trees assume vertical transmission of genetic information. Findings from DNA-sequence studies reveal that DNA sequences are sometimes transmitted by horizontal gene transfer, in which DNA is exchanged between individuals of different species (see Section 7.2). This process is especially common among bacteria, and there are a number of documented cases in which genes have been transferred from bacteria to eukaryotes. The extent of horizontal gene transfer among eukaryotic organisms is controversial, as there are few well-documented cases. Horizontal gene transfer can obscure phylogenetic relationships and make the reconstruction of phylogenetic trees difficult.

CONCEPTS

New genes can evolve through the duplication of genes and through the duplication of whole genomes. Genes can be passed among different organisms through horizontal gene transfer.

CONCEPTS SUMMARY

- Population genetics examines the genetic composition of groups of individuals and how this composition changes with time.
- A Mendelian population is a group of interbreeding, sexually reproducing individuals, whose set of genes constitutes the population's gene pool. A population's genetic composition can be described by its genotypic and allelic frequencies.
- The Hardy–Weinberg law describes the effects of reproduction and Mendel's laws on the allelic and genotypic frequencies of a population. When a population is large, randomly mating, and free from the effects of mutation, migration, and natural selection, the allelic frequencies do not change and the genotypic frequencies stabilize after one generation in the Hardy–Weinberg equilibrium proportions p^2, $2pq$, and q^2, where p and q equal the frequencies of the alleles.
- Nonrandom mating affects the frequencies of genotypes but not those of alleles. Inbreeding increases the frequency of homozygotes while decreasing the frequency of heterozygotes.
- Recurrent mutation eventually leads to an equilibrium, with the allelic frequencies being determined by the relative rates of forward and reverse mutation.

- Migration, the movement of genes between populations, increases the amount of genetic variation within populations and decreases the difference in allelic frequencies between populations.
- Genetic drift is change in allelic frequencies due to chance. Genetic drift arises when a population consists of a small number of individuals, is established by a small number of founders, or undergoes a major reduction in size. Genetic drift changes allelic frequencies, reduces genetic variation within populations, and causes genetic divergence among populations.
- Natural selection is the differential reproduction of genotypes; it is measured by the relative reproductive successes (fitnesses) of genotypes.
- Evolution is genetic change that takes place within a group of organisms. It is a two-step process: **(1)** genetic variation arises and **(2)** genetic variants change in frequency. Anagenesis refers to change within a single lineage; cladogenesis is the splitting of one lineage into two.
- A species can be defined as a group of organisms that are capable of interbreeding with one another and are reproductively isolated from the members of other species.
- Species are prevented from exchanging genes by prezygotic or postzygotic reproductive isolating mechanisms.
- Allopatric speciation arises when a geographic barrier prevents gene flow between two populations. Sympatric speciation arises when reproductive isolation exists in the absence of any geographic barrier.
- A phylogeny can be represented by a phylogenetic tree, consisting of nodes that represent organisms and branches that represent their evolutionary connections. Two different approaches to constructing phylogenetic trees are the distance approach and the parsimony approach.
- Different parts of the genome show different amounts of genetic variation. In general, those parts that have the least effect on function evolve at the highest rates.
- The molecular clock hypothesis proposes a constant rate of nucleotide substitution, providing a means of dating evolutionary events by looking at nucleotide differences between organisms.
- Genome evolution takes place through the duplication of genes to form gene families, whole-genome duplication, and horizontal transfer of genes.

IMPORTANT TERMS

Mendelian population (p. 486)
gene pool (p. 486)
genotypic frequency (p. 487)
allelic frequency (p. 487)
Hardy–Weinberg law (p. 489)
Hardy–Weinberg equilibrium (p. 489)
inbreeding (p. 492)
inbreeding depression (p. 492)
equilibrium (p. 493)
migration (gene flow) (p. 493)
sampling error (p. 495)
genetic drift (p. 495)
effective population size (p. 495)
founder effect (p. 495)
genetic bottleneck (p. 495)
fixation (p. 496)
fitness (p. 497)
selection coefficient (p. 497)
directional selection (p. 498)
overdominance (p. 498)
underdominance (p. 498)
evolution (p. 499)
anagenesis (p. 499)
cladogenesis (p. 499)
species (p. 500)
biological species concept (p. 500)
reproductive isolating mechanism (p. 500)
prezygotic reproductive isolating mechanism (p. 500)
postzygotic reproductive isolating mechanism (p. 500)
speciation (p. 501)
allopatric speciation (p. 501)
sympatric speciation (p. 501)
phylogeny (p. 504)
phylogenetic tree (p. 505)
node (p. 505)
branch (p. 505)
rooted (p. 505)
molecular clock (p. 507)

ANSWERS TO CONCEPT CHECKS

1. a

2. 0.10

3. c

4. d

5. b

6. b

7. Genetic drift can bring about changes in the allelic frequencies of populations and lead to genetic differences among populations. Genetic differentiation is the cause of postzygotic and prezygotic reproductive isolation between populations that leads to speciation.

8. b

COMPREHENSION QUESTIONS

Section 18.1

1. What is a Mendelian population? How is the gene pool of a Mendelian population usually described?

Section 18.2

2. What are the predictions given by the Hardy–Weinberg law?
3. What assumptions must be met for a population to be in Hardy–Weinberg equilibrium?
4. Define inbreeding and briefly describe its effects on a population.

Section 18.3

5. What determines the allelic frequencies at mutational equilibrium?
6. What factors affect the magnitude of change in allelic frequencies due to migration?
7. Define genetic drift and give three ways in which it can arise. What effect does genetic drift have on a population?
8. What is effective population size? How does it affect the amount of genetic drift?
9. Define natural selection and fitness.
10. Briefly describe the differences between directional selection, overdominance, and underdominance. Describe the effect of each type of selection on the allelic frequencies of a population.
11. Compare and contrast the effects of mutation, migration, genetic drift, and natural selection on genetic variation within populations and on genetic divergence between populations.

Section 18.4

12. What are the two steps in the process of evolution?
13. How does anagenesis differ from cladogenesis?

Section 18.5

14. What is the biological species concept?
15. What is the difference between prezygotic and postzygotic reproductive isolating mechanisms?
16. What is the basic difference between allopatric and sympatric modes of speciation?

Section 18.6

17. Briefly describe the difference between the distance approach and the parsimony approach to the reconstruction of phylogenetic trees.

Section 18.7

18. Outline the different rates of evolution that are typically seen in different parts of a protein-encoding gene. What might account for these differences?
19. What is the molecular clock?
20. What is a gene family? What processes produce gene families?
21. Define horizontal gene transfer. What problems does it cause for evolutionary biologists?

For more questions that test your comprehension of the key chapter concepts, go to **LearningCurve** for this chapter.

APPLICATION QUESTIONS AND PROBLEMS

Section 18.1

22. How would you respond to someone who said that models are useless in studying population genetics because they represent oversimplifications of the real world?

*23. Voles (*Microtus ochrogaster*) were trapped in fields in southern Indiana and genotyped for a transferrin locus. The following numbers of genotypes were recorded, where T^E and T^F represent different alleles.

[Tom McHugh/Science Source.]

$T^E T^E$	$T^E T^F$	$T^F T^F$
407	170	17

Calculate the genotypic and allelic frequencies of the transferrin locus for this population.

24. DATA ANALYSIS Jean Manning, Charles Kerfoot, and Edward Berger studied the allelic frequencies at the glucose phosphate isomerase (GPI) locus in the cladoceran *Bosmina longirostris* (a small crustacean known as a water flea). They collected 176 of the animals from a single location in Union Bay in Seattle, Washington, and determined their GPI genotypes by using electrophoresis (J. Manning, W. C. Kerfoot, and E. M. Berger. 1978. *Evolution* 32:365–374).

Genotype	Number
S^1S^1	4
S^1S^2	38
S^2S^2	134

Determine the genotypic and allelic frequencies for this population.

Section 18.2

25. A total of 6129 North American Caucasians were blood typed for the MN locus, which is determined by two codominant alleles, L^M and L^N. The following data were obtained:

Blood type	Number
M	1787
MN	3039
N	1303

Carry out a chi-square test to determine whether this population is in Hardy–Weinberg equilibrium at the MN locus.

26. Assume that the phenotypes of the lady beetles shown in **Figure 18.1b** are encoded by the following genotypes:

Phenotype	Genotype
All black	*BB*
Some black spots	*Bb*
No black spots	*bb*

a. For the lady beetles shown in the figure, calculate the frequencies of the genotypes and frequencies of the alleles.

b. Use a chi-square test to determine if the lady beetles shown are in Hardy Weinberg equilibrium.

***27.** Most black bears (*Ursus americanus*) are black or brown in color. However, occasional white bears of this species appear in some populations along the coast of British Columbia. Kermit Ritland and his colleagues determined that white coat color in these bears results from a recessive mutation (*G*) caused by a single nucleotide replacement in which guanine substitutes for adenine at the melanocortin-1 receptor locus (*mcr1*), the same locus responsible for red hair in humans (K. Ritland, C. Newton, and H. D. Marshall. 2001. *Current Biology* 11:1468–1472). The wild-type allele at this locus (*A*) encodes black or brown color. Ritland and his colleagues collected samples from bears on three islands and determined their genotypes at the *mcr1* locus.

[Wendy Shattil/Alamy.]

Genotype	Number
AA	42
AG	24
GG	21

a. What are the frequencies of the *A* and *G* alleles in these bears?

b. Give the genotypic frequencies expected if the population is in Hardy–Weinberg equilibrium.

c. Use a chi-square test to compare the number of observed genotypes with the number expected under Hardy–Weinberg equilibrium. Is this population in Hardy–Weinberg equilibrium? Explain your reasoning.

28. Genotypes of leopard frogs from a population in central Kansas were determined for a locus (*M*) that encodes the enzyme malate dehydrogenase. The following numbers of genotypes were observed:

Genotype	Number
M^1M^1	20
M^1M^2	45
M^2M^2	42
M^1M^3	4
M^2M^3	8
M^3M^3	6
Total	125

a. Calculate the genotypic and allelic frequencies for this population.

b. What would the expected numbers of genotypes be if the population were in Hardy–Weinberg equilibrium?

***29.** Full color (*D*) in domestic cats is dominant over dilute color (*d*). Of 325 cats observed, 194 have full color and 131 have dilute color.

a. If this population of cats is in Hardy–Weinberg equilibrium for the dilution locus, what is the frequency of the dilute (*d*) allele?

b. How many of the 194 cats with full color are likely to be heterozygous?

30. Tay–Sachs disease is an autosomal recessive disorder. Among Ashkenazi Jews, the frequency of Tay–Sachs disease is 1 in 3600. If the Ashkenazi population is mating randomly for the Tay–Sachs gene, what proportion of the population consists of heterozygous carriers of the Tay–Sachs allele?

***31.** The human MN blood type is determined by two codominant alleles, L^M and L^N. The frequency of L^M in Eskimos on a small Arctic island is 0.80.

a. If random mating takes place in this population, what are the expected frequencies of the M, MN, and N blood types on the island?

b. If inbreeding is present in this population, what effect will it have on the expected numbers of the different blood types?

Section 18.3

***32.** Pikas are small mammals that live at high elevations on the talus slopes of mountains. Most populations located on mountaintops in Colorado and Montana in North America are isolated from one another: the pikas don't occupy the low-elevation habitats that separate the mountaintops and don't venture far from the talus slopes. Thus, there is little gene flow between populations.

Furthermore, each population is small in size and was founded by a small number of pikas. A group of population geneticists proposes to study the amount of genetic variation in a series of pika populations and to compare the allelic frequencies in different populations. On the basis of the biology and distribution of pikas, predict what the population geneticists will find concerning the within- and between-population genetic variation.

*33. Two chromosome inversions are commonly found in populations of *Drosophila pseudoobscura*: Standard (*ST*) and Arrowhead (*AR*). When the flies are treated with the insecticide DDT, the genotypes for these inversions exhibit overdominance, with the following fitnesses:

Genotype	Fitness
ST/*ST*	0.47
ST/*AR*	1
AR/*AR*	0.62

What will the frequencies of *ST* and *AR* be after equilibrium has been reached?

*34. The fruit fly *Drosophila melanogaster* normally feeds on rotting fruit, which may ferment and contain high levels of alcohol. Douglas Cavener and Michael Clegg studied allelic frequencies at the locus for alcohol dehydrogenase (*Adh*) in experimental populations of *D. melanogaster* (D. R. Cavener and M. T. Clegg. 1981. *Evolution* 35:1–10). The experimental populations were established from wild-caught flies and were raised in cages in the laboratory. Two control populations (C1 and C2) were raised on a standard cornmeal–molasses–agar diet. Two ethanol populations (E1 and E2) were raised on a cornmeal–molasses–agar diet to which was added 10% ethanol. The four populations were periodically sampled to determine the frequencies of two alleles at the alcohol dehydrogenase locus, Adh^S and Adh^F. The frequencies of these alleles in the four populations are shown in the graph.

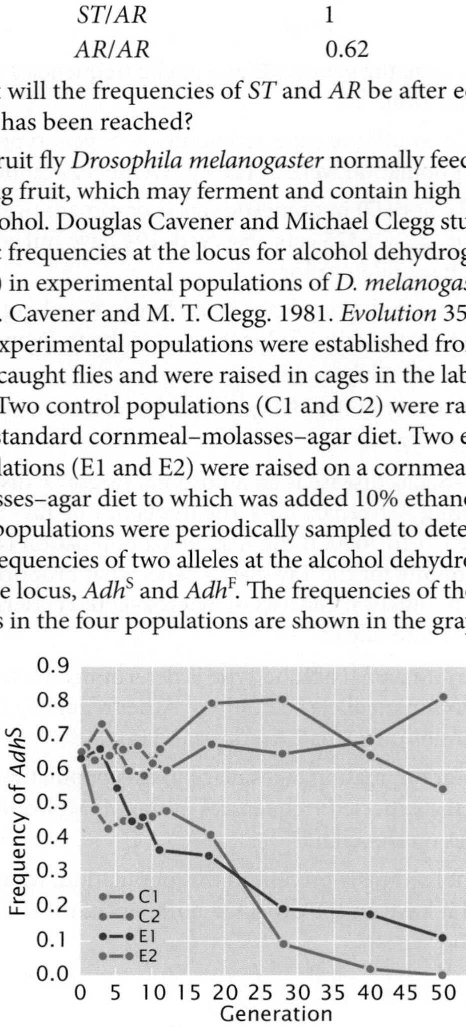

a. On the basis of these data, what conclusion might you draw about the evolutionary forces that are affecting the *Adh* alleles in these populations?

b. Cavener and Clegg measured the viability of the different *Adh* genotypes in the ethanol environment and obtained the following values:

Genotype	Relative viability
Adh^F/Adh^F	0.932
Adh^F/Adh^S	1.288
Adh^S/Adh^S	0.596

Using these relative viabilities, calculate relative fitnesses for the three genotypes.

35. Examine **Figure 18.8**. Which evolutionary forces

a. cause an increase in genetic variation both within and between populations?

b. cause a decrease in genetic variation both within and between populations?

c. cause an increase in genetic variation within populations but cause a decrease in genetic variation between populations?

Section 18.4

*36. The following illustrations represent two different patterns of evolution. Briefly discuss the differences in these two patterns with regard to how evolutionary change (on the *x* axis) occurs with respect to time (one the *y* axis).

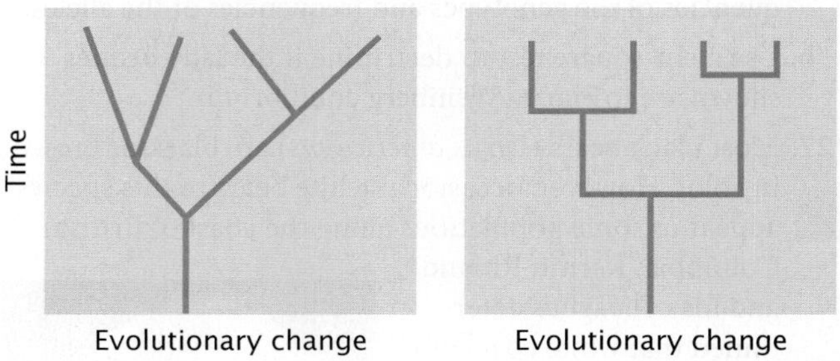

Section 18.6

37. Which of the isolating mechanisms listed in **Table 18.3** have partly evolved between apple and hawthorn host races of *Rhagoletis pomonella*?

38. In Section 18.5, we considered the sympatric evolution of reproductive isolating mechanisms in host races of *Rhagoletis pomonella*, the apple maggot fly. The wasp *Diachasma alloeum* parasitizes apple maggot flies, laying its eggs on the larvae of the flies. Immature wasps hatch from the eggs and feed on the fly larvae. Research by Andrew Forbes and his colleagues (Forbes et al. 2009. *Science* 323:776–779) demonstrated that wasps that parasitize apple host races of *R. pomonella* are genetically differentiated from those that parasitize hawthorn host races of *R. pomonella*. They also found that wasps that prey on the apple host race of the flies are attracted to odors from apples, whereas wasps that prey on the hawthorn host race are attracted to odors from hawthorns.

a. Propose an explanation for how genetic differences might have evolved between the wasps that parasitize the two races of *R. pomonella*.

b. How might these differences lead to speciation in the wasps?

39. Michael Bunce and his colleagues in England, Canada, and the United States extracted and sequenced mitochondrial DNA from fossils of Haast's eagle, a gigantic eagle that was driven to extinction 700 years ago when humans first arrived in New Zealand (M. Bunce et al. 2005. *PLoS Biology* 3:44–46). Using mitochondrial DNA sequences from living eagles and from Haast-eagle fossils, they created the following phylogenetic tree. On this phylogenetic tree, identify (a) all nodes; (b) one example of a branch; and (c) the outgroup.

DATA ANALYSIS

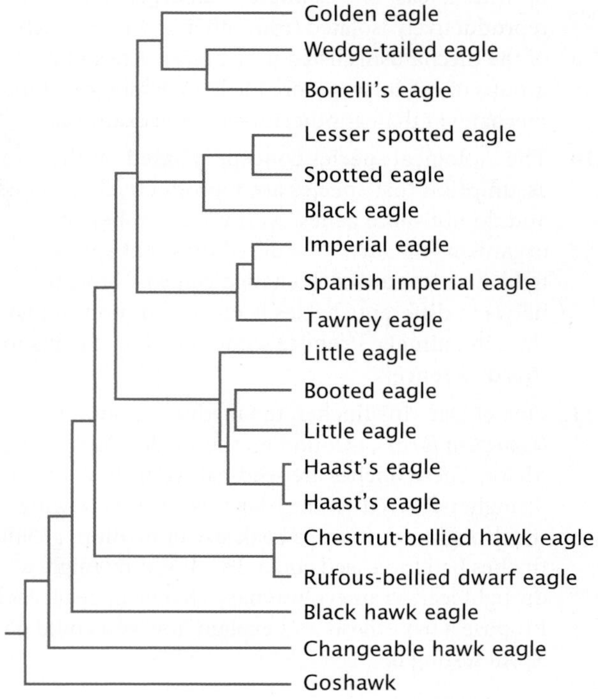

[After M. Bunce et al., PLOS Biology 3:44–46, 2005.]

40. On the basis of the phylogeny of Darwin's finches shown in **Figure 18.12**, predict which two species in each of the following groups will be the most similar genetically.

a. *Camarhynchus parvulus, Camarhynchus psittacula, Camarhynchus pallida*

b. *Camarhynchus parvulus, Camarhynchus pallida, Platyspiza crassirostris*

c. *Geospiza difficilis, Geospiza conirostris, Geospiza scandens*

d. *Camarhynchus parvulus, Certhidea fusca, Pinaroloxias inornata*

41. Assume that one of the genes shown in **Table 18.4** showed similar nonsynonymous and synonymous rates of substitution. What might this suggest about the evolution of this gene?

42. Based on the information provided in **Figure 18.17**, do introns or 3′ untranslated regions of a gene have higher rates of nucleotide substitution? Explain why.

CHALLENGE QUESTIONS

Section 18.3

43. The Barton Springs salamander is an endangered species found in several springs at a single locality in the city of Austin, Texas. There is growing concern that a chemical spill on a nearby freeway could pollute the springs and wipe out the species. To provide a source of salamanders to repopulate the spring in the event of such a catastrophe, a proposal has been made to establish a captive breeding population of the salamander in a local zoo. You are asked to provide a plan for the establishment of this captive breeding population, with the goal of maintaining as much of the genetic variation of the species as possible. What factors might cause loss of genetic variation in the establishment of the captive population? How could loss of such variation be prevented? With the assumption that only a limited number of salamanders can be maintained in captivity, what procedures should be instituted to ensure the long-term maintenance of as much variation as possible?

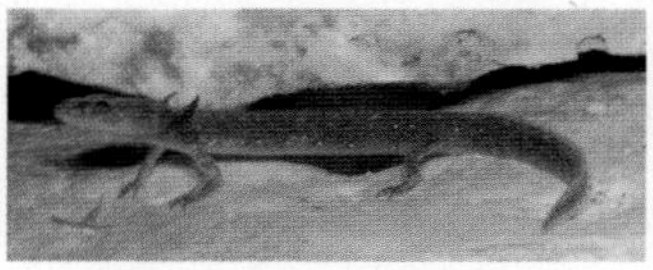

[Dante Fenolio/Getty Images]

Section 18.5

44. Explain why natural selection may cause prezygotic reproductive isolating mechanisms to evolve if postzygotic reproductive isolating mechanisms are already present, but can never cause the evolution of postzygotic reproductive isolating mechanisms.

THINK-PAIR-SHARE QUESTIONS

Introduction

1. Imagine that you are asked to manage a population of wolves that have taken up residence on an island off the coast of Alaska. Because prey resources are limited, the island will support only about 50 wolves at any one time. What steps would you take to prevent inbreeding and genetic drift in the population in the future?
2. Do you think the National Park Service should have brought in new wolves to Isle Royale in an attempt to rescue the population, or do you think that it was correct in allowing natural processes to proceed? Provide some possible reasons for and against attempting genetic rescue.

Section 18.2

3. Miguel says that the Hardy–Weinberg law is only theoretical, of no practical value, and totally worthless because populations will never be large, randomly mating for all traits, and free from all evolutionary forces. Thus he asserts that populations will never be in Hardy–Weinberg equilibrium. Barbara says she knows the Hardy–Weinberg law is important. Who is correct and why?

Section 18.3

4. Figure 18.4 presents a simple mathematical model of migration. What are the assumptions of this model? Write down as many as you can think of. How reasonable are these assumptions?
5. Is migration good or bad for populations? Defend your answer.
6. What historical, social, religious, cultural, and economic factors promote genetic drift in humans? Can you think of some specific human groups in which genetic drift is likely to have occurred?

Section 18.4

7. Consider Theodosius Dobzhansky's remark, "Nothing in biology makes sense except in the light of evolution." Why is evolution so important to the field of biology? Give some specific examples of how the theory of evolution helps us make sense of biology.
8. Evolution is often misunderstood and misinterpreted. What are some common misconceptions about evolution? How are these misconceptions wrong?

Section 18.5

9. Examine **Table 18.3**, which lists different types of reproductive isolating mechanisms. Try to come up with at least one example of an organism that is reproductively isolated from other species by each of the mechanisms listed in the table. Are some groups of organisms more likely to exhibit one type of mechanism than another? Give some examples.
10. The biological species concept is based on the assumption that species are reproductively isolated and do not share genes. And yet a number of organisms that are considered different species hybridize (mate and exchange genes). Hybridization between different species is more common in plants than in animals. Propose some possible reasons for this difference.
11. One of Darwin's finches, the medium ground finch (*Geospiza fortis*), is found on the small island of Daphne Major. These finches are seed-eating birds. A major drought occurred on the island in 1977. Following the drought, the average beak size of medium ground finches had increased about 3%–4%. Why might a drought lead to an evolutionary change in beak size? Propose a hypothesis and explain how you could go about testing it.

Section 18.7

12. In most cases, the rate of synonymous substitution for a gene is higher than the rate of nonsynonymous substitution. Sometimes, however, the rate of nonsynonymous substitution is higher. When would you expect to see this? What might bring it about?
13. What changes, if any, would you predict would occur in the pigmentation of *Drosophila melanogaster* with increased global warming? What type of genetic changes would you expect to see? Be as specific as you can.

Glossary

acentric chromatid Chromatid that lacks a centromere; produced when crossing over takes place within a paracentric inversion. The acentric chromatid does not attach to a spindle fiber and does not segregate in meiosis or mitosis; so it is usually lost after one or more rounds of cell division.

acrocentric chromosome Chromosome in which the centromere is near one end, producing a long arm at one end and a knob, or satellite, at the other end.

addition rule Rule stating that the probability of any of two or more mutually exclusive events occurring is calculated by adding the probabilities of the individual events.

additive genetic variance Component of genetic variance that can be attributed to the additive effect of different genotypes.

adenine (A) Purine base in DNA and RNA.

adenosine-3′,5′-cyclic monophosphate (cAMP) Modified nucleotide that functions in catabolite repression. Low levels of glucose stimulate high levels of cAMP; cAMP then attaches to CAP, which binds to the promoter of certain operons and stimulates transcription.

A-DNA Right-handed helical structure of DNA that exists when little water is present.

allele One of two or more alternative forms of a gene.

allelic frequency Proportion of a particular allele in a population.

allopatric speciation Speciation that arises when a geographic barrier first splits a population into two groups and blocks the exchange of genes between them. *Compare* **sympatric speciation**.

allopolyploidy Condition in which the sets of chromosomes of a polyploid individual are derived from two or more species. *Compare* **autopolyploidy**.

allosteric protein Protein that changes its conformation upon binding with another molecule.

alternative processing Existence of several different pathways by which a single pre-mRNA can be processed to produce different types of mRNA, resulting in the production of different proteins from the same DNA sequence.

alternative splicing Process by which a single pre-mRNA can be spliced in more than one way to produce different types of mRNA.

Ames test Test in which special strains of bacteria are used to evaluate the potential of chemicals to cause cancer.

amino acid Repeating unit of proteins; consists of an amino group, a carboxyl group, a hydrogen atom, and a variable R group.

aminoacyl (A) site One of three sites in a ribosome occupied by a tRNA in translation. All charged tRNAs (with the exception of the initiator tRNA) first enter the A site in translation.

aminoacyl-tRNA synthetase Enzyme that attaches an amino acid to a tRNA. Each aminoacyl-tRNA synthetase is specific for a particular amino acid.

amphidiploidy Type of allopolyploidy in which two different diploid genomes are combined in such a way that every chromosome has one and only one homologous partner and the genome is functionally diploid.

anagenesis Evolutionary change within a single lineage. *Compare* **cladogenesis**.

anaphase Stage of mitosis in which chromatids separate and move toward the spindle poles.

anaphase I Stage of meiosis I in which homologous chromosomes separate and move toward the spindle poles.

anaphase II Stage of meiosis II in which chromatids separate and move toward the spindle poles.

aneuploidy Change from the wild type in the number of individual chromosomes; most often an increase or decrease of one or two chromosomes.

anticodon Sequence of three nucleotides in tRNA that pairs with the corresponding codon in mRNA in translation.

antigenic drift Change in a viral genome that takes place through the continual accumulation of mutations.

antigenic shift Major change in a viral genome that takes place when genetic material from different strains is recombined.

antiparallel A characteristic of the DNA double helix, in which the two polynucleotide strands run in opposite directions: the 5′ end of one polynucleotide strand is opposite the 3′ end of the other strand.

apoptosis Programmed cell death, in which a cell degrades its own DNA, the nucleus and cytoplasm shrink, and the cell undergoes phagocytosis by other cells without leakage of its contents.

archaea One of the three primary divisions of life, consisting of prokaryotic unicellular organisms.

artificial selection Selection practiced by humans.

autopolyploidy Condition in which all the sets of chromosomes of a polyploid individual are derived from a single species. *Compare* **allopolyploidy**.

autosomes Chromosomes that are the same in number and morphology in males and females. *Compare* **sex chromosomes**.

auxotroph Bacterium or fungus that possesses a nutritional mutation that disrupts its ability to synthesize an essential biological molecule; cannot grow on minimal medium, but can grow on minimal medium to which has been added the biological molecule that it cannot synthesize.

backcross Cross between an F_1 individual and one of the parental (P) genotypes.

bacteria One of the three primary divisions of life, consisting of prokaryotic unicellular organisms and including most of the common bacterial species. Also referred to as eubacteria.

bacterial artificial chromosome (BAC) Cloning vector used in bacteria that is capable of carrying DNA fragments as large as 500 kb.

bacteriophage Virus that infects bacterial cells.

Barr body Inactivated X chromosome that appears as a condensed, darkly staining structure in most cells of female placental mammals.

base *See* **nitrogenous base**.

base analog Chemical substance that has a structure similar to that of one of the four standard bases of DNA and may be incorporated into newly synthesized DNA molecules in replication.

base-excision repair DNA repair that first excises modified bases and then replaces the entire nucleotide.

base substitution Mutation in which a single pair of bases in DNA is altered.

B-DNA Right-handed helical structure of DNA that exists when water is abundant; the secondary structure described by Watson and Crick and probably the most common DNA structure in cells.

bidirectional replication Replication at both ends of a replication bubble.

bioinformatics Synthesis of molecular biology and computer science that develops databases and computational tools to store, retrieve, and analyze nucleic-acid and protein-sequence data.

biological species concept Definition of a species as a group of organisms whose members are capable of interbreeding with one another but are reproductively isolated from the members of other species. Because different species do not exchange genes, each species evolves independently. Not all biologists adhere to this concept.

biotechnology Use of biological processes, particularly molecular genetics and recombinant DNA technology, to produce products of commercial value.

bivalent A homologous pair of synapsed chromosomes.

blending inheritance Early concept of heredity proposing that offspring possess a mixture of the traits from both parents.

branch Connection between nodes in a phylogenetic tree representing an evolutionary connection between organisms.

branch point Adenine nucleotide in nuclear pre-mRNA introns that lies from 18 to 40 nucleotides upstream of the 3′ splice site.

broad-sense heritability Proportion of phenotypic variance that can be attributed to genetic variance.

catabolite activator protein (CAP) Protein that functions in catabolite repression. When bound with cAMP, CAP binds to the promoter of certain operons and stimulates transcription.

catabolite repression System of gene control in some bacterial operons in which glucose is used preferentially and the metabolism of other sugars is repressed in the presence of glucose.

cDNA (complementary DNA) library Collection of bacterial or phage colonies containing DNA fragments that have been produced by reverse transcription of cellular mRNA.

cell cycle Stages through which a cell passes from one cell division to the next.

cell theory Theory stating that all life is composed of cells, that cells arise only from other cells, and that the cell is the fundamental unit of structure and function in living organisms.

centiMorgan *See* **map unit**.

central dogma Concept that genetic information passes from DNA to RNA to protein in a one-way information pathway.

centriole Cytoplasmic organelle consisting of microtubules; present at each pole of the spindle apparatus in animal cells.

centromere Constricted region on a chromosome that stains less strongly than the rest of the chromosome; region where spindle microtubules attach to a chromosome.

Chargaff's rules Rules developed by Erwin Chargaff and his colleagues concerning the ratios of bases in DNA.

checkpoint A key transition point at which progression to the next stage in the cell cycle is regulated.

chiasma (pl., **chiasmata**) Point of attachment between homologous chromosomes at which crossing over takes place.

chi-square goodness-of-fit test Statistical test used to evaluate how well a set of observed values fit the expected values. The probability associated with a calculated chi-square value is the probability that the differences between the observed and the expected values is due to chance.

chloroplast DNA (cpDNA) DNA in chloroplasts; has many characteristics in common with eubacterial DNA and typically consists of a circular molecule that lacks histone proteins and encodes some of the rRNAs, tRNAs, and proteins found in chloroplasts.

chromatin Material found in the eukaryotic nucleus; consists of DNA and proteins.

chromatin-remodeling complex Complex of proteins that alters chromatin structure without acetylating histone proteins.

chromosome Structure consisting of DNA and associated proteins that contains genetic information. The cells of each species have a characteristic number of chromosomes.

chromosome deletion Loss of a chromosome segment.

chromosome duplication Mutation that doubles a segment of a chromosome.

chromosome inversion Rearrangement in which a segment of a chromosome has been inverted 180 degrees.

chromosome mutation Difference from the wild type in the number or structure of one or more chromosomes; often affects many genes and has large phenotypic effects.

chromosome rearrangement Change from the wild type in the structure of one or more chromosomes.

chromosome theory of heredity Theory stating that genes are located on chromosomes.

cladogenesis Evolution in which one lineage is split into two. *Compare* **anagenesis**.

clonal evolution Process by which mutations that enhance the ability of cells to proliferate predominate in a clone of cells, allowing the clone to become increasingly rapid in growth and increasingly aggressive in proliferation properties.

cloning vector Stable, replicating DNA molecule to which a foreign DNA fragment can be attached for transfer to a host cell.

cloverleaf structure Secondary structure common to all tRNAs.

coactivator Protein that cooperates with a transcriptional activator protein. In eukaryotic transcriptional control, coactivators often physically interact with transcriptional activators and the basal transcription apparatus.

codominance Type of allelic interaction in which the heterozygote simultaneously expresses traits of both homozygotes.

codon Sequence of three nucleotides that encodes one amino acid in a protein.

coefficient of coincidence Ratio of observed double crossovers to expected double crossovers.

cohesin Molecule that holds the two sister chromatids of a chromosome together. The breakdown of cohesin at the centromeres enables the chromatids to separate in anaphase of mitosis and anaphase II of meiosis.

cohesive end Short, single-stranded overhanging end on a DNA molecule produced when the DNA is cut by certain restriction enzymes; also called a **sticky end**. Cohesive ends are complementary and can spontaneously pair to rejoin DNA fragments that have been cut with the same restriction enzyme.

colinearity Concept that there is a direct correspondence between the nucleotide sequence of a gene and the continuous sequence of amino acids in a protein.

colony Clump of genetically identical bacteria derived from a single bacterial cell that undergoes repeated rounds of division.

comparative genomics Comparative studies of the genomes of different organisms.

competent Capable of taking up DNA from the environment (capable of being transformed).

complementary DNA strands Nucleotide strands of DNA in which each purine on one strand pairs with a specific pyrimidine on the opposite strand (A pairs with T, and G pairs with C).

complementation Exhibition of two different mutations in the heterozygous condition as the wild-type phenotype; indicates that the mutations are at different loci (are nonallelic).

complementation test Test designed to determine whether two different mutations are at the same locus (are allelic) or at different loci (are nonallelic). Two individuals that are homozygous for two independently derived mutations are crossed, producing F_1 progeny that are heterozygous for the mutations. If the mutations are at the same locus, the F_1 will have a mutant phenotype. If the mutations are at different loci, the F_1 will have a wild-type phenotype.

complete dominance Type of dominance in which the heterozygote has the same phenotype as the homozygote. *Compare* **incomplete dominance**.

complete medium Medium used to culture bacteria or other microorganisms that contains all the nutrients required for growth and reproduction, including those normally synthesized by the organism. Nutritional mutants can grow on complete medium. *Compare* **minimal medium**.

compound heterozygote An individual that carries two different alleles at a locus but possesses a recessive phenotype.

concept of dominance Principle of heredity discovered by Mendel stating that when two different alleles are present in a genotype, only one allele may be expressed in the phenotype. The dominant allele is the allele that is expressed, and the recessive allele is the allele that is not expressed.

conditional mutation Mutation that is expressed only under certain conditions.

conditional probability Modification of probability based on additional information. For example the probability of a progeny genotype might be modified by information about its phenotype.

conjugation Mechanism by which genetic material may be exchanged between bacterial cells. During conjugation, two bacteria lie close together, and a cytoplasmic connection forms between them. A plasmid, or sometimes a part of the bacterial chromosome, passes through this connection from one cell to the other.

consanguinity Mating between related individuals.

consensus sequence Sequence that comprises the most commonly encountered nucleotides found at a specific location in DNA or RNA.

constitutive gene Gene that is not regulated and is expressed continually.

constitutive mutation Mutation that causes a structural gene to be continually expressed.

contig Set of overlapping DNA fragments that have been assembled in the correct order to form a continuous stretch of DNA sequence.

continuous characteristic Characteristic that displays a large number of possible phenotypes that are not easily distinguished, such as human height. *Compare* **discontinuous characteristic**.

continuous replication Replication of the leading strand in the same direction as that of unwinding, allowing new nucleotides to be added continuously to the 3′ end of the new strand as the template is exposed.

coordinate induction Stimulation of the simultaneous synthesis of several enzymes by a single environmental factor.

copy-number variation (CNV) Difference among individual organisms in the number of copies of any large DNA sequence (larger than 1000 bp).

core enzyme Set of five subunits at the heart of most bacterial RNA polymerases that, during transcription, catalyzes the elongation of the RNA molecule by the addition of RNA nucleotides.

corepressor Substance that inhibits transcription in a repressible system of gene regulation; usually a small molecule that binds to a repressor protein and alters it so that the repressor is able to bind to DNA and inhibit transcription.

correlation Degree of association between two or more variables.

correlation coefficient Statistic that measures the degree of association between two or more variables. A correlation coefficient can range from $+1$ to -1. A positive value indicates a direct relation between the variables; a negative correlation indicates an inverse relation. The absolute value of the correlation coefficient provides information about the strength of association between the variables.

cosmid Cloning vector consisting of a plasmid packaged in an empty viral protein coat that can be transferred to bacteria by viral infection; can carry large pieces of DNA into bacteria.

cotransduction Process in which two or more genes are transferred together from one bacterial cell to another. Only genes located close together on a bacterial chromosome will be cotransduced.

cotransformation Process in which two or more genes are transferred together during cell transformation.

coupling (cis) configuration Arrangement of linked genes in which two or more wild-type genes are on one chromosome and their mutant alleles are on the homologous chromosome. *Compare* **repulsion (trans) configuration**.

CRISPR-Cas system A molecular tool for precisely cutting DNA, which has been developed in recent years and is a powerful way of editing the genome.

CRISPR RNA (crRNA) Class of small RNA molecules in prokaryotes that assists in the destruction of foreign DNA molecules.

crossing over Exchange of genetic material between homologous but nonsister chromatids.

C-value Haploid amount of DNA found in a cell of an organism.

C-value paradox Variation among eukaryotic species in their C-values that cannot be explained by differences in organismal complexity.

cyclin A key protein in the control of the cell cycle; combines with a cyclin-dependent kinase (CDK). The levels of cyclin rise and fall in the course of the cell cycle.

cyclin-dependent kinase (CDK) A key protein in the control of the cell cycle; combines with cyclin.

cytokinesis Process by which the cytoplasm of a cell divides.

cytoplasmic inheritance Inheritance of characteristics encoded by genes located in the cytoplasm. Because the cytoplasm is usually contributed entirely by only one parent, most cytoplasmically inherited characteristics are inherited from a single parent.

cytosine (C) Pyrimidine base in DNA and RNA.

deamination Loss of an amino group (NH_2) from a base.

degenerate Containing more information than is needed; refers to the fact that in the genetic code, amino acids may be specified by more than one codon.

deletion Mutation in which one or more nucleotides are deleted from a DNA sequence.

deoxyribonucleotide Basic building block of DNA, consisting of deoxyribose, a phosphate group, and a nitrogenous base.

deoxyribose Five-carbon sugar in DNA; lacks a hydroxyl group on the 2′-carbon atom.

depurination Break in the covalent bond connecting a purine base to the 1′-carbon atom of deoxyribose, resulting in the loss of the purine base. The resulting apurinic site cannot provide a template in replication, and a nucleotide with another base may be incorporated into the newly synthesized DNA strand opposite the apurinic site.

dicentric bridge Structure produced when the two centromeres of a dicentric chromatid are pulled toward opposite poles, stretching the dicentric chromosome across the center of the nucleus. Eventually, the dicentric bridge breaks as the two centromeres are pulled apart.

dicentric chromatid Chromatid that has two centromeres; produced when crossing over takes place within a paracentric inversion. The two centromeres of the dicentric chromatid are frequently pulled toward opposite poles in mitosis or meiosis, breaking the chromosome.

dideoxyribonucleoside triphosphate (ddNTP) Special substrate for DNA synthesis used in the Sanger dideoxy-sequencing method; identical with dNTP (the usual substrate for DNA synthesis) except that it lacks a 3′-OH group. The incorporation of a ddNTP into DNA terminates DNA synthesis.

dihybrid cross Cross between two individuals that differ in two characteristics—more specifically, a cross between individuals that are homozygous for different alleles at the two loci (*AA BB* × *aa bb*); also refers to a cross between two individuals that are both heterozygous at two loci (*Aa Bb* × *Aa Bb*).

diploid Possessing two sets of chromosomes.

directional selection Selection in which one trait or allele is favored over another.

direct repair DNA repair in which modified bases are changed back into their original structures.

direct-to-consumer genetic test Test for a genetic condition that can be purchased directly by a consumer, without the participation of a physician or other health-care provider.

discontinuous characteristic Characteristic that exhibits only a few, easily distinguished phenotypes. An example is seed shape in which seeds are either round or wrinkled. *Compare* **continuous characteristic**.

discontinuous replication Replication of the lagging strand in the direction opposite that of unwinding, which means that DNA must be synthesized in short stretches (Okazaki fragments).

displaced duplication Chromosome rearrangement in which the duplicated segment is some distance from the original segment, either on the same chromosome or on a different one.

DNA fingerprinting Technique used to identify individuals by examining their DNA sequences.

DNA gyrase *E. coli* topoisomerase enzyme that relieves the torsional strain that builds up ahead of the replication fork.

DNA helicase Enzyme that unwinds double-stranded DNA by breaking hydrogen bonds.

DNA library Collection of bacterial colonies containing all the DNA fragments from one source.

DNA ligase Enzyme that catalyzes the formation of a phosphodiester bond between adjacent 3′-OH and 5′-phosphate groups in a DNA molecule.

DNA methylation Modification of DNA by the addition of methyl groups to certain positions on the bases.

DNA polymerase Enzyme that synthesizes DNA.

DNA polymerase I Bacterial DNA polymerase that removes and replaces RNA primers with DNA nucleotides.

DNA polymerase III Bacterial DNA polymerase that adds new nucleotides to the 3′ end of a growing DNA strand.

DNA polymerase α Eukaryotic DNA polymerase that initiates replication.

DNA polymerase δ Eukaryotic DNA polymerase that replicates the lagging strand after initiation by DNA polymerase α; also carries out DNA repair and DNA translesion synthesis.

DNA polymerase ε Eukaryotic DNA polymerase that replicates the leading strand during DNA synthesis.

DNA polymerase γ Eukaryotic DNA polymerase that replicates mitochondrial DNA.

DNA sequencing Process of determining the sequence of bases along a DNA molecule.

DNA transposon Transposable element that transposes as DNA (instead of being first copied into RNA). *Compare* **retrotransposon**.

dominance genetic variance Component of genetic variance that can be attributed to dominance (interaction between genes at the same locus).

dominant Refers to an allele or a phenotype that is expressed in homozygotes (*AA*) and in heterozygotes (*Aa*); only the dominant allele is expressed in a heterozygote phenotype. *Compare* **recessive**.

dosage compensation Equalization in males and females of the amount of protein produced by X-linked genes. In placental mammals, dosage compensation is accomplished by the random inactivation of one X chromosome in the cells of females.

Down syndrome (trisomy 21) Human syndrome characterized by variable degrees of intellectual disability, characteristic facial features, some slowing of growth and development, and an increased incidence of heart defects, leukemia, and other abnormalities; caused by the duplication of all or part of chromosome 21.

Edward syndrome (trisomy 18) Human syndrome characterized by severe intellectual disability, low-set ears, a short neck, deformed feet, clenched fingers, heart problems, and other disabilities; results from the presence of three copies of chromosome 18.

effective population size Effective number of breeding adults in a population; influenced by the number of individuals contributing genes to the next generation, their sex ratio, variation between individuals in reproductive success, fluctuations in population size, the age structure of the population, and whether mating is random.

elongation factor G (EF-G) Protein required for movement of the ribosome along the mRNA during translation.

elongation factor Ts (EF-Ts) Protein that regenerates elongation factor Tu in the elongation stage of protein synthesis.

elongation factor Tu (EF-Tu) Protein taking part in the elongation stage of protein synthesis; forms a complex with GTP and a charged tRNA and then delivers the charged tRNA to the ribosome.

engineered nuclease Artificially constructed enzyme that is capable of making double-stranded cuts to DNA at a predetermined DNA sequence.

enhancer Sequence that stimulates maximal transcription of distant genes; affects only genes on the same DNA molecule (is cis acting), contains short consensus sequences, is not fixed in relation to the transcription start site, can stimulate almost any promoter in its vicinity, and may be upstream or downstream of the gene. The function of an enhancer is independent of sequence orientation.

environmental variance Component of phenotypic variance that is due to environmental differences among individual members of a population.

epialleles Genetically identical alleles that produce heritable differences in phenotypes through epigenetic processes.

epigenetic changes (epigenetics) Stable alterations of chromatin structure that can be passed on to descendant cells or individuals.

epigenetic mark Modification of histone proteins that alters chromatin structure and affects the transcription of genes.

epigenetic process A process that affects the expression of genes; often a process that brings about genetic alterations that can be reversed, such as the methylation of DNA.

epigenetics Phenomena due to alterations in DNA that do not include changes in the base sequence; often affect the way in which DNA sequences are expressed. Such alterations are often stable and heritable in the sense that they are passed on to descendant cells or organisms.

epigenome Overall pattern of epigenetic modifications in a genome.

episome Plasmid capable of integrating into a bacterial chromosome.

epistasis Type of gene interaction in which a gene at one locus masks or suppresses the effects of a gene at a different locus.

epistatic gene Gene that masks or suppresses the effect of a gene at a different locus.

equilibrium Situation in which no further change takes place; in population genetics, refers to a situation in which allelic frequencies in a population do not change.

equilibrium density gradient centrifugation Method used to separate molecules or organelles of different density by centrifugation.

euchromatin Chromatin that undergoes condensation and decondensation in the course of the cell cycle. *Compare* **heterochromatin**.

eukaryote One of the three primary divisions of life, consisting of organisms whose cells have a complex structure including a nuclear envelope and membrane-bounded organelles. Eukaryotes include unicellular and multicellular forms.

evolution Genetic change taking place in a group of organisms.

exit (E) site One of three sites in a ribosome occupied by a tRNA. In the elongation stage of translation, the tRNA moves from the peptidyl (P) site to the E site, from which it then exits the ribosome.

exon Coding region of a gene interrupted by introns; remains after transcription and posttranscriptional processing. *Compare* **intron**.

expanding nucleotide repeat Mutation in which the number of copies of a set of nucleotides increases in succeeding generations.

expression vector Cloning vector containing DNA sequences, such as a promoter, a ribosome-binding site, and transcription initiation and termination sites, that allow DNA fragments inserted into the vector to be transcribed and translated.

expressivity Degree to which a trait is expressed.

F_1 (first filial) generation Offspring of the initial parents (P) in a genetic cross.

F_2 (second filial) generation Offspring of the F_1 generation in a genetic cross; the third generation of a genetic cross.

familial Down syndrome Human syndrome caused by a Robertsonian translocation in which the long arm of chromosome 21 is translocated to another chromosome; tends to run in families.

fertilization Fusion of gametes to form a zygote.

F (fertility) factor Episome of *E. coli* that controls conjugation and gene exchange between cells. The F factor contains an origin of replication and genes that enable the bacterium to undergo conjugation.

first polar body One of the products of meiosis I in oogenesis; contains half the chromosomes but little of the cytoplasm.

fitness Reproductive success of a genotype compared with that of other genotypes in a population.

5′ cap Modified 5′ end of eukaryotic mRNA, consisting of an extra, methylated nucleotide and methyl groups at the 2′ position of the ribose sugar in one or more subsequent nucleotides; plays a role in the binding of the ribosome to mRNA and affects mRNA stability and the removal of introns.

5′ end End of a polynucleotide chain at which a phosphate group is attached to the 5′-carbon atom in the sugar in a nucleotide. *Compare* **3′ end.**

5′ splice site The 5′ end of an intron where cleavage takes place in RNA splicing.

5′ untranslated (UTR) region Sequence of nucleotides at the 5′ end of mRNA; does not encode the amino acids of a protein.

fixation Point at which one allele reaches a frequency of 1 in a population. At this point, all members of the population are homozygous for that allele.

flanking direct repeat Short, directly repeated sequence produced on either side of a transposable element when the element inserts itself into DNA.

forward genetics Traditional approach to the study of gene function that begins with a phenotype (a mutant organism) and proceeds to a gene that encodes the phenotype. *Compare* **reverse genetics.**

forward mutation Mutation that alters a wild-type phenotype. *Compare* **reverse mutation.**

founder effect Sampling error that arises when a population is established by a small number of individuals; leads to genetic drift.

fragile site Constriction or gap that appears at a particular location on a chromosome when cells are cultured under special conditions.

fragile-X syndrome A form of X-linked intellectual disability that appears primarily in males; associated with a fragile site that results from an expanding trinucleotide repeat.

frameshift mutation Mutation that alters the reading frame of a gene.

frequency distribution Graphical representation of the frequencies of the different phenotypes found in a group of individuals. Typically, the phenotypes are plotted on the horizontal (x) axis and the numbers (or proportions) of individuals with each phenotype are plotted on the vertical (y) axis.

functional genomics Area of genomics that studies the functions of the genetic information contained within genomes.

G_0 (gap 0) Nondividing stage of the cell cycle.

G_1 (gap 1) Stage in interphase of the cell cycle in which the cell grows and develops.

G_2 (gap 2) Stage of interphase in the cell cycle that follows DNA replication. In G_2, the cell prepares for division.

gain-of-function mutation Mutation that produces a new trait or causes a trait to appear in inappropriate tissues or at inappropriate times in development. *Compare* **loss-of-function mutation.**

gamete Male or female sex cell.

gametophyte Haploid phase of the life cycle in plants.

gel electrophoresis Technique for separating charged molecules (such as proteins or nucleic acids) on the basis of molecular size or charge, or both.

gene An inherited factor that helps determine a trait; often defined at the molecular level as a DNA sequence that is transcribed into an RNA molecule.

gene cloning Insertion of DNA fragments into bacteria in such a way that the fragments will be stable and will be copied by the bacteria.

gene desert A region of the genome that is gene poor—that is, a long stretch of DNA, possibly consisting of hundreds of thousands to millions of base pairs, that is completely devoid of any known genes or other functional sequences.

gene family A set of genes similar in sequence that arose through repeated duplication events and often encode different proteins.

gene flow *See* **migration.**

gene interaction Interaction between genes at different loci that affect the same characteristic.

gene interaction variance Component of genetic variance that can be attributed to gene interaction.

gene mutation Mutation that affects a single gene or locus.

gene pool Total of all genes in a population.

generalized transduction Transduction in which any gene may be transferred from one bacterial cell to another by a virus.

general transcription factor Protein that binds to eukaryotic promoters near the transcription start site and is a part of the basal transcription apparatus that initiates transcription.

gene regulation Mechanisms and processes that control the phenotypic expression of genes.

gene therapy Use of recombinant DNA to treat a disease or disorder by altering the genetic makeup of the patient's cells.

genetic bottleneck Sampling error that arises when a population undergoes a drastic reduction in population size; leads to genetic drift.

genetic drift Change in allelic frequency due to sampling error.

genetic engineering Common term for recombinant DNA technology.

genetic–environmental interaction variance Component of phenotypic variance that results from an interaction between genotype and environment such that genotypes are expressed differently in different environments.

genetic map Map of the relative distances between genetic loci, markers, or other chromosome regions determined by rates of recombination; measured in recombination frequencies or map units. *Compare* **physical map.**

genetic marker Any gene or DNA sequence used to identify a location on a genetic or physical map.

genetic maternal effect Determination of the phenotype of an offspring not by its own genotype, but by the nuclear genotype of the mother.

genetic rescue Introduction of new genetic variation into an inbred population in an effort to improve the health of the population and increase its chances of long-term survival.

genetic variance Component of phenotypic variance that is due to genetic differences among individual members of a population.

genic sex determination Sex determination in which the sexual phenotype is specified by genes at one or more loci, but there are no obvious differences in the chromosomes of males and females.

genome Complete set of genetic instructions for an organism.

genome-wide association study A study that looks for nonrandom associations in a population between the presence of a particular phenotype and alleles at many different loci across the genome.

genomic imprinting Differential expression of a gene that depends on the sex of the parent that transmitted the gene.

genomic library Collection of bacterial or phage colonies containing DNA fragments that constitute the entire genome of an organism.

genomics Study of the content, organization, and function of genetic information in whole genomes.

genotype The set of genes possessed by an individual organism.

genotypic frequency Proportion of a particular genotype in a population.

germ-line mutation Mutation in a germ-line cell (one that gives rise to gametes).

germ-plasm theory Theory stating that cells in the reproductive organs carry a complete set of genetic information.

G_2/M (gap 2/mitotic) checkpoint Important checkpoint in the cell cycle near the end of G_2 after which the cell undergoes mitosis.

G overhang A guanine-rich sequence of nucleotides that protrudes beyond the complementary C-rich strand at the end of a chromosome.

G_1/S (gap 1/synthesis) checkpoint Important checkpoint in the cell cycle after which DNA replicates and the cell is committed to dividing.

guanine (G) Purine base in DNA and RNA.

gyrase *See* **DNA gyrase.**

hairpin Secondary structure formed when sequences of nucleotides on the same strand are complementary and pair with each other.

haploid Possessing a single set of chromosomes.

haploinsufficiency The appearance of a mutant phenotype in an individual cell or organism that is heterozygous for a normally recessive trait.

haploinsufficient gene Gene that must be present in two copies for normal function. If one copy of the gene is missing, a mutant phenotype is produced.

haplotype A specific set of linked genetic variants or alleles on a single chromosome or on part of a chromosome.

Hardy–Weinberg equilibrium Frequencies of genotypes when the conditions of the Hardy–Weinberg law are met.

Hardy–Weinberg law Principle of population genetics stating that in a large, randomly mating population not affected by mutation, migration, or natural selection, allelic frequencies will not change and genotypic frequencies will stabilize after one generation in the proportions p^2 (the frequency of *AA*), $2pq$ (the frequency of *Aa*), and q^2 (the frequency of *aa*), where p equals the frequency of allele *A* and q equals the frequency of allele *a*.

heat-shock protein Protein produced by many cells in response to extreme heat and other stresses that helps cells prevent damage from such stressing agents.

helicase *See* **DNA helicase**.

hemizygosity Possession of a single allele at a locus. Males of organisms with XX-XY sex determination are hemizygous for X-linked loci because their cells possess a single X chromosome.

heritability Proportion of phenotypic variation due to genetic differences. *See also* **broad-sense heritability**, **narrow-sense heritability**.

heterochromatin Chromatin that remains in a highly condensed state throughout the cell cycle; found at the centromeres and telomeres of most chromosomes. *Compare* **euchromatin**.

heteroduplex DNA DNA consisting of two strands, each of which is from a different chromosome.

heterogametic sex The sex (male or female) that produces two types of gametes with respect to sex chromosomes. For example, in the XX-XY sex-determining system, the male produces both X-bearing and Y-bearing gametes. *Compare* **homogametic sex**.

heterozygous Having two different alleles at a locus.

highly repetitive DNA DNA that consists of short sequences that are present in hundreds of thousands to millions of copies; clustered in certain regions of chromosomes.

histone Low-molecular-weight protein found in eukaryotes that complexes with DNA to form chromosomes.

histone code Modifications of histone proteins, such as the addition or removal of phosphate groups, methyl groups, or acetyl groups, that encode information affecting how genes are expressed.

Holliday junction Special structure resulting from homologous recombination that is initiated by single-strand breaks in a DNA molecule.

holoenzyme Complex of an enzyme and other protein factors necessary for complete function.

homogametic sex The sex (male or female) that produces gametes that are all alike with regard to sex chromosomes. For example, in the XX-XY sex-determining system, the female produces only X-bearing gametes. *Compare* **heterogametic sex**.

homologous genes Evolutionarily related genes descended from a gene in a common ancestor.

homologous pair Two chromosomes that are alike in structure and size and that carry genetic information for the same set of hereditary characteristics. One chromosome of a homologous pair is inherited from the male parent and the other is inherited from the female parent.

homologous recombination Exchange of genetic information between homologous DNA molecules.

homozygous Having two identical alleles at a locus.

horizontal gene transfer Transfer of genes from one organism to another by a mechanism other than reproduction.

human papilloma virus (HPV) Virus associated with cervical cancer.

Human Proteome Project Project to identify and characterize all proteins in the human body.

hypostatic gene Gene that is masked or suppressed by the action of a gene at a different locus.

inbreeding Mating between related individuals that takes place more frequently than expected on the basis of chance.

inbreeding depression Decreased fitness arising from inbreeding; often due to the increased expression of lethal or deleterious recessive alleles when inbreeding takes place.

incomplete dominance Type of dominance in which the phenotype of the heterozygote is intermediate between the phenotypes of the two homozygotes. *Compare* **complete dominance**.

incomplete linkage Linkage between genes that exhibit some crossing over; intermediate in its effects between independent assortment and complete linkage.

incomplete penetrance A case in which individuals with a particular genotype do not always express the expected phenotype.

incorporated error Incorporation of damaged, incorrect, or mismatched bases into a DNA molecule.

independent assortment Independent separation of chromosome pairs in anaphase I of meiosis; contributes to genetic variation.

induced mutation Mutation that results from an environmental agent, such as a chemical or radiation.

inducer Substance that stimulates transcription in an inducible system of gene regulation; usually a small molecule that binds to a repressor protein and alters that repressor so that it can no longer bind to DNA and inhibit transcription.

inducible operon Operon or other system of gene regulation in which transcription is normally turned off, so that something must happen for transcription to be induced, or turned on. *Compare* **repressible operon**.

in-frame deletion Deletion of some multiple of three nucleotides, which does not alter the reading frame of the gene.

in-frame insertion Insertion of some multiple of three nucleotides, which does not alter the reading frame of the gene.

inheritance of acquired characteristics Early notion of inheritance proposing that acquired traits are passed to descendants.

initiation codon The codon in mRNA that specifies the first amino acid (fMet in bacterial cells; Met in eukaryotic cells) of a protein; most commonly AUG; also called a **start codon**.

initiation factor 1 (IF-1) Protein required for the initiation of translation in bacterial cells; enhances the dissociation of the large and small subunits of the ribosome.

initiation factor 2 (IF-2) Protein required for the initiation of translation in bacterial cells; forms a complex with GTP and the charged initiator tRNA and then delivers the charged tRNA to the initiation complex.

initiation factor 3 (IF-3) Protein required for the initiation of translation in bacterial cells; binds to the small subunit of the ribosome and prevents the large subunit from binding during initiation.

initiator protein Protein that binds to an origin of replication and unwinds a short stretch of DNA, allowing helicase and other single-strand-binding proteins to bind and initiate replication.

insertion Mutation in which nucleotides are added to a DNA sequence.

insulator DNA sequence that blocks or insulates the effect of an enhancer; must be located between the enhancer and the promoter to have blocking activity; also may limit the spread of changes in chromatin structure.

integrase Enzyme that inserts prophage, or proviral, DNA into a chromosome.

intercalating agent Chemical substance that is about the same size as a nucleotide and may become sandwiched between adjacent bases in DNA, distorting the three-dimensional structure of the helix and causing single-nucleotide insertions and deletions in replication.

interference Degree to which one crossover interferes with additional crossovers.

intergenic suppressor mutation Mutation that suppresses another mutation and occurs in a gene (locus) that is different from the gene containing the original mutation.

interkinesis Period between meiosis I and meiosis II.

interphase Major phase of the cell cycle between cell divisions. In interphase, the cell grows, develops, and prepares for cell division. *Compare* **M phase**.

interspersed repeat sequence Repeated sequence found at multiple locations throughout the genome.

intragenic suppressor mutation Mutation that suppresses another mutation and occurs in the same gene (locus) as the original mutation.

intron Noncoding sequence intervening between coding regions in a eukaryotic gene; removed from the RNA after transcription. *Compare* **exon**.

inverted repeats Sequences on the same strand that are inverted and complementary.

isoaccepting tRNAs Different tRNA molecules with different anticodons that specify the same amino acid.

isotopes Different forms of an element that have the same number of protons and electrons but differ in the number of neutrons in the nucleus.

karyotype The complete set of chromosomes possessed by an organism; usually presented as a picture of metaphase chromosomes lined up in descending order of their size.

kinetochore Set of proteins that assembles on the centromere, providing the point of attachment for spindle microtubules.

Klinefelter syndrome Human syndrome in which cells contain one or more Y chromosomes along with multiple X chromosomes (most commonly XXY, but may also be XXXY, XXXXY, or XXYY). Persons with Klinefelter syndrome are male in appearance but frequently possess small testes, some breast enlargement, and reduced facial and pubic hair; they are often taller than normal and sterile, and most have normal intelligence.

knock-in mice Mouse that carries a foreign sequence inserted at a specific chromosome location.

knockout mice Mouse in which a normal gene has been disabled ("knocked out").

labeling Method for adding a radioactive or chemical label to the ends of DNA molecules.

lagging strand The new DNA strand that is replicated discontinuously. *Compare* **leading strand.**

large ribosomal subunit The larger of the two subunits of a functional ribosome.

lariat Loop-like structure created in the splicing of nuclear pre-mRNA when the 5′ end of an intron is attached to the branch point.

leading strand The new DNA strand that is replicated continuously. *Compare* **lagging strand.**

lethal allele Allele that causes the death of an individual organism, often early in development, so that individuals do not appear in the progeny of a genetic cross. Recessive lethal alleles kill individual organisms that are homozygous for the allele; dominant lethals kill both heterozygotes and homozygotes.

lethal mutation Mutation that causes premature death.

LINE *See* **long interspersed element.**

linkage analysis Gene mapping based on the detection of physical linkage between genes, as measured by the rate of recombination, in the progeny of a cross.

linkage disequilibrium Nonrandom association between genetic variants within a haplotype.

linkage group Genes located together on the same chromosome.

linked genes Genes located on the same chromosome.

linker DNA Stretch of DNA separating two nucleosomes.

locus (pl., **loci**) Position on a chromosome where a specific gene is located.

long interspersed element (LINE) Long DNA sequence repeated many times and interspersed throughout the genome.

long noncoding RNA (lncRNA) Class of relatively long RNA molecules found in eukaryotes that do not code for proteins but provide a variety of other functions, including regulation of gene expression.

loss-of-function mutation Mutation that causes the complete or partial absence of normal function. *Compare* **gain-of-function mutation.**

loss of heterozygosity Inactivation or loss of the wild-type allele in a heterozygote.

Lyon hypothesis Idea proposed by Mary Lyon in 1961 that one X chromosome in each female cell becomes inactivated (a Barr body) and that which X becomes inactivated is random and varies from cell to cell.

lysogenic cycle Life cycle of a bacteriophage in which phage genes first integrate into the bacterial chromosome and are not immediately transcribed and translated.

lytic cycle Life cycle of a bacteriophage in which phage genes are transcribed and translated, new phage particles are produced, and the host cell is lysed.

malignant Consisting of cells that are capable of invading other tissues.

map-based sequencing Method of sequencing a genome in which sequenced fragments are ordered into contigs with the use of genetic or physical maps.

mapping function Mathematical function that relates recombination frequencies to actual physical distances between genes.

map unit (m.u.) Unit of measure for distances on a genetic map; also called a **centiMorgan**. 1 map unit equals 1% recombination.

mass spectrometry Method for precisely determining the mass of a molecule by using the migration rate of an ionized molecule in an electrical field.

mean Statistic that describes the center of a distribution of measurements; calculated by dividing the sum of all measurements by the number of measurements; also called the average.

megaspore One of the four products of meiosis in plants.

megasporocyte Diploid reproductive cell in the ovary of a plant that undergoes meiosis to produce haploid megaspores.

meiosis Process in which the chromosomes of a eukaryotic cell divide to give rise to haploid reproductive cells. Consists of two divisions: meiosis I and meiosis II.

meiosis I First division of meiosis, in which chromosome number is reduced by half.

meiosis II Second division of meiosis. Events in meiosis II are essentially the same as those in mitosis.

Mendelian population Group of interbreeding, sexually reproducing individuals.

meristic characteristic Characteristic whose phenotype varies in whole numbers, such as number of vertebrae.

merozygote *See* **partial diploid**.

messenger RNA (mRNA) RNA molecule that carries genetic information for the amino acid sequence of a protein.

metacentric chromosome Chromosome in which the two chromosome arms are approximately the same length.

metagenomics Area of genomics in which the genome sequences of an entire group of organisms inhabiting a common environment are sampled and determined.

metaphase Stage of mitosis in which chromosomes align in the center of the cell.

metaphase I Stage of meiosis I in which homologous pairs of chromosomes align in the center of the cell.

metaphase II Stage of meiosis II in which individual chromosomes align on the metaphase plate.

metaphase plate Plane in a cell between two spindle poles where chromosomes align in metaphase.

metastasis The separation of cells from malignant tumors and their movement to other sites, where they establish secondary tumors.

microarray Ordered array of DNA fragments fixed to a solid support, where they serve as probes to detect the presence of complementary sequences; often used to assess the expression of genes in various tissues and under different conditions.

microbiome The entire set of microbes that inhabit an organism, including the intestinal tract, skin, and other parts of the body.

microRNA (miRNA) Small RNA molecules, typically 21 or 22 bp in length, that are produced by cleavage of double-stranded RNA arising from small hairpins within RNA that is mostly single stranded. The miRNAs combine with proteins to form a complex that binds (imperfectly) to mRNA molecules and inhibits their translation.

microsatellite Very short DNA sequence repeated in tandem; also called a **short tandem repeat**.

microspore Haploid product of meiosis in plants.

microsporocyte Diploid reproductive cell in the stamen of a plant; undergoes meiosis to produce four haploid microspores.

microtubule Long fiber composed of the protein tubulin; plays an important role in the movement of chromosomes in mitosis and meiosis.

migration Movement of genes from one population to another; also called **gene flow**.

minimal medium Medium used to culture bacteria or other microorganisms that contains only the nutrients required by prototrophic (wild-type) cells—typically, a carbon source, essential elements such as nitrogen and phosphorus, certain vitamins, and other required ions and nutrients. *Compare* **complete medium**.

mismatch repair Process that corrects mismatched nucleotides in DNA after replication has been completed. Enzymes excise incorrectly paired nucleotides from the newly synthesized strand and use the original nucleotide strand as a template when replacing them.

missense mutation Mutation that alters a codon in the mRNA, resulting in a different amino acid in the protein encoded.

mitochondrial DNA (mtDNA) DNA in mitochondria, which has some characteristics in common with eubacterial DNA and typically consists of a circular molecule that lacks histone proteins and encodes some of the rRNAs, tRNAs, and proteins found in mitochondria.

mitosis Process by which the nucleus of a eukaryotic cell divides.

mitotic spindle Array of microtubules that radiate from two poles; moves chromosomes in mitosis and meiosis.

model genetic organism An organism that is widely used in genetic studies because it has characteristics, such as short generation time and large numbers of progeny, that make it well suited to genetic analysis.

moderately repetitive DNA DNA consisting of sequences from 150 to 300 bp in length that are repeated thousands of times.

modified bases Rare bases found in some RNA molecules. Such bases are modified forms of the standard bases (adenine, guanine, cytosine, and uracil).

molecular chaperone Molecule that assists in the proper folding of another molecule.

molecular clock Use of molecular differences to estimate the time of evolutionary divergence between organisms; assumes a roughly constant rate at which one neutral mutation replaces another.

molecular genetics Study of the chemical nature of genetic information and how it is encoded, replicated, and expressed.

monohybrid cross Cross between two individuals that differ in a single characteristic—more specifically, a cross between individuals that are homozygous for different alleles at the same locus ($AA \times aa$); also refers to a cross between two individuals that are both heterozygous for two alleles at a single locus ($Aa \times Aa$).

monosomy Absence of one of the chromosomes of a homologous pair.

M (mitotic) phase The major phase of the cell cycle that encompasses active cell division; includes mitosis (nuclear division) and cytokinesis (cytoplasmic division). *Compare* **interphase**.

multifactorial characteristic Characteristic determined by multiple genes and environmental factors.

multigene family *See* **gene family**.

multiple alleles Presence of more than two alleles at a locus in a group of individuals. Although, within the group, the locus has more than two alleles, each diploid individual in the group has only two of the possible alleles.

multiplication rule Rule stating that the probability of two or more independent events occurring together is calculated by multiplying the probabilities of each of the individual events.

mutagen Any environmental agent that significantly increases the rate of mutation above the spontaneous rate.

mutation Heritable change in genetic information.

mutation rate Frequency with which a gene changes from the wild-type allele to a mutant allele; generally expressed as the number of mutations per biological unit (that is, mutations per cell division, per gamete, or per round of replication).

narrow-sense heritability Proportion of phenotypic variance that can be attributed to additive genetic variance.

natural selection Differential reproduction of individuals with different genotypes.

negative control Gene regulation in which the binding of a regulatory protein to DNA inhibits transcription (the regulatory protein is a repressor). *Compare* **positive control**.

negative supercoiling Tertiary structure that forms when strain is placed on a DNA molecule by underwinding of the helix. *Compare* **positive supercoiling**.

neutral mutation Missense mutation that changes the amino acid sequence of a protein but does not alter the function of that protein.

next-generation sequencing technologies Sequencing methods, such as pyrosequencing, that are capable of simultaneously determining the sequences of many DNA fragments; these technologies are much faster and less expensive than traditional methods of sequencing DNA.

nitrogenous base Nitrogen-containing base that is one of the three parts of a nucleotide.

node A point where the branches split in a phylogeny; represents a past common ancestor that existed before divergence between organisms took place.

nondisjunction Failure of homologous chromosomes or sister chromatids to separate in meiosis or mitosis.

nonhistone chromosomal protein One of a heterogeneous assortment of nonhistone proteins in chromatin.

nonoverlapping genetic code Refers to the fact that in the genetic code, each nucleotide is usually a part of only one codon.

nonreciprocal translocation Movement of a chromosome segment to a nonhomologous chromosome or chromosome region without any (or with unequal) reciprocal exchange of segments.

nonrecombinant (parental) gamete Gamete that contains only the original combinations of genes present in the parents.

nonrecombinant (parental) progeny Progeny that possess only the original combinations of alleles possessed by the parents.

nonreplicative transposition Type of transposition in which a transposable element excises itself from an old site and moves to a new site, resulting in no net increase in the number of copies of the transposable element.

nonsense codon *See* **stop codon**.

nonsense mutation Mutation that changes a sense codon into a stop codon.

nontemplate strand The DNA strand that is complementary to the template strand; not ordinarily used as a template during transcription.

normal distribution Common type of frequency distribution that exhibits a symmetrical, bell-shaped curve; usually arises when a large number of independent factors contribute to a measurement.

Northern blotting Process by which RNA is transferred from a gel to a solid medium.

nuclear envelope Membrane that surrounds the genetic material in eukaryotic cells to form a nucleus; segregates the DNA from other cellular contents.

nucleoid Bacterial DNA confined to a definite region of the cytoplasm.

nucleoside Ribose or deoxyribose bonded to a base.

nucleosome Basic repeating unit of chromatin, consisting of a core of eight histone proteins (two each of H2A, H2B, H3, and H4) and about 146 bp of DNA that wraps around the core about two times.

nucleotide Repeating unit of DNA and RNA made up of a sugar, a phosphate group, and a base.

nucleotide-excision repair DNA repair that removes bulky DNA lesions and other types of DNA damage.

nucleus Organelle found in eukaryotic cells that is enclosed by the nuclear envelope and contains the chromosomes.

nullisomy Absence of both chromosomes of a homologous pair ($2n - 2$).

Okazaki fragment Short stretch of newly synthesized DNA produced by discontinuous replication on the lagging strand; these fragments are eventually joined together.

oncogene Dominant-acting gene that stimulates cell division, leading to the formation of tumors and contributing to cancer; arises from mutated copies of a normal cellular gene (proto-oncogene).

one-gene, one-enzyme hypothesis Idea proposed by Beadle and Tatum that each gene encodes a separate enzyme.

one-gene, one-polypeptide hypothesis Modification of the one-gene, one-enzyme hypothesis; proposes that each gene encodes a separate polypeptide chain.

oogenesis Egg production in animals.

oogonium Diploid cell in the ovary; capable of undergoing meiosis to produce an egg cell.

operator DNA sequence in the operon of a bacterial cell. A regulator protein binds to the operator and affects the rate of transcription of structural genes.

operon Set of structural genes in a bacterial cell that are transcribed together, along with their common promoter and other sequences (such as an operator) that control their transcription.

origin of replication Site where DNA synthesis is initiated.

overdominance Selection in which the heterozygote has higher fitness than either homozygote; also called heterozygote advantage. *Compare* **underdominance**.

ovum Final product of oogenesis.

palindrome Sequence of nucleotides that reads the same on complementary strands; inverted repeat.

pangenesis Early concept of heredity proposing that particles carry genetic information from different parts of the body to the reproductive organs.

paracentric inversion Chromosome inversion that does not include the centromere in the inverted region. *Compare* **pericentric inversion**.

paramutation Epigenetic change in which one allele of a genotype alters the expression of another allele; the altered expression persists for several generations, even after the altering allele is no longer present.

parental gamete *See* **nonrecombinant (parental) gamete**.

parental progeny *See* **nonrecombinant (parental) progeny**.

partial diploid Bacterial cell that possesses two copies of some genes, including one copy on the bacterial chromosome and the other on an extra piece of DNA (usually a plasmid); also called a **merozygote**.

Patau syndrome (trisomy 13) Human syndrome characterized by severe intellectual disability, a small head, sloping forehead, small eyes, cleft lip and palate, extra fingers and toes, and other disabilities; results from the presence of three copies of chromosome 13.

pedigree Pictorial representation of a family history outlining the inheritance of one or more traits or diseases.

penetrance Percentage of individuals with a particular genotype that express the phenotype expected of that genotype.

peptide bond Chemical bond that connects amino acids in a protein.

peptidyl (P) site One of three sites in a ribosome occupied by a tRNA in translation. In the elongation stage of protein synthesis, tRNAs move from the aminoacyl (A) site into the P site.

pericentric inversion Chromosome inversion that includes the centromere in the inverted region. *Compare* **paracentric inversion**.

P (parental) generation First set of parents in a genetic cross.

phage *See* **bacteriophage**.

phenocopy Phenotype produced by environmental effects that is the same as the phenotype produced by a genotype.

phenotype Appearance or manifestation of a characteristic.

phenotypic variance Measure of the degree of phenotypic difference among a group of individuals; composed of genetic, environmental, and genetic–environmental interaction variances.

phosphate group A phosphorus atom attached to four oxygen atoms; one of the three components of a nucleotide.

phosphodiester Molecule containing R–O–P–O–R, where R is a carbon-containing group, O is oxygen, and P is phosphorus.

phosphodiester linkage Phosphodiester bond connecting two nucleotides in a polynucleotide strand.

phylogenetic tree Graphical representation of a phylogeny.

phylogeny The evolutionary relationships among a group of organisms or genes, usually depicted as a family tree or branching diagram.

physical map Map of physical distances between loci, genetic markers, or other chromosome segments; measured in base pairs. *Compare* **genetic map**.

pilus (pl., **pili**) Extension of the surface of some bacteria that allows conjugation to take place. When a pilus on one cell makes contact with a receptor on another cell, the pilus contracts and pulls the two cells together.

Piwi-interacting RNA (pi-RNA) Class of small RNA molecules that plays a role in suppressing the expression of transposable elements in reproductive cells.

plaque Clear patch of lysed cells on a continuous layer of bacteria on the agar surface of a petri plate. Each plaque represents a single original phage that multiplied and lysed many cells.

plasmid Small, circular DNA molecule found in bacterial cells that is capable of replicating independently from the bacterial chromosome.

pleiotropy Ability of a single gene to influence multiple phenotypes.

poly(A) tail String of adenine nucleotides added to the 3′ end of a eukaryotic mRNA after transcription.

polycistronic mRNA Single RNA molecule transcribed from a group of several genes; uncommon in eukaryotes.

polygenic characteristic Characteristic encoded by genes at many loci.

polymerase chain reaction (PCR) Method of amplifying DNA fragments using DNA polymerase.

polynucleotide strand Series of nucleotides linked together by phosphodiester bonds.

polypeptide Chain of amino acids linked by peptide bonds; also called a protein.

polyploidy Possession of more than two haploid sets of chromosomes.

polyribosome Messenger RNA molecule with several ribosomes attached to it.

population genetics Study of the genetic composition of populations (groups of members of the same species) and how their gene pools change with the passage of time.

positional cloning Method that allows for the isolation and identification of a gene by examining the cosegregation of a phenotype with previously mapped genetic markers.

position effect Dependence of the expression of a gene on the gene's location in the genome.

positive control Gene regulation in which the binding of a regulatory protein to DNA stimulates transcription (the regulatory protein is an activator). *Compare* **negative control**.

positive supercoiling Tertiary structure that forms when strain is placed on a DNA molecule by overwinding of the helix. *Compare* **negative supercoiling**.

posttranslational modification Alteration of a protein after translation; can include cleavage from a larger precursor protein, the removal of amino acids, and the attachment of other molecules to the protein.

postzygotic reproductive isolating mechanism Reproductive isolation that arises after gametes from two different species have fused to form zygotes, either because the resulting hybrids are inviable or sterile or because reproduction breaks down in subsequent generations. *Compare* **prezygotic reproductive isolating mechanism**.

preformationism Early concept of inheritance proposing that a miniature adult (homunculus) resides in either the egg or the sperm and increases in size during development, with all traits inherited from the parent that contributes the homunculus.

pre-messenger RNA (pre-mRNA) Eukaryotic RNA molecule that is modified after transcription to become mRNA.

prezygotic reproductive isolating mechanism Reproductive isolation in which gametes from two different species are prevented from fusing and forming a hybrid zygote. *Compare* **postzygotic reproductive isolating mechanism**.

primary Down syndrome Human syndrome caused by the presence of three copies of chromosome 21.

primary oocyte Oogonium that has entered prophase I.

primary spermatocyte Spermatogonium that has entered prophase I.

primary structure of a protein The amino acid sequence of a protein.

primase Enzyme that synthesizes a short stretch of RNA on a DNA template; functions in replication to provide a 3′-OH group for the attachment of a DNA nucleotide.

primer Short stretch of RNA on a DNA template; provides a 3′-OH group for the attachment of a DNA nucleotide at the initiation of replication.

principle of independent assortment (Mendel's second law) Principle of heredity discovered by Mendel that states that genes encoding different characteristics (genes at different loci) separate independently; applies only to genes located on different chromosomes or to genes far apart on the same chromosome.

principle of segregation (Mendel's first law) Principle of heredity discovered by Mendel that states that each diploid individual possesses two alleles at a locus and that these two alleles separate when gametes are formed, one allele going into each gamete.

probability Likelihood of a particular event occurring; more formally, the number of times that a particular event occurs divided by the number of all possible outcomes. Probability values range from 0 to 1.

proband Person with a trait or disease for whom a pedigree is constructed.

probe Known sequence of DNA or RNA that is complementary to a sequence of interest and will pair with it; used to find specific DNA sequences.

prokaryote Unicellular organism with a simple cell structure. Prokaryotes include eubacteria and archaea.

prometaphase Stage of mitosis in which the nuclear membrane breaks down and the spindle microtubules attach to the chromosomes.

promoter DNA sequence to which the transcription apparatus binds so as to initiate transcription; indicates the direction of transcription, which of the two DNA strands is to be read as the template, and the starting point of transcription.

proofreading Process by which DNA polymerases remove and replace incorrectly paired nucleotides in the course of replication.

prophage Phage genome that is integrated into a bacterial chromosome.

prophase Stage of mitosis in which the chromosomes contract and become visible, the cytoskeleton breaks down, and the mitotic spindle begins to form.

prophase I Stage of meiosis I in which chromosomes condense and pair, crossing over takes place, the nuclear membrane breaks down, and the spindle forms.

prophase II Stage of meiosis after interkinesis in which chromosomes condense, the nuclear membrane breaks down, and the spindle forms. Some cells skip this stage.

protein-coding region The part of mRNA consisting of the nucleotides that specify the amino acid sequence of a protein.

protein microarray Large number of different proteins applied to a solid support as a series of spots, each containing a different protein; used to analyze protein–protein interactions.

proteome Set of all proteins found in a cell.

proteomics Study of the proteome, the complete set of proteins found in a given cell.

proto-oncogene Normal cellular gene that controls cell division; when mutated, it may become an oncogene and contribute to cancer progression.

provirus DNA copy of viral DNA or RNA that is integrated into the host chromosome and replicated along with the host chromosome.

pseudoautosomal region Small region of the X and Y chromosomes that contains homologous gene sequences.

pseudodominance Expression of a normally recessive allele owing to a deletion on the homologous chromosome.

Punnett square Shorthand method of determining the outcome of a genetic cross. On a grid, the gametes of one parent are written along the upper edge and the gametes of the other parent are written along the left-hand edge. Within the cells of the grid, the alleles in the gametes are combined to form the genotypes of the offspring.

purine Type of nitrogenous base in DNA and RNA. Adenine and guanine are purines.

pyrimidine Type of nitrogenous base in DNA and RNA. Cytosine, thymine, and uracil are pyrimidines.

pyrimidine dimer Structure in which a bond forms between two adjacent pyrimidine molecules on the same strand of DNA; disrupts normal hydrogen bonding between complementary bases and distorts the normal configuration of the DNA molecule.

quantitative characteristic Continuous characteristic; displays a large number of possible phenotypes, which must be described by a quantitative measurement. Also includes meristic and threshold characteristics that are determined by multiple genetic and environmental factors.

quantitative genetics Genetic analysis of complex characteristics or characteristics influenced by multiple genetic factors.

quantitative trait locus (QTL) Gene or chromosomal region that contributes to the expression of quantitative characteristics.

quaternary structure of a protein Interaction of two or more polypeptides to form a functional protein.

reading frame Particular way in which a nucleotide sequence is read in groups of three nucleotides (codons) in translation; begins with a start codon and ends with a stop codon.

realized heritability Narrow-sense heritability measured from a response-to-selection experiment.

recessive Refers to an allele or phenotype that is expressed only when homozygous (*aa*); the recessive allele is not expressed in a heterozygote phenotype. *Compare* **dominant**.

reciprocal crosses Pair of crosses in which the phenotypes of the male and female parents are reversed. For example, in one cross, a tall male is crossed with a short female and in the other cross, a short male is crossed with a tall female.

reciprocal translocation Reciprocal exchange of segments between two nonhomologous chromosomes.

recombinant DNA technology Set of molecular techniques for locating, isolating, altering, combining, and studying DNA segments.

recombinant gamete Gamete that possesses new combinations of alleles.

recombinant progeny Progeny that possess new combinations of alleles not present in the parents; formed from recombinant gametes.

recombination Sorting of alleles into new combinations.

recombination frequency Proportion of recombinant progeny produced in a cross.

regulator gene Gene associated with an operon in bacterial cells that encodes a protein or RNA molecule that functions in controlling the transcription of one or more structural genes.

regulator protein Protein produced by a regulator gene that binds to another DNA sequence and controls the transcription of one or more structural genes.

regulatory element DNA sequence that affects the transcription of other DNA sequences to which it is physically linked.

regulatory gene DNA sequence that encodes a protein or RNA molecule that interacts with DNA sequences and affects their transcription or translation or both.

relaxed state of DNA Energy state of a DNA molecule when there is no structural strain on the molecule.

release factor Protein required for the termination of translation; binds to a ribosome when a stop codon is reached and stimulates the release of the polypeptide chain, the tRNA, and the mRNA from the ribosome.

repetitive DNA Sequences that exist in multiple copies in a genome.

replicated error Replication of an incorporated error leading to a permanent mutation.

replication Process by which DNA is synthesized from a single-stranded nucleotide template.

replication bubble Segment of a DNA molecule that is unwinding and undergoing replication.

replication fork Point at which a double-stranded DNA molecule separates into two single strands that serve as templates for replication.

replication licensing factor Protein that ensures that replication takes place only once at each origin of replication; required at the origin before replication can be initiated and removed after the DNA has been replicated.

replication origin *See* **origin of replication.**

replicative transposition Type of transposition in which a copy of a transposable element moves to a new site while the original copy remains at the old site; increases the number of copies of the transposable element.

replicon Unit of replication, consisting of DNA from the origin of replication to the point at which replication on either side of the origin ends.

repressible operon Operon or other system of gene regulation in which transcription is normally turned on, so that something must take place for transcription to be repressed, or turned off. *Compare* **inducible operon.**

repressor Regulatory protein that binds to a DNA sequence and inhibits transcription.

reproductive isolating mechanism Any biological factor or mechanism that prevents gene exchange. *See also* **postzygotic reproductive isolating mechanism, prezygotic reproductive isolating mechanism.**

repulsion (trans) configuration Arrangement of linked genes in which each of a homologous pair of chromosomes contains one wild-type (dominant) gene and one mutant (recessive) gene. *Compare* **coupling (cis) configuration.**

response element DNA sequence shared by the promoters or enhancers of several eukaryotic genes to which a regulatory protein can bind to stimulate the coordinate transcription of those genes.

response to selection The amount of change in a characteristic in one generation owing to selection; equals the selection differential times the narrow-sense heritability.

restriction endonuclease *See* **restriction enzyme.**

restriction enzyme Enzyme that recognizes particular base sequences in DNA and makes double-stranded cuts nearby; also called a **restriction endonuclease.**

restriction mapping Physical mapping based on the locations of sites cut by restriction enzymes.

retrotransposon Type of transposable element in eukaryotic cells that possesses some characteristics of retroviruses and transposes through an RNA intermediate. *Compare* **DNA transposon.**

retrovirus RNA virus capable of integrating its genetic material into the genome of its host. The virus injects its RNA genome into the host cell, where reverse transcription produces a complementary, double-stranded DNA molecule from the RNA template. The DNA copy then integrates into the host chromosome to form a provirus.

reverse duplication Duplication of a chromosome segment in which the sequence of the duplicated segment is inverted relative to the sequence of the original segment.

reverse genetics Molecular approach to the study of gene function that begins with a genotype (a DNA sequence) and proceeds to the phenotype by altering the sequence or by inhibiting its expression. *Compare* **forward genetics.**

reverse mutation (reversion) Mutation that changes a mutant phenotype back into the wild type. *Compare* **forward mutation.**

reverse transcriptase Enzyme capable of synthesizing complementary DNA from an RNA template.

reverse transcription Synthesis of DNA from an RNA template.

R factor *See* **R plasmid.**

rho-dependent terminator Sequence in bacterial DNA that requires the presence of the rho factor (a subunit of RNA polymerase) to terminate transcription.

rho (ρ) factor Subunit of bacterial RNA polymerase that facilitates the termination of transcription of some bacterial genes.

rho-independent terminator Sequence in bacterial DNA that does not require the presence of the rho factor to terminate transcription.

ribonucleoside triphosphate (rNTP) Substrate of RNA synthesis; consists of ribose, a nitrogenous base, and three phosphate groups linked to the 5′-carbon atom of the ribose. In transcription, two of the phosphates are cleaved, producing an RNA nucleotide.

ribonucleotide Basic building block of RNA, consisting of ribose, a phosphate group, and a nitrogenous base.

ribose Five-carbon sugar in RNA; has a hydroxyl group attached to the 2′-carbon atom.

ribosomal RNA (rRNA) RNA molecule that is a structural component of the ribosome.

ribozyme RNA molecule that can act as a biological catalyst.

RNA-coding region Sequence of DNA nucleotides that encodes an RNA molecule.

RNA-induced silencing complex (RISC) Complex of a small interfering RNA (siRNA) or microRNA (miRNA) with proteins that can cleave mRNA, leading to its degradation, or affect transcription or repress translation of mRNA.

RNA interference (RNAi) Process in which cleavage of double-stranded RNA produces small RNAs (siRNAs or miRNAs) that bind to mRNAs containing complementary sequences and bring about their cleavage and degradation.

RNA polymerase Enzyme that synthesizes RNA from a DNA template during transcription.

RNA polymerase I Eukaryotic RNA polymerase that transcribes large ribosomal RNA molecules (18S rRNA and 28S rRNA).

RNA polymerase II Eukaryotic RNA polymerase that transcribes pre-messenger RNA, some small nuclear RNAs, and some microRNAs.

RNA polymerase III Eukaryotic RNA polymerase that transcribes transfer RNA, small ribosomal RNAs (5S rRNA), some small nuclear RNAs, and some microRNAs.

RNA polymerase IV RNA polymerase that transcribes small interfering RNAs in plants.

RNA polymerase V RNA polymerase that transcribes RNA that has a role in heterochromatin formation in plants.

RNA replication Process in some viruses by which RNA is synthesized from an RNA template.

RNA splicing Process by which introns are removed from RNA and exons joined together.

Robertsonian translocation Translocation in which the long arms of two acrocentric chromosomes become joined to a common centromere, resulting in a chromosome with two long arms and usually another chromosome with two short arms.

rooted Refers to a phylogenetic tree in which one internal node represents the common ancestor of all other organisms (nodes) in the tree; in a rooted tree, all the organisms depicted have a common ancestor.

R plasmid (R factor) Plasmid having genes that confer antibiotic resistance to any cell that contains the plasmid.

sampling error Deviations from expected ratios due to chance occurrences when the number of events is small.

secondary oocyte One of the products of meiosis I in female animals; receives most of the cytoplasm.

secondary spermatocyte Product of meiosis I in male animals.

secondary structure of a protein Regular folding arrangement of amino acids in a protein. Common secondary structures found in proteins include the alpha helix and the beta pleated sheet.

second polar body One of the products of meiosis II in oogenesis; contains a set of chromosomes but little of the cytoplasm.

segmental duplication Duplication of a chromosome segment larger than 1000 bp.

selection coefficient Measure of the relative intensity of selection against a genotype; equals 1 minus fitness.

selection differential Difference in phenotype between selected individuals and the average of the entire population.

semiconservative replication Replication in which the two nucleotide strands of DNA separate and each serves as a template for the synthesis of a new strand. All DNA replication is semiconservative.

sense codon Codon that specifies an amino acid in a protein.

70S initiation complex Final complex formed in the initiation of translation in bacterial cells; consists of the small and large subunits of the ribosome, mRNA, and initiator tRNA charged with fMet.

sex Sexual phenotype: male or female.

sex chromosomes Chromosomes that differ in number or morphology in males and females. *Compare* **autosomes**.

sex determination Specification of sex (male or female). Sex-determining mechanisms include chromosomal, genic, and environmental sex-determining systems.

sex-determining region Y *(SRY)* gene A gene on the Y chromosome that triggers male development.

sex-influenced characteristic Characteristic encoded by autosomal genes that is more readily expressed in one sex. For example, an autosomal dominant gene may have higher penetrance in males than in females, or an autosomal gene may be dominant in males but recessive in females.

sex-limited characteristic Characteristic encoded by autosomal genes and expressed in only one sex. Both males and females carry genes for sex-limited characteristics, but the characteristics appear in only one of the sexes.

sex-linked characteristic Characteristic determined by a gene or genes on sex chromosomes.

shelterin Multiprotein complex that binds to mammalian telomeres and protects the ends of the DNA from being inadvertently repaired as a double-strand break in the DNA.

Shine–Dalgarno sequence Consensus sequence found in the bacterial 5′ untranslated region of mRNA; contains the ribosome-binding site.

short interspersed element (SINE) Short DNA sequence repeated many times and interspersed throughout the genome.

short tandem repeat (STR) *See* **microsatellite**.

sigma (σ) factor Subunit of bacterial RNA polymerase that allows the RNA polymerase to recognize a promoter and initiate transcription.

silencer Sequence that has many of the properties possessed by an enhancer but represses transcription.

silent mutation Change in the nucleotide sequence of DNA that does not alter the amino acid sequence of a protein.

SINE *See* **short interspersed element**.

single-nucleotide polymorphism (SNP) Single-base-pair difference in DNA sequence between individual members of a species.

single-strand-binding (SSB) protein Protein that binds to single-stranded DNA during replication and prevents it from annealing with a complementary strand and forming secondary structures.

sister chromatids Two copies of a chromosome that are held together at the centromere. Each chromatid consists of a single DNA molecule.

small interfering RNA (siRNA) Single-stranded RNA molecule (usually 21–25 nucleotides in length) produced by the cleavage and processing of double-stranded RNA; binds to complementary sequences in mRNA and brings about the cleavage and degradation of the mRNA. Some siRNAs bring about methylation of DNA.

small nuclear ribonucleoprotein (snRNP) Structure found in the nuclei of eukaryotic cells that consists of small nuclear RNA (snRNA) and protein; functions in the processing of pre-mRNA.

small nuclear RNA (snRNA) Small RNA molecule found in the nuclei of eukaryotic cells; functions in the processing of pre-mRNA.

small nucleolar RNA (snoRNA) Small RNA molecule found in the nuclei of eukaryotic cells; functions in the processing of rRNA and in the assembly of ribosomes.

small ribosomal subunit The smaller of the two subunits of a functional ribosome.

somatic mutation Mutation in a cell that does not give rise to gametes.

SOS system System of proteins and enzymes that allows a cell to replicate its DNA in the presence of a distortion in DNA structure; makes numerous mistakes in replication and increases the rate of mutation.

Southern blotting Process by which DNA is transferred from a gel to a solid medium.

specialized transduction Transduction in which genes near special sites on the bacterial chromosome are transferred from one bacterium to another; requires lysogenic bacteriophages.

speciation Process by which new species arise. *See also* **allopatric speciation, sympatric speciation**.

species **(1)** A particular type of organism to which a unique name has been given. **(2)** An evolutionarily independent group of organisms. *See also* **biological species concept**.

spermatid Immediate product of meiosis II in spermatogenesis; matures to sperm.

spermatogenesis Sperm production in animals.

spermatogonium Diploid cell in the testis; capable of undergoing meiosis to produce sperm.

S (synthesis) phase Stage of interphase in the cell cycle during which DNA replicates.

spindle microtubule Microtubule that moves chromosomes in mitosis and meiosis.

spindle pole Point from which spindle microtubules radiate.

spliceosome Large complex consisting of several RNA molecules and many proteins that splices protein-encoding pre-mRNA; contains five small ribonucleoprotein particles (U1, U2, U4, U5, and U6).

spontaneous mutation Mutation that arises spontaneously from natural changes in DNA structure or from errors in replication.

start codon *See* **initiation codon**.

sticky end *See* **cohesive end**.

stop codon Codon in mRNA that signals the end of translation; also called a **termination codon** or **nonsense codon**. The three common stop codons are UAA, UAG, and UGA.

strand slippage Slipping of the template and newly synthesized strands in replication in which one of the strands loops out from the other and nucleotides are inserted or deleted on the newly synthesized strand.

structural gene DNA sequence that encodes a protein that functions in metabolism or biosynthesis or that has a structural role in the cell.

structural genomics Area of genomics that studies the organization and sequence of information contained within genomes; sometimes used by protein chemists to refer to the determination of the three-dimensional structure of proteins.

submetacentric chromosome Chromosome in which the centromere is displaced toward one end, producing a short arm and a long arm.

supercoiling Coiled tertiary structure that forms when strain is placed on a DNA helix by overwinding or underwinding of the helix. *See also* **positive supercoiling, negative supercoiling**.

suppressor mutation Mutation that hides or suppresses the effect of another mutation at a site that is distinct from the site of the original mutation.

sympatric speciation Speciation arising in the absence of any geographic barrier to gene flow, in which reproductive isolating mechanisms evolve with a single interbreeding population. *Compare* **allopatric speciation.**

synapsis Close pairing of homologous chromosomes.

synonymous codons Different codons that specify the same amino acid.

synthetic biology A field that seeks to design organisms that might provide functions useful to humanity.

tandem duplication Duplication of a chromosome segment that is adjacent to the original segment.

tandem repeat sequences DNA sequences repeated one after another; tend to be clustered at specific locations on a chromosome.

***Taq* polymerase** DNA polymerase commonly used in PCR reactions. Isolated from the bacterium *Thermus aquaticus*, the enzyme is stable at high temperatures, so it is not denatured during the strand-separation step of the cycle.

telocentric chromosome Chromosome in which the centromere is at or very near one end.

telomerase Enzyme made up of both protein and RNA that replicates the ends (telomeres) of eukaryotic chromosomes. The RNA part of the enzyme has a template that is complementary to repeated sequences in the telomere and pairs with them, providing a template for the synthesis of additional copies of the repeats.

telomere Stable end of a chromosome.

telomeric sequence Sequence found at the ends of a chromosome; consists of many copies of short, simple sequences repeated one after the other.

telophase Stage of mitosis in which the chromosomes arrive at the spindle poles, the nuclear membrane re-forms, and the chromosomes relax and lengthen.

telophase I Stage of meiosis I in which chromosomes arrive at the spindle poles.

telophase II Stage of meiosis II in which chromosomes arrive at the spindle poles.

temperate phage Bacteriophage that can undergo either the lytic or the lysogenic cycle; can integrate into the bacterial chromosome and remain there as a prophage. *Compare* **virulent phage.**

temperature-sensitive allele Allele that is expressed only at certain temperatures.

template strand The strand of DNA that is used as a template during transcription. The RNA synthesized during transcription is complementary and antiparallel to the template strand.

−10 consensus sequence (Pribnow box) Consensus sequence (TATAAT) found in most bacterial promoters approximately 10 bp upstream of the transcription start site.

terminal inverted repeats Sequences found at both ends of a transposable element that are inverted complements of one another.

termination codon *See* **stop codon.**

terminator Sequence of DNA nucleotides that causes the termination of transcription.

tertiary structure of a protein Higher-order folding of amino acids in a protein to form the overall three-dimensional shape of the molecule.

testcross Cross between an individual with an unknown genotype and an individual with the homozygous recessive genotype.

tetrad The four products of meiosis; all four chromatids of a homologous pair of chromosomes.

tetrasomy Presence of two extra copies of a chromosome ($2n + 2$).

theta replication Replication of circular DNA that is initiated by the unwinding of the two nucleotide strands, producing a replication bubble. Unwinding continues at one or both ends of the bubble, making it progressively larger. DNA replication on both of the template strands is simultaneous with unwinding until the two replication forks meet.

−35 consensus sequence Consensus sequence (TTGACA) found in many bacterial promoters approximately 35 bp upstream of the transcription start site.

30S initiation complex Initial complex formed in the initiation of translation in bacterial cells; consists of the small subunit of the ribosome, mRNA, initiator tRNA charged with fMet, GTP, and initiation factors 1, 2, and 3.

three-point testcross Cross between an individual heterozygous at three loci and an individual homozygous for recessive alleles at those loci.

3′ end End of a polynucleotide chain at which an OH group is attached to the 3′-carbon atom in the sugar of the nucleotide. *Compare* **5′ end.**

3′ splice site The 3′ end of an intron where cleavage takes place in RNA splicing.

3′ untranslated (UTR) region Sequence of nucleotides at the 3′ end of mRNA; does not encode the amino acids of a protein, but affects both the stability of the mRNA and its translation.

threshold characteristic Characteristic that has only two phenotypes (presence and absence) but whose expression depends on an underlying susceptibility that varies continuously.

thymine (T) Pyrimidine base in DNA but not in RNA.

topoisomerase Enzyme that adds or removes rotations in a DNA helix by temporarily breaking nucleotide strands; controls the degree of DNA supercoiling.

transcription Process by which RNA is synthesized from a DNA template.

transcription activator–like effector nuclease (TALEN) an engineered nuclease in which a protein that normally binds to sequences in promoters is attached to a restriction enzyme.

transcriptional activator protein Protein in eukaryotic cells that binds to consensus sequences in regulatory promoters or enhancers and affects transcription initiation by stimulating or stabilizing the assembly of the basal transcription apparatus.

transcriptional regulator protein Protein in eukaryotic cells that binds to DNA and stimulates or inhibits transcription.

transcription factor Protein that binds to DNA sequences in eukaryotic cells and affects transcription.

transcription start site The first DNA nucleotide that is transcribed into an RNA molecule.

transcription unit Sequence of nucleotides in DNA that encodes a single RNA molecule and the sequences necessary for its transcription; normally contains a promoter, an RNA-coding sequence, and a terminator.

transcriptome Set of all RNA molecules transcribed from a genome.

transducing phage Phage that contains a piece of the bacterial chromosome inside the phage coat. *See also* **generalized transduction**.

transductant Bacterial cell that has received genes from another bacterium through transduction.

transduction Type of gene exchange that takes place when a virus carries genes from one bacterium to another. After it is inside the cell, the newly introduced DNA may undergo recombination with the bacterial chromosome.

transfer RNA (tRNA) RNA molecule that carries an amino acid to the ribosome and transfers it to a growing polypeptide chain in translation.

transformant Cell that has received genetic material through transformation.

transformation Mechanism by which DNA found in the medium is taken up by a cell. After transformation, recombination may take place between the introduced genes and the cellular chromosome.

transforming principle Material responsible for transformation. DNA is the transforming principle.

transgene Foreign gene or other DNA fragment carried in germ-line DNA.

transition Base substitution in which a purine is replaced by a different purine or a pyrimidine is replaced by a different pyrimidine. *Compare* **transversion**.

translation Process by which a protein is assembled from information contained in messenger RNA.

translocation (1) Movement of a chromosome segment to a nonhomologous chromosome or to a region within the same chromosome. (2) Movement of a ribosome along mRNA in the course of translation.

translocation carrier Individual organism heterozygous for a translocation.

transmission genetics Field of genetics that encompasses the basic principles of genetics and how traits are inherited.

transposable element DNA sequence capable of moving from one site to another within the genome through a mechanism that differs from that of homologous recombination.

transposition Movement of a transposable element from one site to another. Replicative transposition increases the number of copies of the transposable element; nonreplicative transposition does not increase the number of copies.

transversion Base substitution in which a purine is replaced by a pyrimidine or a pyrimidine is replaced by a purine. *Compare* **transition**.

triplet code Refers to the fact that three nucleotides encode each amino acid in a protein.

triploidy Possession of three haploid sets of chromosomes ($3n$).

triple-X syndrome Human syndrome in which cells contain three X chromosomes. A person with triple-X syndrome has a female phenotype without distinctive features other than a tendency to be tall and thin; a few such women are sterile, but many menstruate regularly and are fertile.

trisomy Presence of an additional copy of a chromosome ($2n + 1$).

trisomy 8 Presence of three copies of chromosome 8; in humans, results in intellectual disability, contracted fingers and toes, low-set malformed ears, and a prominent forehead.

trisomy 13 Presence of three copies of chromosome 13; in humans, results in Patau syndrome.

trisomy 18 Presence of three copies of chromosome 18; in humans, results in Edward syndrome.

trisomy 21 Presence of three copies of chromosome 21; in humans, results in Down syndrome.

tRNA charging Chemical reaction in which an aminoacyl-tRNA synthetase attaches an amino acid to its corresponding tRNA.

tumor-suppressor gene Gene that normally inhibits cell division. Recessive mutations in such genes often contribute to cancer.

Turner syndrome Human syndrome in which cells contain a single X chromosome and no Y chromosome (XO). Persons with Turner syndrome are female in appearance but do not undergo puberty and have poorly developed female secondary sex characteristics; most are sterile but have normal intelligence.

two-point testcross Cross between an individual heterozygous at two loci and an individual homozygous for recessive alleles at those loci.

unbalanced gamete Gamete that has a variable number of chromosomes; some chromosomes may be missing and others may be present in more than one copy.

underdominance Selection in which the heterozygote has lower fitness than either homozygote. *Compare* **overdominance**.

unequal crossing over Misalignment of the two DNA molecules during crossing over, resulting in one DNA molecule with an insertion and the other with a deletion.

unique-sequence DNA Sequence that is present only once or a few times in a genome.

universal Shared by all organisms; refers to the fact that in the genetic code, particular codons specify the same amino acids in almost all organisms.

uracil (U) Pyrimidine base in RNA but not normally in DNA.

variance Statistic that describes the variability of a group of measurements.

virulent phage Bacteriophage that reproduces only through the lytic cycle and kills its host cell. *Compare* **temperate phage**.

virus Noncellular replicating structure consisting of nucleic acid surrounded by a protein coat; can replicate only within a host cell.

Western blotting Process by which protein is transferred from a gel to a solid medium.

whole-genome shotgun sequencing Method of sequencing a genome in which sequenced fragments are assembled into the correct sequence of contigs by using only the overlaps in sequence.

wild type The trait or allele that is most commonly found in natural (wild) populations.

wobble Base pairing between codon and anticodon in which there is nonstandard pairing, usually at the third (3′) position of the codon; allows more than one codon to pair with the same anticodon.

X-linked characteristic Characteristic determined by a gene or genes on the X chromosome.

X-ray diffraction Method for analyzing the three-dimensional shape and structure of chemical substances. Crystals of a substance are bombarded with X-rays, which hit the crystals, bounce off, and produce a diffraction pattern on a detector. The pattern of the spots produced on the detector provides information about the molecular structure.

Y-linked characteristic Characteristic determined by a gene or genes on the Y chromosome.

Z-DNA Secondary structure of DNA characterized by 12 bases per turn, a left-handed helix, and a sugar–phosphate backbone that zigzags back and forth.

zinc-finger nuclease (ZFN) An engineered nuclease that uses a DNA-binding domain called a zinc finger attached to a restriction enzyme.

Answers to Selected Questions and Problems

Chapter 1

1. In the Hopi culture, people with albinism were considered special and given special status. Because extensive exposure to sunlight could be damaging or deadly, Hopi men with albinism did no agricultural work. Albinism was a considered a positive trait rather than a negative physical condition, which gave people with albinism a mating advantage and thus increased the frequency of the albino mutation. Finally, the small population size of the Hopi tribe may have helped increase the frequency of the albino mutation owing to chance.

16. Evolution is genetic change over time. For evolution to take place, genetic variation must first arise, and then evolutionary forces must change the proportion of genetic variants over time. Genetic variation is therefore the basis of all evolutionary change.

17. (**a**) Transmission genetics; (**b**) population genetics; (**c**) population genetics; (**d**) molecular genetics; (**e**) molecular genetics; (**f**) transmission genetics.

18. Genetics is old in the sense that humans have been aware of hereditary principles for thousands of years and have applied them since the beginning of agriculture. It is very young in the sense that the fundamental principles were not uncovered until Mendel's time, and the structure of DNA and the principles of recombinant DNA were discovered only within the past 60 years.

19. (**a**) Germ-plasm theory; (**b**) preformationism; (**c**) inheritance of acquired characteristics; (**d**) pangenesis.

20. (**a**) Pangenesis postulates that pieces of genetic information travel from all parts of the body to the reproductive organs, and that genetic information is then conveyed to the embryo. According to the germ-plasm theory, gamete-producing cells found within the reproductive organs contain a complete set of genetic information that is passed to the gametes. Pangenesis and the germ-plasm theory are similar in that both propose that genetic information is contained in discrete units that are passed on to offspring. They differ in where that genetic information resides. In pangenesis, it resides in different parts of the body and must travel to the reproductive organs. In the germ-plasm theory, all the genetic information is already in the reproductive cells.
(**b**) Preformationism holds that the sperm or egg contains a miniature preformed adult called a homunculus. In development, the homunculus grows to produce an offspring. Only one parent contributes genetic traits to the offspring. Blending inheritance requires contributions of genetic material from both parents. The genetic contributions from the parents blend to produce the genetic material of the offspring. Having been blended, the genetic material cannot be separated in future generations.
(**c**) The inheritance of acquired characteristics postulates that traits acquired in a person's lifetime alter the genetic material and can be transmitted to offspring. Our modern theory of heredity states that offspring inherit genes located on chromosomes passed to them by their parents. These chromosomes segregate in meiosis in the parent's germ cells and are passed into the gametes.

21. (**a**) Eukaryotic cells have a nucleus containing chromosomal DNA and possess internal membrane-bounded organelles. Prokaryotic cells have neither of these features.
(**b**) A gene is a basic unit of hereditary information, usually encoding an RNA molecule or a protein. Alleles are variant forms of a gene, arising through mutation.
(**c**) The genotype is the set of genes or alleles inherited by an organism from its parent(s). The expression of the genes of a particular genotype, through interaction with environmental factors, produces the phenotype, the observable trait.
(**d**) Both are nucleic acid polymers. RNA contains a ribose sugar, whereas DNA contains a deoxyribose sugar. RNA also contains uracil as one of the four bases, whereas DNA contains thymine. The other three bases are common to both DNA and RNA. Finally, DNA is usually double stranded, consisting of two complementary strands, whereas RNA is single stranded.
(**e**) Chromosomes are structures consisting of DNA and associated proteins. The DNA contains the genetic information.

22.

Type of albinism	Phenotype	Gene mutated
OCA2	Pigment reduced in skin, hair, and eyes, but small amount of pigment acquired with age; visual problems	*OCA2*
OCA1B	General absence of pigment in hair, skin, and eyes, but there may be small amount of pigment; does not vary with age; visual problems	Gene that encodes tyrosinase
OCA1A	Complete absence of pigment; visual problems	Gene that encodes tyrosinase
OCA3	Some pigment present; sun sensitivity and visual problems	Gene that encodes tyrosinase-related protein 1
OASD	Lack of pigment in the eyes and deafness later in life	Unknown
OA1	Lack of pigment in the eyes but normal elsewhere	*GPR143*
ROCA	Bright copper-red coloration in skin and hair of Africans; dilution of color in iris	Gene that encodes tyrosinase-related protein 1
OCA4	Reduced pigmentation	*MATP*

26. All genomes must have the ability to store complex information and to vary. The blueprint for the entire organism must be contained within the genome of each reproductive cell. The information has to be in the form of a code that can be used as a set of instructions for assembling the components of the cells. The genetic material of any organism must be stable, be replicated precisely, and be transmitted faithfully to the progeny, but must be capable of mutating.

28. Legally, she is not required to inform her children or other relatives about her test results, but people may have different opinions about her moral and parental responsibilities. On the one hand, she has the legal right to keep any medical information, including the results of genetic testing, private. On the other hand, her children may be at an increased risk of developing these disorders and might benefit from that knowledge. For example, the risk of colon cancer can be reduced by regular examinations so that tumors can be detected and removed before they become cancerous. Some people might argue that her parental responsibilities include providing her children with information about possible medical problems. Another issue to consider is the possibility that her children or other relatives might not want to know their genetic risk, particularly for a disorder such as Alzheimer disease for which there is no cure.

29. (**a**) Having the genetic test removes doubt about the potential for the disorder: you are either susceptible or not. Knowing about the potential of a genetic disorder enables you to make lifestyle changes that might lessen the effect of the disease or lessen the risk. The types and nature of future medical tests could be guided by the result of genetic testing, thus allowing for early warning and screening for the disease. The knowledge could also enable you to make informed decisions about having children, given the potential of passing the trait to your offspring. Additionally, by knowing what to expect, you could plan your life accordingly. Reasons for not having the test typically concern the potential for testing positive for susceptibility to the genetic disease. If the susceptibility is detected, there is potential for discrimination. Knowledge of the potential future condition could lead to psychological difficulties in coping with the anxiety of waiting for the disease to manifest.
(**b**) There is no "correct" answer, but some of the reasons for wanting to be tested are that the test would remove doubt about the susceptibility, particularly if family members have had the genetic disease; and that either a positive or a negative result would allow for informed planning of lifestyle, medical testing, and family choices in the future.

Chapter 2

20.

Stage	Number of cells counted	Proportion of cells at each stage	Average duration (hours)
Interphase	160	0.80	19.2
Prophase	20	0.10	2.4
Prometaphase	6	0.03	0.72
Metaphase	2	0.01	0.24
Anaphase	7	0.035	0.84
Telophase	5	0.025	0.6
Totals	200	1.0	24

Average duration of M phase is 4.8 hours.

23. (**a**) 12 chromosomes and 24 DNA molecules
(**b**) 12 chromosomes and 24 DNA molecules
(**c**) 12 chromosomes and 24 DNA molecules
(**d**) 12 chromosomes and 24 DNA molecules
(**e**) 12 chromosomes and 12 DNA molecules
(**f**) 6 chromosomes and 12 DNA molecules
(**g**) 12 chromosomes and 12 DNA molecules
(**h**) 6 chromosomes and 6 DNA molecules

25. (**a**) 6 chromosomes
(**b**) First cell: anaphase I of meiosis
Second cell: anaphase of mitosis
Third cell: anaphase II of meiosis
(**c**) First cell: 6 chromosomes and 12 DNA molecules
Second cell: 12 chromosomes and 12 DNA molecules
Third cell: 6 chromosomes and 6 DNA molecules

26.

Stage of mitosis	Amount of DNA per cell
(**a**) G_1	7.3 pg
(**b**) Prophase I	14.6 pg
(**c**) G_2	14.6 pg
(**d**) Following telophase II and cytokinesis	3.7 pg
(**e**) Anaphase I	14.6 pg
(**f**) Metaphase II	7.3 pg

The amount of DNA in the cell will be doubled after the completion of the S phase and prior to cytokinesis in either mitosis or meiosis I. At the completion of cytokinesis following meiosis II, the amount of DNA will be halved.
(**a**) G_1 occurs prior to the S phase and the doubling of the amount of DNA.
(**b**) The amount of DNA doubles in the S phase, so during prophase I of meiosis, the amount of DNA in the cell is twice the amount in G_1.
(**c**) G_2 takes place directly after the completion of the S phase, so the amount of DNA is twice the amount in G_1.
(**d**) Following cytokinesis associated with meiosis II, each daughter cell will contain only one-half the amount of DNA of a cell found in G_1 of interphase.
(**e**) During anaphase I of meiosis, the amount of DNA in the cell is twice the amount in G_1.
(**f**) In metaphase II of meiosis, the amount of DNA in each cell is the same as G_1 because the DNA doubled in the S phase but then was reduced by half in the first meiotic division.

29. The progeny of the organism whose cells contain the larger number of homologous pairs of chromosomes should be expected to exhibit more variation. The number of different combinations of chromosomes that are possible in the gametes is 2^n, where n is equal to the number of homologous pairs of chromosomes. For the fruit fly, which has four pairs of chromosomes, the number of possible combinations is $2^4 = 16$. For the house fly, which has six pairs of chromosomes, the number of possible combinations is $2^6 = 64$.

30. (**a**) Metaphase I

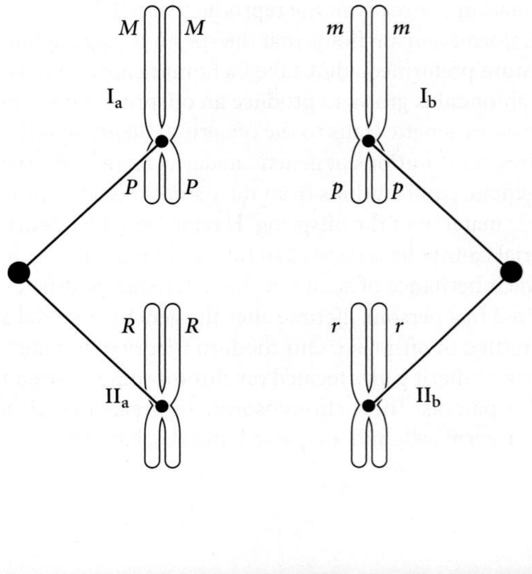

(b) Gametes

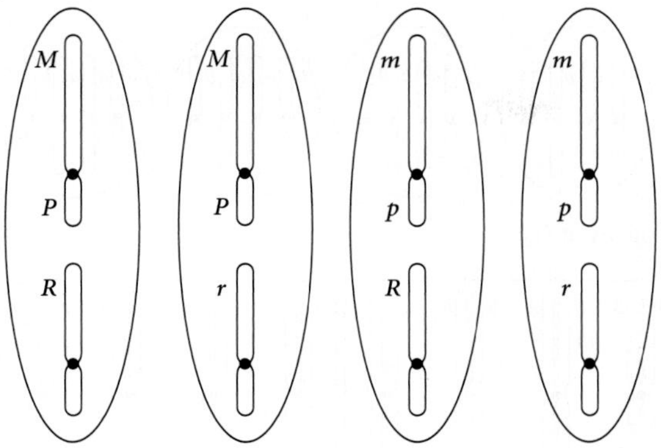

32.

Cell type	Number of chromosomes	Number of DNA molecules
(a) Spermatogonium	64	64

Assuming the spermatogonium is in G_1 prior to the production of sister chromatids in the S phase, the chromosome number will be the diploid number of chromosomes.

| **(b)** First polar body | 32 | 64 |

The first polar body is the product of meiosis I, so it will be haploid; but the sister chromatids have not separated, so each chromosome will consist of two sister chromatids.

| **(c)** Primary oocyte | 64 | 128 |

The primary oocyte has stopped in prophase I of meiosis. So the homologs have not yet separated, and each chromosome consists of two sister chromatids.

| **(d)** Secondary spermatocyte | 32 | 64 |

The secondary spermatocyte is a product of meiosis I and has yet to enter meiosis II. So the secondary spermatocyte will be haploid because the homologous pairs were separated in meiosis I; but each chromosome is still composed of two sister chromatids.

34. **(a)** No, the information is not identical with that found in the secondary oocyte. The first polar body and the secondary oocyte are the result of meiosis I, which produces two nonidentical cells. The first polar body and the secondary oocyte will contain only one member of each original chromosome pair. Additionally, crossing over that took place in prophase I will have generated new and different arrangements of genetic material on those chromatids that participated in crossing over.
(b) No, the information is not identical. The second polar body and the ovum will contain copies of the same members of the homologous pairs of chromosomes that separated in meiosis. However, because of crossing over, the sister chromatids that separated in anaphase II and gave rise to the ovum and second polar body are no longer identical.

35. Most male animals produce sperm by meiosis. Because meiosis takes place only in diploid cells, haploid male bees do not undergo meiosis. Male bees can produce sperm, but only through mitosis. Haploid cells that divide mitotically produce more haploid cells.

Chapter 3

14. Useful characteristics: Easy to grow and maintain; grows rapidly, produces many generations in a short period. Examples of organisms: *Neurospora*, a fungus; *Saccharomyces cerevisiae*, a yeast; *Arabidopsis*, a plant; *Caenorhabditis elegans*, a nematode; *Drosophila melanogaster*, a fruit fly.

18. **(a)** Although the white female gave birth to the offspring, her eggs were produced by the ovary from the black female guinea pig. The transplanted ovary produced only eggs containing the allele for black coat color.
(b) The white male guinea pig contributed a w allele, while the white female guinea pig contributed the W allele from the transplanted ovary. The offspring are thus Ww.
(c) The production of black offspring suggests that the allele for black coat color was passed to the offspring from the transplanted ovary, in agreement with the germ-plasm theory. If pangenesis were correct, the white coat alleles from the female's body would have traveled to the transplanted ovary and then to into the gametes. The absence of any white offspring indicates that pangenesis did not occur.

19. **(a)** Female parent is $i^B i^B$; male parent is $I^A i^B$.
(b) Both parents are $i^B i^B$.
(c) Male parent is $i^B i^B$; female parent is $I^A I^A$ or, possibly, $I^A i^B$, but a heterozygous female in this mating is unlikely to have produced eight blood-type-A kittens owing to chance alone.
(d) Both parents are $I^A i^B$.
(e) Either both parents are $I^A I^A$ or one parent is $I^A I^A$ and the other parent is $I^A i^B$.
(f) Female parent is $i^B i^B$; male parent is $I^A i^B$.

22. **(a)** Sally is Aa, Sally's mother is Aa, Sally's father is aa, and Sally's brother is aa.
(b) $\frac{1}{2}$
(c) $\frac{1}{2}$

24. **(a)** Because 2 is found on only one side of a six-sided die, there is a $\frac{1}{6}$ chance of rolling a 2.
(b) The probability of rolling a 1 on a six-sided die is $\frac{1}{6}$, and the probability of rolling a 2 on a six-sided die is $\frac{1}{6}$. Use the addition rule of probability to determine the probability of rolling a 1 or a 2: $\frac{1}{6} + \frac{1}{6} = \frac{2}{6} = \frac{1}{3}$.
(c) A single die contains three even numbers (2, 4, 6). The probability of rolling any one of these three numbers is $\frac{1}{6}$. Apply the addition rule: $\frac{1}{6} + \frac{1}{6} + \frac{1}{6} = \frac{3}{6} = \frac{1}{2}$.
(d) The number 6 is found on only one side of a six-sided die. The probability of rolling a 6 is therefore $\frac{1}{6}$. The probability of rolling any number but 6 is $(1 - \frac{1}{6}) = \frac{5}{6}$.

25. **(a)** $\frac{1}{18}$; **(b)** $\frac{1}{36}$; **(c)** $\frac{11}{36}$; **(d)** $\frac{1}{6}$; **(e)** $\frac{1}{4}$; **(f)** $\frac{3}{4}$.

27. Parents:

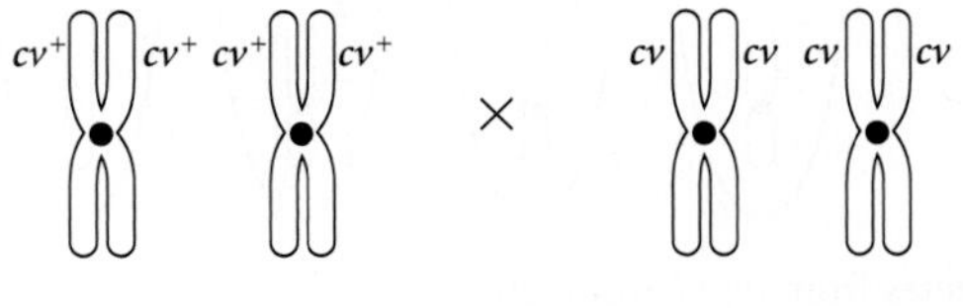

F_1 generation:

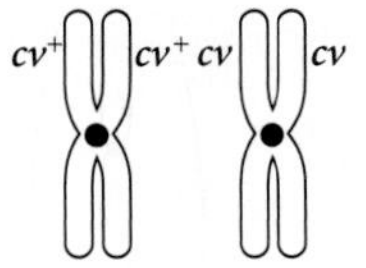

F_2 generation:

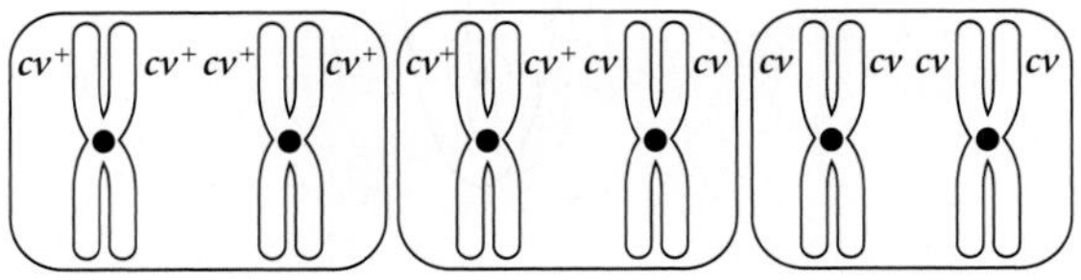

28. (**a**) In the F_1 black guinea pigs (*Bb*), only one of the two homologous chromosomes possesses the black allele, and so the number of copies present at each stage are as follows: G_1, one black allele; G_2, two black alleles; metaphase of mitosis, two black alleles; metaphase I of meiosis, two black alleles; after cytokinesis of meiosis, one black allele but only in half of the cells produced by meiosis.
(**b**) In the F_1 brown guinea pigs (*bb*), both homologs possess the brown allele, and so the number of copies present at each stage are as follows: G_1, two brown alleles; G_2, four brown alleles; metaphase of mitosis, four brown alleles; metaphase I of meiosis, four brown alleles; metaphase II, two brown alleles; after cytokinesis of meiosis, one brown allele.

31. (**a**) The parents are *gg CuCu* × *GG cucu*, and they produce F_1 progeny with genotype *Gg Cucu*. If two F_1 cats mated (*Gg Cucu* × *Gg Cucu*), the phenotypic ratios expected in the F_2 are as follows:
$^9/_{16}$ black with curly ears (*G_ Cu_*)
$^3/_{16}$ black with normal ears (*G_ cucu*)
$^3/_{16}$ gray with curly ears (*gg Cu_*)
$^1/_{16}$ gray with normal ears (*gg cucu*)
(**b**) The mating of an F_1 cat (*Gg Cucu*) with a gray cat with normal ears (*gg cucu*) is expected to produce the following progeny:
$^1/_4$ black with curly ears (*Gg Cucu*)
$^1/_4$ black with normal ears (*Gg cucu*)
$^1/_4$ gray with curly ears (*gg Cucu*)
$^1/_4$ gray with normal ears (*gg cucu*)

32. (**a**) $^1/_2\,(Aa) \times {}^1/_2\,(Bb) \times {}^1/_2\,(Cc) \times {}^1/_2\,(Dd) \times {}^1/_2\,(Ee) = {}^1/_{32}$
(**b**) $^1/_2\,(Aa) \times {}^1/_2\,(bb) \times {}^1/_2\,(Cc) \times {}^1/_2\,(dd) \times {}^1/_4\,(ee) = {}^1/_{64}$
(**c**) $^1/_4\,(aa) \times {}^1/_2\,(bb) \times {}^1/_4\,(cc) \times {}^1/_2\,(dd) \times {}^1/_4\,(ee) = {}^1/_{256}$
(**d**) No offspring have this genotype. The *Aa Bb Cc dd Ee* parent cannot contribute a *D* allele, and the *Aa bb Cc Dd Ee* parent cannot contribute a *B* allele. Therefore, their offspring cannot be homozygous for the *BB* and *DD* gene loci.

34. (**a**) Gametes from *Aa Bb* individual:

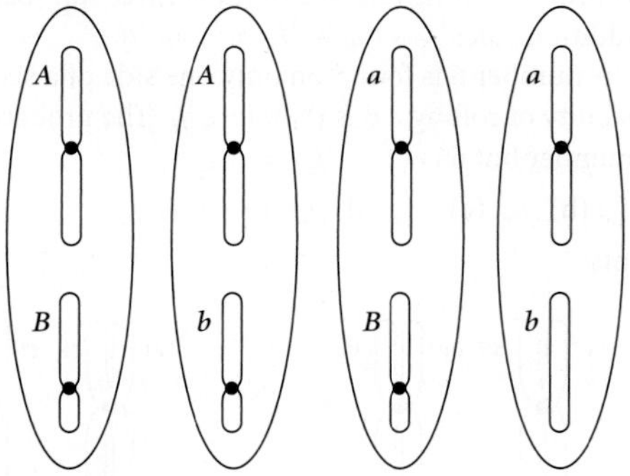

Gametes from *aa bb* individual:

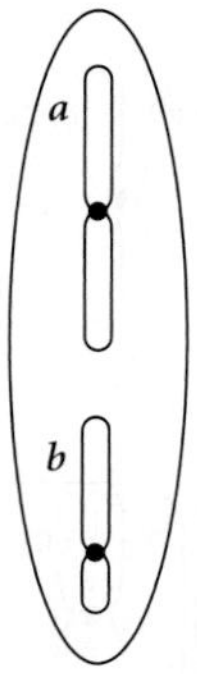

(**b**) Progeny at G_1:

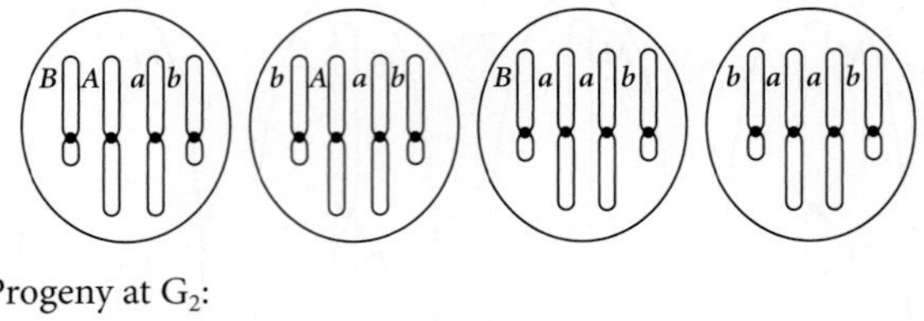

Progeny at G_2:

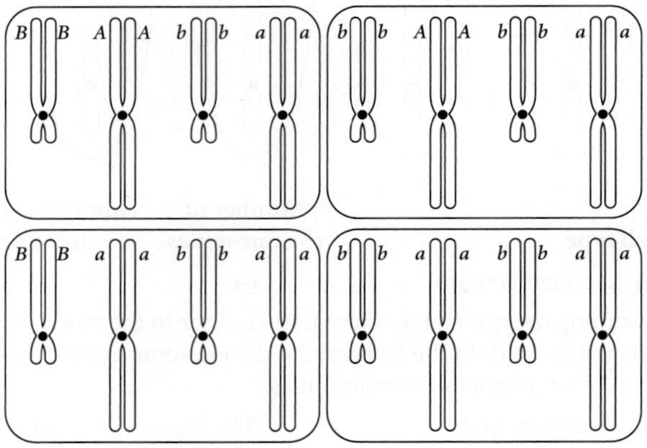

Progeny at metaphase of mitosis:

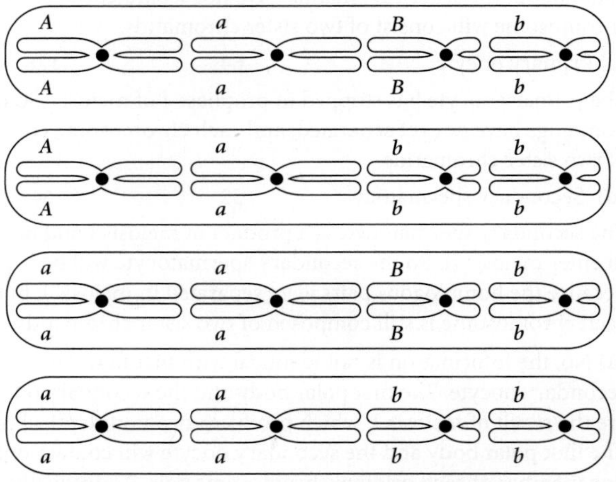

The order of chromosomes on the metaphase plate can vary.

35. (**a**) The burnsi × burnsi cross produced both burnsi and pipiens offspring, indicating that the parents were heterozygous, with each possessing a *burnsi* allele and a *pipiens* allele. This cross indicates that the *burnsi* allele is dominant over the *pipiens* allele. The progeny of the burnsi × pipiens crosses suggest that each of the crosses was between a homozygous recessive frog (pipiens) and a heterozygous dominant frog (burnsi). The results of both crosses are consistent with the burnsi phenotype being dominant to the pipiens phenotype.
(**b**) Let *B* represent the *burnsi* allele and B^+ represent the *pipiens* allele.
burnsi (BB^+) × burnsi (BB^+)
burnsi (BB^+) × pipiens (B^+B^+)
burnsi (BB^+) × pipiens (B^+B^+)
(**c**) For the burnsi × burnsi ($BB^+ \times BB^+$) cross, we would expect a phenotypic ratio of 3 : 1 in the offspring. A chi-square test gives a chi-square value of 2.706 with 1 degree of freedom and a probability between 0.1 and 0.05, indicating that the differences between what we expected and what we observed could have been

generated by chance. For the first burnsi × pipiens ($BB^+ \times B^+B^+$) cross, we expect a phenotypic ratio of 1:1. A chi-square test yields $\chi^2 = 1.78$, df = 1, $P > 0.05$. For the second burnsi × pipiens ($BB^+ \times B^+B^+$) cross, we expect a phenotypic ratio of 1:1. A chi-square test yields $\chi^2 = 0.46$, df = 1, $P > 0.05$. Thus, all three crosses are consistent with the predication that burnsi is dominant over pipiens.

36. (**a**) A heterozygous F_1 plant (*Cc Ff*) is crossed to a homozygous recessive plant (*cc ff*) We expect a 1:1:1:1 phenotypic ratio in the progeny, or $^1/_4$ of each phenotype. A chi-square test yields a chi-square value of 35 with df = 3 and $P < 0.005$.
(**b**) The probability that chance produced the differences between the observed and the expected ratios is very low, indicating that the progeny are not in a 1:1:1:1 ratio.
(**c**) The number of plants with the *cc ff* genotype is much less than expected. Poppies with the *cc ff* genotype may be less viable than other genotypes.

40. The alleles for obesity from both laboratories are recessive, but they are located at different gene loci.
Obese mouse from laboratory 1 ($o_1o_1\ O_2O_2$)
Obese mouse from laboratory 2 ($O_1O_1\ o_2o_2$)

P	$o_1o_1\ O_2O_2 \times O_1O_1\ o_2o_2$
	↓
F_1	All normal ($O_1o_1\ O_2o_2$)

Chapter 4

1. In the XX-XY system, females are homogametic (XX) and males are heterogametic (XY). In the ZZ-ZW system, females are heterogametic (ZW) and males are homogametic (ZZ).

14. (**a**) Yes; (**b**) yes; (**c**) no; (**d**) no.

15. (**a**) F_1: $^1/_2$ X^+Y (gray males), $^1/_2$ X^+X^y (gray females); F_2: $^1/_4$ X^+Y (gray males), $^1/_4$ X^yY (yellow males), $^1/_4$ X^+X^y (gray females), $^1/_4$ X^+X^+ (gray females)
(**b**) F_1: $^1/_2$ X^yY (yellow males), $^1/_2$ X^+X^y (gray females); F_2: $^1/_4$ X^+Y (gray males), $^1/_4$ X^yY (yellow males), $^1/_4$ X^+X^y (gray females), $^1/_4$ X^yX^y (yellow females)

19. Because Bob must have inherited the Y chromosome from his father, and his father has normal color vision, a nondisjunction event in the paternal lineage cannot account for Bob's genotype. Bob's mother must be heterozygous X^+X^c because she has normal color vision, and she must have inherited an X^c chromosome from her color-blind father. For Bob to inherit two X^c chromosomes from his mother, the egg must have arisen from a nondisjunction in meiosis II. In meiosis I, the homologous X chromosomes separate, and so one cell has the X^+ chromosome and the other has X^c. The failure of sister chromatids to separate in meiosis II would then result in an egg with two copies of X^c.

25. (**a**) F_1: All males have miniature wings and red eyes ($X^mY\ s^+s$), and all females have long wings and red eyes ($X^+X^m\ s^+s$). F_2: $^3/_{16}$ male, normal, red; $^1/_{16}$ male, normal, sepia; $^3/_{16}$ male, miniature, red; $^3/_{16}$ male, miniature, sepia; $^3/_{16}$ female, normal, red; $^1/_{16}$ female, normal, sepia; $^3/_{16}$ female, miniature, red; $^1/_{16}$ female, miniature, sepia.
(**b**) F_1: All females have long wings and red eyes ($X^+X^m\ s^+s$), and all males have long wings and red eyes ($X^+Y\ s^+s$). F_2: $^3/_{16}$ male, long wings, red eyes; $^1/_{16}$ male, long wings, sepia eyes; $^3/_{16}$ male, miniature wings, red eyes; $^1/_{16}$ male, miniature wings, sepia eyes; $^3/_8$ female, long wings, red eyes; $^1/_8$ female, long wings, sepia eyes.

26. (**a**) The results of the crosses indicate that cremello and chestnut are pure-breeding traits (homozygous). Palomino is a hybrid trait (heterozygous) that produces a 2:1:1 ratio when palominos are crossed with each other. The simplest hypothesis consistent with these results is incomplete dominance, with palomino as the phenotype of the heterozygotes resulting from chestnuts crossed with cremellos.
(**b**) Let C^B = chestnut, C^W = cremello. The parents and offspring of these crosses have the following genotypes: chestnut = C^BC^B; cremello = C^WC^W; palomino = C^BC^W.

28. To have long earlobes, the child must inherit the dominant allele and also express it. The probability of inheriting the dominant allele is 50%; the probability of expressing it is 30%. The combined probability of both is $0.5 \times 0.3 = 0.15$, or 15%.

29. (**a**) The 2:1 ratio in the progeny of two spotted hamsters suggests lethality, and the 1:1 ratio in the progeny of a spotted hamster and a hamster without spots indicates that spotted is a heterozygous phenotype. If *S* and *s* represent alleles at the locus for white spotting, spotted hamsters are *Ss* and solid-colored hamsters are *ss*. One-fourth of the zygotes expected from a mating of two spotted hamsters are *SS*—embryonic lethal—and missing from the progeny, resulting in the 2:1 ratio of spotted to solid progeny.
(**b**) Because spotting is a heterozygous phenotype, obtaining Chinese hamsters that breed true for spotting is impossible.

32. (**a**) B or AB.
(**b**) No. Many other men have these blood types. The results would have meant only that Chaplin could not be eliminated as a possible father of the child.

33. (**a**) All walnut (*Rr Pp*)
(**b**) $^1/_4$ walnut (*Rr Pp*), $^1/_4$ rose (*Rr pp*), $^1/_4$ pea (*rr Pp*), $^1/_4$ single (*rr pp*)
(**c**) $^9/_{16}$ walnut (*R_ P_*), $^3/_{16}$ rose (*R_ pp*), $^3/_{16}$ pea (*rr P_*), $^1/_{16}$ single (*rr pp*)
(**d**) $^3/_4$ rose (*R_ pp*), $^1/_4$ single (*rr pp*)
(**e**) $^1/_4$ walnut (*Rr Pp*), $^1/_4$ rose (*Rr pp*), $^1/_4$ pea (*rr Pp*), $^1/_4$ single (*rr pp*)
(**f**) $^1/_2$ rose (*Rr pp*), $^1/_2$ single (*rr pp*)

36. Let *A* and *B* represent the two loci. The F_1 heterozygotes are *Aa Bb*. The F_2 are *A_ B_* disc-shaped, *A_ bb* spherical, *aa B_* spherical, *aa bb* long.

38. In genetic maternal effects, the genotype of the mother determines the phenotype of the offspring. Because Martha is sinistral, we know her mother must be genotype *ss*. If Martha's mother is *ss*, Martha must carry at least one *s* allele. We have no information about Martha's father.
(**a**) False. Martha might have inherited an s^+ from her father and therefore could be s^+s.
(**b**) True. Martha must have inherited an *s* allele from her mother and therefore cannot be s^+s^+.
(**c**) False. The phenotype of Martha's offspring will be determined by Martha's genotype, which we do not know. Martha might have inherited an s^+ allele from her father, in which case her genotype would be s^+s and she would produce all dextral offspring.
(**d**) False. Martha's genotype could be s^+s, in which case all her offspring would be dextral.
(**e**) False. Her mother's phenotype is determined by her own mother's genotype, which could have been s^+s.
(**f**) True. Because Martha is sinistral, her mother's genotype is *ss* and all her offspring should be sinistral like Martha.

Chapter 5

9. (**a**) $^1/_4$ wild-type eyes, wild-type wings; $^1/_4$ red eyes, wild-type wings; $^1/_4$ wild-type eyes, white-banded wings; $^1/_4$ red eyes, white-banded wings.
(**b**) The recombinant progeny are the 19 with red eyes, wild-type wings and the 16 with wild-type eyes, white-banded wings.
recombination frequency = recombinants/total progeny × 100%
$= (19 + 16)/879 \times 100\% = 4.0\%$
The distance between the genes is 4 map units.

10. The genes are linked and have not assorted independently.

11. Because the genes for leaf shape and fruit spines are 32.6 m.u. apart, we expect that 32.6% of the progeny of the cross will be recombinants. There will be two types of recombinants (those with heart-shaped leaves and few spines and those with normal-shaped leaves and numerous spines), so each recombinant phenotype will constitute 16.3% of the progeny. The nonrecombinant progeny (those with heart-shaped leaves and numerous spines and those with normal-shaped leaves and few spines) will make up the remainder of the progeny (100% – 32.6% = 67.4%), which will be equally divided between the two nonrecombinant phenotypes (67.4%/2% = 33.7% each).

Heart-shaped, numerous spines	33.7%
Normal-shaped, few spines	33.7%
Heart-shaped, few spines	16.3%
Normal-shaped, numerous spines	16.3%

16.

Genotype	Body color	Eyes	Bristles	Proportion
(**a**) $e^+ ro^+ f^+$	normal	normal	normal	20%
$e^+ ro^+ f$	normal	normal	forked	20%
$e\ ro\ f^+$	ebony	rough	normal	20%
$e\ ro\ f$	ebony	rough	forked	20%
$e^+ ro\ f^+$	normal	rough	normal	5%
$e^+ ro\ f$	normal	rough	forked	5%
$e\ ro^+ f^+$	ebony	normal	normal	5%
$e\ ro^+ f$	ebony	normal	forked	5%
(**b**) $e^+ ro^+ f^+$	normal	normal	normal	5%
$e^+ ro^+ f$	normal	normal	forked	5%
$e\ ro\ f^+$	ebony	rough	normal	5%
$e\ ro\ f$	ebony	rough	forked	5%
$e^+ ro\ f^+$	normal	rough	normal	20%
$e^+ ro\ f$	normal	rough	forked	20%
$e\ ro^+ f^+$	ebony	normal	normal	20%
$e\ ro^+ f$	ebony	normal	forked	20%

18.

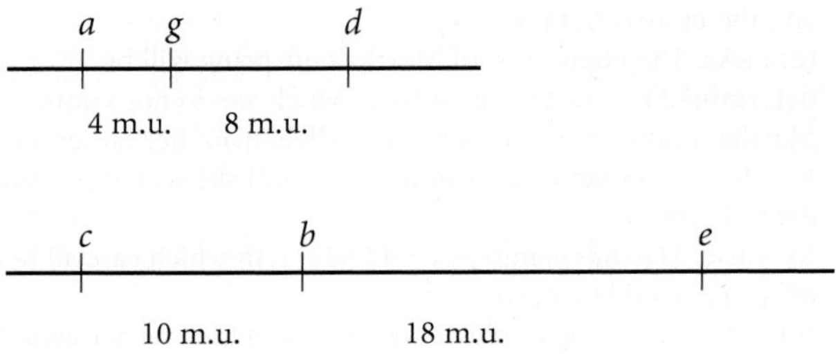

Gene *f* is unlinked to either of these groups; it is on a third linkage group.

20. (**a**) *V* is the middle gene.
(**b**) The *Wx*–*V* distance = 7 m.u. and the *Sh*–*V* distance = 30 m.u. The *Wx*–*Sh* distance is the sum of these two distances, or 37 m.u.
(**c**) Coefficient of coincidence = 0.80; interference = 0.20.

23. (**a**) All the progeny receive $p\ sh\text{-}1\ Hb^2$ from the male parent, shown as the lower chromosome in each progeny type. The upper chromosomes are the products of meiosis in the heterozygous parent, and are identified in the table. The nonrecombinants have $p\ sh\text{-}1\ Hb^2$ or $P\ Sh\text{-}1\ Hb^1$; the double crossovers have $p\ Sh\text{-}1\ Hb^2$ or $P\ sh\text{-}1\ Hb^1$. The two classes differ in the *Sh*-1 locus; therefore, *Sh*-1 is the middle locus.
(**b**) *P* and *Sh*-1: Recombinants have *P sh*-1 or *p Sh*-1.
Recombination frequency = (57 + 45 + 1)/708 = 0.145 = 14.5 m.u.
Sh-1 and *Hb*: Recombinants have $Sh\text{-}1\ Hb^2$ or $sh\text{-}1\ Hb^1$.
Recombination frequency = (6 + 5 + 1)/708 = 0.017 = 1.7 m.u.
(**c**) Expected double crossovers = RF1 × RF2 × total progeny = 0.145(0.017)(708) = 1.7
Coefficient of coincidence = number of observed double crossovers/number of expected double crossovers = 1/1.7 = 0.59
Interference = 1 – coefficient of coincidence = 0.41

24.

Genotype	Body	Eyes	Wings	Proportion
$B^+ pr^+ vg^+$	normal	normal	normal	40.7%
$b\ pr\ vg$	black	purple	vestigial	40.7%
$b^+ pr^+ vg$	normal	normal	vestigial	6.3%
$b\ pr\ vg^+$	black	purple	normal	6.3%
$b^+ pr\ vg$	normal	purple	vestigial	2.8%
$b\ pr^+ vg^+$	black	normal	normal	2.8%
$b^+ pr\ vg^+$	normal	purple	normal	0.2%
$b\ pr^+ vg$	black	normal	vestigial	0.2%

Chapter 6

13. (**a**) Duplications; (**b**) polyploidy; (**c**) deletions; (**d**) inversions; (**e**) translocations.

14. (**a**) Tandem duplication of AB
(**b**) Displaced duplication of AB
(**c**) Paracentric inversion of DEF
(**d**) Deletion of B
(**e**) Deletion of FG
(**f**) Paracentric inversion of CDE
(**g**) Pericentric inversion of ABC
(**h**) Duplication and inversion of DEF
(**i**) Duplication of CDEF, inversion of EF

18. (**a**)

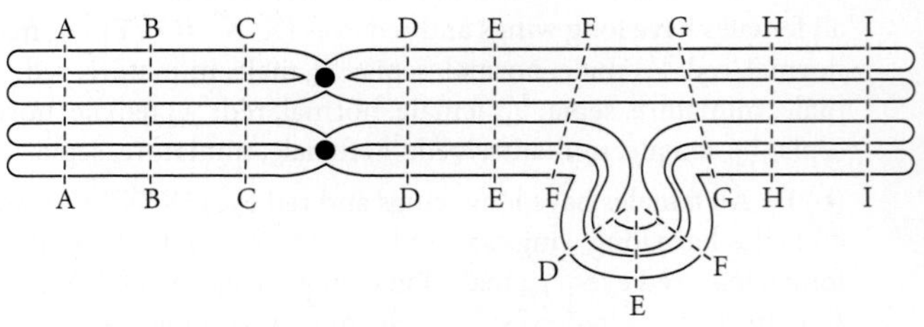

(b)

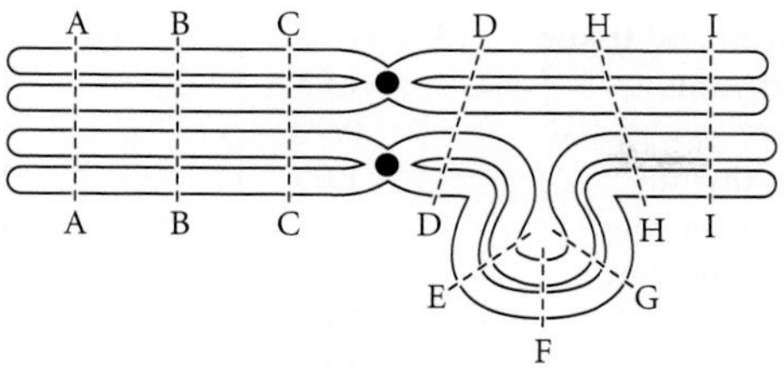

(c)

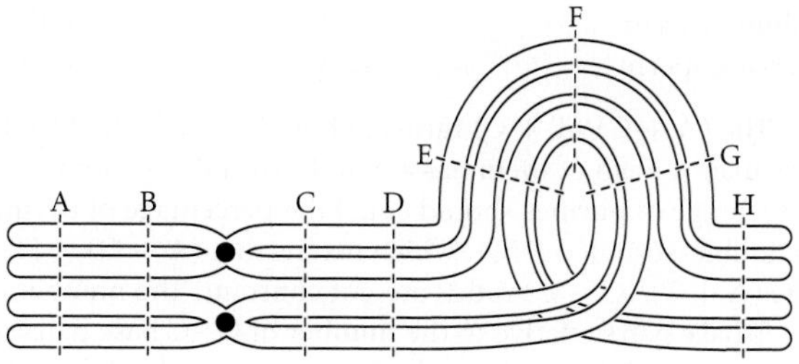

(d)

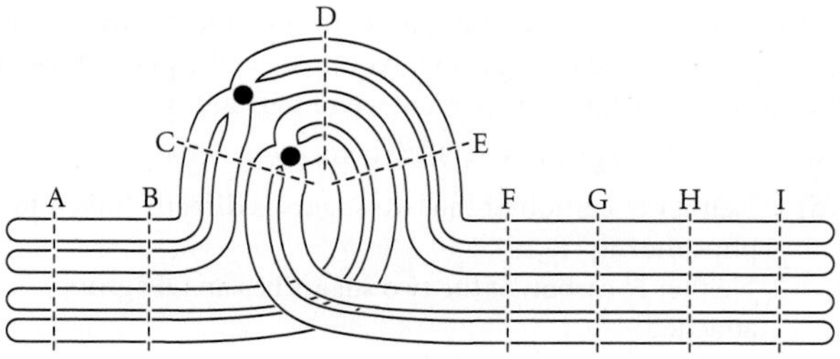

19. Each crossover event results in two recombinant and two nonrecombinant gametes. If one crossover occurs in 100% of meioses, the result would be 50% recombinant gametes. If crossing over occurs within the pericentric inversion at a rate of 26% of meioses, then 13% of the woman's oocytes will have a duplication-containing or deficient chromosome 8. If all these oocytes form viable eggs, and if they do not result in early miscarriage after fertilization, then the probability of the couple having a child with a syndrome caused by the crossing over is 13%.

21. Because the father has normal color vision, the mother must be the carrier for color blindness. The color-blind young man with Klinefelter syndrome must have inherited two copies of the X chromosome with the color-blindness allele from his mother. The nondisjunction event took place during meiosis II of the egg.

22. The high incidence of Down syndrome in Bill's family and among Bill's relatives is consistent with familial Down syndrome, caused by a Robertsonian translocation of chromosome 21. Bill and his sister, who are unaffected, are phenotypically normal carriers of the translocation and have 45 chromosomes. Their children and Bill's brother, who have Down syndrome, have 46 chromosomes. From the information given, there is no reason to suspect that Bill's wife Betty has any chromosome abnormalities. Therefore, statement *d* is most likely correct.

23. In mammals, the higher frequency of sex-chromosome aneuploidy than of autosomal aneuploidy is due to X-chromosome inactivation and the lack of essential genes on the Y chromosome. If fishes do not have X-chromosome inactivation, and if both of their sex chromosomes have numerous essential genes, then the frequency of aneuploids should be similar for both sex chromosomes and autosomes.

27. **(a)** Such allotriploids could have $1n$ from species I and $2n$ from species II for $3n = 18$; alternatively, they could have $2n$ from species I and $1n$ from species II for $3n = 15$.
(b) $4n = 28$
(c) $2n + 1 = 9$
(d) $2n - 1 = 13$
(e) $2n + 2 = 10$
(f) Allotetraploids must have chromosomes from both species, and total $4n$. There are three possible combinations for such allotetraploids: $2n$ from each: $2(4) + 2(7) = 22$; $1n$ from species I + $3n$ from species II: $1(4) + 3(7) = 25$; $3n$ from species I + $1n$ from species II: $3(4) + 1(7) = 19$.

Chapter 7

12. By using low doses of antibiotics for five years, Farmer Smith selected for bacteria that are resistant to the antibiotics. The doses he used killed sensitive bacteria, but not moderately sensitive or slightly resistant bacteria. As time passed, only resistant bacteria remained in his pigs because any sensitive bacteria were eliminated by the low doses of antibiotics.
In the future, Farmer Smith should continue to use the vitamins, but should use the antibiotics only when a sick pig requires them. In this manner, he will not be selecting for antibiotic-resistant bacteria, and the chances of successful treatment of his sick pigs will be greater.

14. The closer genes are to the F factor, the more quickly they will be transferred. The transfer process will occur in a linear fashion. When conjugation is interrupted, the transfer will stop, and the F^- strain will have received only genes carried on the portion of the Hfr strain's chromosome that entered the F^- cell prior to the interruption. From the graph, we can determine when each allele from the Hfr strain was first identified and subsequently approximate the minutes that separate the genes.

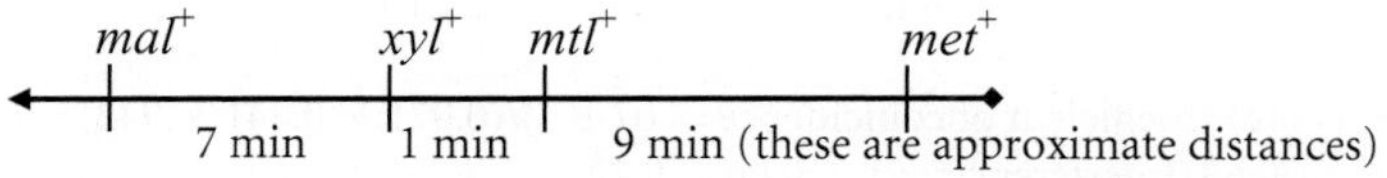

17. **(a)** The observation that the *M* and *S* genes were cotransformed at a high rate suggests that they are located very close to each other on the *Streptococcus pneumoniae* chromosome.
(b) *S* and *F* are cotransformed more frequently (0.22) than are *M* and *F* (0.0058), suggesting that *S* and *F* are located closer together than are *M* and *F*.

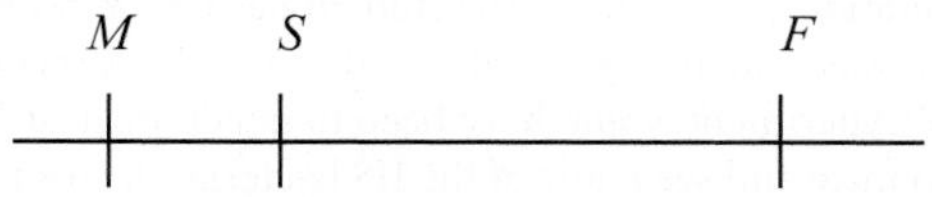

(c) Genes *M* and *F* cotransform infrequently, which is likely due to the physical distance between them. Genes *M* and *S* are more closely linked on chromosome. The relative positions of *M*, *S*, and *F* on the chromosome make it likely that, if *M* and *F* are cotransformed, then *S* will be transformed as well because it is between *M* and *F*.

18. The original infecting phages were wild-type ($a^+ b^+$) and doubly mutant ($a^- b^-$). Any phages that give rise to the $a^+ b^-$ plaque phenotype or the $a^- b^+$ plaque phenotype were produced by recombination between the two types of infecting phage particles.

Plaque phenotype	Number
$a^+ b^+$	2043
$a^+ b^-$	320 (recombinant)
$a^- b^+$	357 (recombinant)
$a^- b^-$	2134
Total plaques	4854

The frequency of recombination is calculated by dividing the total number of recombinant plaques by the total number of plaques (677/4854), which gives a frequency of 0.14, or 14%.

19. **(a)** Transductants were initially screened for the presence of *proC*$^+$. Thus, only *proC*$^+$ transductants were identified.
(b) The wild-type genotypes (*proC*$^+$ *proA*$^+$ *proB*$^+$ *proD*$^+$) represent single transductants of *proC*$^+$. Both the *proC*$^+$ *proA*$^-$ *proB*$^+$ *proD*$^+$ and *proC*$^+$ *proA*$^+$ *proB*$^-$ *proD*$^+$ genotypes represent cotransductants of *proC*$^+$, *proA*$^-$ and *proC*$^+$, *proB*$^-$.
(c) Both *proA* and *proB* were cotransduced with *proC* at about the same rate, which suggests that they are similar in distance to *proC*. The *proD* gene was never cotransduced with *proC*, suggesting that *proD* is more distant from *proC* than *proA* and *proB*. However, there were a small number of all types of cotransductants, and the absence of cotransduction between *proD* and *proC* might be due to chance.

20. Number of plaques produced by $c^+ m^+ = 460$; by $c^- m^- = 460$; by $c^+ m^- = 40$ (recombinant); by $c^- m^+ = 40$ (recombinant).

21. **(a)**

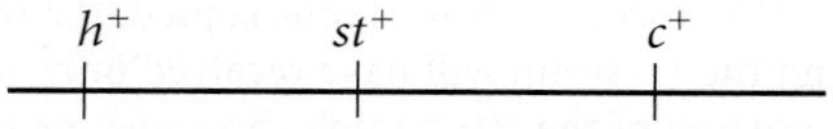

(b)

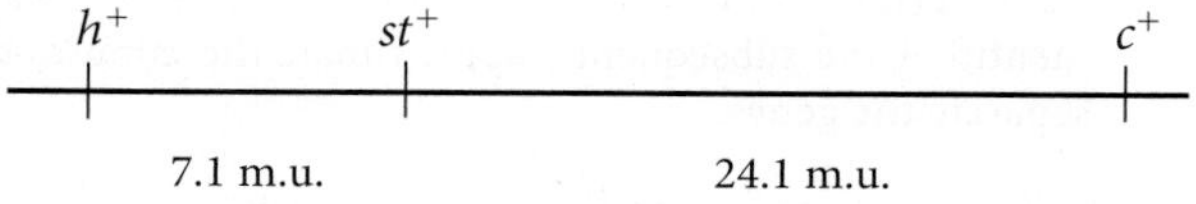

(c) Coefficient of coincidence $= (6 + 5)/(0.071 \times 0.241 \times 942) = 0.68$. Interference $= 1 - 0.68 = 0.32$.

Chapter 8

19. No, the student has not demonstrated that transformation has taken place. A single mutation could convert the IIR strain into the virulent IIS strain. Thus, the student cannot determine whether the conversion from IIR to IIS is due to transformation or to a mutation. Additionally, the student has not demonstrated that the heat was sufficient to kill all the IIS bacteria. A second useful control experiment would have been to inject the heat-killed IIS into mice and see if any of the IIS bacteria survived the heat treatment.

21. The phosphate backbone of DNA molecules typically carries a negative charge, thus making the DNA molecules attractive to the positive pole of the current.

22. **(a)**

Organism and tissue	(A + G)/(C + T)	(A + T)/(C + G)
Sheep thymus	1.03	1.36
Pig liver	0.99	1.44
Human thymus	1.03	1.52
Rat bone marrow	1.02	1.36
Hen erythrocytes	0.97	1.38
Yeast	1.00	1.80
E. coli	1.04	1.00
Human sperm	1.00	1.67
Salmon sperm	1.02	1.43
Herring sperm	1.04	1.29

(b) The (A + G)/(T + C) ratio of about 1.0 is constant for these organisms. Each of them has a double-stranded genome. The percentage of purines should equal the percentage of pyrimidines in double-stranded DNA, which means that (A + G) = (T + C). The (A + T)/(C + G) ratios are not constant. The number of A—T base pairs relative to the number of G—C base pairs is unique to each organism and can vary among the different organisms.
(c) The (A + G)/(T + C) ratio is about the same for the sperm samples, as should be expected. As in part *b*, the percentage of purines should equal the percentage of pyrimidines.

25. Adenine = 15%; guanine = 35%; cytosine = 35%.

27. **(a)** 1. Neither 5′ carbon of the two sugars is directly linked to phosphorus.
2. Neither 5′ carbon of the two sugars has an OH group attached.
3. Neither sugar molecule has oxygen in its ring structure between the 1′ and 4′ carbons.
4. In both sugars, the 2′ carbon has an OH group attached, which does not occur in deoxyribonucleotides.
5. At the 3′ position in both sugars, only hydrogen is attached, as opposed to an OH group.
6. The 1′ carbon of both sugars has an OH group, as opposed to just a hydrogen attached.

(b)

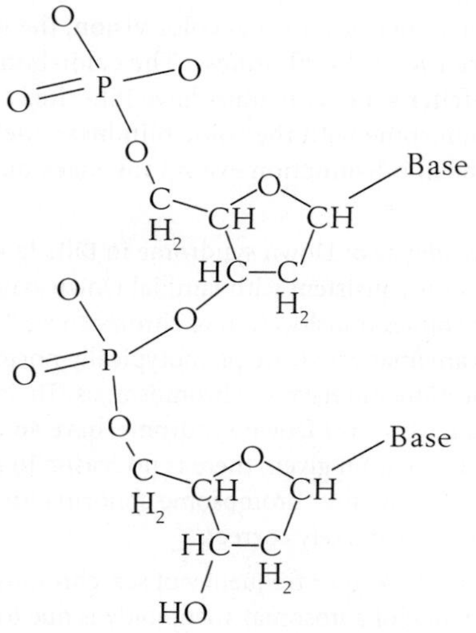

29. No. The flow of information predicted by the central dogma is from DNA to RNA to protein. An exception is reverse transcription, whereby RNA encodes DNA. However, biologists do not currently know of a process in which the flow of information is from proteins to DNA, which is required by the theory of the inheritance of acquired characteristics.

31. Prokaryotic chromosomes are usually circular, whereas eukaryotic chromosomes are linear. Prokaryotic chromosomes generally contain the entire genome, whereas each eukaryotic chromosome has only a part of the genome: the eukaryotic genome is divided into multiple chromosomes. Prokaryotic chromosomes are generally much smaller than eukaryotic chromosomes. Prokaryotic chromosomes are typically condensed into nucleoids, loops of DNA compacted into a dense body. Eukaryotic chromosomes contain DNA packaged into nucleosomes, which are further coiled and packaged into structures of successively higher order. The condensation state of eukaryotic chromosomes varies with the cell cycle.

32. **(a)** 3.2×10^7; **(b)** 2.9×10^8.

33. Although the chemical composition of the genetic material may be different from that of DNA, it more than likely will have properties similar to those of DNA. As stated in this chapter, the genetic material must contain complex information, replicate or be replicated faithfully, and encode the phenotype. Even if the material on the planet is not DNA, it must meet these criteria. Additionally, the genetic material must be stable.

36. Tubes 1, 4, and 5. The DNA of the bacteriophage contains phosphorus and the protein contains sulfur. When the bacteriophages infect a cell, they inject their DNA into the cell, but the protein coats stay on the surface of the cell. The protein coats are sheared off in the blender. After centrifugation, the protein coats remain in the fluid, while the cells containing the phage DNA are at the bottom of the tube. Thus, bacteria infected with ^{35}S-labeled phage will have radioactivity associated with the protein coats, whereas bacteria infected with ^{32}P-labeled phage will have radioactivity associated with the cells.

Chapter 9

18. **(a)**

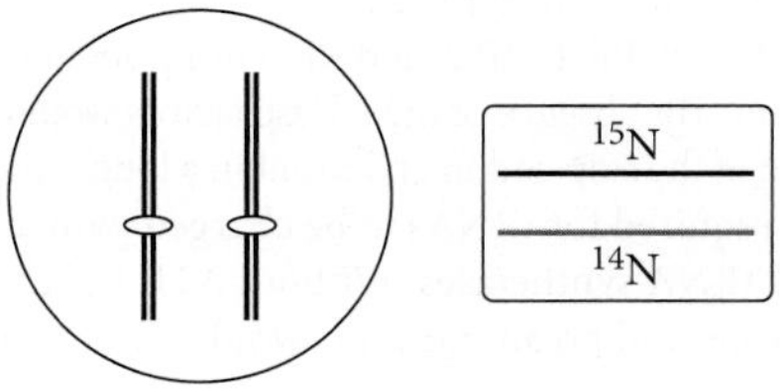

(b)

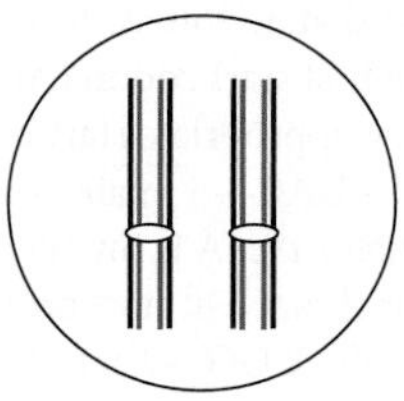

(c)

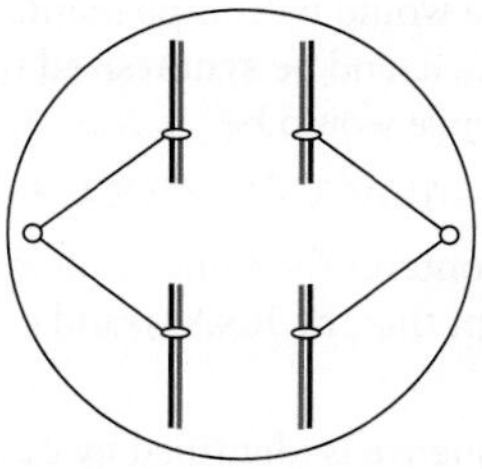

(d)

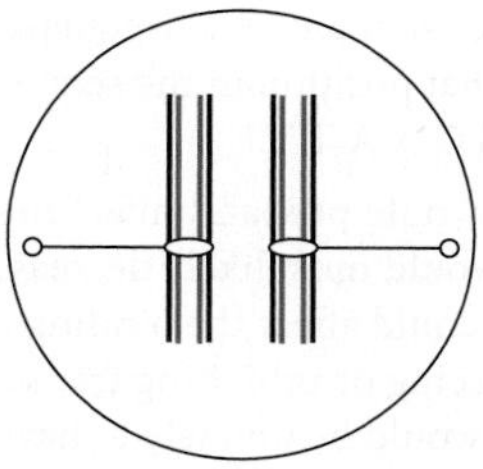

(e)

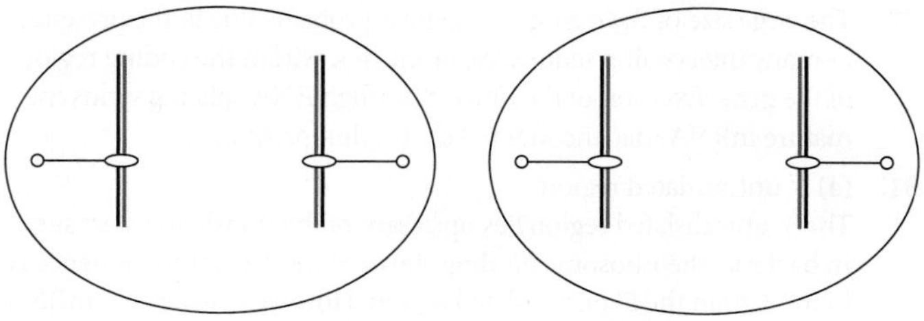

19. 5 minutes.

22.

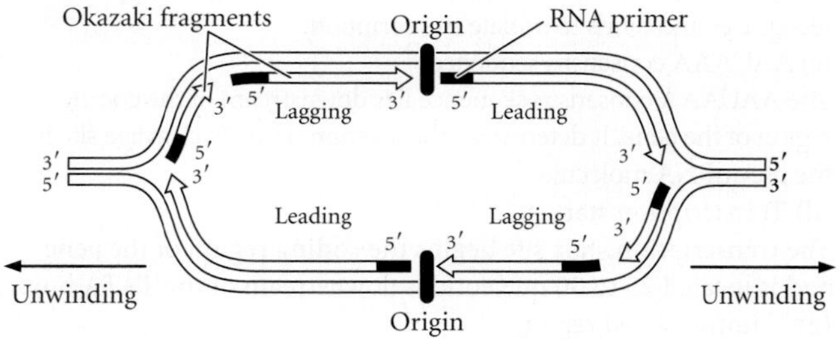

23. **(a)** More errors in replication.
(b) Primers would not be removed.
(c) Primers that had been removed would not be replaced.

27. The RNA part of telomerase is needed to provide the template for synthesizing complementary DNA telomere sequences at the ends of the chromosomes. A deletion of the gene that encodes the RNA part of telomerase would eliminate the template and prevent telomere synthesis at the ends of the chromosomes.

Chapter 10

15. **(a)** The RNA molecule is likely to be single stranded. If the molecule were double stranded, we would expect nearly equal percentages of adenine and uracil and equal percentages of guanine and cytosine.
(b) Because the DNA template strand is complementary to the RNA molecule, we would expect equal percentages of bases in DNA complementary to the RNA bases. Therefore, in the template strand of DNA, we would expect A = 42%, T = 23%, C = 14%, and G = 21%.

16. The RNA molecule would be complementary to the template strand, contain uracil, and be synthesized in an antiparallel fashion. The sequence would be

5′—AUAGGCGAUGCCA—3′

The RNA strand contains the same sequence as the nontemplate DNA strand, except that the RNA strand contains uracil in place of thymine.

25. The consensus sequence is identified by determining which nucleotide is used most frequently at each position. For the two nucleotides that occur at an equal frequency at the first position, both are listed at that position in the sequence and identified by a slash mark: T/A G C A A T T.

26. **(a)** This mutation would probably affect the −10 consensus sequence, which would most likely decrease transcription.
(b) This mutation could affect the binding of the sigma factor to the promoter, reducing or inhibiting transcription.
(c) This mutation would be unlikely to have any effect on transcription.
(d) This mutation would have little effect on transcription.

27. The large size of the dystrophin gene is probably due to the presence of many intervening sequences, or introns, within the coding region of the gene. Excision of the introns through RNA splicing yields the mature mRNA that encodes the dystrophin protein.

31. **(a)** 5′ untranslated region
The 5′ untranslated region lies upstream of the translation start site. In bacteria, the ribosome binding site or Shine-Dalgarno sequence is found within the 5′ untranslated region. However, eukaryotic mRNA does not have the equivalent sequence, and a eukaryotic ribosome binds at the 5′ cap of the mRNA molecule.
(b) Promoter
The promoter is the DNA sequence that the transcription apparatus recognizes and binds to initiate transcription.
(c) AAUAAA consensus sequence
The AAUAAA consensus sequence lies downstream of the coding region of the gene. It determines the location of the 3′ cleavage site in the pre-mRNA molecule.
(d) Transcription start site
The transcription start site begins the coding region of the gene and is located 25 to 30 nucleotides downstream of the TATA box.
(e) 3′ untranslated region
The 3′ untranslated region is a sequence of nucleotides at the 3′ end of the mRNA that is not translated into proteins. However, it does affect the translation of the mRNA molecule as well as the stability of the mRNA.
(f) Introns
Introns are noncoding sequences of DNA that intervene within coding regions of a gene.
(g) Exons
Exons are coding regions of a gene.
(h) Poly(A) tail
A poly(A) tail is added to the 3′ end of the pre-mRNA; it affects mRNA stability and the binding of the ribosome to the mRNA.
(i) 5′ cap
The 5′ cap functions in the initiation of translation and mRNA stability.

Chapter 11

12. **(a)** 1; **(b)** 2; **(c)** 3; **(d)** 3; **(e)** 4.

13. 3^3, or 27 possible codons.

15. **(a)** amino—fMet-Phe-Lys-Phe-Lys-Phe—carboxyl
(b) amino—fMet-Tyr-Ile-Tyr-Ile—carboxyl
(c) amino—fMet-Asp-Glu-Arg-Phe-Leu-Ala—carboxyl
(d) amino—fMet-Gly—carboxyl (The stop codon UAG follows the codon for glycine.)

17. There are two possible sequences:

mRNA:	5′—AUGUGGCAU—3′
DNA template:	3′—TACACCGTA—5′
DNA nontemplate:	5′—ATGTGGCAT—3′
mRNA:	5′—AUGUGGCAC—3′
DNA template:	3′—TACACCGTG—5′
DNA nontemplate:	5′—ATGTGGCAC—3′

21. Initiation factor 3
fMet-$tRNA^{fMet}$
30S initiation complex
70S initiation complex
Elongation factor Tu
Elongation factor G
Release factor 1

24. **(a)** The lack of IF-3 would prevent protein synthesis. IF-3 separates the large and small ribosomal subunits, which is required for the initiation of translation. The absence of IF-3 would mean that translation would not be initiated and no proteins would be synthesized.
(b) No translation would take place. IF-2 is necessary for the initiation of translation. The lack of IF-2 would prevent fMet-$tRNA^{fMet}$ from being delivered to the small ribosomal subunit, thus blocking translation.
(c) Although translation would be initiated by the delivery of methionine to the ribosome–mRNA complex, no other amino acids would be delivered to the ribosome. EF-Tu, which binds to GTP and the charged tRNA, is necessary for elongation. This three-part complex enters the A site of the ribosome. If EF-Tu is not present, the charged tRNA will not enter the A site, thus stopping translation.
(d) EF-G is necessary for the translocation (movement) of the ribosome along the mRNA in the 5′→3′ direction. Once a peptide bond had formed between Met and Pro, the lack of EF-G would prevent the movement of the ribosome along the mRNA, and so no new codons would be read. The formation of the dipeptide Met-Pro does not require EF-G.
(e) Release factors RF-1, RF-2, and RF-3 bring about termination of translation. The absence of the release factors would prevent termination at the stop codon and result in a longer protein produced.
(f) ATP is required for tRNAs to be charged with amino acids by aminoacyl-tRNA synthetases. Without ATP, the charging would not take place, and no amino acids would be available for protein synthesis.
(g) GTP is required for the initiation, elongation, and termination of translation. If GTP is absent, protein synthesis will not take place.

28. **(a)** The results suggest that, to initiate translation, the ribosome scans the mRNA to find the first start codon (AUG) that it encounters. The ribosome finds the appropriate start codon by the anticodon on the initiator tRNA (3′—UAC—5′) pairing with 5′—AUG—3′. If the anticodon on the initiator tRNA is mutated, so that it is now 3′—ACC—5′, the initiator tRNA will start protein synthesis when it pairs with its complement (5′—UGG—3′) in the mRNA.
(b) The initiation of translation in bacteria occurs in a different way—it requires the 16S rRNA of the small ribosomal subunit to interact with the Shine–Dalgarno sequence. This interaction serves

to line up the ribosome over the start codon. If the anticodon had been changed such that the start codon could not be recognized, then protein synthesis would be unlikely to take place.
(**c**) After the anticodon on the initiator tRNA is mutated to 3′—ACC—5′, protein synthesis will be initiated whenever the ribosome encounters 5′—UGG—3′ in the mRNA. If the first UGG codon occurs before the normal AUG start codon, then extra amino acids will be added to the protein. If the first UGG codon occurs after the AUG start codon, then fewer amino acids will be added.

Chapter 12

17. (**a**) Inactive repressor; (**b**) active activator; (**c**) active repressor; (**d**) inactive activator.

18. (**a**) The operon will never be turned off, and transcription will take place all the time.
(**b**) The result will be constitutive expression, and transcription will take place all the time.

20. RNA polymerase will bind the *lac* promoter poorly, significantly decreasing the transcription of the *lac* structural genes.

24.

	Lactose absent		Lactose present	
Genotype of strain	**β-Galactosidase**	**Permease**	**β-Galactosidase**	**Permease**
$lacI^+$ $lacP^+$ $lacO^+$ $lacZ^+$ $lacY^+$	−	−	+	+
$lacI^-$ $lacP^+$ $lacO^+$ $lacZ^+$ $lacY^+$	+	+	+	+
$lacI^+$ $lacP^+$ $lacO^c$ $lacZ^+$ $lacY^+$	+	+	+	+
$lacI^-$ $lacP^+$ $lacO^+$ $lacZ^+$ $lacY^-$	+	−	+	−
$lacI^-$ $lacP^-$ $lacO^+$ $lacZ^+$ $lacY^+$	−	−	−	−
$lacI^+$ $lacP^+$ $lacO^+$ $lacZ^-$ $lacY^+$/ $lacI^-$ $lacP^+$ $lacO^+$ $lacZ^+$ $lacY^-$	−	−	+	+
$lacI^-$ $lacP^+$ $lacO^c$ $lacZ^+$ $lacY^+$/ $lacI^+$ $lacP^+$ $lacO^+$ $lacZ^-$ $lacY^-$	+	+	+	+
$lacI^-$ $lacP^+$ $lacO^+$ $lacZ^+$ $lacY^-$/ $lacI^+$ $lacP^-$ $lacO^+$ $lacZ^-$ $lacY^+$	−	−	+	−
$lacI^+$ $lacP^-$ $lacO^c$ $lacZ^-$ $lacY^+$/ $lacI^-$ $lacP^+$ $lacO^+$ $lacZ^+$ $lacY^-$	−	−	+	−
$lacI^+$ $lacP^+$ $lacO^+$ $lacZ^+$ $lacY^+$/ $lacI^+$ $lacP^+$ $lacO^+$ $lacZ^+$ $lacY^+$	−	−	+	+
$lacI^s$ $lacP^+$ $lacO^+$ $lacZ^+$ $lacY^-$/ $lacI^+$ $lacP^+$ $lacO^+$ $lacZ^-$ $lacY^+$	−	−	−	−
$lacI^s$ $lacP^-$ $lacO^+$ $lacZ^-$ $lacY^+$/ $lacI^+$ $lacP^+$ $lacO^+$ $lacZ^+$ $lacY^+$	−	−	−	−

26. The *lacI* gene encodes the *lac* repressor protein, which can diffuse within the cell and attach to any operator. It can therefore affect the expression of genes on the same molecule or on a different molecule of DNA. The *lacO* gene encodes the operator. It affects the binding of RNA polymerase to DNA, and therefore affects the expression of genes only on the same molecule of DNA.

34. The action of an enhancer is blocked when an insulator is located between the enhancer and the promoter of the gene. It is likely that genes A, B, and C will be stimulated by the enhancer and that gene D will not be stimulated. In the example from the figure, the insulator lies between gene D and the enhancer. The enhancer's effect on genes A, B, and C is not likely to be affected by the insulator, and these genes will be stimulated.

37. The phenotypic differences in traditional genetic traits such as the color and shape of peas that Mendel studied are due to differences in the DNA base sequences within the alleles. In epigenetics, the phenotypic differences are not due to changes in allele DNA base sequences, but are differences in the expression of genes that are passed on to other cells and sometimes to other generations.

38. We would expect to see differences in DNA methylation and histone acetylation that altered expression of genes involved in response to stress. We would also expect that the adults would show increased fear and heightened hormonal response to stress.

Chapter 13

13. (**a**) Leucine, serine, or phenylalanine
(**b**) Isoleucine, tyrosine, leucine, valine, or cysteine
(**c**) Phenylalanine, proline, serine, or leucine
(**d**) Methionine, phenylalanine, valine, arginine, tryptophan, leucine, isoleucine, tyrosine, histidine, glutamine, or a stop codon

15. (**a**) amino—Met-Thr-Gly-**Ser**-Gln-Leu-Tyr-Stop—carboxyl
(**b**) amino—Met-Thr-Gly-Asn-**Stop**—carboxyl
(**c**) amino—Met-Thr-**Ala-Ile-Asn-Tyr-Ile**—carboxyl
(**d**) amino—Met-Thr-Gly-Asn-**His**-Leu-Tyr-Stop—carboxyl
(**e**) amino—Met-Thr-**Thr**-Gly-Asn-Gln-Leu-Tyr-Stop—carboxyl
(**f**) amino—Met-Thr-**Gly**-Asn-Gln-Leu-Tyr-Stop—carboxyl

17. (**a**) A single base-pair substitution resulting in a missense mutation.
(**b**) A single-base substitution resulting in a nonsense mutation.
(**c**) The deletion of a single nucleotide resulting in a frameshift mutation.
(**d**) A six-base pair deletion has occurred, resulting in the elimination of two amino acids (Arg and Leu) from the protein.
(**e**) The insertion of three nucleotides resulting in the addition of a Leu codon.

20. (**a**)

5′–A**G**–3′ to 5′–A**A**–3′
3′–T**C**–5′ 3′–T**T**–5′

(**b**)

5′–**A**G–3′ if A is deaminated, then 5′–**G**G–3′
3′–**T**C–5′ 3′–**C**C–5′

5′–A**G**–3′ if C is deaminated, then 5′–A**A**–3′
3′–T**C**–5′ 3′–T**T**–5′

23. The flanking repeat is in boldface type.
(**a**) 5′—ATTCGAAC**TGAC** [transposable element] **TGAC**CGATCA—3′
(**b**) 5′—ATT**CGAA** [transposable element] **CGAA**CTGACCGATCA—3′

27. The breeder can look for plants that have increased mutation rates in either their germ-line or somatic tissues. If they are defective in DNA repair, they should have higher rates of mutation.

Chapter 14

15. *Ara*I

17. 10

19. (**a**) 460,800; (**b**) 1,036,800; (**c**) 5,120,000.

20.

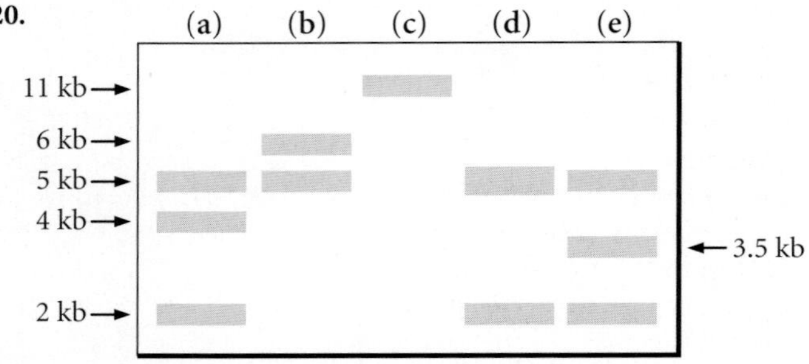

21. (**a**) Plasmid; (**b**) phage λ; (**c**) cosmid; (**d**) bacterial artificial chromosome.

25. 5′—NGCATCAGTA—3′

Chapter 15

14.

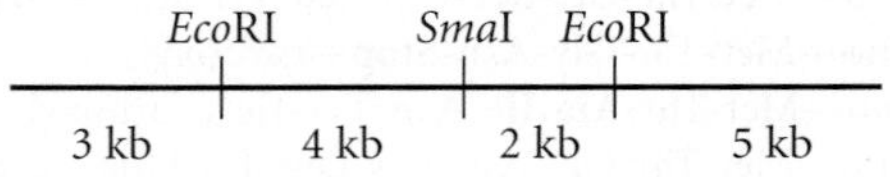

15. 5′—ATTTACCGCAATACCCTAGTTAAAAC—3′

18. Genes 2 and 24 are expressed at far higher levels in the antibiotic-resistant bacteria than in the nonresistant cells. Conversely, genes 4, 17, and 22 are downregulated. All of these genes may be involved in antibiotic resistance. The upregulated genes may be involved in metabolism of the antibiotic or may perform functions that are inhibited by the antibiotic. The downregulated genes may be involved in uptake of the antibiotic or represent a cellular mechanism that accentuates the potency of the antibiotic. Characterization of these genes might lead to information regarding the mechanism of antibiotic resistance, and thus to the design of new antibiotics that can circumvent this resistance mechanism.

21. Any association between genome size and number of protein-encoding genes among the 12 *Drosophila* species is weak at best. This observation is consistent with the pattern seen among eukaryotic organisms in general, in which there is no general association between genome size and number of genes.

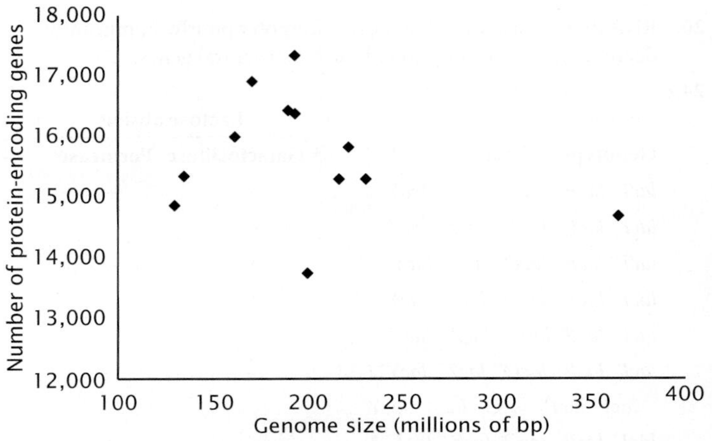

Data from *Drosophila* 12 Genome Consortium, Evolution of genes and genomes in the Drosophila phylogeny, *Nature* 450:203–218, 2007.

22. Proteomes. Generally, the genomes of two cells from the same person are genetically identical. However, the proteins in the two cells are likely to differ widely because different genes are expressed in each cell type.

Chapter 16

17. If the differences in cancer rates are due to genetic differences in the two populations, then people who migrated from Utah or Shanghai to other locations should have rates of prostate cancer incidence similar to those of people who stayed in Utah or Shanghai. If the cancer rates are due to environmental factors, then people who migrated from Utah or Shanghai should have rates of prostate cancer determined by their new locations and not by their place of origin.

18. Because oncogenes promote cell proliferation, they act in a dominant manner. In contrast, mutations in tumor-suppressor genes cause loss of function and act in a recessive manner. When introduced into cells, the mutated *palladin* gene increases cell migration. Such a dominant effect suggests that *palladin* is an oncogene.

Chapter 17

11. (**a**) Discontinuous characteristic because only a few distinct phenotypes are present and alleles at a single locus determine the characteristic.

(**b**) Discontinuous characteristic because there are only two phenotypes (dwarf and normal) and a single locus determines the characteristic.
(**c**) Quantitative characteristic because susceptibility is a continuous trait determined by multiple genes and environmental factors (an example of a quantitative phenotype with a threshold effect).
(**d**) Quantitative characteristic because it is determined by many loci (an example of a meristic characteristic).
(**e**) Discontinuous characteristic because only a few distinct phenotypes are determined by alleles at a single locus.

12. (**a**) All weigh 10 grams.
(**b**) $^1/_{16}$ weighing 16 grams, $^4/_{16}$ weighing 13 grams, $^6/_{16}$ weighing 10 grams, $^4/_{16}$ weighing 7 grams, and $^1/_{16}$ weighing 4 grams.

13. $^1/_{64}$ 22 cm tall; $^6/_{64}$ 20 cm tall; $^{15}/_{64}$ 18 cm tall; $^{20}/_{64}$ 16 cm tall; $^{15}/_{64}$ 14 cm tall; $^6/_{64}$ 12 cm tall; $^1/_{64}$ 10 cm tall.

15. (**a**) 0.38; (**b**) 0.69.

20. The only reasonable conclusion is (**d**). Statement (**a**) is not justified because the heritability value does not apply to absolute height nor to an individual, but to the variance in height among Southwestern undergraduates. Statement (**b**) is not justified because the heritability has been determined only for Southwestern University students; students at other universities, with different ethnic backgrounds and from different regions of the country, may have different heritability for height. Statement (**c**) is again not justified because the heritability refers to the variance in height rather than absolute height. Statement (**e**) is not justified because the heritability has been determined for the range of variation in nongenetic factors experienced by the population under study; environmental variation outside this range (such as severe malnutrition) may have profound effects on height.

23. (**a**) Use the equation $R = h^2 \times S$, where S is the selection differential. In this case, $S = 10\text{ cm} - 4\text{ cm} = 6\text{ cm}$, and we are given that the narrow-sense heritability h^2 is 0.6. Therefore, the response to selection $R = 0.6 \times 6\text{ cm} = 3.6\text{ cm}$.
(**b**) The average wing length of the progeny should be the mean wing length of the population plus R: 4 cm + 3.6 cm = 7.6 cm.

25. 0.75

Chapter 18

23. $f(T^ET^E) = 0.685; f(T^ET^F) = 0.286; f(T^FT^F) = 0.029; f(T^E) = 0.828; f(T^F) = 0.172$.

27. (**a**) $f(A) = (2 \times 42 + 24)/174 = 0.62$

$f(G) = (2 \times 21 + 24)/174 = 0.38$

(**b**) Expected genotypic frequencies:
$f(AA) = (0.62)(0.62) = 0.384$
$f(AG) = 2(0.62)(0.38) = 0.471$
$f(GG) = (0.38)(0.38) = 0.144$

(**c**)

Genotype	Observed	Expected	$O-E$	$(O-E)^2$	$(O-E)^2/E$
AA	42	33	9	81	2.45
AG	24	41	17	289	7.05
GG	21	13	8	64	4.92

$\chi^2 = \Sigma\,(O - E)^2/E = 14.42$

The number of degrees of freedom is the number of genotypes minus the number of alleles = 3 − 2 = 1.
The p value is much less than 0.05; therefore, we reject the hypothesis that these genotypic frequencies may be expected from Hardy–Weinberg equilibrium.

29. (**a**) $f(\text{dilute color}) = f(dd) = q^2 = 131/325 = 0.403; q = 0.635$.
(**b**) If $q = f(d) = 0.635$, then $p = f(D) = 1 - q = 0.365$.
$f(Dd) = 2pq = 2(0.365)(0.635) = 0.464; 0.464 \times 325 =$ 151 heterozygous cats

31. (**a**) 0.64 for M, 0.32 for MN, and 0.04 for N.
(**b**) With inbreeding, the frequencies of the M and N blood types will be higher than expected with random mating, and the frequency of the MN blood type will be lower.

32. The small population sizes and founder effects will cause strong genetic drift. The geneticists will find large variation between populations in allelic frequencies. Within populations, the same factors, coupled with inbreeding, will cause loss of genetic variation and a high degree of homozygosity.

33. The selection coefficient is 1 − fitness, so the selection coefficients for the genotypes are

Genotype	Fitness	Selection coefficient
ST/*ST*	0.47	0.53
ST/*AR*	1.0	0.00
AR/*AR*	0.62	0.38

The chromosome inversions exhibit overdominance: the heterozygote has higher fitness than either homozygote. With overdominance, an equilibrium is reached, at which point the frequency of *AR* (q) will be

$$q = f(AR) = \frac{S_{AR/AR}}{S_{AR/AR} + S_{ST/ST}} = \frac{0.53}{0.53 + 0.38} = 0.58$$

The frequency of *ST* (p) will be $1 - q = 0.42$.

34. (**a**) The populations raised on the ethanol-containing diet appear to be experiencing directional selection in favor of the Adh^F allele and against the Adh^S allele.
(**b**) In the absence of data about relative reproductive rates, we use the relative viability data as a proxy for relative fitness.
$W_{FF} = 0.932/1.288 = 0.724$
$W_{FS} = 1.288/1.288 = 1.0$
$W_{SS} = 0.596/1.288 = 0.463$

36. The illustration on the left shows evolutionary change occurring gradually over time, during both periods of anagenesis and periods of cladogenesis. The illustration on the right shows evolution occurring only during cladogenesis, when one lineage splits into two.

Subject Index

Note: Page numbers followed by f indicate figures; those followed by t indicate tables.

C

D

H

M

N

Q

T